REVIEWS IN MINERALOGY AND GEOCHEMISTRY

Volume 61   2006

SULFIDE MINERALOGY AND GEOCHEMISTRY

EDITOR

David J. Vaug_
University of Manchester
Manchester, United Kingdom

COVER PHOTOGRAPH: Cubo-octahedral crystals of galena (PbS) with calcite (white) on brown crystalline siderite, Neudorf, Harz, Germany. Copyright the Natural History Museum (London); reproduced with permission.

Series Editor: ***Jodi J. Rosso***

GEOCHEMICAL SOCIETY
MINERALOGICAL SOCIETY OF AMERICA

Copyright 2006

MINERALOGICAL SOCIETY OF AMERICA

The appearance of the code at the bottom of the first page of each chapter in this volume indicates the copyright owner's consent that copies of the article can be made for personal use or internal use or for the personal use or internal use of specific clients, provided the original publication is cited. The consent is given on the condition, however, that the copier pay the stated per-copy fee through the Copyright Clearance Center, Inc. for copying beyond that permitted by Sections 107 or 108 of the U.S. Copyright Law. This consent does not extend to other types of copying for general distribution, for advertising or promotional purposes, for creating new collective works, or for resale. For permission to reprint entire articles in these cases and the like, consult the Administrator of the Mineralogical Society of America as to the royalty due to the Society.

REVIEWS IN MINERALOGY AND GEOCHEMISTRY

(Formerly: REVIEWS IN MINERALOGY)

ISSN 1529-6466

Volume 61

Sulfide Mineralogy and Geochemistry

ISBN 0-939950-73-1

Additional copies of this volume as well as others in this series may be obtained at moderate cost from:

THE MINERALOGICAL SOCIETY OF AMERICA
3635 CONCORDE PARKWAY, SUITE 500
CHANTILLY, VIRGINIA, 20151-1125, U.S.A.
WWW.MINSOCAM.ORG

SULFIDE MINERALOGY and GEOCHEMISTRY

Reviews in Mineralogy and Geochemistry

FROM THE SERIES EDITOR

When the first Sulfide volume was published in 1974, by the Mineralogical Society of America with Paul Ribbe as the Editor, little did we forsee what exciting adventure was ahead. But here we are, 33 years and 60 RiMG volumes later. Publishing technologies may have changed (no more typewriters or ditto's), the faces of the scientists (and editors) have grown older (or changed all together), but our enthusiasm for mineralogy and geochemistry hasn't. Let's hope that this present volume on Sulfide Mineralogy and Geochemistry will usher in the next 60 RiMG volumes with as much impact as the first did!

Many thanks to David Vaughan for his dedication to this volume. He was always available when questions arose, was faithful to adhere to our deadlines, and gracefully encouraged his authors to do the same. It was a delight to work with him.

Supplemental material and errata (if any) can be found at the MSA website *www.minsocam.org*.

Jodi J. Rosso, Series Editor
West Richland, Washington
June 2006

PREFACE

The metal sulfides have played an important role in mineralogy and geochemistry since these disciplines first emerged and, for reasons detailed in the first chapter, have achieved an even greater importance in both fundamental and applied mineral sciences in recent years. It is no surprise that Volume 1 in the series, of which this book is Volume 61, was devoted to "Sulfide Mineralogy." For some years there has been the need for an updated volume covering the subject areas of Volume 1 and addressing the new areas which have developed since its publication in 1974.

I am grateful to Paul Ribbe, former RiMG Series Editor, for persistently encouraging me to undertake this new volume on sulfides. I am also grateful to my longstanding collaborator Jim Craig for facilitating acquisition of the copyright of our 1978 book "*Mineral Chemistry of Metal Sulfides*" and for his encouragement in this project. Liane Benning, Jim Craig, Jeff Grossman, Richard Harrison, Michael Hochella, Jon Lloyd, Dave Polya, Wayne Nesbitt, Pat Shanks, Michele Warren and Jack Zussman have all kindly given of their time to review draft versions of the chapters in this volume. I am also grateful to Jodi Rosso, RiMG Series Editor, for her heroic efforts in bringing this book to completion, and to our contributors for all their hard work.

David J. Vaughan
University of Manchester
June 2006

SULFIDE MINERALOGY and GEOCHEMISTRY

61   *Reviews in Mineralogy and Geochemistry*   **61**

TABLE OF CONTENTS

1   Sulfide Mineralogy and Geochemistry: Introduction and Overview

David J. Vaughan

BACKGROUND: WHY STUDY SULFIDES? 1
PREVIOUS WORK: THE RELEVANT LITERATURE 2
THE PRESENT TEXT: SCOPE AND CONTENT 3
REFERENCES 5

2   Crystal Structures of Sulfides and other Chalcogenides

Emil Makovicky

INTRODUCTION 7
    Crystal chemical concepts 7
    Structural classification 9
PRINCIPLES OF THE DESCRIPTION AND CLASSIFICATION OF CRYSTAL
    STRUCTURES RELEVANT TO SULFIDES: A BRIEF OUTLINE 12
    Non-modular categories of similarity 12
    Modular categories of structural similarity 12
CHALCOGENIDES WITH PRONOUNCED ONE-DIMENSIONAL
    BUILDING UNITS 16
    Fibrous sulfides with chains of edge-sharing FeS_4 tetrahedra 16
    Other fibrous sulfides 19
STRUCTURE TYPES COMPOSED OF LAYERS:
    CHALCOGENIDE LAYERS WITH M:S =1:1 19
    Mackinawite and related families 19
    Other M:S=1:1 layer types 22
STRUCTURES COMPOSED OF SESQUICHALCOGENIDE LAYERS 25
STRUCTURES COMPOSED OF DICHALCOGENIDE LAYERS 28
LAYER-LIKE STRUCTURES WITH TWO TYPES OF CHALCOGENIDE LAYERS ... 31
    Cannizzarite and related structures 31
    Derivatives of cannizzarite: combinations of accretion with variable-fit 32

Selected synthetic and natural chalcogenides with layer-misfit structures 35
Cylindrite family: a doubly-noncommensurate example 38
STRUCTURES WITH 3D-FRAMEWORKS 40
    Adamantane structures 40
    Structures based on the sphalerite stacking principle.. 44
    Structures based on the wurtzite stacking principle 45
    Adamantane structures with additional, interstitial tetrahedra 45
    The problem of interstitial cations in *hcp* and related stackings 50
    Defect adamantane structures and more complicated cases.. 52
    Summary of cation-cluster configurations 52
    Tetrahedrite-tennantite... 53
    Structures with anion vacancies 55
    Structures with tetrahedral clusters 55
    Monosulfides with octahedral cations: *ccp* vs. *hcp* stacking of anions 60
    Thiospinels 63
    Disulfides, sulfarsenides and their analogs... 65
CHANNEL AND CAGE STRUCTURES... 69
CATION-SPECIFIC STRUCTURES 69
    Mercury sulfides 69
    Nickel sulfides 70
    Sulfides of palladium and platinum.. 72
IONIC CONDUCTORS AND THE CHALCOGENIDES OF Cu AND Ag 77
LONE ELECTRON PAIR COMPOUNDS 84
    Molecular arsenic sulfides 84
    Ribbon-like arrangements 87
    The bismuthinite–aikinite series of sulfosalts 88
    Layer-like structures 90
LARGE SULFOSALT FAMILIES.. 90
    General features 90
    Lillianite homologous series 92
PAVONITE HOMOLOGOUS SERIES... 94
CUPROBISMUTITE HOMOLOGOUS SERIES 95
    Rod-based sulfosalts.. 97
    Case study: dadsonite 99
    Other sulfosalt families 105
ACKNOWLEDGMENTS.. 107
REFERENCES 107

3   Electrical and Magnetic Properties of Sulfides

Carolyn I. Pearce, Richard A.D. Pattrick, David J. Vaughan

INTRODUCTION 127
THEORY AND MEASUREMENT OF ELECTRICAL AND
    MAGNETIC PROPERTIES... 127
    Electrical properties... 127
    Measurement of electrical properties 128

Magnetic properties 133
Measurement of magnetic properties 139
ELECTRICAL AND MAGNETIC PROPERTIES OF METAL SULFIDES 145
Metal sulfides of sphalerite and wurtzite (ZnS) structure-type 148
Sulfides of pyrite (FeS_2) and related structure-types 150
Sulfides with halite (NaCl) structures 156
Sulfides with niccolite (NiAs)-based structures 157
Mackinawite, smythite, greigite and other thiospinels 163
PROPERTIES OF SULFIDE NANOPARTICLES 164
Synthesis of sulfide nanoparticles 166
Case study: pure and transition metal (TM) ion doped ZnS 166
Sulfide nanoparticles in the environment 169
APPLICATIONS 169
Electronics 169
Paleomagnetic investigations 170
Beneficiation of sulfide ores 171
CONCLUDING COMMENTS 172
ACKNOWLEDGMENTS 172
REFERENCES 172

4 Spectroscopic Studies of Sulfides

Paul L. Wincott and David J. Vaughan

INTRODUCTION 181
ELECTRONIC (OPTICAL) ABSORPTION AND REFLECTANCE SPECTRA 182
Spectra of sphalerite and wurtzite-type sulfides 184
Spectra of pyrite-type disulfides 186
Spectra of the layer sulfides (MoS_2) 187
INFRARED AND RAMAN SPECTRA 188
X-RAY AND ELECTRON EMISSION SPECTROSCOPIES 191
X-ray photoelectron spectroscopy (XPS) and
  Auger electron spectroscopy (AES) 192
X-ray emission spectroscopy 194
Applications to non-transition metal monosulfides: PbS, ZnS, CdS and HgS 195
Applications to the transition metal sulfides:
  FeS, $Fe_{1-x}S$, CuS, $Cu_{1-x}S$, Ag_2S, $CuFeS_2$ and related phases 198
Application to the disulfides: FeS_2, MoS_2 200
X-RAY ABSORPTION SPECTROSCOPIES 201
Extended X-ray absorption fine structure spectroscopy (EXAFS) and
  X-ray absorption near-edge structure spectroscopy (XANES) 201
Applications to binary metal sulfides:
  PbS, ZnS, CdS, HgS, CaS, MgS, MnS, FeS, CoS, NiS and CuS 204
Applications to disulfides: pyrite group (MS_2 where M = Fe, Co, Ni) and MoS_2 206
Ternary and complex sulfides: chalcopyrite, tetrahedrite group 206
MÖSSBAUER STUDIES 208
Pyrite, marcasite, arsenopyrite and related minerals 210

Thiospinel minerals 211
Mackinawite and pentlandite..214
Sphalerite and wurtzite-type sulfide minerals 215
Troilite and pyrrhotite 217
High-pressure Mössbauer spectroscopy of sulfides 218
OTHER SPECTROSCOPIC TECHNIQUES 219
CONCLUDING REMARKS..221
ACKNOWLEDGMENTS...221
REFERENCES 221

5   Chemical Bonding in Sulfide Minerals

David J. Vaughan, Kevin M. Rosso

INTRODUCTION 231
IONIC AND COVALENT BONDS...231
LIGAND FIELD THEORY 232
QUALITATIVE MOLECULAR ORBITAL (MO) AND BAND MODELS........................234
QUANTITATIVE APPROACHES: ATOMISTIC COMPUTATION 237
QUANTITATIVE APPROACHES: ELECTRONIC STRUCTURE CALCULATIONS 238
CHEMICAL BONDING AND ELECTRONIC STRUCTURE IN SOME
    MAJOR SULFIDE MINERALS AND GROUPS 241
       Generalities...241
       Galena (PbS) 243
       Sphalerite and related sulfides (ZnS, Zn(Fe)S, CdS, HgS, CuFeS$_2$) 245
       Transition metal monosulfides (FeS, Fe$_{1-x}$S, CoS, NiS) 248
       Pyrite and the related disulfides (FeS$_2$, CoS$_2$, NiS$_2$, CuS$_2$, FeAsS, FeAs$_2$)..............252
       Other (including complex) sulfides (pentlandites and metal-rich sulfides,
          thiospinels, layer structure sulfides, tetrahedrites) 255
CONCLUDING REMARKS..258
ACKNOWLEDGMENTS...259
REFERENCES 259

6   Thermochemistry of Sulfide Mineral Solutions

Richard O. Sack, Denton S. Ebel

INTRODUCTION 265
SULFIDE THERMOCHEMISTRY 266
       Basic concepts 266
       Dimensionality of systems 271
       Mineral stability and criteria for equilibrium 272
       Experimental methods in sulfide research..273
SULFIDE SYSTEMS IN METEORITES 280
Fe-S...282

Zn-Fe-S 290
    Ag_2S-Cu_2S-Sb_2S_3-As_2S_3 299
    Sulfosalts (Ag_2S-Cu_2S)-Sb_2S_3-As_2S_3 300
    Ag_2S-Cu_2S 307
    As-Sb exchange 310
THE SYSTEM Ag_2S-Cu_2S-ZnS-FeS-Sb_2S_3-As_2S_3 311
    Fahlore thermochemistry..313
    Thermochemical database 320
Bi_2S_3- AND PbS-BEARING SYSTEMS 323
    The Bi_2S_3-Sb_2S_3-$AgBiS_2$-$AgSbS_2$ quadrilateral 324
    The quaternary system Ag_2S-PbS-Sb_2S_3-Bi_2S_3 328
    The system Ag_2S-Cu_2S-ZnS-FeS-PbS-Sb_2S_3-As_2S_3 333
ACKNOWLEDGMENTS..336
REFERENCES 337
APPENDICES 349

7

Phase Equilibria at High Temperatures

Michael E. Fleet

INTRODUCTION 365
Fe, Co AND Ni SULFIDES AND THEIR MUTUAL SOLID SOLUTIONS 365
    Monosulfides of Fe, Co and Ni 370
    Fe-S phase relations..374
    Ni-S phase relations..380
    The system Fe-Co-S 381
    Ternary phase relations in the system Fe-Ni-S 382
Cu-S, Cu-Fe-S AND Cu-Fe-Zn-S SYSTEMS 383
    Phase relations in the system Cu-S..385
    Phase relations in the ternary system Cu-Fe-S 388
    Sulfur fugacity in ternary Cu-Fe-S system 392
    Quaternary System Cu-Fe-Zn-S 392
ZnS AND Fe-Zn-S SYSTEM 395
    ZnS 395
    Ternary Fe-Zn-S system and geobarometry 396
    Calculated phase relations 401
HALITE STRUCTURE MONOSULFIDES 403
    Electronic structure and magnetism 403
    Phase relations...404
ACKNOWLEDGMENTS..405
REFERENCES 405

8 Metal Sulfide Complexes and Clusters

David Rickard, George W. Luther, III

INTRODUCTION	421
Background	421
Metals considered in this chapter	423
Lewis acids and bases	424
Hard A and soft B metals	425
Complexes and clusters	426
Coordination numbers and symmetries	427
Equilibrium constants of complexes	428
METHODS FOR MEASUREMENT OF METAL SULFIDE STABILITY CONSTANTS	430
Theoretical approaches to the estimation of stability constants	432
Experimental approaches to the measurement of metal sulfide stability constants.	434
METHODS USED TO DETERMINE THE MOLECULAR STRUCTURE AND COMPOSITION OF COMPLEXES	442
Chemical synthesis of complexes	443
LIGAND STABILITIES AND STRUCTURES.	444
Molecular structures of sulfide species in aqueous solutions	444
S (−II) equilibria in aqueous solutions	446
Polysulfide stabilities	448
METAL SULFIDE COMPLEXES	450
The first transition series: Cr, Mn, Fe, Co, Ni, Cu	450
Molybdenum	461
Silver and gold	462
Zinc, cadmium and mercury	466
Tin and lead	472
Arsenic and antimony	474
Selected metal sulfide complex stability constants	478
KINETICS AND MECHANISMS OF METAL SULFIDE FORMATION	483
Properties of metal aqua complexes	483
Mechanisms of formation of metal sulfide complexes	485
Metal isotopic evidence for metal sulfide complexes	489
COMPLEXES, CLUSTERS AND MINERALS	489
Zinc sulfide clusters	491
Iron sulfide clusters	493
The relationship between complexes, clusters and the solid phase	494
ACKNOWLEDGMENTS	495
REFERENCES	496

9   Sulfide Mineral Surfaces

Kevin M. Rosso, David J. Vaughan

INTRODUCTION	505
SURFACE PRINCIPLES	506
Surface energy	506
Atomic structure	507
Electronic structure	512
EXAMPLE SURFACES	516
Crystallographic surfaces	516
Fracture surfaces	542
FUTURE OUTLOOK	548
ACKNOWLEDGMENTS	549
REFERENCES	550

10   Reactivity of Sulfide Mineral Surfaces

Kevin M. Rosso, David J. Vaughan

INTRODUCTION	557
ELEMENTARY ADSORPTION/OXIDATION REACTIONS	558
Galena	558
Pyrite	562
AIR/AQUEOUS OXIDATION	565
Galena	565
Pyrite	567
Pyrrhotite	571
Chalcopyrite	571
Arsenopyrite	573
CATALYSIS	576
Laurite	576
Molybdenite	578
FUNCTIONALIZATION AND SELF-ASSEMBLY	580
Collector molecule adsorption	581
METAL ION UPTAKE	592
Galena	592
Pyrite	593
Pyrrhotite and troilite	595
Mackinawite	596
Sphalerite and wurtzite	599
FUTURE OUTLOOK	599
ACKNOWLEDGMENTS	600
REFERENCES	600

11 Sulfide Mineral Precipitation from Hydrothermal Fluids
Mark H. Reed, James Palandri

INTRODUCTION 609
    Calculation method..609
    Data sources 610
SULFIDE MINERAL SOLUBILITY: BASIC CONTROLS 610
SULFIDE MINERAL NUMERICAL EXPERIMENTS..611
    Chloride addition (Run 1)..613
    Simple dilution (Run 2)..614
    Silicate-buffered dilution (Run 3)...618
    Temperature change (Run 4) 619
    pH change (Runs 5, 6, 7, 8) 619
    Simultaneous dilution and cooling by cold water mixing (Run 9)...................625
    Oxidation and reduction of a sulfide-oxide system (Run 10)..........................627
CONCLUSION...629
REFERENCES 630

12 Sulfur Isotope Geochemistry of Sulfide Minerals
Robert R. Seal, II

INTRODUCTION 633
FUNDAMENTAL ASPECTS OF SULFUR ISOTOPE GEOCHEMISTRY 634
ANALYTICAL METHODS 636
REFERENCE RESERVOIRS...637
FACTORS THAT CONTROL SULFUR ISOTOPE FRACTIONATION 638
EQUILIBRIUM FRACTIONATION FACTORS 638
    Experimentally determined fractionation factors 639
    Geothermometry..641
PROCESSES THAT RESULT IN STABLE ISOTOPIC VARIATIONS OF SULFUR 642
    Mass-dependent fractionation processes 642
    Mass-independent fractionation processes 648
GEOCHEMICAL ENVIRONMENTS 650
    Meteorites..650
    Marine sediments 651
    Coal 653
    Mantle and igneous rocks..655
    Magmatic sulfide deposits 657
    Porphyry and epithermal deposits 658
    Seafloor hydrothermal systems 662
    Mississippi Valley-type deposits 666
SUMMARY 668
ACKNOWLEDGMENTS...669
REFERENCES 670

13   Sulfides in Biosystems

Mihály Pósfai, Rafal E. Dunin-Borkowski

INTRODUCTION	679
BIOLOGICAL FUNCTION AND MINERAL PROPERTIES: CONTROLLED MINERALIZATION OF IRON SULFIDES	680
Biologically controlled mineralization in magnetotactic bacteria	681
Sulfide-producing magnetotactic bacteria	682
Structures and compositions of iron sulfide magnetosomes	684
Magnetic sensing with sulfide magnetosomes	685
Mechanical protection: iron sulfides on the foot of a deep-sea snail	689
BIOLOGICALLY INDUCED FORMATION OF SULFIDE MINERALS	692
Microbial sulfate and metal reduction	692
The role of biological surfaces in mineral nucleation	693
Iron sulfides in marine sediments	693
Biogenic zinc sulfides: from mine-water to deep-sea vents	699
BIOLOGICALLY MEDIATED DISSOLUTION OF SULFIDE MINERALS	701
Acid mine drainage	701
Microbial degradation of sulfides in marine environments	703
PRACTICAL APPLICATIONS OF INTERACTIONS BETWEEN ORGANISMS AND SULFIDES	704
Biomimetic materials synthesis	704
Bioremediation	705
Bioleaching of metals	705
IRON SULFIDES AND THE ORIGIN OF LIFE	706
CONCLUDING THOUGHTS	707
ACKNOWLEDGMENTS	708
REFERENCES	708

Sulfide Mineralogy and Geochemistry: Introduction and Overview

David J. Vaughan

School of Earth, Atmospheric and Environmental Sciences
and Williamson Research Centre for Molecular Environmental Science
University of Manchester
Manchester, United Kingdom
e-mail: david.vaughan@manchester.ac.uk

BACKGROUND: WHY STUDY SULFIDES?

The metal sulfides are the raw materials for most of the world supplies of non-ferrous metals and, therefore, can be considered the most important group of ore minerals. Although there are several hundred known sulfide minerals, only a half dozen of them are sufficiently abundant as to be regarded as "rock-forming minerals" (pyrite, pyrrhotite, galena, sphalerite, chalcopyrite and chalcocite; see Bowles and Vaughan 2006). These mostly occur as accessory minerals in certain major rock types, with pyrite being by far the most important volumetrically. It is also important to note that the synthetic analogs of certain sulfide minerals are of interest to physicists and materials scientists because of their properties (electrical, magnetic, optical) and have found uses in various electronic or opto-electronic devices. Recently, this interest has been directed towards the use in such devices of metal sulfide thin films and of sulfide nanoparticles, topics touched upon in Chapter 3 but not discussed in detail in this volume (but see, for example, Fuhs and Klenk 1998; Bernede et al. 1999; Trindade et al. 2001).

The importance of sulfide minerals in ores has long been, and continues to be, a major reason for the interest of mineralogists and geochemists in these materials. Determining the fundamental chemistry of sulfides is key to understanding their conditions of formation and, hence, the geological processes by which certain ore deposits have formed. This, in turn, may inform the strategies used in exploration for such deposits and their subsequent exploitation. In this context, knowledge of structures, stabilities, phase relations and transformations, together with the relevant thermodynamic and kinetic data, is critical. As with many geochemical systems, much can also be learned from isotopic studies.

The practical contributions of mineralogists and geochemists to sulfide studies extend beyond areas related to geological applications. The mining of sulfide ores, to satisfy ever increasing world demand for metals, now involves extracting very large volumes of rock that contains a few percent at most (and commonly less than one percent) of the metal being mined. This is true of relatively low value metals such as copper; for the precious metals commonly occurring as sulfides, or associated with them, the mineable concentrations (*grades*) are very much lower. The "as-mined" ores therefore require extensive processing in order to produce a concentrate with a much higher percentage content of the metal being extracted. Such mineral processing (*beneficiation*) involves crushing and grinding of the ores to a very fine grain size in order to liberate the valuable metal-bearing (sulfide) minerals which can then be concentrated. In some cases, the metalliferous (sulfide) minerals may have specific electrical or magnetic properties that can be exploited to enable separation and, hence, concentration. More commonly, froth flotation is used, whereby the surfaces of particles of a particular mineral phase are rendered water repellent by the addition of chemical reagents and hence are attracted

1529-6466/06/0061-0001$05.00   DOI: 10.2138/rmg.2006.61.1

to air bubbles pulsed through a mineral particle-water-reagent pulp. An understanding of the surface chemistry and surface reactivity of sulfide minerals is central to this major industrial process and, of course, knowledge of electrical and magnetic properties is very important in cases where those particular properties can be utilized.

In the years since the publication of the first ever "Reviews in Mineralogy" volume (at that time called MSA "Short Course Notes") which was entitled "Sulfide Mineralogy" (Ribbe 1974), sulfides have become a focus of research interest for reasons centering on at least two other areas in addition to their key role in ore deposit studies and mineral processing technology. It is in these two new areas that much of the research on sulfides has been concentrated in recent years. The first of these areas relates to the capacity of sulfides to react with natural waters and acidify them; the resulting Acid Rock Drainage (ARD), or Acid Mine Drainage (AMD) where the sulfides are the waste products of mining, has the capacity to damage or destroy vegetation, fish and other aquatic life forms. These acid waters may also accelerate the dissolution of associated minerals containing potentially toxic elements (e.g., As, Pb, Cd, Hg, etc.) and these may, in turn, cause environmental damage. The much greater public awareness of the need to prevent or control AMD and toxic metal pollution has led to regulation and legislation in many parts of the world, and to the funding of research programs aimed at a greater understanding of the factors controlling the breakdown of sulfide minerals. Amongst general references dealing with these topics, which are all part of the emerging field of "Environmental Mineralogy," are the texts of Jambor and Blowes (1994), Vaughan and Wogelius (2000), Cotter-Howells et al. (2000) and Jambor et al. (2003).

The second reason for even greater research interest in sulfide minerals arose initially from the discoveries of active hydrothermal systems in the deep oceans. The presence of life forms that have chemical rather than photosynthetic metabolisms, and that occur in association with newly-forming sulfides, has encouraged research on the potential of sulfide surfaces in catalyzing the reactions leading to assembling of the complex molecules needed for life on Earth (see, for example, reviews by Vaughan and Lennie 1991; Russell and Hall 1997; Hazen 2005). These developments have been associated with a great upsurge of interest in the interactions between microbes and minerals, and in the role that minerals can play in biological systems. In the rapidly growing field of *geomicrobiology*, metal sulfides are of major interest. This interest is related to a variety of processes including, for example, those where bacteria interact with sulfides as part of their metabolic activity and cause chemical changes such as oxidation or reduction, or those in which biogenic sulfide minerals perform a specific function, such as that of navigation in magnetotactic bacteria.

The development of research in areas such as geomicrobiology and environmental mineralogy and geochemistry, is also leading to a greater appreciation of the role of sulfides (particularly the iron sulfides) in the geochemical cycling of the elements at or near the surface of the Earth. For example, the iron sulfides precipitated in the reducing environments beneath the surface of modern sediments in many estuarine areas may play a key role in the trapping of toxic metals and other pollutants. In our understanding of "Earth Systems," geochemical processes involving metal sulfides are an important part of the story.

Thus, interest in sulfide mineralogy and geochemistry has never been greater, given the continuing importance of the long established applications in ore geology and mineral technology, and the new fields of application related to geomicrobiology, the environment, and "Earth Systems."

PREVIOUS WORK: THE RELEVANT LITERATURE

The literature on sulfide mineralogy and geochemistry is spread across a very wide range of journals and related publications. This includes not only journals of mineralogy and geochemistry but also of crystallography, spectroscopy, materials science, pure and applied chemistry and physics, surface science, mineral technology and, more recently, of biogeochemistry and the environmental sciences.

Textbooks devoted specifically to sulfide mineralogy are Ribbe (1974), Vaughan and Craig (1978), and Kostov and Mincheeva-Stefanova (1981). All of these volumes are now out of print, although sometimes still available through library copies. The text by Vaughan and Craig (1978) has recently been made freely available on the internet (*http://www.manchester.ac.uk/ digitallibrary/books/vaughan*). As well as these books dealing with sulfide minerals in general, Mills (1974) provides a wealth of thermodynamic data, and Shuey (1976) reviews the electrical properties of many common sulfides. The books by Barnes (1979, 1997), Garrels and Christ (1965) and Stanton (1972) also contain much of relevance to those interested in the geochemical and petrological aspects of sulfide studies.

As all but a handful of the sulfide minerals are opaque to visible light, the standard method used for initial identification (in a manner analogous to study of a petrographic thin section) and for the characterization of textural and paragenetic relationships, is examination in polished section using the reflected light microscope (commonly known as *ore microscopy*). Here, the texts by Ramdohr (1969, 1980), Uytenbogaart and Burke (1971), Craig and Vaughan (1994), Jambor and Vaughan (1990) and Cabri and Vaughan (1998) all present much information on the ore microscopy and ore petrography of sulfides. Routine ore microscopy involves qualitative observation of a range of properties, and this is often sufficient for the trained observer to identify phases (with techniques such as X-ray powder diffraction and electron probe microanalysis being used to provide confirmation and more detailed characterization). Certain key properties observed qualitatively using the ore microscope can also be measured using a suitably equipped microscope. Quantitative ore microscopy data (reflectance, quantitative color values, indentation microhardness values) for sulfides are provided by Criddle and Stanley (1993).

THE PRESENT TEXT: SCOPE AND CONTENT

The main objective of the present text is to provide an up-to-date review of sulfide mineralogy and geochemistry, chiefly covering the areas previously dealt with in the books by Ribbe (1974) and by Vaughan and Craig (1978). The emphasis is, therefore, on such topics as crystal structure and classification, electrical and magnetic properties, spectroscopic studies, chemical bonding, high and low temperature phase relations, thermochemistry, and stable isotope systematics. In the context of this book, emphasis is on metal sulfides *sensu stricto* where only the compounds of sulfur with one or more metals are considered. Where it is appropriate for comparison, there is brief discussion of the selenide or telluride analogs of the metal sulfides. When discussing crystal structures and structural relationships, the *sulfosalt* minerals as well as the sulfides are considered in some detail (see Chapter 2; also for definition of the term "sulfosalt"). However, in other chapters there is only limited discussion of sulfosalts, in part because there is little information available beyond knowledge of chemical composition and crystal structure.

Given the dramatic developments in areas of research that were virtually non-existent at the time of the earlier reviews, major sections have been added here on sulfide mineral surface chemistry and reactivity, formation and transformation of metal-sulfur clusters and nanoparticles, modeling of hydrothermal precipitation, and on sulfides in biosystems. However, it should be emphasized that the growth in the literature on certain aspects of sulfide mineralogy over the past 20 years or so has been such that comprehensive coverage is not possible in a single volume. Thus, the general area of "sulfides in biosystems" is probably worthy of a volume in itself, and "environmental sulfide geochemistry" (including topics such as oxidative breakdown of sulfides) is another area where far more could have been written. In selecting areas for detailed coverage in this volume, we have been mindful of the existence of other relatively recent review volumes, including those in the RiMG series. So, for example, readers looking for more information on mineral-microbe interactions are referred to Banfield and Nealson (1997), or on nanoparticles to Banfield and Navrotsky (2001). It has also been our intention not to cover any aspects of the natural occurrence, textural or paragenetic relationships involving sulfides, given the coverage already provided by Ramdohr (1969, 1980) and by Craig and Vaughan (1994), in particular. This is published information that, although it may be supplemented by new observations, is likely to remain useful for a long period and largely not be superceded by later work.

In the following chapters, the crystal structures, electrical and magnetic properties, spectroscopic studies, chemical bonding, thermochemistry, phase relations, solution chemistry, surface structure and chemistry, hydrothermal precipitation processes, sulfur isotope geochemistry and geobiology of metal sulfides are reviewed. Makovicky (Chapter 2) discusses the crystal structures and structural classification of sulfides and other chalcogenides (including the sulfosalts) in terms of the relationships between structural units. This very comprehensive survey, using a rather different and complementary approach to that used in previous review volumes, shows the great diversity of sulfide structures and the wealth of materials that remain to be characterized in detail. These materials include rare minerals, and synthetic sulfides that may represent as yet undescribed minerals. Pearce, Pattrick and Vaughan (Chapter 3) review the electrical and magnetic properties of sulfides, discussing the importance of this aspect of the sulfides to any understanding of their electronic structures (chemical bonding) and to applications ranging from geophysical prospecting and mineral extraction to geomagnetic and palaeomagnetic studies. Rapidly developing new areas of interest discussed include studies of the distinctive properties of sulfide nanoparticles. Wincott and Vaughan (Chapter 4) then outline the spectroscopic methods employed to study the crystal chemistry and electronic structures of sulfides. These range from UV-visible through infrared and Raman spectroscopies, to X-ray emission, photoemission and absorption, and to nuclear spectroscopies. Chemical bonding (electronic structure) in sulfides is the subject of the following chapter by Vaughan and Rosso (Chapter 5), a topic which draws on knowledge of electrical and magnetic properties and spectroscopic data as experimental input, as well as on a range of rapidly developing computational methods. Attention then turns to the thermochemistry of sulfides in a chapter by Sack and Ebel (Chapter 6) which is followed by discussion of phase equilibria at high temperatures in the review by Fleet (Chapter 7). Sulfides in aqueous systems, with emphasis on solution complexes and clusters, forms the subject matter of the chapter written by Rickard and Luther (Chapter 8). Sulfide mineral surfaces are the focus of the next two chapters, both by Rosso and Vaughan. The first of these chapters (Chapter 9) addresses characterization of the pristine sulfide surface, its structure and chemistry; the second (Chapter 10) concerns surface reactivity, including redox reactions, sorption phenomena, and the catalytic activity of sulfide surfaces. Reed and Palandri (Chapter 11) show in the next chapter how much can now be achieved in attempting to predict processes of sulfide precipitation in hydrothermal systems. The final chapters deal with two distinctive areas of sulfide mineralogy and geochemistry. Seal (Chapter 12) presents a comprehensive account of the theory and applications of sulfur isotope geochemistry; sulfur isotope fractionation can provide the key to understanding the natural processes of formation of sulfide deposits. In the final chapter, Posfai and Dunin-Borkowski (Chapter 13) review the rapidly developing area of sulfides in biosystems, discussing aspects of both sulfide mineral-microbe interactions and biomineralization processes involving sulfides.

It is inherent in multi-author compilations such as this volume that some relevant topics have been given relatively little attention, or even omitted, and there is also occasional duplication of material between chapters. In effect the reader should view each of the chapters as a "stand alone" account, albeit complemented by the other chapters. It is hoped that readers will appreciate the perspectives offered by the authors in approaching their material from very different standpoints. Having said this, every effort has been made to provide adequate cross referencing between chapters. As noted above, no attempt has been made to review the more "petrological" aspects of sulfide research (natural occurrence, textural and paragenetic relationships of sulfides in rocks and ores). It had been hoped to include more discussion of certain topics, such as the formation and transformation of metal sulfide precipitates, but in this and a number of other areas, that will have to await a future volume. Our hope is that despite its shortcomings, which are entirely the responsibility of the volume editor, this book will be a fitting successor to the 1974 "Sulfide Mineralogy" volume, the book which launched the whole "Reviews in Mineralogy and Geochemistry" series.

REFERENCES

Banfield JF, Navrotsky A (eds) (2001) Nanoparticles and the Environment. Rev Mineral Geochem, Vol. 44. Mineralogical Society of America

Banfield JF, Nealson KH (eds) (1997) Geomicrobiology: Interactions between Microbes and Minerals. Rev Mineral Geochem, Vol. 35. Mineralogical Society of America

Barnes HL (ed) (1979, 1997) Geochemistry of Hydrothermal Ore Deposits.(2nd and 3rd edition) Wiley-Interscience

Bernede JC, Pouzet J, Gourmelon E, Hadouda H (1999) Recent studies on photovoltaic thin films of binary compounds. Synth Met 99:45-52

Bowles JFW, Vaughan DJ (2006) Oxides and Sulphides. Vol. 5A. Rock-Forming Minerals (Deer, Howie, Zussman) Geol Soc London Publication (in press)

Cabri LJ, Vaughan DJ (1998) Modern Approaches to Ore and Environmental Mineralogy. Min Assoc Canada Short Course, Vol. 27, 421 pp.

Cotter-Howells J, Campbell L, Batchelder M, Valsami-Jones E (eds) (2000) Environmental Mineralogy: Microbial Interactions, Anthropogenic Influences, Contaminated Land and Waste Management. Mineral Soc Monog 9

Craig JR, Vaughan DJ (1994) Ore Microscopy and Ore Petrography. (2nd edition) Wiley-Interscience

Criddle AJ, Stanley CJ (eds) 1993) Quantitative Data File for Ore Minerals. Chapman and Hall

Fuhs W, Klenk R (1998) Thin film solar cells. EUR 18656. World Conference on Photovoltaic Solar Energy Conversion 1:381-386

Garrels RM, Christ CL (1965) Solutions, Minerals and Equilibria. Harper and Row

Hazen RM (ed) (2005) Genesis: Rocks, Minerals and the Geochemical Origin of Life. Elements 1:135-161

Jambor JL, Blowes DW (eds) (1994) The Environmental Geochemistry of Sulfide Minewastes. Mineral Assoc Canada Short Course Handbook, Vol. 22

Jambor JL, Vaughan DJ (eds) (1990) Advanced Microscopic Studies of Ore Minerals. Mineral Assoc Canada Short Course, Vol.17

Jambor JL, Blowes DW, Ritchie AIM (eds) (2003) Environmental Aspects of Minewastes. Mineral Assoc Canada Short Course, Vol. 31

Kostov I, Mincheeva-Stefanova J (1981) Sulphide Minerals: Crystal Chemistry, Parageneses and Systematics. Bulgarian Academy of Sciences

Mills KC (1974) Thermodynamic Data for Inorganic Sulphides, Selenides and Tellurides. Butterworth

Ramdohr P (1969, 1980) The Ore Minerals and their Intergrowths. (1st and 2nd English Editions) Pergamon

Ribbe PH (ed) (1974) Sulfide Mineralogy. Mineral Soc Am Short Course Notes. Vol. 1. Mineralogical Society of America

Russell MJ, Hall A (1997). The emergence of life from iron monosulphide bubbles at a submarine hydrothermal redox and pH front. J Geol Soc Lond 154:377-402

Shuey RT (1975) Semiconducting Ore Minerals. Devel in Econ Geol. No.4, Elsevier

Stanton RL (1972) Ore Petrology. McGraw Hill

Trindade T, O'Brien P, Pickett NL (2001) Nanocrystalline semiconductors: synthesis properties, and perspectives. Chem Mater 13:3843-3858

Uytenbogaart W, Burke EAJ (1971) Tables for Microscopic Identification of Ore Minerals. Elsevier

Vaughan DJ, Craig JR (1978) Mineral Chemistry of Metal Sulfides. Cambridge Univ Press, Cambridge [freely available electronically at: *http://www.manchester.ac.uk/digitallibrary/books/vaughan*]

Vaughan DJ, Lennie AR (1991) The iron sulphide minerals: their chemistry and role in nature. Science Progr Edinburgh 75: 371-388

Vaughan DJ, Wogelius RA (2000) Environmental Mineralogy. Notes in Mineralogy Vol. 2. European Mineral Union

Crystal Structures of Sulfides and Other Chalcogenides

Emil Makovicky
*Geological Institute
University of Copenhagen
Øster Voldgade 10, DK1350 Copenhagen, Denmark
e-mail: emilm@geol.ku.dk*

INTRODUCTION

In this chapter, the crystal structures of sulfides and related chalcogenide compounds (involving Se, Te, As and Sb as well as S) are reviewed with emphasis on mineral phases but with some discussion of related synthetic compounds. Before presenting a systematic account of the principal families, relevant crystal chemical concepts are discussed, and structural classification schemes for these phases are reviewed.

Crystal chemical concepts

Chalcogenide ions. Chalcogens differ from oxygen by their moderate electronegativity. The Pauling electronegativity of oxygen is 3.44, whereas that of S is 2.58, Se 2.55, and Te 2.1 (Emsley 1994). The Shannon crystal radius of O^{2-} is 1.40 Å, that of S^{2-} 1.70 Å, Se^{2-} 1.84 Å, and Te^{2-} 2.07 Å (Shannon 1981). We have estimated the radius of As^{3-} as about 1.78 Å, and that of Sb^{3-} as 1.89 Å, based on the structures of selected sulfide-pnictides. The difference in polarizability of chalcogens and oxygen is of great structural importance. The electric polarizability (i.e., the value expressing the displacement of the electron cloud relative to the nucleus) is 3.88×10^{-24} cm^3 for O^{2-}, but it is 10.2×10^{-24} cm^3 for S^{2-}, 10.5×10^{-24} cm^3 for Se^{2-}, and 14.0×10^{-24} cm^3 for Te^{2-} (Pauling 1927). The polarizability values can also be interpreted as expressing covalency trends of anion-cation bonds and anion-anion interactions in the structure.

The difference between oxygen and chalcogens leads to differences in crystal chemistry of oxides and chalcogenides for nearly all element combinations, exceptions being very few. Structural features of certain chalcogenide categories are in common with hydroxides, which have low-charge and polarizable OH^- groups; this is true especially for layer-like structures. There are greater similarities between the above values for S and Se, and a more pronounced difference for Te. However, already the differences between the behaviour of S and Se, and presumably also the possibilities which the larger radius of Se offers for certain cation-cation distances in selenides, lead to important differences between phase diagrams of sulfides and selenides.

With the notable exception of chalcogenides of typical non-metals, a fundamental crystal-chemical division of chalcogenide structures can be based on the difference in electronegativity and bonding schemes between "cations" and "anions," and on Pearson's valence rule,

$$\Sigma/N = 8, \quad \text{where } \Sigma = n_e - b_a - b_c$$

In this expression, n_e is the total number of valence electrons, b_a the number of electrons involved in anion-anion bonds, and b_c number of electrons involved in cation-cation bonds plus those left unshared on cations (i.e., constituting lone electron pairs). In polyanionic compounds, $n_e/N < 8$, in normal-valence compounds, $n_e/N = 8$, and in polycationic compounds, $n_e/N > 8$. As, Sb, and Bi may play either the role of an anion or that of a cation; details are in the section on structures with dichalcogenide groups X_2. Cation-cation interactions may affect the $M:X$ ratio as, e.g., in pentlandite (M_9S_8), but a range of weaker, although obvious cation-cation interactions occur, especially in the chalcogenides of Cu and Ag, which do not alter their stoichiometry. Typically, the interaction distances in these compounds exceed those in a metal; see discussion by Chevrel (1992).

Packing of anions. Another fundamental crystal-chemical principle relevant to the chalcogenides is that of "close packing of anions" (eutaxy). This is an arrangement in which the shortest distances between pairs of anions are as large as possible (O'Keeffe and Hyde 1996). The result, however, is the densest packing of atoms, corresponding to that seen in metals. To the cubic close packing (*ccp*) and the hexagonal close packing (*hcp*) of anions, and their numerous combinations (O'Keeffe and Hyde 1996), is added a body-centered close packing, which has a lesser space-filling efficiency. It is generated, for example, by the steric requirements of silver cations in Ag_2X, especially at elevated temperatures. The third most efficient mode of packing, the icosahedral configuration, cannot propagate as a 3D periodic structure and can occur only as a cluster in other, especially cubic, structures.

It is not necessary that an entire structure exhibits a continuous pattern of close packing. In a large number of sulfosalts, and in some sulfides, layers, rods or even blocks of a close-packed or a distorted close-packed arrangement occur, joined to the adjacent such layers, rods or blocks, by interfaces on which certain cuts through the close-packed arrangement of anions face quite differently oriented cuts through (mostly) the same arrangement. For example, the pseudotetragonal (100) cuts through a PbS archetype (see below) often face (111) cuts through the same archetype. The keys to this arrangement are suitable cations, or lone electron pair cations, capable of spanning and populating this interface.

Cation coordination. The preferred coordination of a cation in a chalcogenide structure depends on its size, charge and electron configuration. The cations with a "noble gas" (sp^3) configuration have coordinations according to their radius. The appropriate "crystal radii" for sulfides were derived by Shannon (1981) and appear to serve well in most instances. A necessary addition to his table is the tetrahedral radius of Cu^{2+}, equal to ~0.51 Å (Makovicky and Karup-Møller 1994), as opposed to the 0.635 Å given for Cu^+ in sulfides by Shannon (1981). Cations with d^{10} configurations prefer tetrahedral coordinations; the same is true for transition metals with maximum valencies. For spin-paired d^8 configurations, such as in Pd^{2+} and Pt^{2+}, square-planar coordinations occur, whereas for such configurations of Ni^{2+}, a square-pyramidal coordination is typical (Jellinek 1968).

The same picture is obtained using the hybridization concept (e.g., Belov et al. 1982). Tetrahedral coordination is explained by the use of sp^3 hybrid orbitals, octahedral by the use of d^2sp^3 orbitals, square planar is based on dsp^2 orbitals, triangular is based on the sp^2 hybridization (e.g., Cu^+), and the linear coordination of Cu^+ and Ag^+ results from an sp hybrid. These concepts are summarized in Table 1. Other crystal chemical concepts relevant to sulfides have been outlined by Prewitt and Rajamani (1974), Vaughan and Craig (1978), and Vaughan and Wright (1998) and are discussed elsewhere in this volume.

Jahn-Teller distortions due to the differential occupancy of individual d orbitals are unusual among sulfides because of electron delocalization (Jellinek 1968) but they do explain the configurations found in Cr^{2+}, Pd^{2+} and Pt^{2+} sulfides.

The coordination properties of the cations need to find their counterpart in the anion packing. The majority of the cations in natural chalcogenides have coordination requirements that fit with the filling of tetrahedral, octahedral or other interstices in the anion arrays. Thus, *ccp* and *hcp* offer coordination octahedra and regular tetrahedra, as well as a plethora of

Table 1. Coordination numbers and hybridization types for the cations commonly occurring in chalcogenides.

CN	Type of Hybridization	Category
2	$sp;\ p^2$	$Cu^+, Cu^{2+}, Ag^+, Hg^{2+}$
3	sp^2	Cu^+, Ag^+
3	p^3	$As^+, Sb^{3+}, Bi^{3+}, Pb^{2+}$
4	sp^3	$Cu^+, Ag^+, Au^+; Zn^{2+}, Cd^{2+}, Hg^{2+}; Ga^{3+}, In^{3+}, Tl^{3+}; Ge^{4+}, Sn^{4+}; As^{5+}, Sb^{5+}; Mn_H^{2+}, Fe_H^{2+}$
4 (sq)	dsp^2	$Fe_H^{3+}, Co_H^{3+}, Ni_H^{2+}, Ni_H^{3+}, Pd^{2+}, Pt^{2+}, Cu^{2+}$
5	dsp^3	Ni_L^{2+}
	p^3d^2	$Sb^{3+}, Bi^{3+}; Pb^{2+}$
6	d^2sp^3	$Fe_L^{2+}, Fe_L^{3+}, Co_L^{3+}, Ni_L^{4+}, Pt_L^{4+}$
	sp^3d^2	Sn^{4+}
	$p^3;\ d^2sp^3$	$Pb^{2+}; Bi^{3+}\ (Sb^{3+})$

triangular coordinations at different structure sites. They preclude quadratic coordinations, whereas linear coordinations, such as S-Cu-S require a distortion of the anion array. Some coordinations, such as the trigonal prismatic coordination of the early $4d$ or $5d$ elements (Mo, W, Nb, Ta) change anion stacking schemes; the van der Waals intervals of layered structures tend to preserve them. The lone electron pair elements, As^{3+}, Sb^{3+}, Bi^{3+}, Sn^{2+}, Ge^{2+}, and to a lesser extent Tl^+ and Pb^{2+}, require an excessive, usually asymmetrically situated structural space for their non-bonding s^2 pair. This volume is not comparable with that of the anion, which is the case in the oxides, with the relatively smaller O^{2-} (Andersson et al. 1973). The degree of stereochemical activity of the lone pair varies with the species (decreasing with increasing Z) and the structure type.

Structural classification

Structural classifications generally employ a combination of the above criteria: the large scale structural configuration (3D/2D/1D character of principal building units), anion packing, bonding characteristics/type of coordination polyhedra of cations, and cation-cation and anion-anion interactions. These factors are used indirectly in the classification of Strunz and Nickel (2001) in which chalcogenides are divided into categories with M:S > 1:1, M:S = 1:1 (and derivatives), M:S = 3:4 and 2:3, and M:S < 1:2. Sulfosalts are classified separately, mainly according to their archetypes. Ross (1957) follows similar principles. Kostov and Minčeva-Stefanova (1982) divide sulfide- and sulfosalt structures into axial, planar, pseudoisometric and isometric types, using the axial ratios of their lattices as an expression for the anisotropy of strongest bonds in the structure. The classification of Povarennykh (1972) is similar, although more "classical" in the choice of categories. Takéuchi (1970) and Parthé (1990) divide chalcogenides into polyanionic, polycationic and normal-valence compounds based on Pearson's valence rule. Belov et al. (1982) discern normal-, cluster-, molecular-, and polyanionic chalcogenides, as well as distinct cation coordinations in the compound. Vaughan and Craig (1978), Robert and Makovicky (1984) and Vaughan and Wright (1998) define chalcogenide groups based on well-known structure types, each of them unifying a number of the above defined structural criteria, in which they include a number of iso-, homeo-, and poly-types as well as interstitial and omission derivatives, eventually even homologues and plesiotypes. An updated overview of the most important groups is given in Table 2.

Table 2. Selection of important sulfide structure families.

Type compound	Principal structural characteristics: Anion array; % of holes filled; stacking; specific properties	Selected isotypes, derivatives and related structures
Sphalerite β-ZnS	ccp, ½ of tetrahedral holes filled, ABC sequence of corner-connected [ZnS$_4$] tetrahedra	CdS, HgTe, GaAs, CuAsS, CuFeS$_2$, Cu$_6$As$_4$S$_9$, Cu$_4$FeS$_8$
Wurtzite α-ZnS	hcp, ½ of tetrahedral holes filled, ABAB sequence of corner-connected [ZnS$_4$] tetrahedra	CdS, ZnTe, SiC, Cu$_3$AsS$_4$, ZnAl$_2$S$_4$, AgInS$_2$, CuFe$_2$S$_3$
Galena PbS	ccp, octahedral holes filled, isotypic with NaCl	MgS, CaS, MnS, PbSe, ZrS, AgSbS$_2$, PbAgAsS$_3$
Niccolite NiAs	hcp, octahedral holes filled, anion has trigonal prismatic coord., cation omission derivatives	FeSe, FeSb, NiS, PtSn, FeS, Fe$_{1-x}$S: from Fe$_{11}$S$_{12}$ to Fe$_7$S$_8$, Cr$_{1-x}$S
Pyrite FeS$_2$	S$_2$ groups instead of anions in ccp, octahedral cation coordination	CoS$_2$, NiS$_2$, MnS$_2$, PtAs$_2$, CoAsS, PtAsS, Cu$_3$FeS$_8$
Marcasite FeS$_2$	dimorph of pyrite, S$_2$ groups and Fe octahedra, distorted hcp	FeSe$_2$, FeAs$_2$, NiSb$_2$, RuSb$_2$, FeAsS, CoAsS, CoSbS, OsAsS
Thiospinels FeNi$_2$S$_4$	ccp, 1/8 of tetr. + ½ of oct. holes filled ; sulphur analogues of normal and inverse spinels	Co$_3$S$_4$, Ni$_3$S$_4$, FeIn$_2$S$_4$, CuCo$_2$S$_4$, ZnAl$_2$S$_4$, HgCr$_2$S$_4$
Pentlandite (Fe, Ni)$_9$S$_8$	ccp, ½ of tetr. + 1/8 of oct. holes filled;8 edge-sharing tetrahedra with M-M bonds	Co$_9$S$_8$, Ag (Fe,Ni)$_8$S$_8$, K$_6$Na(Fe, Cu, Ni)$_{24}$S$_{26}$Cl
Acanthite – Argentite Ag$_2$S	(distorted) bcc , Ag in oct. + ½ of tetr. holes; above 177 °C ionic conductor: Ag mobile	Ag$_2$Se, Ag$_2$Te, ionic conductors Ag$_8$GeSe$_6$, Cu$_2$S, Cu$_9$S$_5$, Cu$_5$FeS$_4$
Mackinawite FeS	*tetragonal layers composed of edge-sharing [FeS$_4$] tetrahedra; 4 M-M bonds per cation	CuTe, TlFeCu$_3$S$_4$, TlFe$_2$S$_2$, KCu$_4$S$_3$, Tl (Cu, Fe) $_{6.35}$ SbS$_4$, FeCuS$_2$, 1.35 [Mg$_{0.666}$Al$_{0.33}$ (OH)$_2$]
Berndtite SnS$_2$	*layers composed of edge-sharing [SnS$_6$] octahedra; the CdI$_2$ (i.e. Cd(OH)$_2$) layer type	NiTe$_2$, TiS$_2$, PtS$_2$, LiCrS$_2$, NaCrS$_2$, AgCrS$_2$, K$_x$TiS$_2$, Ti$_5$S$_8$
Molybdenite MoS$_2$	* layers composed of edge-sharing trigonal [MoS$_6$] prisms; ABAB and ABCABC variants in layer stacking	WS$_2$, NbS$_2$, TaS$_2$, intercalates Ta$_{1+x}$S$_2$, Cu$_x$NbS$_2$, K$_{0.5}$(H$_2$O)$_y$NbS$_2$
Stibnite Sb$_2$S$_3$	A chain structure with quadruple ribbons [Sb$_4$S$_6$]; Sb in trigonal– and square-pyramidal coordination	Bi$_2$S$_3$, Bi$_2$Se$_3$, Sb$_2$S$_2$O. Sn$_2$S$_3$, CuPbBi$_5$S$_9$, CuPbBiS$_3$
synth. KFeS$_2$	A chain structure; single chains of edge-sharing [FeS$_4$] tetrahedra are separated by large cations	TlFeS$_2$, CsFeS$_2$, Ba$_{19}$(FeS$_2$)$_{18}$, Na$_3$Fe$_2$S$_4$, NaFeS$_2$·2H$_2$O
Lillianite Pb$_3$Bi$_2$S$_6$	periodically mirror-twinned slabs of ccp array with octahedral holes filled; trigonal prisms on composition planes; several homologous series	AgPbSbS$_3$, Pb$_6$Bi$_2$S$_9$, Ag$_3$Pb$_8$Bi$_{13}$S$_{30}$; UFeS$_3$, Y$_5$S$_7$, MnY$_2$S$_4$, Cr$_2$Er$_6$S$_{11}$, AgBi$_3$S$_5$, Cu$_{1.6}$Bi$_{4.6}$S$_8$
Cosalite Pb$_2$Bi$_2$S$_5$	zigzag layers of ccp array with oct. holes filled or of SnS-like arrangement; incommensurate interspaces	Pb$_5$In$_9$S$_{17}$, Ce$_{1.3}$Bi$_{3.8}$S$_8$, Pb$_4$Sb$_6$S$_{13}$, FePb$_4$Sb$_6$S$_{14}$
Realgar As$_4$S$_4$	Molecular structure, As–S and As-As pairs or, in As$_4$S$_3$ also As-As triangles	β-As$_4$S$_4$, As$_4$S$_3$, As$_4$S$_5$, [(As,Sb)$_6$S$_9$] [As$_4$S$_5$]
Tetradymite Bi$_2$Te$_2$ S	* double-octahedral layers, anion-anion contacts may be intercalated by Bi-Bi layers; triple-octahedral layers M^{2+}BiTe$_4$ also present	BiTe, Bi$_3$Se$_2$Te, Bi$_2$Te$_3$, Bi$_2$Se$_3$, Bi$_4$Se$_3$, PbBi$_4$Te$_7$, PbBi$_2$Te$_4$
Crookesite TlCu$_7$Se$_4$	a channel structure, Se–Cu and Cu-Cu bonds in channel walls, Tl in channels. Other channel str.: octahedral and square-planar walls	NH$_4$Cu$_7$S$_4$, TeCu$_3$Se$_2$, Rb$_3$Cu$_8$Se$_6$, channel structures K$_2$Hg$_6$S$_7$; K$_{0.3}$Ti$_3$S$_4$, KC$_5$Se$_8$, CsBi$_3$S$_5$

* A layer structure, layers held together by van der Waals forces. When layers become electrically charged, interstitial or intercalation derivatives are formed.
hcp – hexagonal close packing; bcc – body-centered cubic packing; ccp - cubic close packing

The approach adopted in this contribution will highlight a prominent structural principle for each category, sometimes splitting some of the classical description groups, such as sulfosalts. The following divisions are adopted:

Fibrous sulfides:
- with FeS$_4$ tetrahedra
- other cases

Layer-like structures:
Monochalcogenides and related
- mackinawite, its intercalates and derivatives
- covellite merotypes
- valleriite

Sesquichalcogenides
Dichalcogenides
Derivatives with two kinds of chalcogenide layers
- cannizzarite and derivatives
- layer misfit structures

Structures with three-dimensional frameworks:
- tetrahedral *ccp, hcp* and combined "adamantane" structures
- pentlandite group and other structures with tetrahedral clusters
- octahedral *ccp* and *hcp* structures and derivatives
- thiospinels
- dichalcogenides with X_2 groups

Channel- and cage structures

Selected cation-specific structures:
- Ni chalcogenides
- Pd and Pt chalcogenides
- ion conductors and other Cu, Ag chalcogenides
- compounds of lone electron pair cations

Selected larger sulfosalt families

A more detailed account of the concepts involved in this classification is given below. At present, often a number of more or less accurate and complete crystal structure determinations have been made on all important sulfide minerals, investigating not only the structure but also specific substitutions, and structural changes under non-ambient conditions.

For example, in the case of tetrahedrite-tennantite, the pioneering structure determination by Machatschki (1928) was followed by Pauling and Neuman (1934) whose results were refined by Wuensch (1964) and Wuensch et al. (1966). Kalbskopf (1971, 1972) determined structures of Hg-rich and Ag-rich tetrahedrite, respectively. Crystal structures of the natural members of the series, with variable types of mixed substitution, were also determined by Kaplunnik et al. (1980), Johnson and Burnham (1985), as well as Peterson and Miller (1986) and Karanović et al. (2003). Structures of the synthetic, pure copper tetrahedrite varieties were studied by Makovicky and Skinner (1979), Pfitzner (1997) and Pfitzner et al. (1997). The crystal structure of a natural copper-rich tennantite was determined by Makovicky et al. (2005), whereas those of the tellurian members of the tetrahedrite-tennantite solid solution by Kalbskopf (1974), Dmitriyeva et al. (1987) and Pohl et al. (1996). These studies were joined by others, in which relationships were determined between element substitutions and unit-cell parameters (e.g., Hall 1972; Johnson et al. 1987; Makovicky and Karup-Møller 1994; Karup-Møller and Makovicky 2003, 2004). Structurally important Mössbauer and other studies of the valence of iron/copper in these minerals were performed as well (e.g., Vaughan and Burns 1972; Charnock et al. 1989; Makovicky et al. 1990, 2003; Di Benedetto et al. 2002, 2005).

As we can see from this example, it is not even remotely possible to treat, or even quote, all the relevant studies in this review. Therefore, preference will be given to recent, accurate structure determinations and, with our deepest apologies to all those other authors, the reader is advised to consult the half-yearly updated Inorganic Crystal Structure Database (ICSD) released by the Fachinformationszentrum Karlsruhe (Germany)/National Institute of Standards and Technology (U.S. Dept. of Commerce) or the chalcogenide file by Matsushita (2005), and similar sources, where a complete picture of structural knowledge can be obtained for a particular compound or family.

PRINCIPLES OF THE DESCRIPTION AND CLASSIFICATION OF CRYSTAL STRUCTURES RELEVANT TO SULFIDES: A BRIEF OUTLINE

The present section gives a brief outline of principal classification criteria used for inorganic structures. Modular and non-modular (i.e., polyhedral) crystal chemistry are interrelated by so-called configuration levels of structure description, to be found in Makovicky (1997b) and Ferraris et al. (2004).

Non-modular categories of similarity

Definition of *isotypy* and *homeotypy* (Lima-de-Faria et al. 1990) is the most important point in non-modular description.

Two structures are *crystal-chemically isotypic* if they (a) are isoconfigurational, i.e., the *entire configurations* of the two structures are similar: axial ratios, unit-cell angles, values of adjustable x, y, z parameters for corresponding atoms and their coordination polyhedra and (b) the corresponding atoms and corresponding bonds (interactions) have similar physical and chemical characteristics (e.g., electronegativities, radius ratios, electronic states, or bond-strength distribution). Without point (b) they are only configurational isotypes. There are no *a priori* limits on geometric, chemical or physical similarity in this definition. These limits will change according to the type of compounds studied, the physical or chemical properties we concentrate upon, and the purpose of study.

Two structures are *homeotypic* if one or more of the following conditions required for isotypism is relaxed:

(a) instead of an identical (or enantiomorphic) space group they can have a subgroup or a supergroup;

(b) limited variation in axial ratios, interaxial angles, x, y, z values and coordination properties is allowed;

(c) site occupancy limits are relaxed allowing given sites to be occupied by different atomic species in an ordered fashion (splitting of the original Wyckoff position due to (a)).

Bergerhoff et al. (1999) include interstitial derivatives among homeotypes.

Examples are numerous: among site-ordering variants (and omission and interstitial variants) we can quote C (diamond), ZnS (sphalerite), $CuFeS_2$ (chalcopyrite) or Cu_3SbS_4 (famatinite) and also a variety of "adamantane" structures; among distortion derivatives, $PdSe_2$ vs. parent FeS_2. The structures of bismuthinite-aikinite series are a series of homeotypic structures. The ideal, undistorted structure is called an aristotype and the derived distorted structures hettotypes (Megaw 1973).

Modular categories of structural similarity

Structure building operators. Modular structures are composed of modules/fragments (blocks, rods or layers) of simpler, archetypal structures that are recombined in various orientations by the action of structure building operators. The structure-building operators are (mostly planar) defects which produce a definite geometric relationship between the structural portions they join; they are connected with localised chemical (crystal-chemical) changes in the structure.

In the majority of cases, they are various types of unit-cell twinning (Andersson and Hyde 1974). Takéuchi (1997) uses the term chemical twinning for this type of operation. The most frequent type is reflection- (i.e., mirror-reflection) twinning that acts either on a full set of atoms (Andersson and Hyde 1974) or only on a partial set of atoms (Takéuchi 1978). Mirror-reflection twinning is usually connected with the creation of a new type of coordination polyhedron straddling the mirror plane. Examples are the capped trigonal coordination prisms of two distinct kinds on composition planes of reflection twinning of ccp and hcp arrays, respectively (illustrated by the lillianite homologues and the wittichenite group). The next types are glide-reflection twinning (Andersson and Hyde 1974) and cyclic twinning (Hyde et al. 1974). For complex sulfides, noncommensurability between adjacent building blocks comes next in order of importance (Makovicky and Hyde 1981, 1992). It means that the substructures in two adjacent blocks meet according to a vernier principle across the block interface (consult cannizzarite and rod-based structures of sulfosalts). Further categories are antiphase and out-of-phase boundaries, crystallographic shear, the $t/2$-shear (or slip) in the structures with a short pronounced t period (mostly, $t \sim 4$ Å), and the intergrowth of two different structure types on a unit cell scale (Hyde et al. 1974).

The definition of unit cell twinning implies appreciable configurational and chemical changes on the twin composition planes, unlike that of classical twinning. The coordination states and, often, the chemical species of atoms on the block surfaces or in the interfaces differ from those inside the building blocks. The structure-building operators, especially reflection twinning, can apply either to all the details of the structure, including the chemical species, or only to its general topology (or even only to its anion array), something akin to the difference between isotypes and homeotypes.

Archetypes used in the modular description are those simpler structures that display/encompass all fundamental bonding and geometric properties of the structure portions in the interior of building blocks/moduli. For sulfides they are, for example, PbS, SnS and ZnS. Modular structures can be monoarchetypal or diarchetypal (or even triarchetypal, etc...) with distinct types of blocks based on one, two (three, etc...) different archetypes. Many sulfides are monoarchetypal structures (lillianite homologous series, the jordanite-kirkiite pair, rod-based complex sulfides). Diarchetypal examples include complex sulfides of the kobellite homologous series in which blocks of PbS and SnS archetype combine.

Accretional homologous series/polysomatic series. The accretional series is a series of structures in which the type(s) and general shapes of building blocks (rods, layers) as well as the principles that define their mutual relationships (the recombination operators) remain preserved, but the size of these blocks varies incrementally by varying the number of fundamental coordination polyhedra in them in an exactly defined way. The order N of a homologue in this type of series can be defined by the number of coordination polyhedra (polyhedral layers) across a suitably defined diameter of the building block (rod, layer) (see Fig. 1). Every member of the accretional series has its own chemical formula, unit cell parameters and symmetry; a general chemical formula can be devised for the entire series. A given homologue can represent a single compound or a (dis)continuous solid solution which can exsolve into a series of structurally ordered phases with well defined compositions (or compositional ranges) but with the same N. The ideal space group can then be reduced to subgroups in the process of (cation) ordering. It should be stressed that all these ordered compounds are members of the homologous series (Ferraris et al. 2004). The best example is the lillianite homologous series of (primarily) sulfosalts.

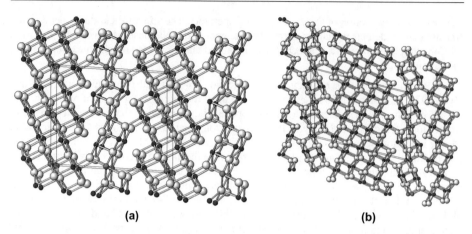

Figure 1. Pavonite (a) and benjaminite (b), the $N = 5$ and 7 members of the pavonite homologous (accretional) series (Makovicky and Mumme 1979). White spheres: S sites, grey spheres: Ag sites, dark spheres: pure Bi and predominantly Bi sites.

Thompson (1978) described accretional homologous series using a different name, that of a polysomatic series and a different approach: the coordination polyhedra on the surfaces and in the interfaces of the blocks (in his case, layers) are treated as one type of (layer) module whereas the incrementally accreting polyhedral layers in the layer (block) interior as another type of layer modules. Thus, all accretional homologues are treated as ordered intergrowths of (usually) two structure types which occur in different proportions in different homologues (polysomes). The two descriptions are equally valid. Takéuchi (1978) denotes the principle behind homologous series as tropochemical twinning.

Some accretional series, for example the lillianite and the pavonite homologues, are extensive series with N varying over a range of values. Besides the members with equal widths ($N_1 = N_2$) across the unit-cell twinning plane or an interface of another kind, those with unequal widths ($N_1 \neq N_2$) can occur. A number of accretional series are limited to only pairs of homologues (N_1, and $N_2 = N_1 + 1$) because of various local or global crystal chemical reasons (e.g., the pair bertrandite – hutchinsonite of Tl-(Pb)-As sulfosalts). These pairs can be extended into combinatorial series, the members of which represent regular intergrowths of the above two accretional homologues: $N_1N_2N_1N_2...$, $N_1N_1N_2N_1N_1N_2...$, $N_1N_2N_2N_2N_1N_2N_2N_2...$, etc., as in the sartorite homologous series or in the MnS-Y_2S_3 system. In some instances, e.g., for the cuprobismutite homologues, only one intermediate member, $N_1N_2N_1N_2...$, is known (Ferraris et al. 2004).

Chemical-composition series. A series of compositions combining formally two sulfide compositions A_mX_n-B_oX_p usually is not a single homologous (polysomatic) series but it may contain one or two homologous series and/or several structurally unrelated compounds. For example, the Cu_2S-Bi_2S_3 "series" contains wittichenite, as well as (idealized) members of the cuprobismutite and the pavonite homologous series.

Variable-fit homologous series and series with a combined character. Variable fit homologous series occur in crystal structures that are composed of two kinds of alternating, mutually noncommensurate layers (rarely of columns or of a matrix/infilling combination). Each kind of layer has its own short-range (sub)periodicity and it takes m periods of one layer and n periods of the other layer before they meet in the same configuration as at the origin. These are so-called semi-commensurate cases for which the two matching (sub)periodicities of the two layers comprise a ratio of two not very large integers. Besides them, incommensurate cases exist for which these short-range periodicities comprise irrational fractions or a ratio of very large integers (these two cases are practically indistinguishable) (Fig. 2a). Noncommensurability of layers may occur in one or two interplanar directions. With minor compositional changes in the ratio or type of cations, the m/n ratio and the M/S ratio vary within certain, rather narrow limits, leading to a "variable-fit" series of closely related compounds (Makovicky and Hyde 1981, 1992). Incommensurate layer structures are also known as vernier compounds (Hyde et al. 1974) and layer-misfit compounds. The best examples are cannizzarite, the cylindrite family and the synthetic sulfide layer-misfit structures. The "infinitely adaptive structures" of the $Ba_{1+x}Fe_2S_4$ series belong here as well; the two matching elements in them are structural rods rather than layers.

The cases which represent a combination of the accretional (separately for each layer) and variable-fit (for the interlayer match) *principles* are discussed in connection with the derivatives of cannizzarite (Fig. 2b) and rod-based structures of Pb-Sb(Bi) sulfosalts.

Merotype and plesiotype series express structural relationships that are geometrically less accurate than the relationships between members of a homologous series (Makovicky 1997b; Ferraris et al. 2004). Merotypic and plesiotypic families rank higher than the homeotypic, ordering- or chemical variations between individual real structures.

Merotypic structures (merotypes; meros = part) are composed of alternating layers (blocks). One set of these building layers (blocks) are common to all merotypes (i.e., they are isotypic, homeotypic or they are mutually related via homologous expansion/ contraction) whereas the layers (blocks) of the other set(s) differ for different merotypes. Structures with a layer set missing may exist as well. Examples are the hutchinsonite merotypes, covellite merotypes, and the complete series of layered derivatives of Bi_2Se_3.

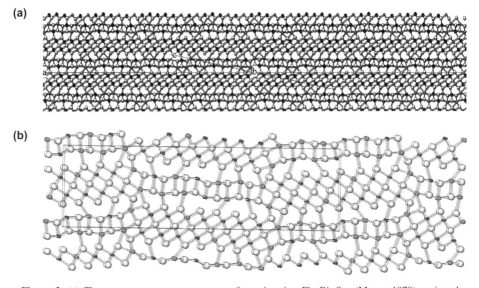

Figure 2. (a) The non-commensurate structure of cannizzarite, $Pb_{46}Bi_{54}S_{127}$ (Matzat 1979) projected upon (010). White circles: S, grey circles: metal atoms (predominant Bi: darker). Double octahedral H layers alternate with two-atoms thick Q, PbS-like layers. A total of 46 primitive Q subcells matches with 27 centered H subcells. (b) Weibullite $Ag_{0.33}Pb_{5.33}Bi_{8.33}(S,Se)_{18}$ (Mumme 1980), a derivative of the cannizzarite structure by means of composition-nonconservative glide planes (100) at $x = ¼$ and $¾$. An example of a structure which combines accretional and variable-fit criteria.

Plesiotypic structures (plesiotypes; plesios = near, close) (a) contain fundamental structural elements (blocks, layers) of the same general type(s), and (b) mutual disposition/ interconnection of these elements in all plesiotypes follows the same general rules. However, unlike the homologous series, (1) the plesiotypic structures may contain additional structural elements that differ from one member of the family to another; and (2) details of fundamental elements, or of their disposition, may differ. For further details, see Ferraris et al. (2004). Rod-layer structures of complex Pb-Sb sulfides and of selected Pb-Bi sulfides belong to this category.

Polytypes and polytype categories. Polytypes are modular structures in which modules (usually layers) are of one or two (in rare cases even more) distinct kinds in strictly regular alternation (Ďurovič 1992). The well-defined geometric relations on the interfaces of two adjacent layers do not determine a unique stacking of modules, i.e., an unambiguously defined three-dimensionally periodic structure.

Ideally, distinct polytypes composed of the same kinds of layers should have the same chemical composition. The strict definition of polytypy stresses the chemical (near) identity of different polytypic modifications. Everyday usage extends these strict criteria. Thus, polytypes will then be defined either on a crystal-chemical or on a configurational level. Polytypes with radically different chemical compositions but with the same layer configurations and layer-stacking principles can be grouped in configurational (= heterochemical) polytypic series (Makovicky 1997).

OD polytypes (order-disorder polytypes) are a family of polytypic structures with a uniquely defined geometric relationship between any two adjacent layers, i.e., with only one kind of layer pairs for each combination of adjacent layers present (Dornberger-Schiff 1966; Ďurovič 1992; Merlino 1990). From the point of view of symmetry, they belong to one groupoid family (ibid.). Polytypes can be described as OD-polytypes only when the appropriate number and types of layers have been selected.

In a family of non-OD polytypes, different stacking principles of layers are active in different members of the family or these principles combine in the observed sequences of layers in a single compound. It means that more than one kind of layer pairs are present for given type(s) of unit layers. Besides these "proper polytypes," with configurationally unmodified layers, there is a spectrum of polytype-like cases (commonly described as polytypes in the literature), in which configurational modifications occur; these were put in the "improper polytype" category with several subdivisions by Makovicky (1997b) and Ferraris et al. (2004). Among our compounds, e.g., the pair stibnite Sb_2S_3 - pääkkönenite Sb_2S_2As (Bonazzi et al. 1995) (see Fig. 3a,b) belong to this category.

CHALCOGENIDES WITH PRONOUNCED ONE-DIMENSIONAL BUILDING UNITS

Fibrous sulfides with chains of edge-sharing FeS_4 tetrahedra

Crystal structures of this plesiotypic family consist of isolated chains of FeS_4 tetrahedra which share edges perpendicular to the chain extension giving, in principle, a unit cell dimension equal to twice the "tetragonal" height of a tetrahedron. These chains are separated by insertion of large, more complex polyhedra of alkali and alkaline earth cations. Structural water is exceptional, compensating for the small size of interstitial sodium. This family contains several minerals (Table 3).

Stoichiometry of these compounds suggests on the one hand trivalent iron, as in $KFeS_2$, $KFeSe_2$ and $BaFe_2S_4$ (Table 3, Fig. 4a), or, on the other hand, mixed-valence iron: 2.5+ in $K_3Fe_2S_4$ and $K_3Fe_2Se_4$, 2.9+ in $Ba_5Fe_9S_{18}$ (Table 3). The three trivalent cases quoted have

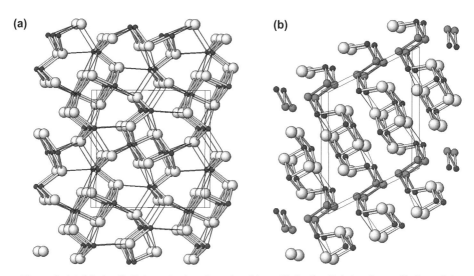

Figure 3. (a) Stibnite Sb_2S_3 in projection along the ribbons Sb_4S_6. Small dark spheres: Sb, large light spheres: S. (b) Pääkkönenite Sb_2S_2As (Bonazzi et al. 1995) with ribbons $Sb_4S_4As_2$; intermediate grey spheres: As-As pairs. (a-b): a pair of pseudopolytypes, with unit layers (100) for stibnite and (10$\bar{1}$) for pääkkönenite. Note the fundamentally different interconnection of unit ribbons in the two structures.

Table 3. Fibrous sulfides with chains of edge-sharing FeS_4 tetrahedra

Compound	Chains	Lattice Parameters				Space Group	Ref.
		a (Å)	b (Å)	c (Å)	β (°)		
$Ba_5Fe_9S_{18}$	single	7.776	7.776	49.860		$P4/ncc$	[1]
$Ba_9Fe_{16}S_{32}$	single	7.776	7.776	44.409		$P4/mnc$	[2]
$\alpha BaFe_2S_4$	single	8.111	8.111	5.59		$I4/mcm$	[3]
$BaFe_2Se_3$	double ∥b	11.878	5.447	9.160		$Pnma$	[4]
$Cs_3Fe_2S_4$	single	7.540	11.168	12.923		$Pnma$	[5]
$CsFeS_2$	single	7.126	11.945	5.420		$Immm$	[6]
$K_3Fe_2S_4$	single sinusoidal	7.157	10.989	11.560		$Pnma$	[5]
KFe_2S_4 rasvumite	single	9.049	11.019	5.431		$Cmcm$	[7]
$Na_3Fe_2S_4$	single twisted	6.633	10.675	11.677		$Pnma$	[8]
$NaFeS_2 \cdot 2H_2O$ erdite	single	10.693	9.115	5.507	92.17	$C2/c$	[9]
$Rb_3Fe_2S_4$	single twisted	7.407	11.141	11.997		$Pnma$	[5]
$RbFeS_2$	single	7.223	11.725	5.430	112.0	$C2/c$	[10]
$RbFeSe_2$	single	7.474	12.091	5.662	112.38	$C2/c$	[10]
$KFeSe_2$	single	7.342	11.746	5.629	113.52	$C2/c$	[10]
$Rb_3Fe_2Se_4$	single	7.691	11.403	12.471		$Pnma$	[11]
$K_3Fe_2Se_4$	single sinusoidal	7.433	11.341	12.016		$Pnma$	[12]
$KFeS_2$	single	7.084	11.303	5.394		$C2/c$	[10]
$TlFe_2S_3$ picotpaulite	single	9.083	10.753	5.411		$Cmcm$	[13]

References: [1] Grey 1975; [2] Hoggins and Steinfink 1977; [3] Boller 1978; [4] Hong and Steinfink 1972; [5] Bronger et al. 1995; [6] Bronger and Müller 1980; [7] Clark and Brown 1980; [8] Klepp and Boller 1981; [9] Konnert and Evans 1980; [10] Bronger et al. 1987; [11] Klepp et al. 2000; [12] Bronger et al. 1999; [13] Balić-Žunić, pers. comm.

Fe-S(Se) distances of 2.231 and 2.237 Å, 2.363 Å, and 2.316 Å, respectively. The uniform Fe-Fe distances, 2.700, 2.815, and 2.800 Å, respectively, indicate metal-metal interactions. The cases with mixed valence of iron display (at least) two types of Fe-S and Fe-Fe distances. Thus, $K_3Fe_2S_4$ and $K_3Fe_2Se_4$ display a range of Fe-X distances equal to 2.282-2.345 Å (average 2.309 Å) and 2.402-2.465 (aver. 2.428 Å). The Fe-Fe distances are 2.78 and 2.90 Å in the sulfide, 2.865 and 3.015 Å in the selenide, and a string... 2.80-2.67-2.67-2.80-2.77-2.77-2.90-2.77 Å in $Ba_5Fe_9S_{18}$.

The sulfide phases have chain (sub)periodicities typically equal to 5.34-5.59 Å, selenides 5.63-5.67 Å (only exceptionally less, Table 3). The chain-and-cation arrangements lead to a limited choice of tetragonal, orthorhombic and monoclinic space groups. Straight chains predominate; the $A_3Fe_2S_4$ phases have sinusoidal or twisted chains, from $Na_3Fe_2S_4$ with the smallest cation to $Rb_3Fe_2S_4$ with the largest one. A double chain is observed in $BaFe_2Se_3$ (Hong and Steinfink 1972). Coordinations of large cations vary from 8 in α–$BaFe_2S_4$ (Fig. 4a) and 9 in $BaFe_2Se_3$ to 6 for sodium in $Na_3Fe_2S_4$ (Klepp and Boller 1981) and in erdite $NaFeS_2\cdot 2H_2O$ (Konnert and Evans 1980) (Figs. 4b,c; Table 3). In $NaFeS_2\cdot 2H_2O$, the coordination octahedra of Na comprise ($4H_2O + 2S$).

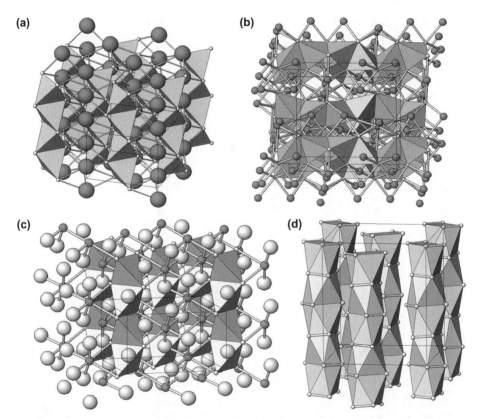

Figure 4. Crystal structures of fibrous sulfides. (a) α–$BaFe_2S_4$ (Boller 1978). Large spheres: Ba; straight chains of edge sharing Fe-S tetrahedra contain Fe^{3+}. (b) $Na_3Fe_2S_4$ (Klepp and Boller 1981). Spheres: Na; folded chains of edge-sharing Fe-S tetrahedra with mixed-valence Fe. (c) $NaFeS_2(H_2O)_2$ (Konnert and Evans 1980) with Na (grey) coordinated by both S (small spheres which participate in tetrahedral chains) and H_2O molecules (large white spheres). (d) Patronite, $V(S_2)_2$ (Kutoglu and Allmann 1972). Chains of "square" VS_8 antiprisms sharing rectangular faces. S-S bonds are indicated by bars.

The variable-fit "infinitely adaptive" structure series, $Ba_{1+x}Fe_2S_4$ (Nakayama et al. 1981), consists of tetrahedral Fe_2S_4 chains and strings of Ba coordination polyhedra which differ in periodicity. It stretches from $Ba_8(Fe_2S_4)_7$ to $Ba_{17}(Fe_2S_4)_{16}$, with a from 7.78 Å to 7.74 Å, and c from 39.03 Å to 82.17 Å (and 104.84 Å for $Ba_{21}(Fe_2S_4)_{19}$, the only phase in which the difference (p-q) in the formula $Ba_p(Fe_2S_4)_q$, in which p,q are believed to be integers, is not one). These phases are rich in Fe^{3+}.

Other fibrous sulfides

An archetypal fibrous structure is that of synthetic SiS_2 (Peters and Krebs 1982) (Table 4), in which single chains of edge-sharing SiS_4 tetrahedra have only van der Waals contacts. Cs_2TiS_3 (Rad and Hoppe 1978) contains yet another type of chain, one composed of square coordination pyramids (TiS_5) sharing two of the edges of the square base with the adjacent pyramids.

The crystal structure of a disulfide of vanadium, patronite $V(S_2)_2$, (Kutoglu and Allmann 1972) consists of isolated chains which are composed of face-sharing distorted (VS_8) tetragonal antiprisms (Fig. 4d). The shared faces are rectangular and not square-shaped, being limited by two parallel S-S covalent pairs each: 2.04 × 3.28 Å and 2.03 × 3.16 Å, respectively. Vanadium – sulfur bonds have the lengths 2.38-2.54 Å. The shortest S-S contacts between chains are 3.24 Å; and along chains are 3.17 Å.

Table 4. Other fibrous structures.

Compound	Space Group	Lattice Parameters				Chain Type	Ref.
		a (Å)	b (Å)	c (Å)	Angle (°)		
SiS_2	$Ibam$	9.545	5.564	5.552		tetrahedral	[1]
$V(S_2)_2$	$I2/c$	6.775	10.42	12.11	100.8	antiprisms	[2]
Cs_2TiS_3	$Cmc2_1$	12.51	9.03	6.55		sq.pyramids	[3]

References: [1] Peters and Krebs 1982; [2] Kutoglu and Allmann 1972; [3] Rad and Hoppe 1978

STRUCTURE TYPES COMPOSED OF LAYERS: CHALCOGENIDE LAYERS WITH M:S =1:1

Mackinawite and related families

Mackinawite. The crystal structure of mackinawite $Fe_{1+x}S$ (possible metal excess $x \leq 0.07$) consists of sheets of edge-sharing FeS_4 tetrahedra parallel to (100) of the tetragonal cell (Table 5, Fig. 5a). Each cation has short metal-metal contacts with four adjacent cations forming an uninterrupted tetragonal net of metal-metal interactions (Fe-Fe = 2.60 Å). The S atoms are bonded to four Fe atoms on one side only (2.26 Å), their opposite side faces the interlayer space with van der Waals bonds (Evans et al. 1964; Lennie et al. 1995). The interlayer S-S distances are 3.55 Å whereas the intralayer distances are 3.67 Å. Layers are stacked in a tetragonal AAA stacking, so that an S atom from the adjacent layer lies above four Fe atoms from the starting layer. Thus, the Fe atoms form strings parallel to [001] but their distance and screening by S layers preclude interactions.

The "metal surplus" in some samples of mackinawite has been identified as a result of anion vacancies. However, sporadically occupied tetrahedra between adjacent layers may also be envisaged. The mackinawite-type layer is also an anti-litharge (γ-PbO) layer (Dickinson and Friauf 1924), with cations and anions interchanged. The mackinawite-type layer is the fundamental structural element of the thalcusite-rohaite series, and the layer-misfit tochilinite series.

Table 5. Mackinawite and tochilinites.

Mineral/Formula	Comp. Layer	Space Group	Lattice parameters (Å,°)						Ref.
			a	b	c	α	β	γ	
Mackinawite $Fe_{1+x}S$	Q	$P4/nmm$	3.674	3.674	5.033				[1]
Tochilinite I[a] $6Fe_{0.9}S \cdot 5[(Mg_{0.7}Fe_{0.3})(OH)_2]$	Q:[b]	$A1$	10.72	15.65	5.37	90	95	90	[2]
	H:[b]	$A1$	10.72	15.65	5.37	90	95	90	
Tochilinite II $6Fe_{0.8}S \cdot 5[(Mg_{0.7}Fe_{0.3})(OH)_2]$	Q:[b]	$P1$	10.74	8.34	8.54	92	92.7	85.5	[3]
	H:[b]	$A1$	10.74	15.65	5.42	90	95	90	

Notes: [a] Selected semicommensurate cases, for layer match see the text.
[b] Q: Pseudotetragonal sulfide component layer. H: Octahedral hydroxide component layer with hexagonal anion patterns on surfaces.

References: [1] Lennie et al. 1995; [2] Organova et al. 1972; [3] Organova et al. 1973

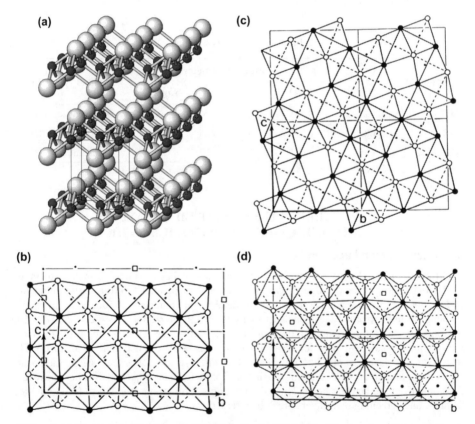

Figure 5. (a) Mackinawite (Lennie et al. 1995). Coordination tetrahedra of Fe share four edges, with four Fe-Fe bonds for each Fe atom. (b-d) Geometry of the sulfide (b, c) and hydroxide (d) layers in tochilinite. Black and white circles: anions at two height levels, dots: cations. Note two different vacancy schemes in the mackinawite-like layers and the Fe^{3+}(Al)-Mg substitution scheme in hydroxide layers; both are indicated by squares.

Mackinawite $Fe_{1-x}S$ is believed to be a metastable polymorph of FeS without a stability field. However, its selenide analog, FeSe, is an important phase in the Fe-Se system, stable up to 457 °C, and equal in importance to the pyrrhotite analog $Fe_{1-x}Se$. Karup-Møller and Makovicky (unpublished) found Cu solubility in mackinawite-like FeSe up to the composition $Cu_{13.6}Fe_{37.2}Se_{49.2}$. This, and the crystal chemistry of thalcusite and its analogs, give credence to the experiments trying to use mackinawite for scavenging metal pollution from reduced sediments.

Tochilinite varieties. Tochilinites (Table 5) are 1:1 intergrowths of tetragonal, mackinawite-like layers with cation vacancies, with a composition between $Fe_{0.91}S$ and $Fe_{0.72}S$, and octahedral "brucite-like" layers $(Mg,Fe)(OH)_2$ or $(Mg,Al)(OH)_2$ (Fig. 5b-d). The initial descriptions of tochilinite (Organova et al. 1972 a,b) concentrated upon exact crystallography with respect to layer match, the later descriptions only relied upon these results and were less detailed in character. The tochilinite problem has been summarized in detail by Organova (1989) as well as by Makovicky and Hyde (1981, 1992).

The two-layer sandwich of tochilinite has a repeat period of 10.7 Å. The sulfide layer has a mackinawite-like 5.2 × 5.4 Å substructure perturbed by diverse patterns of cation vacancies: centered orthorhombic for tochilinite I, tetragonal with a large supercell for tochilinite II (Fig. 5b-d) and tetragonal with 1:1 distribution of partial vacancies in "tochilinites III-VII," or disordered for other samples (including those studied after long standing in water).

The periodicity of the hydroxide layer, i.e., its superstructure, is created by a centered orthorhombic distribution of Fe^{3+}(Al) among the essentially Mg-occupied octahedra of the trioctahedral layer (Fig. 5b-d). If the vacancies are disregarded, stacking of both layer types is monoclinic ($\beta = 95°$); varieties with a two-layer sequence of hydroxide layers mutually rotated by 22° ("Phase 1" of Organova et al. 1974), or with extensive layer orientation disorder, and with twinning of the hydroxide sequence by 90° rotation (the Mg-Al case) have also been observed.

The layers can be denoted according to the anion patterns on their surfaces as the tetragonal (Q) sulfide layers and the hexagonal (H) hydroxide layers. The layer match between the non-commensurate Q and H layers is complex and variable. The commensurate, semicommensurate or incommensurate, approximate matches observed are apparently connected with the distribution of trivalent cations in the hydroxide layer, which interact with the distribution of fractional cation vacancies in the sulfide layer.

In tochilinite I, the semi-commensurate:commensurate match $3b_Q = 5b_H$ and $c_Q = c_H$ indicates the same A1 supercell for the two layers; in tochilinite II, the coincidence mesh is in a complicated relationship to the subcells. The layers undergo a common compositional modulation either based on centered subcells (in tochilinite I) or along the *b* direction (tochilinite II). Further tochilinite types (Organova 1989) have mutually incommensurate layers, in one or both directions, mostly with $5b_H < 3b_Q$ but with variable $c_H:c_Q$ ratios. This situation appears connected with a detailed chess-board pattern of fractional vacancies.

The composition of this remarkable mineral varies with the concentration of vacancies, reflected in the subcell-area match:

$$2Fe_{1-x}S \cdot k(Mg,Fe,Al)(OH)_2$$

where $0.08 < x < 0.28$ and $1.58 < k < 1.75$, based on crystallographic evidence. Hydroxide layers vary from $(Mg_{0.9}Fe_{0.1})(OH)_2$ to $(Mg_{0.7}Al_{0.3})(OH)_2$. Chemical analyses indicate a surplus of hydroxide layers, not observed in the structure; Cu and Ni substitutions in the sulfide layers have been observed (Muramatsu and Nambu 1980). Terrestrial occurrences of tochilinite have been followed by discoveries in carbonaceous chondrites (Zolensky 1984; Tomeoka and Busek 1985) making it an important component in cosmic mineralogy. Both platy and cylindrical varieties of tochilinite are known.

Thalcusite-rohaite series. The thalcusite-rohaite series (Makovicky et al. 1980) with only one known intermediate member, chalcothallite, is a merotypic series. Its structures are composed of regularly occurring layers of coordination cubes of thallium, or of other large cations, which alternate either with quadratic, tetrahedral sulfide layers of mackinawite type (Fig. 6a), fully occupied by (Cu, Fe) in $Tl_2Cu_3FeS_4$ (thalcusite), or with thicker, complex sulfide-antimonide layers in $Tl_2Cu_{8.67}Sb_2S_4$ (rohaite). In chalcothallite, $Tl_2(Cu,Fe)_{6.35}SbS_4$, these two layer types alternate regularly (Fig. 6b). Only Tl and minor K have been found in association with the sulfide-antimonide layers.

In $TlCu_4Se_3$ and $TlCu_6S_4$, the single-tetrahedral, Cu-filled layers between two consecutive Tl layers undergo homologous expansion giving double- and triple-tetrahedral layers (Berger 1987). Other large cations occur in these phases (Table 6) but the multiple tetrahedral layers are known only for Cu. The chemical formulae and the plethora of short Cu-Cu (in one-layer varieties also Fe-Fe and Fe-Cu) contacts imply a complex bonding/valence scheme. Natural compounds exhibit a combination of Cu and Fe close to the 3:1 ratio. Ordered vacancies are present in tetrahedral layers of, e.g., $Rb_2Mn_3S_4$ (Bronger and Böttcher 1972) (Table 6).

Other M:S=1:1 layer types

Covellite merotypes. Covellite, CuS, is a sulfide with a layer structure that is composed of consecutive sheets (0001) of tetrahedrally coordinated copper, trigonally coordinated copper and of covalent S-S dimers that are parallel to the hexagonal c-axis (Fig. 7a). The same structure is adopted by CuSe (klockmannite).

For the sake of comparison with other structures of this small merotypic series, the covellite structure can be divided into composite layers that are limited on both surfaces by the sulfurs of the trigonal planar sheet, and include the tetrahedral sheets adjacent to them, and the central S-S layer. In covellite, these layers share the bounding S atoms; the "exchangeable interlayer" is formed in this case only by the trigonal planar Cu atoms connected to these sulfur atoms.

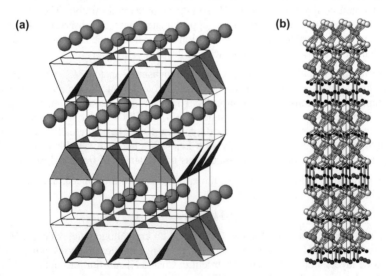

Figure 6. (a) $TlFe_2Se_2$ (Klepp and Boller 1978), a typical structure of the thalcusite family. Large circles: Tl in a cube-like coordination; Fe is in tetrahedral coordination with short Fe-Fe contacts. (b) Chalcothallite $Tl_{1.7}K_{0.2}Cu_{5.5}Fe_{0.7}Ag_{0.1}SbS_4$ (Makovicky et al. 1980). Tl (grey) in 8-fold coordination; Cu (black) in tetrahedral mackinawite-like layers. Cu also participates in triple "antimonide" layers with a partial, $^1/_3$ Cu site (black) and Sb (dark grey) in their central row.

Table 6. Mackinawite and the thalcusite family

Mineral	Formula	Space Group	Lattice Parameters (Å)		Ref.
			a	c	
Mackinawite	FeS	$P4/nmm$	3.674	5.033	[1]
Thalcusite	$Tl_2Cu_3FeS_4$	$I4/mmm$	3.88	13.25	[2]
Murunskite	$K_2Cu_3FeS_4$	$I4/mmm$	3.88	13.10	[3]
Bukovite	$Tl_2Cu_3FeSe_4$	$I4/mmm$	3.98	13.70	[4], [5]
Rohaite	$Tl_2Cu_{8.7}Sb_2S_4$	$Pmmm$ or $P222$	a 7.60 b 3.80	c 20.99	[5]
Chalcothalllite	$Tl_2Cu_{5.7}Fe_{0.7}SbS_4$	$I4/mmm$	a 3.827	c 34.280	[5]
Synthetic	$TlCu_4Se_3$	$P4/mmm$	3.894	9.33	[6]
Synthetic	$TlCu_6S_4$	$I4/mmm$	3.936	24.183	[7]
Synthetic	$Rb_2Mn_3S_4$	$Ibam$	*	6.295	[8]
Synthetic	KCu_4S_3	$P4/mmm$	3.899	9.262	[9], [10]
Synthetic	KCu_4Se_3	$P4/mmm$	4.019	9.720	[11]
Synthetic	$TlFe_2Se_2$	$I4/mmm$	3.890	14.00	[12]
Synthetic	$KCuFeS_2$	$P4/mmm$	3.837	13.384	[13]
Synthetic	$NH_4Cu_4S_3$	$P4/mmm$	3.907	9.453	[14]
Synthetic	$RbCu_4S_3$	$P4/mmm$	3.928	9.43	[10]
Synthetic	$CsCu_4S_3$	$P4/mmm$	3.975	9.689	[15]
Synthetic	$CsCoCuS_2$	$I4/mmm$	3.967	13.833	[16]
Synthetic	$CsCuFeS_2$	$I4/mmm$	3.942	14.238	[17]
Synthetic	$TlCo_2S_2$	$I4/mmm$	3.741	12.956	[18]
Synthetic	$TlCu_2Se_2$	$I4/mmm$	3.80	13.77	[19]
Synthetic	$TlCu_2S_2$	$I4/mmm$	3.777	13.379	[20]

* $a = 7.119, b = 13.140$

References: [1] Lennie et al. 1995; [2] Kovalenker et al. 1976; [3] Dobrovolskaya et al. 1981; [4] Johan and Kvaček 1971; [5] Makovicky et al. 1980; [6] Klepp et al. 1980, Berger 1987; [7] Berger and Eriksson 1990; [8] Bronger and Böttcher 1972; [9] Brown et al. 1980, [10] Ruedorff et al. 1952; [11] Stoll et al. 1999; [12] Klepp and Boller 1978; [13] Ramirez et al. 2001; [14] Boller and Sing 1997; [15] Burschka 1980; [16] Oledzka et al. 1996; [17] Llanos et al. 1996; [18] Huan and Greenblatt 1989; [19] Avilov et al. 1971; [20] Berger 1989

In the first of the known merotypes of covellite (Table 7), $Cu_{3.39}Fe_{0.61}S_4$, nukundamite (Sugaki et al. 1981), the "exchangeable interlayer" consists of edge-sharing tetrahedra, all of which are filled with (Cu, Fe) in a valleriite-like arrangement (for valleriite see the next section) (Fig. 7b). The "exchangeable interlayer" in Cu_4SnS_6 (Chen et al. 1999) (Fig. 7c) and in synthetic $NaCu_4S_4$ (Zhang et al. 1996) contains trigonal planar copper on both surfaces, and octahedrally coordinated Sn^{4+} inside the sandwich.

Valleriite is a two-layer misfit structure composed of two trigonal layers—a sulfide and a hydroxide layer in an incommensurate match (Evans and Allmann 1968). The sulfide component represents a trigonal layer of edge-sharing MS_4 tetrahedra, with the stoichiometry MS (M = Cu and Fe) (Fig. 8). The boundary S layers are 3^6 nets. In the original valleriite (Evans and Allmann 1968), a three-layer rhombohedral sequence of sulfide layers was observed, with $a = 3.792$ Å, $c = 34.10$ Å, component space group $R\bar{3}m$.

The hydroxide component of the original valleriite is a brucite-like trioctahedral layer with Mg^{2+} and Al^{3+}. These layers form a one-layer sequence with $a = 3.070$ Å, $c = 11.37$ Å and a different space group, $P\bar{3}m1$. The a and c axes of the two components coincide, giving an approximate match of 17 a (sulph) = 21 a (brucite), without a visible common modulation.

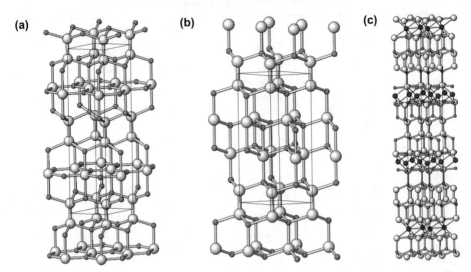

Figure 7. (a) Covellite CuS (Evans and Konnert 1976). Large spheres: S; small grey spheres: Cu. Alternating layers of CuS_3 coordination triangles and CuS_4 coordination tetrahedra. S_2 groups are parallel to [001]. (b) Nukundamite $Cu_{3.4}Fe_{0.6}S_4$ with alternating layers of tetrahedrally coordinated Cu, of edge-sharing (Cu,Fe) tetrahedra and of S_2 groups. (c) Cu_4SnS_6 (Chen et al. 1999). Grey small spheres: Cu; white spheres: S; dark spheres: partly occupied Sn^{4+} positions. These are flanked by partly occupied triangular Cu sites, fully occupied Cu tetrahedra and S_2 groups.

Table 7. Covellite merotypes.

Mineral	Formula	Space Group	Lattice Parameters (Å)		Interlayer	Ref.
			a	c		
Covellite	CuS	$P6_3/mmc$	3.79	16.34	$Cu^{[3]}$	(1)
Klockmannite	CuSe	$P6_3/mmc$	3.94	17.25	$Cu^{[3]}$	(2)
Nukundamite	$Cu_{3.39}Fe_{0.61}S_4$	$P\bar{3}m$	3.78	11.2	$(Cu,Fe)^{[4]}$	(3)
Synthetic	Cu_4SnS_6	$R\bar{3}m$	3.74	32.94	$Cu^{[3]}, Sn^{[6]}$	(4)
Synthetic	$NaCu_4S_4$	$P\bar{3}m1$	3.83	12.07	$Cu^{[3]}, Na^{[6]}$	(5)

References: (1) Evans and Konnert 1976; (2) Effenberger and Pertlik 1981; (3) Sugaki et al. 1981; (4) Chen et al. 1999; (5) Zhang et al. 1996

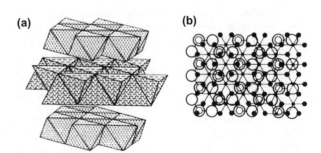

Figure 8. Structural scheme of valleriite. (a) Tetrahedral sulfide layers alternate with octahedral hydroxide layers. (b) In the projection on (0001), the sulfide layer is indicated by large circles, the hydroxide layer by smaller, filled circles.

For valleriite with Mg and Fe^{3+} in brucite-like layers, a one-layer stacking sequence was found for both components by Organova et al. (1973), with ordered Mg and Fe^{3+} in the brucite layer, resulting in $a_{brucite}$ = 5.34 Å, at 30° to the a_{sulph} = 3.79 Å. This appears to be a unique case. In her compilation, Organova (1989) quotes all other known valleriites as having a three-layered sequence of sulfide layers, with an almost constant composition, with Cu:Fe about 1:1. The hydroxide components vary from $Mg_{0.68}Al_{0.32}$ to Mg-Fe combinations (0.1-1.0 Fe), with a increasing from 3.07 Å to 3.20 Å in this process. These changes result in the formula $2MS·1.53-1.37M(OH)_2$. Chemical analyses indicate a surplus of hydroxide component, not explained by the structure investigations. The brucite-like layers display negative charge; compensation by sulfide layers is assumed although less obvious because of the dense net of metal-metal interactions.

Wang and Buseck (1990) suggest that valleriite is triclinic, the 11.47 Å layer pair being quadrupled and with occasional wavy modulation similar to antigorite. This reminds us that Organova (1989) found a regularly interstratified valleriite-serpentinite phase. For the Ni-analog, haapalaite (Huhma et al. 1973), $Fe_{1.26}Ni_{0.74}S_2·1.61$ $[Mg_{0.84}Fe_{0.16}(OH)_2]$, only limited crystallographic data exist. The vanadium analog, yushkinite, $V_{1-x}S·0.53-0.61$ $[(Mg,Al)(OH)_2]$ (Makeyev et al. 1984) with x = 0.38-0.47 has a_{sulph} = 3.21 Å, $a_{brucite}$ =3.06 Å, the stacking axis c = 11.3 Å. The structural scheme by Organova (1989) suggests a slight surplus of metal in the sulfide layer and the above layer ratio equal to 0.55.

STRUCTURES COMPOSED OF SESQUICHALCOGENIDE LAYERS

Tetradymite series (selenides and tellurides of bismuth). Under ambient conditions, the orthorhombic structure of bismuthinite (Bi_2S_3) type is limited to the sulfide of bismuth. At high pressures, however, it is also assumed by the selenide of bismuth. Otherwise, with the polarizable anions Se and Te, bismuth forms a series of layer-like selenide/telluride or more complex structures. On the one hand, these can be extended to subselenides and subtellurides with a portion of Bi-Bi bonds present and, on the other hand, layers of this series may be modified to incorporate suitable divalent cations, originating yet another, parallel structural series (Table 8). The works on the structural classification of this series include Bayliss (1991), Strunz and Nickel (2001), and Cook et al. (in press).

The fundamental unit layer of this series is an octahedral double layer Bi_2X_3, consisting of three hexagonal anion sheets in an ABC stacking sequence, with intermediate Bi atoms in slightly eccentric octahedral coordination (Fig. 9a). A complete sheet- and layer-stacking sequence is AcBaC-BaCbA-CbAcB, with anion sheets symbolized by capitals, cation sheets by lower-case letters, and anion-anion contacts without cation interlayer by hyphenation.

Bismuth-anion bonds to the median sheet of anions are longer than those to the marginal sheets. In Bi_2Se_3 these distances are 3.30 Å and 2.89 Å, respectively; in Bi_2Te_3 they are 3.26 Å and 3.07 Å, in the same sequence (Fig. 9b). Tetradymite, Bi_2Te_2S, accommodates sulfur in the median plane (central anion sheet, with Bi-S equal to 3.06 Å) and tellurium in the outer sheets (Bi-Te 3.13 Å) (Fig. 10a); here the difference between anion radii comes into play. On the other hand, in skippenite, Bi_2Se_2Te, with 10% sulfur substituting for Te, tellurium fills the central plane of the slab (Bi-(Te,S) 3.06 Å), and Se forms its boundary sheets (Bi-Se 2.93 Å).

The anion surfaces of two adjacent double-layers show a eutactic relationship, with the interlayer X-X distance of 3.70–3.65 Å for Te and 3.51 Å for Se. These are van der Waals contacts across planes of excellent cleavage. The Se-Se distances are 4.14 Å and 4.23 Å in the marginal sheets and the interior of a double layer of Bi_2S_3, obviously as a result of the coordination requirements of Bi. This fundamental Bi_2X_3 structure has symmetry $R\overline{3}m$ (Table 8).

At lower anion fugacities (native bismuth is not a rare constituent of ore deposits with

Table 8. Selected structures of the tetradymite family

Mineral	Formula	Space Group	Lattice Parameters (Å,°)			Ref.
			a	c	γ	
tsumoite	BiTe	$P\bar{3}m1$	4.422	24.050	120	[1]
nevskite	BiSe	$P\bar{3}m1$	4.212	22.942	120	[2]
telluronevskite	Bi_3Se_2Te	$P\bar{3}m1$	4.264	23.25	120	[3]
paraguanajuatite	Bi_2Se_3	$R\bar{3}m$	4.143	28.636	120	[4]
kawazulite	Bi_2Te_2Se	$R\bar{3}m$	4.298	29.774	120	[4]
tellurobismutite	Bi_2Te_3	$R\bar{3}m$	4.395	30.440	120	[5]
tetradymite	Bi_2Te_2S	$R\bar{3}$	10.33	—	24.17	[6]
skippenite	Bi_2Se_2Te	$R\bar{3}m$	4.183	29.137	120	[7]
laitakarite	Bi_4Se_3	$R\bar{3}m$	4.27	40.0	120	[8]
rucklidgeite	$PbBi_2Te_4$	$R\bar{3}m$	4.452	41.531	120	[9]
kochkarite	$PbBi_4Te_7$	$P\bar{3}m$	4.44	71.7	120	[10]*

* Shelimova et al. (2004) indicate $P\bar{3}m1$, $a = 4.426$ Å, $c = 23.892$ Å

References: [1] Shimazaki and Ozawa 1978; [2] Gaudin et al. 1995; [3] Řídkošil et al. 2001; [4] Nakajima 1963; [5] Feutelais et al. 1993; [6] Harker 1934; [7] Bindi and Cipriani 2004; [8] Stasova 1968; [9] Zhukova and Zaslavskii 1972; [10] Talybov and Vainshtein 1962

bismuth sulfides and sulfosalts), the Bi_2X_3 layers can be intercalated by trigonal double-layers of covalently bonded bismuth. Such an intercalate in essence is a tightly-bonded double layer from the rhombohedral structure of native bismuth. In Bi_4Se_3, the Bi-Bi bonds in this double-layer are 3.04 Å long, whereas the Bi-Se distances to the surfaces of the Bi_2Se_3 layers are 3.46 Å. Bi_4Se_3 (laitakarite) is the limiting case of the series, in which the Bi_2 interlayer is intercalated at every X–X contact (Fig. 10b). Its presence has a pronounced influence on layer stacking, and it results in a partial sequence...B-a-c-B..., i.e., the lowermost and the uppermost anions of the two adjacent Bi_2X_3 double-layers are situated above one another. The Bi-Se distances in the interior of double-layers do not appear to be modified by this intercalation.

An ordered derivative with a Bi_2 interlayer in every second interlayer space is nevskite, BiSe (Gaudin et al. 1995) (Fig. 10c). Here, the intercalation appears to influence the distortion of the scheme of shorter and longer Bi-Se bonds in the octahedra adjacent to the intercalated double-layer of pure Bi. In telluronevskite, Bi_3Se_2Te (Řídkošil et al. 2001), tellurium forms anion sheets lining the interfaces between the telluroselenide double-layers, whereas the corresponding interspaces, which are occupied by the intercalated bismuth, are lined by selenium.

Incorporation of large divalent cations leads to MBi_2X_4 triple-octahedral layers (Fig. 10d). In the structure of $PbBi_4Te_7$, they alternate regularly with the double layers described above; in $PbBi_2Te_4$ they stand alone. Both structures (Table 8), however, were not sufficiently refined.

At present, at least 26 minerals of this series have been defined; caution should be exercised in the cases when the definition is based on powder data only: the visible reflections are usually limited to relatively low d values when compared to the unit cell c parameter derived from them. Based on inspection of the ranges of different published microprobe investigations, it is this author's opinion that many intermediate, disordered layer sequences may exist, both in the subselenide/subtelluride series and among the M^{2+} containing tellurides.

Both the subchalcogenide and the M^{2+}-containing series can be understood as individual homologous (polysomatic) series. Among the subsulfides, the Bi_2 layers are the unit-size

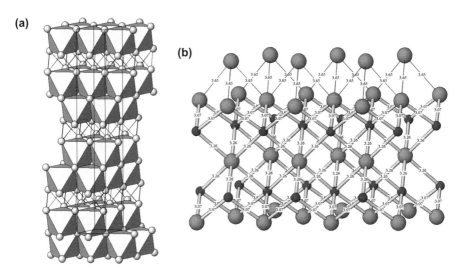

Figure 9. (a) Paraguanajuatite Bi$_2$Se$_3$ (Nakajima 1963). Double layers of coordination octahedra of bismuth are separated by long van der Waals Se-Se interactions. ABC packing of double layers. (b) Configuration details for Bi (black spheres) and Te (large spheres) in tellurobismutite, Bi$_2$Te$_3$ (Feutelais et al. 1993). One double-layer and a contact to an adjacent layer are shown.

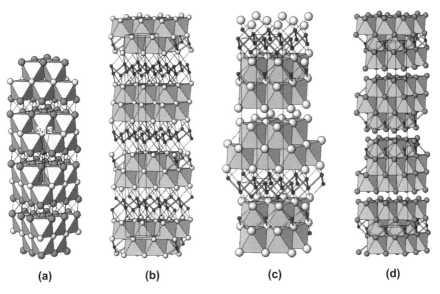

Figure 10. (a) Distribution of sulfur (light spheres) and tellurium (dark spheres) in double-octahedral layers of tetradymite Bi$_2$Te$_2$S (Harker 1934). (b) Laitakarite Bi$_4$Se$_3$ (Stasova 1968). Double-octahedral layers Bi$_2$Se$_3$ alternate with corrugated Bi$_2$ layers. The latter Bi atoms are coordinated to 3 Bi at 3.04 Å and 3 Se at 3.46 Å. (c) Nevskite BiSe (Gaudin et al. 1995). Bi$_2$Se$_3$ double layers are alternately in van der Waals Se-Se contact and intercalated by corrugated Bi$_2$ layers. Tsumoite BiTe is isotypic. (d) PbBi$_2$Te$_4$ (Zhukova and Zaslavskii 1972). Pb is concentrated in the central octahedral layer, Bi in the marginal octahedral layers of a three-layer packet. ABC stacking of layer packets.

portions of one end-member (rhombohedral Bi) and they are combined with one, two or more Bi_2X_3 layers separating them in consecutive polysomes. Pure Bi_2X_3 structures are the other pole of the series. An analogous scheme can be constructed for the combination of Bi_2X_3 and MBi_2S_4 layers. Layers thicker than the triple-octahedral MBi_2X_4 appear largely hypothetical at present (Cook et al. in press).

STRUCTURES COMPOSED OF DICHALCOGENIDE LAYERS

Layer-like MS_2 structures. The layer-like MS_2 structures consist of MS_2 layers in which a sheet of central cations is sandwiched between sheets of anions (S, Se, Te). Adjacent layers adhere via van der Waals contacts between their marginal anion sheets. The electrostatic repulsion of anions on these contacts is reduced by the strongly covalent character of *M-X* bonds, which limits the extent of electron transfer onto anions. Long *M-X* interactions, across the interlayer space, contribute to the stacking modes; each anion resides above three anions of the previous layer.

Besides the structure of berndtite, SnS_2 (Table 9), layer MS_2 structures are typical for transition metals on the left-hand side of each transition series: up to d^1 (Ti^{4+}), d^2 (Mo^{4+}) and d^3(Re^{4+}), respectively. In these cases, the d energy levels are high above the valence band. When we move along the transition series to the right, their energy decreases, until they start overlapping with the S-based valence band and draw electrons from it, giving disulfides such as pyrite instead of layer structures (Rouxel 1991; see also Vaughan and Rosso 2006, this volume).

In their basic form, these layer sulfides have very simple structures—regular stacking of either octahedral (Sn, Ti, high-temperature TaS_2 and $TaSe_2$, ReS_2) or trigonal-prismatic (Nb, Ta, Mo, W, Hf) layers (Figs. 11a-c). Compounds and solid solutions with one type of layer strongly predominate, layer sequences in which octahedral and prismatic layers mix are rare.

The schemes used to describe layer stacking include the standard polytype notation, e.g., 2H- or 3R-TaS_2; sheet stacking notations which are based on the prismatic layers denoted as AbA and octahedral ones as AbC, and the standard sections (11$\bar{2}$0) in which both the layer type and the layer stacking are visible (Figs. 12a,b).

The two fundamental stacking modes of trigonal prismatic layers are 2H and 3R. Their stability differs; e.g., for MoS_2 the 2H polytype (Fig. 11c) is widespread whereas the 3R polytype (Fig. 13a) is assumed to have formed at lower temperatures. For octahedral layers, the 1T stacking (isotypic with CdI_2) or the 3R stacking (isotypic with $CdCl_2$) is present (Table 9).

Dependence on the conditions of formation has been investigated thoroughly for chalcogenides of the d^1 metals Nb and Ta. The octahedral 1T form of $TaSe_2$ is a stable high-temperature form whereas the intermediate, 4H and 6R polytypes with a mixed-layer character, in which there are sequences of regularly alternating octahedral and trigonal-prismatic layers (Fig. 13b), are secured by quenching from about 950 K. The trigonal prismatic 2H form is stable at room temperature (Wilson et al. 1975). Below a transition temperature, the simple structures of the layered polytypes of Ta and Nb become complicated by charge-density waves connected with a formation of superlattices and/or strong diffuse scattering (Wilson et al. 1975). Thus, at room temperature, the $4H_b$ polytype of $TaSe_2$ (a structure with alternating octahedral and trigonal-prismatic layers), forms an $a\sqrt{13} \times a\sqrt{13} \times c$ superstructure by the action of charge density waves in the octahedral layers.

These lamellar structures are eminently suited for intercalation of various cations (Fig. 13d). Besides the large, low-charge alkali cations, many other cations such as Cu or Bi, even H^+ and different organic cations, have been successfully intercalated (Table 9). The intercalated element *A* contributes an electron to the MX_2 layer, changing primarily the valence of the metal in the layer. It is a low-energy process and it is reversible, a subject of "soft chemistry" of practical

Table 9. Selected dichalcogenides with layer structures.

Mineral	Formula	Space group	Lattice Parameters (Å,°)				Ref.
			a	b	c	Angle	
Synth.	NbS$_2$	R3m	3.330	3.330	17.918	γ 120	[1]
Synth.	NbS$_2$	P6$_3$/mmc	3.31	3.31	11.89	γ 120	[2]
Synth.	TaS$_2$-1T	P$\bar{3}$m1	3.365	3.365	5.897	γ 120	[3]
Synth.	TaS$_2$-2H	P6$_3$/mmc	3.314	3.314	12.097	γ 120	[4]
Synth.	TaS$_2$-6R	R$\bar{3}$m	3.335	3.335	35.85	γ 120	[5]
Synth.	Ti$_{1-x}$S$_2$	P$\bar{3}$m1	3.407	3.407	5.695	γ 120	[6]
Synth.	Ti$_{0.5}$Nb$_{0.5}$S$_2$	P$\bar{3}$m1	3.387	3.387	5.784	γ 120	[6]
Synth.	Cu$_{0.7}$TiS$_2$	R$\bar{3}$m	3.439	3.439	18.900	γ 120	[7]
Synth.	Cu$_{0.375}$TiS$_2$	P$\bar{3}$m1	3.415	3.415	5.855	γ 120	[8]
Synth.	Li$_{0.33}$TiS$_2$	P$\bar{3}$m1	3.444	3.444	14.0	γ 120	[9]
Synth.	Ag$_{0.167}$TiS$_2$	P$\bar{3}$m1	3.413	3.413	12.092	γ 120	[10]
Synth.	[NH$_4$]$_{0.22}$TiS$_2$	R$\bar{3}$m	3.418	3.418	41.81	γ 120	[11]
Synth.	RbTiS$_2$	R$\bar{3}$m	3.43	3.43	24.2	γ 120	[12]
Synth.	Na$_{0.3}$TiS$_2$	R$\bar{3}$m	3.406	3.406	38.20	γ 120	[13]
Synth.	LiNbS$_2$	P6$_3$/mmc	3.331	3.331	12.90	γ 120	[14]
Synth.	NaNbS$_2$	P6$_3$/mmc	3.336	3.336	14.52	γ 120	[14]
Synth.	KNbS$_2$	P6$_3$/mmc	3.345	3.345	16.22	γ 120	[14]
Synth.	Nb$_{1.09}$S$_2$	R$\bar{3}$m	3.330	3.330	17.869	γ 120	[15]
Synth.	Cs$_{0.6}$NbS$_2$	P6$_3$/mmc	3.354	3.354	18.324	γ 120	[16]
Synth.	Bi$_{0.64}$NbS$_2$	P6$_3$/mmc	3.313	3.313	17.350	γ 120	[17]
Synth.	H$_{0.76}$NbS$_2$	P6$_3$/mmc	3.34	3.34	12.39	γ 120	[18]
Synth.	Ge$_{0.33}$NbS$_2$	P6$_3$/mmc	5.767	5.767	13.518	γ 120	[19]
Molybdenite 3R	MoS$_2$-3R	R3m	3.163	3.163	18.37	γ 120	[20]
Molybdenite 2H	MoS$_2$-2H	P6$_3$/mmc	3.161	3.161	12.295	γ 120	[20]
Drysdallite 2H	MoSe$_2$-2H	P6$_3$/mmc	3.289	3.289	12.927	γ 120	[21]
Synth.	MoSe$_2$-3R	R3m	3.292	3.292	19.392	γ 120	[22]
Tungstenite 2H	WS$_2$-2H	P6$_3$/mmc	3.153	3.153	12.323	γ 120	[23]
Tungstenite 3R	WS$_2$-3R	R3m	3.158	3.158	18.490	γ 120	[23]
Synth.*	Mo$_{0.5}$Ta$_{0.5}$S$_2$	R$\bar{3}$m	3.252	3.252	18.138	γ 120	[24]
Synth.	Mo$_{0.2}$TaS$_2$	P6$_3$/mmc	3.29	3.29	12.30	γ 120	[25]
Synth.	FeMo$_2$S$_4$	C1c1	11.815	6.550	13.014		[26]
Synth.	LiMoS$_2$	P$\bar{1}$	6.963	6.386	6.250	α 88.60, β 89.07, γ 120.06	[27]
Synth.	In$_{0.25}$MoSe$_2$	P6$_3$/mmc	3.291	3.291	12.929		[28]
Berndtite 4H	SnS$_2$-4H	P6$_3$/mc	3.645	3.645	11.802	γ 120	[29]
Berndtite 2T	SnS$_2$-2T	P$\bar{3}$m1	3.638	3.638	5.88	γ 120	[30]
Synth.	SnSe$_2$-2T	P$\bar{3}$m1	3.811	3.811	6.137	γ 120	[31]
Synth.	SnTaS$_2$	P6$_3$/mmc	3.307	3.307	17.442	γ 120	[32]
Synth.	SnNbS$_2$	P6$_3$/mmc	3.324	3.324	17.37	γ 120	[32]
Synth.	LiSnS$_2$	P$\bar{3}$m1	3.67	3.67	7.90	γ 120	[33]
Synth.	NaSnS$_2$	P$\bar{1}$	3.69	3.69	25.54	γ 120	[33]
Synth.	KSnS$_2$	R3m	3.67	3.67	25.61	γ 120	[33]
Calaverite	AuTe$_2$	C2/m**	7.189	4.407	5.069	β 89.96	[34]
High P	AuTe$_2$	P$\bar{3}$m1	4.078	4.078	5.000	β 120	[34]
Sylvanite	AuAgTe$_4$	P2/c	8.95	4.478	14.62	β 145.35	[35]
Krennerite[+]	(Au,Ag)Te$_2$	Pma2	16.58	8.849	4.464		[35]

Notes: *An example for the phases of the series Mo$_{0.05}$Ta$_{0.95}$S$_2$-Mo$_{0.9}$Ta$_{0.1}$S$_2$. [+] Unit-cell twinned octahedral layer stack.
** A subcell for a non-commensurately modulated structure

References: [1] Morosin 1974; [2] Jellinek et al. 1960; [3] Spijkerman et al. 1997; [4] Meetsma et al. 1990; [5] Jellinek 1962; [6] Furuseth 1992; [7] le Nagard et al. 1974; [8] Kusawake et al. 2000; [9] Hallak and Lee 1983; [10] Burr et al. 1990; [11] McKelvy et al. 1990; [12] Bichon et al. 1973; [13] Bouwmeester et al. 1982; [14] Omloo and Jellinek 1970; [15] Meerschaut and Deudon 2001; [16] Chen et al. 1993; [17] Eppinga and Wiegers 1980; [18] Riekel et al. 1979; [19] Pocha and Johrendt 2002; [20] Schoenfeld et al. 1983; [21] Bronsema et al. 1986; [22] Towle et al. 1966; [23] Schutte et al. 1987; [24] Remmert et al. 1994; [25] Saeki and Onoda 1987; [26] Vaqueiro et al. 2002; [27] Petkov et al. 2002; [28] Shamrai et al. 1986; [29] Guenter and Oswald 1968; [30] Hazen and Finger 1978; [31] Busch et al. 1961; [32] Eppinga and Wiegers 1977; [33] le Blanc and Rouxel 1972; [34] Reithmayer et al. 1993; [35] Pertlik 1984

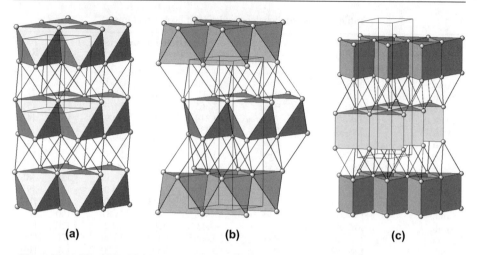

Figure 11. (a) 2H and (b) 4H polytypes of SnS$_2$, respectively, with octahedrally coordinated Sn^{4+}. Thin S-S connections help to outline the configuration of the interlayer space. (c) hexagonal 2H polytype of MoS$_2$ with a trigonal prismatic coordination of molybdenum.

importance as, e.g., in the TiS$_2$ ↔ LiTiS$_2$ cathode in lithium batteries (Rouxel 1991). The influence of selected substitutions on unit cell parameters and stacking of these phases can also be seen in Table 9. This table also contains examples of layer structures slightly "pillared" by their own (or a very similar) cations, e.g., Mo$_{0.2}$TaS$_2$ (Saeki and Onoda 1987). Intercalated forms should be expected in nature.

The crystal structures of gold and gold-silver tellurides (Table 9) are layered MX_2 structures as well. Calaverite, AuTe$_2$, (space group $C2/m$, β 89.96°) consists of octahedral layers which display a non-commensurate modulation in one direction (Reithmayer et al. 1993). In the averaged structure, the octahedrally coordinated Au has two shorter Au-Te bonds (2.67 Å) in a trans-configuration and four longer distances, 2.97 Å. The average Te-Te distances are: one at 3.46 Å, in a trans position to the above Au-Te distance of 2.67 Å, and two at 3.20 Å, opposing the Au-Te distances of 2.97 Å. Apparently the latter distances are the sites of the majority of Te-Te bonds in the real, modulated structure. These bonds solve the problem created by the presence of Au^{3+} (instead of M^{4+}) in a layered MTe$_2$ structure. In the high-pressure form of AuTe$_2$, the layered structure becomes a regular, unmodulated $P\bar{3}m1$ MX_2 structure, which at the pressure of 5130 MPa has Au-Te bonds of 2.77 Å and uniform Te-Te distances of 3.13 Å (Reithmayer et al. 1993).

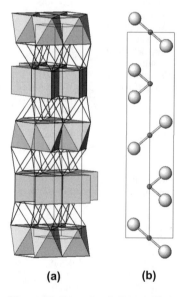

Figure 12. Example of the graphical shorthand used in the literature for the layered structures composed of octahedral and trigonal prismatic layers: (a) the crystal structure of TaSe$_2$ composed of alternating layer types, and (b) its illustration as cation and anion sites in a (11$\bar{2}$0) section of the hexagonal cell.

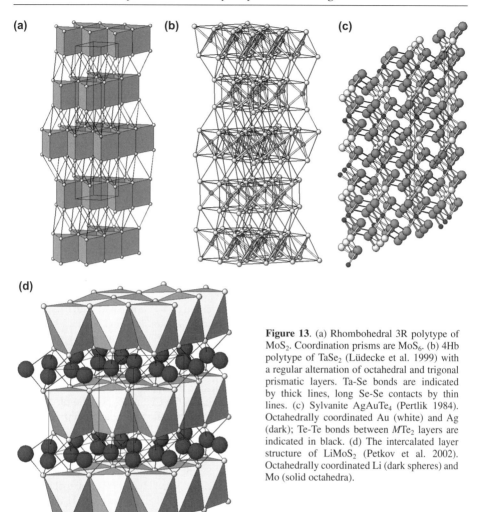

Figure 13. (a) Rhombohedral 3R polytype of MoS_2. Coordination prisms are MoS_6. (b) 4Hb polytype of $TaSe_2$ (Lüdecke et al. 1999) with a regular alternation of octahedral and trigonal prismatic layers. Ta-Se bonds are indicated by thick lines, long Se-Se contacts by thin lines. (c) Sylvanite $AgAuTe_4$ (Pertlik 1984). Octahedrally coordinated Au (white) and Ag (dark); Te-Te bonds between MTe_2 layers are indicated in black. (d) The intercalated layer structure of $LiMoS_2$ (Petkov et al. 2002). Octahedrally coordinated Li (dark spheres) and Mo (solid octahedra).

The mutual shift of MX_2 layers out of the close-packed situation, observed in calaverite, is even more pronounced in sylvanite, $AuAgTe_2$ (Pertlik 1984) (Fig. 13c). The interlayer Te-Te bond is 2.82 Å long. The cation octahedra are asymmetric, 2×2.74 Å, 2×2.93 Å, and 2×3.23 Å for Ag, and close to a square coordination for Au: 2×2.68 Å, 2×2.69 Å, and 2×3.33 Å. The short distances are in a *trans* configuration for both cations. The dimorph of $AuTe_2$, krennerite $(Au_{0.81}Ag_{0.12})Te_2$ (space group *Pma*2) is a unit-cell twinned version of a layered MX_2 structure, twinned on every fourth octahedron of the sequence.

LAYER-LIKE STRUCTURES WITH TWO TYPES OF CHALCOGENIDE LAYERS

Cannizzarite and related structures

Cannizzarite, $\sim Pb_{48}Bi_{54}S_{127}$, (Fig. 2a) is believed to form a true variable-fit series. The structure consists of a regular alternation of (two atomic sheets) thick pseudotetragonal layers and of double-octahedral layers. Both layers are occupied by a mixture of Pb and Bi. The pyramidal sites of the Q layer form distorted capped trigonal prismatic sites and, in some cases, distorted octahedral sites by extending across the interlayer space. The two layers are mutually non-commensurate, $b_H = b_{Qprimitive}$ and $c_H > c_{Qprimitive}$ (e.g., $17c_Q$ approximately corresponds to $10c_H$).

Matzat (1969) found a match of 46 subcells of the pseudotetragonal (Q) layer with 27 centered orthohexagonal subcells of the pseudohexagonal (H) layer (Fig. 3), i.e., practically a non-commensurate structure. There are a number of near-matches between the two layer sets in this model. Should these be fully materialized, the match of the two layer components, expressed as the match coefficient k in the reduced chemical formula $MS·kM_2S_3$ (0.587 for the above case) will alter as follows: for 12Q/7H, the composition would be $Pb_{12}Bi_{14}S_{33}$, i.e., $k = 0.583$; for 17Q/10H, i.e., $Pb_{17}Bi_{20}S_{47}$, $k = 0.588$. For the very simple match, perhaps not attainable by the structure, 5Q/3H, i.e., the composition is $Pb_5Bi_6S_{14}$, and $k = 0.600$. This scheme suggests that very small changes in chemical composition, requiring greater accuracy than provided by microprobe measurements, will lead to a change in the match ratio.

The cannizzarite interlayer match is typical for about equally large average cations in the Q and H layers (a mixture of Pb and Bi in both layer types). The $LaCrS_3$-like match, typical for the layer misfit compounds discussed elsewhere, occurs for the largest Q/H cation radius ratio. The Q layers in these "ABX_3 compounds" are turned by 45° compared with cannizzarite and the match ratios in them differ from cannizzarite (Makovicky and Hyde 1992). The cannizzarite interlayer match is the point of reference for interlayer matches in a number of Cu(or Ag)-Pb-Bi sulfosalts and Pb-Sb sulfosalts, in which different segments of this match occur in the interspaces with Q/H contacts.

In cannizzarite, the H layer is two octahedra thick, in the synthetic phase of Graham et al. (1953) it is presumably three octahedra thick, whereas in the $Cr_2Sn_3Se_7$ polymorph (Jobic et al. 1994) it is a single-octahedral layer, i.e, an accretional homologous relationship. A Q layer of the same thickness occurs in all these phases (Table 10).

Derivatives of cannizzarite: combinations of accretion with variable-fit

Cannizzarite (Fig. 2a) acts as a parent structure for the majority of layer-like structures of sulfosalts with non-commensurate layer contacts (Makovicky and Hyde 1981; Makovicky 1981, 1992; Ferraris et al. 2004) (Table 11). They represent stacks of alternating Q- and H-type layers with different thicknesses (not always the layer thicknesses observed in cannizzarite itself); they have similar match modes to cannizzarite and similar average cation radii. The layers, however, are periodically broken up into strips by different structural adjustments, avoiding the poor fit problems periodically encountered along the non-commensurate direction.

The structure of weibullite $Ag_{0.33}Pb_{5.33}Bi_{8.33}(S,Se)_{18}$ (Mumme 1980) is the most obvious derivative of cannizzarite, obtained by shearing its layers by means of composition non-conservative glide planes (Fig. 2b). Due to the necessity to satisfy both the accretional and the variable-fit principles, the nearest lower homologue of weibullite is structurally very distant from it. It is the short-range structure of galenobismutite, $PbBi_2S_4$. This structure combines pairs of distorted octahedra of Bi, those of "lying" monocapped trigonal prisms of Bi and "standing" bicapped trigonal prisms of Pb; it shows affinity to that of $CaFe_2O_4$ (Mumme 1980). For the second and third cation site, Pinto et al. (in press) indicate mixed (Bi, Pb) and (Pb, Bi) occupancies. In the "weibullite-like" expansion (parallel to a of galenobismutite) the ½ subcell broad Q module in galenobismutite expands to 5½ subcells in weibullite, whereas the ½ H subcell (i.e., 1 octahedron) broad pseudohexagonal module expands to a 3½ H subcells long layer fragment (compare Figs. 14a and 2b).

Table 10. Selected cannizzarite homologues.

Mineral	Formula	Cell Type	Lattice Parameters [Å,°]				Space Group	Ref.
			a	b	c	β		
Cannizzarite	$Pb_{46}Bi_{54}S_{127}$ [a]	Unit cell	189.8	4.09	74.06	11.9	$P2_1/m$	[1]
		Q subcell	4.13	4.09	15.48	98.6	$P2_1/m$	
		H subcell	7.03	4.09	15.46	98.0	$C2/m$	
Synth.	$Pb_{38}Bi_{28}S_{80}$ [a]	Q subcell	4.11	4.08	18.58	93.6	$P2/m$ [b]	[2]
		H subcell	7.03	4.08	27.16	93.6	$F2/m$	
		unit cell n.d.						
Synth.	$Cr_2Sn_3Se_7$	Unit cell	12.77	3.84	11.79	105.2	$P2_1/m$	[3]

Notes: n.d. = not determined; [a] The compositions are derived from the crystal structure; [b] Probably $P2_1/m$ [Matzat 1979].
References: [1] Matzat 1979; [2] Graham et al. 1953; [3] Jobic et al. 1994

Table 11. Selected plesiotypes based on a cannizzarite structure type.

Mineral	Formula	Lattice Parameters (Å)			β (°)	S.G.	Ref.
Galenobismutite	$PbBi_2S_4$	a 11.81	b 14.59	c 4.08	—	$Pnam$	[1]
Weibullite	$Ag_{0.33}Pb_{5.33}Bi_{8.33}(S,Se)_{18}$	a 53.68	c 15.42	b 4.11	—	$Pnma$	[2]
Synth.	$Pb_4In_3Bi_7S_{18}$	a 21.02	b 4.01	c 18.90	971	$P2_1/m$	[3]
Junoite[b]	$Cu_2Pb_3Bi_8(S,Se)_{16}$	26.66	4.06	17.03	127.2	$C2/m$	[4]
Felbertalite[b]	$Cu_2Pb_6Bi_8S_{19}$	27.64	4.05	20.74	131.3	$C2/m$	[5]
Proudite	$Cu_4Bi_{40}Pb_{32}(Se,S)_{94}$	31.81	4.10	36.56	109.3	$C2/m$	[6]
Nordströmite	$CuPb_3Bi_7(S.Se)_{14}$	17.98	4.11	17.62	94.3	$P2_1/m$	[7]
Neyite	$Ag_2Cu_6Pb_{25}Bi_{26}S_{68}$	3753	407	4370	1088	$C2/m$	[8]
Synth.	$Pb_3In_{6.67}S_{13}$ (a)	a 38.13	c 3.87	b 13.81	γ 91.3	$B2/m$	[9]
Synth.	$Bi_3In_5S_{12}$	33.13	3.87	14.41	91.2	$C2/m$	[10]

Notes: (a) 1/3 M is vacant in one metal site. (b) Homologues $N = 1$ and $N = 2$, respectively, of the junoite homologous series.
References: [1] Pinto et al. submitted; [2] Mumme 1980; [3] Krämer and Reis 1986; [4] Mumme 1975; [5] Topa et al. 2000; [6] Mumme, oral comm.; [7] Mumme 1980; [8] Makovicky et al. 2001; [9] Ginderow 1978; [10] Krämer 1980

A simple solution to the mismatch problem occurs when the pseudotetragonal layer fragment A continues laterally as a pseudohexagonal layer fragment B and vice versa. The misfit problems of the pseudotetragonal and pseudohexagonal layers facing one another are solved and compensated for by this periodical reversal (Makovicky and Hyde, 1992; Ferraris et al. 2004). A simple example of this phenomenon is the orthorhombic polymorph of $Cr_2Sn_3Se_7$ (Jobic et al.1995)(Fig. 16a) in which the interior match is identical to that of the above mentioned layer-like polymorph, but it is periodically reversed after a width of three peudotetragonal polyhedra.

Other derivatives of cannizzarite are members of the pair junoite $Cu_2Pb_3Bi_8(S,Se)_{16}$ - felbertalite $Cu_2Pb_6Bi_8S_{19}$ (Mumme 1975 and Topa et al. 2000). These structures are composed of fragments of $(100)_{PbS}$ layer, 2 atomic sheets thick and 3 coordination polyhedra broad (stippled in Figs. 14b,c), alternating with en échelon fragments of $(111)_{PbS}$ octahedral layers which have a thickness of one octahedron (Fig. 14b) or two octahedra (Fig. 14c).

Owing to the differences in the thickness of their octahedral layers, these two compounds were recognized as the first and second homologue of an accretional homologous series (Topa

Figure 14. Structures derived from the structural principles observed in cannizzarite. (a) Galenobismutite $PbBi_2S_4$. S: light spheres, Pb: light grey; Bi: dark grey. (b) Junoite $CuPb_3Bi_7(S,Se)_{14}$ (Mumme 1980), the $N = 1$ member of the junoite-felbertalite homologous pair. (c) Felbertalite $Cu_2Pb_6Bi_8S_{19}$ (Topa et al. 2000), the $N = 2$ homologue of this pair. In (b) and (c), in order of increasing size, circles indicate Cu, Bi, Pb and S. Sheared pseudohexagonal slabs are ruled, pseudotetragonal ribbon-like fragments with pairs of Cu tetrahedra in the step regions are stippled. Shading of all circles indicates two levels of the 4 Å structure, 2 Å apart. (d) Neyite $Ag_2Cu_6Pb_{25}Bi_{26}S_{68}$ (Makovicky et al. 2001), with two types of blocks indicated by different shades of grey, and with the subvertical layers left unshaded. In order of decreasing size, circles describe S, Pb, Bi, Ag (linear coordination) and Cu.

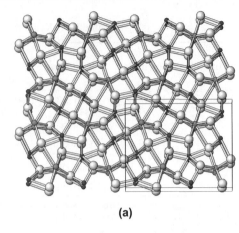

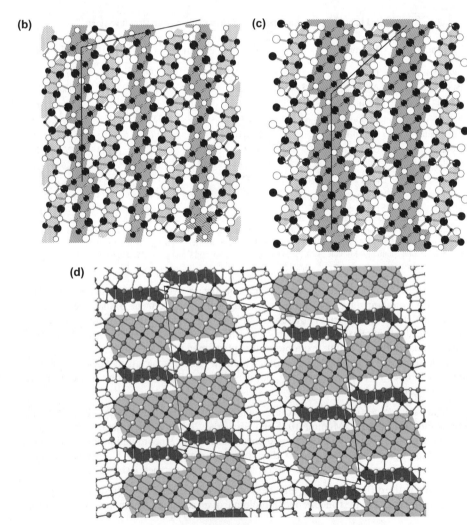

et al. 2000). The next member of the accretional sequence, with triple octahedral layers and (two atomic sheets thick) pseudotetragonal slab fragments was not found as an independent phase but forms a substantial part of the box-work structure of neyite $Ag_2Cu_6Pb_{25}Bi_{26}S_{68}$ (Makovicky et al. 2001) (Fig. 14d). Junoite participates in another homologous series, in which the layer thicknesses are preserved but the length of straight layer portions between two adjacent kinks varies. The shortest member of this series is the pavonite homologue $N = 3$ (Tomeoka et al.1980), the longest one is proudite (Mumme, pers. comm.).

Zero mutual shift of octahedral fragments in the case of felbertalite and the plus-one-octahedron shift for junoite (compare Figs. 14b,c) were both adopted by these sulfosalts in order to create a lone electron pair micelle, i.e., an extended common space for lone electron pairs of bismuth in the overlapped, kinked portions of the $(111)_{PbS}$ layer. For octahedral elements without active lone electron pairs, other shift geometries might be formed instead. This, and other examples (e.g., the pairs $Pb_4Sb_6S_{11}$ (robinsonite)–$Pb_5Sb_6S_{14}$ (synthetic) and $Pb_5Sb_4S_{11}$ (boulangerite)–$Pb_7Sb_4S_{13}$ (synthetic), led to a definition of a "sliding series of structures" described in Ferraris et al. (2004), in which complex structural slabs slide past one another by increments, gradually creating new coordination sites at their stepped interface.

Selected synthetic and natural chalcogenides with layer-misfit structures

The family of synthetic sulfides known as layer-misfit structures or alternatively as "ABX_3" sulfides (or sulfides with alternating incommensurate layers) is composed of pseudotetragonal MX layers which can be described as (100) slices of NaCl/PbS type, interleaved with MX_2 octahedral layers which are (111) slices of NaCl type or, alternatively, with trigonal-prismatic layers (0001) of NbS_2 (Figs. 15a,b). The A cation in the pseudotetragonal layers can be a lanthanide ion, Pb, Sn^{2+}, Bi,.... whereas the B cation is Ti, V, Cr,...in the octahedral layers, and Nb or Ta in the trigonal prismatic layers (Table 12). Crystallography of these compounds implies non-integral valence of (presumably) cations. This may be a real situation (compare with the intercalates of NbS_2 etc.) or it may be an idealized situation in other cases (e.g., Rouxel et al. 1994; Moëlo et al. 1995). Detailed reviews of these topics have been written by Makovicky and Hyde (1992), Wiegers and Meerschaut (1992) and van Smaalen (1995).

The interlayer interactions of the "ABX_3" layer-like structures are weak, van der Waals and charge-transfer in nature, resulting in a weak modulation of the component layers. Still, the resulting mutual adjustment of the two layer sets preserves a high degree of order in the layer stacking (Kuypers et al. 1990) and the equality $b_Q = b_H$.

The bulk of these compounds are composed of two atomic sheets thick pseudotetragonal layers and single octahedral or trigonal-prismatic layers. Examples are $(LaS)_{1.14}NbS_2$ and $(PbS)_{1.14}NbS_2$ with trigonal-prismatic layers and $(PbS)_{1.13}VS_2$ with octahedral layers. Rarely, two trigonal prismatic layers are found sandwiched between adjacent pseudotetragonal layers, e.g., in $PbNb_2S_5$ (Table 12) and three such layers, partly pillared by additional Nb in their van der Waals interspaces, in $(Gd_xSn_{1-x}S)_{1.16}(NbS_2)_3$ (Hoistad et al. 1995). Another complication of the simple basic scheme is observed in the structure of $(Pb_2FeS_3)_{0.58}NbS_2$ (Lafond et al. 1999), with three atomic sheets thick pseudotetragonal layers in which Pb assumes the exterior, half-octahedral positions and iron, split into a five-fold cluster of sub-sites, the interior, octahedral positions (Figs. 15a,b). As well as these non-commensurate cases, commensurate or, semi-commensurate cases exist, with the ratio of periodicities represented by a fairly simple ratio of small values. The 3Q:2H match is observed in the structures of $(SnS)_6(InS_2)_4$ (Adenis et al. 1986) and $Sn_{3.04}Cr_{1.9}Se_7$ (Jobic et al. 1994) (Fig. 15c). Here the interlayer space contains portions of two bicapped trigonal coordination prisms and an octahedron of (primarily) Sn.

Buckhornite, $(Pb_2BiS_3)(AuTe_2)$ (Effenberger et al. 2000) and nagyagite, $Pb(Sb,Pb)S_2(Au,Te)$ (Effenberger et al. 1999) are semi-commensurate composite structures.

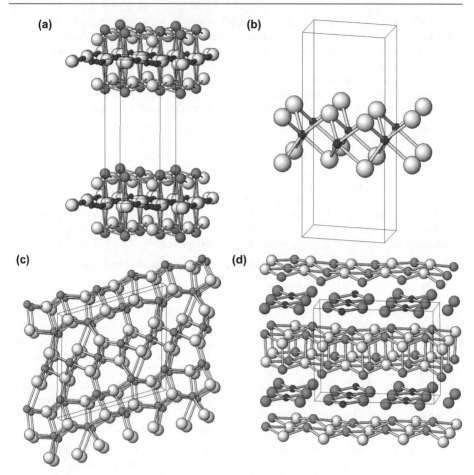

Figure 15. (a, b) Components of the layer-misfit structure of $(Pb_2FeS_3)_{0.58}NbS_2$ (Lafond et al. 1999). (a) The pseudotetragonal layer with Pb on the outside and statistical Fe sites in the interior, and (b) the intervening pseudohexagonal NbS_2 layer of trigonal prisms. (c) $Cr_3Sn_2S_7$ (Jobic et al. 1994). Pseudotetragonal layers are two atomic sheets thick and host primarily Sn^{2+}; pseudohexagonal (octahedral) layers contain primarily Cr^{3+}. Note the semi-commensurate match of the respective subperiodicities. (d) Buckhornite (Effenberger et al. 2000). Pseudotetragonal layers with Pb and Bi (undifferentiated, grey) in a square-pyramidal coordination are intercalated by telluride sheets with square planar Au (dark) coordination in $AuTe_2$ ribbons.

They consist of a combination of: (a) pseudotetragonal sulfide layers, respectively 2 atomic sheets thick and composed of Pb and Bi in the case of buckhornite, and 4 atomic sheets thick in an SnS like arrangement and composed of Pb and Sb in the case of nagyagite, with (b) single-atom thick sheets of tellurium involved in square planar coordination to the atoms of gold. In buckhornite, 3 subcells of the pseudotetragonal layer match with two subperiods of the telluride layer. The latter consist of ribbons of edge-sharing coordination squares of Au (Au-S = 2.70 Å), separated by strips of van der Waals Te-Te interactions (Te-Te = 3.34 Å) (Fig. 15d). Both the four atomic sheets thick sulfide layers of nagyagite, with Sb situated in their interiors, and the intervening Te coordinations, are asymmetric causing the structure to be polar. Both layers, or at least the telluride sheet, are positionally disordered. All the positions observed in the telluride sheet are mixed Te and Au sites.

Crystal Structures of Sulfides & Chalcogenides

Table 12. Selected non-commensurate, two-layer "ABX_3" structures

Compound	Comp. Layer	Space Group	Lattice Parameters (Å,°)			Match Q/H	Ref.*
			a, α	b, β	c, γ		
OCTAHEDRAL H LAYERS							
$(LaS)_{1.20}CrS_2$	Q: LaS	$C\bar{1}$	5.936	5.752	11.040	1.675	[1]
			α 90.33	β 95.30	γ 90.02		
	H: CrS_2		5.936	3.435	11.045		
			α 93.29	β 95.30	γ 90.02		
$(PbS)_{1.12}VS_2$	Q: PbS	$C121$	5.728	5.789	23.939	1.778	[2]
				β 98.95			
	H: VS_2		5.728	3.256	23.939		
				β 98.95			
$(SmS)_{1.25}TiS_2$	Q: SmS	$C211$	5.494	5.818	10.991	1.593	[3]
			α 95.38				
	H: TiS_2	$C2/m11$	3.448	5.821	21.990		
			α 95.40				
$(SnS)_{1.20}TiS_2$	Q: SnS	tricl.	5.833	5.683	11.680	1.666	[4]
			α 94.78	β 95.85	γ 90.03		
	H: TiS_2		5.835	3.412	23.288		
			α 90.30	β 95.86	γ 90.01		
$(SnS)_6(InS_2)_4$	Q: SnS	$P12_1/m1$	11.643	3.784	12.628	1.5	[5]
				β 105.81			
	H: InS_2						
$(Sn_3Se_3)(Cr_{1.9}Sn_{0.04}Se_4)$		$P12_1/m1$	12.765	3.835	11.785	1.5	[6]
				β 105.21			
	H:$CrSe_2$						
TRIGONAL PRISMATIC H LAYERS							
$(LaS)_{1.15}NbS_2$	Q: LaS	$Cm2a$	5.828	5.797	11.512	1.761	[7]
	H: NbS_2	$Fm2m$	3.310	5.797	23.043		
$(CeS)_{1.14}TaS_2$	Q: CeS	$Cm2a$	5.737	5.749	11.444	1.742	[8]
	H: TaS2	$Fm2m$	3.293	5.752	22.892		
$(PbS)_{1.14}NbS_2$	Q: PbS	$Cm2a$	5.834	5.801	11.902	1.761	[7]
	H: NbS_2	$Cm2m$	3.313	5.801	23.807		
$(PbS)_{1.14}(NbS_2)_2$	Q: PbS	$Cmc2_1$	5.829	5.775	35.861	1.752	[9]
	$H_{1,2}$: NbS_2	$Cmc2_1$	3.326	5.776	35.876		
$(SmS)_{1.19}TaS_2$	Q: SmS	$Fm2m$	5.562	5.648	22.56	1.690	[7]
	H: TaS_2	$Fm2m$	3.293	5.679	22.50		
$(YS)_{1.23}NbS_2$	Q: YS	$Fm2m$	5.393	5.658	22.284	1.623	[10]
	H: NbS_2	$C2$	3.322	5.662	11.13		
				β 92.62			
$(BiS)_{1.11}NbS_2$	Q: BiS	$F2mm$	5.752	36.156	23.001	1.809	[11]
	H: NbS_2	$F2mm$	5.750	3.331	23.000		
$(Pb_3FeS_3)_{0.58}NbS_2$	Q: Pb_2FeS_3	$Cmmm$	5.763	5.795	14.081	1.732	[12]
	H: NbS_2	$Cm2m$	3.328	5.795	14.081		
$(BiS)_{1.09}TaS_2$	Q: BiS	$Pm2m$	3.135	2.984	12.174	1.833	[13]
	H: TaS_2	$Fm2m$	3.421	5.970	24.341		

References: [1] Kato 1990; [2] Onoda et al. 1990; [3] Cario et al. 1997; [4] Wiegers et al. 1992; [5] Adenis et al. 1986; [6] Jobic et al. 1994; [7] Wiegers et al. 1990; [8] de Boer et al 1991; [9] Meerschaut et al. 1990; [10] Rabu et al. 1990; [11] Gotoh et al. 1995; [12] Lafond et al. 1999; [13] Petříček et al. 1993

Cylindrite family: a doubly-noncommensurate example

Noncommensurability in two dimensions is typical for the chemically complex structures of the cylindrite-franckeite family of Pb-Sn^{2+}-Sn^{4+}-Sb-Fe sulfides, rather recently broadened to include similar compounds with Bi (and Cu) or As as well as synthetic selenides (Table 13). In this family, layers which are two or four atomic layers thick and based on the SnS archetype, alternate regularly with octahedral, essentially SnS_2 layers. Minor elements, Sb (Bi,As), and Fe (Cu) are believed to be distributed over both layer types.

The type compound, cylindrite, with average composition $FePb_3Sn_4Sb_2S_{14}$, has the coincidence mesh in terms of b and c parameters of pseudotetragonal subcells 19,0/0,13, or in terms of orthohexagonal subcells 30,0/0,12 (see below). Layer stacking proceeds in the a direction (Fig. 16b). Common geometrical and compositional modulation occurs along [001], resulting in the semi-commensurate match 13Q/12H. Along [010], there is incommensurate match, expressed as the empirical ratio 19Q/30H without a manifest modulation, i.e., the layers only have the repetition periods 5.79 Å and 3.67 Å, respectively (Makovicky 1976).

Pseudotetragonal layers are essentially (Pb,Sn^{2+})S with substitutions of minor elements; with the change in the Pb:Sn^{2+} ratio in these layers, a change in the c_Q/c_H ratio occurs, giving matches from 12Q/11H to 16Q/15H; variation in the ratio of b parameters was not systematized. The synthetic, all-Sn selenide cylindrite has a c_Q/c_H ratio of 11.5Q/10.5H, apparently as a result of containing selenium instead of sulfur (Makovicky et al. in prep.; Fig. 16b).

If the ratio of b axes is fixed in a simple way as $7b_Q = 11b_H$, the 12/11 ratio means a composition $(12 \times 7)M_4S_4 + (11 \times 11)M_2S_4$, resulting in an M_Q/M_H ratio of 1.388 and M/S ratio of 0.7048. In the same way, the ratio 16/15 gives an M_Q/M_H ratio equal to 1.358 but an M/S ratio of 0.7022, i.e., a very small change in the resulting valence distribution in this variable-fit series (Makovicky and Hyde 1992). Levyclaudite is a Pb rich, Cu-Bi containing analog of cylindrite.

Franckeite (Table 13) is closely related to cylindrite and displays all the above variations and a similar variable-fit series (Henriksen et al. 2002). It is an accretion homologue of cylindrite: the Q layers are 2 atomic layers thick in the latter, whereas they are 4 such layers thick in franckeite, stacked in a SnS-like fashion. Lengenbachite, approximating $Pb_{182}Ag_{45}Cu_{20}As_{117}S_{390}$ is a closely related structure with the semi-commensurate match fixed

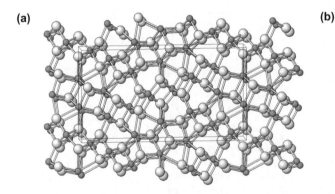
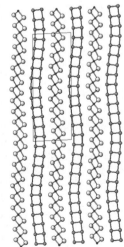
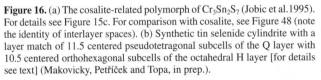

Figure 16. (a) The cosalite-related polymorph of $Cr_3Sn_2S_7$ (Jobic et al.1995). For details see Figure 15c. For comparison with cosalite, see Figure 48 (note the identity of interlayer spaces). (b) Synthetic tin selenide cylindrite with a layer match of 11.5 centered pseudotetragonal subcells of the Q layer with 10.5 centered orthohexagonal subcells of the octahedral H layer [for details see text] (Makovicky, Petříček and Topa, in prep.).

Table 13. Selected crystallographic data on the members of the cylindrite-franckeite series.

Compound		Q component		H component		Coin. data	Ref.
		Subcell	Unit cell	Subcell	Unit cell		
Cylindrite ~$FePb_3Sn_4Sb_2S_{14}$	a	11.73 Å	11.73 Å	11.71 Å	11.71 Å	13Q/12H	[1]
	b	5.79 Å	5.79 Å	3.67 Å	3.67 Å		
	c	5.81 Å	75.53 Å	6.32 Å	37.92 Å		
	α	90°	90°	90°	90°		
	β	92.38°	92.38°	92.58°	92.58°		
	γ	93.87°	93.88°	90.85°	90.85°		
	S.gr.	A1	A1	A1	P1		
Potosiite ~$Pb_{24}Ag_{0.2}Sn_{8.8}$ $Sb_{7.8}Fe_{3.74}S_{55.6}$	a	17.28 Å	17.28 Å	17.28 Å	17.28 Å	16Q/15H	[2]
	b	5.84 Å	5.84 Å	3.70 Å	3.70 Å		
	c	5.88 Å	188.06 Å	6.26 Å	188.06 Å		
	α	90°	90°	90°	90°		
	β	92.2°	92.2°	92.2°	92.2°		
	γ	90°	90°	90°	90°		
	S.gr.	A1 or A$\bar{1}$	P1 or P$\bar{1}$	A1 or A$\bar{1}$	A1 or A$\bar{1}$		
Franckeite natural	a	17.3 Å		17.3 Å		16Q/15H	[3]
	b	5.84 Å		3.68 Å			
	c	5.90 Å		6.32 Å			
	α	91°		91°			
	β	95°		96°			
	γ	88°		88°			
	S.gr	A$\bar{1}$		A$\bar{1}$			
Lévyclaudite subcell	a	11.84 Å		11.84 Å		13Q/12H	[4]
	b	5.83 Å		3.67 Å			
	c	5.83 Å		6.31 Å			
	α	90°		90°			
	β	92.6°		92.6°			
	γ	90°		90°			
	S.gr.	A2/m, A2 or Am		A2/m, A2 or Am			
Langenbachite	a	36.89 Å	36.89 Å	36.89 Å	36.89 Å	12Q/11H	[5]*
	b	5.84 Å	11.68 Å	3.90 Å	11.68 Å		
	c	5.85 Å	70.16 Å	6.38 Å	70.16 Å		
	α	90°	90°	90°	90°		
	β	90°	90°	90°	90°		
	γ	91°	91°	91°	91°		
		A centered	A centered	A centered	A centered		

*Also other variants and stacking disorder of the H component are reported.
References: [1] Makovicky 1976; [2] Wolf et al. 1981; [3] Wang 1989; [4] Moëlo et al. 1990; [5] Makovicky et al. 1994

at 12Q/11H subcells in the c direction and at exactly 2Q/3H subcells in the b direction.

The match of the 13Q subcells tall A-centered "Q" cell with two 6H subcells tall, primitive orthohexagonal cells in cylindrite, and the corresponding 11.5Q/10.5H match in its synthetic Sn selenide analog, results in its OD character with 2 positions of the Q layer, ½ b_Q apart. The OD character becomes dominant in lengenbachite. It is caused by coincidences due to the combination of the $12c_Q/11c_H$ match with the exact $2b_Q/3b_H$ match. These coincidencies result in complete disorder of the H layer in lengenbachite (Makovicky et al. 1994).

STRUCTURES WITH 3D-FRAMEWORKS

Adamantane structures

The structures treated in this section can be considered as homeotypes of the structures of (cubic) diamond and lonsdaleite, the hexagonal diamond polymorph. Consequently, they have been named the adamantane structures by Parthé (1990). The common feature of adamantane structures is tetrahedral coordination of cations and anions. Simple adamantane structures are normal compounds, i.e., when all sites are occupied (C_nA_n) the valence electron concentration (VEC) on cations is four and that on anions, VEC_A is 8 (Parthé 1990). In defect adamantane structures (C_{n-p} vacancy$_p$ A_n), VEC_A is still 8 but VEC has to be larger than four. The anions surrounding the cation vacancy have non-bonding orbitals present. Experience shows that the defect adamantane structures exist for $4 \leq VEC \leq 4.923$, or the maximum number of vacancies are three out of eight potential tetrahedral cation sites (Parthé 1990). Cases with anion vacancies are rather rare and are connected with the presence of non-bonding "lone" electron pairs on some cations (such as Sb and As).

The adamantane compounds (Table 14) can be classified according to several criteria. The universally accepted fundamental categories are: (a) the homeotypes of cubic sphalerite; (b) those of hexagonal wurtzite, and (c) a few structures with combined hexagonal and cubic stacking sequences of cation tetrahedra. The site symmetries for cation tetrahedra in the ZnS aristotypes are $\overline{4}3m$ and $3m$, respectively, imposing different constraints on the occupying cations; these are transferred to a different degree onto cations in the homeotypes.

Across the boundaries of this classification, the adamantane compounds can be divided into:

(i) normal compounds C_nA_n, with both the cation and anion sites fully occupied;
(ii) compounds (omission derivatives) with vacancies in certain tetrahedral sites:
   (a) cation vacancies (e.g., Ga_4GeS_8);
   (b) anion vacancies (e.g., sinnerite $Cu_6As_4S_9$);
(iii) compounds ("stuffed derivatives") with additional cation tetrahedra occupied, in excess of the aristotype composition C_nA_n (e.g., talnakhite, $Cu_{18}Fe_{16}S_{32}$);
(iv) compounds with excess cation tetrahedra and cation vacancies present in one structure;
(v) structures with more complex modifications of certain parts (e.g., tetrahedrite, $Cu_{10}(Fe,Zn)_2Sb_4S_{13}$).

The third classification principle divides the tetrahedral structures into binary, ternary, quaternary, etc., according to the number of distinct cations present in ordered substitution for the Zn of the aristotypes. This replacement results in a "tree" of sub- and supergroup relationships between the ZnS aristotypes and the homeotypes of various compositions. Ordered vacancies enter these relationships as well. Recently, Pfitzner and Reiser (2002), Pfitzner and Bernert (2004), and Bernert and Pfitzner (2005) have stressed the importance of the variability in volumes of different tetrahedra in the tetrahedral compounds (i.e., in atomic radii of the species occupying these sites) for the preference of the structure for a wurtzite or sphalerite stacking principle.

Sphalerite–wurtzite. The two fundamental polytypes, cubic sphalerite and hexagonal wurtzite, differ by possessing *ccp* and *hcp* arrays of anions, respectively. Occupancy of a half of the tetrahedral positions in them, with the coordination tetrahedra of cations sharing only their corners, gives respective stackings of tetrahedral layers which have layer symmetry $P3m1$. When the tetrahedral positions in the initial layer are denoted as A, tetrahedra of the next layer can alternatively be positioned above the triangular voids of the first layer or above the S atoms joining three adjacent tetrahedra in this layer. Stacking according to the first principle results in

Table 14. Adamantane structures and their derivatives.

Mineral	Chemical Formula	Space Group	Lattice Parameters (Å,°)				Ref.
			a	b	c	Angle	
Chalcopyrite	$CuFeS_2$	$I\bar{4}2d$	5.289	5.289	10.423		[1]
Wurtzite	ZnS	$P6_3mc$	3.811	3.811	6.234	γ 120	[2]
Enargite	Cu_3AsS_4	$Pmn2_1$	7.413	6.440	6.158	—	[3]
Synth.	Cu_2MnGeS_4	$Pmn2_1$	7.635	6.527	6.244	—	[4]
Nowackiite	$Cu_6Zn_3As_4S_{12}$	$R3$	13.440	13.440	9.17	γ 120	[5]
Sinnerite	$Cu_6As_4S_9$	$P1$	9.064	9.830	9.078	α 90.00	[6]
						β 109.50	
						γ 107.80	
Talnakhite	$Cu_{18.3}Fe_{15.9}S_{32}$	$I\bar{4}3m$	10.593	10.593	10.593	—	[7]
Haycockite	$Cu_4Fe_5S_8$	$P222$	10.705	10.734	31.630	—	[8]
Mooihoekite	$Cu_9Fe_9S_{16}$	$P\bar{4}2m$	10.585	10.585	5.383	—	[9]
Colusite	$Cu_{24}As_6V_2S_{32}$	$P\bar{4}3n$	10.538	10.538	10.538	—	[10]
Arsenosulvanite	$Cu_{25.1}As_{5.6}V_{2.2}S_{32}$	$P\bar{4}3n$	10.527	10.527	10.527	—	[11]
Germanite	$Cu_{26}Ge_4Fe_4S_{32}$	$P\bar{4}3n$	10.586	10.586	10.586	—	[12]
Renierite	$Cu_{9.6}Fe_{3.6}Zn_{0.8}Ge_{1.5}As_{0.2}S_{16}$	$P\bar{4}2c$	10.623	10.623	10.551	—	[13]
Mawsonite	$Cu_6Fe_2SnS_8$	$P\bar{4}2m$	7.603	7.603	5.358	—	[14]
Sphalerite	ZnS	$F\bar{4}3m$	5.417	5.417	5.417	—	[15]
Stannite	$Cu_2Fe_{0.8}Zn_{0.2}SnS_4$	$I\bar{4}2m$	5.449	5.449	10.757	—	[16]
Kesterite	$Cu_2Zn_{0.7}Fe_{0.3}SnS_4$	$I\bar{4}$	5.427	5.427	10.871	—	[16]
Briartite	Cu_2FeGeS_4	$I\bar{4}2m$	5.325	5.325	10.51	—	[17]
Černyite	Cu_2CdSnS_4	$I\bar{4}2m$	5.487	5.487	10.848	—	[18]
Lenaite	$AgFeS_2$	$I\bar{4}2d$	5.66	5.66	10.3	—	[19]
Gallite	$CuGaS_2$	$I\bar{4}2d$	5.347	5.347	10.474	—	[20]
Synth.	$CuGaSe_2$	$I\bar{4}2d$	5.596	5.596	11.004	—	[20]
Synth.	$CuGaTe_2$	$I\bar{4}2d$	6.023	6.023	11.940	—	[21]
Famatinite	Cu_3SbS_4	$I\bar{4}2m$	5.385	5.385	10.754	—	[22]
Luzonite	Cu_3AsS_4	$I\bar{4}2m$	5.332	5.332	10.57	—	[23]
Stannoidite	$Cu_{16}Fe_{4.3}Zn_{1.7}Sn_4S_{24}$	$I222$	10.767	5.411	16.118	—	[24]
Sulvanite	Cu_3VS_4	$P\bar{4}3m$	5.393	5.393	5.393	—	[25]
Lautite	CuAsS	$Pna2_1$	11.35	5.546	3.749	—	[26]
Synth.	CuAsSe	$Pbcn$	11.75	6.79	19.21	—	[27]
Cubanite	$CuFe_2S_3$	$Pcmn$	6.457	11.093	6.223	—	[28]
Catamarcaite	Cu_6GeWS_8	$P6_3mc$	7.524	7.524	12.390	—	[29]
Kiddcreekite	Cu_6SnWS_8	$F***$	10.856	10.856	10.856	—	[30]
Synth.	$CdGa_2S_4$	$I\bar{4}$	5.553	5.553	10.172	—	[31]

References: [1] Hall and Stewart 1973; [2] Xu and Ching 1993; [3] Karanovic et al. 2002; [4] Bernert and Pfitzner 2005; [5] Marumo 1967; [6] Makovicky and Skinner 1975; [7] Hall and Gabe 1975; [8] Rowland and Hall 1975; [9] Hall and Rowland 1973; [10] Spry et al. 1994; [11] Frank-Kamenetskaya et al. 2002; [12] Tettenhorst and Corbato 1984; [13] Bernstein et al. 1989; [14] Szymanski 1976; [15] Agrawal et al. 1994; [16] Hall et al. 1978; [17] Wintenberger 1979; [18] Szymanski 1978; [19] Zhuze et al. 1958; [20] Abrahams and Bernstein 1973; [21] Leon et al. 1992; [22] Garin et al. 1972; [23] Marumo and Nowacki 1967; [24] Kudoh and Takeuchi 1976; [25] Mujica et al. 1998; [26] Craig and Stephenson 1965; [27] Whitfield 1981; [28] McCammon et al. 1992; [29] Putz et al. 2005; [30] Harris et al. 1984; [31] Gastaldi et al. 1985

layers with parallel orientation and a cubic stacking ABCABC... whereas stacking according to the second principle results in every other layer being in antiparallel orientation (its tetrahedra "reversed" in respect to the preceding layer) and the stacking sequence becomes AB'AB'. We choose to denote the "reversed" layers by priming. These situations are shown in Figs. 17a-b and 17c-d, respectively. Natural representatives are almost always these two polytypes, 3C (3-

layer repeat, cubic sphalerite) and 2H (2-layer repeat, hexagonal wurtzite). In a parallel notation used for polytypes (O'Keeffe and Hyde 1996), they possess stacking schemes *ccc* and *hh* where each symbol *c* signifies formation of a layer pair according to the cubic stacking principle and *h* a corresponding layer pair according to the hexagonal stacking principle. The reader is invited to consider these two notations using the examples given below. Cation site symmetry differs in these two structures, being $\overline{4}3m$ and $3m$, respectively. Thus, the Zn-S distances are 4×2.35 Å in sphalerite, and 3×2.31 Å plus 1×2.32 Å in wurtzite.

At present, 60 structure determinations on ZnS polytypes have been published (ICSD), not counting the repeated investigations on certain polytypes under non-ambient conditions. All these data fall into three categories:

1. Polytypes 2,4,6,8,10,12,14-, and 16H, as well as 20H, which have space group $P6_3mc$. The layer multiplicities are divisible by two, i.e., their layer stacks consist of two equal portions, always joined by an *h* stacking event. Thus, 4H has the stacking sequence *hchc*, 8H is *ccchcch*, 10H (Fig. 18a) is *ccchccch*, and 20H is $(cccccccch)_2$. In terms of tetrahedral layers 4H is ABC′B′, 8H is ABCAB′A′C′B′, and 20H is C(ABC)$_3$A′(C′B′A′)$_3$.

2. Polytypes 15,18,24,30,36,42,48,54,60,66,72,78,84-, and 114R with the space group $R3m$. This group includes the cubic 3C polytype, in the structure of which a rhombohedral cell can be outlined as well. Their layer multiplicities are divisible by 3 and above 15H, also by six. Divisibility by three predicts a structure composed of three equivalent layer stacks in a rhombohedral arrangement which does not explain the additional divisibility by two. The lowest polytype, 3C, represents a *ccc* sequence (ABC) whereas the 15H polytype is an exceptional structure, *hhhhchhhhchhhhc*, i.e.,

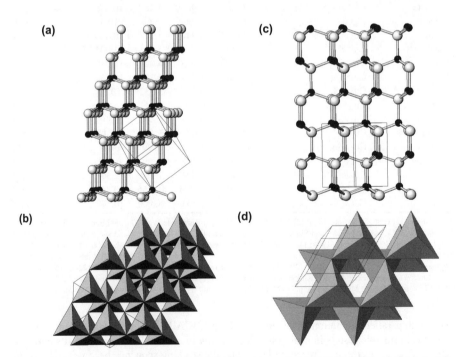

Figure 17. (a, b) Cubic 3C polytype of ZnS, sphalerite (Agrawal et al. 1994). Zn (dark) in tetrahedral coordination; ABC stacking of corner-sharing ZnS$_4$ tetrahedra. (c, d) Hexagonal 2H polytype of ZnS, wurtzite (Xu and Ching 1993). AB′AB′ stacking of ZnS$_4$ tetrahedra.

AB'ABC'BC'BCA'CA'CAB', not known from the next higher homologue, 18H. This latter structure and the higher polytypes are actually composed of two alternating layer stacks with cubic sequences, always separated by a single *h* stacking event. Their multiplicities are in the ratio $n:(n+2)$. Thus, 18R has the stacking sequence *hccchchcccchchcccchc*, and 24R is $(ccccchcch)_3$. In the alternative notation these sequences are AB'A'C'B'ABC'B'A'C'BCA'C'B'A'C, and ABC'B'A'C'B'ABCA'C 'B'A'C'BCAB'A'C'B'A'C. Finally, 36R has the stacking $(cccccchcccch)_3$, starting as ABCABC'B'A'C'B'A'C', and followed by the rhombohedrally displaced variations of this sequence. As a result of the above-defined difference in layer multiplicities, each of the two cubic subsets is arranged in a rhombohedral stacking sequence (Fig. 18b); this difference also explains the divisibility by two.

3. Polytypes 10,12,14,16,18,20,22,24,26,28-, and 44H, with layer multiplicities again divisible by two, but with the terms below the 10-fold stack missing. The multiplicities combined with the space group *P3m* indicate that these structures ought to be composed of two alternating *c*-type layer stacks with generally different multiplicities, always separated by a single *h* event. Indeed, the *P3m*-10H structure (Fig. 18c; BAC'A'B'C'A'B'C'A') is *hchccccccch* and *P3m*-20H is *hchccccchccccccchcc*, as two examples. Both these cases have $P6_3mc$ analogs which were quoted above.

Formation of the short-period polytypes obviously is a result of crystal-chemical factors. A lot of literature deals with the models for long-period polytypes. The ANNNI model, for example (Angel et al. 1985), and others have been summarized by Baronnet (1997). However, when amalgamated, the ZnS polytypes listed here amount to a fairly random incidence of combinations of two different or equal layer-stack thicknesses which, whenever they happen to satisfy a particular multiplicity ratio, produce a certain type of space group symmetry. There might be a human factor enhancing the frequency of observation of the first two categories from the above list in comparison to the last one, for which the structure determination is more difficult. The reader may have noticed the presence of "sphalerite-like blocks" of different thicknesses in the above polytype sequences. Mardix (1986) suggests that these thicknesses are a function of formation temperature. Propagation of these long-period sequences would be a result of growth mechanisms, such as screw dislocations perpendicular to the layer stacks.

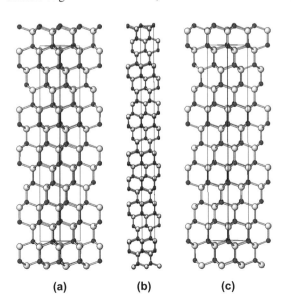

Figure 18. (a) The hexagonal 10H polytype of ZnS, space group $P6_3mc$. Zn: small black spheres. Projection on $(11\bar{2}0)$. The *h* stacking preserves orientation of ZnS_4 tetrahedra, *c* stacking (2 events) reverses it. (b) The rhombohedral 18R polytype of ZnS, space group *R3m*. A regular, repeating combination of *h* and *c* stacking principles. (c) The hexagonal 10H polytype of ZnS with the space group *P3m*1. Only two closely spaced events of *c* stacking (close to the top of the figure) occur in the *h* sequence. Compare this stacking sequence with that in (a); for discussion of all three sequences see the text.

Structures based on the sphalerite stacking principle

Chalcopyrite. The aristotype structure for ternary sulfides $C(1)C(2)A$ is that of chalcopyrite (CuFeS$_2$) (Fig. 19a). The tetragonal unit cell has $a_1 = a_2$ equal to that of the sphalerite aristotype, whereas the c parameter is twice the a parameter of sphalerite. The space group of this structure type is $I\bar{4}2d$. The structure consists of identical layers (001), with alternative cation sites (e.g., 000 and ½½0) occupied by the two cations (Cu and Fe, respectively in chalcopyrite). In the c direction, four such layers follow one another, with origin shifts ½a_1, ½a_2, –½a_1, –½a_2, producing the familiar Fe/Cu distribution in the unit cell of chalcopyrite. Neutron diffraction studies (Donnay et al. 1958) and Mössbauer investigations (summarized by DiGiuseppe et al. 1974) suggest that Cu in chalcopyrite is univalent whereas iron is trivalent. The Fe tetrahedron is very regular (Fe-S= 2.256 Å; tetrahedral angles 109.4°-109.6°), that of Cu is somewhat distorted (Cu-S=2.299 Å; 108.68°-111.06°). Lenaite AgFeS$_2$ (Zhuze et al. 1958) and gallite CuGaS$_2$ (Abrahams and Bernstein 1973) are isostructural, as also are CuGaSe$_2$ (Abrahams and Bernstein 1974), the technologically important CuInS$_2$ (Abrahams and Bernstein 1973), and CuGaTe$_2$ (Leon et al. 1992) (Table 14).

Kesterite and stannite. According to Hall et al. (1978), the normal quaternary adamantanes fall into two categories: the kesterite Cu$_2$ZnSnS$_4$ structure type, with the space group $I\bar{4}$, and the stannite Cu$_2$FeSnS$_4$ structure type, space group $I\bar{4}2m$. In terms of (001) tetrahedral layers, these are 2-layer structures, with two such layer pairs in one c period.

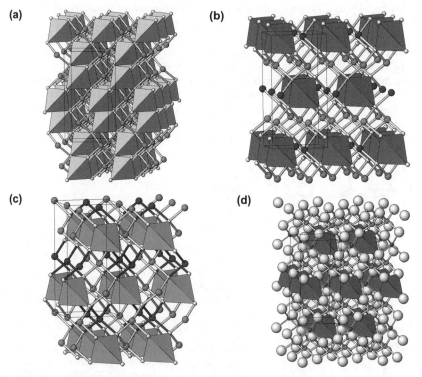

Figure 19. (a) Chalcopyrite CuFeS$_2$ (Hall and Stewart 1973). Fe coordination tetrahedra are shown as solid tetrahedra, the coordination of Cu (medium grey) by Cu-S bonds. (b) Stannite Cu$_2$Fe$_{0.8}$Zn$_{0.2}$SnS$_4$ (Hall et al. 1978). Solid tetrahedra: Fe, grey cations: Cu, black cations: Sn. (c) Kesterite Cu$_2$Zn$_{0.7}$Fe$_{0.3}$SnS$_4$ (Hall et al. 1978). Solid tetrahedra: Zn, grey and black cations: Cu and Sn, respectively. (d) Černyite Cu$_2$CdSnS$_4$ (Szymanski 1978). Solid tetrahedra: Sn, black cations: Cd, grey cations: Cu.

In stannite ($Cu_2Fe_{0.8}Zn_{0.2}SnS_4$, Hall et al. 1978), a layer of Fe and Sn tetrahedra (at $z = 0$ and ½; note I centration), occupied in a chess-board fashion by these two cations, alternates with a layer of copper tetrahedra (at $z = ¼$ and ¾)(Fig. 19b). According to Hall et al. (1978), in kesterite, $Cu_2Fe_{0.3}Zn_{0.7}SnS_4$, both layers have a chess-board occupation scheme: the tetrahedral layer at $z = 0$ and ½ is occupied by Sn and Cu whereas that at ¼ and ¾ by Zn and Cu. Hall et al. suggest that not only the Cu-Sn layer but also the Cu-Zn layer is ordered (Fig. 19c); the latter ordering reduces the space group to $I\bar{4}$.

Bonazzi et al. (2003) examined six crystal structures along the join stannite–kesterite quenched from 750 °C. They confirmed the structure scheme for stannite but found that the space group symmetry $I\bar{4}2m$ is valid for all compositions. As the Zn content increases, Zn moves progressively into the layer which was occupied by pure Cu in stannite and Cu moves into the Fe sites of stannite. For 70 mol% of stannite component, Cu is dominant in the latter site, which is marked by a break in the trend of unit cell parameters. The c parameter is rising whereas the a parameter is slightly decreasing until they "meet" at this point, as $2a \approx c$. The structure becomes and stays pseudocubic, with the above value approximately equal to 10.86 Å. Bonazzi et al.(2003) found that Cu and Zn are statistically distributed in the layers at $z = ¼$ and ¾, giving the space group $I\bar{4}2m$. The question whether there is a difference in the state of ordering between natural, hydrothermal samples and the synthetic products from 750 °C remains open. The same holds for the existence of a synthetic cubic stannite described by several investigators using powder diffraction, and recently refined by Rietveld method by Evstigneeva et al. (2001). For a complete list of stannite-kesterite related investigations see Bonazzi et al. (2003).

Other representatives. Briartite, Cu_2FeGeS_4 (Winterberger 1979) and černyite $Cu(Cd,In,Fe)SnS_4$ (Szymanski 1978) are isostructural with stannite, $I\bar{4}2m$. The alternation of the structural layers with larger cations (Cd and Sn) and those with smaller cations (Cu) leads to a marked flattening of Cu tetrahedra in černyite in the direction along the four-fold inversion axis (Fig. 19d). Famatinite Cu_3SbS_4 (Garin et al. 1972), with the same symmetry, $I\bar{4}2m$, has layers in which Sb tetrahedra alternate with tetrahedra of Cu; these layers are sandwiched by the tetrahedral layers which contain only copper (Fig. 20a). Thus, in spite of its 3:1 cation composition, it follows the same structural principles as the quaternary sulfides. Luzonite, Cu_3AsS_4 (Marumo and Nowacki 1967) has the same structure.

Structures based on the wurtzite stacking principle

Enargite (the latest one of several structure determinations is by Karanović et al. 2002), is a substitution derivative of wurtzite. The mutually identical tetrahedral (Cu_3AsS_4) (0001) layers have tetrahedra pointing in the [00$\bar{1}$] direction. They are trigonal nets of Cu tetrahedra, one quarter of which are periodically replaced by As tetrahedra (Fig. 20b). The latter tetrahedra lie on lines that are one lattice a spacing (or one b spacing) apart; the resulting layer symmetry is Pm.

A more complicated substitution scheme is found in *Cu_2MnGeS_4* (Bernert and Pfitzner 2005), also based on a wurtzite-like stacking. A single layer again has symmetry Pm, with [100] rows of pure Cu tetrahedra alternating with the combined Ge-Mn-Ge-Mn rows. The cited authors use this compound as an example of how the large differences in the volumes of component tetrahedra favour the wurtzite-type of layer stacking.

Adamantane structures with additional, interstitial tetrahedra

Colusite, $Cu_{24}As_6V_2S_{32}$ (Spry et al. 1994), space group, $P\bar{4}3n$, is an example of a compound with an additional tetrahedron, interstitial to the simple tetrahedral motif, and occupied by vanadium (Fig. 21a). It shares edges—and short metal-metal (Cu-V) contacts—with 6 surrounding copper tetrahedra. In a two-layer (001) representation (here in all three directions <100>), the VAs_2Cu_6 layer with symmetry $P\bar{4}2m$, alternates with a Cu_7As_2 layer,

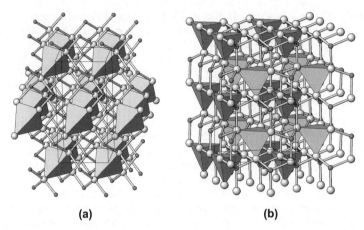

Figure 20. (a) Famatinite Cu_3SbS_4 (Garin et al. 1972). Solid tetrahedra: Sb^{5+}, grey cations: Cu^+. (b) Enargite Cu_3AsS_4 (Karanovic et al. 2002). Solid tetrahedra: As^{5+}, grey cations: Cu^+.

with the same layer group but with the origin shifted; in both layer types the cations other than copper are much "diluted" against the 1:1 situation observed in the chalcopyrite group. In a 3D description, the As tetrahedra form an icosahedron around every V tetrahedron, with edge lengths of 5.33 and 6.52 Å; it also encloses nine Cu atoms (Fig. 21b). The interstitial position of vanadium is best appreciated in the edge-on view of tetrahedral (111) layers. The octahedral metal clusters (VCu_6) are widely dispersed, and represent flat-lying octahedra between adjacent sulfur (111) layers in this view (Fig. 21c).

Colusite is a mineral with a rather flexible composition; the number of (Cu+Fe+Zn) atoms in a formula unit with 32 S atoms varies from 24 (the above described structure scheme) to 28, whereas the number of V atoms stays at about two and the sum of As+Sb+Sn+Ge replacing one another stays equal to 6. Spry et al. (1994), Wagner and Monecke (2005) and Frank-Kamenetskaya et al. (2002) propose substitution mechanisms $(As,Sb)^{5+} \leftrightarrow (Sn,Ge)^{4+} + Cu^+$, $(As,Sb)^{5+} + Cu^+ \leftrightarrow (Sn,Ge)^{4+} + (Fe,Zn,Cu)^{2+}$, and, with some differences in interpretation, $V^{5+} \leftrightarrow V^{4+} + Cu^+$. Two of these mechanisms involve filling of the tetrahedra not occupied in the structure model described here.

Arsenosulvanite is $Cu_{25.1}V_{2.2}As_{5.6}S_{32}$, space group $P\bar{4}3n$ (Frank-Kamenetskaya et al. 2002). Its structure is composed of non-intersecting <100> rods of edge-sharing, alternating V and As tetrahedra, immersed in a matrix of Cu tetrahedra (Fig. 21d). In a description by means of (100) layers, Cu-rich Cu_7As layers of corner-connected tetrahedra alternate with $V_3As_2Cu_7$ layers which display a number of shared tetrahedral edges. Similar to colusite, As tetrahedra form an icosahedron (edges of 5.26 and 6.45 Å). It is centered upon Cu1, which is the Cu atom with 6 close Cu neighbours (Cu-Cu = 2.74 Å). The As-icosahedron has an octahedron of vanadium atoms inscribed, vanadium being positioned on 6 edges of the icosahedron. The metal coordinations around Cu1 and V, respectively, form a perovskite-like framework of corner-connected octahedra (only Cu surrounds Cu1 whereas 2As + 4Cu is situated around V). The unit cell a parameter is twice that of perovskite, and the configuration is I-centered, with Cu1 at 0,0,0. Sulfur decorates every second octahedral face, forming a tetrahedron interstitial to the perovskite-like motif.

The structure contains little cube-like subunits defined by metal-metal contacts along their edges: V-As-V (2 × 2.63 Å), V-Cu-V (2 × 2.63 Å), and V-Cu-Cu (2.52 & 2.74 Å); these contacts are present (the last one twice) in each of the <100> M-M rows. The entire scheme of edge-sharing tetrahedra, leaving only Cu4 as a corner-sharing tetrahedron, is analogous to that

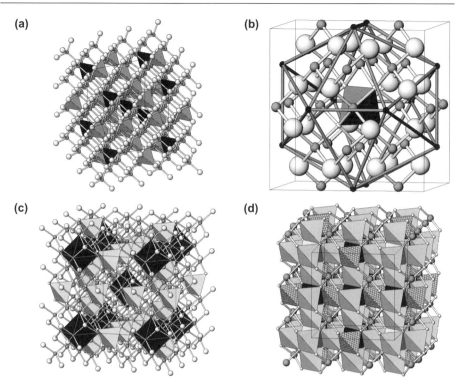

Figure 21. (a) Colusite $Cu_{24}As_6V_2S_{32}$ (Spry et al. 1994). Tetrahedral stacking (As: grey tetrahedra) with interstitial vanadium tetrahedra shown in black. (b) Icosahedral configuration of arsenic (black) around a central coordination tetrahedron of V in colusite; Cu atoms in grey. (c) Octahedral VCu_6 metal clusters (black) in the crystal structure of colusite. Coordination tetrahedra of As are grey. (d) Arsenosulvanite (Frank-Kamenetskaya et al. 2002). Edge-sharing solid tetrahedra form a "scaffolding": grey tetrahedra contain Cu, cross-hatched As, and black V. Interstitial, corner-sharing tetrahedrally coordinated cations: Cu (grey spheres).

observed in sulvanite but is populated by a complex combination of 3 elements. Cu1 and V are the nodes of this 3D framework. Furthermore, the positions of Cu4 are empty in sulvanite; these two phases are therefore quite distinct minerals.

Talnakhite, mooihoekite and haycockite. The crystal structure of talnakhite $Cu_{18.3}Fe_{15.9}S_{32}$, space group $I\bar{4}3m$ (Hall and Gabe 1972), contains eight interstitial tetrahedra in a unit cell of an otherwise regular 3D tetrahedral array with a sphalerite-like stacking. Only one such site is classified as Fe (at 000 and ½½½), the remaining 6 interstitial tetrahedra are Cu. The framework of edge-sharing tetrahedra (i.e., of *M-M* contacts) is three-dimensional, every such tetrahedron sharing edges with two interstitial tetrahedra (Fig. 22a). This results in a perovskite-like framework of corner-sharing cation octahedra (there are 6 Cu-ligands around each interstitial Fe or Cu atom), with S and additional tetrahedral Cu in the cuboctahedral cavities of this framework (Fig. 22b). The latter tetrahedra share only corners with the edge-sharing set of tetrahedra; one such tetrahedron occurs in every large cavity.

The tetragonal structure of mooihoekite $Cu_9Fe_9S_{16}$ ($P\bar{4}2m$) contains interstitial Fe atoms at 000 and ½½0 (Hall and Rowland 1973). There are infinite chains of metal-metal contacts parallel to [001], alternately Fe-Fe-Fe-Fe and Fe-Cu-Fe-Cu. However, the $Fe_{interstitial}Fe_2Cu_4$ and $Fe_{interstitial}Cu_2Fe_4$ clusters share only corners and not edges of common tetrahedra as in talnakhite (Fig. 22c). Octahedral metal-metal clusters share metal atoms along [001] but are isolated in the (001) plane; chains of octahedra result (Fig. 22d). In a layer description, the "defect mackinawite-type (001) layer," with some tetrahedra missing, alternates with a layer which contains only a few tetrahedra. These share their edges with the tetrahedra in the top and bottom layers; the bulk of tetrahedra in this layer are only corner-sharing tetrahedra of Fe and Cu.

Haycockite, $Cu_4Fe_5S_8$, space group $P222$ (Rowland and Hall 1975), is the most complicated of these "stuffed derivatives of chalcopyrite" (Fig. 23). Along [001], planes with interstitial Fe arranged in a chess-board pattern alternate with those arranged in a primitive square mesh. The former yield isolated octahedral clusters $Fe_{interstitial}Fe_4Cu_2$, with additional, corner-sharing tetrahedra (both Cu and Fe) capping the net openings in this layer; the latter are analogous to the (100) layer from talnakhite. However, the cluster composition is identical with the previous layer. The clusters in adjacent layers are shifted by one half of an octahedron height, so that the "wings," i.e., tetrahedra protruding from the two layer types towards the layer of the opposite kind meet only via common S atoms and no 3D pattern of metal-metal contacts is present (Fig. 23).

Stannoidite. In order to compare it with the 2-layer structures, such as those of stannite-kesterite groups, the crystal structure of stannoidite, $Cu_{16}Fe_{4.3}Zn_{1.7}Sn_4S_{24}$, space group $I222$, (Kudoh and Takeuchi 1976), should be considered in terms of tetrahedral (100) layers. In this

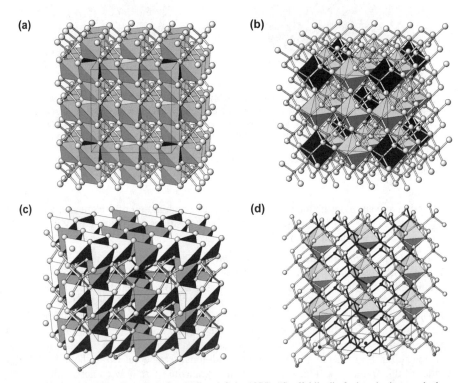

Figure 22. (a) Talnakhite $Cu_{18.3}Fe_{15.9}S_{32}$ (Hall and Gabe 1975). "Scaffolding" of edge-sharing tetrahedra; grey tetrahedra: Cu, black tetrahedra: Fe. Tetrahedral Cu sharing only tetrahedron corners: grey spheres. (b) The "perovskite-like" configuration of octahedral metal clusters in talnakhite. Light octahedra are Cu-centered, dark octahedra are Fe-centered. (c) Mooihoekite $Cu_9Fe_9S_{16}$ (Hall and Rowland 1973). Edge-sharing tetrahedra: solid light-coloured: Cu, solid dark-coloured: Fe. Interstitial, corner-sharing tetrahedra: ball-and-stick representation, Cu grey, Fe black. (d) A layer of the crystal structure of mooihoekite with chains [001] of Fe-centered octahedral metal clusters. Dark cations: Fe, light cations: Cu.

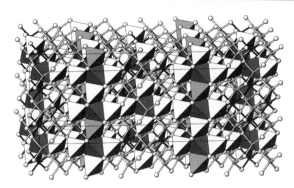

Figure. 23. The structure of haycockite $Cu_4Fe_5S_8$ (Rowland and Hall 1975). Fe: dark atoms and all-grey tetrahedra; Cu: light cations and light tetrahedra. Solid tetrahedra represent groups of edge-sharing tetrahedra (octahedral metal clusters). c axis is horizontal, b axis vertical; note the differences between the (001) layers at $z = 0$ and those at $z = ¼$.

interpretation, the layer at $x = 0$ and ½ contains Fe, Zn, Sn and both "normal" and "interstitial" Cu whereas the layer at $x = ¼$ and ¾ only "normal" tetrahedral Cu (Fig. 24a). In the layer at $x = 0$, Zn and "normal" Cu are surrounded by Sn and Fe. The interstitial tetrahedral Cu has two close Cu neighbours and two Fe neighbours in the $x = 0$ layer (at 2.71 Å and 2.74 Å, respectively) as well as two Cu neighbours in the adjacent Cu layers (at 2.70 Å), forming octahedral clusters interconnected along [010]. The intra-layer configuration of this layer along the [001] direction can be described as a cation sequence Zn-Sn-Fe-(Cu and Cu interstitial)-Fe-Sn-Zn. The layer at $x = ¼$ is a regular net of corner-connected coordination tetrahedra of Cu, positioned over vacant tetrahedra, as well as over the interstitial Cu atom, of the $x = 0$ layer.

Germanite, $Cu_{13}Fe_2Ge_2S_{16}$, space group $P\bar{4}3n$ (Tettenhorst and Corbato 1984) can again be described by means of two alternating <100> layers: (1) a pure Cu layer of corner-connected tetrahedra, with an additional, interstitial Cu tetrahedron inserted in a primitive (2a, 2a) pattern, and (2) a chalcopyrite-like Fe-Cu layer with a chess-board ordering of Fe and Cu. Interstitial Cu has 6 Cu neighbours at 2.72 Å, the resulting Cu_7 octahedral clusters are isolated, in an I-centered motif. Ge and other cations are not differentiated in the extant structure refinements. Each Cu_7 octahedron is surrounded by 8 Fe tetrahedra arranged as corners of a cube; the tetrahedra of Cu are positioned in the cube faces (Fig. 24b).

Renierite, $Cu_{9.56}Fe_{3.56}Zn_{0.8}Ge_{1.46}As_{0.16}S_{16}$ (Bernstein et al. 1989), a tetragonal structure with the space group $P\bar{4}2c$ and a pseudocubic cell is a structure with additional tetrahedral Fe forming octahedral clusters with four Cu and two Fe atoms as ligands (Fig. 24c). The framework of edge-sharing tetrahedra in renierite looks very much like that in arsenosulvanite but only Fe2 forms a tetrahedron present in the intersection of the rods of edge-sharing tetrahedra. As mentioned, it has four Cu and two Fe as close contacts. The corresponding tetrahedron in the other half of rod intersections, surrounded by 2 Zn, 2 Cu and 2 Ge tetrahedra is void, destroying the expected scheme of M-M contacts in these intersections. Out of these surrounding tetrahedra, only 2 copper-containing ones share edges with (i.e., form a part of a cluster around) the Fe2 tetrahedron. The tetrahedra of Zn and Ge do not share edges with any other tetrahedra, always being sandwiched between two empty tetrahedra.

In the (001) layer description, similar to arsenosulvanite, layers densely populated in an open-grid pattern by Fe, Zn, Ge and Cu, alternate with layers that contain corner-sharing Cu and Ge, as well as edge-sharing Cu and Fe.

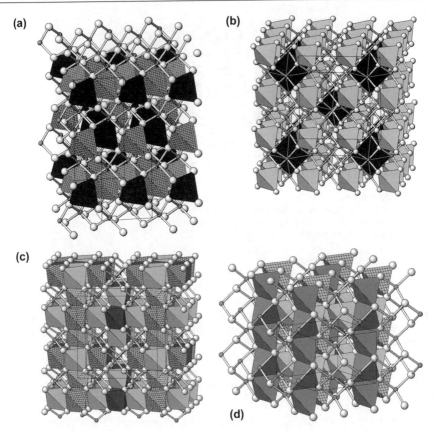

Figure 24. (a) Stannoidite $Cu_{16}Fe_{4.3}Zn_{1.7}Sn_4S_{24}$ (Kudoh and Takeuchi 1976). Fe: black tetrahedra, Sn: cross-hatched tetrahedra, Zn: stippled tetrahedra, Cu: grey spheres. (b) Germanite $Cu_{26}Ge_4Fe_4S_{32}$ (Tettenhorst and Corbato 1984). Copper-centered Cu_7 cation octahedra are black, Fe tetrahedra are solid grey; Cu atoms are grey spheres. (c) Renierite (Bernstein et al. 1989). Edge-sharing coordination tetrahedra: Fe – black, Cu – light grey, Zn – stippled, and Ge – cross-hatched. Corner-sharing tetrahedral cations: Cu –grey spheres. (d) Mawsonite $Cu_6Fe_2SnS_8$ (Szymanski 1976). Fe tetrahedra: dark, cross-hatched tetrahedra: Sn, grey cations: Cu.

Mawsonite, $Cu_6Fe_2SnS_8$, space group $P\bar{4}2m$ (Szymanski 1976) is based on a sphalerite-like packing of tetrahedra, with additional Fe forming interstitial tetrahedra. Columns of edge sharing tetrahedra of Fe define the $a \times a$ mesh, one set "normal," the other interstitial. The central row is built by corner-sharing Sn tetrahedra (Fig. 24d).

The interstitial tetrahedra share edges with four Cu tetrahedra in the (001) layer but these Cu tetrahedra from adjacent Fe-centered clusters share only corners with one another and with the Sn tetrahedra above and below. The remaining Cu tetrahedron shares only corners with adjacent tetrahedra. The octahedral clusters ($FeCu_4Fe_2$) form corner-connected chains parallel to [001], separated by columns of Cu tetrahedra.

The problem of interstitial cations in *hcp* and related stackings

Insertion of an interstitial tetrahedron is not possible in the adamantane structures with a hexagonal packing of anions. The interstitial tetrahedron would share a face with a regular tetrahedron of the wurtzite packing scheme, giving in fact a trigonal bipyramidal coordination. Two ways which nature has used to circumvent these problems are as follows.

Crystal Structures of Sulfides & Chalcogenides

In cubanite, $CuFe_2S_3$ (McCammon et al. 1992), a hexagonal close packing of sulfurs is maintained but the wurtzite-like structure is limited to two-octahedra wide (010) slabs, with Cu tetrahedra situated in the central portions and Fe tetrahedra being marginal (Fig. 25a). Adjacent slabs are related by two-fold rotations about [010] and the zig-zag surfaces of slabs are characterized by short Fe-Fe contacts (2.799 Å) across slab boundaries. This is expressed by edge-sharing of two adjacent Fe tetrahedra. The reversal of an entire wurtzite-like slab instead of an individual tetrahedron avoids the above mentioned problem of exceedingly close cation-cation contacts via a common face of two tetrahedra. Above 200-210 °C, cubanite transforms into "isocubanite," a cubic $F\bar{4}3m$ phase with $a = 5.296$ Å and disordered Cu and Fe, i.e., into an isotype of sphalerite (Cabri et al. 1973).

In catamarcaite Cu_6GeWS_8 (Putz et al. in press), the collision of a "regular" and an "inserted" cation is avoided by the presence of *hchc* stacking instead of the *hh* stacking of wurtzite type. The excess tetrahedron of tungsten is inserted into the bottom layer of a pair of upward-pointing layers stacked according to the sphalerite principle (Fig. 25b). It shares edges, and 2.73 Å contacts, with six surrounding Cu sites, forming an isolated regular metal octahedron. Tungsten is hexavalent; germanium in the structure is tetravalent, and is situated in corner-sharing tetrahedra with no short *M-M* contacts. There is a vacant Cu tetrahedron in

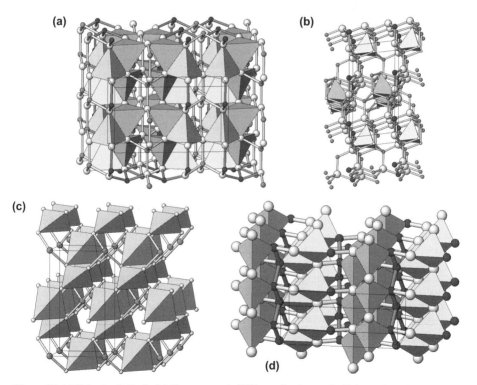

Figure 25. (a) Cubanite $CuFe_2S_3$ (McCammon et al. 1992). small spheres: Cu (light) and Fe (dark), solid tetrahedra: Fe. Note the edge-sharing of Fe tetrahedra. *b* axis is horizontal, *c* axis vertical. Slabs (010) of wurtzite-like arrangement are centered on the planes of Cu tetrahedra. (b) Catamarcaite Cu_6GeWS_8 (Putz et al. 2005). Grey tetrahedrally coordinated cations: Cu, black ones: Ge. Solid interstitial tetrahedra: W. (c) $CdGa_2S_4$ (Gastaldi et al. 1985). Gallium – solid tetrahedra of two kinds; Cd – grey spheres. (d) Lautite CuAsS (Craig and Sephenson 1965). Coordination tetrahedra of copper; anions: S - white spheres, As – black spheres. Covalent As-As bonds (dark).

the structure, below the WS_4 tetrahedron, along the $[00\bar{1}]$ direction. However, it shares only a corner with the WS_4 tetrahedron and the obvious reason for its existence is the bond saturation of the sulfur atom forming this corner. It is bound to one tungsten and three adjacent Cu atoms; a lone electron pair of this sulfur points into the empty tetrahedral cavity. This part of the structure is stacked according to the wurtzite principle.

An alternative stacking of these "sphalerite-like" double-layers in a cubic *cc* fashion leads to a cubic structure with the same type of cation clusters but arranged in such a way as to form an F-centered cubic structure. This obviously is the structure of kiddcreekite Cu_6SnWS_8 (Harris et al. 1984). The difference in stacking is apparently caused by the much larger tetrahedral radius of Sn^{4+} than that of Ge^{4+}.

Defect adamantane structures and more complicated cases

$CdGa_2S_4$. An often quoted example of a defect (omission-) tetrahedral structure is that of $CdGa_2S_4$ (Gastaldi et al. 1985). It displays a "chalcopyrite-like" cell but the symmetry is reduced to $I\bar{4}$. Fully occupied layers with a chess-board pattern of Cd- and Ga-occupied regular tetrahedra (at $z = 0$ and ½) alternate with half-empty layers of flattened Ga tetrahedra in a primitive $a \times a$ arrangement (Fig. 25c). These tetrahedra lie under one half of the empty tetrahedral spaces in the former layer and above such spaces in the next Cd-Ga layer which is displaced by (½a, ½a). The empty space is an irregular, flattened octahedron.

Sulvanite Cu_3VS_4, space group $P\bar{4}3m$ (Mujica et al. 1998) is a combined "interstitial and vacancy structure." In the (001) layer description, "defect mackinawite" VCu_3^j layers alternate with Cu_3^j layers (which are portions of the former layers in the (010) and (100) orientations). The six Cu-V contacts are 2.70 Å and the corner-connected VCu_6 octahedra outline a 3D perovskite-like motif, with S atoms at the vertices of flat three-fold pyramids above every second triangular Cu_3 face.

Lautite. The crystal structure of lautite CuAsS (Craig and Stephenson 1965) is often described as an ordered homeotype of the crystal structure of diamond, because no sphalerite-like distinction into the subsets of anion and cation sites, respectively, can be seen. Copper is coordinated tetrahedrally to three S and one As, whereas each As has two As, one S and one Cu as ligands, and each S is surrounded by three Cu and one As in a tetrahedral coordination. The As-As distances are 2.51 Å, the S-As distance 2.26 Å, whereas the Cu-S bonds vary between 2.25 and 2.36 Å, plus an additional Cu-As distance of 2.43 Å.

In a modular description, (100) slabs of tetrahedral S-Cu arrangement, two atomic layers thick are separated by corrugated (100) layers consisting of parallel zig-zag [001] chains of covalently bound arsenic atoms (Fig. 25d). The Cu-As and Cu-S bonds complete the tetrahedral coordination of all three elements involved. Orientation of Cu-centered tetrahedra in the adjacent (100) slabs reminds one of that in cubanite, i.e., they are oppositely oriented in the [012] and [01$\bar{2}$] directions. Nearly the same structure type is found for CuAsSe (Whitfield 1981). It is a superstructure of lautite in which, in each of the above (100) slabs, the orientation of CuS_3As tetrahedra along [010] is reversed after a period of three tetrahedra. Thus, each slab exhibits a *cch* stacking.

Summary of cation-cluster configurations

Talnakhite and arsenosulvanite are the phases with the largest concentration of (MM_6) octahedra; these form perovskite-like frameworks, with two species of central metals alternating in both phases. Sulvanite has a similar framework, populated only by clusters with central vanadium. Haycockite contains $(100)_{perovskite}$-type planes of interconnected cation octahedra, interleaved by layers of isolated M_7 octahedra between them. Parallel chains of metal octahedral clusters occur in mooihoekite, stannoidite and mawsonite, whereas isolated octahedra occur in colusite, germanite and renierite.

All "octahedral clusters" are clusters of seven edge-sharing tetrahedra; in the perovskite-like frameworks each "ligand" also belongs to an adjacent cluster. Perovskite-like structures can also be presented as 3D openworks composed of these edge-sharing tetrahedra, with corner-sharing tetrahedra filling the interstices. In all these structures we have to rely upon the author's own idea of Cu/Fe distribution among the cation sites; this might be problematic in some larger structures. Edge-sharing involves Cu, Fe, V, W, and (exceptionally) As. Although of fundamental importance for some minerals, Zn, Sn and Ge avoid edge-sharing clusters and their coordination tetrahedra are corner-sharing, as also is a portion of Cu in the structure. The Cu-Fe phases—talnakhite, mooihoekite and haycockite—transform above 165-185 °C into a cubic $F\bar{4}3m$ "intermediate solid solution" with $a \sim 5.36$ Å in which the interstitial cations are statistically distributed.

Tetrahedrite-tennantite

The crystal structure of $(Cu,Ag)_{10}(Fe,Zn,Hg,Cu,....)_2(Sb,As)_4S_{13}$, known in the literature as the tetrahedrite-tennantite isomorphous series (Wuensch 1964, Wuensch et al. 1966 and numerous other works quoted in the introduction), has habitually been interpreted as a derivative of the sphalerite structure with extensive omission, substitution and insertion of new sites. Voids in the tetrahedral framework have the shape of truncated tetrahedra, called also "Laves polyhedra," and are separated by single-tetrahedral walls (Fig. 26a). Each wall consists of a wreath of six tetrahedra around an originally tetrahedral position MS_4, which is always replaced by a trigonal coordination pyramid $(As,Sb)S_3$. Four such pyramids point into the cavity, filling it partly with the lone electron pairs of the metalloids. A further six copper sites of the original framework are altered into trigonal planar sites by coordinating to a new sulfur site placed in the centre of the cavity. Thus, the Laves polyhedron accommodates a "spinner" of six Cu coordination triangles, as well as the lone electron pairs of four metalloids in its walls (Fig. 26b). A pronounced anisotropic displacement ellipsoid of the trigonal planar Cu2 site has been found in all structure determinations. The Cu-S interatomic distances become more credible when a slightly displaced, flat-pyramidal copper position is assumed. The Rietveld refinements of powder neutron diffraction data on Fe-bearing tennantite and tetrahedrite (Andreasen et al. in prep.) suggest that this copper site is indeed flat pyramidal, statistical around the ideal triangular position.

An alternative picture of the structure of tetrahedrite-tennantite (Belov 1976; Johnson et al. 1987) postulates that the tetrahedral framework is a maximally collapsed sodalite-type framework. The large cage openings, hexagonal in sodalite, became reduced, and nearly perfectly trigonalized by clasping of the trigonal coordination pyramids of (As,Sb) by surrounding tetrahedra. This dual image is useful when considering the pure-copper tetrahedrite-tennantite with additional copper atoms in the unit cell (up to the ideal composition $Cu_{14}(Sb,As)_4S_{13}$; Makovicky and Skinner 1979; Pfitzner 1997; Makovicky et al. 2005). These atoms move inside (for As), or between, the cages, assuming new tetrahedral or trigonal-planar positions, respectively, which were not occupied in the basic structure. In the case of Sb, the original tetrahedral Cu sites appear to be involved in copper mobility.

The tetrahedrite-tennantite structure type $Cu_6^{[3]}(Cu,....)_6^{[4]}(Sb,As)_4^{[3]}S_{12}^{[4]}S^{[6]}$ allows extensive substitutions. Nature tries to avoid the presence of formally divalent Cu (i.e., the composition $Cu_{10}^+Cu_2^{2+}Sb_4S_{13}$) by substituting Zn, Fe, Hg, Cd, rarely also Mn, etc., in the tetrahedral sites (up to 2 sites out of 6). Synthetic tetrahedrites can also accommodate up to two atoms p.f.u. of Co, Ni, Mn, as well as partial substitutions by Sn, Ge, Pb, Ga and In (Hall 1972; Makovicky and Karup-Møller 1994; Klünder et al. 2003a,b; Karup-Møller and Makovicky 2004). Bismuth can partly substitute for As and Sb (Breskovska and Tarkian 1994; Klünder et al. 2003) whereas tellurium does so in full, following a substitution principle $Cu_{10}M_2^{2+}(Sb,As)_4S_{13} \rightarrow Cu_{12}Te_2(Sb,As)_2S_{13} \rightarrow Cu_{10}Te_4S_{13}$ (Trudu and Kittel 1998 and references therein; Karup-Møller and Makovicky in prep.). Here, copper vacancies occur in the triangular copper sites

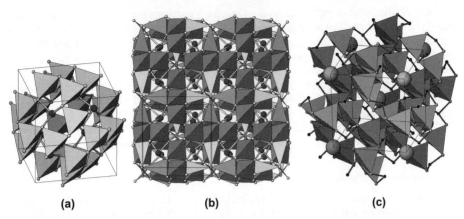

Figure 26. (a) A void Laves polyhedron in the structure of tetrahedrite $Cu_{10}(Zn,Fe)_2Sb_4S_{13}$ (Wuensch 1964). Tetrahedra containing (Cu,Zn,Fe) form the walls of a truncated tetrahedral cavity. Ball-and-stick SbS_3 pyramids close the openings in them. Four Sb atoms have lone electron pairs pointing into the cavity. (b) Tetrahedrite (Wuensch 1964), a complete structure. A sphalerite-like framework of (Cu, Zn, Fe) tetrahedra, with large truncated-tetrahedral cavities in an I-centered cubic arrangement, filled by "spinners" of six triangularly coordinated Cu atoms around a central S atom. A part of the framework tetrahedra are replaced by trigonal pyramidal coordinated Sb. (c) Galkhaite $(Cs,Tl)(Hg,Cu,Zn)_6As_4S_{12}$ (Chen and Szymanski 1981). Oblique projection upon (100). Tetrahedral openwork, with $(Hg,Cu,Zn)S_4$ tetrahedra and (AsS_3) pyramids (As: black small spheres) forming large cages which contain (Cs,Tl) (large spheres).

(Kalbskopf 1974). Electronic processes act as well, e.g., the selenium analog of tetrahedrite exists only as a fully substituted form, $Cu_{10}(Fe,Zn)_2Sb_4Se_{13}$, whereas the partly substituted, as well as the unsubstituted compound, $Cu_{~12}Sb_4Se_{13}$, known for the sulfur analog, is not stable (Karup-Møller and Makovicky 1999).

All these substitutions influence the unit cell parameter a. For example, the substitution by zinc leads to a (Å) = 0.0293 Zn (atoms p.f.u.) + 10.324 for the substitution line $Cu_{12}Sb_4S_{13}$-$Cu_{10}Zn_2Sb_4S_{13}$ and a (Å) = 0.319 Zn (atoms p.f.u.) + 10.448 for the line $Cu_{14}Sb_4S_{13}$-$Cu_{10}Zn_2Sb_4S_{13}$ (Karup-Møller and Makovicky 2004).

For $Cu_{12}Sb_4S_{13}$, copper acts as a mixture of 10 Cu^+ with the Shannon crystal radius of 0.635 Å (Shannon 1981) with two Cu^{2+} with a radius ~0.51 Å (Makovicky and Karup-Møller 1994). Although the mixed valence of Cu appears to be a collective property, this scheme works perfectly for the unit cell size calculations. Some of the older equations (especially Johnson et al. 1987) fail for partly substituted tetrahedrite-tennantite compositions because they do not address this phenomenon. Substitution by Ag is universally assumed to start in the triangular sites and spill over into tetrahedral sites only when the former are (nearly) filled. The presence or absence of the central S atom in the Laves polyhedra of the varieties heavily substituted by Ag is still contested.

Incorporation of iron into tetrahedrite-tennantite starts, at low concentrations, in the form of Fe^{3+}, ideally up to the composition of $Cu_{11}^+Fe^{3+}Sb_4S_{13}$. Afterwards, additional incorporation of Fe proceeds parallel to the conversion of the already incorporated Fe into Fe^{2+}, until the final composition $Cu_{10}^+Fe_2^{2+}Sb_4S_{13}$ is reached (Charnock et al. 1989; Makovicky et al. 1990). For broad compositions about Cu:Fe = 1:1, in tennantite (As) but to some extent also in tetrahedrite (Sb) iron with intermediate valence is observed, indicating strong electron delocalization/charge-transfer via intervening S (and Cu) atoms (Makovicky et al. 2003). The position of Fe in the structure has been contested. Charnock et al. (1989), working with EXAFS, place Fe^{3+} preferentially into trigonal planar sites and Fe^{2+} into tetrahedral sites.

Makovicky et al. (1990, 2003) place all Fe into tetrahedral sites, based on their Mössbauer studies. Finally, Andreasen, Lebech, Makovicky and Karup-Møller (in prep.) place Fe into tetrahedra at all stages of substitution, based on the Rietveld refinements of neutron diffraction data on synthetic tetrahedrite.

Galkhaite: a cage-like sulfide. In galkhaite $(Cs,Tl)(Hg,Cu,Zn)_6As_4S_{12}$ (Chen and Szymanski 1981) the Laves polyhedron of the tetrahedrite-tennantite structure is occupied by a single (rattling) Cs(+Tl) atom, and the size and a charge balance in the structure is achieved by a heavy substitution of tetrahedral Cu^+ by divalent metals, especially Hg^{2+}. It is cubic, a = 10.365(3) Å and the space group is $I\bar{4}3m$. Nearly regular $(Hg, Cu, Zn)S_4$ tetrahedra have M-S = 2.496 Å and the trigonal pyramids AsS_3 have As-S = 2.265 Å. (Cs,Tl) in the cavity centre (Fig. 26c) is 12-coordinated, with Cs-S distances equal to 3.863 Å. There are 19% vacancies in the position of the large cation. Chen and Szymanski (1981) suggest the exchange of 2 Tl for 3 Cs but we interpret their plot as the exchange of 0.8 Tl for 0.8 Cs, with the exception of the case with Cs ≈ 1.0.

Structures with anion vacancies

Nowackiite and sinnerite. A plesiotypic pair of anion-deficient structures consists of nowackiite, $Cu_6Zn_3As_4S_{12}$ (Marumo 1967) (as well as the isostructural aktashite $Cu_6Hg_3As_4S_{12}$) and sinnerite $Cu_6As_4S_9$ (Makovicky and Skinner 1975). Both structures are based on a sphalerite-like cubic stacking of tetrahedral layers. Selected clusters of four adjacent tetrahedra are replaced in them by four AsS_3 pyramids, with lone electron pairs of As^{3+} directed into a common volume with a missing anion (an octet of lone pair electrons). In nowackiite, a rhombohedral structure with a polar c axis (space group $R3$), the (Cu,Zn):As ratio guarantees that the As_4S_{12} clusters are isolated (Fig. 27a). A part of the Cu is replaced by Zn in order to maintain charge balance. In sinnerite, with a lower Cu:As ratio, the As_4S_{12} groups are partly condensed via common S atoms and the resulting As-S configurations are kinked chain fragments As_3S_7 and As_5S_9 (Fig. 27b). The larger, branched fragment is just an extension of the shorter one. Mistakes in the fragment growth, i.e., formation of a larger fragment instead of the smaller one and vice versa in the growing layer, lead to OD phenomena and ubiquitous twinning: the $P1$ structure of sinnerite forms up to 24 twin orientations, restoring the $F\bar{4}3m$ symmetry of the underlying sphalerite-like substructure.

Structures with tetrahedral clusters

Pentlandite and its derivatives. The crystal structure of the parent compound, pentlandite $(Ni,Fe)_9S_8$ or Co_9S_8 (Rajamani and Prewitt 1975) contains isolated cubic clusters of eight metal-sulfur tetrahedra. Tetrahedra in a cluster share edges, yielding 3 direct short metal-metal interactions (2.505 Å for Co-Co, 2.549 Å for Fe-Ni, and 2.670 for argentopentlandite, Table 15) for each cation. Clusters are hinged via common apical S atoms and the cubic space between 6 adjacent clusters is occupied by a cation (Co, Fe and Ni, Ag or PGE in different pentlandites) in octahedral coordination (Fig. 28a). Cages in the pentlandite structure are closed, preventing movement and exchange of atoms between cages.

Djerfisherite is a complex sulfide of iron and copper, containing large cations, K and lesser amounts of Na, as well chlorine in an independent site. Its structure was determined by Dmitrieva et al. (1979) and refined on a synthetic analog $K_6LiFe_{23}S_{26}Cl$ by Tani et al. (1986). Natural djerfisherite, $K_6Na_{0.81}(Fe_{0.84}Cu_{0.16})_{24}S_{26}Cl$, is an iron-copper sulfide but its synthetic analog and the related bartonite $K_{5.68}Fe_{20.37}S_{26.93}$ (Evans and Clark 1981) indicate that copper can be fully exchanged by Fe and, in bartonite and owensite, even Cl is replaced by S. A nickel-barium analog of djerfisherite, $Ba_6Ni_{25}S_{27}$, with S instead of Cl, has been synthetized (Gelabert et al. 1997) (Table 15).

In cubic djerfisherite, pentlandite-like units—cubic cages with 6 surrounding clusters—are interconnected into a loose framework with a channel system <100>, i.e., these channels run

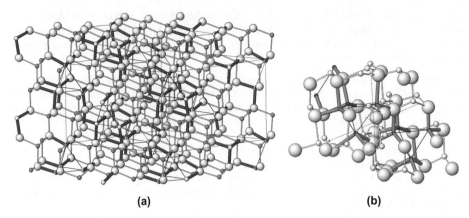

Figure 27. (a) Nowackiite $Cu_6Zn_3As_4S_{12}$ (Marumo 1967). A 3D framework of (Cu,Zn)S_4 tetrahedra (Cu-S bonds as thin lines) with clusters composed of four AsS_3 pyramids (As-S bonds as thick lines) concentrated around a sulfur vacancy. Imaginary tetrahedra, outlined in thin lines, indicate the S vacant sites. (b) A fragment of the crystal structure of sinnerite $Cu_6As_4S_9$ (Makovicky and Skinner 1975). Two adjacent clusters, with four AsS_3 coordination pyramids each, situated around S vacancies of the sphalerite-like framework of Cu tetrahedra. Amalgamation of adjacent AsS_3 groups into larger groups (note an $(As_3S_7)^{5-}$ group in foreground) proceeds via common S atoms. The anion vacancies are outlined as in (a).

Table 15. Pentlandite, djerfisherite and related phases.

Mineral	Formula	Lattice parameters (Å)			Space Group	Ref.
		a	b	c		
Bartonite	$K_{5.68}Fe_{20.368}S_{26.925}$	10.424(1)	—	20.626(2)	$I4/mmm$	[1]
Chlorbartonite	$K_6(Fe,Cu)_{24}S_{26}(Cl,S)$	10.381(8)	—	20.614(2)	$I4/mmm$	[2]
Djerfisherite	$K_6LiFe_{23}S_{26}Cl$	10.353(1)			$Pm\bar{3}m$	[3]
Synthetic	$Ba_6Ni_{25}S_{27}$	10.057(1)			$Pm\bar{3}m$	[4]
Thalfenisite	$Tl_6Fe(Fe,Ni)_{24}S_{26}Cl$	10.92			$Pm\bar{3}m$	[5]
Owensite	$(Ba,Pb)_6(Cu,Fe,Ni)_{25}S_{27}$	10.349(1)			$Pm\bar{3}m$	[6]
Argentopentlandite	$AgFe_8S_8$	10.521(3)			$Fm\bar{3}m$	[7]
Co-pentlandite	Co_9S_8	9.923(1)			$Fm\bar{3}m$	[8]

References: [1] Evans and Clark 1981; [2] Yakovenchuk et al. 2003; [3] Tani et al. 1986; [4] Gelabert et al. 1997; [5] Rudashevskyi et al. 1979; [6] Szymanski 1995; [7] Hall and Stewart 1973; [8] Rajamani and Prewitt 1975

along all edges of the cubic unit cell when the cubic cage, occupied by Na(Li), is situated at ½ ½ ½ (Fig. 28b). Chlorine atoms lie at intersections of the channel systems and are octahedrally surrounded by K atoms; these occur as two cations per each [100] channel interval (Fig. 28b). No constrictions are found in the channels. The Fe-Fe distance in the clusters is 2.731 Å. In thalfenisite (Table 15), K is replaced by Tl and Na (Li) by a divalent octahedrally coordinated metal (as in pentlandite). In owensite the univalent metals in the channels are replaced by Ba and, in the ratio 1:9, also by Pb with an active lone electron pair. Chlorine is replaced by S and Na by octahedrally coordinated metal with the M-S distance equal to 2.50 Å.

Tetragonal bartonite (Fig. 28c) contains intersecting channels [100] and [010] but lacks continuous channels in the [001] direction. The "nuclei" of vertical channels reach only

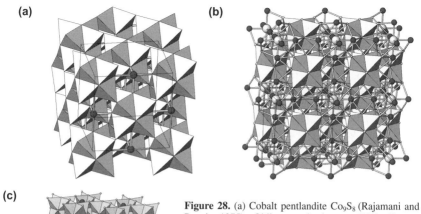

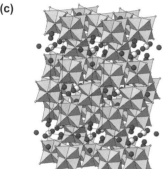

Figure 28. (a) Cobalt pentlandite Co_9S_8 (Rajamani and Prewitt 1975). Oblique projection on (001). Corner-sharing tetrahedral clusters Co_8S_{14} with octahedrally coordinated cobalt atoms (black) in the intervening cages. (b) Synthetic djerfisherite $K_6LiFe_{23}S_{26}Cl$ (Tani et al. 1986). Oblique projection on (100). A framework of corner-sharing Fe_8S_{14} tetrahedral clusters contains cubic cages with Li (hidden from view, at ½ ½ ½) and intersecting channels with K (dark spheres) and Cl (light spheres). (c) Bartonite $K_{5.7}Fe_{20.4}S_{27}$ (Evans and Clark 1981). Oblique projection upon (100). A framework of corner-sharing tetrahedral clusters Fe_8S_{14} and intersecting channels with K (dark) and additional S (light).

one cage up and one cage down from each intersection, with appropriate K atoms again completing an octahedron around the non-framework S atoms in the channel intersection. Lacking enclosed cages, bartonite lacks the smaller cation (Na or Fe, Ni, Cu). The Fe-Fe distance in the clusters is 2.708-2.737 Å.

Djerfisherite and bartonite can be considered two polytypes, where the channel-containing (in terms of cluster interconnection "loosely built") (001) layers are stacked in the AAAA sequence in djerfisherite whereas in bartonite they follow the ABAB stacking sequence. Thus, any occurrence of the "bartonite sequence" in djerfisherite will block the [001] channels. Eight K-S bonds in djerfisherite (constituting a square antiprism) are 4×3.31 and 4×3.44 Å, the single K-Cl bond is 3.10 Å long. The corresponding Ba-S bonds in owensite are 4×3.24 and 4×3.30 Å, plus a single Ba-S bond equal to 3.15 Å.

Bornite. Crystallographic investigations on bornite, ideally Cu_5FeS_4, by means of X-ray diffraction and High Resolution Transmission Electron Microscopy (HRTEM) have been the topic of a series of publications, twenty-five or more, since the first one by Frueh (1950). A complete listing can be found in Ding et al. (2005a,b).

Bornite occurs in three different polymorphs, respectively known under the names of high-, intermediate and low-temperature bornite in the literature (Table 16). The high-temperature form of Cu_5FeS_4 is stable above 265 °C (Pierce and Buseck 1978). It is cubic, with sulfur atoms in an *fcc* arrangement, and six cations in a unit cell distributed over eight tetrahedral sites (Morimoto 1964). Its quenched structure (Fig. 29a) alters only slowly to a low form on standing. This polymorph forms a complete solid solution with high-temperature digenite, $Cu_{7.2+x}S_4$, and also in the direction of excess iron, up to the composition of $Cu_{36.7}Fe_{21.9}S_{41.4}$ at 900 °C (Karup-Møller and Makovicky, unpublished). At room temperature, however, only the composition Cu_5FeS_4 is stable, and all excess Cu or Fe, as well as the contents of minor metals dissolved in bornite-digenite at high temperatures and sulfur fugacities, are exsolved again as metal and/or sulfides.

According to Morimoto and Kullerud (1961), the intermediate form, stable between 265 °C and 200 °C, has the tetrahedral cation vacancies ordered, and the cubic unit cell has a cell edge doubled against that of the F-centered high form (i.e., it is known as a $2a$ form). Its structure was described as an alternation of $1a1a1a$ cubes with antifluorite- and zincblende structure, respectively (see below). On cooling below 200 °C, further ordering of tetrahedral atoms and empty tetrahedra was postulated, leading to the $2a4a2a$ superstructure of the low, stoichiometric bornite (Koto and Morimoto 1975).

Figure 29b shows that the antifluorite cubes actually represent clusters of eight edge-sharing tetrahedra, like those observed in pentlandite and its more complex derivatives. The M-M distances in these clusters vary from 2.76 to 2.97 Å; they are supposed to host a mixture of Cu and Fe. Those cube-like spaces between six clusters, which host M octahedra in pentlandite, host four Cu atoms in slightly convex, three-fold planar coordination in bornite. These copper atoms are strongly eccentric in four corner-sharing tetrahedra of the "sphalerite-like" cube. They have moved into four faces of the empty octahedron which was filled in pentlandite by a cation. Their separation (≥ 3.05 Å) exceeds the distances typical for Cu-Cu interactions. Alternative orientations of these Cu configurations are the basis of the $2a4a2a$ superstructure of the low form.

These observations are in accordance with the refinement of the $2a$ superstructure of the intermediate form by Kanazawa et al. (1978), who described it as an alternation of cubes with half-filled sites, occupied by Cu, and of those with fully occupied sites, hosting Cu and Fe (Fig. 29c). There is, however, ambiguity in the space group of the cubic symmetry; $Fm\overline{3}m$ was chosen. Ding et al. (2005) express a number of doubts about this model, including the questionable homogeneity of bornite at this temperature.

Other compositions of the intermediate-temperature bornite-digenite solid solution produce Na superstructures, with $N = 6$ for the compositions close to digenite, and the value of N decreasing gradually through 5, 4, and 3, to reach 2 for stoichiometric bornite. Samples may present peculiar diffraction patterns: in the $[10\overline{1}]$ axis patterns, only those reciprocal lattice rows which are parallel to $[NNN]$ and pass through the $1a$-substructure reflections are present. In order to explain these, twinning or antiphase domains were invoked but Pierce and Buseck (1978) did not find them. Instead, they suggested, as did Van Dyck (1979,1980) and Conde et al. (1978), vacancy or cation ordering. The corresponding 3D structure models were calculated for

Table 16. Crystallographic data for polymorphs of bornite.

Polymorph	Space Group	Lattice parameters (Å, °)				Temp.	Refs.
		a	b	c	Angle		
High-T bornite	$Fm\overline{3}m$	5.5	5.5	5.5	—	Quenched	[1]
High-T bornite*	$R3m$	6.7	6.7	6.7	rhomb. 33.53°	Quenched	[1]
High-T bornite	$R3m$	3.862	3.862	19.975	hexag. axes 120°	Quenched	[2]
Intermediate bornite	$Fm\overline{3}m$	10.981	10.981	10.981	—	458 K	[3]
Low-T bornite*	$Pbca$	10.950	21.862	10.950	—	Room T	[4]

Notes: * For details of interpretation see the text. Further interpretation/existence problems are discussed in Ding et al. (2005a,b)
References: [1] Morimoto 1964; [2] Daams and Villars 1993; [3] Kanazawa et al. 1978; [4] Koto and Morimoto 1975

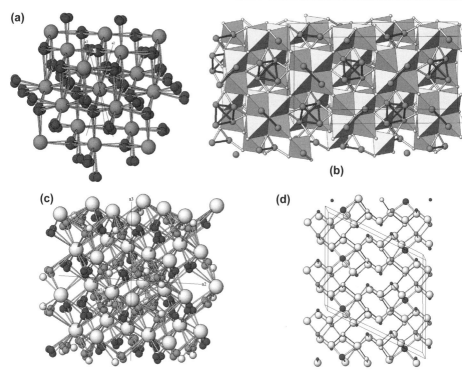

Figure 29. (a) High-temperature bornite (Morimoto 1964). Large spheres: S, dark spheres: disordered Cu sites, refined as clusters of split, statistically distributed atoms. All coordination tetrahedra are occupied. (b) Low-temperature polymorph of bornite according to Koto and Morimoto (1975). Pentlandite-like clusters ("antifluorite cubes") of eight (Cu,Fe) edge-sharing coordination tetrahedra alternating with clusters ("zincblende cubes") of four quasi-planar CuS_3 groups surrounding an empty octahedron. Dark outlined Cu-Cu connections help to define this space. Compare with Ding et al. (2005). (c) Intermediate bornite (Kanazawa et al. 1978). Two distinct, statistically occupied tetrahedral (Cu,Fe) sites (indicated by shades of grey) alternate with clusters of eight statistically occupied trigonal planar sites (white spheres), analogous to the triangular sites in (b). (d) Diaphorite projected along [001]. Large white spheres: S; small white spheres with bonds rendered in dark: Ag; small dark spheres: Sb; large dark spheres: Pb; intermediate grey spheres: positions in which Ag and Pb are mixed due to disorder caused by the OD character of diaphorite.

the cubic $4a$ and $6a$ superstructures (unit cell edges equal to 21.88 Å and 32.82 Å, respectively) by Ding et al. (2005), fitting them to the electron-diffraction and HRTEM data. The space group $Fm\overline{3}m$ was assumed. Calculated occupancies of cation sites (Cu and Fe could not be distinguished) vary from 0.67 to 1.0 for the $4a$ model and from 0.81 to 1.0 for the $6a$ model.

Ding et al. (2005) note that only Pierce and Buseck (1978) claim to have observed the $2a4a2a$ polymorph of bornite by HRTEM whereas Ding et al. (2005) only observed intergrowths of $1a$, $2a$, $4a$, and $6a$ domains in different proportions. Based on these observations and on a number of inconsistencies observed in structural and magnetic studies of bornite (e.g., a changing ratio of Fe^{2+} and Fe^{3+}; summarized in Ding et al. 2005a,b), they propose that the $2a4a2a$ polymorph might actually be a mixture of $2a$ and $4a$ superstructures, simulating the orthorhombic structure. For the $2a$ polymorph, Ding et al. (2005) derived a new structure model, with the space group $F\overline{4}3m$, with (preferentially) Cu and Fe concentrated in the clusters of tetrahedra and Cu in the "zincblende cubes." The resulting chemistry is $Cu_8Fe_4S_8$, however, so it obviously can be only one of the components present in low-temperature bornite. All these models were tested by comparing the calculated and observed HRTEM images. The temperatures of the two polymorphic transformations were refined for the digenite-bornite series by Grguric and Putnis (1998).

Monosulfides with octahedral cations: *ccp* vs. *hcp* stacking of anions

The fundamental difference between the *ccp* and *hcp* variants of the structures with octahedrally coordinated cations is the absence or presence of direct cation-cation interactions. Candidate cations for these two general structure types sharply differ. Occupation of all octahedral positions precludes occupation of tetrahedral voids. Partial vacancies in the *hcp* category, if frequent, can also be considered "pillared" versions of the layered MX_2 structure type, although a smooth transition is virtually unknown (but it is assumed for NiTe – $NiTe_2$ at elevated temperatures).

Phases derived from cubic close packing. The fundamental *ccp* type is represented by galena PbS and its isotypes PbSe, PbTe, as well as the series CaS-MgS-MnS. It goes virtually unnoticed that the octahedral coordination of Pb (Pb-S 2.966 Å) in galena is an exception and not a rule for this lone electron pair element in chalcogenides; its coordination becomes less symmetrical or its coordination number increases whenever possible, i.e., in a less symmetrical environment. Of interest are the homeotypes of PbS, obtained by means of coupled substitutions, such as Ag + Bi = 2Pb, Ag + Sb = 2Pb, which at elevated temperatures work for all substitution proportions. Cubic $AgBiS_2$ and $AgSbS_2$ are limited to higher temperatures (about 200 °C for $AgBiS_2$) and have disordered distribution of cations. Of particular interest are the two monoclinic room-temperature ordered homeotypes which occur on the $PbS-AgSbS_2$ join, the sulfosalts freieslebenite $PbAgSbS_3$ (Ito and Nowacki 1974) and diaphorite $Pb_2Ag_3Sb_3S_8$ (Armbruster et al. 2003). In these structures, Pb forms the largest, trapezoidally deformed octahedra, Ag polyhedra have rhomb-like cross-sections whereas Sb forms distorted square coordination pyramids. Diaphorite (Fig. 29d) appears to be an OD structure (Armbruster et al. 2003). An analog of sinnerite and nowackiite amongst the adamantane structures is the structure of $Tl_4Bi_2S_5$ (Julien-Pouzol et al. 1979). It displays a missing anion, replaced by a lone electron pair micelle of four Tl atoms.

Phases derived from hexagonal close packing. Among the *MX* compounds with octahedral cation coordination, the NiAs structure type occurs for the transition metals which favour the presence of direct metal-metal interactions, especially via the shared octahedral faces in the (0001) planes (Table 17). This is connected with high electrical conductivity, sometimes with variable stoichiometry and the dependence of lattice parameters on electron configuration. For example, for selenides, the *c/a* ratio changes as follows: TiSe (1.68), VSe (1.67), CrSe (1.64), $Fe_{1-x}Se$ (1.64), CoSe (1.46), and NiSe (1.46). For the arsenides and antimonides, rich in valence electrons, the *c/a* ratio is even lower (e.g., 1.39 for NiAs). The ideal *c/a* ratio of hexagonal close packing being 1.633, this substantial reduction of the *c* parameter corresponds to intense *d* orbital metal-metal interaction in this direction (Müller 1996). The anion is surrounded by six cations forming a trigonal coordination prism (Fig. 30a).

The room-temperature modification of stoichiometric FeS, troilite, exhibits triangular Fe_3 clusters parallel to (001) (the Fe-Fe distance being 2.93 Å). They are interconnected by Fe-Fe interactions at 2.99 Å between the Fe_3 triangles stacked above one another, and 2.95 Å for those in different [001] stacks of octahedra (Fig. 30b). The Fe-S bonds range from 2.36 Å to 2.72 Å in length. At 453 K, the high-temperature modification of FeS (Keller-Besrest and Collin 1990) is hexagonal, space group $P6_3mc$. Octahedral bond distances vary between 2.43 Å and 2.54 Å, and there is a continuous string [001] of Fe-Fe interactions, 2.91-2.92 Å long, via the shared octahedral faces. The intermediate modification of FeS (King and Prewitt 1982) has an orthorhombic structure, described as the MnP structure type, with distorted octahedra and a zig-zag pattern of short Fe-Fe interactions parallel to [001]. The Fe-S bonds

Table 17. Selected homeotypes of NiAs.

Mineral	Formula	Space Group	Lattice Parameters (Å, °)				Ref.
			a	b	c	Angle	
Pyrrhotite 6c	$Fe_{11}S_{12}$	$F1d1$	6.895	11.954	34.518	β 90.0	[1]
Pyrrhotite 3c	Fe_7S_8	$P3_121$	6.866	6.866	17.046	γ 120	[2]
Pyrrhotite 4c	Fe_7S_8	$F12/d1$	11.902	6.859	22.787	β 90.4	[3]
Troilite 2H	FeS	$P\bar{6}2c$	5.965	5.965	11.756	γ 120	[4]
Troilite (530 K)*	FeS	$P6_3mc$	6.588	6.588	5.400	γ 120	[5]
Troilite (463 K)+	FeS	$Pnma$	5.825	3.466	6.003	—	[6]
Westerveldite	FeAs	$Pnam$	5.440	6.026	3.371	—	[7]
Modderite	CoAs	$Pnam$	5.286	5.868	3.488	—	[7]
Nickeline	NiAs	$P6_3/mmc$	3.619	3.619	5.025	γ 120	[8]
—	Fe_7Se_8	$P31$	7.21	7.21	17.6	γ 120	[9]
—	Fe_3Se_4	$I12/m1$	6.187	3.525	11.290	β 91.98	[10]
—	CrS	$C12/c1$	3.826	5.913	6.089	β 101.6	[11]
—	Cr_5S_6	$P\bar{3}1c$	5.982	5.982	11.509	γ 120	[12]
Brezinaite	Cr_3S_4	$I12/m1$	5.694	3.428	11.272	β 91.50	[11]
—	Cr_2S_3	$P\bar{3}1c$	5.939	5.939	11.192	γ 120	[11]
—	Cr_2S_3	$R\bar{3}$	5.937	5.937	16.698	γ 120	[11]

Notes: * Pressure 9000 MPa; + Pressure 0.1 MPa; other conditions: Up to 6350 MPa at 294 K
References: [1] Koto et al. 1975; [2] Keller-Besrest and Collin 1983; [3] Tokonami et al. 1972; [4] Keller-Besrest and Collin 1990; [5] Fei et al. 1998; [6] King and Prewitt 1982; [7] Lyman and Prewitt 1984; [8] Yund 1962; [9] Andresen and Leciejewicz 1964; [10] Andresen and van Laar 1970; [11] Jellinek 1957; [12] van Laar 1967

range from 2.44 to 2.50 Å, the Fe-Fe interactions result in a distance of 2.92 Å. All distortions characteristic for the MnP structure type are much more pronounced in westerveldite FeAs (Lyman and Prewitt 1984) (Fig. 30c). The Fe-As distances range from 2.35 to 2.51 Å, and the Fe-Fe interactions along the stacking direction [100] are much shorter than in the sulfide; they are equal to 2.79 Å, in agreement with the a parameter of 5.44 Å.

The omission derivatives of FeS, the family of pyrrhotites with a general formula $Fe_{1-x}S$, has been extensively studied. In the intermediate polymorphs, the cation vacancies undergo partial ordering and the occupied, partly ocupied and least occupied cation sites form occupational waves. These are oriented diagonally to (0001) in NA pyrrhotite (between ~209 and 266 °C, Kissin 1974), and parallel to (0001) in NC pyrrhotite (~100-213 °C). In both cases, non-integral periodicities in the a and c directions result. The low-temperature polymorphs of pyrrhotite (below 100 °C) have ordered cation vacancies in the structure. They occur in every second octahedral (0001) layer, or every second and third layer, and result in unit cells being a multiple of the two-layer cell of high-temperature FeS. The known ordered compositions are: Fe_7S_8 (pyrrhotite 4M, c = 22.88 Å; Fe_9S_{10} (5H, c = 28.67 Å); $Fe_{11}S_{12}$ (6M, c = 34.52 Å); Fe_9S_{10} (7H, c = 40.15 Å); and $Fe_{10}S_{11}$ (11H, c = 63.22 Å). Details of their crystallography, and the structure schemes and phase relations of these phases have been summarized by Scott (1974) and Vaughan and Craig (1978).The structures of the pyrrhotite family can be illustrated using the example of the structure of Fe_7Se_8 obtained from neutron diffraction data by Andresen and Leciejewicz (1964) (Fig. 30d). Every second octahedral layer has 25% of octahedra vacant in a hexagonal pattern; with the adjacent fully occupied layer it gives 7 Fe for every 8 Se atoms. Vacancies in the defect layers are stacked with a shift $1/2 a_i + 1/3 c$, where i = 1, 2 and 3 in the sequence of the layer involved.

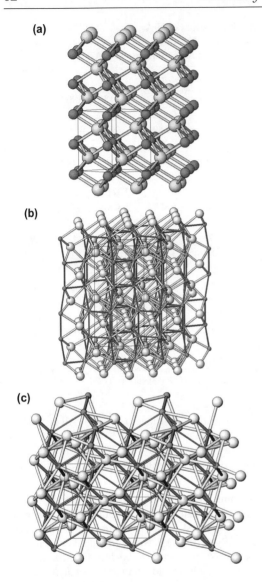
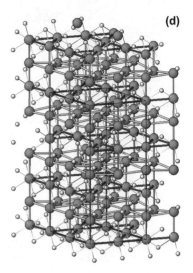

Figure 30. (a) NiAs, the aristotype of the monosulfide structural family with coordination octahedra $NiAs_6$ and coordination prisms $AsNi_6$, respectively. (b) Troilite FeS at ambient temperature. Short Fe-Fe distances are indicated as dark lines (see the text for details). (c) Westerveldite, FeAs (Lyman and Prewitt 1984), a natural analog to the aristotype structure of MnP. Dark small spheres: octahedrally coordinated Fe, light spheres: As. Fe-As bonds are white. Short Fe-Fe contacts are indicated in grey; thinner grey lines indicate longer, second order contacts. Note a pronounced distortion of coordination octahedra. (d) Disposition of Fe atoms (large spheres) in Fe_7Se_8 (Andresen and Leciejewicz 1964). Note the distribution of octahedral vacancies in every second octahedral layer. Sulfur atoms are presented as small white spheres. Fe-Fe connections serve for orientation.

An extreme concentration of vacancies, not attained in the sulfides, is found in the structure of Fe_3Se_4 (Andresen and Laar 1970) where 50% of octahedra in the defect layer are vacant (Fig. 31a). The structure is slightly distorted, with the monoclinic angle $\beta = 91.8°$, and the occupied rows of octahedra run in the [010] direction. Octahedra of the full layer are sizably distorted, the Fe-Se bonds ranging from 2.44 to 2.61 Å. In the Fe-S system, the Fe_3S_4 composition is a thiospinel. The phase closest to this composition, smythite $Fe_{3+x}S_4$ ($0 < x < 0.3$) is rhombohedral, the c parameter is three times that of Fe_3Se_4.

A situation broadly analogous to $Fe_{1-x}S$ exists for the sulfides of chromium. CrS (Jellinek 1957) differs from FeS, however, because the coordination octahedra of the d^4 cation Cr^{2+} undergo Jahn-Teller distortion. Four S atoms at 2.43 Å and two more S atoms at 2.88 Å form an elongated octahedron, and the structure is transitional between NiAs and PtS. Cr_7S_8, Cr_5S_6, Cr_3S_4 and Cr_2S_3 are "defect" structures stable at room temperature (Jellinek 1957). The vacancies in the metal positions are confined to every second layer of octahedra. For the two compounds with a smaller number of vacancies, ordering disappears at about 320 °C; ordering of vacancy-rich compounds persists up to high temperatures (Jellinek 1968).

Thiospinels

The thiospinels and less common selenospinels (AB_2X_4) are the structural analogs of oxide spinels AB_2O_4. The unit cell edge of the cubic cell, based on cubic close packing of anions, has values from about 9.4 Å to 10.6 Å for sulfur-based spinels, i.e., twice the edge length of the F-centered cell of the anion submotif. The space group of nearly all spinels is $Fd\overline{3}m$, subgroups due to tetrahedral cation ordering or distortion are rare (Table 18).The unit cell contains 32 anion atoms with coordinates (u,u,u). This u value is equal to 0.25 for the ideal ccp structure; it ranges from 0.25 to over 0.26 in thiospinels. This range is narrower than observed in oxide spinels, 0.24-0.27 (Lavina et al. 2002). An increase in u produces larger tetrahedral sites at the expense of diminished octahedral sites (Waychunas 1991).

The tetrahedral site has point symmetry $\overline{4}3m$, multiplicity 8, whereas the octahedral site has point symmetry $\overline{3}m$, multiplicity 16. Reduction of symmetry occurs primarily as separation of two distinct kinds of tetrahedral sites. The octahedral M-S distance depends on u as follows:

$$M\text{-}S = a\sqrt{(\tfrac{1}{2}-u)^2 + 2(\tfrac{1}{4}-u)^2}$$

whereas the tetrahedral cation (T)-S distance is

$$T\text{-}S = a\sqrt{3(\tfrac{1}{8}-u)^2}$$

(Lavina et al. 2002).

(a)

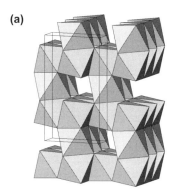

(b)

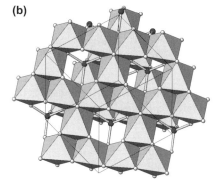

(c)

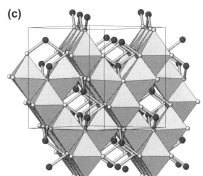

Figure 31. (a) The pyrrhotite-related structure of Fe_3Se_4 (Andresen and Laar 1970) in a polyhedral representation. Note the vacancy distribution. c axis is vertical, a axis points to the right. (b) Violarite $FeNi_2S_4$ (Vaughan and Craig 1985) projected along [111]. Note the dioctahedral layer with the system of three intersecting rows of edge-sharing octahedra. Tetrahedral coordinations are shown in a ball-and-stick presentation. (c) The same structure projected along [110].

Table 18. Selected thiospinels.

Mineral	Formula	Space Group	Lattice Parameters	T-S tetr.	M-S oct.	u	Refs.
Polydymite	Ni_3S_4	$Fd\bar{3}m$	9.457	2.211	2.274	.260	[1]
Linnaeite	Co_3S_4	$Fd\bar{3}m$	9.406	2.185	2.269	.259	[2]
Greigite	Fe_3S_4	$Fd\bar{3}m$	9.880	2.139	2.470	.250	[3]*
			9.876	2.147	2.464	.2505	[4]*
Violarite	$FeNi_2S_4$	$Fd\bar{3}m$	9.465	2.197	2.284	.259	[5]
Carrollite	$CuCo_2S_4$	$Fd\bar{3}m$	9.478	2.265	2.253	.263	[6]*
Cuproirdsite	$CuIr_2S_4$	$Fd\bar{3}m$	9.847	2.303	2.367	.260	[7]
Cuprorhodsite	$CuRh_2S_4$	$Fd\bar{3}m$	9.788	2.272	2.362	.259	[8]
Indite(disord.)	$FeIn_2S_4$	$Fd\bar{3}m$	10.618	2.466	2.562	.2591	[9]
Indite (ordered)	$FeIn_2S_4$	$F\bar{4}3m$	10.618	2.466	2.562	.2591	[9]
Daubréelite	$FeCr_2S_4$	$Fd\bar{3}m$	9.981	2.322	2.406	.2593	[10]*
Synth.(disord.)	$FeCuCr_4S_8$	$Fd\bar{3}m$	9.904	2.282	2.399	.258	[11]*
Synth.(ordered)	$FeCuCr_4S_8$	$F\bar{4}3m$	9.901	Fe 2.349 Cu 2.178	2.362	.252	[12]*
Synth.	$FeRhCrS_4$	$Fd\bar{3}m$	9.944	2.291	2.409	.258	[13]
Synth.	$ZnCr_2S_4$	$Fd\bar{3}m$	9.974	2.325	2.401	.2596	[14]
Rhodostannite	$Cu_2FeSn_3S_8$	$I4_1/a$	a 7.305 c 10.330	2.320	2.535	.2593	[15]
Toyohaite	$Ag_2FeSn_3S_8$	$I4_1/a$	a 7.46 c 10.80	—	—	—	[16]
Synth.	$HgCr_2S_4$	$Fd\bar{3}m$	10.235	2.500	2.406	.266	[17]*

Notes: * powder data (newest ones by using Rietveld refinement)
Hill et al. (1978): parallel refinements as (a) disordered model (mixed In^{3+} and Fe^{2+} in both types of sites) and (b) ordered model with In:Fe about 1:1 in M and one pure In, and one mixed (In,Fe) T site; two distinct T sites in ordered $FeCuCr_4S_8$ as well.
References: [1] Lundqvist 1947; [2] Knop et al. 1968; [3] Uda 1968; [4] Skinner et al.1964; [5] Vaughan and Craig 1985; [6] Williamson and Grimes 1974; [7] Furubayashi et al. 1994; [8] Riedel et al. 1976; [9] Hill et al. 1978; [10] Kim et al. 2002; [11] Riedel et al. 1981; [12] Zaritskii et al. 1986; [13] Riedel and Karl 1980; [14] Wittlinger et al. 1997; [15] Jumas et al. 1979; [16] Yajima et al. 1991; [17] Konopka et al. 1973

Normal spinels have divalent cations in the tetrahedral sites and trivalent cations in the octahedral sites. Inverse spinels have a trivalent cation in the tetrahedral site and a mixture of di-and trivalent cations in the octahedral sites. This division, e.g., $ZnCr_2S_4$ and $CdCr_2S_4$ vs. Fe_3S_4 and $FeIn_2S_4$, is much less obvious in thiospinels than in oxide spinels. Two coexisting valencies are displayed by Fe, Co and Ni; in the case of iron, a temperature dependent gradual transformation of individualized Fe^{2+} and Fe^{3+} in the structure into a completely mixed-valence iron Fe^{n+} was demonstrated for Cu-Fe rhodium and chromium thiospinels by Riedel et al. (1981) using Mössbauer spectroscopy.

Coordination octahedra form dioctahedral layers (111), separated by layers with a set of occupied tetrahedra and "isolated" octahedra (Fig. 31b). This is true for all layers of the {111} set and the "tetrahedral-octahedral" layers of a particular orientation actually contain the octahedra from dioctahedral layers of the other orientations. In this way, each octahedron shares edges with six neighbours. A characteristic result of this is a "pile-up" of octahedral rows [110], and equivalent, separated by interstitial rows containing isolated tetrahedra, in the projection along [001]. Projection of the spinel structure along [110], down the rows of tetrahedra (Fig. 31c), enabled Horiuchi et al. (1981), Navrotsky (1994) and Ferraris et al.(2004) to interpret spinel as a block structure/polytypic structure leading to high-pressure, deep-earth spinelloids in the unary system $(Mg,Fe)_2SiO_4$; we have not (yet) seen a sulfide analog of spinelloids.

In the majority of cases the octahedral cations are transition metals and Al; the choice of tetrahedral cations is wider, including Hg, Cd, Zn, Cu, In and the transition metals. The closest cation-cation distances in the structure are those via the shared edges of two neighbouring octahedra (shared edges are shortened). These Fe-Fe distances in greigite are 3.49 Å, compared with 2.97 Å in magnetite.

Disulfides, sulfarsenides and their analogs

The disulfide S_2^{2-} group with its covalent S-S bond, and the diselenide, sulfarsenide and diarsenide group analogs, are components of a number of highly stable stuctures. The valence of AsS and As_2 groups has been interpreted in a number of ways (e.g., Tossell et al. 1981), often dependent on other assumptions, e.g., that the valence of iron increases along the series FeS_2-FeAsS-$FeAs_2$. Based on the results of Mössbauer spectroscopy and the LCAO calculations, Ioffe et al. (1985) suggest that iron has the same valence state in all three compounds and that the population of the d levels changes little, and never correspons to Fe^{3+}. This is accomplished by back-donation of electrons due to the donor-acceptor interaction between the sp^3 atomic orbitals of the anion and the hybridized d^2sp^3 orbitals of iron. Thus, in FeAsS and $FeAs_2$ the X_2 groups act as divalent anions; they all force Fe into a low-spin state (see Vaughan and Rosso 2006, this volume, for further discussion).

Characteristic X-X distances in the dianion pairs in the structures of sulfides and related substances are listed in Table 19. These pairs are an expression of high sulfur fugacity in the formation environment although, in some of these compounds, they also play a space-filling role, enabling the matching of two distinct structure regions. For example, in $HgSb_4S_8$ (livingstonite, Srikrishnan and Nowacki 1975), the S_2^{2-} groups interconnect double-pyramidal Sb_2S_4 rods into layers $Sb_2S_2(S_2)$, extending the repetition period of these layers in order to match that of the alternating $HgSb_2S_4$ layers. The space-filling role is obvious also in $Cu_4Bi_4S_9$ (Takéuchi and Ozawa 1975; Bente and Kupčík 1984) and $Cu_4Bi_4Se_9$ (Makovicky et al. 2002); S_2^{2-} groups join two different structure portions in $Ba_4Sb_4Se_{11}$ (Cordier et al. 1980) and the triselenide Se_3^{2-} and trisulfide S_3^{2-} groups substitute for interlayers in the rod-based structures of $Sr_4Bi_4Se_9(Se_3)$ (Cook and Schäfer 1982) and $Pb_6Sb_6S_{14}(S_3)$ (moëloite, Orlandi et al. 2002). All of these stuctures are confined to the associations with a high activity of the anion in question; the occurrence of diarsenides and sulfarsenides is also limited by a high activity of sulfur, at which As starts acting as a cation. A somewhat different role is played by the S_2 groups in patronite, $V(S_2)_2$, and the Te_2 groups in sylvanite $AgAuTe_4$, mentioned among the fibrous and layered chalcogenides, respectively.

Pyrites. The cubic disulfide of iron, pyrite FeS_2 is the most widespread sulfide in the Earth's crust. In its structure, three coordination octahedra of low-spin divalent iron meet at each end of the S_2 dumbell, with no edge-sharing present. The characteristic configuration of <100> slabs of pyrite structure is shown in Fig. 32a. The pyrite structure type is found in high-spin d^5 transition elements (e.g., MnS_2), low-spin d^6 (e.g., FeS_2, $FeSe_2$, $FeTe_2$, $PtAs_2$, $PdSb_2$, RuS_2, OsS_2), d^7 (CoS_2, $AuSb_2$), high-spin d^8 (NiS_2, $NiSe_2$), and under high pressure even in d^9 and d^{10} elements (CuS_2, ZnS_2) (Hulliger 1968). These compounds are semiconductors or behave as metals (CoS_2, $PdSb_2$). Modifications of the pyrite structure type are a series of rhodium and iridium chalcogenides known as, e.g., $RhS_{~3}$ and $IrSe_{~3}$, because of the cation vacancies in the pyrite-like structure, and an orthorhombic, pseudotetragonal structure of $PdSe_2$ and PdS_2 (Table 22), in which very elongated octahedra with a [4+2] bond scheme, typical for Pd^{2+}, extend one of the unit cell parameters of the "pyrite cube" substantially.

Distinct electron configurations are the cause of miscibility gaps encountered among the isotypes of pyrite: FeS_2 and NiS_2 exhibit only a limited solubility (Clark and Kullerud 1963 and a number of later studies, e.g., 3.6 atomic% Ni in FeS_2 and 9.6 atomic% Fe in coexisitng NiS_2 at 725 °C; Karup-Møller and Makovicky 1995), and the semiconducting Pt pnictides do not form mixed crystals with the metallic Pd pnictides (Hulliger 1963).

Table 19. X-X bond distances in the covalent X_2 anion groups in sulfides and related substances.

Compound	X-X distance (Å)	Reference
$Cu_4Bi_4S_9$	2.095	Bente and Kupčík 1984
$Cu_4Bi_4Se_9$	2.402	Makovicky et al. 2002
$Ba_4Sb_4Se_{11}$	2.368 2.421	Cordier et al. 1980
FeS_2 cub.	2.140	Will et al. 1984
NiS_2	2.075	Nowack et al. 1991
CoS_2	2.112	Nowack et al. 1991
MnS_2	2.091	Chattopadhyay et al. 1991
$FeSe_2$ orthohomb.	2.550	Pickardt et al. 1975
$HgSb_4S_8$	2.060	Srikrishnan and Nowacki 1975
$Rb_2Sb_8S_{12}(S_2)\cdot 2H_2O$	2.108	Berlepsch et al. 2001
FeAsS	2.346	Fuess et al. 1987
NiAsS	2.299	Foecker and Jeitschko 2001
CoAsS ($Pca2_1$)	2.292	Fleet and Burns 1990
$FeAs_2$	2.488	Lutz et al. 1987
$PtAs_2$	2.409	Szymanski 1979

The sulfarsenides, sulfantimonides and sulfobismutides (as well as the corresponding seleno- and telluro- varieties) are very important homeotypes of pyrite (Table 20). Some have been described as ordered arrangements of chalcogenide-pnictide dumbels (Figs. 32b,c) such as, for example, ullmanite, NiSbS, $P2_13$, and isotypic PdSbSe (Pratt and Bayliss 1980; Foecker and Jeitschko 2001; Paar et al. 2005). For cobaltite CoAsS, both a disordered and an ordered variety is known, $Pa3$ and $Pca2_1$ (Giese et al. 1965; Scott et al. 1976), the orthorhombic symmetry of the latter causing the weak anisotropy of this mineral in reflected light. For gersdorffite, NiAsS, the disordered, $Pa3$ variety and the ordered $P2_13$ and $Pca2_1$ polymorphs have been found (Bayliss 1986). In both ordered structure types of this family, the cation is coordinated by 3S+3As (or equivalents) in an ordered fashion and the S-As dumbells have ordered orientations. The different disposition of "all-S" and "all-As" faces of octahedra in the two ordering schemes is illlustrated in Figs.32b and 32c. This family is typical for d^6 and d^7 elements (examples are CoAsS and IrAsS, and NiAsS, PdAsS and PdBiTe, respectively). Overgrowths of platinum group element (PGE)-rich MAsS phases on Ni-Co sulfarsenides, typically found in PGE deposits are caused by their isotypism.

Marcasites. In the broad marcasite structure type, the coordination octahedra of cations share edges forming chains parallel to the short crystallographic axis (~3.3 Å for FeS_2). Adjacent chains share vertices and the channels comprised by four adjacent chains are reduced to rhomb-like cross-section by the bridging X-X bonds (Fig. 33a). Symmetry constraints are less strict than in pyrite and the variations in octahedron size, elongation and X-X distance are easily adjusted by rotating the chains of octahedra around their axes. The marcasite structure type is related to that of rutile, the latter having no anion-anion bond in the channels; this allows the chain rotation until the channels assume a square cross section. Another relative of the marcasite-type is the $CaCl_2$ structure type, in which the $(100)_{CaCl_2}$ boundaries drawn through anions are planar and not warped as in FeS_2. Using this analogy, marcasite can be described as composed of slightly warped $(100)_{FeS_2}$ layers of octahedra, with every second row of octahedra in each layer emptied and substituted by X-X bonds.

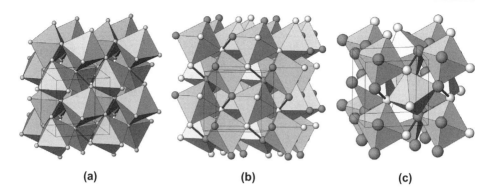

Figure 32. (a) Pyrite FeS_2 in a polyhedral representation. Coordination octahedra (FeS_6) share corners and the covalent S-S bonds (indicated in black). (b) Gersdorfite NiAsS (Foecker and Jeitschko 2001). Light and dark spheres: S and As, respectively. (c) Orthorhombic cobaltite CoAsS (Fleet and Burns 1990). Light and dark spheres: S and As, respectively.

Table 20. Selected disulfides, diarsenides and sulfarsenides.

Mineral	Formula	Space Group	Lattice Parameters (Å,°)				Ref.
			a	b	c	Angle	
Pyrite	FeS_2	$Pa\bar{3}$	5.418	5.418	5.418		[1]
Marcasite	FeS_2	$Pnnm$	4.443	5.424	3.386		[2]
Ferroselite	$FeSe_2$	$Pnnm$	4.804	5.784	3.586		[3]
Loellingite	$FeAs_2$	$Pnnm$	5.300	5.984	2.882		[4]
Sperrylite	$PtAs_2$	$Pa\bar{3}$	5.968	5.968	5.968		[5]
Vaesite	NiS_2	$Pa\bar{3}$	5.677	5.677	5.677		[6]
Cattierite	CoS_2	$Pa\bar{3}$	5.539	5.539	5.539		[6]
Cobaltite	CoAsS	$Pca2_1$	5.583	5.589	5.581		[7]
Gersdorffite	NiAsS	$P2_13$	5.689	5.689	5.689		[8]
Hauerite	MnS_2	$Pa\bar{3}$	6.104	6.104	6.104		[9]
Arsenopyrite	FeAsS	$C112_1/d$	6.546	9.451	5.649	89.84	[10]
Rammelsbergite	$NiAs_2$	$Pnnm$	5.301	5.986	2.882		[11]
Pararammelsbergite	$NiAs_2$	$Pbca$	5.772	5.834	11.421		[12]
Alloclasite	CoAsS	$P2_1$	4.661	5.501	3.411	90.03	[13]

References: [1] Will et al. 1984; [2] Brostigen et al. 1973; [3] Pickardt et al. 1975; [4] Lutz et al. 1987; [5] Szymanski 1979; [6] Nowack et al. 1991; [7] Fleet and Burns 1990; [8] Foecker and Jeitschko 2001; [9] Chattopadhyay et al. 1992; [10] Fuess et al. 1987; [11] Kjekshus et al. 1979; [12] Kjekshus and Rakke 1979; [13] Scott and Nowacki 1976

This model leads directly to an explanation of the topotactic formation of marcasite by mild oxidation of pyrrhotite. This mechanism was proposed for the pyrrhotite-marcasite pair by Fleet (1978) and for the analogous nickeline (NiAs)-rammelsbergite ($NiAs_2$) pair by Karup-Møller and Makovicky (1979). Every second row in the layer of octahedra (ideally (0001)) of the original FeS or NiAs structure is leached out, creating an overall chess-board pattern of occupied octahedral rows, and the charge balance is maintained by partial oxidation of anions which form X_2 pairs. The orthorhombic products form in three orientations on the (ideally) hexagonal substrate. For both pairs of minerals, misfit of the original periodicities

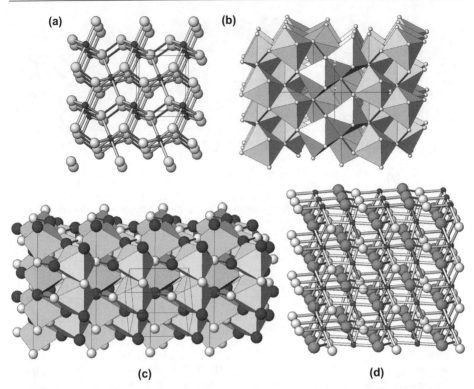

Figure 33. (a) Marcasite FeS$_2$ (Brostigen et al. 1973) in a ball-and-stick representation. Fe (small dark spheres) is octahedrally coordinated, S-S bonds are accentuated in black. (b) Pararammelsbergite, NiAs$_2$ (Kjekshus and Rakke 1979) projected along [100]. Alternating slabs of a marcasite (i.e., rammelsbergite)-like and pyrite-like arrangement are parallel to (001). As-As bonds are indicated in black. (c) Arsenopyrite, FeAsS (Fuess et al. 1987) in a polyhedral representation. Octahedrally coordinated Fe, S and As: white and grey spheres, respectively, with the covalent As-S bond indicated. (d) Alloclasite CoAsS (Scott and Nowacki 1976). Small spheres: Co, dark spheres: As, light spheres: S. As-S bonds: black.

and those of the products is sizable, leading to extremely fine-grained products, unless further ripening or outward growth (common for marcasite) takes place. It is the present author's opinion that much of the marcasite observed in nature is a result of these topotactic processes. For rammelsbergite, the result of such topotaxy is the formation of rammelsbergite far below its stability limits, in the field of stability of pararammelsbergite.

The relationship of the pyrite structure type to the marcasite structure type has long fascinated crystallographers. Schemes in which a composition plane of a pyrite twin on (110) is the thinnest slab of marcasite or the twin contact of two marcasite individuals on {110} forms a slice of the pyrite {001} configuration, are described by Fleet (1970) and Ramdohr and Strunz (1978).

A compound in which twinning of a marcasite array occurs on a unit cell level with maximum frequency is pararammelsbergite (Fleet 1972), the low-temperature polymorph of NiAs$_2$. Two octahedra thick slabs of rammelsbergite-type arrangement, at $x = 0$ and ½, built by edge-sharing of octahedra in their median planes, are joined in a perfectly pyrite-like fashion at the $z =$ ¼ and ¾ planes (Fig. 33b). A single octahedral <100> layer of pyrite configuration (layer symmetry $P2_1/g$) can be stacked with further such layers either by: (1) applying an interlayer glide plane or interlayer two-fold axes, in order to create a pyrite-like layer pair or, alternatively, (2) the next <100> pyrite-like layer can be attached as a translation equivalent of the original one, displaced by ~($\frac{1}{4} a_1 + \frac{1}{4} a_2$)$_{pyrite}$, giving an arrangement seen in marcasite. In the real structure of pararammelsbergite, those halves of octahedra which are situated along the pyrite-like composition plane are an exact copy of the pyrite motif. However, the "outer" portions of these octahedra, facing the marcasite-like stacking, have to be distorted, in order to fit this stacking. Symmetry of the octahedral layer is reduced to *Pg* and the interlayer symmetry elements are reduced as well (to a parallel glide plane). We do not know of a sulfide equivalent of pararammelsbergite, nor do we know structures with thicker marcasite-like slabs.

The general marcasite type structure is formed by transition metal cations d^2-d^9. In the d^6 (low-spin semiconducting) and metallic d^7-d^9 marcasites (e.g., FeS$_2$, FeSe$_2$ CoSe$_2$ and CuSe$_2$), the d_{xy} orbitals are fully occupied, acting repulsively and their *c/a* ratios are about 0.74. The d^2-d^4 marcasites are Jahn-Teller instable, the d_{xy} orbitals are empty and able to overlap along the *c* axis and result in the *c/a* ratios being reduced to 0.47-0.57; these form the loellingites (e.g., CrSb$_2$, FeAs$_2$, RuAs$_2$, and OsAs$_2$) (Hulliger 1967).

Compounds with d^5 lie between these two categories. The half-filled d_{xy} orbitals overlap at every second octahedral contact whereas a non-overlap, i.e., repulsion occurs in the intervening contacts. The resulting structure, the arsenopyrite structure type, has a *c/a* ratio intermediate between marcasites and loellingites and a doubled periodicity. Besides CoAs$_2$, RhAs$_2$ and IrAs$_2$, many ternary compounds belong to this structure type, such as FeAsS (arsenopyrite), FeSbS, RuAsS, OsAsS, etc. Note that Hulliger's interpretation of d electron count on cations differs from that of Ioffe et al. (1985) (See also Vaughan and Rosso 2006, this volume)

Arsenopyrite (Fig. 33c) and its isotypes are a superstructure of the marcasite structure type. In each [001] row of dumbells, the two orientations of S-As pairs alternate. Along the *c* axis, the shared edges of octahedra in the rows are alternately S-S and As-As and the [3As+3S] coordinations in the adjacent octahedra of one row are inversion-related. Cation-cation distances along the row are alternately shorter, 2.82 Å, and longer, 3.62 Å for arsenopyrite FeAsS. All this leads to a distorted derivative of marcasite, with the space group *B*12$_1$/*d*1 instead of the *Pmnn* of the marcasite aristotype. Results of Fuess et al. (1987) who found partly mixed S and As populations in arsenopyrite Fe$_{0.87}$Co$_{0.13}$As$_{0.88}$S$_{1.12}$ can be interpreted by the presence of antiphase domains in the S-As arrangement, as observed by them using HRTEM. According to Scott and Nowacki (1976), alloclasite CoAsS (Fig. 33d) has a pattern of As and S distribution different from arsenopyrite. The stacks of parallel S-As dumbells along the *c* axis have all dumbells oriented in the same way, i.e., pure S and As [001] stacks are present.

CHANNEL AND CAGE STRUCTURES

Chalcogenides form a range of channel and several cage structures, combining framework-forming cations of various kinds with large cations, in most cases M^+ and M^{2+}. A more complete account of channel structures with octahedral walls, cetineites (Sabelli et al. 1988; Wang and Liebau 1998), channel structures with Cu- and Ag-rich walls, channel-like sulfosalts, and cage-like structures of the skutterudite family is given in Makovicky (2005). Compounds belonging to the mineralogically most important category of channel structures are listed in Table 21.

CATION-SPECIFIC STRUCTURES

Mercury sulfides

Cinnabar. The crystal structure of α-HgS (cinnabar, Schleid et al. 1999) is commonly referred to as a distortion derivative of the PbS archetype. The contents of "octahedra" are

Table 21. Compounds with copper-rich partitions and related structures.

Compound	Lattice Parameters (Å,°)				Space Group	Ref.
	a	b	c	β		
TlCu$_{3.99}$Se$_3$	12.431(0)	12.800(0)	3.935(1)	—	Pnnm	[1]
TlCu$_5$Se$_3$	12.900	—	3.968	—	P4$_2$/mnm	[2]
TlCu$_3$Se$_2$	15.213(1)	4.012(0)	8.394(0)	111.70(1)	C2/m	[3]
Tl$_5$Cu$_{14}$Se$_{10}$	18.097(2)	3.958(0)	18.118(2)	116.09(1)	C2/m	[4]
TlCu$_7$Se$_4$	10.448(1)	—	3.968(0)	—	I4/m	[5]
TlCu$_7$S$_4$	10.180(0)	—	3.859(0)	—	I$\bar{4}$	[6]
NH$_4$Cu$_7$S$_4$	10.25(2)	—	3.84(1)	—	I$\bar{4}$	[7]
Rb$_3$Cu$_8$Se$_6$	18.458(6)	4.010(1)	10.212(3)	104.44(2)	C2/m	[8]
Cs$_3$Cu$_8$Se$_6$	19.076(4)	4.078(1)	10.449(3)	106.04(3)	C2/m	[8]
K$_2$Hg$_6$S$_7$	13.805(8)	—	4.080(3)	—	P42$_1$m	[9]
Betekhtinite Pb$_2$(Cu,Fe)$_{21}$S$_{15}$	3.86	14.67	22.80		Immm	[10]
Miharaite Cu$_4$FePbBiS$_6$	10.880	12.003	3.874		Pb2$_1$m	[11]
Synthetic Cu$_3$Bi$_2$S$_3$I$_3$	28.056	4.105	10.580	110.57	C2/m	[12]

Reference: [1] Berger et al. 1995; [2] Berger et al. 1990; [3] Berger 1987; [4] Berger and Meerschaut 1988 [5] Eriksson et al. 1991; [6] Berger and Sobott 1987; [7] Gattow 1957; [8] Schils and Bronger 1979; [9] Kanatzidis 1990; [10] Dornberger-Schiff and Höhne 1959; [11] Petrova et al. 1988; [12] Balić-Žunić et al. 2005

actually covalently bonded S-Hg-S groups (Hg-S = 2.34-2.40 Å), sharing common S atoms. The Hg-S-Hg angle is 173°. Hg coordination is completed by four long Hg-S distances (2 × 3.28 Å and 2 × 3.09 Å) (Fig. 34a). The most interesting view is that along [100] and another along the trigonal axis; they reveal the spiralling system of short Hg-S bonds. Thus, cinnabar is a compound with infinite [001] chains of short Hg-S-Hg-S bonds, with a triple period; these are analogous to the Se-Se-Se chains in native selenium. The dimorph, metacinnabar, is an isotype of sphalerite ($F\bar{4}3m$, a = 5.903 Å).

Nickel sulfides

The sulfides of nickel are distinguished by a number of short metal-metal distances, with lengths comparable to the Ni-Ni distance in the metal. The observed coordinations range from fivefold square-pyramidal to four-fold tetrahedral and flat-tetrahedral or even square planar. At elevated temperatures, NiS has octahedrally coordinated Ni and there is a gradual change in the coordination of nickel from 6- (octahedral) to 5- or 4-fold as the temperature decreases. In addition to the phases described below, pentlandite and a synthetic Ba-Ni sulfide with Ni in tetrahedral coordination and three Ni-Ni bonds belong to this category. The high-temperature phases NiS$_2$ and Ni$_3$S$_4$ with octahedral and tetrahedral nickel coordination are distinct from them.

Heazlewoodite. The rhombohedral, pseudocubic (a = 4.072 Å, α = 89.46°) structure of heazlewoodite, Ni$_3$S$_2$, is characterized by four short Ni-Ni distances for each nickel atom (Parise 1980). Two of the interactions are perpendicular to the c axis and form a part of a Ni$_3$ triangle (Ni-Ni = 2.53 Å). These triangles are stacked in a rhombohedral arrangement, creating Ni spirals parallel to [001] (Ni-Ni = 2.50 Å). Nickel is tetrahedrally coordinated by sulfur at 2 × 2.25 Å and 2 × 2.29 Å. Three tetrahedra share a common edge (cf. the above triangles). Along [001], each of them has two more shared edges (Fig. 34b).

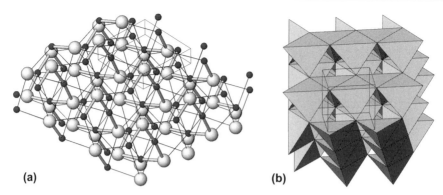

Figure 34. (a) Cinnabar (Schleid et al. 1999). Spirals [001] of short Hg-S bonds are indicated by thick lines, the remaining Hg-S contacts by thin lines. (b) Heazlewoodite, Ni_3S_2 (Parise 1980). Tetrahedrally coordinated nickel with five tetrahedron edges shared with adjacent tetrahedra.

Millerite. The crystal structure of millerite, NiS (Grice and Ferguson 1974) consists of [001] columns of three edge-connected coordination pyramids (NiS_5), with vertices pointing outward and joined to the bases of pyramids in the adjacent columns (Fig. 35a). Pyramidal bases are oriented inward and generate triangles of short Ni-Ni interactions equal to 2.53 Å. In the case of millerite, these triangles do not form a three-dimensional net (Ni-Ni = 3.14 Å along the column direction). Towards the vertex of the NiS_5 coordination pyramid, the Ni-S bond is 2.26 Å long, whereas it is 2.37 Å at the base. Interestingly, the S-centered polyhedra, SNi_5, form the same motif as the NiS_5 polyhedra.

Godlevskite. The structure of orthorhombic godlevskite, Ni_9S_8, space group *C*222 (Fleet 1987), can be described as a sphalerite-like packing of tetrahedra with numerous vacancies in which the tetrahedra are replaced by NiS_5 square pyramids. In the stacking of ($1\bar{1}1$) layers, a two-layer sequence is observed: the more complete one of the two tetrahedral layers has large irregularly-shaped vacancies with interstitial Ni tetrahedra inserted in them; the less fully occupied, rudimentary tetrahedral layer is filled by square coordination pyramids. In the latter, rows of paired pyramids, parallel to [101] occur, sharing edges with one another and with tetrahedra. Ni-Ni interactions run as infinite chains with some interconnections. The distances are 2.62-2.89 Å; for distinct Ni sites the number of metal-metal contacts varies from zero to five.

Godlevskite (Fig. 35b) is an OD structure composed of two types of (001) layers: (a) layers with a zig-zag arrangement of square pyramids and with a layer group $P2_12(2)$ (the direction perpendicular to the layer indicated by round brackets), and (b) layers with "bisphenoid clusters" of tetragonal pyramids and with edge-sharing tetrahedra, with the layer group $P(\bar{4})2m$. There is only a single position for the $P(\bar{4})2m$ layer after the $P2_12(2)$ layer, but there are two alternative positions for the latter layer after the former one. The zig-zag chains of square pyramids of Ni in these two alternatives run at 90° to one another, suggesting the possibility of a tetragonal polytype. For the planes ($1\bar{1}1$), the OD phenomenon manifests itself by the selected [101] strings of polyhedral configurations sliding past one another, assuming two alternative positions, ½ [101] apart.

Maucherite, $Ni_{11}As_8$, is tetragonal, space group $P4_12_12$, (Fleet 1973). It is a true OD structure although it has not yet been described as such. The layer group of the OD layer is $P(\bar{4})m2_1$. The OD layer consists of two sheets formed by rows of $NiAs_5$ square coordination pyramids; these sheets are related by a four-fold rotoinversion operation. It is limited by the arsenic atoms on planes of glide-reflection (Fig. 35c). The interlayer symmetry is $n_{1/2,1}$, parallel to (001), and 2_1 parallel to [010].

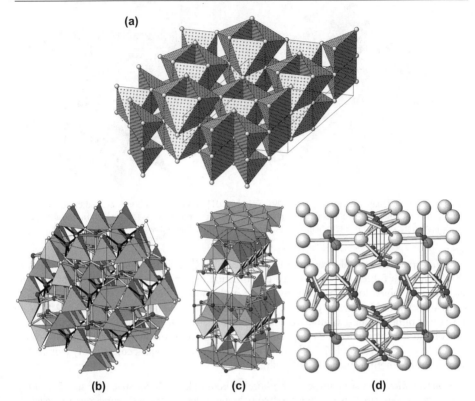

Figure 35. (a) Millerite, NiS (Grice and Ferguson 1994). Columns [001] of square pyramids (NiS$_5$) sharing edges. Nickel forms triangles of short Ni-Ni contacts inside the columns. (b) Godlevskite, Ni$_9$S$_8$ (Fleet 1987). Combination of NiS$_4$ coordination tetrahedra and NiS$_5$ square coordination pyramids. Note the layers (001) (horizontal) with edge-sharing tetrahedra (central tetrahedra of clusters are rendered in a black, ball-and-stick presentation) which alternate with layers rich in square coordination pyramids (bonds outlined in light). OD character: see text. (c) The OD structure of maucherite, Ni$_{11}$As$_8$ (Fleet 1973). Edge- and face sharing coordination pyramids (NiAs$_5$) and a coordination octahedron NiAs$_6$. The OD layer (see text) is composed of the two top layers of pyramids in the figure. (d) Hauchecornite (Kocman and Nuffield 1974). Coordination squares of Ni (shown as small dark spheres), with Ni-Ni interactions as thin black lines; medium grey: Bi, large white spheres: S.

Hauchecornite group. Hauchecornite, Ni$_9$(Bi$_{1.3}$Sb$_{0.7}$)S$_8$, (Kocman and Nuffield 1974) and arsenohauchecornite, Ni$_{18}$AsBi$_3$S$_{16}$ (Grice and Ferguson 1989) are sulfide-bismuthides. Their structures (Fig. 35d) can be understood as a channel structure, with double-columns of accordion-arranged coordination squares of nickel, sharing edges, and forming a very regular ladder-like motif of longitudinal- and cross- Ni-Ni bonds, 2.68 and 2.73 Å for the former, and 2.64 Å for the latter. Ni-S distances lie between 2.26 and 2.32 Å. Square channels created by the accordion-like walls contain Bi in BiNi$_8$ and Ni-Bi-Ni coordinations, respectively. Bi-Ni distances are 2.69 and 2.70 Å, Bi-S only 2.78 Å.

Sulfides of palladium and platinum

Fairly complete structural knowledge exists for the numerous sulfides of palladium (Table 22).

Cation-based polyhedra. In sub-sulfides of Pd, the coordination polyhedra of palladium are very complex. The number of direct Pd-Pd contacts is a function of the Pd:S ratio. In

Pd$_4$S (an unnamed mineral), palladium atoms are 12-coordinated, by 10 palladium atoms up to a distance of 3.12 Å and two S ligands at 2.34 and 2.48 Å, respectively, in an irregular polyhedron. The sulfur atoms are in a near-trans configuration (Fig. 36a). In Pd$_3$S, different Pd sites are surrounded by 7 and 8 Pd atoms, at 2.78-2.95 Å, and again coordinated only to two sulfur ligands (at 2.28 and 2.46 Å) in an approximate trans-configuration.

Vasilite, Pd$_{16}$S$_7$, contains two distinct palladium sites. Pd1 has a five-fold ring of Pd atoms at 2.74-2.79 Å and, in the orientation perpendicular to this ring, a pair of S ligands in a trans-configuration, at 2.27 and 2.32 Å, respectively. Pd2 has six Pd ligands at 2.74-2.95 Å, and three S ligands forming a triangular configuration, all situated at 2.45 Å from Pd2. For comparison, in the cubic F-centered palladium, the Pd-Pd distance is 2.749 Å.

In the crystal structure of its 1:1 sulfide vysotskite, PdS, palladium assumes a square-planar coordination, with S ligands at 2.32-2.35 Å (Fig. 36b). Each Pd atom has zero, one or two very long Pd-Pd distances (minimum value is 3.30 Å). In the projection on (001), the tetragonal stucture imitates a $3^2.4.3.4$ net. One of the [001] square channels is filled by a stack of mutually parallel, horizontal PdS$_4$ coordination squares, whereas the other has walls lined by alternating, vertical PdS$_4$ square configurations. The latter is also true for the triangular channels separating the square ones. This configuration lowers the symmetry to $P4_2/m$.

The structure of PdSe (Fig. 36c) is an $N = 2$ homologue of the structure of PdS. The square piles of "horizontal" coordination PdSe$_4$ squares are doubled, the paired squares being inserted in a network of square- and triangular channels with walls lined by alternating vertical PdSe$_4$ squares. The structure is still tetragonal, space group $P4_2/mbc$, the Pd-Se distances are 2.427-2.468 Å and the closest Pd-Pd distances, in the interior of vertically lined square channels, are 3.20 Å.

A different stacking of square coordination units occurs in PtS (Fig. 37a,b). In this simple tetragonal structure, space group $P4_2/mmc$, only square channels parallel to [001] are present;

Table 22. Palladium and platinum chalcogenides.

Mineral	Formula	Space Group	Lattice Parameters (Å,°)				Ref.
			a	b	c	β	
Vysotskite	PdS	$P4_2/m$	6.429		6.611		[1]
Synthetic	PdS$_2$	$Pbca$	5.460	5.541	7.531		[2]
Synthetic	PdSe	$P4_2/mbc$	11.565		6.998		[3]
Cooperite	PtS	$P4_2/mmc$	3.47		6.110		[4]
Synthetic	PtS$_2$	$P\bar{3}m1$	3.543		5.039		[5]
Synthetic	Pd$_4$S	$P\bar{4}2_1c$	5.115		5.590		[6]
Synthetic	Pd$_3$S	$Ama2$	6.088	5.374	7.453		[7]
Vasilite	Pd$_{16}$S$_7$	$I\bar{4}3m$	8.93				[8]
Palladseite	Pd$_{17}$Se$_{15}$	$Pm\bar{3}m$	10.606				[9]
Prassoite	Rh$_{17}$S$_{15}$	$Pm\bar{3}m$	9.911				[9]
Synthetic	Ni$_{10}$Pd$_7$S$_{15}$	$Pm\bar{3}m$	9.872				[10]
synthetic	Tl$_2$Pt$_5$S$_6$	$P2_1/n$	6.971	6.941	11.088	97.83	[11]
Jaguéite	Cu$_2$Pd$_3$Se$_4$	$P2_1/c$	5.672	9.909	6.264	115.40	[12]
Chrisstanleyite	Ag$_2$Pd$_3$Se$_4$	$P2_1/c$	5.676	10.342	6.341	115.00	[12]

References: [1] Brese et al. 1985; [2] Grønvold and Rost 1957; [3] Ijjaali and Ibers 2001; [4] Grønvold et al. 1960; [5] Furuseth et al. 1965; [6] Grønvold and Rost 1962; [7] Rost and Vestersjö 1968; [8] Matković et al. 1976; [9] Geller 1962; [10] Dubost et al. in press; [11] Klepp 1993; [12] Topa et al. in press

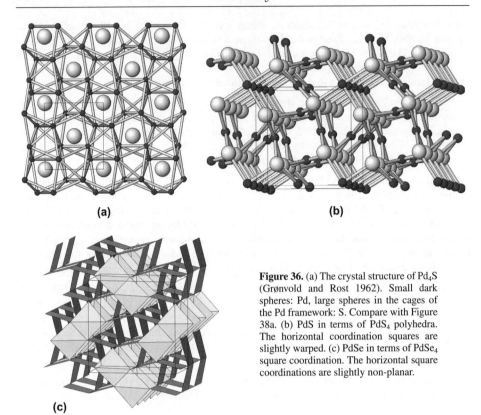

Figure 36. (a) The crystal structure of Pd_4S (Grønvold and Rost 1962). Small dark spheres: Pd, large spheres in the cages of the Pd framework: S. Compare with Figure 38a. (b) PdS in terms of PdS_4 polyhedra. The horizontal coordination squares are slightly warped. (c) PdSe in terms of $PdSe_4$ square coordination. The horizontal square coordinations are slightly non-planar.

all are lined by vertical coordination PtS_4 squares. The alternation of these configurations results in bands <100> of edge-sharing PtS_4 squares. Along [001], the [100] and [010] strips alternate, meeting in common S atoms (Fig. 37b).

PtSe is not known. The structure of palladseite, $Pd_{17}Se_{15}$, stoichiometrically close to the above described PdSe, is one of the most complicated structures in this category. Isotypic are prassoite, $Rh_{17}S_{15}$ and synthetic $Ni_{10}Pd_7S_{15}$ (Table 22), whereas $Pd_{17}S_{15}$ has never been encountered. This complicated cubic structure has four distinct cation sites of widely different type. It contains: (a) a "scallop" consisting of two "shells," i.e., cup-like configurations of four edge-sharing, flattened tetrahedra (Fig. 37c); (b) a coordination octahedron sandwiched between the dorsal portions of six "scallops," and (c) two distinct types of square coordination situated in a <100> system of channels between the above elements. The first type of coordination square occupies all faces of a cube formed in each channel intersection, whereas the second one halves the channels between two intersections (Fig. 37d).

In the ternary sulfide, $Ni_{10}Pd_7S_{15}$, nickel shows preference for the octahedral site (90% Ni, M-S = 2.41 Å), and the flat-tetrahedral site (60% Ni, M-S = 2.25-2.33 Å), whereas the square coordinations forming the cube have 50% Ni and M-S distance equal to 2.27 Å. The other square site is 90% Pd, here M-S = 2.41 Å. The flattened tetrahedra are a hallmark of this structure type. In these sites, the M-S distance varies from 2.26 Å to 2.35 Å in $Rh_{17}S_{15}$; this corresponds to 2.43-2.51 Å in $Pd_{17}Se_{15}$. The trans-configuration angle, which would be 180° for the planar coordination and 109.47° for a regular tetrahedron, is 2 × 163.9° and 2 × 166.5° in the flat tetrahedra of the ternary compound. The only short cation-cation distance

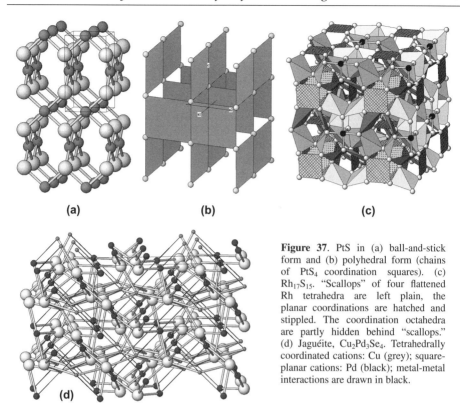

Figure 37. PtS in (a) ball-and-stick form and (b) polyhedral form (chains of PtS$_4$ coordination squares). (c) Rh$_{17}$S$_{15}$. "Scallops" of four flattened Rh tetrahedra are left plain, the planar coordinations are hatched and stippled. The coordination octahedra are partly hidden behind "scallops." (d) Jaguéite, Cu$_2$Pd$_3$Se$_4$. Tetrahedrally coordinated cations: Cu (grey); square-planar cations: Pd (black); metal-metal interactions are drawn in black.

is found between the two square coordinations outside the cube, equal to 2.63 Å in the case of Ni$_{10}$Pd$_7$S$_{15}$, and commensurate with distances in pure metals also for both remaining compounds. Differences in electronic configurations in the four distinct cation sites (metal-metal bonds in the latter case) are solved by the distribution of Ni *vs.* Pd in Ni$_{10}$Pd$_7$S$_{15}$ but must be resolved using the same cation in the two binary compounds.

Square-planar configurations occur also in ternary sulfides/selenides of Pd and Pt. Staggered, variously folded and in one direction also cut-up strips of PtS$_4$ squares occur, e.g., in Tl$_2$Pt$_5$S$_6$ (Klepp 1993), in which their interspaces are occupied by Tl. Another interesting structure with planar coordinations is that of jaguéite, Cu$_2$Pd$_3$Se$_4$, and of the isotypic chrisstanleyite, Ag$_2$Pd$_3$Se$_4$ (Topa et al. in press). In these structures, Pd1 forms isolated square coordinated PdSe$_4$, whereas Pd2 occurs as coordination squares paired via a common edge (Fig. 37d). The Pd-Se distances are 2.49-2.50 Å, and 2.47-2.49 Å, respectively. The Pd1Se$_4$ coordination is perfectly square, whereas that of paired Pd2 squares is slightly puckered (their planarity expressed by the trans-configuration Se-Pd-Se angles of 180.0° and 175.2° ± 0.3°, respectively). The openwork structure created by corner-connection of these square coordinations is stabilized by the presence of elongated CuSe$_4$ (resp. AgSe$_4$) tetrahedra, occuring as edge-sharing pairs, and by linear cation-cation interactions Pd2-Cu(Ag)-Pd1-Cu(Ag)-Pd2, with Cu-Pd distances equal to 2.86 Å for Pd1 and 2.75 Å (plus additional 2.90 Å) for Pd2 (Fig. 37d).

Anion-based coordination polyhedra. The complex structures of lower sulfides of Pd are easier to comprehend when they are presented in terms of S-centered coordination polyhedra, with cations as ligands. In the structure of Pd$_4$S, the anion-based polyhedra are bisdisphenoids (SPd$_8$) which share all eight corners with adjacent bisdisphenoids (Fig. 38a). In Pd$_3$S, the coordination number of sulfur is 6 (Makovicky 2002). Completely different coordinations are observed in vasilite, $Pd_{16}S_7$: tetrahedral clusters of trigonal coordination prisms (SPd_6), radiating from a common Pd_4 tetrahedron, are interspersed by disphenoids (SPd_4) which interconnect free vertices of the trigonal coordination prisms (Fig. 38b). The shortest Pd-S bonds (2.28 Å) occur in these disphenoids.

In both 1:1 sulfides and the selenide, the central anion is tetrahedrally coordinated by the cation. In PdS, there are (S_2Pd_6) clusters composed of two edge-sharing tetrahedra, strung along [001] and interconnected with adjacent strings of such groups, one at the same height, whereas the others are displaced by ½ [001]. The structure can be described as stacking of $\bar{4}2m$ propellers of four tetrahedra along [001] by means of horizontal, inter-propeller m planes and only corner-sharing with adjacent stacks (Fig. 38c). Similar strings of tetrahedra that alternatively share and do not share edges along [001] occur in PdSe. However, the "propeller scheme" is modified compared with that in PdS. In PtS, the S-centered tetrahedra form strings of edge-sharing tetrahedra along [001]; these strings share only tetrahedral corners.

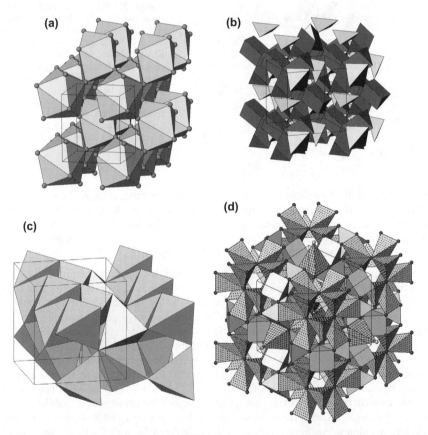

Figure 38. (a) Sulfur-centered bisdisphenoids (SPd_8) in the crystal structure of Pd_4S. Compare with Figure 36a. (b) Sulfur-centered coordination polyhedra in the crystal structure of $Pd_{16}S_7$. Tetrahedral clusters of trigonal coordination prisms (SPd_6) are interspersed by SPd_4 coordination tetrahedra. (c) PdS in terms of edge-sharing SPd_4 tetrahedra. Note the "propellers" of four tetrahedra. (d) S-centered coordination polyhedra in the crystal structure of $Rh_{17}S_{15}$. Cage-like configurations of SRh_4 coordination tetrahedra, two- and three-dimensional "iron crosses" of SRh_5 coordination pyramids. The latter also occur in pentlandite as $S(Fe,Ni)_5$.

In spite of similar unit cell parameters, the $Pd_{17}Se_{15}$ structure type is different from pentlandite; among the sulfur-centered polyhedra, however, three-dimensional "iron crosses" of six tall square coordination pyramids (M_5S) with a common S vertex (Fig. 38d) are prominent in both structure types.

The concept of anion-based coordination polyhedra has only rarely been applied to chalcogenides. It can bring quite a new insight into the crystal chemistry of these compounds.

IONIC CONDUCTORS AND THE CHALCOGENIDES OF Cu AND Ag

Ionic conductors are substances in which cations or anions become mobile above a transition temperature, and become "frozen" in fixed coordination polyhedra below this temperature. They dominate chalcogenide phase systems with copper and silver. They also occur among compounds with cations such as lithium, sodium, and other alkalis. This behaviour has been documented for many channel structures (see Makovicky 2005), some cage structures, and for many layer intercalates. A number of sulfosalts also belong to this category. In the present account we concentrate upon the sulfides and sulfosalts of Cu and Ag.

Compounds of copper. The Cu-S phase diagram has a number of phases belonging to this category, situated in the range Cu_2S-CuS (see Fleet 2006, this volume). Structure analyses exist for chalcocite Cu_2S (Evans 1979; Cava et al. 1981), djurleite $Cu_{1.934-1.965}S$ (Evans 1979) and anilite (Koto and Morimoto 1970). Furthermore, recent structural investigations have been published on high and low digenite, $Cu_{1.8}S$ (Will et al. 2002) and ample material has been published on high and low bornite (Cu_5FeS_4) and its solid solution with digenite (discussed among compounds with clusters of tetrahedra).

At temperatures above 103.5 °C (Roseboom 1966), chalcocite has a small hexagonal cell with symmetry $P6_3/mmc$ (Table 23). According to Buerger and Wuensch (1963) and Cava et al. (1981), its structure consists of an *hcp* arrangement of sulfur anions where copper is highly mobile, found primarily in triangular coordinations of the disordered cation substructure. Cu with anharmonic thermal motion occupies primarily the alternate triangular sites in the hexagonally packed (0002) layers (70% occupancy), and all triangular sites between two consecutive (0002) layers (20% occupancy). These occupancies should be understood as a fractional residence time. Below 103.5 °C, the Cu atoms become immobilized in the sites of a monoclinic $P2_1/c$ structure with 24 fully occupied Cu sites and 12 S sites in a cell (Table 23). The c axis is equal to $4c$ of the hexagonal (sub)cell, the a axis dips across the layers of the hexagonal close packing at an interval of $2c_{hex}$. The structures of the chalcocite – djurleite pair are built along closely related principles. The $P2_1/n$ cell of djurleite contains 62 Cu and 32 S atoms.

The situation in low-temperature chalcocite is illustrated well by the portion of it shown in Figure 39a. With the exception of two Cu atoms with close-to-linear coordination, the Cu atoms are in more or less regular triangular coordination. Only rarely do the Cu atoms deviate more substantially from the plane of the triangle defined by the sulfur ligands. Eight Cu atoms in a unit cell are in the triangles in the (0001) plane (the same for 20 Cu atoms in djurleite); the remaining Cu atoms are at the fixed ~1/3 and ~2/3 positions between two consecutive layers. Cu-Cu distances play an important role in the structure of chalcocite, the closest contacts being 2.52 Å (2.45 Å in djurleite). The most frequent Cu-Cu distances are about 2.77 Å; they are distributed symmetrically between 2.5 and 3.0 Å. Cu-S distances peak at 2.30 Å, the average triangular bond length. They are situated between 2.20 and 2.40 Å, whereas any tendencies towards tetrahedral coordination are very distorted.

Typical features of both structures are the empty coordination octahedra with 3-4 alternate triangular faces occupied by Cu in distinct combinations (Evans 1981). Exceptionally, two of these triangles share an edge. In an ideal octahedron (S-S distance of 3.95 Å), the triangular

Table 23. Copper and silver chalcogenides (ionic conductors).

Phase	Space Group	Lattice Parameters (Å,°)				Notes	Ref.
		a	b	c	Angle		
High-T Cu$_2$S	$P6_3/mmc$	3.985	3.985	6.806	120°	[a]	(1)
Low-T Cu$_2$S chalcocite	$P2_1/c$	15.246	11.884	13.494	β 116.35°	[b]	(2)
Digenite Cu$_{1.8}$S	$Fm\overline{3}m$	5.593	5.593	5.593	—	[c]	(3)
Djurleite Cu$_{1.96}$S	$P2_1/n$	26.897	15.745	13.565	β 90.13°	[d]	(2)
Anilite Cu$_{1.75}$S	$Pnma$	7.89	7.84	11.01	—	[e]	(4)
High-T Ag$_2$S	$Im\overline{3}m$	4.889	4.889	4.889	—	[f]	(5)
Low-T Ag$_2$S	$P2_1/n$	4.23	6.91	7.87	β 99.58°	[g]	(6)
High-T Ag$_2$Te	$Im\overline{3}m$	5.329	5.329	5.329	—	[h]	(7)
Intermediate-T Ag$_2$Te	$Fm\overline{3}m$	6.643	6.643	6.643	—	[i]	(7)
Low-T Ag$_2$Te	$P2_1/c$	8.164	4.468	8.977	β 124.16°	[j]	(8)
Stromeyerite CuAgS	$Cmc2_1$	4.059	6.617	7.967	—	[k]	(9)
Jalpaite Ag$_3$CuS$_2$	$I4_1/amd$	8.671	8.671	11.757	—	[l]	(10)

Notes: [a] 598 K; [b] Stable below 103.5 °C; [c] 673 K; [d] Ambient T; [e] under 75 °C; [f] 533K; [g] Stable below 177 °C; [h] 1123 K; [i] 723 K; [j] 295 K; [k] 298 K; [l] Ambient T
References: (1) Cava et al. 1981; (2) Evans 1979; (3) Will et al. 2002; (4) Koto and Morimoto 1970; (5) Cava et al. 1980; (6) Frueh 1958; (7) Schneider and Schulz 1993; (8) van der Lee and de Boer 1993; (9) Baker et al. 1991; (10) Baker et al. 1992

Cu-S distance is 2.28 Å, linear Cu-S distance is 1.98 Å, whereas Cu-Cu is 2.63 Å for alternate triangles and 3.23 Å across the volume of the octahedron. Thus, appreciable distortion of the sulfur framework is required in order to accommodate the Cu-Cu distances of about 2.77 Å, and the linear Cu-S distances especially (Evans 1981). Thus far, no principles have been defined for the long-range occupational patterns of triangles in these minerals; the two representations in Figures 39a-b illustrate well why this is so. The structure of chalcocite as well as that of djurleite is almost always multiply twinned with reference to their hexagonal substructure.

At Cu:S ratios less than 2:1, and above 435 °C (Roseboom 1966) for stoichiometric Cu$_2$S, a cubic phase, digenite (formally Cu$_9$S$_5$) exists, with a mobile copper array as in high chalcocite. As a point-like approximation to this mobile array, Will et al. (2002) found ~0.25 Cu in all tetrahedral sites, below 0.1 Cu in the octahedral positions of the F-centered cubic cell, and an envelope with a radius of about 0.9 Å of fractional ~0.03 sites around the tetrahedral site. These sites are at a distance of only 1.84 Å from the nearest S atom.

Anilite, Cu$_7$S$_4$ (Koto and Morimoto 1970) is based on *ccp* of sulfur atoms, its unit cell being $a\sqrt{2} \times a\sqrt{2} \times 2a$ of *ccp*. The structure contains rods composed of paired, edge-sharing tetrahedra of copper in an A-centered arrangement. They are actually rows of empty, edge-sharing octahedra with the tetrahedral interspaces occupied (Fig. 39c). The rest of the Cu atoms are present in triangular sites. Cu5 forms rows [010] of triangles, whereas Cu2 and Cu4 occupy three faces of empty coordination octahedra which are enclosed by four adjacent rows of paired tetrahedra. Triangularly coordinated Cu5 interconnects the empty octahedra with the filled tetrahedra of the tetrahedral rows.

An alternative description is that in terms of *ccp* layer stacking. The ABC sequence of layers (011)$_{anilite}$ is overprinted by a two-layer scheme of Cu occupancy, alternating the layers composed exclusively of triangular copper sites with the layers consisting of Cu tetrahedra and only occasional triangles. The latter are in parallel channels [100]. Anilite is stable only to 75 °C (Potter 1977). Obviously it does not survive mobilization of Cu, yielding digenite instead.

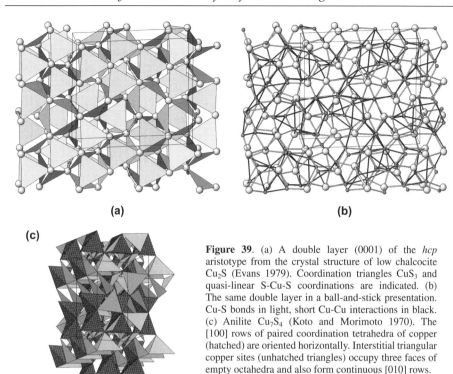

Figure 39. (a) A double layer (0001) of the *hcp* aristotype from the crystal structure of low chalcocite Cu_2S (Evans 1979). Coordination triangles CuS_3 and quasi-linear S-Cu-S coordinations are indicated. (b) The same double layer in a ball-and-stick presentation. Cu-S bonds in light, short Cu-Cu interactions in black. (c) Anilite Cu_7S_4 (Koto and Morimoto 1970). The [100] rows of paired coordination tetrahedra of copper (hatched) are oriented horizontally. Interstitial triangular copper sites (unhatched triangles) occupy three faces of empty octahedra and also form continuous [010] rows.

Among sulfosalts of copper, typical ionic conductors are found in the family of Cu_3XS_3 compounds (X = Sb,Bi; S = S,Se) (Table 24). These structures consist of an *hcp* array of anions which is twinned on $(11\bar{2}2)_{hcp}$ (Andersson and Hyde 1974). The thickness of *hcp* lamellae is $3d(11\bar{2}2)$, i.e., that of one S_6 octahedron. On the composition planes of the unit cell twinning, large trigonal coordination prisms are formed, occupied by the metalloid atom positioned close to one of the prism bases, i.e., in a pyramidal XS_3 coordination. Above the "freezing" temperature, this framework has symmetry *Pnma,* with Cu distributed (in sulfides) statistically over the available trigonal planar sites (Makovicky 1994; Pfitzner 1998). These are both the sites in the triangular walls shared by two tetrahedra and the sites of two kinds in the walls separating an octahedron and a tetrahedron. There are twice as many of these sites as the number of Cu atoms available, making these compounds eminently suitable for ionic conduction. In Cu_3SbSe_3 (Pfitzner 1995) a tetrahedral position straddling the mirror plane of the unit-cell twinning is occupied instead; yet another occupation scheme is found in the related Li_3AsS_3 (Seung et al. 1998).

In wittichenite, Cu_3BiS_3, every second row [001] of the triangles sandwiched between two tetrahedra is occupied by Cu in a continuous fashion (Fig. 40a). It is interspaced by 3D "rotors" composed of other types of occupied triangles. In the adjacent twin lamellae, these schemes alternate, related by 2_1 axes. Above the "freezing transformation" of 118 °C, this $P2_12_12_1$ polymorph has a precursor in the form of a modulated structure, with the **q** vector (parallel to c^*) changing from about 0.32 at the temperature of transformation to almost 0.5 at the point of fading out between 170° and 190 °C (Makovicky 1983) [1/**q** describes a periodicity of non-commensurate modulation in terms of the subcell periodicity in the given lattice direction]. The movement of satellites with temperature exhibits a change in behavior at

Table 24. Wittichenite homeotypes.

Mineral	Formula	Space Group	Lattice Parameters (Å,°)				Ref.
			a	b	c	β	
Wittichenite	Cu_3BiS_3	$P2_12_12_1$	7.72	10.4	6.72		[1]
Intermediate wittichenite	Cu_3BiS_3	modulated	7.66	10.5	6.72[a]		[2]
High wittichenite	Cu_3BiS_3[(2)]	Pnma	7.66	10.5	6.71[b]		[2]
Low skinnerite (γ)	Cu_3SbS_3	$P2_12_12_1$	7.88	10.2	6.62		[3][c]
Intermediate skinnerite (β)	Cu_3SbS_3	$P2_1/c$	7.81	10.2	13.27	90.29	[4][d]
High skinnerite (α)	Cu_3SbS_3	Pnma	7.81	10.3	6.59		[5]
Synth.	Cu_3SbSe_3[(e)]	Pnma	7.99	10.6	6.84		[6]
Synth.	Li_3AsS_3[(e)]	$Pna2_1$	8.05	9.82	6.63		[7]

Notes: (a) at 142 °C; (b) at 350 °C; (c) also Whitfield (1980); (d) also Pfitzner (1994), who gives transformation temperatures as −9 °C and +121 °C, respectively; (e) cation arrangements differ from the Cu_3BiS_3-Cu_3SbS_3 scheme.
References: [1] Kocman and Nuffield 1973; [2] Makovicky 1983; [2]; [3] Pfitzner 1994 [c]; [4] Makovicky and Balić-Žunić 1995; [5] Pfitzner 1998; [6] Pfitzner 1995; [7] Seung et al. 1998

about 135 °C, close to the temperature of 150 °C given by Lugakov et al. (1975) as the onset of intense ionic conduction. Mizota et al. (1998) demonstrate a substantial decrease in resistivity of Cu_3BiS_3 at 100-115° but put the onset of ionic conduction at about 157 °C. The precursor has a substantial hysteresis both in the movement of satellites and in recovering the resistivity value that is appropriate for a given temperature. Several days are needed for a reequilibration from the reduced values of resistivity and larger values of the modulation **q** vector, which were inherited from the higher temperatures, on cooling. This suggests a heavily cooperative mechanism of ion movement in the structure. Models for the precursor structure have been proposed by Makovicky (1994).

Such a precursor has not been seen in Cu_3SbS_3 which has a phase transformation at 120 °C. In the ambient-temperature form of this compound, intermediate skinnerite, the [001] rows of tetrahedra are interrupted after two triangles, and shifted by $a/2$ to the alternate rows in the same twin lamella (Fig. 40b). The other Cu coordination triangles always occur as two opposing triangles in an empty octahedron. The Cu_3XS_3 structures actually are stackings of unit layers (001) of Cu-occupied triangles: the same layer-like enantiomorphs are related by a 2_1 axis perpendicular to them when they are stacked in wittichenite, whereas a two-layer pack of one enantiomorph is always followed by a two-layer pack of the opposite enantiomorph in intermediate skinnerite (Makovicky 1994). Low skinnerite, below −9 °C (Pfitzner 1994), does not have the (expected) symmetry $P2_1/n$, with single layers of enantiomorphs following one another, but it crystallizes in $P2_12_12_1$. The apparent reason is that the wittichenite structure offers the maximum number of structurally favoured short Cu-Cu distances, more than intermediate skinnerite, whereas the $P2_1/n$ form does not (Makovicky and Pfitzner, unpublished).

Another family of ionic conductors are the synthetic and natural copper-bismuth sulfides and selenides of the pavonite homologous series, typified by $Cu_{1.78}Bi_{4.73}Se_8$ (Makovicky et al. 2006). In this structure (Fig. 40c), the octahedral columns of the non-accretional layer contain three fractional copper sites, which add up to 1.5 Cu in one column period: the occupancy of the flattened tetrahedral sites in the upper and lower portions of the empty coordination octahedron is 0.31, that of the sideways positioned tetrahedral Cu is 0.26, whereas the tetrahedral sites between two adjacent octahedra contain 0.16 Cu each. Although

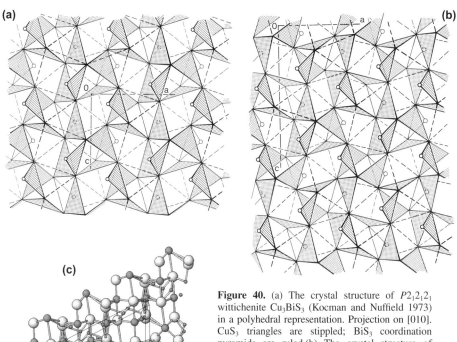

Figure 40. (a) The crystal structure of $P2_12_12_1$ wittichenite Cu_3BiS_3 (Kocman and Nuffield 1973) in a polyhedral representation. Projection on [010]. CuS_3 triangles are stippled; BiS_3 coordination pyramids are ruled. (b) The crystal structure of $P2_1/c$ skinnerite Cu_3SbS_3 (Makovicky and Balić-Žunić 1995). Projection on [010]. Triangular CuS_3 coordinations are presented as stippled triangles, SbS_3 coordination pyramids are ruled. (c) The crystal structure of $Cu_{1.78}Bi_{4.73}Se_8$ (Makovicky et al. 2006), the third member of the pavonite homologous series. A number of statistically occupied copper positions (sites of "frozen" Cu, small grey spheres) were detected in the octahedral columns situated in the thinner structural slabs (at $z = \frac{1}{2}$; c axis vertical) as well as replacing a part of Bi (larger grey circles) in the thicker slabs.

no experimental confirmation is available as yet, the statistical character of Cu distribution along these columns suggests that they have been one-dimensional diffusion paths at the temperatures of formation. The statistically distributed fractional Cu in the accretional layer of Bi octahedra (site occupancies 0.02-0.09) must encounter much more hindrance when percolating through the framework of Bi octahedra. Olivier-Fourcade et al. (1983) prepared an ionic conductor $Li_{3x}Sb_{6-x}S_9$, a pavonite homologue $N = 4$ and Tomeoka et al. (1980) refined the $N = 3$ and 4 pavonite homologues in the system Cu-Bi-S, analogous to the above structure.

Compounds of silver. In the conducting state, the silver chalcogenides Ag_2S and Ag_2Se have body centered S frameworks in which the mobile Ag atoms reside preferentially at regular three-fold, slightly pyramidal sites. For Ag_2S (Cava et al. 1980), the three Ag-S distances are 2.73 Å; central angles are 126.9° and 2 × 101.5°). The Ag-Ag distances in the resulting "sodalite-like cage" of silver sites around a sulfur site (Fig. 41a) are not compatible with full occupancy, being 1.73 Å between all nearest neighbors, and 2.44 Å across the square faces of the cage. Occupancy is 33%. In high-temperature Ag_2Se (Oliveria et al. 1988), the corresponding Ag-Se distances are 2.82 Å, the Ag-Ag distances are 1.78 Å and 2.52 Å, respectively. The structure depicted is that above about 200 °C for Ag_2S, when practically only the tetrahedral positions are occupied by silver atoms. Below this temperature, down to the temperature of a phase transformation at 177 °C, the octahedral positions centered between four such tetrahedral sites also show increasing occupancy. Fourier maps of high-T silver sulfide and selenide show continuous bands <100> of electron density, i.e., the preferred diffusion paths for silver. These are situated along the diameters of the square faces of the cuboctahedra; across the structure, these faces are strung along <100>. According to Wuensch (1993), diffusion in Ag_2S proceeds via crowded [100] second-nearest tetrahedral pairs and octahedral coordinations in a correlated, cooperative motion. The boundary with the other mode of ion conduction, diffusive hops between neighboring tetrahedral sites (such as in AgI), should lie at the $Ag_{1.5}X$ composition. The smooth distribution of Ag along its diffusion paths is a result of relaxation of the closely spaced tetrahedral dimers.

The highest temperature, I-centered polymorph of Ag_2Te (at about 1123 K) is isotypic, with distances Ag-Te = 2.98 Å, and Ag-Ag = 1.88 and 2.66 Å (Schneider and Schulz 1993). The intermediate "high" Ag_2Te polymorph (measured at 523-923 K) is F-centered and has Ag-Te distances of 2.65 Å and 2.84 Å. Ag forms a small cuboctahedron of sites, half-way between S atoms, with site-to-site distances equal to 1.12 and 1.58 Å, and in additional Ag sites at ¼ ¼ ¼ (Ag-Ag distances 1.95 and 2.13 Å); these distances might suggest diffusion paths for Ag. Distances from one cuboctahedron to an adjacent one are 2.47 Å, and less likely for direct diffusion. The low-temperature form of Ag_2Te (van der Lee and de Boer 1993) is composed of mackinawite-like layers (Ag-Te 2.88-3.02 Å) alternating with an openwork layer composed of elongated Ag tetrahedra (Ag-Te 2.84-3.03 Å); these share three edges with adjacent tetrahedra. All six edges of the tetrahedra in the mackinawite-like layer are shared with adjacent tetrahedra, two of these being below and above the layer (Fig. 41b). This leads to Ag-Ag distances between 2.91 and 3.13 Å.

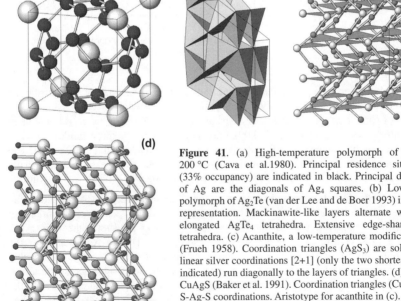

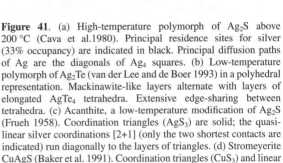

Figure 41. (a) High-temperature polymorph of Ag_2S above 200 °C (Cava et al.1980). Principal residence sites for silver (33% occupancy) are indicated in black. Principal diffusion paths of Ag are the diagonals of Ag_4 squares. (b) Low-temperature polymorph of Ag_2Te (van der Lee and de Boer 1993) in a polyhedral representation. Mackinawite-like layers alternate with layers of elongated $AgTe_4$ tetrahedra. Extensive edge-sharing between tetrahedra. (c) Acanthite, a low-temperature modification of Ag_2S (Frueh 1958). Coordination triangles (AgS_3) are solid; the quasi-linear silver coordinations [2+1] (only the two shortest contacts are indicated) run diagonally to the layers of triangles. (d) Stromeyerite CuAgS (Baker et al. 1991). Coordination triangles (CuS_3) and linear S-Ag-S coordinations. Aristotype for acanthite in (c).

The low-temperature form of Ag_2S, acanthite, consists of (010) layers which are distorted layers of hexagonally packed sulfur atoms, in which alternate triangles are occupied by Ag in trigonal coordination (Ag-S = 2.49, 2.57, and 2.60 Å). This coordination can also be interpreted as a very distorted tetrahedral coordination with the fourth bond of 2.98 Å out of the triangle plane (Fig. 41c). These planes are interconnected by quasi-linear Ag (S-Ag = 2.42 and 2.45 Å), with an additional distance equal to 2.92 Å. The only Ag-Ag interactions of interest are the isolated bonds at 2.94 Å between the two types of Ag; the network of other Ag-Ag interactions starts only at 3.05 Å. The *bcc* cell of the high-temperature form of Ag_2S becomes distorted in the low-temperature form into a pseudo-*bcc* cell with $a_1 = a_2 = 4.89$ Å, $a_3 = 4.76$ Å, $\alpha = 89.1°$, $\beta = 90.9°$, and $\gamma = 89.7°$ (Wuensch 1993). Adjustments between this state and the high-*T* cubic structure proceed first by change in the β angle of the low form, then by means of multiple twinning, followed by the transformation itself (Sadanaga and Sueno 1967).

The undistorted aristotype of acanthite is stromeyerite $CuAgS_2$ (Baker et al. 1991). This structure (Fig. 41d) consists of distorted hexagonal nets (001) of sulfur atoms, with alternate triangles occupied by Cu (2 × 2.25 Å and 2.34 Å). These nets are interconnected by linear S-Ag-S coordination units (2.44 and 2.45 Å). Ag-Ag distances are 3.88 and 3.89 Å whereas a zig-zag net of Ag-Cu distances in (010) planes of the structure are 2.88 Å and 3.45-3.52 Å. Another Cu-Ag sulfide, jalpaite, Ag_3CuS_2 (Baker et al. 1992) is essentially a structure based on linear coordination of both metals with the Cu-S distance equal to 2.18 Å and the Ag-S bonds equal to 2.50 and 2.57 Å. The Ag coordinations are of a quasi-tetrahedral (additional distances 2 × 2.94 Å) and linear type, respectively. In the interpretation based on these polyhedra, parallel and corner connected chains of edge-sharing Ag tetrahedra host linear Cu and Ag in interspaces. The structure is based on a distorted *I*-centered packing of sulfur.

The crystal structure of stephanite, Ag_5SbS_4 (Ribár and Nowacki 1970) can be treated as a distorted version of a unit-cell twinned *hcp* array, with the unit slabs only $1d_{11\bar{2}2}$ of *hcp* array thick, in comparison to wittichenite where they are $3d_{11\bar{2}2}$ thick. The trigonal coordination prisms on composition planes of unit cell twinning again contain Sb pyramids situated at a prism base, but these occur only in alternate [001] rows of prisms. The rows of prisms intervening in the *b* direction are occupied by three-fold coordinated Ag whereas the unit slabs themselves contain two distinct Ag sites, in 3-fold and 3+1 coordination, respectively. The polarity of the structure, space group $Cmc2_1$, is caused by the same orientation of all SbS_3 pyramids, in the $-c$ direction, throughout the entire structure.

In the case of Ag_3AsS_3, proustite, (Engel and Nowacki 1966), the basis of the structure is a rhombohedral arrangement of AsS_3 pyramids, stacked in continuous columns [001], and situated at the bases of trigonal coordination prisms preserving the same orientation of their As vertices (Fig. 42a). Silver is present in intervening trigonal channels [001], with a spiralling system of both S atoms and short S-Ag-S bonds (2.44 Å and 2.45 Å). Every silver atom has an additional S-Ag distance of 2.89 Å, i.e., a 2+1 coordination. The Ag-Ag distances in the channels are 3.25 Å, those between channels 3.48 Å, so there are no cation-cation interactions. A complete coordination polyhedron of Ag is a bipyramid with a rhomb-shaped cross-section in which the short, 2.45 Å, distances combine with long interactions (up to 3.61 Å). Three orientations of such pyramids intermesh in the silver-housing channels.

A remarkable low-temperature ionic conductor is pearceite $(Ag,Cu)_{16}(As,Sb)_2S_{11}$ (Bindi et al. 2006), refined in $P\bar{3}m1$, with $a = 7.388$ Å and $c = 11.888$ Å at room temperature and at 120 and 15 K. This structure consists of two alternating kinds of (0001) slabs, about 6 Å thick. One slab type, $(Ag,Cu)_6 (As,Sb)_2S_7$, has fully occupied trigonal pyramidal (As,Sb) sites, combined with trigonal planar Ag/Cu sites (average M-S=2.417 Å). These are fully occupied, with a certain amount of disorder present. The lone electron pairs of As/Sb point into the slab interior. The alternating types of slabs, Ag_9CuS_4, contain no semimetals, only a

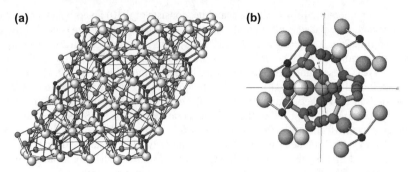

Figure 42. (a) Proustite Ag_3SbS_3 (Engel and Nowacki 1986) with spirals [001] of short S-Ag-S bonds and polar stacks of SbS_3 pyramids in the channels. Thin lines: long Ag-S interactions. (b) Argyrodite-like high-temperature Cu_7PSe_6 (Gaudin et al. 2000). A truncated tetrahedron of Cu diffusion paths with residence sites, surrounded by PSe_4 tetrahedra (partly shown) and additional Se atoms forming a cage which is centered by another Se atom.

linearly coordinated Cu site and mutually interconnected hexagonal anuli of highly smeared silver residence sites sandwiching the Cu position (Cu-Ag distance drops from 2.84 to 2.75 Å on temperature decrease). The Ag sites which are recognized display linear, trigonal and tetrahedral coordination with sulfur. There is a progressive individualization of fractional Ag sites from 300 K to 15 K, especially of the site interconnecting adjacent anuli, but the joint probability function still exhibits strong overlapping of neighbouring sites, i.e., a considerable mobility of Ag even at the lowest temperature measured.

The last example of silver-based ion conductors are the high-temperature phases of the argyrodite group, Ag_8MX_6, M= Ge, Sn, Si...., X=S, Se, Te (Gorochov 1968; Boucher et al. 1993). In the structures of Ag_8GeTe_6 and Ag_8SiTe_6 (Boucher et al. 1993), the M atoms are tetrahedrally coordinated by Te. Triangular walls of four isolated MTe$_4$ tetrahedra, and the combination of these tetrahedra with four independent Te atoms, form large cages with 24 triangular faces. Five faces meet at the corners of MTe$_4$ tetrahedra whereas six faces meet at each independent Te atom. The last Te atom centers the cavity and is surrounded by a truncated tetrahedron (Laves polyhedron) of three distinct Ag residence sites interconnected by clear diffusion paths: (a) the sites at the corners of the tetrahedral shape; (b) those at the long edges of the large tetrahedron, and (c) those above the short edges of the truncating faces. The short interconnections between the first type of residence sites in two adjacent cages serve as diffusion paths between the cages. This structure type is illustrated in Figure 42b using the new structure determination on another argyrodite-type compound, Cu_7PSe_6, by Gaudin et al. (2000).

LONE ELECTRON PAIR COMPOUNDS

Molecular arsenic sulfides

Most arsenic sulfides are typical molecular structures, with As_mS_n basket-like molecules. The lone electron pairs of all atoms in a molecule point outwards, into the intermolecular space. Thus, these molecules may be classified as inverted lone electron pair micelles. Molecules are held together by weak interactions of van der Waals character. The only non-molecular compound of this category, orpiment As_2S_3, and the As–Sb sulfides are discussed elsewhere.

The three-dimensional arrangement of the As_mS_n molecules is difficult to visualize. The As_4S_4 molecule of realgar (Fig. 43a, Table 25) with the ideal point-group symmetry $\bar{4}2m$, has covalent As-S bonds concentrated around the value of 2.24 Å and the As-As bonds of 2.57 Å.

Intermolecular distances start at 3.62 Å in the case of As-As interactions, and at 3.44 for the As-S interactions. Figure 44a shows the arrangement of As_4S_4 molecules in the structure, viewed along approximately [001].

The same molecule occurs in alacranite, β-As_4S_4 (Burns and Percival 2001); the interatomic distances are concentrated at 2.22 Å and 2.60 Å for the As-S and As-As interactions, respectively. For the columns of As_4S_4 molecules parallel to their $\bar{4}$-axes the stacking is similar to that of realgar. Configuration of the *pararealgar* molecule with the ideal point-group symmetry m is very different (Fig. 43b). It has one arsenic atom bound covalently to two As atoms at 2.48 Å and 2.53 Å and to one S atom at 2.19 Å; there are two further As atoms with one As-As and two As-S interactions, and the fourth As atom is bonded to three sulfurs. For the latter three As atoms the As-S distances are 2.23-2.26 Å. Columns of As_4S_4 molecules follow the direction of an approximate three-fold axis of the molecule, which runs through the triangle of As-As interactions (2.48 Å-2.53 Å-3.34 Å) at one pole of the molecule and the AsS_3 group at its opposing pole. This axis is perpendicular to the $(1\bar{1}1)$ layers of molecules in the structure.

Dimorphite, As_4S_3, is a subsulfide of arsenic; it occurs in two modifications, I and II. The As_4S_3 molecules (Fig. 43c) in both modifications have an As_3 triangle (As-As 2.44 Å) opposed by the AsS_3 group (As-S 2.22 Å); the remaining As-S bonds have a length equal to 2.20 Å. The three-fold axes of the molecules (ideal symmetry $3m$) point sideways from their [010] zig-zag columns and are inclined to their [001] rows in dimorphite II; the same is true for the [010] and [110] rows in dimorphite I. The intermolecular As-As distances are 3.58 Å and the As-S distances 3.75 – 3.77 Å in the latter structure. The sulfur-rich member of the As-S family, uzonite, As_4S_5 (Whitfield 1973) contains two sulfur-bonded As atoms (As-S 2.23 – 2.27 Å) and an As-As pair (2.55 Å; its As-S bonds are 2.23-2.25 Å long) in an As_4S_5 molecule with

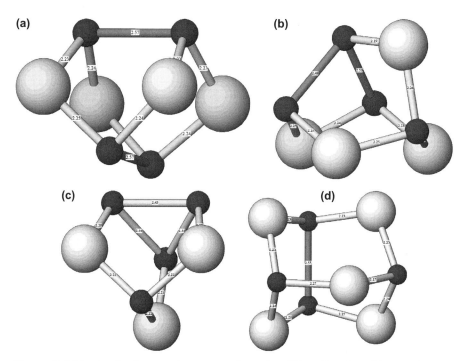

Figure 43. Molecules forming crystal structures of arsenic sulfides: (a) realgar, (b) pararealgar, (c) dimorphite, (d) uzonite. As: small dark spheres, S: large white spheres.

Table 25. Arsenic sulfides with molecular structures.

Mineral	Formula	Space Group	Lattice Parameters (Å,°)				Ref.
			a	b	c	β	
Realgar	$\alpha\text{-}As_4S_4$	$P2_1/n$	9.325	13.571	6.587	106.38	[1]
Pararealgar	As_4S_4	$P2_1/c$	9.909	9.655	8.502	97.29	[2]
Alacranite	$\beta\text{-}As_4S_4$	$C2/c$	9.943	9.366	8.908	102.01	[3]*
Dimorphite I	$\alpha\text{-}As_4S_3$	$Pnma$	9.12	7.99	10.10		[4]
Dimorphite II	$\beta\text{-}As_4S_3$	$Pnma$	11.21	9.90	6.58		[5]
Uzonite	As_4S_5	$P2_1/m$	7.98	8.10	7.14	101.0	[5]
Wakabayshilite	$[(As,Sb)_6S_9][As_4S_5]$	$Pna2_1$	25.262	14.563	6.492		[6]

Note: *high-temperature form: Pertlik (1994)
References: [1] Mullen and Nowacki 1972; [2] Bonazzi et al. 1995; [3] Burns and Percival 2001; [4] Whitfield 1970; [5] Whitfield 1973; [6] Bonazzi et al. 2005

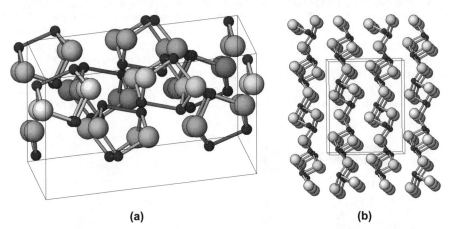

Figure 44. (a) Realgar AsS (Mullen and Nowacki 1972). Cage-like As_4S_4 molecules with lone electron pairs oriented into the intermolecular space. (b) Orpiment As_2S_3 (Mullen and Nowacki 1972). Layers (010) composed of edge-sharing As_2S_4 double chains. As: black spheres, S: large spheres. For OD description see text.

the ideal point-group symmetry $mm2$ (Fig. 43d). Only one mirror plane, that in the As-As dumbell, is preserved in the packing scheme of the molecules. The two-fold axis is nearly perpendicular to the [001] sequence of the molecules. Intermolecular As-As distances start at 3.74 Å; the As-S distances at 3.57 Å.

The remarkable light-induced degradation of realgar has been the subject of many investigations (summarized, e.g., in Kyono et al. 2005). It results in a formation of thin films and spherules of pararealgar on the surface of realgar crystals. Bonazzi et al. (2003) and Bindi et al. (2003) suggest an initial reaction $5As_4S_4+3O_2 \rightarrow 4As_4S_5+2As_2O_3$. The As_4S_5 molecules created by splitting one of the As-As bonds and inserting a sulfur atom are unstable and release another S atom giving a pararealgar molecule instead. According to Kyono et al. (2005), the resulting S radical is re-attached to another realgar molecule and produces an As_4S_5 molecule, repeating the reaction. This cycle is promoted by light and repeated during light exposure, even under low fugacities of oxygen.

An interesting plesiotype of the structures with As_mS_n cages is wakabayshilite, $[(As,Sb)_6S_9][As_4S_5]$ (Bonazzi et al. 2005). This structure contains: (a) hexagonally arranged $(As,Sb)_6S_9$ tubes, which look like a tube-like openwork of corner-connected $As_{0.8}Sb_{0.2}S_3$ pyramids (a tube with hexagonal symmetry, outward oriented As coordination pyramids and inner diameter, defined by S atoms, equal to 4.2 Å), and (b) columns [001] of stacked As_4S_5 cages filling space between the widely spaced tubes. Four out of six such columns have their As-As group oriented according to a six-fold rotation principle around the centering tube whereas two columns, on the opposite sides, break this orientation and conform with that valid for adjacent tubes. This arrangement results in ubiquitous twinning which caused a plethora of different interpretations in crystal structure studies before that of Bonazzi et al. (2005).

Ribbon-like arrangements

The crystal structure of diarsenic trisulfide, orpiment, is a layer-like structure with well expressed lone electron pairs of arsenic. However, as illustrated in Figure 44b, its (010) layers, separated by lone electron pair interspaces (planar micelles) with only very weak interactions, can be interpreted as composed of [001] double-columns of (very elongated) AsS_5 coordination pyramids, making it akin to the Sb and Bi-based structures with a much smaller lone electron eccentricity. The AsS_5 pyramids with elongated bases display three short bonds of 2.24-2.31 Å, with two very long distances (3.22-3.54 Å) in the bases, and distances of 3.48 Å, 3.67 Å or more across the interlayer space. Another interpretation of the (010) layer of orpiment is as an openwork of AsS_3 pyramids; the closest affinities are to the structure of As_2O_3 and the $(As,Sb)_6S_9$ tubes in wakabayshilite.

The structure of orpiment is a typical OD structure. If we follow the second (i.e., right-hand) (010) layer in the unit cell in Figure 44b, we see the double columns of As coordination pyramids which are interconnected via "hinges" of single S atoms. The configuration of the layer is mirror-symmetrical around these hinges. A corner of a double column of pyramids from both the layer below and the layer above protrudes into this symmetrical niche. However, they are not mirror-symmetrical themselves and inspection of Figure 44b shows that the layer below or above can, with equal probability, occur either in the orientation which is shown or in a reversed orientation, i.e., rotated by 180°. Configuration of a single layer pair will be the same in either case—this is the fundamental property of an OD structure. The observed structure has space group $P\,12_1/n1$; the polytype with every other layer rotated would have the space group $P112_1/b$.

Stibnite, Sb_2S_3, and bismuthinite, Bi_2S_3, are isotypic; their structures consist of ribbons of four MS_5 coordination pyramids which share edges (Figs. 3a and 45a). The condensed pyramids belong to two types: a pair of inner pyramids with less eccentric cation positions (Bi-S: 2.58 Å to the pyramid vertex and 2.74 Å, opposed by 2.97 Å, at the pyramid base), flanked by eccentric pyramids (2.69 Å to the vertex and 2.67 Å opposed by 3.06 Å at the base) which constitute margins of the ribbon. The ribbons form corrugated layers (010) via 3.03 Å Bi-S contacts interconnecting the bases of marginal pyramids. For antimony the bond scheme is similar but the latter contacts are longer than the distances in the pyramidal base. The lone electron pair of the inner Bi pyramid is accommodated in a standing trigonal coordination prism situated below its base (i.e., in the inter-ribbon space) (Bi-S 3.32 Å); that of the marginal pyramid is accommodated in the volume of a trigonal coordination prism below that pyramid and oriented at 90° to the 4 Å axis. In the latter case, Bi-S distances in the volume of the prism are 3.03 Å and 3.40 Å. Behavior of lone electron pairs in Bi_2S_3 and Sb_2S_3 at high pressure was described by Lundegaard et al. (2003, 2005). The resulting space group is $Pbnm$.

Nearly isotypic are U_2S_3 and Dy_2S_3 but without lone electron pairs, and homeotypic is Sn_2S_3 in which the marginal pyramidal/lying prism coordinations are replaced by octahedra of Sn^{4+} whereas the central pyramids contain Sn^{2+} and resemble those of Sb^{3+}. In Sb_2S_2As,

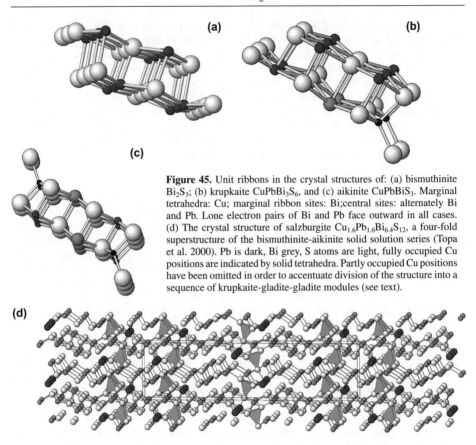

Figure 45. Unit ribbons in the crystal structures of: (a) bismuthinite Bi_2S_3; (b) krupkaite $CuPbBi_3S_6$, and (c) aikinite $CuPbBiS_3$. Marginal tetrahedra: Cu; marginal ribbon sites: Bi; central sites: alternately Bi and Pb. Lone electron pairs of Bi and Pb face outward in all cases. (d) The crystal structure of salzburgite $Cu_{1.6}Pb_{1.6}Bi_{6.4}S_{12}$, a four-fold superstructure of the bismuthinite-aikinite solid solution series (Topa et al. 2000). Pb is dark, Bi grey, S atoms are light, fully occupied Cu positions are indicated by solid tetrahedra. Partly occupied Cu positions have been omitted in order to accentuate division of the structure into a sequence of krupkaite-gladite-gladite modules (see text).

pääkkonnenite, (Bonazzi et al. 1995), the antimony and sulfur positions in the ribbons of this structure correspond to those in stibnite but the marginal sulfurs of stibnite are replaced by arsenic which is covalently bonded to Sb. The arsenic atoms from adjacent ribbons form covalent As-As pairs (2.47 Å) and the ribbons are arranged in layers (Fig. 3b) instead of the herringbone pattern of Sb_2S_3 (Fig. 3a). The resulting space group is $C2/m$. Although the overall configuration of the structure of kermesite Sb_2S_2O (Kupčík 1967; Bonazzi 1987; Ďurovič and Hybler 2006) is similar to pääkkonenite, the marginal Sb in the ribbons is coordinated primarily to oxygen whereas the central Sb is coordinated exclusively to sulfur.

The bismuthinite–aikinite series of sulfosalts

Stibnite is compositionally isolated, the only known solid solution is $(Sb,Bi)_2S_3$ (Kyono and Kimata 2004). Bismuthinite, Bi_2S_3, however, gives rise to an extensive solid solution series Bi_2S_3–$CuPbBiS_3$, which is continuous above ~300 °C (Springer 1971). At ambient temperatures, this series contains a number of derivatives with ordered structures, separated by exsolution gaps (Table 26). Some of them appear metastable, for example, salzburgite and paarite convert gradually into a gladite–krupkaite intergrowth (Topa et al. 2002). All ordered derivatives of the bismuthinite – aikinite solid solution series have the a and c axes in common (~4 Å and ~11.5 Å) whereas the b axis represents multiples of ~ 11.2 Å (Table 26). The space group is $Pmcn$ for the end-members with b~11.2 Å and for the phases with b equal to odd multiples of 11.2 Å. It is $Pmc2_1$ for those with even multiples of b and for $CuPbBi_3S_6$, i.e., the central member of the series.

The crystal structures of this series are formed by ordered two-and-two combinations of three types of ribbons: bismuthinite-like ribbons, Bi_4S_6, krupkaite-like ribbons, $CuPbBi_3S_6$ and aikinite-like ribbons, $Cu_2Pb_2Bi_2S_6$ (Figs. 45a-c). They are named after the minerals in which they occur alone (Ohmasa and Nowacki 1970; Mumme et al. 1976). Thus, gladite $CuPbBi_5S_9$ is a 1:1 combination of bismuthinite- and krupkaite-like ribbons and hammarite is a corresponding combination of krupkaite- and aikinite-like ribbons. Ordered ribbon combinations (Fig. 45d) lead to a system of "supercells" arranged symmetrically around krupkaite, $CuPbBi_3S_6$, a phase half-way between bismuthinite and aikinite. In this sequence, the multiples of the bismuthinite-like subcell along the b direction are: 1 (= bismuthinite), 3, 3, 4, 5, 1 (= krupkaite), 5, 4, 3, 3, 1 (= aikinite).

For the entire series, substitution of a Bi position by Pb, connected with the occupation of an adjacent tetrahedral vacancy by Cu, involves the inner Bi sites of the ribbon and leaves the marginal Bi sites untouched. The tetrahedral vacancy involved is attached laterally to the ribbon (Figs. 45b-d). The filled tetrahedra (Cu sites) are organized into slightly warped (010) planes. Below the level of 100% substitution by Pb and Cu which results in the composition $Cu_2Pb_2Bi_2S_6$, some of the (010) planes in the structure are planes with all tetrahedra vacant. This leads to another way of looking at the ordered structures of the bismuthinite-aikinite series as a combination of three types of modules: (1) gladite-like modules (G) with copper-occupied planes 1½ subcells apart; (2) krupkaite-like modules (K) with such planes 1 subcell period apart; and (3) aikinite modules (A), with these planes ½ subcell period apart; in pekoite there may be a need for a still larger module.

One subcell spacing is equal to 11.2 Å. In their type structures, these modules occur in simple sequences (e.g., gladite is GGGG), but in the structures intermediate to these, 1:1, 1:2 or more complicated sequences of modules occur (Makovicky et al. 2001; Topa et al. 2002). Thus, in salzburgite (Fig. 45d) the sequence of modules is gladite-gladite-krupkaite, i.e., GGKGGK…; in paarite GKGK, in lindströmite KKAKKA, and in emilite KKAKAKKAKA… (Ferraris et al. 2004).

The majority of authors prefer ideal, simple stoichiometric formulae for the minerals of the bismuthinite–aikinite series., The a and c dimensions (as well as the subcell b dimensions) increase, however, parallel with the extent of Pb and Cu substitution so that a combination of two different modules of distinct types involves strain in the (010) planes

Table 26. Selected members of the bismuthinite-aikinite series.

Mineral	Formula	Lattice Parameters (Å)			Space Group	Ref.
		a	b	c		
Bismuthinite	Bi_2S_3	3.985	11.163	11.314	$Pmcn$	[1]
Gladite	$CuPbBi_5S_9$	4.004	33.575	11.48	$Pmcn$	[1]
Salzburgite	$Cu_{1.6}Pb_{1.6}Bi_{6.4}S_{12}$	4.007	44.81	11.513	$Pmc2_1$	[2]
Paarite	$Cu_{1.7}Pb_{1.7}Bi_{6.3}S_{12}$	4.007	55.998	11.512	$Pmcn$	[3]
Krupkaite	$CuPbBi_3S_6$	4.013	11.208	11.56	$Pmc2_1$	[1]
Lindströmite	$Cu_3Pb_3Bi_7S_{15}$	4.018	56.141	11.578	$Pmcn$	[1]
Emilite	$Cu_{10.7}Pb_{10.7}Bi_{21.3}S_{48}$	4.029	44.986	11.599	$Pmc2_1$	[4]
Hammarite	$Cu_2Pb_2Bi_4S_9$	4.025	33.773	11.595	$Pmcn$	[1]
Aikinite	$CuPbBiS_3$	4.042	11.339	11.652	$Pmcn$	[1]

References relate to the newest refinements of unit cell parameters.
References: [1] Topa et al. 2002; [2] Topa et al. 2000; [3] Makovicky et al. 2001; [4] Balić-Žunić et al. 2002

of "module composition." As might be expected, the strain becomes alleviated by a small fractional occupation of empty tetrahedral sites in the less-substituted set of modules and by a small number of Cu vacancies in the nominally full Cu sites. Both of these processes are accompanied by a corresponding shift in the amount of Pb in the relevant sites housing the large cations and cause, in some cases, deviations from simple stoichiometries. The possible degree of deviation from the ideal stoichiometry varies for different members of the series. For krupkaite, with 50% substitution, the amount of additional Cu and Pb substitution may reach 8 mol%, the undersubstitution in aikinite is quite common and may reach 17 mol%, hence even at 83% substitution there may be no signs of ordering (Topa et al. in prep.) or there may be only weak continuous streaks instead of superstructure reflections in the b^* direction of the reciprocal lattice (Ohmasa and Nowacki 1970).

The aikinite-bismuthinite series is just a member $N = 2$ of the meneghinite homologous series of sulfosalts (Makovicky 1989). In this series, the copper containing configurations on glide planes are identical for all members of the series, whereas the SnS-like slabs become broader with growing homologue order N. The structure of jaskolskiite $Cu_{0.2}Pb_{2.2}(Sb,Bi)_{1.8}S_5$ (Makovicky and Norrestam 1985) is a member with $N = 4$ and that of meneghinite $CuPb_{13}Sb_7S_{24}$ (Euler and Hellner 1960; Moëlo et al. 2002) is $N = 5$.

Layer-like structures

Herzenbergite. The crystal structure of SnS (herzenbergite) (Chattopadhayay et al. 1986) is a layered structure, with tightly-bonded double layers separated by spaces accommodating lone electron pairs of Sn^{2+}. The bonding scheme of Sn^{2+} is reminiscent of As^{3+} and Sb^{3+}, with three strong short bonds (Sn-S = 2.62-2.66 Å)(Fig. 46a), accompanied by two considerably longer, intralayer Sn-S distances (3.29 Å), and by still longer distances across the interlayer space (distances up to 3.6 Å are shown in Fig. 46a). The SnS_5 coordination pyramids with rectangular bases are illustrated in this figure.

SnS is one of the archetypes used by Makovicky (1985) for the definition and description of families of sulfosalts. Some references interpret it as a "distortion derivative" of the PbS archetype, although such an interpretation obscures the decisive role of stereochemically active lone electron pairs of Sn in this archetype. A modification of this motif is present in $TlSbS_2$ (Takéuchi 1997; Balić-Žunić and Makovicky 1995).

LARGE SULFOSALT FAMILIES

General features

According to the bond-valence concept, sulfosalts or, rather, chalcogenosalts are complex sulfides, selenides or tellurides containing a subset of cations with high valence, such as As^{3+}, Te^{4+}, Ge^{4+}, As^{5+}, and W^{6+} (Moëlo and Makovicky in press). Because the bulk of natural thioarsenates, thiostannates, thiotungstates, etc. are homeotypes of simple sulfides, the usual definition of sulfosalts is restricted to the compounds containing a subset of elements with elevated valence and, at the same time, with stereochemically active lone electron pairs. This restricted definition is a result of the defining influence of lone electron pair configurations upon the configuration of the resulting crystal structure.

Thus, sulfosalts are complex sulfides/selenides in which formally trivalent lone-electron pair metalloids As^{3+}, Sb^{3+} and/or Bi^{3+} are combined with different cations. In a half of natural sulfosalt species this cation is Pb^{2+}; otherwise Cu, Ag, Fe, Tl, Mn, Sn and other, less frequently occurring elements. Rare natural finds and a growing number of synthetic sulfosalts have enlarged the range of compositions to include sulfosalts of Li, Na, K, Rb, Cs, as well as those of alkaline earths and lanthanoids, and with a number of sulfosalts that contain $(NH_4)^+$ and

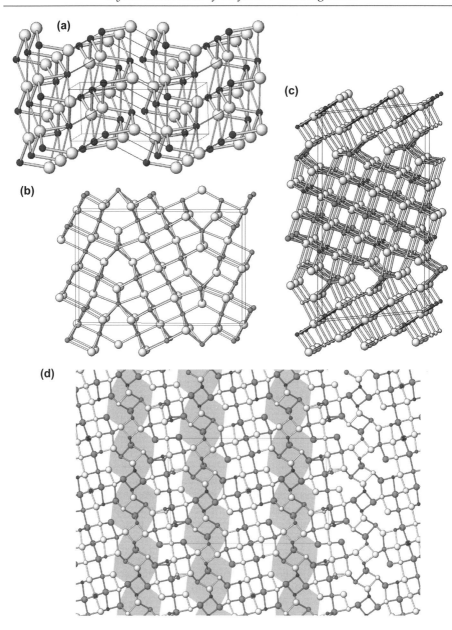

Figure 46. (a) An oblique edge-on view of a stack of double-layers in the structure of herzenbergite, SnS. Each Sn atom has three short Sn-S bonds (thick lines) and two longer bonds (thinner lines) inside the layer (i.e., forming coordination pyramids SnS$_5$), and longest distances (thin lines) across the lone electron pair interspace. (b) Lillianite, Pb$_3$Bi$_2$S$_6$ (Takagi and Takeuchi 1972). Pb and the mixed (Pb,Bi) positions are grey, S white. Number of octahedra in a diagonal edge-sharing chain across a unit slab (i.e., homologue order N) is four. (c) Vikingite Pb$_8$Ag$_5$Bi$_{13}$S$_{30}$ (Makovicky et al. 1992). White cations: Ag, dark cations: Bi, medium grey cations: Pb. Unit slabs are alternatively 4 and 7 octahedra wide when counted along diagonals. (d) Crystal structure of hodrushite Cu$_{7.6}$Ag$_{0.4}$Fe$_{0.4}$Bi$_{11.6}$S$_{22}$ (Topa et al. 2003), $N_{1,2} = 1,2$ homologue of the cuprobismutite combinatorial series. Spheres in order of decreasing size indicate S, Bi (dark) and Cu. Note the PbS-like portions (331)$_{PbS}$ interleaved by thinner, Cu-Bi based layers (accentuated in grey). Shading of atoms indicates two height levels, 2 Å apart.

organic cations of various sizes. Li emulates Cu in sulfosalts (resulting in ionic conductivity); Na is a fairly small cation as also is Ca. Potassium can either be a "channel-building" element in sulfosalts, similar to Rb or Cs, or it can replace Pb^{2+} in the structures by means of coupled substitution, with Sb^{3+} (Bi^{3+}) or a lanthanoid ion replacing another Pb atom. In its channel- and interlayer-building role, K^+ also replaces Tl^+.

A more detailed outline of the crystal chemistry of sulfosalts is to be found in Makovicky (1989, 1997a) and Ferraris et al. (2004). However, it is important to mention the difference between the stereochemical activity of the lone electron pair (LEP) in As^{3+} and Sb^{3+} (a pronounced activity resulting in capped trigonal coordination prisms of these elements and an SnS-like archetypal configuration of structural units) and Bi^{3+} (a limited stereochemical activity of LEP, yielding octahedral, distorted octahedral as well as capped trigonal prismatic coordinations, and a PbS-like archetypal arrangement of atoms in structural blocks/layers). Other configurations occur in the structures with Cu or Ag as principal cations. Especially in these structures, when the concentration of metalloids (primarily As^{3+}, Sb^{3+}, and Te^{4+}) is low, the metalloids form MX_3 trigonal coordination pyramids, with a large space for the lone electron pair above the metalloid atom, away from the ligands. Pb^{2+} and Tl^+ have close to insignificant LEP activity and can, therefore, be replaced by alkali metals or alkaline earths.

As in the silicates, a large number of natural sulfosalts with very different crystal structures are formed from combinations of a small number of the same chemical elements in various proportions. Prominent examples are the combinations Pb-Bi-Cu-Ag-S and Tl-As-S. A very large proportion of sulfosalts are amenable to modular classification principles and form homologous series or plesiotypic series of various kinds and degrees of generalization. Several such families are described in this review; more complete listings are in Makovicky (1989, 1997a), Ferraris et al. (2004), as well as in Moëlo and Makovicky (in press). A number of other sulfosalts have already been discussed in the sections on layer-like structures, PbS-related or adamantane structures as well as on the ionic conductors.

Lillianite homologous series

The extensive accretional series of lillianite homologues (Makovicky and Karup-Møller 1977a, b) is produced by reflection twinning on a unit-cell level. These are sulfides with *ccp* anion arrays (portions of PbS archetype) in which all coordination octahedra are filled. These arrays are cut parallel to and twinned on $(311)_{PbS}$ with corresponding coordination changes on twin planes. The most prominent members of this series are Pb-Bi-Ag sulfosalts (Fig. 46b). The overlapping octahedra of the adjacent, mirror-related layers are replaced on planes of unit-cell twinning by bicapped trigonal coordination prisms PbS_{6+2}, with the Pb atoms positioned on the mirror planes (Otto and Strunz 1968; Takéuchi and Takagi 1974, etc.).

Distinct homologues differ in the thickness of the PbS-like layers, expressed as the number N of octahedra in the chain of octahedra that runs diagonally across an individual archetypal layer and is parallel to $[011]_{PbS}$ (Fig. 46b). A lillianite homologue is denoted as $^{N1,N2}L$ where N_1 and N_2 are the (not necessarily equal) values of N for the two alternating sets of layers (Fig. 46c). Its chemical formula is $Pb_{N-1-2x}Bi_{2+x}Ag_xS_{N+2}$ (Z = 4) where $N = (N_1 + N_2)/2$ and x is the coefficient of the Ag + Bi = 2Pb substitution in the coordination octahedra of the PbS-like layers. If the trigonal coordination prisms of Pb cannot be substituted by Ag and Bi (which is very close to the real situation), $x_{max} = (N - 2)/2$. This structure type is also quite frequently found outside the Pb-Bi-Ag compositional space, e.g., for a number of complex lanthanide sulfides; the general formula then becomes $M^{2+}_{N-1}M^{3+}_2S_{N+2}$. The rare instance of $TlSb_3S_5$ ($N = 3$) leads to the formula $M^+_{(N-1)/2}M^{3+}_{(N+3)/2}S_{N+2}$.

The existence of lillianite homologues (Table 27) depends on suitable sizes of coordination polyhedra (trigonal prisms vs. octahedra), satisfactory local valence balance and the feasibility of close-to-regular octahedral (i.e., *ccp* or PbS-like) arrays. The cases with $N = 1$

Table 27. Selected homologues of lillianite[(4)]

Compound	Homologue	Lattice Parameters (Å, °)				S.G.	Note	Ref.
NdYbS$_3$	[1,1]L	a 12.55	b 9.44	c 3.85		$B22_12$		[1]
UFeS$_3$	[1,1]L	a 11.63	c 8.72	a 3.80		$Cncm$		[2]
CuEu$_2$S$_3$	[1,1]L	b 12.86	a 10.35	c 3.95		$Pnam$	Cu tetrahedral	[3]
MnEr$_2$S$_4$[(1)]	[2,2]L	b 12.60	c 12.75	a 3.79		$Cmc2_1$		[4]
CrEr$_2$S$_4$	[2,2]L	a 12.56	b 12.48	c 7.54		$Pb2_1/a$	$c' = c/2$	[5]
FeHo$_4$S$_7$	[1,2]L	a 12.57	c 11.35	b 3.78	β 105.7	$C2/m$		[6]
MnEr$_4$S$_7$[(1)]	[1,2]L	a 12.54	c 11.44	b 3.76	β 105.4	$C2/m$		[7]
TlSb$_3$S$_5$	[3,3]L	a 7.23	b 15.55	c 8.95	β 113.6	$P2_1/c$	$c' = c/2$	[8]
Pb$_3$Bi$_2$S$_6$ (lillianite)	[4,4]L	a 13.54	b 20.45	c 4.10		$Bbmm$	minor Ag	[9]
Pb$_3$Bi$_2$S$_6$ (xilingolite)	[4,4]L	a 13.51	c 20.65	b 4.09	β 92.2	$C2/m$	cation ordering	[10]
PbAgBi$_3$S$_6$ (gustavite)	[4,4]L	a 7.08	b 19.57	c 8.27	β 107.2	$P2_1/c$	$c' = c/2$	[11]
Pb$_8$Ag$_5$Bi$_{13}$S$_{30}$[(2)] (vikingite)	[4,7]L	a 7.10	c 25.25	b 8.22	α 90.0; β 95.36; γ 106.8	$P\bar{1}$	$c' = c/2$	[12][13]
Pb$_6$Sb$_{11}$Ag$_3$S$_{24}$ (ramdohrite)	[4,4]L	b 13.08	a 19.24	c 8.73	β 90.3	$P2_1/n$	$c' = c/2$	[14]
AgMnPb$_3$Sb$_5$S$_{12}$ (uchucchacuaite)	[4,4]L	a 12.67	b 19.34	c 4.38(3)			pseudoorthorhomb.	[15]
Pb$_{18}$Ag$_{15}$Sb$_{47}$S$_{96}$ (andorite IV = quatrandorite)	[4,4]L	a 13.04	b 19.18	c 17.07	γ 90.0	$P2_1/a$	$c' = c/4$	[16]
Pb$_{24}$Ag$_{24}$Sb$_{72}$S$_{144}$ (andorite VI = senandorite)	[4,4]L	a 13.02	b 19.18	c 25.48		$Pn2_1/a$	$c' = c/6$	[16][17]
Pb$_6$Bi$_2$S$_9$ (heyrovskyite)	[7,7]L	a 13.71	b 31.21	c 4.13		$Bbmm$		[18]
Pb$_{5.92}$Bi$_{2.06}$S$_9$ (aschamalmite)	[7,7]L	a 13.71	c 31.43	b 4.09	β 91.0	$C2/m$	cation ordering	[19]
Pb$_{3.36}$Ag$_{1.32}$Bi$_{3.32}$S$_9$ (Ag-Bi-heyrovskyite)	[7,7]L	b 13.60	c 30.49	a 4.11		$Cmcm$		[20]
Ag$_7$Pb$_{10}$Bi$_{15}$S$_{36}$ (eskimoite)	[5,9]L	a 13.46	b 30.19	c 4.10	β 93.4	$B2/m$		[12]
Pb$_4$Ag$_3$Bi$_5$S$_{13}$ (ourayite)	[11,11]L	a 13.49	b 44.17	c 4.05		$Bbmm$		[22]

Notes: (1) Complex homologues Mn$_2$Er$_6$S$_{11}$ ([2,2,2,1]L) and Mn$_4$Er$_8$S$_{15}$ ([2,2,1,2,2,1]L) occur in this system (Landa-Canovas and Otero-Diaz 1992); the equivalent phases, Mn$_2$Y$_2$S$_4$, MnY$_4$S$_7$, and Mn$_2$Y$_6$S$_{11}$ occur in the MnS-Y$_2$S$_3$ system (Bakker and Hyde 1978). Random complex intergrowths [4,7–2]L occur in the PbS-Bi$_2$S$_3$-Ag$_2$S system ('schirmerite'; Makovicky and Karup-Møller 1977b). (2) The 4 Å subcell of vikingite has a 13.60 Å, b 4.11 Å, c 25.25 Å and β 95.6°, space group $C2/m$. (3) Determined from powder data. (4) Synthetic homologues [8,8]L, [7,8]L, [4,5]L were observed in the system Ag$_2$S-Bi$_2$S$_3$-PbS at high temperatures, by HRTEM (Skowron and Tilley 1990).

References: [1] Carré and Laruelle 1974; [2] Noël and Padiou 1976; [3] Lemoine et al. 1986; [4] Landa-Canovas and Otero-Diaz 1992; [5] Tomas and Guittard 1980; [6] Adolphe and Laruelle 1968; [7] Landa-Canovas and Otero-Diaz 1992; [8] Gostojić et al. 1982; [9] Takagi and Takéuchi 1972; [10] Berlepsch et al. 2002; [11] Harris and Chen 1975; [12] Makovicky and Karup-Møller 1977b; [13] Makovicky et al. 1992; [14] Makovicky and Mumme 1983; [15] Moëlo et al. 1984; [16] Moëlo et al. 1988; [17] Sawada et al. 1987; [18] Takéuchi and Takagi 1974; [19] Mumme et al. 1983; [20] Makovicky et al. 1991; [12] Makovicky and Karup-Møller 1977b; [22] Makovicky and Karup-Møller 1984 to 3 cannot accommodate lone electron pairs of Bi^{3+} or Sb^{3+} that enlarge selected volumes of individual layers while leaving the coordination pyramids (half-octahedra) of these elements unchanged. Therefore, $PbBi_2S_4$ is not a lillianite homologue, although about a half of its structure approximates the $N = 2$ configuration.

Sulfides with $N = 1$ to 3 present both the cases with ideal "aristotype" symmetry and those with subgroup symmetry (Table 27); the reduction of symmetry is caused either by distortions of coordination polyhedra or by the asymmetric position of cations in the trigonal coordination prisms. The tetrahedral voids on the mirror planes of unit cell twinning are occupied only in exceptional cases (Eu_2CuS_3, $^{1,1}L$). Bakker and Hyde (1978) found that the homologous pair MnY_2S_4 ($N = 2$) and "$MnYS_3$" ($N = 1$) (which occurs only as a layer in MnY_4S_7) form a combinatorial series that comprises MnY_4S_7 ($^{1,2}L$), $Mn_2Y_6S_{11}$ ($^{1,2,2}L$), $Mn_4Y_{10}S_{19}$($^{1,2,2,2,2}L$), etc.

Starting at $N = 4$ (although already at $N = 3$ for $TlSb_3S_5$, perhaps because of the very large size of the Tl^+ ion), the higher homologues allow more pronounced departures from the galena-like array, especially in the form of locally "inflated" interspaces that accommodate lone electron pairs of quasi-octahedral Bi or Sb (a common volume for lone electron pairs of several adjacent cations was named a lone electron pair micelle by Makovicky and Mumme 1983). Only $^{4,4}L$ (lillianite) and $^{7,7}L$ (heyrovskyite) are known for the Ag-free subsystem Pb-Bi-S. Reduction of symmetry from the usual orthorhombic to monoclinic, caused by different Pb/Bi ordering in the two mirror-related slabs, was observed for $^{4,4}L$ (xilingolite, Berlepsch et al. 2001) and $^{7,7}L$ (aschamalmite, Mumme and Makovicky unpublished). With the Ag + Bi = 2Pb substitution active in the octahedral layers, not only these combinations but also the cases $^{4,7}L$ (Fig. 46c), $^{4,8}L$, $^{5,9}L$ and $^{11,11}L$ (Makovicky and Karup-Møller 1977) are found (Table 27), as well as disordered combinations based on $N_1 = 4$ and $N_2 = 7$ in different proportions (ibid., Skowron and Tilley 1990). The structures with close-to-ideal PbS-like arrays and those with extensive lone electron pair micelles either do not mix as continuous solid solutions or they become separated by exsolution at low temperatures (e.g., the pair $Pb_3Bi_2S_6$-$PbAgBi_3S_6$ with $N = 4$).

A good match of coordination polyhedra in the octahedral layers of Pb-Ag-Bi sulfosalts results in an extensive accretional series. In the parallel system Pb-Ag-Sb-S, the size and shape mismatch of Pb, Ag and Sb coordination polyhedra appears serious; only members with $N = 4$ were found and they are known only for > 50% of (Ag + Sb) substitution for Pb. They form a string of intermediate (often mutually exsolved) phases with different spatial distribution of lone electron pair micelles in the structure and with different superperiods (2-, 4-, and 6-fold) of the 4 Å dimension. These superperiods result from complex distribution patterns of short and long Sb-S bonds in the walls of the lone electron pair micelles. For phases with substitution close to 50%, incorporation of smaller M^{2+} ions instead of a portion of octahedral Pb appears critical for the formation of viable unit layers; e.g., $AgPb_3MnSb_5S_{12}$ (Moëlo et al. 1984).

Symmetry and size of the trigonal prismatic site (its occupation by a symmetrically coordinated Pb atom or by an asymmetric, sideways oriented Bi configuration) is critical for the distinction between the lillianite and the pavonite homologous series in the Pb-Bi-Ag-Cu system (Makovicky et al. 1977; Makovicky 1981, 1989).

PAVONITE HOMOLOGOUS SERIES

The pavonite homologous series was defined by Makovicky et al. (1977) as an extensive series of complex sulfides in the Cu-Ag-(Pb)-Bi-S system. It has seen a vigorous growth in numbers of reported phases, the relatively new acquisitions being Cu-Bi selenides, sulfides of cadmium, lead and indium, as well as several hitherto unknown homologues.

Topology of the pavonite homologous series is similar to that of lillianite homologues with

$N_1 \neq N_2$, i.e., it is a case of contracted-set reflection twinning (Takeuchi 1997) and, to certain degree, of a heterochemical homologous series. In the pavonite homologues (Figs. 1a,b), all members have $N_1;N_2 = 1;N_{pav}$ where the order of pavonite homologue, N_{pav}, is 2 to 8, and possibly even higher (Table 28). The trigonal coordination prisms on the planes of contracted-set unit-cell twinning are distorted and are occupied by square-pyramidal Bi and its lone electron pair. Bi is displaced towards those prism caps that form a part of the thin $N = 1$ layers. The extensive PbS-like portions (those with N_2) contain quasi-octahedrally to octahedrally coordinated Bi combined with, or partly replaced by, Ag, Cu and some Pb, and occasionally Cd and In. The sole, skewed octahedron in the narrow portions ($N_1 = 1$) represents AgS_{2+4} or this octahedral column contains three- and four-coordinated Cu situated in the interior of the octahedra or in the tetrahedral spaces between them (see the description of $Cu_{1.78}Bi_{4.73}S_8$, Fig. 40c, in the section on ion conductors). Only for $N_{pav} = 2$ are these octahedra substantially occupied by Pb; in the cases with $N = 3$ they also accommodate Cd, Hg or Mn (Table 28).

The primarily synthetic Cu-Bi pavonite homologues exhibit a statistical substitution of Bi by copper in the thick layers. This copper is accommodated in the walls of, and adjacent to, the vacated Bi polyhedra (Fig. 40c). Synthetic phases (sulfides and selenides) with $N_{pav} = 3$ and 4 are constructed according to this principle. For natural phases with $N_{pav} = 4, 5$, and 8, this substitutional-and-interstitial solid solution exsolves into an intimate lamellar intergrowth of two phases with the same N_{pav}, one Cu-Pb poor and the other with a substantial substitution by Cu-Pb; the nature of the Cu-Pb rich phase is discussed further below. Natural pavonite homologues with $N_{pav} = 5$ and 7 prefer Ag in the isolated skewed octahedra but do not exclude Cu; the natural $N = 8$ homologue has mixed occupancy, and $N = 4$ is known both as a Cu-rich and as an Ag-rich variety (Table 28).

Sb alone does not form pavonite homologues because of its inability to assume near-regular octahedral coordination in the interior of the thicker slabs. The only exceptions are triclinic $Li_{3x}Sb_{6-x}S_9$ ($N = 4$) with large lone-electron pair micelles and partial Li for Sb substitution, and $MnSb_2S_4$ ($N = 3$), in which the central site of the accreting layer is formed by octahedral Mn.

Two ways of bridging the differences between the lillianite and pavonite accretional series have been observed. The first one is the "Phase V homologous series" investigated and summarized by Takéuchi (1997). We can describe this series, defined as $Pb_{1-x}Bi_{2x/3}S \cdot 2Bi_2S_3$ by Takéuchi, as a combinatorial series of ordered intergrowths of lillianite $^{2,2}L$ and pavonite $^{1,2}L$ modules; the latter are present as the only component in the phase V-1 (Takéuchi et al. 1974). In this case, the lillianite and pavonite-like composition planes of unit-cell twinning are spatially separated, with varying alternation frequencies. The second way has recently been observed in cupromakovickyite ($N_{1,2} = 1,4$) (Topa et al., in press.) and apparently relates also to cupropavonite ($N_{1,2} = 1,5$) and, possibly, other "cupro-" varieties. In "classical" pavonite homologues, the trigonal prisms, asymmetrically occupied by Bi, share edges of their caps across the thin, $N = 1$, layer, resulting in a column of paired Bi square pyramids. In the cupro-varieties, one Bi polyhedron from such a pair is exchanged with a symmetrically occupied trigonal prism hosting lead. This leads to an asymmetric occupation, by statistically distributed Cu atoms, of the adjacent octahedral column in the thin layer. The ordering pattern of these substitutions results in a doubled c parameter (Table 28).

CUPROBISMUTITE HOMOLOGOUS SERIES

Cuprobismutite homologues are copper-bismuth sulfosalts, with minor contents of Fe^{3+}, Pb and/or Ag. The crystal structures of the cuprobismutite homologous series (Fig. 46d) represent a regular 1:1 intergrowth, on a unit-cell scale, of two types of slabs: (a) $(331)_{PbS}$ slabs of a galena-like structure with two known thicknesses, typified by the layers in kupčíkite

Table 28. Selected pavonite homologues.

Mineral/Formula	Homologue	Lattice Parameters (Å)			β (°)	Space Group	Ref.
Synth. V-I $Pb_{1.46}Bi_{8.36}S_{14}$ (~ $PbBi_4S_7$)	^2P	a 13.25	b 4.03	c 12.04	105	$C2/m$	[1]
Synth. $CdBi_4S_7$	^2P	a 13.11	b 4.00	c 11.77	105.2	$C2/m$	[2]
Synth. $Cd_{2.8}Bi_{8.1}S_{15}$	$^{2,3}P$	a 13.11	b 3.99	c 24.71	97.8	$C2/m$	[2]
Synth. $Cd_2Bi_6S_{11}$	$^{2,2,3}P$	a 13.11	b 4.00	c 35.84	90.4	$C2/m$	[2]
Synth. $CdBi_2S_4$	^3P	a 13.10	b 3.98	c 14.61	116.3	$C2/m$	[2]
Synth. $Cu_{1.57}Bi_{4.57}S_8$	^3P	a 13.21	b 4.03	c 14.09	115.6	$C2/m$	[3][4]
Synth. $HgBi_2S_4$	^3P	a 14.17	b 4.06	c 13.99	118.3	$C2/m$	[5]
Synth. $Cu_{3.21}Bi_{4.79}S_9$	^4P	a 13.21	b 3.99	c 14.81	100.2	$C2/m$	[6]
Synth. $Cu_2Pb_{1.5}Bi_{4.5}S_9$	^4P	a 13.45	b 4.03	c 14.99	99.8	$C2/m$	[7]
Makovickyite $Cu_{1.12}Ag_{0.81}Pb_{0.27}Bi_{5.35}S_9$	^4P	a 13.37	b 4.05	c 14.71	99.5	$C2/m$	[8][9]
Cupromakovickyite $Cu_{1.85}Ag_{0.60}Pb_{0.70}Bi_{4.40}S_9$	$^4P(2)$	a 13.40	b 4.01	c 29.93	100.07	$C2/m$	[10]
Ag-rich makovickyite $Cu_{0.11}Ag_{0.80}Pb_{0.86}Bi_{5.01}S_9$	^4P	a 13.83	b 4.04	c 14.72	97.50	$C2/m$	[8][9]
Synth. $Li_{3x}Sb_{6-x}S_9$ (x = 1/3)	^4P	b 6.68	a 4.09	c 14.70	(1)	$P\bar{1}$	[11]
Synth. $AgBi_3S_5$	^5P	a 13.31	b 4.04	c 16.42	94.0	$C2/m$	[12]
Pavonite $Cu_{0.27}Ag_{0.78}Pb_{0.33}Bi_{2.78}S_5$	^5P	a 13.42	b 3.99	c 16.39	94.3	$C2/m$	[13]
Cupropavonite $Cu_{0.9}Ag_{0.5}Pb_{0.6}Bi_{2.5}S_5$	$^5P(2)$	a 13.45	b 4.02	c 33.06	93.5	$C2/m$	[13]
Benjaminite $Ag_3Bi_7S_{12}$	^7P	a 13.25	b 4.05	c 20.25	103.1	$C2/m$	[14]
Benjaminite $Cu_{0.5}Ag_{2.3}Pb_{0.4}Bi_{6.8}S_{12}$	^7P	a 13.30	b 4.07	c 20.21	103.3	$C2/m$	[15]
Mummeite $Cu_{0.58}Ag_{3.11}Pb_{1.10}Bi_{6.65}S_{13}$	^8P	a 13.47	b 4.06	c 21.63	92.9	$C2/m$	[16][8]

Notes: Pavonite homologues are denoted as ^NP where N is the order of the homologue as defined in the text. (1) α 96.84 β 90.28 γ 107.70. (2) Transitional between the pavonite and lillianite series.

References: [1] Takéuchi et al. 1974, 1979; [2] Choe et al. 1997; [3] Ohmasa and Nowacki 1973; [4] Tomeoka et al. 1980; [5] Mumme and Watts 1980; [6] Ohmasa 1973; [7] Mariolacos et al. (oral. comm.); [8] Mumme 1990; [9] Žák et al. 1994; [10] Topa et al. in prep.; [11] Olivier-Fourcade et al. 1983; [12] Makovicky et al. 1977; [13] Karup-Møller and Makovicky 1979; [14] Herbert and Mumme 1981; [15] Makovicky and Mumme 1979; [16] Karup-Møller and Makovicky 1992

(N = 1) and in cuprobismutite (N = 2) (Table 29), and (b) slabs composed of columns of paired BiS_5 pyramids and paired CuS_4 coordination tetrahedra (shaded in Fig. 46d), identical for all homologues.

The first kind of slab consists of regular BiS_6 octahedra, BiS_5 pyramids flanking them, and the fitting portions of the trigonal coordination bipyramids of Cu(Fe). In kupčíkite

($N = 1$), coordination pyramids of Bi attach themselves to the opposite sides of the central Bi octahedron (Topa et al. 2003a). In cuprobismutite, $N = 2$, there are pairs of such pyramids attached; the inner one can be completed as a distorted octahedron, the outer one as a capped trigonal prism analogous to the sole pyramid in kupčíkite. Hodrushite (Fig. 46d) is a regular 1:1 combination of these two slab thicknesses, $N = 1;2$ (Kupčík and Makovicky 1968).

The regular coordination octahedron of Bi is typical for this homologous series. Pure Cu-Bi kupčíkite requires some Cu^{2+} in the structure; but in natural samples this is accommodated by the presence of Fe^{3+} in some of the trigonal bipyramidal → distorted tetrahedral Cu positions, resulting in a composition $Cu_{3.4}Fe_{0.6}Bi_5S_{10}$ (Topa et al. 2003a). The $N = 2$ member has the opposite valence problem, solved by the substitution of some of the octahedral Bi by silver, $Cu_8AgBi_{13}S_{24}$ and also by its partial substitution by Pb (Topa et al. 2003b). In synthetic, admixture-free cuprobismutite, this problem is solved by extensive heterotopic substitution of Bi by Cu (Ozawa and Nowacki 1975). Owing to its $N = 1;2$ character, hodrushite (ideally $Cu_8Bi_{12}S_{22}$) displays all these substitutions, located in the appropriate slabs of the structure.

In the structures of $Cu_4Bi_4S_9$ (Takéuchi and Ozawa 1975; Bente and Kupčík 1984) and $Cu_4Bi_4Se_9$ (Makovicky et al. 2002), related to this homologous series, the thin galena-like portions contain paired Bi octahedra instead of the isolated ones whereas the intervening layers are more complicated than in the true cuprobismutite homologues. These structures contain covalently bonded X-X pairs. Another plesiotype is padčraite, $Cu_7((Cu,Ag)_{0.33}Pb_{1.33}Bi_{11.33})_{\Sigma 13}S_{22}$ (Mumme 1986; Topa and Makovicky in press). A complex modular description of this structure was offered by Mumme (1986); a different modular description was given by Topa and Makovicky (in press) as a regular 1:1 intergrowth of kupčíkite-like slabs with thick slabs of a complex structure containing Bi_4S_6 ribbons and an independent, prismatic Pb site.

Rod-based sulfosalts

The plesiotype family of rod-based sulfosalts includes complex sulfides of Pb and Sb, Pb and Bi, Sn and Sb, as well as a number of synthetic sulfides of alkalis, alkaline earths and lanthanides combined with Sb or Bi. Minor elements, such as Cu, Ag, Fe or Mn, rarely also Sn^{4+}, are confined to specific roles or structures. Further references are Makovicky (1993), Ferraris et al. (2004), and Makovicky (2005).

Table 29. Cuprobismutite homologues and plesiotypes.

Mineral (Formula)	N1;N2	Lattice Parameters (Å,°)				Space Group	Ref.
Synth. ($Cu_4Bi_5S_{10}$)	1;1	a 17.54	b 3.93	c 12.85	β 108.0	C2/m	[1]
Kupčíkite ($Cu_{3.3}Fe_{0.7}Bi_5S_{10}$)	1;1	a 17.51	b 3.91	c 12.87	β 108.57	C2/m	[2]
Hodrushite[3] ($Cu_{8.12}Fe_{0.29}Bi_{11.54}S_{22}$)	1;2	c 17.58	b 3.94	a 27.21	β 92.15	A2/m	[3]
Synth.[1] ($Cu_{10.4}Bi_{12.6}S_{24}$)	2;2	a 17.52	b 3.93	c 15.26	β 100.2	C2/m	[4]
Cuprobismutite ($Cu_8AgBi_{13}S_{24}$)	2;2	a 17.65	b 3.93	c 15.24	β 100.5	C2/m	[5]
Synth. ($Cu_4Bi_4S_9$)	plesio[2]	b 11.66	c 3.97	a 31.68	—	Pnam	[6]
Synth. ($Cu_4Bi_4Se_9$)	plesio	c 12.20	b 4.12	a 32.69	—	Pnma	[7]
Paderaite ($Cu_{5.9}Ag_{1.3}Pb_{1.2}Bi_{11.2}S_{22}$)	plesio	a 28.44	b 3.90	c 17.55	β 106.0	P2$_1$/m	[8]

Notes: (1) Synthetic cuprobismutite (Ozawa and Nowacki 1975). (2) Plesiotypes of cuprobismutite homologues. (3) Hodrushite from Felbertal (Austria) is $Cu_{7.55}Ag_{0.39}Fe_{0.36}Bi_{11.58}S_{22.11}$ (Topa 2001).
References: [1] Mariolacos et al. 1975; [2] Topa et al. 2003; [3] Kodera et al. 1970; [4] Ozawa and Nowacki 1975; [5] Nuffield 1952; [6] Bente and Kupčík 1984; [7] Makovicky et al. 2002; [8] Mumme 1986

The rod-based sulfosalts (Figs. 47-49) contain rods of simpler configuration, based on a PbS or SnS archetype and generally with a lozenge-like cross section, which are several PbS/SnS subcells (i.e., cation polyhedra) wide and two or more (commonly four) atomic sheets thick. Their surfaces are alternately of a pseudotetragonal type (like the (100) surface of PbS) and of a (sheared or non-sheared) pseudohexagonal type (i.e., a hexagonal $(111)_{PbS}$ or a sheared-hexagonal $3^2.4^2$ arrangement of anions) [the symbol of the sheared hexagonal net indicates that two triangular and two square mesh meet at each node]. Typically, the former surfaces face the latter in a complicated non-commensurate match across rod interfaces, so that the structure is not based on a single, continuous close-packed or sheared close-packed arrangement of anions.

The reader is asked to note that when we outline individual structural elements (layers, rods, blocks, ribbons) in these structures, we do not select only the groups based on strong bonds but we consider parts of the structure in their entirety. For example, a part of the structure based on the SnS archetype will contain both the tightly-bonded double-layers and the weakly-bonded interspaces (lone electron pair micelles) which are no less important for the entire configuration than the strongly bonded ones. Even if the tightly-bonded parts contribute most of the inner energy, these are common to many structures in a certain group of compounds and it is the weak interactions which determine the modular crystal chemistry of an individual compound. We draw boundaries between structural blocks (rods, etc.) on non-commensurate interfaces, i.e., between individual blocks of archetypal structure.

All rod-based structures show a distinct accumulation of Sb and Bi in the rod interiors, forming lone electron pair micelles, and of Pb and the above-mentioned large cations on the pseudotetragonal surfaces; the interiors of larger rods can contain mixed Pb-Sb and Sn-Sb positions. In synthetic products, the presence of univalent large cations is outweighed by the presence of trivalent cations, e.g., $KBi_{6.33}S_{10}$ (Kanatzidis et al. 1996) is homeotypic with $Pb_4Sb_4Se_{10}$ (Skowron and Brown 1990). These substitutions take place on rod surfaces (Makovicky 2005). The plesiotypic family of rod-based structures can be divided into four subfamilies according to the pattern which the rod-like elements compose as follows.

(a) Rod-layer structures (Table 30) in which rods are arranged in parallel rows and joined into layers via common polyhedra/atoms at their acute corners (Figs. 47a, 47b, and 48a). Their make-up will be explained in detail below using the example of dadsonite. Individual structures in this family differ in rod dimensions, number of rod types (one or two), and the modes of their interconnection (Makovicky 1993). This is not a homologous series but a more complicated, plesiotype, series; it contains only a very limited number of homologous pairs.

(b) Chess-board structures in which the rods form a chess-board pattern with non-commensurate interfaces. In the kobellite homologous series of Pb-Sb-Bi sulfides that belong to this category, the PbS-like rods can have complicated cross sections (Fig. 48b).

(c) Structures with a cyclic (six-fold) arrangement of rods; rod interfaces can either be of noncommensurate kind or they resemble contacts of archetype slabs in the lillianite series. An example is zinkenite (Fig. 49a). In the nearest lower homologue of zinckenite, $Bi_{0.67}Bi_{12}S_{18}Hal_2$, where *Hal* is either iodine or bromine, trigonal channels are formed which accommodate individual halogen anions and the lone electron pairs of Bi (Miehe and Kupčík 1971).

(d) Large-scale, doubly non-commensurate structures of a box-work type. They contain three types of rods, two of which form a box-work with pseudotetragonal surfaces, and the third one occurs inside the channels of this box-work and displays primarily pseudohexagonal surfaces (Table 31, Fig. 49b). Layer match problems encountered in several of these structures are mostly solved by incorporation of oxygen into the coordination polyhedra of selected Sb atoms; chlorine may participate as well.

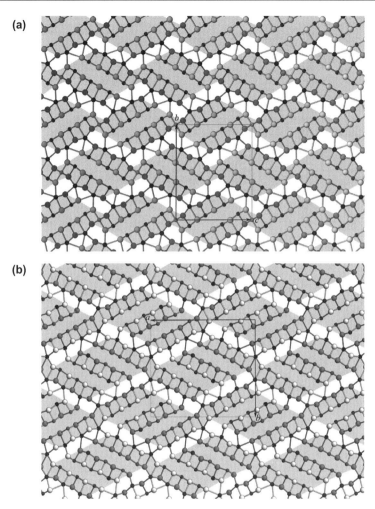

Figure 47. (a) Jamesonite FePb$_4$Sb$_6$S$_{14}$ (Matsushita and Ueda 2003). In order of decreasing size, spheres represent S, Pb (dark), Sb (dark), and Fe (light grey, situated in the octahedra joining adjacent rods in rod-layers). Rod-layers Type 2 of Makovicky (1993) are shaded, non-commensurate interlayer spaces are left unshaded. Dark and light spheres represent atoms at two levels, 2 Å apart. (b) Boulangerite (Skowron and Brown 1990). In order of decreasing size, spheres represent S, Pb, and Sb (as well as the mixed Sb-Pb sites). Heights of the light and dark spheres differ by 2 Å. Rod-layers of Type 1 (Makovicky 1993) are shaded, non-commensurate interspaces have been left unfilled.

The close similarity of rod types and identity of match modes found in these four categories justifies joining all these plesiotypic categories into a unique, higher plesiotype family. In all categories it is mostly only homologous *pairs* which are found because of their combined, accretional and variable-fit character. The pair FePb$_4$Sb$_6$S$_{14}$ (jamesonite)- Pb$_4$Sb$_4$S$_{11}$ or the pair Ba$_9$Bi$_{18}$S$_{36}$-Ba$_{12}$Bi$_{24}$S$_{48}$ are examples.

Case study: dadsonite

The crystal structure of the Pb-Sb sulfosalt dadsonite (Table 30; Makovicky et al. in press) can be used to highlight the majority of features exhibited by rod-based sulfosalts. Dadsonite was originally defined as Pb$_{11}$Sb$_{12}$S$_{29}$ (Jambor 1969) and later as Pb$_{23}$Sb$_{25}$ClS$_{60}$ (Moëlo 1979

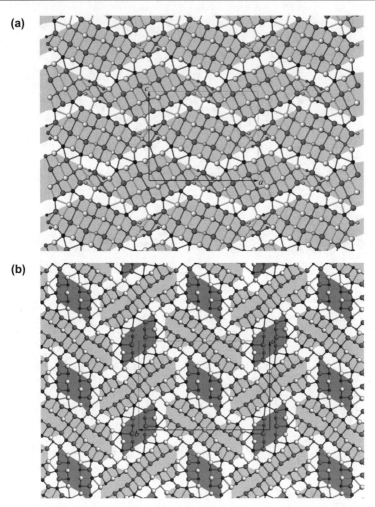

Figure 48. (a) Cosalite, Cu-and-Ag substituted $Pb_2Bi_2S_5$ (Topa et al. in prep.). Projection along the 4 Å axis (heights of light and dark spheres differ by 2 Å). In order of decreasing size, spheres represent S, (predominantly) Pb, (predominantly) Bi,otherwise Ag and Cu. Trigonal planar coordinations of Cu in cosalite were also found by Macíček (1986). Rod layers Type 3 of Makovicky (1993) (shaded) have rod interconnections consisting of two edge-sharing octahedra. Non-commensurate interspaces are uncoloured. (b) Kobellite, $(Cu,Fe)_2Pb_{12}(Bi,Sb)_{14}S_{35}$ (Miehe 1972). In order of decreasing size, spheres represent S, Pb, Sb or Bi, Cu and Fe. Light shading: Bi-enriched rods based on PbS archetype; dark shading: Sb-enriched rods based on SnS archetype. These are two pyramidal widths broad so that kobellite is a member of the series with $N = 2$.

and Cervelle et al. 1979). The crystal structure of the 4 Å subcell of dadsonite was determined by Makovicky and Mumme (1984).

Rod-layers. The structure consists of rods of SnS-like arrangement, infinite along [001] of the SnS archetype. These rods, three coordination pyramids of Pb or Sb wide, and four combined sulfur-and-metal sheets thick, are organized into two alternating, distinct types of layers (shaded in Fig. 50a), the so-called rod-layers of Makovicky (1993). Their inner portions are occupied by square coordination pyramids of (primarily) Sb, which form the caps of the central row of "standing" trigonal coordination prisms hosting the lone electron pairs of Sb.

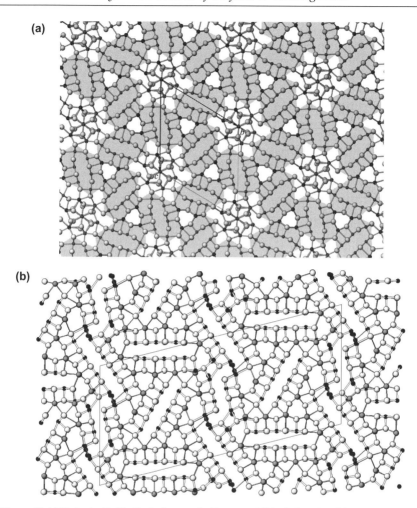

Figure 49. (a) Zinkenite Pb$_9$Sb$_{22}$S$_{42}$ (refinement by Topa, unpublished). In order of decreasing size, spheres are S, Pb and Sb. The SnS-based rods are shaded; triangular and hexagonal (Sb-filled) channels are left unshaded. (b) Scainiite Pb$_{14}$Sb$_{30}$S$_{54}$O$_5$ (Moëlo et al. 2000). a axis pointing to the right, c axis downward. Complex wavy layers (100) consisting of two types of amalgamated rods are interconnected by N = 3 rods at z = 0, forming a box-work arrangement that encloses another type of N = 3 rods at z = ½. Central portions of rods are occupied by Sb (dark), their surfaces primarily by Pb (grey). Oxygen is shown as small black circles situated very close to some Sb sites.

The outer surfaces consist of square coordination pyramids of Pb, forming parts of variously capped trigonal coordination prisms. The Pb prisms span the zigzag interspaces between adjacent rod layers. Each straight interval in these interspaces (left unshaded in Fig. 50a) represents a non-commensurate fit between the pseudotetragonal, Pb-rich surface (Q) of one rod and the purely anionic, sheared hexagonal $3^2 \cdot 4^2$ surface (H) of the opposing rod. The observed matches are 2Q:1½H and 2½Q:1½H in terms of a primitive pseudotetragonal and a centered orthohexagonal sub-mesh; they can be traced in Figure 50a. The match values found for the entire family of rod-based sulfosalts range from 1½Q:¾H to 4Q:2½H (Makovicky 1993). Coordinations spanning the interspaces in this family range from variously distorted capped trigonal prisms to occasional octahedra.

Table 30. Selected sulfosalts with rod-based layer structures.

Mineral	Formula	Layer Type	Rod Type	N	N'	Lattice Parameters[1] (Å, °)			Space Group	Ref.
Jamesonite	$FePb_4Sb_6S_{14}$	2	[001]SnS	3	4	a 15.57	b 18.98	c 4.03	β 91.8	$P2_1/a$	[1]
Boulangerite	$Pb_5Sb_4S_{11}$	1	[001]SnS	3	6	b 23.51	a 21.24	c 4.04	β 100.71	$Pbnm$[3]	[2]
						b 23.54	a 21.61	c 8.08		$P2_1/a$	[3]
Synth.	$Sn_3Sb_2S_6$	4	[001]SnS	5	6	c 34.91	a 23.15	b 3.96	—	$Pbnm$	[3]
Robinsonite	$Pb_4Sb_6S_{13}$	4	[001]SnS	3	4	b 17.69	a 16.56	c 3.98	α 96.5; β 97.8; γ 91.1	$Pbnm$	[4]
		4	[001]SnS	2	4						
Synth.	$Sn_4Sb_6S_{13}$	1	[001]SnS	3	4	a 24.31[2]		c 23.49	β 94.05	$I2/m$	[5]
Synth.	$Pb_{12.65}Sb_{11.35}S_{28.35}Cl_{2.65}$	4	[001]SnS	2	6	c 35.13	a 19.51	b 4.05	β 96.34	$I2/m$	[6]
		1	[001]SnS	2	4						
		5	[001]SnS	2	4						
Dadsonite	$Pb_{23}Sb_{25}S_{60}Cl$	3	[001]SnS	3	4	c 17.39	b19.51	a 8.28	α 83.53; β 77.88; γ 89.22	$P\bar{1}$	[7]
		5	[001]SnS	3	4						
Synth.	$Pb_5Sb_6S_{14}$	6	[001]SnS	4	4	a 28.37[2]	c 22.04	b 4.02	α 89.59; β 92.28; γ 89.93	$I\bar{1}$	[8]
		4	[001]SnS	4	4						
Synth.	$Pb_7Sb_4S_{13}$	6	[001]SnS	4	6	a 23.67	b 25.55	c 4.00	—	$Pnam$	[9]
Cosalite	$Pb_2Bi_2S_5$	3	[011]PbS	4	4	b 23.87	a 19.13	c 4.06	—	$Pbnm$	[10]
Synth.	$KBi_{6.33}S_{10}$	3	[011]PbS	4	4	a 24.05	b 19.44	c 4.10	—	$Pbnm$	[11]
Synth.	$Pb_4Sb_4Se_{10}$	3	[011]PbS	4	4	a 24.59	b 19.58	c 4.17	—	$Pbnm$	[12]
Synth.	$Ce_{1.25}Bi_{3.78}S_8$	3	[011]PbS	3	4	c 21.52	a 16.55	b 4.05	—	$Pbnm$	[13]
Synth.	$KLa_{1.28}Bi_{3.72}S_8$	4	[011]PbS	3	4	c 21.59	a 16.65	b 4.07	—	$Pbnm$	[14]
Moëloite	$Pb_6Sb_6S_{14}(S_3)$	10	[001]SnS	3	4	c 23.05	a 15.33	b 4.04	—	$P2_12_12_1$	[15]
Synth.	$BaBiSe_3$	10	[011]PbS	2	4	a 17.24	b 16.00	c 4.37	—	$P2_12_12_1$	[16]
Synth.	$SrBiSe_3$	10	[011]PbS	4	4	a 33.55	b 15.76	c 4.26	—	$P2_12_12_1$	[17]
Synth.	$Sr_6Sb_6S_{17}$	10	[001]PbS	3	4	c 22.87	b 15.35	c 8.29	—	$P2_12_12_1$	[18]

Notes: N denotes number of coordination polyhedra along, N' number of atom planes across the rod-like element.
[1] When not indicated otherwise, the first parameter (or the relevant vector d) is parallel to the periodicity of the rod-layer which has rods infinite along the 4 Å direction.; [2] Unit-cell vectors were selected diagonal to the direction and stacking of rod-layers.; [3] Presumed ordering variants for boulangerite. Structure refinements were published for $Pb_4Sb_6S_{13}$ in $P1$ (Petrova et al. 1978) and $I2/m$ (Skowron and Brown 1990a) and for $Pb_5Sb_4S_{11}$ in $Pnam$ (Skowron and Brown 1990b, Topa et al. in prep.).; [4] Variable amounts of substitution by Cu and Ag

References: [1] Niizeki and Buerger 1957; [2] Petrova et al. 1978; [3] Mumme 1989; [3] Smith 1984; [4] Parise et al. 1984; [5] Jumas et al. 1980; [6] Kostov and Macíček 1995; [7] Makovicky, Topa and Mumme, in press; [8] Skowron et al. 1992 ; [9] Skowron and Brown, written comm.; [10] Topa, pers. comm.; [11]Kanatzidis et al. 1996; [12] Skowron and Brown 1990c; [13] Ceolin et al. 1977; [14] Iordanidis et al. 1999; [15] Orlandi et al. 2002; [16] Volk et al. 1980; [17] Cook and Schäfer 1982; [18] Choi and Kanatzidis 2000

Table 31. Box-work type structures of complex sulfides.

Mineral	Formula	Lattice Parameters (Å,°)				Space Group	Ref.
Neyite	$Ag_2Cu_6Pb_{25}Bi_{26}S_{68}$	a 37.53	b 4.07	c 43.7	β 108.80	$C2/m$	[1]
Pillaite[1]	$Pb_9Sb_{10}S_{23}ClO_{0.5}$	a 49.49	b 4.13	c 21.83	β 99.62	$C2/m$	[2]
Scainiite[1]	$Pb_{14}Sb_{30}S_{54}O_5$	a 52.00	b 8.15	c 24.31	β 104.09	$C2/m$	[3]
Synth.	$Er_9La_{10}S_{27}$	c 21.83	b 3.94	a 29.71	β 122	$C2/m$	[4]

Notes: [1] Quasi-homologues; related to cyclically twinned structures of $Ba_{12}Bi_{24}S_{48}$ type
References: [1] Makovicky et al. 2001; [2] Meerschaut et al. 2001; [3] Moëlo et al. 2000; [4] Carré and Laruelle 1973.

The two rod-layers differ in the interconnection of the rods. Those at $y = \frac{1}{2}$ in Figure 50a have rods interconnected via a single row of anions. These anions are underbonded, a site assumed by Cl in the structure. This layer is the rod-layer Type 5 of Makovicky (1993), also known from the structure of $Pb_{12.7}Sb_{11.4}S_{28.4}Cl_{2.7}$ (Kostov and Macíček 1995). The rod-layers at $y = 0$ have adjacent rods interconnected via a pair of standing, distorted monocapped coordination prisms of antimony. A similar interconnection in cosalite ~$Pb_2Bi_2S_5$ (Srikrishnan and Nowacki 1974) (Fig. 48a with a new interpretation of this structure portion) contains coordination octahedra. A different interconnection in boulangerite, $Pb_5Sb_4S_{11}$ (Type 1 layer) is shown in Figure 47b.

Ribbons of Sb coordination pyramids. The triple [100] ribbons of the square coordination pyramids of antimony in Type 3 layers are polar, with all short Sb-S bonds oriented in the +[100] direction. These ribbons are very similar to those in jamesonite (Niizeki and Buerger 1957). The approximate correspondence between the two consecutive Sb sites in each [100] row of pyramids means that Type 3 layer has, to a good approximation, two repetition periods within one ~8 Å long a period of the dadsonite lattice. The Type 5 layer at $y = \frac{1}{2}$ has a completely different configuration. The triple ribbon of Sb pyramids (one of the pyramids being substituted by Pb) consists of trapezoidally distorted "square" coordination pyramids, with the pair of short distances in a base of the pyramid facing sideways. They have approximate mirror planes perpendicular to the ribbon extension. Sb sites form solitary SbS_3 groups of short bonds and one Sb_2S_4 pair. The non-crystallographic symmetry (i.e., in excess of the symmetry elements of the space group) in both layers is important for a potential order-disorder in the structure of dadsonite, discussed below.

The trapezoidally distorted bases of MX_5 coordination pyramids are a common configuration in sulfosalts, e.g., they occur for Sb in ramdohrite $Pb_6Sb_{11}Ag_3S_{24}$ (Makovicky and Mumme 1983), andorite $PbAgSb_3S_6$ (Sawada et al. 1987), zinkenite (Topa, unpublished), and for As in the sartorite group of Pb-As sulfosalts (Berlepsch et al. 2001). Partial substitution of Sb by Pb or a regular Sb-Pb alternation producing an ~8 Å periodicity occurs in, e.g., boulangerite $Pb_5Sb_4S_{11}$ and robinsonite $Pb_4Sb_6S_{13}$ (details in Mumme 1989 and Makovicky et al. 2004).

Coordination of Sb and its complications. There are two distinct types of coordination polyhedra of antimony in dadsonite. In the first group of polyhedra three short Sb-S bonds (2.42-2.59 Å) are combined with three-to-four long Sb-S distances, starting at 3.09 Å or even at 3.44 Å. Nearly all of these Sb atoms form typical [SbS_5] coordination pyramids with trapezoidal bases in the Type 5 layer. The two longest Sb-S distances span the space hosting lone electron pairs. The other set of polyhedra has three regularly increasing short empirical Sb-S distances, starting at 2.41-2.45 Å and increasing to 2.60-2.72 Å. These polyhedra contain split Sb sites, which in the unsplit refinement exhibited augmented U_{iso} values, distinctly elongated displacement ellipsoids, and also an average bond length atypical for Sb. Such split half-atoms will have a nearly equal bond length to the vertex of the [SbS_5] coordination pyramid in which they are situated (e.g., in a selected example it is 2.42 Å and 2.41 Å,

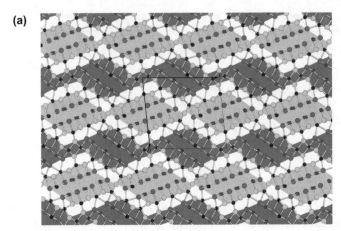

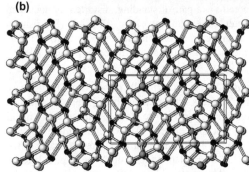

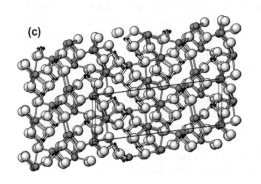

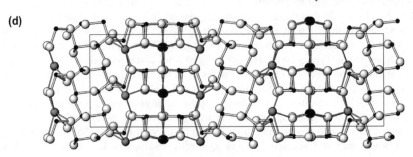

Figure 50. (a) Projection of dadsonite along [100]. The b axis points downward, c axis is horizontal. In order of decreasing size, spheres represent: sulfur (and chlorine) (white), lead (black), and (predominantly) antimony (grey). Light and dark spheres are strings of atoms differing in height along [100] by ~ 2 Å. Two shades of infill indicate rods constituting rod-layers (010) of two types, at $y = 0$ and $y = \frac{1}{2}$, respectively. (b) Jaskolskiite $Cu_{0.2}Pb_{2.2}$ (Sb, Bi)$_{1.8}$ S$_5$, the $N = 4$ member of the meneghinite homologous series. Light spheres denote S, grey spheres Pb, dark spheres Sb and Bi; tetrahedral Cu. Atoms at $z = 0$ and $z = \frac{1}{2}$. (c) Baumhauerite $Pb_{4.6}As_{15.7}Ag_{0.6}S_{36}$ (Engel and Nowacki 1969), the $N_{1,2} = 3,4$ member of the combinatorial sartorite series (note different thicknesses of alternating slabs separated by zig-zag planes populated by Pb). Cation count includes both the As (dark) coordination polyhedra and those of Pb (grey). (d) Hutchinsonite TlPbAs$_3$S$_9$ (Takéuchi et al. 1965), the $N_{1,2} = 2,3$ member of a hutchinsonite series of merotypes. Four height levels are present in the ~8.8 Å structure. Large light spheres: S, black: Pb, grey: Tl, small dark spheres: As. In the $N_{1,2} = 2,2$ homologue, bernardite TlAs$_5$S$_8$ (Pašava et al. 1989), the Pb position is missing and the two portions are joined, one of them having been rotated by 180°.

respectively). They will also have one more-or-less common, short Sb-S distance to one S atom in the square base of the pyramid (2.48 Å and 2.41 Å), and one long Sb-S distance that opposes this bond (3.18 Å and 3.28 Å, respectively). The other short Sb-S bond of each split Sb half-atom of antimony always overlaps the long Sb-S distance of its counterpart (2.70 Å and 2.69 Å, respectively, overlap with the distances 2.96-2.99 Å). Splitting of the two positions is never complete because also their S ligands are an average of two positions. The Sb-S bond distances should be investigated by means of a bond length hyperbola established for Sb by Berlepsch et al. (2001) ; the data for split positions will deviate from the normal hyperbolic trend. Data points belonging to the split/mixed cation sites also yield unrealistic bond valence values. Splitting of Sb sites inside a "square" coordination pyramid appears to be a fairly common and important phenomenon in the Pb-Sb sulfosalts; the same is true in the Pb-As sulfosalts of the sartorite group (Berlepsch et al. 2001).

Order-disorder phenomena. When we preserve the labelling of crystallographic axes used for the unit cell of dadsonite (Table 30), the Type 3_{SnS} layer has an ideal layer symmetry $P\bar{1}$ with the a' axis equal to one half of the a value of the unit cell. The Type 5 layer has an ideal layer symmetry $B2/m11$, with the c parameter doubled against the triclinic unit cell. After every Type 5 layer with its horizontal mirror planes, the triclinic stacking of layers ($\gamma = 89.2°$) can proceed in the $+\vec{a}$ or in the $-\vec{a}$ direction. Furthermore, the halved repetition period $a/2$ of the Type 3 layer means that the consecutive Type 5 layer can attach itself either at (practically) the same x height as the preceding Type 5 layer or it can attach itself at ($x + 0.5$) instead. Both the double-periodicity of the Type 3 layer and the B2/m symmetry of the Type 5 layer are only approximate, violated to a small but always present extent. Therefore, we deal with a desymmetrized OD structure in which the OD phenomena are present only as an occasional twinning or antiphase-boundary event (Ďurovič 1979). Among sulfosalts, OD phenomena cause twinning in sinnerite $Cu_6As_4S_9$ (Makovicky and Skinner 1975), imhofite $Tl_3As_{7.66}S_{13}$ (Balić-Žunić and Makovicky 1993), gillulyite $Tl_2(As,Sb)_8S_{13}$ (Makovicky and Balić-Žunić 1999) as well as structural disorder in owyheeite $Ag_3Pb_{10}Sb_{11}S_{28}$ (Makovicky et al. 1998), vurroite $Pb_{20}Sn_2(Bi,As)_{22}S_{54}Cl_6$ (Pinto et al. in prep.) and diaphorite $Pb_2Ag_3Sb_3S_8$ (Armbruster et al. 2002).

8 Å superstructure vs. 4 Å substructure. Many lead-antimony sulfosalts exhibit a pronounced ~4 Å substructure, producing a subset of strong X-ray reflections, and an 8 Å superstructure which modifies the 4 Å submotif by ordered substitutions, occupation of alternative split positions in a coordination polyhedron or even vacancies, producing additional weak to very weak reflections. The intensity of the generally weak 8 Å reflections varies with the degree of modification present in the structure. Some sulfosalts, e.g., jamesonite $FePb_4Sb_6S_{14}$ (Niizeki and Buerger 1957) or robinsonite $Pb_4Sb_6S_{13}$ (Skowron and Brown 1990a) have only the 4 Å periodicity, confirmed by the latest refinements (e.g., Leone et al. 2003; Matsushita and Ueda 2003; Franzini et al. 1992; Makovicky et al. 2004). Others, like boulangerite $Pb_5Sb_4S_{11}$, might have both the ordered variants (Mumme 1986) and disordered (synthetic) variants (e.g., Petrova et al. 1978; Skowron and Brown 1990b) of the Pb-Sb substitution schemes. Finally, for zinkenite ~ $Pb_9Sb_{22}S_{42}$, both ordered and disordered 8 Å variants are known. The remarkable box-work oxy-sulfosalt structures from Bucca della Vena (Italy) show very weak 8 Å levels; they were found to be too weak to be further analyzed (e.g., scainiite $Pb_{14}Sb_{30}S_{54}O_5$, Moëlo et al. 2000).

Other sulfosalt families

Glide-plane twinned accretional series have coordination polyhedra on composition planes of adjacent slabs different from those in the slab interior. Such are the meneghinite series of sulfosalts of Cu, Pb, Sb or/and Bi (Fig. 50b; the bismuthinite-aikinite series dealt with above are $N = 2$ members of the meneghinite series) and the sartorite homologous series (sulfosalts of Pb and As). These respectively represent twinning of SnS-like arrays on $(501)_{SnS}$ and $(301)_{SnS}$ (Makovicky 1985).

Sartorite homologues. Members of the sartorite homologous series are Pb-As sulfosalts which can be obtained by a glide-plane type of unit cell twinning of a (modified) SnS like archetype on the planes $(301)_{SnS}$. A simplified formula of the series is $Pb_{4N-8}As_8S_{4N+4}$. The zigzag composition planes are occupied by tricapped coordination prisms of lead whereas the slabs by coordination pyramids of arsenic, replaced in central portions by Pb and also some Ag or Tl. The match problems expected between the As pyramids and the inserted multiple Pb polyhedra in the potential higher homologues of sartorite limit this series to combinations of $N = 3$ and $N = 4$ slabs in different proportions: $N = 3,3; 3,4$ (Fig. 50c); $3,4,4; 4,3,4,3,4,4,3,4,3,4$; etc....; and, finally, $4,4$; members with $N > 4$ would already have entire slabs of Pb polyhedra inserted into their SnS-based layers. Thus, the sartorite homologous series remains a combinatorial series. This is apparently valid also for its Ba-Sb analogs. This series has been treated in detail by Makovicky (1985), Berlepsch et al. (2001, 2003), Pring (2001) and Ferraris et al. (2004).

Axial unit-cell twins. The plagionite homologous series (Takéuchi 1997) consists of four Pb-Sb sulfosalts: füloppite $Pb_3Sb_8S_{15}$, plagionite $Pb_5Sb_8S_{17}$, heteromorphite $Pb_7Sb_8S_{19}$ and semseyite $Pb_9Sb_8S_{21}$. They differ from other Pb-Sb sulfosalts by not being rod-based structures. Instead, they consist of continuous slabs of SnS archetype (or $TlSbS_2$ archetype, Takeuchi 1997), cut diagonally and joined together on composition planes via the action of 2_1 axes. Takéuchi (1997) gives a detailed analysis and description of both the local and the modular aspects of this series; he used a distorted PbS archetype as a basis. Plagionite homologues form only a subset of all the known compounds in the system Pb-Sb-S, i.e., only those which have a restricted formula $Pb_{3+2x}Sb_8S_{15+2x}$ with $x = 0, 1, 2$ or 3 (Takéuchi 1997).

Hutchinsonite merotypes (Makovicky 1997) are a group of complex sulfides combining As or Sb with large uni- and divalent cations (Tl^+, Pb^{2+}, Na^+, Cs^+, NH_4^+, and others, including organic cations). The structures of these sulfosalts are regular 1:1 intergrowths of slabs (A) which can be described as $(010)_{SnS}$ cut-outs of different widths from the SnS-archetype, or as $(110)_{PbS}$ cut-outs from the PbS-archetype, with layers (B) of variable thickness and configuration. The (B) layers contain primarily MS_3 pyramids (M = As, Sb) with active lone electron pairs, mostly combined with coordination polyhedra of large (even of organic) cations. Slabs A and B share certain S atoms in common. In this family of merotypes the A slabs are built according to the common principles in all these structures, whereas the B slabs differ, being always adapted to the requirements of different large cations (Makovicky 1997).This series has been treated in detail by Ferraris et al. (2005) and only a single example will be given here.

The phases $TlPbAs_3S_9$ (hutchinsonite, *Pbca*, Takéuchi et al. 1965) - $TlAs_5S_8$ (bernardite, $P2_1/c$, $TlAs_5S_8$, Pašava et al. 1989) are examples of a pair of homologues. Both structures are composed of alternating slabs (A) of SnS-like configuration, with As, Tl ± Pb, and slabs (B) of a complex spiral configuration, accommodating only As with its active lone electron pairs (Fig. 50d). The similarity between the two structures, in nearly all coordination polyhedra and layer configurations, is striking. The only difference is that in bernardite the SnS-like slab is 2 coordination pyramids of As (As/Tl in the case of marginal pyramids) broad, i.e., a 2,2-homologue when considering the two faces of a tightly-bonded fragment in the SnS-like slab, whereas in hutchinsonite this tightly-bonded fragment is diagonally split, half of it rotated 180° about [010], and an additional Pb pyramid is inserted into only one face of the fragment. This operation defines it as a 2,3-homologue of the series(Fig. 50d). It is easy to imagine a 3,3-homologue with two Pb pyramids inserted. However, this has not been found, apparently because of the match problems between the large Pb polyhedra and the As pyramids.

ACKNOWLEDGMENTS

This paper profitted from the secretarial assistance of Mrs. Camilla Sarantaris and Mrs. Lisbeth Skjoldager; Dr. Dan Topa, Mrs Brita Munch and the deceased Mr. Ole Bang Berthelsen helped to edit the figures. Critical reading by the chief editor Prof. D.J. Vaughan and the kind assistance of Dr. J. J. Rosso improved this contribution and the patience and understanding of my wife created the necessary conditions for it. Partial support by the National Science Foundation project 21-03-0519 and the Carlsberg Foundation project ANS 1185/10 is gratefully acknowledged.

REFERENCES

Abrahams SC, Bernstein JL (1973) Piezoelectric nonlinear optic $CuGaS_2$ and $CuInS_2$ crystal structure: Sublattice distortion in A(I) B(III) C(VI)2 and A(II) B(IV) C(V)2 type chalcopyrites. J Chem Phys 59: 5415-5420

Abrahams SC, Bernstein JL (1974) Piezoelectric nonlinear optic $CuGaSe_2$ and $CdGe As_2$: Crystal structure, microhardness, and sublattice distortion. J Chem Phys 61:1140-1146

Adenis C, Olivier-Fourcade J, Jumas JC, Philippot E (1986) Etude structurale de $In_2Sn_3S_7$ par spectrometrie moessbauer de 119Sn et diffraction des rayons X. Rev Chimie minerale 23:7335-745

Adolphe C, Laruelle P (1968) Structure cristalline de $FeHo_4S_7$ et de certains composés isotypes. Bull Soc Franc Minér Crist 91:219-232

Agrawal BK, Yadav PS, Agrawal S. (1994) Ab initio calculation of the electronic, structural and dynamical properties of Zn-based semiconductors. Phys Rev B – Cond Matt 50:14881-14887

Andersson S, Åström A, Galy J, Meunier G (1973) Simple calculations of bond lengths and bond angles in certain oxides, fluorides and oxide fluorides of Sb^{3+}, Te^{4+} and Pb^{2+}. J Solid State Chem 6:187-190

Andersson S and Hyde BG (1974) Twinning on the unit cell level as a structure-building operation in the solid state. J Solid State Chem 9:92-101

Andresen AF, Leciejewicz J (1963) A neutron diffraction study of Fe_7Se_8. J Phys (Paris) 25:574-578

Andresen AF, van Laar R (1970) The magnetic structure of Fe_3Se_4. Acta Chim Scand 24:2435-3439

Angel RJ, Price GD, Yeomans J (1985) The energetics of polytypic structures: Further applications of the ANNNI model. Acta Cryst B41:310-319

Armbruster T, Hummel W (1987) (Sb, Pb) ordering in sulfosalts: Crystal-structure refinement of a Bi-rich izoklakeite. Am Mineral 72:821-831

Armbruster T, Makovicky E, Berlepsch P, Sejkora J (2003) Crystal structure, cation ordering and polytypic character of diaphorite $Pb_2Ag_3Sb_3S_8$, a PbS based structure. Eur J Mineral 15:137-146

Aurivilius B (1983) The crystal structures of two forms of $BaBi_2S_4$. Acta Chem Scand A37:399-407

Avilov AS, Imamov RM, Pinsker ZG (1971) Electron diffraction study of the Cu_2TlSe_2 phase. Kristallografiya 16:635-636

Baker CL, Lincoln FJ, Johnson AWS (1991) A low-temperature structural phase transformation in CuAgS. Acta Cryst 47:891-899

Baker CL, Lincoln FJ, Johnson AWS (1992) Crystal structure determination of Ag_3CuS_2 from powder X-ray diffraction. Australian J Chem 45:1441-1449

Bakker M, Hyde BG (1978) A preliminary electron microscope study of chemical twinning in the system MnS + Y_2S_3, an analog of the mineral system PbS + Bi_2S_3 (galena + bismuthinite). Phil. Mag A38:615-628

Balić-Žunić T, Makovicky E (1993) Contributions to the crystal chemistry of thallium sulphosalts. I. The O-D nature of imhofite. N Jahrb Mineral Abh 165:317-330

Balić-Žunić T, Topa D, Makovicky E (2002) The crystal structure of emilite, $Cu_{10.7}Pb_{10.7}Bi_{21.3}S_{48}$, the second 45 Å derivative of the bismuthinite-aikinite solid-solution series. Can Mineral 40:239-245

Balić-Žunić T, Makovicky E, Moelo Y (1995) Contributions to the crystal chemistry of thallium sulphosalts. III. The Crystal structure of lorandite $TlAsS_2$ and its relation to weissbergite $TlSbS_2$. N Jahrb Mineral Abh 168:213-235

Balić-Žunić T, Mariolacos K, Friese K, Makovicky E (2005) Structure of a synthetic halogen sulfosalt $Cu_3Bi_2S_3I_3$. Acta Cryst B 61:239-245

Baronett A (1997) Equilibrium and kinetic processes for polytype and polysome generation. EMU Notes in Mineralogy 1:119-152

Bayliss P. (1968) The crystal structure of disordered gersdorffite. Am Mineral 53:290-293

Bayliss P (1991) Crystal chemistry and crystallography of some minerals in the tetradymite group. Am Mineral 76:257-265

Belov NV, Godovikov AA, Bakakin VV (1982) Ocherki po Teoreticheskoy Mineralogii (Essays on Theoretical Mineralogy). Publ House Nauka, Moscow

Belov NV (1976) Ocherki po strukturnoy mineralogii (Essays on Structural mineralogy). Publ House Nedra, Moscow

Bente K, Kupčík V (1984) Redetermination and refinement of the structure of tetrabismuth tetracopper enneasulphide, $Cu_4Bi_4S_9$. Acta Cryst C40:1985-1986

Bergerhoff G, Berndt M, Brandenburg K, Degen T (1999) Concerning inorganic crystal structure types. Acta Crystallogr B55:147-156

Berger R. (1987) A phase-analytical study of the Tl-Cu-Se system. J of Solid State Chem 70:65-70

Berger R. (1989) Synthesis und characterization of a layered metal $TlCu_2S_2$. J Less-Common Metals 147:141-148

Berger R, Eriksson L (1990) Crystal structure and properties of $TlCu_6S_4$. J Less-Common Metals 161:165-173

Berger R, Eriksson L (1990) Crystal structure refinement of monoclinic $TlCu_3Se_2$. J of Less-Common Metals 161:101-108

Berger R, Eriksson L, Meerschaut A (1990) The crystal structure of $TlCu_5Se_3$. J Solid State Chem 87:283-288

Berger R, Meerschaut A (1988) The crystal structure of $Tl_5Cu_{14}Se_{10}$. Eur Solid State Inorg Chem 25:279-288

Berger RA, Sobott RJ (1987) Characterization of $TlCu_7S_4$, a crookesite analog. Mh für Chem und verw Teile anderer Wissenschaften 118:967-972

Berger R, Tergenius L-E, Noren L, Eriksson L (1995) The crystal structure of room-temperature synthesized orthorhombic $TlCu_4Se_3$ from direct methods on X-ray powder data. J Alloys Comp 224:171-176

Berger R, van Bruggen CF (1984) $TlCu_2Se_2$: A p-type metal with a layer structure. J Less-Common Metals 99:113-123

Berlepsch P, Armbruster T, Makovicky E, Hejny C, Topa D, Graeser S (2001) The crystal structure of (001) twinned xilingolite, $Pb_3Bi_2S_6$, from Mittal-Hohtenn, Valais, Switzerland. Can Mineral 39:1653-1663

Berlepsch P, Armbruster T, Makovicky E, Topa D (2003) Another step toward understanding the true nature of sartorite: Determination and refinement of a ninefold superstructure. Am Mineral 88:450-461

Berlepsch P, Armbruster T, Topa D (2003) Structural and chemical variation in rathite, $Pb_8Pb_{4-x}(Tl_2As_2)_x(Ag_2As_2)As_{16}S_{40}$: modulations of a parent structure. Z Kristallogr 217:581-590

Berlepsch P, Makovicky E, Balić-Žunić T (2001) Crystal chemistry of sartorite homologues and related sulfosalts. N Jahrb Mineral Abh 176:45-66

Berlepsch P, Makovicky E and Balić-Žunić T (2001) Crystal chemistry of meneghinite homologues and related sulfosalts. N Jahrb Mineral Mh 2001:115-135

Berlepsch P, Miletich R, Makovicky E, Balić-Žunić T, Topa D (2001) The crystal structure of synthetic $Rb_2Sb_8S_{12}(S_2)\cdot 2(H_2O)$ a new member of the hutchinsonite family of merotypes. Z Kristallogr 216:272-277

Berner RA (1962) Tetragonal iron sulfide. Science, 137:669-669

Bernert T, Pfitzner A (2005) $Cu_2MnM^{IV}S_4$ (M^{IV} = Si, Ge, Sn)-analysis of crystal structures and tetrahedra volumes of normal tetrahedral compounds. Z Kristallogr 220:968-972

Bernstein LR, Reichel DG, Merlino S (1989) Renierite crystal structure refined from Rietveld analysis of powder neutron-diffraction data. Am Mineral 74:1177-1181

Bichon J, Davot M, Rouxel J (1973) Systematique structurale pour les series d'intercalaires M_xTiS_2 (M = Li, Na, K, Rb, Cs). Comptes Rendus Acad Sci, Ser C, Sci Chim 276:1283-1286

Bindi L, Popova V, Bonazzi P (2003) Uzonite, As_4S_5, from the type locality: single-crystal X-ray study and effects of exposure to light. Can Mineral 41:1463-1468

Bindi L, Cipriani C (2004) The crystal structure of skippenite, Bi_2Se_2Te, from the Kochkar deposit, southern Urals, Russian Federation Can Miner 42:835-840

Bindi L, Evain M, Menchetti S (2006) Temperature dependence of the silver distribution in the crystal structure of natural pearceite, $(Ag, Cu)_{16} (As, Sb)_2 S_{11}$. Acta Crystallogr 62:212-219

Boller H (1978) Faserförmige Erdalkali–thioferrate. Mh Chem 109:975-985

Boller H, Sing M (1997) On the formation and thermal stability of $NH_4Cu_7S_4$ and $NH_4Cu_4S_3$. Topotactic decomposition of $NH_4Cu_7S_4$. Solid State Ionics 101: 1287-1291

Bonazzi P, Borrini D, Mazzi F, Olmi F (1995) Crystal structure and twinning of $Sb_2 AsS_2$, the synthetic analog of pääkkönenite. Am Mineral 80:1054-1058

Bonazzi P, Menchetti S, Sabelli C (1987) Structure refinement of kermesite: Symmetry, twinning, and comparison with stibnite. N Jahrb Mineral Mh 1987:557-567

Bonazzi P, Bindi L, Bernardini GP, Menchetti S (2003) A model for the mechanism of incorporation of Cu, Fe and Zn in the stannite – kësterite series, $Cu_2FeSnS_4 – Cu_2ZnSnS_4$. Can Mineral 41: 639-647

Bonazzi P, Bindi L, Olmi F, Menchetti S (2003) How many alacranites do exist? A structural study of non-stoichiometric As_8S_{9-x} crystals. Eur J Mineral 15:283-288

Bonazzi P, Lampronti GI, Bindi L, Zanardi S (2005) Wakabayshilite, [$(As,Sb)_6S_9$][As_4S_5]: Crystal structure, pseudosymmetry, twinning, and revised chemical formula. Am Mineral 90:108-1114
Bonazzi P, Menchetti S, Pratesi G (1995). The crystal structure of pararealgar. Am Mineral 80:400-403
Boucher F, Evain M, Brec R (1993) Distribution and ionic diffusion path of silver in gamma-Ag_8GeTe_6: a temperature dependent anharmonic single crystal structure study. J Solid State Chem 107:332-346
Bouwmeester HJM, Dekker EJP, Bronsema KD, Haange RJ, Wiegers GA (1982) Structures and phase relations of compounds Na_xTiS_2 and Na_xTiSe_2. Rev Chim Minérale 19:333-342
Brandt G, Raeuber A, Schneider J (1973) ESR and X-ray analysis of the ternary semiconductors $CoAlS_3$,$CuInS_2$, $AgGaS_2$. J Solid State Chem12:481-483
Brese NE, Squattrito PJ, Ibers JA (1985) Reinvestigation of the structure of PdS. Acta Crystallogr C41:1829-1830
Breskovska V, Tarkian M (1994) Compositional variations in Bi-bearing fahlores. N Jb Miner Mh 1994:230-240
Bronger W, Böttcher P (1972) Über Thiomanganate und - kobaltate der schweren Alkalimetalle: $Rb_2Mn_3S_4$, $Cs_2Mn_3S_4$, $Rb_2Co_3S_4$, $Cs_2Co_3S_4$. Z anorg allg Chem 390:1-96
Bronger W, Kyas A, Muller P (1987) The antiferromagnetic structures of $KFeS_2$, $RbFeS_2$, $KFeSe_2$ and $RbFeSe_2$ and the correlation between magnetic moments and crystal field calculations. J Solid State Chem 70: 262-270
Bronger W, Genin HS, Mueller P (1999) K_3FeSe_3 und $K_3Fe_2Se_4$. Zwei neue Verbindungen im System K/Fe/Se. Z anorg allg Chem 625:274-278
Bronger W, Müller P (1980) Low-spin Anordnungen in Tetraederstrukturen von Eisensulfiden: Untersuchungen zum System $CsGa_{1-x}Fe_xS_2$. J Less-Common Metals 70:253-262
Bronger W, Ruschewitz V, Müller P (1995) New ternary iron sulphides $A_3Fe_2S_4$(A = K, Rb, Cs): syntheses and crystal structures. J Alloys Compounds 218:22-27
Bronsema KD, de Boer JL, Jellinek F (1986) On the structure of molybdenum diselenide and disulfide. Z anorg allg Chem 540:15-17
Brostigen G, Kjekshus A, Romming C (1973) Compounds with the marcasite type crystal structure VIII, Redetermination of the prototype. Acta Chem Scand 27:2791-2796, Acta Chem Scand 24:1925-1940
Brouwer R, Jellinek F (1979) Modulation of the intergrowth structures of $A_{1-p}Cr_2X_{4-p}$ (A = Ba, Sr, Eu, Pb; x = S, Se; p ≈ 0.29). Am Inst Phys Conf Proceedings 53:114-116
Brown DB, Zubieta J, Vella PA, Wrobleski JT, Watt T, Hatfield WE, Day P (1980) Solid-State and Electronic Properties of a Mixed-Valence Two-Dimensional Metal, KCu_4S_3. Inorg Chem 19:1945-1950
Buerger MJ, Wuensch BJ (1963) Distribution of atoms in high chalcocite, Cu_2S. Science 141:276-277
Burns PC, Percival JB (2001) Alacranite As_4S_4: a new occurrence, new formula and determination of the crystal structure. Can Mineral 39:809-818
Burr GL, Young VG Jr, McKelvy MJ, Glaunsinger WS, von Dreele RB (1990) A structural investigation of $Ag_{0.167}TiS_2$ by time-of-flight neutron powder diffraction. J Solid State Chem 84:355-364
Burschka C (1980) $CsCu_4S_3$ und $CsCu_3S_2$: Sulfide mit tetraedrisch und linear koordiniertem Kupfer. Z anorg allg Chem 463:65-71
Busch G, Froehlich C, Hullinger F (1961) Struktur, elektrische und thermoelektrische Eigenschaften von $SnSe_2$. Helvet Phys Acta 34:359-368
Cabri LJ, Hall SR, Szymanski JT, Stewart JM (1973) On the transformation of cubanite. Can Mineral 12: 33-38
Cario L, Meerschaut A, Moëlo Y, Nader A, Rouxel J (1997) Structure determination and electrical properties of a new misfit layer compound $(SmS)_{1.25}$ TiS_2. Eur J Solid State Inorg Chem 34:913-924
Carré D, Laruelle P (1973) Structure cristalline du sulfure d'erbium et de lanthane, $Er_9La_{10}S_{27}$. Acta Crystallogr B29:70-73
Carré D, Laruelle P (1974) Structure cristalline du sulfure de néodyme et d'ytterbium, $NdYbS_3$. Acta Crystallogr B30:952-954
Cava RJ, Reidinger F, Wuensch BJ (1980) Single-crystal neutron diffraction study of the fast-ion conductor beta-Ag_2S between 186 and 325 C. J Solid State Chem 31:69-80
Cava RJ, Reidinger F, Wuensch BJ (1981) Mobile ion distribution and anharmonic thermal motion in fast ion conducting Cu_2S. Solid State Ionics 5:501-504
Chattopadhyay TK, von Schnering HG, Stansfield RFD, McIntyre GJ (1992) X-ray and neutron diffraction study of the crystal structure of MnS_2. Z Kristallogr 199:13-24
Chattopadhyay TK, Pannetier J, von Schnerring HG (1986) Neutron diffraction study of the structural phase transition in SnS. J Phys Chem Solids 47:879-885
Ceolin R, Tofoli P, Khodadad P, Rodier N (1977) Structure cristalline du sulfure mixte de cerium et de bismuth $Ce_{1.25}Bi_{3.78}S_8$. Acta Crystal B33:2804-2806
Cervelle BD, Cesbron FP, Sichère MC (1979) La chalcostibite et la dadsonite de Saint-Pons, Alpes de Haute Provence, France. Can Mineral 17:601-605

Charnock JM, Garner CD, Pattrick RAD, Vaughan DJ (1989) EXAFS and Mössbauer spectroscopic study of Fe-bearing tetrahedrites. Miner Mag 53:193-199

Chen BH, Eichhorn BW, Peng JL, Greene RL (1993) Superconductivity in the A_xNbS_2 intercalation compounds (A = Cs, Rb). J Solid State Chem 103:307-313

Chen TT, Szymanski JT (1981) The structure and chemistry of galkhaite, a mercury sulfide containing Cs and Tl. Can Mineral 19:571-581

Chen X-A, Wada H, Sato A (1999) Preparation, crystal structure and electrical properties of Cu_4SnS_6. Mater Res Bull 34:239-247

Chevrel R (1992) Cluster Solid State Chemistry: a frontier discipline between metallurgy and molecular chemistry. Parthé Eeditor Modern Perspectives in Inorganic Crystal Chemistry. NATO ASI Series C382: 17-26

Choe W, Lee S, O'Connel P, Covey A (1997) Synthesis and structure of new Cd-Bi-S homologous series: a study in intergrowth and the control of twinning patterns. Chem Mater 9:2025-2030

Choi K-S, Kanatzidis MG (2000) Sulfosalts with alkaline earth metals. Centrosymmetric vs acentric interplay in $Ba_3Sb_{4.66}S_{10}$ and $Ba_{2.62}Pb_{1.38}Sb_4S_{10}$ based on the Ba/Pb/Sb ratio. Phases related to the arsenosulfide minerals of the rathite group and the novel polysulfide $Sr_6Sb_6S_{17}$. Inorg Chem 39:5655-5662

Clark R, Brown GE (1980) Crystal structure of rasvumite, KFe_2S_3. Am Mineral 65:477-482

Conde C, Manolikas C, Van Dyck D, Delavignette P, Van Landuyt J, Amelinckx S (1978) Electron microscopic study of digenite-related phases ($Cu_{2-x}S$). Mater Res Bull 13:1055- 1063

Cook R, Schaefer H (1982) Darstellung und Kristallstrukur von $SrBiSe_3$. Rev chimie Minerale 19:19-27

Cook NJ, Ciobanu CL, Wagner T, Stanley CL (in press) Minerals of the system (Pb)-Bi-Te-Se-S related to the tetradymite archetype: Review of classification and compositional variation. Can Mineral in press

Cordier G, Cook R, Schäfer H (1980) Isolierte Selenoantimonat (III) anionen in $Ba_4Sb_4Se_{11}$. Revue Chim Miner 17:1-6

Craig DC, Stephenson NC (1965) The crystal structure of lautite, CuAsS. Acta Crystallogr 19:543-547

Daams JLC, Villars P (1993) Atomic environment classification of the rhombohedral "intermetallic" structure types. J Alloys Compd 201:11-16

de Boer JL, Meetsma A (1991) Structures of misfit layer compounds $(LaS)_{1.13}TaS_2$ "$LaTaS_3$" and $(CeS)_{1.15}TaS_2$ "$CeTaS_3$." Acta Crystallogr C47:924-930

di Benedetto F, Bernardini GP, Borrini D, Emiliani C, Cipriani C, Danti C, Caneschi A, Gatteschi D, Romanelli M (2002) Crystal chemistry of tetrahedrite solid solution: EPR and magnetic investigations. Can Mineral 40:837-847

di Benedetto F, Bernardini GP, Cipriani C, Emiliani C, Gatteschi D, Romanelli M (2003) The distribution of Cu(II) and the magnetic properties of the synthetic analog of tetrahedrite: $Cu_{12}Sb_4S_{13}$. Phys Chem Min 32:155-164

Dickinson RG, Friauf JB (1924) The crystal structure of tetragonal lead monoxide. J Am Chem Soc 46:2457-2462

Digiuseppe M, Steger J, Wold A, Kostiner E (1974) Preparation and characterization of the system $CuGa_{1-x}Fe_xS_2$. Inorg Chem 13:1828-1831

Ding Y, Veblen DR, Prewitt CT (2005) High-resolution transmission electron microscopy (HRTEM) study of the 4a and 6a superstructure of bornite Cu_5FeS_4. Am Mineral 90:1256-1264

Ding Y, Veblen DR, Prewitt CT (2005) Possible Fe/Cu ordering schemes in the 2a superstructure of bornite (Cu_5FeS_4). Am Mineral 90:1265-1269

Divjaković V, Nowacki W (1976) Die Kristallstruktur von Imhofit, $Tl_{5.6}As_{15}S_{25.3}$. Z Kristallogr 144:323-333

Dmitrieva MT, Ilyukhin VV, Bokii GB (1979) Close packing and cation arrangement in the djerfisherite structure. Kristallografiya 24:1193-1197

Dmitrieva MT, Yefremov VA, Kovalenker VA (1987) Crystal structure of As goldfieldite Dokl Acad Sci USSR, Earth Sci Sect 297:141-144

Dobrovolskaya MG, Tsepin AI, Evstigneyeva TL, Vyaltsov LN, Zaozerina AO (1981) Murunskite, $K_2Cu_3FeS_4$, a new sulfide of potassium, copper and iron. Zap Vses Mineral Obshch 110:468-473

Donnay G, Donnay JDH, Elliott N, Hastings JM (1958) Symmetry of magnetic structures: Magnetic structure of chalcopyrite. Phys Rev 112:1917-1923

Dornberger-Schiff K, Hoehne E (1959) Die Kristallstruktur des Betechtinit $Pb_2(CuFe)_{21}S_{15}$. Acta Crystallogr 12:646-651

Dornberger-Schiff K (1966) Lehrgang über OD-Strukturen. Akademie-Verlag, Berlin.

Dubost V, Balić-Žunić T, Makovicky E (in press) The crystal structure of $Ni_{10}Pd_7S_{15}$. Can Miner in press

Ďurovič S (1979) Desymmetrization of OD structures. Kristall und Technik 14:1047-1053

Ďurovič S (1992) Layer stacking in general polytypic structures. International Tables for X-ray Crystallography, Vol. C, chapter 9.2.2., 667-680. Kluwer Acad. Publ.

Ďurovič S, Hybler J (2006) OD Structures in crystallography – basic concepts and suggestions for practice. Z Kristallogr 221:63-76

Effenberger H, Pertlik F (1981) Ein Beitrag zur Kristallstruktur von α-CuSe (Klockmannit). N Jahrb Mineral Mh 1981:197-205
Effenberger H, Paar WH, Topa D, Culetto FJ, Giester G (1999) Toward the crystal structure of nagyagite (Pb (Pb, Sb) S$_2$) (Au,Te)). Am Mineral 84:669-676
Effenberger H, Culetto FJ, Topa D, Paar WH (2000) The crystal structure of synthetic buckhornite (Pb$_2$BiS$_3$) (AuTe$_2$). Z Kristallogr 215:10-16
Emsley J (1994) The Elements, 2nd ed. Oxford Univ Press
Engel P, Nowacki W (1969) Die Kristallstruktur von Baumhauerit. Z Kristallogr 129:178-202
Engel P, Nowacki W (1966) Die Verfeinerung der Kristallstruktur von Proustit, Ag$_3$AsS$_3$ und Pyrargyrit, Ag$_3$SbS$_3$. N Jahrb Mineral, Mh 1966:181-184
Eppinga R, Wiegers GA (1977) The crystal structure of the intercalates SnTaS$_2$ and SnNbS$_2$. Mater Res Bull 12:1057-1062
Eppinga R, Wiegers GA (1980) A generalized scheme for niobium and tantalum dichalcogenides intercalated with post-transition elements. Physica B and C (Netherland) 99:121-127
Eriksson L, Werner P-E, Berger R, Meerschaut A (1991) Structure refinement of TlCu$_7$Se$_4$ from X-ray powder profile data. J Solid State Chem 90:61-68
Euler R, Hellner E (1960) Ueber komplex zusammengesetzte sulfidische Erze VI. zur Kristallstruktur des Meneghinits CuPb$_{13}$Sb$_7$S$_{24}$ 113:345-372
Evans HT Jr (1979) The crystal structures of low chalcocite and djurleite. Z Kristallogr 150:299-320
Evans HT Jr, Allmann R (1968) The crystal structure and crystal chemistry of valleriite. Z Kristallogr 127: 73-93
Evans HT Jr, Konnert JA (1976) Crystal structure refinement of covellite. Am Mineral 61:996-1000
Evans HT Jr, Clark JR (1981) The crystal structure of bartonite, a potassium iron sulfide, and its relationship to pentlandite and djerfisherite. Am Mineral 66:376-384
Evstigneeva TL, Kabalov YuK (2001) Crystal structure of the cubic modification of Cu$_2$FeSnS$_4$. Kristallografiya 46:418-422 (in Russian)
Fei Y, Prewitt CT, Frost DJ, Parise JB, Brister K (1998) Structures of FeS polymorphs at high pressure and temperature. Rev High Pressure Sci Technol 7:55-58
Ferraris G, Makovicky E, Merlino S (2004) Crystallography of Modular Materials. IUCr Monographs on Crystallography 15. Oxford Sci Publ
Feutelais Y, Legendre B, Rodier N, Agafonov V (1993) A study of phases in the bismuth – tellurium system. Materials Res Bull 28:591-596
Fleet ME (1972) The crystal structure of pararammelsbergite NiAs$_2$. Am Mineral 57:1-9
Fleet ME (1973) The crystal structure of maucherite. Am Mineral 58:203-210
Fleet ME (1987) Structure of godlevskite Ni$_9$S$_8$. Acta Crystallogr 43:2255-2257
Fleet ME (1970) Structural aspects of the marcasite-pyrite transformation. Can Mineral 10:224-231
Fleet ME (1970) Refinement of the crystal structure of cubanite and polymorphism of CuFe$_2$S$_3$. Z Kristallog 132:276-287
Fleet ME (2006) Phase equilibria at high temperatures. Rev Mineral Geochem 61:365-419
Fleet ME, Burns PC (1990) Structure and twinning of cobaltite. Can Mineral 28:719-723
Foecker AJ, Jeitschko W (2001) The atomic order of the pnictogen and chalcogen atoms in equiatomic ternary compounds TPnCh (T=Ni, Pd; Pn=P, Sb ; Ch=S, Se,Te). J Solid State Chem 162:69-78
Foit FF, Robinson PD, Wilson JR (1995) The crystal structure of gillulyite, Tl$_2$ (As, Sb)$_8$ S$_{13}$, from the Mercur gold deposit, Tooele County, Utah, U.S.A. Am Mineral 80:394-399
Frank-Kamenetskaya OV, Rozhdenstvenskaya IV, Yanulova LA (2002) New data on the crystal structures of colusites and arsenosulvanites. J Struct Chem (USSR) 43:89-100
Franzini M, Orlandi P, Paserp M (1992) Morphological, chemical and structural study of robinsonite (Pb$_4$Sb$_6$S$_{13}$) from Alpi Apuane, Italy. Acta Volcanologica 2:231-235
Frueh AJ Jr (1950) Disorder in the mineral bornite Cu$_5$FeS$_4$. Am Mineral 35:185-192
Frueh AJ jr. (1958) The crystallography of silver sulfide Ag$_2$S. Z Kristallogr 110:136-144
Fuess H, Kratz T, Töpel-Schadt J, Miehe G (1987) Crystal structure refinement and electron microscopy of arsenopyrite. Z Kristallogr 179:335-346
Furubayashi T, Matsumoto T, Hagino T, Nagata S (1994) Structural and magnetic properties of metal-insulator transition in thiospinel CuIr$_2$S$_4$. J Phys Soc Japan 63:3333-3339
Furuseth S (1992) Structural properties of the Ti$_{1-x}$S$_2$ phase. J Alloys Comp 178:211-215
Furuseth S, Selte K, Kjekhus A (1965) Redetermined crystal structures of NiTe$_2$, PtS$_2$, PtTe$_2$. 19:257-258
Gaines RV, Skinner HCW, Foord EE, Mason B, Rosenzweig A (1997) Dana's New Mineralogy, 8th Ed. John Wiley and Sons
Garin J, Parthe E, Oswald HR (1972) The crystal structure of Cu$_3$PSe$_4$ and other ternary normal tetrahedral structure compounds with composition 13 5 64. Acta Crystallogr B 28:3672-3674

Gastaldi L, Simeone MG, Viticoli S (1985) Cation ordering and crystal structures in AGa_2X_4 compounds ($CoGa_2S_4$, $CdGa_2S_4$, $CdGa_2Se_4$, $HgGa_2Se_4$, $HgGa_2Te_4$). Solid State Comm 55:605-607

Gattow G (1957) Die Kristallstruktur von $NH_4Cu_7S_4$. Acta Crystallogr 10:549-553

Gaudin E, Jobic S, Evain M, Brec R, Rouxel J (1995) Charge balance in some Bi_xSe_y phases through atomic structure determination and band structure calculations. Mater Res Bull 30:549-561

Gaudin E, Brocher F, Patricek V, Taulelle F, Evain M (2000) Structures and phase transition of the $A_7 P Se_6$ (A=Ag, Cu) argyrodite-type ionic conductors. II. beta-and gamma $–Cu_7PSe_6$. Acta Crystallogr B56:402-408

Gelabert MC, Ho MH, Malik A-S, DiSalvo FJ, Deniard P (1997) Structure and properties of $Ba_6Ni_{25}S_{27}$. Chem Eur J 3:1884-1889

Geller S (1962a) The crystal structure of $Pd_{17}Se_{15}$. Acta Crystallogr 15:713-721

Geller S (1962b) The crystal structure of the superconductor $Rh_{17}S_{15}$. Acta Crystallogr 15:1198-1201

Giese RF, Kerr DF (1965) The crystal structure of ordered and disordered cobaltite. Am Mineral 50:1002-1014

Ginderow D (1978) Structures cristallines de $Pb_4In_9S_{17}$ et $Pb_3In_{6.67}S_{13}$. Acta Crystallogr B34:1804-1811

Gorochov O (1968) Les composés Ag_8MX_6 (M=Si,Ge,Sn et X= S,Se,Te). Bull Soc Chim France 1968:2263-2275

Gostojić M, Nowacki W, Engel P (1982) The crystal structure of synthetic $TlSb_3S_5$, Z Kristallogr 159:217-224

Gotoh Y, Akimoto J, Goto M, Oosawa Y, Onoda M (1995) The layered composite crystal structure of the ternay sulfide $(BiS)_{1.11}NbS_2$. J Solid State Chem 116:61-67

Graham AR, Thompson RM, Berry LG (1953) Studies of mineral sulphosalts: XVII -Cannizzarite. Am Mineral 38:536-544

Grey IE (1975) The structure of $Ba_5Fe_9S_{18}$. Acta Crystallogr B31:45-48

Grguric BA, Putnis A (1998) Compositional controls on phase-transition temperatures in bornite: a differential scanning calorimetry study. Can Mineral 36:215-227

Grice JD, Ferguson RB (1974) Crystal structure refinement of millerite (beta-NiS). Can Mineral 12:248-252

Grice JD, Ferguson RB (1989) The crystal structure of arsenohauchecornite. Can Mineral 27:137-142

Grønvold F, Haraldsen H, Kjekshus A, (1960) On the sulfides, selenides and tellurides of platinum. Acta Chem Scand 14:1879-1893

Grønvold F, Rost E (1957) The crystal structures of $PdSe_2$ and PdS_2. Acta Crystallogr 10:329-331

Grønvold F, Rost E (1962) The crystal structures of Pd_4Se and Pd_4S. Acta Crystallogr 15:11-13

Guinier A, Bokij GB, Boll-Dornberger K, Cowley JM, Ďurovič S, Jagodzinski M, Krishna P et al (1984) Nomenclature of polytype structures. Report of the IUCr Ad-hoc Committee on the Nomenclature of Disordered, Modulated and Polytype Structures. Acta Crystallogr A40:399-404

Guenter JR, Oswald (1968) Neue polytype Form von Zinn (IV)-Sulfid. Naturwissen-schaften 55:177

Hall AJ (1972) Substitution of Cu by Zn, Fe and Ag in synthetic tetrahedrite. Bull Soc franç Minér Crist 99: 152-158

Hall SR, Gabe EJ (1972) The crystal structure of talnakhite $Cu_{18}Fe_{16}S_{32}$. Am Mineral 57:368-380

Hall SR, Rowland JF (1973) The crystal structure of synthetic mooihoekite, $Cu_9Fe_9S_{16}$. Acta Crystallogr B29: 2365-2372

Hall SR, Stewart JM (1973) The crystal structure of argentian pentlandite (Fe $Ni)_8AgS_8$, compared with the refined structure of pentlandite (Fe $Ni)_9S_8$.Can Mineral 12:61-65

Hall SR, Stewart JM (1973) The crystal structure refinement og chalcopyrite, $CuFeS_2$. Acta Crystallogr 1329: 579-585

Hall SR, Szymanski JT, Stewart JM (1978) Kesterite, $Cu_2(Zn,Fe)SnS_4$, and stannite, $Cu_2(Fe,Zn)SnS_4$, structurally similar but distinct minerals. Can Mineral 16: 131-137

Hallak H, Lee P (1983) Lithium ordering in Li_xTiS_2: a superlattice structure for $Li_{0.33}TiS_2$. Solid State Comm 47:503-505

Harker D (1934) The crystal structure of the mineral tetradymite Bi_2Te_2S. Z Kristallogr 89:175-181

Harris DC, Chen TT (1975) Gustavite - two Canadian occurrences. Can Mineral 13:411-414.

Harris DC, Roberts AC, Thorpe RI, Criddle AJ, Stanley CJ (1984) Kiddcreekite, a new mineral species from the Kidd Creek Mine, Timmins, Ontario and from the Campbell orebody, Bisbee, Arizona. Can Mineral 22:227-232

Hazen RM, Finger LW (1978) The crystal structures and compressibilities of layer minerals at high pressure. I. SnS_2, berndtite. Am Mineral 63:289-292

Henriksen RB, Makovicky E, Stipp SLS, Nissen C, Eggleston CM (2002) Atomic-scale observations of franckeite surface morphology. Am Mineral 87:1273-1278

Herbert HK, Mumme WG (1981) Unsubstituted benjaminite from the A W Mine, NSW: a discussion of metal substitutions and stability. N Jahrb Mineral Mh 1981:69-80

Hill RJ, Craig JR, Gibbs GV (1978) Cation ordering in the tetrahedral sites of the thiospinel FeIn$_2$S$_4$ (indite). J Phys Chem Solids 39:1105-1111

Hoistad LM, Meerschaut A, Bonneau P, Rouxel J (1995) Structure determination of a trilayer misfit compound (Gd$_\epsilon$ Sn$_{1-\epsilon}$S)$_{1.16}$(NbS$_2$)$_3$. J Solid State Chem 114:435-441

Hoggins J, Steinfink H (1977) Compounds in the infinitely adaptive series Ba$_P$(Fe$_2$S$_4$)$_Q$: Ba$_9$(Fe$_2$S$_4$)$_8$. Acta Crystallogr B33:673-678

Horiuchi H, Akaogi M, Sawamoto H (1982) Crystal structure studies on spinel-related phases. *In:* High-presssure Research in Geophysics. Akimoto S, Manghnani MH (eds) Reidel Publ Co, p 391-403

Hong HYP, Steinfink H (1972) The crystal chemistry of phases in the Ba-Fe-S and Se systems. J Solid State Chem 5:93-104

Huan G, Greenblatt M (1989) Antiferromagnetic-to-ferromagnetic transition in metallic Tl$_{1-x}$K$_x$Co$_2$Se$_2$ (0 ≤ x ≤ 1.0) with ThCr$_2$Si$_2$-type structure. J Less-Common Metals 156:247-257

Huhma M, Vuorelainem Y, Häkli TA, Papunen H (1973) Haapalaite, a new nickel-iron sulphide of the valleriite type from East Finland. Bull geol Soc Finland 45:103-106

Hulliger F (1963) Marcasite-type semiconductors. Nature 198:1081-1082

Hulliger F (1968) Crystal chemistry of chalcogenides and pnictides of the transition elements. Struct Bonding 4:83-229

Hyde BG, Bagshaw AN, Andersson S, O'Keeffe MO (1974) Some defect structures in crystalline solids. Ann Rev Mat Sci 4:43-92

Iitaka Y, Nowacki W (1962) A redetermination of the crystal structure of galenobismutite, PbBi$_2$S$_4$. Acta Crystallogr 15: 691-698

Ijaali I, Ibers JA (2001) Crystal structure of palladium selenide, PdSe. Z Kristallogr New Cryst St, 216: 485-486

Ioffe PA, Tsemekham LSh, Parshukova LN, Bobkovskii AG (1985) The chemical state of the iron atoms in FeS$_2$, FeAsS and FeAs$_2$. Russian J Inorg Chem 30:1566-1567.

Iordanidis L, Schindler JL, Kannewurf CR, Kanatzidis MG (1999) ALn$_{1+x}$Bi$_{4+x}$S$_8$ (A = K, Rb; Ln = La, Ce, Pr, Nd): New semiconducting quaternary bismuth sulfides. J Solid State Chem 143:151-162

Ito T, Nowacki W (1974) The crystal structure of freieslebenite, PbAgSbS$_3$. Z Kristallogr 139:85-102

Jambor JL (1969) Dadsonite (minerals Q and QM), a new lead sulphantimonide. Mineral Mag 37:437-441

Jellinek F (1957) The structure of chromium sulfides. Acta Crystallogr 10:620-628

Jellinek F (1968) Sulphides. *In:* Inorganic Sulphur Chemistry. Nicless G (ed) Elsevier, p 669-747

Jellinek F, Brauer G, Mueller H (1960) Molybdenum and niobium sulphides. Nature 185:376-377

Jellinek F (1962) The system tantalum-sulfur. J Less-Common Metals 4:9-15

Jobic S, Le Boterf, P, Brec R, Ouvrard G (1994) Structural determination and magnetic properties of a new mixed valence tin chromium selenide: Cr$_2$Sn$_3$Se$_7$. J. Alloys Comp 205:139-145

Jobic S, Bodenan F, Ouvrard G, Elkaim E, Lauriat JP (1995) Structural determination and magnetic properties of a new orthorhobic chromium seleno stannate Cr$_2$Sn$_3$Se$_7$.J Solid State Chem 155:165-173

Johan Z, Kvaček M (1971) La bukovite Cu$_{3+x}$ Tl Fe Se$_{4-x}$, une nouvelle espèce minérale. Bull Soc fr Minéral 94:529-533

Johnson ML, Burnham CW (1985) Crystal structure refinement of an arsenic bearing argentian tetrahedrite. Am Mineral 70:165-170

Johnson NE, Craig JR, Rimstidt JD (1987) Substitutional effects on the cell dimension of tetrahedrite. Can Miner 25:237-244

Julien-Pouzol M, Jaulmes S, Laruelle P (1979) Structure cristalline du sulfure de bismuth et thallium Tl$_4$Bi$_2$S$_5$. Acta Crystallogr B35:1313-1315

Jumas JC, Olivier-Fourcade J, Philippot E, Maurin M (1980) Sur le système SnS-Sb$_2$S$_3$: Etude structurale de Sn$_4$Sb$_6$S$_{13}$. Acta Crystallogr B36:2940-2945

Jumas JC, Philippot E, Maurin M (1979) Structure du rhodostannite synthetique. Acta Crystallogr 35:2195-2197

Kalbskopf R (1972) Strukturverfeinerung des Freibergits. Tschermaks Mineral Petrogr Mitt 18:147-155

Kalbskopf R (1974) Synthese und Kristallstruktur von Cu $_{12-x}$Te$_4$S$_{13}$, dem Tellur-Endglied der Fahlerze. TMPM Tschermaks Mineral Petrogr Mitt 21:1-10

Kalbskopf R (1971) Die Koordination des Quecksilbers im Schwazit. Tschermaks Mineral Petrogr Mitt 16: 173-175

Kanatzidis MG, McCarthy TJ, Tanzer TA, Chen L-H, Iordanidis L, Hogan T, Kannewurf CR, Uher C, Chen B (1996) Synthesis and thermoelectric properties of the new ternary bismuth sulfides KBi$_{6.33}$S$_{10}$ and K$_2$Bi$_8$S$_{13}$. Chem Materials 8:1465-1474

Kanatzidis MG (1990) Molten alkali-metal polychalcogenides as reagents and solvents for the synthesis of new chalcogenide materials. Chem Materials 2:353-363

Kanazawa Y, Koto K, Morimoto N (1978) Bornite (Cu$_5$FeS$_4$): Stability and crystal structure of the intermediate form. Can Mineral 16:397-404

Kaplunnik LN, Pobedimskaya EA, Belov NV (1980) The crystal structure of schwazite ($Cu_{4.4}Hg_{1.6}$) $Cu_6Sb_4S_{12}$. 253:105-107

Karanovic Lj, Cvetkovic Lj, Poleti D, Balić-Žunić T, Makovicky E (2002) Crystal and absolute structure of enargite from Bor (Serbia). N Jahrb Mineral Mh 2002:241-253

Karanović L, Cvetković L, Poleti D, Balić-Žunić T, Makovicky E(2003) Structural and optical properies of schwazite from Dragodol (Serbia). N Jb Miner Mh 2003:503-520

Karup-Møller S, Makovicky E (1979) On pavonite, cupropavonite, benjaminite and "oversubstituted" gustavite. Bull Minéral 102:351-367

Karup-Møller S, Makovicky E (1979) Topotactic replacement of niccolite by rammelsbergite; new data on alloclasite, $Co_{0.56}Ni_{0.45}Fe_{0.01}As_{1.18}S_{0.80}$. N Jb Miner Abh 136:310-325

Karup-Møller S, Makovicky E (1992) Mummeite - A new member of the pavonite homologous series from Alaska Mine, Colorado. N Jahrb Mineral Mh 1992:555-576

Karup-Møller S, Makovicky E (1995) The phase system Fe-Ni-S at 725 °C. N Jb Min Mh 1995:1-10

Karup-Møller S, Makovicky E (1999) Exploratory studies of element substitutions in synthetic tetrahedrite. Part II. Selenium and tellurium as anions in Zn-Fe tetrahedrites. N Jb Miner Mh 1999:385-399

Karup-Møller S, Makovicky E (2003) Exploratory studies of element substitutions in synthetic tetrahedrite Part V. Mercurian tetrahedrite. N Jb Miner Abh 179:73-83

Karup-Møller S, Makovicky E (2004) Exploratory studies of the solubility of minor elements in tetrahedrite VI. Zinc and the combined zinc-mercury and iron-mercury substitutions. N Jb Miner Mh 2004:508-524

Kato K, Kawada I, Takahashi T (1977) Die Kristallstruktur von $LaCrS_3$. Acta Crystallogr B33:3437-3493

Kawada I and Hellner E (1971) Dei Kristallstruktur der Pseudozell (subcell) von Andorit VI (Ramdohrit). N Jahrb Mineral Mh 1971:551-560

Keller-Besrest F, Collin G (1983) Structure and planar faults in the defective NiAs-type compound 3c, Fe_7S_8. Acta Crystallogr B39:296-303

Keller-Besrest F, Collin G (1990) Structural aspects of the alpha transitition in stoichiometric FeS. Identification of the high-temperature phase. J Solid State Chem 84:194-210

Kim S-J, Kim WC, Kim CS (2002) Neutron diffraction and Mössbauer studies on $Fe_{1-x}Cr_2S_4$ (x=0.0, 0.04,.0.08). J Appl Phys 91:7935-7937

King HEjr, Prewitt CT (1982) High-pressure and high-temperature polymorphism of iron sulfide. Acta Crystallogr B38:1877-1887

Kissin SA (1974) Phase relations in a portion of the Fe-S system. PhD Dissertation, Univ Toronto

Kjekshus A, Peterzons PG, Rakke T, Andresen AF (1979) Compounds with the marcasite type crystal structure. XIII. Structural and magnetic properties of $Cr_tFe_{1-t}As_2$, $Cr_tFe_{1-t}Sb_2$, $Fe_{1-t}Ni_tAs_2$ and $Fe_{1-t}Ni_tSb_2$. Acta Chem Scand A33:469-480

Kjekshus A, Rakke T (1979) Structural transformations in $Co_tNi_{1-t}As_2$, $NiAs_{2-x}S_x$, $NiAs_{2-x}Se_x$, and $CoAs_{1-x}Se_{1+x}$. Acta Chem Scand A33:609-615

Klepp KO (1993) $Tl_2Pt_5S_6$ - a new thioplatinate with a channel-type structure. J Alloys Comp 196:25-28

Klepp KO, Boller H (1978) Ternäre Thallium-Übergangsmetall-Chalkogenide mit $ThCr_2Si_2$ Struktur. Mh Chemie 109:1049-1057

Klepp KO, Boller H, Völlenkle H (1980) Neue Verbindungen mit KCu_4S_3-Struktur. Mh Chemie 111:727-733

Klepp KO, Boller H (1981) $Na_3Fe_2S_4$ ein Thioferrat mit gemischt valenter (FeS_2) – kette. Mh Chemie und verw Teile anderer Wiss 112:83-89

Klepp KO, Pantschov S, Boller H (2000) Crystal structure of mixed-valent trirubidium tetraselenidoferrate $Rb_3Fe_2Se_4$. Z Kristallogr – New cryst struct 215:5-6

Klünder-Hansen M, Makovicky E, Karup-Møller S (2003) Exploratory studies on substitutions in tetrahedrite-tennantite solid solution. Part IV. Substitution of germanium and tin. N Jb Miner Abh 179:43-71

Klünder-Hansen M, Karup-Møller S, Makovicky E (2003) Exploratory studies on substitutions in tetrahedrite-tennantite solid solution. Part III. The solubility of bismuth in tetrahedrite –tennantite containing iron and zinc. N Jb Miner Mh 2003: 153-175

Knop O, Reid KIG, Sutarno R, Nakagawa Y (1968) Chalkogenides of the transition elements VI X-Ray neutron and magnetic investigation of the spinels Co_3O_4, $NiCo_2O_4$, Co_3S_4, $NiCo_2S_4$. Can J Chem 46:3463-3476

Kocman V, Nuffield EW (1973) The crystal structure of wittichenite, Cu_3BiS_3. Acta Crystallogr B29:2528-2535

Kocman V, Nuffield EW (1974) Crystal structure of antimonian hauchecornite from Westphalia. Can Mineral 12:269-274

Koděra M, Kupčik V, Makovicky E (1970) Hodrushite - a new sulphosalt. Mineral Mag 37:641-648

Konnert JA, Evans HT jr. (1980) The crystal structure of erdite $NaFeS_2$ (H_2O)$_2$. Am Mineral 65:516-521

Konopka D, Kozlowska I, Chelkowski A (1973) X-ray investigations of spinel structure compounds of the $HgCr_2$ (Se_x S_{1-x})$_4$ type. Phys Lett A44:289-290

Kostov VV, Macíček J (1995) Crystal structure of synthetic $Pb_{12.65}Sb_{11.35}S_{28.35}Cl_{2.65}$ - A new view of the crystal chemistry of chlorine-bearing lead-antimony sulphosalts. Eur Mineral 7:1007-1018

Kostov I, Minčeva Stefanova J (1982) Sulphide Minerals. Crystal Chemistry, Parageneses and Systematics. E.Schweizerbartsche Verlagsbuchhandlung.

Koto K, Morimoto N (1975) Superstructure investigation of bornite Cu_5FeS_4 by the modified partial Patterson function. Acta Crystallogr B31:2268-2273

Koto K, Morimoto N (1970) The crystal structure of anilite. Acta Crystallogr 26:915-924

Koto K, Morimoto N, Gyobu A (1975) The superstructure of the intermediate pyrrhotite. I. Partially disordered distribution of metal vacancy in the 6C type, $Fe_{11}S_{12}$. Acta Crystallogr B31:2759-2769

Kovalenker VA, Laputina IP, Yevstigneyeva TL, Izoitko VM (1976) Thalcusite, $Cu_{3-x}Tl_2Fe_{1-x}S_4$, a new sulfide of thallium from copper – nickel ores of the Talnakh Deposit. Zap Vsez Min Obshch 105:202-206

Krämer V (1980) Structure of bismuth indium sulphide $Bi_3In_5S_{12}$. Acta Crystallogr B 36:1922-1923

Krämer V (1983) Lead indium bismuth chalcogenides III. Structure of $Pb_4In_2Bi_4S_{13}$. Acta Crystallogr C42: 1089-1091

Krämer V, Reis I (1986) Lead indium bismuth chalcogenides. II. Structure of $Pb_4In_3Bi_7S_{18}$. Acta Crystallogr C42:249-251

Kudoh Y, Takeuchi Y (1976) The superstructure of stannoidite. Z Kristallogr 144:145-160

Kupčík V (1967) Die Kristallstruktur des Kermesits, Sb_2S_2O. Naturwissenschaften 54:114-115

Kupčík V, Makovicky E (1968) Die Kristallstruktur des Minerals (Pb, Ag, Bi) $Cu_3Bi_5S_{11}$. N Jahrb Mineral Mh 236-237.

Kupčík V, Steins M (1991) Verfeinerung der Kristallstruktur von Gustavit $Pb_{1.5}Ag_{0.9}Bi_{2.5}Sb_{0.1}S_6$. Berichte Deutsch Mineral Gesellschaft 1990/2. 151

Kusawake T, Takahashi Y, Oshima K-I (2000) Structural analysis of the layered compounds Cu_xTiS_2. Molec Cryst Liquid Cryst 341:93-98

Kutoglu A, Allmann R (1972) Strukturverfeinerung des Patronits, $V(S_2)_2$. N Jahrb Mineral Mh 1972:339-345

Kyono A, Kimata M, Matsuhisa M, Miyashita Y, Okamoto K (2002) Low-temperature crystal structures of stibnite implying orbital overlap of Sb $5is^2$ inert pair electrons. Phys Chem Mineral 29:254-260

Kyono A, Kimata M (2004) Structural variations induced by difference of the inert pair effect in the stibnite – bismuthinite solid solution series $(Sb,Bi)_2S_3$. Am Mineral 89:932-940

Kyono A, Kimata M, Hatta T (2005) Light-induced degradation dynamics in realgar: in situ-structural investigation using single-crystal X-ray diffraction study and X-ray photoelectron spectroscopy. Am Mineral 90:1563-1570

Lafond A, Deudon C, Meerschaut A, Palvadeau P, Moelo Y, Briggs A (1999) Structure determination and physical properties of the misfit layered compound Pb_2FeS_3 0.58 NbS_2. J Solid State Chem142:461-469

Landa-Canovas A R, Otero-Diaz L C (1992) A trasmission electron microscopy study of the $MnS-Er_2S_3$ System. Austral J Chem 45:1473-1487

Lavina B, Salviulo G, Della Giusta A (2002) Cation distribution and structure modelling of spinel solid solutions. Phys Chem Mineral 29:10-18

le Blanc A, Rouxel J (1972) Sur les types structuraux des composés intercalaires $MSnS_2$ (M = Li, Na, K, Rb). Comptes Rendus Acad Sci, Serie C, Sci Chimi 274:786-788

Lemoine P, Carré D, Guittard M (1986) Structure de sulfure d'europium et de cuivre Eu_2CuS_3. Acta Crystallogr 42: 390-391

Lemoine P, Carré D, Guittard M (1986) Structure du sulfure d'europium et de bismuth $Eu_{1.1}Bi_2S_4$. Acta Crystallogr C42:259-261

le Nagard N, Gorochov O, Collin G (1974) Structure cristalline et proprietes physiques de Cu_xTiS_2. Materials Res Bull 10:1287-1296

Lennie AR, Redfern SAT, Schofield PF, Vaughan DJ (1995) Synthesis and Rietveld crystal structure refinement of mackinawite, tetragonal FeS. Mineral Mag 59:677-683.

Leon M, Merino JM, de Vidales JLM (1992) Crystal structure of synthesized $CuGaTe_2$ determined by X-ray powder diffraction using the Rietveld method. J Materials Sci 27:4495-4500

Léone P, Le Leuch L-M, Palvadeau P, Molinié P, Moëlo Y (2003) Single crystal structures and magnetic properties of two iron- or manganese-lead-antimony sulfides: $MPb_4Sb_6S_{14}(M:Fe,Mn)$. Solid State Sci 5: 771-776

Lima-de-Faria J, Hellner E, Liebau F, Makovicky E, Parthé E (1990) Nomenclature of inorganic structure types. Report of the IUCr Commission on Crystallographic Nomenclature, Subcommittee on the Nomenclature of Inorganic Structure Types. Acta Crystallogr A46:1-11

Llanos J, Tapia M, Mujica C, Oro-Sole J, Gomez-Romero P (2000) A new structural modification of stannite. Bol Soc Chilena Quimica 45:605-609

Luedecke J, van Smaalen S, Spijkerman A, de Boer JL, Wiegers GA (1999) Commensurately modulated structure of 4Hb- (Ta Se2) determined by X-ray crystal structure refinement. Phys Rev, Serie 3. B - Condensed Matter 59:6063-6071

Lugakov NF, Movchanskiy EA, Pokrovskiy II (1975) Self-diffusion of copper and ion conductivity in Cu_3BiS_3. Izv Akad Nauk BSSR, Ser Khim 3:42-44

Lundegaard LF, Miletich R, Balić-Žunić T, Makovicky E (2003) Equation of state and crystal structure of Sb_2S_3 between 0 and 10 GPa. Phys Chem Min 30:463-468

Lundegaard LF, Makovicky E, Boffa-Ballaran T, Balić-Žunić T (2005) Crystal structure and cation lone electron pair activity of Bi_2S_3 between 0 and 10 GPa. Phys Chem Min 32:578-584

Lundqvist D (1947) X-ray studies on the binary system Ni-S. Arkiv foer Kemi, Mineral Geol 24:1-12

Lutz HD, Jung M, Waeschenbach G (1987) Kristallstrukturen des Loellingits $FeAs_2$ und des Pyrits $RuTe_2$. Z anorg allg Chem 554:87-91

Lyman PS, Prewitt CT (1984) Room- and high-pressure crystal chemistry of CoAs and FeAs. Acta Crystallogr B40:14-20

Machatschki F (1928) Formel und Kristallstruktur des Tetraedrites. Norsk Geol Tidsskr 10:23

Machatschki F (1928) Praezisionmessungen der gitterkonstanten verschiedener Fahlerze Formel und Struktur derselben. Z Kristallogr 68:204-222

Maciček J (1986) The crystal chemistry of cosalite. Coll Abstr Xth Eur Cryst Meeting Wroclaw, 260

Makeyev AB, Evstigneeva TL, Troneva NV, Vyalsov LN, Gorshkov AI, Trubkin NV (1984) Yushkinite $V_{1-x}S.n[(Mg, Al)(OH)_2]$ – a new mineral. Mineral Zhurnal 6:91-97

Makovicky E (1976) Crystallography of cylindrite. I. Crystal lattices of cylindrite and incaite. N Jahrb Mineral Abh 126:304-326

Makovicky E (1981) The building principles and classification of bismuth-lead sulphosalts and related compounds. Fortschr Mineral 59:137-190

Makovicky E (1985) The building principles and classification of sulphosalts based on the SnS archetype. Fortschr Mineral 63:45-89

Makovicky E (1989) Modular classification of sulphosalts - current status. Definition and application of homologous series. N Jahrb Mineral Abh 160:269-297

Makovicky E (1993) Rod-based sulphosalt structures devided from the SnS and PbS archetype. Eur J Mineral 5:545-591

Makovicky E (1994) Polymorphism in Cu_3SbS_3 and Cu_3BiS_3: The ordering schemes for copper atoms and electron microscope observations. N Jahrb Mineral Abh 168:185-212

Makovicky E (1997a) Modular crystal chemistry of sulphosalts and other complex sulphides. Europ Mineral Union Notes Mineral 1:237-271

Makovicky E (1997b) Modularity - different types and approaches. Europ Mineral Union Notes Mineral 1: 315-343

Makovicky E, Balić-Žunić T (1993) Contributions to the crystal chemistry of thallium sulphosalts. II. $TlSb_3S_5$ - the missing link of the lillianite homologous series. N Jahrb Mineral Abh 165:331-344

Makovicky E, Balić-Žunić T (1995) The crystal structure of skinnerite, $P2_1/c$-Cu_3SbS_3, from powder data. Can Mineral 33:655-663

Makovicky E, Balić-Žunić T(1999) Gillulyite $Tl_2(As, Sb)_8S_{13}$: Reinterpretation of the crystal structure and order-disorder phenomena. Am Mineral 84:400-406

Makovicky E, Balić-Žunić T, Topa D (2001) The crystal structure of neyite, $Ag_2Cu_6Pb_{25}Bi_{26}S_{68}$. Can Mineral 39:1365-1376

Makovicky E (2002) Experimental studies of palladium containing systems and compounds. Bol Soc Españ Miner 25:5-37

Makovicky E (2005) Micro and mesoporous sulfide and selenide structures. Rev Mineral Geochem 57:403-433

Makovicky E, Mumme WG, Topa D (in press) The crystal structure of dadsonite. Can Mineral, in press

Makovicky E, Balić-Žunić T, Karanović L, Poleti D, Pršek J (2004) Structure refinement of natural robinsonite, $Pb_4Sb_6S_{13}$: Cation distribution and modular description. N Jahrb Mineral 2004:49-67

Makovicky E, Forcher K, Lottermoser W, Amthauer D (1990) The role of Fe^{2+} and Fe^{3+} in synthetic Fe-substituted tetrahedrite. Miner Petrol 43:73-81

Makovicky E, Hyde BG (1981) Non-commensurate (misfit) layer structures. Struct Bonding 46:101-170

Makovicky E, Hyde BG (1992) Incommensurate, two-layer structures with complex crystal chemistry: minerals and related synthetics. *In*: Incommensurate Misfit Sandwiched Layered Compounds. Meerschaut A (ed) Materials Sci Forum 100-101:1-100 Trans. Tech Publ. Ltd.

Makovicky E, Karanović L, Poleti D, Balić-Žunić T (2005) Crystal structure of copper-rich unsubstituted tennantite, $Cu_{12.5}As_4S_{13}$. Can Miner 43:679-688

Makovicky E, Johan Z, Karup-Møller S (1980) New data on bukovite, thalcusite, chalcothallite and rohaite. N Jahrb Mineral Abh 138:122-146

Makovicky E, Karup-Møller S (1977 a) Chemistry and crystallography of the lillianite homologous series. I. General properties and definitions. N Jahrb Mineral Abh 130:264-287

Makovicky E, Karup-Møller S (1977 b) Chemistry and crystallography of the lillianite homologous series. Part II: Definition of new minerals: eskimoite, vikingite, ourayite and treasurite. Redefintion of schirmerite and new data on the lillianite - gustavite solid solution series. N Jahrb Mineral Abh 131:56-82

Makovicky E, Karup-Møller S (1984) Ourayite from Ivigtut, Greenland. Can Mineral 22:565-575
Makovicky E, Karup-Møller S (1994) Exploratory studies on substitution of minor elements in synthetic tetrahedrite. Part I. Substitution by Fe, Zn, Co, Ni, Mn, Cr, V and Pb. Unit-cell parameter changes on substitution and the structural role of "Cu^{2+}." N Jb Miner Abh 167:89-123
Makovicky E, Leonardsen E, Moëlo Y (1994) The crystallography of lengenbachite, a mineral with the non-commensurate layer structure. N Jahrb Mineral Abh 166:169-191
Makovicky E, Mumme WG, Watts JA (1977) The crystal structure of synthetic pavonite, $AgBi_3S_5$ and the definition of the pavonite homologous series. Can Mineral 15:339-348
Makovicky E, Mumme WG (1979) The crystal structure of benjaminite $Cu_{0.50}Pb_{0.40}Ag_{2.30}Bi_{6.80}S_{12}$. Can Mineral 17:607-618
Makovicky E, Mumme WG (1983) The crystal structure of ramdohrite, $Pb_6Sb_{11}Ag_3S_{24}$ and its implications for the andorite group and zinckenite. N Jahrb Mineral Abh 147:58-79
Makovicky E, Mumme WG (1984) The crystal structures of izoklakeite, dadsonite and jaskolskiite. 15th Int. Congr Crystallog., Coll Abstr, Hamburg: C-246
Makovicky E, Mumme WG (1986) The crystal structure of isoklakeite, $Pb_{51.3}Sb_{20.4}Bi_{19.5}Ag_{1.2}Cu_{2.9}Fe_{0.7}S_{11.4}$. The kobellite homologous series and its derivatives. N Jahrb Mineral Abh 153:121-148
Makovicky E, Mumme WG, Hoskins BF (1991) The crystal structure of Ag-Bi bearing heyrovskyite. Can Mineral 29:553-558
Makovicky E, Mumme WG, Madsen IC (1992) The crystal structure of vikingite. N Jahrb Mineral Mh 454-468
Makovicky E, Norrestam R (1985) The crystal structure of jaskolskiite, $Cu_x Pb_{2+x}$ (Sb, Bi)$_{2-x}$ S_5 (x ≈ 0.2), a member of the meneghinite homologous series. Z Kristallogr 171:179-194
Makovicky E, Skinner BJ (1975) Studies of the sulfosalts of copper. IV. Structure and twinning of sinnerite, $Cu_6As_4S_9$. Am Mineral 60:998-1012
Makovicky E, Skinner BJ (1979) Studies of the sulfosalts of copper. VII. Crystal structures of the exsolution products $Cu_{12.3}Sb_4S_{13}$ and $Cu_{13.8}Sb_4S_{13}$ of unsubstituted synthetic tetrahedrite. Can Miner 17:619-634
Makovicky E (1993) Rod-based sulphosalt structures derived from the SnS and PbS archetype. Eur J Miner 5:545-591
Makovicky E (1983) The phase transformations and thermal expansion of the solid electrolyte Cu_3BiS_3 between 25 and 300 °C. J Solid State Chem 49:85-92
Makovicky E, Søtofte I, Karup-Møller S (2002): The crystal structure of $Cu_4Bi_4Se_9$. Z Kristallogr 217:597-604
Makovicky E, Søtofte I, Karup-Møller S (2006) The crystal structure of $Cu_{1.78}Bi_{4.73}S_8$, an N=3 pavonite homologue with a Cu-for-Bi substitution. Z Kristallogr 221:122-127
Makovicky E, Tippelt G, Forcher K, Lottermoser W, Karup-Møller S, Amthauer G (2003) Mössbauer study of Fe-bearing synthetic tennantite. Can Mineral 41:1125-1134
Makovicky E, Topa D, Balić-Žunić T (2001) The crystal structure of paarite, the newly discovered 56 Å derivative of the bismuthinite-aikinite solid-solution series. Can Mineral 39:1377-1382
Mardix S (1986) Polytypism: A controlled thermodynamic phenomenon. Phys Rev Ser 3B Cond Matter 33: 8677-8684
Mariolacos K, Kupčík V, Ohmasa M, Miehe G (1975) The crystal structure of $Cu_4Bi_5S_{10}$ and its relation to the structures of hodrushite and cuprobismutite. Acta Crystallogr B31:703-708
Marumo F (1967) The crystal structure of nowackiite, $Cu_6Zn_3As_4S_{12}$. Z Kristallogr 124:352-368
Marumo F, Nowacki W (1967) A refinement of the crystal structure of luzonite, Cu_3AsS_4. Z Kristallogr 124: 1-8
Matković P, El-Boragy, M, Schubert K (1976) Kristallstruktur von $Pd_{16}S_7$. J Less Common Metal, 50:65-176
Matsushita Y, Ueda Y (2003) Structure and physical properties of 1D magnetic chalcogenide, jamesonite $FePb_4Sb_6S_{14}$. Inorg Chem 42:7830-7838
Matsushita Y (2005) Chalcogenide structure data base version 4.3M. (downloadable database)
Matzat E (1972) Die Kristallstruktur des Wittichenits, Cu_3BiS_3. Tschermaks Mineral-Petrogr Mitt 18:312-316
Matzat E (1979) Cannizzarite. Acta Crystallogr B35:133-136
Matsushita Y, Takéuchi Y (1994) Refinement of the crystal structure of hutchinsonite, $TlPbAs_5S_9$. Z Kristallogr 209:475-478
McCammon C, Zhang J, Hazen RM, Finger LW (1992) High-pressure crystal chemistry of cubanite, $CuFe_2S_3$. Am Mineral 77:937-944
McKelvy MJ, Wiegers GA, Dunn JM, Young VG Jr, Glaunsinger W.S. (1990): Structural investigation of the ammonnium intercalates of titanium and niobium disulfides. Solid State Ionics 38:163-170
Meerschaut A, Deudon C (2001) Crystal structure studies of the 3R-$Nb_{1.09}S_2$ and the 2H-$NbSe_2$ compounds: correlation between nonstoichiometry and stacking type (= polytypism). Mater Res Bull 36:1721-1727

Meerschaut A, Guemas L, Auriel C, Rouxel J (1990) Preparation structure determination and transport properties of a new misfit layer compound $PbS_{1.14}(NbS_2)_2$. Eur J Solid State Inorg Chem 27:557-570

Meerschaut A, Palvadeau P, Moëlo Y, Orlandi P (2001) Lead-antimony sulfosalts from Tuscany (Italy). IV. Crystal structure of pillaite, $Pb_9Sb_{10}S_{23}ClO_{0.5}$, an expanded monoclinic derivative of hexagonal $Bi(Bi_2S_3)_9I_3$, from the zinkenite group. Eur J Mineral 13:779-790

Meetsma A, Wiegers GA, Haange RJ, de Boer JL (1990) Structure of $2H$-TaS_2. Acta Crystallogr C46:1598-1599

Megaw HD (1973) Crystal Structures: A Working Approach. W.B. Saunders Co.

Merlino S (1988) Average and real structures in minerals. Z Kristallogr 185:13-14

Merlino S, Pasero M (1997) Polysomatic approach in the crystal chemical study of minerals. EMU Notes Mineral 1:297-312

Miehe G (1971) Crystal structure of kobellite. Nature Phys Sci 231:133-134

Miehe G, Kupčík V (1971) Die Kristallstruktur des $Bi(Bi_2S_3)_9J_3$. Naturwiss 58:219-220

Mizota T, Inove A, Yamada T, Nakatsuka A, Nakayama N (1998) Ionic conduction and thermal nature of synthetic Cu_3BiS_3. Min J (Japan) 20:81-90

Moëlo Y, Palvadeau P, Meisser N, Meerschaut A (2002) Structure cristalline d'une meneghinite naturelle pauvre en cuivre $Cu_{0.58}Pb_{12.72}(Sb_{7.04}Bi_{0.24})S_{24}$. Comptes Rendus Geosci 334:529-536

Moëlo Y (1979) Quaternary compounds in the system Pb-Sb-S-Cl: dadsonite and synthetic phases. Can Mineral 17:595-600

Moëlo Y (1983) Contribution à l'étude des conditions naturelles de formation des sulfures complexes d'antimoine et plomb (sulfosels de Pb/Sb). Signification métallogènique. Série "Documents du B.R.G.M.," Orléans 55 : 624 p

Moëlo Y, Makovicky E, Karup-Møller S (1988) Sulfures complexes plombo-argentifères: Minéralogie et cristallochimie de la série andorite-fizelyite (Pb, Mn,Fe, $Cd,Sn)_{3-2x}$ $(Ag,Cu)_x(Sb,Bi,As)_{2+x}(S,Se)_6$. Documents BRGM (Orlèans) 167:107 pp

Moëlo Y, Makovicky E, Karup-Møller S, Corvelle B, Maurel C (1990) La lévyclaudite, $Pb_8Sn_7Cu_3(Bi,Sb)_3S_{28}$, une nouvelle espèce à structure incommensurable, de la série de la cylindrite. Eur J Mineral 2:711-723

Moëlo Y, Meerschaut A, Rouxel J, Auriel C (1995) Precise analytical characterization of incommensurate sandwiched layered compounds $[(Pb,Sn)S]_{1+x}[(Nb,Ti)S_2]_m$ ($0.08<x<0.28$), $m=1-3$). Role of cationic coupling on the properties and the structural modulation. Chem Mater 7:1759-1771

Moëlo Y, Meerschaut A, Orlandi P, Palvadeau P (2000) Lead-antimony sulfosalts from Tuscany (Italy): II- Crystal structure of scainiite, $Pb_{14}Sb_{30}S_{54}O_5$, an expanded monoclinic derivative of $Ba_{12}Bi_{24}S_{48}$ hexagonal sub-type (zinkenite group). Eur J Mineral 12:835-846

Moëlo Y, Oudin E, Picot P, Caye R (1984) L'uchucchacuaite, $AgMnPb_3Sb_5S_{12}$, une nouvelle espèce minérale de la série de l'andorite. Bull Minéral 107:597-604

Moëlo Y, Makovicky E (in press) Revision of sulfosalt definition and nomenclature. Report COM IMA

Morimoto N, Kullerud G (1961) Polymorphism in bornite. Am Mineral 46:1270-1282

Morimoto N (1964) Structures of two polymorphic forms of Cu_5FeS_4. Acta Crystallogr 17:351-360

Morimoto N, Koto K (1970) Phase relations of the Cu-S system at low temperatures: stability of anilite. Am Mineral 55:106-117

Morosin B (1974) Structure refinement on NbS_2. Acta Crystallogr B30:551-552

Mujica C, Carvajal G, Llanos J, Wittke O (1998) Redetermination of the crystal structure of copper (I) tetrathiovanadate (sulvanite), Cu_3VS_4. Z Kristallogr – New Crystal Struct 213:p 12

Mujica C, Paez J, Llanos J (1994) Synthesis and crystal structure of layered chalcogenides $KCuFeS_2$ and $KCuFeS_2$. Mater Res Bull 29:263-268

Mullen DJE, Nowacki W (1972) Refinement og the crystal structures of realgar AsS and orpiment As_2S_3. Z Kristallogr 136:48-65

Müller U (1996) Anorganische Strukturchemie. B.G. Teubner

Mumme WG (1975) Junoite, $Cu_2Pb_3Bi_8(S,Se)_{16}$, a new sulfosalt from Tennant Creek, Australia: Its crystal structure, and relationship with other bismuth sulfosalts. Am Mineral 60:548-558

Mumme WG (1980) Weibullite, $Ag_{0.32}Pb_{5.02}Bi_{8.55}Se_{6.08}S_{11.92}$ from Falun, Sweden. A higher homologue of galenobismutite. Can Mineral 18:1-18

Mumme WG (1980c) The crystal structure of nordströmite, $CuPb_3Bi_7(S,Se)_{14}$ from Falun, Sweden: A member of the junoite homologous series. Can Mineral 18:343-352

Mumme WG (1986) The crystal structure of paderaite, a mineral of the cuprobismutite series. Can Mineral 24:513-521

Mumme WG (1989) The crystal structure of $Pb_{5.05}(Sb_{3.75}Bi_{0.28})Se_{10.72}Se_{0.28}$: boulangerite of near ideal composition. N Jahrb Mineral Mh 1989:498-512

Mumme WG (1990) A note on the occurrence, composition and crystal structures of pavonite homologous series members ^4P, ^6P, and ^8P. N Jahrb Mineral Mh 1990:193-204

Mumme WG, Niedermayr G, Kelly PR, Paar WH (1983) Aschamalmite, $Pb_{5.92}Bi_{2.06}S_9$, from Untersulzbach Valley in Salzburg, Austria - "monoclinic heyrovskyite." N Jahrb Mineral Mh 1983:433-444

Mumme WG, Watts JA (1980) $HgBi_2S_4$: Crystal structure and relationship with the pavonite homologous series. Acta Crystallogr B 36:1300-1304

Mumme WG, Welin E, Wuensch BJ (1976) Crystal chemistry and proposed nomenclature for sulfosalts in the system bismuthinite-aikinite (Bi_2S_3-$CuPbBiS_3$). Am Mineral 61:15-20

Nakajima S (1963) The crystal structure of $Bi_2Te_{3-x}Se_x$. J Phys Chem Solids 24:479-485

Nakayama N, Kosuge K, Kachi S (1981) Studies on the compounds in Ba-Fe-S system (III). Phase relations of $Ba_{1+x}Fe_2S_4$ with infinitely adaptive structure. J Solid State Chem 36:9-19

Navrotsky A(1994) Physics and Chemistry of Earth Materials. Cambridge Univ Press

Niizeki W, Buerger MJ (1957) The crystal structure of jamesonite, $FePb_4Sb_6S_{14}$. Z Kristallogr 109:161-183

Noël H, Padiou J (1976) Structure cristalline de $FeUS_3$. Acta Crystallogr B32:1593-1595

Nowack E, Schwarzenbach D, Hahn T (1991) Charge densities in CoS_2 and NiS_2 pyrite structure. Acta Crystallogr B47:650-659

Nowacki W (1982) Isotypy in aktashite $Cu_3Hg_3As_4S_{12}$ and nowackiite $Cu_6Zn_3As_4S_{12}$. Kristallografiya 27: 49-50

Nuffield EW (1952) Studies of mineral sulpho-salts: XVI-cuprobismutite. Am Mineral 37:447-452

Nuffield EW (1980) Cupropavonite from Hall's Valley, Park Country, Colorado. Can Mineral 18:181-184

O'Keefe M, Hyde BG (1996) Crystal Structures. I. Patterns and Symmetry. Mineralogical Society of America

Ohmasa M (1973) The crystal structure of $Cu_{2+x}Bi_{6-x}S_9$ (x = 1.21). N Jahrb Mineral Mh 1973:227-233

Ohmasa M, Nowacki W (1970) A redetermination of the crystal structure of aikinite [$BiS_2|S|Cu^{IV}Pb^{VIII}$]. Z Kristallogr 137:422-432

Ohmasa M, Nowacki (1973) The crystal structure of synthetic $CuBi_5S_8$. Z Kristallogr 137:422-432

Oledzka M, Lee J-G, Ramanujacheri KV, Greenblatt M (1996) Synthesis and characterization of quaternary sulfides with $ThCr_2Si_2$-type structures: $KCo_{2-x}Cu_xS_2$ (0.5 L × L 1.5) and $ACoCuS_2$ (A = K, Rb, Cs). J Solid State Chem 127:151-160

Oliveria M, McMullan RK, Wuensch BJ (1988) Single crystal neutron diffraction analysis of cation distribution in the high-temperature phases alpha-$Cu_{2-x}S$, alpha-$Cu_{2-x}Se$ and alpha Ag_2Se. Solid State Ionics 28-30: 1332-1337

Olivier-Fourcade J, Maurin M, Philippot E (1983) Étude cristallochimique du système Li_2S-Sb_2S_3. Revue Chim Minérale 20:196-217

Omloo WP, Jellinek F(1970) Intercalation compounds of alkali metals with niobium and tantalum dichalcogenides. J Less-Common Metals 20:121-129

Onoda M, Kato K, Gotoh Y, Oosawa Y (1990) Structure of the incommensurate composite crystal $PbS_{1.12}VS_2$. Acta Crystallogr 46:487-492

Organova NI, Drits VA, Dmitrik AL (1972) Strutural study of tochilinite Part I The isometric variety. Kristallografiya 17:761-767

Organova NI, Drits VA, Dmitrik AL (1973) Strutural study of tochilinite PartII Acicular variety. Unusual diffraction patterns. Kristallografiya 18:966-972

Organova NI, Drits VA, Dmitrik AL (1974) Selected area diffraction study of a type II valleriite like mineral. Am Mineral 59:190-200

Organova NI (1989) Crystal Chemistry of Incommensurate and Modulated Mixed-layer Minerals. Nauka, Moscow

Orlandi P, Meerschaut A, Palvadeau P, Merlino S (2002) Lead-antimony sulfosalts from Tuscany (Italy). V. Definition and crystal structure of moëloite, $Pb_6Sb_6S_{14}(S_3)$, a new mineral from the Ceragiola marble quarry. Eur J Mineral 14:599-606

Otto HH, Strunz H (1968) Zur Kristallchemie synthetischer Blei-Wismut-Spiessglanze, N Jahrb Mineral Abh 108:1-19

Ozawa T, Nowacki W (1975) The crystal structure of, and the bismuth-copper distribution in synthetic cuprobismutite. Z Kristallogr 142:161-176

Ozawa T, Tachikawa O (1996) A transmission electron microscope observation of 138 Å period in Pb-As-S sulfosalts. Mineral J 18:97-101

Ozawa T, Takéuchi Y (1993) X-ray and electron diffraction study of sartorite - A periodic antiphase boundary structure and polymorphism. Mineral J 16:358-370

Paar WH, Topa D, Makovicky E, Culetto FY (2005) Milotaite, PbSbSe, a new palladium mineral species from Předbořice, Ccech Republic. Can Mineral 43:689-694

Parise JB (1980) Structure of Heazlewoodite Ni_3S_2. Acta Crystallogr 36:1179-1180

Parise JB, Smith PPK, Howard CJ (1984) Crystal structure refinement of $Sn_3Sb_2S_6$ by high-resolution neutron powder diffraction. Mater Res Bull 19:503-508

Parthé E (1990) Elements of Inorganic Structural Chemistry, A Course on Selected Topics. Publ K Sutter Parthé, Petit-Lancy

Pašava J, Pertlik F, Stumpfl EF, Zemann J (1989) Bernardite, a new thallium arsenic sulphosalt from Allchar, Macedonia, with a determination of the crystal structure. Mineral Mag 53:531-538

Pattrick RAD, Hall AJ (1983) Silver substitution into synthetic zinc, cadmium and iron tetrahedrites Mineral Mag 47:441-450

Pauling L(1927) Electronic polarizabilities. Proc Roy Soc (London) A114:181

Pauling L, Neumann EW (1934) The crystal structure of binnite $(Cu,Fe)_{12}As_4S_{13}$, and the chemical composition and structure of minerals of the tetrahedrite group. Z Kristallogr 88:54-62

Pertlik F (1984) Kristallchemie natuerlicher Telluride I Verfeinerung der Kristallstruktur des Sylvanits $AuAgTe_4$.TMPM. Tscher Mineral Petrogr Mitt 33:203-212

Pertlik F (1984) Crystal chemistry of natural tellurides II Redetermination of the crystal structure of Krennerite $Au_{1-x}Ag_x Te_2$ with x about 0.2. v TMPM. Tscher Mineral Petrogr Mitt 33:253-262

Pertlik F (1994) Kristallstrukturbestimmung der monoklinen Hochtemperaturmodi-fikation von AsS (alpha-AsS), Oesterreische Akad Wiss, Math-Naturwiss Klasse, Sitzungsberichte 131:3-5

Peters J, Krebs B (1982) Silicon disulphide and silicon diselenide: A reinvestigation. Acta Crystallogr B38: 1270-1272

Peterson RC, Miller I (1986) Crystal structure and cation distribution in freibergite and tetrahedrite. Min Magazine 50:717-721

Petkov V, Billinge SJL, Larson P, Mahanti SD, Vogt T, Rangan KK, Kanatzidis MG (2002) Structure of nanocrystalline materials using atomic pair distribution function analysis: Study of $LiMoS_2$. Phys Rev 3B Cond Matt 65:0921051-0921054

Petříček V, Cisařová I, de Boer JL, Zhou W, Meetsma A, Wiegers A, van Smaalen S (1993) The modulated structure of the commensurate misfit-layer $(BiSe)_{1.09} TaSe_2$. Acta Crystallogr B49:258-266

Petrova IV, Kaplunnik LN, Bortnikov NS, Pobedimskaya YeA, Belov NV (1978) The crystal structure of synthetic robinsonite. Dokl Akad Nauk SSSR 241:88-90

Petrova IV, Bortnikov NS, Pobedimskaya Ye A, Belov NV (1979) The crystal structure of a new synthetic Pb,Sb-sulphosalt. Dokl Akad Nauk SSSR 244:607-609

Petrova IV, Pobedimskaya EA, Brygzgalov IA (1988) Crystal structure of miharaite $Cu_4FePbBiS_6$. Dokl Akad NaukSSSR 299:123-127- Soviet Physics Doklady 33:157-159

Petrova I V, Kuznetsov A I, Belokoneva Ye L, Simonov M A, Pobedimskaya Ye A, Belov NV (1978) On the crystal structure of boulangerite (in Russ.). Dokl Akad Nauk SSSR 242:337-340

Pfitzner A (1994) Cu_3SbS_3: Zur Kristallstruktur und Polymorphie. Z anorg allg Chemie 620:1992-1997

Pfitzner A (1995) Cu_3SbSe_3: Synthese und Kristallstruktur. Z anorg allg Chemie 621: 685-688

Pfitzner A (1997) Die Präparative Anvendung der Kupfer (I)-halogenid-Matrix zur Synthese neuer Materialen. Habilitationschrift Universität Siegen.

Pfitzner A (1998) Disorder of Cu^+ in Cu_3SbS_3: Structural investigations of the high- and low-temperature modification. Z Kristallogr 213:228-236

Pfitzner A, Bernert T (2004) The system Cu_3AsS_4-Cu_3SbS_4 and investigations on normal tetrahedral structures. Z Kristallogr 219:20-26

Pfitzner A, Evain M, Petříček V (1997) $Cu_{12}Sb_4S_{13}$ a temperature-dependent structure investigation. Acta Crystallogr 53:337-345

Pfitzner A, Reiser S (2002) Refinement of the crystal structures of Cu_3PS_4 and Cu_3SbS_4 and a comment on normal tetrahedral structures. Z Kristallogr 217:51-54

Pickardt J, Reuter B, Riedel E, Soechtig J (1975) On the formation of $FeSe_2$ single crystals by chemical transport reactions. J Solid State Chem 15:366-368

Pierce L, Buseck PR (1978) Superstructuring in the bornite-digenite series: a high resolution electron microscopy study. Am Mineral 63:1-16

Pocha R, Johrendt D (2002) Kristallstrukturen und elektronische Eigenschaften von $Ge_{½}NbS_2$ und $Ge_{¼}NbS_2$. Z Naturforsch B. Anorg Org Chemie 57:1367-1374

Pohl D, Liessmann W, Okrugin VM (1996) Rietveld analysis of selenium-bearing goldfieldites. N Jahrb Mineral. Mh 1996:1-8

Portheine JC, Nowacki W (1975) Refinement of the crystal structure of zinckenite, $Pb_6Sb_{14}S_{27}$. Z Kristallogr 141:79-96

Potter RW (1977) An electrochemical investigation of the system Cu-S. Econ Geol 72:1524-1542

Povarennykh AS (1963) Grundsätze einer kristallchemischen Klassifikation der Sulfide. Geologie 12:377-400

Pratt JL, Bayliss P (1980) Crystal structure refinement of a cobaltian ullmannite. Am Mineral 65:154-156

Prewitt CT, Rajamani V (1974) Electron interaction and chemical bonding in sulfides. In: Sulfide Mineralogy, MSA Short Course Notes 1. Ribbe P (ed) Mineral Soc Am, p 1-41

Pring A (1990) Disordered intergrowths in lead-arsenic sulfide minerals and the paragenesis of the sartorite-group minerals. Am Mineral 75:289-294

Pring A (2001) The crystal chemistry of the sartorite group minerals from Lengenbach, Binntal, Switzerland - a HRTEM study. Schweiz Mineral-Petrogr Mitt 81:69-87

Pring A, Graeser S (1994) Polytypism in baumhauerite. Am Mineral 79:302-307
Pring A, Jercher M, Makovicky E (1999) Disorder and compositional variation in the lillianite homologous series. Mineral Mag 63:(6) 917-926
Pring A, Williams T, Withers R (1993) Structural modulation in sartorite: An electron microscope study. Am Mineral 78:619-626
Putnis A, Grace J (1976) The transformation behavior of bornite. Contrib Mineral Petrol 55:311-315
Putz H, Paar WH, Topa D, Makovicky E, Roberts AC (in press): Catamarcaite, Cu_6GeWS, a new germanium – tungsten sulfide from Capillitas, Catamaren, Argentina: description, paragenesis and crystal structure. Can Mineral in press.
Rabu P, Meerschaut A, Rouxel J, Wiegers GA (1990) The crystal structure of the misfit layer compound $(YS)_{1.23}NbS_2$. J Solid State Chem 88:451-458
Rad HD, Hoppe R (1978) Über thiotitanate (IV): Synthese und Struktur von $Cs_2(TiS_3)$. Z Naturforschung B33: 1184-1185
Rajamani V, Prewitt CT (1975) Refinement of the structure of Co_9S_8. Can Mineral 13:75-78
Ramdohr P, Strunz H (1978) Klockmanns Lehrbuch der Mineralogie, 16th edition. Ferdinand Enke Verlag
Ramirez R, Mujica C, Buljan A & Llanos J (2001) A family of new compounds derived from chalcopyrite. Common patterns in their electronic and crystal structures. Bol Soc Chilena Química 46:235-245
Reithmayer K, Steurer W, Schulz H, de Boer JL (1993) High-pressure single-crystal structure study on calaverite $AuTe_2$. Acta Crystallogr 49:6-11
Remmert P, Fischer E, Hummel HV (1994) Phasenuntersuchungen im System $2Ha-TaS_2-2Hc-MoS_2$. Z Naturforsch B, Anorg Org Chemie 49:1175-1178
Ribar B, Nowacki W (1970) Die Kristallstruktur von Stephanit. Acta Crystallogr 26:201-207
Řídkošil T, Skala R, Johan Z, Srein V (2001) Telluronevskite Bi_3TeSe_2 a new mineral. Eur J Mineral 13:177-185
Riedel E, Karl R, Rackwitz R (1981) Moesbauer studies of thiospinels V system $Cu_{1-x}Fe_xMe_2S_4$ Me=Cr, Rh and $Cu_{1-x}Fe_xCr_2(S_{.7}Se_{.3})_4$. J Solid State Chem 40:255-265
Riedel E, Karl R (1980) Moesbauer studies of thiospinels I The system $FeCr_2S_4$- Fe Rh_2S_4. J Solid State Chem 35:77-82
Riedel E, Pickardt J, Soechtig J (1976) Roentgenographische Untersuchung des Spinellsystems $CuRh_2(S_{1-x}Se_x)_4$. Z Anorg Allg Chem 419:63-66
Riekel C, Reznik HG, Schoellhorn R, Wright CJ (1979) Neutron diffraction study on formation and structure of D_xTaS_2 and H_xNbS_2. J Chem Physics 70:5203-5212
Robert JL, Makovicky E (1984) Cristallochimie Minérale. Encyclop Univer Paris: 747-753
Roseboom EH (1966) An investigation of the system Cu-S and some natural copper sulfides between 25° and 700 °C. Econ Geol 61:641-672
Ross V (1957) Geochemistry, crystal structure and mineralogy of the sulfides. Econ Geol 52:755-774
Rost E, Vesterjö E (1968) The crystal structure of the high temperature phase Pd_3S. Acta Chem Scand 22: 819-826
Rouxel J (1991) Des solides à moins de trois dimensions. Pour la Science 165:64-73
Rouxel J, Moëlo Y, Lafond A, DiSalvo FJ, Meerschaut A, Roesky R (1994) Role of vacancies in misfit layered compounds: The case of gadolinium chromium sulfide compound. Inorg Chem 33:3358-3363
Rowland JF, Hall SR (1975) Haycockite, $Cu_4Fe_5S_8$: a superstructure in the chalcopyrite series. Acta Crystallogr B 57:689-708
Rudashevski NS, Karpenkov AM, Shipova GS, Shishkin NN, Ryabkin VA (1979) Thalfenisite, the thallium analog of djerfisherite. Vses Mineralog Obsch Zapiski 108:696-701 (in Russ.)
Ruedorff W, Schwarz HG, Walter MC (1952) Strukturuntersuchungen an Alkalithiocupraten. Z Anorg Allg Chemie 269:141-152
Saeki M, Onoda M (1987) Preparation of molybdenum intercalated tantalum disulphide Mo_xTaS_2. Chem Lett 1987:1353-1356
Sawada H, Kawada I, Hellner E, Tokonami M (1987) The crystal structure of senandorite (andorite VI): $PbAgSb_3S_6$. Z Kristallogr 180:141-150
Schils H, Bronger W (1979) Ternäre Selenide des Kupfers. Z anorg allg Chemie 456:187-193
Schleid T (1990) Das System Na_zGdClH_x/S. II Einkristalle von Gd_2S_3 im U_2S_3-Typ. Zeitschrift für Anorgan Allgemeine Chemie 590:111-119
Schleid T, Lauxmann P, Schneck C (1999) Roentgenographische Einkristalluntersuchungen an alpha-HgS (zinnober). Z Kristallogr 16:95-95
Schneider J, Schultz H (1993) X-ray powder diffraction of Ag_2Te at temperatures up to 1123K. Z Kristallogr 203:1-15
Schoenfeld B, Huang JJ, Moss SC (1983) Anisotropic mean-square displacement (MSD) in single crystals of 2H- and 3R-MoS_2. Acta Crystallogr B39:404-407

Schutte WJ, de Boer JL, Jellinek F (1987) Crystal structures of tungsten disulfide and diselenide. J Solid State Chem 70:207-209
Scott JD (1974) Experimental methods in sulfide synthesis *In:* Sulfide Mineralogy, MSA short Course Notes 1. Ribbe P (ed) Mineral Soc Am, p. S1-S38
Scott JD, Nowacki W (1976) The crystal structure of alloclasite, CoAsS, and the alloclasite-cobaltite transformation. Can Mineral 14:561-566
Seung DY, Gravereau P, Trut L, Levasseur A (1988) Li_3AsS_3. Acta Crystallogr C54: 900-902
Shamrai VF, Leitus GM, Meshcheryakov VN, Faustov NI (1986) Study of the ternary chalcogenides of layer structure. Izv Akad Nauk SSSR, Metally 1986:213-219
Shannon RD (1981) Bond distances in sulfides and a preliminary table of sulfide crystal radii. Struct Bond Crystals 2:53-70
Shelimova LE, Karpinski OG, Svechnikova TE, Avilov ES, Kretova MA, Zemskov VS (2004) Synthesis and structure of layered compounds in the $PbTe-Bi_2Te_3$ and $PbTe-Sb_2Te_3$ systems. Neorg Mat 40:1440-1447
Shimazaki H, Ozawa T (1978) Tsumoite, Bi Te, a new mineral from Tsumo mine, Japan. Am Mineral 63: 1162-1165
Skinner BJ, Erd RC, Grimaldi FS (1964) Greigite the thio-spinel of iron; a new mineral. Am Mineral 49:543-555
Skowron A, Brown ID, Tilley RJD (1992) A single-crystal X-ray and high resolution microscope study of $Pb_5Sb_6S_{14}$. J Solid State Chem 97:199-211
Skowron A, Brown ID (1990a) Refinement of the structure of robinsonite, $Pb_4Sb_6S_{13}$. Acta Crystallogr C46: 527-531
Skowron A, Brown ID (1990b) Refinement of the structure of boulangerite, $Pb_5Sb_4S_{11}$. Acta Crystallogr C46: 531-534
Skowron A, Brown ID (1990c) Structure of antimony lead selenide, $Pb_4Sb_4Se_{10}$, a selenium analog of cosalite. Acta Crystallogr C46:2287-2291
Skowron A, Tilley RJD (1990) Chemically twinned phases in the $Ag_2S-PbS-Bi_2S_3$ system. Part 1. Electron microscope study. J Solid State Chem 85:235-250
Smith PPK (1984) Structure determination of diantimony tritin hexasulphide, $Sn_3Sb_2S_6$, by high-resolution transmission electron microscopy. Acta Crystallogr C40:581-584
Spijkerman A, de Boer JL, Meetsma A, Wiegers GA, van Smaalen S (1997) X-ray crystal structure refinement of the nearly commensurate phase of $1T-(TaS_2)$ in (3+2) dimensional superspace. Phys Rev 3B - Cond Matter 56:13757-13767
Springer G (1971) The synthetic solid-solution series $Bi_2S_3-BiCuPbS_3$ (bismuthinite-aikinite). N Jb Miner Mh 1971:13-24
Spry PG, Merlino S, Wang S, Zhang XM, Buseck PR (1994) New occurrences and refined crystal chemistry of colusite, with comparisons to arsenosulvanite. Am Mineral 79:750-762
Srikrishnan T, Nowacki, W (1974) A redetermination of the crystal structure of cosalite, $Pb_2Bi_2S_5$. Z Kristallogr 140:114-136
Srikrishnan T, Nowacki W (1975) A redetermination of the crystal structure of livingstonite $HgSb_4S_8$. Z. Kristallogr 141:174-192
Stasova MM (1968) The crystal structure of bismuthum selenide Bi_4Se_3. Izvestiya Akad Nauk SSSR Neorg Materialy 4:28-31
Stoll P, Näther C, Jess I, Bensch W (1999) KCu_4Se_3. Acta Crystallogr C55:286-288.
Strunz H, Nickel EH (2001) Strunz Mineralogical Tables. E. Schweizerbart'sche Verlagsbuchhandlung
Sugaki A, Shima H, Kitakaze A, Mizota T (1981) Hydrothermal synthesis of nukundamite and its crystal structure. Am Mineral 66:398-402
Szymanski JT (1995) The crystal structure of owensite $(Ba,Pb)_6(Cu,Fe,Ni)_{25}S_{27}$, a new member of the djerfisherite group. Can Mineral 33:671-677
Szymanski JT (1974) The crystal structure of high-temperature $CuFe_2S_3$. Z Kristallogr 140:240-248
Szymanski JT (1976) The crystal structure of mawsonite, $Cu_6Fe_2SnS_8$. Can Mineral 3:79-86
Szymanski JT (1978) The crystal structure of černyite, Cu_2CdSnS_4, a cadmium analog of Stannite. Can Mineral 16:147-151
Szymanski JT (1979) The crystal structure of platarsite, $Pt(As,S)_2$, and a comparison with sperrylite, $PtAs_2$. Can Mineral 17:117-123
Takagi J, Takeuchi Y (1972) The crystal structure of lillianite. Acta Crystallogr B 28:649-651
Takeuchi Y (1957) The absolute structure of ullmannite NiSbS. Mineral J 2:90-102
Takeuchi Y, Ozawa T (1975) The structure of $Cu_4Bi_5S_9$ and its relation to the structures of covellite CuS and bismuthinite Bi_2S_3. Z Kristallogr 141:217-232
Takeuchi Y (1970) On the crystal chemistry of sulphides and sulphosalts. *In:* Volcanism and Ore Deposits. Tstsumi T (ed), Univ of Tokyo Press, p. 395-420
Takéuchi Y (1997) Tropochemical cell-twinning. Terra Scientific Publishing Co

Takeuchi Y, Ghose S, Nowacki W (1965) The crystal structure of hutchinsonite, $(Tl,Pb)_2 As_5S_9$. Z Kristallogr 121:321-348

Takeuchi Y, Ozawa T, Takagi J (1974) Structural characterization of the high-temperature phase V on the PbS - Bi_2S_3 join. Z Kristallogr 140:249-272

Takeuchi Y, Ozawa T, Takagi J (1979) Tropochemical cell-twinning and the 60 Å structure of phase V in the PbS - Bi_2S_3 system. Z Kristallogr 150:75-84

Takeuchi Y, Takagi I (1974) The crystal structure of heyrovskyite ($6PbS\text{-}Bi_2S_3$). Proc Japan Acad 50:75-79

Talybov AG, Vainshtein BK (1962) An electron diffraction study of the second superlattice in $Pb\ Bi_4\ Te_7$. Kristallografiya 7:43-50

Tani B, Mrazek F, Faber J jr; Hitterman RL (1986) Neutron diffraction study of electrochemically synthesized djerfisherite. J Electrochemi Soci 133:2644-2649

Tettenhorst RT, Corbato CE (1984) Crystal structure of germanite, $Cu_{26}Ge_4Fe_4S_{32}$, determined by powder X-ray diffraction. Am Mineral 69:943-947

Thompson JB Jr. (1978) Biopyriboles and polysomatic series. Am Mineral 58:239-249

Tokonami M, Nishiguchi K, Morimoto N (1972) Crystal structure of a monoclinic pyrrhotite (Fe_7S_8). Am Mineral 57:1066-1080

Tomas A, Guittard M (1980) Cristallochimie des sulfures mixtes de chrome et d'erbium. Mater Res Bull 15: 1547-1556

Tomeoka K, Ohmasa M, Sadanaga R (1980) Crystal chemical studies on some compounds in the copper-bismuth sulfide ($Cu_2S\text{-}Bi_2S_3$) system. Mineral J 10:57-70

Tomeoka K, Buseck PR (1985) Indicators of aqueous alteration in CM carbonaceous chondrites: Microtextures of a layered mineral containing Fe,S,O and Ni. Geochem Cosmochim Acta 49:2149-2164

Topa D (2001) Mineralogy, Crystal Structure and Crystal chemistry of the Bismuthinite- Aikinite Series from Felbertal, Austria. Ph D Tesis, Inst Mineral Univ Salzburg, Austria.

Topa D, Balić-Žunić T, Makovicky E (2000) The crystal structure of $Cu_{1.6}Pb_{1.6}Bi_{6.4}S_{12}$, a new 44.8 Å derivative of the bismuthinite-aikinite solid-solution series. Can Mineral 38:611-616

Topa D, Makovicky E, Balić-Žunić T, Berlepsch P (2000) The crystal structure of $Cu_2Pb_6Bi_8S_{19}$. Eur J Mineral 12:825-833

Topa D, Makovicky E, Criddle A, Paar WH, Balić-Žunić T (2001) Felbertalite, $Cu_2Pb_6Bi_8S_{19}$, a new mineral species from Felbertal, Salzburg Province, Austria. Eur J Mineral 13:961-972

Topa D, Makovicky E, Balić-Žunić T (2002) The structural role of excess Cu and Pb in gladite and krupkaite based on new refinements of their structure. Can Mineral 40:1147-1159

Topa, D., Makovicky, E. and Paar, W.H. (2002): Composition ranges and exsolution pairs for the members of the bismuthinite-aikinite series from Felbertal, Austria. Can Mineral 40:549-869

Topa D, Makovicky E, Balić-Žunić T, Paar WH (2003) Kupčíkite, a new Cu-Bi sulfosalt from Felbertal, Austria and its crystal structure. Can Mineral 41:1155-1166

Topa D, Makovicky E, Balić-Žunić T (in press) The crystal structure of jaguéite, $Cu_2Pd_3Se_4$ and chrisstanleyite, $Ag_2Pd_3Se_4$. Can Miner in press

Topa D, Makovicky E (in press) The crystal structure of paderaite $Cu_7 (X_{0.33}Pb_{1.32}Bi_{11.33}) S_{11}$, with x = Cu or Ag. Can Mineral, in press

Tossell JA, Vaughan DJ, Burdett JK (1981) Pyrite, marcasite and arsenopyrite type minerals: Crystal chemical and structural principles. Phys Chem Minerals 7:177-184

Towle LC, Oberbeck V, Brown BE, Stajdohar RE (1966) Molybdenum diselenide: rhombohedral high pressure - high temperature polymorph. Science 154:895-896

Trudu AG, Kittel U (1998) Crystallography, mineral chemistry and chemical nomenclature of goldfieldite, the tellurian member og the tetrahedrite solid-solution series. Can Mineral 36:1115-1137

Uda M (1968) Synthesis of magnetic Fe_3S_4. Scientific Papers of the Institute of Phys Chem Research 62:14-23

Uda M (1968) The structure og tetragonal FeS. Z Anorg Allg Chem 361:94-98

Van der Lee A, de Boer JL (1993) Redetermination of the structure of hessite, Ag_2Te-III. Acta Crystallogr C49: 1444-1446

Van Dyck D, Conde C, Amelinckx S (1979) The diffraction pattern of crystals presenting a digenite type of disorder I. Phys Status Solidi 56:327-334

Van Dyck D, Conde-Amiano C, Amelickx S (1980) The diffraction pattern of crystals presenting a digenite type of disorder II. Phys Status Solidi 58:451-468

van Laar V (1967) Ferrimagnetic and anti-ferromagnetic structures of Cr_5S_6. Phys Rev 156:654-662

van Smaalen S (1995) Incommensurate crystal structures. Cryst Rev 4, 79-202

Vaughan DJ, Burns RG (1972) Mössbauer spectroscopy and bonding in sulfide minerals containing four-coordinated iron. Proc 24th Int Geol Congr 14:156-167

Vaughan DJ, Wright KV (1998) Crystal chemistry of ore minerals. *In:* Modern Approaches to Ore and Environmental Mineralogy. Cabri LJ, Vaughan DJ (eds) Min Soc Canada Short Course Ser 27:75-109

Vaughan DJ, Craig JR (1978) Mineral Chemistry of Metal Sulfides. Cambridge Univ Press
Vaughan DJ, Craig JR (1985) The crystal chemistry of iron-nickel spinels. Am Mineral 70:1036-1043
Vaughan DJ, Rosso KM (2006) Chemical bonding in sulfide minerals. Rev Mineral Geochem 61:231-264
Vaqueiro P, Kosidowski ML, Powell AV (2002) Structural distortions of the metal dichalcogenide units in AMo_2S_4 (A = V, Cr, Fe, Co) and magnetic and electrical properties. Chem Mater 14:1201-1209
Volk K, Cordier G, Cook R, Schäfer H (1980) $BaSbTe_3$ und $BaBiSe_3$. Verbindungen mit BiSe-bzw. SbTe-Schichtverbänden. Z Naturforsch 35B: 136-140
Volk K, Schäfer H (1979) $Cs_2Sb_8S_{13}$, ein neuer Formel- und Strukturtyp bei Thioantimoniten. Z Naturforsch 34b: 1637-1640
Wagner T, Monecke T (2005) Germanium-bearing colusite from the Waterloo volcanic-rock- hosted massive sulfide deposit, Australia: Crystal chemistry and formation of colusite-group minerals. Can Mineral 43: 655-669
Wang X (1989) Transmission electron microscope study of the minerals of the franckeite family. PhD Thesis (in Chinese). Chinese Geological University,
Waychunas GA (1991) Crystal chemistry of oxides and oxyhydroxides. Reviews Mineral 25:11-61
Whitfield HJ (1980) Polymorphism in skinnerite, Cu_3SbS_3. Solid State Comm 33:747-748
Whitfield HJ (1981): The crystal structure of hcc-CuAsSe. J Solid State Chem 39:209-214
Whitfield HJ (1970) The crystal structure of tetraarsenic trisulfide. J Chem Soc A Inorg Phys Theor Chem 1970:1800-1803
Whitfield HJ (1973) Crystal structure of the beta-form of tetraarsenic trisulphide. J Chem Soc Dalton Trans Inorg Chem 1973:1737-1738
Wiegers GA, Meerschaut A (1992) Structures of misfit layer compounds $(MS)_nTS_2$ (M=Sn, Pb, Bi, rare earth metals; T=Nb,Ta, Ti, V, Cr; 1.08<n<1.23). J Alloys Comp 178:351-368
Wiegers GA, Meetsma A, Haange RJ, de Boer JL (1990) Structure, electrical transport and magnetic properties of the misfit layer compound $SmS_{1.19}TaS_2$ ($SmTaS_3$). J Less-Common Metals 168:347-359
Wiegers GA, Meetsma A, Haange RJ, van Smaalen S, de Boer JL, Meerschaut A, Rabu P, Rouxel J (1990) The incommensurate misfit layer structure of $PbS_{1.14}NbS_2$ ($PbNbS_3$) and $LaS_{1.14}NbS_2$ ($LaNbS_3$) an x-ray diffraction study. Acta Crystallogr 46:324-332
Will G, Hinze E, Abdel RAM, (2002) Crystal structure analysis and refinement of digenite ($Cu_{1.8}S$) in the temperature range 20 to 500 C under controlled sulfur partial pressure. Eur J Mineral 14:591-598
Will G, Lauterjung J, Schmitz H, Hinze E (1984) The bulk moduli of 3d-transition element pyrites measured with synchrotron radiation in a new belt-type apparatus. Mater Res Soc Symp Proc 22:49-52
Williamson DP, Grimes NW (1974 +1973) An X-ray diffraction investigation of sulphide spinels. J Phys D Appl Phys 7:1-6 Mater Res Bull 8:973-982
Wilson JA, Di Salvo FJ, Mahajan S (1975) Charge-density waves and superlattices in the metallic layered transition metal dichalcogenides. Adv Phys 24:117-200
Wintenberger M (1979) Etude de la structure cristallographique et magnetique de Cu_2FeGeS_4 et remarque sur la structure magnetique de Cu_2MnSnS_4. Mater Res Bull 14:1195-1202
Wittlinger J, Werner S, Schultz H (1997) On the amorphisation of $ZnCr_2S_4$ spinel under high pressure x-ray diffraction studies. Phys Chem Mineral 24:597-600
Wolf M, Hunger H-J, Bewilogua K (1981) Potosiit - ein neues Mineral der Kylindrit-Franckeit-Gruppe. Freiberg Forsch Hefte C364:113-133
Wuensch BJ (1964) The crystal structure of tetrahedrite, $Cu_{12}Sb_4S_{13}$. Z Kristallogr 119:437-453
Wuensch BJ, Takeuchi Y, Nowacki W (1964) Refinement of the crystal structure of binnite, $Cu_{12}As_4S_{13}$. Z Kristallogr 123:1-20
Wuensch BJ (1993) Cation distributions, bonding and transport behavious in silver and copper fast-ion conductors with simple anion packings. Mat Sci Eng B18:186-200
Xu YN, Ching WY (1993) Electronic, optical, and structural properties of some wurtzite crystals. Phys Rev Cond Matter 48:4335-4351
Yajima J, Okta E, Kanazawa Y (1991) Toyohaite, $Ag_2 FeSn_3S_8$, a new mineral. Mineral J (Japan) 15:222-232
Yakovenchuk VN, Pakhomovsky YA, Men'shikov YP, Ivanyuk GYu, Krivorichev SV, Burns PC (2003) Chlorbartonite, $K_6Fe_{24}S_{26}(Cl,S)$ a new mineral species from a hydrothermal vein in the Khikina massif, Kola Peninsula, Russia: description and crystal structure. Can Mineral 41:503-511.
Yund RA (1962) Phase relations in the system Ni-As. Econ. Geology 56:1273-1296
Žák L, Frýda J, Mumme WG, Paar WM (1994) Makovickyite, $Ag_{1.5}Bi_{5.5}S_9$, from Baita Bihorului, Romania: The ^4P natural mineral member of the pavonite series. N Jahrb Mineral Abh 168:147-169
Zaritskii VN, Sadykov RA, Kostyuk YaI, Sizov RA, Aminov TG, Gubaidullin RK, Safin ShR (1986) Superlattice and cation distribution in the $Fe_{1-x} Cu_x Cr_2S_4$ system. Fiz Tverd Tela Solid State Physics 28: 3293-3298
Zhang X, Kanatzidis MG, Hogan T, Kannewurf CR(1996) $NaCu_4S_4$, a simple new low-dimensional, metallic copper polychalcogenide, structurally related to CuS. J Am Chem Soc 118:693-694

Zhukova TB, Zaslavskii AI (1972) Crystal structures of the compounds $PbBi_4Te_7$ $PbBi_2T_4$ $SnBi_4Te_7$ $SnBi_2Te_4$ $SnSb_2Te_4$ and $GeBi_4Te_7$. Kristallografiya 16:918-922.

Zhuze VP, Sergeeva VM, Strum EL (1958) New semiconductor compounds. Zh Tekhn Fiziki 3:208-211

Zolensky ME (1984) Hydrothermal alteration of CM carbonaceous chondrites: Implications of the identification of tochilinite as one type of meteoritic PCP. Meteoritics 19:346-347

Electrical and Magnetic Properties of Sulfides

Carolyn I. Pearce, Richard A.D. Pattrick, David J. Vaughan

School of Earth, Atmospheric and Environmental Sciences, and Williamson Research Centre for Molecular Environmental Science
University of Manchester
Manchester, United Kingdom
e-mail: carolyn.pearce@manchester.ac.uk

INTRODUCTION

The metal sulfides exhibit a great diversity of electrical and magnetic properties with both scientific interest and practical applications. These properties apply major constraints on models of the electronic structure (or chemical bonding) in sulfides (Vaughan and Rosso 2006, this volume). The pure and doped synthetic equivalents of certain sulfide minerals have actual or potential applications in the electronics industries (optical devices, photovoltaics, photodiodes and magnetic recording devices). Sulfides are also components of many thin film devices and have been extensively investigated as part of the nanotechnology revolution. Certain electrical and magnetic properties of sulfide minerals mean they contribute to geomagnetism and paleomagnetism, and provide the geophysical prospector with exploration tools for metalliferous ore deposits. To the mineral technologist, these same properties provide methods for the separation of the metal-bearing sulfides from associated waste minerals after mining and milling and before extraction of the metal by pyrometallurgical or hydrometallurgical treatment.

In this chapter, the theory and measurement of electrical and magnetic properties are outlined along with spectroscopic and diffraction studies that can provide insights into magnetic behavior are discussed. A brief review of electrical and magnetic studies of major sulfide minerals includes some examples of the applications of sulfide electrical and magnetic properties, including special consideration of the properties of sulfide nanoparticles. Most of the available data presented are for pure synthetic binary and ternary sulfides, as very small concentrations of impurities can dramatically affect electrical properties leading to problems of interpreting data from natural samples. Although data for several of the commonly found sulfides are discussed in this chapter, no attempt is made at a comprehensive coverage.

The section below on theory and measurement of electrical and magnetic properties draws on the account given in Vaughan and Craig (1978) to which readers are referred for further details. It is useful to provide this overview as the electrical and magnetic properties of sulfides provide critically important information for understanding them as materials, as well as giving rise to important applications. Also, the theoretical background is not commonly dealt with in texts concerning mineralogy or geochemistry, hence its inclusion here.

THEORY AND MEASUREMENT OF ELECTRICAL AND MAGNETIC PROPERTIES

Electrical properties

Metals, in which characteristically high conductivity is associated with overlapping valence and conduction bands or a partly filled band (see below and Vaughan and Rosso 2006), have room-temperature conductivities which are largely independent of impurities or lattice defects. Generally, the electrical conductivity (σ) is given by the expression:

$$\sigma = \frac{Ne^2\tau}{m} \qquad (1)$$

where N = concentration of "free" (conduction) electrons, e = electron charge (1.6×10^{-19} coulombs), m = electron mass, and τ = relaxation time (approximate mean time between collisions for the electron) This generalized expression derives from the "free-electron model" for metals. It shows that conductivity is mainly influenced by the number of conduction electrons (i.e., position of the metal in the periodic table) and the mean time between collisions which scatter the electrons. The lattice vibrations causing the scattering clearly dominate at room temperature but would be expected to die out (i.e., τ would go to infinity) at 0 K and the conductivity becomes infinite in a perfect crystal. Since no real crystal is perfect, resistivity does not fall to zero and, at lower temperatures, impurities and defects have a marked effect. By contrast, the resistivity of metals at relatively high temperatures varies only slowly with temperature. The quantity $e\tau/m$ is defined as the mobility (μ) and is the drift velocity per unit electric field. The "free-electron model" from which these expressions are derived is a classical model which does not involve quantum mechanics and, although adequate in describing conductivity, breaks down when applied to other properties such as magnetic susceptibility.

Semiconductors characteristically exhibit conductivity which increases rapidly as a function of temperature over certain ranges. In the pure intrinsic semiconductor a filled valence band is separated by a narrow energy gap from the vacant conduction band, although this situation is only really applicable at 0 K. As the temperature is raised, the Fermi surface becomes "fuzzy" and electrons are excited above this level by thermal energies of amount kT,[†] and as more and more of them receive enough energy to cross the forbidden gap and become effectively "free," the conductivity increases. Impurities which act as donors or acceptors of electrons complicate the situation by having electronic energies that fall in the forbidden energy gap. The donor levels provide electrons for the conduction band and such (negative) electron conduction is called n-type, whereas in p-type conduction, the excitation of electrons from the valence band into acceptor levels results in (positive) holes in the valence band. Both are examples of extrinsic semiconduction and both may occur in the same sample. If so, donor levels may empty into acceptor levels and when equal in concentration, effectively compensate for each other. Since intrinsic and extrinsic semiconduction are both temperature-dependent properties, many real examples exhibit temperature ranges over which either intrinsic or extrinsic mechanisms dominate. The concentrations of holes and of electrons in both the intrinsic and extrinsic ranges can be treated as problems in establishing chemical equilibria.

It is important to emphasize that occupation of energy states by electrons in the valence band and conduction band of semiconductors is statistical, and dependent on the number of available states and on the absolute temperature. In an intrinsic semiconductor, it can be shown that above 0 K, the probability factor is one half of the energy between occupied and empty states. This defines the Fermi energy (E_F) and the position of the Fermi level, see Figures 1a and 1b. This is also the activation energy (E_a) required for electrical conduction, which in this simple case is half the energy of the forbidden (or band) gap (i.e., $E_a = \frac{1}{2} E_g$).

Measurement of electrical properties

A wide range of techniques is available for the measurement of the electrical properties of materials, but only a small number need be considered here. The main properties of interest to

---

[†] Boltzmann's constant, K, is given by R/N_A where R is the gas constant and N_A is Avogadro's number. It has the value of 1.380×10^{-16} erg K^{-1}.

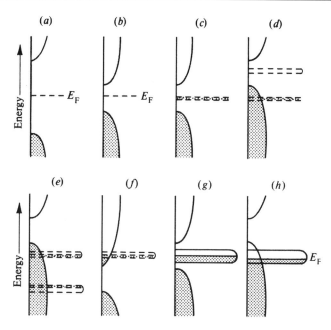

Figure 1. Simplified band models for main group sulfides (at 0 K). Energy (E) is plotted versus the density of states (i.e., statistical occupancy of the energy levels). Occupied states are shaded and empty states are open; localized d levels are shown with broken lines. (a) main group element insulator with wide energy gap (E_F = Fermi level); (b) Main group element semi-conductor (intrinsic); (c) and (d) transition element semiconductors with paramagnetism or diamagnetism; (e) and (f) transition element metallic conductors (p- and n-type respectively) with paramagnetism; (g) and (h) transition metal conductors with Pauli paramagnetism. After Jellinek (1972).

the sulfide mineralogist are those which characterize the material as a metal or semiconductor, give the band gap for the latter, and provide information on conduction mechanisms. As shown above for metals, conductivity is a function both of carrier (electrons or holes) concentration and carrier mobility; unfortunately, more than just conductivity must be known to determine these.

Values of conductivity (or its reciprocal, resistivity) range over about 25 orders of magnitude for all known compounds. Ranges of resistivity for a number of natural sulfide minerals are given in Table 1 and plots of resistivity against reciprocal temperature in Figure 2. Resistivity is normally measured in terms of resistance, which is a function of sample geometry, so that careful orientation and measurement of samples is required, although single crystals are not necessary in many cases because it is possible to achieve good electrical connectivity in polycrystalline materials. Conductivity relates to the current density and the electric field:

$$J = \sigma E \tag{2}$$

where J = current density in amperes cm^{-2}, σ = conductivity in Ω^{-1} cm^{-1}; and E = electrical field in V cm^{-1}. The current density (J) for a hole charge-carrier system is given by:

$$J = p\, e V_\alpha \tag{3}$$

where p = number of hole carriers per cm^3, e = electronic charge (1.6 × 10^{-19} coulombs), V_α = average velocity of carriers in cm s^{-1}. When Equation (3) is combined with Equation (2):

$$\sigma = pe\frac{V_\alpha}{E} \tag{4}$$

Table 1. Electrical resistivity of certain sulfide ores and minerals. After Parasnis (1956).

Mineral	Formula	Resitivity (ohm cm) Ore	Resitivity (ohm cm) Mineral
Pyrite	(FeS$_2$)	0.01-1000	0.005-5
Chalcopyrite	(CuFeS$_2$)	0.01-10	0.01-0.07
Pyrrhotite	(Fe$_{1-x}$S)	0.001-0.1	0.001-0.005
Arsenopyrite	(FeAsS)	0.1-10	0.03
Löllingite	(FeAs$_2$)	—	0.003
Cobaltite	(CoAsS)	—	1-5
Galena	(PbS)	1-30000	0.003-0.03

This is often written:

$$\sigma = pe\mu_p \tag{5}$$

where μ_p is the carrier mobility of holes in cm^2 V^{-1} s^{-1}. This equation shows the dependence of conductivity on both carrier concentration and mobility. A similar equation applies to electrons as charge carriers, and where both are involved then:

$$\sigma = e(\mu_n n + \mu_p p) \tag{6}$$

where n and p are the electron and hole concentrations and μ_n, μ_p their respective mobilities. Techniques for conductivity measurement are described by van der Pauw (1958), Fischer et al. (1961), and Baleshta and Keys (1968).

Hall effect measurements are a method of determining the mobility of charge carriers. The Hall effect is observed when a magnetic field is applied at right-angles to a conductor carrying a current (Fig. 3). The magnetic field deflects the current carriers and a restoring force (the Hall potential) is generated in order to maintain equilibrium so the current can continue flowing. This equilibrium can be expressed as:

$$B\,eV = eE_H \tag{7}$$

where B = magnetic flux density in gauss, e = electronic charge (1.6 × 10^{-19} coulombs), V = velocity of the carriers in cm s^{-1}, E_H = Hall field in v cm^{-1}. The average velocity of the charge carriers can be expressed in terms of the current density (J) and the conduction electron density (n), since $J = n\,eV$. Then, substituting in Equation (7) gives:

$$E_H = \left(\frac{I}{ne}\right)JB = RJB \tag{8}$$

where R = the Hall constant = I/ne, which gives the carrier concentration directly. What is experimentally determined is the Hall voltage (V_H), which is the potential actually measured across the sample in the direction normal to the current and magnetic field directions. It is dependent on sample geometry ($E_H = V_H/w$ where w = sample width; also $J = I/wt$ where I = current flowing through sample, t = sample thickness). Thus Equation (8) becomes:

$$R = \frac{I}{ne} = \frac{V_H t}{IB} \tag{9}$$

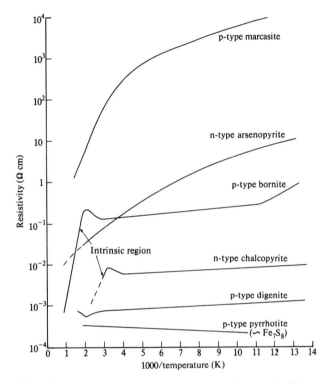

Figure 2. Resistivity versus reciprocal temperature for several sulfides. Modified after Baleshta and Dibbs (1969).

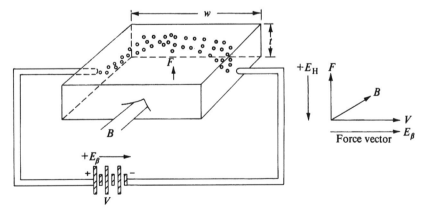

Figure 3. The Hall effect experiment – the effect of a magnetic field on positive charge-carriers in a crystal.

and the carrier concentration is given from the measured Hall voltage and known applied current and magnetic field. The sign of the Hall voltage indicates whether the carriers are holes (+) or electrons (−). As shown in Equation (5), carrier mobility is a function of electron or hole concentration (expressed as n and p respectively) and conductivity. Thus, from combined Hall effect and conductivity measurements, the mobility of electron (μ_n) or hole (μ_p) carriers may be determined; that is:

$$\mu_n = R\sigma \text{ for electrons (where } R = 1/ne)$$
$$\mu_p = R\sigma \text{ for holes (where } R = 1/pe) \tag{10}$$

This applies to cases where the conduction is overwhelmingly due to either electrons or holes. Where both electrons and holes play an important role, as in intrinsic semiconductors:

$$R = \frac{p - nb^2}{(p + nb)^2 e} \tag{11}$$

Where $b = \mu_n/\mu_p$. It is possible for the sign of the Hall voltage to change with temperature in such samples. The relationship between n and p shown in Equation (11) has to be established by further experiments involving varying the sample composition and temperature of measurement. Data for mobility as a function of temperature in some galena and pyrite samples are shown in Figure 4.

Thermoelectric power measurements show the tendency of mobile charge carriers to move from the hot end to the cold end of a sample which is placed in a temperature

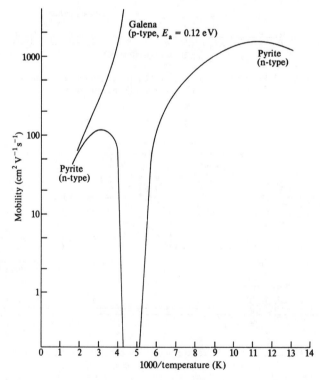

Figure 4. Mobility versus reciprocal temperature for a galena sample (hole mobility) and an *n*-type pyrite (electron mobility). Modified after Baleshta and Dibbs (1969).

gradient. Thus, if two contacts are placed on areas of the sample maintained at different temperatures, the potential difference measured between them is called the Seebeck voltage (normally expressed as V deg K^{-1}). The polarity of this thermoelectric voltage can be used to determine the sign of the majority carrier in a semiconductor, since electrons give rise to negative Seebeck voltages, whereas hole carriers result in a positive potential. The variation of thermoelectric power with temperature gives information regarding the activation energy and the position of the Fermi level:

$$\Phi_T = -\frac{k}{e}\left[A - \frac{(E_c - E_F)}{kT}\right] \quad (12)$$

Where Φ_T = thermoelectric power (V K^{-1}) at temperature T (K), K = Boltzmann constant, e = electronic charge, A = constant (≈ 2), E_c = energy of bottom of conduction band, E_F = Fermi energy. The Seebeck coefficients measured for pyrite-type disulfides, for example, range from −500 to +311 and can be correlated with conduction mechanism (Table 2). Plots of Φ_T against temperature are shown for several sulfides in Figure 5.

Magnetic properties

All materials can be categorized into a number of general classes according to their response to magnetic fields. When a magnetization is induced which opposes the external magnetic field, the material is diamagnetic. When the induced magnetization is parallel to the external field, the material is paramagnetic. Paramagnetism results from the presence of atoms with permanent magnetic dipoles, and commonly occurs in atoms or molecules with unpaired electrons. This is because a single electron can be regarded, in terms of a classical model, as a magnet formed by the spinning of the negatively charged particle on its axis. Also, an electron traveling in a closed path around the nucleus produces a magnetic moment. The magnetic moments of atoms, ions and molecules are expressed as Bohr magnetons (BM or μ_B):

$$1\mu_B = \frac{eh}{4\pi mc} \quad (13)$$

where e = electronic charge, m = electron mass, h = Planck's constant, c = speed of light.

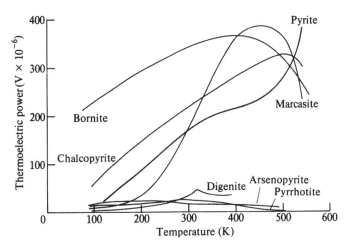

Figure 5. Thermoelectric power versus temperature for several sulfides. Modified after Baleshta and Dibbs (1969).

Table 2. Electrical and Magnetic properties of pyrite-type sulfides.

Compound	Electrical Properties[a]			Magnetic Properties[b]		
	Resistivity at 25 °C (ohm cm)	E_a (eV)	Seeback Coefficient (μ V deg^{-1})	Susceptibility at 25 °C ($\times 10^6$ c.g.s. units mole^{-1})	Magnetic Moment (μ_{eff}) (μ_B)	T_N, T_c or θ
MnS$_2$ (hauerite)	—	Semiconducting (1)	—	Paramagnetic/ antiferromagnetic (1)	~6.3 (3)	T_N = 48.2 K (4)
FeS$_2$ (pyrite)	1.74 (2)	Semiconducting 0.20 (298 K) (2) 0.46 (500 K)	−500 (2)	Diamagnetic ~ 10 (9)	0 (2)	—
CoS$_2$ (cattierite)	2 × 10^{-4} (2) (for CoS$_{1.99}$) 7.6 × 10^{-6} at 4.2 K (8)	Metallic (2)	−28 (2)	Paramagnetic/ ferromagnetic 4000 (9)	~1.84 (2)	T_c = 110 K (2) θ = 193 K
NiS$_2$ (vaesite)	6 × 10^{-1} (2) (for NiS$_{1.99}$)	Semiconducting 0.25 (500 K) (5) 0.12 (298 K) (2) 0.32 (500 K)	+311 (2)	Paramagnetic 700 (9)	~3.2 (6) ~2.48 (9)	θ = −1500 K (6) −740 K (9)
CuS$_2$	1.5 × 10^{-4} (2)	Metallic (2) Superconducting <1.56 K (7)	+3 (2)	0.29 (2) Pauli paramagnetic ~40 (9)	—	—
ZnS$_2$	1.0 × 10^6 (2)	Semiconducting ~2.5	—	Diamagnetic −0.36 emu g^{-1} (2)	—	—

References: (1) Hulliger (1968); (2) Bither et al. (1968); (3) Hastings et al. (1959); (4) Lin and Hacker (1968); (5) Hulliger (1959); (6) Benoit (1955); (7) Bither et al. (1966); (8) Butler and Bouchard (1971); (9) Adachi et al (Adachi et al. 1969).

Notes: [a] E_a, activation energy (eV).
[b] μ_{eff}, magnetic moment (in Bohr magnetons; μ_B); T_N Néel temperature; T_C, Curie temperature; θ, Weiss constant

However, this is not the magnetic moment of a single electron, since the complete quantum mechanical treatment is more complex. In fact, the magnetic moment μ_s of a single electron is given from wave mechanics as:

$$\mu_s \text{ (in } \mu_B) = g\sqrt{s(s+1)} \tag{14}$$

where s is the spin quantum number and g is the gyromagnetic ratio which relates the angular momentum of the electron, $[s(s+1)]^{1/2}$, to the magnetic moment ($g = 2$ for a spin only moment). From Equation (14), the spin magnetic moment of one electron is calculated to be $1.73\mu_B$ from the electron spin alone. This may be reduced or augmented by an orbital contribution. What is significant to this discussion is that knowledge of the magnetic moment (μ_{eff}) of a paramagnetic ion or atom in a compound provides information on the number of unpaired electrons on that atom or ion. Magnetic moments cannot be directly measured; the material property normally measured is the magnetic susceptibility χ where:

$$M = \chi H \tag{15}$$

in which M is the intensity of magnetization and H is the applied field intensity. In the simplest case, the magnetic moment is related to the susceptibility:

$$\mu_{eff} = \left(\frac{3k}{N_A\mu_B^2}\right)^{1/2} (\chi_A T)^{1/2} = 2.828(\chi_A T)^{1/2} \tag{16}$$

where K = Boltzmann's constant; N_A = Avogadro's number; μ_B = Bohr magneton; χ_A = magnetic susceptibility per gram atom; T = temperature (K). The measurement of magnetic susceptibility will be further discussed below.

An important aspect of the study of magnetic properties is their variation as a function of temperature. Diamagnetism (for which χ is negative) does not vary in magnitude with temperature, whereas the qualitative temperature dependence of magnetic susceptibility for a simple paramagnetic is shown in Figure 6(a). Curie demonstrated that paramagnetic susceptibilities depend inversely on temperature and may follow the simple law:

$$\chi_m^{corr} = \frac{C}{T} \tag{17}$$

where χ_m^{corr} is paramagnetic susceptibility per mole corrected for the diamagnetism due to closed shells; T is the absolute temperature; and C is a characteristic constant for the substance (Curie constant). If T is plotted against $1/\chi_m^{corr}$ for a paramagnet which obeys this Curie law

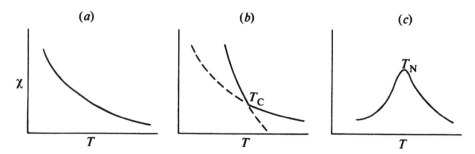

Figure 6. The qualitative temperature (T) dependence of magnetic susceptibility (χ) in (a) a simple paramagnet, (b) a ferromagnet and (c) an antiferromagnet. T_c, Curie temperature. T_N, Néel temperature. Modified after Cotton and Wilkinson (1972).

(Eqn. 17), a straight line of slope C should be obtained which intersects the origin. Although many paramagnetic substances do show this behavior, there are many others for which this line does not go through the origin but cuts the temperature axis above or below 0 K. Such behavior can be represented by a modification of Equation (17):

$$\chi_m^{corr} = \frac{C}{T-\theta} \qquad (18)$$

in which θ is the temperature at which the line cuts the T axis. This equation expresses the Curie-Weiss law and θ is the Weiss constant. This complication arises because the magnetic dipoles of individual atoms or ions are not independent of each other. It also results in a more complex relationship between susceptibility and magnetic moment than shown in Equation (16).

In addition to simple paramagnetism, there are four other forms of paramagnetism, several of which are particularly important in the metal sulfides. These are ferromagnetism, antiferromagnetism, ferrimagnetism, and Pauli paramagnetism, all of which arise through the interaction of unpaired electrons on neighboring "paramagnetic" ions. A spontaneous alignment of magnetic moments in the same direction occurs in ferromagnetic materials and is retained even after the external magnetic field has been removed. The resulting permanent magnetization (as in metallic iron) is attributed to a quantum mechanical exchange interaction between the electrons and adjacent atoms. The efficiency of this interaction (and therefore the value of the susceptibility) decreases with increasing temperature, breaking down at the Curie temperature (T_c) above which such materials exhibit simple paramagnetic behavior (Fig. 6b).

In antiferromagnets, the moments spontaneously align themselves but are antiparallel on adjacent atoms, so that they cancel and result in no net moment. Such antiferromagnetic coupling frequently takes place between two paramagnetic metal cations via an anion intermediary. A number of mechanisms have been proposed (see Fig. 7) to account for antiferromagnetism and ferromagnetism in ionic solids such as metal sulfides:

(1) *Superexchange*. For example, one of the two $3p$ valence electrons of S^{2-} could be transferred to the half-filled shell of a transition metal ion (M, e.g., Mn^{2+} with five d electrons) and according Hund's rule its spin would be antiparallel to the spins of the five d electrons. The second sulfur p electron would have its spin antiparallel to the first (because of the Pauli exclusion principle) and would remain localized on the sulfur ion. Such a configuration can also be formed with the opposite metal ion, antiparallel alignment of unpaired electrons on the two metal ions shown in Figure 7 results. The same mechanism predicts ferromagnetic coupling for cations with less than half-filled shells. The superexchange interaction clearly depends on orbital overlap and is strongest for a linear M-S-M arrangement and weakest when the M-S bonds are at right-angles.

(2) *Indirect exchange* is similar to superexchange but transfer (or "promotion") of a sulfur p electron to the d shell of the overlapping cation is not involved. Here the two $3p$ electrons of the sulfur atom simultaneously participate in antiferromagnetic coupling by remaining in their orbits but occupying areas near the metal ions to which they are coupled in antiparallel alignment. The coupling is always antiferromagnetic.

(3) *Double exchange* is suggested for systems containing mixed valence states and involves a simultaneous transfer of electrons from cation to sulfur and sulfur to cation. Ferromagnetism always results.

(4) *Semi-covalent exchange* is a modification of superexchange which assumes appreciable covalent character in the metal-anion bond; distribution of electrons amongst hybridized orbitals leads to formation of electron pair bonds between the central anion and adjacent cations which are thus antiferromagnetically coupled.

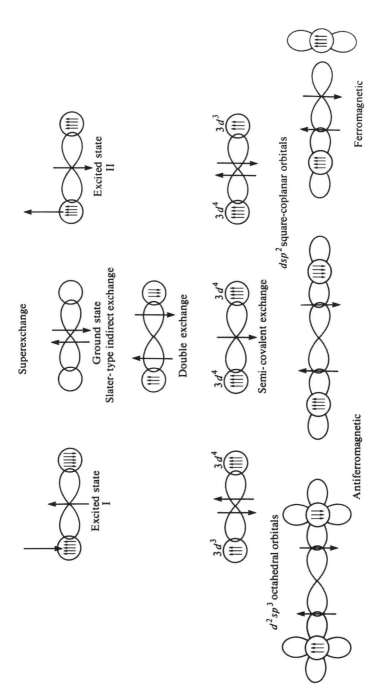

Figure 7. Schematic representation of different types of exchange mechanisms. After Schull and Wollan (1956).

Examples of exchange interaction in a sulfide system are afforded by the magnetic structures of the three polymorphs of MnS—the NaCl structure-type (alabandite) and the sphalerite and wurtzite analogs. In each case, the Mn^{2+} cation has 12 nearest Mn^{2+} neighbors but different numbers of S^{2-} anion neighbors. In alabandite, Mn^{2+} is bonded to next-nearest-neighbor Mn^{2+} ions via 180° sulfur p orbitals, whereas in the other polymorphs the linkage is tetrahedral. The magnetic structures determined by neutron diffraction, after Corliss et al. (1956) are shown in Figure 8. In the alabandite structure, the nearest Mn^{2+} neighbors are antiferromagnetically coupled by superexchange via the sulfur intermediaries, whereas in the sphalerite and wurtzite structure-types, next-nearest-neighbors are coupled. The magnetic structure of chalcopyrite ($CuFeS_2$) is closely related to sphalerite-type MnS.

It is characteristic of antiferromagnetic materials that their magnetic susceptibility increases with temperature over the range of antiferromagnetic behavior up to the Néel temperature (T_N), above which normal paramagnetic behavior is exhibited with decreasing susceptibility as temperature increases (see Fig. 6c).

From the discussion of exchange mechanisms giving rise to ferromagnetism and antiferromagnetism, it can be seen that the possibility exists of materials in which both types of interaction occur simultaneously. Such materials do occur and are known as ferrimagnets. The best known material exhibiting this phenomenon is magnetite (Fe_3O_4). Greigite (Fe_3S_4), the sulfur analogue of magnetite, is also ferrimagnetic. Another important example is provided by monoclinic pyrrhotite (Fe_7S_8), in which metal atom vacancies on one of the magnetic sublattices result in imbalance and consequent ferrimagnetism.

The last type of paramagnetism, Pauli paramagnetism, is found only in metallic materials in which the outermost electrons are extensively delocalized. Here, in a partly filled band, application of an external field causes imbalance between spin-up and spin-down electrons and hence a net magnetic moment. Such susceptibility values are very small and are independent of temperature.

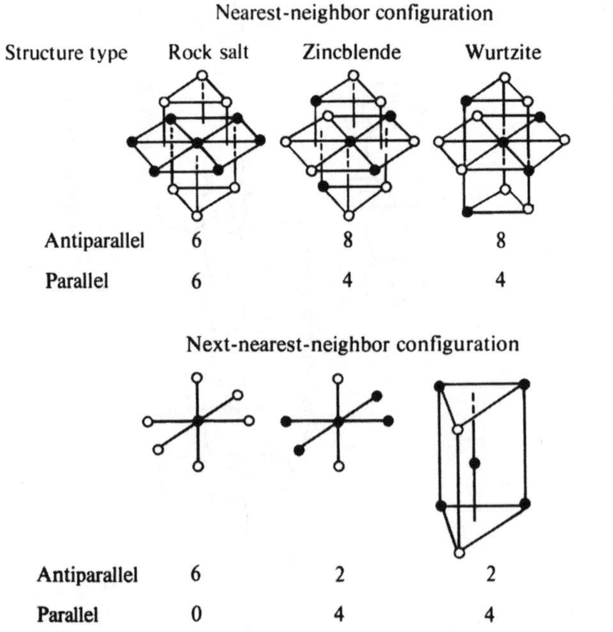

Figure 8. Magnetic structures of the three polymorphic forms of MnS showing nearest- and next-nearest-neighbor spin configurations. Only metal atoms are shown, and open and closed circles represent up and down spins respectively. [Used with permission of APS from Corliss et al. (1956), *Physical Review*, Vol. 104, Fig. 2, p. 925.]

Structure type	Rock salt	Zincblende	Wurtzite
Nearest-neighbor configuration			
Antiparallel	6	8	8
Parallel	6	4	4
Next-nearest-neighbor configuration			
Antiparallel	6	2	2
Parallel	0	4	4

Measurement of magnetic properties

The measurement of magnetic susceptibility is the most important technique in characterizing a substance magnetically. Conventional methods depend on measuring the force exerted on the sample when it is placed in an inhomogeneous magnetic field. For example, in the Guoy balance, a cylindrical sample is suspended between the pole pieces of an electromagnet and the weights of the sample in zero field and in a known magnetic field are recorded. A paramagnetic sample will be attracted by the field (or diamagnetic sample repelled) resulting in an apparent change in the mass of the sample (Δm) then:

$$\Delta m = \tfrac{1}{2} \kappa H^2 A \qquad (19)$$

where H is the field strength at the center (in Oesteds); A is the cross-sectional area of the sample and κ is its volume susceptibility. κ is converted to the susceptibility per gram by dividing by the sample density, and then to the susceptibility per mole by multiplying by the molecular weight. Other magnetic balances apply similar principles and by the addition of Dewar and heating coil systems, susceptibility can be studied as a function of temperature and Curie and Néel points can be determined.

In the case of ferri- or ferromagnetic materials, the intensity of magnetization (I) increases markedly with increase of the applied field up to a certain value at which saturation magnetization is reached. Determination of the saturating field and the study of magnetization as a function of applied field is another important application of the magnetic balance, particularly in relation to paleomagnetic work (McElhinny 1973).

A very sensitive commercially available laboratory instrument for measuring anisotropy of magnetic susceptibility and bulk magnetic susceptibility, at high and low temperatures and in weak variable magnetic fields (field range from 2 A/m to 700 A/m, peak values), particularly in the area of rock magnetism, is the Kappa Bridge. This system works by measuring the change in the induction properties of a bridge circuit with and without the sample (the sample is automatically lifted in and out of the sensor unit).

SQUID magnetometry. Currently the SQUID (Superconducting QUantum Interference Device) magnetometer is the most sensitve magnetic sensor available to study the magnetic properties of solids. The sensitivity of the SQUID magnetic sensor makes it possible to measure the magnetic response of a material (in the 10^{-15} tesla range) in its proper field without significant disturbance of the magnetic moments by external magnetic fields. The SQUID magnetometer measures magnetic field variations using a thin electrically resistive junction (the Josephson junction) between two superconducting electromagnets (Fig. 9) (O'Connor 1982). The SQUID sensing coil surrounding the sample is cooled using liquid helium down to its superconducting temperature and a current is passed through while the voltage across the junction is monitored. The magnetization of the sample changes the magnetic flux through the sensing coil which acts as an amplifier and is recorded as a variation in the output signal. The high sensitivity, fast response time and high field of the superconducting magnets, along with the ability to examine samples over a temperature range, mean it is an excellent tool for examining mineral samples with small magnetic moments as they pass through various structural transformations, or with variations in composition. In dilute magnetic semiconductors (such as transition metal doped ZnS) where small amounts of paramagnetic species are present, this sensitivity is particularly important. Drawbacks include sensitivity to outside noise and vibrations, expense, and the fact the SQUID measures variations in field so that samples need to move in and out of the coils to establish a baseline. To date, SQUID investigations of sulfides have been relatively few, but see Brun del Re et al. (1992) and Di Benedetto et al. (2002).

Electron paramagnetic resonance (EPR) can be employed with a SQUID detection system to quantitatively measure the change in magnetic moment of a specimen at different values of

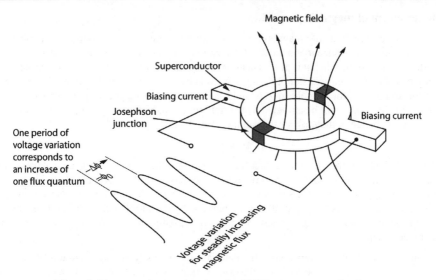

Figure 9. Diagrammatic representation of a SQUID magnetometer showing a superconducting loop interrupted by Josephson junctions. After Nave (2000).

applied magnetic field. EPR measurements of metal sulfides are briefly discussed elsewhere in this volume (Wincott and Vaughan 2006, this volume).

X-ray magnetic circular dichroism (XMCD). XMCD is a recently developed analytical technique that utilizes the differences in X-ray photoabsorption that occurs near X-ray absorption edges when circularly polarized light is incident on a magnetized sample, and either the alignment of the light helicity or the magnetization direction is reversed (Thole et al. 1985; van der Laan et al. 1986; Schutz et al. 1997). It can be used to provide information about the electronic and magnetic structure of ferromagnetic and ferrimagnetic solids by tuning the X-ray energy through the absorption edges that excite electrons to the valence band responsible for magnetism. Although it has largely been employed in the study of magnetic thin films and multilayers, it can just as easily be used on mineral powders, slices or orientated samples. In particular, it has been used extensively on magnetic oxides (such as the ferrite spinels and garnets), dilute magnetic semiconductors and magnetic nanostructures (van der Laan et al. 1986; van der Laan and Thole 1991; Chen et al. 1995; Stöhr 1995; Schutz et al. 1997). Even though little used on sulfides so far, the search for new magnetic materials, including nanomaterials and thin film multilayers containing chalcogenides, means it is becoming an increasing important technique for the mineral physicist (Pearce et al. 2006b).

XMCD utilizes the tunability of synchrotron sources and the element specific property of X-ray absorption spectroscopy (XAS) that allows the investigator to focus on the contribution of specific elements in chemically complex materials. It can potentially provide information about the oxidation state (including mixed states), site symmetry, spin state and crystal-field splitting of the absorbing $3d$ transition metal ions. XMCD is also sensitive to the direction and size of the local magnetic moments. In the $3d$ transition metals the magnetic moments (spin and orbital) are almost entirely derived from the $3d$ electrons and, therefore, the most effective way to examine magnetic materials using XAS is by observing the excitation of $2p$ core electrons to unfilled $3d$ states by analysis of the $L_{2,3}$ edge absorption.

In samples subjected to circularly polarized light, the polarized photons transfer angular momentum to the excited electron and, because of the usually large spin-orbit coupling, this is transferred to both the spin and orbital components of the electron. At the L_3 absorption edge ($2p_{3/2}$) light polarized in one direction will preferentially excite electrons parallel (arbitrarily "spin up") to the polarized photons compared to those antiparallel ("spin down"), whereas at the L_2 absorption edge ($2p_{1/2}$) in the initial state, spin down electrons will be preferentially excited (Fig. 10). If either the light polarization or the magnetic field is reversed, it has the opposite effect on the $2p_{3/2}$ and $2p_{1/2}$ electrons; the difference between the two spectra is determined by the empty states available in the valence band and can be related to the spin and orbital components of the magnetism (see Thole and van der Laan 1988 and van der Laan and Thole 1988 for details).

The sum rule, developed by Thole et al. (1992), states that the XMCD signal integrated over the $2p$ absorption edge is proportional to the orbital part of the $3d$ magnetic moment per hole. The weighted difference of the integrated XMCD signals of the L_3 and L_2 edge provides the spin moment per hole. Therefore, if the number of holes in the electronic state is known, or can be estimated, the absolute magnetic moments for individual atoms can be derived (van der Laan and Thole 1991).

Since the cross-section for soft X-ray absorption is very high, samples in the form of very thin films would be required for transmission measurement. Instead, measuring the photoabsorption in the soft X-ray regime is better done by recording the total electron yield (TEY) as the photon energy is scanned across the absorption edge (Fig. 11a,b). However this method is inherently surface sensitive due to the limited escape depth of the electrons

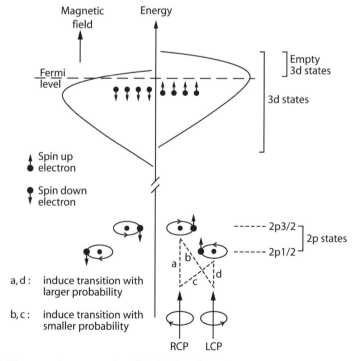

Figure 10. Diagrammatic representation of the XMCD effect. The arbitrarily assigned 'spin up' and 'spin down' electrons in a mineral containing unpaired electrons are not equal and if the applied magnetic field or light helicity are in the 'up' direction there is a high probability that the 'up spin' electrons will be excited from $2p$ to $3d$. If the orbital motion of the $2p$ electrons in the same sense as circular motion of light the transition is larger and the L_3 peak is enhanced; change the magnetization direction or the polarization of the light, and the effect is reversed. [Reproduced with kind permission of Ian Kirkman.]

(typically ~20-50 Å for the energy region of the transition metal L-edges). It is, therefore, vital that surface oxidation/contamination is avoided in order for meaningful results to be obtained. Also the measurement will give the absorption spectrum averaged over all contributions within the sampling depth, and even a small surface oxidation will contribute to the total spectrum. The fluorescence yield (FY) signal can also be used to measure the XAS and can be collected simultaneously. This signal is less surface sensitive; however, the fluorescence rate in the soft X-ray region is rather low and FY shows a lower signal to noise ratio than TEY. Magnetic reversal is achieved by "flipping" the magnetization of the sample using an electromagnet (Pattrick et al. 2002) and computer controlled action is undertaken for every energy step across the absorption edge. The two resulting spectra are normalized and then the difference spectrum produces the XMCD spectrum. Spectra can also be derived scanning firstly with one magnetization direction and then in the reverse direction, but variations in beam characteristics can lead to loss of fine structure and normalization problems.

The effectiveness of XMCD for sulfide minerals is well demonstrated by greigite (Fe_3S_4). Figure 12 (Letard et al. 2005) shows the isotropic Fe $L_{2,3}$ absorption spectra of natural greigite collected by reversing the applied 2 Tesla magnetic field. The sample need not be aligned along any specific direction and, although the strength of the applied field changes the shape of the spectra measured with positive and negative magnetization, it does not change the shape of the XMCD, only its magnitude. In the case of greigite, the three peaks relate to the contributions from Fe in the tetrahedral lattice site and Fe in two oxidation states in the octahedral lattice site. To separate the contributions of Fe^{2+} and Fe^{3+} in the octahedral and tetrahedral sites an approach similar to that employed for the oxide, magnetite, is used (van der Laan and Kirkman 1992; Pattrick et al. 2002). However, in iron sulfides the covalence of the bonds is stronger than in oxides, implying that the multiplet structure will be less visible and the selectivity of site and oxidation state will be less precise (Letard et al. 2005).

Neutron diffraction. As neutrons interact more strongly with atomic nuclei than X-rays, they are better than either diffracted X-rays or electrons in determining atomic positions. They also interact less with solid materials and, as a result, cause much less damage than X-rays. In magnetic studies, neutron diffraction is particularly valuable because the neutrons have a small magnetic moment and, therefore, interact with the spin and orbital magnetic moments of unpaired electrons in solid phases, and they are sensitive to magnetic ordering. Neutrons are being increasingly used by mineralogists, although the requirement for relatively large sample sizes (a few grams) and for sample homogeneity have meant using synthetic phases. Excellent reviews can be found in Harrison (2006) and Dove (2002). New generations of neutron instruments (e.g., GEM at ISIS, UK) and neutron sources e.g., Spallation Neutron Source (SNS) at Oakridge, Tennessee <*http://www.sns.gov/*> will mean that smaller samples can be used, with powder diffraction performed on samples of the order 10-100 mg.

A beam of neutrons can be diffracted by interaction with crystalline materials. Neutrons are either produced continuously by nuclear reactors or by both continuous and pulsed spallation sources. The world's most powerful neutron source is the SNS, where negatively charged hydrogen ions comprising a proton and two electrons are injected into a linear accelerator and accelerated up to energies of 1 G eV. The ions are stripped off the electrons as they pass through a foil and the resulting protons enter a ring where they form bunches. These proton bunches are ejected from the ring as a pulse to strike a liquid mercury target, producing neutron pulses by the resulting spallation. The neutron pulses are slowed by a "moderator" and guided into beamlines where neutrons of specific energies can be selected for a range of experiments. Water moderators are used to produce room temperature neutrons and liquid nitrogen to produce cold (20 K) neutrons.

The neutrons hitting a sample have a range of energies and wavelengths and neutrons are diffracted when atomic distances and wavelengths match. Magnetic neutron diffraction occurs

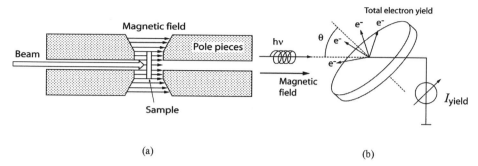

Figure 11. Diagrams showing (a) XMCD sample set up and (b) measurement of total electron yield.

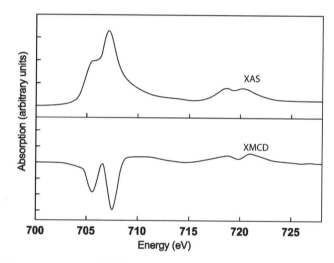

Figure 12. Fe $L_{3,2}$ absorption and XMCD spectra of natural greigite. The spectra were collected in a reversible 0.6 Tesla magnetic field, and the resulting XMCD difference spectrum is shown below (intensity scale enhanced by a factor of three). (Letard et al. 2005).

as a result of the interaction between the magnetic moment of the neutron and the magnetic moment of the target atom (μ). The magnetic moment of the atom is derived from the unpaired electron "cloud" and interferences from contributions from different parts of the cloud mean that magnetic diffraction reduces with an increasing angle to the scattering vector (Q). Thus, magnetic contributions are seen in diffraction spectra at low values of 2θ. Neutron diffraction is only sensitive to $\mu_\perp$ which is the component of the atomic magnetic moment perpendicular to the scattering vector (Q). Unlike the nuclear (crystal) structure contribution, which remains constant as a function of Q, the magnetic contribution to a neutron diffraction pattern is restricted to small Q. Hence, the magnetic contribution to a Bragg diffraction peak will be exactly zero if the moments are normal to the diffracting planes (Fig. 13).

Multi-collectors are used to detect the intensity of the diffracted neutrons as a function of angle for fixed wavelength experiments, or intensity as a function of time-of-flight for spallation source experiments (time of flight is related to the velocity, which is related to the energy), thus generating a diffraction pattern of peaks and intensities (Rodríguez-Carvajal 1993). The "magnetic" atoms in the (mineral) phase form a magnetic lattice that contributes

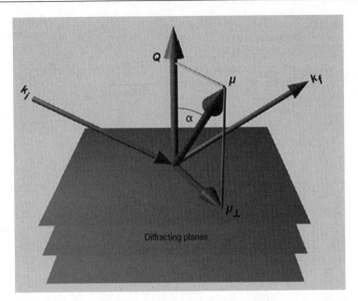

Figure 13. Diagrammatic representation of the angular relationship between scattering vector and magnetic interaction vector for magnetic neutron scattering. The incident and scattered neutron beams are described by the wave vectors k_i and k_f. The scattering vector, Q, is defined as $Q = k_f - k_i$. For Bragg diffraction, Q points normal to the set of diffracting planes. The magnetic moment of the atom is defined by the vector μ. The scattering amplitude is determined by the component of μ perpendicular to Q ($\mu_\perp$). (Harrison 2006).

to the diffraction pattern. In general, the periodicity of the magnetic structure will be different from the nuclear structure, leading to the presence of magnetic superlattice peaks. The magnetic and nuclear reflections coincide, however, when the nuclear and magnetic unit cells are equal. Figure 14 shows the neutron diffraction pattern of α-MnS taken at 4.2 K and a wavelength of 1.064 Å. The diffraction peaks labeled "nuclear" are the fundamental peaks and persist unchanged above the Néel temperature. The superstructure peaks are entirely magnetic in origin and disappear when the Néel temperature is exceeded (Corliss et al. 1956). These magnetic superstructure peaks are evident in the neutron diffraction patterns for all three polymorphs of MnS, indicating three distinct antiferromagnetic structures. The nearest and next-nearest neighbor configurations for the three forms of MnS are shown in Figure 8. The magnetic diffraction by the individual atoms is sensitive to the relative orientation of the neutron magnetic moment, the atomic magnetic moment, and the scattering vector (Harrison 2006). The diffraction trace will, thus, comprise both nuclear structure and magnetic structure components and, for unpolarized neutrons, it is the sum of the nuclear and magnetic contributions. By using polarized neutrons, the magnetic component can be separated from the nuclear component. When the polarization direction is set parallel to the scattering vector, magnetic diffraction will always flip the polarization of the incident beam whereas nuclear diffraction preserves the polarization of the beam. By analyzing the spin-flip versus non-spin-flip intensities, the two signals can be separated.

Elastic scattering arises from the static component of a crystal structure and therefore has zero angular frequency but is not identical to the scattering function. The main information obtained from elastic studies is the magnetic structure including the direction and the magnitude of the magnetic moments for each magnetic atom. The vibration of atoms about their mean positions, as in a crystalline solid, results in inelastic scattering. Inelastic neutron

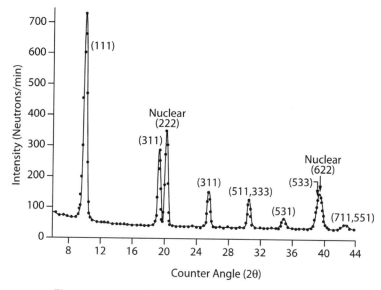

Figure 14. Neutron diffraction pattern of MnS - rock salt structure (Temperature = 4.2 K, wavelength = 10.423 Å) (Corliss et al. 1956).

scattering can be used to study magnetic dynamics or spin waves, which give an indication of the strength and nature of the exchange forces between neighboring spins (Dove 2002).

ELECTRICAL AND MAGNETIC PROPERTIES OF METAL SULFIDES

All sulfides of main group (i.e., non-transition) elements (including Zn, Cd, Hg) are diamagnetic insulators or semiconductors. This is because, in terms of a simple band structure model, the valence and conduction bands are, respectively, filled and empty and do not overlap. Such situations are shown in Figure 1a,b. Generally, the valence band in such compounds should be largely made up of S $3p$ and $3s$ type orbitals and the conduction band of metal s and p orbitals. Galena (PbS) and sphalerite (ZnS) are the two most important examples of this group of sulfides.

The sulfides of the transition metals can be considered as intermediate between the transition metal oxides, whose properties are determined by strong electronic correlation effects, and the transition metal selenides, showing a variety of electronically induced structural phase transitions. Transition metal sulfides have many technological applications, ranging from lubricants to catalysts, motivating the need to develop an understanding of their properties at an atomistic level (Hobbs and Hafner 1999). In the transition metal sulfides, the electronic structures, and hence electrical and magnetic properties, are complicated by the presence of the d electrons (Jellinek 1972). The implications are best examined by simple molecular orbital/band model approach including the d electrons (for further discussion, see Vaughan and Rosso 2006). In Figure 1c-h, valence and conduction band energy levels comprised of metal and sulfur s and p orbitals are shown with d orbital energy levels added. Separations in the range of one to several electron volts are expected between s-p valence and conduction bands, but the relative energy of the d orbital levels may vary widely as shown in the figure. Also, most importantly, the d orbitals may overlap with sulfur orbitals or even with the d orbitals of adjacent metals to form a band (or bands), or may not overlap at all; i.e., remain localized. When overlap produces delocalization of the d electrons, the resulting bands are usually narrow. In Figure 1c, localized

d electrons occur between the s-p valence and conduction bands in energy, so the material is a semiconductor. If the d electrons are completely paired, the material will be diamagnetic (e.g., pyrite, FeS_2); if not, then it will exhibit temperature-dependent paramagnetism (e.g., hauerite, MnS_2) and may be magnetically ordered over a certain temperature range. The d levels in Figure 1d,e are also localized but below the top of the valence band in energy. If Figure 1d applies, the material will again exhibit semiconduction and diamagnetism or paramagnetism. However, if the energy of the d orbitals of the same cation (T^{M+}) but in a different oxidation state ($T^{(M-1)+}$) also fall below the top of the valence band, as in Figure 1e, then the cation will be reduced by the removal of an electron from the valence band. The holes created in the valence band permit p-type metallic conduction. Thus, the material will be metallic but exhibit paramagnetism. Similar properties may be shown by a sulfide of the type illustrated in Figure 1f where the localized d levels overlap the empty conduction band. When an electric field is applied, transfer of electrons to the conduction band occurs, producing n-type metallic conductivity (CoS_2, cattierite, exhibits behavior akin to this). Finally, in the sulfides illustrated in Figure 1g,h, the d levels are delocalized to form a narrow band near the top of the s-p valence band or in the band gap. If such a narrow band is partly filled, the sulfide will exhibit metallic conductivity and Pauli paramagnetism or diamagnetism (Co_3S_4, linnaeite, and a number of metal-rich sulfides exhibit such behavior).

The above discussion offers a simple model to explain the diversity of sulfide magnetic and electrical properties. However, it should be noted that in this discussion the d electrons have been regarded as degenerate (equal in energy). Of course, in all sulfides the degeneracy of the d orbitals is partly removed by the ligand field of the surrounding sulfur atoms (Vaughan and Rosso 2006, this volume). This leads to further complexities; e.g., the t_{2g} levels of an octahedrally coordinated transition metal ion may remain localized and the e_g levels overlap to form a narrow band. Such situations will be discussed in following sections, but they do not invalidate the broad categories outline above, which were originally organized into four classes of transition metal sulfide by Jellinek (1972):

(A) Semiconductors with paramagnetism or diamagnetism i.e., localized d electrons, see Figure 1c,d.

(B) Metallic conductors of p-type with paramagnetism i.e., localized d electrons, see Figure 1e.

(C) Metallic conductors of n-type with paramagnetism i.e., localized d electrons, see Figure 1f.

(D) Metallic conductors with Pauli paramagnetism i.e., delocalized d electrons, see Figure 1g,h.

The magnetic and electrical data available for several major sulfides and groups of sulfides will now be considered, beginning with the sphalerite- and wurtzite-type metal sulfides, including ZnS. The effect of adding a transition metal to such a material will then be considered using data on (Zn,Fe)S, and other sulfides of this structure type will also be discussed. The series of pyrite-type sulfides provides a good example of an isostructural series involving successive addition of electrons and the NiAs-type sulfides an opportunity to consider a group of materials with complex magnetic ordering behavior. Also discussed are galena (PbS) and alabandite (α-MnS) with their simple halite-type structures, and a number of other phases (makinawite, smythite and greigite) of interest because of their geochemical or rock magnetic importance. A summary of the magnetic and electrical properties of these important sulfides is provided in Table 3. Shuey (1975) has also reviewed the electrical properties of many mineral sulfides and discussed their interpretation in detail. This includes sulfides with a range of different properties from sphalerite (ZnS), which is a notoriously poor electrical conductor with a wide band gap (~3.5 eV); to covellite (CuS), which has very high

Table 3. Electrical and magnetic characteristics of some common sulfides. After Corry (2005) and Vaughan and Craig (1978).

Mineral	Formula	Structure	Resistivity (ohm m)	T_C (or T_N where shown) in °C	Magnetic Property; Magnetic susceptibility in $k \times 10^{-6}$ cgs
Acanthite	Ag_2S	Ortho.	semiconductor n-type; 1.5×10^{-3} to 1.5	179	diamagnetic; $K = -30$
Alabandite	MnS	Cubic	semiconductor	T_N –121	antiferromagnetic; $K = 5,600$
Arsenopyrite	FeAsS	Ortho.	semiconductor mixed type; 10^{-5} to 20	<500; (Trans. to Hex at 103)	diamagnetic or paramagnetic; $K = 240$
Bismuthinite	Bi_2S_3	Ortho.	semiconductor n-type; 3 to 5.7×10^4	50	diamagnetic; $K = -123$
Bornite	Cu_5FeS_4	Ortho.	semiconductor p-type; 10^{-5} to 1.6	<228; T_N –265	paramagnetic; antiferromagentic (1) below –197
Chalcocite	Cu_2S	Mono.	large decrease at $T \sim 228$ semiconductor p-type; 10^{-4} to 2.3×10^3	103 to 130; (Trans. to Hex. at 103)	diamagnetic
Chalcopyrite	$CuFeS_2$	Tetrag.	semiconductor n-type ; 10^{-5} to 150; large decrease at $T \sim 330$°C	<550; T_N 277	antiferromagnetic; k~32
Cinnabar	HgS	Hex.	semiconductor; 10^9 to 10^{11}	<344	diamagnetic; $K = -55$
Covellite	CuS	Hex.	p-type metal; 10^{-7} to 10^{-3}	<500	diamagnetic; $K = -2$
Enargite	Cu_3AsS_4	Ortho.	semiconductor p-type; 2×10^{-4} to 0.9	<383	diamagnetic
Galena	PbS	Cubic	semiconductor mixed type; 6.8×10^{-6} to 17.5 (z mean 2×10^{-3})	1127	diamagnetic; $K = -3$ to $+84$
Greigite	Fe_3S_4	Cubic	metal?		ferrimagnetic
Mackinawite	$Fe_{1+x}S$	Tetrag.	metal		pauli paramagnetic
Marcasite	FeS_2	Ortho.	semiconductor p-type; 10^{-3} to 10^4. large decrease at T~200°C,	<520	diamagnetic $k \sim 18,720$
Metacinnabar	HgS	Cubic	semiconductor; 10^{-6} to 10^{-2}		diamagnetic; $K = -55$
Molybdenite	MoS_2	Hex.	semiconductor mixed type; 10^{-36}	<1675	diamagnetic; $K = -63$ to -77
Pentlandite $(Fe,Ni)_9S_8$			metal		Pauli paramagnetic
Pyrrhotite	$Fe_{7/8}S_8$	Mono.	p-type metal; 10^{-6} to 10^{-1}	315	ferrimagnetic; ($k = 10^2$ to 5×10^5 mean 1.25×10^5)
Pyrrhotite	$Fe_{1-x}S$	Hex.	metal	T_N 315	antiferromagnetic
Sphalerite	ZnS	Cubic	semiconductor; 2.7×10^{-3} to 10^{12}	1020	diamagnetic; $K = -25$ to -60
Stibnite	Sb_2S_3	Ortho	semiconductor; 10^5 to 10^{12} (mean 5×10^6)	147 to 217	diamagnetic; $K = -86$
Tennantite	$(Cu,Fe,Zn,Ag)_{12}As_4S_{13}$	Cubic	semiconductor; 7×10^{-4} to 0.4		paramagnetic
Tetrahedrite	$(Cu,Fe,Zn,Ag)_{12}Sb_4S_{13}$	Cubic	semiconductor; 0.3 to 3×10^4		paramagnetic
Troilite	FeS	Hex.	p-type metal; 10^{-6} to 10^{-1}	T_N 315	antiferromagnetic; $K = 5,187$
Wurtzite	ZnS	Hex.	semiconductor ; 3.5×10^2		diamagnetic; $K = -25$

(1) Collins et al. (1981)

conductivity as a result of the presence of "charge carrier" holes, formed when one electron per three CuS is removed to give electrical neutrality, due to the S-S pairs in the structure that reduce the charge on the sulfur anions (Tossell 1978).

Metal sulfides of sphalerite and wurtzite (ZnS) structure-type

Pure zinc sulfide is a diamagnetic semiconductor, although the band gap is sufficiently large that it is sometimes described as an insulator. The measured specific magnetic susceptibilities of sphalerite and wurtzite are -0.262×10^{-6} and -0.290×10^{-6} emu g^{-1} respectively (Larach and Turkevich 1955), and are in good agreement with the theoretical value calculated from the susceptibilities of the zinc and sulfur ions of -0.30×10^{-6} emu g^{-1} (International Critical Tables 1929). The forbidden energy gap in ZnS has been report between 3.2 and 3.9 eV at 300 K, depending on the method of measurement (Cheroff and Keller 1958; Fok 1963). Pressure reduces the forbidden energy gap of ZnS. Samara and Drickamer (1962) observed that sphalerite undergoes a semiconductor-to-metal transition to become electrically conducting at very high pressures, probably due to a transformation to an NaCl-type structure with large numbers of defects (Rooymans 1963). Substantial discrepancies in the magnitude of the transformation pressure for ZnS reported in the literature occur due to factors such as the different techniques used to identify the transition and the grain size (Jiang et al. 1999). The energy gap is also reduced by cadmium or mercury substitution. The mechanism of intrinsic semiconduction in ZnS may be p- or n-type, depending on the stoichiometry. Sphalerite and wurtzite are now known not to be true polymorphs; in fact, sphalerite is zinc-deficient and wurtzite sulfur-deficient compared to the stoichiometric compound. Thus p-type semiconduction is expected in sphalerite and n-type in wurtzite. Studies of the electrical properties of ZnS heated in air and in excess sulfur vapor support these observations (Morehead 1963; Morehead and Fowler 1962).

The presence of transition metal impurities in ZnS has a pronounced effect on the magnetic and electrical properties. Iron (Spokes and Mitchell 1958) and manganese (Brummage et al. 1964) render sphalerite paramagnetic, and paramagnetic copper centers have been observed in ZnS by Holton et al. (1969). Di Benedetto et al. (2002) used electron paramagnetic resonance spectroscopy to study synthetic sphalerite doped with Mn and suggested that two different Mn^{2+} species are present. Brummage et al. (1964) observed that ZnS containing substantial manganese departs from Curie behavior above 77 K. This was attributed to antiferromagnetic coupling between adjacent manganese ions in the structure. Such impurities also have a pronounced effect on the value of the forbidden energy gap and on the electrical transport properties. This is well illustrated by the detailed study of iron-containing zinc sulfide by Keys et al. (1968). This study is of particular mineralogical interest because iron, by far the most important substituent in natural sphalerite, reaches concentrations of ≈20 atomic%. Measurements of the electrical conductivity of a sphalerite with 12.4 atomic% iron showed typical semiconductor behavior with a forbidden energy gap reduced to 0.49 eV. The absence of a measurable Hall effect and positive sign for the thermoelectric power suggested "hopping" of holes in a band of largely d orbital character as the conduction mechanism. The model requires the holes to be generated by the presence of Fe^{3+} ions (with one less d electron than Fe^{2+}) although only 1 in 800 iron atoms need be Fe^{3+}. Magnetic susceptibility measurements on the samples showed paramagnetic behavior at lower iron concentrations and a linear variation in susceptibility with iron content. At higher iron concentrations, the variation becomes non-linear and antiferromagnetic coupling is suggested (see Fig. 15). More recently, Nikolic et al. (1999) have studied the electronic transport properties of natural (Zn,Fe)S single crystals containing 12.2 wt% Fe and determined the mobility of (photogenerated) free carrier holes (μ_p) to be ~428 cm^2 V^{-1} s^{-1}. Deulkar et al. (2003) studied the electrical properties of (Zn,Fe)S samples, performing resistivity and thermoelectric power measurements; the latter show p-type (semi)conductivity over a wide range of compositions. In this case, samples were prepared by sintering ZnS and FeS powders made by precipitation from aqueous solution

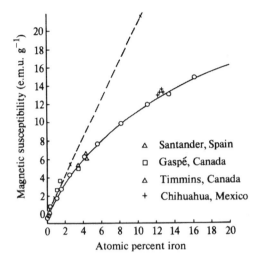

Figure 15. Magnetic susceptibility plotted against iron content for natural and synthetic (plotted as O) iron-bearing sphalerites. In the dashed curve, the scale is expanded ten-fold; this shows the linear relationship between susceptibility and iron content at low temperatures. The diamagnetism due to ZnS has been subtracted from this curve. The solid curve shows the deviation from linear behavior at higher iron concentrations—the diamagnetic contribution of ZnS has not been subtracted in this case. After Keys et al. (1968).

which, although yielding a sphalerite structure solid solution single phase in most cases, at the highest concentrations produced some contamination by other phases (probably FeS). The sphalerite structure cadmium and mercury sulfides are also diamagnetic semiconductors, and there is an extensive literature on their electrical properties (Jellinek 1972). The CdS and HgS polymorphs probably also exhibit deviations from stoichiometry which can be linked with intrinsic semiconduction mechanisms.

Chalcopyrite ($CuFeS_2$). In $CuFeS_2$, the zinc atoms of the sphalerite structure are replaced equally by copper and iron is antiferromagnetically ordered up to a Néel temperature of 823 K (Donnay et al. 1958; Teranishi 1961). The magnetic structure of chalcopyrite has been determined by neutron diffraction (Donnay et al. 1958) which shows that the spins on the two iron atoms, tetrahedrally bonded to a common sulfur atom, are opposed and directed along the c-axis. Chalcopyrite behaves as a typical semiconductor and exhibits intrinsic behavior above ~623 K (see Fig. 2) and extrinsic behavior below 223 K (Teranishi 1961), although as chalcopyrite is heated it loses sulfur with a consequent irreversible increase in conductivity (Frueh 1959). Room-temperature determinations on natural chalcopyrite nearly all show n-type conduction and band gap values of 0.5 eV, 0.33 eV and 0.6 eV are proposed by Austin et al. (1956), Baleshta and Dibbs (1969) and Teranishi et al. (1974). The metal enriched "stuffed" derivatives of chalcopyrite, talnakhite ($Cu_{18}Fe_{16}S_{32}$) and mooihoekite ($Cu_9Fe_9S_{16}$), have been studied by Townsend et al. (1971a). Talnakhite exhibits zero ferromagnetism and is antiferromagnetically ordered at room temperature, with a small paramagnetic component due to partial disorder. Mooihoekite shows weak ferromagnetism of approximately 0.2 μ_B/$Cu_8Fe_9S_{16}$ formula unit (Townsend et al. 1971a).

Stannite (Cu_2FeSnS_4). The antiferromagnetic coupling observed in chalcopyrite is no longer so readily possible in Cu_2FeSnS_4 because half of the iron atoms are replaced by tin. Stannite exhibits normal paramagnetic behavior in the temperature range 77–206 K (Eibschutz et al. 1967) and the susceptibility data indicate oxidation states of Cu_2^{2+} Fe^{2+} Sn^{4+} S_4^{2-}; a result confirmed by Mössbauer data. However, SQUID magnetometer measurements by Bernardini et al. (2000) of pure synthetic stannite show a transition to an antiferromagnetically ordered structure at 8 K (see Fig. 16). These authors also studied magnetic susceptibility as a function of temperature for iron-bearing samples of the stannite – kesterite (Cu_2ZnSnS_4) solid solution series, demonstrating paramagnetic (Curie-Weiss law) behavior attributed to the tetrahedrally coordinated Fe^{2+}, and determining Curie and Weiss constants. The isostructural mineral

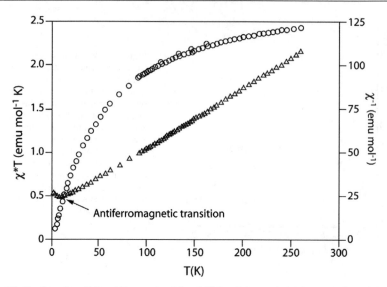

Figure 16. Graph to show X^{-1} vs. T (open triangle) and X^*T vs. T (open circle) for synthetic stannite. After Bernardini et al. (2000). [Used with kind permission of Springer Science and Business Media]

briartite (Cu_2FeGeS_4) also exhibits a transition to an antiferromagnetically ordered mineral at very low temperature ($T_N = 12$ K) (Allemand and Wintenberger 1970).

Cubanite ($CuFe_2S_3$). Cubanite is another mineral of this group which exhibits interesting magnetic and electrical properties; natural samples have been studied by Takeno et al. (1972). At room temperature, these authors, along with Sleight and Gillson (1973) observed semiconducting behavior in natural cubanite. Mössbauer data suggest formal valence states of $Cu^+(Fe^{2+}Fe^{3+})S_3$ in cubanite, but with electron hopping taking place between Fe^{2+} and Fe^{3+} ions across shared edges of the iron tetrahedra. Such hopping must be restricted to isolated pairs of iron atoms which must also be antiferromagnetically coupled to explain the ordered magnetism. However, weak ferromagnetism is exhibited by cubanite, and Fleet (1970) has suggested "canting" of the antiferromagnetic spins on the adjacent iron atoms as the origin; a mechanism similar to that suggested by Dzyaloshinsky (1958) to account for ferromagnetism in α-Fe_2O_3 which involves an imperfect antiparallel alignment of the spins.

Sulfides of pyrite (FeS_2) and related structure-types

The series of pyrite-structure disulfides from MnS_2 to ZnS_2 exhibits a wide range of electrical and magnetic properties as shown in Table 2. Hauerite (MnS_2) can be considered a "Class A" sulfide in terms of Jellinek's (1968) subdivisions. The localized d electrons of MnS_2 probably require excitation up into the σ^* band for conduction to occur. MnS_2 is cubic above the Néel temperature (295 K), and consists of an NaCl-like arrangement of Mn^{2+} and $(S_2)^{2-}$ ions, with the axes of the $(S_2)^{2-}$ groups directed along the various body diagonals of the cell. The magnetic data (Hastings et al. 1959) show that MnS_2 is paramagnetic and has five unpaired d electrons—a half-filled d shell which represents a particularly stable configuration. This sulfide undergoes a first-order phase transition from the paramagnetic phase to an antiferromagnetic phase at $T_N = 48.2$ K (Hastings and Corliss 1976). Chattopadhyay et al. (1991, 1995) investigated the magnetic diffuse scattering from neutron diffraction. They found that this first-order antiferromagnetic phase transition of MnS_2 could not be explained purely by symmetry considerations. The center of the diffuse scattering does not coincide with the commensurate superlattice point (1, ½, 0) but corresponds to an incommensurate position $K = (1, k_y, 0)$, which is weakly temperature dependant (Fig. 17). This suggests that the antiferromagnetic phase transition in MnS_2 can be understood as a first-order lock-in transition from the incommensurate short-range order to the commensurate phase at T_N. Commensurate and incommensurate structures are represented schematically in Figure 18 (see also Makovicky 2006, this volume).

Pyrite (FeS_2). FeS_2 is diamagnetic and must therefore have its six d electrons paired and completely filling the t_{2g} orbitals. The low-spin configuration of Fe^{2+} in pyrite is an indication of the strength of the ligand field due to the disulfide anions. Figure 19 shows a schematic band model for the pyrite-type disulfides (such qualitative MO/band-model energy level diagrams for pyrite-type disulfides are discussed in greater detail by Vaughan and Rosso (2006, this volume). Semiconducting properties are observed for pyrite in which the e_g^* levels are expected to overlap to form a band. The observed energy gap (≈0.9 eV) and large negative Seebeck

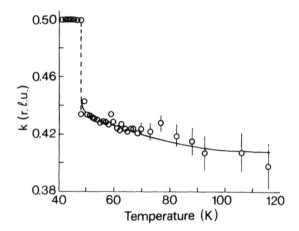

Figure 17. Temperature variation of the incommensurate component of the vector at which the diffuse magnetic neutron scattering is centered in MnS_2. [Reprinted with permission of American Physical Society, from Chattopadhyay et al. (1991), *Physical Review B*, Vol. 44, Fig. 5, p. 7396.]

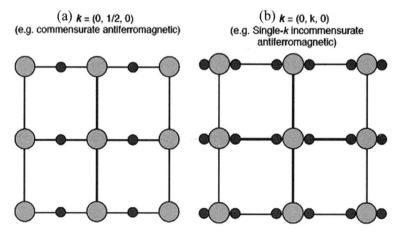

Figure 18. Schematic diagram of the relationship between nuclear (light grey) and magnetic (dark grey) diffraction peaks for (a) a commensurate antiferromagnet where the magnetic cell is a supercell of the nuclear structure and pure magnetic reflections appear as superlattice peaks in between the nuclear reflections, and (b) a single-k incommensurate antiferromagnet, where satellite peaks are present at position $+k$ and $-k$ from each nuclear reflection (Harrison 2006).

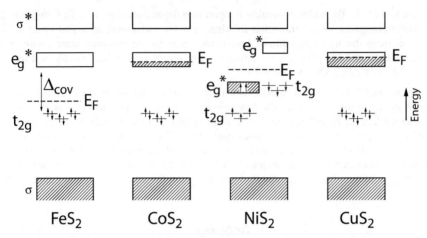

Figure 19. Schematic band-model energy level diagrams for the pyrite-type transition metal disulfides. E_F are the approximate Fermi levels; filled bands are shown shaded.

coefficient (Table 2) support a conduction process via electrons excited from the localized t_{2g} levels in the e_g^* band. Willeke et al. (1992) found that high purity single crystals prepared by chemical vapor transport are dominantly n-type conductors with carrier concentration (n) of 10^{16} cm^{-3} on the basis of Hall effect studies. Very similar results were obtained by Schieck et al. (1990) who also noted difficulties in synthesizing p-type single crystals. Harada (1998), who studied Hall effect along with conductivity and thermoelectric power of pure synthetic powder samples, showed that the sign of the dominant carriers in the intrinsic region is negative.

There has long been interest in reported deviations in the stoichiometry of pyrite from the ideal 1:2 ratio and its impact on properties. Birkholz et al. (1991) claimed from detailed X-ray diffraction measurements that synthetic pyrite can exhibit sulfur deficiency in the "few at% range." However, Ellmer and Hopfner (1997) undertook a detailed evaluation of all earlier work on the extent of pyrite nonstoichiometry and, combining this with theoretical arguments concerning the energetics of vacancy formation, concluded that deviations from the ideal stoichiometry do not exceed 1 at%. This would still be sufficient compositional variation to have a significant impact upon electrical properties. The considerable interest in the electrical properties of pyrite has arisen partly because of its possible use in photoelectrochemical and photovoltaic applications, particularly in solar cells. It is not our intention to discuss here the considerable literature dealing with this potential application, excellent reviews are provided elsewhere, for example, by Ennaoui and Tributsch (1984). One problem with using pyrite semiconductors in technical applications is the difficulty in obtaining high photopotentials with n-type or p-type material. The conductivity type can be adjusted by doping the crystals with P, As, Sb (p-type) or Co, Au, Cl, Br (n-type). High photo-effects are correlated with a low density of dislocations and low concentrations of impurities (Abd Al Halim et al. 2002). Ni-doped pyrite (Fe$_{0.99}$Ni$_{0.01}$S$_2$ – Fe$_{0.9}$Ni$_{0.1}$S$_2$) has been studied by Ho et al. (2004) and Ni shown to be an n-type dopant causing increasing conductivity with increasing concentration. Abd Al Halim et al. (2002) investigated the electrodeposition of Co and Cu on n-type pyrite slices. The free energy of interaction of a few monolayers of Co with pyrite was much higher than for the interaction of electrodeposited Cu with pyrite. This is because the Co atom was able to donate electrons into the conduction band of pyrite whereas the Cu atom is exchanging electrons with the valence band. The difference in free energy per electron corresponds to the width of the forbidden energy gap of pyrite (1 eV per electron). Not surprisingly, given the above findings, natural pyrite exhibits both n- and p-type semiconduction, sometimes within the same single crystal. Figure 20 indicates the variation in the conductivity of pyrite and a review of compositional, textural and electrical variations in pyrite is provided in Abraitis et al. (2004). It has also been suggested that natural pyrite formed at lower temperatures tends to be *p*-type (iron-deficient) whereas high-temperature pyrite tends to be *n*-type (sulfur-deficient). Electron mobility in natural pyrite is highly variable (Fig. 4) but generally is about two orders of magnitude greater than hole mobility (Shuey 1975). As far as is known, marcasite exhibits a similar range of electrical and magnetic properties (Figs. 2 and 5).

Cattierite (CoS_2). The magnetic data for cattierite (Table 2) indicate a *d* electron configuration of $(t_{2g})^6(e_g)^1$ and the metallic *n*-type conductivity observed supports conduction via the e_g^* band as shown in Figure 20. CoS_2 is very unusual amongst metal sulfides because it exhibits itinerant ferromagnetism below the Curie temperature of ~120 K, a phenomenon related to the very narrow e_g^* band width which also results in temperature-dependence of the paramagnetism above the Curie temperature. According to Adachi et al. (1969) the saturated magnetic moment for Co atoms in CoS_2 is 0.85μ_B. In the pyrite-cattierite solid solution $Fe_{1-x}S_2$-Co_xS_2, as Co is added, electrons occupy the previously empty e_g orbitals (in FeS_2) and, even with *x* values as low as 0.03, ferromagnetism is seen (Ramesha et al. 2004) although pure FeS_2 is a diamagnetic semiconductor.

Muro et al. (1996) undertook a detailed XMCD study of CoS_2 looking at both Co and S absorption edges. The Co 2*p* spectra and resulting XMCD are shown in Figure 21. The X-ray absorption spectrum (XAS) shows the Co $2p_{3/2}$ and $2p_{1/2}$ absorption peaks and the difference between the reversed magnetization directions is clear. The XMCD spectrum shows a positive peak corresponding to the $2p_{3/2}$ XAS and a negative peak at the $2p_{1/2}$ peak with the change

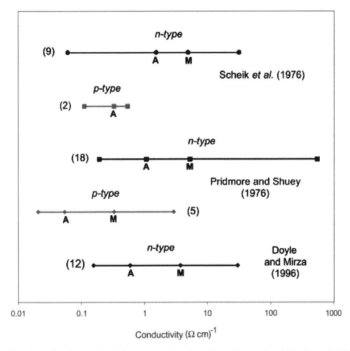

Figure 20. Variations in the conductivities of pyrite, based on the work of Doyle and Mirza (1996), Pridmore and Shuey (1976) and Schieck et al. (1990). Note the logarithmic conductivity scale. Maximum and minimum values are shown in each case. (A denotes the mean value, M denotes the median and the bracketed Figure nest to each bar denotes the number of measurements).

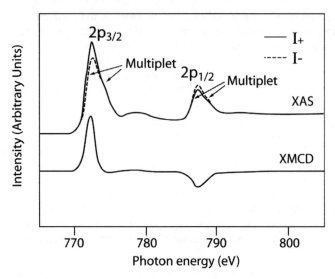

Figure 21. Co $L_{3,2}$ XAS absorption and XMCD spectra of synthetic CoS_2. The absorption spectra for the two magnetization directions are shown, with the XMCD difference spectra - XMCD intensity enhanced by a factor of three. [Used with permission of S. Suga and the American Physical Society, from Muro et al. (1996), *Physcial Review B*, Vol. 53, Fig. 2b, p. 7057.]

in sign across the two edges due to the different effect on the spin up/spin down electrons. Multiplet structure is observed in the main absorption peaks (indicated in Fig. 21) and the resulting XMCD is dependent on the individual multiplet components; in this case, the main contribution comes from the multiplet on the lower energy side of the main XAS peaks with a decrease in XMCD on the high energy side of the $2p_{3/2}$ peak. This dependence reveals the importance of the electron-core hole exchange and Coulomb interactions for the Co $2p - 3d$ excitation. Using the sum rules and the XMCD spectrum, the integrated intensity ratio of the Co $2p_{3/2}$ and $2p_{1/2}$ is 1.5:1.0 and the orbital magnetic moment, $<L_z>$, is evaluated as $0.060\mu_B$ and the $<L_z>/<S_z>$ is 0.18 ($<S_z>$ is the spin magnetic moment). The high orbital moment (>10% of the total moment) suggests localized 3d orbitals and the presence of multiplet structure in the spectrum supports localization. There are also satellite peaks 6 eV above the XAS and XMCD main peaks which are due to hybridization of the Co $3d$ and S states and Muro et al. (1996) also recorded the XMCD spectrum from the S $2p$ XAS. This confirms the S $2p$ hybridization with the Co $3d$ state and, therefore, a slight magnetic moment on the S atom. As a result of these magnetic properties, CoS_2 has potential use in the rapidly developing area of spintronics (Néel and Benoit 1953; Jarrett et al. 1968; Muro et al. 1996; Ramesha et al. 2004).

The magnetic properties of CoS_2 have also been examined using neutron diffraction by a number of groups (Ohsawa et al. 1976a, 1976b; Panissod et al. 1979; Brown et al. 2005). Panissod et al. (1979) used neutron diffraction in conjunction with NMR in their study of the $CoS_{2-x}Se_x$ solid solution which indicated that the magnetic properties are determined by an inhomogeneous distribution of moments on Co atoms. The neutron diffraction spectra contained overlapping magnetic components and nuclear components below the T_C of 120 K (Fig. 22). The NMR data confirm the presence of both ferromagnetic and paramagnetic Co atoms in $CoS_{2-x}Se_x$ with the appearance of the non-magnetic atoms relating to Se in the inner coordination sphere. This can be explained by the narrower band width of the antibonding $3p$-states of S in CoS_2 than the $4p$-states of Se in $CoSe_2$ (Yamada et al. 1998). The moment on the magnetic Co atoms was determined as $0.9\mu_B$, as in pure CoS_2 and at values $x < 0.28$

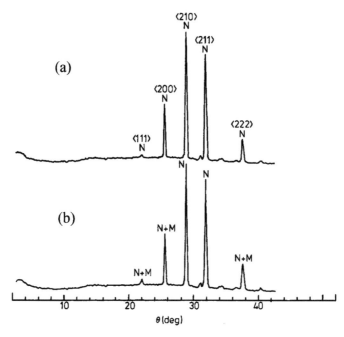

Figure 22. Neutron diffraction patterns ($2 \leq \theta \leq 52°$) for $CoS_{1.675}Se_{0.325}$ at 140 K (a) and 4 K (b) [Used with permission of IOP Publishing Limited, from Panissod et al. (1979).]

in $CoS_{2-x}Se_x$, almost all Co atoms are ferromagnetic. Polarized neutron diffraction has also revealed a small magnetic moment distributed around S atoms in CoS_2. The magnetic moment distribution is aspherical around the Co nuclei and results from a covalent bonding effect between $3d_{eg}$ orbitals of Co atoms and $3p\pi$ orbitals of S atoms (Ohsawa et al. 1976b).

Investigating the possibility that CoS_2 is a half-metallic ferromagnet, Brown et al. (2005) used unpolarised neutrons on a single crystal to define the crystal structure, atomic positions and a magnetic moment per cobalt atom of $0.882\mu_B$ at a temperature of 20 K. Using a 9.6 T applied field, the flipping ratio was determined for reflections of the ferromagnetic phase at 1.8 K and the paramagnetic phase at 150 K. The magnetization associated with the Co atoms in full cubic symmetry (h,k,l = all odd or all even) coincide and contribute to the crystal lattice reflections, while those associated with mixed index reflections are either magnetic in origin and associated with the non-cubic Co atoms or with the S atoms. The orbital contribution to the moment was 15% at 150 K, but was small at 1.8 K and 300 K, a factor of two less than that determined from XMCD measurements. The data suggest that the magnetization distribution around the Co^{2+} ions in the ferromagnetic phase at 1.8 K is nearly spherical as a result of e_g, t_{2g} mixing, with a small magnetic moment of approximately 2% of that on the Co atoms residing on the S atoms. At 150 K, just above the T_c the ordered paramagnetism has the aspericity characteristics of e_g electrons. At higher temperatures (300 K), the band with the highest density of states lies sufficiently close to the Fermi energy for thermal excitations to create holes in it, resulting in orbitals with t_{2g} character contributing more to the magnetization. The magnetization distribution observed in CoS_2 indicates that, in the ferromagnetic phase, both e_g and t_{2g} bands contain unpaired electrons. Therefore, CoS_2 is not a half-metallic ferromagnet (Brown et al. 2005).

Vaesite (NiS_2). This mineral exhibits unusual properties as a result of unpaired d electrons which result in the splitting of the t_{2g} and e_g^* energy levels into two sets containing electrons of opposite spin, *spin-up* and *spin-down* (Fig. 19). NiS$_2$ is a Mott-Hubbard semiconductor because the splitting is greater than the width of the e_g^* band and the resulting band gap (~0.27 eV at 300 K) (Kautz et al. 1972) is the energy required to excite an electron into the empty *spin-down* e_g^* band. The insulating ground state is only present when there is no vacancy on the Ni sites and disappears progressively with the introduction of vacancies or the substitution of S by Se to form a metallic phase (Adachi et al. 1969). These metal-insulator transitions of the Mott-Hubbard type play a fundamental role in the physical transformation of the 3*d* electronic states from localized atomic states on the insulating side into itinerant band states on the metallic side; no intermediate state is possible. Magnetic ordering effects in NiS$_2$ have been repeatedly investigated though magnetization and neutron diffraction studies. Neutron diffraction showed that magnetic ordering occurs in two steps: (1) a transition to the antiferromagnetic state takes place (40-60 K), with the Néel point depending on the stoichiometry, and (2) a second transition at 30 K to a weak ferromagnetic state which coexists with the first state. The magnetic moments in the two configurations are 1.0 and 0.6μ_B respectively (Honig and Spalek 1998).

CuS$_2$ and ZnS$_2$. The transition metal disulfides CuS$_2$ and ZnS$_2$ can be synthesized at high pressures but are not found in nature. In CuS$_2$ (d^9), the e_g^* band results in temperature-independent Pauli paramagnetism. This material is an example of "class D" compound of Jellinek (1968). Electrical resistivity data $\rho(T)$ for CuS$_2$ show a metallic behavior which is consistent with band-structure calculations. The Hall coefficient is weakly temperature dependant and positive, implying that the charge carriers and, therefore, the magnetic properties, are dominated by holes in the anion *p* band (Ueda et al. 2002). Krill et al. (1976) found a high *p* and *d* density of states at the Fermi level which is indicative of a relatively narrow conduction band. A sudden decrease of the bulk susceptibility was observed below 160 K. In ZnS$_2$, the *d* levels are completely filled with electrons so that the material is no longer a transition metal sulfide. Properties akin to ZnS are observed; i.e., it is a diamagnetic semiconductor with a large band gap.

Sulfides with halite (NaCl) structures

Galena (PbS). PbS is a diamagnetic semiconductor, the properties of which have been reviewed by Dalven (1969). The forbidden energy gap at 300 K is 0.41 eV and values for the isostructural PbSe (clausthalite) and PbTe (altaite) of 0.27 and 0.31 eV have been reported (Zemel et al. 1965). Figure 1b is a simplistic representation of the type of band structure found in galena, although a great deal more detailed information is actually known about the band structure of PbS (Vaughan and Rosso 2006, this volume). The conduction mechanisms and carrier concentrations, as in all such compounds, are very sensitive to precise stoichiometry. Slightly metal-rich galenas with *n*-type semiconductivity or sulfur-rich *p*-type semiconductors are readily synthesized. Most natural galena samples exhibit *n*-type conductivity, although carrier concentrations are strongly influenced by the presence of silver, bismuth and antimony in the samples (Kravchenko et al. 1966). Shuey (1975) observed that galena exhibiting *p*-type conductivity occurs particularly from limestone-lead-zinc ("Mississippi Valley type") mineral deposits and silver-rich hydrothermal deposits. Conductivities and transport phenomena in synthetic samples are discussed by Dalven (1969). Conductivities range over several orders of magnitude. The plot of hole mobility against reciprocal temperature in Figure 4 is for a synthetic *p*-type galena, although other *n*- and *p*-type galenas exhibit similar behavior over this temperature range. Nimtz and Schlicht (1983) studied a series of narrow-gap semiconductors based on PbS over a range of temperatures and pressures, including Sn-bearing sulfides, selenides and tellurides. In particular, optical properties, band structures and transport properties were investigated.

Alabandite (α-MnS). This sulfide actually exists in three crystalline modifications, the halite structure (alabandite, α-MnS) and the sphlalerite and wurzite structures, but α-MnS is the most stable form. Neutron diffraction measurements indicate that, below its Néel temperature (150 K), α-MnS is a high spin antiferromagnet (Corliss et al. 1956). Experimentally, Mn moments of $\mu = 4.54\mu_B$ have been reported. According to the Hall measurements, the conductivity in α-MnS is a result of holes in the 3d band of the manganese ions (Aplesnin et al. 2004), with a hole concentration per Mn ion of $n \sim 0.1$ at $T = 435$ K (Aplesnin et al. 2005).

The electronic structure of α-MnS has been studied by Tappero and Lichanot (1998), Tappero et al. (2001) and Youn et al. (2004), using quantum mechanical calculations and it provides an example of the differences between various electronic structure models. Tappero and Lichanot (1998) showed that the predicted electronic structure of α-MnS is quite different when calculated using a Hartree-Fock (HF) compared to a Density Functional Theory (DFT) formalism (see Vaughan and Rosso 2006, this volume). In the HF scheme, α-MnS is a highly ionic compound with a magnetic moment per Mn^{2+} of $4.90\mu_B$ corresponding to a high spin electronic configuration $t_{2g}^3 e_g^2$. It has strong insulator character as shown by the band structure. Using DFT, α-MnS is a less ionic compound, the Mn d shell is populated with spin down electrons and the magnetic moment is $4.40\mu_B$. Comparisons with the experimental data ($\mu = 4.54\mu_B$) indicate that the DFT scheme gives a better description of the electronic structure of α-MnS. However, recent simulations of the neutron diffraction spectra of α-MnS have shown that the use of the exact HF exchange potential is more suitable for an accurate determination of electron charge and spin densities (Tappero et al. 2001). Youn et al. (2004) applied local density approximation band calculations, where the strong Mn d correlation effect is taken into consideration, to obtain an enhanced energy gap and magnetic moment for α-MnS, more comparable with the experimental data. α-MnS is a band insulator even though correlation effect between the d electrons is large, and can be considered as a "crossroads" material between a charge transfer and a band insulator.

MnS can also be doped with other transition metal (*TM*) ions to form the disordered system of $Mn_{1-x}TM_xS$ (*TM* = Cr, Fe, V) in which metal-insulator transitions, connected with electron localization at the edge of the conduction band have been observed. The disordered solid solutions form two antiferromagnetic 3d metal monosulfides at intermediate compositions, exhibiting magnetic properties different from the original MnS (Petrakovskii et al. 1995). The diluted magnetic semiconductor $Mn_{1-x}Fe_xS$ exhibits colossal magnetoresistance, i.e., a strong resistivity decrease in an applied magnetic field.

Rare earth sulfides. A number of the rare-earth sulfides, including europium, gadolinium and lanthanum sulfide, have simple NaCl structures. They are magnetic semiconductors, exhibiting pure spin magnetism with magnetic ordering varying from ferromagnetic (EuS) to antiferromagnetic (GaS) and diamagnetic (LaS) (Hauger et al. 1976; Klein et al. 1976). These sulfides have potential applications in electroluminescent devices as a result of their electro-optical characteristics.

Sulfides with niccolite (NiAs)-based structures

The pyrrhotites are metal-deficient iron monosulfides, $Fe_{1-x}S$, and form a structurally complex series between monoclinic Fe_7S_8 and hexagonal FeS (Makovicky 2006, this volume). Troilite is the low temperature (<140 °C) stoichiometric form of FeS. The pyrrhotites are based on the NiAs-type structure, and their ferri- and antiferromagnetic magnetic properties have been of particular interest to mineralogists, geologists and mineral technologists.

Troilite (FeS). This has alternating Fe and S layers parallel to 001 with all the metal sites filled. The magnetic properties of FeS have been studied by a number of authors (Haraldsen 1941a, 1941b; Benoit 1952; Murakami and Hirahara 1958; Schwarz and Vaughan 1972; Kruse 1990; Li and Franzen 1996; Li et al. 1996). The change in magnetization of troilite during heating is shown in Figure 23a and indicates that above the so-called "β-transition" paramagnetic behavior is observed. An effective magnetic moment of 5.24 μ_B has been

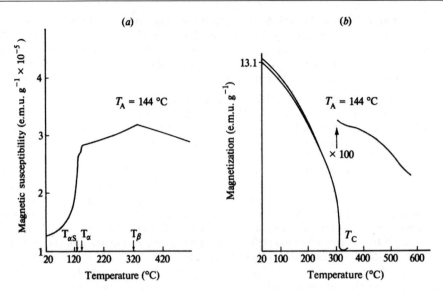

Figure 23. Thermomagnetic curves showing the thermal change in susceptibility or magnetization (J, applied field, 5350 Oe) for: (a) troilite and (b) monoclinic pyrrhotite; T_A is annealing temperature of synthetic sample, T_α, T_β and $T_{\alpha S}$ are transition temperatures and T_c is the Curie temperature. After Schwarz and Vaughan (1972).

reported from high-temperature susceptibility measurements. The β-transition, which extends to metal-deficient compositions but still occurs at ~315 °C, has been attributed by some authors to a first-order crystallographic transformation. However, Hirone et al. (1954) attribute the β-transition to the Néel temperature over the whole range of troilite–pyrrhotite compositions, a theory supported by specific heat measurements in this temperature range which agree with predicted values for such a mechanism. In FeS, below T_β, antiferromagnetic behavior is observed, but two further transitions occur. The α-transition represents the change from the high-temperature NiAs structure to the troilite superstructure. The second transition, in some reports occurring at the same temperature as T_α (140 °C), is attributed to a change in orientation of the antiferromagnetically ordered spins. This "Morin transition" (at temperature $T_{\alpha S}$) is associated with the structural transformation at T_α and has been studied by neutron diffraction and Mössbauer spectroscopy (Andresen and Torbo 1967). In an X-ray diffraction and Mössbauer study of phase transitions in synthetic FeS and the associated kinetics, Kruse (1990) found that the α-transition is spread over ~40 °C but is complete at 140 °C and the transition involving a change in spin orientation is completed by 159 °C in these synthetic samples. However, the later effect was not complete at 156 °C after 23 days when troilite from a meteorite was used.

Determination of its electrical properties shows that troilite is a semiconductor which exhibits considerable electrical anisotropy and a sharp change in c-axis conductivity at T_α (Hirahara and Murakami 1958). Townsend et al. (1976) have observed a semiconductor-metal transition in FeS at T_α from electrical, magnetic and Mössbauer measurements on single crystals. A semiconductor-metal transition has also been observed by Kobayashi et al. (2005) at high pressure (2.9 GPa) and low temperature (100 K). Troilite (i.e., FeS below T_α) is a ferroelectric material, a property observed and discussed by van den Berg et al. (1969).

Pyrrhotite **(Fe_7S_8).** The Fe_7S_8 endmember has a monoclinic "defect" NiAs structure, in which cation-deficient layers alternate with fully occupied cation layers. The vacancies are in alternate sites, in every other row of cations in the layer, and the ABCD stacking of the vacancy layers leads to a 4C notation for the classic pyrrhotite structure. Increased iron in the structure filling the vacancies ($Fe_{7+x}S_8$) leads to less regular metal ordering in the layers and more complex stacking of the metal deficient layers and most commonly a hexagonal symmetry (Lotgering 1956; Makovicky 2006, this volume).

Monoclinic pyrrhotite is a very highly conducting material (see Fig. 2), although there appears to be some dispute regarding the actual mechanism of the conductivity. Synthetic samples show a room-temperature resistivity of $\approx 3 \times 10^{-4}$ Ωcm which increases with decreasing temperature as in semiconductors, whereas natural samples exhibit a temperature dependence characteristic of metals. Theodossiou (1965) made resistivity and Hall effect measurements on natural samples; p-type conduction was observed both normal and parallel to the c-axis direction at room temperature (although n-type conduction was found perpendicular to the c-axis below −108 °C). Synthetic hexagonal pyrrhotite has a room-temperature resistivity of ~10^{-2}–10^{-3} Ω cm (Perthel 1960) and thermoelectric power measurements indicate p-type conduction below 225 °C and n-type above this temperature (Kamigaichi 1956). Method of preparation strongly influences the electrical, magnetic and structural properties of these "intermediate" pyrrhotites.

The magnetic properties of pyrrhotites with compositions between FeS and Fe_7S_8 ($Fe_{0.875}$ S) can best be visualized with reference to the magnetic phase diagram shown in Figure 24. Here, synthesized initial compositions (annealed at 144 °C) are plotted along the abscissa, and the ordinate combines a temperature scale and a susceptibility/magnetization scale. Solid lines on the diagram are magnetic phase transitions with reference to the temperature scale, and dashed lines represent susceptibility in the region where antiferromagnetic behavior is observed. Dotted lines occur in the ferromagnetic region of the diagram and represent the estimated saturation

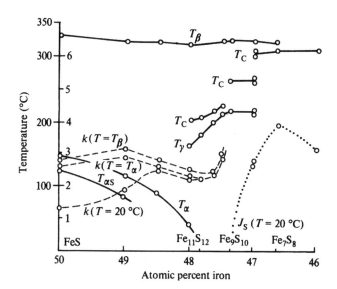

Figure 24. Magnetic phase diagram for synthetic pyrrhotite samples annealed at 144 °C. Initial compositions are plotted along the abscissa and the ordinate combines a temperature/magnetization scale. The solid lines are magnetic phase transitions with reference to the temperature scale and the dashed lines represent susceptibility in the region where antiferromagnetic behavior is observed (scale of $k \times 10^5$ emu g^{-1}). The dotted line appears in the ferromagnetic region and represents the estimated saturation magnetization J_s (on this scale $J_s \times 1/5$ emu g^{-1}) at 20 °C. The susceptibility is recorded at 20 °C and at the β-transition and α-transition temperatures. After Schwarz and Vaughan (1972).

magnetization (J_s). Susceptibility and saturation magnetization are shown at 20 °C and at the α- and β-transition temperatures. Compositions containing ≥48 at% iron ($Fe_{0.92}S$) exhibit antiferromagnetic behavior. The origins of the β-transition, α-transition and Morin transition ($T_{\alpha s}$) are as discussed for troilite, although with increasing metal deficiency the α-transition corresponds to the solvus between the "troilite + hexagonal pyrrhotite" two-phase field and the hexagonal pyrrhotite field. The region of 48.0 to 47.0 at% iron (~$Fe_{0.92}S$ to $Fe_{0.89}S$) (containing the "intermediate" or "low-temperature hexagonal" pyrrhotites) is of particular interest. Such compositions, at the so called γ-transition or anti-Curie point of Haraldsen (1941a,b), exhibit a very sharp increase in susceptibility which is retained over only a short temperature range (see Fig. 24). This apparent ferrimagnetism disappears at a "Curie temperature" well below T_β. At compositions <47 at% Fe, the pyrrhotite is ferrimagnetic (note that Fe_7S_8 contains 46.67 at% Fe). Thermomagnetic curves for natural monoclinic pyrrhotite (Fig. 23b) show characteristic behavior, i.e., initially high magnetization which gradually decreases on heating to the Curie temperature (315 °C).

Various theories of both vacancy and spin order = disorder transitions have been put forward to explain the magnetization in pyrrhotites (Hirone et al. 1954; Lotgering 1956; Kawiaminami and Okazaki 1970; Ward 1970; Schwarz and Vaughan 1972). According to Néel (1953), the spontaneous room temperature ferrimagnetic behavior of monoclinic pyrrhotite is due to the ferrimagnetic alignment of cations (Fe) within layers, with an antiparallel arrangement of the spins on the iron atoms on successive cation layers in the (001) plane (Bertaut 1953). There is antiferromagnetic coupling between layers and, as there are uncompensated magnetic moments, this results in the ferrimagnetic behavior. Kruse (1990) undertook a Mössbauer investigation of hexagonal pyrrhotites and noted that repulsion may inhibit the development of monoclinic pyrrhotite at room temperature. Marusak and Mulay (1980, 1979) used data from X-ray diffraction, magnetic and Mössbauer studies, and arguments based on reaction rate theory, to support a vacancy ordering model. Here, the observed ferrimagnetism above the gamma transition (Fig. 24) is attributed to a vacancy-ordered 4C superstructure and the antiferromagnetism below it to a different (5C) superstructure (Fleet 2006, this volume; Makovicky 2006, this volume). Novakova and Gendler (1995) also used Mössbauer spectroscopy to observe hexagonal and monoclinic nonstoichiometric pyrrhotites with varied degrees of vacancy ordering and disordering during pyrite oxidation.

Li and Franzen (1996) and Li et al. (1996) undertook a detailed thermometric study of the structural and magnetic transitions in the "$Fe_{0.875}S$ – FeS" series (Fe_7S_8, pyrrhotite – $Fe_{1.0}S$, troilite). They confirmed the previous structural changes, noting also a transformation in Fe_7S_8 to a trigonal 3C stacking between 200-240 °C and, in the range of $Fe_{0.88}S$ to $Fe_{0.92}S$ a magnetic "λ-transition" takes place with an increase and then decrease in magnetization over the range 200-250 °C, consistent with the anti-Curie point behavior recorded by Haraldsen (see above). A large enthalpy change at 315 °C determined using differential thermal analysis (DTA) of Fe_7S_8 is coincident with the magnetic transition and is associated with the disordering of the vacancies and offers further confirmation that this is the Curie temperature of monoclinic Fe_7S_8 (Li and Franzen 1996; Li et al. 1996).

A very detailed, time of flight powder neutron diffraction study of synthetic Fe_7S_8 in the temperature range 11-773 K by Powell et al. (2004) further confirmed the existing magnetic models. The powder diffraction data produced (Fig. 25) fitted well with the assumed monoclinic unit cell, but there were major discrepancies between calculated and observed intensities, especially the strongest reflection at 5.689 Å. Although this represents $(001)_{NiAs}$ there is no calculated reflection at this angle of 2θ and this is also the case for reflections at 2.62 Å. These additional reflections/intensities are magnetic in origin and coincident with the potential structural reflections. Thermometric studies revealed the $(001)_{NiAs}$ magnetic reflection decreases with increasing temperature and disappears at 325 °C, with little hysteresis

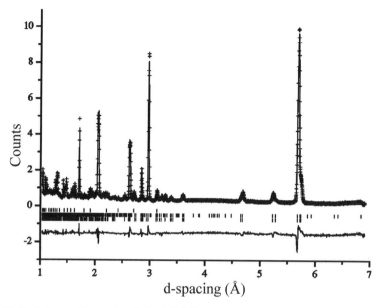

Figure 25. Final observed (crosses), calculated (full line) and difference (lower full line) neutron profiles for Fe_7S_8 collected at 11 K. [Used with permission of A. Powell and the American Physical Society, from Powell et al. (2004), *Physical Review B*, Vol. 70, Fig. 2, p. 014415-3.]

on cooling (Fig. 26). This disappearance is coincident with the loss of the monoclinic 4C structure and the formation of a hexagonal, cation-deficient NiAs structure, in which the vacancies are statistically distributed between all the layers, demonstrating the link between magnetic and structural properties. The ordered magnetic moment was measured as $3.16\mu_B$ at 11 K and $2.993\mu_B$ at 298 K.

An XMCD study of monoclinic (4C) pyrrhotite was recently carried out by Pearce et al. (2006a). The results have not yet been analysed in terms of the absolute magnetic moments for the individual atoms but the XMCD spectrum, produced due to the net magnetic moments in the cation deficient layers, is shown in Figure 27.

A SQUID study of selenium-bearing pyrrhotites (Ericsson et al. 2004, 1997) revealed a spin-flip transition due to the re-orientation of the spins from perpendicular (parallel to *a,b*) to parallel to the c axis that is not present in S-rich members of the Fe_7S_8-Fe_7Se_8 series. By varying the applied fields the relationship between composition and vacancy ordering was established with superstructures (of 3C and 4C) close to the end-member compositions ($y < 0.15$; $0.85 < y$ in $Fe_7(S_{1-y}Se_y)_8$. Figure 28 shows the magnetization versus applied field curves measured at 150 K for four different compositions that illustrate the different types of behavior. The transition temperature is lowered below the 130 K found in Fe_7Se_8 with increasing S-content.

CoS and NiS. The cobalt and nickel monosulfides with the NiAs-type structure can be metastably retained on quenching from high temperature and have been studied to determine their magnetic and electrical properties. CoS is antiferromagnetically ordered with $T_N = 358$ K; the effective magnetic moment on the cobalt atom is $1.7\mu_B$ (Benoit 1955). Metallic conductivity has been reported. NiS is extremely interesting; Sparks and Komoto (1968) observed a phase transition from a low-temperature antiferromagnetic semiconductor to a paramagnetic metallic phase at close to room temperature ($T_N \approx 263$ K). This transition temperature is strongly dependent on the sulfur content (NiS – $NiS_{1.03}$ is the approximate range of compositions) and

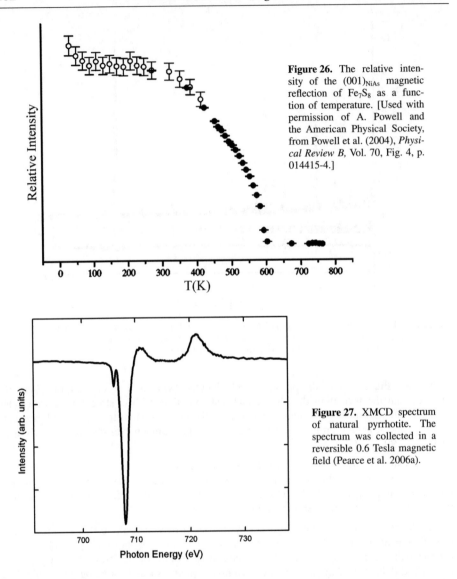

Figure 26. The relative intensity of the $(001)_{NiAs}$ magnetic reflection of Fe_7S_8 as a function of temperature. [Used with permission of A. Powell and the American Physical Society, from Powell et al. (2004), *Physical Review B*, Vol. 70, Fig. 4, p. 014415-4.]

Figure 27. XMCD spectrum of natural pyrrhotite. The spectrum was collected in a reversible 0.6 Tesla magnetic field (Pearce et al. 2006a).

detailed electrical, magnetic and Mössbauer studies of this transition have been performed (Townsend et al. 1971b; Coey and Brusetti 1975). The low-temperature form of NiS (millerite) is reported to be metallic and Pauli paramagnetic (Hulliger 1968), properties attributed to the metal-metal bonding which stabilizes this structure (Grice and Ferguson 1974; Rajamani and Prewitt 1974; Vaughan and Rosso 2006, this volume). McCammon and Price (1982) have studied the magnetic behavior of $(Fe,Co)S_{1-x}$ solid solutions using Mossbauer spectroscopy; samples with >16% Fe show magnetic ordering but, below that concentration, no ordering is observed down to 4.2 K. The absence of magnetic ordering at the lower Fe concentrations is attributed to a substantial increase in electron delocalization towards the ligands as the M-S distance decreases, rather than a high-spin to low spin transition which is ruled out by the Mossbauer (isomer shift) parameters.

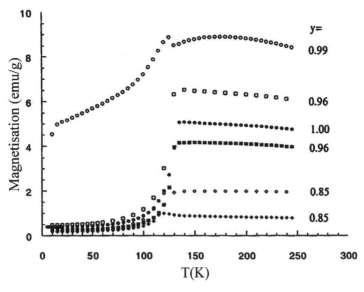

Figure 28. Magnetization measured at increasing temperatures in an external field of 1 kOe for zero field cooled $Fe_7(S_{1-y}Se_y)_8$ samples. The abrupt change in magnetization taking place at 130 K in Fe_7Se_8 is due to a spin-flip. Open symbols indicate samples annealed at 423 K; filled symbols indicate samples annealed at 873 K. [Used by permission of E. Schweizerbart'sche Verlagsbuchhandlung <http://www.schweizerbart.de>, from Ericsson et al. (1997), Eur. J. Mineral. Vol. 9, p. 1131-1146.] (Ericsson et al. 1997).

Mackinawite, smythite, greigite and other thiospinels

Three other iron sulfide minerals should be mentioned because of their geochemical importance or potential relevance to magnetic studies. The tetragonal iron monosulfide, mackinawite (FeS) is the first formed sulfide in many modern sedimentary environments; a range of transformations of this mineral can occur, leading ultimately to pyrite or pyrrhotite (Fleet 2006, this volume; Rickard and Luther 2006, this volume), but in some cases involving greigite (Fe_3S_4). The apparently much rarer mineral smythite (probable composition ~Fe_9S_{11}), like the other two phases, appears only to occur in very low temperature hydrothermal (Krupp 1994) or sedimentary environments. These minerals only occur as fine particles and cannot be synthesized except via precipitation from aqueous solution. Mackinawite has a layer structure, greigite has a spinel structure (as the sulfide analog of magnetite) and smythite appears to be structurally related to pyrrhotite but with an even higher concentration of vacant sites than is found in monoclinic pyrrhotite (Makovicky 2006, this volume).

The lack of single crystal material for all of these phases, and the metastable nature of mackinawite and greigite, in particular, has limited attempts to define their electrical and magnetic properties. From their structures and compositions, all three are assumed to be highly conducting (although a mackinawite crystal might be expected to show lower conductivity normal to the layers in its structure). Unlike electrical properties, magnetic properties can be measured in bulk, and inferences drawn from techniques such as Mössbauer spectroscopy (Wincott and Vaughan 2006, this volume). Mackinawite shows no magnetic ordering down to very low temperatures (<4.2 K) and is believed to be a Pauli paramagnet. Both greigite and smythite show evidence of being magnetically ordered up to (and probably above) room temperature. The Curie temperature of synthetic greigite has been reported at between 297 and 527 °C based on various extrapolations, but several authors urge caution over such data as greigite probably begins to break down at ~200 °C (Roberts 1995; Snowball and Torii 1999). Smythite has a net magnetic moment and Mössbauer parameters in line with the kind of ferrimagnetism seen in monoclinic pyrrhotite (Hoffmann 1993). Greigite is also regarded as a ferrimagnet, with this behavior arising in the same way as for magnetite (Fe_3O_4), although the saturation magnetization per formula unit is 2.5× less than Fe_3O_4 and it shows no Verwey transition. An XMCD study of greigite indicated the presence of iron vancancies in both natural and synthetic Fe_3S_4 leading to a structure similar to that for the lacunary iron oxide maghemite γ-Fe_2O_3 (Letard et al. 2005).

Such departures from stoichiometry have been found to affect the magnetic and transport properties in a range of thiospinels. In the semiconducting and ferromagnetic chromium chalcogenide spinels ($A^{2+}Cr^{3+}_2S_4$), such as $CdCr_2S_4$, $CoCr_2S_4$ and $FeCr_2S_4$, the defects appear as sulfur vacancies, affecting the Curie point and the saturation magnetization. The type of conduction (n-type or p-type) depends on the nature of the defects and the ion in the spinel lattice (Gibart et al. 1976). Natural and synthetic thiospinels of the series polydymite (Ni_3S_4)-violarite ($FeNi_2S_4$)-greigite (Fe_3S_4) have been studied using Mössbauer spectroscopy to show low-spin Fe^{2+} in the octahedral sites (Vaughan and Craig 1985). Ni_3S_4 is metallic and it exhibits itinerant electron ferrimagnetism as a result of strong covalence (Manthiram and Jeong 1999). Certain physical properties of these and other thiospinel minerals including carrollite ($CuCo_2S_4$), linnaeite (Co_3S_4) and siegenite [(Co, Ni)$_3S_4$], which occur in ore deposits and daubreelite [(Fe,Mn,Fe)Cr_2S_4], which is present in meteorites, have been interpreted using molecular orbital and band theories (Vaughan et al. 1971).

Only in the last decade has it become clear that, in fine grained marine and terrestrial sediments, greigite is an important carrier of the magnetic information needed by paleomagnetists. In particular, the neoformation of greigite in sediments can complicate or compromise studies of environmental magnetism and geomagnetic field behavior using such rocks (Roberts 1995; Jiang et al. 2001; Roberts and Weaver 2005). Dekkers et al. (2000) provide detailed information on the rock magnetic properties of greigite, and this topic is discussed further below.

PROPERTIES OF SULFIDE NANOPARTICLES

This chapter has been chiefly concerned with the electrical and magnetic properties of bulk sulfide materials. In the past two decades there has been considerable interest in nanoparticles of sulfide-based semiconductors with critical dimensions in the range of 1-20 nm, many of which are "quantum dots." These small particles can be viewed as having properties intermediate between a molecule and a bulk solid. The optical, magnetic, electrical and chemical properties of such nanoparticles are often strongly composition, size, and shape dependant, resulting in novel uses in a range of optical or electronic devices, as well as in catalysts. Two fundamental factors, related to the size of the individual semiconductor nanoparticles, are responsible for their unique properties. The first is the large surface to volume ratio. As the particle becomes smaller, the ratio of the number of surface atoms to those in the interior increases, e.g., in a typical approximately spherical 1nm nanoparticle ca 80% of the atoms are at the surface. The surface plays a very significant role in the properties of the nanomaterial. The second factor is the relationship between the electronic properties of the material and the critical dimensions of the particle. In semiconductors, these size dependant properties appear when the radius of the particle is comparable to the excitonic (Bohr) radius of the bulk material. The band gap increases in energy with decreasing particle diameter, as a result of a quantum size effect due to confinement of the electron and hole in a small volume. This phenomenon gives rise to discrete energy levels rather than the continuous band found in the corresponding bulk material (Pickett and O'Brien 2001). This produces a shift in the band gap, and light absorption moves towards the higher energy end of the spectrum accompanied by the appearance of a strong peak, excitonic in origin rather than the band edge associated with a bulk material. The manipulation of particle size, as well as chemical composition, leads to control over the chemical and physical behavior of the nanoparticles (Kulkarni et al. 2001). The variation of band gap with size can be modeled using the particle in the box model as first described by Brus (1984).

Chalcogenide semiconductor nanomaterials have been studied in detail, especially those of zinc, cadmium, mercury and lead, and there is some work on transition metal sulfides e.g., those of copper and nickel. Potential applications are anticipated in color change materials such as phosphors, and in photovoltaics. The quantum size effects described above have been demonstrated in a study of quantum dot sized particles of CdS with different and well-defined size distributions (Vossmeyer et al. 1994). A shift in the excitonic peak to energies as high as 4.8 eV was observed for nanoparticles with a mean diameter of 0.64 nm, compared to bulk CdS, which has a 2.42 eV band gap (Fig. 29). Substitutional doping of these chalcogenide semiconductor nanomaterials with paramagnetic transition-metal ions can also produce magnetic materials termed dilute magnetic semiconductors (DMS). Figure 30 shows a comparison of the photoluminescence and photoluminescence excitation data for nanoscale and bulk ZnS:Mn. The large shift of 1215.5 nm in the excitation spectrum is attributed to an increase in the value of the s-p electron band gap in the ZnS nanocrystal from quantum confinement. Reducing the particle size will also affect the magnetic properties of the phases. For example, α-NiS nanoplates show paramagnetism with weak ferromagnetic interactions, unlike the bulk counterpart which is antiferromagnetic (Zhang et al. 2005a). The possibility of using sulfide-bearing nanoparticles as magnetic devices remains largely unexplored, although the magnetic properties of a range of hybrid materials containing NiS has been assessed by Lui et al (2005).

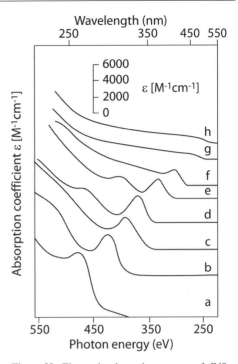

Figure 29. Electronic absorption spectra of CdS nanoparticles with different mean diameters (nm): (a) 0.64, (b) 0.72, (c) 0.8, (d) 0.93, (e) 1.16, (f) 1.94, (g) 2.8 and (h) 4.8. [Redrawn with permission from Vossmeyer et al. Copyright 1994 American Chemical Society.]

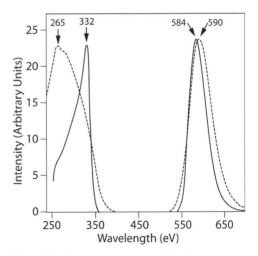

Figure 30. Photoluminescence and photoluminescence excitation spectra of bulk (solid lines) and nanocrystalline (dashed lines) ZnS:Mn. [Used with permission of American Physical Society, from Bhargava et al. (1994), *Physical Review Letters,* Vol. 72, Fig. 1, p. 4162.]

Synthesis of sulfide nanoparticles

Extensive studies have been reported of preparative methods for sulfide nanomaterials, and a range of morphologies including terapodal structures, rods, wires, plates, oblate proloids and spheres are known, as reviewed in Trindade et al. (2001). Fabrication of ordered assemblies of these nanoparticles can be achieved through "top down" processes involving lithography and etching. However, this approach usually only yields semiconductor nanostructures of good structural and optical quality for sizes larger than 50 nm. Below this size, these fabrication processes result in surface damage which has a detrimental effect on the structural quality and makes it difficult to differentiate between intrinsic and extrinsic properties in the study of electronic and magnetic properties as a function of nanoparticle size. There are also a number of "bottom up" fabrication techniques that rely on nucleation and precipitation processes, either in glass matrices including sol-gel methods, or in solution including the trioctyl-phosphine/ trioctylphosphine oxide (TOP/TOPO) solvent system (Brieler et al. 2005; Trindade et al. 2001). Chemical synthesis of nanoparticles can also be achieved by using a size-limiting "host" such as a surfactant in a nonpolar solvent to form reverse micelles, or the use of microporous substances such as zeolites (Wang and Herron 1987). Several types of nanoparticles synthesized using biomimetric processes have been reported, including bio-inorganic complexes of CdS-ferritin (Wong and Mann 1996; Trindade et al. 2001). In a number of cases, nanoparticles need to be "capped" to inhibit surface reactions, and to avoid agglomeration and Ostwald ripening. An example is the production of CdS from cadmium sulfate and ammonium sulfide, where styrene is used as a capping agent (Boxall and Kelsall 1991).

Complex nanocomposites can be made by successive reactions, for instance reacting relatively soluble HgS nanoparticles with dissolved Cd to form HgS with a CdS coating (Häesselbarth et al. 1993); an early example of a "core-shell" particle. Advances in the development of sulfide-polymer nanocomposites for optical, magnetic, and electronic uses is driven by new fabrication methods which are devised to produce emitting diodes, photodiodes, solar cells and gas sensors (Godovsky 2000). These often comprise conductive polymer matrices containing embedded semiconducting nanoparticles of sulfides such as ZnS, CdS, PbS or CuS. Light emitting diodes and lasers exploit the electroluminescence properties of nanocomposites that result from the radiative recombination of holes in the valence band and electrons in the conduction band (Godovsky 2000). A very efficient electroluminescent nanocomposite comprises relatively high band gap CdS surrounding a lower band gap CdSe core (with a semi-conducting polymer and a dodeylamine capping agent). Nanocomposite devices are much more efficient than those employing only CdSe, exhibiting enhanced properties such as low operating voltages and voltage dependent color emission (Schlamp et al. 1997).

It is not possible to provide comprehensive coverage of the sulfide nanoparticle literature in this chapter due to the vast amount of information available. The pertinent concepts will, therefore, be covered in a case study of ZnS in both pure and doped form.

Case study: pure and transition metal (*TM*) ion doped ZnS

ZnS is a wide band gap II-IV semiconductor and is a good host material as a result of its band gap characteristics (~3.6 eV). *TM* ion doping in ZnS nanocrystals (ZnS:*TM*) enhances the optical transition efficiency of the charge carriers (electrons or holes) and increases the number of optically active sites (Karar et al. 2004). ZnS:*TM* may also exhibit ferromagnetism, allowing the potential for bifunctional devices with luminescent and magnetic properties. Manganese is one of the most common *TM* dopants because: (1) it can be incorporated into a (II, IV) semiconductor host in large proportions without altering the crystal structure; (2) it has a relatively large magnetic moment and (3) it is electrically neutral in a (II, IV) host, thus avoiding the formation of any acceptor or donor impurities in the crystal (Bandaranayake et al. 1997). Manganese usually substitutes in the Zn-lattice sites as a divalent ion with a tetrahedral crystal field (Brieler et al. 2005). Figure 31 shows a TEM micrograph of doped chalcogenide semiconductor nanoparticles. The excitation and decay of the Mn^{2+} ion produces an intense yellow luminescence at ~585 nm as a result of the electronic interaction between the Mn^{2+} ions d-electron states and the s-p electronic states of the host ZnS nanostructure (Bhargava et al. 1994). The incorporation of *TM* ions into the ZnS nanostructures is revealed in the photoluminescence spectra (Singh et al. 2004). The luminescence of ZnS at 425 nm is affected by *TM* dopants in terms of the decay time and the intensity (Kulkarni et al. 2001). Yuan et al. (2004) found that the intensity of the ZnS-related emission decreased with increasing Mn concentrations, while the Mn^{2+} emission intensity increased with an increasing concentration of Mn^{2+} incorporated in the nanostructure at lower Mn concentrations. However, the decrease in Mn^{2+} emission at higher Mn contents was possibly due to the presence of Mn^{3+} or Mn^{4+} ions, which act as efficient quenchers of the emission due to the low position of the charge transfer absorption band of these ions in the sulfide. Figure 32 shows a schematic diagram of the complex optical processes observed in wide gap (II, Mn) VI semiconductors. The magnetic properties of the ZnS:Mn system also provide information about the relationship between the s-p electron hole of the host (ZnS) and the d-electrons of the impurity (Mn). Magnetization is often found to be higher in ZnS:Mn nanostructures as compared with their bulk counterparts (Yuan et al. 2004). Schuler et al. (2005) used X-ray emission and absorption spectroscopy to examine the electronic structure of $Zn_{1-x}Mn_xS$ and found that these compounds exhibit similar behavior to that of MnS within 4 eV of the Fermi energy, with the Zn s and p states providing little influence on the Mn $3d$ bonding states. A band gap of 3 eV (±1.0 eV) was found.

TM doped ZnS nanoparticles can be synthesized using various methods such as solid-state reaction, sol-gel processes and hydrothermal techniques (Table 4). The first synthesis of nanoscale Mn-doped ZnS employed a precipitation reaction and was reported by Bhargava et al (1994). They maintained particle separation with a surfactant, which resulted in an increase in the intensity of the yellow Mn^{2+} emission as a result of surface passivation of the ZnS:Mn particles. Nanoscale TM doped ZnS is used commercially as a phosphor and in electroluminescent devices such as electronic displays (Trindade et al. 2001). Correct treatment of the doped ZnS nanoparticle surface also allows for selective attachment to specific bacterial cell walls for

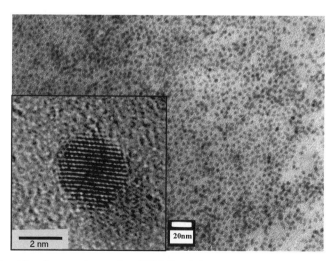

Figure 31. Transition electron micrograph and High Resolution-transition electron microbgraph showing doped chalcogenide semiconductor nanoparticles. The lattice fringes in the HRTEM micrograph are indicative of the crystalline nature of the particles. [Reproduced with kind permission of Paul O'Brien.]

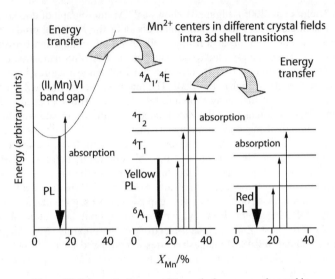

Figure 32. Schematic diagram of the optical processes observed in wide-gap (II,Mn)VI semiconductor nanoparticles (Brieler et al. 2005).

Table 4. Routes for the preparation of ZnS and ZnS:Mn nanoparticles.

Synthetic Method	Particle Size Range (nm)	References
Solutions of polymeric aducts of zinc alkyls + H_2S	2-4	Li et al. 1994
Arrested precipitation in aqueous and methanolic solutions	<2	Rossetti et al. 1985
Photodegradation of 3 nm particles or precipitation in phosphate containing solution	1.7	Weller et al. 1984
Treatment of zinc containing copolymers with H_2S	<3	Sankaran et al. 1993
Thermolysis of single molecule precursors in TOPO	3.5-4.2	Ludolph and Malik 1998
Precipitation by reacting diethylzinc with hydrogen sulfide in toluene with addition of diethylmanganese in a tetrahydrofuran solvent. Coated with methacrylic acid.	3.5-7.5	Bhargava et al. 1994
Chemical Bath Deposition at 85°C - ammonium hydroxide and hydrazine with aqueous $ZnSO_4$ thiourea.	4	Singh et al. 2004
Addition of sodium sulfide solution to $Zn(CH_3COO)_2$ and $Mn(CH_3COO)_2$ solutions in iso-propanol medium with polyvinylpyroledone to control the particle size	2-4	Karar et al. 2004
Precipitation with $ZN(CH_3COO)_2$ and NaS - Reaction performed under microwave irradiation.	5	Yang et al. 2005
Metal nanocluster-catalyzed chemical vapor deposition for nanowire growth	20	Radovanovic et al. 2005
NaS added to aqueous solution containing $Zn(CH_3COO)_2$ and $Mn(CH_3COO)_2$ with sodium polyphosphate as stabilizing agent.	60-80	Warad et al. 2005

biological labeling (Warad et al. 2005). These sulfide-based diluted magnetic semiconductors also have potential application in the novel area of spin-transport electronics (or spintronics), in which the spin of charge carriers (electrons or holes) is exploited to provide new functionality for microelectric devices (Pearton et al. 2003). However, the ability to tailor their magnetic and electronic properties will be required in much the same way that band gaps are engineered in conventional semiconductors. Research in this area is extensive and ongoing but, as yet, no systematic understanding of the relationship between properties, such as Curie temperature and band gap, exists for these novel magnetic semiconductors (Erwin and Žutic 2004). The synthesis and properties of TM doped ZnS have been recently reviewed by Hu and Zhang (2006).

Sulfide nanoparticles in the environment

Sulfide nanoparticles can be formed in natural aqueous environments due to the gross insolubility of many sulfides in aqueous media (e.g., CdS has a K_{sp} value of ca. 10^{-28}). ZnS typically forms 2-3 nm diameter particles in natural systems as it is fairly insoluble with a low barrier for nucleation (Gilbert and Banfield 2005). CuS nanoparticles, with reported radii of 3-26 nm, have also been reported in oxic waters where they account for substantial fractions of filterable Cu and reduced sulfur (Rozan et al. 2000; Cigleneèki et al. 2005). Transition metal sulfide minerals are often formed as products of the metabolism of sulfate-reducing bacteria in anoxic environments (Posfai and Dunin-Borkowski 2006, this volume). These bacteria can couple the oxidation of carbon to the reduction of aqueous sulfate ions to produce HS$^-$, which in the presence of a suitable counter-ion, will precipitate as a metal sulfide. Labrenz et al. (2000) have shown that the first sulfide mineral formed by bacteria in groundwater associated with an abandoned Pb-Zn ore deposit is ZnS, with 3-5 nm particles formed in proximity to the bacteria. However, biogenic sulfide nanoparticles can also include mackinawite, pyrite, marcasite, greigite and pyrrhotite, with the product being dependant on solution chemistry (Benning et al. 2000). Microbial sulfide production rates reported by Sahm et al. (1999) correspond to production of fraction of a milligram quantities of nanocrystalline metal sulfides per cubic centimeter per year, which is significant over geological time scales. For further information on sulfide nanoparticles in the environment, see Banfield and Zhang (2001).

APPLICATIONS

Electronics

One of the major areas of potential technological exploitation of sulfides is as photovoltaics in solar energy conversion devices, using thin films on various substrates in conjunction with various organic polymers (Fuhs and Klenk 1998). In particular, chalcopyrite-like structures/stoichiometries such as $CuInS_2$ (the mineral roquesite), which has an ideal band gap of 1.5 eV, have been investigated. Production of vapor-deposited thin films with efficiencies in excess of 10% make commercial exploitation a genuine possibility and assessment continues (Klenk et al. 2005). CdTe also has a band gap of 1.5 eV and photovoltaic devices with outputs of 50W are commercially available while potential CdTe/CdS heterojunction thin film solar cells have yielded efficiencies of 16.5% on coated glass substrates (Albin et al. 2002).

Nanoparticle fabrication has boosted potential developments, especially when nanoparticles are used in conjunction with polymers or other semiconductors. PbS is a good photovoltaic because it has mobile electrons and holes, the lifetime of excited photoelectrons allows them to be "collected" and, importantly, quantum dot sized PbS nanoparticles (~4 nm) have a wide spectral absorption range (Plass et al. 2002; Watt et al. 2005). Nanocomposites involving oleic acid and octylamine capped PbS in poly[2-methoxy-5-(2′-ethylhexyloxy p-phenylenevinylene)] (MEH-PPV) have shown particularly good photovoltaic response with the octylamine-capped PbS showing response in the infrared region (Zhang et al. 2005b).

MoS$_2$ and WS$_2$ are two other sulfides that also have potential uses as photovoltaics in solar cells (Jaeger-Waldau et al. 1993; Bernede et al. 1999). A further use of CuIn$_{1-x}$Ga$_x$S$_2$ thin films is in water splitting photoelectrochemical cells using a multi-band gap process employing RuO$_2$ and RuS$_2$ as photocatalysts (Dhere and Jahagirdar 2005).

Takayama and Takagi (2006) have proposed the use of the magnetic transition in pyrrhotite in phase-change memory devices. They demonstrated the thermally controlled switching of the two magnetic states of pyrrhotite in Fe$_{0.92}$S (Fe$_{0.74}$S$_8$) by continual thermal cycling over the ferrimagnetism transition temperature at 580 K. Switching between the ferrimagnetic and the superparamagnetic states takes place with total recovery of the ferrimagnetic state, due to the ordering of the cation vacancies. The authors propose development of pyrrhotite thin films.

Paleomagnetic investigations

Mineral phases which have remanent magnetism are essential for paleomagnetic studies. Although magnetite (including maghemite and titanomagnetite) is the predominant mineral for studying rock magnetism, pyrrhotite and greigite can be the dominant contributors in sediments, and pyrrhotite in magmatic and hydrothermal ore deposits. In clastic sediments, detrital allogenic magnetite predominates whereas sulfides, which are easily broken down during transport, are usually authigenic or syndiagenetic in origin. However, in young sediments where sedimentation rates have been high, pyrrhotite of detrital origin has been recognized (Horng and Roberts 2006). Authigenic Fe-sulfides, such as pyrrhotite and greigite, form in anoxic conditions during diagenesis; this typically involves the bacterial reduction of sulfate to provide a source of H$_2$S and the release of Fe from detrital iron oxides. As greigite is relatively unstable (Benning and Barnes 1998), and does not survive the oxidation reactions that characterize many geological processes (and often convert it to pyrite), it is only present in younger sedimentary sequences. Pyrrhotite can retain original magnetic signatures over long time spans, but even low oxygen fugacities at 500 °C can cause the conversion of pyrrhotite to magnetite (Bina and Daly 1994) with consequences for the resetting of the rock magnetic signature in metamorphosed rocks. In studies of ore formation, paleomagnetic analysis of ores and host rocks have been employed as a means of age dating mineralization where suitable radiometric minerals are absent; it is particularly useful for low temperature hydrothermal mineralization (Rochette et al. 2001).

The increasing use of greigite in paleomagnetic studies of younger sedimentary rocks relates to the increasing awareness of its presence in these sediments (Horng et al.; Thompson and Cameron 1995). The greigite in these rocks typically forms in early diagenesis so its paleomagnetic information relates closely to the time of deposition and, if the sediment remains unoxidized, it can be used a reliable paleomagnetic indicator. For example, the Pleistocene (Pastorian) estuarine clays of the Norfolk (UK) coast have proved highly ferrimagnetic and X-ray diffraction and scanning electron microscopy revealed 1-2 μm grains of greigite to be the only magnetic phase present (Thompson and Cameron 1995). The greigite was formed by reduction of iron by organic matter during diagenesis in the clays. Natural remanent magnetization measurements made using high sensitivity magnetometers on oriented cores exposed a magnetic polarity reversal during the "blue" clay deposition while oxidation has affected the sulfides in other "brown" clays in the sequence. Greigite is also responsible for stable characteristic remanent magnetism in Neogene sediments in north western New Zealand (Rowan and Roberts 2006). These sediments were characterized by high accumulation rates that restricted concentrations of organic matter and allowed incomplete pyritization to take place and, as a result, authigenic greigite to be preserved. The source of iron was detrital magnetite dissolved by H$_2$S released by bacteria activity in anoxic conditions. Some of the greigite is thought to have formed >1my after sedimentation, signaling caution when interpreting authigenic sulfides in paleomagnetic studies (Rowan and Roberts 2006). In deltaic sequences, diagenetic greigite is a major contributor to rock magnetism in the clayey lithologies (prodelta, delta front marine clays, lagoonal clays) while magnetite is dominant in the fluvio-deltaic sands. In oilfield borehole samples of a deltaic succession, greigite–bearing samples have been used to determine stratigraphic correlation and sedimentation rates (Thompson and Cameron 1995).

The recognition that the pyrrhotites, and monoclinic pyrrhotite in particular, can be important contributors to the magnetic properties of rocks has led to a significant number of publications dealing with parameters of interest in geomagnetism and paleomagnetism. Readers are referred to articles by Dekkers (1988, 1989), Menyeh and O'Reilly (1995, 1997) and O'Reilly et al. (2000) as examples of such studies. Magnetic remanence was determined from pyrrhotite-bearing sulfide in core samples from the active 26N° TAG hydrothermal field associated with the North Atlantic mid-oceanic ridge. As the deposits are formed in the last 140 ka, no magnetic reversal was apparent but the data were surprisingly complex, with highly variable inclinations over relatively short distances (Zhao et al. 1998). Multistage mineralization, resetting during repeated hydrothermal episodes (including paragenetically late veining), local tectonic activity and the potential effect of magnetic excursions all may have contributed to the small scale complexity. These magnetic excursions have also been recognized in sediments bordering the Atlantic using paleomagnetic studies (Hall et al. 1997) allowing paleomagnetic stratigraphic correlation over long distances.

Pyrrhotite concentrations are common in many ancient sulfide-bearing mineral/ore deposits and have been used to provide paleomagnetic information. In particular paleomagnetic investigations of the low temperature, epigenetic hydrothermal base-metal Mississippi Valley Type (MVT) deposits have been used to help with age dating of the mineralization using the determination of paleopoles during mineralizing episodes. Sphalerite-dominated epigenetic MVT mineralization hosted by Lower Ordovician carbonates at Madison-Jefferson City, Tennessee, contains pyrrhotite (with minor magnetite) that gave characteristic remanent magnetism that indicated a paleopole at 128.7°E, 34.0°N, demonstrating that it acquired its primary chemical remanence at 316 ± 8 Ma. This places the mineralization after the peak orogenesis (Alleghanian) and the tectonic context suggests that formation is consistent with models of topographically or thrust-driven driven fluid flow. Paleomagnetic analysis of host rocks and mineralization in Tri-State MVT district show that the remanence of the samples is carried mostly by single to pseudosingle domain pyrrhotite and/or magnetite. The mineralization in the boulders yields a characteristic remanent magnetism with an inclination of 7.4° ± 1.9°, which constrains ore genesis to an equatorial paleolatitude, which for the Tri-State area has only been true in the Middle Pennsylvanian to the Middle Triassic. Two genetic models for ore deposition are constrained by the paleomagnetic data that support a single major epigenetic MVT ore-forming event during either the Late Pennsylvanian (304 ± 6 Ma) or the Late Permian (257 ± 10 Ma) (Symons and Stratakos 2002). Paleomagnetic analysis of sulfide ore components, dolomitic carbonate host rocks, altered igneous rocks and later mineralization established an Early Permian age of ~272 ± 18 Ma origin for the Sherman-type deposits of the Colorado mineral belt, supporting a Mississippi-Valley type origin for those deposits (Symons et al. 2000).

Beneficiation of sulfide ores

Due to its ferromagnetic properties and its wide occurrence in Ni-Cu-Fe magmatic ores, such as those associated with ultramafic intrusions, pyrrhotite is the most common sulfide that has been magnetically separated during mineral processing (Wells et al. 1997; Yalcin et al. 2000; Hu and Sun 2001). In this type of ore deposit it is often the main sulfide present and is associated and intergrown with the main mineral of economic interest, pentlandite $(Fe,Ni)_9S_8$. Pyrrhotite can be nickeliferous and contain inclusions of platinum group elements but is usually of little economic value. However, its separation prior to smelting of the pentlandite (and chalcopyrite) is of enormous economic and environmental benefit (Wells et al. 1997). The different magnetic properties of monoclinic (ferromagnetic) and hexagonal (antiferromagnetic) forms of pyrrhotite requires different approaches to their separation (Schwarz and Vaughan

1972). In addition to the magmatic ores, the magnetic separation of "waste" pyrrhotite from complex base-metal and oxide ores has been achieved successfully (Hudson 1967; Shoji et al. 1974). Recent mineral processing innovations have focused on new flotation methodologies to separate the pyrrhotite (Wells et al. 1997) as well as a combination of techniques by applying magnetic fields to the flotation froth as it leaves the flotation cells (Yalcin et al. 2000).

Secondary methods of magnetic separation of sulfides are sometimes employed. Partially roasting or microwaving and oxidation of Fe-bearing sulfides ores can lead to the development of magnetic surface species ($Fe_{1-x}S$, Fe_3O_4, γ-Fe_2O_3) that improve magnetic separation efficiency. In this way partially oxidized pyrite can be separated from completely oxidized Fe_2O_3 in iron ores, while maghemite forms on the surface of chalcopyrite such that it can be magnetically separated from other copper and base-metal sulfides (Jirestig and Forssberg 1994; Lovas et al. 2003; Uslu et al. 2003). The thermal oxidation of pyrite in coal to pyrrhotite allows its magnetic separation, and sulfides containing magnetic mineral inclusions (magnetite or pyrrhotite) can be selectively separated (Kim et al. 1985; Weng and Wang 1992).

The relative susceptibilities of the paramagnetic sulfides are often employed on the laboratory scale (by using magnetic fields of differing strengths) to separate and concentrate sulfides (Jirestig and Forssberg 1994). On the industrial scale the relatively high magnetic susceptibility of chalcopyrite in a strong applied field means it can be separated from other less susceptible base-metal sulfides, such as galena, (Kim et al. 1995). Zinc recovery at the Boliden complex, Sweden, has been improved by recovering sphalerite from the galena concentrate using a high gradient magnetic separator (Jirestig and Forssberg 1994).

CONCLUDING COMMENTS

The great diversity of electrical and magnetic properties exhibited by the metal sulfides has long been of interest to mineralogists and been exploited in geophysical exploration and mineral extraction technology. Recent decades have seen an even greater awareness amongst geophysists of the importance of sulfide minerals in geomagnetism and paleomagnetism, and considerable growth in interest from materials scientists in the potential for applications of sulfide thin films or nanoparticles. These areas of research are likely to undergo substantial further development, and mineralogists and geochemists are well place to contribute to their development, as well as benefiting from the insights into natural materials afforded by experiments on synthetic samples and the application of new techniques.

ACKNOWLEDGMENTS

The authors are very grateful for the comments received on the manuscript from Richard Harrison, Gerrit van der Laan and Paul O'Brien. We would also like to thank Richard Hartley for his sterling work on the Figures and Helen Weedon for her assistance.

REFERENCES

Abd Al Halim AM, Fiechter S, Tributsch H (2002) Control of interfacial barriers in n-type FeS_2 (pyrite) by electrodepositing metals (Co, Cu) forming isostructural disulfides. Electrochim Acta 47:2615-2623

Abraitis P K, Pattrick RAD, Vaughan DJ (2004) Variations in the compositionsal, textural and electrical properties of natural pyrite: a review. Int J Miner Process 74:41-59

Adachi K, Sato K, Takeda M (1969) Magnetic properties of cobalt and nickel dichalcogenide compounds with pyrite stucture. J Phys Soc Japan 26:631-638

Albin D, Dhere R, Wu X, Gessert T, Romero MJ, Yan Y, Asher S (2002) Perturbation of copper substitutional defect concentrations in CdS/CdTe heterojunctions solar cells devices. Meter Res Soc Spring Meeting, Proc Symp F

Allemand J, Wintenberger M (1970) Propriétés structurales et magnetiques de quelques composes du type stannite. Bull Soc Fr Mineral 93:341-396
Andresen AF, Torbo P (1967) Phase transitions in Fe_xS (x = 0.90 - 1.00) studied by neutron diffraction. Acta Chem Scand 21:2841-2848
Aplesnin SS, Petrakovskii GA, Ryabinkina LI, Abramova GM, Kiselev NI, Romanova OB (2004) Influence of magnetic odering on the resisteivity anisotropy of α–MnS single crystal. Solid State Comm 129:195-197
Aplesnin SS, Ryabinkina LI, Abramova GM, Romanova OB, Vorotynov AM, Velikanov DA, Kiselev NI, Balaev AD (2005) Conductivity, weak ferromagnetism , and charge instability in an α-MnS single crystal. Phys Rev B 71:125204
Austin IG, Goodman CHL, Pengelly A (1956) New semiconductors with the chalcopyrite structure. J Electrochem Soc 103:218-219
Baleshta TM, Dibbs HP (1969) An introduction to the theory, measurement and application of semiconductor transport properties of minerals. Mines Branch Technical Bulletin TB 106, Ottawa, Canada
Baleshta TM, Keys JD (1968) Single electrometer method of measuring transport properties of high-resistivity semiconductors. Am J Phys 36:23-26
Bandaranayake RJ, Lin JY, Jiang HX, Sorensen CM (1997) Synthesis and properties of $Cd_{1-x}Mn_xS$ diluted magnetic semiconductor ultrafine particles. J Magn Magn Mat 169:289-302
Banfield JF, Zhang H (2001) Nanoparticles in the environment. Rev Mineral 44:41-51
Benning LG, Barnes HL (1998) *In situ* determination of the stability of iron monosulfides and kinetics of pyrite formation. Mineral Mag 62A:151-152
Benning LG, Wilkin RT, Barnes HL (2000) Reaction pathways in the Fe-S system below 100 °C. Chem Geol 167:25-51
Benoit R (1952) Sur le paramagnetisme des sulfures de fer. Compt Rend 234:2174
Benoit R (1955) Etude paramagnetique des composés binaires. J Chim Phys 52:119-132
Bernardini GP, Borrini D, Caneschi A, Di Benedetto F, Gatteschi D, Ristori S, Romanelli M (2000) EPR and SQUID magnetometry study of Cu_2FeSnS_4 (stannite) and Cu_2ZnSnS_4 (kesterite). Phys Chem Min 27: 453-461
Bernede JC, Pouzet J, Gourmelon E, Hadouda H (1999) Recent studies on photoconductive thin films of binary compounds. Synth Met 99:45-52
Bertaut EF (1953) Contributions a l'étude des structures lacunaires: la pyrrhotine. Acta Cryst 6:557-561
Bhargava RN, Gallagher D, Hong X, Nurmikko A (1994) Optical properties of manganese-doped nanocrystals of ZnS. Phys Rev Lett 72:416-419
Bina M, Daly L (1994) Mineralogical change and self-reversed magnetizations in pyrrhotite resulting from partial oxidation; geophysical implications. Phys Earth Planet Inter 85:83-99
Birkholz M, Fiechter S, Hartmann A, Tributsch H (1991) Sulfur deficiency in iron pyrite (FeS_{2-x}) and its consequences for band-structure models. Phys Rev B 43:11926
Bither TA, Bouchard RJ, Cloud WH, Donohue PC, Siemons WJ (1968) Transition metal pyrite chalcogenides. High pressure synthesis and correlation of properties. Inorg Chem 7:2208-2220
Bither TA, Prewitt CT, Gillson JL, Bierstedt PG, FLippen RB, Young HS (1966) New transition metal dichalcogenides formed at high pressure. Solid State Comm 4:533
Boxall C, Kelsall GH (1991) Photoelectrophoresis of colloidal semiconductors. 1. The technique and its applications. J Chem Soc Faraday Trans 87:3537-3545
Brieler FJ, Grundmann P, Fröba M, Chen L, Klar PL, Heimbrodt W, von Nidda H-A K, Kurz T, Loidl A (2005) Comparison of the magnetic and optical properties of wide-gap (II,Mn)VI nanostructures confined in mesoporous silica. Eur J Inorg Chem:3597-3611
Brown PJ, Neumann K-U, Simon A, Ueno F, Ziebeck KRA (2005) Magnetization distribution in CoS_2; is it a half metallic ferromagnet? J Phys: Cond Matter 17:1583-1592
Brummage WH, Yarger CR, Lin CL (1964) Effect of the exchange coupling of Mn^{++} ions on the magnetic susceptibilities of ZnS:MnS crystals. Phys Rev 133:765-767
Brun del Re R, Lamarche G, Woolley JC (1992) Magnetic behavior of $CuIn_{1-z}Fe_zS_2$. J Phys: Cond Matt 4: 8221-8232
Brus LE (1984) Electron-electron and electron-hole interactions in small semiconductor crystallites: The size dependence of the lowest excited electronic state. J Chem Phys 80:4403-4409
Butler SR, Bouchard RJ (1971) Single crystal growth of pyrite solid solutions. J Cryst Growth 10:163-169
Chattopadhyay T, Brükel T, Burlet P (1991) Spin correlation in the frustrated aniferromagnet MnS_2 above the Néel temperature. Phys Rev B 44:7394-7402
Chattopadhyay T, Brükel T, Holwein D, Sonntag R (1995) Magnetic diffuse scattering from the frustrated antiferromagnet MnS_2. J Magn Magn Mat 140-144:1759-1760
Chen CT, Idzerda YU, Kao C-C, Tjeng LH, Lin H-J, Meigs G (1995) Recent progress in soft-X-ray magnetic circular dichroism. J China Univ Sci Tech 25:1-10

Cheroff G, Keller SP (1958) Optical transmission and photoconductive and photovoltaic effects in activated and unactivated single crystals of ZnS. Phys Rev 111:98-102

Cigleneèki I, Krznariæ D, Helz GR (2005) Voltammetry of copper sulfide particles and nanoparticles: investigation of the cluster hypothesis. Environ Sci Technol 39:7492-7498

Coey JMD, Brusetti R (1975) Hest capacity of nickel sulfide and its semimetal-metal transition. Phys Rev B 11:671-677

Collins MF, Longworth G, Townsend MG (1981) Magnetic structure of bornite, Cu_5FeS_4. Can J Phys 59:535-539

Corliss L, Elliot N, Hastings J (1956) Magnetic structures of the polymorphic forms of manganous sulfide. Phys Rev 104:924-928

Corry CE (2005) <http://www.zonge.com/ferro/proper_1.htm>

Cotton FA, Wilkinson G (1972) Advanced Inorganic Chemistry, Interscience, New York

Dalven R (1969) A review of semiconductor properties of PbTe, PbSe, PbS and PbO. Infrared Phys 9:141-184

Dekkers M (1988) Magnetic properties of natural pyrrhotite Part I: Behavior of initial susceptibility and saturation-magnetization-related rock-magnetic parameters in a grain-size dependent framework. Phys Earth Planet Int 52:376-393

Dekkers M (1989) Magnetic properties of natural pyrrhotites. II. High- and low-temperature behavior of J_{rs} and TRM as function of grain size. Phys Earth Planet Int 57:266-283

Dekkers M, Passier HF, Schoonen MAA (2000) Magnetic properties of hydrothermally synthesized greigite (Fe_3S_4)-II. High- and low-temperature characteristics. Geophys J Int 141:809-819

Deulkar SH, Bhosale CH, Sharon M, Neumann-Spallart M (2003) Preparations of non-stoichiometric (Zn, Fe)S chalcogenides and evaluation of their thermal, optical and electrical properties. J Phys Chem Solids 64: 539-544

Dhere NG, Jahagirdar AH (2005) Photoelectrochemical water splitting for hydrogen production using combination of CIGS2 solar cell and RuO_2 photocatalyst. Thin Solid Films 480-81:462-465

Di Benedetto F, Bernardini GP, Caneschi A, Cipriani C, Danti C, Pardi L, Romanelli M (2002) EPR and magnetic investigations on sulphides and sulphosalts. Eur J Mineral 14:1053-1060

Donnay G, Corliss LM, Donnay JDH, Elliot N, Hastings JM (1958) Symmetry of magnetic structures: magnetic structure of chalcopyrite. Phys Rev 112:1917-1923

Dove MT (2002) An introduction to the use of neutron scattering methods in mineral sciences. Eur J Mineral 14:203-224

Doyle FM, Mirza AH (1996) Electrochemical oxidation of pyrite samples with known compositions and electrical properties. Electrochem Proc 96:203-214

Dzyaloshinsky I (1958) A thermodynamic theory of 'weak' ferromagnetism of antiferromagnetics. J Phys Chem Solids 4:241-255

Ellmer K, Hopfner C (1997) On the stoichiometry of the semiconductor pyrite FeS_2. Phil Mag A 75:1129-1151

Ennaoui A, Tributsch H (1984) Iron sulphide solar cells. Solar Cells 13:197-200

Ericsson T, Amcoff Ö, Nordblad P (1997) Superstructure formation and magnetism of synthetic selenian pyrrhotites of $Fe_7(S_{1-y}Se_y)_8$, y<1 composition. Eur J Mineral 9:1131-1146

Ericsson T, Amcoff Ö, Nordblad P (2004) Vacancy ordering in Fe_7Se_8-Fe_7S_8 solid solution studied by Mössbauer, X-ray and magnetization techniques. Hyperfine Interact 90:515-520

Erwin SC, Žutic I (2004) Tailoring ferrimagnetic chalcopyrites. Nature Mater 3:410-414

Fischer G, Grieg D, Mooser E (1961) Apparatus for the measurement of galvanomagnetic effects in high resistance semiconductors. Rev Sci Instrum 32:842-848

Fleet ME (1970) Refinement of the crystal structure of cubanite and polymorphism of $CuFe_2S_3$. Z Krist 132: 276-287

Fleet ME (2006) Phase equilibria at high temperatures. Rev Mineral Geochem 61:365-419

Fok MV (1963) Forbidden band width and effective charge of ions in the crystal lattice of ZnS. Sov Phys Solid State 5:1085-1088

Frueh AJ (1959) The use of zone theory in problems of sulfide mineralogy. II. The resistivity of chalcopyrite. Am Mineral 44:1010-1019

Fuhs W, Klenk R (1998) Thin-film solar cells - overview. EUR 18656, World Conference on Photovoltaic Solar Energy Conversion 1:381-386

Gibart P, Goldstein L, Brossard L (1976) Non stoichiometry in $FeCr_2S_4$. J Magn Magn Mater 3:109-116

Gilbert B, Banfield JF (2005) Molecular-scale processes involving nanoparticulate minerals in biogeochemical systems. Rev Mineral Geochem 59:109-146

Godovsky DY (2000) Device applications of polymer-nanocomposites. Adv Polym Sci 153:163-205

Grice JD, Ferguson RB (1974) Crystal structure refinement of millerite (β-NiS). Can Mineral 12:248-252

Häesselbarth A, Eychmueller A, Eichberger R, Giersig M, Mews A, Weller H (1993) Chemistry and photophysics of mixed cadmium sulfide/mercury sulfide colloids. J Phys Chem 97:5333-5340

Hall F, Cisowski S, John S (1997) Environmental rock-magnetic evidence of authigenic-magnetic mineral formation/preservation (Amazon Fan). Proc Ocean Drilling Prog: Scientific Results 155:251-270

Harada T (1998) Transport properties of iron dichalcognenides FeX_2 (X = S, Se and Te). J Phys Soc Jap 67: 1352-1358

Haraldsen H (1941a) Über die eisen (II) sulfid Mischkristalle. Z Anorg Chem 246:169-194

Haraldsen H (1941b) Über die hochtemperaturum – wandlungen der eisen (II) sulfid Mischkristalle. Z Anorg Chem 246:195-226

Harrison RJ (2006) Neutron diffraction of magnetic materials. Rev Mineral Geochem 62: in press

Hastings JM, Corliss LM (1976) First-order anitferromagnetic phase transition in MnS_2. Phys Rev B 14:1995-1996

Hastings JM, Elliot N, Corliss LM (1959) Antiferromagnetic structures of MnS_2, $MnSe_2$ and $MnTe_2$. Phys Rev 115:13-17

Hauger R, Kaldis E, von Schulthess G, Wachter P, Zürcher C (1976) Electrical resistivity of Gd and La monochalcogenides as a function of stoichiometry. J Magn Magn Mat 3:103-108

Hirahara E, Murakami M (1958) Magnetic and electrical anisotropies of iron sulfide single crystals. J Phys Chem Solids 7:281-289

Hirone T, Maeda S, Chiba S, Tsuya N (1954) Thermal analysis of iron sulfides at the temperature range of the β-transformation. J Phys Soc Jap 9:500-502

Ho CH, Huang CE, Wu CC (2004) Preparation and characterization of Ni-incorporated FeS_2 single crystals. J Cryst Growth 270:535-541

Hobbs D, Hafner J (1999) Magnetism and magneto-structural effects in transition metal-sulphides. J Phys: Cond Matt 11:8197-8222

Hoffmann V (1993) Mineralogical, magnetic and Mössbauer data of smythite (Fe_9S_{11}). Studia Geophy Geodaet 37:366-381

Holton WC, DeWit M, Watts R K, Estle TL, Schneider J (1969) Paramagnetic copper centres in ZnS. J Phys Chem Solids 30:963-977

Honig JM, Spalek J (1998) Electronic properties of $NiS_{2-x}Se_x$ single crystals: from magnetic Mott-Hubard insulators to normal metals. Chem Mater 10:2910-2929

Horng C-S, Roberts AP (2006) Authigenic or detrital origin of pyrrhotite in sediments?: resolving a paleomagnetic conundrum. Earth Planet Sci Lett 241:750-762

Horng C-S, Torii M, Shea K-S, Kao S-J (1998) Inconsistent magnetic polarities between greigite - and pyrrhotite/magnetite-bearing marine sediments from the Tsailiao-chi section, southwestern Taiwan. Earth Planet Sci Lett 164:467-481

Hu H, Zhang W (2006) Synthesis and properties of transition metals and rare-earth metals doped ZnS nanoparticles. Opti Mater 28:536-550

Hu Z, Sun C (2001) Separating pure minerals from Jinchuan copper-nickel mine. Nonferrous Met 53:73-75

Hudson SB (1967) High-intensity wet magnetic separation of a zinc cleaner tailing from Broken Hill, New South Wales. Ore Dressing Investigations 672:3

Hulliger F (1959) Über den zusammenhang zwischen magnetismus und elektrischer leitfähigkeit von verbindungen mit ubergangselementen. Helv Phys Acta 32:615-654

Hulliger F (1968) Crystal chemistry of the chalcogenides and pnictides of the transition elements. Struct Bonding 4:83-229

Jaeger-Waldau A, Lux-Steiner MC, Bucher E, Jaeger-Waldau G (1993) WS_2 thin films: a new candidate for solar cells. In: Conference Record of the IEEE Photovoltaic Specialists Conference. 597-602

Jarrett HS, Cloud WH, Bouchard RJ, Butler SR, Frederick CG, Gillson JL (1968) Evidence for itinerant d-electron ferromagnetism. Phys Rev Lett 21:617-620

Jellinek F (1972) Sulfides, Selenides and Tellurides of the Transition Elements. MTP International Review of Science. Series 1, vol. 5, Butterworth

Jiang JZ, Gerward L, Frost D, Secco R, Peyronneau J, Olsen JS (1999) Grain-size effect on pressure-induced semiconductor-to-metal transition in ZnS. J Appl Phys 88:6608-6610

Jiang W-T, Horng C-S, Roberts AP, Peacor DR (2001) Contradictory magnetic polarities in sediments and variable timing of neoformation of authigenic greigite. Earth Planet Sci Lett 193:1-12

Jirestig JA, Forssberg KSE (1994) Magnetic separation in sulfide processing. Trans Soc Min Metall Exploration 294:176-181

Kamigaichi T (1956) Electrical conductivity and thermoelectric power of FeS_x (pyrrhotite). J Sci Hiroshima Univ Ser A 19:499-505

Karar N, Singh F, Mehta BR (2004) Structure and photoluminescence studies of ZnS:Mn nanoparticles. J Appl Phys 95:656-660

Kautz RL, Dresselhaus MS, Adler D, Linz A (1972) Electrical and optical properties of NiS_2. Phys Rev B 6: 2078-2082

Kawiaminami M, Okazaki A (1970) Neutron diffraction study of Fe_7S_8. II. J Phys Soc Jap 29:649-655

Keys JD, Horwood JL, Baleshta TM, Cabri LJ, Harris DC (1968) Iron-iron interaction in iron-containing zinc sulfide. Can Mineral 9:453-467

Kim YS, Fujita T, Hashimoto S, Shimoiizaka J (1985) The removal of copper sulfide minerals from lead flotation concentrate of black ore by high-gradient magnetic separation. Compt Rend 15:381-390

Klein UF, Wortman G, Kalvius GM (1976) High-pressure Mössbauer study of hyperfine interactions in magnetically ordered europium chalcogenides: EuO, EuS, EuTe. J Magn Magn Mat 3:50-54

Klenk R, Klaer J, Scheer R, Lux-Steiner MC, Luck I, Meyer N, Ruehle U (2005) Solar cells based on $CuInS_2$ - an overview. Thin Solid Films 480-481:509-514

Kobayashi H, Kamimura T, Ohishi Y, Takeshita N, Mori N (2005) Structural and electrical properties of stoichiometric FeS compounds under high pressure at low temperature. Phys Rev B 71:014110

Kravchenko AF, Timchenko A K, Godovikov AA (1966) Electrophysical properties of galena from different deposits. Dokl Akad Nauk SSSR 167:74-77

Krill G, Panissod P, Lapierre MF, Gautier F, Robert C, Nassr Eddine M (1976) Magnetic properties and phase transitions of the metallic CuX_2 dichalcogenides (X = S, Se, Te) with pyrite structure. J Phys C 9:1521-1533

Krupp RE (1994) Phase-transitions and phase-transformations between the low-temperature iron sulfides mackinawite, gregite, and smythite. Eur J Mineral 6:265-578

Kruse O (1990) Mössbauer and X-ray study of the effects of vacancy concentration in synthetic hexagonal pyrrhotites. Am Mineral 75:755-763

Kulkarni S K, Winkler U, Deshmukh N, Borse PH, Fink R, Umbach E (2001) Investigations on chemically capped CdS, ZnS and ZnCdS nanoparticles. App Surf Sci 169-170:438-446

Labrenz M, Druschel G K, Thomsen-Ebert T, Gilbert B, Welch SA, Kemner K, Logan GA, Summons R, De Stasio G, Bond PL, Lai B, Kelly SD, Banfield JF (2000) Natural formation of sphalerite (ZnS) by sulfate-reducing bacteria. Science 290:1744-1747

Larach S, Turkevich J (1955) Magnetic properties of zinc sulfide and cadmium sulfide phosphors. Phys Rev 98:1015-1019

Letard I, Sainctavit P, Menguy N, Valet J-P, Isambert A, Dekkers M, Gloter A (2005) Mineralogy of greigite Fe_3S_4. Physica Scripta T115:489-491

Li F, Franzen HF (1996) Ordering, incommensuration, and phase transition in pyrrhotite. Part II: a high-temperature X-ray powder diffraction and thermomagnetic study. J Solid State Chem 126:108-120

Li F, Franzen HF, Kramer MJ (1996) Ordering, incommensuration, and phase transition in pyrrhotite. Part I: a TEM study of Fe_7S_8. J Solid State Chem 124:264-271

Li X, Fryer JR, Cole-Hamilton DJ (1994) A new, simple and versitile method for the production of nano-scale particles of semiconductors. Chem Comm:1715-1716

Lin MS, Hacker H (1968) Antiferromagnetic transitions in MnS_2 and $MnTe_2$. Solid State Comm 6:687-689

Liu W-J, He W-D, Wang Y-M, Wang D, Zhang Z-C (2005) New approach to hybrid materials: functional sub-micrometer core/shell particles coated with NiS clusters by γ-irradiation. Polymer 46:8366-8372

Lotgering FK (1956) Ferrimagnetism of sulfides and oxides. Philips Res Rep 11:190-217, 213-249, 337-350

Lovas M, Murova I, Mockovciakova A, Rowson N, Jakabsky S (2003) Intensification of magnetic separation and leaching of Cu ores by microwave radiation. Separat Purif Tech 31:291-299

Ludolph B, Malik MA (1998) Novel single molecule precursor routes for the direct synthesis of highly monodispersed quantum dots of cadmium or zinc sulfide or selenide. Chem Comm 17:1849-1850

Makovicky E (2006) Crystal structures of sulfides and other chalcogenides. Rev Mineral Geochem 61:7-125

Manthiram A, Jeong YU (1999) Ambient temperature synthesis of spinel Ni_3S_4: an itinerant electron ferrimagnet. J Solid State Chem 147:679-681

Marusak LA, Mulay LN (1979) Mössbauer and magnetic study of the antiferro to ferrimagnetic phase transition in Fe_9S_{10} and the magnetokinetics of the diffusion of iron atoms during the transition. J Appl Phys 50:1865-1867

Marusak LA, Mulay LN (1980) Polytypism in the cation-deficient iron sulfide, Fe_9S_{10}, and the magnetokinetics of the diffusion process at temperatures about the antiferro- to ferrimegnetic (λ) phase transition. Phys Rev B 21:238

McCammon CA, Price DC (1982) A Mössbauer effect investigation of the magnetic behavior of (iron, cobalt) $sulfide_{1+x}$ solid solutions. J Phys Chem Solids 43:431-437

McElhinny MW (1973) Paleomagnetism and Plate Tectonics, Cambridge University Press

Menyeh A, O'Reilly W (1995) The coercive force of fine particles of monoclinic pyrrhotite (Fe_7S_8) studied at elevated temperature. Phys Earth Planet Int 89:51-62

Menyeh A, O'Reilly W (1997) Magnetic hysteresis properties of fine particles of monoclinic pyrrhotite Fe_7S_8. J Geomag Geoelec 49:965-976

Morehead FF (1963) A Dember effect study of shifts in the stoichiometry of ZnS. J Electrochem Soc 110:285-288

Morehead FF, Fowler AB (1962) The Dember effect in ZnS-type materials. J Electrochem Soc 109:688-695

Murakami M, Hirahara E (1958) A certain anomalous behavior of iron sulfides. J Phys Soc Jap 13:1407
Muro T, Shishidou T, Oda F, Fukawa T, Yamada H, Kimura A, Imada S, Suga S, Park SY (1996) Magnetic circular dichroism of the S 2p, Co 2p, and Co 3p core absorption and orbital angular momentum of the Co 3d state in low-spin CoS_2. Phys Rev B 53:7055-7058
Nave R (2000) SQUID Magnetometer, HyperPhysics <http://hyperphysics.phy-astr.gsu.edu/Hbase/hframe.html>
Néel L (1953) Some new results on antiferromagnetism and ferromagnetism. Rev Mod Phys 25:58-63
Néel L, Benoit R (1953) Magnetic properties of certain disulfides. Compt Rend 237:444-447
Nikolic PM, Uric S, Todorovic DM, Blagojevic V, Urosevic D, Mihajlovic P, Bojicic AI, Radulovic KT, Vasiljevic-Radovic DG, Elazar J, Dimitrijevic M (1999) Some thermal and electronic transport properties of mineral marmatite (ZnS:Fe). Mat Res Bull 34:2247-2261
Nimtz G, Schlicht B (1983) Narrow-Gap Semiconductors, Springer-Verlag, Berlin
Novakova AA, Gendler TS (1995) Metastable structural-magnetic transformations in sulfides in course of oxidation. J Radioanal Nucl Chem 190:363-368
O'Connor CJ (1982) Magnetochemistry - Advances in theory and experimentation. Prog Inorg Chem 29:203
Ohsawa A, Yamaguchi Y, Watanabe H, Itoh H (1976a) Polarized neutron diffraction study of cobalt disulfide. II. Covalent magnetic moment around the sulfur atoms. J Phys Soc Jap 40:992-995
Ohsawa A, Yamaguchi Y, Watanabe H, Itoh H (1976b) Polarized neutron diffraction study of cobalt disulfide. I. Magnetic moment distribution of cobalt disulfide. J Phys Soc Jap 40:986-991
O'Reilly A, Hoffman V, Chouker AC, Soffel HC, Manyeh A (2000) Magnetic properties of synthetic analogues of pyrrhotite ore in the grain size range 1-24 μm. Geophys J Int 142:669-683
Panissod P, Krill G, Lahrichi M, Lapierre-Ravet MF (1979) Magnetic properties of the $CoS_{2-x}Se_x$ compounds II. Neutron diffraction and NMR investigation of the magnetism and the 'metamagnetic' transition. J Phys C 12:4281-4294
Parasnis DS (1956) The electrical resistivity of some sulfide and oxide minerals and their ores. Geophys Pros 4:249-278
Pattrick RAD, van der Laan G, Henderson CMB, Kuiper P, Dudzik E, Vaughan DJ (2002) Cation site occupancy in spinel ferrites studied by X-ray magnetic circular dichroism: developing a method for mineralogists. Eur J Mineral 14:1095-1102
Pearce CI, Coker VS, van der Laan G, Pattrick RAD (2006a) Magnetic properties of ferrimagnetic sulfides as defined by X-ray magnetic circular dichroism: troilite (FeS) to pyrrhotie (Fe_7S_8). (in preparation)
Pearce CI, Henderson CMB, Pattrick RAD, van der Laan G, Vaughan DJ (2006b) Direct determination of cation site occupancies in natural ferrite spinels by $L_{2,3}$ X-ray absorption spectroscopy and X-ray magnetic circular dichroism. Am Mineral 91:880-893
Pearton SJ, Abernathy CR, Overberg ME, Thaler GT, Norton DP, Theodoropoulou N, Hebard AF, Park YD, Ren F, Kim J, Boatner LA (2003) Wide band gap ferromagnetic semiconductors and oxides. J Appl Phys 93:1-13
Perthel R (1960) Über den Ferrimagnetismus nichtstöchiometrischer Eisensulfide. Ann Physik 7:273-295
Petrakovskii GA, Loseva GV, Ryabinkina LI, Aplesnin SS (1995) Metal-insulator transitions and magnetic properties in disordered systems of solid solutions $Me_xMn_{1-x}S$. J Magn Magn Mat 140-144:147-148
Pickett NL, O'Brien P (2001) Syntheses of semiconductor nanoparticles using single-molecular precursors. Chem Rec 1:467-479
Plass R, Pelet S, Krueger J, Graetzel M, Bach U (2002) Quantum dot sensitization of organic-inorganic hybrid solar cells. J Phys Chem B 106:7578-7580
Pósfai M, Dunin-Borkowski RE (2006) Sulfides in biosystems. Rev Mineral Geochem 61:679-714
Powell AV, Vaqueiro P, Knight KS, Chapon LC, Sánchez RD (2004) Structure and magnetism in synthetic pyrrhotite Fe_7S_8: A powder neutron-diffraction study. Phys Rev B 70:014415
Pridmore DF, Shuey RT (1976) The electrical resistivity of galena, pyrite and chalcopyrite. Am Mineral 61:248-259
Radovanovic PV, Barrelet CJ, Gradeèak S, Quian F, Lieber CM (2005) General synthesis of manganese-doped II-VI and III-V semiconductor nanowires. Nano Lett 5:1407-1411
Rajamani V, Prewitt CT (1974) The crystal structure of millerite. Can Mineral 12:253-257
Ramesha K, Seshadri R, Ederer C, He T, Subramanian MA (2004) Experimental and computational investigation of structure and magnetism in pyrite $Co_{1-x}Fe_xS_2$: Chemical bonding and half-metallicity. Phys Rev B 70:214409
Rickard D, Luther GW III (2006) Metal sulfide complexes and clusters. Rev Mineral Geochem 61:421-504
Roberts AP (1995) Magnetic properties of sedimentary greigite (Fe_3S_4). Earth Planet Sci Lett 134:227-236
Roberts AP, Weaver R (2005) Multiple mechanisms of remagnetization involving sedimentary greigite. Earth Planet Sci Lett 231:263-277
Rochette P, Lorand J-P, Fillion G, Sautter V (2001) Pyrrhotite and the remanent magnetization of SNC meteorites: a changing perspective on Martian magnetism. Earth Planet Sci Lett 190:1-12

Rodríguez-Carvajal J (1993) Recent advances in magnetic structure determination by neutron powder diffraction. Physica B 192:55-69

Rooymans CJM (1963) A phase transformation in the wurtzite and zinc blende lattice under pressure. J Inorg Nucl Chem 25:253-255

Rossetti R, Hull R, Gibson JM, Brus LE (1985) Excited electronic states and optical spectra of ZnS and CdS crystallites in the 15 to 50 Å size range: Evolution from molecular to bulk semiconducting properties. J Chem Phys 82:552-559

Rowan CJ, Roberts AP (2006) Magnetite dissolution, diachronous greigite formation, and secondary magnetizations from pyrite oxidation: unravelling complex magnetizations in Neogene marine sediments from New Zealand. Earth Planet Sci Lett 241:119-137

Rozan TF, Lassman ME, Ridge DP, Luther GW, III (2000) Evidence for iron, copper and zinc complexation as multinuclear sulphide clusters in oxic rivers. Nature 406:879-882

Sahm K, MacGregor BJ, Jorgensen BB, Stahl DA (1999) Sulphate reduction and vertical distribution of sulphate-reducing bacteria quantified by rRNA slot-blot hybridization in a coastal marine sediment. Environ Microbiol 1:65-74

Samara GA, Drickamer HG (1962) Pressure induced phase transitions in II-VI compounds. J Phys Chem 23:457-461

Sankaran V, Yue J, Cohen RE, Schrock RR, Silbey RJ (1993) Synthesis of zinc sulfide clusters and zinc particles within microphase-separated domains of organometallic block copolymers. Chem Mater 5:1133-1142

Schieck R, Hartmann A, Fiechter S, Konenkamp R, Wetzel H (1990) Electrical properties of natural and synthetic pyrite (FeS_2) crystals. J Mater Res 5:1567-1572

Schlamp MC, Peng X, Alivisatos AP (1997) Improved efficiencies in light emitting diodes made with CdSe(CdS) core/shell type nanocrystals and a semiconducting polymer. J Appl Phys 82:5837-5842

Schuler TM, Stern A, NcNorton R, Willoughby SD, Maclaren JM, Ederer DL, Perez-Dieste V, Himpsel FJ, Lopez-Rivera SA, Callcott TA (2005) Electronic structure of the dilute magnetic semiconductor $Zn_{0.90}Mn_{0.10}S$ obtained by soft X-ray spectroscopy and first principles calculations. Phys Rev B 72:045211

Schull C, Wollan E (1956) Applications of neutron diffraction to solid state problems. Solid State Phys 2:137-217

Schutz G, Fischer P, Attenkofer K, Ahlers D (1997) X-ray magnetic circular dichroism. *In:* Roentgen Centennial. Haase A, Landwehr G, Umbach E (eds) p. 341-363

Schwarz EJ, Vaughan DJ (1972) Magnetic phase relations of pyrrhotite. J Geomag Geoelec 24:441-458

Shoji K, Takamura Y, Kuroda A, Shimoiizaka J (1974) Mineral processing of low-grade bismuth-tungsten ore at Akagane mine, Japan. *In:* Proc Int Miner Process Congr. London. Jones M (ed) p. 667-679

Shuey RT (1975) Semiconducting Ore Minerals. Elsevier Scientific Publishing Company

Singh SP, Perales-Perez OJ, Tomar MS, Mata OV (2004) Synthesis and characterization of nanostructured Mn-doped ZnS thin films and nanoparticles. Phys Stat Solidi C 1:811-814

Sleight AW, Gillson JL (1973) Electrical resistivity of cubanite: $CuFe_2S_3$. J Solid State Chem 8:29-30

Snowball IF, Torii M (1999) Incidence and significance of ferrimagnetic iron sulphides in Quaternary studies. *In:* Quaternary Climates and Magnetism. Maher BA, Thompson R (eds) Cambridge University Press, p. 199-230

Sparks JT, Komoto T (1968) Metal to semiconductor transition in hexagonal nickel sulfide. Rev Mod Phys 40:752-754

Spokes EM, Mitchell DR (1958) Relation of magnetic susceptibility to mineral composition. Min Eng 60:373-379

Stöhr J (1995) X-ray magnetic circular dichroism spectroscopy of transition metal thin films. J Elect Spect Relat Phenom 75:253-272

Symons DTA, Lewchuk MT, Taylor CD, Harris MJ (2000) Age of the Sherman-type Zn-Pb-Ag deposits, Mosquito Range, Colorado. Econ Geol 95:1489-1504

Symons DTA, Stratakos KK (2002) Paleomagnetic dating of Alleghanian orogenesis and mineralization in the Mascot-Jefferson City zinc district of East Tennessee, USA. Tectonophys 348:51-72

Takayama T, Takagi H (2006) Phase-change magnetic memory effect in cation-deficient iron sulfide $Fe_{1-x}S$. Appl Phys Lett 88:012512

Takeno S, Masumoto K, Kasamatsu Y, Kamigaichi T (1972) Electrical and magnetic properties of cubanite. J Sci Hiroshima Univer Ser C 7:11-19

Tappero R, Lichanot A (1998) A comparative study of the electronic structure of α-MnS (alabandite) calculated by the Hartree-Fock and Density Functional levels of theory. Chem Phys 236:97-105

Tappero R, Wolfers P, Lichanot A (2001) Electronic, magnetic structures and neutron diffraction in B_1 and B_2 phases of MnS: a density functional approach. Chem Phys Lett 335:449-457

Teranishi T (1961) Magnetic and electrical properties of chalcopyrite. J Phys Soc Jap 16:1881-1887

Teranishi T, Sato K, Kondon K (1974) Optical properties of a magnetic semiconductor: chalcopyrite $CuFeS_2$. I. Absorption spectra of $CuFeS_2$ and Fe-doped $CuAlS_2$ and $CuGaS_2$. J Phys Soc Jap 36:1618-1624

Theodossiou A (1965) Measurement of the Hall effect and resistivity in pyrrhotite. Phys Rev 137:1321-1326

Thole BT, Carra P, Sette PF, van der Laan G (1992) X-ray circular dichroism as a probe of orbital magnetization. Phys Rev Lett 68:1943-1946

Thole BT, van der Laan G (1988) Linear relation between X-ray absorption branching ratio and valence-band spin-orbit expectation value. Phys Rev A 39:1943-1947

Thole BT, van der Laan G, Sawatzky GA (1985) Strong magnetic dichroism predicted in the $M_{4,5}$ X-ray absorption spectra of magnetic rare earth materials. Phys Rev Lett 55:2086-2088

Thompson R, Cameron TDJ (1995) Palaeomagnetic study of Cenozoic sediments in North Sea boreholes: an example of a magnetostratigraphic conundrum in a hydrocarbon-producing area. Geol Soc Sp Pub 98:223-236

Tossell JA (1978) Theoretical studies of the electronic structure of copper in tetrahedral and triangular coordination with sulfur. Phys Chem Min 2:225-236

Townsend MG, Gosselin JR, Tremblay RJ, Webster AH (1976) Semiconductor to metal transition in iron(2+) sulfide. J de Physique, Colloque 4:11-16

Townsend MG, Horwood JL, Hall SR, Cabri LJ (1971a) Mössbauer, magnetic susceptibility and crystallographic investigations of $Cu_4Fe_4S_8$, $Cu_{18}Fe_{16}S_{32}$ and $Cu_9Fe_9S_{16}$. AIP Magnetic Materials Conf Proc 5:887-991

Townsend MG, Tremblay RJ, Horwood JL, Ripley J (1971b) Metal-semiconductor transition in single crystal hexagonal nickel sulfide. J Phys C 4:598-606

Trindade T, O'Brien P, Pickett NL (2001) Nanocrystalline semiconductors: synthesis, properties, and perspectives. Chem Mater 13:3843-3858

Ueda H, Nohara M, Kitazawa K, Takagi H, Fujimori A, Mizokawa T, Yagi T (2002) Copper pyrites CuS_2 and $CuSe_2$ as anion conductors. Phys Rev B 65:155104

Uslu T, Atalay U, Arol AL (2003) Effect of microwave heating on magnetic separation of pyrite. Colloids Surf A 225:161-167

van den Berg CB, van Delden JE, Bouman J (1969) α-transition in FeS- a ferroelectric transition. Phys Status Solidi 36:K89-K93

van der Laan G, Kirkman IW (1992) The 2p absorption spectra of 3d transition metal compounds in tetrahedral and octahedral symmetry. J Phys: Condens Matter 4:4189-4204

van der Laan G, Thole BT (1988) Local probe for spin-orbit interaction. Phys Rev Lett 60

van der Laan G, Thole BT (1991) Strong magnetic X-ray dichroism in 2p absorption spectra of 3d transition metal ions. Phys Rev B 43:13401-13411

van der Laan G, Thole BT, Sawatzky GA, Goedkoop JB, Fuggle JC, Esteva JM, Karnatak RC, Remeika JP, Dabkowska HA (1986) Experimental proof of magnetic X-ray dichroism. Phys Rev B 34:6529-6531

van der Pauw LJ (1958) A method of measuring specific resistivity and Hall effect of discs of arbitrary shape. Philips Res Rep 13:1-9

Vaughan DJ, Burns RG, Burns VM (1971) Geochemistry and bonding of thiospinel minerals. Geochim Cosmochim Acta 35:365-381

Vaughan DJ, Craig JR (1978) Mineral Chemistry of Metal Sulfides, Cambridge Universtiy Press, London

Vaughan DJ, Craig JR (1985) The crystal chemistry of iron-nickel thiospinels. Am Mineral 70:1036-1043

Vaughan DJ, Rosso KM (2006) Chemical bonding in sulfide minerals. Rev Mineral Geochem 61:231-264

Vossmeyer T, Katsikas L, Giersig M, Popovic G, Diesner K, Chemseddine A, Eychmüller A, Weller H (1994) CdS nanoclusters: synthesis, characterization, size dependant oscillator strength, temperature shift of the excitonic transition energy, and reversible absorbance shift. J Phys Chem 98:7665-7673

Wang Y, Herron N (1987) Chemical effects on the optical properties of semiconductor particles. J Phys Chem 91:5005-5008

Warad HC, Ghosh SC, Hematanon B, Thanachayanout C, Dutta J (2005) Luminescent nanoparticles of Mn doped ZnS passivated with sodium hexametaphosphate. Sci Technol Ad Mat 6:296-301

Ward JC (1970) The structure and properties of some iron sulfides. Rev Pure Appl Chem 20:175-206

Watt A, Eichmann T, Rubinsztein-Dunlop H, Meredith P (2005) Carrier transport in PbS nanocrystal conducting polymer composites. Appl Phys Lett 87:253109

Weller H, Koch U, Gutierrez M, Henglein A (1984) Photochemistry of colloidal metal sulfides. 7. Absorption and fluorescence of extremely small ZnS particles. Ber Bunsen-Ges Pys Chem 88:649

Wells PF, Kelebek S, Burrows MJ, Suarez DF (1997) Pyrrhotite rejection at Falconbridge's Strathcona Mill Falconbridge Ltd., Onaping, Ontario, Canada. In: Proceedings of the UBC-McGill Bi-Annual International Symposium on Fundamentals of Mineral Processing. Ontario. Finch J, Rao S, Holubec I, (eds) p. 51-62

Weng S, Wang J (1992) Mössbauer spectroscopy study of enhanced magnetic separation of pyrite from coal by microwave irradiation. J Fuel Chem Tech 20:368-374

Willeke G, Blenk O, Kloc C, Bucher E (1992) Preparation and electrical transport properties of pyrite (FeS_2) crystals. J Alloys Comp 178:181-191

Wincott PL, Vaughan DJ (2006) Spectroscopic studies of sulfides. Rev Mineral Geochem 61:181-229

Wong KKW, Mann S (1996) Biomimetic synthesis of cadmium sulfide-ferritin nanocomposites. Adv Mater 8: 928-932

Yalcin T, Sabau A, Wells P (2000) Magnetoflotation of nickel ore. CIM Bulletin 93:97-102

Yamada H, Terao K, Aoki M (1998) Electronic structure and magnetic properties of CoS_2. J Magn Magn Mat 177-181:607-608

Yang H, Huang C, Su X, Tang A (2005) Microwave-assisted synthesis and luminescent properties of pure and doped ZnS nanoparticles. J Alloys Comp 402:274-277

Youn SJ, Min BI, Freeman AJ (2004) Crossroads electronic structure of MnS, MnSe, and MnTe. Phys Stat Solidi B 241:1411-1414

Yuan HJ, Yan XQ, Zhang ZX, Liu DF, Zhou ZP, Cao L, Wang JX, Gao Y, Song L, Liu LF, Zhao XW, Dou XY, Zhou WY, Xie SS (2004) Synthesis, optical and magnetic properties of $Zn_{1-x}Mn_xS$ nanowires grown by thermal evaporation. J Cryst Growth 271:403-408

Zemel JN, Jensen JD, Schoolar RB (1965) Electrical and optical properties of epitaxial films of PbS, PbSe, PbTe, and SnTe. Phys Rev A 140:330-342

Zhang HT, Wu G, Chen XH (2005a) Synthesis and magnetic properties of NiS_{1+X} nanocrystallites. Mat Lett 59: 3728-3731

Zhang S, Cyr PW, McDonald SA, Konstantatos G, Sargent EH (2005b) Enhanced infrared photovoltaic efficiency in PbS nanocrystal/semiconducting polymer composites. 600-fold increase in maximum power output via control of the ligand barrier. Appl Phys Lett 87:233101

Zhao X, Housen B, Solheid P, Xu W (1998) Magnetic properties of leg 158 cores: the origin of remanence and its relation to alteration and mineralization of the active TAG mound. Proc Ocean Drilling Prog: Scientific Results 158:337-351

Spectroscopic Studies of Sulfides

Paul L. Wincott and David J. Vaughan

*School of Earth, Atmospheric and Environmental Sciences, and
Williamson Research Centre for Molecular Environmental Science
University of Manchester
Manchester M13 9PL, United Kingdom
e-mail: paul.wincott@manchester.ac.uk*

INTRODUCTION

Spectroscopic methods are the most direct and powerful means of obtaining experimental information on the electronic structures of materials. They also provide crucial information on aspects of crystal structure, crystal chemistry, solution chemistry and chemical speciation in mineralogical or geochemical systems. Many advanced methods for the elemental analysis of solids, surfaces and solutions are also based on various spectroscopies.

In this chapter, we review applications of spectroscopic techniques to sulfides, particularly to the bulk solids, but also noting the importance of spectroscopic studies to the investigation of surfaces, fine particle solids, and solutions. Certain of these latter topics, particularly the study of mineral surfaces, are discussed in greater detail in the chapters that follow and will be discussed here only briefly. Not all of the spectroscopic methods employed by mineralogists and geochemists have proved useful in studying sulfides, and the following account focuses on those which have provided the greatest insights. Where we have considered it appropriate, some background information on the principles of a spectroscopic method are provided, as well as references to more comprehensive accounts.

We begin with a discussion of the spectroscopies involving the interaction of radiation in the near-infrared–visible–near-ultraviolet region of the electromagnetic spectrum with sulfide minerals, and specifically by discussing optical (electronic) absorption and reflectance spectra. This is followed by an outline of infrared and Raman (vibrational) spectroscopies and their applications. Interactions with higher energy forms of radiation are then considered with the use of X-rays in photoemission, X-ray emission and X-ray absorption spectroscopies. At still higher energies, we consider the absorption of γ-rays in Mössbauer spectroscopy. Finally, a number of methods which have been hitherto rather limited in their application to sulfides (Nuclear Magnetic Resonance, Electron Spin (or paramagnetic) Resonance, Rutherford Backscattering, Secondary Ion Mass Spectroscopy) are briefly discussed.

The interpretation of spectra commonly requires an electronic structure (chemical bonding) model, whether qualitative or quantitative, for the material being studied. It is particularly helpful if this is a quantitative model derived from *ab initio* computation. Modeling electronic structures of sulfides is described elsewhere in this volume (Vaughan and Rosso 2006); however, to facilitate discussion of spectroscopic data in this chapter, reference is made to such modeling and the results of calculations. It is important to appreciate the reciprocal relationship between spectroscopy and computation, whereby theory can be used to interpret experiment and experiment used to validate a theoretical model.

ELECTRONIC (OPTICAL) ABSORPTION AND REFLECTANCE SPECTRA

Many minerals are translucent, but also colored due to the absorption of light over particular regions of the visible range. Measurement of the intensity of light (including near-ultraviolet and near-infrared radiation) transmitted through the mineral compared with the incident beam produces data for plotting an absorption spectrum. A common cause of such absorption is the excitation of electrons between transition metal $3d$ orbital energy levels which have been split by the ligand field of the anions. In addition to such d-d transitions, light can also be absorbed by charge transfer transitions where the energy is used in exciting electrons from filled (or partly filled) orbitals on the metal (cation) to empty orbitals on an adjacent ligand (anion) or on an adjacent metal.

For any particular transition metal ion in a particular site in a crystal, the absorption spectrum can provide information on the electronic configuration (oxidation state, spin-state), the ligand (crystal) field splitting parameter (Δ), distortion of the coordination site and covalence of the metal-anion bond. The theory, experimental methods and literature data on optical absorption spectral studies of minerals have been comprehensively reviewed by Burns (1993). A number of metal sulfides have been studied using this method; for example, the results on iron-bearing sphalerites are discussed below. However, most metal sulfides are not translucent and incident light is partly absorbed and partly reflected by them. Spectral techniques which involve measuring this reflected light must therefore be used. A few sulfide minerals which are opaque in the visible range do not absorb radiation in the near-infrared region of the spectrum, and their absorption spectra can be studied in this region.

Measurements of the light reflected from solids are of two types. Light, including near-infrared and near-ultraviolet radiation, reflected from a flat polished surface of the material is the specular component and a plot of its intensity with varying wavelength gives a specular reflectance spectrum. If the same material is ground to a fine powder, there will be a diffuse component in addition to the specular component. This arises from radiation which has penetrated the crystals (rather as in an absorption spectrum) and reappeared at the surface after multiple scatterings. The distinction between the two components is familiar to the mineralogist as the difference between the color of an opaque mineral in a hand specimen and its "streak." In measuring the diffuse reflectance spectrum, the specular component of the reflectance of a powder is suppressed by grinding it with a fine white material such as MgO.

Whereas it is possible to measure directly an absorption spectrum for a translucent sulfide such as (Zn,Fe)S, specular reflectance measurements do not yield equivalent information directly. This is because in the reflecting process, the absorption coefficient is only partly responsible for the reflectance R:

$$R(\%) = \frac{[n-1]^2 + k^2}{[n+1]^2 + k^2} (\times 100) \tag{1}$$

where n is the relative refractive index or ratio of the velocities of light in the two adjoining media and k is the absorption coefficient. This is the Fresnel equation for the special case of normal incidence; the Fresnel equations for non-normal incidence are much more complex. Using specular reflectance data taken at different angles of incidence, the values of n and k as a function of wavelength (λ) can be obtained through a Kramers-Kronig calculation (see Wendlandt and Hecht 1966). The plot of k against λ approximates more closely the information obtained from a transmission measurement on a translucent solid. Examples of n and k values for sulfide minerals are shown by the data for disulfides discussed below.

In diffuse reflectance, an alternative approach to obtaining the equivalent of an electronic absorption spectrum is adopted (see Vaughan and Craig 1978; Wood and Strens 1979). Here, the Kubelka-Munk function $f(r)$ plotted against photon energy gives a good approximation to the absorption spectrum. This is derived from the measured reflectance (r) by the expression:

$$f(r) = \frac{(1-r)^2}{2r} \approx \frac{k}{s} \qquad (2)$$

where k and s are absorption and scattering coefficients. Wood and Strens (1979) have reported diffuse reflectance spectra over the wavelength range 200-2500 nm for sphalerite, cinnabar, alabandite, chalcopyrite, bornite, orpiment, stibnite, bismuthinite, enargite and pyrargyrite, plus eight pyrite structure-type and four NiAs structure-type phases. For some of these phases, spectral assignments have been made, and general trends in reflectance and its dispersion have also been discussed. Vaughan and Craig (1978) also report diffuse reflectance data for the pyrite-type disulfides. A particular application of diffuse reflectance spectroscopy is in measuring the optical band gaps of semiconducting solids, i.e., the energy separation between the valence band which is filled with electrons and the empty conduction band. Boldish and White (1998) have used this approach to measure the optical band gaps of binary and ternary sulfide minerals; typical absorption edge spectra for a number of binary sulfides and several other typical semiconductors are shown in Figure 1. Comparisons with other methods suggest that this technique can produce values that are accurate to within ±0.1 eV.

In the following section, the spectra of sphalerite and wurtzite samples containing transition metal impurities, chalcopyrite, the pyrite-type disulfides, and molybdenite are discussed as representative of translucent and of opaque sulfides. In addition to these examples, both transmittance and reflectance spectra of PbS (galena), PbSe (clausthalite) and PbTe (altaite) have been reported and discussed in detail in the context of their electronic structures (Cardona and Greenaway 1964; Schoolar and Dixon 1965). In some cases, specular

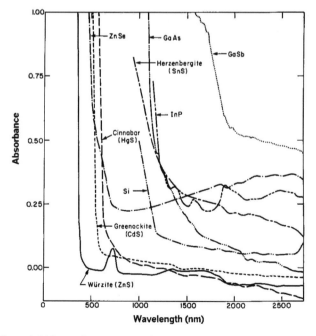

Figure 1. Diffuse reflectance absorption edge spectra for some binary sulfides and other semiconductors (after Boldish and White 1998).

reflectance spectra have also been used to study sulfide mineral surfaces which have become tarnished on exposure to air and thus to gain insights into the products of air oxidation (Remond et al. 1985; Vaughan et al. 1987).

Spectra of sphalerite and wurtzite-type sulfides

Sphalerite. Pure sphalerite (β-ZnS) is a semiconductor with a band gap which has been estimated as 3.6 eV (see also Pearce et al. 2006; this volume). The location of the absorption edge in the ultraviolet region (~3400 Å or 29400 cm^{-1}) results in pure sphalerite being transparent and colorless. However, natural sphalerites are never pure, and usually contain considerable amounts of transition metal ions, especially Fe^{2+}, substituting for Zn; the resulting colors range from pale yellow to dark brown and some specimens appear nearly opaque. Thus, impure sphalerites provide an opportunity to study the electronic absorption spectra of transition metal ions in tetrahedral coordination to sulfur.

As iron is the most common substituent for Zn in natural sphalerites, its spectrum has been extensively studied. For example, Low and Weger (1960) observed an absorption band in the infrared (at 30000 Å or 3500 cm^{-1}) which they attributed to Fe^{2+} in tetrahedral sites. This would be the single spin-allowed transition between the e and t_2 levels of the $3d$ orbitals (distinguished in the spectroscopic nomenclature as $^5E \rightarrow {}^5T_2$) and a measure of the ligand field splitting (Δ). Marfunin et al. (1968) also obtained a value for Δ of 3500 cm^{-1} and reported further splitting of this allowed transition band into three peaks which they tentatively attributed to the Jahn-Teller effect. Their absorption spectrum for ZnS:Fe^{2+} (7% Fe) is shown in Figure 2. Jahn-Teller coupling was also proposed to explain the infrared absorption of ZnS:Fe^{2+} in a detailed analysis by Ham and Slack (1971). However, Mössbauer data do not support the occurrence of Jahn-Teller distortion of the Fe^{2+} sites in sphalerite. Data for the visible and near-ultraviolet regions are also shown in Figure 2 and these spectral features have been variously attributed to spin-forbidden d-d transitions and to Fe-S charge transfer. Manning (1967) attributed features in similar spectra to Fe^{3+} in interstitial octahedral sites but this has not been substantiated by Mössbauer spectroscopy or by density determinations (see Cabri 1969; see also Table 3 in Vaughan and Rosso 2006, this volume).

Platonov and Marfunin (1968) have studied the absorption spectra of natural sphalerites and identified features which they attributed to the fundamental absorption ("band edge"), to crystal field and charge transfer transitions in Mn^{2+}, Fe^{2+}, Co^{2+} and Ni^{2+}, and to donor and acceptor levels related to minor amounts of Cu^+, Ag^+, In^{3+}, As^{3+}, Sb^{3+} and Sn^{4+}. Their composite diagram of the absorption spectra of natural sphalerites is shown in Figure 3. The color of natural sphalerites is principally determined by the iron content, low-iron sphalerites

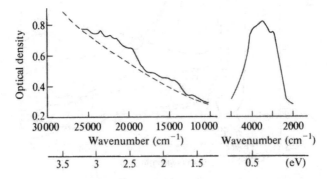

Figure 2. Electronic absorption spectrum of iron-bearing sphalerite (after Marfunin et al. 1968). Spectral energies are shown in wavenumbers (cm^{-1}) and electron volts (eV) and the absorption as optical density.

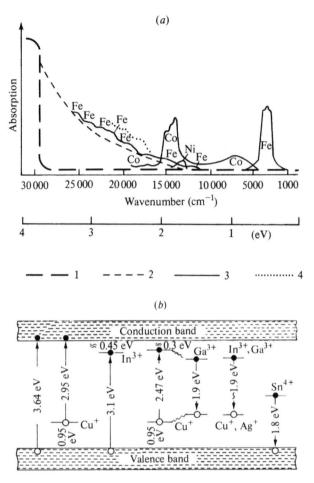

Figure 3. (a) Composite diagram of the electronic absorption spectrum of natural sphalerites containing various impurities: 1, fundamental absorption (band edge) region; 2, charge transfer bands; 3, crystal field bands; 4, donor-acceptor level bands. (b) Position of donor and acceptor levels in the forbidden band of ZnS, based on the work of various authors (after Platonov and Marfunin 1968).

having a yellow color and higher iron concentrations eventually producing the black, opaque varieties. When the iron content is less than 1 atom%, very small amounts of Co^{2+} can produce green-blue colors, and the M^+, M^{3+}, M^{4+} impurities giving rise to donor and acceptor level produce orange and red colors. The positions of these donor and acceptor levels in the forbidden band of ZnS are shown in Figure 3. The d-d optical absorption spectra of transition metal impurities in ZnS and related semiconductors have also been investigated theoretically by Mizokawa and Fujimori (1993).

Wurtzite. The absorption edge of wurtzite is at a slightly higher energy (shorter wavelength) than that of sphalerite. For example, Beun and Goldsmith (1960) give a value of 3.73 eV (30120 cm^{-1}) parallel to the c-axis and 3.69 eV (29760 cm^{-1}) normal to the c-axis of wurtzite. The absorption spectra of transition metal impurity ions in wurtzite are very similar to the sphalerite spectra. Also, the spectra obtained for CdS doped with Fe^{2+} are similar to those observed in ZnS:Fe^{2+} (Pappalardo and Dietz 1961) although the value of Δ is closer to 3000 cm^{-1}. In the

ZnS-CdS solid solution series, the position of the fundamental absorption edge shifts from ~3.6 eV for ZnS to ~2.4 eV for CdS (Kikuchi and Iijima 1960). The introduction of mercury into these systems also reduces the energy of the band gap (Kawai et al. 1971).

Chalcopyrite. The ternary sulfide chalcopyrite ($CuFeS_2$) which is isostructural with sphalerite, has been studied using a range of spectroscopic methods including optical reflectance (Fujisawa et al. 1994). The reflectivity spectrum over the range 0.2 to 25 eV (see Fig. 4) from which other parameters such as the dielectric constants have been derived, is one of the sources of information on the electronic structure of this material input into the models to be discussed in later chapters (see Vaughan and Rosso 2006).

Spectra of pyrite-type disulfides

Bither et al. (1968) studied the specular reflectance of the pyrite-structure sulfides MX_2 where M = Mn, Fe, Co, Ni, Cu, and X = S, Se, Te. Specular reflectance spectra were measured from 0.5 to 5.0 eV and optical constants calculated using a Kramers-Kronig analysis. Plots of n and k for FeS_2, CoS_2, NiS_2 and CuS_2 are

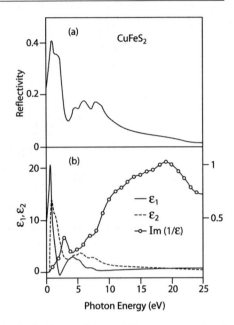

Figure 4. Specular reflectivity spectrum of chalcopyrite (a) and data for the dielectric constants (ε_1 and ε_2) and a loss function ($I_m(-1/\varepsilon)$) obtained by Kramers-Kronig analysis of the reflectivity data (after Fujisawa et al. 1994; reprinted with permission from the American Physical Society).

shown in Figure 5. Bither et al. (1968) interpret the optical properties of these disulfides in terms of a qualitative band model discussed elsewhere in this volume (see Pearce et al. 2006, Vaughan and Rosso 2006, and figures in these articles). As magnetic and electrical evidence has shown, the t_{2g} levels of the iron $3d$ orbitals in FeS_2 are filled with electrons and the e_g levels are empty; i.e., pyrite contains low-spin Fe^{2+}. Bither et al. (1968) regard the metallic and semiconducting properties of the disulfides as not due to metal-metal interactions but to metal-sulfur-metal overlap by the e_g orbitals which thus form an e_g^* band (as confirmed by later experimental and computational studies). The t_{2g} orbitals of the transition metals are still regarded, however, as "localized" on the cations. The absorption edge in the diffuse reflectance spectrum of pyrite and E_1 in Figure 5, correspond to the onset of $t_{2g} \to e_g^*$ transitions. The energy of this optical band gap (0.9 ± 0.1 eV) agrees with the observed activation energy of electrical conduction (Bither et al. 1968).

The next available empty energy levels above the band are e_g^* formed by the s, p and t_{2g} antibonding orbitals and constitute the σ^* band. Bither et al. (1968) concluded from examining ZnS_2 (in which the σ^* band provides the lowest energy vacant levels) that these must be at least 2.5 eV above the d levels, and that E_2, E_3 and E_4 are also due to $t_{2g} \to e_g^*$ transitions. However, as they point out, some of these peaks may be due to σ (valence band) $\to e_g^*$ transitions, and it was also noted that a constant e_g^*–σ^* or σ separation is by no means certain. In this regard, the shift of the absorption bands (of which E_2 is the major) to lower energies with increasing atomic weight of the cation (Fig. 5) may be attributed to increasing stability of the d levels from iron to Zn, such that the t_{2g} levels may be falling below the top of the σ band. The observed decrease in Δ would then be only an apparent one. The electronic structure of pyrite and related disulfides is further discussed later in this volume (see Vaughan and Rosso 2006).

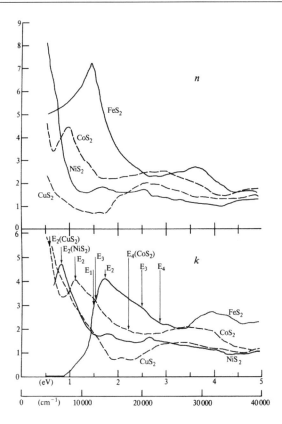

Figure 5. Plots of the refractive index (n) and the absorption coefficient (k), the real and imaginary parts of the complex refractive index (N), versus energy (in cm^{-1} and eV) obtained by Kramers-Kronig analysis of reflectivity data for the pyrite structure disulfides FeS$_2$, CoS$_2$, NiS$_2$ and CuS$_2$. (after Bither et al. Inorg Chem 7, 2111 ©1968 American Chemical Society).

The partial filling of the e_g^* band in CoS$_2$ and CuS$_2$ correlates with their metallic rather than semiconducting properties and consequent disappearance of the band edge in their optical spectra. NiS$_2$ presents the complexities of spin-splitting of the d orbitals whereas MnS$_2$ shows the maximum tendency to localized $3d$ electrons, all five being unpaired and in the high-spin state. An absorption edge around 18000 cm^{-1} in MnS$_2$ may correspond to a $d \rightarrow \sigma^*$ transition causing the onset of electrical conduction.

In addition to the classic study of pyrite-type disulfides by Bither et al. (1968), there have been studies which combine electronic structure calculations with data from optical (and other) spectra (e.g., Lauer et al. 1984) and very detailed measurements of reflectivity down to low temperature (~6 K; see Yamamoto et al. 1999).

Spectra of the layer sulfides (MoS$_2$)

The transition metal sulfides with layer structures, of which molybdenite (MoS$_2$) is the most significant mineralogically, have been examined in detail by solid state physicists because of their interesting electrical properties. A MoS$_2$ absorption spectrum for the range 10000–40000 cm^{-1} (1.25–5 eV), reported in a detailed review by Wilson and Yoffe (1969), is shown in Figure 6, together with a simple band model used in the interpretation of the spectrum. MoS$_2$ contains molybdenum in trigonal prismatic coordination, and by the arguments of ligand field theory, the molybdenum d orbital energy levels are split by the coordinating anions. In this treatment, the d_{yz} and d_{xz} orbitals are considered as being involved in σ bonding and forming part of the σ (valence) band. The $d_{x^2-y^2}$ and d_{xy} orbitals mix somewhat with neighboring metal p_x and p_y states and form a d/p nonbonding band above and separate from a nonbonding band

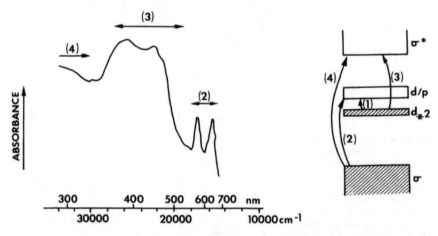

Figure 6. Electronic absorption spectrum of MoS$_2$ (2H polytype) and a simple band model interpretation (after Wilson and Yoffe 1969).

based on d_z^2 which overlaps very little with near-neighbor metals and forms a distinct narrow band. In MoS$_2$, sufficient electrons are available to fill the σ and d_z^2 bands only. The lowest energy transition occurs between this d_z^2 band and the empty nonbonding band above, and accounts for the semiconducting properties of MoS$_2$. This transition has been identified as an absorption edge in the infrared (≈0.2 eV) not shown in Figure 6. This transition across the "d band gap" determines the whole character of electrical properties of the group VI semiconductors (Wilson and Yoffe 1969). The higher energy transitions observed in the spectrum were assigned as shown in Figure 6.

Studies of the reflectivity of MoS$_2$ and related synthetic layer structure dichalcogenides have been conducted over a very wide energy range using synchrotron radiation (Mamy et al. 1977) or using single crystals (Hughes and Liang 1974). There are also numerous studies of molybdenite using other spectroscopic techniques (see below) and a large number of calculations of its electronic structure. In some cases, these calculations include predictions of optical properties and reflectivity spectra (e.g., Reshak and Auluck 2005). In an intriguing development of chalcogenide materials science, Frey et al. (1998) have reported the optical absorption spectra of nanoparticle "fullerene-like" MoS$_2$ (and WS$_2$) and compared it with the bulk material. Although MoS$_2$ retains its semiconductivity, the band gap is decreased compared to the bulk when the number of layers is >6, although there is an increase when the number of layers is very small (<5). The authors discuss these observations in terms of the possible changes in crystal and electronic structure (see also Pearce et al. 2006, this volume, for discussion of sulfide nanoparticles).

INFRARED AND RAMAN SPECTRA

Electromagnetic radiation in the infrared (IR) region of the spectrum can be absorbed by matter through the vibration of bonded atoms or (in a gaseous system, for example) rotations of the bonded atoms in molecular units. In infrared spectroscopy in its simplest forms, infrared light is passed through a sample (gas, liquid, powdered solid or single crystal) and its selective absorption measured as a function of the wavelength of that light. Reflection of infrared light from a powder or polished crystal surface may also be studied as a function of wavelength. Modern instruments employ lasers and Fourier transform manipulation to achieve much improved resolution over older designs (in Fourier Transform Infrared Spectroscopy, FTIR; see McMillan and Hofmeister 1988). IR spectra can provide information on lattice dynamics, bond strengths, coordination of atoms and linkage between polyhedral units in complex structures. A proper analysis of such spectra involves the application of group theory.

Raman spectroscopy arises from the Raman Effect which also involves interactions of electromagnetic radiation with matter that are associated with vibrational and rotational effects. In this case, monochromatic light in the visible or UV range is directed through the sample (gas, liquid, crystal or powder). In modern instruments, this light is from a laser source and may be polarized. Light scattered at 90° or 180° is collected, analyzed and recorded. The Raman effect involves shifts in the frequency of the scattered light relative to that from the source. Raman spectra can be used to identify particular molecular units and polyhedra in matter, and the linkages between them.

The theoretical and experimental aspects of IR and Raman spectroscopies and their applications in mineralogy and geochemistry have been extensively reviewed (e.g., Karr 1975; McMillan 1985) and will not be discussed here. These are techniques that have proved less useful in studying sulfides than many other mineral groups, partly as a consequence of the fact that many sulfide minerals are opaque to IR and visible light. However, there have been various systematic investigations of the IR and Raman spectra of sulfides along with a number of more focused studies, as discussed below.

Soong and Farmer (1978) measured the IR spectra of thirty sulfide minerals (as finely ground samples dispersed in polyethylene discs) in the wavelength region 420-90 cm^{-1}. Twenty four of these gave spectra with absorption peaks that range from quite well-defined (cinnabar, realgar, orpiment, sphalerite, wurtzite, stibnite, bismuthinite, chalcopyrite, stannite, pyrite, marcasite, molybdenite, cobaltite, arsenopyrite, enargite, bournonite, tetrahedrite, pyrargyrite, proustite, boulangerite) to broader features (galena, jamesonite, plagionite, alabandite). Six exhibiting metallic conductivity showed featureless spectra, with almost total absorption of IR radiation (troilite, pyrrhotite, covellite, chalcocite, bornite, millerite). The emphasis of this work was on the potential uses of IR spectroscopy for the rapid identification of these minerals, rather than a detailed analysis of the origins of spectral features. Typical examples of their spectra are shown in Figure 7a.

The IR spectra of a wide compositional range of pyrite, marcasite, loellingite and arsenopyrite structure phases were studied by Lutz et al. (1983). This included MX_2, MY_2 and MXY compounds where M = Fe, Co, Ru, Rh, Os and Ir; X = S, Se and Te, and Y = As, Sb and P. Data were analyzed in terms of a group theoretical treatment and discussed in the context of the strengths of MX and MY bonds. Both IR and Raman spectra of realgar (AsS) and orpiment (As_2S_3) were studied by Forneris (1969) and related to structures of these phases. The infrared reflectivity spectra of cinnabar (α-HgS) using polarized light have been studied in detail by Marqueton et al (1973); all the vibration modes predicted by group theory were observed and analyzed in terms of oscillators.

A major area of application of FTIR spectroscopy is in work on sulfide mineral surfaces, a topic explored in greater detail elsewhere in this volume (see Rosso and Vaughan 2006a,b). One of the advantages of IR spectroscopies is the ability to perform *in situ* experiments; i.e., to analyze directly surfaces undergoing reaction in contact with a fluid. The form of infrared reflection spectroscopy known as Internal Reflection Spectroscopy (IRS) or Attenuated Total Reflection (ATR), where the IR light is introduced into an optically transparent solid medium at angles above the critical angle and undergoes multiple internal reflections from the experimental surface, has proved particularly effective. Amongst sulfide surface studies employing FTIR spectroscopies has been work on galena before and after treatments related to flotation with xanthates (Cases and De Donato 1991) and after electrochemical oxidation

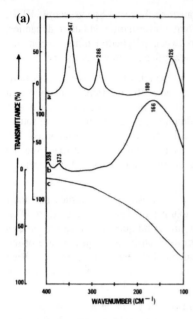
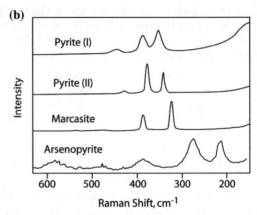

Figure 7. Examples of vibrational spectra of sulfide minerals: (a) IR spectra of cinnabar [a], galena [b] and pyrrhotite [c] (after Soong and Farmer 1978; reproduced with the publishers permission), and (b) Raman spectra of pyrite, marcasite and arsenopyrite. [Used with permission of Elsevier, from Mernagh and Trudu (1993), *Chemical Geology*, Vol. 103, p. 113-127.]

(Chernyshova 2001). The surface oxidation of pulverized galena, sphalerite, pyrite and chalcopyrite (Godocikova et al. 2002), galena and pyrite oxidation in aqueous solution (Chernyshova 2003), pyrite oxidation with water and dissolved oxygen (Usher et al. 2004), and the complex effects of interactions between the different components in a mixture of minerals on their surface properties (Mielczarski and Mielczarski 2005) have also been studied.

As with FTIR, Raman spectroscopy has been developed using laser and microbeam techniques for application as a method of mineral identification and characterization of phases including sulfides (see, for example, Mernagh and Trudu 1993; Coleyshaw et al. 1994; Bouchard and Smith 2003; Hope et al. 2001) (see Fig. 7b). More focused studies of sulfides have included work on cinnabar (HgS), realgar (AsS) and orpiment (As_2S_3) by Frost et al. (2002), where peaks were specifically assigned to lattice vibrations and to stretching and bending vibrations. Measurements on galena single crystals over the temperature range 80-373 K also led to detailed peak assignments (Smith et al. 2002), which were further refined by the study of synthetic PbS samples of varying stoichiometry or doped with ^{34}S (Sherwin et al. 2005). The vibrational spectra (Raman and IR) of pyrite have been well studied under ambient conditions (e.g., Verble and Wallace 1969); the Raman spectrum has also been studied at pressures up to 55 GPa by Kleppe and Jephcoat (2004), and Blanchard et al. (2005) have used electronic structure (DFT) calculations to predict the vibrational spectra up to 150 GPa. Their studies suggest a nonlinear pressure dependence of all vibrational frequencies.

Raman spectroscopy is also having an impact in studies of sulfide mineral surfaces; again, this is partly because Raman spectroscopy can be used as an *in situ* technique for the direct observation of reacting surfaces. For example, Turcotte et al. (1993) obtained Raman spectra from electrochemically oxidized pyrite and galena surfaces. Using this technique, elemental sulfur was identified at the galena surface after it had been held at high positive potentials; pyrite surfaces held at positive potentials (0.42 and 1.0 V vs. a Saturated Calomel Electrode) showed peaks characteristic of elemental sulfur, together with a broad peak that probably arises from some or all of the possible polysulfides species (S_3^{2-} to S_6^{2-}). Despite criticism by these authors of the most widely used surface analysis method (XPS) because it requires *ex situ* measurements (i.e., with the sample in UHV), Buckley and Woods (1994) have presented strong arguments for their XPS results on oxidized pyrite and galena being in line with those obtained using Raman spectroscopy. Other examples of the use of Raman spectroscopy in sulfide surface studies include its use in combination with other methods to identify products at the mineral surface during the leaching of copper and copper-iron sulfides (Parker et al. 2003), and in microbially mediated sulfide mineral (pyrite, marcasite, arsenopyrite) dissolution (McGuire et al. 2001).

X-RAY AND ELECTRON EMISSION SPECTROSCOPIES

When matter is irradiated with high-energy photons, electrons can be ejected from their orbitals. If monoenergetic photons are used to cause the ejection of electrons from a material then, to a first approximation, the kinetic energy of the emitted electrons will equal the energy of the incident radiation less the binding energy. In X-ray photoelectron spectroscopy (XPS) the binding energies of electrons in solids may be obtained by measuring the kinetic energies of electrons ejected by a monochromatic beam of X-rays (see Fig. 8).

Also, when an incident high-energy photon ejects an electron from one of the inner orbitals of an atom in the material being bombarded, the resulting vacancy can be filled by an electron from an outer orbital dropping into the inner orbital hole, with the energy lost by this electron being emitted as an X-ray photon (Fig. 8). The energy of this X-ray emission corresponds to the difference in ionization energies between the two orbitals. Measurement of the energies and intensities of the emitted X-rays (in X-ray emission spectroscopy, XES) can therefore provide information on the energy separation of inner orbitals and less tightly bound orbitals, including the valence orbitals. It is also possible for the energy lost by an outer electron undergoing "relaxation" so as to fill a core orbital vacancy, to be internally converted. Here, the energy is used

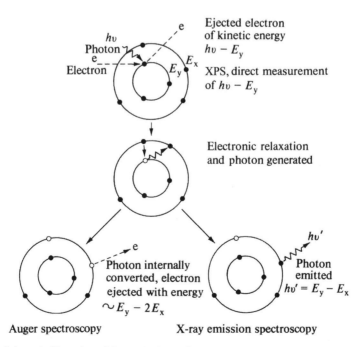

Figure 8. Schematic illustration of the mechanisms of electron and X-ray generation in X-ray emission, X-ray photoelectron and Auger electron spectroscopies (redrawn after Urch 1971).

to eject one of the outer electrons rather than being emitted as X-rays (Fig. 8). Measurement of the kinetic energies of these electrons is the basis of Auger Electron Spectroscopy (AES).

These X-ray and electron spectroscopies have contributed substantially to our understanding of the mineral chemistry and geochemistry of sulfides. Because these techniques can directly provide information about the energies of the electrons in the atomic and molecular orbitals (or energy bands) of materials, they have been used in attempts to "dissect" the chemical bond. It is important to note that whereas XPS features arise from all orbitals in a system, XES peaks are subject to selection rules that relate to transitions involving electrons of particular atomic orbital character. The capacity to provide information on electronic structure is therefore much enhanced when such spectroscopies are used together, and particularly when combined with quantum mechanical calculations, in which case the calculations may aid the interpretation of spectra; conversely, spectroscopic data may provide evidence to support a model derived from computation. In this context, both cluster models based on a molecular orbital (MO) approach and band models have given useful insights. In the following sections, further discussion of these spectroscopic methods is provided, along with key examples of applications to sulfides. Although reference is made to quantum mechanical calculations of electronic structure, these are dealt with in more detail in a later chapter of this volume (Vaughan and Rosso 2006).

X-ray photoelectron spectroscopy (XPS) and Auger electron spectroscopy (AES)

If a sample is bombarded with radiation of frequency v, then photoemission will occur for electrons with ionization energies of less than hv (where h is Planck's constant). If the binding energy of the electron is E_b, then the ejected photoelectron will have a kinetic energy of E_k, where $E_k = hv - E_b$ for a free atom in the gas phase (subject to a very minor correction for the recoil energy of the positive ion). In a solid, further corrections are required in the form of a work function (C), which is a constant for the sample and spectrometer concerned. This takes account of effects such as positive charging of the surface following removal of photoelectrons; this charge will attract the emerging photoelectrons and degrade their kinetic energies. Thus we have:

$$E_k = hv - E_b - C \tag{3}$$

An X-ray photoelectron spectrum is a plot of electron binding energies against intensities, as shown in Figure 9. This valence-region spectrum of PbS (galena) reported by McFeely et al. (1973) can be interpreted in terms of band-structure calculations and calculations based on molecular-orbital methods, as further discussed by Vaughan and Rosso (2006) later in this volume. The peaks labeled 1 and 1' at the top of the valence band arise from electrons in orbitals of essentially sulfur $3p$ character (nonbonding molecular orbitals). Below these in energy lie the main bonding orbitals (peak 2) of chiefly Pb $6s$ plus sulfur $3p$ character. Peak 3 arises from the sulfur $3s$ orbitals, which are not involved in bonding. The intense double peak at ~20 eV binding energy comes from electrons in the Pb $5d$ orbitals. These more tightly bound, compact, and "core-like" orbitals give rise to much more intense photoemission.

As can be seen from this example, XPS is a powerful method for directly probing the electronic structure of a material. In addition, a major application of XPS arises from the very limited escape depth of the electrons from a solid sample (a few nm depending on the material and electron energy). This makes XPS the workhorse technique for the analysis of surfaces. It is used for quantitative elemental surface analysis because XPS peaks are associated with specific elements and have intensities proportional to abundances, and for determining chemical speciation because precise peak positions are sensitive to oxidation state, coordination, and the nature of the surrounding ligands. Conventional photoelectron spectrometers involve a monoenergetic X-ray source (usually Al_{K_α} or Mg_{K_α}) used to bombard the sample, with the ejected electrons being collected in an energy analyzer to measure their kinetic energies, and the whole system operating under UHV. It should still be emphasized, however, that in such a system the bulk of the XPS signal (~50-90%) is derived from the subsurface.

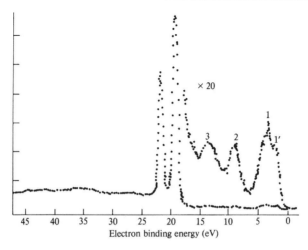

Figure 9. Valence region photoelectron spectrum of galena (PbS) (redrawn after McFeely et al. 1973).

In recent years, increasing use has been made of synchrotron X-ray sources for XPS studies; here the incident photon energy can be selected to give maximum resolution and sensitivity, whether for analysis of the surface or of the bulk. For the latter, sources tunable to very high photon energies (2000-8000 eV) are becoming available (see, for example, Sekiyama and Suga 2004), enabling enhancement of the bulk signal to ~99% of the total and permitting definitive XPS analysis of the bulk electronic structure. At the other extreme of the energy scale, photoelectron studies of the valence region can also employ conventional or synchrotron ultraviolet light sources (in Ultraviolet Photoelectron Spectroscopy, UPS).

Auger electrons will also be detected in the same spectrometer used for X-ray photoelectron spectroscopy. The ejection of a second electron by the process of internal conversion (Fig. 8) leaves a doubly ionized atom. The energy of this second ejected electron, the Auger electron, will be approximately:

$$E(\text{Auger}) = E_b - E_c - E_d \qquad (4)$$

where E_b is the binding energy of the original photoelectron (as in Eqn. 3), E_c is the binding energy of the electron involved in the process of relaxation and generation of an X-ray photon, and E_d is the binding energy for an electron from orbital d in the atom with an atomic number that is one greater than the atom under consideration. In Figure 10 is shown an LMM Auger spectrum of PbS, the interpretation of which is more fully discussed below (and see below for more about the terminology used to describe Auger spectra). Vaughan and Tossell (1986) also discuss the interpretation of AES of sulfide minerals in general; for example, pointing out that for many sulfides the LVV Auger spectrum is essentially a self-convolution of the valence band density of states weighted by the amount of S$3p$ character. Auger electron spectroscopy is also used for the elemental analysis of solid surfaces and in a "scanning mode" for the mapping of surfaces in terms of compositional variations on the micron scale.

More detailed accounts of the principles and technical aspects of XPS, AES and related techniques can be found in Briggs and Seah (1996), Briggs and Grant (2003) and the earlier mineralogical applications are reviewed by Hochella (1988) and Perry (1990). In fact, the applications of XPS to the study of sulfide minerals have passed through a number of stages of development since the XPS technique was first described in the late 1960s. The core and valence level XPS data for many important sulfides were first reported ion the 1970s and

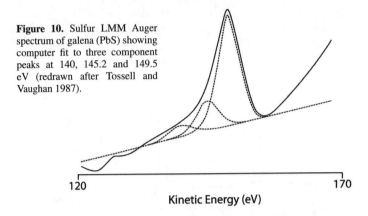

Figure 10. Sulfur LMM Auger spectrum of galena (PbS) showing computer fit to three component peaks at 140, 145.2 and 149.5 eV (redrawn after Tossell and Vaughan 1987).

1980s, particularly through the work of Buckley and collaborators. This work included key observations, such as that concerning the ability to resolve distinct core level signals for structurally different S atoms in complex sulfides (Buckley et al. 1988). From the 1990s onwards, many of the earlier studies were revisited and new investigations undertaken using improved instruments and methods, and with the aim of improving understanding of both bulk and surface electronic properties. For example, Nesbitt and Muir (1994) and Pratt et al. (1994) were able to demonstrate that the Fe2p spectra can be used to determine the spin state of iron in sulfides as well as oxidation state and the ratio of iron (Fe^{2+}: Fe^{3+}) in different oxidation states. Yin et al. (1995) showed the unusual nature of the Fe2p spectrum of chalcopyrite ($CuFeS_2$) in relation to its tetrahedral coordination. The most recent stage of development concerns the use of synchrotron radiation (SR) sources for XPS studies, and arises from such pioneering studies as those of Pettenkofer et al. (1991) and Leiro et al. (1998) in showing the greatly improved resolution and sensitivity obtainable using tunable SR sources. These methods make it possible to study bulk and surface properties separately or in combination, and to analyze in detail the impact of surface phenomena on the electrons in core levels. Many examples of the application of XPS to the study of sulfide mineral surfaces are provided in separate chapters later in this volume (Rosso and Vaughan 2006a,b). Below, some key examples of XPS studies of the bulk electronic structures of sulfides are discussed, after a brief account of X-ray emission spectroscopy (XES), as XPS and XES data have frequently been combined in presenting a model for the electronic structure of a particular material.

X-ray emission spectroscopy

In X-ray emission spectroscopy, the energy lost by relaxation of an outer electron (binding energy E_b) is emitted as an X-ray photon of energy hv'. Thus, the frequency (v) of the emitted X-radiation is:

$$v = \frac{E_b - E_c}{h} \tag{5}$$

X-ray emission spectra are subject to an electric-dipole selection rule that requires that the orbital angular momentum quantum number (l) changes only one unit during the transition; hence:

$$\Delta l = \pm 1 \tag{6}$$

Permitted X-rays ("diagram lines") will only be generated by the transitions: $s \leftarrow p$, $p \leftarrow s$ or d, $d \leftarrow p$ or f, $f \leftarrow d$ or g. Although this is essentially an atomic selection rule, it can be applied to transitions involving molecular orbitals in the valence band. Thus, X-ray spectra can provide valuable information about the atomic contributions to molecular orbitals. X-ray emission peaks are classified according to the orbital (or "shell" – K, L, M etc.) in which the initial vacancy was created, a familiar system summarized in Figure 11.

X-ray emission spectra are generated from solids in a number of instruments widely used for elemental analysis. In the X-ray fluorescence spectrometer, the source of ionizing radiation is X-rays from an X-ray tube and in the electron probe microanalyzer, the source is a beam of electrons. Both instruments can be used to record X-ray emission spectra, although the former is usually employed. In practice, the use of spectrometers for chemical bonding studies requires higher resolution than for routine analysis, and this may require some modifications to commercial instruments.

More detailed accounts of X-ray emission spectroscopy can be found in, for example, Urch (1985). One aspect of the study of X-ray emission spectra is the application of such measurements (particularly of peak positions, intensities and shapes) to determining the valence states of elements, particularly first row transition metals, in solids. Despite numerous studies, this has not become a routine procedure. An incisive review of the problems inherent in such measurements has been published by Armstrong (1999).

Applications to non-transition metal monosulfides : PbS, ZnS, CdS and HgS

Galena. The valence-region X-ray photoelectron spectrum of galena (PbS) reported by McFeely et al. (1973) is shown in Figure 9. A substantial number of MO and band structure calculations have been carried out on galena (e.g. Rabii and Lasseter 1974; Hemstreet 1975; Tossell and Vaughan 1987, 1992; Mian et al. 1996; Gurin 1998; Gerson and Bredow 2000; Muscat and Klauber 2001) which aid in the interpretation of the spectra. The basis of these and similar calculations are reviewed in a later chapter (Vaughan and Rosso 2006). It is possible, therefore, to compare the results of various calculations and evaluate their agreement with the experimental data, as illustrated in Figure 12. Hence, as noted above, in examining the spectra alongside the calculations, we can see that the peaks labeled 1 and 1' at the top of the valence band are essentially S p orbital in character (and therefore nonbonding molecular

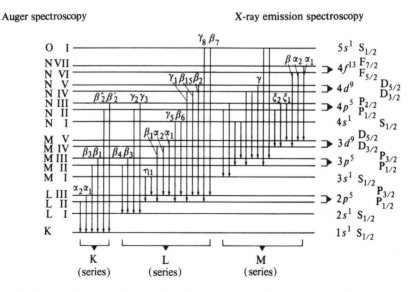

Figure 11. Permitted X-ray emission transitions. Electron configurations and corresponding spectroscopic state are indicated on the right (redrawn after Urch 1971).

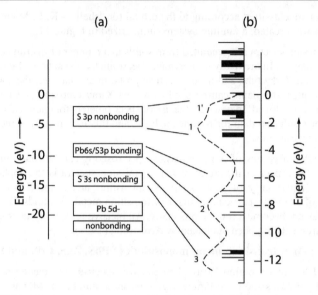

Figure 12. XPS data and electronic structure models for galena: (a) simple band structure representation based upon spectroscopic data; (b) molecular orbital energy levels calculated for a $SPb_6S_{12}Pb_8$ cluster using the multiple scattering (SCF) Xα method (after Hemstreet 1975) with the XPS data superimposed (after McFeely et al 1973; peaks labeled as in Fig. 7).

orbitals). Below these lie the main bonding orbitals of the system, which are chiefly Pb6s/Sp in character. Peak 3 represents S3s orbitals that are not involved in bonding, and the intense double peak below this (at ~20 eV binding energy) arises from the Pb5d orbitals.

Leiro et al. (1998) have studied PbS using synchrotron radiation XPS, and were the first to identify a separate S2p core level signal associated with surface species shifted in energy (−0.30 eV ± 0.02 eV) by comparison with the signal from the bulk These and other aspects of the surface electronic structure of PbS are further discussed in later chapters of this volume (Rosso and Vaughan 2006a,b).

The Auger electron spectrum of PbS (Fig. 10) has been fitted to three peaks at kinetic energies of 149.5, 145.2 and 140eV; comparison with XPS data and calculations suggests that these peaks arise from S3p nonbonding orbitals, Pb6s-S3p nonbonding orbitals, and the dominantly S3s nonbonding orbitals, respectively. PbS was also one of a large number of binary sulfides for which the sulfur $K_β$ X-ray emission spectra were recorded by Sugiura et al. (1997). These spectra, arising from a transition from a S3p to a S1s state, reflect the partial S3p density of states for the valence band and support the above interpretation. In Figure 13 are shown the S $K_β$ and S $L_{2,3}$ emission spectra along with the XPS valence band spectra on a common energy scale (after Sugiura et al. 1997).

Sphalerite. The structure of the valence band in sphalerite structure ZnS has also been much investigated using XPS and XES (Ley et al. 1974; Sugiura et al. 1974; Domashevskaya et al. 1976; Sugiura 1994; Laihia et al. 1996, 1998). Combination of these two spectroscopic methods aids in the assignment of features in the spectra. Furthermore, sphalerite provides a good example of the insights that can be gained regarding the electronic structure of a material from both cluster (MO) computations and band theory models. Thus, in Figure 14 we can suggest that peak I seen in the XPS and XES data arises from the same orbitals. Since this the S $K_β$ main peak in the XES, we can assign it in both spectra to orbitals that are dominantly S3p in character. Using similar arguments, peaks II, III and IV may be assigned to orbitals with Zn and S s and p and some Zn$3d$ character (II), to dominantly Zn$3d$ orbitals, and to orbitals of dominantly S s character (IV). Quantum mechanical calculations on sphalerite have been performed using an MO method (the MS-SCF-$X\alpha$ cluster method; Tossell, 1977a) and employing a ZnS$_4^{6-}$ cluster. The results, in the form of MO energy levels labeled according to the irreducible representations of the T_d symmetry group, are also shown in Figure 14. Results are given here for both the *ground-state* (GS) electronic configuration and using the *transition-state* (TS) procedure, which incorporates electronic relaxation effects occurring during ionization (see Vaughan and Rosso 2006). As seen in Figure 14, the latter show quite good agreement with the XPS and XES results. Of course, as well as providing information on the order and relative separation of MO energy levels, the calculations also provide information on the composition of particular molecular orbitals in terms of the contribution of atomic orbitals of metal or sulfur. Thus, the $4t_2$, $1t_2$, and $2e$ molecular orbitals are dominantly sulfur $3p$ in character and nonbonding, whereas the $3t_2$ and $2a_1$ orbitals are the main metal – sulfur bonding orbitals. The $1e$ and $2t_2$ orbitals are the dominantly metal $3d$ ("crystal-field")-type orbitals in this system, and the $1t_2$, $1a_1$ orbitals are S$3s$ nonbonding in character.

For comparison, Figure 15 shows the ZnS valence band XPS and both Zn and sulfur XES data aligned and compared with a band model (the theoretical density of states calculated using a

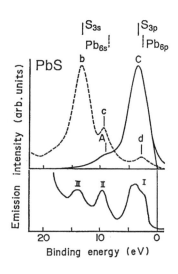

Figure 13. X-ray emission and XPS data for galena (PbS). Top: SK$_\beta$ emission and SL$_{2,3}$ emission (broken line to left). Bottom: XPS valence band spectrum. Data are aligned on a common energy scale where the zero is at the top of the valence band. Peaks have been associated with particular Pb and S states as shown. [Redrawn with permission of The Physical Society of Japan, from Sugiura et al. (1997), *J. Physical Society of Japan*, Vol. 66, Fig. 3, p. 504.]

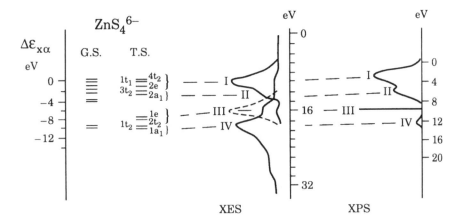

Figure 14. Valence region XPS and XES data for ZnS (after Ley et al. 1974; Sugiura et al. 1974; Domashevskaya et al. 1976) compared to MO calculations on a ZnS$_4^{6-}$ cluster for both the ground state (G.S.) and transition state (T.S.) (after Tossell 1977a).

pseudopotential plane-wave method by Martins et al. 1991). These spectroscopic data of Sugiura (1994), together with the analysis in terms of the band model, are in good general agreement with the above "dissection" of sphalerite chemical bonding, adding further details to the picture derived from an MO approach.

Hawleyite and metacinnabar. The phases CdS (hawleyite) and metacinnabar (HgS) are isostructural with sphalerite. The close similarities of the valence region XPS and X-ray emission spectra (see, for example, S K_β spectra reported by Sugiura et al. 1997) point to similarities in electronic structures of ZnS, CdS and HgS in the valence region which are supported by both MO and band structure calculations (Zunger and Freeman 1978; Farberovich et al. 1980; Tossell and Vaughan 1981; see further discussion by Tossell and Vaughan 1992, and Vaughan and Rosso 2006, in this volume).

Applications to the transition metal sulfides: FeS, $Fe_{1-x}S$, CuS, $Cu_{1-x}S$, Ag_2S, $CuFeS_2$ and related phases

Troilite and the pyrrhotites. A variety of spectra have been studied of FeS and the pyrrhotites ($Fe_{1-x}S$). For example, S K_β and S K_α X-ray emission spectra (Marusak and Tongson 1979; Sugiura et al. 1997), Fe K_β X-ray emission spectra (Reuff et al. 1999; Gamblin and Urch 2001) and XPS (Skinner et al. 2004). Band structure and cluster calculations have also been undertaken, as well as more qualitative speculation on electronic structure based on observed physical properties (e.g., Goodenough 1967; Wilson 1972). Tossell (1977b)

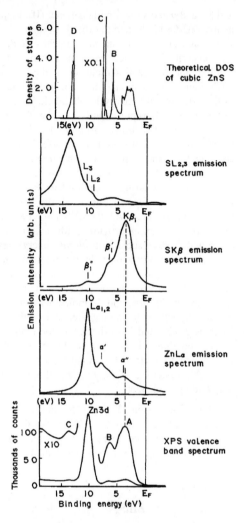

Figure 15. Comparison of sphalerite valence region XPS and XES data (Zn L_α, SK_β, $SL_{2,3}$) and the theoretical density of states (DOS) calculated using a band structure model and aligned on a common energy scale. [Used with permission of The Physical Society of Japan, from Sugiura (1994), *J. Physical Society of Japan*, Vol. 63, Fig. 6, p. 3768.]

performed molecular orbital (MS-SCF-$X\alpha$) calculations on an FeS_6^{10-} cluster with Fe^{2+} in the high-spin state (quintet) as a model for NiAs-structure-type FeS phases. The result of this calculation is shown in Figure 16. In the quintet state, the $2a_{1g}$, $2t_{1u}$, and $1t_{2g}$ orbitals are the main Fe-S bonding orbitals and are followed by a group of essentially S3p nonbonding orbitals ($2e_g$, $1t_{2u}$, $3t_{1i}$, and $1t_{1g}$). Above these lie the $2t_{2g}$ and $3e_g$ antibonding crystal-field orbitals of Fe3d and S3p character. Calculated energies are in reasonable agreement with the data available from X-ray emission and absorption spectra as can be seen in Figure 16 which also shows an energy band model for pyrrhotite proposed by Sakkopoulos et al. (1984) and based on the cluster calculations of Tossell (1977b). This model

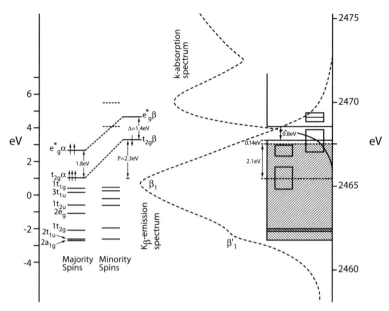

Figure 16. MO energy level diagram for the FeS$_6^{10-}$cluster (after Tossell 1977b) in a high spin Fe^{2+} (quintet) state and showing spin-up and spin-down energy levels, along with sulfur K$_\beta$ emission and K absorption spectra and a simplified band model for pyrrhotite (redrawn after Sakkopoulos et al. 1984).

consists of a S3p valence band, 5.4 eV wide, separated by a 0.8 eV gap from the conduction band. The spin-pairing energy (2.3 eV) deduced from spectroscopic data, puts the one-third-filled $t_{2g}\beta$ band in such a position that the 0.8 eV gap is almost bridged. The occupancy of the $t_{2g}\beta$ band determines the position of the highest occupied level (E_F) and leaves an empty strip about 0.2 eV wide, at the top of the valence band.

In very detailed studies of aspects of the crystal chemistry of FeS, Rueff et al. (1999) used synchrotron radiation to study the Fe K$_\beta$ X-ray emission spectrum as a function of pressure and to demonstrate a reversible high-spin to low-spin transition in Fe^{2+} at pressures of ~6.5 Gpa (see also discussion of high pressure Mössbauer spectra later in this chapter). Skinner et al. (2004) studied S2p and Fe2p XPS of clean surfaces of both FeS and Fe$_7$S$_8$, identifying metal-like Fe states in troilite due to increased occupancy of Fe lattice sites and increased Fe-Fe 3d orbital interaction in the structure.

Copper sulfides, chalcopyrite. The XPS data (Cu2p, S2p) for a very large number of Cu minerals were reported by Nakai et al. (1978) and used to support the presence of monovalent copper in all of the sulfides studied (including CuFeS$_2$, CuCo$_2$S$_4$ and CuS). Perry and Taylor (1986) also reported only Cu^{1+} in both Cu$_2$S and CuS from XPS studies. Sugiura and Gohshi (1981) report S K$_\beta$ and K absorption spectra of a Cu$_{1.8}$S phase and, subsequently, Sugiura et al. (1988) undertook comparative studies of Ag$_2$S using the same techniques. The spectra show the presence of an energy gap in semiconducting Ag$_2$S compared with the metallic Cu$_{1.8}$S, where the K$_\beta$ band edge overlaps with the K absorption edge. Kurmaev et al. (1998) studied both XPS and SL emission spectra of CuS, the latter using a synchrotron source, and they also report the same range of spectra for CuFeS$_2$, a phase earlier studied in detail by Tossell et al. (1982) using a comprehensive range of spectra (S K$_\beta$, S L, Cu K$_\beta$, Cu L, Fe K$_\beta$, Fe L and XPS) and involving comparison with molecular orbital calculations on relevant polyhedral units. Detailed discussion of these data are provided by Tossell et al (1982) and summarized in Tossell and Vaughan (1992).

Application to the disulfides: FeS_2, MoS_2

Pyrite and related phases. Pyrite (FeS_2) has been the subject of numerous spectroscopic studies including photoemission and X-ray emission measurements (e.g., Li et al. 1974; Ohsawa et al. 1974; Sugiura et al. 1976; van der Heide et al. 1980; Folmer et al. 1988; Fujimori et al. 1996; Bocquet et al. 1996) and associated electronic structure calculations (see Vaughan and Rosso 2006, this volume). More recent X-ray emission and photoelectron studies have employed both conventional and synchrotron radiation sources in order to achieve the best possible resolution. Thus, Kurmaev et al. (1998) examined S $L_{2,3}$, S K_β, Fe L_α, and valence region XPS spectra of pyrite, and Prince et al. (2005) both core and valence region XPS spectra in a fundamental study of electronic structure that has further refined previous models. The essential features of the pyrite spectra are seen in the valence region UV photoelectron spectrum which shows an intense peak at low binding energy arising from the six spin-paired electrons in the t_{2g} levels (Fig. 17). Less pronounced features arise from the other valence-band electrons. The photoelectron spectrum can be aligned with X-ray emission spectra using more deeply buried core orbitals and, as shown from Figure 17, provide further experimental data on the composition of the valence region. Thus, the Fe K_β spectrum shows the contribution from orbitals that are predominantly Fe4p in character, the S K_β spectrum shows the S3p contribution, and the SL spectrum arises from orbitals with S3s character. Interpretation of these spectroscopic data is, therefore, facilitated by comparison with the results of calculations.

As noted above, pyrite has been the subject of a large number of molecular orbital and band structure calculations. Earlier MO (MS-SCF-$X\alpha$) calculations on an FeS_6^{10-} cluster were performed by Li et al. (1974), Tossell (1977b), Harris (1982), and Braga et al. (1988) and the data of Tossell (1977b) are also shown in Figure 17 and compared against the spectroscopic data. The calculation places the filled $2t_{2g}$ levels at 2.7 eV above the top of the main valence band orbitals and shows them to be localized nonbonding orbitals of almost entirely Fe3d character. The empty $3e_g$ orbitals have, by contrast, appreciable S3p character, whereas the $2t_{1u}$, $1t_{2g}$ and $2a_{1g}$ orbitals are the main bonding orbitals of the system with substantial iron and sulfur character. About 10 eV below the sulfur nonbonding orbitals are a set of orbitals ($1a_{1g}$, $1t_{1g}$ and $1e_g$) that are essentially S3s nonbonding orbitals.

In addition to the work on pyrite itself, there have been spectroscopic studies of the pyrite structure phases CoS_2, NiS_2 and CuS_2 (and their Se and Te analogs) and computational studies on these phases which are further discussed in later articles in this volume (see Vaughan and

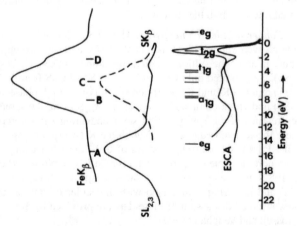

Figure 17. UV photoelectron (labeled ESCA) data and XES (SK$_\beta$, SL$_{2,3}$ and FeK$_\beta$) data for pyrite together with calculated MO energy levels (based on Tossell 1977b). [After Tossell and Vaughan 1992. By permission of Oxford Univeristy Press.]

Rosso 2006). An example of such a spectroscopic study using both XPS and UPS techniques is that by Krill and Amamou (1980).

Molybdenite. The optical absorption spectrum of MoS_2 has already been discussed (see Fig. 6) and a simple band model described in the context of the optical spectra. Many band structure calculations have been performed on MoS_2, and much of the earlier work has been reviewed by Calais (1977). A study using self-consistent augmented-spherical-wave calculations was undertaken by Coehoorn et al. (1987). MO cluster calculations using the MS-SCF-$X\alpha$ method have also been undertaken by de Groot and Haas (1975) and by Harris (1982). For example, in Figure 18 are shown the energy levels for various (MoS_6 and MoS_6^{8-}) clusters calculated by de Groot and Haas (1975), along with the X-ray photoelectron spectrum of MoS_2 from Wertheim at al. (1973). The calculation shows an electronic structure which is broadly the same as that proposed above on qualitative grounds. Above the sulfur s band are a group of levels largely of S p character but with some molybdenum character in the states of symmetry A_1', E' and E'' (see Fig. 18). The highest occupied level is of symmetry A_1', and this has mainly molybdenum d_z^2 character. The lowest unoccupied levels (E', E'') are of molybdenum d with some sulfur character. Agreement between calculation and the XPS data is reasonable, particularly when relaxation effects are incorporated (as in Fig. 18c).

Because large oriented single-crystal surfaces of molybdenite are readily prepared, X-ray emission spectra can be studied as a function of angular dependence and using polarized X-rays (Haycock et al. 1978; Yarmoshenko et al. 1983; Simunek and Wiech 1984). Such studies indicate that S3p and Mo4d character are extensively mixed in the valence-band region (Haycock et al. 1978). The polarized X-ray emission spectra (S $K_{\beta 1}$ and Mo $L_{\beta 2}$) obtained from a single crystal of MoS_2(2H) as a function of angle of incidence (Yarmonshenko et al. 1983) suggest mixing of S3p and Mo4d states, and specifically of S3p_x and 3p_y orbitals with Mo4d. Angular dependence of the Mo $L_{\beta 2}$ spectrum indicates that the Mo4d_z^2 orbital is among the least tightly bound. These angle resolved XES data have been followed by detailed measurements of angle resolved photoelectron spectra by Park et al. (1996) and by Böker et al. (2001; see Fig. 19), the latter complemented by *ab initio* calculations. These studies have provided an even more detailed picture of the valence band in MoS_2.

X-RAY ABSORPTION SPECTROSCOPIES

Extended X-ray absorption fine structure spectroscopy (EXAFS) and X-ray absorption near-edge structure spectroscopy (XANES)

If an intense X-ray source that can scan through a range of energies, such as a synchrotron source, is available then the absorption of incident X-rays as a function of energy (wavelength) can be readily measured in X-ray absorption spectroscopy (XAS). In this case, given the critical energy associated with an absorption edge, electrons from core or valence level orbitals are excited into higher energy vacant orbitals or the continuum. The fine structure associated with an absorption edge spectrum is the basis of X-ray Absorption Near-Edge Structure spectroscopy (XANES), also known as Near-Edge X-ray Absorption Fine Structure spectroscopy (NEXAFS). The fine structure at energies above the absorption edge provides the data used in Extended X-ray Absorption Fine Structure spectroscopy (EXAFS).

When the X-ray photon energy (E) is tuned to the binding energy of some core level of an atom in the material, an abrupt increase in the absorption coefficient occurs (giving an absorption edge). For an isolated atom, a spectrum of X-ray absorption as a function of energy around and beyond the absorption edge shows little or no fine structure; after the sharp increase at the edge, the absorption coefficient decreases monotonically. For atoms in a molecule or in a condensed phase, the variations of absorption coefficient around the absorption edge and

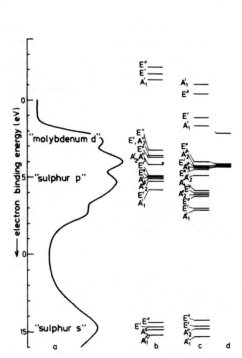

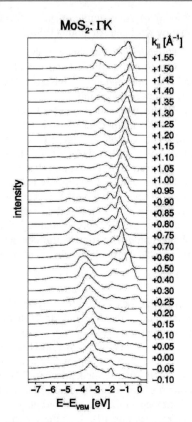

Figure 18. The XPS spectrum (a) of MoS$_2$ and the results of MO calculations (using the multiple scattering Xα method) on: (b) an MoS$_6$ cluster using a non-self consistent calculation; (b) an MoS$_6^{8-}$ cluster using a self-consistent calculation; (c) an MoS$_6^{8-}$ cluster using a self-consistent calculation including relaxation effects for the five highest energy orbitals (after de Groot and Hass 1975; reproduced with the permission from Elsevier).

Figure 19. Polarized angle resolved photoemission spectra for an MoS$_2$ single crystal recorded along the high symmetry line of the structure (after Böker et al. 2001; reproduced with permission of the American Physical Society).

above the absorption edge display complex fine structure. The former is studied in XANES and the latter in EXAFS. The energy regions studied in each of these techniques are shown in Figure 20. Near or below the edge, there also generally appear absorption peaks due to excitation of core electrons to some bound states [e.g., $1s \rightarrow nd$, $(n+1)s$, or $(n+1)p$ orbitals for the K edge]. The processes occurring here, which are akin to those studied in XES, can provide valuable bonding information, such as the energetics of virtual orbitals and the electronic configuration.

EXAFS is a final-state interference effect involving scattering of the outgoing photoelectron from the neighboring atoms. Hence, whereas in a monatomic gas the photoelectron ejected by absorption of an X-ray photon will travel outwards unimpeded and the X-ray absorption curve simply decays smoothly, in the presence of neighboring atoms this outgoing photoelectron can be backscattered, producing an incoming wave that can interfere either constructively or destructively with the outgoing wave near the origin. The result is oscillatory behavior of the absorption rate, and the amplitude and frequency of this modulation depend on the type and bonding of the neighboring atoms and their distances from the absorber. The distinction between XANES and EXAFS can be simply pictured as shown in Figure 21. The XANES

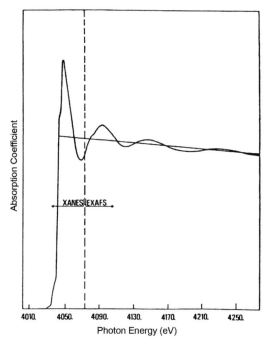

Figure 20. A typical X-ray absorption (XAS) spectrum showing the XANES and EXAFS regions.

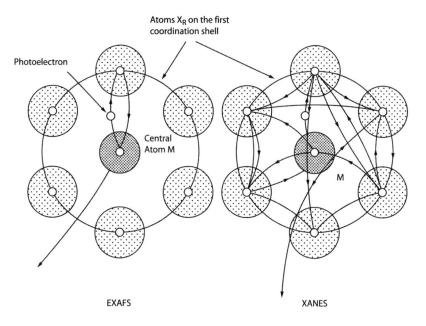

Figure 21. The processes involved in EXAFS and XANES spectroscopies: the scattering processes of the excited internal photoelectron determining the EXAFS oscillations (single scattering regime) and the XANES (multiple scattering regime).

are determined by multiple-scattering resonances where the final-state wave function in the continuum is localized inside the molecule; EXAFS oscillations are a product of scattering processes in the single-scattering regime.

Because the information from EXAFS concerns the local structure surrounding a particular atom, it is especially useful in determining the site occupancy of atoms in a complex structure (as illustrated below for the tetrahedrites), or in studying materials lacking long range order such as solution complexes, amorphous materials, fine particle precipitates or sorbates. Certain of these applications are discussed in more detail elsewhere in this volume. The XANES spectra provide a powerful technique for investigating electronic structure, particularly when combined with other spectroscopies and with *ab initio* calculations. As observed by Fleet (2005), the XANES spectrum can provide a "map" of the unoccupied partial density of states above the valence band in a material. The XANES can also be used to probe the chemical states of particular atoms in a mineral, precipitate, solution or surface species. Some examples of the uses of XAS methods in studying sulfides are discussed below. An excellent overview of the applications of XAS in mineralogy is provided by Brown et al. (1988). A review of NEXAFS (or XANES) investigations of transition metal compounds including sulfides has been published by Chen (1997), a range of S K- and L-edge XAS data published by Li et al. (1995), and an excellent review of XANES spectra of sulfur in Earth materials including sulfide minerals published by Fleet (2005). O'Day et al. (2004) have published a very useful set of XANES and EXAFS data for 27 common Fe-bearing reference compounds, data of value in distinguishing the mineral components in fine grained soils or sediments.

Applications to binary metal sulfides: PbS, ZnS, CdS, HgS, CaS, MgS, MnS, FeS, CoS, NiS and CuS

XANES of galena, sphalerite and related phases. Sulfur K-edge XANES spectra have been reported for galena (PbS) and sphalerite (ZnS) by von Oertzen et al. (2005a,b). These spectra represent transitions (subject to the $\Delta l = \pm 1$ selection rule) from the S$1s$ state to the conduction band comprized from Sp-like orbitals, so that the shape of the absorption spectrum reflects the shape of the density of states for the relevant conduction bands. The authors compare their experimental spectra with results of *ab initio* calculations using four different methods, as illustrated in Figure 22 for the example of sphalerite. The calculations show generally good agreement with experiment; peaks labeled a – d in the spectrum can be identified with peaks in the Sp density of states, supporting their being due to transitions to Sp-like states. This interpretation is also in good agreement with the earlier S K-edge XANES studies of ZnS by Sainctavit et al. (1987), Pong et al. (1994) and Li et al. (1994a,b). The latter authors also studied both the K and L-edge spectra of sulfur in ZnS, CdS and HgS, including polymorphic forms. The positions of the SK-edge and L-edge features shift to lower energy by ~2 eV through this series of monosulfides. Figure 23 reproduces the K and L-edge spectra of Li et al. (1994a) for which they proposed qualitative assignments based upon proposed MO/energy band structures of these monosulfides (see later in this volume, Vaughan and Rosso 2006). In general, K-edge features are assigned to transitions from S$1s$ to S$3p$-like states, and L-edge features assigned to S$2p$ to various S$3s$ and S$3d$-like states. At higher energies above the edge (i.e., the EXAFS region) the features observed can be interpreted using a multiple scattering model and hence related to the crystal structure.

The XANES spectra have also been used to probe the local environment and bonding characteristics of transition metals (Mn, Fe, Co, Ni) as dopants in ZnS (e.g., Lawniczak-Jablonska et al. 1996), and the nature of coupled substitutions (e.g., Cu/In) in the sphalerite lattice (Patttrick et al. 1998). These metals are involved in direct replacement of the Zn, with evidence for some local site distortion and clustering in Cu/In and Cd-doped materials. It can also be noted that the S K-edge XANES of wurtzite (the hexagonal form of ZnS) is essentially identical to that of the cubic (sphalerite) form (Li et al. 1994a).

Spectroscopic Studies of Sulfides 205

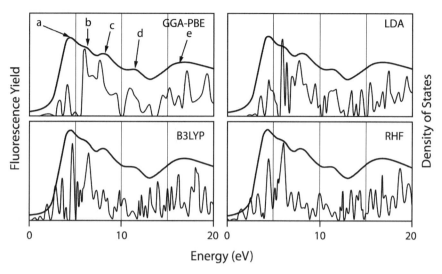

Figure 22. Sphalerite XANES spectrum (the heavy line at top of each figure) with features labeled a-e, and the conduction band Sp density of states (the lighter line below in each figure) calculated using four different computational methods (GGA-PBE, LDA, B3LYP, RHF). [Redrawn from von Oertzen et al. 2005a, with kind permission of Springer Science and Business Media.]

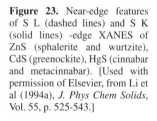

Figure 23. Near-edge features of S L (dashed lines) and S K (solid lines) -edge XANES of ZnS (sphalerite and wurtzite), CdS (greenockite), HgS (cinnabar and metacinnabar). [Used with permission of Elsevier, from Li et al (1994a), *J. Phys Chem Solids*, Vol. 55, p. 525-543.]

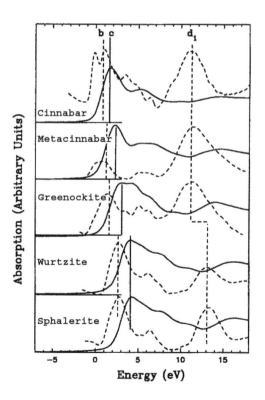

XANES of transition metal sulfides. Turning to the transition metals, there have been XANES studies of FeS (Sugiura 1984; Kitamura et al. 1988; Womes et al. 1997) and of the NiAs-structure monosulfide series FeS-CoS-NiS (Soldatov et al. 2004). Farrell and Fleet (2001) also studied the S K-edge XANES spectra of ternary solid solutions in the system $Fe_{0.923}S$-$Co_{0.923}S$-$Ni_{0.923}S$ (the nonstoichiometric composition being used to avoid precipitation of a Co pentlandite phase) and the S K-edge and L-edge XANES of synthetic endmembers and binary solid solutions of halite (NaCl) structure monosulfides in the CaS-MgS-MnS-FeS system (Farrell and Fleet 2000; Farrell et al. 2002; Kravtsova et al. 2004). For many of these phases the XANES spectra have been simulated using multiple scattering calculations (see Fleet 2005 for an overview). The S K-edge XANES of transition metal-bearing monosulfides generally show anomalous absorption consistent with hybridization of the final $S3p\sigma^*$ antibonding states with empty $3d$ orbitals on the metal atoms (Fleet 2005).

CuS and other binary copper sulfides such as Cu_2S and $Cu_{1.8}S$ have also been studied (Sugiura and Gohsi 1981; Li et al. 1994c). Copper provides a good example of the use of Cu L-edge spectra in determining oxidation states. For example, van der Laan et al. (1992) investigated a large number of copper minerals and show that Cu^{2+} gives a narrow single peak due to excitation into the empty $3d$ density of states, whereas Cu^+ gives a broad band at higher energy due to the transitions to the empty metal s and ligand states. The data are used to demonstrate the presence of monovalent copper not only in Cu_2S but also in $CuFeS_2$, Cu_5FeS_4, Cu_9S_5 and even in CuS.

Applications to disulfides: pyrite group (MS_2 where M = Fe, Co, Ni) and MoS_2

XANES of pyrite and related phases. As for galena and sphalerite, von Oertzen et al. (2005a,b) have reported the $S1s$ XANES spectra for pyrite (FeS_2) and compared this with *ab initio* computations as shown in Figure 24. In pyrite, the conduction band is composed of Sp states mixed with Fep and d states; the shoulder b in the XANES spectrum can be explained by a transition to predominantly Sp states and the peak a by a transition to predominantly Fep and d states. This agrees with peak assignments made by Mosselmans et al. (1995) in an earlier study of the $FeS_2 - CoS_2 - NiS_2 - CuS_2$ series of disulfides and in which a complete series of metal and sulfur S and L-edge spectra were reported and interpreted with reference to earlier band structure calculations of Bullet (1982) and of Temmerman et al. (1993). For the three disulfides FeS_2, CoS_2 and NiS_2 the XANES at the metal $L_{3,2}$ edges have also been recorded by Charnock et al. (1996) and compared with simulations using an atomic multiplet approach; this is shown reasonably to reproduce the observed experimental data.

XANES of molybdenite. The optical absorption, XPS and XES spectra of MoS_2 have all been discussed above; this phase has been a focus of particular interest because of its properties (see also Abrams and Wilcoxon 2005 for review). Ohno et al. (1983a,b) have reported the S K-edge and the Mo L_{II}-edge X-ray absorption spectra of MoS_2 and a series of other layered dichalcogenides. These studies suggest that the metal d orbitals are strongly mixed with the Sp orbitals, in line with ideas on the electronic structure discussed elsewhere in this chapter.

Ternary and complex sulfides: chalcopyrite, tetrahedrite group

Chalcopyrite. Chalcopyrite has been studied using XAS by a number of workers including Sainctavit et al. (1986), Petiau et al. (1988) and Li et al. (1994b). The investigation by Li et al. (1994b) of SK and L-edge spectra also involved the isostructural phase stannite (Cu_2FeSnS_4) and, as discussed above, sphalerite (including (Zn,Fe)S). The SK-edge spectra of this series of minerals (Fig. 25) clearly illustrate that, for energies more than ~15 eV above the edge (peaks d, e, f, g) where multiple scattering dominates, the structural similarities between these minerals leads to very similar spectral features. Closer in energy to the edge, the differences in electronic structure are reflected in very different spectra. These peaks have again been qualitatively assigned using MO/energy band structure models (as further

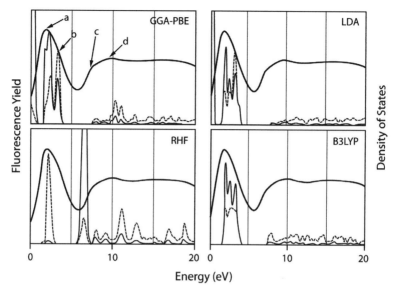

Figure 24. Pyrite SK-edge XANES (the heavy line at the top of each figure) with features labeled a – d, and the calculated conduction band Sp density of states (dashed line) and sum of Fep and d density of states (solid line) shown below; each figure shows comparison of the XANES data with the results of different methods of calculation (see Fig. 22). [Redrawn wih permission of Elsevier, from von Oertzen et al. (2005b), *J. Electron Spect. Rel. Phenomena*, Vol. 144-147, p. 1245-1247.]

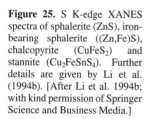

Figure 25. S K-edge XANES spectra of sphalerite (ZnS), iron-bearing sphalerite ((Zn,Fe)S), chalcopyrite (CuFeS$_2$) and stannite (Cu$_2$FeSnS$_4$). Further details are given by Li et al. (1994b). [After Li et al. 1994b; with kind permission of Springer Science and Business Media.]

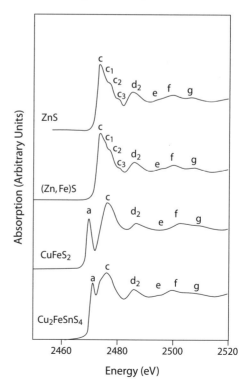

discussed by Vaughan and Rosso 2006, in this volume). Using an identical approach, Li et al. (1994c) have also discussed S K- and L-edge spectra of bornite, cubanite, enargite and tetrahedrite.

Tetrahedrite group. The complex substitutions in the tetrahedrite group (also known as "fahlore") minerals (see Makovicky, 2006, this volume) have been investigated using EXAFS spectroscopy (Charnock et al. 1988, 1989a) in a study that well illustrates the power of this technique for determining the local environment around an element of interest. These researchers collected data on the Cu, Ag, Fe, Cd and Sb K-edges of a series of natural and synthetic tetrahedrite group minerals. The tetrahedrite endmember, of ideal composition $Cu_{12}Sb_4S_{13}$, has a structure with half of the Cu atoms in tetrahedral and half in trigonal sites, and Sb bonded to three S atoms. Twelve of the S atoms are in tetrahedral coordination and the thirteenth S atom in octahedral coordination. However, substitutions are extensive, allowing general compositions such as $(Cu,Ag)_{10}(Cu,Fe,Zn,Cd)_2Sb_4S_{13}$, but the location of the substituting cations in the structure cannot be determined by traditional methods such as X-ray diffraction. It was shown by EXAFS, for example, that silver goes into trigonal sites (see Fig. 26) and that iron occupies mainly tetrahedral sites, as does cadmium. Combined EXAFS and ^{57}Fe Mössbauer studies (Charnock et al. 1989b) show that

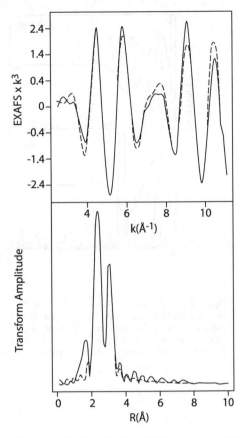

Figure 26. The k^3 weighted Ag EXAFs spectrum (solid line, top) and Fourier transform (solid line, bottom) of a natural tetrahedrite. Broken lines represent best fit theoretical simulation. [Used by permission of Elsevier from Charnock et al. (1989a), *J. Solid State Chemistry*, Vol. 82, p. 279-289.]

the $Fe^{2+}:Fe^{3+}$ ratio plays a part in maintaining charge balance when these complex substitutions occur. Pattrick et al. (1993) also used L-edge XAS spectra to determine the oxidation state and electronic configuration of copper in the tetrahedrites.

Thiospinels. The thiospinel minerals are another example of a more complex structure type where EXAFS has provided insights into the substitutions that occur, in this case into the tetrahedral and octahedral sites of the spinel structure (Charnock et al. 1990). These studies confirm that daubréelite ($FeCr_2S_4$) and carrollite ($CuCo_2S_4$) are normal spinels, and that in the series $Ni_3S_4 - FeNi_2S_4$, Fe substitutes for Ni in the octahedral sites.

MÖSSBAUER STUDIES

The recoilless emission and resonant absorption of nuclear γ-rays in solids, known as the Mössbauer effect after its discoverer Rudolf Mössbauer, is the basis of a routine spectroscopic method (Mössbauer 1962). Many excellent introductions to the theory and experimental methods of Mössbauer spectroscopy have been published, including several with an emphasis on mineralogical applications (e.g. Bancroft 1973; Maddock 1985; Hawthorne 1988). Almost all of the studies relevant to this work concern the Mössbauer effect in the nucleus of ^{57}Fe (≈2% natural abundance), so the discussion can be confined to this isotope.

The decay of the radioactive isotope ^{57}Co (illustrated in Fig. 27a) results in the formation of the stable isotope ^{57}Fe. It is accompanied by the emission of gamma-rays of various energies as the decaying ^{57}Co passes through various excited states of ^{57}Fe characterized by

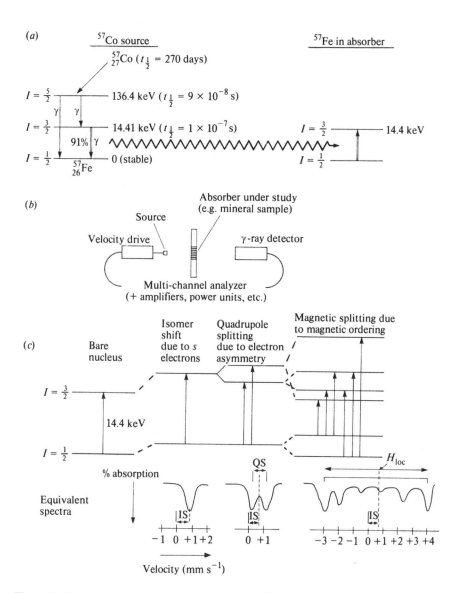

Figure 27. Mössbauer spectroscopy: (a) nuclear transitions giving rise to the Mössbauer effect in ^{57}Fe; (b) principles of the Mössbauer spectrometer; (c) Mössbauer resonant absorption of iron in different crystal environments and resulting spectral types.

different nuclear spin quantum numbers (I). These are very short-lived species with half-lives ($t_{1/2}$) of 10^{-7} s or less. The 14.4 keV γ-rays emitted when ^{57}Fe in the first excited state ($I = 3/2$) passes to the ground state ($I = 1/2$) are used in the Mössbauer effect experiment. When γ-rays of this energy are emitted without recoil, which is possible only from a solid material, they may be absorbed resonantly (again without recoil) by ^{57}Fe atoms in an absorber such as a mineral. However, the precise energy at which this absorption occurs and the nature of the absorption is influenced by the environment of the ^{57}Fe nucleus, and thus a range of γ-ray energies must be provided by the source. This is achieved by making use of the Doppler effect, i.e., by vibrating the source through a range of velocities. The experimental system is shown in Figure 27b. The sample lies between the source, mounted on its vibrating system, and the γ-ray detector. As the source sweeps through a range of velocities, the counts from the detector pass into a range of counting cells in a multi-channel analyzer or computer counting system. The resulting spectrum is a plot of γ-ray energies (expressed as velocity of the source in mm s^{-1}) versus absorption (see Fig. 27c).

The influence which the environment of the nucleus exerts on the energies of the ground and first excited states of ^{57}Fe makes the Mössbauer effect a powerful tool in the study of the chemistry of iron-bearing solids. The effects of the perturbations on the energy levels and their corresponding spectra are summarized in Figure 27c. Differences in the densities of s electrons at the nucleus between source and absorber result in shifting of the relative energies of ground and excited states. This *isomer shift* (IS) varies from one substance to another because covalency, oxidation state, coordination number and type of ligand all influence the s electron density. Isomer shift (normally referred to a standard such as pure iron) can, therefore, provide information on oxidation state, spin state, coordination and degree of covalence of the iron atoms in a substance. If the electronic environment of the iron nucleus is not spherically symmetrical (i.e., if the electrons are not symmetrically distributed amongst the orbitals, or if the coordination site is distorted) the excited state energy level will be split (its degeneracy will be partially removed). As a result of this *quadrupole splitting* (QS), two transitions can occur producing a two-peak spectrum. Quadrupole splitting values can indicate spin states, oxidation states and distortion of coordination sites. The final perturbation is that produced by magnetic ordering which further splits the energy levels, between which six transitions are now possible, giving a six-peak spectrum. The separation of the outermost peaks can be used to obtain the value of the internal magnetic field at the nucleus (H_{loc}).

Hence, the Mössbauer effect can be used to study the oxidation states, spin states, coordination, magnetic state and covalence of iron atoms in minerals. When iron atoms occur in more than one site in a mineral structure, their distribution between sites can be determined. Applications to the study of iron in sulfide minerals will now be considered. It is important to bear in mind that many minerals were studied in the decades immediately following Mössbauer's discovery in 1958. These earlier studies are briefly reviewed here, along with more recent investigations; more detailed accounts of the earlier work can be found in Vaughan and Craig (1978) and Mitra (1992). Representative Mössbauer data for most of the iron-bearing sulfides are provided in Table 1, along with references to the sources of the data.

Pyrite, marcasite, arsenopyrite and related minerals

Like pyrite, all of these minerals contain dianion groups in their structures and all have Mössbauer spectra comprising a single quadrupole split doublet (see Fig. 28). At room temperature, the isomer shifts of all of these compounds lie in the range 0.25–0.47 mm s^{-1}, whereas the quadrupole splittings cover the range 0.32–1.68 mm s^{-1} (Table1). The magnetic and electrical properties of pyrite show that the iron 3d electrons are in the low-spin state. The Mössbauer spectrum of iron in pyrite supports this conclusion. The small isomer shift is characteristic of a low-spin species and the quadrupole splitting is a consequence only of the departure of the iron site from cubic symmetry. The small isomer shifts of iron in all of these

Spectroscopic Studies of Sulfides 211

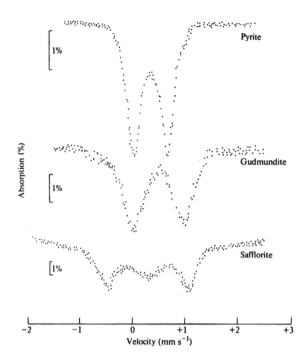

Figure 28. Mössbauer spectra of pyrite (FeS$_2$), gudmundite (FeSbS) and safflorite (Co, Fe)As$_2$ at room temperature. The zero of the velocity scale in this and succeeding spectra is the center of the spectrum of iron at room temperature (Craig and Vaughan 1978).

dianion compounds are compatible with models involving maximum spin-pairing of electrons (see Vaughan and Rosso 2006, in this volume) although the decrease in iron 3d electron populations predicted in the series FeS$_2$ → FeAsS → FeAs$_2$ by some models is not indicated by the isomer shifts.

Studies of members of the pyrite-structured (Fe,Cu)S$_2$ solid solution series (Schmid-Beurmann and Lottermoser 1993) show that the low-spin Fe^{2+} configuration is retained; a minimum in the quadrupole splitting and a change in isomer shift (~0.30 to ~0.36 mm s^{-1}) at around 30 mol% substitution of Fe by Cu correlates with the onset of metallic conductivity. Similar behavior is seen in the pyrite-structured (Fe,Co)S$_2$ solid solution (Nishihara and Ogawa 1980). Mössbauer parameters for pyrite and Fe-doped isostructural MS_2 phases (M = Co, Ni, Cu, Zn) have also been successfully calculated quantum mechanically by Lauer et al. (1984).

Thiospinel minerals

The iron-bearing thiospinel mineral end-members, greigite (Fe$_3$S$_4$), daubréelite (FeCr$_2$S$_4$), violarite (FeNi$_2$S$_4$) and indite (FeIn$_2$S$_4$), have all been studied (see Table 1). The room-temperature spectrum of daubréelite is a simple doublet which may be assigned to high-spin Fe^{2+} ions in the regular tetrahedral sites. Below 192 K, daubréelite is antiferromagnetically ordered and a six-line spectrum is observed. These parameters contrast with the indite data which indicate high-spin Fe^{2+} in the octahedral sites.

Greigite spectra have only been obtained for fine particle synthetic materials. In such materials, magnetic reversals due to thermal motion take place in a shorter time than the Mössbauer transition at room temperature (i.e., they are *superparamagnetic*); therefore spectra

Table 1. Representative Mössbauer parameters for iron in sulfide minerals.

Mineral	Temp of Spectrum (K)	Isomer Shift [a] (mm s^{-1})	Quadrupole Splitting (mm s^{-1})	Hyperfine Magnetic Field (H_{loc}; Kgauss)	Reference
Disulfides and related phases:					
Pyrite (FeS$_2$)	300	0.31	0.61	—	Temperley and Lefevre (1966)
	81	0.40	0.62	—	"
Marcasite (FeS$_2$)	300	0.273	0.507	—	"
	81	0.369	0.504	—	"
Arsenopyrite (FeAsS)	300	0.25	1.05	—	Imbert et al. (1963)
Gudmundite (FeSbS)	79	0.49	0.94	—	"
Safflorite ((Co,Fe)As$_2$)	300	0.28	1.48	—	Vaughan and Craig (1978)
Thiospinel minerals:					
Daubréelite (FeCr$_2$S$_4$)	295	0.60	0.00	—	Greenwood and Whitfield (1968)
[High spin tet. Fe^{2+}]	77	0.72	0.17	206	"
Indite (FeIn$_2$S$_4$)	295	0.88	3.27	—	Yagnik and Mathur (1967)
[High spin oct. Fe^{2+}]	79	0.81	3.12	—	"
Violarite (FeNi$_2$S$_4$)	300	0.30	0.60	—	Vaughan and Craig (1985)
[Low spin oct. Fe^{2+}]	77	0.31	0.64	—	"
Greigite (Fe$_3$S$_4$)					
(i) High spin oct. Fe^{2+}	4.2	0.70	0.30	322	Vaughan and Ridout (1971)
(ii) High spin oct. Fe^{3+}	4.2	0.45	0.40	465	"
(iii) High spin tet. Fe^{3+}	4.2	0.40	0.00	486	"
Metal-enriched sulfides:					
Mackinawite (Fe$_{1+x}$S)	4.2	0.20	0	—	Vaughan and Ridout (1971)
site type 1 (52%)	300	0.42	0	—	Mullet et al. (2002)
site type 2 (19%)	0.13	0.38	—		"
site type 3 (29%)	0.41	0.64	—		"
Pentlandite (Fe,Ni)$_9$S$_8$					
(i) octahedral site	300	0.649	0	—	Knop et al. (1970)
(ii) tetrahedral site	300	0.356	0.298	—	

	T (K)	IS	QS	H	Reference
Sphalerite/wurtzite and related phases:					
Sphalerite (ZnS; 1-5% Fe)					Gerard et al. (1971)
- single peak	300	0.66	0	—	"
- doublet	300	0.66	0.60	—	"
	77	0.66	2.40	—	"
	4.2	0.66	3.19	—	"
Wurtzite (ZnS; 1-5% Fe)					"
- single peak	300	0.69	0	—	"
- doublet	300	0.69	0.56	—	"
	77	0.69	2.38	—	"
	4.2	0.69	3.19	—	"
Chalcopyrite ($CuFeS_2$)	295	0.23	−0.012	356	Greenwood and Whitfield (1968)
	77	0.37	0.17	206	"
Stannite (Cu_2FeSnS_4)	295	0.57	2.90	—	Vaughan and Burns (1972)
Germanite ($Cu_3(Fe,Ge)S_4$)	300	0.34	0.35	—	"
	77	0.37	0.88	—	"
Bornite (Cu_5FeS_4)	300	0.39	0.16	—	"
Cubanite ($CuFe_2S_3$)	300	0.40	0.27	333	"
Sterbergite ($AgFe_2S_3$)	300	0.38	2.10	266	"
Argentopyrite ($AgFe_2S_3$)					"
six peak spectrum (1)	300	0.49	2.36	293	"
six peak spectrum (2)	300	0.35	2.21	279	"
Pyrrhotite and related phases:					
Troilite (FeS)	300	0.81	−0.32	315	Hafner and Kalvius (1966)
	77	0.81	−0.32	320	"
Monoclinic pyrrhotite (Fe_7S_8)					Vaughan and Ridout (1971)
six peak spectrum (1)	77	0.81	−0.36	345	"
six peak spectrum (2)	77	0.79	0.60	311	"
six peak spectrum (3)	77	0.79	0.12	267	"
six peak spectrum (4)	77	0.77	0.32	229	"
two peak spectrum (5)	77	0.39	0.25	0	"

Note: a – relative to metallic iron

have to be recorded at very low temperature to overcome this effect. Vaughan and Ridout (1971) studied greigite, synthesized by different methods, at temperatures down to 1.4 K (see Fig. 29). They observed ferrous and ferric iron, magnetically ordered in both tetrahedral and octahedral sites, and suggested greigite to be the sulfur analog of magnetite; i.e., $Fe^{3+}(Fe^{2+},Fe^{3+})S_4$. The Fe^{2+}/Fe^{3+} ratio showed variation with method of preparation. Coey et al. (1970) studied the spectra at low temperatures and in external magnetic fields and also concluded that greigite is the sulfur analog of magnetite. They observed only a single species in the octahedral sites and suggested electron hopping as in magnetite. The Mössbauer spectrum of violarite at 77 K consists of a doublet with a small isomer shift and quadrupole splitting, indicating low-spin Fe^{2+} in the octahedral sites. Compositions in the solid solution series $Fe_{0.25}Ni_{2.75}S_4 - FeNi_2S_4$ show this same doublet attributed to octahedral Fe^{2+}, but with the small isomer shift increasing with increasing iron (0.23-0.53 mm s^{-1}), although with a possible small contribution (~18%) from iron in tetrahedral coordination (Vaughan and Craig 1985).

Given the interest in synthetic spinel-structure compounds because of their diverse magnetic and electrical properties, it is not surprising that ^{57}Fe Mossbauer spectroscopy has been used, along with X-ray diffraction and various electrical and magnetic measurements, to study the detailed effects of various substitutions (e.g., work on $Fe_{1-x}Cu_xCr_2S_4$ phases by Kim et al. 2004)

Mackinawite and pentlandite

Mackinawite (FeS) and pentlandite [(Fe,Ni)$_9$S$_8$] both contain iron predominantly in tetrahedral coordination, and with short metal-metal distances between the tetrahedral site metals in certain directions in the structures. At 300 K, the Mössbauer spectrum of pentlandite is an asymmetric doublet which has been interpreted as a small unsplit peak due to the octahedral iron with a superimposed quadrupole doublet from iron in the distorted tetrahedral sites (Knop et al. 1970; Vaughan and Ridout 1971). Parameters are given in Table 1. Cooling of pentlandite

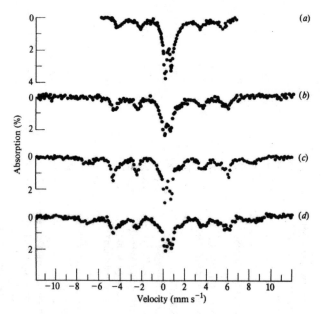

Figure 29. Mössbauer spectra of greigite (Fe$_3$S$_4$): (a) at 300 K; (b) at 77 K; (c) at 4.2 K; (d) at 1.4 K [Used with permission of Elsevier, from Vaughan and Ridout (1971), *J. Inorganic and Nuclear Chemistry*, Vol. 33, p. 741-746.]

to 4.2 K produces only a slight increase in isomer shift and quadrupole splitting, and in an external magnetic field of 30 kgauss, no internal field contribution is observed. This suggests the absence of a paramagnetic moment on at least the tetrahedral site iron atoms, and together with the small isomer shift, suggests metal-metal bonding and 3d electron delocalization. The formal oxidation state of iron in the tetrahedral sites is therefore less than +2. In the octahedral sites, high- or low-spin Fe^{2+} could be present, although Rajamani and Prewitt (1973) favor high-spin Fe^{2+} on the basis of interatomic distances. In natural pentlandite, the iron atoms show a slight preference for the octahedral sites relative to a random distribution. Knop et al. (1970) noted, however, that on annealing at 200 °C a migration of the iron atoms to the tetrahedral sites took place, such that the iron atoms then showed a slight preference for the tetrahedral sites relative to a random distribution.

Mackinawite has a layer structure and very short metal-metal distances between tetrahedral sites within the layers, but large separation of metal atoms between layers (see Makovicky 2006, this volume). Reported Mössbauer spectra of mackinawite (Bertaut et al. 1965; Vaughan and Ridout 1971) showed no evidence of magnetic ordering down to 1.4 K, and this was confirmed by neutron diffraction studies (Bertaut et al. 1965). Vaughan and Ridout (1971) report a single-line spectrum with a very small isomer shift (0.20 mm s^{-1}). Application of an external magnetic field of 30 kgauss showed no internal field to be present, suggesting that the iron in mackinawite is involved in the same type of metal-metal interaction as that proposed in pentlandite. In their Mössbauer study, Mullet et al. (2002) confirmed the dominance of this single peak with a small isomer shift (in this case, reported to be ~0.4 mm s^{-1}) again retained to low temperature (11 K) and attributed by them to tetrahedrally coordinated low spin Fe^{2+}. However, they also fitted their room temperature spectrum with two lower intensity doublets attributed to tetrahedrally coordinated low spin Fe^{2+} and low spin Fe^{3+} species and suggested to arise from a "weathered thin layer covering the bulk material that consists of both Fe(II) and Fe(III) bound to S(−II) atoms and in a less extent polysulfide and elemental sulfur." Mackinawite is a highly reactive metastable phase, and this "weathered thin layer" may represent an initial stage in the oxidative breakdown to form greigite, as in the process postulated by Lennie et al. (1997).

Sphalerite and wurtzite-type sulfide minerals

Gerard et al. (1971) studied synthetic sphalerite, wurtzite and greenockite (CdS) containing iron as an impurity and observed only a single peak for very low iron concentrations (≤6 mol% Fe) but a single peak and superimposed quadrupole doublet for higher concentrations (<31%). They interpret the single peak as due to iron atoms without other iron impurity atoms as nearest neighbors, and the doublet as the effect of one or more other iron atoms as nearest neighbors. The singlet and doublet are temperature-independent but composition-dependent in their intensity. Gerard et al. (1971) used these results to argue against static Jahn-Teller distortions in Fe^{2+}-containing ZnS. These Mössbauer data indicate that iron sphalerites (and wurtzites) contain Fe^{2+} ions substituting for Zn in regular tetrahedral sites. Lepetit et al. (2003) synthesized Fe-containing sphalerites over a wide range of Zn:Fe ratios, temperature and sulfur fugacity conditions, and confirmed the findings of Gerard et al. (1971). However, their study focused on the conditions leading to the segregation of a second phase (pyrrhotite or, in the presence of Cu, chalcopyrite) at high iron content, and showed that Fe^{3+} can occur at higher sulfur fugacities. (Within single phase solid solutions of 20 mol% FeS, Fe^{3+} increases from 0 to 7.3 mol% with increasing sulfur fugacities, up to conditions of the $Fe_{1-x}S/FeS_2$ buffer). Most recently, Di Benedetto et al. (2005a) used Mössbauer spectroscopy and magnetic susceptibility measurements on iron-bearing sphalerites to demonstrate the presence of clustering of Fe even at low concentrations.

The Mössbauer spectrum of chalcopyrite ($CuFeS_2$) consists of a single six-line hyperfine magnetic spectrum arising from magnetically ordered iron in one position of the structure (see, for example, Vaughan and Burns 1972; Boekema et al. 2004). The isomer shift (Table 1) is small and no quadrupole splitting is observed. The observed parameters are consistent with magnetically coupled high-spin Fe^{3+} in the tetrahedral sites. The question of the valencies of copper and iron in chalcopyrite has long been the subject of debate, but these Mössbauer data confirm the formal scheme $Cu^+Fe^{3+}S_2$. The Mössbauer spectra of the related phases $Cu_{18}Fe_{16}S_{32}$ (talnakite) and $Cu_9Fe_9S_{16}$ (mooihoekite) have been reported by Townsend et al. (1972). Interpretation of the talnakite spectrum shows it to be antiferromagnetic with the additional metals in interstitial sites. Mooihoekite was also studied by Kalinowski et al. (2000) who observed four sextets and a low magnetic field component at room temperature; they proposed a statistical distribution of Fe and Cu atoms in the interstitial sites with these additional atoms, as well as those in normal chalcopyrite sites, involved in an antiferromagnetically ordered structure.

Mössbauer spectra for a number of iron-bearing sulfides structurally related to sphalerite and wurtzite have been reported (see Table 1). The main contributions of these Mössbauer studies has been determination of the oxidation state of the iron (and by charge balance arguments, of the other cations present). Thus, formal valence schemes of $Cu_2^+Fe^{2+}Sn^{4+}S_4$ for stannite, $Cu_5^+Fe^{3+}S_4$ for bornite, and (assuming Ge^{4+}) of $(Cu^+,Cu^{2+})_3(Fe^{3+},Ge^{4+})(S^{2-},As^{3-})_4$ for germanite, are indicated by the reported Mössbauer data. In the case of stannite, the Mössbauer effect in ^{119}Sn was also studied and confirmed the presence of tetravalent tin (Eibschutz et al. 1967), and Di Benedetto et al. (2005c) also studied both nuclei in the stannite–kesterite (Cu_2ZnSnS_4) solid solution. High-spin states and an absence of magnetic ordering characterizes the iron in these phases, although bornite orders antiferromagnetically at ~62 K. The series of spectra recorded by Collins et al. (1981), and shown in Figure 30, are a good example of the changes observed in a Mössbauer spectrum on cooling through a magnetic ordering transition. The authors fit the low temperature spectra to a six-line pattern, and the higher temperature spectra to a quadrupole doublet with a small splitting (~0.2 mm s^{-1}) and isomer shift (~0.38 mm s^{-1}).

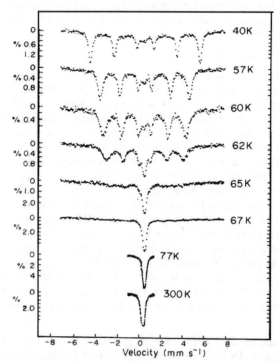

Figure 30. Mössbauer spectra of bornite at temperatures between 40 K and 300 K (after Collins et al. 1981).

The Mössbauer spectrum of cubanite, on the other hand, shows antiferromagnetic ordering at room temperature, and the six-line spectrum has been attributed to rapid electron exchange between ferrous and ferric iron in $Cu^+Fe^{2+}Fe^{3+}S_3$ (Greenwood and Whitfield 1968; Vaughan and Burns 1972). It has been proposed that the sternbergite spectrum indicates similar rapid electron exchange between ferrous and ferric iron as in cubanite (i.e., $AgFe^{2+}Fe^{3+}S_3$), in this case with the magnetic coupling within a pair of crystallographically identical iron atoms being ferromagnetic (Wintenberger et al. 1990). Although the Mössbauer parameters for the argentopyrite form of $AgFe_2S_3$ again suggest electron hopping, the presence of a second six-line subspectrum shows that iron occurs in two different positions in the argentopyrite structure (Vaughan and Burns 1972).

Troilite and pyrrhotite

FeS with the troilite structure has a room-temperature Mössbauer spectrum comprising a single six-peak hyperfine set (Hafner and Kalvius 1966; Thiel and van den Berg 1968; Kruse and Ericsson 1988). This arises from antiferromagnetically ordered iron in one type of position. Mössbauer parameters for troilite are given in Table 1. In troilite, interest has centred around the nature of two observed transitions. One is the transition from the troilite-type superstructure to the NiAs-type structure (the "α-transition") at ≈ 140 °C; the second is the change in spin alignment of the antiferromagnetically ordered electrons on adjacent iron atoms ("spin flip" or "Morin" transition). Mössbauer measurements made on a single crystal of FeS by Horita and Hirahara (1966) show that above the Morin transition, the antiferromagnetically aligned spins are perpendicular to the c-axis and below it they are parallel to the c-axis. In a very detailed Mössbauer study of a natural meteoritic troilite, Kruse and Ericsson (1988) were able to follow the transitions from the high temperature NiAs-type structure (~ 483 K) to an intermediate MnP-type structure and then to the troilite structure (~ 413 K). They propose that the magnetic spin flip of 90° occurs only in the MnP phase.

The metal-deficient monoclinic pyrrhotite (Fe_7S_8) has been studied using the Mössbauer effect by Hafner and Kalvius (1966), Levinson and Treves (1968) Vaughan and Ridout (1970) and Jeandey et al. (1991). The vacancy ordering model proposed by Bertaut (1953) in which iron atom vacancies are confined to alternate cation layers in (001) plane and to every other position in alternate rows in these layers, generates four non-equivalent iron atom positions. Vaughan and Ridout (1970) fitted four magnetically split subspectra and an additional weak non-magnetic doublet (see Fig. 31 and Table 1). In all of these studies the spectra were discussed in terms of the Bertaut model and used to support his scheme of vacancy ordering which has been confirmed by the structural refinement of Tokonami et al. (1972). None of the authors found evidence to suggest the presence of Fe^{3+}, despite the original dependence of Bertaut's model on a formal valence scheme of the type $Fe_5^{2+}Fe_2^{3+}S_8^{2-}$. Levinson and Treves (1968) suggested a valence model of the type $Fe_7^{2.286+}S_8^{2-}$ from the reduced isomer shift values (relative to FeS). In more recent work, Xie et al. (2001) synthesized low dimensional platelike Fe_7S_8 microcrystals and characterized them using a range of techniques including Mössbauer spectroscopy. The room temperature spectrum shows a broad sextet from magnetically ordered iron in the structure and a nonmagnetic doublet (~23% of the intensity) believed to arise from a superparamagnetic (very fine particle) component.

As regards the so-called intermediate (hexagonal) pyrrhotites, there have been a number of Mössbauer studies (Goncharov et al. 1970; Vaughan et al. 1971; Schwarz and Vaughan 1972; Marusak and Mulay 1979; Townsend et al. 1979; Igaki et al. 1981, 1982; Kruse 1990). These have attempted to relate the spectra to various vacancy-containing or vacancy ordered structures lying in composition between troilite and monoclinic pyrrhotite. For example, Kruse (1990) reported a series of room temperature spectra with increasing vacancy concentrations (x in $Fe_{1-x}S$ phases; see Fig. 32) interpreted as overlapping sextets arising from magnetically ordered iron atoms in different nearest neighbor iron atom environments. Thus,

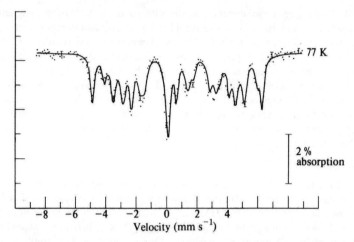

Figure 31. Mössbauer spectra of monoclinic pyrrhotite (Fe$_7$S$_8$) at 77 K. [Redrawn with permission of Elsevier, from Vaughan and Ridout (1970), *Solid State Communications*, Vol. 8, p. 2165-2167.]

for compositions where $0.004 < x < 0.042$, three sextets were assigned respectively to Fe atoms in triangular (troilite-type) clusters without any vacancy amongst the eight neighbors, cluster members with one neighboring vacancy, and Fe atoms which are not cluster members. For $0.054 < x < 0.079$, two of the sextets corresponded to atoms with zero or one vacancy amongst the Fe neighbors. For $0.106 < x < 0.143$ three sextets corresponded to atoms with zero, one, or two vacancies amongst the Fe atom neighbors. These different overlapping sextets are regarded as responsible for the spectra shown in Figure 32.

In the context of the "pyrrhotites," the apparently metastable low temperature phase smythite (~ Fe$_9$S$_{11}$) should be mentioned. Although Mössbauer spectra have been reported for this phase (e.g., Makarov et al. 1969; Hoffmann 1993) there remain uncertainties as to the characterization of these materials. Smythite appears to be magnetically ordered at room temperature and to show similar behavior to monoclinic pyrrhotite. It should also be noted that a very useful review of the mineral chemistry of the pyrrhotites has been published by Wang and Salveson (2005).

High-pressure Mössbauer spectroscopy of sulfides

The high-pressure spectra of a number of sulfides have been studied, including pyrite, troilite, loellingite, iron-doped hauerite, chalcopyrite, monoclinic pyrrhotite and FeSb$_2$. In many cases, systematic changes in Mössbauer parameters have been observed but in others there are more dramatic effects. For example, an elegant study of the effect of pressure on 2% ^{57}Fe substituting for manganese in MnS$_2$ (hauerite) was performed by Bargeron et al. (1971). At 1 atm, the iron was predominantly in the high-spin state (≈10% low-spin). The primary effect of pressure was the conversion of Fe^{2+} from the high-spin to the low-spin state. A measurable change was observed at 40 kbar. The conversion was reversible with only slight hysteresis.

Troilite (FeS) and chalcopyrite (CuFeS$_2$) are both minerals which are magnetically ordered (antiferromagnetic) at room temperature and exhibit six-peak spectra. The ferrimagnetic monoclinic pyrrhotite (Fe$_7$S$_8$; see above) has a much more complex magnetic spectrum at room temperature. Kasper and Drickamer (1968) report that the six-peak troilite spectrum is transformed to a single peak by 36 kbar with a reduction in isomer shift from 0.69 mm s^{-1} to 0.29 mm s^{-1}. Kobayashi et al. (2001) also studied FeS at high pressure using synchrotron-based Mössbauer spectroscopy and obtained results broadly consistent with conventional Mössbauer studies. Vaughan and Tossell (1973) report such transitions from magnetically ordered to disordered states in chalcopyrite and monoclinic pyrrhotite over the pressure range 5 to 16 kbar. In chalcopyrite, the magnetic spectrum is replaced by a doublet with essentially the same isomer shift (0.23 mm s^{-1}) and a fairly large quadrupole splitting (0.72 mm s^{-1}). In pyrrhotite, it is replaced by a single peak with a much reduced isomer shift ($\approx$ 0.80 mm s^{-1} at 1 atm → 0.36 mm s^{-1} at 16 kbar). The spectra of pyrrhotite over a range of pressures are shown in Figure 33. The changes in these spectra can be interpreted as evidence of transitions from an antiferromagnetic to a paramagnetic state for pyrrhotite. The transitions are totally reversible (see Fig. 33) and both ordered and disordered species appear to coexist over a range of pressures. The large changes in isomer shift observed in pyrrhotite and quadrupole splitting in chalcopyrite at the magnetic transitions are more likely to be high-spin → low-spin transitions than structural changes. Takele and Hearne (2001) have studied both FeS and Fe$_7$S$_8$ at high pressure using Mössbauer spectroscopy and electrical resistance measurements. They report that the high pressure (>7 GPa) (monoclinic) FeS phase adopts a magnetically quenched low-spin state and non-metallic behavior, and that Fe$_7$S$_8$ is magnetic-metallic at lower pressures (<5 GPa) and diamagnetic-metallic at high pressures.

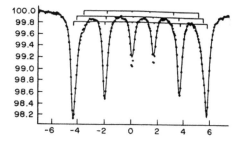
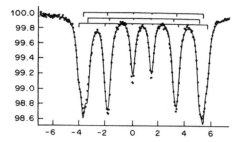
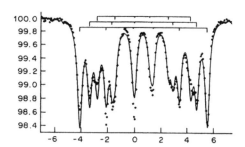

Figure 32. Mössbauer spectra of hexagonal pyrrhotites (Fe$_{1-x}$S): top $x = 0.037$ at 283 K; middle $x = 0.079$ at 286 K; bottom $x = 0.143$ at 283 K. Sextets are indicated by bar diagrams and the strongest sextet in the top spectrum is absent in the lower two (after Kruse 1990).

These high pressure spectroscopic studies need to be considered in the context of the known pressure-induced transformations in phases in the Fe-S system which are reviewed elsewhere in this volume (Fleet 2006).

OTHER SPECTROSCOPIC TECHNIQUES

There are a number of other spectroscopic methods that have been employed in sulfide mineralogy and geochemistry, but to a more limited extent.

Nuclear magnetic resonance. In Nuclear Magnetic Resonance (NMR) spectroscopy, resonant absorption of radio energy radiation by transitions of the nuclei of certain isotopes from ground to various excited states is measured. Information can be obtained from NMR spectra on the local environment of particular atoms in a molecule or crystal structure, and

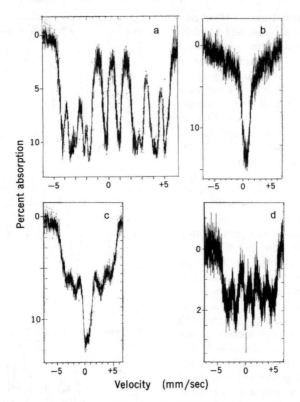

Figure 33. Mössbauer spectra of monoclinic pyrrhotite(Fe$_7$S$_8$): (a) at 1 atm; (b) at 16 kbar; (c) at 5 kbar (after release of pressure); (d) after pellet used in experiment has been reground. [Reprinted with permission from Vaughan and Tossell (1973). Copyright (1973) AAAS.]

on aspects of bonding. General reviews of the applications of NMR in mineralogy have been published by McWhinnie (1985) and Kirkpatrick (1988). Amongst nuclides of use or potential use in NMR studies of sulfide minerals and related species are ^{33}S, ^{63}Cu, ^{67}Zn, ^{95}Mo, ^{109}Ag, ^{113}Cd, ^{119}Sn, ^{195}Pt, ^{199}Hg and ^{207}Pb. For example, Eckert and Yesinowski (1986) studied a variety of metal sulfides using ^{33}S solid state NMR and found a large range in NMR (chemical shift) parameters which they interpreted in relation to trends in ionicities and bond polarities. Luther et al. (2002) used ^{63}Cu NMR in studying aqueous copper sulfide clusters in solution as precursors of the formation of copper sulfide minerals. Tossell and Vaughan (1993) predicted structures and properties and calculated NMR spectra for bisulfide complexes of Zn and cadmium in aqueous solution using *ab initio* methods.

Electron spin resonance. In Electron Spin Resonance (ESR) or Electron Paramagnetic Resonance (EPR) spectroscopy, absorption of microwave radiation by molecules, atoms or ions possessing electrons with unpaired spins is measured in the presence of an external magnetic field. The technique enables fingerprint identification of species such as paramagnetic impurity ions in a diamagnetic material and provides information on the stereochemistry and bonding of such species. It is best suited to systems where the paramagnetic centers are a very dilute component in the diamagnetic matrix. Mineralogical applications in general are reviewed by Poole et al. (1977), McWhinnie (1985) and Calas (1988). EPR has been applied to sulfide (and sulfosalt) minerals in relatively few cases. Seehra and Srinivasan (1982) studied Mn^{2+} ions in the pyrite lattice and Siebert et al (1989,1990,1992) studied Cr^{3+}, Cr^{+} and Ni^{2+} in the same host. Zhou and Rudowicz (1996) studied Fe^{2+} in $FeSb_2S_4$ and Cr^{2+} impurities in ZnS and CdS. Most commonly, the technique has been used to determine the valence state of Cu in complex sulfide and sulfosalt minerals. Examples include work on the Cu-Fe-Bi-Pb-Sn sulfides (Bente 1987), stannite and kesterite (Bernardini et al. 2000; Di Benedetto et al. 2002b, 2005c), and the tetrahedrites (Di Benedetto et al. 2002a,b, 2005b).

Other surface spectroscopies. A number of spectroscopic methods, in addition to the XPS and AES techniques discussed above, are available for determining information on the chemical composition of the surfaces of solids such as sulfide minerals. These include Rutherford Backscattering Spectroscopy (RBS) and Secondary Ion Mass Spectrometry (SIMS). The latter is proving particularly powerful as a result of recent technical developments, such as the use of ^{60}C sources which provide unprecedented sensitivity and both depth and lateral resolution. Applications of SIMS in ore mineralogy have been reviewed by McMahon and Cabri (1998) and in studies of sulfide mineral surfaces by Smart et al. (2003). The chemistry of sulfide mineral surfaces is discussed in detail later in this volume (Rosso and Vaughan 2006a,b).

CONCLUDING REMARKS

The last thirty years have seen remarkable developments in both the range and power of spectroscopic methods available to researchers working in the field of sulfide mineralogy and geochemistry, such that in a single article like this, it is possible only to provide the most cursory review. These developments look set to continue and, in some cases, to accelerate over the coming decade. On the one hand, laboratory scale instruments are becoming more widely available and offering both greater resolution (e.g., small spot scanning XPS machines, nanoscale SIMS instruments) and the advantages of combining several techniques on one instrument. On the other hand, the large scale facilities required for experiments employing synchrotron radiation (e.g., XAS, reflectivity, diffraction and scattering experiments) are becoming more widely available, and a new generation of synchrotron laboratories is under construction. The mineralogy and geochemistry of sulfides will surely provide many interesting problems to address with these new facilities.

ACKNOWLEDGMENTS

The authors thank Helen Weedon and Richard Hartley for help in preparation of the manuscript, and acknowledge the support of the Natural Environment Research Council and the Engineering and Physical Sciences Research Council in undertaking research on sulfide systems using spectroscopic methods.

REFERENCES

Abrams B, Wilcoxon J (2005) Nanosize semiconductors for photooxidation. Crit Rev Solid State and Mater Sci 30:153-182
Armstrong JT (1999) Determination of chemical valence state by X-ray emission analysis using electron beam instruments: pitfalls and promises. Anal Chem 71:2714-2724
Bancroft GM (1973) Mössbauer Spectroscopy: An Introduction for Inorganic Chemists and Geochemists. McGraw-Hill
Bargeron CB, Avinor M, Drickamer HG (1971) Effect of pressure on the spin state of iron(II) in manganese(IV) sulfide. Inorg Chem 10:1338-1339
Bente K (1987) Stabilization of copper-iron-bismuth-lead-tin-sulfides. Mineral Petrol 36:205-17
Bernardini GP, Borrini D, Caneschi A, Di Benedetto F, Gatteschi D, Ristori S, Romanelli M (2000) EPR and SQUID magnetometry study of Cu_2FeSnS_4 (stannite) and Cu_2ZnSnS_4 (kesterite). Phys Chem Minerals 27: 453-461

Bertaut F, Burlet P, Chappert J (1965) The absence of magnetic order in the tetragonal form of FeS. Solid State Comm 3:335-338

Bertaut F (1953) Defect structures; pyrrhotite. Acta Crystallogr 6:557-561

Beun JA, Goldsmith G J (1960) Electrical and optical properties of zinc sulphide crystals in polarized light. Helvet Phys Acta 33:508-513

Bither TA, Bouchard RJ, Cloud, WH, Donohue PC, Siemons WJ (1968) Transition metal pyrite dichalcogenides. High-pressure synthesis and correlation of properties. Inorg Chem 7:2208-2220

Blanchard M, Alfredsson M, Broholt J, Price GD, Wright K, Catlow CRA (2005)Electronic structure study of the high pressure vibrational spectrum of FeS_2 pyrite. J Phys Chem B 109:22067-22073

Bocquet AE, Mamiya K, Mizokawa T, Fujimori A, Miyadai T, Takahashi H, Mori MSuga S (1996) Electronic structure of 3d transition metal pyrites MS_2 (M = Fe,Co or Ni) by analysis of the M 2p core-level photoemission spectra. J Phys Cond Matt 8:2389-2400

Boekema C, Krupski AM, Varasteh M, Parvin K, van Til F, van der Woude F, Sawatzky GA (2004) Cu and Fe valence states in $CuFeS_2$. J Mag Magn Mater 272-276:559-561

Böker Th, Severin R, Muller A, Janowitz C, Manzke R, Voss D, Kruger P, Mazur A, Pollmann J (2001) Band structure of MoS_2, $MoSe_2$, and α-$MoTe_2$: Angle-resolved photoelectron spectroscopy and *ab initio* calculations. Phys Rev B 64: 235305/1-235305/11.

Boldish SI, White WB (1998) Optical band gaps of selected ternary sulfide minerals. Am Mineral 83:865-871

Bouchard M, Smith DC (2003) Catalogue of 45 reference Raman spectra of minerals concerning research in art history or archaeology, especially on corroded metals and colored glass. Spectrochim Acta 59A:2247-2266

Braga M, Lie SK, Taft CA, Lester WA Jr. (1988) Electronic structure, hyperfine interaction, and magnetic properties for iron octahedral sulfides. Phys Rev B 38:10837-10851

Briggs D, Grant JT (eds) (2003) Surface Analysis by Auger and X-ray photoelectron Spectroscopy. IM Publications

Briggs D, Seah MP (eds) (1996) Practical Surface Analysis. Vol.1, Auger and X-ray Photoelectron Spectroscopy. 2nd ed, Wiley

Brown GE Jr, Calas G, Waychunas GA, Petiau J (1988) X-ray absorption spectroscopy and its applications in mineralogy and geochemistry. Rev Mineral 18:431-512

Buckley AN, Wouterland HJ, Cartwright PS, Gilbuckley RD (1988) Core electron binding energies of platinium and rhodium polysulfides. Inorg Chim Acta 143:77-80

Buckley AN, Woods R (1994) On the characterization of sulfur species on sulphide mineral surfaces by X-ray photoelectron spectroscopy and Raman spectroscopy. J Electroanal Chem 370:295-296

Bullett DW (1982) Electronic structure of 3d pyrite- and marcasite-type sulfides. J Phys C 15:6163-6174

Burns RG (1993) Mineralogical Applications of Crystal Field Theory. (2nd ed), Cambridge University Press

Cabri LJ (1969) Density determinations: accuracy and application to sphalerite stoichiometry. Am Mineral 54: 539-548

Calais JL (1977) Band structure of transition metal compounds. Adv Phys 26:847-885

Calas G (1988) Electron paramagnetic resonance. Rev Mineral 18:513-571

Cardona M, Greenaway DL (1964) Optical properties and band structure of group IV- VI and group V materials. Phys Rev 133:1685-1697

Cases JM, De Donato P (1991) FTIR analysis of sulfide mineral surfaces before and after collection: galena. Int J Mineral Process 33:49-65

Charnock JM, Garner CD, Pattrick RAD, Vaughan DJ (1988) Investigation into the nature of copper and silver sites in argentian tetrahedrites using EXAFS spectroscopy. Phys Chem Mineral 15:296-299

Charnock JM, Garner CD, Pattrick RAD, Vaughan DJ (1989a) Coordination sites of metals in tetrahedrite minerals determined by EXAFS. J Solid State Chem 82:279-289

Charnock, JM, Garner CD, Pattrick RAD, Vaughan DJ (1989b) EXAFS and Mössbauer spectroscopic study of iron-bearing tetrahedrites. Mineral Mag 53:193- 199

Charnock J, Garner CD, Pattrick RAD, Vaughan DJ (1990) An EXAFS study of thiospinel minerals. Am Mineral 75:247-255

Charnock JM, Henderson CMB, Mosselmans JFW, Pattrick RAD (1996) 3d transitional metal L-edge X-ray absorption studies of the dichalcogenides of Fe, Co and Ni. Phys Chem Mineral 23:403-408

Chen JG (1997) NEXAFS investigations of transition metal oxides, nitrides, carbides, sulfides and other interstitial compounds. Surf Sci Rep 30:1-152

Chernyshova IV (2001) Anodic oxidation of galena (PbS) studied FTIR- spectroelectrochemically. J Phys Chem B 105:8178-8184

Chernyshova, IV (2003) An *in situ* FTIR study of galena and pyrite oxidation in aqueous solution. J Electroanal Chem 558:83-98

Coehoorn R, Haas C, Dijkstra J, Flipse CJF, De Groot RA, Wold A (1987) Electronic structure of molybdenum diselenide, molybdenum disulfide and tungsten diselenide. I.Band-structure calculations and photoelectron spectroscopy. Phys Rev B 35:195-202

Coey JMD, Spender, MR, Morrish AH (1970) Magnetic structure of the spinel Fe_3S_4. Solid State Comm 8: 1605-1608

Coleyshaw EE, Griffith WP, Bowell R J (1994) Fourier-transform Raman spectroscopy of minerals. Spectrochim Acta 50A:1909-1918

Collins MF, Longworth G, Townsend MG (1981) Magnetic structure of bornite,Cu_5FeS_4. Can J Phys 59:535-539

de Groot RA, Hass C (1975) Multiple scattering theory-X_α calculations on molybdenum disulfide and some related compounds. Solid State Comm 17: 887- 890

Di Benedetto F, Bernardini GP, Caneschi A, Cipriani C, Danti C, Pardi L, Romanelli M (2002a) EPR and magnetic investigations on sulphides and sulphosalts. Euro J Mineral 14:1053-1060

Di Benedetto F, Bernardini GP, Borrini D, Emiliani C, Cipriani C, Danti C, Caneschi AG, Dante RM (2002b) Crystal chemistry of tetrahedrite solid- solution: EPR and magnetic investigations. Can Mineral 40:837-847

Di Benedetto F, Andreozzi GB, Bernardini GP, Borgheresi M, Caneschi A, CiprianiC, Gatteschi D, Romanelli M (2005a) Short-range order of Fe^{2+} in sphalerite by^{57}Fe Mössbauer spectroscopy and magnetic susceptibility. Phys Chem Mineral 32:339-348

Di Benedetto F, Bernardini GP, Cipriani C, Emiliani C, Gatteschi D, Romanelli M (2005b) The distribution of Cu(II) and the magnetic properties of the synthetic analogue of tetrahedrite: $Cu_{12}Sb_4S_{13}$. Phys Chem Mineral 32:155-164.

Di Benedetto F, Bernardini GP, Borrini D, Lottermoser W, Tippelt G, Amthauer G (2005c) ^{57}Fe- and ^{119}Sn-Mössbauer study on stannite (Cu_2FeSnS_4)-kesterite (Cu_2ZnSnS_4) solid solution. Phys Chem Mineral 31: 683-690

Domashevskaya EP, Terekhov, VA, Marshakova LN, Ugai Ya A, Nefedov VI, Sergushin NP (1976) Participation of d-electrons of metals of groups I, II, and III in chemical bonding with sulfur. J Elect Spectr Rel Phenom 9:261-267

Eckert H, Yesinowski JP (1986) Sulfur-33 NMR at natural abundance in solids. J Am Chem Soc 108:2140-2146

Eibschutz M, Hermon E, Shtrikman S (1967) Determination of cation valencies in Cu_2FeSnS_4 by Moessbauer effect and magnetic susceptibility measurements. J Phys Chem Solids 28:1633-1636

Farberovich OV, Kurganskii SI, Domashevskaya EP (1980) Problems of the OPW method. II. Calculation of the band structure of zinc sulfide and cadmium sulfide. Phys Stat Solidi B 97:631-640

Farrell SP, Fleet ME (2000) Evolution of local electronic structure in cubic $Mg_{1-x}Fe_xS$ by S K-edge XANES spectroscopy. Solid State Comm 113: 69-72

Farrell SP, Fleet ME (2001) Sulfur K-edge XANES study of local electronic structure in ternary monosulfide solid solution [(Fe,Co, Ni)$_{0.923}$S]. Phys Chem Mineral 28: 17- 27

Farrell SP, Fleet ME, Stekhin IE, Kravtsova A, Soldatov AV, Liu X (2002) Evolution of local electronic structure in alabandite and niningerite solid solutions [(Mn,Fe)S, (Mg,Mn)S, (Mg,Fe)S using sulfur K and L-edge XANES spectroscopy. Am Mineral 87: 1321-1332

Folmer JCW, Jellinek F, Calis GHM (1988) The electronic structure of pyrites, particularly CuS_2 and $Fe_{1-x}Cu_xSe_2$: an XPS and Moessbauer study. J Solid State Chem 72:137-144

Fleet ME (2005) XANES spectroscopy of sulfur in Earth materials. Can Mineral 43: 1811-1838

Fleet ME (2006) Phase equilibria at high temperatures. Rev Mineral Geochem 61:365-419

Forneris R (1969) Infrared and Raman spectra of realgar and orpiment. Am Mineral54:1062-1074

Frey GL, Elani S, Homyonfer M, Feldman Y, Tenne R (1998) Optical –absorption spectra of inorganic fullerenelike MS_2 (M= Mo,W). Phys Rev B 57: 6666-6671

Frost RL, Martens WN, Kloprogge JT (2002) Raman spectroscopic study of cinnabar (HgS), realgar (As_4S_4), and orpiment (As_2S_3) at 298 and 77 K. Neues Jahrb Mineral Monats 10:469-480

Fujimori A, Mamiya K, Mizokawa T, Miyadai T, Sekiguchi T, Takahashi H, Mori N, Suga S (1996) Resonant photoemission study of pyrite-type NiS_2, CoS_2 and FeS_2. Phys Rev B 54:16329-16332

Fujisawa M, Suga S, Mizokawa T, Fujimori A, Sato K (1994) Electronic structures of $CuFeS_2$ and $CuAl_{0.9}Fe_{0.1}S_2$ studied by electron and optical spectroscopies. Phys Rev B 49:7155-7164

Gamblin SD, Urch DS (2001) Metal K_β X-ray emission spectra of first row transition metal compounds. J Elect Spectr Rel Phenom 113:179-192

Gerard A, Imbert P, Prange H, Varret F, Wintenberger M (1971) Fe^{2+} impurities, isolated and in pairs, in zinc sulfide and cadmium sulfide studied by the Mössbauer effect. J Phys Chem Solids 32:2091-2100

Gerson AR, Bredow T (2000) Interpretation of sulfur 2p XPS spectra in sulphide minerals by means of *ab initio* calculations.Surf and Interf Anal 29:145-150

Godocikova E, Balaz P, Bastl Z, Brabec L (2002) Spectroscopic study of the surface oxidation of mechanically activated sulphides. Appl Surf Sci 200:6-47

Goncharov GN, Ostanevich YuM, Tomilov SB, Cser L (1970) Mössbauer effect in the FeS1+π system. Phys Stat Solidi 37:141-150

Goodenough JB (1967) Description of transition metal compounds. Application to several sulfides. Colloques Internationaux du Centre National de la Recherche Scientifique 157:263-290
Greenwood NN, Whitfield HJ (1968) Mössbauer effect studies on cubanite ($CuFe_2S_3$) and related iron sulfides. J Chem Soc A 7:1697-1699
Gurin VS (1998) Observation and simulation of PbS nanocrystal formation at the initial steps. Macromolecular Symp 136:13-16
Hafner S, Kalvius GM (1966) The Mössbauer resonance of iron-57 in troilite and pyrrhotite. Z Kristallogr 123: 443-458
Ham FS, Slack GA (1971) Infrared absorption and luminescence spectra of Fe^{2+} in cubic zinc sulfide. Role of the Jahn-Teller coupling. Phys Rev B 4:777-798
Harris S (1982) Study of the electronic structure of first and second row transition metal sulfides using SCF-SW-Xα cluster calculations. Chem Phys 67:229-237
Hawthorne FC (1988) Mössbauer spectroscopy. Rev Mineral 18:255-340
Haycock DE, Casrai M, Urch DS (1978) Electronic structure of molybdenum disulphide: angle resolved X-ray spectroscopy. Jap J Appl Phys 17: 138-140
Hemstreet LA Jr (1975) Cluster calculations of the effect of single vacancies on the electronic properties of lead(II) sulfide. Phys Rev B 11:2260-2270
Hochella MF Jr (1988) Auger electron and X-ray photoelectron spectroscopies. Rev Mineral 18:573-637
Hoffmann V (1993) Mineralogical, magnetic and Moessbauer data of smythite. Stud Geophys Geodet 37: 366-381
Hope GA, Woods R, Munce CG (2001) Raman microprobe mineral identification. Mineral Eng 14:1565-1577
Horita H, Hirahara E (1966), Mössbauer measurements of single-crystal FeS. J Phys Soc Jap 21:1447-1447
Hughes HP, Liang WY (1974) Vacuum ultraviolet reflectivity spectra of the molybdenum and tungsten dichalogenides. J Phys 7:1023-1032
Igaki K, Sato M, Shinohara T (1981) Mössbauer study on the iron vacancy distribution in iron sulfide $Fe_{1-x}S$ ($0.083 \leq x \leq 0.125$). Trans Jap Inst Metals 22:627-632
Igaki K, Sato M, Shinohara T (1982), Mössbauer study on the distribution of iron vacancies in iron sulfide $Fe_{1-x}S$. Trans Jap Inst Metals 23:221-228
Imbert P, Gerard A, Wintenberger M (1963) Study of naturally occurring sulfides, sulfarsenides, and arsenides of iron by Mössbauer effect. Compt Rend 256:4391-4393
Jeandey C, Oddou JL, Mattei JL, Fillion G (1991) Mössbauer investigation of the pyrrhotite at low temperature. Solid State Comm 78:195-198
Kalinowski M, Kalska B, Szymanski K, Dobrzynski L (2000) Mössbauer spectroscopy applied to synthetic $Fe_{9+x}Cu_{9-x}S_{16}$. Physica B 275:328-335
Karr C (1975) Infrared and Raman Spectra of Lunar and Terrestrial Minerals. Academic Press
Kasper H, Drickamer HG (1968) High-pressure Mössbauer resonance studies of compounds of iron with GroupV and Group VI elements. Proc Nat Acad Sci 60:773-775
Kawai S, Kiriyama R, Nakahara F (1971) Optical and electrical properties of (zinc, cadmium, mercury) sulfide and cadmium (sulfide, selenide) solid solutions. Mem Inst Sci Indust Res Osaka Univ 28:101-112
Kikuchi M, Iijima S (1960) Photoconductivity of cadmium-zinc sulfide mixed crystal. J Phys Soc Jap 15:357-537
Kim SJ, Son BS, Lee BW, Kim CS (2004) Mössbauer studies of dynamic Jahn-Teller relaxation on the Cu-substituted sulfur spinel. J Appl Phys 95:6837-6839
Kirkpatrick RJ (1988) MAS NMR spectroscopy of minerals and glasses. Rev Mineral 18: 341-403
Kitamura, M, Sugiura C, Muramatsu S (1988) Multiple-scattering calculation of sulfur K X-ray absorption spectra for iron monosulfide, cobalt monosulfide and nickel monosulfide. Solid State Comm 67:313-316
Kleppe AK, Jephcoat AP (2004) High-pressure Raman spectroscopic studies of FeS_2 pyrite. Mineral Mag 68: 433- 441
Knop O, Huang C-H, Woodhams FWD (1970) Chalcogenides of the transition elements. VII. Moessbauer study of pentlandite. Am Mineral 55:1115-1130
Kobayashi H, Yoda Y, Kamimura T (2001) Moessbauer spectroscopic study of FeS under pressure using synchrotron radiation. J Phys Soc Jap 70: 1128-1132
Kravtsova AN, Stekhin IE, Soldatov AV, Liu X, Fleet ME (2004) Electronic structure of MS (M = Ca,Mg,Fe,Mn): X-ray absorption analysis. Phys Rev B 69:134109/1-134109/12
Krill G, Amamou A (1980) XPS and UPS studies of pyrite compounds ($CoX_2 \rightarrow CuX_2$, X = S;Se). Investigation of the insulator-metal transition in the $NiS_{2-x}Se_x$ system. J Phys Chem Solids 41:531-538
Kruse O (1990) Mössbauer and X-ray study of the effects of vacancy concentration in synthetic hexagonal pyrrhotites. Am Mineral 75:755-763
Kruse O, Ericsson TA (1988) Mössbauer investigation of natural troilite from the Agpalilik meteorite. Phys Chem Mineral 15: 509-513

Kurmaev EZ, van Ek J, Ederer DL, Zhou, L, Callcott TA, Perera RCC, Cherkashenko VM, ShaMineral SN, Trofimova VA, Bartkowski S, Neumann M, Fujimori A, Moloshag VP (1998) Experimental and theoretical investigation of the electronic structure of transition metal sulfides: CuS, FeS_2 and $FeCuS_2$. J Phys Cond Mat 10:1687-1697

Laihia R, Leiro JA, Kokko K, Mansikka K (1996) The X-ray $K\beta_{2,5}$ – emission band and the electronic structure of Zn, ZnS and ZnSe crystals. J Phys Cond Matter 8: 6971-6801

Laihia R, Kokko K, Hergert W, Leiro JA (1998) K-emission spectra of Zn, ZnS, and ZnSe within dipole and quadrupole approximations. Phys Rev B 58:1272-1278

Lauer S, Trautwein AX, Harris FE (1984) Electronic-structure calculations, photoelectron spectra, optical spectra, and Mössbauer parameters for the pyrites MS_2 (M = Fe, Co, Ni, Cu, Zn). Phys Rev B 29:6774-6783

Lawniczak-Jablonska K, Iwanowski RJ, Golacki Z, Traverse A, Pizzini S, Fontaine A, Winter I, Hormes J (1996) Local electronic structure of ZnS and ZnSe doped by Mn, Fe, Co, and Ni from X-ray-absorption near-edge structure studies. Phys Rev B 53:1119-28

Leiro JA, Laajalehto K, Kartio I, Heinonen MH (1998) Surface core-level shift and phonon broadening in PbS (100). Surf Sci 412-413:L918-923

Lennie AR, Redfern SAT, Champness PE, Stoddart CP, Schofield PF, Vaughan DJ (1997) Transformation of mackinawite to greigite : an in situ X-ray powder diffraction and transmission electron microscope study. Am Mineral 82:302-309

Lepetit P, Bente K, Doering T, Luckhaus S (2003) Crystal chemistry of Fe- containing sphalerites. Phys Chem Mineral 30:185-191

Levinson LM, Treves D (1968) Moessbauer study of the magnetic structure of Fe_7S_8. J Phys Chem Solids 29: 2227-2231

Ley L, Pollak RA, McFeely FR, Kowalczyk SP, Shirley DA (1974) Total valence- band densities of states of [Groups] III-V and II-VI compounds from X-ray photoemission spectroscopy. Phys Rev B 9:600-621

Li EK, Johnson KH, Eastman DE, Freeouf JL (1974) Localized and bandlike valence-electron states in iron sulfide (FeS_2) and nickel sulfide (NiS_2). Phys Rev Lett 32:470-472

Li, D, Bancroft GM, Kasrai M, Fleet ME, Feng XH, Tan KH, Yang BX (1994a) Sulfur K- and L-edge XANES and electronic structure of zinc, cadmium and mercury monosulfides: a comparative study. J Phys Chem Solids 55:535-543

Li D, Bancroft GM, Kasrai M, Fleet ME, Yang BX, Feng X. H, Tan K, Peng M (1994b) Sulfur K- and L-edge X-ray absorption spectroscopy of sphalerite, chalcopyrite and stannite. Phys Chem Mineral 20:489-499

Li D, Bancroft GM, Kasrai M, Fleet M, Feng XH, Yang BX, Tan KH (1994c) S K- and L-edge XANES and electronic structure of some copper sulfide minerals. Phys Chem Mineral 21: 317-324

Li D, Bancroft GM, Kasrai M, Fleet ME, Feng X, Tan K (1995) S K- and L-edge X-ray absorption spectroscopy of metal sulfides and sulfates: applications in mineralogy and geochemistry. Can Mineral 33:949-960

Low W, Weger M (1960) Paramagnetic resonance and optical spectra of bivalent iron in cubic fields. II. Experimental results. Phys Rev 118:1130-1136

Luther GW III, Theberge SM, Rozan TF, Rickard D, Rowlands CC, Oldroyd A (2002) Aqueous copper sulfide clusters as intermediates during copper sulphide formation. Env Sci Tech 36:394-402

Lutz HD, Schneider G, Kliche G (1983) Chalcides and pnictides of Group VIII transition metals: far-infrared spectroscopic studies on compounds MX_2, MXY, and MY_2 with pyrite, marcasite, and arsenopyrite structure. Phys Chem Mineral 9:109-114

Maddock AG (1985) Moessbauer spectroscopy in mineral chemistry. In: Chemical Bonding and Spectroscopy in Mineral Chemistry. Berry FJ, Vaughan DJ (eds) Chapman & Hall, p 141-208

Makarov EF, Marfunin AS, Mkrtchyan AR, Nadzharyan G, Povitskii V, Stukan RA (1969) Moessbauer spectroscopic study of magnetic properties of Fe_3S_4. Sov Phys Solid State 11:391-392

Makovicky E (2006) Crystal structures of sulfides and other chalcogenides. Rev Mineral Geochem 61:7-125

Manning PG (1967) Absorption spectra of iron(III) in octahedral sites in sphalerite. Can Mineral 9:57-64

Mamy R, Thieblemont B, Martin L, Pradal F (1977) Reflectivity of layer-type transition metal dichalcogenides from 6eV to 40eV. Nuovo Cimento 38B:196-205

Marfunin AS, Mkrtchyan AR (1968) Mössbauer effect on tin-119 nuclei in stannite. Geokhimiya 4:498-500

Marqueton Y, Ayrault B, Decamps EA, Toudic Y (1973) Infrared reflectivity of synthetic cinnabar (α - mercury(II) sulfide). Physica Stat Solidi B 60:809-820

Martins JL, Troullier N, Wei SH (1991) Pseudopotential plane -wave calculations for zinc sulfide. Phys Rev B 43:2213-2217

Marusak LA, Mulay LN (1979) Mössbauer and magnetic study of the antiferro to ferrimagnetic phase transition in iron sulfide (Fe_9S_{10}) and the magnetokinetics of the diffusion of iron atoms during the transition. J Appl Phys 50:1865-1867

Marusak LA, Tongson LL (1979) Soft X-ray emission and Auger electron spectroscopic study of FeS, $Fe_{0.9}S$, $Fe_{0.875}S$, and $Fe_{0.5}S$. J Appl Phys 50:4350- 4355

McFeely FR, Kowalczyk S, Ley L, Pollak RA, Shirley DA (1973) High-resolution X-ray-photoemission spectra of lead sulfide, lead selenide, and lead telluride valence bands. Phys Rev B 7:5228-5237
McGuire MM, Edwards KJ, Banfield JF, Hamers RJ (2001) Kinetics, surface chemistry, and structural evolution of microbially mediated sulfide mineral dissolution. Geochim Cosmochim Acta 65:1243-1258
McMahon G, Cabri LJ (1998) The SIMS technique in ore mineralogy. *In:* Modern Approaches to Ore and Environmental Mineralogy. Cabri LJ, Vaughan DJ (eds) Mineral Assoc Can Short Course Ser Vol 27, p 199-240
McMillan P (1985) Vibrational spectroscopy in the mineral sciences. Rev Mineral 14:9-63
McMillan PF, Hofmeister AM (1988) Infrared and Raman spectroscopy. Rev Mineral 18:99-159
McWhinnie WR (1985) Electron spin resonance and nuclear magnetic resonance applied to minerals. *In:* Chemical Bonding and Spectroscopy in Mineral Chemistry. Berry FJ, Vaughan DJ (eds) Chapman & Hall, p 209-249
Mernagh TP, Trudu AG (1993) A laser Raman microprobe study of some geologically important sulfide minerals. Chem Geol 103:113-127
Mian M, Harrison NM, Saunders VR, Flavell WR (1996) An *ab initio* Hartree-Fock investigation of galena (PbS). Chem Phys Lett 257:627-632
Mielczarski E, Mielczarski JA (2005) Infrared spectroscopic studies of galvanic effect influence on surface modification of sulfide minerals by surfactant adsorption. Environ Sci Technol 39:6117-6122
Mitra S (1992) Applied Moessbauer Spectroscopy. Pergamon Press
Mizokawa T, Fujimori A (1993) Electronic structure of 3d transition-metal impurities in semiconductors. Jap J Appl Phys 32:417-418
Mössbauer RL (1962) Recoilless nuclear resonance absorption of γ-radiation. Science 137:731-738
Mosselmans JFW, Pattick RAD, van der Laan G, Charnock JM, Vaughan DJ, Henderson CMB, Garner CD (1995) X-ray absorption near-edge spectra of transition metal disulfides FeS_2 (pyrite and marcasite), CoS_2, NiS_2 and CuS_2, and their isomorphs FeAsS and CoAsS. Phys Chem Mineral 22:311-317
Mullet M, Boursiquot S, Abdelmoula M, Genin J-M, Ehrhardt J-J (2002) Surface chemistry and structural properties of mackinawite prepared by reaction of sulfide ions with metallic iron. Geochim Cosmochim Acta 66:829-836
Muscat J, Klauber C (2001) A combined *ab initio* and photoelectron study of galena (PbS). Surf Sci 491:226-238
Nesbitt HW, Muir IJ (1994) X-ray photoelectron spectroscopic study of a pristine pyrite surface reacted with water vapour and air. Geochim Cosmochim Acta 58:4667-4679.
Nakai I, Sugitani Y, Nagashima K, Niwa Y (1978) X-ray photoelectron spectroscopic study of copper minerals. J Inorg Nucl Chem 40:789-791
Nishihara Y, Ogawa S (1980) Mössbauer study of cobalt thioferrate ($Co_{0.99557}Fe_{0.005}S_2$) in external magnetic fields: magnetic structure of cobalt disulfide and the quadrupole splitting of the iron disulfide-cobalt disulfide-nickel disulfide system. Phys Rev B 22:5453-5459
O'Day PA, Rivera N Jr., Root R, Carroll SA (2004) X-ray absorption spectroscopic study of Fe reference compounds for the analysis of natural sediments. Am Mineral 89:572-585
Ohno Y, Hirama K, Nakai S, Sugiura C, Okada S (1983a) X-ray absorption spectroscopy of layered 4d transition-metal dichalcogenides. J Phys C 16: 6695-6701
Ohno Y, Hirama K, Nakai S, Sugiura C, Okada S (1983b) X-ray absorption spectroscopy of layer transition-metal disulfides. Phys Rev B 27: 3811-3820
Ohsawa A, Yamamoto H, Watanabe H (1974) X-ray photoelectron spectra of valence electrons in iron disulfide, cobalt disulfide, and nickel disulfide. J Phys Soc Jap 37:568-568
Pappalardo R, Dietz RE (1961) Absorption spectra of transition ions in CdS crystals. Phys Rev 123:1188-1203
Park KT, Richards-Babb M, Hess JS, Weiss J, Klier K (1996) Valence-band electronic structure of MoS_2 and Cs/MoS_2 (0002) studied by angle-resolved x- ray photoemission spectroscopy. Phys Rev B 54:5471-5479
Parker GK, Woods R, Hope GA (2003) Raman investigation of sulfide leaching. Proceedings Electrochemical Society 2003-18 (Electrochemistry in Mineral and Metal Processing VI), p 181-192
Pattrick RAD, van der Laan G, Vaughan DJ, Henderson CMB (1993) Oxidation state and electronic configuration determination of copper in tetrahedrite group minerals by L-edge X-ray absorption spectroscopy. Phys Chem Mineral 20:395-401
Pattrick RAD, Mosselmans JFW, Charnock JM (1998) An X-ray absorption study of doped sphalerites. Euro J Mineral 10: 239-249
Pearce CI, Pattrick RAD, Vaughan DJ (2006) Electrical and Magnetic properties of sulfides. Rev Mineral Geochem 61:127-180
Perry DL (ed) (1990) Instrumental Surface Analysis of Geologic Materials. VCH
Perry DL, Taylor JA (1986) X-ray photoelectron and Auger spectroscopic studies of Cu_2S and CuS. J Mater Sci Lett 5:384-386
Petiau J, Sainctavit P, Calas G (1988) K X-ray absorption spectra and electronic structure of chalcopyrite $CuFeS_2$. Mater Sci Eng B1:237-249

Pettenkofer C, Jaegermann W, Bronold M. (1991) Site specific surface interaction of electron donors and acceptors on FeS$_2$ (100) cleavage planes. Ber Buns Gesell Physik Chem 95:560-565

Platonov AN, Marfunin AS (1968) Optical absorption spectra of sphalerites. Geochem Internat 5:245-259

Pong WF, Mayanovic RA, Wu KT, Tseng PK, Bunker BA, Hiraya A, Watanabe M (1994) Influence of transition metal type and content on local-order properties of $Zn_{1-x}M_xS$ (M = Mn,Fe,Co) alloys studied using XANES spectroscopy. Phys Rev B 50: 7371-7377

Poole CP Jr., Farach HA, Bishop TP (1977) Electron spin resonance of minerals. Part I. Nonsilicates. Magn Res Rev 4:137-195

Pratt AR, Muir IJ, Nesbitt HW (1994) X-ray photoelectron and Auger electron spectroscopic studies of pyrrhotite and mechanism of air oxidation. Geochim Cosmochim Acta 58: 827-841

Prince KC, Matteuci M, Kuepper K, Chiuzbaian SG, Bartkowski S, Neumann M (2005) Core-level spectroscopic study of FeO and FeS$_2$. Phys Rev B 71: 085102/1-085102/9

Rabii S, Lasseter RH (1974) Band structure of lead polonide and trends in the lead chalcogenides. Phys Rev Lett 33:703-705

Rajamani V, Prewitt CT (1973) Crystal chemistry of natural pentlandites. Can Mineral 12:178-187

Remond G, Holloway PH, Kosakevitch A, Ruzakowski P, Packwood RH, Taylor JA (1985) X-ray spectrometry, electron spectroscopies and optical microreflectometry applied to the study of zinc sulfide tarnishing in polished sulfide ore specimens. Scan Elect Micros 4:1305-1326

Reshak AH, Auluck S (2005) Band structure and optical response of 2H-MoX$_2$ compounds (X = S, Se, and Te). Phys Rev B 71:155114/1-155114/6

Rosso KM, Vaughan DJ (2006a) Reactivity of sulfide mineral surfaces. Rev Mineral Geochem 61:557-607

Rosso KM, Vaughan DJ (2006b) Sulfide mineral surfaces. Rev Mineral Geochem 61:505-556

Rueff J-P, Kao C-C, Struzhkin VV, Badro J, Shu J, Hemley RJ, Mao HK (1999) Pressure-induced high-spin to low-spin transition in FeS evidenced by X-ray emission spectroscopy. Phys Rev Lett 82:3284-3287

Sainctavit P, Calas G, Petiau J, Karnatak R, Esteva JM, Brown GE Jr. (1986) Electronic structure from X-ray K-edges in iron-doped zinc sulfide and copper iron sulfide (CuFeS$_2$). J Phys Colloq 8:411-414

Sainctavit P, Petiau J, Calas G, Benfatto M, Natoli CR (1987) XANES study of sulfur and zinc K-edges in zincblende: experiments and multiple scattering calculations. J Physique 48:C9-1109- C9-1112

Sakkopoulos S, Vitoratos E, Argyreas T (1984) Energy-band diagram for pyrrhotite. J Phys Chem Solid 45: 923-928

Schoolar RB, Dixon JR (1965) Optical constants of lead sulfide in the fundamental absorption edge region. Phys Rev 137:667-670

Schmid-Beurmann P, Lottermoser W (1993) Iron-57-Moessbauer spectra, electronic and crystal structure of members of the CuS$_2$-FeS$_2$ solid solution series. Phys Chem Mineral 19:571-577

Schwarz EJ, Vaughan DJ (1972) Magnetic phase relations of pyrrhotite. J Geomag Geoelect 24: 441-458

Seehra MS, Srinivasan G (1982) Electron spin resonance from impurities in coal- derived pyrites. Fuel 61: 396-398

Sekiyama A, Suga S (2004) High-energy bulk-sensitive angle-resolved photoemission study of strongly correlated systems. J Elect Spect Rel Phenom 137-140: 681-685

Siebert D, Dahlem J, Fiechter S, Hartmann A (1989) An ESR investigation of synthetic pyrite crystals. Z Naturforsch A 44:59-66

Siebert D, Dahlem J, Fiechter S, Miller R (1992) EPR investigation of monovalent chromium centers in synthetic pyrite crystals. Phys Stat Solidi B 171:K93-K96

Siebert D, Miller R, Fiechter S, Dulski P, Hartmann A (1990) An EPR investigation of chromium(3+) and nickel(2+) in synthetic pyrite crystals. Z Naturforsch A 45:1267-1272

Sherwin R, Clark RJH, Lauck R, Cardona M. (2005) Effect of isotope substitution and doping on the Raman spectrum of galena (PbS). Solid State Comm 134:565- 570

Skinner WM, Nesbitt HW, Pratt AR (2004) XPS identification of bulk hole defects and itinerant Fe 3d electrons in natural troilite (FeS). Geochim Cosmochim Acta 68: 2259-2263

Simunek A, Wiech G (1984) Angle-dependent X-ray sulfur K-emission bands and electronic structure of tin disulfide and molybdenum disulfide. Phys Rev B 30:923-930

Smart RSC, Amarantidis J, Skinner WM, Prestidge CA, La Vanier L, Grano SR (2003) surface analytical studies of oxidation and collector adsorption in sulfide mineral flotation. Topics Appl Phys 85: 3-60

Soldatov AV, Kravtsova AN, Fleet ME, Harmer SL (2004) Electronic structure of MeS (Me= Ni,Co,Fe): X-ray absorption analysis. J Phys Cond Matt 16: 7545-7556

Soong R, Farmer VC (1978) The identification of sulfide minerals by infrared spectroscopy. Mineral Mag 42: M17-M20

Smith GD, Firth S, Clark RJH, Cardona M (2002) First- and second-order Raman spectra of galena (PbS). J Appl Phys 92:4375-4380

Sugiura C (1984) Iron K X-ray absorption-edge structures of iron(II) sulfide and iron disulfide. J Chem Phys 80:1047-1049

Sugiura C (1994) Lα X-ray emission spectra and electronic structures of zinc and its compounds. J Phys Soc Jap 63:3763-3774

Sugiura C, Gohshi Y, Suzuki I (1974) Sulfur Kβ X-ray emission spectra and electronic structures of some metal sulfides.Phys Rev B 10:338-343

Sugiura C, Gohshi Y (1981) Sulfur Kβ emission and K absorption spectra and electronic structure of copper sulfide: $Cu_{1.8}S$. J Chem Phys 74:4204-4205

Sugiura C, Kitamura M, Muramatsu S, Shoji S, Kojima S, Tada Y, Umezu I, Arai T (1988) Sulfur-K X-ray spectra and electronic structure of a semiconductor silver sulfide (Ag_2S). Jap J Appl Phys 27:1216-1219

Sugiura C, Suzuki I, Kashiwakura J Gohshi Y (1976) Sulfur Kβ X-ray emission bands and valence-band structures of transition-metal disulfides. J Phys Soc Jap 40:1720-1724

Sugiura C, Yorikawa H, Muramatsu S (1997) Sulfur Kβ X-ray emission spectra and valence-band structures of metal sulfides. J Phys Soc Jap 66:503-504

Takele S, Hearne GR (2001) Magnetic-electronic properties of FeS and Fe_7S_8 studied by ^{57}Fe Moessbauer and electrical measurements at high pressure and variable temperatures. J Phys C 13: 10077-10088

Temmerman WM, Durham PJ, Vaughan DJ (1993) The electronic structures of the pyrite-type disulfides (MS_2, where M = manganese, iron, cobalt, nickel, copper, zinc) and the bulk properties of pyrite from local density approximation (LDA) band structure calculations. Phys Chem Mineral 20:248-254

Temperley AA, Lefevre HW (1966) The Mössbauer effect in marcasite-structure iron compounds. J Phys Chem Solids 27:85-92

Thiel RC, van den Berg CB (1968) Temperature dependence of hyperfine interactions in near-stoichiometric ferrous sulfide. I. Experiment. Phys Stat Solidi 29:837-846

Tokonami M, Nishiguchi K, Morimoto N (1972) Crystal structure of a monoclinic pyrrhotite (Fe_7S_8). Am Mineral 57:1066-1080

Tossell JA (1977a) Theoretical studies of valence orbital binding energies in solid zinc sulfide, zinc oxide, and zinc fluoride. Inorg Chem 16:2944-2949

Tossell JA (1977b) SCF-Xα scattered wave MO studies of the electronic structure of ferrous iron in octahedral coordination with sulfur. J Chem Phys 66:5712-5719

Tossell JA, Urch DS, Vaughan DJ, Wiech G (1982) The electronic structure of $CuFeS_2$, chalcopyrite, from X-ray emission and X-ray photoelectron spectroscopy and Xα calculations. J Chem Phys 77:77-82

Tossell JA, Vaughan DJ (1981) Relationships between valence orbital binding energies and crystal structures in compounds of copper, silver, gold, zinc, cadmium, and mercury. Inorg Chem 20:3333-3340

Tossell JA, Vaughan DJ (1987) Electronic structure and the chemical reactivity of the surface of galena. Can Mineral 25:381-392

Tossell JA, DJ Vaughan (1992) Theoretical Geochemistry: Applications of Quantum Mechanics in the Earth and Mineral Sciences. Oxford University Press

Tossell JA, Vaughan DJ (1993) Bisulfide complexes of zinc and cadmium in aqueous solution: calculation of structure, stability, vibrational, and NMR spectra, and of speciation on sulfide mineral surfaces. Geochim Cosmochim Acta 57:1935-1945

Townsend MG, Horwood JL, Hall SR, Cabri LJ (1972) Moessbauer, magnetic susceptibility and crystallographic investigations of $CuFeS_2$, $Cu_9Fe_8S_{16}$ and $Cu_9Fe_9S_{16}$. AIP Conf Proc 5:887-891

Townsend MG, Webster AH, Horwood JL, Roux-Buisson H (1979) Ferrimagnetic transition in $Fe_{0.9}S$: magnetic, thermodynamic and kinetic aspects. J Phys Chem Solids 40: 183-189

Turcotte SB, Benner RE, Riley AM, Li J, Wadsworth ME, Bodily D (1993) Application of Raman spectroscopy to metal-sulfide surface analysis. Appl Optics 32:930-934

Urch DS (1985) X-ray spectroscopy and chemical bonding in minerals. In: Chemical Bonding and Spectroscopy in Mineral Chemistry. Berry F, Vaughan DJ (eds) Chapman & Hall, p 31-61

Usher CR, Cleveland CA Jr, Strongin DR, Schoonen MA (2004) Origin of oxygen in sulfate during pyrite oxidation with water and dissolved oxygen: An in situ horizontal attenuated total reflectance infrared spectroscopy isotope study. Env Sci Tech 38:5604-5606.

van der Heide H, Hemmel R, van Bruggen CF, Haas C (1980) X-ray photoelectron spectra of 3d transition metal pyrites. J Solid State Chem 33:17-25

van der Laan G, Pattrick RAD, Henderson CMB, Vaughan DJ (1992) Oxidation state variations in copper minerals studied with Cu 2p X-ray absorption spectroscopy. J Phys Chem Solids 53:1185-1190

Vaughan DJ, Burns RG (1972) Moessbauer spectroscopy and bonding in sulphide minerals containing four-coordinated iron. Int Geol Congr (Rep Sess 24th) 14:158-167

Vaughan DJ, Burns RG, Burns VM (1971) Geochemistry and bonding of thiospinel minerals. Geochim Cosmochim Acta 35:365-381

Vaughan DJ, Craig JR (1978) Mineral Chemistry of Metal Sulfides. Cambridge Univ Press

Vaughan DJ, Craig JR (1985) The crystal chemistry of iron-nickel thiospinels. Am Mineral 70:1036-1043

Vaughan DJ, Ridout MS (1970) Moessbauer study of pyrrhotite. Solid State Comm 8:2165-2167

Vaughan DJ, Ridout MS (1971) Mössbauer studies of some sulfide minerals. J Inorg Nucl Chem 33:741-746

Vaughan DJ, Rosso KM (2006) Chemical bonding in sulfide minerals. Rev Mineral Geochem 61:231-264
Vaughan DJ, Schwarz EJ, Owens DR (1971) Pyrrhotites from the Strathcona Mine, Sudbury, Canada. Thermomagnetic and mineralogical study. Econ Geol 66:1131-1144
Vaughan DJ, Tossell JA (1973) Magnetic transitions observed in sulfide minerals at elevated pressures and their geophysical significance. Science 179:375-377
Vaughan DJ, Tossell JA (1986) Interpretation of the Auger electron spectra (AES) of sulfide minerals. Phys Chem Mineral 13:347-350
Vaughan DJ, Tossell JA, Stanley CJ (1987) The surface properties of bornite. Mineral Mag 51:285-293
Verble JL, Wallis RF (1969) Infrared studies of lattice vibrations in iron pyrite. Phys Rev 182:783- 789
von Oertzen GU, Jones RT, Gerson AR (2005a) Electronic and optical properties of Fe, Zn and Pb sulfides. Phys Chem Mineral 32:255-268
von Oertzen GU, Jones RT, Gerson AR (2005b) Electronic and optical properties of Fe, Zn and Pb sulfides. J Elect Spectr Rel Phenom 144-147:1245-1247
Wang H, Salveson I (2005) A review on the mineral chemistry of the non- stoichiometric iron sulphide, $Fe_{1-x}S$ ($0<x<0.125$): polymorphs, phase relations and transitions, electronic and magnetic structures. Phase Trans 78:547-567
Wendlandt WW, Hecht HG (1966) Reflectance Spectroscopy. Wiley Interscience
Wertheim GK, DiSalvo FJ, Buchanan DNE (1973) Valence bands of layer structure transition-metal chalcogenides. Solid State Comm 13:1225-1228
Wilson JA (1972) Systematics of the breakdown of Mott insulation in binary transition metal compounds. Adv Phys 21:143-198
Wilson JA, Yoffe AD (1969) Transition metal dichalcogenides. Discussion and interpretation of the observed optical, electrical, and structural properties. Adv Phys 18:193-335
Wintenberger MAG, Perrin M, Garcin C, Imbert P (1990) Magnetic structure and Moessbauer data of sternbergite ($AgFe_2S_3$), an intermediate valence iron compound. J Mag Magn Mater 87:123-129
Womes M, Karnatak RC, Esteva JM, Lefebvre I, Allan G, Olivier-Fourcade J, Jumas JC (1997) electronic structures of FeS and FeS_2: X-ray absorption spectroscopy and band structure calculations. J Phys Chem Solids 58: 345-352
Wood BJ, Strens RGJ (1979) Diffuse reflectance spectra and optical propertiesof some sulfides and related minerals. Mineral Mag 43:509-518
Xie Y, Zhu L, Jiang X, Lu J, Zheng X, Wei H, Li Y (2001) Mild hydrothermal- reduction synthesis and Moessbauer study of low-dimensional iron chalcogenide microcrystals and single crystals. Chem Mater 13:3927-3932
Yagnik CM, Mathur HB (1967) Moessbauer spectra of sulfo-spinels containing Fe^{2+} ions. Solid State Comm 5: 841-844
Yamamoto R, Machida A, Moritomo Y, Nakamura A (1999) Reconstruction of the electronic structure in half-metallic CoS_2. Phys Rev B 59:R7793-R7796
Yarmoshenko YM, Cherkashenko VM, Kurmaev EZ (1983) The anisotropy of X-ray emission spectra for 2H-molybdenum(IV) sulfide single crystals. J Elect Spectr Rel Phenom 32:103-112
Yin Q, Kelsall GH, Vaughan DJ, England KER (1995) Atmospheric and electrochemical oxidation of the surface of chalcopyrite ($CuFeS_2$). Geochim Cosmochim Acta 59:1091-1100
Zhou Y-Y, Rudowicz C (1996) Crystal field and EPR analysis for 5D ($3d^4$ and $3d^6$) ions at tetragonal sites: applications to Fe^{2+} ions in minerals and Cr^{2+} impurities in semiconductors. J Phys Chem Solids 57:1191-1199
Zunger A, Freeman AJ (1978) Local-density self-consistent energy-band structure of cubic cadmium sulfide. Phys Rev B 17:4850-4863

Chemical Bonding in Sulfide Minerals

David J. Vaughan

*School of Earth Atmospheric and Environmental Sciences
and Williamson Research Centre for Molecular Environmental Science
University of Manchester
Manchester, United Kingdom
e-mail: david.vaughan@manchester.ac.uk*

Kevin M. Rosso

*Chemical Sciences Division
and W. R. Wiley Environmental Molecular Sciences Laboratory
Pacific Northwest National Laboratory, P. O. Box 999, MSIN K8-96
Richland, Washington, 99352, U.S.A.
e-mail: kevin.rosso@pnl.gov*

INTRODUCTION

An understanding of chemical bonding and electronic structure in sulfide minerals is central to any attempt at understanding their crystal structures, stabilities and physical properties. It is also an essential precursor to understanding reactivity through modeling surface structure at the molecular scale. In recent decades, there have been remarkable advances in first principles (*ab initio*) methods for the quantitative calculation of electronic structure. These advances have been made possible by the very rapid development of high performance computers. Several review volumes that chart the applications of these developments in mineralogy and geochemistry are available (Tossell and Vaughan 1992; Cygan and Kubicki 2001).

An important feature of the sulfide minerals is the diversity of their electronic structures, as evidenced by their electrical and magnetic properties (see Pearce et al. 2006, this volume). Thus, sulfide minerals range from insulators through semiconductors to metals, and exhibit every type of magnetic behavior. This has presented problems for those attempting to develop bonding models for sulfides, and also led to certain misconceptions regarding the kinds of models that may be appropriate.

In this chapter, chemical bonding and electronic structure models for sulfides are reviewed with emphasis on more recent developments. Although the fully *ab initio* quantitative methods are now capable of a remarkable degree of sophistication in terms of agreement with experiment and potential to interpret and predict behavior with varying conditions, both qualitative and more simplistic quantitative approaches will also be briefly discussed. This is because we believe that the insights which they provide are still helpful to those studying sulfide minerals. In addition to the application of electronic structure models and calculations to solid sulfides, work on sulfide mineral surfaces (Rosso and Vaughan 2006a,b) and solution complexes and clusters (Rickard and Luther 2006) are discussed in detail later in this volume.

IONIC AND COVALENT BONDS

As with all crystalline solids, representation using a simple ionic model in which the ions are considered to be charged spheres of a particular radius has proved useful in gaining a qualitative understanding of sulfide crystal structures. As discussed in more detail elsewhere in this volume (Makovicky 2006), the Shannon crystal radius of the S^{2-} ion (1.70 Å; Shannon 1981) which dominates in sulfide minerals is significantly greater than that of the O^{2-} ion of oxides and silicates, and the sulfide ion is more polarizable. Nevertheless, concepts such as the (cubic or hexagonal) close packing of anion spheres are also useful in understanding sulfide crystal structures. However, such approaches do not yield information on electronic structure or the subtleties of chemical bonding and will not be discussed in detail in this chapter. For further discussion, readers are referred to Vaughan and Craig (1978) and the very detailed analysis of sulfide structures and crystal chemistry provided by Makovicky (2006; this volume).

The qualitative theories of covalent bonding such as the hybridization of atomic orbitals have also been applied to sulfides, emphasizing the directionality of metal-sulfur bonds and significant delocalization of the electrons involved in bonding. As with the purely ionic approach, the usefulness of concepts derived from valence bond theory has largely been in gaining an understanding of crystal structures. Again, in the detailed discussion of crystal structures elsewhere in this volume (Makovicky 2006) use is made of these concepts in understanding structural principles. Also, fuller accounts of such approaches are provided by Vaughan and Craig (1978).

As discussed in detail below, both qualitative and quantitative modeling of bonding and electronic structure now center on molecular orbital and band theories, although for transition metal sulfides, the ligand (or crystal) field theories have also provided useful insights. Atomistic modeling, a quantitative approach developed from the ionic view has also, perhaps surprisingly, proved useful for certain applications, as described below.

LIGAND FIELD THEORY

The ligand field and crystal field theories have played an important role in the development of geochemistry (see Burns 1993 for a full account). In crystal field theory, the effects of the anions surrounding a transition metal cation on the energies of its d-electrons are modeled. Whereas in crystal field theory, these anions (S^{2-} ions in most sulfides) are treated as point charges, in ligand field theory, account is taken of the covalent contribution to metal-anion bonding through overlap between metal and anion orbitals. These theories therefore provide descriptions of only the d-electrons in the system being considered. As the transition metal sulfides are amongst the most important of this mineral group, the ligand field approach has yielded some useful insights, although a further limitation is that the d-electrons are treated as *localized* on the cations, whereas some sulfides exhibit considerable d-electron *delocalization*.

Hulliger (1968) and Nickel (1968, 1970) applied a ligand field model to explain the structural stabilities of certain disulfide and diarsenide minerals (see also Vaughan and Craig 1978; Burns 1993 for review). The transition metal disulfides containing S-S (and S-As or As-As) dimers exist in four principal structure-types characterized by pyrite, marcasite, arsenopyrite and loellingite (see Makovicky 2006). In each of these structures the metals are octahedrally coordinated to six anions which are, in turn, linked to other anions. However, in pyrite the octahedra only share corners, whereas in arsenopyrite, loellingite and marcasite, they share edges lying in the (001) plane. Relative to marcasite, loellingite has a compressed c-axis whereas arsenopyrite has alternate short and longer metal-metal distances along c (Fig. 1). The outer electron distribution in iron ($3d^6\,4s^2$) indicates that in FeS_2, the d-electrons are not involved in metal-sulfur bonding (in terms of a simple ionic or covalent model). However, replacement of a sulfur by arsenic as in FeAsS results in a deficit of one electron per formula unit. If cation-anion electron pair bonds are to be maintained, a d-electron must be used in bonding. In $FeAs_2$, two such d-electrons will be required to maintain the electron pair bonds.

Chemical Bonding in Sulfide Minerals 233

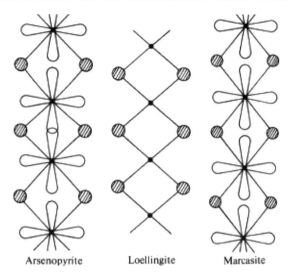

Figure 1. Proposed interactions of the t_{2g} orbitals (Fe $3d$) along the direction parallel to the c-axis in arsenopyrite (FeAsS) and in marcasite (FeS$_2$); in loellingite (FeAs$_2$) this orbital would be empty. (Redrawn after Nickel 1970).

Magnetic and Mössbauer data indicate that pyrite and marcasite contain low-spin Fe^{2+} (t_{2g}^6) (see Pearce et al. 2006, this volume). However, in marcasite the octahedral edges are shared, making interaction between adjacent iron atoms along the c-axis possible, particularly as this is the direction of one of the t_{2g} orbital sets (Fig. 1). In fact, the angle subtending the shared octahedral edge (82°) indicates mutual repulsion of the filled d-electron clouds. In FeAs$_2$, there are four d-electrons not involved in bonding and they cannot achieve spin pairing if placed into the three t_{2g} orbitals. If the four electrons are paired into two of these orbitals (Pearson 1965), complete spin pairing will be achieved. The orbital parallel to the c-axis is the obvious choice for the empty orbital. This could explain the contraction along the c-axis (Fe-Fe distance 2.85 Å in loellingite, 3.38 Å in marcasite) and the low magnetic moment reported for iron in loellingite (Wintenberger 1962). In FeAsS, the iron atom has five d-electrons not involved in metal-anion bonding and spin-pairing can only be achieved if an unpaired electron in a t_{2g} orbital on one atom is paired with an equivalent electron on the adjacent metal atom across the shared octahedral edge. The alternate short (2.89 Å) and long (3.53 Å) metal-metal distances along the c-axis in arsenopyrite result. Again magnetic evidence supports this model (Wintenberger 1962). These structural relationships are all illustrated in Figure 1.

Various iron, cobalt and nickel chalcogenides having the same or very similar structures can be considered using the same models (see Table 1). For example, CoAsS and NiAs$_2$ (cobalt having a $3d^7 4s^2$ configuration, nickel $3d^8 4s^2$) are isoelectronic with FeS$_2$ and have a modified pyrite and a marcasite structure, as the minerals cobaltite (CoAsS) and rammelsbergite (NiAs$_2$) respectively. CoAs$_2$ is isoelectronic with FeAsS and does, indeed, crystallize with the arsenopyrite structure. Compounds with seven (CoS$_2$, NiAsS) and eight (NiS$_2$) nonbonding d-electrons crystallize with a pyrite or closely related structure as the model would predict. The additional electrons go into the two e_g orbitals which are proximal to the ligands and tend to have a repulsive effect, increasing metal-sulfur distances and the overall unit cell dimensions.

The stability of the marcasite structure, despite the repulsion of metal atoms across the shared edge, has been suggested as being due to a complex bonding scheme involving second-nearest-neighbor sulfur atoms (Pearson 1965). The anion pairs in this structure tend to form

Table 1. The crystal structures of the dichalcogenides in relation to d-electron configuration (modified from Nickel 1970).

Composition	Mineral name	# of d-electrons	Structure-type[a]
$FeAs_2$	Loellingite	4	Loellingite
$FeSb_2$	—	4	Loellingite
FeAsS	Arsenopyrite	5	Arsenopyrite
FeSbS	Gudmundite	5	Arsenopyrite
$CoAs_2$	Safflorite	5	Arsenopyrite
FeS_2	Pyrite	6	Pyrite
CoAsS	Cobaltite	6	Pyrite
CoSbS	—	6	Pyrite
FeS_2	Marcasite	6	Marcasite
FeS_2	Ferroselite	6	Marcasite
$FeTe_2$	Frohbergite	6	Marcasite
$NiAs_2$	Rammelsbergite	6	Marcasite
CoS_2	Cattierite	7	Pyrite
$CoSe_2$	Trogtalite	7	Pyrite
NiAsS	Gersdorffite	7	Pyrite
NiAbS	Ullmannite	7	Pyrite
$CoSe_2$	Hastite	7	Marcasite
$CoTe_2$	—	7	Brucite
NiS_2	Vaesite	8	Pyrite
$NiSe_2$	Penroseite	8	Pyrite
$NiTe_2$	Melonite	8	Brucite

[a] Ignoring minor structural distortions.

a ladder-like arrangement parallel to the c-axis; similar arrangements are found in loellingite and arsenopyrite but not in pyrite. FeS_2 appears to represent the marginal case in which energy differences between the pyrite and marcasite structures are very small, repulsion between cations and attraction between anion pairs being balanced.

The dichalcogenides, therefore, are a good example of the insight provided by a ligand field approach to understanding a family of sulfide minerals. Nickel (1970) used a similar approach in discussing bonding in the skutterudites ((Co,Ni,Fe)As_{3-x}), explaining observed magnetic data and rationalizing solid solution limits in terms of the d-electron configurations. However, as detailed below, the ligand field model for dichalcogenides has been criticized and alternative theories proposed to explain structural relationships.

QUALITATIVE MOLECULAR ORBITAL (MO) AND BAND MODELS

The molecular orbital (MO) theory approach to bonding in sulfides has been a particularly powerful one, given the degree of "covalence" exhibited by these materials. Both qualitative and quantitative MO models provide great insights into the small molecules and clusters important in the solution chemistry of sulfides (and, in some cases, such clusters interacting with surfaces). Although metal-sulfur clusters have obvious limitations for the description of crystalline solids, they have provided powerful ways of modeling certain properties and of modeling the behavior dominated by localized electrons (e.g., many types of spectra in the visible-UV and X-ray regions). Also, in terms of qualitative models, the familiar molecular orbital energy level diagram for a single metal-sulfur cluster forms the basis (through "overlap" of large numbers of adjacent clusters) for the transition to a simple qualitative band theory (or "MO/band theory") model (and the so-called "one-electron" energy band scheme).

Pyrite and the related disulfides and dichalcogenides provide a good example with which to illustrate the MO and MO/band theory approach. An MO energy level diagram for pyrite based on Bither et al. (1968), Burns and Vaughan (1970) and Burns (1993) is shown in Figure 2. The tetrahedral coordination of the sulfur atoms in pyrite to three metals and another sulfur suggests involvement of $3s$ and $3p$ orbitals (sp^3 hybridized) in forming σ-bonds. One hybrid sp^3 orbital from each of the six sulfurs forms six σ-bonds with d^2sp^3 hybrid orbitals of the central transition metal. The d^2sp^3 hybrids consist of the two e_g orbitals ($d_{x^2-y^2}$ and d_{z^2}), the one $4s$ orbital and the three $4p$ orbitals. Bither et al. (1968) assumed that the three t_{2g} orbitals (d_{xy}, d_{yz}, d_{xz}) of the transition metal remain *nonbonding*. The energy separation between nonbonding t_{2g} and antibonding $e_g{}^*$ orbitals is designated $\Delta_{cov-\sigma}$ and is analogous to the ligand field stabilization energy. Burns and Vaughan (1970) suggest that the paired electrons in the nonbonding t_{2g} orbitals may form π bonds with vacant t_{2g}-type $3d$ orbitals of the sulfur atoms. This would result in the increased energy separation between nonbonding t_{2g} and antibonding $e_g{}^*$ levels shown in Figure 2, i.e., $\Delta_{cov-\pi} > \Delta_{cov-\sigma}$.

Overlapping of molecular orbitals between Fe(S-S)$_6$ "clusters" in an FeS$_2$ crystal would cause broadening into bands. The "one-electron" energy band scheme for pyrite shown in Figure 3a originates from this qualitative approach (after Bither et al. 1968; Goodenough 1972). The main bonding molecular orbitals now form the filled σ band and the corresponding antibonding orbitals the empty σ* band, constituting the main valence and conduction bands

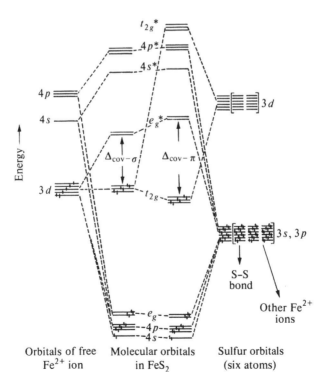

Figure 2. Molecular orbital energy level diagram for pyrite, FeS$_2$. (redrawn after Burns and Vaughan 1970).

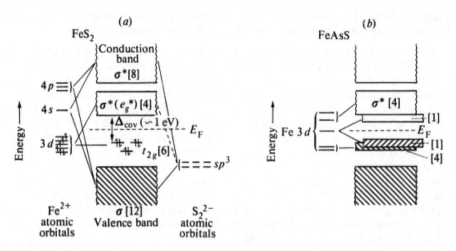

Figure 3. Schematic "one electron" energy band diagrams for: (a) pyrite (FeS_2) and (b) arsenopyrite (FeAsS). Numbers in square brackets refer to states per molecule; E_F = Fermi level; hatched bands are filled with electrons. (Redrawn after Bither et al. 1972; Goodenough 1972.)

respectively. The iron $3d$ orbitals then lie between these in energy, the t_{2g} orbitals being regarded as essentially localized on the cation but the e_g orbitals forming a band through overlap via sulfur intermediaries. Thus, conduction in pyrite would occur when electrons are excited into the band formed from e_g^* orbitals.

Using this approach to the electronic structures of the pyrite-, arsenopyrite- and loellingite-type disulfides, Goodenough (1972) criticized the ligand field model of Hulliger (1968) (the basis for the ligand field model described in the section above). Essential features of Goodenough's energy band model for FeAsS (and $CoAs_2$) are shown in Figure 3b. The splitting of the metal $3d$ orbital energy levels by the ligand field in the arsenopyrite structure differs markedly from that in regular octahedral coordination and is such as to raise the energy of the t_{2g}-type orbital which is parallel to the c-axis. The other t_{2g}-type orbitals remain nonbonding orbitals and form a filled narrow band (cf. the localized t_{2g} electrons in FeS_2); the e_g-type orbitals are the empty antibonding orbitals forming a conduction band. The unique t_{2g}-type orbital is split by a bonding interaction into a lower-energy filled band and higher-energy empty band. However, the important difference between this model and the ligand field model outlined above is that the interaction (and hence the structural deformation) is attributed to metal-anion bonding not metal-metal bonding. Goodenough (1972) argues that the arsenopyrite structure represents an expansion, not a contraction, of alternate metal-metal separations along the c-axis, and that electron density is concentrated in the regions of *greater* separation.

The pyrite and arsenopyrite disulfides serve to emphasize the difference between the localized electron approach of ligand field theory and the collective electron MO/band theory in describing the d orbitals in sulfides. Further illustrations can be provided by comparing the electronic structures of the pyrite-structured MnS_2 (hauerite) and NiS_2 (vaesite) with pyrite itself. In MnS_2, the $3d$ electrons are localized on the cation, partly because of the large intra-atomic stabilization energy gained by the high-spin d^5 configuration. Effectively, the five single d-electrons are more tightly bound. The intra-atomic exchange energy (Δ_{ex}) required to produce pairing of the d-electrons is a maximum value and can be represented as in the $3d$ orbital energy band diagram of Figure 4. Expressed another way, the d orbital energy levels are split into a spin-up (α) and spin-down (β) set. It is unnecessary to invoke such splitting to explain the properties of FeS_2 shown in Figure 4. The series of pyrite-structure

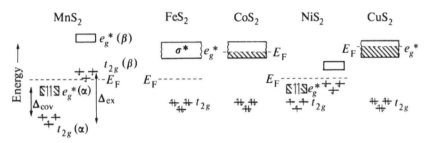

Figure 4. Schematic energy level diagram for the 3d orbitals in the pyrite-type disulfides (only 3d orbital energy levels are shown). The effects of intra-atomic exchange are shown in MnS_2 and NiS_2.

compounds FeS_2-CoS_2-NiS_2-CuS_2 sees successive addition of electrons into the e_g^* band and the progressive changes in properties outlined elsewhere in this volume (see, for example, Pearce et al. 2006). The energy band diagrams for just the 3d orbitals are shown in Figure 4 and accord with the metallic conductivity of CoS_2 and CuS_2, although in CoS_2 the e_g^* band is sufficiently narrow that ferromagnetism is observed at low temperatures. In NiS_2 the intra-atomic exchange energy again becomes important, splitting the energy levels into spin-up and spin-down states. The localized magnetism and semiconductivity of NiS_2 can be viewed as consequences of this interaction. The MO/band approach can, therefore, describe the complete range of properties exhibited by the pyrite-type disulfides, from the localized 3d electron behavior of MnS_2 to the broad band (metallic) delocalized 3d electron behavior of CuS_2.

The same MO/band theory approach has been used to describe the bonding in other major sulfide mineral groups, including the NiAs-type monosulfides, thiospinels, and the pentlandites. Some of these examples are discussed below (but see also Vaughan and Craig 1978, for further discussion of the earlier work).

QUANTITATIVE APPROACHES: ATOMISTIC COMPUTATION

One of the quantitative approaches that has been employed to model chemical bonding in minerals is the use of so-called *atomistic* calculations based on the Born Model of solids. Here, in a classical rather than quantum mechanical formalism, the atoms comprising the crystalline solid are regarded simply as ions, and the (non-directional) forces between them modeled using interatomic potentials. Short range interactions are generally modeled using a Buckingham potential:

$$U_{ij} = \sum_{ij} A_{ij} \exp\left[r_{ij}/\rho_{ij}\right] - C_{ij}r_{ij}^{-6} \qquad (1)$$

where i and j are two ions with separation r, A and ρ are constants describing short range repulsion and C is a term that takes dispersion effects into account. Long range contributions from Coulombic interactions are also incorporated in these models in various ways. Also, rather than having to regard the ions involved purely as rigid spheres, the effects of the polarizability of the ion (and hence distortion of charge distribution in its environment in the crystal) can be incorporated, and additional potential terms added to account for the directionality of bonds. The actual potential parameters are most commonly derived by empirical fitting to accurately known experimental data, although an alternative approach is to use parameters derived *ab initio* using quantum mechanical methods. A full account of atomistic computational methods can be found in references such as Catlow (1997), and a good introduction for the mineralogist is provided by Gale (2001).

High quality atomistic calculations can be successfully used to model crystal structure, crystal morphology, energetics, lattice dynamics and properties such as elastic and dielectric constants. They have also been used to simulate the structure and stability of the surfaces of crystalline solids and to model defects, and behavior related to the presence of defects, such as diffusion. Of course, they provide no information on electronic structure or on properties (electrical, magnetic, optical, etc.) that can only be understood in terms of electronic structure.

Although atomistic methods have mainly been used to model materials such as oxides, silicates and carbonates, there have been successful applications to sulfides, despite the limited extent to which sulfides can be regarded simply as "ionic" solids. Wright and Jackson (1995) used this approach to simulate the structure and defect properties of ZnS. Their potentials reproduce the structure and elastic constants to within a few percent of the experimental values, and their calculations suggest that Zn diffusion in ZnS takes place via an interstitial mechanism, whereas S diffuses by way of vacancies in the lattice. Wright et al. (1998) went on to use the same potentials to simulate the structure and stability of sphalerite surfaces. Hamad et al. (2002) generated a new set of Zn-S and S-S potential parameters and used them to model both the sphalerite and wurtzite forms of ZnS. This work included modeling of the relaxed surface geometry of the sphalerite (110) surface and the $(10\bar{1}0)$ surface of wurtzite. Calculated surface energies were used to predict the most stable crystal morphologies of both dimorphs. In the case of sphalerite, this is a dodecahedron comprised of only the (110) surface and its equivalents; in wurtzite, it is a highly anisotropic cylindrical-like shape. Recently, new interatomic potentials have been presented by Wright and Gale (2004) and used to model the structures and stabilities of both sphalerite and wurtzite polytypes of ZnS and CdS. These potentials reproduce many of the properties of all four minerals to within a few percent of experimental values. In contrast to the majority of previous calculations, the relative stabilities of the cubic and hexagonal phases are correctly predicted, with the cubic form being more stable for ZnS and the hexagonal form for CdS. In $Zn_xCd_{1-x}S$ solid solutions, the transition from hexagonal to cubic is predicted to occur at $x = 0.6$.

More surprising than the success in using atomistic simulations to model particular properties of ZnS has been the work on pyrite and marcasite by Sithole et al. (2003). Here, interatomic potential parameters derived at simulated temperatures of 0 K and 300 K (referred to as P1 and P2, respectively) were used to predict structures and elastic properties as a function of pressure up to 44 GPa. Predicted pyrite structures were within 1% of those determined experimentally and the calculated bulk modulus was within 7%. As illustrated in Figure 5, the calculated equation of state (EOS) for pyrite gives good agreement with experimental data (such as that of Merkel et al. 2002). The calculations show that Fe-S bonds shorten more rapidly than S-S dimer bonds, and that marcasite shows very similar behavior to pyrite at high pressure. The vibrational spectrum of pyrite has also been modeled with some success using lattice dynamics atomistic simulations (Lutz and Zminscher 1996).

QUANTITATIVE APPROACHES: ELECTRONIC STRUCTURE CALCULATIONS

These days, the electronic structure of sulfide minerals is most commonly discussed from first-principles terms using a quantitative approach based on either molecular orbital theory or band theory. As noted above, this has largely to do with the continually improving efficiency and sophistication of electronic structure calculations. In the present view, we concern ourselves generally with the energies and spatial distribution of electronic states traveling throughout the solid. The electronic structure depends on all possible interactions between particles in the material. We approach the problem using the Schrödinger equation:

$$H\Psi = E\Psi \qquad (2)$$

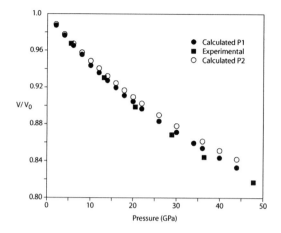

Figure 5. Atomistic calculation of the equation of state (EOS) of pyrite compared with experimental data (after Sithole et al 2003).

where H is the Hamiltonian operator, Ψ is the wavefunction, and E is the total energy of the system. A usual non-relativistic Hamiltonian is written:

$$H = H_e + H_{eN} + H_N \tag{3}$$

where the electron-electron interaction term is:

$$H_e = \sum_i -\frac{\hbar^2}{2m}\nabla_i^2 + \frac{1}{2}\sum_{i,j\neq}\frac{e^2}{|\mathbf{r}_i - \mathbf{r}_j|} \tag{4}$$

the nuclear-nuclear interaction term is:

$$H_N = \sum_I -\frac{\hbar^2}{2M_I}\nabla_I^2 + \frac{1}{2}\sum_{I,J\neq}\frac{Z_I Z_J e^2}{|\mathbf{R}_I - \mathbf{R}_J|} \tag{5}$$

and the electron-nuclear interaction term is:

$$H_{eN} = -\sum_{iI} Z_I \frac{e^2}{|\mathbf{R}_I - \mathbf{r}_i|} \tag{6}$$

where $\mathbf{r}$ and $\mathbf{R}$ are position vectors for electrons and nuclei, respectively, m and M are electron and nuclear masses, respectively, and Z is a nuclear charge.

In a first principles calculation one seeks a solution to the Schrödinger equation. To accomplish this requires many approximations. Some of the main ones relevant to this discussion are: (1) the Born-Oppenheimer approximation involving separation of electronic and nuclear motions; (2) the one-electron or mean-field approximation where each electron is treated as traveling on a periodic potential arising from the nuclear charges modified by the average potential contribution of all the other electrons, and (3) Koopmans' approximation, which allows one to impart physical significance (i.e., ionization potentials) to the one-electron eigenvalues by assuming electron removal does not affect the energies of other one-electron states. In general, other approximations are specific to the approach for implementing the calculation in a practical manner.

For example, the Hartree-Fock (HF) approach is an attempt to treat the many-body problem directly by keeping track of the coordinates of all the electrons. HF describes the wavefunction in terms of a single Slater determinant, a determinant of a matrix of one-electron "spin" orbitals. The interactions between the electrons are expressed as Coulomb and exchange integrals. The task is to determine the lowest energy wavefunction for the entire system of interacting electrons, where a change in the spin orbital for one electron influences the behavior of other electrons due to coupling of the electronic motions. In HF, this is done by focusing on a single electron in a spin orbital interacting with the fixed field of nuclei and the fixed field of other electrons. But since the solution for one electron affects the remaining electrons, an iterative scheme called the self-consistent field (SCF) approach is used to find the overall solution. Electron exchange is a short range electron-electron interaction that excludes the possibility of electrons of like spin from occupying the same orbital (the Pauli exclusion principle). In HF this is a built-in property of the Slater determinant (antisymmetry). In the HF equations, the exchange term modifies electron-electron repulsion only for electrons of like spin. The effect is to spatially separate electrons of like spin in the calculation, which gives rise to a slight reduction in the total energy called the exchange energy. A strength of the HF approach is its exact expression for the electron exchange energy. However, the exchange interaction does not fully describe the tendency of electrons to avoid each other. The difference between the HF energy and the exact energy is the correlation energy. This deficiency leads to overbinding of electrons to nuclei, and consequently very poor prediction of bond energies and band gaps. The usefulness of the HF approach for understanding the electronic structure of "strongly correlated" systems such as sulfide minerals is very constrained.

Density functional theory (DFT) is a very different approach that does provide for a treatment of both electron exchange and correlation. The main idea of DFT is to replace the many-body electronic wavefunction with the electron density as the central quantity. An early implementation of DFT that was heavily used for solid-state electronic structure calculations was the multiple-scattering SCF X_α method. MS-SCF-X_α is a molecular orbital method wherein the one-electron Schrödinger equation is set up for a so-called "muffin-tin" approximation of the true potential, spherically symmetrical within spheres surrounding the various nuclei and constant in the region between the spheres. It uses a statistical approximation for exchange-correlation. This method was often successful for describing sulfide mineral electronic structure and is mentioned here because this early work is still useful, some of which is reviewed below (for more detailed information on this method and results for sulfides see Tossell and Vaughan 1992). Presently, however, it is typical for reports of sulfide mineral electronic structure to be based on so-called "modern" density functional theory (DFT) (Hohenberg and Kohn 1964; Kohn and Sham 1965). In this Kohn-Sham approach, the ground state wave function and energy are expressed as functionals of the electron density distribution. Exchange-correlation interaction is included in terms of an empirically parameterized functional. Incremental improvements to the accuracy of DFT historically have come in the form of improved exchange-correlation functionals. Early functionals were based on the limiting behavior of the uniform electron gas in what is known as the local density approximation (LDA). More refined exchange-correlation functionals include terms involving the spatial gradient of the charge density in what is known as the generalized gradient approximation (GGA). For sulfide minerals, the BLYP (Becke 1998; Lee et al. 1988), PW (Perdew and Wang 1992), and PBE (Perdew et al. 1996) GGA functionals are in common use. Becke's three parameter hybrid functional (B3LYP)(Becke 1993), based in part on a prescribed amount of HF exact exchange, is also in common use. While DFT performs better than HF, it is not without problems. Unlike HF, where electron interaction with itself is completely canceled, this self-interaction is only partially cancelled in typical DFT implementations (Perdew and Zunger 1981). Therefore, recently, improvements to DFT have been made by incorporating a self-interaction correction (Svane et al. 2004), by an amount that is sometimes empirical. Other recent improvements to DFT have been based on incorporating exact exchange (Stadele et al. 1999). These have led to substantially more accurate band gaps for instance.

DFT band structure calculations for crystals may be implemented using periodic boundary conditions at the unit cell edges. Basis sets (one-electron trial functions) could have the form of either local functions (typically Gaussians) such as in the Crystal code (Saunders et al. 2003) or continuous (e.g., planewave) functions such as in the NWChem code (Apra et al. 2003). The solution is a set of wavefunctions and energy eigenvalues for each one-electron state. The one-electron wavefunctions are Bloch functions, which have the form of planewaves, modulated by a function whose periodicity is the same as that of the crystal lattice, as given by:

$$\psi_k(\mathbf{r}) = e^{i\mathbf{k}\cdot\mathbf{r}} u_k(\mathbf{r}) \quad (7)$$

where $\psi_k(\mathbf{r})$ is the Bloch function, $u_k(\mathbf{r})$ is the periodic function, $\mathbf{r}$ is a position in the unit cell, and $\mathbf{k}$ is the propagation vector. The energy level structure for these states consists of groupings (or bands) of allowed energy levels (where $\mathbf{k}$ is real), separated by energy gaps where no electronic states are allowed (where $\mathbf{k}$ is complex). The band structure is displayed as energy vs. $\mathbf{k}$, where $\mathbf{k}$ vectors are chosen along high symmetry directions in the Brillouin zone. The Fermi level (E_F) lies in the band gap between the occupied (valence) and unoccupied (conduction) bands, designating the energy at which the chemical potential of electrons is zero.

DFT works well for describing electrons in metallic or small band gap sulfide minerals where very strong interatomic interactions exist, such as covellite, millerite, and cattierite. The band structure calculations can provide useful insights into how atomic orbitals are combined into crystalline orbitals comprising bands. In many sulfide minerals, band energies are typically closely related to the energies of the free atom states. One example is where metal cations lead to partially filled d-bands, such as in many of the pyrite-type disulfide minerals where the cation is a first-row transition metal atom. In this case, band widths can be relatively narrow and band gaps are non-zero. Various methods are available for decomposing the wavefunction for the crystal in terms of the individual atomic wavefunctions (e.g., projected densities of states, crystal orbital overlap population analysis, etc.). This allows bands to be spoken of in terms of their atomic orbital character (e.g., s-p band, d-band, etc.). Close spacing and high degree of atomic orbital hybridization between the atoms is the basis for covalent bonding, which gives bands that are strongly mixed in their atomic orbital character. Hybridization is most significant for valence atomic orbitals, producing a mixed band which is smoothly varying across atom types in the crystal and allowing for electron delocalization.

Lastly, it is worth noting that most DFT studies of sulfide minerals so far have primarily been "static" calculations, intrinsically for a temperature of zero Kelvin. *Ab initio* molecular dynamics, particularly using the Car and Parrinello scheme (Car and Parrinello 1985) (CPMD) is beginning to be applied to sulfide minerals. In CPMD, the minimization of the total energy with respect to the total wavefunction and atomic coordinates is solved simultaneously and molecular dynamics is performed at only a slight increase in computational expense. A planewave basis set is used, often with pseudopotentials to mimic the scattering properties of core electrons and explicit description only of the valence electrons. CPMD is highly efficient quantum mechanical molecular dynamics at the DFT level of theory. It has provided a means to study sulfide mineral properties and reactions that depend on both the electronic and nuclear motion at temperatures of interest.

CHEMICAL BONDING AND ELECTRONIC STRUCTURE IN SOME MAJOR SULFIDE MINERALS AND GROUPS

Generalities

The development of quantitative bonding models for sulfide minerals has lagged behind the work in this field on oxides and silicates. This has been partly because of the more limited geological abundance and perceived importance of sulfides, but chiefly because of the challenge to modeling posed by the diversity of electronic structure types found in this mineral group. In what follows, our current knowledge of bonding in some major sulfide minerals and groups is reviewed, noting the usefulness of both qualitative and quantitative cluster (MO) and periodic (band model) approaches. A number of more wide ranging computational studies of sulfides have been published in recent years, in addition to relevant material in the review volumes already mentioned, and these are discussed here.

Gibbs et al. (1999) calculated electron density distributions $\rho(\mathbf{r})$ for a large number of model sulfide molecules and their oxide equivalents, and applied the concepts developed by Bader and coworkers to characterizing the differences between sulfide and oxide bonds. Bader (1990, 1998) has argued that atomic interactions in "molecular" systems can be identified and characterized by the topological properties, the gradient, and the Laplacian $\nabla^2\rho(\mathbf{r})$ of their electron density distributions (e.g., Bader 1990, 1998). In their study, Gibbs et al. (1999) provide a quantitative demonstration that bonded interactions in sulfides are more directional for a given cation compared with oxides. They also show that the value of the electron density distribution at the bond critical point, and the length of the M-S bond, are reliable measures of a bonded interaction; the greater the accumulation of electron density at the bond critical point and shorter the bond, the greater the covalent interaction. Laplacian maps of the electron density distribution for sulfur-containing molecules in comparison with the same molecules containing oxygen are illustrated in Figure 6. Here it can be seen that the valence shell charge concentration (VSCC) of the sulfide anion (dashed line contours) is highly polarized and extends into the internuclear region of the M-S bonds. In contrast, the corresponding oxide anion tends to be less polarized and more locally concentrated in the vicinity of its valence shell.

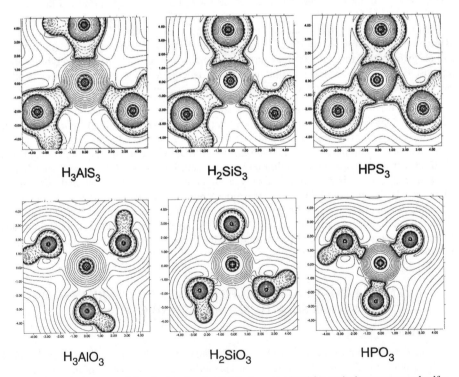

Figure 6. Laplacian maps of the electron density distribution calculated for equivalent oxygen and sulfur containing molecules. A metal cation is located at the center of each molecule and H atoms are attached to the oxide or sulfide anions to achieve electrical neutrality. (Reproduced from Gibbs et al. 1999; see this reference for further details.)

Similarly, analyses of electron density distributions have given insights into how various types of bonded interactions present in sulfides can be related to electronic properties such as electrical conductivity, a macroscopic observable. For example, electron density distributions were computed and analyzed for the bonded interactions comprising the nickel sulfide minerals millerite, vaesite, heazlewoodite, and also Ni metal (Gibbs et al. 2005). These analyses identified Ni-Ni bond paths restricted to isolated Ni_3S_9 clusters in millerite (a $Sp - Nid$ charge transfer conductor), and were found to form contiguous highly branched networks in heazlewoodite (a Ni $3d$ metallic conductor). No Ni-Ni bond paths were found in vaesite (an insulator). Electron transport in Ni metal and heazlewoodite was pictured as occurring along the Ni-Ni bond paths, which behave as networks of atomic-size wires that radiate in a contiguous circuit throughout the two structures. In contrast, electron transport in millerite is pictured as involving a cooperative hopping of the d-electrons from the Ni_3 rings comprising Ni_3S_9 clusters to Ni_3 rings in adjacent clusters via the p orbitals on the interconnecting S atoms.

A number of theoretical studies have been aimed at calculating the electronic structures of a series or large group of sulfides. In earlier work, for example, Harris (1982) used cluster calculations (MS_6^{n-} clusters) to study the electronic structures of first and second row transition metal sulfides, calculating trends in energy levels and charge distributions, and concluding that the bonding in second row transition metal sulfides is more covalent than in the first row due to increased metal-sulfur d-$p\pi$ interactions.

Certain first row transition metal sulfides have also featured in systematic studies aimed at exploring different theoretical approaches or the subtleties of electrical or magnetic properties (Bauschlicher and Maitre 1995; Saitoh et al. 1995; Rohrbach et al. 2003). In a series of calculations using DFT, Raybaud et al. (1997a,b) investigated the structural and cohesive properties and the electronic structures of more than thirty transition metal sulfides, including the monosulfides of V, Cr, Mn, Fe, Co, Ni, Pd and Pt; disulfides of Mn, Fe, Co, Ni, Mo, Ru, Pd, Re, Os, Ir, and more unusual phases such as Co_9S_8 and Ni_3S_2. Although initial calculations using the LDA tended to overestimate the strength of bonding, using GGA the structures and cohesive energies were more accurately predicted. Calculated cohesive energies are shown in Table 2. The general conclusions regarding electronic structure were that it is determined by short range interactions in the S $3p$ – metal d band complex, with the ligand field splitting of the metal d states by the surrounding S atoms determining the structure of the d band. The authors were able to predict electrical properties, and to comment on factors influencing the catalytic activity of the various transition metal sulfides studied. Correlations between catalytic activity and both metal-sulfur bond strength and the character of the highest occupied states (so-called "frontier orbitals") were noted. This formidable series of calculations provides a good illustration of how much can now be achieved using *ab initio* computation to study sulfide minerals. In addition, Hobbs and Hafner (1999) have extended the work of Raybaud et al. (1997a,b) by using the DFT approach to calculate the magnetic properties of key transition metal sulfides (CrS, MnS, FeS, CoS, NiS, MnS_2, FeS_2, CoS_2, NiS_2).

Galena (PbS)

As noted elsewhere in this volume (Mackovicky 2006; Pearce et al. 2006; Wincott and Vaughan 2006), the rocksalt structure PbS is a small bandgap (~ 0.4 eV) intrinsic semiconductor, although natural samples may show either n- or p-type behavior dependant on the presence of defects and impurities. Both bulk and surface electronic structure and properties have been the subject of numerous spectroscopic and computational studies (surface studies are discussed in this volume by Rosso and Vaughan 2006a,b). Amongst experimental data available are optical reflectance and transmittance spectra (Cardona and Greenaway 1964; Schoolar and Dixon 1965), detailed X-ray and UV photoemission and inverse photoemission spectra (McFeely et al. 1973; Grandke et al. 1978; Santoni et al. 1992; Ollonqvist et al. 1995; Schedin et al. 1997; Leiro et al. 1998; Muscat and Klauber 2001), Auger electron spectra

Table 2. Cohesive energies of the transition-metal sulfides (in eV/atom).

Structure	E_{exp}	E_{GGA}	E_{GGA}/E_{exp}	E_{LDA}/E_{exp}
3d				
VS	5.61	5.02	0.90	1.12
Cr_2S_3	4.79	3.81	0.80	1.06
CrS	—	4.09	—	—
MnS	4.03	4.03	1.00	1.12
FeS(NiAs)	4.14	4.42(6)	1.07	—
FeS (troilite)	—	4.43(5)	—	—
FeS_2 (pyrite)	3.92	4.29	1.08	—
FeS_2 (marcasite)	3.92	4.31	1.10	—
Co_9S_8	4.23	4.91	1.15	1.44
CoS	4.10	4.68	1.14	—
Ni_3S_2	4.30	4.54	1.06	—
NiS (millerite)	4.12	4.39	1.07	—
NiS (NiAs)	—	4.32	—	—
NiS_2	—	4.02	—	—
4d				
NbS	—	6.16	—	—
NbS_2	5.68	5.54	0.98	—
MoS_2	5.18	5.11	0.99	1.21
RuS_2	4.88	5.05	1.03	1.28
Rh_2S_3	4.61	4.84	1.05	1.29
PdS	3.80	3.76	0.99	1.26
PdS_2	—	3.60	—	—
5d				
TaS_2	5.87	6.04	1.03	—
WS_2	5.78	1.00	1.00	1.20
ReS_2	5.23	5.35	1.02	—
OsS_2	5.19	5.34	1.03	1.26
Ir_2S_3	5.06	5.37	1.06	—
IrS_2	4.77	5.07	1.06	—
PtS	4.83	4.69	0.97	1.20

(Tossell and Vaughan 1987), and a range of X-ray emission (Sugiura et al. 1997) and X-ray absorption spectra (von Oertzen et al. 2005). These spectroscopic studies have been reviewed elsewhere in this volume (Wincott and Vaughan 2006).

A wide variety of cluster (MO) and periodic (band structure) computational approaches have been applied to galena (Tung and Cohen 1969; Rabii and Lasseter 1974; Hemstreet 1975; Tossell and Vaughan 1987, 1992; Mian et al. 1996; Gurin 1998; Gerson and Bredow 2000; Satta et al. 2000; Muscat and Klauber 2001; Ma et al. 2004; Zeng et al. 2005). As noted by Rosso (2001), accurate computation of the electronic structure of galena is more difficult than might first appear because proper treatment of the heavy element Pb requires dealing with relativistic effects, and an all-electron treatment has to include the f electrons. Despite these complications, the computed values for band energies, band widths and densities of states of the valence band are in excellent agreement with experiments. Wincott and Vaughan (2006; this volume) illustrate how earlier cluster and band structure calculations inform interpretation of photoemission and X-ray emission and absorption spectra of galena. The composition of the valence region is illustrated in Figure 7 using results of a recent calculation of the density

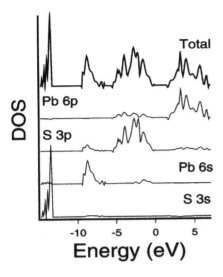

Figure 7. Calculated total and projected densities of states (DOS) for galena using data from Becker and Rosso (2001). The Fermi level (zero on the energy scale) is arbitrarily located at the valence band edge. The top of the valence band is seen here as being comprised of predominantly S 3p states and the bottom of the conduction band as comprised of predominantly Pb 6p states. (Reproduced from Rosso 2001.)

of states (see Rosso 2001; Becker and Rosso 2001). As can be seen, the conduction band is dominated by Pb 6p states with a minor admixture of S 3p states. The top of the valence band consists of non-bonding S 3p states, overlying a Pb 6s – S 3p bonding band. It is interesting to note that calculation of electron density maps and of other indictors of the nature of bonding, such as orbital overlaps, show that the bonding in galena can be described as ionic with minor covalent character (Mian et al. 1996).

Sphalerite and related sulfides (ZnS, Zn(Fe)S, CdS, HgS, CuFeS$_2$)

The cubic (sphalerite) and hexagonal (wurtzite) forms of ZnS are classic crystal structure types, both containing Zn and S in regular tetrahedral coordination. Both are also diamagnetic semiconductors with large band gaps (sphalerite ~ 3.6 eV; see Pearce et al. 2006, this volume for details). Sphalerite and wurtzite, along with the two isostructural CdS species (hawleyite and greenockite) have also been the subject of numerous experimental and computational studies of bonding and electronic structure, partly because of their interest to solid state physicists. This includes investigations of sphalerite using X-ray photoelectron and X-ray emission spectroscopies (e.g., Ley et al. 1974; Sugiura et al. 1974; Domashevskaya et al. 1976; Sugiura 1994; Laihia et al. 1996, 1998) and X-ray absorption spectroscopies (Li et al. 1994a,c; von Oertzen et al. 2005) reviewed elsewhere in this volume (Wincott and Vaughan 2006).

Calculations have been performed on ZnS using both MO cluster methods (e.g., Tossell 1977) and a large variety of band structure methods (e.g., Stukel et al. 1969; Pantelides and Harrison 1975; Faberovich et al. 1980; Bernard and Zunger 1987; Martins et al. 1991; Schroer et al. 1993; Edelbro et al. 2003). As noted above, ZnS has also been successfully studied using atomistic methods. The calculations of Edelbro et al. (2003) using an DFT approach are typical of the level of detail now attainable and in Figure 8 are shown the band structure (Fig. 8a) and the calculated density of states (DOS) with Zn and S contributions (Fig. 8b) obtained by these authors for sphalerite ZnS. As can be seen from Figure 8b, the valence band is calculated to be a mixture of Zn and S orbitals and to lie in the interval from −5.2 to 0 eV, in good agreement with experiment (such as the XPS data of Ley et al. 1974). Below these levels (at −6.5 eV) are the Zn 3d bands which are around 3 eV higher in energy than determined by experiment. The band at −11.7 to −12.9 eV comprises the S 3s electrons according to the calculations. The calculated band gap is somewhat underestimated compared with experiment

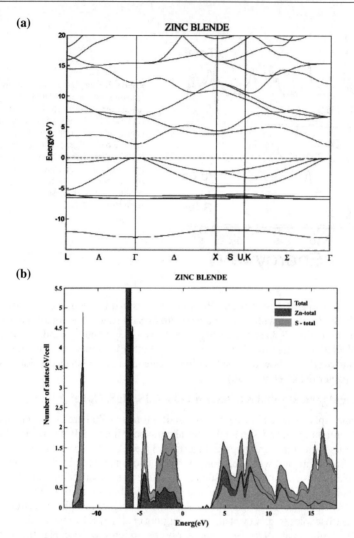

Figure 8. The electronic structure of ZnS (sphalerite): (a) calculated band structure with the zero of energy set at the highest occupied state; (b) the calculated density of states (DOS) showing the Zn and S contributions. [Used with permission of Elsevier, from Edelbro et al. (2003), *Applied Surface Science*, Vol. 206, Figs. 7 and 8, pp. 306 and 307.]

(2.23 eV vs. 3.6 eV), a well known problem with DFT calculations. Both these calculations and earlier studies aid in interpretation of the large amount of available experimental data, and show that the Zn 3d orbitals in sphalerite are non-bonding and located below the valence band, which has strong bonding orbital character.

In a study using MO cluster calculations applied to MS_4^{6-} tetrahedral units (using the MS-SCF-X_α method), Tossell and Vaughan (1981) were able to compare ZnS, CdS and HgS. As discussed elsewhere in this volume (Wincott and Vaughan 2006), the calculations on the ZnS_4^{6-} cluster can be used to interpret XPS and XES data for sphalerite, the generally good agreement between experiment and theory providing support for these models (which are applicable to both sphalerite and wurtzite structure types). As seen in Figure 9, the calculations suggest an overall similarity in the electronic structure of the valence region in ZnS, CdS and HgS, with S $3p$ non-bonding orbitals at the top of the valence band, below them the main metal-sulfur bonding orbitals, then non-bonding metal $3d$ orbitals and at around 12 eV below the Fermi level, the S $3s$ non-bonding orbitals. The order of molecular orbitals is the same in each case, but there are significant differences, notably the trend towards decreasing stabilization of the main metal-sulfur bonding orbitals moving across the series ZnS-CdS-HgS. In Figure 9, for comparison, a calculation is also shown of a HgS_2^{2-} linear cluster, the basic unit found in the cinnabar form of HgS.

The well known substitution of Fe^{2+} for Zn in the ZnS structure has also been modeled using MO cluster calculations (Vaughan et al. 1974), in particular to calculate the optical absorption spectra of Fe-doped sphalerite (see Wincott and Vaughan 2006; this volume, for a figure showing this spectrum). The calculated and experimental absorption features are given in Table 3 and generally show good agreement; these data also demonstrate the importance of spin polarization due to the four unpaired $3d$ electrons on the Fe^{2+} cation, which splits the MO energy levels into spin-up and spin-down groups (see discussion in earlier section of this chapter). These calculations used a transition-state (TS) procedure which takes into account relaxation effects associated with electron transitions. Other transition metals (Ti, Mn, Co, Ni) can substitute for the Zn in ZnS, and the electronic structures of these dopants (as well as Fe) have also been studied using X-ray absorption spectroscopy (XANES) by a number of researchers (Lawniczak-Jablonska et al. 1996; Pattrick et al. 1998; Perez-Dieste et al. 2004).

Chalcopyrite ($CuFeS_2$) has a sphalerite-type structure, and is one of the relatively few ternary sulfides to be the subject of detailed studies of its electronic structure. Tossell et al.

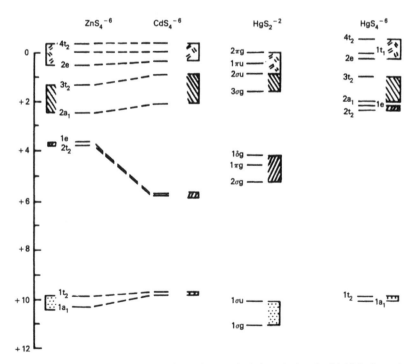

Figure 9. Energy level diagrams derived from cluster calculations (using the Multiple Scattering X_α method) on tetrahedral ZnS_4^{6-}, CdS_4^{6-}, HgS_4^{6-} clusters and the linear HgS_2^{2-} cluster. The makeup of particular MO energy levels in terms of atomic orbital character are shown by the boxes on the diagram. (After Tossell and Vaughan 1981.)

Table 3. Experimental and calculated optical spectra for the FeS_4^{6-} unit. Transition energies, ΔE, are all in cm^{-1}.

Type of transition	Assignment	Calc. ΔE (TS)	Experimental ΔE
Spin-allowed d-d	$3e\downarrow \rightarrow 10t_2\downarrow$	2194	3850, 3500, 2950, 3700, 2850
Spin-forbidden	$10t_2\uparrow \rightarrow 3e\downarrow$	13,149	12,120, 13,000
Spin-flip (very weak)	$10t_2\uparrow \rightarrow 10t_2\downarrow$	14,279	14,500
	$3e\uparrow \rightarrow 3e\downarrow$	17,586	16,950
	$3e\uparrow \rightarrow 10t_2\downarrow$	18,586	19,600
Ligand-metal charge transfer	$2t_1\downarrow \rightarrow 3e\downarrow$	14,279	14,500
	$2t_1\downarrow \rightarrow 10t_2\downarrow$	13,100	
	$9t_2\downarrow \rightarrow 3e\downarrow$	17,586	16,950
	$9t_2\downarrow \rightarrow 10t_2\downarrow$	19,764	19,600

Source: After Vaughan et al. 1974, who provide details of the experimental data.

(1982) reported a fairly complete series of XES and XPS spectra, and used the results of MO calculations on CuS_4^{7-} and FeS_4^{5-} clusters to assign the features in these spectra. Mikhlin et al. (2005) studied the X-ray absorption (XANES) spectra of Fe, Cu and S (L-edges) in chalcopyrite and used a qualitative molecular orbital approach to interpret their data. Edelbro et al. (2003), following on from earlier band structure calculations (e.g., Hamajima et al. 1981), used an *ab initio* DFT approach to calculate the band structure of chalcopyrite, but their calculation predicted metallic conductivity, contrary to experimental findings. Lavrentyev et al. (2004) compared a combination of their own and other workers XES, XAS and XPS spectra for chalcopyrite with their calculated density of states using a cluster model. In Figure 10, a comparison of theoretical and experimental data from their study is shown. These data suggest that the main peak maximum (B) in the XPS arises from the copper d-electrons which, together with the d states of iron, also produce the shoulder A in the XPS. The shoulder C is attributed to Cu and Fe d states and a substantial contribution from sulfur p states. It is noteworthy that the $3d$ electrons of Cu appear to participate in bonding in this system, even though spectroscopic and magnetic evidence clearly shows that Cu is monovalent with a nominally filled $3d$ shell (see Pearce et al. 2006, Wincott and Vaughan 2006; both in this volume). Fujisawa et al. (1994), in reporting a series of detailed spectroscopic studies of $CuFeS_2$, state that the Cu $2p$ core XPS spectrum reveals a mixing of the d^9 ("Cu^{2+}") configuration into the formally monovalent Cu and interpret this as due to Cu $3d$-Fe $3d$ hybridization mediated by the S $3sp$ valence states. There is further discussion of the electronic structure of chalcopyrite, particularly in the context of surface structure, in a later chapter of this volume (Rosso and Vaughan 2006b).

Transition metal monosulfides (FeS, $Fe_{1-x}S$, CoS, NiS)

The monosulfides FeS, CoS and NiS have the nickel arsenide structure at elevated temperatures but undergo distortion, structural transformation or breakdown on cooling (see Makovicky 2006; Fleet 2006, both in this volume). In the case of FeS, troilite (stable below 140°C) is a distorted form of the NiAs structure in which triangular clusters of iron atoms form in the basal plane. There is also the $Fe_{1-x}S$ omission solid solution of the pyrrhotites, with its complex series of vacancy ordered superstructures, the most important of which is that of monoclinic pyrrhotite (Fe_7S_8). CoS undergoes dissociation forming cobalt pentlandite (Co_9S_8; see the separate section below) and NiS transforms to the millerite structure (see below).

As discussed, in part, elsewhere in this volume (Wincott and Vaughan 2006), a wide variety of spectra of these monosulfides have been studied. This includes S K_β and S K_α X-ray

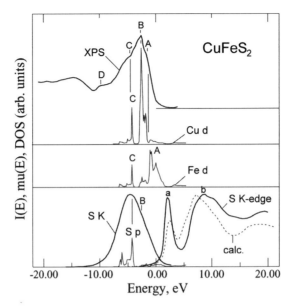

Figure 10. Comparison of experimental and theoretical data for chalcopyrite (CuFeS$_2$): Experimental XPS and S K$_\beta$ emission spectra from Tossell et al. (1983); experimental S K absorption spectrum from Petiau et al. (1988); calculated densities of states for Cu 3d, Fe 3d and S 3p electron states (cluster calculation, Lavrentyev et al. 2004). [Used with permission of Elsevier, from Lavrentyev et al. (2004), *J. Electron Spectroscopy and Related Phenomenon*, Vol. 137-140, Fig. 4, p. 497.]

emission spectra (Sugiura et al. 1974; Marusak and Tongson 1979), Fe K$_\beta$ X-ray emission spectra (Reuff et al. 1999; Gamblin and Urch 2001), XPS and UPS (Gopalakrishnan et al. 1979; Krishnakumar et al. 2002; Skinner et al. 2004), sulfur K and L$_{2,3}$ XANES and other XAS (Womes et al. 1997; Zajdel et al. 1999; Soldatov et al. 2004; Lavrentyev et al. 2004). The discussions of electronic structure have ranged from qualitative MO/band models based on observed properties (Goodenough 1967; Wilson 1972) to cluster calculations (e.g., Tossell 1977; Soldatov et al. 2004; Lavrentyev et al. 2004) and band structure calculations (Raybaud et al. 1997a,b; Krishnakumar et al. 2002).

The NiAs structure forms of FeS, CoS and NiS have been the subject of both experimental and computational studies. (Although high temperature forms, they can be retained at room temperature by quenching). Raybaud et al. (1997a) for example, using DFT band structure calculations, were able to make quite accurate predictions of the equilibrium crystal structures and of structural parameters such as cell volume and $c:a$ axial ratios. In the companion paper, Raybaud et al. (1997b) the same computational approaches were used to model the electronic structures of the monosulfides and in Figure 11 shown their calculated electronic densities of states for FeS, CoS and NiS. Here the total densities of states (DOS) are shown and the s, p and d partial DOS for each phase. The bands can be arranged into four distinct groups: (1) S 3s at ~ −15 to −14 eV, (2) S 3p at ~ −8 to −3 eV overlapping with (3) the transition metal 3d bands around the Fermi level, and (4) the conduction band at > 4 eV made up from transition metal 4s, p and sulfur 3d states. The same authors also studied the low temperature forms, troilite, millerite and Co$_9$S$_8$ using the same computational methods. In comparison with the NiAs form, they found slight broadening of all the bands and increased S 3p – Fe 3d overlap in the case of troilite. They attribute the stabilization of the troilite phase over the NiAs-type to an increase in the binding energies of some of the Fe 3d (e_g) states associated with the closer Fe-Fe coordination. The

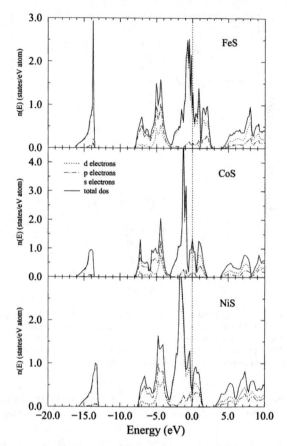

Figure 11. Calculated densities of states of the NiAs-structure FeS, CoS and NiS; solid line: total density of states; dotted, dashed and chain lines are *s*, *p* and *d* partial densities of states. [Used with permission of IOP Publishing Limited, from Raybaud et al. (1997), *J. Phys. Condens. Matter*, Vol. 9, Fig. 13, p. 11125.]

energetic stabilization of the millerite structure was similarly attributed to a slight shift to lower energies of certain Ni 3*d* (t_{2g}) states arising from the metal-metal interactions in the structure.

In considering other studies of the monosulfides, the results of calculations on an FeS_6^{10-} cluster (octahedrally coordinated high-spin Fe^{2+}; Tossell 1977) are presented and discussed elsewhere in this volume (Wincott and Vaughan 2006; see their Fig. 16), as is the simple band model for pyrrhotite based on these calculations and proposed by Sakkopoulos et al. (1984). Such a model is also used in the interpretation of spectroscopic data for a reacted pyrrhotite surface (Mikhlin et al. 1998) as discussed by Rosso and Vaughan (2006b, this volume; see their Fig. 33). Soldatov et al. (2004) used calculations on small clusters (19-37 atoms) employing a multiple scattering approach to compare with the results of their experiments recording the sulfur K and $L_{2,3}$ X-ray absorption near-edge structure (XANES) spectra of FeS, CoS and NiS. As can be seen from Figure 12 the agreement between experiment and theory was good in these studies. The electronic structures of the monosulfides were discussed on the basis of the calculations, with conclusions similar to those discussed above from the work of Raybaud et al. (1997b). For example, a systematic decrease in the separation between the sulfur 3*p* and the transition metal 3*d* bands in the series FeS-CoS-NiS (separations estimated at 3.52 eV (FeS),

3.50 eV (CoS) and 2.9 eV (NiS) by Soldatov et al. 2004) is regarded as a measure of increasing covalency across the series.

As noted above, NiS transforms on cooling below ~379 °C from the NiAs-type structure to the unique millerite structure in which nickel is in a NiS_5 square pyramidal coordination and sulfur is also in five-fold coordination (see Makovicky 2006; this volume). Millerite, with its very short metal-metal distances (~2.5 Å) has attracted interest as regards its electronic structure, in addition to that discussed above. Goodenough (1997) notes the metallic behavior and Pauli paramagnetism of millerite at room temperature, and the occurrence of another transition (T_N) at 264 K, below which millerite is a semimetallic antiferromagnet. Ikoma et al. (1995) and others have performed XPS experiments below this temperature (at 130 K). As discussed by Goodenough (1997), this must involve a transition from itinerant to more localized electron behavior. This transition is likely to be highly pressure dependant; indeed, millerite has been shown experimentally to be highly compressible compared with the NiAs-structure form (Sowa et al. 2004).

Krishnakumar et al. (2002) studied the electronic structure of millerite using XPS and UPS measurements and (DFT) band structure calculations (using the LMTO method and atomic sphere approximation). It was also found necessary to perform cluster calculations (using an NiS_5 cluster) that included electron correlation effects in order to model certain spectral features. Figure 13 shows the experimental valence band XPS spectrum from this work, along

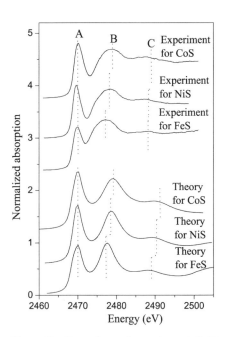

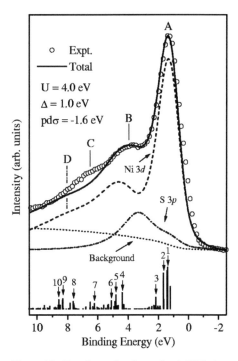

Figure 12. Comparison of experimental S K XANES spectra for FeS, CoS and NiS with theoretical spectra calculated using a cluster method. [Used with permission by IOP Publishing Ltd., from Soldatov et al. (2004), *J. Phys. Cond. Matt*, Vol. 16, Fig. 3, p. 7551.]

Figure 13. Experimental valence band XPS data (open circles) for millerite (NiS), along with the calculated spectrum (solid line) and the calculated Ni 3d and S 3p contributions and final states of the cluster calculation. [Reprinted with permission from Krishnakumar et al. (2002), *Phys. Rev. B*, Vol. 66, Fig. 5, p. 115105-5. © 2002 American Physical Society.]

with the calculated XPS spectrum and contributions to this spectrum from the Ni $3d$ and S $2p$ components. The detailed analysis of electronic structure provided by Krishnakumar et al. (2002) was summarized by them in describing millerite as a "highly covalent pd metal."

Pyrite and the related disulfides (FeS_2, CoS_2, NiS_2, CuS_2, FeAsS, $FeAs_2$)

Pyrite, with its low-spin Fe^{2+} octahedrally coordinated to S atoms in their dianion pairs, has been exhaustively studied experimentally and computationally. Much of the experimental evidence relevant to understanding its electronic structure is described elsewhere in this volume (Pearce et al. 2006; Wincott and Vaughan 2006). The data, which commonly extends to the isostructural MS_2 phases (where M is Co, Ni, Cu, and sometimes Zn) include optical reflectance spectra (Bither et al. 1968; Schlegel and Wachter 1976; Suga et al. 1983; Sato 1984; Ferrer et al. 1990; Huang et al 1993), photoemission spectra (Li et al. 1974; Ohsawa et al. 1974; van der Heide et al. 1980; Folmer et al. 1988; Fujimori et al. 1996; Bocquet et al. 1996) and X-ray emission and absorption spectra (Sugiura et al. 1976; Matsukawa et al. 1978; Mosselmans et al. 1995; Charnock et al. 1996; Lavrentyev et al. 2004; Prince et al. 2005).

A detailed discussion of qualitative MO and band models for pyrite and the isostructural Co, Ni and Cu disulfides has been presented earlier in this chapter. A large number of *ab initio* cluster and periodic calculations have been used to elucidate the electronic structure of these materials (Li et al. 1974; Tossell 1977; Bullett 1982; Folkerts et al. 1987; Temmerman et al. 1993; Fujimori et al. 1996; Raybaud et al. 1997; Eyert et al. 1998; Rosso et al. 1999; Gerson and Bredow 2000; Muscat et al. 2002; Edelbro et al. 2003). These calculations are generally in agreement with the each other and with the qualitative MO/band models for pyrite outlined above and illustrated in Figures 3a and 4. The pyrite band structure and total and partial densities of states from an LDA calculation of Eyert et al. (1998) are also illustrated here, in Figure 14. Thus, the top of the valence region is comprised of a narrow band of non-bonding Fe $3d$ t_{2g} electron states which lies a little above the main bonding band comprised of mixed S $3p$ and Fe $3d$ states (more specifically σ, π and π* S_2^{2-} $3p$ states and e_g Fe $3d$ states). The high degree of mixing found here between cation and anion states is indicative of the strongly covalent bonding interactions in pyrite. As regards the bottom of the conduction band, both the qualitative models and most computational studies attribute it to a mixed S $3p$ – Fe $3d$ band composed of σ* S $3p$ and e_g* Fe $3d$ orbitals. However, in their calculation, Eyert et al. (1998) suggested that the bottom of the conduction band comprises exclusively S $3p$ states, this being attributed to a larger splitting of S $3p$ states arising from strong S-S σ interaction. This study also found a weak π-bonding component between the Fe $3d$ t_{2g} and S $3p$ orbitals supporting an earlier proposal by Burns and Vaughan (1970) that the t_{2g} states are not completely non-bonding.

The band gap which separates the top of the valence band from the bottom of the conduction band in this semiconducting sulfide, and which has a well established experimental value of 0.9-0.95 eV, is not well reproduced computationally. Typically the value is underestimated (e.g., a calculated value of 0.6 eV; Raybaud et al. 1997) although Eyert et al. (1998) predicted a value of 0.9 eV in what is undoubtedly a fortuitous result. In a detailed evaluation of different computational approaches to the electronic properties of pyrite, Muscat et al. (2002) attribute the discrepancy between experimental and computational values of the band gap to large deviations from stoichiometry in this mineral. However, there is no evidence for non-stoichiometry in pyrite on the scale proposed by them, and these discrepancies are more likely due to well known failings of the present levels of theory used to perform the calculations, as described above.

The series of pyrite-structure disulfides FeS_2-CoS_2 (cattierite)-NiS_2 (vaesite)-CuS_2-ZnS_2 show interesting variations in electrical and magnetic properties (Pearce et al. 2006, this volume) and have been extensively studied experimentally and using various computational methods. Most of the references to experimental studies cited at the beginning of this section include work on these other pyrite-structure phases. Qualitative models of electronic structure for these phases are also outlined earlier in this chapter (see also Fig. 4). Elsewhere in this volume (Rosso and Vaughan 2006b) there is also some discussion of the electronic structure of the pyrite-structure RuS_2 (laurite), noting that its band structure is very similar to that of pyrite and describing, in some detail, the electronic structures of the (100), (111) and (210) surfaces of this important catalyst.

Qualitative models have also been discussed above for the marcasite (FeS_2) - arsenopyrite (FeAsS) – loellingite ($FeAs_2$) series of minerals, including ligand field and MO/band model approaches. Tossell et al. (1981) and Tossell (1984) have suggested that these models give

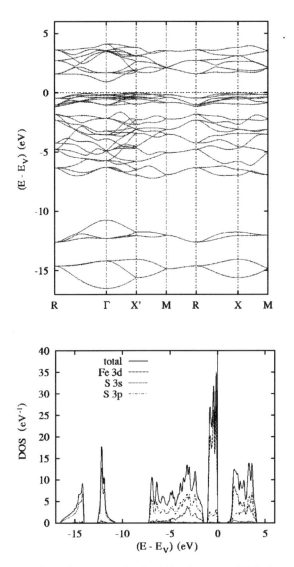

Figure 14. The calculated electronic structure of pyrite: (a) band structure of FeS_2 along selected symmetry lines within the first Brillouin zone of the simple cubic lattice; (b) total and partial densities of states. [Used with permission from Eyert et al. (1998), *Phys. Rev. B*, Vol. 57, Figs. 4 and 5, p. 6353. © 1998 American Physical Society.]

inadequate attention to the influence of the dianion electron distribution on the structures adopted in these compounds; they used MO calculations on the dianion units along with qualitative MO arguments to provide another interpretation. Here, attention is focused on the highest occupied MO of the electron donor and lowest unoccupied MO of the electron acceptor and their energies and overlaps. For example, the calculated MO scheme for S_2^{2-} has 14 electrons filling orbitals up to a $1\pi_g^*$ antibonding orbital (see Fig. 15) which can mix with metal d orbitals of σ symmetry to generate two pairs of orbitals, one oriented in the xz plane and the other in the yz plane (where z is the internuclear axis direction). Each pair consists of a metal-sulfur bonding orbital (π_b) stabilized relative to $1\pi_g^*$ and a destabilized antibonding orbital (π^*). In FeS_2, all of the π_b orbitals would be filled and all π^* empty. In $FeAs_2$, spectroscopic evidence suggests the iron may still be divalent and the dianion therefore As_2^{2-}, which would mean only 12 valence electrons, so that only one component of the π_b orbital set would be filled leading to the more distorted metal-anion coordination which is observed (see Fig. 15). Such arguments can be extended to other related dichalcogenides (listed in Table 1); for example, in FeAsS, the As end of the AsS group is effectively a "12 electron system" resulting in alternately greater or less distortion of the coordination of metals around the anion.

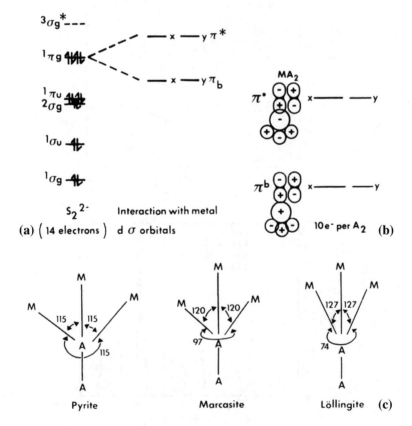

Figure 15. Models for the bonding in disulfides (after Tossell et al. 1981; Tossell and Vaughan 1992): (a) molecular orbital energy level diagram for the S_2^{2-} dianion and splitting of the highest energy orbital containing electrons on interaction with $M\,d\sigma$ orbitals; (b) perturbed molecular orbitals formed by mixing of $A_2\,\pi_g^*$ and $M\,d\sigma$; (c) geometries of M atoms about A in pyrite, marcasite, and loellingite structures. [Figures used by permission of Springer Verlag and Oxford Univeristy Press.]

Other (including complex) sulfides (pentlandites and metal-rich sulfides, thiospinels, layer structure sulfides, tetrahedrites)

The chemical bonding in a number of other sulfide minerals and mineral groups has been analyzed using both qualitative and quantitative models. Generally these minerals have been studied both because of their mineralogical importance, and because they present interesting challenges for theorists.

Pentlandites ((Ni,Fe)$_9S_8$ and Co_9S_8). The minerals pentlandite and cobalt pentlandite form a solid solution series and are interesting examples of metal-rich sulfides. In pentlandite, a cube cluster of tetrahedrally coordinated cations has very short metal-metal distances (e.g., Co-Co = 2.505 Å); 32 of the 36 metal atoms in the unit cell occupy the tetrahedral sites of the cube clusters, with the other four in octahedral sites between the clusters (see Makovicky 2006, this volume). Tossell and Vaughan (1992) reviewed qualitative MO/band models for Co_9S_8 and discussed the reported metallic conductivity and Pauli paramagnetism used to make inferences about electronic structure.

Burdett and Miller (1987) pointed out the inadequacies of a ligand field analysis of the pentlandites, going on to analyze the cube clusters using MO calculations. Chauke et al. (2002) performed *ab initio* calculations using DFT. They were able to rationalize the high stability of the Co_9S_8 (and $Fe_5Ni_4S_8$) stoichiometries in terms of their Fermi levels falling in a pseudo(band)gap corresponding to an average number of electrons per atom of 7.58. They also performed calculations on Co_8S_8 which demonstrated the importance of the additional (octahedral) metal atoms to the stability of the structure, and calculated equilibrium lattice parameters and heats of formation in good agreement with experiment (the former being 1% smaller than experiment, the latter being calculated as −91.5 kJ/mole versus an experimental value of −85.1 kJ/mole). Raybaud et al. (1997b) in their *ab initio* DFT calculations on Co_9S_8 also found that the Fermi level falls at a deep pseudogap between the lower part of the Co $3d$ band (which has only weak interaction with the S $3p$ band) and the upper part which shows appreciable hybridization with the S $3p\sigma^*$ states.

Other metal-rich sulfides (Ni_3S_2, Cu_2S, Ag_2S). A number of sulfides in addition to pentlandite form phases with a metal:sulfur ratio > 1. Heazlewoodite, Ni_3S_2, is a metallic conductor which exhibits Pauli paramagnetism, the properties of which have been studied in detail by Metcalf et al.(1994). As noted above, Gibbs et al. (2005) computed electron density distributions in Ni_3S_2 and analyzed them for bonded interactions, showing that Ni-Ni bond paths form contiguous, highly branched networks in this "Ni $3d$ metallic conductor". Raybaud et al. (1997) also studied this phase and report calculated (total, local and partial) densities of states. Their electronic structure shows an overall similarity to that of the millerite and NiAs-structure forms of NiS, with the Fermi level separating the Ni orbitals interacting with the S $3p\sigma^*$, π band complex from the Ni $3d$ band.

The binary copper sulfides are more complex structurally than their formulae suggest (see Makovicky 2006, this volume); in chalcocite (Cu_2S) copper occurs in two kinds of triangular coordination. Experimental investigations of the electronic structure of Cu_2S include studies of valence region X-ray emission and photoelectron spectra (Domashevskaya et al. 1976; Nakai et al. 1978; Folmer and Jellinek 1980) and of S K and SL edge XANES (Li et al. 1994). These experimental data have been interpreted using MO cluster calculations (Tossell and Vaughan 1981) which suggest that the top of the valence band is comprised of Cu $3d$ states, with the S $3p$ non-bonding levels being significantly more tightly bound.

The binary silver sulfide Ag_2S (acanthite) is also structurally complex (see Makovicky 2006, this volume); half of the silver atoms occur in a twofold, nearly linear, coordination and half in a distorted tetrahedral coordination. X-ray emission and XPS data are available for Ag_2S (e.g., Domashevskaya et al. 1976) and MO calculations have been performed for the AgS_2^{3-} and AgS_4^{7-} clusters (Tossell and Vaughan 1981). The calculations suggest that, in contrast to Cu_2S, the top of the valence band is comprised of S $3p$ non-bonding orbitals with, below them, the main Ag-S bonding orbitals, and the Ag $3d$ orbitals "buried"~ 4 eV beneath the top of the valence band. For further discussion of acanthite, chalcocite and related phases see Tossell and Vaughan (1981, 1992).

Thiospinels. The thiospinels, a group of sulfide minerals and synthetic compounds with the spinel structure have, therefore, both tetrahedrally and octahedrally coordinated metals in phases of the type $A^{tet}(B_2^{oct})S_4$ and with the possibilities of "formally" divalent and trivalent cations in both A and B sites. However, many of the thiospinels exhibit metallic behavior with valence electrons delocalized between tetrahedral and octahedral cations, so that formal valencies are not applicable. These materials have been well characterized as regards their electrical and magnetic properties but spectroscopic data are limited. In early qualitative work on bonding, Goodenough (1969) and Vaughan et al. (1971) were able to rationalize those properties and the solid solution limits found in the thiospinels using MO/band models. Such models were developed further by Vaughan and Tossell (1981) using MO calculations on appropriate clusters. The magnetic and Mössbauer parameters of greigite (Fe_3S_4) have also been discussed on the basis of similar (MS-SCF-X_α) MO calculations by Braga et al. (1988). A more detailed account of work on bonding in thiospinels is given in Tossell and Vaughan (1992).

Layer structure sulfides (MoS_2, CuS, FeS). Molybdenite (MoS_2), a diamagnetic, moderate band gap semiconductor, has been much studied by physicists because of its interesting properties and, as seen elsewhere in this volume, has been subjected to a wide range of spectroscopic techniques, both as a bulk material and with respect to its surface chemistry (see Wincott and Vaughan 2006; Rosso and Vaughan 2006b; this volume). A simple band model for MoS_2 has already been presented and discussed in the context of understanding the optical absorption spectrum (Wincott and Vaughan 2006). Many band structure calculations have been performed on MoS_2 and the earlier work has been reviewed by Calais (1977); later studies have involved both cluster (de Groot and Haas 1975; Harris 1982) and band structure (Coehoorn et al. 1987) calculations, some of which have also already been presented in the context of interpreting spectroscopic data. Molybdenite was one of the large number of sulfides studied using DFT calculations by Raybaud et al. (1997) who also provide a good summary of previous experimental and computational work. Their results confirm the importance of the ligand field splitting (in trigonal-prismatic coordination) of the transition metal d states, resulting from covalent Mo d – S p bonding interactions. The bulk, as well as the surface electronic structure of MoS_2 is discussed in some detail elsewhere in this volume, and the data of Raybaud et al. (1997) for the DOS of bulk MoS_2 are illustrated (Rosso and Vaughan 2006b; see their Fig. 38).

Covellite (CuS) has a more complex structure than its formula suggests, with Cu in both tetrahedral and triangular coordination and disulfide units in the structure. It is also a metallic conductor (see Pearce et al. 2006, this volume). Investigations of the electronic structure of covellite have been performed experimentally using K-edge and L-edge XANES (Li et al, 1994b), S K_β and S $L_{2,3}$ X-ray emission spectroscopy (Sugiura et al. 1974, Kurmaev et al. 1998), and X-ray photoelectron spectroscopy (Nakai et al. 1978; Folmer et al. 1980; Kurmaev et al. 1998). The spectroscopic data on CuS are discussed elsewhere in this volume (Wincott and Vaughan 2006). These spectra have been interpreted using MO calculations on appropriate cluster units, with the calculations then used to propose a simple band model for covellite (Vaughan and Tossell 1980, 1981); see Figure 16. Although calculations were initially performed on CuS_3^{4-} and CuS_4^{7-}, the metallic conductivity and spectroscopic evidence for essentially monovalent copper being present, led to the suggestion that charge should flow from the $4t_2$ orbital on the tetrahedral "Cu^+" to the $4e$ orbital on the triangular "Cu^{2+}". This was modeled by performing a cluster calculation on $CuS_4^{6.5-}$ and using this with the CuS_3^{5-} calculation to give the composite "one-electron" band model shown in Figure 16. More recent calculations using extended Hückel tight binding

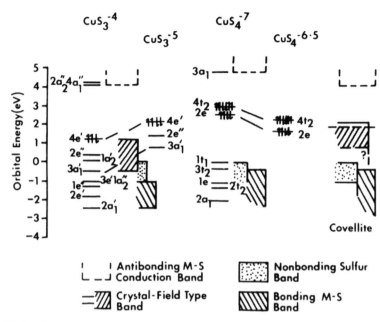

Figure 16. An electronic structure model for covellite (CuS) based on calculations using an MO cluster method (multiple scattering X_α). Discrete energy levels are shown for the clusters CuS_3^{4-}, CuS_3^{5-}, CuS_4^{7-}, $CuS_4^{6.5-}$ and a composite "one electron" band model energy level diagram for the mineral. (After Vaughan and Tossell 1980.)

methods (Liang and Whangbo 1993) and HF calculations on the periodic covellite structure (Rosso and Hochella 1999) largely confirm this model. One point of apparent difference is that these latter studies suggest that the top of the valence band is comprised predominantly of S $3p$ states, in contrast to the crystal field-like Cu $3d$ states suggested by the MO calculations. Formal charges in CuS appear to be best described by the formula $(Cu^+)_3(S^{2-})(S^{2-})$, where Cu in both tetrahedral and trigonal planar sites is approximately monovalent, the S^{2-} sites are the trigonal planar ligands, and the S_2^- sites are three of the four tetrahedral ligand sites (Nakai et al. 1978; Liang and Whangbo 1993; Rosso and Hochella 1999).

Mackinawite (FeS), the layer structured tetragonal iron monosulfide, is of great environmental importance because it is the first sulfide formed in many low T aqueous environments (see elsewhere in this volume; e.g., Rickard and Luther 2006). Data on many of the properties of this phase are lacking, as it occurs only as fine particles. However, Welz and Rosenberg (1987) performed band structure calculations on this form of FeS using a DFT (LMTO) method suggesting it to be metallic with conduction bands of mainly d-electron character. The calculations suggest that direct Fe-Fe interactions across the edge-sharing FeS_4 tetrahedra in the structure are responsible for a reduced density of states at the Fermi level and absence of magnetic ordering, even at very low temperature (see Wincott and Vaughan 2006; this volume).

Tetrahedrites. The minerals forming the complete solid solution series from tetrahedrite ($Cu_{12}Sb_4S_{13}$) to tennantite ($Cu_{12}As_4S_{13}$) are economically and environmentally important because of the numerous substitutions that can occur for Cu in the structure (this includes Zn, Fe, Cd, Hg, Ag). The crystal chemistry of this mineral family is complex. The structure of $Cu_{12}Sb_4S_{13}$ has half of the Cu atoms in tetrahedral and half in trigonal sites, Sb bonded to three S atoms with a "lone pair" of electrons extending in a fourth tetrahedral direction, and twelve of the S atoms in tetrahedral and the thirteenth S atom in octahedral coordination (see Makovicky,

2006, this volume). Considering just this three component endmember, there is also a variation in stoichiometry which accommodates Cu-rich phases with compositions bounded by the line $Cu_{14}Sb_4S_{13}$ – $Cu_{12}Sb_{4.67}S_{13}$ in the Cu-Sb-S system (Johnson and Jeanloz 1983). The substitutions in this mineral group have been extensively investigated using X-ray diffraction, and X-ray absorption and ^{57}Fe Mössbauer spectroscopies (see Wincott and Vaughan 2006, this volume). Despite the complexity of these materials, there have been qualitative and quantitative models presented to describe the electronic structure of compositions within the Cu-Sb-S system. Johnson and Jeanloz (1983) used concepts developed for the description of alloys to predict correctly compositional limits and electrical properties. In this work, the arguments center on phases being stable at a particular electron-to-atom ratio due to the filling of Brillouin zones with electrons. Bullett and Dawson (1986) and Bullett (1987) used non-empirical atomic orbital based techniques to calculate the band structure of Cu-Sb-S system tetrahedrites (see Fig. 17), showing that $Cu_{12}Sb_4S_{13}$ is intrinsically electron deficient. Calculations on the composition $Cu_{14}Sb_4S_{13}$ showed a filled valence band with an energy gap of 0.9 eV between highest filled and lowest empty states. It is suggested that in this "copper rich" variant, the two additional Cu atoms per formula unit are accommodated by displacing two tetrahedrally coordinated Cu atoms into a third kind of interstitial site. The calculated density of states for the different kinds of Cu atoms and displaced coppers are shown in Figure 17.

CONCLUDING REMARKS

Recent decades have seen remarkable advances in our attempts to calculate the electronic structures of minerals and related materials. It is now possible to use *ab initio* quantum mechanical calculations to predict structures and properties that, in many cases, show excellent agreement with experiment, and such calculations will surely be at the forefront of future work.

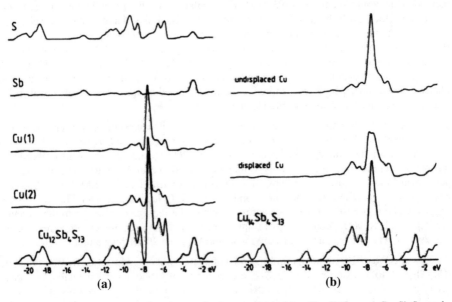

Figure 17. The calculated densities of states in the two tetrahedrites $Cu_{12}Sb_4S_{13}$ and $Cu_{14}Sb_4S_{13}$ and its local site projections for S, Sb, and various types of Cu atom. In the copper-rich phase the copper contribution for undisplaced and displaced (interstitial) Cu sites are compared. [Used with permission of Springer-Verlag from Bullet (1987), *Phys. Chem. Min.*, Vol. 14, Fig. 1, p. 486.]

There have been parallel advances in experimental methods, particularly those spectroscopic methods that can provide direct information on electronic structure, such as photoemission and X-ray absorption spectroscopies (see Wincott and Vaughan 2006, this volume). As the discussions presented in this chapter show, the sulfides are a diverse and complex group of materials in terms of their electronic structures. A range of both quantitative and qualitative approaches has proved valuable in our attempts to the understand structure and bonding in sulfides, and this is likely to continue. Qualitative models will continue to be of value in providing conceptual advances in understanding. However, these will be guided by new concepts, or by methods of analysis that provide much more rigorous definitions of older concepts such as "charge" or "covalency" used to describe sulfide minerals. Quantitative calculations using *ab initio* methods are being consistently incrementally improved, but larger leaps forward in this regard are already on the horizon. In the future, vastly improved accuracy is likely to come in the form of multiconfiguration or excited state electronic structure calculations.

ACKNOWLEDGMENTS

DJV acknowledges the contributions of the Natural Environment Research Council and the Engineering and Physical Sciences Research Council in supporting his research in the mineral sciences. DJV also wishes to thank Jack Tossell and Michele Warren for valuable discussions regarding sulfide electronic structure studies. Helen Weedon and Richard Hartley are thanked for help in the preparation of the manuscript. KMR acknowledges the support of the U.S. Department of Energy (DOE), Office of Basic Energy Sciences, Geosciences Division, and the Stanford Environmental Molecular Sciences Institute (EMSI) jointly funded by the National Science Foundation and the DOE Office of Biological and Environmental Research (OBER). The W. R. Wiley Environmental Molecular Science Laboratory (EMSL) at Pacific Northwest National Laboratory (PNNL) is a national scientific user facility sponsored by the OBER. PNNL is operated for the DOE by Battelle Memorial Institute under contract DE-AC06-76RLO 1830.

REFERENCES

Apra E, Bylaska EJ, de Jong W, Hackler MT, Hirata S, Pollack L, Smith DMA, Straatsma TP, Windus TL, Harrison RJ, Nieplocha J, Tipparaju V, Kumar M, Brown E, Cisneros G, Dupuis M, Fann GI, Fruchtl H, Garza J, Hirao K, Kendall R, Nichols JA, Tsemekhman K, Valiev M, Wolinski K, Anchell J, Bernholdt D, Borowski P, Clark T, Clerc D, Dachsel H, Deegan M, Dyall K, Elwood D, Glendening E, Gutowski M, Hess A, Jaffe J, Johnson B, Ju J, Kobayashi H, Kutteh R, Lin Z, Littlefield R, Long X, Meng B, Nakajima T, Niu S, Rosing M, Sandrone G, Stave M, Taylor H, Thomas G, van Lenthe J, Wong A, Zhang Z (2003) NWChem: A computational chemistry package designed to run on high-performance parallel supercomputers, version 4.5. Pacific Northwest National Laboratory

Bader RFW (1990) Atoms in Molecules. Oxford Science Publications

Bader RFW (1998) A bond path: a universal indicator of bonded interactions. J Phys Chem A 102:7314-7323

Bauschlicher CW Jr., Maitre P (1995) Theoretical study of the first transition row oxides and sulfides. Theorchim Acta 90:189-203

Becke AD (1993) A new mixing of Hartree-Fock and local density functional theories. J Chem Phys 98:1372-1377

Becke AD (1998) A new inhomogeneity parameter in density-functional theory. J Chem Phys 109:2092-2098

Becker U, Rosso KM (2001) Step edges on galena (100): Probing the basis for defect driven surface reactivity at the atomic scale. Am Mineral 86:862-870

Bernard JE, Zunger A (1987) Electronic structure of ZnS, ZnSe, ZnTe and their pseudobinary alloys. Phys Rev B 36:3199-3228

Bither TA, Bouchard RJ, Cloud WH, Donohue PC, Siemons WJ (1968) Transition metal pyrite dichalcogenides. High pressure synthesis and correlation of properties. Inorg Chem 7:2208-2220

Bocquet AE, Mamiya K, Mizokawa T, Fujimori A, Miyadai T, Takahashi H, Mori M, Suga S (1996) Electronic structure of 3d transition metal pyrites MS_2 (M = Fe, Co or Ni) by analysis of the M 2p core-level photoemission spectra. J Phys Condens Matt 8:2389-2400

Braga M, Lie SK, Taft CA, Lester WA Jr. (1988) Electronic structure, hyperfine interaction, and magnetic properties for iron octahedral sulfides. Phys Rev B 38:10837-10851
Bullett DW (1982) Electronic structure of 3d pyrite- and marcasite-type sulfides. J Phys C 15:6163-6174
Bullet DW (1987) Applications of atomic-orbital methods to the structure and properties of complex transition-metal compounds. Phys Chem Mineral 14: 485-491
Bullet DW, Dawson WG (1986) Bonding relationships in some ternary and quaternary phosphide and tetrahedrite structures ($Ag_6M_4P_{12})M_6$', $Cu_{12+x}Sb_4S_{13}$, $Cu_{14-x}Sb_4S_{13}$, $Ln_6Ni_6P_{17}$. J Phys C 19:5837-5847
Burdett JK, Miller GJ (1987) Polyhedral clusters in solids: the electronic structure of pentlandite. J Am Chem Soc 109:4081-4091
Burns RG (1993) Mineralogical Applications of Crystal Field Theory. (2nd edition) Cambridge Univ. Press
Burns RG, Vaughan DJ (1970) Interpretation of the reflectivity behavior of ore minerals. Am Mineral 55:1576-1586
Car R, Parrinello M (1985) Unified approach for molecular dynamics and density functional theory. Phys Rev Lett 55:2471-2244
Calais JL (1977) Band structure of transition metal compounds. Adv Phys 26:847-885
Catlow CRA (1997) Need and scope of modelling techniques. *In* Computer Modelling in Inorganic Crystallography. Catlow CRA (ed). Academic Press, p. 1-22
Cardona M, Greenaway DL (1964) Optical properties and band structure of group IV-VI and group V materials. Phys Rev 133:1685-1697
Charnock JM, Henderson CMB, Mosselmans JFW, Pattrick RAD (1996) 3d transition metal L-edge X-ray absorption studies of the dichalcogenides of Fe, Co and Ni. Phys Chem Mineral 23:403-408
Chauke HR, Nguyen-Manh D, Ngoepe PE, Pettifor DG, Fries SG (2002) Electronic structure and stability of the pentlandites Co_9S_8 and $(Fe,Ni)_9S_8$. Phys Rev B 66:155105
Coehoorn R, Haas C, Dijkstra J, Flipse CJF, De Groot RA, Wold A (1987) Electronic structure of molybdenum diselenide, molybdenum disulfide and tungsten diselenide. I. Band-structure calculations and photoelectron spectroscopy. Phys Rev B 35:195-202
Cygan RT, Kubicki JD (eds) (2001) Molecular Modeling Theory: Applications in the Geosciences. Reviews in Mineralogy and Geochemistry. Vol. 42. Mineralogical Society of America
de Groot RA, Hass C (1975) Multiple Scattering theory-$X\alpha$ .Calculations on molybdenum disulfide and some related compounds. Solid State Comm 17:887-890
Domashevskaya EP, Terekhov, V.A, Marshakova LN, Ugai Ya A, Nefedov VI, Sergushin NP (1976) Participation of d-electrons of metals of groups I, II, and III in chemical bonding with sulfur. J Elec Spec Rel Phen 9: 261-267
Edelbro R, Sandstrom A, Paul J (2003) Full potential calculations on the electron bandstructures of sphalerite, pyrite and chalcopyrite. Appl Surf Sci 206: 300-313
Eyert V, Hock K-H, Fiechter S, Tributsch H (1998) Electronic structure of FeS_2: The crucial role of electron-lattice interaction. Phys Rev B 57:6350-6359
Farberovich, OV, Kurganskii S I Domashevskaya EP (1980) Problems of the OPW method. II. Calculation of the band structure of zinc sulfide and cadmium sulfide. Phys Status Solidi B 97:631-640
Ferrer IJ, Nevskaia DM, de las Herras C, Sanchez C (1990) About the band gap nature of FeS_2 as determined from optical and photoelectrochemical measurements. Sol State Comm 74:913-916
Fleet ME (2006) Phase equilibria at high temperatures. Rev Mineral Geochem 61:365-419
Folkerts W, Sawatzky GA, Haas C, de Groot RA, Hillebrecht FU (1987) Electronic structure of some 3d transition-metal pyrites. J Phys C 20:4135-4144
Folmer JCW, Jellinek F (1980) The valence of copper in sulfides and selenides: An X-ray photoelectron spectroscopy study. J Less Comm Met 76:153-162
Folmer JCW, Jellinek F, Calis GHM (1988) The electronic structure of pyrites, particularly copper disulfide and iron copper selenide ($Fe_{1-x}Cu_xSe_2$): An XPS and Mössbauer study. J Sol Stat Chem 72:137-144
Fujimori A, Mamiya K, Mizokawa T, Miyadai T,.; Sekiguchi T, Takahashi H, Mori N, Suga S (1996) Resonant photoemission study of pyrite-type NiS_2, CoS_2 and FeS_2. Phys Rev B 54:16329-16332
Fujisawa M, Suga S, Mizokawa T, Fujimori A, Sato K (1994) Electronic structures of $CuFeS_2$ and $CuAl_{0.9}Fe_{0.1}S_2$ studied by electron and optical spectroscopies. Phys Rev B 49:7155-7164
Gale JD (2001) Simulating the crystal structures and properties of ionic materials from interatomic potentials. Rev Mineral Geochem 42:37-62
Gamblin SD, Urch DS (2001) Metal $K\beta$ X-ray emission spectra of first row transition metal compounds. J Elect Spect Rel Phenom 113:179-192
Gerson AR, Bredow T (2000) Interpretation of sulfur 2p XPS spectra in sulfide minerals by means of ab initio calculations. Surf Int Anal 29:145-150
Gibbs GV, Tamada O, Boisen MB Jr, Hill FC (1999) Laplacian and bond critical point properties of the electron density distributions of sulfide bonds: A comparison with oxide bonds. Am Mineral 84:435-446

Gibbs GV, Downs RT, Prewitt CT, Rosso KM, Ross NL, Cox DF (2005) Electron density distributions calculated for the nickel sulfides millerite, vaesite, and heazlewoodite and nickel metal: A case for the importance of Ni-Ni bond paths for electron transport. J Phys Chem B 109:21788-21795

Goodenough JB (1967) Description of transition metal compounds. Application to several sulfides. Coll Int Centre Nat Recherche Sci 157:263-90, discussion 290-292

Goodenough JB (1969) Descriptions of outer d-electrons in thiospinels. J Phys Chem Solids 30:261-280

Goodenough JB (1972) Energy bands in TX_2 compounds with pyrite, marcasite and arsenopyrite structures. J Solid State Chem 5:144-152

Goodenough JB (1997) Localized-itinerant electronic transitions in oxides and sulfides. J Alloys Comp 262: 1-9

Gopalakrishnan J, Murugesan T, Hegde MS, Rao CNR (1979) Study of transition metal monosulfides by photoelectron spectroscopy. J Phys C 12:5255-5261

Grandke T, Ley L, Cardona M (1978) Angle resolved UV photoemission and electronic band structures of the lead chalcogenides. Phys Rev B 18:3847-3871

Gurin VS (1998) Observation and simulation of PbS nanocrystal formation at the initial steps. Macromolecular Symposia 136, 2nd International Conference on Chemistry of Highly-Organized Substances and Scientific Principles of Nanotechnology, 13-16

Hamad S, Cristol S, Catlow CRA (2002) Surface structures and crystal morphology of ZnS: A computational study. J Phys Chem 106:11002-11008

Hamajima T, Kambara T, Gondaira KI, Oguchi T(1981) Self-consistent electronic structures of magnetic semiconductors by a discrete variational X_α calculation III. Chalcopyrite $CuFeS_2$. Phys Rev B 24:3349-3353

Harris S (1982) Study of the electronic structure of first and second row transition metal sulfides using SCF-SW-$X\alpha$ cluster calculations. Chem Phys 67:229-237

Hemstreet LA Jr. (1975) Cluster calculations of the effect of single vacancies of the electronic properties of lead(II) sulfide. Phys Rev B 11:2260-2270

Hobbs D, Hafner J (1999) Magnetism and magneto-structural effects in transition-metal sulfides. J Phys Cond Matt 11:8197-8222

Huang YS, Huang JK, Tsay MY (1993) An electroreflectance study of FeS_2. J Phys Cond Matt 5: 7827-7836

Hulliger F (1968) Crystal chemistry of the chalcogenides and pnictides of the transition elements. Struct Bond 4:83-229

Ikoma H, Matoba M, Mikami M, Anzai S (1995) Effect of 4d transition metal atom Rh doping on thermoelectric power, magnetic susceptibility, thermal expansion and X-ray photoemission spectra in the charge transfer type non-metallic state of NiS. J Phys Soc Japan 64:2600-2608

Johnson ML, Jeanloz R (1983) A Brillouin-zone model for compositional variation in tetrahedrite. Am Mineral 68:220-226

Kjekshus A, Nicholson DG (1971) The significance of π back bonding in compounds with the pyrite, marcasite and arsenopyrite type structures. Acta Chem Scand 25:866-876

Krishnakumar SR, Shanthi N, Sarma DD (2002) Electronic structure of millerite NiS. Phys Rev B 66:115105

Kurmaev EZ, van Ek J, Ederer DL, Zhou L, Callcott TA, Perera RCC, Chernashenko VM, Shamin SN, Tromifova VA, Bartkowski S, Neumann M, Fujimori A, Moloshag VP (1998) Experimental and theoretical investigation of the electronic structure of transiotion metal sulphides: CuS, FeS_2 and $FeCuS_2$. J Phys Cond Matt 10:1687-1697

Lavrentyev AA, Gabrelian BV, Nikiforov IYa, Rehr JJ, Ankudinov AL (2004) The electron energy structure of some sulfides of iron and copper. J Elec Spec Rel Phenom 137-140:495-498

Lawniczak-Jablonska K, Iwanowski RJ, Golacki Z, Traverse A, Pizzini S, Fontaine A, Winter I, Hormes J (1996) Local electronic structure of ZnS and ZnSe doped by Mn, Fe, Co, and Ni from X-ray absorption near-edge structure studies. Phys Rev 53:1119-1128

Laihia R, Leiro, JA, Kokko K, Mansikka K (1996) The X-ray $K\beta_{2,5}$ emission band and the electronic structure of Zn, ZnS and Znse crystals. J Phys Cond Matt 8:6791-6801

Laihia R, Kokko K, Hergert W, Leiro JA (1998) K-emission spectra of Zn, ZnS, and ZnSe within dipole and quadrupole approximations. Phys Rev B 58:1272-1278

Lee CT, Yang WT, Parr RG (1988) Development of the Colle-Salvetti correlation energy formula into a functional of the electron density. Phys Rev B 37:785-789

Leiro JA, Laajalehto K, Kartio I, Heinonen MH (1998) Surface core-level shift and phonon broadening in PbS (100). Surf Sci 412/413, L918-L923

Ley L, Pollak RA, McFeely FR, Kowalczyk SP, Shirley DA (1974) Total valence-band densities of states of [Groups] III-V and II-VI compounds from X-ray photoemission spectroscopy. Phys Rev B 9:600-621

Li EK, Johnson KH, Eastman DE, Freeouf JL (1974) Localized and bandlike valence-electron states in iron sulfide (FeS_2) and nickel sulfide (NiS_2). Phys Rev Lett 32:470-472

Li, D, Bancroft GM, Kasrai M, Fleet ME, Feng XH, Tan KH, Yang BX (1994a) Sulfur K- and L-edge XANES and electronic structure of zinc, cadmium and mercury monosulfides: a comparative study. J Phys Chem Solids 55:535-543

Li D, Bancroft GM, Kasrai M, Fleet ME, Feng XH, Yang BX, Tan KH (1994b) S K-edge and L-edge XANES and electronic structure of some copper sulfide minerals. Phys Chem Mineral 21:317-324

Li D, Bancroft GM, Kasrai M, Fleet ME, Yang BX, Feng XH, Tan K, Peng M (1994c) Sulfur K- and L-edge X-ray absorption spectroscopy of sphalerite, chalcopyrite and stannite. Phys Chem Mineral 20:489-499

Liang W, Whangbo MH (1993) Conductivity anisotropy and structural phase transition in covellite CuS. Solid State Comm 85:405-408

Lutz HD, Zminscher J (1996) Lattice dynamics of pyrite FeS_2 polarizable-ion model. Phys Chem Mineral 23:497-502

Ma J-X, Jia Y, Song Y-L, Liang E-J, Wu L-K, Wang F, Wang X-C, Hu X (2004) The geometric and electronic properties of the PbS, PbSe and PbTe (001) surfaces. Surf Sci 551:91-98

Makovicky E (2006) Crystal structures of sulfides and other chalcogenides. Rev Mineral Geochem 61:7-125

Martins JL, Troullier N, Wei SH (1991) Pseudopotential planewave calculations for zinc sulfide. Phys Rev B 43:2213-2217

Marusak LA, Tongson LL (1979) Soft X-ray emission and Auger electron spectroscopic study of iron(II) sulfide, iron sulfides $Fe_{0.9}S$, $Fe_{0.875}S$, and $Fe_{0.5}S$. J Appl Phys 50:4350-4355

Matsukawa T, Obashi M, Nakai S, Sugiura C (1978) The K absorption spectra of FeS_2, CoS_2 and NiS_2. Jap J Appl Phys 17 (Suppl 17-2) 184-186

McFeely FR, Kowalczyk S, Ley L, Pollak RA, Shirley DA (1973) High-resolution X-ray photoemission spectra of lead sulfide, lead selenide, and lead telluride valence bands. Phys Rev B 7:5228-2237

Merkel S, Jephcoat AP, Shu J, Mao H-K, Gillet P, Hemley RJ (2002) Equation of state, elasticity and shear strength of pyrite under high pressure. Phys Chem Mineral 29:1-9

Mian M, Harrison NM, Saunders VR, Flavell WR (1996) An ab initio Hartree-Fock investigation of galena (PbS). Chem Phys Lett 257:627-632

Mikhlin Y, Tomashevich Y, Pashkov GL, Okotrub AV, Asanov IP, Mazalov LN (1998) Electronic structure of the non-equilibrium iron-deficient layer of hexagonal pyrrhotite. Appl Surf Sci 125:73-84

Mikhlin Y, Tomashevich Y, Tauson V, Vyalikh D, Molodtsov S, Szargan R (2005) A comparative X-ray absorption near-edge structure study of bornite Cu_5FeS_4, and chalcopyrite, $CuFeS_2$. J Elec Spec Rel Phenom 142:83-88

Mosselmans JFW, Pattick RAD, van der Laan G, Charnock JM, Vaughan DJ, Henderson CMB, Garner CD (1995) X-ray absorption near-edge spectra of transition metal disulfides FeS_2 (pyrite and marcasite), CoS_2, NiS_2 and CuS_2, and their isomorphs FeAsS and CoAsS. Phys Chem Mineral 22:311-317

Muscat J, Klauber C (2001) A combined ab initio and photoelectron study of galena (PbS). Surf Sci 491:226-238

Muscat J, Hung A, Russo S, Yarovsky I (2002) First-principles studies of the structural and electronic properties of pyrite FeS_2. Phys Rev B 65:054107

Nakai I, Sugitani Y, Nagashima K, Niwa Y (1978) X-ray photoelectron spectroscopic study of copper minerals. J Inorg Nucl Chem 40:789-791

Nickel EH (1968) Structural stability of minerals with the pyrite, marcasite, arsenopyrite and loellingite structures. Can Mineral 9:311-321

Nickel EH (1970) The application of ligand field concepts to an understanding of the structural stabilities and solid solution limits of sulfides and related minerals. Chem Geol 5:233-241

Ohsawa A Yamamoto H, Watanabe H (1974) X-ray photoelectron spectra of valence electrons in iron disulfide, cobalt disulfide, and nickel disulfide. J Phys Soc Jap 37:568

Ollonqvist T, Kaurila T, Isokallio M, Punkkinen M, Vayrynen J (1995) Inverse photoemission and photoemission spectra of the PbS (001) surface. J Elec Spec Rel Phenom 76:729-734

Pantelides S, Harrison WA (1975) Structure of the valence band of zincblende-type semiconductors. Phys Rev B 11:3006-3011

Pattrick RAD, Mosselmans JFW, Charnock JM (1998) An X-ray absorption study of doped sphalerites. Euro J Mineral 10:239-249

Pearce CI, Pattrick RAD, Vaughan DJ (2006) Electrical and Magnetic properties of sulfides. Rev Mineral Geochem 61:127-180

Pearson WD (1965) Compounds with the marcasite structure. Z Kristallogr 121:449-462

Perdew JP, Wang Y (1992) Accurate and simple analytic representation of the electron gas correlation energy. Phys Rev B 45:13244-13249

Perez-Dieste V, Crain JN, Kirakosian A, McChesney JL, Arenholz E, Young AT, Denlinger JD, Ederer DL, Callcott TA, Lopez-Rivera SA, Himpsel FJ (2004) Unoccupied orbitals of 3d transition metals in ZnS. Phys Rev B 70:085205

Prince KC, Matteucci M, Kuepper K, Chiuzbaian SG, Bartkowski S, Neumann M (2005) Core-level spectroscopic study of FeO and FeS_2. Phys Rev B 71:085102.

Rabii S, Lasseter RH (1974) Band structure of PbPo and trends in the Pb chalcogenides. Phys Rev Lett 33:703-704

Raybaud P, Kresse G, Hafner J, Toulhoat H (1997) Ab initio density functional studies of transition metal sulfides; I. Crystal structure and cohesive properties. J Phys Cond Matt 9:11085-11106
Raybaud P, Hafner J, Kresse G, Toulhoat H (1997) Ab initio density functional studies of transition metal sulfides; II. Electronic structure. J Phys Cond Matt 9:11107-11140
Rickard D, Luther GW III (2006) Metal sulfide complexes and clusters. Rev Mineral Geochem 61:421-504
Rohrbach A, Hafner J, Kresse G (2003) Electronic correlation effects in transition-metal sulfides. J Phys Cond Matt 15:979-996
Rosso KM (2001) Structure and reactivity of semiconducting mineral surfaces: Convergence of molecular modeling and experiment. Rev Mineral Geochem 42:199-271
Rosso KM, Becker U, Hochella MF (1999) A UHV STM/STS and ab initio investigation of covellite (001) surfaces. Surf Sci 423:364-374
Rosso KM, Hochella MF (1999) Atomically resolved electronic structure of pyrite {100} surfaces: An experimental and theoretical investigation with implications for reactivity. Am Mineral 84:1535-1548
Rosso KM, Vaughan DJ (2006a) Reactivity of sulfide mineral surfaces. Rev Mineral Geochem 61:557-607
Rosso KM, Vaughan DJ (2006b) Sulfide mineral surfaces. Rev Mineral Geochem 61:505-556
Rueff J-P, Kao C-C, Struzhkin VV, Badro J, Shu J, Hemley RJ, Mao H K (1999) Pressure-induced high-spin to low-spin transition in FeS evidenced by X-ray emission spectroscopy. Phys Rev Lett 82:3284-3287
Saitoh T, Bocquet AE, Mizokawa T, Fujimori A (1995) Systematic variation of the electronic structure of 3d transition-metal compounds. Phys Rev 52:7934-7938
Sakkopoulos S, Vitoratos E, Argyreas T (1984) Energy-band diagram for pyrrhotite. J Phys Chem Solids 45: 923-928
Santoni A, Paolucci G, Santoro G, Prince KC, Christensen NE (1992) Band structure of lead sulphide. J Phys Cond Matt 4:6759-6768
Satta A, de Gironcoli S (2000) Surface structure and core-level shift in lead chalcogenide (001) surfaces. Phys Rev B 63:033402
Saunders VR, Dovesi R, Roetti C, Orlando R, Zicovich-Wilson CM, Harrison NM, Doll K, Civalleri B, Bush IJ, D'Arco P, Llunell M (2003) CRYSTAL03. University of Torino
Schoolar RB, Dixon JR (1965) Optical constants of lead sulfide in the fundamental absorption edge region. Phys Rev 137:667-670
Schroer P, Kruger P, Pollmann J (1993) First-principles calculation of the electronic structure of the wurtzite semiconductors ZnO and ZnS. Phys Rev 47:6971-6980
Schlegel A, Wachter P(1976) Optical properties, phonons, and electronic structure of iron pyrite (FeS_2). J Phys C 9: 3363-3369
Shannon RD (1981) Bond distances in sulfides and a preliminary table of sulphide crystal radii. In: Structure and Bonding in Crystals II. O'Keefe M, Navrotsky A (eds) Academic Press, p. 53-70
Sithole HM, Ngoepe PE, Wright K (2003) Atomistic simulation of the structure and elastic properties of pyrite (FeS_2) as a function of pressure. Phys Chem Mineral 30:615-619
Skinner WM, Nesbitt HW, Pratt AR (2004) XPS identification of bulk hole defects and itinerant Fe 3d electrons in natural troilite (FeS). Geochim et Cosmochim Acta 68:2259-2263
Soldatov AV, Kravtsova AN, Fleet ME, Harmer SL (2004) Electronic structure of MeS (Me = Ni, Co, Fe): X-ray absorption analysis. J Phys Cond Matt 16:7545-7556
Sowa H, Ahsbahs H, Schmitz W (2004) X-ray diffraction studies of millerite NiS under non-ambient conditions. Phys Chem Mineral 31:321-327
Stadele M, Moukara M, Majewski JA, Vogl P, Gorling A (1999) Exact exchange Kohn-Sham formalism applied to semiconductors. Phys Rev B 59:10031-10043
Stukel DJ, Euwema RN, Collins TC, Herman F, Kortum RL (1969) Self-consistent orthogonalised planewave and empirically refined orthogonalised planewave energy band models for cubic ZnS, ZnSe, CdS and CdSe. Phys Rev 179:740-751
Suga S, Inoue K, Taniguchi M, Shin S, Seki M, Sato K, Teranishi T (1983) Vacuum ultraviolet reflectance spectra and band structures of pyrites (FeS_2, CoS_2 and NiS_2) and NiO measured with synchrotron radiation. J Phys Soc Jap 52:1848-1856
Sugiura C (1994) Lα x-ray emission spectra and electronic structures of zinc and its compounds. J Phys Soc Jap 63:3763-3774
Sugiura C, Gohshi Y, Suzuki I (1974) Sulfur K_β X-ray emission spectra and electronic structures of some metal sulfides. Phys Rev B 10:338-343
Sugiura C, Suzuki I, Kashiwakura J Gohshi Y (1976) Sulfur K_β X-ray emission bands and valence-band structures of transition-metal disulfides. J Phys Soc Jap 40:1720-1724
Sugiura C, Yorikawa H, Muramatsu S (1997) Sulfur K_β X-ray emission spectra and valence-band structures of metal sulfides. J Phys Soc Jap 66:503-504
Svane A, Santi G, Szotek Z, Ternmerman WM, Strange P, Horne M, Vaitheeswaran G, Kandiana V, Petit L, Winter H (2004) Electronic structure of Sm and Eu chalcogenides. Phys Stat Sol B 241:3185-3192

Temmerman WM, Durham PJ, Vaughan DJ (1993) The electronic structures of the pyrite-type disulfides (MS_2, where M = manganese, iron, cobalt, nickel, copper, zinc) and the bulk properties of pyrite from local density approximation (LDA) band structure calculations. Phys Chem Mineral 20:248-54

Tossell JA (1977a) Theoretical studies of valence orbital binding energies in solid zinc sulfide, zinc oxide, and zinc fluoride. Inorg Chem 16:2944-2949

Tossell JA (1977b) SCF-X_α scattered wave MO studies of the electronic structure of ferrous iron in octahedral coordination with sulfur. J Chem Phys 66:5712-5719

Tossell JA (1984) A reinterpretation of the electronic structures of $FeAs_2$ and related minerals. Phys Chem Mineral 11:75-80

Tossell JA, Vaughan DJ (1981) Relationships between valence orbital binding energies and crystal structures in compounds of copper, silver, gold, zinc, cadmium, and mercury. Inorg Chem 20:3333-3340

Tossell JA, Vaughan DJ (1987) Electronic structure and the chemical reactivity of the surface of galena. Can Mineral 25:381-392

Tossell JA, Vaughan DJ (1992) Theoretical Geochemistry: Application of Quantum Mechanics in the Earth and Mineral Sciences. Oxford Univ Press

Tossell JA, Vaughan DJ, Burdett JK (1981) Pyrite, marcasite and arsenopyrite type minerals: Crystal chemical and structural principles. Phys Chem Mineral 7:177-184

Tossell JA, Urch DS, Vaughan DJ, Wiech G (1982) The electronic structure of $CuFeS_2$, chalcopyrite, from x-ray emission and x-ray photoelectron spectroscopy and X_α calculations. J Chem Phys 77:7-82

Tung YW, Cohen ML (1969) Relativistic band structure and electronic properties of tin telluride, germanium telluride, and lead telluride. Phy Rev 180:823-826

van der Heide H, Hemmel R, Van Bruggen CF, Haas C (1980) X-ray photoelectron spectra of 3d transition metal pyrites. J Solid State Chem 33:17-25

Vaughan DJ, Craig JR (1978) Mineral Chemistry of Metal Sulfides. Camb Univ Press

Vaughan DJ, Tossell JA (1980) The chemical bond and the properties of sulphide minerals: I. Zn, Fe and Cu in tetrahedral and triangular coordinations with sulfur. Can Mineral 18:157-163

Vaughan DJ, Tossell JA (1981) Electronic structure of thiospinel minerals: Results from MO calculations. Am Mineral 66:1250-1253

Vaughan DJ, Burns RG, Burns VM (1971) Geochemistry and bonding of thiospinel minerals. Geochim Cosmochim Acta 35:365-381

Vaughan DJ, Tossell JA, Johnson KH (1974) The bonding of ferrous iron to sulfur and oxygen: A comparative study using SCF-X_α scattered wave molecular orbital calculations. Geochim Cosmochim Acta 38:993-1005

von Oertzen GU, Jones RT, Gerson AR (2005) Electronic and optical properties of Fe, Zn and Pb sulfides. Phys Chem Mineral 32:255-268

Welz D, Rosenberg M (1987) Electronic band structure of tetrahedral iron sulfides. J Phys C 20:3911-3924

Wilson JA (1972) Systematics of the breakdown of Mott insulation in binary transition metal compounds. Adv Phys 21:143-198

Wincott PL, Vaughan DJ (2006) Spectroscopic studies of sulfides. Rev Mineral Geochem 61:181-229

Wintenberger M (1962) Etude électrique et magnétique de composés sulfurés et arséniés d'éléments de transition. III Propriétés électriques et magnétiques et liaisons dans l'arsenopyrite, la cobaltite et la loellingite. Bull Soc Fr Mineral 85:107-119

Womes M, Karnatak RC, Esteva JM, Lefebvre I, Allan G, Olivier-Fourcade J, Jumas JC (1997) Electronic structures of FeS and FeS_2: X-ray absorption spectroscopy and band structure calculations. J Phys Chem Solids 58: 345-352

Wright KV, Jackson RA (1995) Computer simulation of the structure and defect properties of zinc sulfide. J Mat Chem 5:2037-2040

Wright KV, Gale JD (2004) Interatomic potentials for the simulation of the zinc-blende and wurtzite forms of ZnS and CdS: bulk structure, properties, and phase stability. Phys Rev B 70:035211

Wright KV, Watson GW, Parker SC, Vaughan DJ (1998) Simulation of the structure and stability of sphalerite (ZnS) surfaces. Am Mineral 83:141-146

Zadjel P, Kisiel A, Zimnal-Starnawska M, Lee PM, Boscherini F, Giriat W (1999) XANES study of sulphur K edges of transition metal (V, Cr, Mn, Fe, Co, Ni) monosulphides: experiment and LMYO numerical calculations. J Alloys Comp 286:66-70

Zeng H, Schelly ZA, Ueno-Noto K, Marynick DS (2005) Density functional study of the structure of lead sulfide clusters $(PbS)_n$ ($n = 1 - 9$). J Phys Chem A 109:1616-1620.

Thermochemistry of Sulfide Mineral Solutions

Richard O. Sack

OFM Research
28430 NE 47th Place
Redmond, Washington, 98053-8841, U.S.A.
e-mail: fahlore@centurytel.net or rosack@ofm-research.org

Denton S. Ebel

Department of Earth and Planetary Sciences
American Museum of Natural History
Central Park West at 79th Street
New York, New York, 10024-5192, U.S.A.
e-mail: debel@amnh.org or dsebel@ofm-research.org

INTRODUCTION

In the sulfide mineral assemblages commonly found in terrestrial rocks, the major elements are S with the metals Fe, Zn, Cu, Pb, and Ag and semimetals As and Sb. Minor elements include Te, Se; the metals Ni, Cd, Co, Mn and the semimetals Hg and Bi. Particular ore deposits can be characterized using specific subsystems containing these elements (e.g., McKinstry 1963). Previous reviews have summarized in detail the phase relations in binary, ternary, and higher-order systems. These include Barton (1970), the first volume in this series (Craig and Scott 1976), excellent chapters by Barton and Skinner (1967, 1979), and Vaughan and Craig (1997) in Barnes' valuable compendium, and particularly the comprehensive volume by Vaughan and Craig (1978, chapter 8). Characteristics of a few important subsystems are briefly presented here, but our emphasis is on new results since 1980, with an emphasis on equations of state suitable for computer-assisted calculation of phase relations applicable to terrestrial ore deposits, and generally to assemblages quenchable from natural systems. Most of these new results focus on the properties of sulfide mineral solutions and phase relations in portions of the supersystem Ag_2S-Cu_2S-ZnS-FeS-PbS-Sb_2S_3-As_2S_3-Bi_2S_3-Au-S_2.

Solid solutions of a few minerals in this multisystem constitute a significant fraction of many hydrothermal ores, particular when they are initially deposited. Some of these solid solutions are not quenchable in phase (e.g., $(Ag,Cu)_2S$ solid solutions) or composition (e.g., PbS-$AgSbS_2$-$AgBiS_2$ galena solid solution), their instability with cooling giving rise to a multiplicity of more nearly stoichiometric sulfides and/or an exceeding diverse suite of modular sulfosalts (e.g., Skinner 1966; Hall and Czamanske 1972; Anthony et al. 1990; Makovicky 1997; see also in this volume Makovicky 2006). Here we summarize activity-composition relations for the solid solutions $(Zn,Fe)S$ sphalerite (*sph*), $(Ag,Cu)_{16}(Sb,As)_2S_{11}$ polybasite-pearceite (*plb-prc*), $(Ag,Cu)_3SbS_3$ pyrargyrite-skinnerite (*prg-skn*), $Ag_3(Sb,As)S_3$ *prg*-proustite (*prg-prs*), body-centered cubic (*bcc*)-face-centered-cubic (*fcc*)-, and hexagonally-closed packed (*hcp*)-$(Ag,Cu)_2S$ solid solutions, $(Cu,Fe)_{10}(Fe,Zn)_2(Sb,As)_4S_{13}$ fahlore (*fah*) (also known as "tetrahedrites"; Tatsuka and Morimoto 1977; Makovicky and Skinner 1978, 1979; Spiridonov 1984; Sack et al. 2005), $Ag(Sb,Bi)S_2$ members of the solid solution series miargyrite (*mia*) – aramayoite (*arm*) – matildite (*mat*), $(Bi,Sb)_2S_3$ bismuthinites-stibnites (*bs*), and PbS-$AgSbS_2$-$AgBiS_2$ galena (*gn*). But first we review such relations for $Fe_{1-x}S$ pyrrhotite (*po*) (Toulmin and Barton 1964). To illustrate the efficacy of these models in analysing primary and retrograde ore-forming processes, we compare multicomponent phase diagrams calculated from these activities, and thermodynamic data for sulfide and sulfosalt end-members, with phase assemblage data from laboratory experiments and from petrological studies of polymetallic sulfide ore deposits from metallogenic provinces near the western margins of North and South America. Finally, we use the database and natural assemblages to constrain the thermodynamic properties of some sulfosalts formed by retrograde reactions during cooling and construct some key diagrams for evaluating resource potential of selected ores.

A discussion of phase relations and solid solution thermochemistry requires a brief review of experimental techniques. We rely on pointers to previous, exhaustive treatments (e.g., Vaughan and Craig 1978, chapter 7), and exemplary papers illustrating applications. It is absolutely essential that the modern student of sulfide petrology, who may acquire most references through the internet, be aware that a great deal of the essential literature is only available in non-digital hard copy editions. Details of experimental methods and results are crucial to their evaluation, and much of the primary literature predates the internet and is not adequately reviewed in digitally available literature.

Much of the primary, classic literature in dry sulfide systems, particularly on the Fe-Cu-Ni-Zn-S subsystems, predates the electron microprobe and automated XRD, and substantial detail is missing regarding methods and individual data points. Perhaps for this reason, the uncertainty and incorrect nature of some, perhaps a substantial, fraction of the phase relations repeated in the review literature is unappreciated. For example, the diagram of ternary Fe-Ni-S relations at 650 °C by Kullerud (1963) is reprinted in Kullerud et al. (1969), Barton and Skinner (1967, their Fig. 7.13; 1979, their Fig. 7.18), Vaughan and Craig (1978; their Fig. 8.32), and Vaughan and Craig (1997; their Fig. 8.23b). Yet Karup-Møller and Makovicky (1995), and Sugaki and Kitakaze (1998) demonstrate how very incorrect the earlier pentlandite phase relations were, and Ebel and Naldrett (1996, 1997) determined correct mss-liquid tielines above 1000 °C. There is a lot of research to be done in these systems, to verify and correct the original work of Kullerud, Toulmin, Barton, Skinner, Scott and other pioneers.

SULFIDE THERMOCHEMISTRY

Basic concepts

The laws of thermodynamics are set forth in many texts (e.g., Guggenheim 1967; Kern and Weisbrod 1967; Kubaschewski et al. 1967; Ehlers 1972; Wood and Fraser 1977; Denbigh 1981; Callen 1985; deHeer 1986; Anderson and Crerar 1993; Krauskopf and Bird 1994) and as chapters in review volumes (e.g., Vaughan and Craig 1978; Navrotsky 1987, 1994; Ganguly 2001). Thermodynamic properties, or equations of state, describe the Gibbs Free Energy (referred to in the rest of the chapter simply as "Gibbs energy") of crystalline or amorphous substances, including liquids. More than a century of work in many fields has resulted in constraints that allow thermodynamic properties for many pure substances (e.g., quartz, pyrite) to be tabulated (e.g., Hultgren et al. 1963; Mills 1974; Robie et al. 1978; Barton and Skinner 1979; Chase et al. 1985; Knacke et al. 1991; Barin 1991; Chase 1998). Tabulated equations of state are related to equations describing reaction equilibria through equilibrium constants. A chemical reaction of species A, B, M, etc. with stoichiometric coefficients α, β, etc. :

$$\alpha A + \beta B + ... = \mu M + \nu N + ... \qquad (1)$$

has a standard state Gibbs energy of:

$$\Delta G°_{rxn} = \bar{G}°_{products} - \bar{G}°_{reactants} = (\mu G°_M + \nu G°_N ...) - (\alpha G°_A + \beta G°_B ...) \qquad (2)$$

where all the species are described by some standard state convention (cf. Anderson and Crerar 1993, p. 155) at fixed temperature and pressure, and deviations of the species from their standard states in concentration, internal state of order, etc. may be accounted for in the product of the gas constant R, temperature, and an activity product

$$K = \frac{[a_M]^\mu [a_N]^\nu}{[a_A]^\alpha [a_B]^\beta} \quad (3)$$

equal in magnitude, but opposite in sign, to $G°_{rxn}$ when reactant and product species are in equilibrium. For pure substances, $a = 1$, and for gases at low pressure, the activity of a component i in the gas is equivalent to its fugacity in the mixture: $a_i = f_i$. For example, an H_2S+H_2 mixture might be used to fix the sulfur fugacity in an experiment, where the proportion of H_2 controls the degree of dissociation of H_2S by the reaction:

$$2H_2S = 2H_2 + S_2 \quad (4)$$

The Gibbs energy of each gaseous molecule in the mixture can be written as, for example,

$$G_{H_2S} = G°_{H_2S} + RT\ln(a_{H_2S}) \quad (5)$$

where R is the gas constant (8.3143 J/K·mol), and T is the absolute temperature (Kelvin). The expression containing the activity disappears for a pure substance ($a = 1$), and so describes the deviation from ideality of the molecule in the mixture from that of the molecule as a pure, ideal, substance. The Gibbs energy of reaction is then:

$$\begin{aligned} G_{rxn} &= G_{S_2} + 2G_{H_2} - 2G_{H_2S} \rightarrow \\ &= \left(G°_{S_2} + 2G°_{H_2} - 2G°_{H_2S}\right) + RT\left[\ln(a_{S_2}) + 2\ln(a_{H_2}) - 2\ln(a_{H_2S})\right] \end{aligned} \quad (6)$$

At chemical equilibrium among the gaseous molecules in the mixture, no reaction takes place, and the reaction energy goes to zero so the expression containing the $G°$ must, at equilibrium ($G_{rxn} = 0$), be equal to the negative of the expression in activities. From equations of state for pure gaseous species (e.g., Chase 1998) one can obtain all the necessary values for $G°$, and write:

$$\left(G°_{S_2} + 2G°_{H_2} - 2G°_{H_2S}\right) = G°_{rxn} = -RT\ln(K) = -RT\ln\left[\frac{(a_{S_2})(a_{H_2})^2}{(a_{H_2S})^2}\right] \quad (7)$$

Depending on which database is consulted, one can obtain a set of values for K as a function of temperature, for example, as done by Adachi and Morita (1958):

$$\log_{10} K = \log_{10}(a_{S_2}) + 2\log_{10}(a_{H_2}) - \log_{10}(a_{H_2S})^2 = \frac{-9480}{T(K)} + 5.16 \quad (8)$$

which allows the calculation of a_{S_2} for the assumption that:

$$\frac{a_{H_2}}{a_{H_2S}} = \frac{[H_2]}{[H_2S]} \quad (9)$$

Energies for many reactions of the type "metal + sulfur = metal sulfide" have been determined in the laboratory, and summarized in tables (e.g., Vaughan and Craig 1978, their Appendix II; Barton and Skinner 1979). Reactions of this type are *univariant* because at constant temperature and pressure they depend only on the fugacity (activity, at low pressure) of sulfur, f_{S_2}.

The equations of state for pure sulfide phases, and sulfide solid solutions must be determined from experimental data. For solutions, the mixing properties of endmember components are described by equations describing the Gibbs energy of mixtures which can be related to activities of endmember components [$a_i = \exp+((\mu_i - \mu_i°)/RT)$ where μ_i and $\mu_i°$ are the chemical potentials of component i in the mixture and in an endmember reference state, usually

$G°_i$, Appendix 1] through the equilibrium constant. Here we focus on mixing of sulfide/sulfosalt endmembers with different cationic species. In contrast to the nearly ideal mixing exhibited by Pb(S,Se) and Zn(S,Se) anionic mixtures (e.g., Barton and Skinner 1967, their Fig. 7.2), cationic substitutions relating sulfide/sulfosalt endmembers are invariably nonideal (i.e., $a_i \neq X_i$). There are many sources of this nonideality, including: (1) differences in radii and bonding of the substituent elements (e.g., Ag^+ and Cu^+, and As^{3+}, Sb^{3+} and Bi^{3+}); (2) long-range-ordering of atoms between different sublattice sites (LRO) or short-range-ordering (clustering) of cations within these sites (SRO); (3) incompatibilities between pairs of substituent elements in, for example, a solid solution with two cation substitutions, A for B, and M for N, and (4) defects.

Differences in ionic radii may correlate directly with nonideality of substitutions on cation sites, as in silicates and oxides (e.g., Ghiorso and Sack 1991), and lead to immiscibility, or wide two-phase regions when relative differences in cation radii are large enough (e.g., $[r_{Sb^{3+}}/r_{As^{3+}}]$ ~ 1.3 in *plb-prc*, Fig. 26; Lawson 1947). However, similar radii are not sufficient to insure ideal mixing, as for example, Fe and Ni do not mix ideally in sulfides even at high temperatures (e.g., Hsieh et al. 1987a,b). Also, the cations of several "isomorphous" substitutions are known to exhibit different stereochemical behavior which may lead to compositional dependence in site geometries and atomic displacements (e.g., Ag^+ and Cu^+ in *prg-prs*). Multiple sites of mixing are also a complicating factor in many of the sulfides considered here (e.g., *bcc-*, *fcc-*, and *hcp-*(Ag,Cu)$_2$S solid solutions, *iss, plb-prc, fah, skn, arm, gn*). On the one hand, the substituting cations may exhibit nearly random distributions on individual sites but be strongly ordered between crystallographically distinct sites, with this ordering contributing to negative deviations from ideal mixing and stabilizing intermediate compositions (e.g., *skn*, see below). Alternatively, LRO may potentially control the location of miscibility gaps where the ordering exhibits strong compositional dependence (e.g., *fah*, Figs. 40, 43). Additionally, the cations may not be randomly distributed on a given site, exhibiting instead a tendency to form clusters (e.g., *sph*, see below).

Despite these, and other, complexities several general rules apply to activity-composition relations in binary solid solutions. These include the generalizations (Barton and Skinner 1967, p. 279-280; Vaughan and Craig 1978, p. 215) that:

(1) in dilute solutions, the activity of the solvent (the major component) tends to behave perfectly—i.e., it closely approximates its mole fraction;

(2) in dilute solutions, the activity of the solute (the minor component) is proportional to its mole fraction, but the proportionality constant is usually not unity;

(3) the greater the extent of the solid solution, the greater the compositional range to which dilute solution rules apply;

(4) substitutional solid solutions are more likely to follow dilute solution rules than are interstitial and omission types of solid solutions.

Substitutional solid solution and rules (1)–(3) may be illustrated by FeS and ZnS activities in (Zn,Fe)S *sph* (Fig. 20). In dilute solutions of these components, both of these activities are greater than mole fraction. Also both of the activities of these components are greater than their mole fractions when they are the solvent, but the activity of ZnS resembles the ZnS mole fraction for a greater range in its concentrated region ($a_{ZnS} \sim X_{ZnS}$ for $1.0 \geq X_{ZnS} > 0.6$). Omission type solid solutions may be illustrated by reference to activities of Cu$_2$S, CuFeS$_2$, and FeS in (Cu,□)$_2$S (*dig*) (boxes indicate vacancies), (CuFe□)$_2$, Cu$_2$Fe$_2$)S$_2$ (*iss*), and (Fe□,)S (*po*) at 300, 600 and 700 °C, respectively (Barton 1973, Rau 1967; Toulmin and Barton 1964; Barton and Skinner 1979, their Fig. 7.2, p. 343). Each of these declines roughly linearly with mole fraction, but with a proportionality constant of about 4 (i.e., $a_i \sim 4X_i - 3$, for $X_i > 0.8$) rather than the slope of unity for an ideal solution (e.g., Eqn. 19).

The concept of an ideal solution becomes increasingly anachronistic when we consider mixtures capable of multiple substitutions such as a $(A,B)_\alpha(N,M)_\beta S_x$ solution. For this solution to be ideal, not only must the individual substitutions be ideal, but the endmember components related by the reciprocal reaction:

$$A_\alpha N_\beta S_x + B_\alpha M_\beta S_x = A_\alpha M_\beta S_x + B_\alpha N_\beta S_x \qquad (10)$$

must also be coplanar in G-X space (i.e., $\Delta G = 0$). The latter condition is rarely approximated in nature, but $(Ag,Cu)_3(Sb,As)S_3$ prg-prs appear to be an exception to this rule. Gibbs energies of reciprocal reactions in prg-prs, plb-prc and fah have been established from constraints on isotherms for cation exchange between these phases and a medium setting the potential of the exchanging cations (i.e., the exchange potential). This medium was $(Zn,Fe)S$ sph at various Fe/Zn in the case of Fe-Zn exchange with fah (e.g., Raabe and Sack 1984; Sack and Loucks 1985; O'Leary and Sack 1987) and Ebel and Sack (1989, 1991) employed the multiphase assemblage pyr + cpy + elec to fix the Ag for Cu, $Ag(Cu)_{-1}$, exchange potential, $\mu_{Ag(Cu)-1}$, in Ag-Cu exchange experiments with fah. The absolute values of the reciprocal energies range between 0.7 and 3.9 kJ/gfw on a one sulfur formula basis and are, therefore, significant sources of nonideality. The largest of these energies expresses the incompatibily between Ag and As in fah, and it is sufficiently large to to have a major impact on the distribution of fah compositions in nature (e.g., Fig. 38).

Although many *solid state* reactions have been experimentally explored in determining the solid solution properties of sulfides and sulfosalts, in natural systems, fluids or vapors interact with solids. In nature, a few "pure" phases (e.g., pyrite [*pyr*]) and common solid solutions (e.g., sphalerite [*sph*]) can act as buffers to control the assemblage in equilibrium with fluid. The resulting multiphase equilibria in polymetallic systems can be quite complex. In some cases, simpler subsystems may involve only a few elements. For example, much experimental effort has been directed toward understanding the effect of sulfur fugacity and oxygen fugacity on the stability relations of pure phases and simple binary solid solutions such as *sph*.

Equilibria among sulfides, oxides, and sulfates can be represented on diagrams as stability fields that are functions of the activities of S_2 and O_2, calculated from equations like:

$$2Fe + S_2 = 2FeS \qquad (11)$$

where, at equilibrium:

$$G_{rxn} = -RT\ln(K) = RT\ln(a_{S_2}) \qquad (12)$$

because a_{Fe} and a_{FeS} are unity as they are assumed to be pure phases. The reaction:

$$Fe + S_2 = FeS_2, \qquad (13)$$

yields an analogous boundary curve. Another is from

$$3FeS + 2O_2 = Fe_3O_4 + \tfrac{3}{2}S_2 \qquad (14)$$

through the expression

$$G_{rxn} = G_{Fe_3O_4} - 3G_{FeS} = -RT\ln(K) = RT\{\tfrac{3}{2}\ln(a_{S_2}) - 2\ln(a_{O_2})\} \qquad (15)$$

The combination of many such reactions, and some simple algebra, allows the construction of an activity-activity diagram describing the stability of the pure phases as functions of a_{S_2} and a_{O_2} at a particular temperature (Fig. 1; cf. Vaughan and Craig 1978).

Although experience shows that such approximations are useful in interpreting the broad features of many natural mineral assemblages, they depend in detail on the assumptions that phases are pure endmembers, that Gibbs energy data and their dependence on temperature are accurate, and that the natural assemblages have not re-equilibrated and are not metastable. For example, in systems with high halogen content (e.g., Meridiani Planum on Mars), substitution

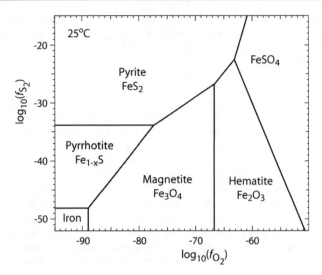

Figure 1. $a(S_2)$-$a(O_2)$ *po-mte-pyr-hm* equilibria at 25 °C, calculated as by Vaughan and Craig (1978).

of chlorine into crystalline structures will change the boundaries on Figure 1. Previous reviews have emphasized evaluation of sulfur fugacity from mineral reactions and phase relations in systems of three or fewer components. Very little has been added to the experimental literature in these areas, as emphasis has shifted to understanding the thermochemistry of complex solid solutions which are useful petrogenetic indicators in natural systems. Further experimental work is necessary to constrain such effects on mineral stability, so that rigorous expressions for activity can be used to perform the multi-variable calculations necessary to understand the vapor-fluid-mineral interactions observed in nature (Lichtner 1985, 1988, 1996). Even in an apparently simple system such as shown in Figure 1, the interpretation of real paragenetic sequences requires some knowledge of the kinetics of retrograde reactions.

Graphical presentation of multivariate thermodynamic data can occur in a great variety of formats. In interpreting data presented by others, it helps to have a thorough understanding of the underlying relationships between activity, temperature, pressure, Gibbs energy, entropy and enthalpy. For example, activity of the gas species S_2, a proxy for the propensity of the mineral environment to sulfidize phases, is frequently plotted as "$\log_{10}(a_{S_2})$" versus "$1000/T(K)$," due to the way that the equilibrium constant is defined (eqn. above). The logarithms of oxygen and hydrogen fugacities are plotted for the same reason, and Eh-pH diagrams derive from the same basic relationships. Conventional presentation includes temperature-composition (*T-X*), pressure-temperature (*P-T*), activity-composition (*a-X*), activity-temperature, activity-activity, and Eh-pH diagrams, all of which are reviewed comprehensively in Vaughan and Craig (1978), section 7.6), and in the texts referred to above (e.g., Anderson and Crerar 1993).

Some of the diagrams presented later in this chapter will be unfamiliar. The algebraic formulations of reaction equilibria and equations of state dictate the choices of axes in every case. Quite simply, the goal in many two-dimensional representations is to place the known quantities on the left side of an equation, represented on the vertical axis of the figure. The horizontal axis usually represents a variable, frequently a composition parameter, that controls an unknown parameter. Experimental data are substituted into the algebraic formulation, and plotted in the chosen dimensions. In this way, values for the unknown parameters can be chosen to best fit the experimental values. Because the goal is to derive equations of state for solid solutions, rather than pure phases, the algebra is more formidable. Compensation is at hand in the common availability of computers, spreadsheet applications, and high-level programming languages.

Thermochemical analysis of experimental results has been reviewed elsewhere with rigor (e.g., Vaughan and Craig 1978, chapter 7; Anderson and Crerar 1993). Data are presented in the literature in many forms. Some of the older work can be formidable to the twenty-first century researcher, who can encounter: Gibbs energies of reactions (e.g., Barton and Skinner 1979, their Table 7.2); EMF measurements (e.g., Lusk and Bray 2002), or equations for $\log(f_{S_2})$ in univariant equilibria (e.g., Vaughan and Craig 1978, their Appendix II). Direct EMF measurements (millivolts, mV) are converted to sulfur fugacity using cell equations, from sulfur polymerization models calibrated against the zero of the electrochemical cell (e.g., Lusk and Bray 2002). The algebra used by Sack (2005), for example, goes beyond the basic thermochemical concepts reviewed by Vaughan and Craig (1978); texts such as Anderson and Crerar (1993) or Denbigh (1981) may be consulted.

Dimensionality of systems

The great strides in ore deposit petrology for most of the twentieth century were mainly due to the detailed study of binary, ternary and the occasional quaternary systems. Thus Vaughan and Craig's (1997, p.368) objective was to provide a modern account of "major systems," and to apply mineral stability data to understanding natural deposits. They state that "... the stabilities of ore minerals in the major systems can be described or presented in the form of graphs or diagrams. Nevertheless, much uncertainty remains, particularly at low temperatures, even in the most important binary and ternary systems." The vision was to constrain complex systems through a thorough understanding of mineral stability relations in the boundary systems, which contain the major elements present in the complex systems of nature. The interplay of degree of equilibration and temperature (i.e., kinetics), in specific systems, was explicitly recognized. How far have we come since then?

Mineral stability is a function of equilibration. Equilibrium relations are sometimes preserved in nature, but not always. So also in the laboratory. Graphs and diagrams illustrate equilibria in the two or three dimensions we can observe on a printed page. Sulfidation equilibria, for example, may be trivially computed from the equations of, for example, Barton and Skinner (1979, their Table 7.2). But the curves describing equilibria in these dimensions stem from algebraic expressions that may apply to systems of many more than three components. The computational equipment available in recent times makes it possible to project, or otherwise illustrate, these many algebraic relationships in an infinite variety of binary and ternary systems printable on a page. The algebraic relationships are represented by the equations of state of the pure endmembers, and solid solutions, of the sulfides and sulfosalts. We can coherently illustrate equilibrium phase relationships on the computer screen in three dimensions, but remain limited to a small number of compositional and intensive (P,T) variables. With an internally consistent database containing equations of state of many sulfides and sulfosalts, such illustrations can be generated and even animated to allow visualization of the phase relations in previously inaccessible composition-temperature-pressure spaces.

Such a view is deceptively simple. The profound understanding of phase relations and thermochemistry achieved by previous scholars working on sulfides remains requisite to the modern sulfide petrologist possessing modern tools. The roots of knowledge continue to lie in experimental work and investigation of natural assemblages informed by sound thermodynamics, with results interpreted in the largest possible context. The modern experimentalist has the advantage in being able to rapidly calculate known equilibria in the context of a proposed experiment, and to constrain intensive parameters (e.g., chemical potentials of cation exchange reactions) of buffer subsystems based on the careful application of known equations of state, including solid solution thermochemistry. The modern petrologist can analyze chemical compositions and crystal structures of natural assemblages *in situ*, at high spatial resolution. The present chapter offers the experimentalist and investigator of natural systems a comprehensive overview, and detailed algebraic treatment, of many such chemical systems of direct relevance to ore deposits. An advantage here is the ability to compute equilibria in systems with dimensions well beyond those expressible on a two-dimensional paper sheet.

Mineral stability and criteria for equilibrium

Much of the data discussed in this chapter results from the identification of sulfide phases and analysis of their chemical compositions, when they coexist in chemical equilibrium. At equilibrium, the chemical potential energy of a system is at a minimum, so the system remains as it is. The chemical potential energy of a sulfide assemblage, represented by the state function G, the Gibbs energy, is minimized when all phases are at constant pressure and temperature, and all the chemical potentials (μ) of all independently variable components in the assemblage are the same in each phase. Equilibrium is more difficult to assess in nature than in the laboratory, where tests can be applied such as: (1) results are the same for experiments of very different duration; (2) duplication of the experiment yields the same result; (3) the same result is obtained using different combinations of reactants, or pressure-temperature paths, or (4) the result is obtained reversibly, by observed transitions across a reaction boundary. Even when these tests are met, such "apparent equilbria" may be metastable, and not the true stable assemblage. This is a particular danger at low temperatures.

In nature, disequilibrium is expressed in crystal zoning, reaction rims on minerals, or the presence of more phases than should coexist based on the phase rule (Anderson and Crerar 1993). Equilibrium between minerals in nature can be assessed by comparison with phase diagrams and by calculations like those presented in this chapter. Most of the data on which these are based comes from "reversed" experiments, in which equilibrium is approached from more than one direction (criterion 3 and 4, above). There are many examples in the literature (e.g., Sack and Loucks 1985; Ebel and Sack 1989; Harlov and Sack 1995b; Andrews and Brenan 2002; Bockrath et al. 2004). Reversed experiments can provide tight brackets on equilibria, if the phase under scrutiny reacts on laboratory timescales, but not so fast that compositions are unquenchable.

Many of the sulfides (e.g., Ag_2S-Cu_2S) and sulfosalts (e.g., *plb*, *prs*, *skn*) discussed here react rapidly. Others, for example *sph*, *pyr*, and arsenopyrite, react much more slowly, but can be reacted at higher temperatures. Fluxes such as ammonium chloride can be used to hasten reactions in the laboratory (Kretschmar and Scott 1976; Moh and Taylor 1971). Equilibrium is also approached more rapidly in hydrothermal experiments, because of the presence of water, but quenching is more difficult. Sulfides in nature have time, and fluxing agents, which cause them to re-equilibrate to low temperatures. As Vaughan and Craig (1978) point out, the refractory sulfides that preserve their high temperature equilibrium state in nature are most difficult to study in the laboratory, and the more easily a sulfide is studied in the laboratory, the less applicable it will be to natural occurrences. The results presented here break through this barrier in specific systems which *can* be addressed at low temperature in the lab. Some minerals (e.g., *fah*, also known as "tetrahedrite") are refractory enough to record conditions in nature, but can be studied on laboratory timescales.

Another refractory phase useful in recording natural conditions is arsenopyrite (*asp*, FeAsS), which dominates the system Fe-As-S. This system was mapped by Clark (1960) and reevaluated, with the component binaries, by Barton (1969). Kretschmar and Scott (1976) examined in detail the range of *asp* compositions and its phase relations, and their dependence on temperature: the arsenopyrite geothermometer (cf. Fig. 2). Because arsenopyrite is slow to re-equilibrate at low temperatures, it can be used as a geothermometer in limited circumstances. Barton and Skinner (1979) concluded: "Until an experimental breakthrough permits successful

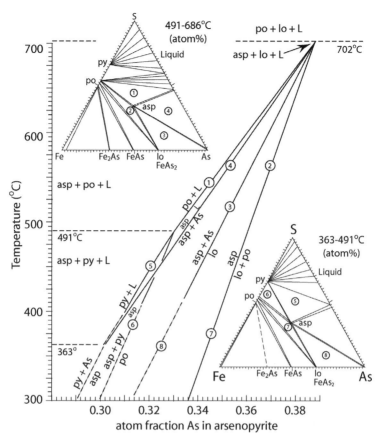

Figure 2. Temperature-composition projection of the system Fe-As-S. The pseudobinary projection shows *asp* composition as a function of temperature, as equilibrium assemblages vary about the *asp* solid solution field. Phases are: *asp*, *pyr*, *po*, *lo* = loellingite, L = liquid. Binary tielines are tentative (adapted from Vaughan and Craig 1997, after Kretschmar and Scott 1976).

experimentation at lower temperatures... the arsenopyrite "geothermobarometer" will remain an interesting, but unproven tool." Sharp et al. (1985) addressed this issue in the most recent critical assessment. They found that pressure effects on the *asp* geothermometer must be taken into account in deposits metamorphosed above lower amphibolite grade, it does not apply consistently to low-temperature hydrothermal deposits, but it appears valid for most other deposits for which alternative thermometric data exist (cf. Vaughan and Craig 1997).

Experimental methods in sulfide research

Many techniques from chemistry and silicate experimental petrology have been adapted for sulfide investigations. Allen and Lombard (1917), Rosenqvist (1954), and Dickson et al. (1962) used the dew-point method to measure sulfur fugacity over sulfide assemblages. Merwin and Lombard (1937) mapped *iss* stability in H_2S gas, using graphite crucibles. Most synthesis of sulfides, and phase equilibrium evidence, however, results from evacuated silica tube experiments (Fig. 3), as described by Kullerud (1971), and by Vaughan and Craig (1978, chapter 8). For most purposes, high-purity fused quartz (>99% SiO_2) glass tubes are required, due to the reactivity of sulfur. Scott (1976) provides a detailed explanation of these techniques

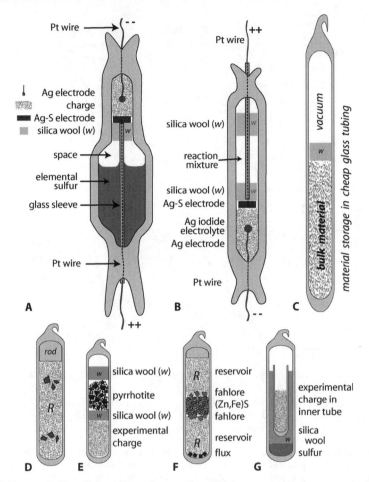

Figure 3. Evacuated silica tube experimental geometries. All tubes are evacuated on a vacuum line prior to and during welding at the top end. A) Calibration apparatus used by Lusk and Bray (2002) for the Ag/AgI/Ag$_{2+x}$S, $f_{S_2(vapor)}$ electrochemical cell (cf. Schneeberg 1973). Glass capillaries are collapsed onto conductor wires. B) Cell used by Lusk and Bray (2002) for measurement of f_{S_2} in reactions. Ag electrode is a silver pellet crimped onto Pt wire. C) Long-term storage configuration for reagents. Glass wool and empty space allow opening and re-sealing of the tube. D) Simple reaction configuration for exchange of target phase (large pieces) with buffer reservoir. Glass rod minimizes vapor space during the experiment. E) Pyrrhotite technique for f_{S_2} measurement. Silica wool allows equilibration of sulfur vapor. F) Reaction configuration for multiply-saturated fahlore-(Zn,Fe)S-reservoir (R) experiments, where R = chalcopyrite, pyrite, electrum assemblage (Ebel and Sack 1989). Fluxing agent NH$_4$Cl dissociates to vapor at reaction temperature. G) Tube-in-tube method for S-fugacity control by a buffer assemblage.

(cf. Kullerud 1971). Glass is welded using an oxygen-gas torch (H$_2$ gas is preferred; acetylene is too slow), with the top end of the tube attached to a vacuum pump. Welding goggles are required. A manifold with several valves facilitates sealing of several charges at one sitting, and judicious placement of valves enables regulation of the force of the pump on tube contents. Powdered charges will erupt up the tube upon sudden evacuation. Tubes of appropriate size are sealed at one end and reactants placed inside sequentially. For synthesis work, the tube is weighed after each addition, but for many types of work this is not required (e.g., Fig. 3D,F). A vibrating engraver and/or a static electricity dissipator (wand with radioactive particles in it) are useful in removing powders from tube walls. A fused quartz rod and/or silica wool is inserted to protect the reactants and minimize vapor space, and a vacuum applied. The evacuated tube is welded shut around the rod or by "necking down," with reactants and researcher protected by a wrapping of water-soaked paper.

Silica tubes are placed in the hot spot of a furnace at up to 1200 °C, and quenched intact directly into water due to the very low thermal expansion of the glass. The ambient pressure within the tube cannot be measured during the experiment, but tube-in-tube techniques can be used to buffer sulfur fugacity (Fig. 3G; Kullerud 1971; Peregoedova et al. 2004). Thermal gradients in tubes can be used to grow large crystals of simple sulfides, aided by the inclusion of halide salts (e.g., NH_4Cl) as transport agents. Salt fluxes can also be used to speed up sluggish reactions (e.g., Boorman 1967; Scott 1976; Ebel 1993; cf. Vaughan and Craig 1978, their Table 8.1).

In addition to direct equilibration of fixed bulk compositions at fixed temperature, silica tube techniques have been central to many other important investigations. Precise synthesis in silica tubes is critical for spectroscopic analysis of sulfide phases (Vaughan and Craig 1978, Chapter 4). Differential thermal analysis (DTA) is useful for establishing the temperatures of exothermic and endothermic reactions, including crystallographic inversions (Kullerud and Yoder 1959; Dutrizac 1976; Sugaki and Kitakaze 1998). The electrum-tarnish method developed by Barton and Toulmin (1964; cf. Barton 1980) allows the measurement of sulfur activity by observing the onset of tarnish formation on Ag-Au alloy (Fig. 4a). Toulmin and Barton (1964) used the method to calibrate the relationships between *po* composition, $\log(f_{S_2})$ and temperature above 300 °C (Fig. 4). Electrochemical cells, encapsulated in silica, have been used many workers (e.g., Kiukkola and Wagner 1957; Schneeberg 1973; Vaughan and Craig 1978, their Fig. 7.11; Lusk 1989; Lusk and Bray 2002) to measure f_{S_2} in particular reactions as a function of temperature. Lusk and Bray (2002) provide a recent example of this difficult technique, with many details (Fig. 3A, B). Interdiffusion and kinetics are also studied using evacuated silica tubes, for example the "diffusion induced segregation" (also known as "chalcopyrite disease") in chalcopyrite (*cpy*) (Bente and Doering 1993), and kinetics of pentlandite dissolution from monosulfide solid solution (*mss*) (Etschmann et al. 2004). Vaughan and Craig (1978, Section 7.3) review electrum (*elec*) tarnish, pyrrhotite (*po*) buffer, dew point, gas mixing, and electrochemical cell techniques for sulfur activity measurement in some detail.

Sulfide assemblages containing several phases in abundance (multiple saturatation) can fix the chemical potential energy of known exchanges, allowing the calibration of unknown reaction equilibria. A simple example is the fixing of sulfur fugacity by including a large mass of *po* powder of known composition in an isothermal experiment (Figs. 4b, 3E). For exploration of more complex systems, careful choice of reservoirs is necessary. Ebel and Sack (1989, 1991) fixed $Ag[Cu]_{-1}$ and $Fe[Zn]_{-1}$ exchange potentials with $(Zn,Fe)S - CuFeS_2 - (AgAu)$ mixtures of fixed X_{FeS} and X_{Ag} compositions in *sph* and *elec*, respectively (Fig. 3F). The simple sulfide phase equilibria reviewed here, and in previous studies, are critical to designing experiments in more complex systems where control of exchange potentials is required.

Gas mixing furnaces provide an alternative for the control of sulfur fugacity. Because SO_2(vapor) forms readily, oxygen-bearing sulfide systems are difficult to investigate in silica tubes (Naldrett 1969). In the gas mixing furnace, internal vapor pressure is not a problem; however, measurement of the partial pressures of vapor phase species (e.g., S_2) is difficult. For given input fractions of pure gases, the speciation of elements among molecules in the vapor phase at a particular temperature and 1 bar pressure can be calculated from the equations of state of the molecules, by minimization of the free energy of the vapor mixture (White et al. 1958; Kress 2003; Kress et al. 2004; Ebel 2006). This is a form of solution equilibrium, dependent on the accuracy of tabulated equation of state data (Lodders 2004). The electrochemical cell technique is similarly dependent upon calculated vapor phase equilibria: "Choosing the most accurate $\log(f_{S_2})$-T data for inclusion in a working cell equation is a vexed problem" (Lusk

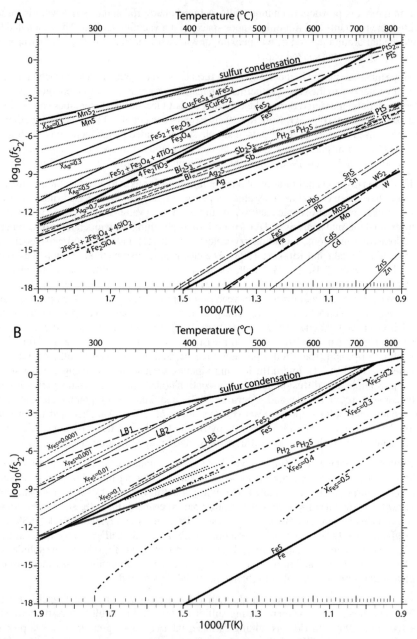

Figure 4. Sulfidation curves for selected systems. A) Metal/metal sulfide buffers and several oxide-sulfide reactions tabulated by Barton and Skinner (1979), and *elec*-Ag_2S equilibrium contours (tarnish curves) with $0.1 < X_{Ag} < 1.0$ in electrum (equation in Barton 1980, p 314). B) FeS content of *sph* and *po* in Fe-Zn-S assemblages. In the *pyr* stability field, solid curves (Barton & Toulmin 1966), and dotted curves (Scott and Barnes 1971) describe X_{FeS} in coexisting *sph* and dashed curves are reactions LB1: *pyr*+idaite (*id*) =*bn*(c)+S, LB2: cpy=*bn*(c)+*pyr*, and LB3: *pyr*+*cpy*= cubanite (*cb*)+S reported by Lusk and Bray (2002). Below the FeS=FeS_2 line, dot-dash curves describe X_{FeS} in *sph* (Scott and Barnes 1971), and dotted curves are results of Lusk and Bray (2002) for X_{FeS}=0.9625, 0.9545, 0.9480 and 0.9464 in *po*, from lowest to highest $f(S_2)$. Curves are truncated to reflect range of experimental data.

and Bray 2002). The relative abundances of the many sulfur-bearing molecular species vary with temperature and total sulfur pressure, and this is also a difficulty in the furnace. Recent work on sulfides using gas mixing furnaces includes the comprehensive approach to the Fe-S-O system by Kress (1997, 2000), and olivine-sulfide liquid and *mss*-sulfide liquid element partitioning studies by Brenan (2002, 2003).

Recent work in the Fe-Ni-S system demonstrates the importance of a very rapid quench in sulfide experiments of all types. For high-temperature silica tube work where sulfide liquids occur, a vertical arrangement dropping samples into water provides a consistent and rapid quench (Ebel and Naldrett 1997). For investigation of Fe-O-S liquid-solid equilibria, Kress (1997, 2000) has developed an extremely rapid quench method using cooled rollers to freeze molten material dropped from the gas mixing furnace, to retain the S and O contents of liquids at experimental conditions. At lower temperatures, horizontal furnaces are typically employed, and samples are pulled out using metal tongs, then plunged into water. This arrangement seems to retain equilibria in most cases. Karup-Møller and Makovicky (1995) were able to reconstruct unquenchable high-temperature pentlandite solid solution equilibria from textural analysis, careful microprobe work, and interpretation of quench products, with important consequences for the Fe-Ni-S system above 610 °C (see below). Although ternary phase compositions cannot be determined precisely using X-ray diffraction (XRD; e.g., Kullerud et al. 1969), unquenchable phases can be explored using XRD at high temperatures, as demonstrated by Sugaki and Kitakaze (1998; cf. Kissin and Scott 1979) in confirming and extending the earlier work on pentlandite solutions.

Preparation of starting materials is difficult for sulfide research because of the tendency of sulfides to react with air and water vapor. Metals commonly used in research rapidly develop tarnish in transit, and must be fire-polished under vacuum, or reduced under H_2 gas, before weighing as reactants in experiments. For example, Sb lumps or Ag shot should be heated on a flame, in a silica tube under vacuum, and H_2 should be passed over Fe powder at >700 °C for some time. Pure metals should be stored in a vacuum dessicator with no sulfides present, while sulfides should also be stored in vacuum. Inexpensive evacuated sodium-glass tubes (e.g., pyrex) serve to keep sulfide and sulfosalt reactants stable for long periods (Fig. 3C).

Analysis of experimental product compositions is critical to accurate understanding of phase relations. Until the widespread use of electron probe microanalysis (EMPA) in the late 1960s (e.g., Cabri 1973), X-ray diffraction was the primary method for analysis of experimental product compositions (e.g., Kullerud and Yoder 1959), supplemented by wet-chemical methods. Phase equilibrium studies require more careful use of EMPA than day-to-day surveys of ore minerals. Many of the common Ag- and Cu-sulfides and sulfosalts are mobile-ion semiconductors, in which atoms move under the electron beam (cf. Ebel 1993). These require special techniques for analysis (e.g., Harlov 1995). Modern techniques open new windows for analysis, particularly of trace elements in sulfide systems. Laser ablation inductively-coupled plasma mass spectrometry (LA-ICPMS) is seeing increasing application in this area (e.g., Ebel and Campbell 1998; Albarède 2004; Mungall et al. 2005), as is the ion microprobe (Chryssoulis et al. 1989; Fleet et al. 1993; Simon et al. 2000) also known as Secondary Ion Mass Spectrometry, SIMS (see Wincott and Vaughan 2006, this volume).

The importance of experimental techniques can be illuminated by comparison of modern work with older work in the Fe-Ni-S system. Minerals in this system dominate opaque sulfide assemblages associated with mafic rocks, including meteorites (Ramdohr 1973, 1980). The hexagonal monosulfides $Fe_{1-x}S$, $Ni_{1-x}S$ form a solid solution (*mss*) that is the primary crystallization product of magmatic sulfide liquids (Naldrett 1989, 1997; Naldrett et al. 1997). Unfortunately, much of the work that is still accepted in the geological community predates the electron microprobe. Many of those phase relations were established by XRD and DTA in silica tubes, where compositions are uncertain, or by XRD of quenched experiments where both Ni and S content of *mss* vary (e.g., Lundqvist 1947; Kullerud and Yund 1962; Kullerud 1963; Klemm 1965; Naldrett et al. 1967). More recent experimental work has substantially improved our understanding of these systems. Chang and Hsieh (1987) computed the Fe-Ni-S phase diagram down to 700 °C using associated solution models for liquid, a statistical thermodynamic model for *mss*, a quasi-subregular model for metal alloy, and a pseudobinary solution model for the disulfide, all based on their experimental work and that in the literature (cf. Chuang 1983; Hsieh 1983; Hsieh et al. 1982, 1987a,b; Sharma and Chang 1979; cf. Kress 2003). Chang's group generated new gravimetric data, and employed the "Calphad" approach commonly used to model alloy systems (Chang et al. 2004). Ebel and Naldrett (1996, 1997) compared these model results, and previous *mss*-liquid phase relations with their experimental results in the Fe-Ni-S system. New experimental work in the quaternary Fe-Ni-Cu-S has also been reported by Fleet and Pan (1994).

Perhaps the most profound revision of the older Fe-Ni-S phase relations regards pentlandite. Vaughan and Craig (1978, p. 306, and 1997, p. 405; following Naldrett et al. 1967) reported that the *mss* field narrows upon cooling, and pentlandite solid solution (*pnt*, $(Fe,Ni)_9S_8$) begins to exsolve from *mss* (± metal alloy) at 610 °C, exhibiting a wide composition range. Sugaki and Kitakaze (1998), however, found the sulfur-deficient, high-temperature *pnt* with atomic Fe = Ni forms at 865 ± 3 °C by a pseudoperitectic reaction of *mss* with liquid (Fig. 5a). They used high-temperature X-ray diffraction to study the unquenchable high-*pnt*. This form coexists with γ (gamma) iron below 746 ± 3 °C, and persists to 584 ± 3 °C, with composition limits of $Fe_{5.07}Ni_{3.93}S_{7.85}$ to $Fe_{3.61}Ni_{5.39}S_{7.85}$ at 850 °C (Fig. 5a). The *pnt* solution field widens with decreasing temperature, reaching the Ni-S join below 806 °C, and forming a continuous solution with $Ni_{3-x}S_2$ (or Ni_4S_3) at 800 and 650 °C. These results are consistent with Fedorova and Sinyakova (1993) and Karup-Møller and Makovicky (1995). The equilibrium compositions reported by Karup-Møller and Makovicky (1995; Fig. 5b) also cast into doubt the entirety of disulfide solid solution limits reported by Klemm (1965). A great deal of research has established Fe-Ni-S phase relations above about 200 °C, but reactions are sluggish at lower temperatures (Craig and Scott 1976; Misra and Fleet 1974). The effects of composition, and cooling rate on *pnt* exsolution from *mss*, and their textural expression, have been studied by Durazzo and Taylor 1982; Kelly and Vaughan (1983). Vaughan and Craig (1997, p. 409) note that "ultimately, such textures in natural ores and their synthetic equivalents can be highly informative. However, their complexity requires more detailed work...".

It is likely that similar difficulties are present in the Cu-Fe-S ternary subsystem, that contains minerals abundant in terrestrial, lunar, and meteoritic rocks, and is especially important in hydrothermal systems. Vaughan and Craig (1978) stated what remaines true: "In spite of intensive investigation into the minerals of this system, many relationships, especially those at low temperature, remain enigmatic. This is because of the large number of phases, extensive solid solutions, non-quenchable phases and persistent metastability." Ebel and Naldrett (1996, 1997) mapped pyrrhotite solid solution (po_{ss})-liquid tielines between 950 and 1150 °C, but by 600 °C solid solutions of bornite (bn_{ss},), intermediate solid solution (iss), and po_{ss} dominate the central portion of the system. The ternary diagrams describing the system between 900 and 300 °C stem primarily from work in the 1960s and early 1970s (e.g., Kullerud et al. 1969; Barton and Skinner 1979; Vaughan and Craig 1997). It is certain that the binary tielines published in these diagrams are as inaccurate as those drawn into the Fe-Ni-S system (Kullerud et al. 1969; cf. Ebel and Naldrett 1996; Naldrett, pers. comm.). With decreasing temperatures, a myriad of Cu-Fe-S phases become stable or metastable (Fig. 6), many poorly understood (e.g., mooihoekite, Cabri and Hall 1972). At 300 °C, the po_{ss} field is very small, but the bn_{ss} and iss fields remain extensive. The thermodynamic properties of these solid solutions are not known, and textures seen in nature may result from many subsolidus processes. As pointed out by Vaughan and Craig (1997), order-disorder transformations often mask true stability relations in this system. And as noted above, the high-temperature solid solutions, of interstitial and omission type rather

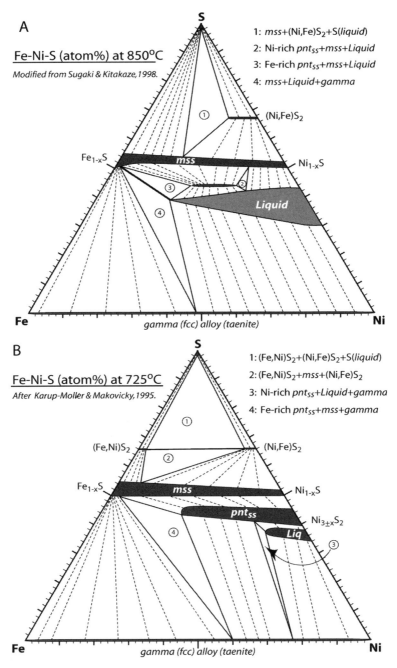

Figure 5. Non-quenchable high-*pnt* stability fields. A) Results at 850 °C, adapted from Sugaki and Kitakaze (1998, their Fig. 6). The *pnt* composition limits from their text are used, not the limits expressed in their diagram. The $(Ni,Fe)S_2$ solution limit is taken from Klemm (1965, his Fig. 8) and cannot be correct. B) Results at 725 °C, after Karup-Møller and Makovicky (1995, their Fig. 1). Disulfide compositions are from their experiment #21.

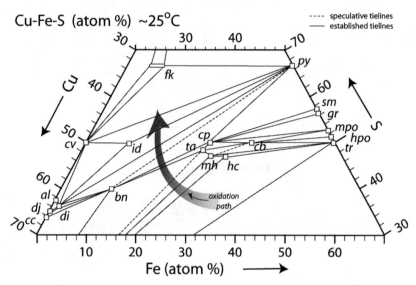

Figure 6. Possible phase relations at 25 °C (atom %), in the central portion of the system Cu-Fe-S, after Vaughan and Craig (1997, their Figs. 8.17 and 8.21). Arrow shows general trend of ores during supergene weathering, as Fe and then Cu are removed. Abbreviations: al = anilite; bn = bornite; cb = cubanite; cc = chalcocite; cp = *cpy*; cv = covellite; di = digenite; dj = djurleite; fk = fukuchilite; gr = greigite; hc = haycockite; hpo = hexagonal *po*; id = idaite; mh = mooihoekite; mpo = monoclinic *po*; py = *pyr*; sm = smythite; ta = talnakhite; tr = troilite.

than simple substitutions, may be highly non-ideal. Upon cooling, they transform through structural readjustments that may persist metastably to lower temperatures. Recently, Ding et al. (2005) confirmed earlier suggestions that structural heterogeneity is a common characteristic of bornite, by exploring low-temperature Fe/Cu ordering using high-resolution transmission electron micrxcoscopy (HRTEM) and magnetic structure calculations. In recent years, attention has turned to exploration of trace element partitioning in the ternary and quaternary Fe-Ni-Cu-S systems at high temperatures (e.g., Li et al. 1996; Fleet et al. 1993; Ballhaus et al. 2001; Brenan 2002). The field is open for modern confirmation of the major-element phase relations, and particularly for modeling the equations of state of the solid solutions in these systems.

SULFIDE SYSTEMS IN METEORITES

The cubic (*Fm3m*) monosulfides oldhamite (CaS), alabandite (MnS), and MgS (niningerite; Keil and Snetsinger 1967) are rare on Earth, but common as solid solutions (with troilite, FeS) in the enstatite chondrite and aubrite meteorites (Keil 1968; Wadhwa et al. 1997). They probably originate as stable solid grains in equilibrium with a reducing vapor in space, at temperatures above 1070 °C (Ebel 2006, his Fig. 4). Various degrees of solid solution exist among these phases. Skinner and Luce (1971) explored their phase relations between 600 and 1000 °C in silica tubes, using both XRD and the electron microprobe. They found complete MgS-MnS solution, extensive FeS solution in both MgS and MnS, but ≤1 atom% solubility of FeS in CaS, and <10% MgS and MnS solubility in CaS (Fig. 7). They also investigated the effect of cooling rate on reequilibration of alabandite and niningerite solutions. Fogel (2005) derived temperature-dependent Margules parameters for modeling the asymmetric CaS-MnS solvus, incorporating new and previously published high temperature data. He concluded that the data for $T > 900$ °C are likely flawed, and called for new experimental work.

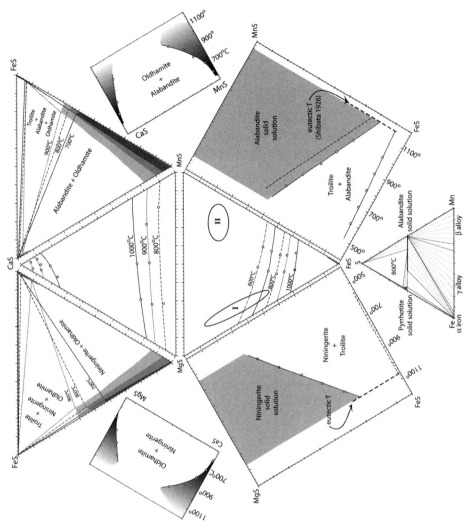

Figure 7. Solvi in the system CaS-FeS-MgS-MnS. Compositions of coexisting monosulfides at 800, 900 and 1000 °C in the CaS-MgS-MnS (upper), and FeS-MgS-MnS (lower) ternaries. Ternary and binary phase fields at 700, 800 and 900 °C are shown in the CaS-MgS-FeS, and CaS-MnS-FeS systems (upper left and right). Temperature-composition relations of solvi in associated binaries are illustrated in central and lower right and left. The Fe-Mn-S system at 800 °C (below) illustrates the general form of Fe-X-S systems where X=Ca, Mg, Mn. In the FeS-MgS-MnS ternary, regions **I** and **II** indicate compositions of niningerite in type I and intermediate (EH3 to EH5), and alabandite in type II (EL6) enstatite chondrites (Keil 1968, with CaS plotted with MgS). Equilibration temperatures of some type I EC reach nearly 800 °C, but retention of compositions requires very rapid cooling. All diagrams after Skinner and Luce (1971).

The MnS-CaS solvus (e.g., Lodders 1996) and the extent of other binary and ternary solvi can be used as geothermometers for enstatite chondrites, because the FeS content of alabandites and niningerites is not significantly affected by CaS (Fig. 7). The temperature-composition relations of binary solvi can be directly compared with natural composition data (recast in the quaternary for the assumption that Cr has insignificant effect) to determine the temperatures of final equilibration of monosulfide assemblages in meteorites (Skinner and Luce 1971; Ehlers and El Goresey 1988; Zhang and Sears 1996; Kimura and Lin 1999). Coexisting troilites contain negligible Mn and other elements (Keil 1968), but may represent exsolution products, so temperature estimates represent minima. Skinner and Luce (1971) described subtleties in this approach, and note the hazard in linear extrapolation of solvi below 500 °C. The Mn contents of daubréelite ($Fe^{2+}Cr_2S_4$; El Goresy and Kullerud 1969) and sphalerite have been used to estimate f_{O_2} and f_{H_2S} in the enstatite chondrites (Lin and El Goresy 2002).

Fe-S

The system Fe-S (see Fleet 2006, Fig. 1a; this volume; Fig. 8) contains many petrologically important iron sulfides, most notably pyrite (*pyr*), the most common iron sulfide in the Earth's crust, and pyrrhotite (*po*). With the important exception of high-temperature (>308 °C) hexagonal pyrrhotite (*po*), most of these minerals are nearly stoichiometric and/or metastable or low-temperature minerals. At temperatures above 400 °C, pyrite (*pyr*) in the system Fe-S (see Fleet 2006, Fig. 1a; this volume) is for all intents and purposes stoichiometric FeS_2 (Kullerud and Yoder 1959) and it coexists with either *po* or sulfur at temperatures below its maximum thermal stability, at 742 ± 1 °C at one bar and at ~14 °C higher T per kbar (Kullerud and Yoder 1959). On the other hand, high-temperature *po*, ~$Fe_{1-x}S$, encompasses a composition range

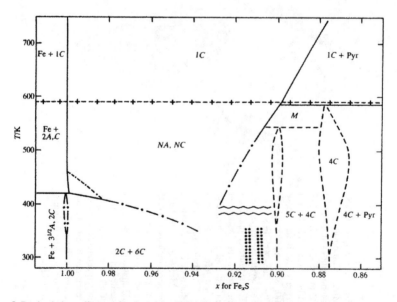

Figure 8. Revised phase diagram for the region FeS - Fe_7S_8. Phase designations as in text. Solid curves are macroscopic phase boundaries, dashed and dotted lines are suggested phase boundary extensions, dashed lines are tentative phase boundaries for Fe_9S_{10} (5C) and Fe_7S_8 (4C), dotted lines are possible existence ranges for $Fe_{11}S_{12}$ (6C) and $Fe_{10}S_{11}$ (11C), crosses separated by two dashes are antiferromagnetic to paramagnetic transitions, and dots separated by two dashes is a spin-flip transition. [Used by permission of Elsevier, from Grønvold and Stølen (1992), *J Chem Thermodyn*, Vol. 24, Fig. 5, p. 927]

0 < x < 0.19, with nearly stoichiometric *po* coexisting with Fe metal at temperatures below, and with sulfur-rich iron liquid above, 988 °C (Brett and Bell 1969), and with the contents of vacancies substituting for Fe in this hexagonal, NiAs-type structure (Hägg and Sucksdorff 1933) rising steadily with temperature where *po* and *pyr* coexist (Arnold 1962; Toulmin and Barton 1964). At temperatures above *pyr* maximum thermal stability, *po* coexists with sulfur-rich liquid until 1083 °C where it reacts with liquid to produce an iron-rich sulfide liquid. *Po* melts congruently at a temperature of about 1192 °C.

The composition of hexagonal *po* is related to the position of the strongest reflection in its X-ray diffraction pattern, with $d_{(102)}$ decreasing with increasing vacancy content. Yund and Hall (1969) obtained the following relationship between these variables:

$$\text{Atomic \% Fe} = 45.212 + 72.86[d_{(102)} - 2.0400] + 311.5[d_{(102)} - 2.0400]^2 \quad (16)$$

This relationship is consistent with earlier calibrations of Arnold (1962) and Toulmin and Barton (1964), calibrations that allowed them to precisely define the compositional dependence of *po* coexisting with *pyr* from their experimental data. Toulmin and Barton (1964) also determined the relationships between sulfur fugacity and temperature for various *po* compositions by utilizing the electrum tarnish method (Barton and Toulmin 1964) to bracket sulfur fugacity for temperatures above 308 °C, the temperature at which hexagonal 1C *po* undergoes ordering to antiferromagnetic superstructures with unit cells that are multiples of those of the 1C NiAs-type structure (see Makovicky 2006, this volume; Fig. 8). Based on previously published data and the sulfur fugacity-temperature brackets they obtained for their *po* compositions, Toulmin and Barton (1964) obtained the following equation relating sulfur fugacity, temperature and *po* composition:

$$\log_{10} f_{S_2} = (70.03 - 85.83 N_{FeS})\left(\frac{1000}{T} - 1\right) + 39.30(1 - 0.9981 N_{FeS})^{\frac{1}{2}} - 11.91 \quad (17)$$

where (N_{FeS} is the mole fraction of FeS in the system FeS-S_2) which they used to contour the *po* field on a $\log_{10}(f_{S_2})$–$1/T$ diagram (see Fleet 2006, Fig. 5; this volume). They then utilized the differential form of Equation (17) when they integrated the Gibbs-Duhem equation for *po*,

$$N_{FeS} d\log_{10} a_{FeS} = -(1 - N_{FeS}) d\log_{10} f_{S_2} \quad \text{(constant } P,T) \quad (18)$$

for the boundary condition $a_{FeS} = 1$ at $N_{FeS} = 1$ to obtain the following relation between activity of FeS (a_{FeS}), temperature, and N_{FeS}:

$$\log_{10} a_{FeS} = 85.83\left(\frac{1000}{T} - 1\right)(1 - N_{FeS} + \ln N_{FeS}) - 39.3(1 - 0.9981 N_{FeS})^{\frac{1}{2}}$$
$$-39.23 \tanh^{-1}(1 - 0.9981 N_{FeS})^{\frac{1}{2}} - 0.002 \quad (19)$$

In addition, Toulmin and Barton (1964) derived estimates for the Gibbs energy of formation of *pyr* from the elements, examined the effect of pressure on a_{FeS} and *pyr-po* equilibria, and attempted to extrapolate their results to lower temperatures. They concluded that a_{FeS} is virtually independent of pressure, because the volume of *po* decreases linearly with N_{FeS}. They also derived and evaluated expressions for the pressure dependence of N_{FeS}, a_{FeS}, and f_{S_2} for *po* in equilibrium with *pyr*.

Since the pioneering study of Toulmin and Barton (1964) there have been several attempts to refine the thermodynamic properties of 1C *po* solid solution (e.g., Burgmann et al. 1968; Scott and Barnes 1971; Ward 1971; Libowitz 1972; Rau 1976; Froese and Gunter 1976, 1978; Chuang et al. 1985; Martin and Gil 2005). In general, these refinements have resulted in calibrations for a_{FeS} and a_S in *po* in good agreement with those of Toulmin and Barton (1964) for *po* with S/Fe ratios greater than 1.02 (cf. Barker and Parks 1986). In several of these calibrations the more physically realistic components FeS and $\square$S ($\square$ denotes a vacancy of Fe sites) have been chosen as reference components rather than FeS and S_2 (e.g., Froese and Gunter 1976; Martin and Gil 2005), with the result that activity composition-relations may be successfully reproduced using an asymmetric solution model to describe the excess Gibbs energy of *po*. In these formulations the activity coefficients of FeS and $\square$S components (box indicates vacancy on Fe site) are related to *po* composition by the familiar expressions:

$$RT \ln \gamma_{FeS} = (1 - X_{FeS})^2 \left[W_{FeS} + 2(W_{\square S} - W_{FeS}) X_{FeS} \right] \quad (20)$$

and

$$RT \ln \gamma_{\square S} = (X_{FeS})^2 \left[W_{\square S} + 2(W_{FeS} - W_{\square S})(1 - X_{FeS}) \right] \quad (21)$$

where $X_{FeS} \equiv$ (mol% FeS)/(mol% FeS + mol% S) rather than the *po* composition measure employed by Toulmin and Barton, $N_{FeS} \equiv$ (mol% FeS)/(mol% FeS + mol% S_2), and W_{FeS} and $W_{\square S}$ are the Margules parameters characterizing the excess partial molar Gibbs energies of FeS and $\square$S components at infinite dilution (e.g., Thompson 1967). Values inferred for these parameters depend on the f_{S_2}–T–X_{FeS} dataset examined and the values employed or inferred for the Gibbs energies of the end-member reactions:

$$Fe + \tfrac{1}{2}S_2 = FeS \quad (22)$$

and

$$\square S \text{ (in } po\text{)} = \tfrac{1}{2}S_2 \text{ (in vapor)} \quad (23)$$

Representative values deduced for these parameters and the standard state Gibbs energy of Reaction (22), $\Delta \bar{G}°_{22}$, range between W_{FeS} = −592.23 + 0.53062T and −464.79 + 0.3122T, $W_{\square S}$ = −392.92 + 0.2479T and −332.31 + 0.1603T, and $\Delta \bar{G}°_{22}$ = −392.92 + 0.14570T and −212.68 + 0.0544T kJ/gfw in the studies of Froese and Gunter (1976) and Martin and Gil (2005), respectively.

The main advantage of describing *po* as an asymmetric solution is simplicity. It expedites integrating of the Gibbs-Duhem equation, finding simple expressions for activity coefficients and calculating compositions of *po* in equilibrium with *pyr* (e.g., Froese and Gunter 1976). The latter may be readily determined by iterative solution of the conditions of equilibrium for Reaction (23) and the reaction:

$$Fe + \tfrac{1}{2}S_2 = FeS \quad (24)$$

$$RT \left[\tfrac{1}{2} \ln f_{S_2} - \ln(1 - X_{FeS}) - \ln \gamma_{\square S} \right] = \Delta \bar{G}°_{22} = \Delta \bar{H}°_{22} - T \Delta \bar{S}°_{22} + P \bar{V}°_{\square S} \quad (25)$$

and

$$RT \left[\tfrac{1}{2} \ln f_{S_2} - \ln(X_{FeS}) - \ln \gamma_{FeS} \right] = \Delta \bar{G}°_{23} = \Delta \bar{H}°_{23} - T \Delta \bar{S}°_{23} + P \left(\bar{V}°_{FeS_2} - \bar{V}°_{FeS} \right) \quad (26)$$

at a given T and P. The main disadvantage of describing *po* as an asymmetric solution is that it does not correctly describe the properties of *po* near the limit of its saturation with respect to Fe metal where, for example, the sulfur fugacity has a finite value rather than the value of −∞ implied by condition of Equilibrium (25).

To remedy this deficiency and correctly describe the thermodynamic properties of stoichiometric FeS, it is necessary to make explicit provision for both interacting iron vacancies and their complementary defects. Although vacancies are the predominant defect in *po* deviating significantly from FeS stoichiometry, iron vacancies must be balanced by a complementary defect of equal abundance in stoichiometric FeS. This complementary defect may be one of three types: vacancies on sulfur sites, interstitial Fe atoms, or Fe atoms substituting for

S atoms on S sites (e.g., Kröger 1974). Ward (1971) and Libowitz (1972) assumed that the complementary defects in stoichiometric *po* are of Frenkel type (iron vacancies balanced by iron interstituals) and used the equations of Lightstone and Libowitz (1969) to deduce the energy of the interaction of vacancies in *po* from analysis of the data of Rosenqvist (1954), Niwa and Wada (1961), Toulmin and Barton (1964), Burgmann et al. (1968) and Turkdogan (1968). Lightstone and Libowitz (1969) derived activity-composition relations for simple solutions based on the changes in configurational entropy due to point defects developed by Libowitz (1969) and Libowitz and Lightstone (1967) assuming Bragg-Williams type pair-wise interactions between defects and the random distribution of these defects. The equations of Lightstone and Libowitz (1969) for activities of *po* components derived for the assumption of Frenkel defects (iron vacancies and iron interstituals) may be written as follows:

$$RT \ln a_S = \left(\Delta \bar{G}^f_{FeS} + g_{\square Fe}\right) + 4n_\square (2 - n_\square) \zeta_{\square Fe} + RT \ln \left[\frac{n_\square}{4}(2 - n_I)^2\right] \quad (27)$$

and

$$RT \ln a_{Fe} = -g_{\square Fe} - 8\zeta_{\square Fe} n_\square + RT \ln \left(\frac{1 - n_\square}{n_\square}\right) \quad (28)$$

where $a_S = \sqrt{f_{S_2}}$ because it is referred to a state for which diatomic sulfur has a fugacity of one atmosphere, $\Delta \bar{G}^f_{FeS}$ and $g_{\square Fe}$ are the Gibbs energy of formation of stoichiometric FeS and of iron vacancies with respect to iron, $\zeta_{\square Fe}$ is the pairwise interaction energy between iron vacancies, $n_\square$ and n_I are the concentrations of iron vacancies and interstituals, and where $X_{Fe}/X_S = (1 - X_S)/X_S = 1 - n_\square + n_I$. In these equations concentrations of defects are governed by the condition of homogeneous equilibrium:

$$RT \ln \left[\left(\frac{n_\square}{1 - n_\square}\right)\left(\frac{n_I}{2 - n_I}\right)\right] = -\left(g_{IFe} + g_{\square Fe} + 8\zeta_{\square Fe} n_\square\right) \quad (29)$$

where g_{IFe} is the Gibbs energy of formation of iron interstituals.

For $n_\square > 0.02$, n_I may be neglected in Equation (27) and it may be simplified to

$$\ln\left(\frac{a_S r}{r - 1}\right) = \frac{\left(\Delta \bar{G}^f_{FeS} + g_{\square Fe}\right)}{RT} + 4\zeta_{\square Fe}\left(\frac{r^2 - 1}{r^2}\right) \quad (30)$$

or

$$\ln\left(\frac{a_S}{\delta}\right) = \frac{\left(\Delta \bar{G}^f_{FeS} + g_{\square Fe}\right)}{RT} + 4\zeta_{\square Fe}\delta(2 - \delta) \quad (31)$$

where $r \equiv X_S/X_{Fe}$ and $\delta \equiv n_\square$. Libowitz (1972) demonstrated that the isotherms of these equations computed from the data of Rosenqvist (1954), Niwa and Wada (1961), Toulmin and Barton (1964), Burgmann et al. (1968) and Turkdogan (1968) are linear on plots of $\ln[a_S r/(r - 1)]$ vs. $[(r^2 - 1)/r^2]$ or $\ln(a_S/\delta)$ vs. $\delta(2 - \delta)$ and Barker and Parks have updated these plots to include data obtained subsequently by Rau (1976) (Figs. 9, 10). As noted by Libowitz (1972), the positive energies, or repulsive interactions, between iron vacancies deduced for these isotherms are consistent with the fact that *po* forms ordered superstructures based on iron vacancies at temperatures below 308 °C. Libowitz (1972) calculated that *po* contains concentrations of Frenkel defects at the stoichiometric composition (intrinsic defect concentration) greater than 10^{-2} at 1100 °C and the parameters of Libowitz's model have been reoptimized by Chuang et al. (1985) to reflect more data than that considered by Libowitz,

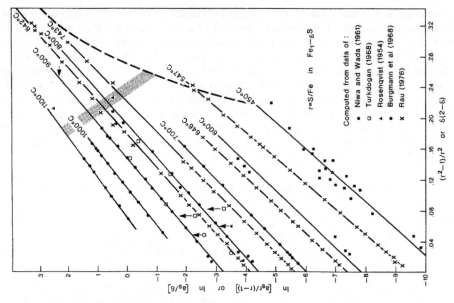

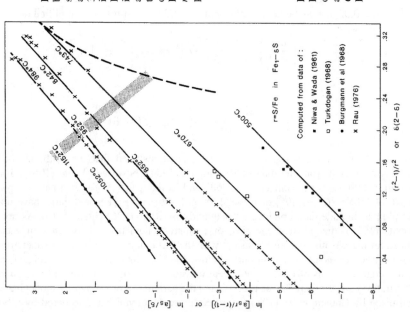

Figure 9 (*left*). Plots of Libowitz relationships between sulfur activity, temperature and *po* composition (eqn. 30 and 31). After Libowitz (1962) but updated by Barker and Parks (1985) to include results of Rau (1976 and private communication). [Used by permission of Elsevier, from Barker and Parks (1986), *Geochim Cosmochim Acta*, Vol. 50, Fig. 2, p. 2187.]

Figure 10 (*right*). Same as Fig. 9. [Used by permission of Elsevier, from Barker and Parks (1986), *Geochim Cosmochim Acta*, Vol. 50, Fig. 3, p. 2188.]

including the very accurate data of Rau (1976). Chuang et al. (1985) obtained the following values for model energetic parameters: $\zeta_{\square Fe} = 26.35395 + 0.1630785T - 0.0196967T(\ln T)$, $g_{\square Fe} = 24.62335 + 0.0263674T$, $g_{IFe} = 111.229 - 0.0334838T$, and $\Delta \bar{G}_{FeS}^f = -129.0028 - 0.154763T + 0.0267165T(\ln T)$ kJ/mol (where γ-Fe and ½S$_2$ are the standard states).

General agreement has been achieved between calculated and predicted a_S within the *po* homogeneity region in the Fe-S system (see Fig. 1a, Fleet 2006; this volume) by employing the assumption of interacting iron vacancies as the prevailing defect with interstitial iron as the complementary defect. However, Rau (1976) demonstrated that the alternative assumption that the complementary defect is iron substituting on sulfur sites (i.e., clusters of metallic iron in the lattice), rather than interstitial iron, provides more satisfactory agreement between calculated a_S and his very accurate measurements of a_S near the stoichiometric limit of the *po* homogeneity region. For the alternative assumption of iron substituting on sulfur sites the appropriate equations given by Libowitz (1969) and Lightstone and Libowitz (1969) are:

$$r = \frac{\left(1 - n_{Fe(S)}\right)}{\left(1 + n_{Fe(S)} - n_\square\right)} \tag{32}$$

$$a_S = \left[\left(1 - n_{Fe(S)}\right)\left(n_\square\right)\right] \exp\left[\frac{\Delta \bar{G}_{FeS}^f + g_{\square Fe}}{RT}\right] \exp\left[\frac{4n_\square \left(2 - n_\square\right)\zeta_{\square Fe}}{RT}\right] \tag{33}$$

$$a_S^2 = \left[\frac{\left(1 - n_{Fe(S)}\right)^2 \left(1 - n_\square\right)}{n_{Fe(S)}}\right] \exp\left[\frac{\Delta \bar{G}_{FeS}^f - g_{Fe(S)}}{RT}\right] \tag{34}$$

where $n_{Fe(S)}$ is the concentration of iron on sulfur sites, $g_{Fe(S)}$ is the Gibbs energy of formation of iron on sulfur sites and Equation (33) simplifies to Equations (30) or (31) for $n_\square > 0.02$. Rau (1976) obtained the following values for energetic parameters: $\zeta_{\square Fe} = 46.262 + 0.006577T$, $g_{\square Fe} = 14.405 + 0.034765T$, $g_{Fe(S)} = 363.906 - 0.173991T$, and $\Delta \bar{G}_{FeS}^f = -150.44 - 0.05264T$ kJ/mol. Calculations of $\log(f_{S_2}) - a_{FeS}$ relations (Barker and Parks 1985) using this calibration of Rau (1976) are in good agreement with determinations for *pyr-po* from other sources (e.g., Gilletti et al. 1968; Schneeberg 1973; Kissin and Scott 1979), with the exception that they are about 0.3 $\log(f_{S_2})$ units lower than those obtained by extrapolation of the results of Scott and Barnes (1971) to the lowest temperatures at which hexagonal *po* has a disordered distribution of vacancies.

At present there is no other basis besides the experimental data for discriminating between interstitial iron or iron substituting on sulfur sites as the complementary defect in *po* near its stoichiometric limit; the Mössbauer investigation cited in Rau (1976) is inconclusive due to the very low concentrations of these defects, even at high temperatures, and it is impossible to quench these defects, or even the 1C *po* structure, to room temperature (e.g., Corlett 1968; Nakazawa and Morimoto 1970, 1971). However, based on evidence to be summarized for (Zn,Fe)S sphalerite in the next section, it appears likely that clusters of metallic iron in the lattice are the phantom complimentary defect in *po*. As noted elsewhere in this volume (see Makovicky 2006; Fleet 2006) all *po* acquire superstructures below 308 °C, the most common of them based on ordering of iron vacancies in alternate layers parallel to the *c*-axis and distinguished by the integral multiple of the *c* dimension of their 1C NiAs-type unit cells. These include the superstructures labeled 4C, 5C, 11C, 6C, that are commonly present at room temperature and approximate the limiting compositions Fe$_7$S$_8$, Fe$_9$S$_{10}$, Fe$_{10}$S$_{11}$, and Fe$_{11}$S$_{12}$, respectively (e.g., Nakazawa and Morimoto 1971; see also Vaughan and Craig 1978). The most commonly observed of these is the 4C superstructure, monoclinic *po* with space group F2/d, for which the ordering of iron vacancies in alternate layers parallel to the c axis was first proposed and determined (e.g., Bertaut 1953; Tokonami et al. 1972). In addition to these superstructures observed at room temperature, there are various incommensurate *po* superstructures with nonintegral repeats of X-ray reflections along hexagonal a^* and c^* axes of the subcell that bridge the gap between 1C *po* with disordered distribution of iron vacancies and the superstructures with complete ordering of iron vacancies in alternate layers. These have compositions between FeS and Fe_7S_8, are stable at temperatures between room temperature and 308 °C (e.g., Fig. 8; see Fleet 2006, Fig. 5; this volume), and have been labeled NA, NC, and MC (or M) by Nakazawa and Morimoto (1971). Finally, high temperature, 1C stoichiometric FeS (γ-FeS) is formed by a second-order phase transition from a 2A, C superstructure of the NiAs structure with space group $P6_3mc$ (β-FeS) at about 317 °C (e.g., Keller-Besrest and Collin 1990; Grønvold and Stølen 1992). This 2A, C superstructure transforms to the closely related troilite structure (space group $P\bar{6}2c$, α-FeS), a $3^{1/2}$A, C superstructure, at about 147 °C, but the 2A, C superstructure at this transition does not appear to achieve the stoichiometric FeS composition where it coexists with α-Fe, extending to limiting composition of only about $Fe_{0.993}S$ (Keller-Besrest and Collin 1990).

Phase and textural relations involving *po* superstructures at low temperatures can be exceedingly complex as they are complicated by metastability of phases and sluggish reaction rates (e.g., Nakazawa and Morimoto 1970, 1971; Pierce and Buseck 1974; Putnis 1975; Kissin and Scott 1982), and by the magnetic origin of some of the phase transitions (e.g., Grønvold and Stølen 1992; Fig. 8). Although it is impossible to unambiguously establish stable, "equilibrium" phase relations under these conditions by conventional techniques, it appears that troilite and monoclinic *po*, 4C (~Fe_7S_8), are stable with respect to other *po* phases, except possibly 6C (~$Fe_{11}S_{12}$), at low temperatures, based on TEM studies (e.g., Pierce and Buseck 1974; Putnis 1975; Putnis and McConnell 1980) and thermodynamic analysis of calorimetric measurements for FeS, $Fe_{0.98}S$, Fe_9S_{10}, $Fe_{0.89}S$ and Fe_7S_8 (Grønvold et al. 1991; Grønvold and Stølen 1992). Based on hydrothermal recrystallization experiments Kissin and Scott (1982; Fig. 11) concluded that the monoclinic *po* + *pyr* assemblage is stable down to below 115 °C and tentatively speculated that this assemblage becomes unstable with respect to smythite (~Fe_9S_{11}; a pseudorhombohedral mineral with a structure possibly related to monoclinic *po*) + *pyr* and smythite + monoclinic *po* assemblages at about 75 °C based on experimental results of Scott and Kissin (1973) in the system Fe-Zn-S (Fig. 11). This inference has been questioned by Krupp (1994) who inferred that troilite + smythite is the stable assemblage in Fe-rich compositions below 65 °C based on observations of natural assemblages. However, Taylor and Williams (1972), Nickel (1972), and Bennett et al. (1972) have observed that Ni is at least a minor constituent of natural smythites, and Ni might be responsible for the apparent stability of smythite assemblages inferred by Krupp (1994). Other mineral phases that have compositions closely conforming to the system Fe-S include marcasite (a well known polymorph of *pyr*), greigite, and mackinawite. Despite their widespread occurrence in various sediments, all of these minerals are metastable with respect to *pyr* and *po* phases. Both greigeite (a thiospinel, ~Fe_3S_4) and mackinawite are found in low temperature sedimentary environments and are recognized as metastable precursors of *pyr* (e.g., Berner 1964, 1971). Based on thermodynamic arguments Grønvold and Westrum (1976) have demonstrated that marcasite is metastable to *pyr* at all pressures and temperatures.

One final issue regarding high-temperature phase relations involving 1C *po* is the pressure dependence of *po* composition in the *po* + *pyr* assemblage. Toulmin and Barton (1964) developed relevant expressions for evaluating the derivative of *po* composition with pressure. Scott (1973) conducted a thorough thermodynamic analysis using these expressions but concluded that the pressure dependence of *po* composition is roughly less than half that suggested by comparison of the results of his hydrothermal experiments at 2.5 and 5 kbar with the 1 bar isobar obtained by a polynomial fit to the data of Arnold (1962) and Barton and Toulmin (1966). On this basis Scott (1973) speculated that most of his *po* had re-equilibrated to more Fe-rich compositions during quenching, but this inference is not easy reconciled with

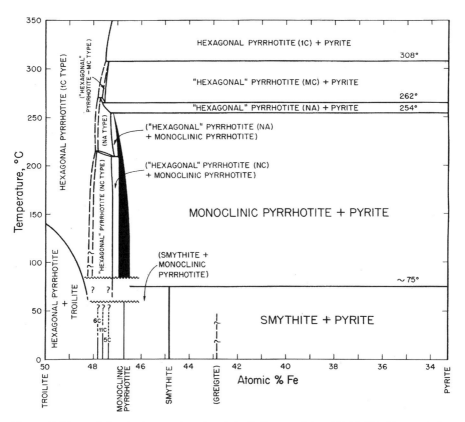

Figure 11. Phase relations in the central portion of the Fe-S system below 350 °C. [Used by permission of the Society of Economic Geologists, Inc., from Kissin and Scott (1982), *Econ Geol*, Vol. 77, Fig. 3, p. 1746.]

the 1 kbar hydrothermal experiments of Udodov and Kashayev (1970) and Chernyshev et al. (1968) which produced *po* compositions very similar to those of Arnold (1962) and Barton and Toulmin (1966). More recently, however, Balabin and Sack (2000) independently calculated isobars for the *po* + *pyr* assemblage at 2.5 and 5 kbar that are in close agreement with 10 of the 14 data points obtained by Scott (1973) and these isobars are displayed in Figure 12. Balabin and Sack (2000) utilized the following expression for the concentration of Fe in *po*:

$$\left(\frac{\partial N_{Fe}^{po}}{\partial P}\right)_{po+pyr,T} = \left(\frac{1-N_{Fe}^{po}}{1-3N_{Fe}^{po}}\right) \times \frac{\left(\bar{V}_{FeS_2}^{pyr} - \bar{V}_{Fe}^{po} - 2\bar{V}_{S}^{po}\right)}{\left(\frac{\partial \mu_{Fe}^{po}}{\partial N_{Fe}^{po}}\right)_{T,P}} \quad (35)$$

which they obtained from the condition of equilibrium between non-stoichiometric *po* and stoichiometric *pyr*,

$$\mu_{Fe}^{po} + 2\mu_{S}^{po} = \bar{G}_{FeS_2} \quad (36)$$

by differentiating Equation (36) with respect to *P* at constant *T*, rearrangement and use of the Gibbs-Duhem relation. Equation (35) was numerically integrated from zero pressure to 5 kbar using a polynomial fit to the data of Arnold (1962) and Barton and Toulmin (1966)

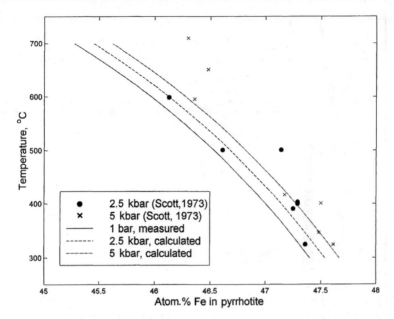

Figure 12. Calculated compositions of 1C *po* in equilibrium with *pyr* at pressures between 1 bar and 5 kbar compared with experimental data. 1 bar isobar is a polynomial fit to the data of Arnold (1962) and Toulmin and Barton (1964). [Used by permission of the Mineralogical Society of Great Britain and Ireland, from Balabin and Sack (2000), *Mineral Mag*, Vol. 64, Fig. 3, p. 934.]

to obtain initial compositions. Estimates of the derivative $(\partial \mu_{Fe}^{po} / \partial N_{Fe}^{po})_{T,P}$ were provided by the *po* model of Chuang et al. (1985). Molar volumes of *pyr* were obtained from the room temperature value of 23.943 cm³/mol (Toulmin et al. 1991) and thermal expansion and bulk modulus corrections of Skinner (1962) and Benbattouche et al. (1989). Molar volumes of nonstoichiometric *po* (Fe$_{1-\delta}$S) were given by the relation

$$\overline{V}_{Fe_{1-\delta}S} \text{cm}^3 = 18.366 - 6.231\delta \tag{37}$$

corrected assuming equivalent thermal expansion and compressibility of Fe$_{1-\delta}$S and FeS, and using the equation of state for stoichiometric *po* that was derived by Balabin and Urusov (1995) from the experimental data of Taylor (1969), Anzai and Ozawa (1974), Novikov et al. (1982), and King and Prewitt (1982). Finally, Equation (37) was obtained by least-squares fitting of volume data of Fleet (1968), Novikov et al. (1982), and Kruse (1990) for *po* with $0.01258 \leq \delta \leq 0.05$, as these are probably the only data representative of the NiAs structure with disordered, "free" vacancies (Novikov et al. 1982; Kruse 1990).

Zn-Fe-S

(Zn,Fe)S sphalerite (*sph*) is a derivative of the diamond structure possessing $F\overline{4}3m$ space group symmetry. It is by far the most petrologically important mineral in the system Zn-Fe-S; it is a common constituent of polymetallic base metal sulfide deposits and is sometimes found in iron meteorites. Due to its refractory character, widespread occurrence and extensive capacity for substitution of Fe for Zn, *sph* has been the object of numerous experimental and petrological studies aimed at defining its thermodynamic properties and using its composition as a petrogenetic indicator of ore-forming processes. At low pressures phase relations involving *sph* may be illustrated by the ternary phase diagram for 600 °C given by Balabin and Sack

(2000; Fig. 13). The topology of phase relations at temperatures between about 200 and 850 °C are identical to those shown in this figure providing allowance is made for the breakdown of *pyr* above 742 °C, the appearance of monoclinic (4C) and other *po* superstructures below 308 °C (Kissin and Scott 1982; Fig. 11), and changes in the intermediate Fe-Zn alloys. At low pressures and temperatures between 400 and 850 °C *sph* spans the composition range from the ZnS end-member to maximum values of mole fraction of FeS given by the relation:

$$X^{sph}_{FeS} = 0.4409 + 0.000125T \text{ (K)} \tag{38}$$

for *sph* coexisting with Fe-metal and C1 *po* (Balabin and Urusov 1995). As illustrated in Figure 14 this maximum bound on the mole fraction of FeS in *sph* has been demonstrated by Hutchison and Scott (1983) to be highly dependent on pressure, decreasing by roughly 40% by 5 kbar; calibrations for the *sph* composition in this assemblage have been used as a cosmobarometer to estimate the size of planetestimals (e.g., Schwarz et al. 1975; Kissin et al. 1986; Kissin 1989; Hutchison and Scott 1983; Balabin and Urusov 1995). For bulk compositions in the FeS-ZnS-S ternary and mole fractions of FeS in *sph* below the maximum bound defined by *sph* in the *po* + Fe-metal assemblage, *sph* coexists with *po* until the *sph* + *po* assemblage becomes saturated with *pyr* with decreasing FeS mole fraction. The mole fraction of FeS in *sph* coexisting with *po* and *pyr* has also been demonstrated to be systematically dependent on pressure (e.g., Fig. 15), but the exact functional dependence of FeS content of *sph* on pressure and temperature has been a matter of some debate (e.g., Boorman 1967; Chernyshev and Anfigolov 1968; Boorman et al. 1971; Scott and Barnes 1971; Scott 1973; Toulmin et al. 1991) and has impacted the efficacy of the use of *sph* composition in the *sph* + *po* + *pyr* assemblage as a geobarometer. At 600 °C *sph* with several mol% FeS coexists with *pyr* and sulfur, and the FeS-content of *sph* in this assemblage increases to roughly 13 mol% by 742 °C when *pyr* decomposes. Finally, Fe-poor *sph* is stable relative to various polytypes of its hexagonally close packed polymorph wurtzite under the $T-f_{S_2}$ conditions prevailing in most environments of ore deposition (e.g., Barton 1970; Scott and Barnes 1972). Such *sph* have S/Zn ratios that are barely detectably higher than wurtzites and the boundary between the stability fields of *sph* and 2H, 4H, and 6H wurtzite polytypes in the system Zn-S appears to intersect the usual $T-f_{S_2}$ conditions of ore deposition and the curve separating the *pyr* and *po* stability fields only at temperatures less than 200 °C (e.g., Craig and Scott 1976). Wurtzites are also reported to be stable relative to *sph* at high temperatures, where Allen and Crenshaw (1912) reported a first order transition between *sph* and wurtzite at 1020 °C in the Zn-S subsystem, and Kullerud (1953) and Barton and Toulmin (1966) produced polytypes in dry synthesis experiments down to 850 °C in the most Fe-rich (Zn,Fe)S compositions.

Barton and Toulmin (1966) provided the first comprehensive experimental study and thermodynamic analysis of *sph*. Using dry synthesis and reaction techniques (evacuated silica tube experiments) they equilibrated *sph* and *po* at temperatures between 580 and 1000 °C, obtaining tight reversal brackets on the compositions of *sph*

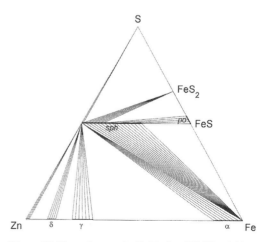

Figure 13. Phase diagram for Fe-Zn-S at 600 °C and 1 bar. [Used by permission of the Mineralogical Society of Great Britain and Ireland, from Balabin and Sack (2000), *Mineral Mag*, Vol. 64, Fig. 1, p. 924.]

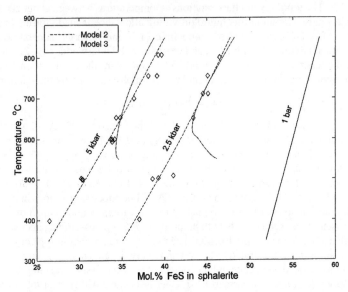

Figure 14. Calculated solubility of FeS in *sph* coexisting with 1C *po* and Fe metal at 2.5 and 5 kbar compared with experimental data of Hutchison and Scott (1983) and 1 bar isobar from Balabin and Urusov (1995). Isobars calculated using Model 3 are truncated at the maximum temperature of spinodal ordering. [Used by permission of the Mineralogical Society of Great Britain and Ireland, from Balabin and Sack (2000), *Mineral Mag*, Vol. 64, Fig. 4, p. 935.]

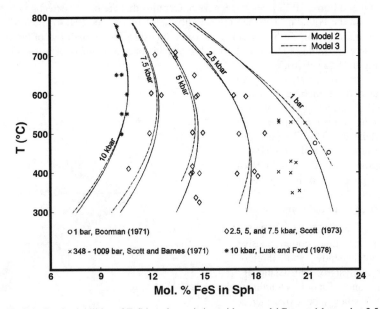

Figure 15. Calculated solubilities of FeS in *sph* coexisting with *pyr* and 1C *po* at 1 bar and at 2.5, 5, 7.5, and 10 kbar compared with experimental data. Isobars from Model 3 are truncated at the temperatures of spinodal ordering. [Used by permission of the Mineralogical Society of Great Britain and Ireland, from Balabin and Sack (2000), *Mineral Mag*, Vol. 64, Fig. 5, p. 937.]

coexisting with *po* for *sph* covering the composition ranges of roughly 20-55 and 10-55 mol% FeS at 850 and 700 °C, respectively (Fig. 16). Using *po* compositions in these experiments, they obtained f_{S_2} and a_{FeS}^{pyr} for these *sph* from Equations (17) and (19) and observed that $a_{FeS}^{pyr}(X_{FeS}^{sph})$ is nearly independent of temperature above 580 °C, a condition that Scott and Barnes demonstrated continues down to at least 340 °C. Barton and Toulmin (1966) also constructed a plot in which *sph* isopleths are superimposed on a T–f_{S_2} diagram (Fig. 17) and they obtained an expression for the *sph* unit cell dimension,

$$a_0 = 5.4093 + 0.05637\left(X_{FeS}^{sph}\right) - 0.0004107\left(X_{FeS}^{sph}\right)^2 \tag{39a}$$

subsequently extended by Barton and Skinner (1967) to account for non-(Zn,Fe)S components in *sph*:

$$a_0 = 5.4093 + 0.0546 X_{FeS}^{sph} + 0.424 X_{CdS}^{sph} + 0.202 X_{MnS}^{sph}$$
$$+ 0.0700 X_{CoS}^{sph} + 0.2592 X_{ZnSe}^{sph} - 0.3 X_{ZnO}^{sph} \tag{39b}$$

However, Equation (39a) for the lattice spacing parameter a_0 does not apply to *sph* with $X_{FeS}^{sph} > 0.3$ synthesized under the low sulfur fugacities of the Fe + FeS buffer; these have values of a_0 distinctly less than the curve defined by Equation (39a) and departures from this curve become more pronounced with increasing temperature of synthesis (Fig. 18; Barton and Toulmin 1966, their Fig. 3). Equation (39a) also accounts for most other cell edge data for synthetic *sph* with $X_{FeS}^{sph} < 0.25$ (e.g., van Aswegen and Verlanger 1960; Chernyshev et al. 1969; Sorokin et al. 1970; Osadchii and Sorokin 1989; Dicarlo et al. 1990) but not for greater values of X_{FeS}^{sph}, where there is a noticeable spread of the data, particularly for *sph* synthesized hydrothermally at lower temperatures (e.g., Chernyshev et al. 1969; Sorokin et al. 1970).

The first attempt to extract the mixing properties of (Zn,Fe)S *sph* from the experimental data for *sph* coexisting with *po* was that of Fleet (1975), who performed a graphical integration

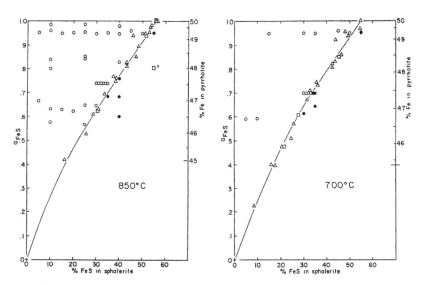

Figure 16. 700 and 850 °C activities of FeS as a function of *sph* composition. Triangles indicate reversed runs and circles indicate bulk compositions of appearance of phase runs; open circles = no *po*, filled circles = + *po*. [Used by permission of the Society of Economic Geologists, Inc., from Barton and Toulmin (1966), *Econ Geol*, Vol. 61, Figs. 12 & 13, p. 837.]

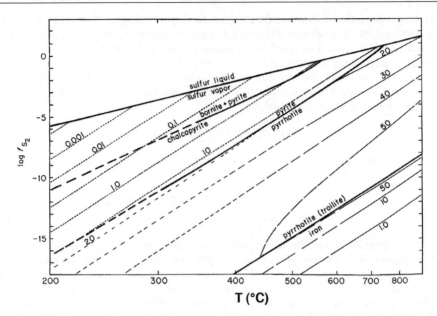

Figure 17. *Sph* composition in the assemblage *sph* + *pyr* + *po* as a function of fugacity of S$_2$ and temperature. The *cpy* + bornite + *pyr* curve is from Barton and Toulmin (1964). [Used by permission of the Society of Economic Geologists, Inc., from Barton and Toulmin (1966), *Econ Geol*, Vol. 61, Fig. 17, p. 842.]

of Barton and Toulmin's (1966) 850 °C data. Based on this analysis Fleet (1975) inferred that (Zn,Fe)S *sph* displays slight positive deviations from ideal mixing over its entire composition range, these deviations are not grossly inconsistent with strictly regular solution behavior, and the ratios of the activities of FeS in coexisting *po* and *sph* are roughly constant with

$$k = \frac{a^{po}_{FeS}}{a^{sph}_{FeS}} = \exp\left(\frac{\mu^{\circ sph}_{FeS} - \mu^{\circ po}_{FeS}}{RT}\right) = 1.62 \tag{40}$$

Subsequently, Hutcheon (1978, 1980) attempted to extract the temperature dependence of *sph* mixing properties from a wider selection of experimental constraints on *sph* + *po* assemblages, including the data of Barton and Toulmin (1966), but excluding their data at 1104, 1001, 1000, and 925 °C (wurtzite present in some runs) and at 555 °C (difficulty in obtaining equilibrium) and all but run 42A of Scott and Barnes (1971). Hutcheon (1978, 1980) assumed *sph* could be described as a simple asymmetric solution (e.g., Margules 1895; Carlson and Colburn 1942; Thompson 1967) in which:

$$RT \ln \gamma^{sph}_{FeS} = \left(X^{sph}_{ZnS}\right)^2 \left[W^{sph}_{G\,FeS} + 2\left(W^{sph}_{G\,ZnS} - W^{sph}_{G\,FeS}\right)X^{sph}_{FeS}\right] \tag{41a}$$

$$RT \ln \gamma^{sph}_{ZnS} = \left(X^{sph}_{FeS}\right)^2 \left[W^{sph}_{G\,ZnS} + 2\left(W^{sph}_{G\,FeS} - W^{sph}_{G\,ZnS}\right)X^{sph}_{ZnS}\right] \tag{41b}$$

He used Equation (41a) in the condition of equilibrium for the reaction FeS (*po*) → FeS (*sph*):

$$\Delta \bar{H}^{\circ}_{po \to sph} - T\Delta \bar{S}^{\circ}_{po \to sph} = RT \ln\left(\frac{a^{po}_{FeS}}{X^{sph}_{FeS}\gamma^{sph}_{FeS}}\right) - \bar{V}^{\circ}_{po \to sph}(P-1) \tag{42}$$

and Equation (19) to constrain values of $\Delta \bar{H}^{\circ}_{po \to sph}$, $\Delta \bar{S}^{\circ}_{po \to sph}$, $W^{sph}_{H\,ZnS}$, $W^{sph}_{S\,ZnS}$ ($W^{sph}_{G\,ZnS} = W^{sph}_{H\,ZnS}$

$-TW^{sph}_{S\,ZnS}$), $W^{sph}_{H\,FeS}$, and $W^{sph}_{S\,FeS}$ ($W^{sph}_{G\,FeS} = W^{sph}_{H\,FeS} - TW^{sph}_{S\,FeS}$) from these data. Hutcheon (1978) obtained a result very similar to that of Fleet (1975) at 850 °C and found that 99.76% of the variation in $RT\ln(a^{po}_{FeS}/X^{sph}_{FeS})$ could be explained assuming $\ln\gamma^{sph}_{FeS}$ is not a function of temperature with $\Delta\bar{S}^{\circ}_{po\rightarrow sph} = 3.515$, $W^{sph}_{S\,ZnS} = 2.686$, and $W^{sph}_{S\,FeS} = 4.845$ J/K-mol, and with $\Delta\bar{H}^{\circ}_{po\rightarrow sph} = 1000$ and $W^{sph}_{H\,ZnS} = W^{sph}_{H\,FeS} = 0$ (J/mol). As will be discussed subsequently, this assumption of temperature independence to $\ln\gamma^{sph}_{FeS}$ (and $\ln\gamma^{sph}_{ZnS}$) was also shown to be a plausible first approximation by Sack and Loucks (1985) in describing Fe-Zn partitioning between sph and $Cu_{10}(Fe,Zn)_2(Sb,As)_4S_{13}$ fah.

There have been further attempts to refine the model of Hutcheon (1978, 1980) using a wider spectrum of data from both low and high pressures, notably that of Martin and Gil (2005). In this most recent treatment a subset of the experimental data of Barton and Toulmin (1966),

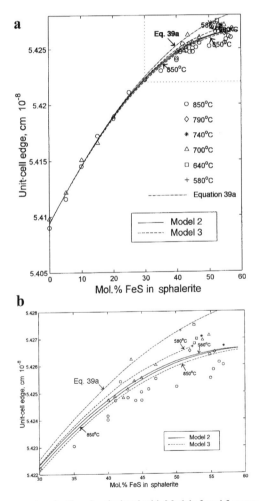

Figure 18. a) Lattice parameters for Fe-sph calculated with Models 2 and 3 compared with experimental results of Barton and Toulmin (1966) for Fe-*sph* prepared under the Fe/FeS buffer (points) and Fe-*sph* synthesized at higher sulfur pressures, but which presumably re-ordered on quench (dotted curve, eqn. 39a). b) Enlarged view of the upper right portion of Figure 18a. [Used by permission of the Mineralogical Society of Great Britain and Ireland, from Balabin and Sack (2000), *Mineral Mag*, Vol. 64, Fig. 6, p. 938.]

Boorman (1967), Boorman et al. (1971), Scott and Barnes (1971), Scott (1973), Hutchison (1978), Lusk and Ford (1978), Hutchison and Scott (1981, 1983), Bryndia et al. (1988, 1990), Lusk et al. (1993), and Balabin and Urusov (1995) were employed along with temperature and pressure dependent Margules parameters (i.e., $W_{G\,FeS}^{sph} = W_{H\,FeS}^{sph} - TW_{S\,FeS}^{sph} + PW_{V\,FeS}^{sph}$ and $W_{G\,ZnS}^{sph} = W_{H\,ZnS}^{sph} - TW_{S\,ZnS}^{sph} + PW_{V\,ZnS}^{sph}$). The prediction of Martin and Gil (2005) is broadly similar to, but less temperature dependent than, that of Hutcheon (1978) in describing the excess Gibbs energy of sph at 1 bar ($W_{S\,ZnS}^{sph} = 2.224$ and $W_{S\,FeS}^{sph} = 3.122$ J/K-mol, $W_{H\,ZnS}^{sph} = 3809.4$ and $W_{H\,FeS}^{sph} = 3694.7$ J/mol). As with Hutcheon (1978), Martin and Gil's (2005) inferred mixing properties for sph at 850 °C are similar to those inferred by Fleet (1975). On closer inspection, however, it is apparent that the temperature dependence of the excess Gibbs energy of sph:

$$\bar{G}^{ex} = X_{FeS}^{sph} RT \ln \gamma_{FeS}^{sph} + X_{ZnS}^{sph} RT \ln \gamma_{ZnS}^{sph} \tag{43}$$

inferred by either of these studies is too large to satisfy various other constraints, including experimental constraints on heat capacities (e.g., Pankratz and King 1965) and experimental and natural constraints on phase relations in ZnS- and FeS-bearing supersystems (e.g., Sack 2000, 2005). Pankratz and King (1965) have determined that the molar heat capacity of Fe-sph is nearly a linear function of composition, restricting the excess entropy of mixing to less than 0.25-0.3 J/K-mol (Balabin and Sack 2000); Sack (2000, 2005) has shown that, in detail, the assumption that $\ln \gamma_{FeS}^{sph}$ (and $\ln \gamma_{ZnS}^{sph}$) are temperature independent does not satisfy a multiplicity of phase relations involving fah and various other sulfosalts in the Ag_2S-Cu_2S-ZnS-FeS-Sb_2S_3-As_2S_3 supersystem. These discrepancies, as well as the decrease in unit cell volume with annealing temperaure (Fig. 18b), may be accounted for by considering short-range-ordering (SRO) of Fe^{2+} in the sph lattice.

SRO, or clustering of Mn^{2+} and Fe^{2+} in nearest-neighbor metal sites, is responsible for low-temperature magnetic behavior, and gives rise to long-range magnetic interactions in Mn- and Fe-substituted sph (e.g., Spalek et al. 1986; Furdyna 1988; Twardowski 1990; Di Benedetto et al. 2002; Bernardini et al. 2004). Different states of SRO ("metastable" states) may account for contradictory experimental results, because multiple "metastable" states are likely. The experiments of Scott and Barnes (1971) may have produced such "metastable" states. In these experiments homogenous patches of a "metastable" phase (up to 8.7 mol% richer in FeS than matrix) were produced during hydrothermal recrystallization. These became more like the "matrix" in Fe/(Fe+Zn) with increasing temperature, becoming identical by 530 °C. Similar phenomena have been reported elsewhere, and it is possible that the apparent experimental temperature independence of sph FeS-content in the $sph+po+pyr$ assemblage at 1 bar, and 5 and 7.5 kbar and between 340 and 520 °C (Boorman 1967; Boorman et al. 1971; Scott 1973) is also a manifestation of "metastable" equilibria. Models based on the assumption of random distribution of metals between tetrahedral sites, such as those cited above, cannot account for these experimental results and are inconsistent with constraints from natural assemblages (e.g., Toulmin et al. 1991).

Balabin and Sack (2000) developed cluster variation method (CVM) models for the analysis of thermodynamic mixing properties of sph which make explicit provision for SRO of metal atoms in the fcc cationic sublattice. These CVM models employ cuboctahedron and octahedron basis clusters and provide accurate descriptions for the enthalpy and configuration entropy, and correctly predict a decrease in the configurational volumes of Fe-sph annealed at high temperatures relative to lower temperatures (Fig. 18b). An innovative and unique feature of these models is their explicit formulation of the linear equations relating different cluster configurations, developing the principles formulated by Hijmans and de Boir (1955). This methodology, rather than the use of the so-called correlation polynomials typically employed as independent variables in CVM calculations (e.g., Sanchez and de Fontaine 1981; Sanchez et al. 1984; de Fontaine 1994), affords dramatic improvements in computational efficiencies.

Models of increasing intricacy in their formulation for internal energy were calibrated to reveal the interactions prevailing in (Zn,Fe)S solutions.

The simplest model that proved adequate contained three empirical interaction parameters for next to nearest neighbor (***nnn***) pair interactions and for many-bodied interactions associated with nearest neighbor (***nn***) equilateral triangles. Within experimental uncertainty it accounts for all experimental data on Fe-solubility in *sph* coexisting with *po* and *pyr* and with *po* and Fe-metal. This model predicts relatively small positive deviations from ideality (Fig. 19) with activity-composition relations qualitatively similar to those obtained by Fleet (1975) by Gibbs-Duhem integration for 850 °C (Fig. 20), and has an excess entropy permitted by the calorimetric data of Pankratz and King (1965). However, it indicates that *sph* undergoes long-range ordering to lower symmetry structures at temperatures only slightly below those investigated experimentally. Moreover, more realistic models prescribe that such ordering occurs at even higher temperatures. This inference, obtained from calculated curves for spinodal ordering (curves marking the appearance of an ordered phase or an assemblage of the disordered and ordered phases; e.g., de Fontaine 1994; Fig. 21), appears to be independently corroborated by: (1) abrupt changes and discontinuities in the electrochemical potential of FeS in (Zn,Fe)S *sph* at high temperatures (Osadchii and Lunin 1994); (2) a Mössbauer investigation of synthetic, hydrothermal *sph* indicating that Fe occurs on two distinct sites (Sorokin et al. 1975); (3) the composition data for natural (Zn,Fe)S phases (Balabin and Sack 2000; Di Benedetto et al. 2005), and (4) widespread metastable phenomena in Fe-*sph* in experiments below 550 °C.

For ease in computation of activities, Balabin and Sack (2000) approximate the Gibbs energy for the three parameter model using a Guggenheim type polynomial (Guggenheim 1937):

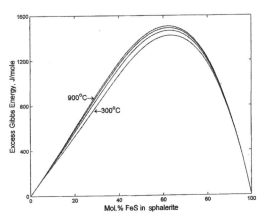

Figure 19. Excess Gibbs energies of (Zn,Fe)S *sph* at 300, 500, 700, and 900 °C calculated using Model 2. [Used by permission of the Mineralogical Society of Great Britain and Ireland, from Balabin and Sack (2000), *Mineral Mag*, Vol. 64, Fig. 7, p. 938.]

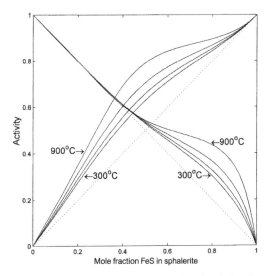

Figure 20. Activities of FeS and ZnS in sph calculated at 300, 500, 700, and 900 °C using Model 2. [Used by permission of the Mineralogical Society of Great Britain and Ireland, from Balabin and Sack (2000), *Mineral Mag*, Vol. 64, Fig. 8, p. 939.]

$$\bar{G}^{ex} = X_{FeS}^{sph} X_{ZnS}^{sph} \Big[A_0 + A_1 \left(X_{ZnS}^{sph} - X_{FeS}^{sph} \right) + A_2 \left(X_{ZnS}^{sph} - X_{FeS}^{sph} \right)^2$$
$$+ A_3 \left(X_{ZnS}^{sph} - X_{FeS}^{sph} \right)^3 + A_4 \left(X_{ZnS}^{sph} - X_{FeS}^{sph} \right)^4 \Big] \quad (44)$$

where the coefficients A_i are expressed as polynomials in T (K) (Appendix 3). Expressions for the activities of FeS and ZnS components are given in Appendix 1; these may be corrected for pressure using Equation (12) of Hutchison and Scott (1983) for the mixing volume of (Zn,Fe)S *sph*. In the CVM model on which these activities are based, ordering in *sph* is driven by a tendency to minimize the concentration of identical atoms in the second, rather than the first coordination sphere in the cationic sublattice (***nnn*** interchange energy parameter $w^{nnn} = \varepsilon_{AA}^{nnn} - 2\varepsilon_{AB}^{nnn} + \varepsilon_{BB}^{nnn}$ >> triangular (***tr***) interchange energy parameters $w_1^{tr} = \varepsilon_{AAA}^{tr} - 3\varepsilon_{ABB}^{tr} + 2\varepsilon_{BBB}^{tr}$ and $w_2^{tr} = \varepsilon_{AAB}^{tr} - 2\varepsilon_{ABB}^{tr} + \varepsilon_{BBB}^{tr}$). Although the activities should be accurate for many temperatures of interest, some common regions of T-X_{FeS}^{sph} space lie below representative ordering spinodals for more comprehensive models (Type 3 Models, Fig. 21) in which the internal energy is formulated in terms of exchange energies for both centered square (***csq***) and triangular (***tr***) clusters. Even though little confidence can be placed on the exact positioning of these spinodals, due to their acute sensitivity to energy parameters, they may offer potentially helpful insights into the origin of several phenomena, including polytypism (e.g., Hollenbaugh and Carlson 1983; Chao and Gault 1998) and repetitive *sph* banding (e.g., Barton et al. 1977; L'Heureux 2000; Loucks 1984; Di Benedetto et al. 2005).

Polytypism in *sph* may arise from the substitution of Fe for Zn due to strong interactions between ***nnn*** atoms. Wurtzite and *sph* polytypes might be stabilized at intermediate Fe/(Fe+Zn), if the exchange energies between ***nnn*** atoms are smaller in these polytypes. Such polytypes usually differ in their Fe/(Fe+Zn) where multiple varieties are found together (e.g., Hollenbaugh and Carlson 1983; Chao and Gault 1998). The ordering of *sph* into various polytypes, which coexist stably at low temperatures (cf. curves of spinodal ordering in Fig. 21) might provide a convenient explanation for repetitive *sph* banding in some hydrothermal vein-type sulfide deposits, including "Black Smokers" and Mississippi Valley-type deposits. Otherwise, it is difficult to reconcile the preservation of sharp compositional zoning to likely cooling rates and temperatures of formation of some of these deposits, given experimental constraints on diffusion rates (Mitzuta 1988).

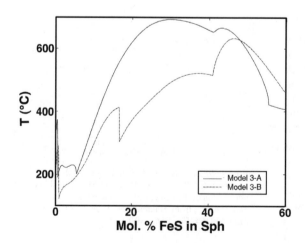

Figure 21. Ordering spinodals marking possible ordering of sph to lower symmetry structures and assemblages. Parameters for these spinodals are given in Balabin and Sack (2000, p. 940). These are not experimentally verified, but their existence is strongly suggested by available data. [Used by permission of the Mineralogical Society of Great Britain and Ireland, from Balabin and Sack (2000), *Mineral Mag*, Vol. 64, Fig. 9, p. 940.]

Ag_2S-Cu_2S-Sb_2S_3-As_2S_3

At temperatures greater than about 119 °C and below melting, the system Ag_2S-Cu_2S is comprised principally of three solid solutions, body-centered cubic (*bcc*)-$(Ag,Cu)_2S$, face centered cubic (*fcc*)-$(Ag,Cu)_2S$ and hexagonally close packed (*hcp*)-$(Cu,Ag)_2S$ (Fig. 22; Cava et al. 1980, 1981). The Ag_2S endmember of the *bcc*-$(Ag,Cu)_2S$ solid solution (space group *Im3m*) forms by a second-order transition from monoclinic α-Ag_2S. Just above the transition temperature of 177 °C this electronic and cation disordered fast-ion conductor has density maxima located on tetrahedral and octahedral sites with about one-fifth of the Ag occupying octahedral interstices, but by 325 °C the density of Ag on octahedral sites is negligible (Cava et al. 1980). Heat capacity decreases with temperature by about 2.4 J/K-mol over this temperature interval (Grønwold and Westrum 1986). The Cu_2S endmember of the *hcp*-$(Cu,Ag)_2S$ (space group *P6₃/mmc*) is also a fast-conductor and exhibits an even more pronounced decrease in heat capacity, of about 8.9 J/K-mol over its 103 to 437 °C range of stability (Grønwold and Westrum 1987). The *fcc*-$(Ag,Cu)_2S$ appears to have an antifluorite structure type (Ferrante et al. 1981) and forms a continuous solid solution above 592 °C with its solid solution range becoming progressively more restricted by the *bcc-fcc* $(Ag,Cu)_2S$ and *hcp-fcc* $(Cu,Ag)_2S$ transition loops until the field of *fcc*-$(Ag,Cu)S$ is terminated by the intersection of these loops at 119 °C (Skinner 1966; Fig. 22).

Although the enthalpies and entropies of the transitions between *bcc*- and *fcc*-Ag_2S and between *hcp*- and *fcc*-Cu_2S are well established at the transition temperatures of 592 and 437 °C (cf. Appendix 3; Grønwold and Westrum 1986, 1987), the differences in heat capacities between the transitional phases have not been established at temperatures below or above these transitions due to the non-quenchability of these phases. At room temperature, the sequence of minerals (and their synthetic equivalents) acanthite/argentite, jalpaite, mckinstryite, stromeyerite, and chalcocite are observed in natural specimens and experimental products spanning the Ag_2S-Cu_2S binary, these minerals initially crystallizing at temperatures from a high of 177 °C (acanthite) to a low of 80 °C (stromeyerite) (cf. Figs. 23, 24; Skinner

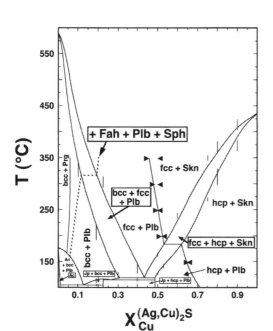

Figure 22. Phase relations in the Ag_2S-Cu_2S system between 100 and 600 °C and compositions of *bcc*-, *fcc*-, and *hcp*-$(Ag,Cu)_2S$ solid solutions in the assemblages *bcc* + *prg* + *plb*, *fcc* + *plb* + *skn*, and *hcp* + *plb* + *skn* in the system Ag_2S-Cu_2S-Sb_2S_3 and of *bcc*- and *fcc*-$(Ag,Cu)_2S$ coexisting with *fah*, *plb*, and *sph* in the system Ag_2S-Cu_2S-ZnS-Sb_2S_3. Vertical lines are experimental constraints on the *bcc*- to *fcc*- and *hcp*- to *fcc*-$(Ag,Cu)_2S$ transition loops and on phase relations involving acanthite (*ac*) and jalpaite (*jp*) are from Skinner (1966). [Used by permission of Elsevier Ltd, from Sack (2005), *Geochim Cosmochim Acta*, Vol. 69, Fig. 4, p. 1161.]

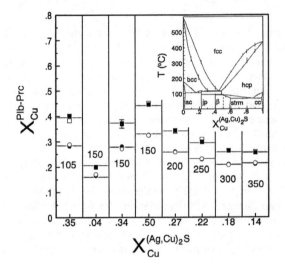

Figure 23. Molar Cu/(Cu+Ag) ratios of *plb* (circles) and *prc* (squares) for various temperatures (°C) and X_{Cu} of coexisting *bcc-fcc-*, or *hcp-*$(Ag,Cu)_2S$ solid solution; insert gives Ag_2S-Cu_2S phase relations from Skinner (1966) where the β phase (β), stromeyerite (*strm*) and chalcocite (*cc*) appear below 100 °C. [Used by permission of Elsevier Ltd, from Harlov and Sack (1994), *Geochim Cosmochim Acta*, Vol. 58, Fig. 2, p. 4368.]

Figure 24. Molar Cu/(Cu+Ag) of plb (circles), plb-prc (diamonds) and prc (squares) in Ag-Cu exchange equilibrium with one or two of the $(Ag,Cu)_2S$ phases acanthite (*ac*), jalpaite (*jp*), β phase (β), stromeyerite (*strm*), hcp-$(Cu,Ag)_2S$ (*hcp*), and chalcocite (*cc*) at 75 °C. Error bar indicates representative uncertainty (± 3%). [Used by permission of Elsevier Ltd, from Harlov and Sack (1994), *Geochim Cosmochim Acta*, Vol. 58, Fig. 1, p. 4367.]

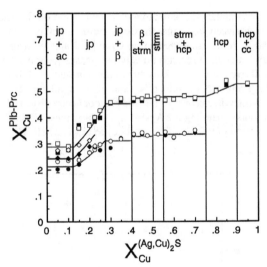

1966). High-temperature brackets on the *bcc-* and *fcc-* and *hcp-* and *fcc*-transition loops (Skinner 1966) provide the remaining constraints on thermodynamic properties of these $(Ag,Cu)_2S$ solid solutions (Fig. 22). By themselves, the totality of constraints within the $(Ag,Cu)_2S$ system are insufficient to define the thermodynamic mixing properties of *bcc-*, *hcp-* and *fcc-*$(Ag,Cu)_2S$ solid solutions, but this may be done by examining Ag-Cu partitioning data involving sulfosalt solutions which coexist with them in the systems Ag_2S-Cu_2S-Sb_2S_3-As_2S_3, polybasite-pearceite (*plb-prc*) [~$(Ag,Cu)_{16}((Sb,As)_2S_{11})$], pyrargyrite-proustite (*prg-prs*) [~$(Ag,Cu)_3(Sb,As)S_3$], and high-skinnerite (*skn*) [~$(Ag,Cu)_3SbS_3$] (cf. Fig. 22).

Sulfosalts $(Ag_2S$-$Cu_2S)$-Sb_2S_3-As_2S_3

Harlov and Sack (1994, 1995a,b) have determined Ag-Cu partitioning relations between these sulfosalt solutions and Ag_2S-Cu_2S phases over the temperature range 75-350 °C. These results and the experimental data of Hall (1966), Keighin and Honea (1969), Skinner et al.

(1972) and Maske and Skinner (1971) permit the construction of the 150-350 °C Ag$_2$S-Cu$_2$S-AgSbS$_2$-CuSbS$_2$ and Ag$_2$S-Cu$_2$S-AgAsS$_2$-CuAsS$_2$ quadrilaterals given in Fig. 25. Harlov and Sack (1994) used brackets on the difference in Ag-Cu partitioning between *plb* and *prc* with identical Ag$_2$S-Cu$_2$S assemblages (Fig. 24) or (Ag,Cu)$_2$S solution compositions (Fig. 23), experimental limits on the extent of a *plb-prc* miscibility gap at 75 °C (Fig. 26), and natural *plb-prc* compositions (Fig. 27) to constrain the thermodynamic mixing properties of *plb-prc* independently of those of (Ag,Cu)$_2$S sulfides. Harlov and Sack (1995a) demonstrated that the

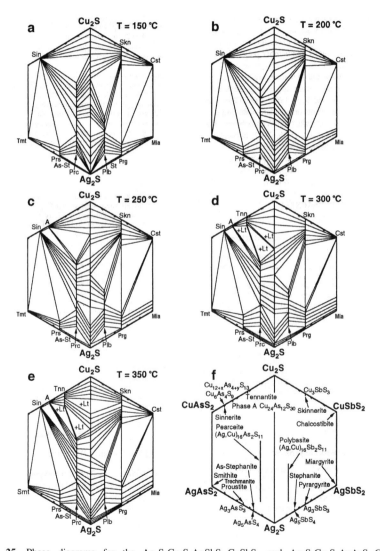

Figure 25. Phase diagrams for the Ag$_2$S-Cu$_2$S-AgSbS$_2$-CuSbS$_2$ and Ag$_2$S-Cu$_2$S-AgAsS$_2$-CuAsS$_2$ quadrilaterals of the systems Ag$_2$S-Cu$_2$S-Sb$_2$S$_3$ and Ag$_2$S-Cu$_2$S-As$_2$S$_3$ at 150, 200, 250, 300, and 350 °C (a-e) and mineral index (f). Mineral abbreviations: stephanite (*st*), ~(Ag,Cu)$_5$SbS$_4$; lautite (*lt*), ~ (Cu,Ag)AsS; tennantite (*tnn*), ~ (Cu,Ag)$_{12+x}$As$_{4+y}$S$_{13}$; phase A (*A*), ~ (Cu,Ag)$_{24}$As$_{12}$S$_{30}$; sinnerite (*sin*), ~ (Cu,Ag)$_6$As$_4$S$_9$; smithite (*smt*) ~ (Ag,Cu)AsS$_2$; trechmannite (tmt), ~ (Ag,Cu)AsS$_2$; arsenostephanite (*As-st*), ~ (Ag,Cu)$_5$AsS$_4$. [Used by permission of Elsevier Ltd, from Harlov and Sack (1995b), *Geochim Cosmochim Acta*, Vol. 59, Fig. 2, p. 4353.]

Figure 26. BSE images of unmixing of plb-prc produced during Ag-Cu exchange experiments with $(Ag,Cu)_2S$ sulfides at 75 °C. Light areas are Sb-rich domains; scale bar = 200 μm in **a**, and = 500 μm in **b**, **c**, and **d**. [Used by permission of Elsevier Ltd, from Harlov and Sack (1994), *Geochim Cosmochim Acta*, Vol. 58, Fig. 4, p. 4370.]

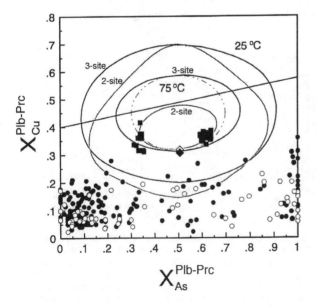

Figure 27. Miscibility gaps calculated for $(Ag,Cu)_{16}(Sb,As)_2S_{11}$ *plb-prc* employing two-site and three-site models for mixing of Cu and Ag; circles are *plb-prc* reported in the literature (open = Se-*plb-prc*). [Used by permission of Elsevier Ltd, from Harlov and Sack (1994), *Geochim Cosmochim Acta*, Vol. 58, Fig. 3, p. 4368.]

properties inferred for the Gibbs energy of *plb* permit description of the partitioning of Ag and Cu between *plb, prg,* and *skn* (Figs. 28, 29), this being accomplished under the banner of the physically plausible, first-approximation assumptions of zero values for the standard-state entropies of Ag-Cu exchange reactions and of homogeneous reactions within solid solutions. Continuing these assumptions Harlov and Sack (1995b) then examined additional Ag-Cu partitioning experiments to determine the mixing properties of *bcc*-, *hcp*-, and *fcc*-(Ag,Cu)$_2$S. Ghosal and Sack (1995) constrained the As-Sb exhange reactions between *plb-prc* and *prg-prs* (Fig. 30), and between *prg-prs*, miargyrite (*mia*) and smithite (*smt*) (Fig. 31), the non-ideality associated with the As-Sb substitution in *prg-prs*, *mia* and *smt* (Fig. 32), and calculated a phase diagram for the AgSbS$_2$-AgAsS$_2$ binary (Fig. 33). Finally, Sack (2000) combined the results of these activities, made minor adjustments to parameters and assembled a database for calculating reaction equilibria involving *bcc*-, *hcp*-, and *fcc*-(Ag,Cu)$_2$S, *plb-prc*, *prg-prs*, *mia-smt*, and trechmanite (*tmt*) (+*skn* in As-free subsystem) at temperatures greater than 119 °C

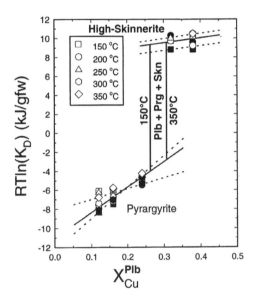

Figure 28. Experimental brackets on the apparent Gibbs energies of the Ag-Cu exchange reactions between *prg* and *plb*, $RT \ln[(X_{Cu}^{prg}/X_{Ag}^{prg})(X_{Ag}^{plb}/X_{Cu}^{plb})]$, and $RT \ln[(X_{Cu}^{skn}/X_{Ag}^{skn})(X_{Ag}^{plb}/X_{Cu}^{plb})]$ between *skn* and *plb*, at 150, 200, 250, 300, and 350 °C. Slanted solid lines are least squares fits to these brackets and vertical lines indicate molar Cu/(Cu+Ag) ratios of *plb* coexisting with *skn* + *prg* at 150 and 350 °C. [Used by permission of Elsevier Ltd, from Harlov and Sack (1995a), *Geochim Cosmochim Acta*, Vol. 59, Fig. 1, p. 869.]

Figure 29. Comparisons between the calibration and brackets on the *prg-skn* miscibility gap limbs deduced from Ag-Cu exchange experiments. [Used by permission of Elsevier Ltd, from Harlov and Sack (1995a), *Geochim Cosmochim Acta*, Vol. 59, Fig. 2, p. 870.]

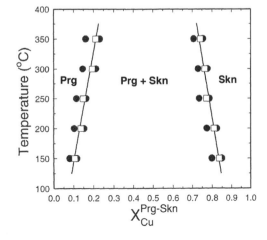

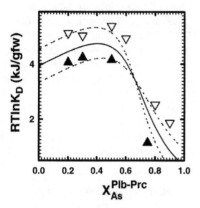

Figure 30. 350 °C experimental brackets on the apparent Gibbs energy of the As-Sb exchange reaction between *plb-prc* (Pbp) and *prg-prs* (Ppr), $RT \ln[(X_{As}^{Pbp}/X_{Sb}^{Pbp})(X_{Sb}^{Ppr}/X_{As}^{Ppr})]$, compared with calibration (dotted lines indicate 1σ uncertainty). [Used by permission of Elsevier Ltd, from Ghosal and Sack (1995), *Geochim Cosmochim Acta*, Vol. 59, Fig. 1, p. 3577.]

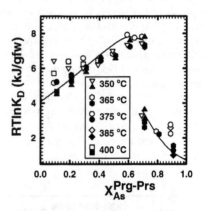

Figure 31. Calculated apparent Gibbs energy of the As-Sb exchange reaction between *prg-prs*, *mia*, and *smt*, $RT \ln[(X_{As}^{Pbr}/X_{Sb}^{Pbr})(X_{Sb}^{\Phi}/X_{As}^{\Phi})]$ (Φ = α-*mia*, β-*mia*, or *smt*) compared with experimental brackets. [Used by permission of Elsevier Ltd, from Ghosal and Sack (1995), *Geochim Cosmochim Acta*, Vol. 59, Fig. 2, p. 3577.]

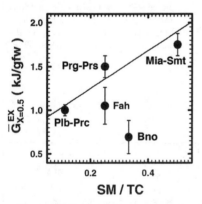

Figure 32. Excess Gibbs energies of As-Sb mixing of *plb-prc*, *prg-prs*, *fah*, *mia-smt*, and *bno* on a one semimetal formula unit basis; *bno* = bournonite-seligmannite, SM = semi-metal ions, TC = total metal plus semi-metal ions. [Used by permission of Elsevier Ltd, from Ghosal and Sack (1995), *Geochim Cosmochim Acta*, Vol. 59, Fig. 5, p. 3579.]

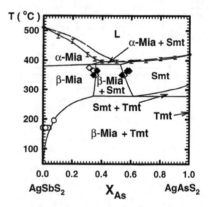

Figure 33. Phase diagram for the AgSbS$_2$-AgAsS$_2$ binary compared with experimental brackets on the *mia-smt* miscibility gap (diamonds), epithermal *mia* (open circles) and half reversals on solidus curves. [Used by permission of Elsevier Ltd, from Ghosal and Sack (1995), *Geochim Cosmochim Acta*, Vol. 59, Fig. 3, p. 3577.]

(e.g., 22, 41). Expressions for the activities of mineral components, solution parameters, and Gibbs energies of formation of the minerals in the database are given in Appendices 1–4.

Polybasite-pearceite. The most complicated of the models for Ag_2S-Cu_2S-Sb_2S_3-As_2S_3 solutions employed by Harlov and Sack (1994, 1995a,b), Ghosal and Sack (1995), and Sack (2000) is that for *plb-prc*. In this model (Harlov and Sack 1994) the vibrational Gibbs energy (G^*) is assumed to conform to a second degree Taylor expansion in two composition [X_3 = As/(As+Sb), X_4 = Cu/(Cu+Ag)] and two Ag-Cu ordering variables, $s = X_{Cu}^A - X_{Cu}^B$, and $t = X_{Cu}^A - X_{Cu}^C$, where atom fractions of Cu and Ag on A, B, and C sites are related to composition and ordering variables by the expressions: $X_{Cu}^A = X_4 + \frac{1}{4}s + \frac{1}{2}t$, $X_{Ag}^A = 1 - X_4 - \frac{1}{4}s - \frac{1}{2}t$, $X_{Cu}^B = X_4 - \frac{3}{4}s + \frac{1}{2}t$, $X_{Ag}^B = 1 - X_4 + \frac{3}{4}s - \frac{1}{2}t$, $X_{Cu}^C = X_4 + \frac{1}{4}s - \frac{1}{2}t$, and $X_{Ag}^C = 1 - X_4 - \frac{1}{4}s + \frac{1}{2}t$. As and Sb are inferred to occupy only one type of site. The expression for molar Gibbs energy is obtained by combining its expressions for vibrational ($\bar{G}^*$) and configurational ($-T\bar{S}^{IC}$) components ($\bar{G} = \bar{G}^* - T\bar{S}^{IC}$):

$$\bar{G} = \bar{G}_1^\circ(1 - X_3 - X_4) + \bar{G}_3^\circ(X_3) + \bar{G}_4^\circ(X_4) + \Delta\bar{G}_{34}^\circ(X_3)(X_4) + W_{AsSb}^{SM}(X_3)(1 - X_3) \quad (45)$$

$$+ \left[W_{AgCu}^A + W_{AgCu}^B + W_{AgCu}^C - \frac{1}{2}\left(\Delta\bar{G}_{4s}^* + \Delta\bar{G}_{4t}^* + \Delta\bar{G}_{st}^*\right)\right](X_4)(1 - X_4)$$

$$+ \frac{1}{4}\left(\frac{3}{2}\Delta\bar{G}_{4s}^* - \frac{1}{2}\Delta\bar{G}_{4t}^* - \frac{1}{2}\Delta\bar{G}_{st}^* + \Delta\bar{G}_s^* + W_{AgCu}^A - 3W_{AgCu}^B + W_{AgCu}^C\right)(s)$$

$$+ \frac{1}{4}\left(\Delta\bar{G}_{4t}^* - \Delta\bar{G}_{4s}^* - \Delta\bar{G}_{st}^* + \Delta\bar{G}_t^* + 2W_{AgCu}^A + 2W_{AgCu}^B - 2W_{AgCu}^C\right)(t)$$

$$+ \frac{1}{16}\left(-\frac{7}{2}\Delta\bar{G}_{4s}^* + \frac{1}{2}\Delta\bar{G}_{4t}^* + \frac{1}{2}\Delta\bar{G}_{st}^* - W_{AgCu}^A - 9W_{AgCu}^B - W_{AgCu}^C\right)(s^2)$$

$$+ \frac{1}{4}\left(\frac{1}{2}\Delta\bar{G}_{4s}^* - \frac{3}{2}\Delta\bar{G}_{4t}^* + \frac{1}{2}\Delta\bar{G}_{st}^* - W_{AgCu}^A + W_{AgCu}^B - W_{AgCu}^C\right)(t^2)$$

$$+ \frac{1}{4}\left(\frac{1}{2}\Delta\bar{G}_{4s}^* + \frac{3}{2}\Delta\bar{G}_{4t}^* - \frac{3}{2}\Delta\bar{G}_{st}^* - W_{AgCu}^A + 3W_{AgCu}^B + W_{AgCu}^C\right)(s)(t)$$

$$+ \frac{1}{4}\Delta\bar{G}_{3s}^*(X_3)(s) + \frac{1}{4}\left(\Delta\bar{G}_{3t}^* - \frac{3}{2}\Delta\bar{G}_{34}^*\right)(X_3)(t)$$

$$+ \frac{1}{2}\left(-\frac{3}{2}\Delta\bar{G}_{4s}^* + \frac{1}{2}\Delta\bar{G}_{4t}^* + \frac{1}{2}\Delta\bar{G}_{st}^* - W_{AgCu}^A + 3W_{AgCu}^B - W_{AgCu}^C\right)(X_4)(s)$$

$$+ \left(\frac{1}{2}\Delta\bar{G}_{4s}^* - \frac{1}{2}\Delta\bar{G}_{4t}^* + \frac{1}{2}\Delta\bar{G}_{st}^* - W_{AgCu}^A - W_{AgCu}^B + W_{AgCu}^C\right)(X_4)(t)$$

$$+ RT\begin{cases}4X_{Cu}^A \ln X_{Cu}^A + 4(1 - X_{Cu}^A)\ln(1 - X_{Cu}^A) + 4X_{Cu}^B \ln X_{Cu}^B \\ + 4(1 - X_{Cu}^B)\ln(1 - X_{Cu}^B) + 8X_{Cu}^C \ln X_{Cu}^C + 8(1 - X_{Cu}^C)\ln(1 - X_{Cu}^C) \\ + 2X_{As}^{SM} \ln X_{As}^{SM} + 2(1 - X_{As}^{SM})\ln(1 - X_{As}^{SM})\end{cases}$$

contains parameters of four types: (1) Gibbs energies of end-member components ($\bar{G}_1^\circ$, $\bar{G}_3^\circ$, $\bar{G}_4^\circ$), (2) Gibbs energies of reciprocal and reciprocal-ordering reactions ($\Delta\bar{G}_{34}^\circ$, $\Delta\bar{G}_{3s}^*$, $\Delta\bar{G}_{4s}^*$, $\Delta\bar{G}_{3t}^*$, $\Delta\bar{G}_{4t}^*$, and $\Delta\bar{G}_{st}^*$), (3) Gibbs energies of Ag-Cu ordering reactions ($\Delta\bar{G}_s^*$, $\Delta\bar{G}_t^*$), and (4) regular-solution parameters (W_{AsSb}^{SM}, W_{AgCu}^A, W_{AgCu}^B, and W_{AgCu}^C). Parameters of the last three types are assumed to be temperature independent, and the formulation is similar to the multi-site formulation for thermodynamic properties of Fe-Mg cummingtonite solid solutions of Ghiorso et al. (1995). Expressions for the chemical potentials of the end-member *plb-prc* components $Ag_{16}Sb_2S_{11}$, $Cu_{16}Sb_2S_{11}$, $Ag_{16}As_2S_{11}$, and $Cu_{16}As_2S_{11}$ (Appendix 1) may be obtained by applying the Darken equation (Tangent Intercept Rule) (cf. Sack 1992):

$$\mu_j = \bar{G} + \sum_i n_{ij}(1 - X_i)\left(\frac{\partial\bar{G}}{\partial X_i}\right)_{X_k/X_1} - s\left(\frac{\partial\bar{G}}{\partial s}\right)_{X_3, X_4, t} - t\left(\frac{\partial\bar{G}}{\partial t}\right)_{X_3, X_4, s} \quad (46)$$

where n_{ij} is the stoichiometric coefficient giving the number of moles of component i in one mole of component j. Finally, the ordering variables s and t are evaluated by solving the conditions of homogeneous equilibria

$$\left(\partial \bar{G}/\partial s\right)_{X_3,X_4,t} = \left(\partial \bar{G}/\partial t\right)_{X_3,X_4,s} = 0 \qquad \text{(see Appendix 2)}$$

Harlov and Sack (1994) justified the complexity of this model on several grounds. First, they noted that their assumption that there are three metal sites with multiplicities of 4(*A*), 4(*B*) and 8(*C*) is the simplest Ag-Cu ordering scheme consistent with the equipoint population groups of the *C2/m* space group, the symmetry inferred for end-member *plb* and *prc* (Peacock and Berry 1947; Frondel 1963). They also found that simpler models, such as one in which Ag and Cu are disordered ($s = t = 0$), could not be justified in light of their constraints on compositions of *plb* and *prc* in osmotic Ag-Cu exchange equilibrium over the temperature range 75-350 °C (Figs. 23, 24) or on the extent of miscibility gaps in *plb-prc* at 75 and 25 °C (Fig. 27). Even a simpler model with only one ordering variable ($s = t$, *B* and *C* metal sites combined, *A*:*C* = 1:3) proved inadequate; solutions that satisfied osmotic Ag-Cu exchange constraints (Figs. 23, 24) were characterized by thermal dependencies to the extent of the miscibility gap too large to satisfy both the 75 °C experimental data and the *plb-prc* compositions from nature (cf. Fig. 27). Finally, Harlov and Sack (1994) note that their three-site Ag-Cu ordering model applies strictly only to regions of temperature-composition space characterized by the single unit cell (Hall 1967). In low-Cu *plb-prc* with Cu/(Cu+Ag) ratios less than those investigated by Harlov and Sack (1994), *plb-prc* solutions exhibit three distinct sets of unit cell dimensions (1:1:1, 2:2:1, and 2:2:2) (e.g., Frondel 1947; Peacock and Berry 1947; Harris et al. 1965; Hall 1966; Barrett and Zolensky 1987), potentially requiring models of even more complexity to make explicit provision for convergent ordering transformations associated with multiple to single cell transitions.

Pyrargyrite-skinnerite. Harlov and Sack (1995a) utilized their 150-350 °C reversed experimental brackets on Ag-Cu partitioning between *plb*, *prg* and *skn* and the thermodynamic calibration of Harlov and Sack (1994) to constrain the thermodynamic mixing properties of *prg* and *skn* and the miscibility gap between them. The latter is readily inferred from their exchange experiments (cf. Fig. 28). Thermodynamic properties were constrained using a one-site Ag-Cu asymmetric mixing model for *prg* and a two-site 2:1 ordered Ag-Cu mixing model for *skn*. A one-site Ag-Cu model was adopted for *prg*, because crystallographic studies (e.g., Harker 1936) indicate that Ag atoms occupy only one type of site in a structure exhibiting $R\bar{3}c$ space group symmetry. Ag-Cu mixing in *prg* is far from ideal, because there is a strong composition dependence to the distribution coefficient for Ag-Cu exchange between *plb* and *prg* (Fig. 28) and Ag-Cu mixing is nearly ideal in *plb*. Symmetric Ag-Cu mixing behavior can also be readily ruled out, because this would imply metastable *prg-prg* miscibility gaps wider than the observed, stable *prg-skn* gap. Accordingly, Harlov and Sack (1995a) adopted an asymmetric Ag-Cu mixing model, but they found it was also necessary to make explicit provision for site distortion associated with the Cu^+ substitution for Ag^+, given the radically different ionic radii and stereochemistry of these monovalent cations. They did this following the example given by Sack and Ghiorso (1994) by assuming that the standard state properties of Ag-*prg* and Cu-*prg* end-members are linearly dependent on Cu/(Cu+Ag) ratio. In their resulting expression of molar Gibbs energy:

$$\begin{aligned}
\bar{G} = {} & \bar{G}^{\circ\,prg}_{Ag_3SbS_3}\left(1 - X^{prg}_{Cu}\right) + \left(\bar{G}^{\circ\,skn}_{Cu_3SbS_3} - \Delta\bar{G}^{\circ}_{Cu\,skn-prg}\right)\left(X^{prg}_{Cu}\right) \\
& + \left[W^{prg}_{AgCu}\left(X^{prg}_{Cu}\right) + X^{prg}_{CuAg}\left(1 - X^{prg}_{Cu}\right)\right]\left(X^{prg}_{Cu}\right)\left(1 - X^{prg}_{Cu}\right) \\
& + \left(\Delta\bar{H}^{\circ}_{DAg\,prg\text{-}Ag\,prg} - T\Delta\bar{S}^{\circ}_{DAg\,prg\text{-}Ag\,prg}\right)\left(X^{prg}_{Cu}\right)\left(1 - X^{prg}_{Cu}\right) \\
& + \left(\Delta\bar{H}^{\circ}_{DCu\,prg\text{-}Cu\,prg} - T\Delta\bar{S}^{\circ}_{DCu\,prg\text{-}Cu\,prg}\right)\left(X^{prg}_{Cu}\right)\left(X^{prg}_{Cu}\right) \\
& + 3RT\left[X^{prg}_{Cu}\ln X^{prg}_{Cu} + \left(1 - X^{prg}_{Cu}\right)\ln\left(1 - X^{prg}_{Cu}\right)\right]
\end{aligned} \qquad (47)$$

only the Gibbs energy terms expressing the compositional dependence of Ag-*prg* and Cu-*prg* end-members ($\Delta \bar{G}°_{DAg\,pyr\text{-}Ag\,pyr} = \Delta \bar{H}°_{DAg\,prg\text{-}Ag\,prg} - T\Delta \bar{S}°_{DAg\,prg\text{-}Ag\,prg}$ and $\Delta \bar{G}°_{DCu\,prg\text{-}Cu\,prg} = \Delta \bar{H}°_{DCu\,prg\text{-}Cu\,prg} - T\Delta \bar{S}°_{DCu\,prg\text{-}Cu\,prg}$) were considered temperature dependent, and $\Delta \bar{G}°_{Cu\,skn\text{-}prg}$, and the analogous $\Delta \bar{G}°_{Ag\,prg\text{-}skn}$, are differences in molar Gibbs energies of formation between end-member Ag and Cu *prg* and *skn*:

$$\Delta \bar{G}°_{Cu\,skn\text{-}prg} = \bar{G}°^{skn}_{Cu_3SbS_3} - \bar{G}°^{prg}_{Cu_3SbS_3} \tag{48}$$

$$\Delta \bar{G}°_{Ag\,prg\text{-}skn} = \bar{G}°^{prg}_{Ag_3SbS_3} - \bar{G}°^{skn}_{Ag_3SbS_3} \tag{49}$$

both of which are negative and were assumed to be temperature independent.

For high-skinnerite Harlov and Sack (1995a) adopted a two site Ag-Cu mixing model, motivated by the equipoint population groups of the space group *Pnma*, which suggests that the metal cations occupy two different sites in a 2:1 (*B:A*) ratio. The molar Gibbs energy was described with a second degree Taylor expansion in the composition variable X^{skn}_{Cu} = Cu/(Cu+Ag) and the ordering variable $s = X^A_{Cu} - X^B_{Cu}$ [$X^A_{Cu} = X^{skn}_{Cu} + 2/3s$; $X^A_{Ag} = (1 - X^{skn}_{Cu}) - 2/3s$; $X^B_{Cu} = X^{skn}_{Cu} - 1/3s$; $X^B_{Ag} = (1 - X^{skn}_{Cu}) + 1/3s$]. Expressions for chemical potentials evaluated for Ag_3SbS_3 and Cu_3SbS_3 *skn* components derived from application of the Darken equation to the expression for *skn* molar Gibbs energy:

$$\bar{G} = \left(\bar{G}°^{prg}_{Ag_3SbS_3} - \Delta\bar{G}°_{Ag\,prg\text{-}skn}\right)\left(1 - X^{skn}_{Cu}\right) + \left(\bar{G}°^{skn}_{Cu_3SbS_3}\right)\left(X^{skn}_{Cu}\right) \tag{50}$$
$$+ \left(W^{A\,skn}_{AgCu} + W^{B\,skn}_{AgCu} - \Delta\bar{G}°^*_{xs}\right)\left(X^{skn}_{Cu}\right)\left(1 - X^{skn}_{Cu}\right)$$
$$- \left(\tfrac{1}{2}\Delta\bar{G}°^*_s + \tfrac{1}{6}\Delta\bar{G}°^*_{xs} - \tfrac{2}{3}W^{A\,skn}_{AgCu} + \tfrac{1}{3}W^{B\,skn}_{AgCu}\right)(s)$$
$$- \left(\tfrac{2}{9}\Delta\bar{G}°^*_{xs} + \tfrac{4}{9}W^{A\,skn}_{AgCu} + \tfrac{1}{9}W^{B\,skn}_{AgCu}\right)(s^2)$$
$$+ \left(\tfrac{1}{3}\Delta\bar{G}°^*_{xs} - \tfrac{4}{3}W^{A\,skn}_{AgCu} + \tfrac{2}{3}W^{B\,skn}_{AgCu}\right)\left(X^{skn}_{Cu}\right)(s)$$
$$+ RT\left(X^A_{Cu}\ln X^A_{Cu} + X^A_{Ag}\ln X^A_{Ag} + 2X^B_{Cu}\ln X^B_{Cu} + 2X^B_{Ag}\ln X^B_{Ag}\right)$$

are given in Appendix 1. In Appendix 3 we give the parameter values Harlov and Sack (1995a) derived from analyzing the Ag-Cu exchange reactions between *prg* and *plb*, *skn* and *plb* (Fig. 28) and between *prg* and *skn* (Fig. 29) utilizing these expressions and those for the Cu and Ag end-member components in *plb* and *prg*. Both *prg* and *skn* are very non-ideal with respect to the substitution of Cu for Ag, they have Cu/(Cu+Ag) values less than 0.25, and greater than 0.70 at 400 °C, values that approach 0 and 1 roughly linearly with temperature until at least 150 °C (Fig. 28).

Ag_2S-Cu_2S

Harlov and Sack (1995b) utilized the calibrations for *plb* and *skn* energies of Harlov and Sack (1994, 1995a) and their reversed brackets on the Ag/Cu ratios of $(Ag,Cu)_2S$ phases coexisting with *plb* and *skn* (Fig. 34) to define the Ag-Cu exchange properties of *bcc-*, *fcc-*, and *hcp-* $(Ag,Cu)_2S$ using the expressions:

$$\bar{Q}^{plb\text{-}i}_{Cu(Ag)_{-1}} = \frac{1}{16}RT\ln\left(\frac{a^{plb}_{Cu_{16}Sb_2S_{11}}}{a^{plb}_{Ag_{16}Sb_2S_{11}}}\right) - \frac{\alpha_i}{2}RT\ln\left(\frac{X^i_{Cu}}{X^i_{Ag}}\right) \tag{51}$$
$$= \Delta\bar{G}°^{plb\text{-}i}_{Cu(Ag)_{-1}} + \frac{1}{2}W^i_{AgCu}\left(1 - 2X^i_{Cu}\right)$$

and

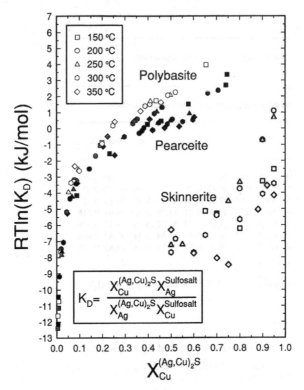

Figure 34. Experimental brackets on the 150–350 °C apparent Gibbs energies of exchange reactions between *bcc*-, *fcc*-, and *hcp*-(Ag,Cu)$_2$S solutions and *plb-prc* and *skn* sulfosalts, $RT \ln[(X_{Cu}^{(Ag,Cu)_2S}/X_{Ag}^{(Ag,Cu)_2S})(X_{Ag}^{sulfo}/X_{Cu}^{sulfo})]$. [Used by permission of Elsevier Ltd, from Harlov and Sack (1995b), *Geochim Cosmochim Acta*, Vol. 59, Fig. 3, p. 4362.]

$$\bar{Q}_{Cu(Ag)_{-1}}^{skn-i} = \frac{1}{3} RT \ln\left(\frac{a_{Cu_3SbS_3}^{skn}}{a_{Ag_3SbS_3}^{skn}}\right) - \frac{\alpha_i}{2} RT \ln\left(\frac{X_{Cu}^i}{X_{Ag}^i}\right) \quad (52)$$

$$= \Delta \bar{G}_{Cu(Ag)_{-1}}^{\circ\,skn-i} + \frac{1}{2} W_{AgCu}^i \left(1 - 2X_{Cu}^i\right)$$

where $a_{Cu_{16}Sb_2S_{11}}^{plb}$ and $a_{Ag_{16}Sb_2S_{11}}^{plb}$ and $a_{Cu_3SbS_3}^{skn}$ and $a_{Ag_3SbS_3}^{skn}$ are the activities of Cu$_{16}$Sb$_2$S$_{11}$ and Ag$_{16}$Sb$_2$S$_{11}$ in *plb*, and of Cu$_3$SbS$_3$ and Ag$_3$SbS$_3$ in *skn*, the terms $\Delta\bar{G}_{Cu(Ag)_{-1}}^{\circ\,plb-i}$ and $\Delta\bar{G}_{Cu(Ag)_{-1}}^{\circ\,skn-i}$ are the standard state Gibbs energies of the Cu-Ag exchange reactions

$$\frac{1}{16} \text{Cu}_{16}\text{Sb}_2\text{S}_{11} + \frac{1}{2} \text{Ag}_2\text{S} = \frac{1}{16} \text{Ag}_{16}\text{Sb}_2\text{S}_{11} + \frac{1}{2} \text{Cu}_2\text{S} \quad (53)$$
$$\text{plb} \qquad i\text{-(Ag,Cu)}_2\text{S} \qquad \text{plb} \qquad i\text{-(Ag,Cu)}_2\text{S}$$

and

$$\frac{1}{3} \text{Cu}_3\text{SbS}_3 + \frac{1}{2} \text{Ag}_2\text{S} = \frac{1}{3} \text{Ag}_2\text{SbS}_3 + \frac{1}{2} \text{Cu}_2\text{S} \quad (54)$$
$$\text{skn} \qquad i\text{-(Ag,Cu)}_2\text{S} \qquad \text{skn} \qquad i\text{-(Ag,Cu)}_2\text{S}$$

and the W_{AgCu}^i and α_i refer to the regular solution parameter, and site multiplicity of Cu-Ag mixing in the ith (Ag,Cu)$_2$S phase, *bcc*-, *fcc*-, or *hcp*-(Ag,Cu)$_2$S (cf. Fig. 35). Differences between values of $\Delta\bar{G}_{Cu(Ag)_{-1}}^{\circ\,plb-i}$ and $\Delta\bar{G}_{Cu(Ag)_{-1}}^{\circ\,skn-i}$ of the *i*-(Ag,Cu)$_2$S phases were further constrained to be consistent with values of $\Delta\bar{G}_{Cu(Ag)_{-1}}^{\circ\,bcc-fcc}$ and $\Delta\bar{G}_{Cu(Ag)_{-1}}^{\circ\,hcp-fcc}$ inferred from a similar analysis of the

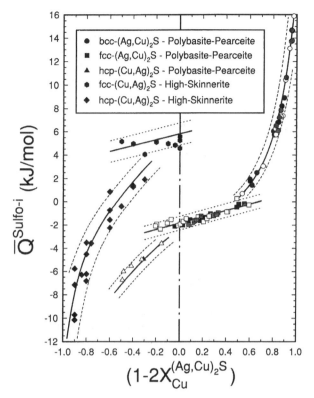

Figure 35. Calibration for Ag-Cu exchange reactions between $(Ag,Cu)_2S$ solutions and *plb-prc* and *skn* sulfosalts. (Eqns. 51 and 52, dotted curve indicate 1σ uncertainty). [Used by permission of Elsevier Ltd, from Harlov and Sack (1995b), *Geochim Cosmochim Acta*, Vol. 59, Fig. 4, p. 4362.]

Ag-Cu exchange reactions:

$$\tfrac{1}{2} Cu_2S + \tfrac{1}{2} Ag_2S = \tfrac{1}{2} Ag_2S + \tfrac{1}{2} Cu_2S \quad (55)$$
$$\text{bcc} \quad \text{fcc} \quad \text{bcc} \quad \text{fcc}$$

$$\tfrac{1}{2} Cu_2S + \tfrac{1}{2} Ag_2S = \tfrac{1}{2} Ag_2S + \tfrac{1}{2} Cu_2S \quad (56)$$
$$\text{hcp} \quad \text{fcc} \quad \text{hcp} \quad \text{fcc}$$

utilizing the brackets of Skinner (1966; cf. Fig. 22) (e.g., $\Delta\bar{G}^{\circ\,hcp\text{-}fcc}_{Cu(Ag)_{-1}} = \Delta\bar{G}^{\circ\,plb\text{-}fcc}_{Cu(Ag)_{-1}} - \Delta\bar{G}^{\circ\,plb\text{-}hcp}_{Cu(Ag)_{-1}}$), the temperature independent $\Delta\bar{G}^{\circ\,i\text{-}j}_{Cu(Ag)_{-1}}$ values. Finally, Harlov and Sack (1995b) attempted to render solutions satisfying Ag-Cu partitioning constraints simultaneously consistent with constraints on the conditions of equilibria for the net transport reactions

$$Ag_2S = Ag_2S \quad (57)$$
$$\text{bcc} \quad \text{fcc}$$

and

$$Cu_2S = Cu_2S \quad (58)$$
$$\text{hcp} \quad \text{fcc}$$

$$\Delta \bar{G}^{\circ\, bcc\text{-}fcc}_{Ag_2S} = \Delta \bar{H}^{\circ\, bcc\text{-}fcc}_{T_r,\, Ag_2S} - T\Delta \bar{S}^{\circ\, bcc\text{-}fcc}_{T_r,\, Ag_2S} + \int_{T_r}^{T} \Delta \bar{C}^{bcc\text{-}fcc}_{p\, Ag_2S} dT - \int_{T_r}^{T} \Delta \bar{C}^{bcc\text{-}fcc}_{p\, Ag_2S} d\ln T \quad (59)$$

$$= RT \ln \left(\frac{\alpha_{bcc} X^{bcc}_{Ag_2S}}{\alpha_{fcc} X^{fcc}_{Ag_2S}} \right) + W^{bcc}_{Ag\text{-}Cu} \left(X^{bcc}_{Cu} \right)^2 - W^{fcc}_{Ag\text{-}Cu} \left(X^{fcc}_{Cu} \right)^2$$

and

$$\Delta \bar{G}^{\circ\, hcp\text{-}fcc}_{Cu_2S} = \Delta \bar{H}^{\circ\, hcp\text{-}fcc}_{T_r,\, Cu_2S} - T\Delta \bar{S}^{\circ\, hcp\text{-}fcc}_{T_r,\, Cu_2S} + \int_{T_r}^{T} \Delta \bar{C}^{hcp\text{-}fcc}_{p\, Cu_2S} dT - \int_{T_r}^{T} \Delta \bar{C}^{hcp\text{-}fcc}_{p\, Cu_2S} d\ln T \quad (60)$$

$$= RT \ln \left(\frac{\alpha_{bcc} X^{hcp}_{Cu_2S}}{\alpha_{fcc} X^{fcc}_{Cu_2S}} \right) + W^{hcp}_{Ag\text{-}Cu} \left(1 - X^{hcp}_{Cu} \right)^2 - W^{fcc}_{Ag\text{-}Cu} \left(1 - X^{fcc}_{Cu} \right)^2$$

provided by the (Ag,Cu)$_2$S phase diagram constraints of Skinner (1966; Fig. 22) and the calorimetric constraints of Grønvold and Westrum (1986, 1987) on the differences in enthalpies, entropies, and heat capacities between *bcc*- and *fcc*-Ag$_2$S and between *hcp*- and *fcc*-Cu$_2$S at their transformation temperatures ($\Delta \bar{H}^{\circ\, bcc\text{-}fcc}_{T_r,\, Ag_2S}$, $\Delta \bar{S}^{\circ\, bcc\text{-}fcc}_{T_r,\, Ag_2S}$, and $\Delta \bar{C}^{\circ\, bcc\text{-}fcc}_{p\, Ag_2S}$, and $\Delta \bar{H}^{\circ\, hcp\text{-}fcc}_{T_r,\, Cu_2S}$, $\Delta \bar{S}^{\circ\, hcp\text{-}fcc}_{T_r,\, Cu_2S}$, and $\Delta \bar{C}^{\circ\, hcp\text{-}fcc}_{p\, Cu_2S}$, respectively; cf. Appendix 3)

From their experimental $\bar{Q}^{sulfo\text{-}i}_{Cu(Ag)_{-1}}$ constrains Harlov and Sack (1995b) inferred that *fcc*-(Ag,Cu)$_2$S exhibits symmetric regular solution behavior ($W^{fcc}_{Ag\text{-}Cu}$ = 5.4 ± 0.2 kJ/mol with α_{fcc} = 2), but that there is a strong composition dependence to the regular solution parameters $W^{bcc}_{Ag\text{-}Cu}$ and $W^{hcp}_{Ag\text{-}Cu}$, if they are assumed temperature independent (cf. Fig. 35). As an alternative, Sack (2000) employed temperature dependent, but composition independent, regular solution parameters $W^{bcc}_{Ag\text{-}Cu}$ and $W^{hcp}_{Ag\text{-}Cu}$, because the strong compositional dependence to these parameters inferred by Harlov and Sack (1995b) is untenable (c.f. Sack 2000, p. 3806-7). This resulted in an improved fit to the $\bar{Q}^{plb\text{-}bcc}_{Cu(Ag)_{-1}}$ and $\bar{Q}^{skn\text{-}hcp}_{Cu(Ag)_{-1}}$ constraints, and transition loops that satisfying the constraints on the (Ag,Cu)$_2$S phase diagram (Fig. 22) calculated using Equations (59) and (60) for values of α_{bcc} = α_{hcp} = 3 and the assumptions that $\Delta \bar{C}^{hcp\text{-}fcc}_{p\, Cu_2S}$ and $\Delta \bar{C}^{bcc\text{-}fcc}_{p\, Ag_2S}$ are constant, all with minimal changes to values of other parameters deduced by Harlov and Sack (1995b). Both Harlov and Sack (1995b) and Sack (2000) noted that values of α_{bcc} and α_{hcp} > 2 are required, the multiplicities of Ag-Cu site mixing greater than 2 presumably reflecting the behavior of *bcc*-(Ag,Cu)$_2$S and *hcp*-(Cu,Ag)$_2$S as ionic (fast-ion) conductors. They also used the brackets of Harlov and Sack (1995b) on the compositions of phases in the three phase assemblages *plb* + *prg* + *bcc*-(Ag,Cu)$_2$S and *plb* + *skn* + *fcc*- or *hcp*-(Cu,Ag)$_2$S and the constraints of Verduch and Wagner (1957) to derive a value for the Gibbs energy for formation of Ag$_{16}$Sb$_2$S$_{11}$ and Cu$_{16}$Sb$_2$S$_{11}$ *plb* (Appendix 4).

As-Sb exchange

Finally, Ghosal and Sack (1995) found that their 350-400 °C As-Sb exchange data for *plb-prc*, *prg-prs*, and *mia-smt* (Figs. 30, 31) are readily satisfied if *prg-prs* and *mia-smt* exhibit symmetric regular solution behavior, if As and Sb mixing occurs on only one type of site, and the nonideality of As-Sb mixing increases progressively with semimetal/total cation ratio in these sulfosalts (Fig. 32, Appendix 3) [terms for As-Sb mixing have been added to the expressions for *prg* and *mia* in Appendix 1 and parameter values appear in Appendix 3]. They speculated that the nonideality in As-Sb mixing in *prg-prs* (W = 6.0 kJ/mol), while only large enough to produce metastable unmixing more than 100 °C below the transitions of *prg* to xanthoconite and *prs* to pyrostilpnite at 192 °C (Hall 1966; Keighin and Honea 1969), is apparently large enough to account for the paucity of *prg-prs* with intermediate As/(As+Sb) relative to As- and Sb-rich varieties in epithermal, Ag-bearing vein-type deposits (Fig. 36).

They also constructed a phase diagram for the AgSbS$_2$-AgAsS$_2$ binary based on determinations of solidus temperatures and *mia-smt* miscibility gap pairs, the calorimetric constraints of Brynzdia and Kleppa (1988, 1989) on the *smt* to trechmanite (*tmt*) reaction, and the assumption of negligible As-solubility in *tmt* (Fig. 33). They demonstrated that their calculated curve for the solubility of As in *mia* coexisting with *tmt* is consistent with the data for Japanese epithermal *mia* of Motomura (1990) (Fig. 33). A more thorough demonstration of the accuracy of this prediction was given by Cambrubí et al. (2001), who determined As-contents of *mia* and fluid inclusion temperatures for various stages of silver mineralization in epithermal veins from the La Guitarra Ag-Au Deposit (Fig. 37) located in the southern most part of the Mexican silver belt. Ghosal and Sack (1995) also noted several discrepancies between the results of their reversed phase equilibrium experiments and the synthesis experiments of Chang et al. (1977). They synthesized, and reversibly reacted *mia* and *smt* with considerably wider ranges of As/(As+Sb) than those reported by Chang et al. (1977) for experiments at 300-400 °C. They also demonstrated the metastability of the phase with an intermediate composition Ag(As$_{0.3}$Sb$_{0.7}$)S$_2$ and the aramayoite (*arm*) structure, reported to occur at 350 and 400 °C by Chang et al. (1977). Finally, Sack (2000) demonstrated that the Ghosal and Sack (1999) calibration of the As-Sb exchange reaction between *prg-prs* + *plb-prc* is compatible with data from nature (e.g., Gemmell et al. 1989).

THE SYSTEM Ag$_2$S-Cu$_2$S-ZnS-FeS-Sb$_2$S$_3$-As$_2$S$_3$

Of the multitude of minerals that have compositions nearly contained within the Ag$_2$S-Cu$_2$S-ZnS-FeS-Sb$_2$S$_3$-As$_2$S$_3$ system, the sulfosalt fahlore (*fah*), ~(Cu,Ag)$_{10}$(Fe,Zn)$_2$(Sb,As)$_4$S$_{13}$ (also known as the tetrahedrite-tennantite group; Tatsuka and Morimoto 1977; Makovicky and Skinner 1978, 1979; Spiridonov 1984; Sack et al. 2005), exhibits the greatest latitude for solid solution (cf. Fig. 38). Most natural *fah* conform closely to this formula (e.g., Johnson and Jeanloz 1983) but may: (1) contain minor amounts of Cd, Hg, and Mn substituting for Fe and Zn, Bi substituting for Sb and As, or Se substituting for S (e.g., Springer 1969a; Charlat and Levy 1974; Pattrick 1978),

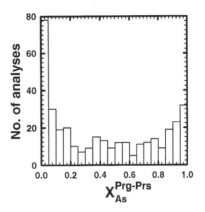

Figure 36. Composite of molar As/(As+Sb) histograms for Inakuraishi-type manganese and other epithermal deposits from Japan (Motomura 1990). [Used by permission of Elsevier Ltd, from Ghosal and Sack (1995), *Geochim Cosmochim Acta*, Vol. 59, Fig. 4, p. 3578.]

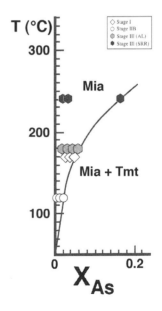

Figure 37. *Mia* deposited during stages I-III in epithermal veins from the La Guitarra Ag-Au deposit compared with the curve defining maximum As in *mia* of Ghosal and Sack (1995). [Used by permission of the Society of Economic Geologists, Inc., from Cambrubí et al. (2001), New Mines and Discoveries in Mexico and Central America, Special Publication 8, Fig. 14, p. 149.]

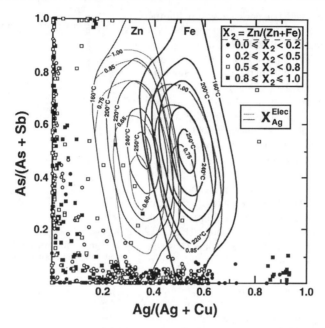

Figure 38. Molar Ag/(Ag+Cu) and As/(As+Sb) ratios of *fah* from nature compared to calculated phase equilibria. Fah compositions are grouped in four categories by molar Zn/(Zn+Fe). Curves correspond to calculated miscibility gaps for $(Cu,Ag)_{10}Fe_2(Sb,As)_4S_{13}$ and $(Cu,Ag)_{10}Zn_2(Sb,As)_4S_{13}$ *fah* and isopleths of *elec* on these gaps for the assemblage *elec* + *fah* + *fah* + *pyr* + *cpy* (equation 69). [Used by permission of the Mineralogical Society of Great Britain and Ireland, from Sack and Ebel (1993), *Mineral Mag*, Vol. 57, Fig. 3, p. 640.]

and (2) depart from the stoichiometry of this formula due to the coupled substitutions $2Fe^{2+} \leftrightarrow CuFe^{3+}$ and $\square Fe^{2+} \leftrightarrow 2Cu$ in low-Ag varieties not in equilibrium with *sph* (e.g., Charnock et al. 1989, Makovicky et al. 1990; Makovicky and Karup-Møller 1994; Sack and Goodell 2002) and possibly one in high-Ag varieties involving the substitution for vacancies for sulfur (e.g., Johnson 1986). As discovered by Pauling and Neumann (1934), the *fah* structure ($I\bar{4}3m$ space group symmetry with two formula units per unit cell) is a complex derivative of that of *sph* in which the metals are equally distributed between two sites which are three- and four-fold coordinated by sulfurs, and semimetals replace Zn atoms at the midpoints of the half-diagonals (¼, ¼, ¼, etc.), eight sulfurs are removed at ⅛, ⅛, ⅛, etc., and sulfurs are added to the center and corners of the unit cell (e.g., Wuensch 1964, Wuench et al 1966). The resulting structural formula for Ag_2S-Cu_2S-ZnS-FeS-Sb_2S_3-As_2S_3 *fah*, $^{III}(Ag,Cu)_6{}^{IV}[(Ag,Cu)_{2/3}(Fe,Zn)_{2/3}]_6[^{III}(Sb,As)^{IV}S_3]_4{}^{IV}S$, (e.g., Johnson 1986) conveniently illustrates that their thermodynamic description requires three composition variables and, potentially, one ordering variable describing the distribution of Cu or Ag between trigonal (III) and tetrahedral (IV) sites. Indeed, spectroscopic studies have been interpreted as indicating that Ag strongly prefers III-fold coordination, at least in Fe-rich *fah* with Ag/(Ag+Cu) < 0.4 at room temperature (e.g., Wuensch 1964, Wuench et al. 1966, Johnson and Burnham 1985; Peterson and Miller 1986; Charnock et al. 1988, 1989), but the temperature and composition dependence of this ordering is not well known. Curiously, volumes of natural Fe-rich *fah* are a sigmoid function of Ag/(Ag+Cu) with a pronounced local maximum at an Ag/(Ag+Cu) of about 0.35 (Indolev et al. 1971; Riley 1974), but volumes of equivalent *fah* synthesized at 400 °C are roughly linear in Ag/(Ag+Cu) (e.g., Pattrick and Hall 1983; Ebel 1988). Noting parallels to phenomena in ferrichromite spinels and K-nephelines (e.g., Francombe 1957; Robbins et al. 1971; Sack and Ghiorso 1991, 1998), it is tempting to attribute the collapsed unit cell in natural *fah* to a change in Ag site preference with Ag/(Ag+Cu) and attribute the nearly linear volume–Ag/(Ag+Cu) behavior of synthetic *fah* to higher temperature disordering of Ag and Cu between trigonal and tetrahedral sites, but alternative explanations have been proposed (e.g., Johnson and Burnham 1985; Peterson and Miller 1986; Charnock et al 1989).

Here we focus on *fah*, because its thermochemical properties have been the subject of some interest since the publication of the last Reviews in Mineralogy (and Geochemistry) volume on sulfides; it is a widely distributed and is often a primary mineral in polymetallic, hydrothermal deposits; its capacity for solid solution make it a potentially ideal petrogenetic indicator of ore-forming processes and, most importantly, it provides a bridge between the Ag_2S-Cu_2S-ZnS-FeS-Sb_2S_3-As_2S_3 system and more complicated systems, such as those containing PbS and Bi_2S_3. Following the currently popular chemical system enumeration approach where chemical systems are discussed in order of increasing complexity, we would most logically first consider *fah* within the simple system Ag_2S-Cu_2S-ZnS-Sb_2S_3 (i.e., $(Cu,Ag)_{10}Zn_2Sb_4S_{13}$ *fah*). This approach has the advantage that it affords graphical illustration of the key net-transport reactions that fix the Ag/(Ag+Cu) ratio in $(Cu,Ag)_{10}(Fe,Zn)_2(Sb,As)_4S_{13}$ *fah* in four-phase assemblages,

$$Ag_{10}Zn_2Sb_4S_{13} = AgSbS_2 + 3\ Ag_3SbS_3 + 2\ ZnS \qquad (60)$$
$$\textit{fah} \qquad\qquad \textit{mia} \qquad \textit{prg} \qquad\quad \textit{sph}$$

$$Ag_{10}Zn_2Sb_4S_{13} + 11\ Ag_2S = 2\ Ag_{16}Sb_2S_{11} + 2\ ZnS \qquad (61)$$
$$\textit{fah} \qquad\qquad \textit{bcc or fcc} \qquad \textit{plb} \qquad\quad \textit{sph}$$

$$5\ Ag_{10}Zn_2Sb_4S_{13} + Ag_{16}Sb_2S_{11} = 22\ Ag_3SbS_3 + 10\ ZnS \qquad (62)$$
$$\textit{fah} \qquad\qquad \textit{plb} \qquad\qquad \textit{prg} \qquad\quad \textit{sph}$$

on the familiar Ag_2S-Cu_2S-Sb_2S_3 plane (Fig. 39). We may then add FeS and As_2S_3 to illustrate how the Ag/(Ag+Cu) ratios in $(Cu,Ag)_{10}(Fe,Zn)_2(Sb,As)_4S_{13}$ fah in these four-phase assemblages vary with Fe/(Fe+Zn) and As/(As+Sb) ratios (e.g., Figs. 40, 41). Finally, we could consider how *fah* may be used as a petrogenetic indicator in systems of higher dimensionality, recording, for example, the Ag-content of primary galena solid solution in Ag-Pb-Zn ore deposits, or how equilibria involving *fah* may be used to evaluate Ag and Au resources. Before proceeding in this manner, we will first briefly review constraints on *fah* thermochemistry in the Fe- and As-bearing supersystems of the Ag_2S-Cu_2S-ZnS-Sb_2S_3 system.

Fahlore thermochemistry

Recent experimental and petrological studies have documented that the principal chemical substitutions in Ag_2S-Cu_2S-ZnS-FeS-Sb_2S_3-As_2S_3 system *fah* [$(Cu,Ag)_{10}(Fe,Zn)_2(Sb,As)_4S_{13}$], Ag for Cu, Zn for Fe, and As for Sb, are coupled energetically as a result of incompatibilities between Ag and As, Zn and As, and Ag and Zn in the *fah* crystal lattice. These incompatibilities are expressed as positive Gibbs energies of the reciprocal reactions

$$Cu_{10}Zn_2Sb_4S_{13} + Cu_{10}Fe_2As_4S_{13} = Cu_{10}Fe_2Sb_4S_{13} + Cu_{10}Zn_2As_4S_{13} \qquad (63)$$

$$Cu_{10}Zn_2Sb_4S_{13} + Ag_{10}Fe_2Sb_4S_{13} = Cu_{10}Fe_2Sb_4S_{13} + Ag_{10}Zn_2Sb_4S_{13} \qquad (64)$$

$$Cu_{10}Fe_2As_4S_{13} + Ag_{10}Fe_2Sb_4S_{13} = Cu_{10}Fe_2Sb_4S_{13} + Ag_{10}Fe_2As_4S_{13} \qquad (65)$$

The magnitude of the first of these reactions, Reaction (63), was deduced by Raabe and Sack (1984), Sack and Loucks (1985), and O'Leary and Sack (1987) from experimental and petrological studies of partitioning of Zn and Fe between *sph* and Ag-poor *fah* (cf. Fig. 42). These workers found that this Zn-Fe partitioning could be described simply with constant values of the standard state Gibbs energies of Reaction (63) and the Fe-Zn exchange reaction

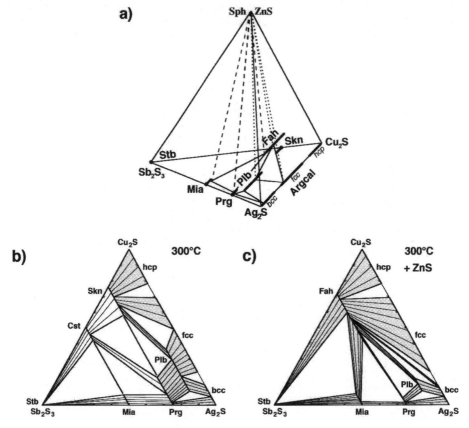

Figure 39. Comparison of phase equilibria in the system Ag_2S-Cu_2S-Sb_2S_3 (**a, b**) with equilibria for ZnS *sph* saturated assemblages in the system Ag_2S-Cu_2S-ZnS-Sb_2S_3 (**a, c**). [Used by permission of Elsevier Ltd, from Sack (2005), *Geochim Cosmochim Acta*, Vol. 69, Fig. 1, p. 1158.]

$$ZnS + \tfrac{1}{2}\, Cu_{10}Fe_2Sb_4S_{13} = FeS + \tfrac{1}{2}\, Cu_{10}Zn_2Sb_4S_{13} \quad (66)$$
$$\text{sph} \qquad \text{fah} \qquad \text{sph} \qquad \text{fah}$$

using an expression of the kind

$$RT \ln K_D + RT \ln\left(\frac{\gamma_{ZnS}^{sph}}{\gamma_{FeS}^{sph}}\right) = \Delta\bar{H}_{(66)}^\circ + \tfrac{1}{2}\Delta\bar{H}_{(63)}^\circ \left(\frac{As}{As+Sb}\right)^{fah} \quad (67)$$

where $K_D = (n_{Fe}^{fah}/n_{Zn}^{fah})(n_{Zn}^{sph}/n_{Fe}^{sph})$ where $\gamma_{ZnS}^{sph}/\gamma_{FeS}^{sph}$ was assumed to be temperature independent. However, O'Leary and Sack (1987) found Equation (67) inadequate, because values of the left-hand-side, rather than being linear in *fah* Ag/(Ag+Cu), display a sigmoid dependence on Ag/(Ag+Cu) (cf. Fig. 43), reminiscent of that displayed by volumes.

O'Leary and Sack (1987) and Sack et al. (1987) interpreted this sigmoid dependence as a manifestation of a change in Ag site preference, and formulated a model for the Gibbs energy of $(Cu,Ag)_{10}(Fe,Zn)_2(Sb,As)_4S_{13}$ *fah* based on three composition and an ordering variable: X_2 = Zn/(Zn+Fe), X_3 = As/(As+Sb), X_4 = Zn/(Zn+Fe), $s = X_{Ag}^{TRG} - \tfrac{3}{2} X_{Ag}^{TET}$. They expanded the vibrational Gibbs energy in a second-degree Taylor series in these variables and obtained the following expression for the molar Gibbs energy:

$$\bar{G} = \bar{G}^*_{Cu_{10}Fe_2Sb_4S_{13}}(X_1) + \bar{G}^*_{Cu_{10}Zn_2Sb_4S_{13}}(X_2) + \bar{G}^*_{Cu_{10}Fe_2As_4S_{13}}(X_3) + \bar{G}^*_{Ag_{10}Fe_2Sb_4S_{13}}(X_4) \quad (68)$$

$$+\bar{G}^*_s(s) + \Delta\bar{G}^\circ_{23}(X_2)(X_3) + \Delta\bar{G}^\circ_{24}(X_2)(X_4) + \Delta\bar{G}^\circ_{34}(X_3)(X_4) + \Delta\bar{G}^\circ_{2s}(X_2)(s)$$

$$+\Delta\bar{G}^\circ_{3s}(X_3)(s) + \tfrac{1}{10}\left(\Delta\bar{G}^*_{4s} + 6W^{TET}_{AgCu} - 4W^{TRG}_{AgCu}\right)(2X_4 - 1)(s)$$

$$+W^{TET}_{FeZn}(X_2)(1-X_2) + W^{SM}_{AsSb}(X_3)(1-X_3) + \left(\Delta\bar{G}^*_{4s} + W^{TET}_{AgCu} + W^{TRG}_{AgCu}\right)(X_4)(1-X_4)$$

$$+\tfrac{1}{25}\left(6\Delta\bar{G}^*_{4s} - 9W^{TET}_{AgCu} - 4W^{TRG}_{AgCu}\right)(s^2)$$

$$+RT\begin{Bmatrix} 2X_2 \ln\left(\dfrac{X_2}{3}\right) + 2(1-X_2)\ln\left(\dfrac{1-X_2}{3}\right) + 4X_3 \ln X_3 \\ +4(1-X_3)\ln(1-X_3) + 6(1-X_4-\tfrac{2}{5}s)\ln(1-X_4-\tfrac{2}{5}s) \\ +6(X_4+\tfrac{2}{5}s)\ln(X_4+\tfrac{2}{5}s) + 6(\tfrac{2}{3}X_4-\tfrac{2}{5}s)\ln(\tfrac{2}{3}X_4-\tfrac{2}{5}s) \\ +6\left[\tfrac{2}{3}(1-X_4)+\tfrac{2}{5}s\right]\ln\left[\tfrac{2}{3}(1-X_4)+\tfrac{2}{5}s\right] \end{Bmatrix}$$

where the mole fractions of atoms on sites are given by the relations: $X^{TET}_{Zn} = \tfrac{1}{3}X_2$, $X^{TET}_{Cu} = \tfrac{1}{3}(1-X_2)$, $X^{SM}_{As} = X_3$, $X^{SM}_{Sb} = (1-X_3)$, $X^{TRG}_{Ag} = X_4 + \tfrac{2}{5}s$, $X^{TRG}_{Cu} = (1-X_4) - \tfrac{2}{5}s$, $X^{TET}_{Ag} = \tfrac{2}{3}X_4 - \tfrac{2}{5}s$, $X^{TET}_{Cu} = \tfrac{2}{3}(1-X_4) + \tfrac{2}{5}s$, and the parameters are: (1) the vibrational Gibbs energies of $Cu_{10}Fe_2Sb_4S_{13}$, $Cu_{10}Zn_2Sb_4S_{13}$, $Cu_{10}Fe_2As_4S_{13}$ and $Ag_{10}Fe_2Sb_4S_{13}$, end-members ($\bar{G}^*_{Cu_{10}Fe_2Sb_4S_{13}}$, $\bar{G}^*_{Cu_{10}Zn_2Sb_4S_{13}}$, $\bar{G}^*_{Cu_{10}Fe_2As_4S_{13}}$, $\bar{G}^*_{Ag_{10}Fe_2Sb_4S_{13}}$); (2) regular-solution parameters for the non-ideality associated with the substitution of Ag for Cu on trigonal-planar and tetrahedral metal sites (W^{TRG}_{AgCu} and W^{TET}_{AgCu}), and the substitutions of Zn for Fe and As for Sb on tetrahedral metal and semimetal sites

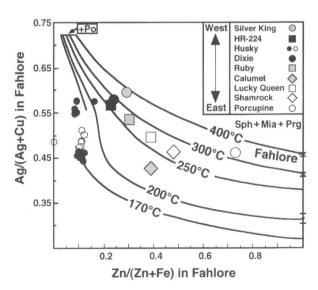

Figure 40. Molar Ag/(Ag+Cu) and Zn/(Zn+Fe) ratios of $(Cu,Ag)_{10}(Fe,Zn)_2Sb_4S_{13}$ *fah* in the *sph* + *mia* + *prg* assemblage (Sack 2005, solid curves) compared with average ratios for *fah* from mines in the Keno Hill Ag-Pb-Zn district, Yukon, Canada (Lynch 1989a, larger symbols) and with the compositions of coexisting *fah* deduced from unmixing features in a sample from the Husky mine, Keno Hill (Fig. 4.8, Sack et al. 2003, small filled circles are light and dark areas in *fah* in Fig. 45b). Error bars on the vertical axis for Zn/(Zn+Fe) = 1.00 represent the experimental constraints of Ebel and Sack (1994) at 200, 300 and 400 °C. Modified from Sack (2005). [Used by permission of Elsevier Ltd, from Sack (2005), *Geochim Cosmochim Acta*, Vol. 69, Fig. 2, p. 1158.]

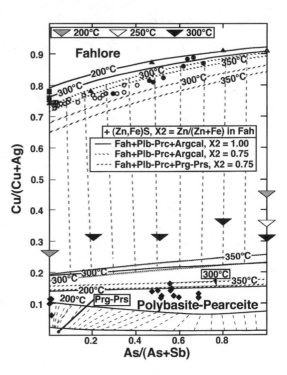

Figure 41. Molar Cu/(Cu+Ag) and As/(As+Sb) ratios for *fah* and *plb* coexisting with *argcal* + *sph* and *prg-prs* + *sph* in the system $Ag_2S-Cu_2S-ZnS-FeS-Sb_2S_3-As_2S_3$ and the FeS-free subsystem (Sack 2005) and for coexisting *prg-prs* and *plb-prc* in *argcal*-bearing assemblages in the simple system $Ag_2S-Cu_2S-Sb_2S_3-As_2S_3$ (Sack 2000). Arrows are experimental half-brackets of Sack (2000), squares are *fah* produced in 300 °C experiments reported by Ebel (1993), open and filled circles and diamonds represent *fah* and *plb-prc* in inclusions in quartz in a sample from the La Colorada Ag-Pb-Zn deposit, Zacatecas, Mexico (Chutas and Sack 2004), and emboldened tie-lines between prg-prs and plb-prc are from mineralization stages I-III of the Santo Nino vein, also in Zacatecas (Gemmell et al. 1989). Modified from Sack (2005). [Used by permission of Elsevier Ltd, from Sack (2005), *Geochim Cosmochim Acta*, Vol. 69, Fig. 3, p. 1159.]

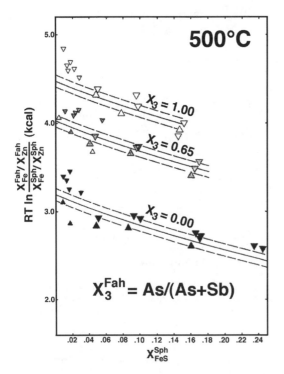

Figure 42. 500 °C experimental brackets on the apparent Gibbs energy of the Fe-Zn exchange reaction between *sph* and Ag-free *fah* (~ $Cu_{10}(Fe,Zn)_2(Sb,As)_4S_{13}$), $RT \ln[(X_{Fe}^{fah}/X_{Zn}^{fah})(X_{Zn}^{sph}/X_{Fe}^{sph})]$, expressed as a function of X_{Fe}^{sph} and compared with a calibration of Equation (67) and 1σ uncertainty of experimental data. From Sack and Loucks (1985)

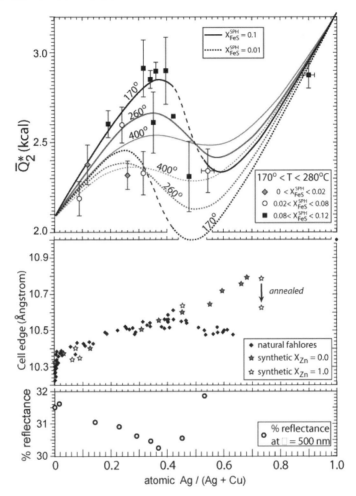

Figure 43. The "fahlore effect" on crystal chemistry. Comparison of the variation with Ag/(Ag+Cu) in *fah*, on (a) the Fe-Zn exchange reaction between *sph* and As-free *fah* as measured by $\bar{Q}_2^* = RT \ln[(X_{Fe}^{fah}/X_{Zn}^{fah})(a_{Zn}^{sph}/a_{Fe}^{sph})]$, estimated from assemblages from nature by O'Leary and Sack (1987), (b) cubic *fah* cell dimensions in natural samples (Riley 1974; Shimada and Hirowatari 1972; Indolev et al. 1971; Charlat and Levy 1975; Peterson and Miller 1986; Timoveyevskiy 1967; Petruk and staff 1971; Criddle, pers. comm 1992), and *fah* synthesized at ~400 °C (Hall 1972; Pattrick and Hall 1983; Ebel 1993), and (c) reflectance data ($\lambda = 500$ nm) reported by Imai and Lee (1980). Cell edge of one sample decreased as shown upon annealing at ~250 °C. Details in Ebel (1993, his Fig. 3.8).

($W_{Fe,Zn}^{TET}$ and $W_{As,Sb}^{SM}$); (3) standard state Gibbs energies of reciprocal Reactions (63)–(65) ($\Delta \bar{G}_{23}^{\circ}$, $\Delta \bar{G}_{24}^{\circ}$, and $\Delta \bar{G}_{34}^{\circ}$), and (4) vibrational Gibbs energies of ordering and reciprocal-ordering reactions ($\Delta \bar{G}_s^*$, and $\Delta \bar{G}_{2s}^*$, $\Delta \bar{G}_{3s}^*$, and $\Delta \bar{G}_{4s}^*$) (cf. Sack 1992). Expressions for the chemical potentials of the end-member vertices of the *fah* cube (Fig. 38) and for their activity-composition relations are readily derived from the extended tangent intercept rule (Eqn. 46), and these are given here in Appendix 1.

Using this model, the previous results of Raabe and Sack (1984) and Sack and Loucks (1985), and their Fe-Zn exchange data from nature constrained by fluid inclusion temperatures, O'Leary and Sack (1987) obtained a calibration for $(Cu,Ag)_{10}(Fe,Zn)_2Sb_4S_{13}$ *fah*, noting that all their solutions required a miscibility gap below 191 °C (cf. lower horizontal join of Fig. 44). Since then high-Ag and low-Ag *fah* miscibility gap pairs have been discovered (Sack et al. 2003) which confirm the predictions of O'Leary and Sack (1987). The exsolution pairs Sack et al. (2003) report for *fah* from the Husky mine, Keno Hill Ag-Pb-Zn district, Yukon, Canada (Fig. 45, X_{4A} = 0.560 ± 0.014 and X_{2A} = 0.104 ± 0.033 for the high-Ag *fah*, and X_{4B} = 0.449 ± 0.008 and X_{2B} = 0.113 ± 0.010 for the low-Ag *fah*) correspond to compositions calculated for a temperature of about 170 °C, a temperature consistent the upper bound on temperature of T < 197 ± 5 °C implied by the presence of stephanite (*st*) (Keighin and Honea 1969) in the Husky sample depicted in Figure 45.

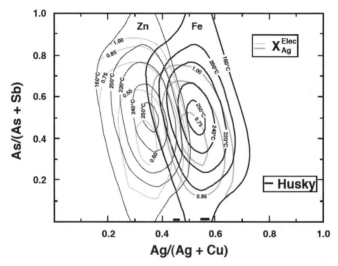

Figure 44. Husky mine fah miscibility gap pairs (X_2 = Zn/(Zn+Fe) ~ 0.11, Fig. 45b) compared with predictions (Fig. 38). [Used by permission of the Mineralogical Society of Great Britain and Ireland, from Sack et al. (2003), *Mineral Mag*, Vol. 67, Fig. 6, p. 1033.]

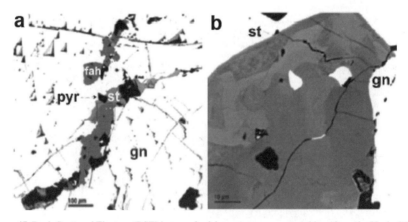

Figure 45. Back-Scattered Electron (BSE) image of a *fah* + *st* + *pyr* + *gn* assemblage from the Husky Mine, Keno Hill Ag-Pb-Zn district, Yukon, Canada (**a**, scale bar = 100 μm) in which the fah has unmixed to Ag-rich (brighter) and Ag-poor (darker) domains (**b**, scale bar = 10 μm). [Used by permission of the Mineralogical Society of Great Britain and Ireland, from Sack et al. (2003), *Mineral Mag*, Vol. 67, Fig. 4, p. 1029.]

Ebel and Sack (1989, 1991) and Sack and Ebel (1993) extended and refined the calibrations of O'Leary and Sack (1987) and Sack et al. (1987) for $(Cu,Ag)_{10}(Fe,Zn)_2(Sb,As)_4S_{13}$ *fah*. Sack and Ebel (1993) experimentally constrained the non-ideality associated with As-Sb mixing in $Cu_{10}(Zn,Fe)_2(Sb,As)_4S_{13}$ and bournonite (*bno*)-seligmannite solid solutions [$CuPb(Sb,As)S_3$] by determining As-Sb partitioning between them at 400 °C and found their study in accord with the As-Sb partitioning data obtained by Mishra and Mookherjee (1986) for *fah* + *bno*-seligmannite pairs in the Rajpura-Dariba polymetallic deposit, Poona, India. Ghosal and Sack (1995) noted that the non-idealities inferred to be associated with the As-Sb substitutions in *fah* and *bno* are less than those established for Ag_2S-$(Sb,As)_2S_3$ sulfosalts based on semi-metal/total cation ratio (Fig. 32), this increased ideality presumably reflecting the dividend of increased energetic simplicity afforded by increasing chemical and structural complexity.

Ebel and Sack (1989, 1991) constrained the Gibbs energy of reciprocal Reaction (65), $\Delta \bar{G}_{34}^\circ$, by obtaining reversed brackets on the Ag/(Ag+Cu) ratios of *fah* with varying As/(As+Sb) ratios coexisting with a matrix assemblage of *pyr*, chalcopyrite (*cpy*), *sph* ($X_{FeS}^{sph} \sim 0.05$), and electrums (*elec*) with 10, 20 and 30 mol% Ag at 400 °C (Fig. 46) and with 20 and 30 mol% at 300 °C. The *elec* + *cpy* + *pyrite* assemblage defines the Ag/(Ag+Cu) of *fah* (X_4) through the reaction:

$$Ag + \tfrac{1}{10} Cu_{10}Fe_2Sb_4S_{13} + FeS_2 = \tfrac{1}{10} Ag_{10}Fe_2Sb_4S_{13} + CuFeS_2 \qquad (69)$$
elec   *fah*   *pyr*   *fah*   *cpy*

Ebel and Sack (1989) employed the activity-composition relations for Ag in *elec* of White et al. (1957), the previous constraints on *fah* solution parameters and the assumptions that As enters only *fah* and that *cpy* and *pyr* are stoichiometric, extracting a value for $\Delta \bar{G}_{34}^\circ$ from the condition of equilibrium for Reaction (69) and their experimental brackets at 400 °C. This value for $\Delta \bar{G}_{34}^\circ$ has only recently been revised downwards by about 1.8% by Sack (2005) (cf. Appendix 3). Ebel and Sack (1989) also concluded that, to first order, the standard state entropy of Reaction (69) should be equal to that of the reaction:

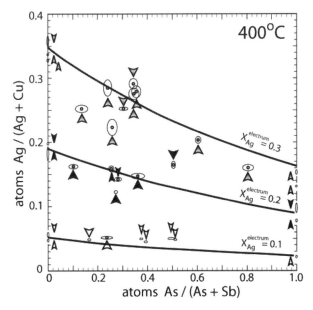

Figure 46. Reversed experimental brackets on molar Ag/(Ag+Cu) of *fah* coexisting with *pyr*, *cpy*, *sph* with 5 mol% FeS, and *elec* with 10, 20, and 30 mol% Ag at 400 °C compared with calibration of reaction (69). Modified from Ebel and Sack (1989).

$$Cu + FeS_2 = CuFeS_2 \tag{70}$$

a reaction which has a large entropy for a solid-solid reaction (43.43 J/K-mol, Barton and Skinner 1979) because it involves the reordering of metals between different sites. Reaction (69) can be thought as the sum of Reaction (70) and the Ag-Cu exchange reaction:

$$\tfrac{1}{10}Cu_{10}Fe_2Sb_4S_{13} + Ag° \leftrightarrow Cu° + \tfrac{1}{10}Ag_{10}Fe_2Sb_4S_{13} \tag{71}$$

and the standard state entropies of such exchange reactions can typically be ignored in geochemical calculations (e.g., Sack and Ghiorso 1989). Ebel and Sack (1991) demonstrated that this assumption leads to a correct prediction of their experimental results at 300 °C.

Sack and Brackebusch (2004) have shown that a calibration for Reaction (69) based on $\Delta \overline{S}° =$ 43.34 J/K-mol and the experimental brackets of Ebel and Sack (1989, 1991) is consistent with the composition-temperature data for coexisting *elec* and *fah* from a variety of precious metal deposits (Fig. 47). They demonstrated that cooling induced zoning of *fah* to lower, and of *elec* to higher, Ag-contents in gold-quartz veins from the Coeur d'Alene mining district (Fig. 48) and illustrated this thermal dependence with curves for the composition of *elec* coexisting with the "primary" (highest Ag) *fah* from these veins (Fig. 49). By matching compositions of *elec* grains isolated as inclusions of *asp* with $T\text{-}X_{Au}^{elec}$ curves, Sack and Brackebusch (2004) demonstrated that isolated *elec* in the High Grade Prospect had their compositions "frozen-in" within 40 °C of the temperature established as the "Hydrothermal Mineralization" temperature from fluid inclusion studies (cf. Arkadakskiy 2000), whereas free *elec* grains in the Coleman vein were substantially enriched in Ag, achieving Ag-contents appropriate to temperatures below 130 °C (Fig. 49). Finally, Sack and Brackebusch (2004) contrasted this solid-state Ag enrichment in *elec* with the Ag depletion in *elec* from environments of weathering and supergene enrichment where Ag is soluble relative to Au (e.g., Krupp and Wieser 1992; Greffié 2002).

Thermochemical database

A thermochemical database for $Ag_2S\text{-}Cu_2S\text{-}ZnS\text{-}Sb_2S_3$ *fah* and the sulfides/sulfosalts with which it coexists (e.g., Sack 2000, 2005) may be generated from (1) the experimental constraints of Ebel (1993), Ebel and Sack (1994) and Sack (2000) on Reactions (60) and (61), and on Ag-Cu partitioning between *fah* and $(Ag,Cu)_2S$ sulfides; (2) the activity-composition

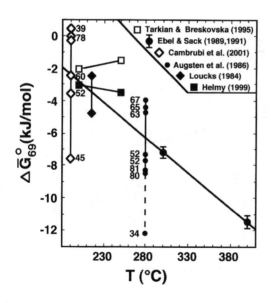

Figure 47. Predicted standard state Gibbs energy of Reaction (69) compared with constraints from nature and from Ebel and Sack (1989, 1991). Numbers adjacent to symbols are Ag-contents of *elec*. [Figure published in the *CIM Bulletin*, Vol. 97, No. 1081. Reproduced with permission of the Canadian Institute of Mining, Metallurgy and Petroleum.]

Thermochemistry of Sulfide Mineral Solutions 321

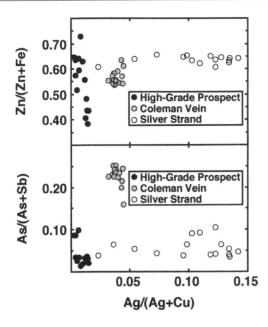

Figure 48. Molar Zn/(Zn+Fe), As/(As+Sb), and Ag/(Ag+Cu) of *fah* from three gold-quartz veins from the Coeur d'Alene mining district, Idaho that are prospects of the New Jersey Mining Company (NJMC). [Figure published in the *CIM Bulletin*, Vol. 97, No. 1081. Reproduced with permission of the Canadian Institute of Mining, Metallurgy and Petroleum.]

Figure 49. Mole fraction of Au in *elec*, X_{Au}^{elec}, defined by the calibration of Reaction (69) for the most Ag-rich *fah* in Figure 48. [Figure published in the *CIM Bulletin*, Vol. 97, No. 1081. Reproduced with permission of the Canadian Institute of Mining, Metallurgy and Petroleum.]

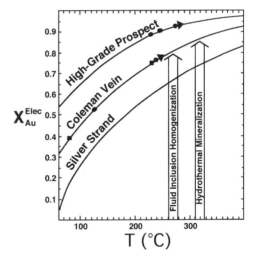

relations summarized above (Appendix 1); (3) the Gibbs energy of formation of the sulfosalts AgSbS$_2$ and Ag$_3$SbS$_3$ at 400 and 275 °C (Schenck et al. 1939; Schenck and von der Forst 1939; Verduch and Wagner 1957); and (4) the database for the simple system Ag$_2$S-Cu$_2$S-Sb$_2$S$_3$ discussed above. Sack (2000) assembled the first such database and examined the extensions of Reactions (60) and (61) into FeS- and As$_2$S$_3$-bearing systems, respectively. Finally, Sack (2005) implemented the activity-composition relations of Balabin and Sack (2000) for (Zn,Fe)S *sph* and revised several *fah* solution parameters (e.g., $\Delta \bar{G}_{24}^{\circ}$ and $\Delta \bar{G}_{s}^{*}$ revised downwards by about 6.5 and 17.3%, respectively) and the $\Delta \bar{H}_i^{\circ}$'s = $\Delta \bar{G}_i^{\circ}$'s of Ag-Cu reactions between *fah* and Ag$_2$S-Cu$_2$S-Sb$_2$S$_3$ sulfides and sulfosalts (e.g., $\Delta \bar{H}_{Cu(Ag)_{-1}}^{\circ \, fah\text{-}fcc}$ revised upwards by 10%) to achieve a database in closer agreement than that of Sack (2000) with the results of experimental investigations of Ebel (1993) and Sack (2000), and with petrological studies (e.g., Sack 2002; Sack et al. 2002, 2003, 2005; Sack and Goodell 2002; Chutas and Sack 2004) (cf. Appendix 3). Some calculations using the database of Sack (2005) are shown in Figures 39, 40, 41, and 22.

The first of these figures (Fig. 39) shows the location of some of the phases of interest in the Ag_2S-Cu_2S-ZnS-Sb_2S_3 tetrahedron (Fig. 39a), phase relations for the Ag_2S-Cu_2S-Sb_2S_3 base of this tetrahedron at 300 °C (Fig. 39b) and for assemblages containing ZnS *sph* at 300 °C projected onto this composition plane (Fig. 39c). Among other things, this diagram illustrates that the presence of *fah* substantially reduces or eliminates the stability of *plb* and *skn* in *sph*-bearing assemblages. We may extend this analysis by drawing a series of such diagrams for varying temperatures and different FeS-contents of *sph*. However, a more efficient, alternate approach is to examine the temperature dependence of the composition variables of low-variance assemblages such as those composed of *fah* + *mia* + *prg* + *sph*, *fah* + *prg* + *plb* + *sph*, and *fah* + *plb* + $(Ag,Cu)_2S$ + *sph* (cf. Figs. 40, 41). Thus, for example, we may display the Ag/(Ag+Cu) ratio of $(Ag,Cu)_2S$ solutions in the Ag_2S-Cu_2S-ZnS-Sb_2S_3 assemblage *fah* + *plb* + $(Ag,Cu)_2S$ + *sph* on the Ag_2S–Cu_2S binary (Fig. 22) to show the temperature at which *fcc*–$(Ag,Cu)_2S$ transforms to *bcc*–$(Ag,Cu)_2$ and to further illustrate the reduction of the *plb* stability field relative to that in the Ag_2S-Cu_2S-Sb_2S_3 system (Fig. 22)

In Figure 40 we show isotherms for the molar Zn/(Zn+Fe) and Ag/(Ag+Cu) ratios of $(Cu,Ag)_{10}(Fe,Zn)_2Sb_4S_{13}$ *fah* calculated between 170 and 400 °C for the assemblage defining the maximum solubility of Ag in *fah*: *fah* + *mia* + *prg* + *sph*. These isotherms are compared with the 200, 300 and 400 °C experimental data of Ebel and Sack (1994) for the Ag_2S-Cu_2S-ZnS-Sb_2S_3 subsystem. They are also compared with average compositions of *fah* coexisting with *sph*, *prg*, galena (*gn*) and siderite from mines in the Keno Hill Ag-Pb-Zn district, Yukon, Canada (Boyle 1965; Lynch 1989a), where Ag is concentrated in fault- and fracture-controlled sulfide veins in the Mississippian Keno Hill graphitic quartzite, which constitutes the distal portion of a laterally zoned, fossil hydrothermal system which extends over 40 km from the Cretaceous Mayo Lake Pluton, and in which *fah* appears to display zoning on the scale of tens of kilometers (Lynch 1989a, Lynch et al. 1990). It is not surprising that the 200, 300, and 400 °C isotherms for *fah* compositions in the *fah* + *mia* + *prg* + *sph* assemblage reproduce Ebel and Sack's (1994) constraints, as these were employed in constructing the database. What is surprising, and was originally noticed by Sack (2002), is that for most mines the average compositions of *fah* determined by Lynch (1989a) lie between the 250 and 300 °C isotherms calculated from the revised database of Sack (2005) (solid curves); 300 °C is at the upper end of the range in fluid inclusion temperatures determined by Lynch (1989b) for quartz and siderite in these ores (250-310 °C).

Sack (2002) interpreted the parallel between isotherms and average *fah* compositions for Keno Hill fah displayed in Fig. 40 as indicating that many of the *fah* were rapidly quenched following hydrothermal mineralization, leaving behind a signature of the incompatibility between Ag and Zn expressed on a district-wide scale. Rapid quenching following mineralization is exactly what was argued by Lynch (1989b) and Lynch et al. (1990) based on stable isotope, mineralogical and fluid inclusion studies. They concluded that the bulk of hydrothermal mineralization accompanied boiling in the 310-250 °C range, that rapid cooling was accomplished by mixing with meteoric water, and that Ag mineralization was a consequence of Ag-ligand complex destabilization due to reduction in oxygen fugacity caused by interaction between hydrothermal fluids and graphite as well as a rise in pH through boiling and loss of CO_2 in fluids infiltrating quartzite. In a further petrological study, however, Sack et al. (2003) concluded *fah* provides a more continuous record of thermal history, finding that temperatures as high as 400 °C are recorded by some isolated Ag- and Zn-rich *fah* compositions and by the As-content of arsenopyrites (*asp*) coexisting with *pyr*, *po* and *sph*, and that the most Fe-and Ag-rich *fah* are the products of retrograde Fe-Zn exchange with *sph*, unmixing or crystallization from late-stage fluids which produced *plb*, stephanite, acanthite and wire silver.

Sack (2000, 2005) also extended the analysis of net-transport Reactions (61) and (62) into FeS- and As_2S_3-bearing systems utilizing Ghosal and Sack's (1995) brackets on the As-Sb exchange reactions between *plb-prc* and *prg-prs* and an approximation for the standard state enthalpy of the As-Sb exchange reaction between *fah* and *plb-prc* (cf. Fig. 41). The revised database of Sack (2005) also provides a better accounting for the experimental constraints of Ebel (1993) and Sack (2000) than that of Sack (2000), correctly reproducing the Cu/(Cu+Ag) ratios of *fah* produced in experiments starting with the ZnS *sph* + *plb-prc* assemblage, and the half-brackets on *plb-prc* compositions (cf. Fig. 41). The revised database also provides an excellent accounting for the compositions of *fah* (0 ≤ As/(As+Sb) < 0.7) coexisting with *plb* and acanthite in inclusions in quartz and *sph* in polymetallic ores from the La Colorado Ag-Pb-Zn deposit in the Mexican silver belt (Chutas and Sack 2004) for the assumption that the ores initially crystallized at temperatures between 325 and 350 °C, a temperature range they established based on compositions of *fah* coexisting with *prg, sph* and *gn*, and on fluid inclusion temperatures of Albinson et al. (2001) interpreted in light of the reconstructed paleosurface (e.g., Albinson 1988). As in the Keno Hill district, the fingerprint of retrograde re-equilibration on *fah* compositions is everywhere evident. *Fah* compositions indicate lower temperatures where *fah* was not protected from reaction with secondary fluids by encapsulation, highlighting the need to pay close attention to petrological details when using *fah* as a geothermometer and petrogenetic indicator.

Bi_2S_3- AND PbS-BEARING SYSTEMS

Based on the experimental studies and analyses of Ghosal and Sack (1999) and Chutas (2004) we may extend the database for *fah*, and the minerals with which it coexists, to Bi_2S_3- and PbS-bearing systems, incorporating the minerals aramayoite (*arm*), matildite (*mat*), bismuthinite-stibnite solid solution (*bs*) and *gn* into the database. From this database we may develop composition diagrams for *gn* + *fah* + *sph* coexistences in the Ag_2S-Cu_2S-ZnS-FeS-PbS-Sb_2S_3-As_2S_3-Bi_2S_3 system and constrain formation energies of some sulfosalts observed in hydrothermal Ag-Pb-Zn deposits: diaphorite [~$Pb_2Ag_3Sb_3S_8$], freieslebinite [~$PbAgSbS_3$], bismuthian diaphorite [~$Pb_2Ag_3(Bi,Sb)_3S_8$], and bournonite [~$PbCuSbS_3$]. Consideration of many other of the 100+ sulfosalt species and solid solutions introduced by adding Bi_2S_3 and PbS to the list of simple sulfides (e.g., Gaines et al. 1997) is premature at this point, as there are few thermochemical and phase equilibrium data on most of these species. Craig and Scott (1976) summarized most of these data in the previous Reviews in Mineralogy (and Geochemistry) volume on sulfides, and they include: pilot thermochemical data on simple sulfosalt ternaries (e.g., Schenck et al. 1939; Schenck and von der Forst 1939; Verduch and Wagner 1957), phase equilibria for several simple constituent subsystems (e.g., Van Hook 1960; Craig et al. 1973; Hoda and Chang 1975), and stability analyses of ternary and quaternary sulfosalt systems based on assemblages from nature (e.g., Goodell 1975).

Shenck and co-workers (1939) determined H_2/H_2S ratios over multiphase assemblages in simple ternary sulfosalt systems Ag-Sb-S, Pb-Sb-S, and Cu-Sb-S at 400 °C and Ag-Bi-S, Pb-Bi-S, and Cu-Bi-S at 510 °C. Verduch and Wagner (1957), Craig and Lees (1972), and Craig and Barton (1973) have extracted Gibbs energies of formation of sulfosalts on the constituent simple system binaries (Ag_2S-Sb_2S_3, PbS-Sb_2S_3, Cu_2S-Sb_2S_3, Ag_2S-Bi_2S_3, PbS-Bi_2S_3, and Cu_2S-Bi_2S_3) under the assumptions that "solid solution effects are absent or negligible and that the semi-metal is pure" (Craig and Lees 1972. p. 375). Sack (2000) noted the prediction for the Gibbs energy of formation of the Cu_2S-Sb_2S_3 sulfosalt *skn* (~Cu_3SbS_3) is not negative enough satisfy the phase equilibrium constraints summarized above, and we advise against using these data for more than "rough" calculations in other such systems where solid solution effects are suspected or reaction times longer than those employed by Shenck and co-workers are required to achieve stable equilibria. We advise similar skepticism with regard to algorithms for estimating Gibbs energies (e.g., Craig and Barton 1973); numbers created by such generalizations are no substitute for those obtained by thermodynamic analysis of phase equilibrium, calorimetric, and structural constraints (see Makovicky 2006, this volume). Here, we will restrict our attention to subsystems of the quinary system Ag-Pb-Sb-Bi-S that contain portions of the Bi_2S_3-Sb_2S_3-$AgBiS_2$-$AgSbS_2$ quadrilateral and that include the main components of hydrothermal galena: PbS, $AgSbS_2$, and $AgBiS_2$.

The Bi_2S_3-Sb_2S_3-$AgBiS_2$-$AgSbS_2$ quadrilateral

To clarify the energetics of the Bi for Sb substitution, Ghosal and Sack (1999) investigated Bi–Sb partitioning between sulfosalts with compositions approximating the $Ag(Sb,Bi)S_2$ binary and bismuthinite-stibnites (*bs*) in the ternary Ag_2S-Sb_2S_3-Bi_2S_3 at temperatures between 300 and 450 °C (Fig. 50). Experiments were limited to values of $X_{Bi}^{bs} < 0.6$ at 450 °C and $X_{Bi}^{bs} < 0.9$ at 300 °C due to the formation of pavonite (~$Ag(Bi,Sb)_3S_5$). Ghosal and Sack (1999) constrained Bi-Sb energetics and constructed a *T-X* diagram for the $AgSbS_2$-$AgBiS_2$ subsystem (Fig. 51) from their reversed brackets on Bi-Sb partitioning, temperature and calorimetric constraints on the enthalpies of transitions of end–member *mia* and *mat* to the face-centered cubic *gn* (i.e., rock salt) structure (Bryndzia and Kleppa 1988b, 1989), and composition brackets on the width of the transition loop at 300 °C defined by both their synthesis experiments and those of Chang et al. (1977). In their analysis Ghosal and Sack (1999) assumed that both cubic $Ag(Sb,Bi)S_2$ and orthorhombic *bs* solid solution exhibit symmetric regular solution behavior with random mixing of Bi and Sb on the two sites per formula unit in *bs* and with random mixing of Ab, Bi, and Ag on the two sites per formula unit in cubic $Ag(Sb,Bi)S_2$.[†] For these assumptions, the following expressions for the molar Gibbs energies of *bs* and cubic $Ag(Sb,Bi)S_2$ may be readily derived:

$$\bar{G}^{bs} = \bar{G}^{\circ bs}_{Sb_2S_3}\left(1 - X^{bs}_{Bi}\right) + \bar{G}^{\circ bs}_{Bi_2S_3}\left(X^{bs}_{Bi}\right) + W^{bs}_{Bi\text{-}Sb}\left(X^{bs}_{Bi}\right)\left(1 - X^{bs}_{Bi}\right) \quad (72)$$
$$+ 2RT\left[X^{bs}_{Bi}\ln X^{bs}_{Bi} + \left(1 - X^{bs}_{Bi}\right)\ln\left(1 - X^{bs}_{Bi}\right)\right]$$

and

$$\bar{G}^{\alpha} = \bar{G}^{*\alpha}_{AgSbS_2}\left(1 - X^{\alpha}_{Bi}\right) + \bar{G}^{*\alpha}_{AgBiS_2}\left(X^{\alpha}_{Bi}\right) + W^{\alpha}_{Bi\text{-}Sb}\left(X^{\alpha}_{Bi}\right)\left(1 - X^{\alpha}_{Bi}\right) \quad (73)$$
$$+ 2RT\left\{½\ln(½) + ½X^{\alpha}_{Bi}\ln\left(½X^{\alpha}_{Bi}\right) + ½\left(1 - X^{\alpha}_{Bi}\right)\ln\left[½\left(1 - X^{\alpha}_{Bi}\right)\right]\right\}$$

From these expressions we may readily obtain the following expression for the condition of equilibrium for the Bi–Sb exchange reaction between *bs* and α-$Ag(Sb,Bi)S_2$:

$$½\, Bi_2S_3 + AgSbS_2 = AgBiS_2 + ½\, Sb_2S_3 \quad (74)$$

$$RT\ln\left[\frac{\left(1 - X^{\alpha}_{Bi}\right)X^{bs}_{Bi}}{\left(1 - X^{bs}_{Bi}\right)X^{\alpha}_{Bi}}\right] + ½W^{bs}_{Bi\text{-}Sb}\left(1 - 2X^{bs}_{Bi}\right) = \Delta\bar{G}^{\circ\, bs\text{-}\alpha}_{BiSb} + W^{\alpha}_{Bi\text{-}Sb}\left(1 - 2X^{\alpha}_{Bi}\right) \quad (75)$$

where $\Delta\bar{G}^{\circ\, bs\text{-}\alpha}_{BiSb} = (\bar{G}^{*\alpha}_{AgBiS_2} - \bar{G}^{*\alpha}_{AgSbS_2}) - ½(\bar{G}^{\circ\, bs}_{Bi_2S_3} - \bar{G}^{\circ\, bs}_{Sb_2S_3})$.

Formulating a thermodynamic model for the *mia-arm-mat* series proves more problematic, because the structural systematics of these minerals are not known in detail unambiguously, even at room temperature (e.g., Graham 1951; Knowles 1964; Mullen and Nowacki 1974;

[†] The random mixing approximation for Bi and Sb in *bs* is not strictly correct as Kyono and Kimata (2004) have demonstrated that *Pnma bs* exhibits ordering of Sb and Bi between two crystallographically distinct *M* sites. However, this ordering is sufficiently slight that it does not have a material impact on the results summarized here.)

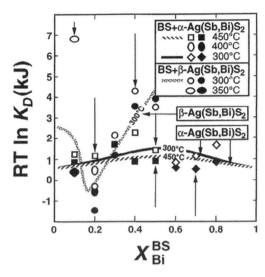

Figure 50. Experimental brackets on the apparent Gibbs energies of the Sb-Bi exchange reaction between *bs* and α- and β-Ag(Sb,Bi)S$_2$ solutions, $RT \ln[(X_{Sb}^{\Phi}/X_{Bi}^{\Phi})(X_{Bi}^{BS}/X_{Sb}^{BS})]$, compared with calibrations. [Used by permission of the Mineralogical Society of Great Britain and Ireland, from Ghosal and Sack (1999), *Mineral Mag*, Vol. 63, Fig. 2, p. 725.]

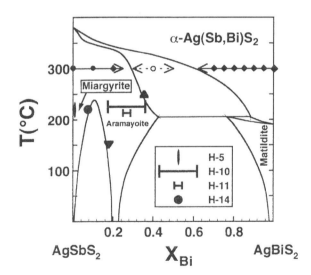

Figure 51. Calculated *T-X* diagram for the AgSbS$_2$-AgBiS$_2$ system (Ghosal and Sack 1999) compared with phase equilibrium constraints and the Bi/(Bi+Sb) ratios of *mia* and *arm* from Julcani, Peru (Sack and Goodell 2002). Filled circles and diamonds at 300 °C represent homogeneous compositions established by Chang et al. (1977) and Ghosal and Sack (1999), open circle is the bulk composition of an initially homogeneous material (X_{Bi} = 0.4) which unmixed at 300 °C to compositions as disparate as indicated by the tips of the accompanying arrows, inverted and upright triangles are natural *arm* compositions reported by Graham (1951, X_{Bi} = 0.17, *T* > 170 °C) and Borodayev et al. (1986, X_{Bi} ~ 0.35, *T* ~ 250 °C), and Julcani samples (H-5, H-10, H-11 and H-14) are plotted for an arbitrary temperature of 220 °C. [Used by permission of the Mineralogical Society of Great Britain and Ireland, from Sack and Goodell (2002), *Mineral Mag*, Vol. 66, Fig. 9, p. 1055.]

Smith et al. 1977; Effenberger et al. 2002). What is known for certain is that each of these minerals is a *gn* derivative based on alternate layers of Ag-Sb-Bi and S atoms with the semimetal and Ag atoms ordered into separate sites and can be described in terms of pseudo-cubic sublattices with nearly identical cell edge lengths (Graham 1951). In addition, there is sufficient phase equilibrium and crystallographic evidence that in *arm* (triclinic, space group $P1$ or $P\bar{1}$) semimetals are further ordered into two or more distinct sites at room temperature, and structural formulas with Sb/Bi ratios from 5:1 to 1.875:1 have been proposed for a fully ordered *arm* end-member (e.g., Graham 1951; Mullen and Nowacki 1974; Chang et al. 1977; Effenberger et al. 2002). Based on phase equilibrium considerations, such as the minimum in the distribution coefficient for Sb-Bi exchange between *bs* and Ag(Sb,Bi)S$_2$ solutions at 300 °C displayed in Fig. 50, Ghosal and Sack (1999) adopted an intermediate value of 4:1 for this ratio and the structural formula $Ag_5Sb_4^I Bi^{II} S_{10}$ for the fully ordered *arm* end-member. They also made the simplifying assumption that, at the temperatures of ore deposition, the *mia-arm-mat* series may be treated as a continuous solid solution for which strong Sb-Bi ordering and nonideality reduce the ranges of mutual solubility between *mia* and *arm* and between *arm* and *mat* with decreasing temperature. For these assumptions, the choice of $X_{Bi}^\beta \equiv Bi/(Bi+Sb)$ and $s^\beta \equiv X_{Bi}^{II\beta} - X_{Bi}^{I\beta}$ as composition and ordering variables, and the additional assumption that a Taylor expansion of second degree adequately describes the vibrational Gibbs energy, Ghosal and Sack (1999) obtained the following expression for the molar Gibbs energy of members of the β-Ag$_5$(Sb,Bi)$_4^I$(Bi,Sb)^{II}S$_{10}$ series:

$$\bar{G}^\beta = \bar{G}^{\circ\beta}_{Ag_5Sb_5S_{10}}\left(1-X_{Bi}^\beta\right) + \bar{G}^{\circ\beta}_{Ag_5Bi_5S_{10}}\left(X_{Bi}^\beta\right) \quad (76)$$

$$+\left(\Delta\bar{G}_x^{\circ\beta} + W_{Bi\text{-}Sb}^{I\beta} + W_{Bi\text{-}Sb}^{II\beta}\right)\left(X_{Bi}^\beta\right)\left(1-X_{Bi}^\beta\right)$$

$$+\left(\tfrac{3}{10}\Delta\bar{G}_x^{\circ\beta} + \tfrac{1}{2}\Delta\bar{G}_s^{\circ\beta} - \tfrac{1}{5}W_{Bi\text{-}Sb}^{I\beta} + \tfrac{4}{5}W_{Bi\text{-}Sb}^{II\beta}\right)\left(s^\beta\right)$$

$$+\left(\tfrac{4}{25}\Delta\bar{G}_x^{\circ\beta} - \tfrac{1}{25}W_{Bi\text{-}Sb}^{I\beta} - \tfrac{16}{25}W_{Bi\text{-}Sb}^{II\beta}\right)\left(s^\beta\right)^2 + \left(-\tfrac{3}{5}\Delta\bar{G}_x^{\circ\beta} + \tfrac{2}{5}W_{Bi\text{-}Sb}^{I\beta} - \tfrac{8}{5}W_{Bi\text{-}Sb}^{II\beta}\right)\left(X_{Bi}^\beta\right)\left(s^\beta\right)$$

$$+RT\left[\begin{array}{l}4\left(1-X_{Bi}^\beta + \tfrac{1}{5}s^\beta\right)\ln\left(1-X_{Bi}^\beta + \tfrac{1}{5}s^\beta\right) + 4\left(X_{Bi}^\beta - \tfrac{1}{5}s^\beta\right)\ln\left(X_{Bi}^\beta - \tfrac{1}{5}s^\beta\right) \\ +\left(1-X_{Bi}^\beta - \tfrac{4}{5}s^\beta\right)\ln\left(1-X_{Bi}^\beta - \tfrac{4}{5}s^\beta\right) + \left(X_{Bi}^\beta + \tfrac{4}{5}s^\beta\right)\ln\left(X_{Bi}^\beta + \tfrac{4}{5}s^\beta\right)\end{array}\right]$$

where the mole fractions of atoms on sites are given by the relations: $X_{Sb}^{I\beta} = 1 - X_{Bi}^\beta + \tfrac{1}{5}s^\beta$, $X_{Bi}^{I\beta} = X_{Bi}^\beta - \tfrac{1}{5}s^\beta$, $X_{Sb}^{II\beta} = 1 - X_{Bi}^\beta - \tfrac{4}{5}s^\beta$, $X_{Bi}^{II\beta} = X_{Bi}^\beta + \tfrac{4}{5}s^\beta$, and the parameters are (1) the molar Gibbs energies of Ag$_5$Sb$_5$S$_{10}$ and Ag$_5$Bi$_5$S$_{10}$ end-members ($\bar{G}^{\circ\beta}_{Ag_5Sb_5S_{10}}$ and $\bar{G}^{\circ\beta}_{Ag_5Bi_5S_{10}}$), (2) regular-solution parameters for the non-ideality associated with the substitution of Bi for Sb ($W_{Bi\text{-}Sb}^{I\beta}$ and $W_{Bi\text{-}Sb}^{II\beta}$), and (3) vibrational Gibbs energies of ordering and reciprocal-ordering reactions ($\Delta\bar{G}_s^{\circ\beta}$ and $\Delta\bar{G}_x^{\circ\beta}$) (cf. Ghosal and Sack 1999). From these expressions Ghosal and Sack (1999) readily obtained the following expression for the condition of equilibrium for the Bi–Sb exchange reaction between *bs* and β-Ag(Sb,Bi)S$_2$

$$\tfrac{1}{2}Bi_2S_3 + \tfrac{1}{5}Ag_5Sb_5S_{10} = \tfrac{1}{5}Ag_5Bi_5S_{10} + \tfrac{1}{2}Sb_2S_3 \quad (77)$$

$$RT\ln\left\{\left[\frac{\left(1-X_{Bi}^\beta + \tfrac{1}{5}s^\beta\right)^4 \left(1-X_{Bi}^\beta - \tfrac{4}{5}s^\beta\right)}{\left(X_{Bi}^\beta - \tfrac{1}{5}s^\beta\right)^4 \left(X_{Bi}^\beta + \tfrac{4}{5}s^\beta\right)}\right]^{1/5} \frac{X_{Bi}^{bs}}{\left(1-X_{Bi}^{bs}\right)}\right\} - \Delta\bar{G}_{BiSb}^{\circ bs\text{-}\beta} + \tfrac{1}{2}W_{Bi\text{-}Sb}^{bs}\left(1-2X_{Bi}^{bs}\right) \quad (78)$$

$$= \tfrac{1}{5}\left(\Delta\bar{G}_x^{\circ\beta} + W_{Bi\text{-}Sb}^{I\beta} + W_{Bi\text{-}Sb}^{II\beta}\right)\left(1-2X_{Bi}^\beta\right) + \tfrac{1}{25}\left(-3\Delta\bar{G}_x^{\circ\beta} + 2W_{Bi\text{-}Sb}^{I\beta} - 8W_{Bi\text{-}Sb}^{II\beta}\right)\left(s^\beta\right)$$

where

$$\Delta \overline{G}_{\text{BiSb}}^{\circ\, bs\text{-}\beta} = \tfrac{1}{5}\left(\overline{G}_{\text{Ag}_5\text{Bi}_5\text{S}_{10}}^{\circ\, \beta} - \overline{G}_{\text{Ag}_5\text{Sb}_5\text{S}_{10}}^{\circ\, \beta}\right) - \tfrac{1}{2}\left(\overline{G}_{\text{Bi}_2\text{S}_3}^{\circ\, bs} - \overline{G}_{\text{Sb}_2\text{S}_3}^{\circ\, bs}\right)$$ (79)

$$= \Delta \overline{G}_{\text{BiSb}}^{\circ\, bs\text{-}\alpha} + \Delta \overline{G}_{\text{AgSbS}_2}^{\circ\, \beta\to\alpha} - \Delta \overline{G}_{\text{AgBiS}_2}^{\circ\, \beta\to\alpha}$$

and the ordering variable s^β is evaluated by setting $(\partial \overline{G}/\partial s^\beta)=0$ (cf. Appendix 2). Finally, Ghosal and Sack (1999) derived an expression for the condition of equilibrium for the Sb-Bi exchange reaction between the α- and β-Ag(Sb,Bi)S$_2$ solid solutions of:

$$\text{AgBiS}_2 + \tfrac{1}{5}\text{Ag}_5\text{Sb}_5\text{S}_{10} = \tfrac{1}{5}\text{Ag}_5\text{Bi}_5\text{S}_{10} + \text{AgSbS}_2$$ (80)

by subtracting Equations (78) and (75).

Ghosal and Sack (1999) obtained a calibration for thermodynamic parameters based on a remarkably simple set of assumptions, including: (1) zero entropies of Bi-Sb exchange, ordering and reciprocal-ordering reactions; (2) identical non-ideality associated with the Bi for Sb substitution on the two Bi-Sb sites in β-Ag(Sb,Bi)S$_2$ solid solutions, and (3) a value of 2ln(2) for the standard state entropies of the β→α transitions in AgSbS$_2$ and AgBiS$_2$ end-members ($\Delta \overline{S}_{\text{AgSbS}_2}^{\circ\, \beta\to\alpha}$ and $\Delta \overline{S}_{\text{AgBiS}_2}^{\circ\, \beta\to\alpha}$) with values of $\Delta \overline{H}_{\text{AgSbS}_2}^{\circ\, \beta\to\alpha}$ and $\Delta \overline{H}_{\text{AgBiS}_2}^{\circ\, \beta\to\alpha}$ given by the relation $\Delta \overline{H}^{\circ\, \beta\to\alpha} = T\Delta \overline{S}^{\circ\, \beta\to\alpha}$ where T~380 °C for AgSbS$_2$ and T~190 °C for AgBiS$_2$. The latter assumption is entirely consistent with the calorimetric constraints of Bryndzia and Kleppa (1988b, 1989). Ghosal and Sack (1999) found that a calibration that satisfies all extant phase equilibrium constraints is readily achieved providing the standard state Gibbs energy of the ordering reaction:

$$\tfrac{3}{5}\text{Ag}_5\text{Sb}_5\text{S}_{10} + \text{Ag}_5(\text{Bi})_4^{\text{I}}(\text{Sb})^{\text{II}}\text{S}_{10} = \text{Ag}_5(\text{Sb})_4^{\text{I}}(\text{Bi})^{\text{II}}\text{S}_{10} + \tfrac{3}{5}\text{Ag}_5\text{Bi}_5\text{S}_{10}$$ (81)

is strongly negative (< −10 kJ/gfw; cf. Appendix 3), with the additional simplifying assumption of zero Gibbs energy of the reciprocal-ordering reaction:

$$\text{Ag}_5\text{Sb}_5\text{S}_{10} + \text{Ag}_5\text{Bi}_5\text{S}_{10} = \text{Ag}_5(\text{Sb})_4^{\text{I}}(\text{Bi})^{\text{II}}\text{S}_{10} + \text{Ag}_5(\text{Bi})_4^{\text{I}}(\text{Sb})^{\text{II}}\text{S}_{10}$$ (82)

Finally, Ghosal and Sack (1999) obtained values for the non-idealities of the Bi for Sb substitution in *bs* and the Ag(Sb,Bi)S$_2$ sulfides subject to the limiting boundary conditions that *bs* apparently exhibits complete solid solution at temperatures ≥200 °C (Springer and Laflamme 1971), and that there are no miscibility gaps in β-Ag(Sb,Bi)S$_2$ sulfides at 300 °C. These solutions were also subject to minimum bounds based on plausibility, size mismatch considerations (e.g., Ghiorso and Sack 1991) and the observation that the paucity of natural *bs* solid solutions with intermediate Bi/(Bi+Sb) ratios relative to nearly end-member compositions (Fig. 52) probably reflects non-ideality in the Bi for Sb substitution, as was inferred to be the case for the As for Sb substitution in *prg-prs* (Fig. 36).

As a further test of the adequacy of the calculated phase diagram for Ag(Sb,Bi)S$_2$ sulfides, Sack and Goodell (2002) examined *fah* + *arm* + *prg* + *sph* assemblages from a prominently zoned ore deposit at Julcani, Peru (e.g., Goodell 1970; Goodell and Petersen 1974; Deen et al. 1994). These assemblages are found in a distal bonanza silver zone of this ore

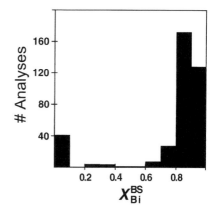

Figure 52. Histogram of molar Bi/(Bi+Sb) ratios of *bs* based on compilations of Springer (1969b) and Leuth (1988). [Used by permission of the Mineralogical Society of Great Britain and Ireland, from Ghosal and Sack (1999), *Mineral Mag*, Vol. 63, Fig. 1, p. 724.]

deposit where *arm* formed as a result of decomposition of Ag-rich gn with cooling (Fig. 53). Sack and Goodell (2002) determined equilibration temperatures in the range of 215 to 220 °C from the Ag/(Ag+Cu) of *fah* and the Bi/(Bi+Sb) of *arm* (cf. Fig. 54), based on the assumption that the T-X displacements of net-transport Reaction (60) from those in the Ag_2S-Cu_2S-ZnS-Sb_2S_3 system are due solely to the substitution of Bi for Sb in *mia-arm* solid solutions, and may be accounted for using the models for thermodynamic properties of $Ag(Sb,Bi)S_2$ solutions of Ghosal and Sack (1999). These assumptions are reasonable in light of microprobe data which indicates that there are only minor amounts of Bi in *prg* and *fah* that the Julcani *sph* and sulfosalts in the *fah* + *arm* + *prg* + *sph* assemblage are virtually Fe- and As-free, respectively. It is especially noteworthy that the 215 to 220 °C temperature range Sack and Goodell (2002) inferred from *fah* Ag/(Ag+Cu) ratio in the *fah* + *arm* + *prg* + *sph* assemblage is exactly that which would be inferred by comparison of Bi/(Bi+Sb) ratios of $Ag(Sb,Bi)S_2$ sulfosalts for the entire Julcani deposit with the $Ag(Sb,Bi)S_2$ phase diagram calculated by Ghosal and Sack (1999; cf. Fig. 51). This demonstration of internal consistency is fully compatible with the ranges of Bi/(Bi+Sb) ratios recently reported for *arm* from similar Bolivian, Argentinan and Austrian deposits by Effenberger et al. (2002), 0.199-0.245, 0.235-0.279, 0.185-0.410, and 0.400, provided it is recognized that these ratios represent similar temperatures of retrograde re-equilibration. The same cannot be stated for the structures determined for *arm*. These are clearly low temperature structures of metastable compositions and they may well have little relevance to the discussion here, as it is unlikely that they are derived directly from the parent rocksalt (i.e., *gn*) structure (e.g., Ivantchev et al. 2000).

The quaternary system Ag_2S-PbS-Sb_2S_3-Bi_2S_3

Various experimental studies within the quaternary system Ag_2S-PbS-Sb_2S_3-Bi_2S_3 have confirmed that *gn* has a composition restricted to near the ternary plane PbS-$AgSbS_2$-$AgBiS_2$ at temperatures of 300-500 °C (e.g., Van Hook 1960; Wernick 1960; Hoda and Chang 1975; Nenasheva 1975; Amcoff 1976; Chutas 2004; Figs. 55-59). This appears to be literally the case for *gn* in Ag_2S-PbS-Sb_2S_3 subsystem over the temperature range 300-500 °C where *gn* virtually conform to the $AgSbS_2$-PbS edge of the PbS-$AgSbS_2$-$AgBiS_2$ plane (e.g., Hoda and Chang 1975; Fig. 55). In the Ag_2S-PbS-Bi_2S_3 subsystem, however, compositions of *gn* may depart noticeably from the $AgBiS_2$-PbS edge of this plane, deviating from it by up to about

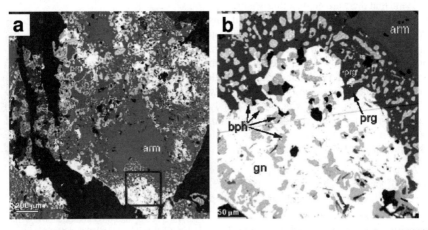

Figure 53. BSE images of *arm*- and *bph*-bearing sample H-11 from Julcani, Peru (Sack and Goodell 2000). (a) *Arm* + *bph* + *gn* + *prg* assemblage; scale bar = 200 μm. (b) Enlarged view of portion of (a) enclosed in rectangle; scale bar = 50 μm. [Used by permission of the Mineralogical Society of Great Britain and Ireland, from Sack and Goodell (2002), *Mineral Mag*, Vol. 66, Fig. 8, p. 1054.]

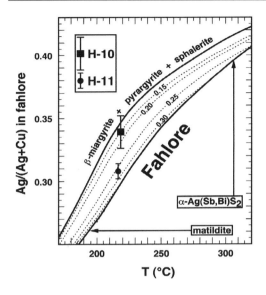

Figure 54. Comparison of molar Ag/(Ag+Cu) of *fah* and Bi/(Bi+Sb) of *arm* for samples H-10 and H-11 from Julcani, Peru (Sack and Goodell 2002) with calculated phase equilibria. Solid curves represent *fah* coexisting with *sph*, *mia*, and *prg* in the system Ag_2S-Cu_2S-ZnS-Sb_2S_3 and with this assemblage plus $AgBiS_2$-rich *gn* or *mat* in the system Ag_2S-Cu_2S-ZnS-Sb_2S_3-Bi_2S_3. Dotted curves between these solid curves are 0.15, 0.20, 0.25, and 0.30 molar Bi/(Bi+Sb) (X_{Bi}) isopleths for *mia-arm* and are calculated for the simplifying assumption that Bi does not enter *fah*, *prg*, or *sph*. [Used by permission of the Mineralogical Society of Great Britain and Ireland, from Sack and Goodell (2002), *Mineral Mag*, Vol. 66, Fig. 10, p. 1056.]

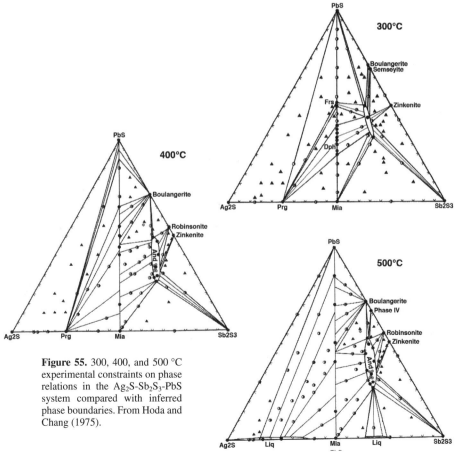

Figure 55. 300, 400, and 500 °C experimental constraints on phase relations in the Ag_2S-Sb_2S_3-PbS system compared with inferred phase boundaries. From Hoda and Chang (1975).

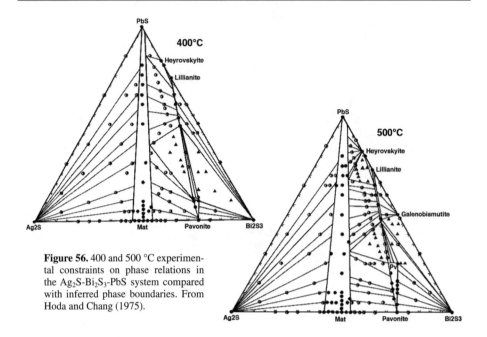

Figure 56. 400 and 500 °C experimental constraints on phase relations in the Ag_2S-Bi_2S_3-PbS system compared with inferred phase boundaries. From Hoda and Chang (1975).

16 mol% Ag_2S and about 8 mol% Bi_2S_3 where *gn* coexists with Ag_2S or with pavovite at 500 °C (Hoda and Chang 1975; Fig. 56). Above 450 °C *gn* may coexist with a variety of sulfosalts, including boulangerite (~$5PbS \cdot 2Sb_2S_3$) and andorite and heyrovskyite (~$82PbS \cdot 18Bi_2S_3$), as well as with lillianite and pavonite solid solutions in the ternary systems Ag_2S-PbS-Sb_2S_3 and Ag_2S-PbS-Bi_2S_3, respectively (Figs. 55, 56). Between 450 and 300 °C there is a miscibility gap in PbS-rich *gn* solid solutions. This gap is most extensive in the PbS-$AgSbS_2$ subsystem and has not been encountered experimentally in the PbS-$AgBiS_2$ subsystem (e.g., Van Hook 1960). Most recently, Chutas (2004) obtained reversed determinations for compositions of *gn* on this gap at 400 and 350 °C (Fig. 57) and demonstrated that the binodes could be adequately described for the assumption that *gn* is an asymmetric ternary solution (e.g., Thompson 1967) with its molar Gibbs energy described by the expression:

$$\bar{G}^{gn} = \bar{G}^{\circ\, gn}_{Pb_2S_2}(X_1) + \bar{G}^{\circ\, gn}_{AgSbS_2}(X_2) + \bar{G}^{\circ\, gn}_{AbBiS_2}(X_3) \quad (83)$$
$$+ 2RT(X_1 \ln X_1 + X_2 \ln X_2 + X_3 \ln X_3)$$
$$+ W_1(X_1)(X_2) + W_2(X_1)(X_3) + W_3(X_2)(X_3) + W_4(X_1)^2(X_2) + W_5(X_1)(X_2)^2$$
$$+ W_6(X_1)^2(X_3) + W_7(X_1)(X_3)^2 + W_8(X_2)^2(X_3) + W_9(X_2)(X_3)^2 + W_{10}(X_1)(X_2)(X_3)$$

where $X_1 = 2PbS/(2PbS + AgSbS_2 + AgBiS_2)$, $X_2 = AgSbS_2/(2PbS + AgSbS_2 + AgBiS_2)$ and $X_3 = (1 - X_1 - X_2)$, and values of $W_1, W_2, \ldots W_{10}$ are given in Appendix 3. At temperatures of about 380, 350 and 340 °C in the system PbS-$AgSbS_2$, $AgSbS_2$ *gn* (α-*mia*) transforms to β-*mia*, and *dph* (e.g., Armbruster et al. 2003) and *frs* (e.g., Ito and Nowacki 1974) become stable relative to *gn* of their own end-member compositions (cf., Fig. 58). $AgSbS_2$-rich *gn* are not stable below 300 °C (Hoda and Chang 1975; Nenasheva 1975).

We may readily calculate the Gibbs energies of formation of *dph* and *frs* from the simple sulfides Ag_2S, PbS, and Sb_2S_3 from Equation (83) and the coincidences in composition between *gn* and *dph*, and *gn* and *frs* at 350 and 340 °C. The Gibbs energies of formation of

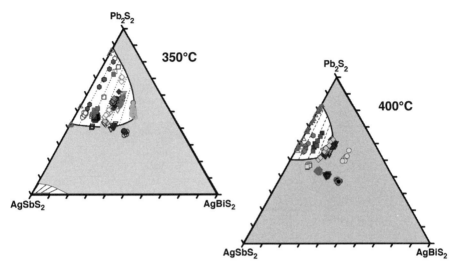

Figure 57. Calculated 350 and 400 °C *gn* miscibility gaps in the AgSbS$_2$-PbS-AgBiS$_2$ system compared with the results of unmixing (filled symbols) and homogenization (open symbols) experiments. [Used by permission of the author, from Chutas (2004), The Solubility of Silver in Galena, PhD Dissertation, Univ. of Washington.]

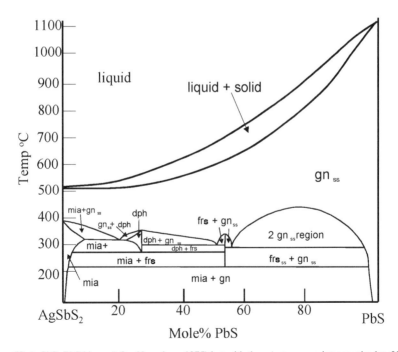

Figure 58. AgSbS$_2$-PbS binary (after Nenasheva 1975) but with the *mia* + *gn* coexistence raised to 220 °C and with the plotting coordinates of *frs* (mol% PbS = 50) and *dph* (mol% PbS = 40) adjusted to artificially PbS-enriched and PbS-depleted compositions to illustrate their phase equilibria. [Used by permission of the author, from Chutas (2004), The Solubility of Silver in Galena, PhD Dissertation, Univ. of Washington.]

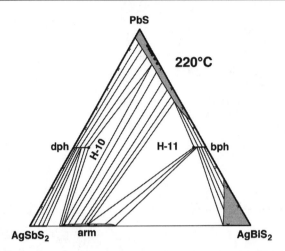

Figure 59. Schematic phase diagram for the AgSbS$_2$-PbS-AgBiS$_2$ system at 220 °C. Circles represent compositions of *arm*, *bph*, *dph*, and *gn* from *arm*, *dph*, and *gn* and *arm*, *bph*, and *gn* assemblages in samples H-10 and H-11 from Julcani, Peru (Sack and Goodell 2002). The diagram assumes that more AgSbS$_2$-rich *gn* are stable at 220 °C in these assemblages at 220 °C, has been constructed on the premise that both end-member Sb-*dph* and end-member Bi-*bph* are stable, ignores *frs* stability, and utilizes phases relations for the system AgSbS$_2$-AgBiS$_2$ from Ghosal and Sack (1999). [Used by permission of the Mineralogical Society of Great Britain and Ireland, from Sack and Goodell (2002), *Mineral Mag*, Vol. 66, Fig. 11c, p. 1057.]

dph and *frs* from the simple sulfides are equal to 4 and ³⁄₂ times the Gibbs energy of formation of *gn* at the *dph* and *frs* stoichiometries. Under these conditions Equation (83) becomes

$$\bar{G}^{gn} = \bar{G}^{°\,gn}_{AgSbS_2}(X_2) + 2RT(X_1 \ln X_1 + X_2 \ln X_2) \qquad (84)$$
$$+ W_1(X_1)(X_2) + W_4(X_1)^2(X_2) + W_5(X_1)(X_2)^2$$

where $X_2 = 3/4$ and $T = 623.15$ K for *dph* and $X_2 = 2/3$ and $T = 613.15$ K for *dph*. Utilizing the values of the parameters given in Appendices 3 and 4 one may readily calculate the Gibbs energies of formation from the simple sulfides to be −47.1212 and −17.3496 kJ/mol for *dph* at 350 °C and *frs* at 340 °C, respectively, or −2.4951 and −2.3916 kJ/mol on an equivalent basis (each sulfur atom in the formula unit equals two equivalents). This estimate of −2.4434 ± 0.0732 kJ/mol per equivalent for the energy of formation of Ag$_2$S-Sb$_2$S$_3$-PbS sulfosalts from the simple sulfides Ag$_2$S, Sb$_2$S$_3$, and PbS are roughly ½ kJ more negative than the average Gibbs energy of formation of binary sulfosalts of simple sulfides inferred by Craig and Lees (1972), −1.8828 ± 1.3598 kJ/mol per equivalent. From a DTA study, Nenasheva (1975) inferred that *dph* breaks down to *mia* and *frs* at about 270 °C and that *frs* in turn breaks down to *mia* and *gn* around 220 °C (Fig. 59). The stability of the *gn* + *mia* assemblage at low temperatures is well known (e.g., Boyle 1965). However, a lower limit of 270 °C for the stability of *dph* may be considered at least 50 °C too high based on petrological data (e.g., Lueth et al. 2000; Sack et al. 2002; Sack and Goodell 2002).

Sack and Goodell (2002) reported *dph*, and its bismuthian analog *bph* (Fig. 53), in their Julcani (Peru) samples in which Ag/(Ag+Cu) ratios in *fah* "equilibrated" down to 215-220 °C (Fig. 54), and they used compositions of *arm*, *dph* and *bph* to construct a schematic phase diagram for the system PbS-AsSbS$_2$-AgBiS$_2$ at 220 °C (Fig. 59). In this rendition of the composition plane the Julcani *arm*, *dph* and *bph* define two vertices of the three-phase triangles for the assemblages *dph* + *arm* + *gn* and *bph* + *arm* + *gn*; the *gn* composition field and tielines are schematic, as primary *gn* compositions are not quenched (e.g., Sack et al. 2002; Sack and

Goodell 2002; Sack et al 2003; Chutas and Sack 2004). This drawing is also based on the assumptions that *dph* and *bph* are stable on the PbS-AgSbS$_2$ and PbS-AgBiS$_2$ edges of the composition plane, respectively, and that *frs* is not stable on the PbS-AgSbS$_2$ edge. This is, of course, just one possibility. The assumed stability of *dph* with respect to *gn* + *mia* or *frs* + *mia* assemblages at 220 °C is consistent with its presence in assemblages which underwent slow cooling to 190 °C (e.g., Sack et al. 2002) and the fact that *dph* is much more common than *frs* in nature (e.g., Goodell 1975; Sharp and Buseck 1985; Lueth 2000; Sack et al. 2002; Sack and Goodell 2002). The proposed stability of *bph* is purely conjectural, as *bph* has not been reported for the system PbS-AgBiS$_2$ either from laboratory synthesis or nature, but a field of *bph* stability is suspected based on a reported range in X_{Bi} for *bph* from 0.95 for *bph* from Julcani to as low as 0.90-0.86 for *bph* obtained from three drill cores from Pirquitas, Peru (W.H. Paar pers. comm.; Effenberger et al. 2002), a range which might well extend to the PbS-AgBiS$_2$ binary. These and other issues must be resolved before it will be productive to proceed with this analysis.

The system Ag$_2$S-Cu$_2$S-ZnS-FeS-PbS-Sb$_2$S$_3$-As$_2$S$_3$

Notwithstanding lingering uncertainties regarding low-temperature phase relations, it is still possible to construct composition diagrams of general use to mining geologists in resource evaluation, and in analyzing the petrogenesis of polymetallic, hydrothermal ore deposits, and evaluating the thermodynamic properties of common Ag-Cu-Pb sulfosalts. Examples can be generated by integrating the model for the thermodynamic properties of *gn* outlined above into the database for *fah*, and the minerals with which it coexists, and using the resulting database to calculate composition diagrams for the *gn* which coexists with *fah*, *sph*, and other sulfides and sulfosalts in the system Ag$_2$S-Cu$_2$S-ZnS-FeS-PbS-Sb$_2$S$_3$-As$_2$S$_3$ or the Ag$_2$S-Cu$_2$S-ZnS-FeS-PbS-Sb$_2$S$_3$-As$_2$S$_3$-Bi$_2$S$_3$ supersystem. A logical place to start is with the subsystem Ag$_2$S-Cu$_2$S-ZnS-FeS-PbS-Sb$_2$S$_3$ and, to consider relations between Ag in *gn* ($X^{gn}_{AgSbS_2}$) and *fah* Ag/(Ag+Cu), X^{fah}_4.

In the system Ag$_2$S-Cu$_2$S-ZnS-FeS-PbS-Sb$_2$S$_3$ the composition of AgSbS$_2$-PbS *gn* is uniquely determined by the coexisting Ag$_2$S-Cu$_2$S-ZnS-FeS-Sb$_2$S$_3$ mineral assemblage. In mineral assemblages containing *fah*, the presence of *sph* and another sulfosalt or sulfide fixes the chemical potential of AgSbS$_2$, and hence galena composition, at a given temperature. For these assemblages the AgSbS$_2$-content of *gn* decreases monotonically with declining X^{fah}_4, (at a fixed X^{fah}_2), with the curves of $X^{gn}_{AgSbS_2}$ (X^{fah}_4) broken into a series of curvilinear segments for each of the different two-phase regions on the composition plane Ag$_2$S-Cu$_2$S-Sb$_2$S$_3$ (Fig. 39c). Figure 60 displays curves of this sort for low-Ag *gn* ($X_{AgSbS_2} < 0.05$) and *fah* with Zn/(Zn+Fe) = 1.0 and 0.4 coexisting with *sph* and one of the following sulfosalts or sulfides: *mia*, *prg*, *plb*, or *bcc*-, *fcc*-, or *hcp*- (Ag,Cu)$_2$S; the companion figure for high-Ag *gn* may be found in Sack (2005, his Fig. 6a). Sack (2005) used these systematics to infer an initial temperature of hydrothermal deposition of 320-350 °C from constraints on initial *gn* and *fah* compositions for Coeur d'Alene Ag-Pb-Zn deposits (Sack et al. 2002, 2005), a temperature estimate in accord with previous determinations (e.g., Arkadakskiy 2000). Besides such geothermometric applications, these systematics may be used to reconstruct primary compositions of minerals, where they are not quenched (e.g., *gn*).

With the addition of *gn* to the database, it becomes fairly straightforward to evaluate the thermodynamic properties of many of the more common Pb-Ag-Cu sulfosalts, given phase equilibrium constraints. An example is the Cu-Pb-Sb sulfosalt bournonite (*bno*, ~CuPbSbS$_3$), a common mineral in Ag-Pb-Zn deposits which may be primary and/or secondary. Sack et al. (2002, 2005), for example, showed that most, if not all, of the *bno* in the Coeur d'Alene Ag-Pb-Zn deposits is secondary, and was produced by the Ag for Cu exchange reaction:

$$\tfrac{1}{10}Cu_{10}(Zn,Fe)_2Sb_4S_{13} + AgSbS_2 + PbS = CuPbSbS_3 + \tfrac{1}{10}Ag_{10}(Zn,Fe)_2Sb_4S_{13} \quad (85)$$
Cu-*fah*   in *gn*   *gn*   *bno*   Ag-*fah*

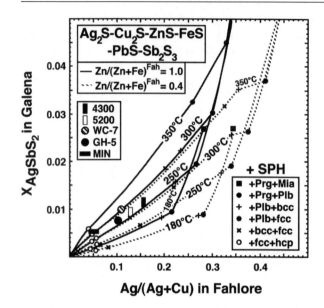

Figure 60. Mole fraction of AgSbS$_2$ in *gn* and Ag/(Ag+Cu) in *fah* in assemblages composed of *fah* + *gn* + *sph* plus one or two of the phases *mia*, *prg*, *plb*, or *bcc*-, *fcc*-, or hcp-(Ag,Cu)$_2$S solid solution in the system Ag$_2$S-Cu$_2$S-ZnS-FeS-PbS-Sb$_2$S$_3$. Solid lines are for Zn/(Zn+Fe) of *fah* = 1.0; dashed lines are for Zn/(Zn+Fe) of *fah* = 0.4. Objects labeled 4300, 5200, WC-7, GH-5, and MIN are estimates of original compositions of *gn* and *fah* in Coeur d'Alene (Idaho) ores. From Sack et al. (2005).

which removed Ag from *gn* and enriched *fah* in Ag (Fig. 61) during post-depositional cooling (cf. Fig. 62). The operation of this reaction resulted in a population of high-Ag *fah* in ores with high proportions of *gn* relative to *fah*, producing the most Ag-rich *fah* in the Coeur d'Alene mining district in the most *gn*-rich samples (e.g., Hackbarth 1984; Sack et al. 2002, 2005). The Gibbs energy of formation of *bno* may be readily estimated from the condition of equilibrium corresponding to Reaction (85),

$$\bar{G}^{\circ\, bno}_{CuPbSbS_3} = \tfrac{1}{10}\left(\mu^{fah}_{Cu_{10}Zn_2Sb_4S_{13}} - \mu^{fah}_{Ag_{10}Zn_2Sb_4S_{13}}\right) + \mu^{gn}_{AgSbS_2} + \tfrac{1}{2}\mu^{gn}_{Pb_2S_2} \qquad (86)$$

where we have treated *bno* as a stoichiometric compound. Substituting our expressions for chemical potentials into Equation (86) we obtain:

$$\bar{G}^{\circ\, bno}_{CuPbSbS_3} = \tfrac{1}{10}\left(\bar{G}^{\circ\, fah}_{Cu_{10}Zn_2Sb_4S_{13}} - \bar{G}^{\circ\, fah}_{Ag_{10}Zn_2Sb_4S_{13}}\right) + \bar{G}^{\circ\, gn}_{AgSbS_2} \qquad (87)$$

$$+ RT\ln\left[\left(1-X^{gn}_{AgSbS_2}\right)\left(X^{gn}_{AgSbS_2}\right)^2\right]$$

$$-\tfrac{1}{10}\left\{\begin{array}{l} 6RT\ln\left(\dfrac{X^{fah}_4 + \tfrac{2}{5}s}{1-X^{fah}_4 - \tfrac{2}{5}s}\right) + 4RT\ln\left[\dfrac{\tfrac{2}{3}X^{fah}_4 - \tfrac{2}{5}s}{\tfrac{2}{3}\left(1-X^{fah}_4\right)+\tfrac{2}{5}s}\right] \\ +\Delta\bar{G}^{\circ}_{24}\left(X^{fah}_2\right)+\Delta\bar{G}^{\circ}_{34}\left(X^{fah}_3\right)+\left(\Delta\bar{G}^{*}_{4s}+W^{TET}_{AgCu}+W^{TRG}_{AgCu}\right)\left(1-2X^{fah}_4\right) \\ +\tfrac{1}{5}\left(\Delta\bar{G}^{*}_{4s}+6W^{TET}_{AgCu}-4W^{TRG}_{AgCu}\right)(s) \end{array}\right\}$$

$$+W_1\left[1-2X^{gn}_{AgSbS_2}+\tfrac{3}{2}\left(X^{gn}_{AgSbS_2}\right)^2\right]+W_4\left[1-4X^{gn}_{AgSbS_2}+6\left(X^{gn}_{AgSbS_2}\right)^2-3\left(X^{gn}_{AgSbS_2}\right)^3\right]$$

$$+W_5\left[2X^{gn}_{AgSbS_2}-\tfrac{9}{2}\left(X^{gn}_{AgSbS_2}\right)^2+3\left(X^{gn}_{AgSbS_2}\right)^3\right]$$

where the term $\tfrac{1}{2}\bar{G}^{\circ\, gn}_{Pb_2S_2}$ has been eliminated because $\bar{G}^{\circ\, gn}_{Pb_2S_2} = 2\bar{G}^{\circ\, gn}_{PbS} = 0$ by definition.

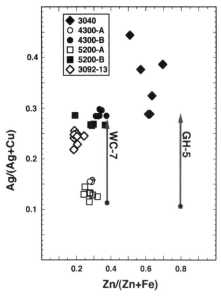

Figure 61 (left). Enrichment of *fah* in molar Ag/(Ag+Cu) in Coeur d'Alene ores. Tips of arrows indicate present Ag/(Ag+Cu) of *fah* from WC-7 and GH-5; asterisks at tails of arrows indicate original Ag/(Ag+Cu) inferred by mass balance from reaction (85). Open squares and rectangles are *fah* in control samples lacking *gn*; filled squares are *fah* from nearby samples containing significant *gn*. From Sack et al. (2005)

Figure 62 (below). BSE and X-ray images from *gn*-rich sample WC-7 from the West Chance vein of the Sunshine Mine, Kellogg, Idaho. Scale bar = 100 μm. [Used by permission of the Mineralogical Society of Great Britain and Ireland, from Sack et al. (2002), *Mineral Mag*, Vol. 66, Fig. 3, p. 221.]

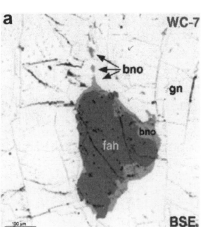
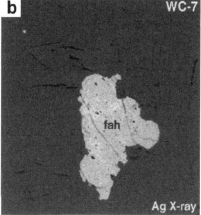

We may use Equation (87) and the mineralogic data for the ore sample displayed in Figure 62 (WC-7) to calculate an estimate for the Gibbs energy of formation of *bno* from the simple sulfides, $\overline{G}^{\circ\, bno}_{CuPbSbS_3}$, at the temperature at which the *fah* + *gn* assemblage became saturated with *bno*. A value of $\overline{G}^{\circ\, bno}_{CuPbSbS_3} = -27.12 \pm 0.84$ kJ/gfw for 310 °C is obtained from Equation (87) assuming this temperature is between 320 and 300 °C and using the initial compositions of *fah* ($X_2^{fah} = 0.383$, $X_3^{fah} = 0.039$, $X_4^{fah} = 0.110$) and *gn* ($X_{AgSbS_2}^{gn} = 0.0100$ to 0.0108) estimated by Sack et al. (2002, 2005) by mass balance. Given the corresponding entropy of formation of *bno* from the simple sulfides, $\overline{S}^{\circ\, bno}_{CuPbSbS_3}$, we may establish the plausibility of this estimate by determining how close to ambient temperature is the temperature calculated from Equation (87) for the present compositions of *gn* and *fah*. Close agreement between these temperatures is required, given that the unquenchability of natural high Ag-*gn* is well established (e.g., Hall and Czamanske 1972; Sharp and Buseck 1993; Lueth et al. 2000; Sack et al. 2002, 2005; Sack and Goodell 2002; Chutas and Sack 2004).

An entirely plausible temperature of about 48 °C is calculated from Equation (87) and present compositions of *fah* ($X_2^{fah} = 0.383$, $X_3^{fah} = 0.039$, $X_4^{fah} = 0.277$; Sack et al. 2002) and *gn* ($0.0022 < X_{AgSbS_2}^{gn} < 0.0030$; Knowles 1983; Hardy 2002; Sack et al. 2002; Sack 2005) for the assumption that $\bar{S}_{CuPbSbS_3}^{\circ\, bno}$ (25.225 J/K-gfw) is given by the algorithm of Craig and Barton (1973):

$$\Delta S = -(1.2 \pm 0.8) R \sum_i N_i \ln N_i \tag{88}$$

where N_i are the mole fractions of simple sulfides with one sulfur atom in their formula unit and ΔS is the entropy of formation of the sulfosalt from the simple sulfides per gram-atom of sulfur. Accordingly, we might adopt this calibration for the Gibbs energy of *bno* ($\bar{H}_{CuPbSbS_3}^{\circ\, bno}$ = −12.41 ± 0.84 kJ/gfw and $\bar{S}_{CuPbSbS_3}^{\circ\, bno}$ =25.225 J/W-gfw) and incorporate it into the thermodynamic database as a first approximation, much as oxide and silicate solids are continually added to the popular MELTS program (Ghiorso and Sack 1995; Ghiorso 2004) to more completely simulate magmatic processes. We note that are the average estimates for the Gibbs energy of formation of *bno* from the simple sulfides at 400 °C, −29.39 ± 0.84 kJ/gfw or −4.90 ± 0.14 kJ/equivalent, is distinctly more negative than the range of formation energies for binary sulfosalt mixtures obtained experimentally, −1.8828 ± 1.3598 kJ/equivalent. This increased stability of *bno* is undoubtedly a reflection of its ternary character. The Gibbs energy of *bno* is more directly comparable with formation energies deduced for Cu- and Sb-endmember *fah* $Cu_{10}Zn_2Sb_4S_{13}$ and $Cu_{10}Fe_2Sb_4S_{13}$ at 400 °C (−4.43 and −3.92 kJ/equivalent) and roughly twice as negative as the formation energies of *dph* and *frs* (−2.5 and −2.37 kJ/equivalent). To this calibration for the Gibbs energy of formation of *bno* we may add the regular solution parameter accounting for the As for Sb substitution ($W_{AsSb}^{bno} = 2.76 \pm 0.75$ kJ/gfw, Sack and Ebel 1993) so that retrograde Reaction (85) may be modeled in As-rich ores.

An approach similar to that employed using Equation (87) might be advocated for characterizing the Gibbs energies of many other sulfosalts formed by retrograde processes. Indeed, for many of these sulfosalts an alternative may not be feasible due to the difficulty in successfully synthesizing and reacting them on realistic laboratory time-scales. Accordingly, it is clear that, in addition to phase equilibria and thermochemical studies, careful characterization of retrograde processes in polymetallic, hydrothermal ores is required to elucidate fundamental relationships between stability, structure and composition for the diverse suite of sulfosalts in the Ag_2S-Cu_2S-ZnS-FeS-PbS-Sb_2S_3-As_2S_3-Bi_2S_3 system. However, we emphasize that the present thermochemical database is quite adequate for modeling primary hydrothermal processes in many polymetallic ore deposits where the solid solutions *po*, *sph*, *plb-prc*, *prg-prs*, *mia-smt*, *skn*, *bcc-*, *fcc-*, and *hcp*-$(Ag,Cu)_2S$ solutions, *bs*, *arm*, *mat*, *gn*, *tmt*, *dph*, and *bph*, and stoichiometric *pyr*, account for most of the sulfides precipitated. By incorporating these solution models into codes that describe transport and reaction in porous media (e.g., FLOWTRAN, Lichtner 2001) and explicitly make provision for solid solutions (Lichtner and Carey 2006), it will become possible to model primary metal zoning in polymetallic sulfide deposits. These models will define formats for further evaluation of the thermodynamics of the retrograde reactions which produce a diverse collection of sulfosalts. They will afford new approaches to mineral exploration and resources evaluation which are based on realistic thermochemical models for solid solutions of sulfides and sulfosalts.

ACKNOWLEDGMENTS

Support given by OFM Research (ROS) and NASA Grant NAG5-12855 (DSE), and assistance, comments, encouragement and gentle prodding provided by Jodi Rosso, David Vaughan, and Phillip Goodell are much appreciated.

REFERENCES

Adachi A, Morita Z (1958) The equilibrium between sulfur and molten iron and H_2–H_2S gas mixture. Techn Rep Osaka Univ 8:385-394

Albinson FT (1988) Geological reconstruction of paleosurfaces in the Sombrete, Colorado, and Fresnillo districts, Zacatecas State, Mexico. Econ Geol 83: 1647–1667

Albinson FT, Norman DI, Cole D, Chomiak B (2001) Controls on formation of low-sulfidation epithermal deposits in Mexico: constraints from fluid inclusions and stable isotope data. *In*: New Mines and Discoveries in Mexico and Central America. Albinson T, Nelson CE (ed), Society of Economic Geologists Special Publication 8:1–27

Albarède F (2004) The stable isotope geochemistry of copper and zinc. Rev Mineral Geochem 55:409-428.

Allen ET, Crenshaw JL (1912) The sulfides of zinc, cadmium and mercury; their crystallographic forms and genetic conditions. Am J Sci 34: 341-396

Allen ET, Lombard RH (1917) A method for the determination of dissociation pressures of sulphides and its application to covellite (CuS) and pyrite FeS_2. Am J Sci 43: 175-195

Amcoff O (1976) The solubility of silver and antimony in galena. Neues Jahrbuch für Mineralogie, Monatschefte 6: 247–261

Anderson GM, Crerar D (1993) Thermodynamics in Geochemistry. Oxford University Press

Andrews DRA, Brenan JM (2002) The solubility of ruthenium in sulfide liquid: implications for platinum group mineral stability and sulfide melt-silicate melt partitioning. Chem Geol 192:163-181

Anthony JW, Bideaux RA, Bladh K, Nichols MC (1990) Handbook of Mineralogy: Volume I. Elements, Sulfides, Sulfosalts. Mineral Data Publishing

Anzai S, Ozawa K (1974) Effect of pressure on the Neel and ferromagnetic Curie temperatures for $FeS_{1+\delta}$. Physica Status Solidi 24:K31-34

Arkadakskiy SV (2000) Fluid inclusion and stable isotopic study of precious and base metal veins from the Coeur d'Alene Ag-Pb-Zn district, Idaho, U.S.A. MSc. Dissertation, University of Alberta, Edmondton, Alberta, Canada

Armbruster T, Makovicky E, Berlepsch P, Sejkora J (2003) Crystal structure, cation ordering, and polytypic character of diaphorite, $Pb_2Ag_3Sb_3S_8$, a PbS-based structure. Eur J Mineral 15:137–146

Arnold RG (1962) Equilibrium relations between pyrrhotite and pyrite from 325° to 743 °C. Econ Geol 57: 72–90

Augsten BEK, Thorpe RI, Harris DC, Fedikow MAF (1986) Ore mineralogy of the Agassiz (MacLellan) gold deposit in the Lynn Lake region, Manitoba. Can Mineral 24:369–377

Balabin AI, Urusov VS (1995) Recalibration of the sphalerite cosmobarometer: experimental and theoretical treatment. Geochim Cosmochim Acta 59:1401–1410

Balabin AI, Sack RO (2000) Thermodynamics of (Zn,Fe)S sphalerite: a CVM approach with large basis clusters. Mineral Mag 64:923-943

Ballhaus C, Tredoux M, Spath A (2001) Phase relations in the Fe-Ni-Cu-PGE-S system at magmatic temperature and application to massive sulphide ores of the Sudbury Igneous Complex. J Petrol 42:1911-1926

Barin I (1991) Thermochemical Data of Pure Substances, 2nd Edition. 2 volumes. Verlag Stahleisen

Barker WW, Parks TC (1986) The thermodynamic properties of pyrrhotite and pyrite; a reevaluation. Geochim Cosmochim Acta 50: 2185–2194

Barrett RA, Zolensky ME (1987) Compositional-crystallographic relations for polybasite. Geol Soc Am Abstr Prog 19:535

Barton MD (1980) The Ag–Au–S system. Econ Geol 75:303–316

Barton PB Jr (1969) Thermochemical study of the system Fe-As-S. Geochim Cosmochim Acta 33: 841-857

Barton PB Jr (1970) Sulfide petrology. Mineralogical Society of America Special Paper 3:187-198

Barton PB Jr (1973) Solid solution in the system Cu-Fe-S. Part I: The Cu-S and CuFe-S joins. Econ Geol 68: 455–465

Barton PB Jr, Skinner BJ (1967) Sulfide mineral stabilities. *In*: Geochemistry of Hydrothermal Ore Deposits. Barnes HL (ed) Holt Rinehart and Winston, p 236-333

Barton PB Jr, Skinner BJ (1979) Sulfide mineral stabilities: *In*: Geochemistry of Hydrothermal Ore Deposits, 2nd edition. Barnes HL (ed) Wiley, p 278-403

Barton PB Jr, Toulmin P (1964) The electrum-tarnish method for the determination of the fugacity of sulfur in laboratory sulfide systems. Geochim Cosmochim Acta 28: 619–640

Barton PB Jr, Toulmin P (1966) Phase relations involving sphalerite in the Fe–Zn–S system. Econ Geol 61: 815–849

Barton PB Jr, Bethke PM, Roedder E (1977) Environment of ore deposition in the Creede mining district, San Juan Mountains, Colorado: III. Progress toward interpretation of the chemistry of the ore-forming fluid for the OH vein. Econ Geol 72: 1–25

Benbattouche N, Saunders GA, Lambsom EF, Hönle W (1989) The dependence of the elastic stiffness moduli and the Poisson ratio of natural iron pyrites FeS_2 upon pressure and temperature. J Phys D: Appl Phys 22: 670-675

Bennett CEG, Graham J, Thornber MR (1972) New observations on natural pyrrhotites. Part I – Mineralographic techniques. Am Mineral 57:445-462

Bente K, Doering T (1993) Solid-state diffusion in sphalerites: an experimental verification of the "chalcopyrite disease". Eur J Mineral 5:465-478

Bernardini GP, Borgheresi M, Cipriani C, Di Benedetto F, Romanelli M (2004) Mn distributions in sphalerite an EPR study. Phys Chem Mineral 31:80-84

Berner RA (1964) Iron sulfides formed from aqueous solution at low temperatures and atmospheric pressure. J Geol 72:293-306

Berner RA (1971) Principles of Chemical Sedimentology. McGraw-Hill

Bertaut EF (1953) Contribution à l'étude des structures lacunaries: La Pyrrhotine Acta Crystallogr 6:537-561

Bertaut EF (1956) Structure de FeS stoichiometrique. Bull Soc Franc Minéralogie et Cristallographie 79:276-292

Bockrath C, Ballhaus C, Holzheid A (2004) Stabilities of laurite RuS_2 and monosulfide liquid solution at magmatic temperatures. Chem Geol 208:265-271

Boorman RS (1967) Subsolidus studies in the $ZnS-FeS-FeS_2$ system. Econ Geol 62: 614–631

Boorman RS, Sutherland JK, Chernyshev LV (1971) New data on the sphalerite–pyrrhotite–pyrite solvus. Econ Geol 66:670–675

Borodayev YS, Nenasheva SN, Gamyanin, GN, Mozgova NN (1986) First find of aramayoite-galena-matildite exsolution textures. Doklady Akademii Nauka SSR 290: 192-195

Boyle RW (1965) Keno Hill-Galena Hill lead-zinc-silver deposits. Geol Survey Can Bull 111

Brenan JM (2002) Re-Os fractionation in magmatic sulfide melt by monosulfide solid solution. Earth Planet Sci Lett 199:257-268

Brenan JM (2003) Effects of fO_2, fS_2, temperature and melt composition on Fe-Ni exchange between olivine and sulfide liquid: implications for natural olivine-sulfide assemblages. Geochim Cosmochim Acta, 67: 2663-2681

Brett PR, Bell PM (1969) Melting relations in the Fe-rich portion of the system Fe-FeS at 30 kb pressure. Earth Planet Sci Lett 6:479–482

Bryndzia, LT, Kleppa OJ (1988a) High-temperature reaction calorimetry of solid and liquid phases in the quasi-binary system $Ag_2S-Sb_2S_3$. Geochim Cosmochim Acta 52:167–176

Bryndzia, LT, Kleppa OJ (1988b) Standard enthalpies of formation of sulfides and sulfosalts in the Ag–Bi–S system by high temperature, direct synthesis calorimetry. Econ Geol 83:174–181

Bryndzia, LT, Kleppa OJ (1989) Standard molar enthalpies of formation of sulfosalts in the Ag–As–S system and thermochemistry of the sulfosalts of Ag with As, Sb, and Bi. Am Mineral 74:243–249

Bryndzia, LT, Scott SD, Spry PG (1988) Sphalerite and hexagonal pyrrhotite geobarometer: Experimental calibration and application to the metamorphosed sulfide ores of Broken Hill, Australia. Econ Geol 83: 1193–1204

Bryndzia, LT, Scott SD, Spry PG (1990) Sphalerite and hexagonal pyrrhotite geobarometer: Correction in calibration and application. Econ Geol 85:408–411

Burgmann W Jr, Urbain G, Frohberg MG (1968) Contribution à l'étude du système fer-soufre limité au domaine du mono-sulfure de fer (pyrrhotine). Mémoires Scientifiques de la Revue de Métallurgie 65:567-578

Cabri LJ (1973) New data on phase relations in the Cu-Fe-S system. Econ Geol 68:443-454

Cabri LJ, Hall SR (1972) Mooihoiekite and haycockite, two new copper-iron sulfides and their relationship to chalcopyrite and talnakite. Am Mineral 57:689-708

Callen HB (1985) Thermodynamics and an Introduction to Thermostatistics, 2nd Edition. Wiley

Cambrubí A, Canals A, Cardellach E, Prol–Ledesma, RA, Rivera R (2001) The La Guitarra Ag–Au low-sulfidation epithermal deposit, Temascaltepec district, Mexico; vein structure, mineralogy, and sulfide-sulfosalt chemistry. *In*: New Mines and Discoveries in Mexico and Central America. Albinson T, Nelson CE (eds), Society of Economic Geologists Special Publication 8:133–158

Carlson, HC, Colburn AP (1942) Vapor-liquid equilibrium of nonideal solutions. Indust Eng Chem 34:581-589

Cava RJ, Reidenger F, Wuensch BJ (1980) Single-crystal neutron diffraction study of the fast-ion-conductor β-Ag_2S between 186 and 325 °C. J Solid State Chem 31:69-80

Cava RJ, Reidenger F, Wuensch BJ (1981) Mobile ion distribution and anharmonic thermal motion in fast-ion conducting Cu_2S. J Solid State Ionics 5:501-504

Chao GY, Gault RA (1998) The occurrence of two rare polytypes of wurtzite, 4H and 8H, at Mont Saint-Hilaire, Quebec. Can Mineral 36:775-778

Chang LLY, Knowles CR, Tzong TC (1977) Phase relations in the systems $Ag_2S-Sb_2S_3-Bi_2S_3$, $Ag_2S-As_2S_3-Sb_2S_3$ and $Ag_2S-As_2S_3-Bi_2S_3$. Memoir Geol Soc China 2:229-237

Chang YA, Hsieh K-C (1987) Thermochemical description of the ternary iron-nickel-sulphur system. Can Metall Q 26:311-327
Chang YA, Chen S, Zhang F, Yan X, Xie F, Schmid-Fetzer R, Oates WA (2004) Phase diagram calculation: past, present and future. Prog Mater Sci 49:313-345
Charlat M, Levy C (1974) Substitutions multiples dans la série tennantite-tétrahedrite. Bull Soc Fr Mineral Cristall 97:241-250
Charlat M, Levy C (1975) Influence principales sur le parametre cristallin dans la série tennantite-tétrahedrite. Bull Soc Fr Mineral Cristall 98:152-158
Charnock JM, Garner CD, Pattrick RAD, Vaughan DJ (1988) EXAFS and Mössbauer spectroscopic study of Fe–bearing tetrahedrites. Phys Chem Mineral 15:296-299
Charnock JM, Garner CD, Pattrick RAD, Vaughan DJ (1989) EXAFS and Mössbauer spectroscopic study of Fe–bearing tetrahedrites. Mineral Mag 53:193–199
Chase MW Jr (1998) NIST-JANAF thermochemical tables, fourth edition. Journal of Physical and Chemical Reference Data, Monograph No. 9, published for the National Institute of Standards and Technology by the American Chemical Society and the American Institute of Physics. (CD-ROM)
Chase MW Jr, Davies CA, Downey JR, Frurip DJ, McDonald RA and Syverud AN (1985) JANAF thermochemical tables, 3rd edition. J Phys Chem Ref Data 14, Supplement 1. American Chemical Society and American Institute of Physics
Chernyshev LV, Anfigolov VN (1968) Subsolidus relations in the ZnS-FeS-FeS$_2$ system. Econ Geol 63:841-847
Chernyshev LV, Anfigolov VN, Pastushkova TM, Suturina TA (1968) Hydrothermal investigation of the system Fe-Zn-S. Geologiya Rudnykh Mestorozhdeniy 3: 50-64 (in Russian)
Chernyshev LV, Anfigolov VN, Berestennikov MI (1969) Cell-edge dimensions of iron-bearing sphalerites synthesized under hydrothermal conditions. Geologiya Rudnykh Mestorozhdeniy 6:85-89 (in Russian)
Chryssoulis SL, Cabri LJ, Lennard W (1989) Calibration of the ion microprobe for quantitative trace precious metal analysis of ore minerals. Econ Geol 84:1684-1689
Chuang Y-Y (1983) The Thermodynamics and Phase Relationships of the Cu-Ni-Fe-S Quaternary System and its Subsystems. PhD Dissertation, University of Wisconsin (Madison)
Chuang Y-Y, Hsieh, K-C, Chang YA (1985) Thermodynamics and phase relationships of transitional metal-sulfur systems: Part V. A reevaluation of the Fe-S system using and associated solution model for the liquid phase. Metall Trans B:277-285
Chutas NI (2004) The solubility of silver in galena. PhD Dissertation, University of Washington, Seattle, Washington
Chutas NI, Sack RO (2004) Ore genesis at La Colorada Ag-Zn-Pb deposit in Zacatecas, Mexico. Mineral Mag 68:923–937
Clark LA (1960) The Fe-As-S system: Phase relations and applications. Econ Geol 55:1345-1381, 1631-1652
Corlett M (1968) Low-iron polymorphs in the pyrrhotite group. Z Kristallogr 126:124-132
Craig JR, Barton PB Jr (1973) Thermochemical approximations for sulfosalts. Econ Geol 68:493-506
Craig JR, Lees WR (1972) Thermochemical data for sulfosalt ore minerals: Formation from simple sulfides. Econ Geol 67 373-377
Craig JR, Scott SD (1976) Sulfide phase equilibria. In: Sulfide Mineralogy. Short Course Notes 1. Ribbe PH (ed) Mineralogical Society of America, p CS104-110
Craig JR, Chang LY, Lees WR (1973) Investigations in the Pb-Sb-S system. Can Mineral 12:199-206
Deen JA, Rye RO, Munoz JL, Drexler JW (1994) The magmatic hydrothermal system at Julcani, Peru: evidence from fluid inclusions and hydrogen and oxygen isotopes. Econ Geol 89:1924–1938
de Fontaine D (1994) Cluster approach in order-disorder transformations in alloys. Solid State Phys 47:33-176
deHeer J (1986) Phenomenological Thermodynamics, with Application to Chemistry. Prentice-Hall
Denbigh K (1981) The Principles of Chemical Equilibrium, 4th Edition. Cambridge University Press
Di Benedetto F, Bernardini GP, Caneschi A, Cipriani C, Danti C, Pardi L, Romanelli M (2002) EPR and magnetic investigations on sulphides and sulphosalts. Eur J Mineral 14:1053-1060
Di Benedetto F, Bernardini GP, Costagliola P, Plant D, Vaughan DJ (2005) Compositional zoning in sphalerite crystals. Am Mineral 90:1384-1392
Dicarlo J, Albert M, Dwight K, Wold A (1990) Preparation and properties of iron-doped II-VI chalcogenides. J Solid State Chem 87:443-448
Dickson FW, Shields LD, Kennedy GC (1962) A method for the determination of equilibrium sulfur pressures of metal sulfide reactions. Econ Geol 112:1917-1923
Ding Y, Veblen DR, Prewitt CT (2005) Possible Fe/Cu ordering schemes in the $2a$ superstructure of bornite (Cu$_5$FeS$_4$). Am Mineral 90:1265-1269.
Durazzo A, Taylor LA (1982) Exsolution in the mss-pentlandite system: textural and genetic implications for Ni-sulfide ores. Mineralium Deposita 17:79-97
Dutrizac JE (1976) Reactions in cubanite and chalcopyrite. Can Mineral 14:172-181

Ebel DS (1988) Argentian Zinc-Iron Tetrahedrite-Tennantite Thermochemistry. MS Dissertation, Purdue University, West Lafayette, IN
Ebel DS (1993) Thermochemistry of fahlore (tetrahedrite) and biotite mineral solutions. PhD Dissertation, Purdue University, West Lafayette, IN
Ebel DS (2006) Condensation of rocky material in astrophysical environments. In: Meteorites and the Early Solar System II. Lauretta D et al. (ed) University Arizona, p. 253-277
Ebel DS, Campbell AJ (1998) Rhodium and palladium partitioning between copper-nickel-pyrrhotite and sulfide liquid. Geol Soc Am Abst Prog 30A:318
Ebel DS, Naldrett AJ (1996) Fractional crystallization of sulfide ore liquids at high temperatures. Econ Geol 91:607-621
Ebel DS, Naldrett AJ (1997) Crystallization of sulfide liquids and the interpretation of ore composition. Can J Earth Sci 34: 352-365
Ebel DS, Sack RO (1989) Ag–Cu and As–Sb exchange energies in tetrahedrite–tennantite fahlores. Geochim Cosmochim Acta 53:2301–2309
Ebel DS, Sack RO (1991) As–Ag incompatibility in fahlore. Mineral Mag 55:521–528
Ebel DS, Sack RO (1994) Experimental determination of the free energies of formation of freibergite fahlores. Geochim Cosmochim Acta 58:1237–1242
Effenberger H, Paar WH, Topa D, Criddle AJ, Fleck M (2002) The new mineral baumstarkite and a structural reinvestigation of aramayoite and miargyrite. Am Mineral 87:753?764
Ehlers EG (1972) The Interpretation of Geological Phase Diagrams. W.H. Freeman and Company
Ehlers K, El Goresy A (1988) Normal and reverse zoning in niningerite: A novel key parameter to the thermal histories of EH-chondrites. Geochim Cosmochim Acta 53: 877-887
El Goresy A, Kullerud G (1969) Phase relations in the system Cr-Fe-S. In: Meteorite Research. Millman PM (ed) D. Reidel, p 638-656
Etschmann B, Pring A, Putnis A, Benjamin A, Grguric BA, Studer A (2004) A kinetic study of the exsolution of pentlandite (Ni, Fe)$_9$S$_8$ from the monosulfide solid solution (Fe,Ni)S. Am Mineral 89:39-50
Fedorova ZN, Sinyakova EF (1993) Experimental investigation of physicochemical conditions of pentlandite formation. Geologiya i Geofizika 34: 84–92 (in Russian)
Ferrante MJ, Stuve JM, Pankratz LB (1981) Thermodynamic properties of cuprous and cupric sulfides. High Temp Sci 14:77-90
Fleet ME (1968) On the lattice parameters and superstructures of pyrrhotites. Am Mineral 53:1846-1855
Fleet ME (1975) Thermodynamic properties of (Zn,Fe)S solid solutions at 850 °C. Am Mineral 60:466-470
Fleet ME (2006) Phase equilibria at high temperatures. Rev Mineral Geochem 61:365-419
Fleet ME, Pan Y (1994) Fractional crystallization of anhydrous sulfide liquid in the system Fe-Ni-Cu-S, with application to magmatic sulfide deposits. Geochim Cosmochim Acta 58:3369-3377
Fleet ME, Chryssoulis SL, Stone WE, and Weisener CG (1993) Partitioning of platinum-group elements and Au in the Fe-Ni-Cu-S system: experiments on the fractional crystallization of sulfide melt. Contrib Mineral Petrol 115:36-44
Fogel RA (2005) An analysis of the solvus in the CaS-MnS system. Lunar Planet Sci Conf XXXVI:2395
Francombe MH (1957) Lattice changes in spinel-type iron chromites. J Phys Chem Solids 3:37-43
Froese E, Gunter AE (1976) A note on the pyrrhotite-sulfur vapor equilibrium. Econ Geol 71:1589-1594
Froese E, Gunter AE (1978) A note on the pyrrhotite-sulfur vapor equilibrium – a discussion. Econ Geol 73: 286
Frondel C (1963) Isodimorphism of the polybasite and pearceite series. Am Mineral 48:565-572
Fryklund VC Jr (1964) Ore deposits of the Coeur d'Alene district, Shoshone County, Idaho. U.S. Geol Survey Prof Paper 445
Furdyna JK (1988) Diluted magnetic semiconductors. J Appl Phys 64:R29-R64
Gaines RV, Skinner HCW, Foord EE, Mason B, Rosenzweig A, King VT, Dowty E (1997) Dana's New Mineralogy, 8th ed. John Wiley & Sons
Ganguly J (2001) Thermodynamic modelling of solid solutions. In: Solid Solutions in Silicate and Oxide Systems. C.A. Geiger CA (ed) EMU Notes in Mineralogy 3:37-70
Gemmell JB, Zantop H, Birnie RW (1989) Sulfosalt geochemistry of the Santa Nino Ag-Pb-Zn vein, Fresnillo district, Mexico. Can Mineral 27:401-428
Ghiorso MS (2004) An equation of state for silicate melts. III. Analysis of stoichiometric liquids at elevated pressure: shock compression data, molecular dynamics simulations and mineral fusion curves. Am J Sci 304:752-810
Ghiorso MS, Sack RO (1991) Thermochemistry of the oxide minerals. Rev Mineral 25:221–264
Ghiorso MS, Sack RO (1995) Chemical mass transfer in magmatic processes: IV. A revised and internally consistent model for the interpolation and extrapolation of liquid–solid equilibria in magmatic systems at elevated temperatures and pressures. Contrib Mineral Petrol 119:197–212

Ghiorso MS, Evans BW, Hirschmann MM, Yang H (1995) Thermodynamics of the amphiboles: Fe-Mg cummingtonite solid solution. Am Mineral 80:502-519
Ghosal S, Sack RO (1995) As–Sb energetics in argentian sulfosalts. Geochim Cosmochim Acta 59:3573–3579
Ghosal S, Sack RO (1999) Bi–Sb energetics in sulfosalts and sulfides. Mineral Mag 63:723–733
Gilletti BJ, Yund RA, Lin TJ (1968) Sulfur vapor pressure of pyrite-pyrrhotite (abstract). Econ Geol 63:702
Goodell PC (1970) Zoning and paragenesis in the Julcani district, Peru. Ph.D. thesis, Harvard University, Cambridge, Massachusetts
Goodell PC (1975) Binary and ternary sulphosalt assemblages in the Cu_2S-Ag_2S-PbS-As_2S_3-Sb_2S_3 system. Can Mineral 13:27-42
Goodell PC, Petersen U (1974) Julcani mining district, Peru: a study of metal ratios. Econ Geol 69:347-361
Graham AR (1951) Matildite, aramayoite, miagyrite. Am Mineral 36:436-439
Greffié C, Bailly L, Milési J-P (2002) Supergene alteration of primary ore assemblages from low-sulfidation Au-Ag epithermal deposits at Pongkor, Indonesia and Nazareño, Peru. Econ Geol 97:561-571
Grønvold F, Stølen S (1992) Thermodynamics of iron sulfides. II. Heat capacity and thermodynamic properties of FeS and $Fe_{0.875}S$ at temperatures from 298.15 K to 1000 K, and of $Fe_{0.89}S$ from 298.15 K to 650 K. Thermodynamics of formation. J Chem Thermodyn 24:913–936
Grønvold F, Westrum EF JR (1976) Heat capacities from iron disulfides. Thermodynamic of marcasite from 5 to 700K and the transition of marcasite to pyrite. J Chem Thermodyn 8:1039–1048
Grønvold F, Westrum EF JR (1986) Silver(I) sulfide: Ag_2S. Heat capacities from 5 to 1000 K, thermodynamic properties, and transitions. J Chem Thermodyn 18:381–401
Grønvold F, Westrum EF JR (1987) Thermodynamics of copper sulfides. I. Heat capacities and thermodynamic properties of copper (I) sulfide, Cu_2S, from 5 to 950 K. J Chem Thermodyn 19:1183–1198
Grønvold F, Stølen S, Labban AK, Westrum EF JR (1991) Thermodynamics of iron sulfides. I. Heat capacitiy and thermodynamic properties of Fe_9S_{10}, at temperatures from 5 K to 740 K. J Chem Thermodyn 23: 261–272
Guggenheim EA (1937) Theoretical basis of Raoult's law. Trans Faraday Soc 33:151-179
Guggenheim EA (1967) Thermodynamics, 5th edition. North Holland
Hackbarth CJ (1984) Depositional modeling of tetrahedrite in the Coeur d'Alene district. PhD Dissertation, Harvard University, Cambridge, Massachusetts
Hägg G, Sucksdorf I (1933) Die kristallstruktur von troilite und magnetkies. Zietschrift für physikalische Chemie B22:444-452
Hall HT (1966) The systems Ag-Sb-S, Ag-As-S, and Ag-Bi-S: phase relations and mineralogical significance. PhD Dissertation, Brown University, Providence, Rhode Island
Hall HT (1967) The pearceite and polybasite series. Am Mineral 52:1311–1321
Hall HT (1972) Substitution of Cu by Zn, Fe, and Ag in synthetic tetrahedrite. Bull Soc Franc Mineral Petrol 95:583-594
Hall WE, Czamanske GK (1972) Mineralogy and trace-element content of the Wood River lead-silver deposit, Blaine County, Idaho. Econ Geol 67:350-361
Hardy LS (2002) Characterization of silver in Coeur d'Alene District ores: the missing component. In: Coeur d'Alene District Symposium. Duff JK, Laidlaw RO (ed) Northwest Mining Association 108th Annual Meeting, Exposition abd Short Courses, Spokane, WA
Harker D (1936) The application of the three-dimensional Paterson method and the crystal structures of proustite, Ag_3AsS_3, and pyrargyrite, Ag_3SbS_3. J Chem Phys 4:381-390
Harlov DE (1995) Thermochemistry of minerals in the system Ag_2S-Cu_2S-Sb_2S_3-As_2S_3. PhD Dissertation, Purdue Univ, West Lafayette, IN
Harlov DE, Sack RO (1994) Thermochemistry of polybasite–pearceite solutions. Geochim Cosmochim Acta 58:4363–4375
Harlov DE, Sack RO (1995a) Ag–Cu exchange equilibria between pyrargyrite, high–skinnerite, and polybasite solutions. Geochim Cosmochim Acta 59:867–874
Harlov DE, Sack RO (1995b) Thermochemistry of Ag_2S–Cu_2S sulfide solutions: constraints derived from coexisting Sb_2S_3–bearing sulfosalts. Geochim Cosmochim Acta 59:4351–4365
Harris DC, Nuffield EW, Frohberg MH (1965) Studies of mineral sulpho-salts: XIX selenian polybasite. Can Mineral 8:172-184
Helmy HM (1999) The UM Saminuki volcanogenic Zn-Cu-Pb-Ag deposit, eastern-desert, Egypt: A possible new occurrence of cervelleite. Can Mineral 37:143-153
Hijmans J, de Boir J (1955) An approximation method for order-disorder problems I. Physica 21:471-484
Hoda SN, Chang LLY (1975) Phase relations in the system PbS–Ag_2S–Sb_2S_3 and PbS–Ag_2S–Bi_2S_3. Am Mineral 60:621–633
Hollenbaugh DW, Carlson EH (1983) The occurrence of wurtzite polytypes in eastern Ohio. Can Mineral 21: 697-703

Hsieh K-C (1983) The Thermodynamics and Phase Equilibria of the Fe-Ni-S-O System. PhD Dissertation, University of Wisconsin, Madison, WI

Hsieh K-C, Chang YA, Zhong T (1982) The Fe-Ni-S system above 700 °C (iron-nickel-sulfur). Bull Alloy Phase Diagrams 3:165-172

Hsieh K-C, Vlach KC, Chang YA (1987a) The Fe-Ni-S system I. A thermodynamic analysis of the phase equilibria and calculation of the phase diagram from 1173 to 1623 K. High Temp Sci 23:17-37

Hsieh K-C, Schmid R, Chang YA (1987b) The Fe-Ni-S system II. A thermodynamic model for the ternary monosulfide phase with the nickel arsenide structure. High Temp Sci 23:39-52

Hultgren R, Orr RL, Anderson PD, Kelley KK (1963) Selected Values of the Thermodynamic Properties of Metals and Alloys. John Wiley and Sons

Hutcheon I (1978) Calculation of metamorphic pressure using the sphalerite-pyrrhotite-pyrite equilibrium. Am Mineral 63: 87-95

Hutcheon I (1980) Calculated phase relations for sphalerite-pyrrhotite-pyrite: Correction. Am Mineral 65: 1063-1064

Hutchison MN, Scott SD (1981) Sphalerite geobarometry in the Cu-Fe-Zn-S system. Econ Geol 76:143-153

Hutchison MN, Scott SD (1983) Experimental calibration of the sphalerite cosmobarometer. Geochim Cosmochim Acta 47:101–108

Indolev LN, Nevoysa IA, Bryzgalov IA (1971) New data on the composition of stibnite and the isomorphism of copper and silver. Doklady Akademii Nauk SSSR 199:115-118

Ito T, Nowacki W (1974) The crystal structure of freieslebenite, $PbAgSbS_3$. Z Kristallogr 139:85-102

Ivantchev S, Kroumova E, Madariaga G, Perez-Mato JM, Aroyo MI (2000) SUBGROUPGRAPH - a computer program for analysis of group-subgroup relations between space groups. J Appl Chrystallogr 33:1190-1191

Johnson ML, Burnham CW (1985) Crystal structure refinement of an arsenic-bearing argentian tetrahedrite. Am Mineral 70:165–170

Johnson ML, Jeanloz R (1983) A brillouin–zone model for compositional variation in tetrahedrite. Am Mineral 68:220–226

Johnson NE (1986) The crystal chemistry of tetrahedrite. PhD Dissertation, Virginia Polytechnic Institute and State University, Blacksburg, Virginia

Karup-Møller S, Makovicky E (1995) The phase system Fe-Ni-S at 725 °C. N Jahr Mineral Mon 1–10

Keighin CW, Honea RM (1969) The system Ag-Sb-S from 600 °C to 200 °C. Mineralium Deposita 4:153–171

Keil K (1968) Mineralogical and chemical relationships among enstatite chondrites. J Geophys Res 73:6945-6976

Keil K, Snetsinger KG (1967) Niningerite, a new meteorite sulfide. Science 155:451-453

Keller-Besrest T, Collin G (1990) Structural aspects of the α-transition in stoichiometric FeS: Identification of the high-temperature phase. J Solid State Chem 84:194-210

Kelly DP, Vaughan DJ (1983) Pyrrhotite-pentlandite ore textures: a mechanistic approach. Mineral Mag 47: 453-463

Kern R, Weisbrod A (1967) Thermodynamics for Geologists. Freeman, Cooper, and Company

Kimura M, Lin Y (1999) Petrological and minearlogical study of enstatite chondrites with reference to their thermal histories. Antarctic Meteorite Res 12:1-18

King HE Jr, Prewitt CT (1982) High-pressure and high-temperature polymorphism of iron sulfide (FeS). Acta Crystallogr B38:1877-1886

Kissin SA (1989) Application of the sphalerite cosmobarometer to the enstatite chondrites. Geochim Cosmochim Acta 53:1649–1655

Kissin SA, Scott SD (1979) Device for the measurement of sulfur fugacity mountable on the precession camera. Am Mineral 64:1306-1310

Kissin SA, Scott SD (1982) Phase relations involving pyrrhotite below 350 °C. Econ Geol 77:1739–1754

Kissin SA, Schwarcz HP, Scott SD (1986) Application of the sphalerite cosmobarometer to IAB iron meteorites. Geochim Cosmochim Acta 50:371–378

Kiukkola K, Wagner C (1957) Galvanic cells for the determination of the standard molar free energy of formation of metal halides, oxides and sulfides at elevated temperatures. J Electrochem Soc 104:308-316

Klemm DD (1965) Synthesen und Analysen in den Dreickdiagrammen FeAsS-CoAsS-NiAsS und FeS_2-CoS_2-NiS_2. N Jahr Mineral Abhand 103:205-255

Knacke O, Kubaschewski O, Hesselmann K (1991) Thermochemical Properties of Inorganic Substances, 2nd Edition. 2 volumes. Springer-Verlag

Knowles CR (1964) A redetermination of the structure of miargyrite, $AgSbS_2$. Acta Crystallogr 17:846–851

Knowles CR (1983) A microprobe study of silver ore in northern Idaho. *In:* Microbeam Analysis. Gooley, R (ed) San Francisco Press, p 61–64

Krauskopf KB, Bird DK (1994) Introduction to Geochemistry, 3rd Edition. McGraw Hill

Kress VC (1997) Thermochemistry of sulfide liquids. I. The system O-S-Fe at 1 bar. Contrib Mineral Petrol 127:176-186

Kress VC (2000) Thermochemistry of sulfide liquids. II. Associated solution model for sulfide liquids in the system O-S-Fe. Contrib Mineral Petrol 139:316-325

Kress VC (2003) On the mathematics of associated solutions. Am J Sci 203:708-722

Kress VC, Ghiorso MS, Lastuka C (2004) Microsoft EXCEL spreadsheet-based program for calculating equilibrium gas speciation in the C-O-H-S-Cl-F system. Computers & Geosciences 30:211-214

Kretschmar U, Scott SD (1976) Phase relations involving arsenopyrite in the system Fe–As–A and their application. Can Mineral 14:364–386

Kröger FA (1974) The Chemistry of Imperfect Crystals: Volume 2, Imperfections Chemistry of Crystalline Solids. North-Holland

Krupp R (1994) Phase relations and phase transformations between the low temperature iron sulfides mackinawite, greigite and smythite. Eur J Mineral 6:265-278

Krupp RE, Weiser T (1992) On the stability of gold-silver alloys in the weathering environment. Mineralium Deposita 27:268-275

Kruse O (1990) Mossbauer and X-ray study of the vacancy concentration in synthetic hexagonal pyrrhotites. Am Mineral 75:755-763

Kubaschewski O, Evans EL, Alcock CB (1967) Metallurgical Thermochemistry, 4th edition. Pergamon Press

Kullerud G (1953) The FeS–ZnS system, a geologic thermometer. Norsk Geologisk Tidsskrift 32:61–147

Kullerud G (1963) Thermal stability of pentlandite. Can Mineral 7:353-366

Kullerud G (1971) Experimental techniques in dry sulfide research. *In*: Research Techniques for High Pressure and High Temperature. Ulmer GC (ed) Springer-Verlag, p 299-315

Kullerud G, Yoder HS (1959) Pyrite stability relations in the Fe-S system. Econ Geol 54:533-572

Kullerud G, Yund RA (1962) The Ni-S system related minerals. J Petrol 3:126-175

Kullerud G, Yund RA, Moh GH (1969) Phase relations in the Cu-Fe-S, Cu-Ni-S, and Fe-Ni-S systems. Econ Geol Monograph 4:323-343

Kyono A, Kimata M (2004) Structural variations induced by difference of the inert pair effect in the stibnite-bismuthinite solid solution series $(Sb,Bi)_2S_3$. Am Mineral 89:932–940

L'Heureux I (2000) Origin of banded patterns in natural sphalerite. Phys Rev E 62:3234–3245

Lawson AW (1947) On binary solid solutions. J Chem Phys 15:831-842

Li C, Barnes S-J, Makovicky E, Rose-Hansen J, Makovicky M (1996) Partitioning of nickel, copper, iridium, rhenium, platinum, and palladium between monosulfide solid solution and sulfide liquid: effects of composition and temperature. Geochim Cosmochim Acta 60:1231-1238

Libowitz GG (1969) Thermodynamic properties and defect structure of nonstoichiometric compounds. J Solid State Chem 1:50-58

Libowitz GG (1972) Energetics of defect formation and interaction in non-stoichiometric pyrrhotite. *In*: Reactivity of Solids. Anderson JS, Roberts MW, Stone FS (ed) Chapman and Hall, p 107-115

Libowitz GG, Lightstone JB (1967) Characterization of point defects in nonstoichiometric compounds from thermodynamic considerations. J Phys Chem Solids 28:1145-1154

Lichtner PC (1985) Continuum model for simultaneous chemical reaction and mass transport in hydrothermal systems. Geochim Cosmochim Acta 49:779-800

Lichtner PC (1988) The quasi-stationary state approximation to coupled mass transport and fluid-rock interaction in a porous medium. Geochim Cosmochim Acta 52:143-166

Lichtner PC (1996) Continuum formulation of multicomponent-multiphase reactive transport. Rev Mineral 34:1-81

Lichtner PC (2001) FLOWTRAN User Manual. Los Almos National Laboratory report LA-UR-01-2349

Lichtner PC, Carey JW (2006) Incorporating solid solutions in geochemical reactive transport models using a kinetic discrete-composition approach. Geochim Cosmochim Acta 70:1356-1378

Lightstone JB, Libowitz GG (1969) Interaction between point defects in nonstoichiometric compounds. J Phys Chem Solids 30:1025-1036

Lin Y, El Goresy A (2002) A comparative study of opaque phases in Qingzhen (EH3) and MacAlpine Hills 88136 (EL3): Representatives of EH and EL parent bodies. Meteo Planet Sci 37: 577-599

Lodders K (1996) Oldhamite in enstatite achondrites (aubrites). Proc NIPR Symp Antarctic Meteorites 9:127-142

Lodders K (2004) Revised and updated thermochemical properties of the gases mercapto (HS), disulfur monoxide (S_2O), thiazyl (NS), and thioxophosphino (PS). J Phys Chem Ref Data 33:357-367

Loucks RR (1984) Zoning and Ore Genesis at Topia, Durango, Mexico. PhD Dissertation, Harvard University, Cambridge, MA

Lueth VW (1988) Studies of the Geochemistry of the Semimetal Elements: Arsenic, Antimony, and Bismuth. PhD Dissertation, University of Texas, El Paso, TX

Lueth VW, Megaw PKM, Pingatore NE, Goodell RC (2000) Systematic variation in galena solid solution at Santa Eulalia, Chihuahua. Mexico. Econ Geol 95:1673–1687

Lundqvist D (1947) X-ray studies on the ternary system Fe-Ni-S. Ark. Kemi Mineral Geol 24A, No. 22, 12 pp.

Lusk J (1989) An electrochemical method for determining equilibration temperatures for sulfide minerals. Econ Geol 84:1663-1670

Lusk JD, Bray M (2002) Phase relations and the electrochemical determination of sulfur fugacity for selected reactions in the Cu–Fe–S and Fe–S systems at 1 bar and temperatures between 185 and 460 °C. Chem Geol 192:227–248

Lusk J, Ford CE (1978) Experimental extension of the sphalerite geobarometer to 10 kbar. Am Mineral 78: 516–519

Lusk J, Scott SD, Ford CE (1993) Phase relations in the Fe-Zn-S system to 5 kbars and temperatures between 325 °C and 150 °C. Econ Geol 88:1880–1903

Lynch JVG (1989a) Large-scale hydrothermal zoning reflected in the tetrahedrite-freibergite solid solution, Keno Hill Ag-Pb-Zn district, Yukon. Can Mineral 27:383-400

Lynch JVG (1989b) Hydrothermal zoning in the Keno Hill Ag–Pb–Zn vein system: a study in structural geology, mineralogy, fluid inclusions and stable isotope geochemistry. PhD Dissertation, University of Alberta, Alberta, Canada

Lynch JVG, Longstaffe FJ, Nesbitt BE (1990) Stable isotopic and fluid inclusion indicators of hydothermal paleoflow, boiling and fluid mixing in the Keno Hill Ag-Pb-Zn district, Yukon territory, Canada. Geochim Cosmochim Acta 54:1045-1059

Makovicky E (1997) Modular crystal chemistry of sulphosalts and other complex sulphides. In: Modular Aspects of Minerals, European Mineralogical Union Notes in Mineralogy 1. Merlino S (ed) Eötvös University Press, p 237-271

Makovicky E (2006) Crystal structures of sulfides and other chalcogenides. Rev Mineral Geochem 61:7-125

Makovicky E, Karup–Møller S (1994) Exploratory studies on substitution of minor elements in synthetic tetrahedrite Part I. Substitution by Fe, Zn, Co, Ni, Mn, Cr, V, and Pb. Unit-cell parameter changes on substitution and the structral role of "Cu^{2+}". N Jahr Mineral Abhand 167:247–261

Makovicky E, Skinner BJ (1978) Studies of the sulfosalts of copper. VI. Low temperature exsolution in synthetic tetrahedrite solid solution, $Cu_{12+x}Sb_{4+y}S_{13}$. Can Mineral 16:611–623

Makovicky E, Skinner BJ (1979) Studies of the sulfosalts of copper. VII. Crystal structures of the exsolution products $Cu_{12.3}Sb_4S_{13}$ and $Cu_{13.3}Sb_4S_{13}$ of unsubstituted synthetic tetrahedrite. Can Mineral 17:619–634

Makovicky E, Forcher K, Lottermoser W, Amthauer G (1990) The role of Fe^{2+} and Fe^{3+} in synthetic Fe–substituted tetrahedrite. Mineral Petrol 43:73–81

Margules M (1895) Uber die Zummensetzung der gesattigten Dampfe von Mischungen. Sitzungber, Akademie der Wissenschaften in Wien 104:1243-1278

Martin JD, Gil ASI (2005) An integrated thermodynamic mixing model for sphalerite geothermometry from 300 to 850 °C and up to 1 Gpa. Geochim Cosmochim Acta 69:995-1006

Maske S, Skinner BJ (1971) Studies of the sulfosalts of copper. I. Phases and phase relations in the system Cu-As-S. Econ Geol 66:901-918

McKinstry H (1963) Mineral assemblages in sulfide ores: The system Cu-Fe-As-S. Econ Geol 58:483-505

Merwin HE, Lombard RH (1937) The system Cu-Fe-S. Econ Geol 32:203-284

Mills KC (1974) Thermodynamic Data for Inorganic Sulphides, Selenides and Tellurides. Butterworths

Mishra B, Mookherjee A (1986) Analytical formulation of phase equilibrium in two observed sulfide-sulfosalt assemblages in the Rajpura-Dariba polymetallic deposit. Econ Geol 81:627-639

Misra KC, Fleet ME (1974) Chemical composition and stability of violarite. Econ Geol 69:391-403

Moh GH, Taylor LA (1971) Laboratory techniques in experimental sulfide petrology. N Jahr Mineral Mon 450-459

Motomura Y (1990) Solid solution ranges for pyrargyrite-proustite series minerals in the Inakuraishi-type manganese deposits of Japan. Ganko: Ganseki Kobutsu Kosho Gakkai shi 85:502-513

Mullen DJE, Nowacki W (1974) the crystal structure of aramayoite $Ag(Sb,Bi)S_2$. Z Kristallogr 139:54-69

Mungall JE, Andrews DR, Cabri L, Sylvester PJ, Tubrett M (2005) Partitioning of Cu, Ni, Au and Platinum Group Elements between monosulfide solid solution and sulfide melt under controlled oxygen and sulfur fugacities. Geochim Cosmochim Acta 69: 4349-4360

Nakazawa H, Morimoto N (1970) Pyrrhotite phase relations below 320 °C. Japanese Acad Proc 34:678-683

Nakazawa H, Morimoto N (1971) Phase relations and superstructures of pyrrhotite $Fe_{1-x}S$. Mater Res Bull 6: 345-358

Naldrett AJ (1969) A portion of the system Fe–S–O between 900 and 1080 °C and its application to sulfide ore magmas. J Petrol 10:171-201

Naldrett AJ (1989) Magmatic Sulfide Deposits. Oxford University Press

Naldrett AJ (1997) Key factors in the genesis of Noril'sk, Sudbury, Jinchuan, Voisey's Bay and other world-class Ni-Cu-PGE deposits: implications for exploration. Australian J Earth Sci 44:283-315

Naldrett AJ, Craig JR, Kullerud G (1967) The central portion of the Fe-Ni-S system and its bearing on pentlandite exsolution in iron nickel sulfide ores. Econ Geol 62:826–847

Naldrett AJ, Ebel DS, Asif M, Morrison G, Moore CM (1997) Fractional crystallization of sulfide melts as illustrated at Noril'sk and Sudbury. Eur J Mineral 9:365-377

Navrotsky A (1987) Models of Crystalline Solutions. Rev Mineral 17:35-69

Navrotsky A (1994) Physics and Chemistry of Earth Materials. Cambridge University Press

Nenasheva SN (1975) Eksperimental'noye issledovaniye priody primesey serebra, sur'my i vismuta v galenite. Nauka, Sibirskoye Otdeleniye Instituta Geologii i Geofiziki, Novosibirsk, USSR, p. 124

Nickel EH (1972) Nickeliferous smythite from some Canadian occurrences. Can Mineral 11:514-519

Niwa K, Wada T (1961) Thermodynamic studies of pyrrhotite. Metall Soc Conf 8:945-961

Novikov GV, Sokolov JuA, Sipavina LV (1982) The temperature dependence of the unit-cell parameters of pyrrhotite $Fe_{1-x}S$. Geochem Internat 19:184-190

O'Leary MJ, Sack RO (1987) Fe-Zn exchange reaction between tetrahedrite and sphalerite in natural environments. Contrib Mineral Petrol 96:415-425

Osadchii EG, Lunin S (1994) Determination of activity-composition relations in the $(Fe_xZn_{1-x})S$ solid solution at 820-1020 K by solid-state EMF measurements. Exp Geosci 3:48-50

Osadchii EG, Sorokin VI (1989) Stannite-Containing Sulfide Systems. Nauka, Moscow (in Russian).

Pankratz LB, King EG (1965) High-temperature heat contents and entropies of two zinc sulfides and four solid solutions of zinc and iron sulfides. Bureau Mines Rep Invest 6708:1-8

Pattrick RAD, Hall AJ (1978) Microprobe analyses of cadmium-rich tetrahedrites from Tyndrum, Perthshire. Mineral Mag 42:286-288

Pattrick RAD, Hall AJ (1983) Silver substitution into synthetic zinc, cadmium, and iron tetrahedrites. Mineral Mag 47:441-451

Pauling L, Neuman EW (1934) The crystal structure of binnite, $(Cu,Fe)_{12}As_4S_{13}$, and the chemical composition and structure of minerals of the tetrahedrite group. Z Kristallogr 88:54-62

Peacock MA, Berry LG (1947) Studies of mineral sulfo-salts: XIII – polybasite and pearceite. Mineral Mag 28: 1-13

Peregoedova A, Barnes S-J, Baker DR (2004) The formation of Pt-Ir alloys and Cu-Pd-rich sulfide melts by partial desulfurization of Fe-Ni-Cu sulfides: results of experiments and implications for natural systems. Chem Geol 208:247-264.

Peterson RC, Miller I (1986) Crystal structure refinement and cation distribution in freibergite and tetrahedrite. Mineral Mag 50: 717-721

Petruk W, staff (1971) Characteristics of the sulfides. Can Mineral 11:196-231

Pierce L, Buseck PR (1974) Electron imaging of pyrrhotite superstructures. Science 186:1209-1212

Putnis A (1975) Observations on coexisting pyrrhotite phases by transmission electron microscopy. Contrib Mineral Petrol 52:307-313

Putnis A, McConnell JDC (1980) Principles of Mineral Behavior. Elsevier

Raabe KC, Sack RO (1984) Growth zoning in tetrahedrite from the Hock Hocking mine. Can Mineral 22: 577–582

Ramdohr P (1973) The Opaque Minerals in Stony Meteorites. Elsevier

Ramdohr P (1980) The Ore Minerals and their Intergrowths, 2nd English edition. Pergamon

Rau HT (1967) Defect equilibria in cubic high temperature copper sulfide (digenite). J Phys Chem Solids 28: 902–916

Rau HT (1976) Energetic of defect formations and interactions in pyrrhotite, $Fe_{1-x}S$, and its homogeneity range. J Phys Chem Solids 37:307–313

Riley JF (1974) The tetrahedrite-freibergite series, with reference to the Mount Isa Pb-Zn-Ag ore body. Mineralium Deposita 9:117-124

Robbins M, Wertheim GK, Sherwood RC, Buchanan DNE (1971) Magnetic properties and site distributions in the system $FeCr_2O_4$-Fe_3O_4 $(Fe^{2+}Cr_{2-x}Fe_x^{3+}O_4)$. J Phys Chem Solids 32:717-729

Robie RA, Hemingway BS, Fisher JR (1978) Thermodynamic Properties of Minerals and Related Substances at 298.15 K and 1 Bar (105 Pascals) Pressure and at Higher Temperatures. USGS Bulletin 1452

Rosenqvist T (1954) A thermodynamic study of of the iron, cobalt, and nickel sulfides. J Iron Steel Inst 176: 37-57

Sack RO (1992) Thermochemistry of tetrahedrite-tennantite fahlores. In: The Stability of Minerals. Ross NL, Price GD (ed) Chapman and Hall, p 243-266

Sack RO (2000) Internally consistent database for sulfides and sulfosalts in the system Ag_2S-Cu_2S-ZnS-Sb_2S_3-As_2S_3. Geochim Cosmochim Acta 64:3803-3812

Sack RO (2002) Note on "Large-scale hydrothermal zoning reflected in the tetrahedrite-freibergite solid solution, Keno Hill Ag-Pb-Zn District, Yukon" by J.V. Gregory Lynch. Can Mineral 40:1717-1719

Sack RO (2005) Internally consistent database for sulfides and sulfosalts in the system Ag_2S-Cu_2S-ZnS-FeS-Sb_2S_3-As_2S_3: Update. Geochim Cosmochim Acta 69:1157-1164

Sack RO, Brackebusch FW (2004) Fahlore as an indicator of mineralization temperature and gold fineness. CIM Bulletin 97:78-83

Sack RO, Ebel DS (1993) As-Sb exchange energies in tetrahedrite-tennantite fahlores and bournonite-seligmannite solutions. Mineral Mag 57:633-640

Sack RO, Ghiorso MS (1989) Importance of considerations of mixing properties in establishing an internally consistent thermodynamic database: thermochemistry of minerals in the system Mg_2SiO_4–$Fe_2SiO_4SiO_2$. Contrib Mineral Petrol 102:41–68

Sack RO, Ghiorso MS (1991) Chromian spinels as petrogenetic indicators: Thermodynamics and petrological applications. Am Mineral 76:827–847

Sack RO, Ghiorso MS (1994) Thermodynamics of multicomponent pyroxenes: II. Phase relations in the quadrilateral. Contrib Mineral Petrol 116:287-300

Sack RO, Ghiorso MS (1998) Thermodynamics of feldspathoid solutions. Contrib Mineral Petrol 130:256-274

Sack RO, Goodell PC (2002) Retrograde reactions involving galena and Ag-sulphosalts in a zoned ore deposit, Julcani, Peru. Mineral Mag 66:1043-1062

Sack RO, Loucks RR (1985) Thermodynamic properties of tetrahedrite-tennantites: Constraints on the interdependence of the Ag↔Cu, Fe↔Zn, Cu↔Fe, and As↔Sb exchange reactions. Am Mineral 70: 1270–1289

Sack RO, Ebel DS, O'Leary MJ (1987) Tennahedrite thermochemistry and zoning. *In:* Chemical Transport in Metasomatic Processes. Helgeson HC (ed) D. Reidel, p 701-731

Sack RO, Kuehner SM, Hardy LS (2002) Retrograde Ag-enrichment in fahlores from the Coeur d'Alene mining district, Idaho. Mineral Mag 66:215-229

Sack RO, Lynch JGV, Foit FF Jr (2003) Fahlore as a petrogenetic indicator: Keno Hill Ag-Pb-Zn district, Yukon, Canada. Mineral Mag 67:1023-1038

Sack RO, Fredericks R, Hardy LS, Ebel DS (2005) Origin of high-Ag-fahlores from the Galena Mine, Wallace, Idaho, U.S.A. Am Mineral 90:1000-1007

Sanchez JM, de Fontaine D (1981) Theoretical prediction of ordered superstructures in metallic alloys. *In*: Structure and Bonding in Crystals. Vol. II. Okeeffe M, Navrotsky A (ed) Academic Press Inc., p 117-132

Sanchez JM, Ducastelle F, Gratias D (1984) Generalized cluster description of multicomponent systems. Physica A 128:334-350

Schenck R, Hoffmann I, Knepper W, Vogler H (1939) Gleichewichtsstudien über erzbinende sulfide. I. Z Anorg Allg Chem 240: 173–197

Schenck R, von der Forst P (1939) Gleichewichtsstudien über erzbinende sulfide. II. Z Anorg Allg Chem 241: 145-157

Schneeberg EP (1973) Sulfur measurements with the electrochemical cell Ag | AgI | $Ag_{2+x}S$, fS_2. Econ Geol 68: 507-517

Schwarz HP, Scott SD, Kissin SA (1975) Pressures of formation of iron meteorites from sphalerite compositions. Geochim Cosmochim Acta 39: 1457-1466

Scott SD (1973) Experimental calibration of the sphalerite geobarometer. Econ Geol 68: 466–474

Scott SD (1976) Application of the sphalerite geobarometer to regionally metamorphosed terraines. Am Mineral 61: 661–670

Scott SD, Barnes HL (1971) Sphalerite geothermometry and geobarometry. Econ Geol 66: 653–569

Scott SD, Barnes HL (1972) Sphalerite-wurtzite equilibria and stoichiometry. Geochim Cosmochim Acta 36: 1275-1295

Scott SD, Kissin SA (1973) Sphalerite composition in the system Zn-Fe-S system below 300 °C. Econ Geol 68: 475–479

Sharp TG, Buseck PR (1993) The distribution of Ag and Sb in galena: Inclusions versus solid solution. Am Mineral 78: 85-95

Sharp ZD, Essene EJ, Kelly WC (1985) A re–examination of the arsenopyrite geothermometer: pressure considerations and applications to natural assemblages. Can Mineral 23: 517–534

Sharma RC, Chang YA (1979) Thermodynamics and phase relationships of transition metal-sulfur systems: Part III. Thermodynamic properties of the Fe-S liquid phase and the calculation of the Fe-S phase diagram. Metall Trans B 10B: 103-108

Shimada N, Hirowatari F (1972) Argentian tetrahedrites from the Taishu-Shigekuma mine, Tsushima Island, Japan. Mineral J 7:77-87

Simon G, Kesler SE, Essene EJ (2000) Gold in porphyry copper deposits: Experimental determination of the distribution of gold in the Cu-Fe-S system at 400° to 700 °C. Econ Geol 95:259-270

Skinner BJ (1966) The system Cu-Ag-S. Econ Geol 61:1–26

Skinner BJ (1962) Thermal expansion of ten minerals. USGS Professional Paper 450-D:109-112

Skinner BJ, Luce FD (1971) Solid solutions of the type (Ca, Mg, Mn, Fe)S and their use as geothermometers for the enstatite chondrites. Am Mineral 56:1269-1296

Skinner BJ, Luce FD, Makovicky E (1972) Studies of the sulfosalts of copper III. Phases and phase relations in the system Cu-Sb-S. Econ Geol 67:924–938
Smith JV, Pluth JJ, Saho-Xu Han (1977) Crystal structure refinement of miargyrite, $AgSbS_2$. Mineral Mag 61: 671-675
Sorokin BI, Gruzdev VS, Shorygin VA (1970) Variation of the a_o parameter with the iron content in sphalerite obtained under hydrothermal conditions. Geochem Inter 7:361–363
Sorokin BI, Novikov VK, Egotov VK, Popov VI, Dipavina LV (1975) An investigation of Fe-sphalerites by means of Mössbauer spectroscopy. Geochimiya 9:1329–1335
Spalek J, Lewicki A, Tarnawski Z, Furdyna JK, Galazka RR, Obuszko Z (1986) Magnetic susceptibility of semimagnetic semiconductors in the high temperature regime and the role of superexchange. Phys Rev B 33:3407-3418
Spiridonov EM (1984) Species and varieties of fahlore (tetrahedrite-tennantite) minerals and their rational nomenclature. Doklady Akademii Nauka SSR 279:166-172
Springer G (1969a) Electron microprobe analyes of tetrahedrites. N Jahr Mineral Mon 24-32
Springer G (1969b) Naturally occurring compositions in the solid solution series Bi_2S_3-Sb_2S_3. Mineral Mag 37: 294-296
Springer G, Laflamme JHG (1971) The system Bi_2S_3–Sb_2S_3. Can Mineral 10:847–853
Sugaki A, Kitakaze A (1998) High form of pentlandite and its thermal stability. Am Mineral 83:133–140
Tarkian M, Breskovska V (1995) Mineralogy and fluid inclusion study of the Zidarova copper polymetallic deposit, eastern Bulgaria. N Jahr Mineral Abhand 168:283-298
Tatsuka K, Morimoto N (1977) Tetrahedrite stability relations in the Cu-Fe-Sb-S system. Am Mineral 62:1101-1109
Taylor LA (1969) Low-temperature phase relations in the Fe-S system. Carnegie Institution of Washington Yearbook 62:175-189
Taylor LA, Williams KL (1972) Smythite, $(Fe, Ni)_{3.25}S_4$? A redefinition. Am Mineral 57:1571-1577
Thompson JB Jr (1967) Thermodynamic properties of simple solutions. In: Researches in Geochemistry. Vol 2. Abelson PH (ed) John Wiley, p 340-361
Timoveyevskiy DA (1967) First find of Ag-rich freibergite in the USSR. Doklady Akademii Nauk SSR 176: 1388-1391
Tokanami T, Nishiguchi K, Morimoto N (1972) Crystal structure of a monoclinic pyrrhotite (Fe_7S_8). Am Mineral 57:1066-1080
Toulmin P III, Barton PB Jr (1964) A thermodynamic study of pyrite and pyrrhotite. Geochim Cosmochim Acta 28: 641-671
Toulmin P III, Barton PB Jr, Wiggins LB (1991) Commentary on the sphalerite geothermometry. Am Mineral 76:1038-1051
Turkdogan ET (1968) Iron-sulfur system. Part I: Growth of ferrous sulfide in iron and diffusivities of iron in ferrous sulfide. Trans Am Inst Mining Metall Petroleum Eng 242:641-671
Twardowski A (1990) Magnetic properties of Fe-based diluted magnetic semiconductors. J Appl Phys 67:5108-5113
Udodov YN, Kashayev AA (1970) An isothermal section (400°) of the state diagram of pyrrhotite. Transactions (Doklady) USSR. Academy of Sciences: Earth Sciences Section 187:103-105
Van Aswegen JTS, Verleger H (1960) Röntgenographishe untersuchung des system ZnS-FeS. Die Naturwissenschaften 47:131
Van Hook HJ (1960) The ternary system Ag_2S–Bi_2S_3–PbS. Econ Geol 55:759–788
Vaughan DJ, Craig JR (1978) Mineral Chemistry of Metal Sulfides. Cambridge University Press
Vaughan JR, Craig JR (1997) Sulfide ore mineral stabilities, morphologies, and intergrowth textures. In: Geochemistry of Hydrothermal Ore Deposits. Barnes HL (ed) Wiley, p 367-434
Verduch AG, Wagner C (1957) Contributions to the thermodynamics of the systems PbS–Sb_2S_3, Cu_2S–Sb_2S_3, Ag_2S–Sb_2S_3, and Ag–Sb. J Phys Chem 61:558–562
Wadhwa M, Zinner EK, Crozaz G (1997) Mn-Cr systematics in sulfides of unequilibrated enstatite chondrites. Meteoritics 32:281-292
Ward JC (1971) Interaction between cation vacancies in pyrrhotite. Solid State Comm 9:357-359
Wernick JH (1960) Constitution of the $AgSbS_2$-PbS, $AgBiS_2$-PbS, and $AgBiS_2$-$AgBiSe_2$ systems. Am Mineral 45:591-598
White JL, Orr RL, Hultgren R (1957) The thermodynamic properties of silver–gold alloys. Acta Metallurgica 5:747–760
White WB, Johnson SM, Dantzig GB (1958) Chemical equilibrium in complex mixtures. J Chem Phys 28: 751-755
Wincott PL, Vaughan DJ (2006) Spectroscopic studies of sulfides. Rev Mineral Geochem 61:181-229
Wood BJ, Fraser DG (1977) Elementary Thermodynamics for Geologists. Oxford University Press
Wuensch BJ (1964) The crystal structure of tetrahedrite, $Cu_{12}Sb_4S_{13}$. Z Kristallogr 119:437-53

Wuensch BJ, Takeuchi Y, Nowacki W (1966) Refinement of the crystal structure of binnite. Z Kristallogr 123: 1-20
Yund RA, Hall HT (1969) Hexagonal and monoclinic pyrrhotites. Econ Geol 64:420–423
Zhang Y, Sears DWG (1996) The thermometry of enstatite chondrites: A brief review and update. Meteo Planet Sci 31:647-655

On the following pages ....

APPENDICES

Appendix 1 — pages 86-93
Appendix 2 — pages 94-95
Appendix 3 — pages 96-99
Appendix 4 — page 100

Appendix 1. Expressions for chemical potentials of components of $(Zn,Fe)S$ sph, $(Ag,Cu)_{16}(Sb,As)_2S_{11}$ plb-prc, $(Ag,Cu)_3(Sb,As)S_3$ prg-prs, $(Cu,Ag)_3SbS_3$ skn, $(Cu,Ag)_{10}(Fe,Zn)_2(Sb,As)_4S_{13}$ fah, $Ag(Sb,Bi)S_2$ β-mia, arm and mat, and Pb_2S_2–$AgSbS_2$–$AgBiS_2$ gn solutions after Balabin and Sack (2000), Harlov and Sack (1994), Harlov and Sack (1995a), Ghosal and Sack (1995), Sack et al. (1987), Ghosal and Sack (1999) and Chutas (2004).

$$\mu^{sph}_{ZnS} = \mu^{o\,sph}_{ZnS} + RT\ln(X_{ZnS}) + (X_{FeS})^2\left[A_o + A_1(3 - 4X_{FeS}) + A_2(5 - 16X_{FeS} + 12X_{FeS}^2) + A_3(7 - 36X_{FeS} + 60X_{FeS}^2)\right.$$
$$\left. - 32X_{FeS}^3\right) + A_4(9 - 64X_{FeS} + 168X_{FeS}^2 - 192X_{FeS}^3 + 80X_{FeS}^4)\right]$$

$$\mu^{sph}_{FeS} = \mu^{o\,sph}_{FeS} + RT\ln(X_{FeS}) + (X_{ZnS})^2\left[A_o - A_1(3 - 4X_{ZnS}) + A_2(5 - 16X_{ZnS} + 12X_{ZnS}^2) - A_3(7 - 36X_{ZnS} + 60X_{ZnS}^2)\right.$$
$$\left. - 32X_{ZnS}^3\right) + A_4(9 - 64X_{ZnS} + 168X_{ZnS}^2 - 192X_{ZnS}^3 + 80X_{ZnS}^4)\right]$$

$$\mu^{plb\text{-}prc}_{Ag_{16}Sb_2S_{11}} = \mu^{o\,plb}_{Ag_{16}Sb_2S_{11}} + RT\left[4\ln(X^A_{Ag}) + 4\ln(X^B_{Ag}) + 8\ln(X^C_{Ag}) + 2\ln(1 - X_3)\right] - \Delta\bar{G}^o_{34}(X_3)(X_4) + W^{SM}_{AsSb}(X_3^2) + \left[W^A_{AgCu} + W^B_{AgCu}\right.$$
$$+ W^C_{AgCu} - \frac{1}{2}\Delta\bar{G}^*_{4s} - \frac{3}{2}\Delta\bar{G}^*_{4t} - \frac{1}{2}\Delta\bar{G}^*_{st}\right](X_4^2) - \left[-\frac{7}{32}\Delta\bar{G}^*_{4s} + \frac{1}{32}\Delta\bar{G}^*_{4t} + \frac{1}{32}\Delta\bar{G}^*_{st} - \frac{1}{16}W^A_{AgCu} - \frac{9}{16}W^B_{AgCu} - \frac{1}{16}W^C_{AgCu}\right](s^2)$$
$$- \left[\frac{1}{8}\Delta\bar{G}^*_{4s} - \frac{3}{8}\Delta\bar{G}^*_{4t} + \frac{1}{8}\Delta\bar{G}^*_{st} - \frac{1}{4}W^A_{AgCu} - \frac{1}{4}W^B_{AgCu} - \frac{1}{4}W^C_{AgCu}\right](t^2) - \left[\frac{1}{8}\Delta\bar{G}^*_{4s} + \frac{3}{8}\Delta\bar{G}^*_{4t} + \frac{1}{8}\Delta\bar{G}^*_{st} - \frac{1}{4}W^A_{AgCu} + \frac{3}{4}W^B_{AgCu}\right]$$
$$+ \frac{1}{4}W^C_{AgCu}\right](s)(t) - \frac{1}{4}\Delta\bar{G}^*_{3s}(X_3)(s) - \frac{1}{4}\left[\Delta\bar{G}^*_{3t} + \Delta\bar{G}^*_{34}\right](X_3)(t) - \left[-\frac{3}{4}\Delta\bar{G}^*_{4s} + \frac{1}{4}\Delta\bar{G}^*_{4t} + \frac{1}{4}\Delta\bar{G}^*_{st} - \frac{1}{2}W^A_{AgCu} + \frac{3}{2}W^B_{AgCu}\right]$$
$$- \frac{1}{2}W^C_{AgCu}\right](X_4)(s) - \left[-\frac{1}{2}\Delta\bar{G}^*_{4s} + \frac{1}{2}\Delta\bar{G}^*_{4t} + \frac{1}{2}\Delta\bar{G}^*_{st} - W^A_{AgCu} - W^B_{AgCu} + W^C_{AgCu}\right](X_4)(t)$$

Appendix 1. *continued*

$$\mu_{Ag_{16}As_2S_{11}}^{plb\text{-}prc} = \mu_{Ag_{16}As_2S_{11}}^{o\,prc} + RT\left[4\ln(X_{Ag}^A) + 4\ln(X_{Ag}^B) + 8\ln(X_{Ag}^C) + 2\ln(X_3)\right] + \Delta\bar{G}_{34}^o(1-X_3)(X_4) + W_{AsSb}^{SM}(1-X_3)^2 + \left[W_{AgCu}^A + W_{AgCu}^B\right]$$

$$+ W_{AgCu}^C - \frac{1}{2}\Delta\bar{G}_{4s}^* - \frac{3}{8}\Delta\bar{G}_{4t}^* + \frac{1}{8}\Delta\bar{G}_{st}^*\right](X_4^2) - \left[-\frac{7}{32}\Delta\bar{G}_{4s}^* + \frac{1}{32}\Delta\bar{G}_{4t}^* + \frac{1}{32}\Delta\bar{G}_{st}^* - \frac{1}{16}W_{AgCu}^A - \frac{9}{16}W_{AgCu}^B - \frac{1}{16}W_{AgCu}^C\right](s^2)$$

$$- \left[-\frac{1}{8}\Delta\bar{G}_{4s}^* - \frac{3}{8}\Delta\bar{G}_{4t}^* + \frac{1}{8}\Delta\bar{G}_{st}^* - \frac{1}{4}W_{AgCu}^A - \frac{1}{4}W_{AgCu}^B - \frac{1}{4}W_{AgCu}^C\right](t^2) - \left[\frac{1}{8}\Delta\bar{G}_{4s}^* + \frac{3}{8}\Delta\bar{G}_{4t}^* - \frac{1}{8}\Delta\bar{G}_{st}^* - \frac{1}{4}W_{AgCu}^A + \frac{3}{4}W_{AgCu}^B\right]$$

$$+ \frac{1}{4}W_{AgCu}^C\right](s)(t) + \frac{1}{4}\Delta\bar{G}_{3s}^*(1-X_3)(s) + \frac{1}{4}\left[\Delta\bar{G}_{3t}^* + \Delta\bar{G}_{34}^*\right](1-X_3)(t) - \left[-\frac{3}{4}\Delta\bar{G}_{4s}^* + \frac{1}{4}\Delta\bar{G}_{4t}^* + \frac{1}{4}\Delta\bar{G}_{st}^* - \frac{1}{2}W_{AgCu}^A + \frac{3}{2}W_{AgCu}^B\right]$$

$$- \frac{1}{2}W_{AgCu}^C\right](X_4)(s) - \left[-\frac{1}{2}\Delta\bar{G}_{4s}^* + \frac{1}{2}\Delta\bar{G}_{4t}^* + \frac{1}{2}\Delta\bar{G}_{st}^* - W_{AgCu}^A - W_{AgCu}^B + W_{AgCu}^C\right](X_4)(t)$$

$$\mu_{Cu_{16}Sb_2S_{11}}^{plb\text{-}prc} = \mu_{Cu_{16}Sb_2S_{11}}^{o\,plb} + RT\left[4\ln(X_{Cu}^A) + 4\ln(X_{Cu}^B) + 8\ln(X_{Cu}^C) + 2\ln(1-X_3)\right] + \Delta\bar{G}_{34}^o(X_3)(1-X_4) + W_{AsSb}^{SM}(X_3^2) + \left[W_{AgCu}^A + W_{AgCu}^B\right]$$

$$+ W_{AgCu}^C - \frac{1}{2}\Delta\bar{G}_{4s}^* - \frac{3}{8}\Delta\bar{G}_{4t}^* + \frac{1}{8}\Delta\bar{G}_{st}^*\right](1-X_4^2) - \left[-\frac{7}{32}\Delta\bar{G}_{4s}^* + \frac{1}{32}\Delta\bar{G}_{4t}^* + \frac{1}{32}\Delta\bar{G}_{st}^* - \frac{1}{16}W_{AgCu}^A - \frac{9}{16}W_{AgCu}^B - \frac{1}{16}W_{AgCu}^C\right](s^2)$$

$$- \left[-\frac{1}{8}\Delta\bar{G}_{4s}^* - \frac{3}{8}\Delta\bar{G}_{4t}^* + \frac{1}{8}\Delta\bar{G}_{st}^* - \frac{1}{4}W_{AgCu}^A - \frac{1}{4}W_{AgCu}^B - \frac{1}{4}W_{AgCu}^C\right](t^2) - \left[\frac{1}{8}\Delta\bar{G}_{4s}^* + \frac{3}{8}\Delta\bar{G}_{4t}^* - \frac{1}{8}\Delta\bar{G}_{st}^* - \frac{1}{4}W_{AgCu}^A + \frac{3}{4}W_{AgCu}^B\right]$$

$$+ \frac{1}{4}W_{AgCu}^C\right](s)(t) - \frac{1}{4}\Delta\bar{G}_{3s}^*(X_3)(s) - \frac{1}{4}\left[\Delta\bar{G}_{3t}^* + \Delta\bar{G}_{34}^*\right](X_3)(t) + \left[-\frac{3}{4}\Delta\bar{G}_{4s}^* + \frac{1}{4}\Delta\bar{G}_{4t}^* + \frac{1}{4}\Delta\bar{G}_{st}^* - \frac{1}{2}W_{AgCu}^A + \frac{3}{2}W_{AgCu}^B\right]$$

$$- \frac{1}{2}W_{AgCu}^C\right](1-X_4)(s) + \left[-\frac{1}{2}\Delta\bar{G}_{4s}^* + \frac{1}{2}\Delta\bar{G}_{4t}^* + \frac{1}{2}\Delta\bar{G}_{st}^* - W_{AgCu}^A - W_{AgCu}^B + W_{AgCu}^C\right](1-X_4)(t)$$

Appendix 1. continued

$$\begin{aligned}
\mu_{Cu_{16}As_2S_{11}}^{plb\text{-}prc} =\ & \mu_{Cu_{16}As_2S_{11}}^{o\,prc} + RT\left[4\ln(X_{Cu}^A) + 4\ln(X_{Cu}^B) + 8\ln(X_{Cu}^C) + 2\ln(X_3)\right] - \Delta\bar{G}_{34}^o(1-X_3)(1-X_4) + W_{AsSb}^{SM}(1-X_3)^2 + \left[W_{AgCu}^A\right.\\
& + W_{AgCu}^B + W_{AgCu}^C - \tfrac{1}{2}\Delta\bar{G}_{4s}^* - \tfrac{1}{2}\Delta\bar{G}_{4t}^* - \tfrac{1}{2}\Delta\bar{G}_{st}^*\left](1-X_4^2) - \left[-\tfrac{7}{32}\Delta\bar{G}_{4s}^* + \tfrac{1}{32}\Delta\bar{G}_{4t}^* + \tfrac{1}{32}\Delta\bar{G}_{st}^* - \tfrac{1}{16}W_{AgCu}^A - \tfrac{9}{16}W_{AgCu}^B\right.\\
& \left. - \tfrac{1}{16}W_{AgCu}^C\right](s^2) - \left[\tfrac{1}{8}\Delta\bar{G}_{4s}^* - \tfrac{3}{8}\Delta\bar{G}_{4t}^* + \tfrac{1}{8}\Delta\bar{G}_{st}^* - \tfrac{1}{4}W_{AgCu}^A - \tfrac{1}{4}W_{AgCu}^B - \tfrac{1}{4}W_{AgCu}^C\right](t^2) - \left[\tfrac{1}{8}\Delta\bar{G}_{4s}^* + \tfrac{3}{8}\Delta\bar{G}_{4t}^* - \tfrac{3}{8}\Delta\bar{G}_{st}^* - \tfrac{1}{4}W_{AgCu}^A\right.\\
& \left. + \tfrac{3}{4}W_{AgCu}^B + \tfrac{1}{4}W_{AgCu}^C\right](s)(t) - \tfrac{1}{4}\Delta\bar{G}_{3s}^*(X_3)(s) - \tfrac{1}{4}\left[\Delta\bar{G}_{3t}^* + \Delta\bar{G}_{34}^*\right](X_3)(t) +\\
& + \tfrac{1}{4}W_{AgCu}^C\right](s)(t) + \tfrac{1}{4}\Delta\bar{G}_{3s}^*(1-X_3)(s) + \tfrac{1}{4}\left[\Delta\bar{G}_{3t}^* + \Delta\bar{G}_{34}^*\right](1-X_3)(t) + \left[-\tfrac{3}{4}\Delta\bar{G}_{4s}^* + \tfrac{1}{4}\Delta\bar{G}_{4t}^* + \tfrac{1}{4}\Delta\bar{G}_{st}^* - \tfrac{1}{2}W_{AgCu}^A + \tfrac{3}{2}W_{AgCu}^B\right.\\
& \left. - \tfrac{1}{2}W_{AgCu}^C\right](1-X_4)(s) + \left[-\tfrac{1}{2}\Delta\bar{G}_{4s}^* + \tfrac{1}{2}\Delta\bar{G}_{4t}^* + \tfrac{1}{2}\Delta\bar{G}_{st}^* - W_{AgCu}^A - W_{AgCu}^B + W_{AgCu}^C\right](1-X_4)(t)
\end{aligned}$$

$$\begin{aligned}
\mu_{Ag_3SbS_3}^{prg\text{-}prs} =\ & \mu_{Ag_3SbS_3}^{o\,prg} + RT\ln\left[(1-X_{Cu}^{prg\text{-}prs})^3(1-X_{As}^{prg\text{-}prs})\right] + W_{AsSb}^{prg\text{-}prs}(X_{As}^{prg\text{-}prs})^2 + \left[W_{CuAg}^{prg\text{-}prs} + \Delta W(1-2X_{Cu}^{prg\text{-}prs})\right](X_{Cu}^{prg\text{-}prs})^2 \\
& + \left[(\Delta\bar{H}_{DAgprg\text{-}Agprg}^o - T\Delta\bar{S}_{DAgprg\text{-}Agprg}^o) - (\Delta\bar{H}_{DCupyr\text{-}Cupyr}^o - T\Delta\bar{S}_{DCuprg\text{-}Cuprg}^o)\right](X_{Cu}^{prg\text{-}prs})^2
\end{aligned}$$

Appendix 1. *continued*

$$\mu^{prg\text{-}prs}_{Cu_3SbS_3} = [\mu^{o\,skn}_{Cu_3SbS_3} - \Delta\overline{G}^o_{Cuskn\text{-}prg}] + RT\ln[(X^{prg\text{-}prs}_{Cu})^3(1 - X^{prg\text{-}prs}_{As})] + W^{prg\text{-}prs}_{AsSb}(X^{prg\text{-}prs}_{As})^2$$

$$+ [W^{prg\text{-}prs}_{CuAg} - 2\,\Delta W(X^{prg\text{-}prs}_{Cu})](1 - X^{prg\text{-}prs}_{Cu})^2 + (\Delta\overline{H}^o_{DAgprg-Agprg} - T\Delta\overline{S}^o_{DAgprg-Agprg})(1 - X^{prg\text{-}prs}_{Cu})^2$$

$$+ (\Delta\overline{H}^o_{DCupyr-Cupyr} - T\Delta\overline{S}^o_{DCuprg-Cuprg})[2\,X^{prg\text{-}prs}_{Cu} - (X^{prg\text{-}prs}_{Cu})^2]$$

$$\mu^{prg\text{-}prs}_{Ag_3AsS_3} = \mu^{o\,prg}_{Ag_3AsS_3} + RT\ln[(1 - X^{prg\text{-}prs}_{Cu})^3(X^{prg\text{-}prs}_{As})] + W^{prg\text{-}prs}_{AsSb}(1 - X^{prg\text{-}prs}_{As})^2 + [W^{prg\text{-}prs}_{CuAg} + \Delta W(1 - 2\,X^{prg\text{-}prs}_{Cu})](X^{prg\text{-}prs}_{Cu})^2$$

$$+ [(\Delta\overline{H}^o_{DAgprg-Agprg} - T\Delta\overline{S}^o_{DAgprg-Agprg}) - (\Delta\overline{H}^o_{DCupyr-Cupyr} - T\Delta\overline{S}^o_{DCuprg-Cuprg})](X^{prg\text{-}prs}_{Cu})^2$$

$$\mu^{prg\text{-}prs}_{Cu_3AsS_3} = [\mu^{o\,skn}_{Cu_3SbS_3} - \Delta\overline{G}^o_{Cuskn\text{-}prg} + \mu^{o\,prg}_{Ag_3AsS_3} - \mu^{o\,prg}_{Ag_3SbS_3}] + RT\ln[(X^{prg\text{-}prs}_{Cu})^3(X^{prg\text{-}prs}_{As})] + W^{prg\text{-}prs}_{AsSb}(1 - X^{prg\text{-}prs}_{As})^2$$

$$+ [W^{prg\text{-}prs}_{CuAg} - 2\,\Delta W(X^{prg\text{-}prs}_{Cu})](1 - X^{prg\text{-}prs}_{Cu})^2 + (\Delta\overline{H}^o_{DAgprg-Agprg} - T\Delta\overline{S}^o_{DAgprg-Agprg})(1 - X^{prg\text{-}prs}_{Cu})^2$$

$$+ (\Delta\overline{H}^o_{DCupyr-Cupyr} - T\Delta\overline{S}^o_{DCuprg-Cuprg})[2\,X^{prg\text{-}prs}_{Cu} - (X^{prg\text{-}prs}_{Cu})^2]$$

Appendix 1. *continued*

$$\mu_{Cu_3SbS_3}^{skn} = \mu_{Cu_3SbS_3}^{o\,skn} + RT\left[\ln(X_{Cu}^{A\,skn}) + 2\ln(X_{Cu}^{B\,skn})\right] + (W_{AgCu}^{A\,skn} + W_{AgCu}^{B\,skn} - \Delta\bar{G}_{xs}^*)(1 - X_{Cu}^{skn})^2 + (\tfrac{2}{9}\Delta\bar{G}_{xs}^* + \tfrac{4}{9}W_{AgCu}^{A\,skn} + \tfrac{1}{9}W_{AgCu}^{B\,skn})(s)^2$$

$$+ (\tfrac{1}{3}\Delta\bar{G}_{xs}^* - \tfrac{4}{3}W_{AgCu}^{A\,skn} + \tfrac{2}{3}W_{AgCu}^{B\,skn})(1 - X_{Cu}^{skn})(s)$$

$$\mu_{Ag_3SbS_3}^{skn} = \mu_{Ag_3SbS_3}^{o\,prg} - \Delta\bar{G}_{Agprg\text{-}skn}^o + RT\left[\ln(1 - X_{Cu}^{A\,skn}) + 2\ln(1 - X_{Cu}^{B\,skn})\right] + (W_{AgCu}^{A\,skn} + W_{AgCu}^{B\,skn} - \Delta\bar{G}_{xs}^*)(X_{Cu}^{skn})^2$$

$$+ (\tfrac{2}{9}\Delta\bar{G}_{xs}^* + \tfrac{4}{9}W_{AgCu}^{A\,skn} + \tfrac{1}{9}W_{AgCu}^{B\,skn})(s)^2 - (\tfrac{1}{3}\Delta\bar{G}_{xs}^* - \tfrac{4}{3}W_{AgCu}^{A\,skn} + \tfrac{2}{3}W_{AgCu}^{B\,skn})(X_{Cu}^{skn})(s)$$

$$\mu_{Cu_{10}Fe_2Sb_4S_{13}}^{fah} = \mu_{Cu_{10}Fe_2Sb_4S_{13}}^{o\,fah} + RT\ln\left[(1 - X_4 - \tfrac{2}{5}s)^6(\tfrac{2}{3}[1 - X_4] + \tfrac{2}{5}s)^4(\tfrac{3}{5})^4(1 - X_2)^2(1 - X_3)^4\right] - \Delta\bar{G}_{23}^o(X_2)(X_3) - \Delta\bar{G}_{24}^o(X_2)(X_4)$$

$$- \Delta\bar{G}_{34}^o(X_3)(X_4) + W_{FeZn}^{TET}(X_2)^2 + W_{AsSb}^{SM}(X_3)^2 + (\Delta\bar{G}_{4s}^* + W_{AgCu}^{TET} + W_{AgCu}^{TRG})(X_4)^2 - \Delta\bar{G}_{2s}^*(X_2)(s) - \Delta\bar{G}_{3s}^*(X_3)(t)$$

$$- \tfrac{1}{5}(\Delta\bar{G}_{4s}^* - 4W_{AgCu}^{TRG})(X_4)(s) - \tfrac{1}{25}(6\Delta\bar{G}_{4s}^* - 9W_{AgCu}^{TET} - 4W_{AgCu}^{TRG})(s)^2$$

$$\mu_{Cu_{10}Zn_2Sb_4S_{13}}^{fah} = \mu_{Cu_{10}Zn_2Sb_4S_{13}}^{o\,fah} + RT\ln\left[(1 - X_4 - \tfrac{2}{5}s)^6(\tfrac{2}{3}[1 - X_4] + \tfrac{2}{5}s)^4(\tfrac{3}{5})^4(X_2)^2(1 - X_3)^4\right] + \Delta\bar{G}_{23}^o(1 - X_2)(X_3) + \Delta\bar{G}_{24}^o(1 - X_2)(X_4)$$

$$- \Delta\bar{G}_{34}^o(X_3)(X_4) + W_{FeZn}^{TET}(1 - X_2)^2 + W_{AsSb}^{SM}(X_3)^2 + (\Delta\bar{G}_{4s}^* + W_{AgCu}^{TET} + W_{AgCu}^{TRG})(X_4)^2 + \Delta\bar{G}_{2s}^*(1 - X_2)(s) - \Delta\bar{G}_{3s}^*(X_3)(s)$$

$$- \tfrac{1}{5}(\Delta\bar{G}_{4s}^* - 4W_{AgCu}^{TRG})(X_4)(s) - \tfrac{1}{25}(6\Delta\bar{G}_{4s}^* - 9W_{AgCu}^{TET} - 4W_{AgCu}^{TRG})(s)^2$$

Appendix 1. *continued*

$$\mu^{fah}_{Cu_{10}Fe_2As_4S_{13}} = \mu^{o\,fah}_{Cu_{10}Fe_2As_4S_{13}} + RT\ln\left[(1-X_4-\tfrac{2}{5}s)^6(\tfrac{6}{4}[1-X_4]+\tfrac{2}{5}s)^{\tfrac{4}{6}}(\tfrac{3}{2})^{\tfrac{4}{5}}(1-X_2)^2(X_3)^4\right] + \Delta\bar{G}^o_{23}(X_2)(1-X_3) - \Delta\bar{G}^o_{24}(X_2)(X_4)$$

$$+ \Delta\bar{G}^o_{34}(1-X_3)(X_4) + W^{TET}_{FeZn}(X_2)^2 + W^{SM}_{AsSb}(1-X_3)^2 + (\Delta\bar{G}^*_{4S} + W^{TET}_{AgCu} + W^{TRG}_{AgCu})(X_4)^2 - \Delta\bar{G}^*_{2S}(X_2)(s) + \Delta\bar{G}^*_{3S}(1-X_3)(s)$$

$$- \tfrac{1}{5}(\Delta\bar{G}^*_{4S} + 6\,W^{TET}_{AgCu} - 4\,W^{TRG}_{AgCu})(X_4)(s) - \tfrac{1}{25}(6\,\Delta\bar{G}^*_{4S} - 9\,W^{TET}_{AgCu} - 4\,W^{TRG}_{AgCu})(s)^2$$

$$\mu^{fah}_{Ag_{10}Fe_2Sb_4S_{13}} = \mu^{o\,fah}_{Ag_{10}Fe_2Sb_4S_{13}} + RT\ln\left[(X_4+\tfrac{2}{5}s)^6(\tfrac{2}{3}X_4-\tfrac{2}{5}s)^{\tfrac{4}{6}}(\tfrac{3}{2})^{\tfrac{4}{5}}(1-X_2)^2(1-X_3)^4\right] - \Delta\bar{G}^o_{23}(X_2)(X_3) + \Delta\bar{G}^o_{24}(X_2)(1-X_4)$$

$$+ \Delta\bar{G}^o_{34}(X_3)(1-X_4) + W^{TET}_{FeZn}(X_2)^2 + W^{SM}_{AsSb}(X_3)^2 + (\Delta\bar{G}^*_{4S} + W^{TET}_{AgCu} + W^{TRG}_{AgCu})(1-X_4)^2 - \Delta\bar{G}^*_{2S}(X_2)(s) - \Delta\bar{G}^*_{3S}(X_3)(s)$$

$$+ \tfrac{1}{5}(\Delta\bar{G}^*_{4S} + 6\,W^{TET}_{AgCu} - 4\,W^{TRG}_{AgCu})(1-X_4)(s) - \tfrac{1}{25}(6\,\Delta\bar{G}^*_{4S} - 9\,W^{TET}_{AgCu} - 4\,W^{TRG}_{AgCu})(s)^2$$

$$\mu^{fah}_{Cu_{10}Zn_2As_4S_{13}} = \mu^{o\,fah}_{Cu_{10}Zn_2As_4S_{13}} + RT\ln\left[(1-X_4-\tfrac{2}{5}s)^6(\tfrac{6}{4}[1-X_4]+\tfrac{2}{5}s)^{\tfrac{4}{6}}(\tfrac{3}{2})^{\tfrac{4}{5}}(X_2)^2(X_3)^4\right] - \Delta\bar{G}^o_{23}(1-X_2)(1-X_3) + \Delta\bar{G}^o_{24}(1-X_2)(X_4)$$

$$+ \Delta\bar{G}^o_{34}(1-X_3)(X_4) + W^{TET}_{FeZn}(1-X_2)^2 + W^{SM}_{AsSb}(1-X_3)^2 + (\Delta\bar{G}^*_{4S} + W^{TET}_{AgCu} + W^{TRG}_{AgCu})(X_4)^2 + \Delta\bar{G}^*_{2S}(1-X_2)(s)$$

$$+ \Delta\bar{G}^*_{3S}(1-X_3)(s) - \tfrac{1}{5}(\Delta\bar{G}^*_{4S} + 6\,W^{TET}_{AgCu} - 4\,W^{TRG}_{AgCu})(X_4)(s) - \tfrac{1}{25}(6\,\Delta\bar{G}^*_{4S} - 9\,W^{TET}_{AgCu} - 4\,W^{TRG}_{AgCu})(s)^2$$

Appendix 1. *continued*

$$\mu^{fah}_{Ag_{10}Zn_2Sb_4S_{13}} = \mu^{o\,fah}_{Ag_{10}Zn_2Sb_4S_{13}} + RT\ln\left[(X_4 + \tfrac{2}{5}s)^6(\tfrac{2}{5}X_4 - \tfrac{2}{5}s)^4(\tfrac{3}{5})^4(X_2)^2(1-X_3)^4\right] + \Delta\bar{G}^o_{23}(1-X_2)(X_3) - \Delta\bar{G}^o_{24}(1-X_2)(1-X_4)$$

$$+ \Delta\bar{G}^o_{34}(X_3)(1-X_4) + W^{TET}_{FeZn}(1-X_2)^2 + W^{SM}_{AsSb}(X_3)^2 + (\Delta\bar{G}^*_{4s} + W^{TET}_{AgCu} + W^{TRG}_{AgCu})(1-X_4)^2 + \Delta\bar{G}^*_{2s}(1-X_2)(s) - \Delta\bar{G}^*_{3s}(X_3)(s)$$

$$+ \tfrac{1}{5}(\Delta\bar{G}^*_{4s} + 6W^{TET}_{AgCu} - 4W^{TRG}_{AgCu})(1-X_4)(s) - \tfrac{1}{25}(6\Delta\bar{G}^*_{4s} - 9W^{TET}_{AgCu} - 4W^{TRG}_{AgCu})(s)^2$$

$$\mu^{fah}_{Ag_{10}Fe_2As_4S_{13}} = \mu^{o\,fah}_{Ag_{10}Fe_2As_4S_{13}} + RT\ln\left[(X_4 + \tfrac{2}{5}s)^6(\tfrac{2}{5}X_4 - \tfrac{2}{5}s)^4(\tfrac{3}{5})^4(1-X_2)^2(X_3)^4\right] + \Delta\bar{G}^o_{23}(X_2)(1-X_3) + \Delta\bar{G}^o_{24}(X_2)(1-X_4)$$

$$- \Delta\bar{G}^o_{34}(1-X_3)(1-X_4) + W^{TET}_{FeZn}(X_2)^2 + W^{SM}_{AsSb}(1-X_3)^2 + (\Delta\bar{G}^*_{4s} + W^{TET}_{AgCu} + W^{TRG}_{AgCu})(1-X_4)^2 - \Delta\bar{G}^*_{2s}(X_2)(s) + \Delta\bar{G}^*_{3s}(1-X_3)(s)$$

$$+ \tfrac{1}{5}(\Delta\bar{G}^*_{4s} + 6W^{TET}_{AgCu} - 4W^{TRG}_{AgCu})(1-X_4)(s) - \tfrac{1}{25}(6\Delta\bar{G}^*_{4s} - 9W^{TET}_{AgCu} - 4W^{TRG}_{AgCu})(s)^2$$

$$\mu^{fah}_{Ag_{10}Zn_2As_4S_{13}} = \mu^{o\,fah}_{Ag_{10}Zn_2As_4S_{13}} + RT\ln\left[(X_4 + \tfrac{2}{5}s)^6(\tfrac{2}{5}X_4 - \tfrac{2}{5}s)^4(\tfrac{3}{5})^4(X_2)^2(X_3)^4\right] - \Delta\bar{G}^o_{23}(1-X_2)(1-X_3) - \Delta\bar{G}^o_{24}(1-X_2)(1-X_4)$$

$$- \Delta\bar{G}^o_{34}(1-X_3)(1-X_4) + W^{TET}_{FeZn}(1-X_2)^2 + W^{SM}_{AsSb}(1-X_3)^2 + (\Delta\bar{G}^*_{4s} + W^{TET}_{AgCu} + W^{TRG}_{AgCu})(1-X_4)^2 + \Delta\bar{G}^*_{2s}(1-X_2)(s)$$

$$+ \Delta\bar{G}^*_{3s}(1-X_3)(s) + \tfrac{1}{5}(\Delta\bar{G}^*_{4s} + 6W^{TET}_{AgCu} - 4W^{TRG}_{AgCu})(1-X_4)(s) - \tfrac{1}{25}(6\Delta\bar{G}^*_{4s} - 9W^{TET}_{AgCu} - 4W^{TRG}_{AgCu})(s)^2$$

Appendix 1. *continued*

$\mu_{AgSbS_2}^{\beta} = \mu_{AgSbS_2}^{o\,\bar{\beta}\text{-}mia} + \frac{1}{5} RT \ln \left[(1 - X_{Bi}^{\beta} + \frac{1}{5} s^{\beta})^4 (1 - X_{Bi}^{\beta} - \frac{4}{5} s^{\beta}) \right] + \frac{1}{5} (\Delta \bar{G}_x^{o\,\beta} + W_{Bi\text{-}Sb}^{I\,\beta} + W_{Bi\text{-}Sb}^{II\,\beta})(1 - X_{Bi}^{\beta})^2$

$\quad - \frac{1}{5} (-\frac{3}{5} \Delta \bar{G}_x^{o\,\beta} + \frac{2}{5} W_{Bi\text{-}Sb}^{I\,\beta} - \frac{8}{5} W_{Bi\text{-}Sb}^{II\,\beta})(X_{Bi}^{\beta}) s^{\beta} - \frac{1}{5} \cdot \frac{4}{5} (\frac{1}{25} \Delta \bar{G}_x^{o\,\beta}) s^{\beta} - \frac{1}{5} W_{Bi\text{-}Sb}^{I\,\beta} - \frac{16}{25} W_{Bi\text{-}Sb}^{II\,\beta})$

$\mu_{AgBiS_2}^{\beta} = \mu_{AgBiS_2}^{o\,\bar{\beta}\text{-}mat} + \frac{1}{5} RT \ln \left[(X_{Bi}^{\beta} - \frac{1}{5} s^{\beta})^4 (X_{Bi}^{\beta} + \frac{4}{5} s^{\beta}) \right] + \frac{1}{5} (\Delta \bar{G}_x^{o\,\beta} + W_{Bi\text{-}Sb}^{I\,\beta} + W_{Bi\text{-}Sb}^{II\,\beta})(X_{Bi}^{\beta})^2$

$\quad + \frac{1}{5} (-\frac{3}{5} \Delta \bar{G}_x^{o\,\beta} + \frac{2}{5} W_{Bi\text{-}Sb}^{I\,\beta} - \frac{8}{5} W_{Bi\text{-}Sb}^{II\,\beta})(1 - X_{Bi}^{\beta}) s^{\beta} - \frac{1}{5} (\frac{4}{25} \Delta \bar{G}_x^{o\,\beta} - \frac{1}{25} W_{Bi\text{-}Sb}^{I\,\beta} - \frac{16}{25} W_{Bi\text{-}Sb}^{II\,\beta})$

$\mu_{Pb_2S_2}^{gn} = \mu_{Pb_2S_2}^{o\,gn} + 2 RT \ln (X_{Pb_2S_2}^{gn}) + W_1 (X_2 - X_1 X_2) + W_2 (X_3 - X_1 X_3) - W_3 (X_2 X_3) - 2 W_4 (X_1 X_2 - X_1^2 X_2) + W_5 (X_2^2 - 2 X_1 X_2^2)$

$\quad + 2 W_6 (X_1 X_3 - X_1^2 X_3) + W_7 (X_3^2 - 2 X_1 X_3^2) - 2 W_8 (X_2^2 X_3) - 2 W_9 (X_2 X_3^2) + W_{10} ([1 - 2 X_1] X_2 X_3)$

$\mu_{AgSbS_2}^{gn} = \mu_{AgSbS_2}^{o\,\alpha\text{-}mia} + 2 RT \ln (X_{AgSbS_2}^{gn}) + W_1 (X_1 - X_1 X_2) - W_2 (X_1 X_3) - W_3 (X_3 - X_2 X_3) + W_4 (X_1^2 - 2 X_1^2 X_2)$

$\quad + 2 W_5 (X_1 X_2 - X_1 X_2^2) - 2 W_6 (X_1^2 X_3) - 2 W_7 (X_1 X_3^2) + 2 W_8 (X_2 X_3 - X_2^2 X_3) + W_9 (X_3^2 - 2 X_2 X_3^2) + W_{10} ([1 - 2 X_2] X_1 X_3)$

$\mu_{AgBiS_2}^{gn} = \mu_{AgBiS_2}^{o\,\alpha\text{-}mat} + 2 RT \ln (X_{AgBiS_2}^{gn}) - W_1 (X_1 X_2) + W_2 (X_1 - X_1 X_3) + W_3 (X_2 - X_2 X_3) - 2 W_4 (X_1^2 X_2) - 2 W_5 (X_1 X_2^2)$

$\quad + W_6 (X_1^2 - 2 X_1^2 X_3) + 2 W_7 (X_1 X_3 - X_1 X_3^2) + W_8 (X_2^2 - 2 X_2^2 X_3) + 2 W_9 (X_2 X_3 - X_2 X_3^2) + W_{10} ([1 - 2 X_3] X_1 X_2)$

Appendix 2. Conditions of homogeneous equilibrium in $(Ag,Cu)_{16}(Sb,As)_2S_{11}$ *plb-prc*, $(Cu,Ag)_{10}(Fe,Zn)_2(Sb,As)_4$ S_{13} *fah*, and $Ag(Sb,Bi)S_2$ β-*mia*, *arm* and *mat* solutions after Harlov and Sack (1994, 1995a), Ghosal and Sack (1995), and Ghosal and Sack (1999).

$$\left(\frac{\partial \bar{G}}{\partial s}\right)_{X_3, X_4, t} = 0 = RT\left[\ln(X_{Cu}^A/X_{Ag}^A) + 3\ln(X_{Ag}^B/X_{Cu}^B) + 2\ln(X_{Cu}^C/X_{Ag}^C)\right] + \frac{1}{4}\left[\frac{3}{2}\Delta\bar{G}_{4s}^* - \frac{1}{2}\Delta\bar{G}_{st}^* + \Delta\bar{G}_s^* + W_{AgCu}^A - 3\,W_{AgCu}^B$$

$$+ W_{AgCu}^C\right] + \frac{1}{8}\left[-\frac{7}{2}\Delta\bar{G}_{4s}^* + \frac{1}{2}\Delta\bar{G}_{4t}^* + \frac{1}{2}\Delta\bar{G}_{st}^* - W_{AgCu}^A - 9\,W_{AgCu}^B - W_{AgCu}^C\right](s) + \frac{1}{4}\left[\frac{1}{2}\Delta\bar{G}_{4s}^* + \frac{3}{2}\Delta\bar{G}_{4t}^* - \frac{3}{2}\Delta\bar{G}_{st}^*\right.$$

$$\left. - W_{AgCu}^A + 3\,W_{AgCu}^B + W_{AgCu}^C\right](t) + \frac{1}{4}\Delta\bar{G}_{3s}(X_3)(s) + \frac{1}{2}\left[-\frac{3}{2}\Delta\bar{G}_{4s}^* + \frac{1}{2}\Delta\bar{G}_{4t}^* + \frac{1}{2}\Delta\bar{G}_{st}^* - W_{AgCu}^A + 3\,W_{AgCu}^B - W_{AgCu}^C\right](X_4)$$

$$\left(\frac{\partial \bar{G}}{\partial t}\right)_{X_3, X_4, s} = 0 = RT\left[2\ln(X_{Cu}^A/X_{Ag}^A) + 2\ln(X_{Cu}^B/X_{Ag}^B) + 4\ln(X_{Ag}^C/X_{Cu}^C)\right] + \frac{1}{4}\left[\Delta\bar{G}_{4t}^* - \Delta\bar{G}_{4s}^* - \Delta\bar{G}_{st}^* + \Delta\bar{G}_t^* + 2\,W_{AgCu}^A + 2\,W_{AgCu}^B\right.$$

$$\left. - 2\,W_{AgCu}^C\right] + \frac{1}{2}\left[\frac{1}{2}\Delta\bar{G}_{4s}^* - \frac{3}{2}\Delta\bar{G}_{4t}^* + \frac{1}{2}\Delta\bar{G}_{st}^* - W_{AgCu}^A - W_{AgCu}^B - W_{AgCu}^C\right](t) + \frac{1}{4}\left[\frac{1}{2}\Delta\bar{G}_{4s}^* + \frac{3}{2}\Delta\bar{G}_{4t}^* - \frac{3}{2}\Delta\bar{G}_{st}^* - W_{AgCu}^A\right.$$

$$\left. + 3\,W_{AgCu}^B + W_{AgCu}^C\right](s) + \frac{1}{4}\left[\Delta\bar{G}_{3t}^* - \frac{3}{2}\Delta\bar{G}_{34}^*\right](X_3) + \left[\frac{1}{2}\Delta\bar{G}_{4s}^* - \frac{1}{2}\Delta\bar{G}_{4t}^* + \frac{1}{2}\Delta\bar{G}_{st}^* - W_{AgCu}^A - W_{AgCu}^B + W_{AgCu}^C\right](X_4)$$

$$\left(\frac{\partial \bar{G}}{\partial s}\right)_{X_{Cu}} = 0 = RT\left[\ln(X(X_{Cu}^{skn} + \frac{2}{3}s)/(1 - X_{Cu}^{skn} - \frac{2}{3}s)\right] + \ln\left[(1 - X_{Cu}^{skn} + \frac{1}{3}s)/(X_{Cu}^{skn} - \frac{1}{3}s)\right] - \frac{3}{2}\left[\frac{1}{2}\Delta\bar{G}_s^* + \frac{1}{6}\Delta\bar{G}_{xs}^* - \frac{2}{3}W_{AgCu}^{A\,skn}\right.$$

$$\left. + \frac{1}{3}W_{AgCu}^{B\,skn}\right](s) - 3\left[\frac{2}{9}\Delta\bar{G}_{xs}^* + \frac{4}{9}W_{AgCu}^{A\,skn} + \frac{1}{9}W_{AgCu}^{B\,skn}\right](s) + \frac{3}{2}\left[\frac{1}{3}\Delta\bar{G}_{xs}^* - \frac{4}{3}W_{AgCu}^{A\,skn} - \frac{2}{3}W_{AgCu}^{B\,skn}\right](X_{Cu}^{skn})$$

Appendix 2. *continued*

$$\left(\frac{\partial \bar{G}}{\partial s}\right)_{X_2, X_3, X_4} = 0 = RT\left[\ln(X_{Ag}^{TRG}/X_{Ag}^{TET}) + \ln(X_{Cu}^{TET}/X_{Cu}^{TRG})\right] + \frac{5}{12}\Delta\bar{G}_s^* + \frac{5}{12}\Delta\bar{G}_{2s}^*(X_2) + \Delta\bar{G}_{3s}^*(X_3)$$
$$+ \frac{1}{24}(\Delta\bar{G}_{4s}^* + 6\,W_{AgCu}^{TET} - 4\,W_{AgCu}^{TRG})(2X_4 - 1) + \frac{1}{30}(6\,\Delta\bar{G}_{4s}^* - 9\,W_{AgCu}^{TET} - 4\,W_{AgCu}^{TRG})(s)$$

$$\left(\frac{\partial \bar{G}}{\partial s^\beta}\right)_{X_{Bi}} = 0 = RT\left[\ln\left[(1 - X_{Bi}^\beta + \tfrac{1}{5}s^\beta)(1 - X_{Bi}^\beta - \tfrac{4}{5}s^\beta)\right] + \ln\left[(X_{Bi}^\beta + \tfrac{4}{5}s^\beta)(X_{Bi}^\beta - \tfrac{1}{5}s^\beta)\right]\right] + \frac{5}{4}\left(\frac{3}{10}\Delta\bar{G}_x^{o\,\beta} + \frac{1}{2}\Delta\bar{G}_s^{o\,\beta} - \frac{1}{5}W_{Bi\text{-}Sb}^{I\beta} + \frac{4}{5}W_{Bi\text{-}Sb}^{II\beta}\right)$$
$$+ \frac{5}{2}\left(\frac{4}{25}\Delta\bar{G}_x^{o\,\beta} - \frac{1}{25}W_{Bi\text{-}Sb}^{I\beta} - \frac{16}{25}W_{Bi\text{-}Sb}^{II\beta}\right)s^\beta + \frac{5}{4}\left(-\frac{3}{5}\Delta\bar{G}_x^{o\,\beta} + \frac{2}{5}W_{Bi\text{-}Sb}^{I\beta} - \frac{8}{5}W_{Bi\text{-}Sb}^{II\beta}\right)(X_{Bi}^\beta)$$

Appendix 3. Thermodynamic parameters.

Parameter	Value (kJ/mol)	Parameter	Value (kJ/mol) or (kJ/K mol)
	1) *sph* Solution Parameters		
A_0	$3.464954 + 0.0049152\,T - 0.0040522(T^2/10^3)$ $+ 0.00120078(T^3/10^6)$	A_1	$-4.86433 + 0.0039523\,T - 0.00350847(T^2/10^3)$ $+ 0.00108473(T^3/10^6)$
A_2	$0.350802 + 0.00151804\,T - 0.00157187(T^2/10^3)$ $+ 0.00052146(T^3/10^6)$	A_3	$2.32679 - 0.0056524\,T + 0.00510041(T^2/10^3)$ $- 0.00159157(T^3/10^6)$
A_4	$3.54139 - 0.00836153\,T + 0.00744021(T^2/10^3)$ $- 0.00230296(T^3/10^6)$		
		2) *plb-prc* Solution Parameters	
W^A_{AgCu}	8.00	W^B_{AgCu}	0.00
W^C_{AgCu}	0.00	W^{SM}_{AsSb}	8.00
$\Delta \bar{G}^o_{34}$	−20.00	$\Delta \bar{G}^*_s$	4.50
$\Delta \bar{G}^*_{3s}$	0.00	$\Delta \bar{G}^*_{4s}$	24.00
$\Delta \bar{G}^*_t$	4.50	$\Delta \bar{G}^*_{3t}$	−12.00
$\Delta \bar{G}^*_{4t}$	0.00	$\Delta \bar{G}^*_{st}$	0.00
		3) *prg-prs* Solution Parameters	
$\Delta \bar{H}^o_{DAgprg-Agprg}$	0.00	$\Delta \bar{H}^o_{DCuprg-Cuprg}$	0.00
$\Delta \bar{S}^o_{DAgprg-Agprg}$	0.0155	$\Delta \bar{S}^o_{DCuprg-Cuprg}$	0.0125

Appendix 3. *continued*

Parameter	Value (kJ/mol)	Parameter	Value (kJ/mol) or (kJ/K mol)
W^{prg}_{CuAg}	11.0	W^{prg}_{AgCu}	21.0
ΔW	10.0	$W^{prg-prs}_{AsSb}$	6.0
$W^{\beta-mia}_{AgCu}$	9.0		
4) *skn* Solution Parameters			
$W^{A\,skn}_{AgCu}$	12.0	$W^{B\,skn}_{AgCu}$	9.0
$\Delta \bar{G}^{*}_{s}$	10.0	$\Delta \bar{G}^{*}_{XS}$	0.0
5) *bcc-, fcc-, and hcp-* Solution and Transition Parameters			
W^{bcc}_{AgCu}	$25.55612 - 0.039788\,(T) + 0.00000692761 6(T)^2$	W^{fcc}_{AgCu}	5.4
W^{hcp}_{AgCu}	$144.3949 - 1.078516(T) + 0.003321395(T)^2 - 0.000004485141(T)^3 + 0.000000002287694(T)^4$		
$\Delta \bar{H}^{o\,bcc-fcc}_{T_t,\,Ag_2S}$ (865 K)	0.784	$\Delta \bar{H}^{o\,hcp-fcc}_{T_t,\,Cu_2S}$ (710.1 K)	1.187
$\Delta \bar{S}^{o\,bcc-fcc}_{T_t,\,Ag_2S}$ (865 K)	0.000906	$\Delta \bar{S}^{o\,hcp-fcc}_{T_t,\,Cu_2S}$ (710.1 K)	0.001672
$\Delta \bar{C}^{o\,bcc-fcc}_{p\,Ag_2S}$ (865 K)	-0.003858	$\Delta \bar{C}^{o\,hcp-fcc}_{p\,Cu_2S}$ (710.1 K)	-0.003563
6) *fah* Solution Parameters			
W^{TET}_{FeZn}	0.0	W^{TRG}_{AgCu}	0.0
W^{TRG}_{AgCu}	29.0077	W^{SM}_{AsSb}	16.736
$\Delta \bar{G}^{o}_{23}$	10.8366	$\Delta \bar{G}^{o}_{24}$	9.00
$\Delta \bar{G}^{o}_{34}$	50.34266	$\Delta \bar{G}^{*}_{s}$	-1.6736

Appendix 3. *continued*

Parameter	Value (kJ/mol)	Parameter	Value (kJ/mol) or (kJ/K mol)
$\Delta \bar{G}_{2s}^{*}$	9.00	$\Delta \bar{G}_{3s}^{*}$	0.00
$\Delta \bar{G}_{4s}^{*}$	−10.8784		
7) *BS* and β-Ag(Sb,Bi)S$_2$ *mia*, *arm* and *mat* Solution and Transition Parameters			
W_{Bi-Sb}^{BS}	12.0	W_{Bi-Sb}^{α}	6.0
$W_{Bi-Sb}^{I\beta}$	34.0	$W_{Bi-Sb}^{II\beta}$	8.50
$\Delta \bar{G}_{s}^{o\,\beta}$	−11.0	$\Delta \bar{G}_{x}^{o\,\beta}$	0.00
$\Delta \bar{H}_{AgSbS_2}^{o\,\beta \to \alpha}$	7.53	$\Delta \bar{H}_{AgBiS_2}^{o\,\beta \to \alpha}$	5.34
$\Delta \bar{S}_{AgSbS_2}^{o\,\beta \to \alpha}$	0.0115	$\Delta \bar{S}_{AgBiS_2}^{o\,\beta \to \alpha}$	0.0115
8) Pb$_2$S$_2$-AgSbS$_2$-AgBiS$_2$ *gn* Solution Parameters			
W_1	8.5	W_2	11.7
W_3	6.0	W_4	18.6
W_5	2.0	W_6	3.4
W_7	0.5	W_8	0.1
W_9	8.5	W_{10}	45.3

Appendix 3. *continued*

Parameter	Value (kJ/mol)	Parameter	Value (kJ/mol) or (kJ/K mol)
		9) Ag-Cu Exchange Energies	
$\Delta \bar{G}^{o\ bcc\text{-}fcc}_{Cu(Ag)_{-1}}$	−0.500	$\Delta \bar{G}^{o\ hcp\text{-}fcc}_{Cu(Ag)_{-1}}$	0.450
$\Delta \bar{G}^{o\ plb\text{-}bcc}_{Cu(Ag)_{-1}}$	−2.35	$\Delta \bar{G}^{o\ plb\text{-}fcc}_{Cu(Ag)_{-1}}$	−1.85
$\Delta \bar{G}^{o\ skn\text{-}fcc}_{Cu(Ag)_{-1}}$	5.67	$\Delta \bar{G}^{o\ fah\text{-}fcc}_{Cu(Ag)_{-1}}$	9.15
$\Delta \bar{G}^{o\ plb\text{-}prg}_{Cu(Ag)_{-1}}$	2.79	$\Delta \bar{G}^{o\ prg\text{-}mia}_{Cu(Ag)_{-1}}$	−0.239
		10) As-Sb Exchange Energies	
$\Delta \bar{G}^{o\ fah\text{-}plb}_{As(Sb)_{-1}}$	8.75	$\Delta \bar{G}^{o\ prg\text{-}plb}_{As(Sb)_{-1}}$	0.40
		11) Sb-Bi Exchange Energies	
$\Delta \bar{G}^{o\ BS\text{-}\alpha}_{BiSb}$	2.70	$\Delta \bar{G}^{o\ BS\text{-}\beta}_{BiSb}$	4.89
		12) Reaction (69)	
$\Delta \bar{H}^{o}$	17.725	$\Delta \bar{S}^{o}$	0.04343

* Calculated for site multiplicities $\alpha_{bcc} = \alpha_{hcp} = 3$ with $119 \geq T \geq 500$ °C and $119 \geq T \geq 400$ °C for bcc-$(Ag,Cu)_2S$ and hcp-$(Cu,Ag)_2S$, respectively.

References: 1) Balabin and Sack (2000); 2) *Harlov* and Sack (1984); 3) *Harlov* and Sack (1995a), Ghosal and Sack (1995); 4) *Harlov* and Sack (1995b); 5) Skinner (1966), Grønvold and Westrum (1986, 1987), *Harlov* and Sack (1995b), Sack (2000); 6) Raabe and Sack (1984), Sack and Loucks (1985), O'Leary and Sack (1987), Sack et al. (1987), Ebel and and Sack (1989, 1991), Sack (1992), Sack and Ebel (1993), Sack (2000, 2005); 7) Ghosal and Sack (1999); 8) Ghosal and Sack (1999), Chutas (2004); 9) Skinner (1966), *Harlov* and Sack (1995a, 1995b), Sack (2000); 10) Ghosal and Sack (1995), Sack (2000); 11) Ghosal and Sack (1999); 12) Ebel and and Sack (1989, 1991), Sack and Brackebusch (2004).

Appendix 4. Enthalpies and entropies of formation of Ag_2S-Cu_2S-ZnS-FeS-Sb_2S_3 sulfosalts from the simple sulfides at 1 bar.

		$\Delta \bar{H}_f$ (kJ/gfw)	$\Delta \bar{S}_f$ (J/K-gfw)
$Ag_{16}Sb_2S_{11}$	Ag-*plb*	−27.8188	6.273
$Cu_{16}Sb_2S_{11}$	Cu-*plb*	−12.3114	−17.42
$Ag_{10}Zn_2Sb_4S_{13}$	Ag-*fah*	−19.5393	7.880
$Cu_{10}Zn_2Sb_4S_{13}$	Cu-*fah*	−119.8472	−6.958
Ag_3SbS_3	Ag-*prg*	−13.3101	8.234
Cu_3SbS_3	Cu-*prg*	−18.7724	3.792
Ag_3SbS_3	Ag-*skn*	−0.1101	8.234
Cu_3SbS_3	Cu-*skn*	−19.7724	3.792
$AgSbS_2$	α-*mia*	−4.9207	6.066
$AgSbS_2$	β-*mia*	−12.4483	−5.459
$Cu_{10}Fe_2Sb_4S_{13}$	Ag-*fah*	−51.6719	49.72
$Ag_{10}Fe_2Sb_4S_{13}$	Cu-*fah*	39.6360	64.53
$CuPbSbS_3$	*bno*	−12.41	25.225

Phase Equilibria at High Temperatures

Michael E. Fleet

Department of Earth Sciences
University of Western Ontario
London, Ontario N6A 5B7, Canada
e-mail: mfleet@uwo.ca

INTRODUCTION

The wide variety of metal sulfide structures and their accommodation of atomic substitution, non-stoichiometry and metal-metal (M-M) and ligand-ligand interactions allows for diverse physical, chemical and electronic properties. The energy band structure of $3d$ transition-metal sulfides, in particular, is strongly influenced by the covalence of metal-S bonds, which results in hybridization of S $3p$ and metal $3d$ bonding states and direct or indirect M-M bonding interactions in favorable cases. Differences in the phase relations of isostructural metal sulfides are often attributable to subtle changes in electronic states. The literature on metal sulfide phase relations relevant to the earth sciences is very extensive and could not possibly be summarized in a single chapter. Therefore, following Craig and Scott (1974), this chapter focuses on the base metal (Fe, Co, Ni, Cu, and Zn) sulfides, with the literature for other metal chalcogenides and pnictides and some sulfosalts summarized in a single table (Table 1). The relevant phase relations of the platinum-group element (PGE) chalcogenides and pnictides have been comprehensively reviewed in Makovicky (2002). A section on the halite structure sulfides (niningerite, alabandite and oldhamite) is also included in this chapter. The material presented here relates closely to that discussed in other chapters, in particular the chapter on sulfide thermochemistry (Sack and Ebel 2006). The importance of understanding phase equilibria in the context of electronic and magnetic properties and, hence, electronic structure is also emphasized, thereby reinforcing material presented in several other chapters (Pearce et al. 2006; Vaughan and Rosso 2006).

Abbreviations used for mineral/phase names include: tr- troilite; po- pyrrhotite; hpo- hexagonal pyrrhotite; mpo- monoclinic pyrrhotite; py- pyrite; pn- pentlandite; hpn- high pentlandite; vs- vaesite; *mss*- (Fe,Ni) monosulfide solid solution; cv- covellite; al- anilite; ya- yarrowite; dg- digenite; cc- chalcocite; bn- bornite; nk- nukundamite; cp- chalcopyrite; tal- talnakhite; mh- mooihoekite; hc- haycockite; cb- cubanite; *iss*- intermediate solid solution; sp- sphalerite; wz- wurtzite; and *nss*- niningerite solid solution.

Fe, Co AND Ni SULFIDES AND THEIR MUTUAL SOLID SOLUTIONS

The sulfides of the ferrous metal triad (Fe, Co and Ni) include pyrite (the most common sulfide in the crust), troilite (the most common sulfide in the solar system), pyrrhotite (important in magmatic and massive sulfide ore bodies), and the ternary Fe-Ni-S monosulfide solid solution (*mss*) (important in the genesis of magmatic sulfide ores). The numerous low-temperature iron sulfide phases, which are of particular importance in aqueous and environmental geochemistry, are discussed in detail in other chapters. The progression of solid state properties and similarity in crystal structures and phase relations necessitates some degree of parallel discussion for equivalent phases in the Fe-S, Co-S and Ni-S binary systems.

Table 1. Summary of studies on sulfide phase relations relevant to Earth sciences.

System	Compound	Mineral/(Notes)	Stability, T_{max} (°C)	References
Ag-S	Ag_2S	acanthite	177	Kracek (1946); Roy et al. (1959);
	Ag_2S	argentite	586-622	Taylor (1969); Bell and Kullerud
	Ag_2S		838	(1970)
As-S	AsS	realgar	307	Clark (1960a,b); Barton (1969);
	AsS		265	Kirkinskiy et al. (1967); Hall (1966);
	As_2S_3	orpiment	315	Nagao (1955)
	As_2S_3		170	
	As_4S_3	dimorphite		
	AsS_2	(a)		
Bi-S	Bi_2S_3	bismuthinite	760	Van Hook (1960); Cubicciotti (1962,
	BiS_2	(high P)		1963); Silverman (1964); Craig et al. (1971)
Ca-S	see text			
Cd-S	CdS	hawleyite		Corrl (1964); Miller et al. (1966);
	CdS	greenockite	1475	D'sugi et al. (1966): Trail and Boyle (1955); Yu and Gidisse (1971); Boer and Nalesnik (1969)
Co-S	Co_4S_3		780-930	Rosenqvist (1954); Curlook and
	Co_9S_8	cobalt pentlandite	832	Pidgeon (1953); Kuznetsov et al.
	$Co_{1-x}S$	jaipurite	460-1181	(1965)
	Co_3S_4	linnaeite	660	
	CoS_2	cattierite	950	
Cr-S	CrS		330	Bunch and Fuchs (1969); Jellinek
	CrS		570	(1957); Hansen and Anderko (1958)
	CrS		~1650	
	Cr_3S_4	brezinaite		
	$Cr_{2.1}S_3$		~1100	
Cu-S	see text			
Fe-S	see text			
Hg-S	HgS	(high P)		Dickson and Tunell (1959); Potter
	$Hg_{1-x}S$	cinnabar	316-345	and Barnes (1971); Kullerud (1965b);
	$Hg_{1-x}S$	metacinnabar	572	Mariano and Warekois (1963); Scott and Barnes (1969); Barnes (1973); Potter (1973)
Mg-S	see text			
Mn-S	MnS	alabandite	1610	Skinner and Luce (1971); Keil and
	MnS	(high P)		Snetsinger (1967)
	MnS_2	haverite	423	
Mo-S	Mo_2S_3		>610	Bell and Herfert (1957); Morimoto
	MoS_2	molybdenite (h)	~1350	and Kullerud (1962); Clark (1970b,
	MoS_2	molybdenite (r)	~1350	1971); Zelikman and Belyaerskaya
	MoS_2	jordisite (a)		(1956); Graeser (1964)
Ni-S	see text			
Os-S	OsS_2	erlichmanite		Snetsinger (1971); Sutarno et al. (1967); Ying-chen and Yu-jen (1973)
Pb-S	PbS	galena	1115	Kullerud (1969); Bloem and Kroger (1956); Stubbles and Birchenall (1959)
Pt-S	PtS	cooperite		Richardson and Jeffes (1952);
	PtS_2			Hansen and Anderko (1958); Cabri (1972); Grønvold et al. (1960)
Ru-S	RuS_2	laurite		Ying-chen and Yu-jen (1973); Leonard et al. (1969); Sutarno et al. (1967)

table continued on following page

Table 1. *continued from previous page*

System	Compound	Mineral/(Notes)	Stability, T_{max} (°C)	References
Sb-S	Sb_2S_3	stibnite	556	Pettit (1964); Barton (1971); Clark (1970d)
	Sb_2S_{3-x}	meta-stibnite (a)		
Sn-S	β-SnS	herzenbergite	600	Moh (1969); Albers and Schol (1961); Karakhanova et al. (1966); Rau (1965); Moh and Berndt (1964); Mootz and Puhl (1967)
	α-SnS		880	
	δ-Sn_2S_3	ottemannite	661-675	
	δ-Sn_2S_3		710-715	
	β-Sn_2S_3		744-753	
	α-Sn_2S_3		760	
	β-SnS_2	berndtite	680-691	
	α-SnS_2		865	
Ti-S	Ti_2S			Franzen et al. (1967); Viaene and Kullerud (1971); Wiegers and Jellinek (1970)
	TiS		>800	
	Ti_8S_9			
	Ti_4S_5			
	Ti_3S_4			
	TiS_2			
	TiS_3		>610	
V-S	V_3S			Chevreton and Sapet (1965); Baumann (1964); Shunk (1969)
	VS			
	V_3S_4			
	~V_2S_3			
	VS_4	patronite		
Zn-S	See text			
Ag-As-S	Ag_7AsS_6	As-billingsleyite	575	Toulmin (1963); Hall (1966, 1968); Roland (1968, 1970); Wehmeier et al. (1968)
	Ag_5AsS_4	As-stephanite	361	
	Ag_3AsS_3	proustite	495	
	Ag_3AsS_3	xanthocanite	192	
	δ-$AgAsS_2$	smithite	415	
	δ-$AgAsS_2$	trechmannite	225	
	α-$AgAsS_2$		415-421	
Ag-Au-S	Ag_3AuS_2		181	Graf (1968); Barton and Toulmin (1964)
Ag-Bi-S	β-$AgBiS_2$	matildite	195	Schenck et al. (1939); Van Hook (1960); Craig (1967); Karup-Møller (1972)
	α-$AgBiS_2$		195-801	
	$AgBi_3S_5$	pavonite	732	
Ag-Cu-S	$Cu_{0.45}Ag_{1.55}S$	jalpaite	117	Djurle (1958); Skinner (1966); Graf (1968); Skinner et al. (1966); Suhr (1965); Krestovnikov et al. (1968); Valverde (1968); Werner (1965)
	$Cu_{0.8}Ag_{1.2}S$	mckinstiyite	94	
	$AgCu_{1+x}S$	stromeyerite	93	
Ag-Cu-Pd-Bi-S		pavonite, benjaminite[1]		Chang et al. (1988)
Ag-Fe-S	$AgFe_2S_3$	argentopyrite	152	Czamanske (1969); Czamanske and Larson (1969); Taylor (1970a,b)
	$AgFe_2S_3$	sternbergite	152	
	$Ag_3Fe_7S_{11}$	argyropyrite		
	$Ag_2Fe_5S_8$	frieseite		
Ag-Pb-S				Vogel (1953); Van Hook (1960); Craig (1967).
Ag-Sb-S	Ag_7SbS_6	Sb-billingsleyite	475	Barstad (1959); Somanchi (1963); Toulmin (1963); Chang (1963); Cambi and Elli (1965); Hall (1966, 1968); Keighin and Honea (1969); Wehmeier et al. (1968)
	Ag_5SbS_4	stephanite	197	
	Ag_3SbS_3	pyrargyrite	485	
	Ag_3SbS_3	pyrostilpnite	192	
	$AgSbS_2$	miargyrite	380	
	β-$AgSbS_2$		510	
Ag-Sn-S				Sugaki et al. (1985)
As-Co-S	CoAsS	cobaltite		Bayliss (1969); Gammon (1966); Klemm (1965)

table continued on following page

Table 1. *continued from previous page*

System	Compound	Mineral/(Notes)	Stability, T_{max} (°C)	References
As-Cu-S	Cu_3AsS_4	luzonite	275-320	Maske and Skinner (1971); Gaines (1957)
	Cu_3AsS_4	enargite	671	
	$Cu_{24}As_{12}S_{31}$		578	
	$Cu_6As_4S_9$	sinnerite	489	
	$Cu_{12}AsS_{13}$	tennantite	665	
	CuAsS	lautite		
As-Fe-S	FeAsS	arsenopyrite	702	Clark (1960a,b); Morimoto and Clark (1961); Barton (1969); Kretschmar and Scott (1976)
As-Ni-S	NiAsS	gersdorffite	>700	Yund (1962); Klemm (1965); Bayliss (1968)
As-Pb-S	$Pb_9As_4S_{15}$	gratonite	250	Rosch and Hellner (1959); Le Bihan (1963); Kutoglu (1969); Burkart-Baumann et al. (1972); Chang and Bever (1973)
	$Pb_9As_4S_{15}$	jordanite	549	
	$Pb_2As_2S_5$	dufrenoysite	485?	
	$Pb_{19}As_{26}S_{58}$	rathite II	474	
	$PbAs_2S_4$	sartorite	305	
As-Sb-S	$AsSbS_3$	getchellite	345	Weissberg (1965); Moore and Dixon (1973); Craig et al. (1974); Dickson et al. (1974); Radtke et al. (1973)
	$(As,Sb)_{11}S_{18}$	wakabayashilite		
	$AsSb_2S_2$		538	
As-Tl-S	Tl_3AsS_3			Canneri and Fernandes (1925); Graeser (1967); Radtke et al. (1974a,b)
	$Tl_4As_2S_5$			
	$Tl_6As_4S_9$			
	$TlAsS_2$	lorandite	~300	
Bi-Cu-S	Cu_9BiS_6		~375-650	Vogel (1956); Buhlmann (1965); Sugaki and Shima (1972); Godovikov and Ptitsyn (1968); Godovikov et al. (1970); Buhlmann (1971); Sugaki et al. (1972)
	Cu_3BiS_3	wittichenite	527	
	$Cu_6Bi_4S_9$			
	$Cu_{24}Bi_{26}S_{51}$	emplectite	~360	
	$Cu_{24}Bi_{26}S_{51}$	cuprobismutite	474	
	$Cu_3Bi_5S_9$		442-620	
	$CuBi_3S_5$		649	
	$Cu_3Bi_3S_7$		~498	
Bi-Fe-S	$FeBi_4S_7$		608-719	Urazov et al. (1960); Sugaki et al. (1972); Ontoev (1964)
Bi-Mo-S				Stemprok (1967)
Bi-Ni-S	$Ni_3Bi_2S_2$	parkerite	>400	Schenck and von der Forst (1939); Fleet (1973); DuPreez (1945); Peacock and McAndrew (1950); Brower et al. (1974)
Bi-Pb-S	$Pb_{10}AgBi_5S_{18}$	heyrovskyite	829	Schenck et al. (1939); Van Hook (1960); Craig (1967); Salanci (1965); Otto and Strunz (1968); Salanci and Moh (1969); Klominsky et al. (1971); Chang and Bever (1973)
	$Pb_3Bi_2S_6$	lillianite	816	
	$PbBi_2S_4$	galenobismutite	750	
	$Pb_2Bi_2S_5$	cosalite	<450	
	$Pb_5Bi_4S_{11}$	bursaite		
	$PbBi_4S_7$	bonchevite		
	$PbBi_6S_{10}$	ustarasite		
Bi-Sb-S	$(Bi,Sb)_2S_3$		>200	Hayase (1955); Springer (1969); Springer and LaFlamme (1971)
Bi-Se-S	$Bi_4(S,Se)_3$	ikunolite		Kato (1959); Markham (1962); Godovikov and Il'yasheva (1971)
Bi-Te-S	$Bi_2Te_{1.5+x}S_{1.5-x}$	δ-tetradymite		Beglaryan and Abrikasov (1959); Godovikov et al. (1970); Yusa et al. (1979)
	$Bi_2Te_{2+x}S_{1-x}$	β-tetradymite		
	$Bi_8Te_7S_5$			
	$Bi_{18}(TeS_3)_3$	joseite-C		
	$Bi_9(Te_2S_2)$			
	$Bi_{15}(TeS_4)$			

table continued on following page

Table 1. *continued from previous page*

System	Compound	Mineral/(Notes)	Stability, T_{max} (°C)	References
Ca-Fe-S	see text			
Ca-Mg-S	see text			
Ca-Mn-S	see text			
Cd-Mn-S	(Pb,Cd)S			Bethke and Barton (1961, 1971)
Cd-Zn-S	(Zn,Cd)S			Hurlbut (1957); Bethke and Barton (1971)
Co-Cu-S	$CuCo_2S_4$	carrollite	>600	Williamson and Grimes (1974); Clark (1974)
Co-Fe-S	see text			
Co-Ni-S	$(Co,Ni)_{1-x}S$ $(Co,Ni)_3S_4$ $(Co,Ni)S_2$			Klemm (1962, 1965); Delafosse and Can Hoang Van (1962); Bouchard (1968)
Co-Sb-S	CoSbS CoSbS	paracostibite costibite	876 <100?	Lange and Schlegel (1951); Cabri et al. (1970a,b)
Cr-Fe-S	Cr_2FeS_4 $(Cr,Fe)_{1-x}S$	daubreelite[2]	>740	El Goresy and Kullerud (1969); Vogel (1968); Bell et al. (1970)
Cu-Fe-S	see text			
Cu-Fe-Bi-S				Sugaki et al. (1984)
Cu-Fe-Sn-S				Ohtsuki et al. (1981a,b)
Cu-As-Sb-S				Sugaki et al. (1982)
Cu-Fe-Zn-Sn-S				Moh (1975)
Cu-Ga-S	$CuGaS_2$	gallite		Strunz et al. (1958); Ueno et al. (2005)
Cu-Ge-S				Wang (1988)
Cu-Mo-S				Grover and Moh (1969)
Cu-Ni-S	$CuNi_2S_6$	villamaninite	503	Kullerud et al. (1969); Bouchard (1968)
Cu-Pb-S	$Cu_{14}Pb_2S_{9-x}$		486-528	Schuller and Wohlmann (1955); Craig and Kullerud (1968); Clark and Sillitoe (1971)
Cu-Sb-S	Cu_3SbS_4 $CuSbS_2$ $Cu_{12+x}Sb_{4+y}S_{13}$ Cu_3SbS_3	famatinite chalcostibite tetrahedrite skinnerite	627 553 543 607	Avilov et al. (1971); Skinner et al. (1972); Tatsuka and Morimoto (1973); Karup-Møller and Makovicky (1974)
Cu-Sn-S	Cu_2SnS_3			Wang (1974); Roy-Choudhury (1974)
Cu-V-S	Cu_3VS_4	sulvanite		Dolanski (1974)
Cu-W-S				Moh (1973)
Cu-Zn-S	$\sim Cu_3ZnS_4$			Craig and Kullerud (1973); Clark (1970c)
Cu-Zn-Cd-Sn-S		kesterite, černyite, mohite, sphalerite, wurtzite and greenockite		Osadchii (1986)
Fe-Cr-S		daubréelite		Balabin et al. (1986)
Fe-Ga-S		c, h, t ternary phases		Ueno and Scott (1994)
Fe-Ge-S	Fe_2GeS_4		>800	Viaene (1968, 1972)
Fe-Mg-S	see text			
Fe-Mn-S	see text			
Fe-Mo-S	$FeMo_3S_4$			Kullerud (1967a); Grover and Moh (1966); Lawson (1972); Grover et al. (1975)
Fe-Ni-S	see text			
Fe-Os-S				Karup-Møller and Makovicky (2002)
Fe-Pb-S				Brett and Kullerud (1967)
Fe-Pb-Bi-S				Chang and Knowles (1977)

table continued on following page

Table 1. continued from previous page

System	Compound	Mineral/(Notes)	Stability, T_{max} (°C)	References
Fe-Pb-Sb-S	$Pb_8Fe_2Sb_{12}S_{28}$	jamesonite, etc.		Chang and Knowles (1977); Bortnikov et al. (1981)
Fe-Sb-S	$FeSb_2S_4$	berthierite	563	Barton (1971)
	FeSbS	gudmundite	280	Clark (1966)
Fe-Sn-S				Stemprok (1971)
Fe-Ti-S	$FeTi_2S_4$		>1000	Plovnick et al. (1968); Hahn et al. (1965); Keil and Brett (1974); Viaene and Kullerud (1971)
	$FeTi_2S_4$		540	
	$Fe_{1+x}Ti_2S_4$	heideite		
Fe-W-S				Stemprok (1971); Grover and Moh (1966)
Fe-Zn-S	see text			
Fe-Zn-Ga-S	$(Zn,Fe,Ga)_{1-x}S$	sphalerite/wurtzite phases		Ueno et al. (1996)
Hg-Sb-S	$HgSb_4S_8$	livingstonite	451	Learned (1966); Tunell (1964); Learned et al. (1974)
Pb-Sb-S				Wang, (1973); Kitakaze et al. (1995)
Pb-S-Se-Te				Liu and Chang (1994)
Pd-Co-S				Karup-Møller (1985)
Pd-Pt-Sn				Shelton et al. (1981)

Notes: 1. three other quinary solid solutions; 2. >14 kbar; 3. c- cubic, t- tetragonal, h- hexagonal, r- rhombohedral, a- amorphous

The phase diagrams for these three systems are similar, differing principally in the presence of metal-excess sulfides and the greater thermal stability of the low-spin pyrite phases in the Co-S and Ni-S systems (Fig. 1). This section is organized into the following topics: monosulfides of Fe, Co and Ni, binary phase relations in the systems Fe-S and Ni-S, the system Fe-Co-S, and ternary phase relations in the system Fe-Ni-S.

Monosulfides of Fe, Co and Ni

Electronic configuration and magnetism. The mineral phases of end-member (or near end-member) composition are troilite (FeS) and jaipurite ($Co_{1-x}S$) (both with NiAs-type or derivative structures) and millerite (NiS). The NiAs-type phase α-$Ni_{1-x}S$ encountered in the Ni-S system has not been reported as a mineral: in synthesis experiments, it inverts to millerite at the stoichiometric composition, but persists as a metastable phase at room temperature on quenching. The crystal structures, electrical and magnetic properties and energy band diagrams of NiAs-type and derivative phases have been extensively studied using conventional laboratory techniques (e.g., Vaughan and Craig 1978; Gopalakrishnan et al. 1979; King and Prewitt 1982; Anzai 1997; Raybaud et al. 1997; Hobbs and Hafner 1999), but details remain controversial. The NiAs-type substructure (space group $P6_3/mmc$) of metal monosulfides has the metal (M) and S atoms in six-fold coordination to nearest neighbors; the metal atom is in octahedral coordination with S (site symmetry $\bar{3}m$), and S is in trigonal prismatic coordination with the metal ($\bar{6}m2$). The metal and S atoms form alternate layers normal to the c-axis, with the S layers hexagonal closest packed and metals in octahedral interstices. A key feature of the NiAs-type structure is the sharing of MS_6 octahedral faces along the c-axis which permits both a direct M-M interaction via either metal $3d(t_{2g})$ or metal $3d(e_g)$ orbitals and an indirect M-S-M interaction via hybridized S $3p$ (or S $3d$) and metal $3d(e_g)$ orbitals. At high-temperature, metal-deficient monosulfides ($M_{1-x}S$) have vacancies in metal positions which are randomly distributed in individual layers of metal atoms normal to c-axis. The $3d$ electron

Phase Equilibria at High Temperatures 371

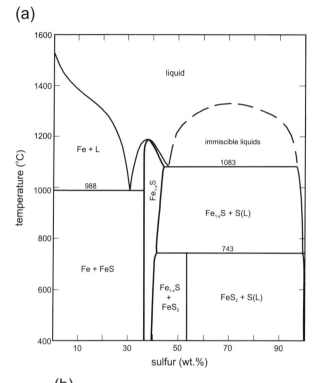

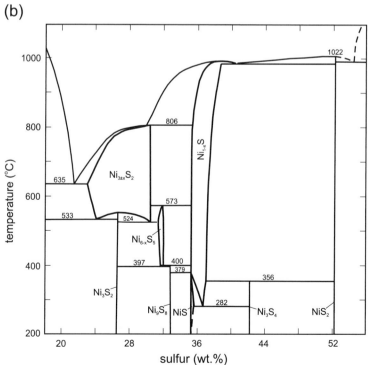

Figure 1. (a) Phase relations in the Fe-S system above 400 °C (after Kullerud et al. 1969); and (b) phase relations in the Ni-S system above 200 °C and from 18 to 56 wt% sulfur (after Kullerud and Yund 1962; Fleet 1988). All phases and phase assemblages coexist with vapor.

configurations of divalent Fe, Co and Ni monosulfides with the NiAs-type or derivative structure are well understood from the magnetic behavior of these compounds. Stoichiometric FeS is antiferromagnetic at room temperature (*RT*; troilite), but paramagnetic above 147 °C. Iron has the high spin t_{2g}^4 - e_g^2 configuration with the majority spin (↑) t_{2g}^α and e_g^α bands filled, minority spin (↓) t_{2g}^β band half filled and minority spin (↓) e_g^β band empty. $Co_{1-x}S$ is Pauli paramagnetic at room temperature, and Co has the low spin t_{2g}^6 - e_g^1 electronic configuration, with majority spin (↑) t_{2g}^α and minority spin (↓) t_{2g}^β bands filled, majority spin (↑) e_g^α band half filled and minority spin (↓) e_g^β band empty. α-NiS is paramagnetic at room temperature and antiferromagnetic below -13 °C (White and Mott 1971; Townsend et al. 1971), and Ni has the high spin t_{2g}^6 - e_g^2 configuration, with majority spin (↑) t_{2g}^α and e_g^α and minority spin (↓) t_{2g}^β bands filled, and the minority spin (↓) e_g^β band empty.

Stoichiometric FeS (troilite) is a small band gap semiconductor, but the nature of this band gap and of the metallic behavior above 147 °C is controversial (Marfunin 1979; King and Prewitt 1982; Sakkopoulos et al. 1984; Tossell and Vaughan 1992; Shimada et al. 1998), because electron delocalization through overlap of the half-filled t_{2g}^β and empty e_g^β bands is prohibited by symmetry. Goodenough (1967) recognized that trigonal distortion of the face-shared FeS_6 octahedra could split the t_{2g}^β levels below 147 °C into narrow bands parallel (Γ_I) and normal (Γ_{II}) to the *c*-axis, and thus directly account for the energy gap; partial overlap of the (Γ_I) and (Γ_{II}) bands at the transition temperature would explain the metallic behavior at higher temperature. Although recent density of states (DOS) studies invoke extensive hybridization of S 3*p* and metal 3*d* states and attribute the metallic behavior to *M-S-M* π bonding. Metallic behavior in $Co_{1-x}S$ and NiS is generally attributed to delocalization of e_g^α electrons through overlap of the e_g^α and e_g^β bands. Fujimori et al. (1990) indicated that a shift to lower X-ray photoelectron (XPS) binding energy should be expected as the metal-nonmetal transition is approached, consistent with closure of the energy band gap. Coey and Roux-Buisson (1979) pointed to a sudden decrease in the Mössbauer hyperfine field and isomer shift for $(Ni_{1-x}Fe_x)S$ (x = 0.1-0.2) at the temperature of the metal-nonmetal transition of NiS and discounted the possibility of this being a high-spin to low-spin transition. Nakamura et al. (1993) invoked Ni(3*d*)-Co(3*d*) charge transfer across the shared MS_6 octahedral faces to explain the contrasting effects of substitution by Co and Fe on the metal-nonmetal transition (at –13 °C) in NiS. Impurities and vacancies in these compounds tend to lower the metal-nonmetal transition temperature due to broadening of the S 3*p* bandwidth with decreasing S-S bond distances (Matoba et al. 1994). The effect of metal vacancies on magnetic ordering and metallic behavior of $(Fe,Ni)_{1-x}S$ solid solutions has been studied by Vaughan and Craig (1974). Most studies show that metallic character increases in the sequence $Co_{1-x}S$ > NiS > FeS, and is associated with progressive increase in covalence and decrease in metal 3*d* electron interaction energy. This trend correlates nicely with the progressive downward shift in the metal-nonmetal transition, from 147 °C for FeS, to –13 °C for NiS, and not detected at cryogenic temperatures for CoS.

FeS. The phase relations of stoichiometric FeS are complex. The phase stable at room temperature and pressure is troilite, which has a NiAs-type derivative structure, with space group $P\bar{6}2c$ and $a = (3)^{½}A$, $c = 2C$ (Bertaut 1956), where A and C are the unit-cell parameters for the hexagonal NiAs-type subcell and are approximately 3.4 and 5.9 Å, respectively. Troilite is stable up to 147 °C at 1 bar. The first order transition (known as the α transition) is well defined by changes in entropy (Robie et al. 1978), electrical conductivity (Gosselin et al. 1976a), and magnetic susceptibility (Horwood et al. 1976). The α transition has been investigated by X-ray powder diffraction (Haraldsen 1941; Grønvold and Haraldsen 1952; Taylor 1970c), neutron powder diffraction (Andresen 1960), and selected area electron diffraction (Putnis 1974). All of these researchers concluded that troilite transforms to FeS with NiAs-derivative structure; Putnis (1974) reporting a 2A,1C superstructure. On the other hand, single-crystal X-ray diffraction study revealed instead a MnP-type phase with space group *Pnma* and a diffraction pattern very similar to that of a NiAs-type phase, corresponding to $a = C$, $b = A$

and $c = (3)^{1/2}A$ (cf. Tremel et al. 1986). However, this tritwinned MnP structure was apparently excluded by detailed study of the α transition in Keller-Besrest and Collin (1990a,b). *In situ* single-crystal X-ray diffraction pointed to a 2A,1C structure in space group $P6_3mc$. Later, Kruse (1992) investigated the α transition in a natural troilite with ~0.17 wt% Cr using *in situ* X-ray diffraction and Mössbauer spectroscopy, and assuming the high-temperature phase to be of MnP type. Like Keller-Besrest and Collin (1990b), he found that the transformation of troilite takes place over a wide interval in temperature, of at least 40 °C: also, the reverse transition was sluggish. Above the α transition, a second order, spin-flip transition occurs at about 167 °C (Andresen and Torbo 1967; Gosselin et al. 1976b; Horwood et al. 1976), and is marked by change in the direction of the magnetic spins from parallel to normal to c-axis. Kruse (1992) observed that this magnetic spin flip transition required several weeks for completion at 156 °C. The final transition of FeS is to the ideal NiAs-type structure with the 1A,1C subcell. This first-order transition coincides with the Néel transition at about 327 °C. Thus, the loss of correlation of spin states appears to be associated with complete disorder and relaxation of the Fe and S sub-lattices. Unit-cell parameters for the subcell structure from *in situ* cooling neutron powder diffraction are $a = 3.5308(2)$, $c = 5.779(10)$ Å, measured just above the Néel transition (at 340 °C; Tenailleau et al. 2005). The 1A,1C subcell structure persists up to the low-pressure melting point. Melting of $Fe_{1-x}S$ begins at about 1080 °C for the stoichiometric composition, and is complete at 1190 °C at 52 at% S (Figs. 1a and 2). The Néel temperature (327 °C at 1 bar) increases linearly with increase in pressure by 3.2 ± 0.1 °C/kbar, measured up to 8.5 kbars (Anzai and Ozawa 1974), whereas the α transition temperature (147 °C at 1 bar) decreases by −0.8 ± 0.1 °C/kbar up to 11 kbars (Ozawa and Anzai 1966). Note that, elsewhere, the three low-pressure FeS polymorphs [troilite, 2A,1C (or MnP) and 1A,1C] may be referred to as the α, β and γ phases, respectively: β and γ are also used to discriminate between low-temperature, ordered (or partially ordered) hexagonal pyrrhotite and high-temperature, disordered 1A,1C pyrrhotite.

FeS at high pressure. The high-pressure phase relations of FeS are actively researched, largely because of the interest in predicting the sulfide mineralogy and melting in the core of Mars and Earth. Two structural phase transitions have been observed experimentally with increasing pressure at room temperature (Karunakaran et al. 1980; King and Prewitt 1982; Fei et al. 1995; Kusaba et al. 1997; Nelmes et al. 1999; Marshall et al. 2000; Kobayashi et al. 2005). Troilite is stable to 3.4 GPa (Pichulo 1979) where it transforms to an MnP-type structure (King and Prewitt 1982). The next transition is at 6.7 GPa to a monoclinic phase ("FeS III" in Urakawa et al. 2004, and other studies) with a 24 atom unit cell (Nelmes et al. 1999) and b-axis parallel to a-axis of troilite and c-axis of MnP. This MnP

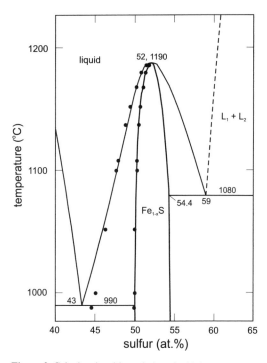

Figure 2. Calculated melting relations for high-temperature iron monosulfide ($Fe_{1-x}S$): dots are experimental values of Burgmann et al. (1968) (after Chuang et al. 1985).

FeS III transition is marked by a volume reduction of 9% (6% in Kusaba et al. 1997), a large decrease in c/a for the equivalent NiAs-type subcell (i.e., C/A), and the disappearance of the magnetic moment (King and Prewitt 1982; Rueff et al. 1999). Kobayashi et al. (1997) suggested that collapse of the magnetic moment in the Fe atom is related to electron delocalization associated with a distinct change in the $3d$ electron configuration of the Fe atoms. Takele and Hearn (1999) confirmed that the troilite and MnP structures have a high-spin configuration whereas the monoclinic phase adopts a magnetically quenched low-spin state. Above 7 GPa, FeS III transforms with increasing temperature first to a 2A,1C hexagonal superstructure (labeled FeS IV and perhaps equivalent to the low-pressure 2A,1C phase of Putnis 1974), and then to the NiAs-type subcell 1A,1C structure (FeS V; Fei et al. 1995; Kusaba et al. 1998). A cubic CsCl-type phase (with Fe and S in eight-fold coordination) was predicted to be stable at the very high pressures of the Earth's inner core (>330 GPa; Sherman 1995; Martin et al. 2001). Temperature-pressure phase diagrams are given in Kavner et al. (2001) and Urakawa et al. (2004; present Fig. 3): the former study used a laser-heated diamond anvil cell and the latter a large volume multianvil apparatus, and both studies were made *in situ* using synchrotron X-ray diffraction. Note that Urakawa et al. (2004) did not encounter the MnP (FeS II) phase, which was earlier reported to be stable in between troilite and FeS III at room temperature (e.g., King and Prewitt 1982; Fei et al. 1995; Marshall et al. 2000).

Fe-S phase relations

Numerous diverse studies have contributed to our understanding of the Fe-S system (e.g., Jensen 1942; Rosenqvist 1954; Kullerud and Yoder 1959; Kullerud 1961, 1967a; Arnold 1962; Carpenter and Desborough 1964; Toulmin and Barton 1964; Clark 1966b; Burgmann et al. 1968; Nakazawa and Morimoto 1971; Rau 1976; Sugaki and Shima 1977; Sugaki et al.1977; Kissin and Scott 1982; Barker and Parks 1986; Chuang et al. 1985; Fig. 1a). The central portion of the phase diagram is dominated by the broad field of pyrrhotites ($Fe_{1-x}S$) which extends from complete melting at 1190 °C and 52 at% S to below room temperature, and from 50 at% S to a maximum width at pyrite breakdown at 55 at% S ($x = 0.18$; Toulmin and Barton 1964). There are no Fe-excess sulfides above the low-temperature region (at low

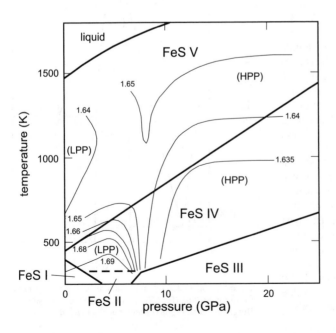

Figure 3. High-pressure phase relations of FeS obtained by *in situ* X-ray diffraction: isopleths are C/A ratio for equivalent NiAs-type subcell; FeS I is troilite; FeS III is monoclinic phase; FeS IV is 2A,1C phase; and FeS V is disordered NiAs-type FeS (*e.g.*, 1A,1C, FeS^{HT}); LPP and HPP are low- and high-pressure phases, respectively, of earlier studies (after Urakawa et al. 2004).

pressure), and the eutectic in the Fe-FeS portion (for coexisting α-iron + FeS + liquid) is at 990 °C and 43 at% S. Details of the melting relationships in this central region are shown in Figure 2. The composition of $Fe_{1-x}S$ coexisting with pyrite (i.e., the pyrrhotite-pyrite solvus) is defined by the experiments of Arnold (1962) and Toulmin and Barton (1964; present Fig. 4). The breakdown of pyrite to $Fe_{1-x}S + S_2$ occurs at 743 °C (742 °C in Chuang et al. 1985).

Sulfur fugacity and activity of FeS. In what has become a classic study, Toulmin and Barton (1964) determined the $f(S_2)$ *versus* temperature (T) curve for the univariant assemblage pyrrhotite + pyrite + vapor from 743-325 °C. The fugacity of S_2 was measured using the electrum tarnish method, which is based on the reaction:

$$4Ag^{electrum} + S_2 = 2Ag_2S^{argentite} \tag{1}$$

Note that their data were only in moderate agreement with later electrochemical measurements at 185-460 °C (Schneeberg 1973; Lusk and Bray 2002), when extrapolated into the lower-temperature range. Toulmin and Barton (1964) also developed the following equation to interrelate the composition of pyrrhotite, $f(S_2)$ and T:

$$\log f(S_2) = (70.03 - 85.83 N_{FeS})(1000/T - 1) + 39.30(1 - 0.9981 N_{FeS})^{1/2} - 11.91 \tag{2}$$

where N_{FeS} is the mole fraction of FeS in the system FeS-S_2. They then mapped the field for pyrrhotite with isopleths of N_{FeS} in a plot of $\log f(S_2)$ vs. (1000/T) (Fig. 5). The activity of FeS in pyrrhotite was extracted by application of the Gibbs-Duhem equation. Note, however, that here the activity of FeS in $Fe_{1-x}S$ relates to mixing of FeS and S_2, but the primary interaction leading to nonideality of the (Fe,)S solid solution (and departure from Raoult's law) is clearly that of Fe atoms and vacancies on the Fe sublattice of the NiAs-type structure. Chuang et al. (1985) handled this problem reasonably well in their thermodynamic analysis of the Fe-S system by considering mixtures of FeS, S, Fe vacancies, and Fe interstitials. Their calculated values for the activity of S in high-temperature pyrrhotite up to 1101 °C compared favorably with values from several experimental studies, including Burgmann et al. (1968) and Rau (1976).

Low-temperature iron sulfides. The low-temperature phase relations for bulk compositions between FeS (or Fe-excess FeS; $Fe_{1+x}S$) and FeS_2 are complex and little understood due to inconsistency between laboratory products and natural assemblages. The

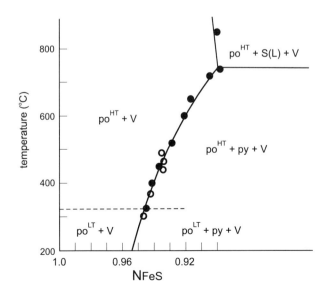

Figure 4. Pyrrhotite-pyrite solvus from low-pressure experiments; filled circles are Arnold (1962); open circles are Toulmin and Barton (1964); N_{FeS} is mole fraction of FeS in the system FeS-S_2 (after Toulmin and Barton 1964).

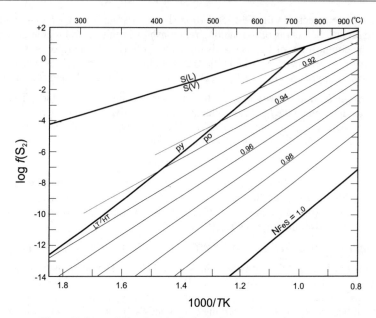

Figure 5. Composition of pyrrhotite in the Fe-S system as a function of temperature and sulfur fugacity, $f(S_2)$ (after Toulmin and Barton 1964).

latter are generally assumed to better reflect mineral stability, whereas synthesis experiments are plagued by sluggish reactions and metastability. Troilite and hexagonal and monoclinic pyrrhotites all have structures based on the NiAs-type structure.

The literature on the structure, phase relations, physical properties and mineralogy of the low-temperature NiAs-type phases is very extensive (e.g., Fleet 1968, 1971; Ward 1970; Scott and Kissin 1973; Craig and Scott 1974; Kissin 1974; Morimoto et al. 1975a,b; Power and Fine 1976; Sugaki et al. 1977; Vaughan and Craig 1978; Kissin and Scott 1982; Pósfai and Dódony 1990; Lusk et al. 1993; Li and Franzen 1996; Li et al. 1996; Farrell and Fleet 2001; Powell et al. 2004; Tenailleau et al. 2005). Superstructure phases of the 1A,1C NiAs-type substructure arise by the clustering of Fe atoms in the troilite structure and the ordering of Fe atoms and vacancies in individual (001) layers in the more metal-deficient compositions. As reviewed above, troilite forms spontaneously on quenching below 147 °C at the stoichiometric composition (FeS), by the triangular clustering of Fe atoms (Bertaut 1953, 1956), giving a hexagonal unit cell with $a = A(3)^{\frac{1}{2}}$, $c = 2C$.

There appear to be numerous low-temperature superstructures based on the commensurate and incommensurate ordering of Fe atoms and vacancies in $Fe_{1-x}S$ compositions. Although there is some correlation between superstructure type and Fe-S composition, there are significant discrepancies between laboratory products and natural hexagonal and monoclinic pyrrhotites. Most of these pyrrhotite superstructures have a unit cell with $a = 2A$, or an incommensurate modulation of 2A, and various patterns of order in the c-axis direction. Experimental products were classified by Nakazawa and Morimoto (1971) into the following superstructure types:

NC for intermediate pyrrhotite compositions below about 200 °C, and characterized by a hexagonal 2A,NC unit cell, where N is generally non-integral with values from 3.0 to 6.0. Superstructure reflections occur as satellites to the subcell reflections as well as in reciprocal lattice rows reflecting doubling of the A dimension.

NA for more S-rich intermediate compositions near and above 200 °C, and characterized additionally by non-integral spacings in the *a*-axis direction, with N varying from about 40 to 90.

MC limited to S-rich compositions near 300 °C, and characterized by a similar diffraction pattern to NC, with N varying from 3.0 to 4.0, but without the satellite reflections.

4C monoclinic pyrrhotite of near Fe_7S_8 composition.

Nakazawa and Morimoto (1971) additionally recognized a field of 1C phases (undifferentiated high-temperature 1A,1C and low-temperature 2A,1C phases) and a 2C phase (troilite). The NC, NA and MC superstructure types have hexagonal diffraction symmetry. Their characteristic single-crystal X-ray diffraction patterns are depicted in Nakazawa and Morimoto (1971) and Craig and Scott (1974), and a representative phase diagram is given in the present Figure 6. Only structures with the ideal Fe_7S_8 composition have been determined in any detail; i.e., synthetic hexagonal 3C (Fleet 1971; a quench product, and broadly equivalent to NA type) and natural monoclinic 4C (Tokonami et al. 1972).

Integral NC superstructures are generally restricted to natural assemblages. As representative natural pyrrhotites, Lennie and Vaughan (1996) list ~Fe_7S_8 (monoclinic 4C), Fe_9S_{10} (intermediate hexagonal pyrrhotite 5C), $Fe_{10}S_{11}$ (intermediate hexagonal pyrrhotite 11C), $Fe_{11}S_{12}$ (intermediate hexagonal pyrrhotite 6C), and ~Fe_9S_{10}-$Fe_{11}S_{12}$ (intermediate hexagonal pyrrhotite non-integral, or incommensurate, NC). These five pyrrhotite superstructure types occur as important minerals in sulfide ores of "magmatic" origin and as accessory minerals in ultramafic and mafic rocks, as well as in metamorphosed massive sulfides and some hydrothermal ores.

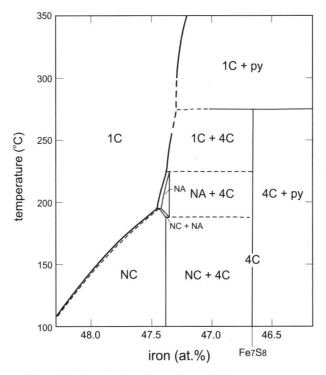

Figure 6. Schematic phase diagram for pyrrhotite superstructures encountered in laboratory studies (after Sugaki et al. 1977).

Pyrrhotite phase relations below 350 °C have been investigated systematically by Nakazawa and Morimoto (1971), Scott and Kissin (1973), Kissin (1974), Sugaki et al. (1977), and Kissin and Scott (1982). The observations of Nakazawa and Morimoto (1971) were based on *in situ* high-temperature X-ray diffraction, whereas those of Kissin (1974), Scott and Kissin (1973) and Kissin and Scott (1982) were based on room-temperature measurements of crystals grown by hydrothermal recrystallization both *in situ* and via transport reactions in 5 *m* aqueous solutions of NH_4I. Sugaki et al. (1977) grew crystals by the thermal gradient hydrothermal transport method, but based their phase diagram (Fig. 6) largely on *in situ* X-ray diffraction measurements.

The stability fields of the pyrrhotite superstructures lie, very generally, on the S-rich side of the low-temperature extension of the hexagonal pyrrhotite-pyrite solvus (cf. Figs. 4 and 6). The field of the 1C phases extends down to near, or below, 100 °C at about 48.6 at% Fe. Interestingly the onset of vacancy ordering in laboratory equilibrated pyrrhotites is anticipated by discontinuities in A and C subcell parameter plots for $Fe_{1-x}S$ compositions quenched from 700 °C (Fleet 1968). Between 48.6 at% Fe and stoichiometric FeS, the literature phase diagrams show a wedge-like field of troilite + 1C. There is overall inter-laboratory agreement on the pyrrhotite phase relations. Kissin and Scott (1982) found that a discontinuity in the hexagonal pyrrhotite-pyrite solvus at 308 °C marked the upper stability of the MC superstructure. The onset of the NA superstructure occurred further down the solvus at 262 °C, and monoclinic 4C pyrrhotite became stable at 254 °C through the peritectic reaction:

hexagonal NA(47.30 at% Fe) + pyrite = monoclinic 4C(47.25 at% Fe)   (3)

At lower temperatures, hexagonal pyrrhotite is separated from the field of monoclinic 4C pyrrhotite by a narrow miscibility gap. Hexagonal NC pyrrhotite appeared at and below 209 °C. The integral 6C, 11C and 5C superstructures of intermediate pyrrhotite were thought to form in nature only below about 60 °C, and the field of coexisting troilite + 1C hexagonal pyrrhotite extended from 147 °C at the stoichiometric FeS composition to about 60-70 °C at 48.3 at% Fe. Sugaki et al. (1977) and Sugaki and Shima (1977) had earlier reported similar results for assemblages of troilite and hexagonal NC and NA and monoclinic 4C pyrrhotites, but did not observe a field of stability for MC pyrrhotite (Fig. 6). Sugaki et al. (1977) found that the monoclinic 4C phase reacted out at a slightly higher temperature (275 ± 5 °C) than in Kissin and Scott (1982). Revising Kissin and Scott (1982), the monoclinic 4C phase was not detected by Lusk et al. (1993) in experiments on phase relations in the Fe-Zn-S system between 325-150 °C. Therefore, they concluded that in nature monoclinic pyrrhotite may become stable somewhere around 140 °C.

Troilite and hexagonal NC phases are antiferromagnetic (Schwarz and Vaughan 1972), monoclinic 4C pyrrhotite is ferrimagnetic (Nakazawa and Morimoto 1971), and the high-temperature 1A,1C subcell phase is paramagnetic (e.g., Vaughan and Craig 1978; Kruse 1990). Magnetic ordering (i.e., the Néel temperature) and ordering of vacancies in $Fe_{1-x}S$ both occur near 250-300 °C (Power and Fine 1976) on cooling, but these ordering events are clearly composition-dependent, and their details remain obscure (e.g., Li and Franzen 1996). Hysteresis of magnetic and electrical properties and phase transitions and transformations is commonly encountered in heating-cooling experiments with nonstoichiometric phases (Nakazawa and Morimoto 1971; Schwarz and Vaughan 1972; McCammon and Price 1982). Li and Franzen (1996) attributed hysteresis in their synthetic and natural pyrrhotites heated above 550 °C to loss of S. Powell et al. (2004) recently confirmed that, at the ideal Fe_7S_8 composition, the magnetic transition is accompanied by a structural transformation from vacancy-ordered monoclinic 4C to vacancy disordered hexagonal 1C.

Of interest here is that a vacancy- and magnetically-ordered hexagonal 3C pyrrhotite is obtained by quenching Fe_7S_8 and more S-rich compositions from well within the field of

1A,1C pyrrhotite (e.g., 500-700 °C; Fleet 1968, 1982). This 3C phase is clearly metastable in the Fe-S system. However, the hexagonal 3C type superstructure is stable in the Fe-Se system at the Fe_7Se_8 composition (Okazaki 1959, 1961). Both hexagonal 3C Fe_7Se_8 and triclinic 4C Fe_7Se_8 (which is equivalent to monoclinic 4C Fe_7S_8) are vacancy ordered and ferrimagnetic and have nearly identical Curie temperatures (189 °C for 3C and 187 °C for 4C). The low-temperature 4C selenide transforms to 3C over the temperature interval 240 to 298 °C, and magnetic and vacancy disordering of 3C occurs over the interval 360 to 375 °C. A similar 4C→3C→1C transformation sequence was suggested for pyrrhotites of Fe_7S_8 composition by Li and Franzen (1996) but this was not supported by later study (Powell et al. 2004).

The other low-temperature binary Fe-S phases are, in the main, either metastable or of questionable stability (Lennie and Vaughan 1996) and include:

1. synthetic iron-pentlandite (Fe_9S_8; a = 10.5 Å; Nakazawa et al. 1973)

2. cubic $Fe_{1+x}S$ (x = 0.03-0.10; sphalerite-type, a = 5.425 Å; Takeno et al. 1970; Murowchick and Barnes 1986a)

3. mackinawite ($Fe_{1+x}S$; x = 0.01-0.08; tetragonal, PbO-type, a = 3.676, c = 5.032 Å; Berner 1962, 1964; Nozaki et al. 1977; Takeno et al. 1982; Lennie et al. 1995b)

4. amorphous FeS/disordered mackinawite/"dorite" (Lennie and Vaughan 1996; Wolthers et al. 2003; Ritvo et al. 2003)

5. smythite (~Fe_9S_{11}, ideally $Fe_{13}S_{16}$; rhombohedral, a = 3.47, c = 34.50 Å; Erd et al. 1957; Rickard 1968; Fleet 1982)

6. greigite (Fe_3S_4; cubic, spinel-type, a = 9.876 Å; Skinner et al. 1964)

7. synthetic Fe_3S_4 (monoclinic, NiAs-derivative, a = 5.93, b = 3.42, c = 10.64 Å; Fleet 1982)

8. marcasite (FeS_2; a = 4.436, b = 5.414, c = 3.381 Å; Murowchick and Barnes 1986b)

9. pyrite (FeS_2; a = 5.417 Å).

The relative stability of these phases has been studied variously by precipitation, sulfidation and replacement in aqueous or hydrothermal media and dry heating and quenching, as well as in natural assemblages: e.g., conversion of mackinawite to greigite (Lennie et al. 1977; Taylor et al. 1979) and mackinawite to hexagonal pyrrhotite (Lennie et al. 1995a). Lennie and Vaughan (1996) concluded that the predominant sulfidation sequence at low temperature is: (cubic FeS/amorphous FeS) → mackinawite → greigite → marcasite/pyrite. On the other hand, Krupp (1994) studied textural relationships of smythite, greigite and mackinawite from the low-temperature hydrothermal Moschellandsberg mercury deposit in south-western Germany, and concluded that greigite and mackinawite were probably metastable phases, smythite was stable, and pyrrhotites may be metastable below 65 °C. Fleet (1982) found that S-rich $Fe_{1-x}S$ compositions spontaneously precipitated smythite ($Fe_{13}S_{16}$) and monoclinic Fe_3S_4 when quenched under dry conditions from the pyrrhotite solvus and also from just above pyrite decomposition (743 °C). Monoclinic Fe_3S_4 is isomorphous with monoclinic Fe_3Se_4 and Cr_3S_4. It also seems appropriate to note here that the cubic (Mg,Fe)S solid solution extends beyond 70 mol% FeS at the highest temperature investigated (Skinner and Luce 1970; Farrell and Fleet 2000), allowing extrapolation to a = 5.07 Å for the unstable halite-type FeS. Both monoclinic Fe_3S_4 and halite-type FeS are potential metastable products of low-temperature experiments with Fe-S compositions.

High-pressure phases. The high-pressure phase relations in the system Fe-FeS are of current interest in connection with the composition and mineralogy of planetary cores, for which S is a leading candidate for the light element component (e.g., Svendsen et al. 1989; Boehler 1992; Sherman 1995; Liu and Fleet 2001; Sanloup et al. 2002; Secco et al. 2002; Lin et al. 2004;

Sanloup and Fei 2004). Earlier research focused on the melting behavior, with Boehler (1992) demonstrating strong depression of melting in an Fe-rich core for mixtures of Fe and FeS. A surprising development has been the discovery of the Fe-S alloys Fe_3S, Fe_2S, and Fe_3S_2. Fe_3S was predicted as a probable core phase based on crystal chemical reasoning and first-principles density functional calculations (Sherman 1995). Subsequently, Fei et al. (1997, 2000) synthesized three Fe-S alloy phases (Fe_3S_2, Fe_3S and Fe_2S) at subsolidus temperatures and high pressures (Fig. 7). Fe_3S has a tetragonally distorted $AuCu_3$-type structure. It forms at 21 GPa and 1027 °C and is stable in the tetragonal structure up to 42.5 GPa. Lin et al. (2004) studied magnetic and elastic properties of Fe_3S up to 57 GPa at room temperature, observing a magnetic collapse at 21 GPa consistent with the high-spin to low-spin electronic transition at 20-25 GPa reported earlier. Whereas S was thought formerly to concentrate in the outer liquid core of Earth, the existence of high-pressure Fe-S alloys as probable host phases for S argues for significant amounts of S in the inner solid core (e.g., Sanloup et al. 2002). Indeed, Lin et al. (2004) concluded that the non-magnetic Fe_3S phase is a dominant component of the Martian core.

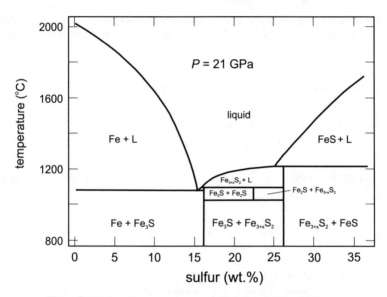

Figure 7. Melting relations in the system Fe-FeS at 21 GPa; Fe_3S, Fe_2S and $Fe_{3+x}S_2$ are new alloy phases (after Fei et al. 2000).

Ni-S phase relations

The phase diagram for the binary Ni-S system (Fig. 1b) is based on the experiments of Rosenqvist (1954) and Kullerud and Yund (1962), with additional details from Arnold and Malik (1974), Craig and Scott (1974), Sharma and Chang (1980), Lin et al. (1978), and Fleet (1988). The system includes five minerals: heazlewoodite (Ni_3S_2), godlevskite (Ni_9S_8), millerite (NiS), polydymite (Ni_3S_4), and vaesite (NiS_2); and three high-temperature fields of solid solution: α-$Ni_{3\pm x}S_2$, $Ni_{6-x}S_5$ (elsewhere known as known as α-Ni_7S_6), and α-$Ni_{1-x}S$. End-member heazlewoodite is stable up to 565 °C, where it inverts to α-$Ni_{3\pm x}S_2$, Ni_9S_8 (elsewhere known as β-Ni_7S_6; godlevskite) is stable up to about 400 °C, NiS (millerite) up to 379 °C, Ni_3S_4 (polydymite) up to 356 °C, and NiS_2 (vaesite) up to 1022 °C, the congruent melting point. The field of the NiAs-derivative phase α-$Ni_{1-x}S$ extends from 999 °C down to 282 °C, being replaced at lower temperature by the assemblage millerite + polydymite.

The structures and detailed phase relations of the two metal-excess sulfides located in between NiS and Ni_3S_2 proved to be troublesome in earlier studies. The orthorhombic (space group *Bmmb*) structure of the high-temperature solid solution $Ni_{6-x}S_5$ was determined by Fleet (1972) for a sample of composition Ni_7S_6. Surprisingly, the structure has an ideal 6:5 stoichiometry; the commonly reported 7:6 stoichiometry being accommodated by partial occupancy of the Ni positions. Three of the Ni positions are disordered, and various superstructures (2*a*, 2*b*, 2*c*; 2*a*, 2*b*, 3*c*; and 2*a*, 2*b*, 4*c*; Putnis 1976; Parise and Moore 1981) have been reported. Fleet (1987, 1988) also showed that godlevskite [$(Ni_{8.7}Fe_{0.3})S_8$] and β-Ni_9S_8 were the same phase and the stoichiometry was pentlandite-like (9:8). However, unlike pentlandite, $(Fe,Ni)_9S_8$, which is a true ternary phase, their crystal structures contain three-fold clusters (Ni_3) of short Ni-Ni interactions. The greater metallic character of Ni and Co and their tendency to form *M-M* bonds, compared with Fe, results in metal-excess sulfides in both of the Ni-S and Co-S binary systems. The activity of sulfur in the metal-excess Ni-S phases has been determined using a gas equilibration technique (Lin et al. 1978). This study predated resolution of the phase relations of β-Ni_9S_8, and also separated the broad field of α-$Ni_{3\pm x}S_2$ solid solution into two phases, which they labelled β'-Ni_3S_2 and $β_2$-Ni_4S_3.

The system Fe-Co-S

Low-pressure phase relations and physical properties in the systems Co-S and Fe-Co-S have been fairly extensively studied (e.g., Lamprecht and Hanus 1973; Rau 1976; Wyszomirski 1976, 1977, 1980; Lamprecht 1978; Terukov et al. 1981; McCammon and Price 1982; Wieser et al. 1982; Barthelemy and Carcaly 1987; Collin et al. 1987; Raghavan 1988; Vlach 1988; Farrell and Fleet 2002). Rau (1976) and Vlach (1988) reviewed the binary Co-S phase relations and calculated equilibrium sulfur pressures. There is some inconsistency in precise temperatures between these two studies. Using Vlach (1988), the field of $Co_{1-x}S$ extends from 1181 °C down to only 460 °C, being replaced at lower temperature by the assemblage cobalt pentlandite + Co_3S_4. Below 785 °C, the bulk-composition CoS yields cobalt pentlandite + $Co_{1-x}S$ to 460 °C, and cobalt pentlandite + Co_3S_4 below 460 °C. Thus, stoichiometric CoS cannot be preserved at room temperature using conventional quenching procedures. Incidentally, Wyszomirski (1976) considered the existence of jaipurite, the mineral of ideal $Co_{1-x}S$ composition, to be doubtful. Cobalt pentlandite is stable up to 834 °C, reacting to Co_4S_3 + $Co_{1-x}S$, and the pyrite structure phase CoS_2 is stable up to 1027 °C, reacting to $Co_{1-x}S$ + S_2.

The *mss* phase in the ternary system, (Fe,Co)-*mss*, is continuous between $Fe_{1-x}S$ and $Co_{1-x}S$ down to at least 500 °C (Wyszomirski 1976; Raghavan 1988; present Fig. 8). Of interest here is that, for experiments at and above 500 °C: (1) the range in nonstoichiometry (*x*) in $Co_{1-x}S$ is similar to that of $Fe_{1-x}S$ and to *mss* in the system Fe-Ni-S; (2) (Fe,Co)-*mss* near FeS composition is metal-rich [>50 at% (Fe,Co)], and analogous to the area of Fe-rich *mss* reported by Misra and Fleet (1973); and, (3) the composition field of cobalt pentlandite is restricted to <0.2 atomic Fe/(Fe + Co). The Fe content of natural cobalt pentlandite is also very low in the absence of Ni, even though there is extensive solid solution between synthetic (Fe-Ni) pentlandite and cobalt pentlandite (Knop and Ibrahim 1961; Geller 1962). Misra and Fleet (1973) found that the Co content of natural pentlandite is highly variable, ranging from almost Co-free pentlandite with less than 0.1 at% Co to an Fe-free pentlandite of composition $Co_{6.9}Ni_{1.3}S_8$, and a Ni-free pentlandite of composition $Co_{9.1}Fe_{0.2}S_8$. For quenched stoichiometric ($Fe_{1-x}Co_xS$) compositions, the troilite phase extends to about *x* = 0.17, although this composition limit varies somewhat with heat treatment and from study to study (Terukov et al. 1981; Wieser et al. 1982; Barthelemy and Carcaly 1987; Collin et al. 1987). The more Co-rich compositions exhibit hysteresis of magnetic and electrical properties and phase behavior. McCammon and Price (1982) investigated the magnetism of $(Fe,Co)_{1-x}S$ solid solutions quenched from 1000 °C using Mössbauer spectroscopy, observing an antiferromagnetic = paramagnetic transition between 0.69 and 0.50 atomic Fe/(Fe + Co) at 25 °C and 0.50 and 0.16

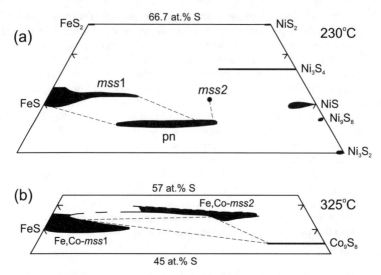

Figure 8. Partial phase relations at low temperature showing incompletely unmixed monosulfide solid solution (*mss*) in: (a) the system Fe-Ni-S at 230 °C (after Misra and Fleet 1973; Fleet 1988); and (b) the system Fe-Co-S at 325 °C (after Farrell and Fleet 2002).

atomic Fe/(Fe + Co) at 4.2 K. Thus, Fe-rich (Fe,Co)-*mss* compositions are antiferromagnetic like troilite and hexagonal pyrrhotites (Schwarz and Vaughan 1972) in the system Fe-S, but Co-rich (Fe,Co)-*mss* compositions are paramagnetic.

Farrell and Fleet (2002) observed that (Fe,Co)-*mss* unmixes abruptly below 425 °C, giving co-existing phases of Fe-rich (Fe,Co)-*mss*1, with the 1A,1C NiAs-type structure, and Co-rich (Fe,Co)-*mss*2, with the hexagonal 2A,3C structure. For bulk compositions with 52.0 at% S, the equilibrium (Fe,Co)-*mss*1 solvus is at about 0.83 atomic Fe/(Fe + Co) at 400 °C and progressively diverges toward the $Fe_{1-x}S$ end-member composition with decrease in temperature to 0.98 atomic Fe/(Fe + Co) at 105 °C. At 400 °C, the equilibrium (Fe,Co)-*mss*2 solvus is at about 0.37 atomic Fe/(Fe + Co) and does not appear to vary significantly with decrease in temperature. There is a metastable solvus within the equilibrium miscibility gap with a critical temperature at 400 °C between 0.45 and 0.50 atomic Fe/(Fe + Co), and a narrow field of spontaneous phase-separation is centered at 0.75 atomic Fe/(Fe + Co), and results in satellite reflections to $h0l$ NiAs-type subcell reflections in (Fe,Co)-*mss* quenched from high temperature (800 °C).

Ternary phase relations in the system Fe-Ni-S

The phase relations in the Fe-Ni-S system have been extensively researched largely because of their important application to Ni-Cu magmatic sulfides (e.g., Lundqvist 1947; Clarke and Kullerud 1963; Naldrett and Kullerud 1966; Naldrett et al. 1967; Kullerud et al. 1969; Shewman and Clark 1970; Craig 1971, 1973; Misra and Fleet 1973, 1974; Vaughan and Craig 1974; Mandziuk and Scott 1977; Coey and Roux-Buisson 1979; Vollstädt et al. 1980; Hsieh and Chang 1987; Naldrett 1989; Karup-Møller and Makovicky 1995, 1998; Sugaki and Kitakaze 1998; Ueno et al. 2000; Etschmann et al. 2004; Kitakaze and Sugaki 2004; Wang et al. 2005). Magmatic sulfide assemblages consist principally of pyrrhotite, chalcopyrite and pentlandite, but interest in the Fe-Ni-(Cu)-S phase relations begins with the high-temperature ternary liquid, because of the putative role of immiscible sulfide liquid in the formation of magmatic sulfide orebodies.

At 1200 °C the system is composed entirely of liquids, except for alloys along the Fe-Ni join. At high subsolidus temperatures, pyrrhotite and pentlandite in magmatic sulfide ores are represented by (Fe,Ni) monosulfide solid solution (*mss*) (Fig. 9). $Fe_{1-x}S$ appears on the Fe-S join at 1190 °C, and *mss* extends to the Ni-S join at 999 °C. For a number of years it was believed that the metal-excess phase pentlandite appeared only on cooling to 610 °C (Kullerud 1963). However, in a series of presentations extending from Sugaki et al. (1982) to Sugaki and Kitakaze (1998) it was demonstrated instead that a high form of pentlandite was stable between 584 ± 3 °C and 865 ± 3 °C and showed limited solid solution from $Fe_{5.07}Ni_{3.93}S_{7.85}$ to $Fe_{3.61}Ni_{5.39}S_{7.85}$, at 850 °C, including the composition point for ideal $Fe_{4.50}Ni_{4.50}S_{8.00}$. High pentlandite is cubic with $a = 5.189$ Å (at 620 °C) corresponding to $a/2$ of pentlandite. Kitakaze and Sugaki (2004) recently showed that the unit-cell of high pentlandite is the subcell of (low) pentlandite, and that the inversion of the low- and high-form solid solution in the quaternary system Fe-Ni-Co-S is of the order-disorder type. The revised phase relations of Sugaki and Kitakaze (1998) in the system Fe-Ni-S at 850 °C and a pseudobinary temperature *versus* composition section with Fe = Ni through the pentlandite and *mss* fields are given in Figures 9 and 10. The latter shows that high pentlandite crystallizes from metal-rich liquid between 865 °C and 746 °C: note that pentlandite in magmatic sulfide ores is generally understood to form by segregation or phase separation from *mss* in the subsolidus (e.g., Francis et al. 1976; Etschmann et al. 2004). The composition limits of pentlandite solid solution are $Fe_{5.95}Ni_{3.07}S_{8.00}$ to $Fe_{2.81}Ni_{6.55}S_{8.00}$ and $Fe_{6.57}Ni_{2.82}S_{8.00}$ to $Fe_{3.79}Ni_{5.45}S_{8.00}$ at 400 °C and 500 °C, respectively (Ueno et al. 2000).

Low-temperature phase relations in the *mss* region of the system Fe-Ni-S were investigated by Misra and Fleet (1973) and Craig (1973). Misra and Fleet (1973) reported that *mss* is continuous between $Fe_{1-x}S$ and $Ni_{1-x}S$ to below 400 °C, but is discontinuous at 300 °C, with a region of solid-solution extending from $Fe_{1-x}S$ to about 25 at% Ni (*mss*1), a second *mss* phase (*mss*2) at about 33 at% Ni, and millerite solid-solution containing about 5 at% Fe. At 230 °C, the composition of *mss*2 had not changed appreciably, but the maximum Ni content of *mss*1 had diminished to 17 at% (Fig. 8a). Craig's (1973) results were similar, except that he found the critical temperature for unmixing to be 263 ± 13 °C, and the field of *mss*2 extended further toward the Ni-S join. Misra and Fleet (1973) noted that the Ni content near the *mss*1 solvus at 230 °C is still appreciably greater than the range in Ni content of 0.2-0.7 at% for intergrown hexagonal and monoclinic pyrrhotite coexisting with pentlandite in magmatic sulfides. These findings suggest that either the progressive chemical readjustment in magmatic sulfides persists to very low temperatures or there is a discontinuous change from *mss*1 to pyrrhotite at some lower temperature. Indeed, the initial phase separation of (Fe,Co)*mss* is discontinuous in the analogue system Fe-Co-S (Farrell and Fleet 2002). This occurs between 425-400 °C and, thereafter, the miscibility gap in (Fe,Co)-*mss* is generally similar to that for the phase separation of *mss* in the system Fe-Ni-S.

Violarite (ideally $FeNi_2S_4$) appears in the subsolidus at 461 °C (Craig 1971) and solid solution with the end-member thiospinel polydymite (Ni_3S_4) is thought to be complete at 356 °C. There remains some uncertainty as to whether the common association of violarite with pentlandite represents stable equilibrium (e.g., Misra and Fleet 1974) or merely metastable juxtaposition (Misra and Fleet 1973; Craig 1973): the phase relations of Craig (1973), in particular, exclude the establishment of violarite-pentlandite tie-lines.

Structural and electrical properties of *mss* in the Fe-Ni-S system have been investigated at high pressures and temperatures by Vollstädt et al. (1980) and Kraft et al. (1982).

Cu-S, Cu-Fe-S AND Cu-Fe-Zn-S SYSTEMS

The ternary system Cu-Fe-S is the road map for understanding the geology and geochemistry of Cu sulfide deposits and has been studied systematically for more than 70 years, beginning with Merwin and Lombard (1937). This early experimental study was followed by Schlegel and Schüller (1952), Hiller and Probsthain (1956), Roseboom and

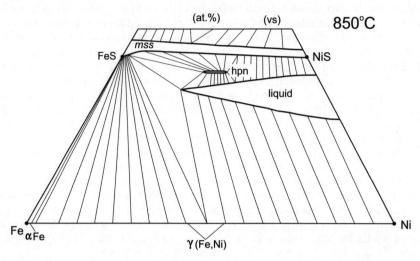

Figure 9. Phase relations in the metal-rich portion of the Fe-Ni-S system at 850 °C, showing extensive field of liquid alloy centered near the Ni_3S_2 composition point, a continuous field of *mss*, and early appearance of high pentlandite (after Sugaki and Kitakaze 1998).

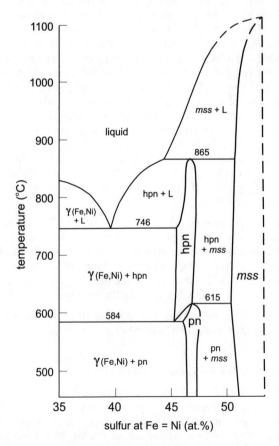

Figure 10. Partial temperature *versus* sulfur section at atomic Fe = Ni through Fe-Ni-S system, showing relationship of new high pentlandite phase to coexisting liquid, *mss* and pentlandite (after Sugaki and Kitakaze 1998).

Kullerud (1958), Yund (1963), Brett (1964), Morimoto and Kullerud (1966), Roseboom (1966), Yund and Kullerud (1966), Cabri (1967, 1973), Kullerud et al. (1969), Morimoto and Koto (1970), Mukaiyama and Izawa (1970), Cabri and Harris (1971), Morimoto and Gyobu (1971), Cabri and Hall (1972), MacLean et al. (1972), Barton (1973), and Sugaki et al. (1975). Since then there have been a number of studies on, variously: electrochemistry of the Cu-S join (Potter 1977) and ternary Cu-Fe-S system (Lusk and Bray 2002); digenite (Grønvold and Westrum 1980; Pósfai and Buseck 1994), bornite (Putnis and Grace 1976; Kanazawa et al. 1978; Grguric and Putnis 1998; Grguric et al. 1998) and the low-temperature phase relations of digenite-bornite solid solution (Grguric and Putnis 1999; Grguric et al. 2000), the ternary system (Wang 1984), the effects of addition of Zn to the ternary system (Wiggins and Craig 1980; Kojima and Sugaki 1984, 1985; Lusk and Calder 2004), and phase relations in the system $CuS-FeS-H_2S-H_2SO_4-HCl-H_2O$ at magmatic temperatures and pressures (McKenzie and Helgeson 1985). However, the present state of knowledge is based largely on the flurry of studies in the sixties and seventies. Therefore, this section builds on the review of Craig and Scott (1973) and the subsequent detailed experimental work of Sugaki et al. (1975).

Phase relations in the system Cu-S

The original list of eleven minerals and phases in the Cu-S system in Craig and Scott (1974) is now expanded to fourteen (Tables 2 and 3) with the addition of the new minerals roxbyite, geerite, and spionkopite. Also, blaubleibender covellite is now recognized as the mineral yarrowite, which has a 9:8 Cu:S stoichiometry and, therefore, is distinct from covellite, and I have included villamininite as the mineral form of synthetic CuS_2, even though this pyrite-group mineral contains significant amounts of Ni, Co and Fe. Most Cu-

Table 2. Minerals and phases in the system Cu-S.

Mineral/Phase	Formula	Symmetry	Stability	References
chalcocite	Cu_2S	monoclinic	to 103 °C	1, 2, 3
hexagonal Cu_2S	Cu_2S	hexagonal	103 ° to ~435 °C	1, 4, 5
cubic Cu_2S	Cu_2S	cubic	~435 ° to 1129 °C	6, 7
tetragonal Cu_2S	Cu_2S	tetragonal	>1 kbar	8, 9
high digenite	$Cu_{1.80+x}S_5$	cubic	83 ° to 1129 °C	6,7
djurleite	$Cu_{1.97}S$	orthorhombic	to 93 °C	1, 3, 10, 11
digenite	$Cu_{1.80}S$	cubic	76 ° to 83 °C	1, 7, 12
roxbyite	$Cu_{1.78}S$	monoclinic		13
anilite	$Cu_{1.75}S$	orthorhombic	to 76 °C	3, 11, 14, 15
geerite	$Cu_{1.60}S$	cubic		16
spionkopite	$Cu_{1.40}S$	hexagonal-R		17
yarrowite	Cu_9S_8	hexagonal-R	to 157 °C	11, 17, 18, 19
covellite	CuS	hexagonal	to 507 °C	3, 20, 21
villamaninite	CuS_2	cubic		22, 23, 24

References: 1. Roseboom (1966); 2. Evans (1971); 3. Potter and Evans (1976); 4. Buerger and Buerger (1944); 5. Wuensch and Buerger (1963); 6. Jensen (1947); 7. Morimoto and Kullerud (1963); 8. Skinner (1970); 9. Janosi (1964); 10. Morimoto (1962); 11. Potter (1977); 12. Grønvold and Westrum (1980); 13. Mumme et al. (1988); 14. Morimoto and Koto (1970); 15. Morimoto et al. (1969); 16. Goble and Robinson (1980); 17. Goble (1980); 18. Moh (1964); 19. Rickard (1972); 20. Kullerud (1965a); 21. Berry (1954); 22. Munson (1966); 23. Taylor and Kullerud (1971, 1972); 24. Bayliss (1989)

Notes: 1. cubic Cu_2S, stable between ~435-1129 °C in solid solution with high digenite, $Cu_{1.80}S$; 2. yarrowite –formerly blaubleibender covellite; 3. villamaninite must be the mineral form of synthetic cubic CuS_2 since both are pyrite structure compounds, even though the mineral contains high amounts of Ni, Co and Fe; 4. hexagonal-R is hexagonal-rhombohedral; 5. covellite- $a = 3.7938$, $c = 16.341$ Å; yarrowite- $a = 3.800$, $c = 67.26$ Å; villamaninite- $a = 5.694$ Å

Table 3. Crystal data for some Cu-excess Cu-S minerals and phases.

Mineral	Formula	Space Group	a (Å)	b (Å)	c (Å)	β (°)	Z
chalcocite	Cu_2S	$P2_1/c$	15.235	11.885	13.496	116.26	48
h chalcocite	Cu_2S	$P6/mmc$	3.95		6.75		2
djurleite	$Cu_{1.97}S$	$P2_1/c$	26.896	15.745	13.565	90.13	128
digenite	$Cu_{1.80}S$	$Fm3m$	5.57				4
roxbyite	$Cu_{1.78}S$	$C2/m$	53.79	30.90	13.36	90.0	512
anilite	$Cu_{1.75}S$	$Pnma$	7.89	7.84	11.01		16
geerite	$Cu_{1.60}S$	F-$43m$	5.410				4
spionkopite	$Cu_{1.40}S$	$P3m1$	22.962		41.429		504

Notes: 1. after Gaines et al. (1997); 2. h- hexagonal

and Cu-Fe-sulfides have crystal structures based on closest-packed arrays of S atoms, with hexagonal closest packed tending to be favored in low-temperature phases and cubic in higher-temperature phases. There is a tendency for tetrahedral coordination to predominate in structures of the ternary phases. This preference is driven by the golden rule that the number of bonding electrons available is four times the number of atoms (e.g., Wuensch 1974). The predominant 1+ oxidation state of Cu requires occupancy of interstitial metal positions for ternary compositions with atomic (Cu,Fe) > S and Cu > Fe. Along the Cu-S join, the stability of chalcocite (Cu_2S) is a direct reflection of the 1+ oxidation state (cf., the analogue formulae of Na_2S, Ag_2S and Au_2S), but now Cu is predominantly in three-fold coordination to minimize the distant Cu-Cu interactions. Covellite (CuS) accommodates the 1+ oxidation state by forming S-S bonds. All of these sulfides are metallic compounds (or small band gap semiconductors). Complexity in the compositions of the low-temperature phases in the vicinity of Cu_2S stoichiometry arises because of their alloy nature and similarity in energy of the 1+ and 2+ oxidation states of Cu for these minerals/compounds. Although not evident in laboratory experiments, there are a total of seven minerals (eight including binary digenite) with compositions extending from Cu_2S (chalcocite) to $Cu_{1.40}S$ (spionkopite).

The binary Cu-S phase diagram (Roseboom 1966; Barton 1973) is complicated by an extensive field of high-digenite($Cu_{9+x}S_5$)-high-chalcocite solid solution extending from 83 °C to 1129 °C and some uncertainty in the precise stability of digenite, as well as numerous phases of very-low-temperature (<103 °C) stability in this same composition region. The detailed phase relations are shown in Figure 11 which is after Barton (1973) and Craig and Scott (1974). There remains considerable uncertainty on the precise temperatures of phase transitions in the very-low-temperature region, with literature values varying markedly with the method of investigation. Also, there have been no studies on the stability of the new minerals roxbyite, geerite and spionkopite, and their investigation could be very demanding. As concluded from the minimal amounts of Ni, Co and Cu dissolved in pyrrhotites (discussed elsewhere), it seems that alloy-like sulfides continue to re-equilibrate toward ambient temperatures on exhumation of orebodies. Morimoto and Koto (1970) suggested that the mineral digenite (Cu_9S_5) was not stable in the binary Cu-S system, but invariably contained a small amount of Fe (~1%). Cubic low-temperature digenite only becomes a stable phase on the Cu-S join above about 70 °C. At a slightly higher, and composition dependent, temperature (76-83 °C), binary Cu-S digenite inverts to cubic high digenite ($Cu_{9+x}S_5$) which is isostructural with cubic high chalcocite (Morimoto and Kullerud 1963; Roseboom 1966). Chalcocite is one of four polymorphs of Cu_2S. The mineral is monoclinic (Evans 1971), and inverts to hexagonal Cu_2S at 103 °C. The cubic high form appears at 435 °C, forming continuous solid solution with cubic $Cu_{9+x}S_5$. A low-temperature, high-pressure polymorph has tetragonal symmetry and is stable only above

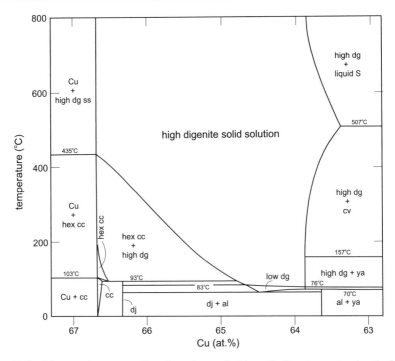

Figure 11. Partial temperature-composition phase diagram for binary Cu-S system centered on the field of high digenite solid solution: see INTRODUCTION for mineral/phase abbreviations (after Barton 1973).

~0.8 kbars (Skinner 1970). Djurleite ($Cu_{1.96}S$) is stable at and below 93 °C: it was discovered as a synthetic compound by Djurle (1958) and described as a mineral by Roseboom (1962) and Morimoto (1962). Anilite ($Cu_{1.75}S$) is stable at and below about 70 °C. The remaining Cu-S phase of intermediate composition, covellite (CuS), exhibits no solid solution and is stable on its composition point up to 507 °C where it melts incongruently to $Cu_{9+x}S_5$ + liquid sulfur.

The Cu-S phase equilibria have since been investigated using electrochemical cells from 0 °C to 250 °C over the composition range Cu/S = 0.95-2.10 (Potter 1977). The electrolytes were vapor-saturated aqueous cupric sulfate and cuprous chloride solutions. Two blaubleibender covellites (e.g., Moh 1971), corresponding in composition to "ideal" yarrowite and spionkopite (Table 2), low digenite, tetragonal Cu_2S, and a protodjurleite were found to be metastable. The temperatures of stable invariant points were in general agreement with literature values, as follows:

1. covellite = high digenite + liquid sulfur (507 ± 2 °C)
2. high chalcocite + Cu = high digenite + Cu (435 ± 8 °C)
3. low chalcocite + Cu = high chalcocite + Cu (103.5 ± 0.5 °C)
4. djurleite = high digenite + high chalcocite (93 ± 2 °C)
5. low chalcocite = djurleite + high chalcocite (90 ± 2 °C)
6. anilite = covellite + high digenite (75 ± 3 °C)
7. high digenite = anilite + djurleite (72 ± 3 °C)

Contradicting earlier results for anilite decomposition, Grønvold and Westrum (1980), using calorimetry and X-ray diffraction, found that the anilite to digenite transition occurred at 37 °C: coexisting covellite was not detected. The transition was rapid in the forward direction, but sluggish and initiated at 17 °C in reversal.

Phase relations in the ternary system Cu-Fe-S

There are numerous ternary Cu-Fe-S phases, and most of them lie within a broad swath of composition space between pyrrhotite on the Fe-S join and high digenite solid solution on the Cu-S join. To the list of eighteen minerals and confirmed phases in Craig and Scott (1974), I have added the new minerals putoranite ($Cu_{1.1}Fe_{1.2}S_2$), isocubanite and nukundamite, and revised the formula for idaite (Clark 1970a; Inan and Einaudi 2002) (Table 4).

Craig and Scott (1974) remarked that, although more time and effort had been expended on the Cu-Fe-S system than any other ternary sulfide system, many relationships remained enigmatic, obscured by extensive solid solutions, non-quenchable phases, and metastability. Since that time, detail has been added to the low-temperature phase relations (e.g., Sugaki et al. 1975; Wang 1984; Grguric et al. 2000; Inan and Einaudi 2002), $f(S_2)$ for selected reactions has been determined by electrochemistry (Lusk and Bray 2002), and the phase relations under hydrothermal conditions have been calculated (McKenzie and Helgeson 1985, 1987; Einaudi 1987). The phase relations in the central part of the Cu-Fe-S system at 600 °C and 350 °C are shown in Figures 12 and 13. As with the binary Cu-S join, the phase relations of many ternary Cu-Fe-S minerals remain unclear; e.g., the metal-excess chalcopyrite-like minerals

Table 4. Minerals and phases in the system Cu-Fe-S.

Mineral/Phase	Formula	(Symmetry)/Unit cell (Å, °)	Stability	Refs.
digenite	$(Cu,Fe)_9S_5$	(c) $a = 5.57$	37 ° to 77 °C	1
high digenite	$\sim(Cu,Fe)_9S_5$	(c) $a = 27.85$	77 ° to 1129 °C	1
bornite	Cu_5FeS_4	(t) $a = 10.95$, $c = 21.86$	to 228 °C	2
Cu_5FeS_4	Cu_5FeS_4	(c) $a = n(5.5)$		2
Cu_5FeS_4	Cu_5FeS_4	(c) $a = 5.5$	228 ° to ~1100 °C	2
x-bornite	$Cu_5FeS_{4.05}$	(t) $a = 10.44$, $c = 21.88$	to 125 °C	3, 4
idaite	Cu_3FeS_4	(h) $a = 3.90$, $c = 16.95$	metastable to 265 °C	5, 6
nukundamite	$Cu_{3.38}Fe_{0.62}S_4$	(r) $a = 3.782$, $c = 11.187$	224 ° to 501 °C	6, 7, 8, 9
fukuchilite	Cu_3FeS_8	(c) $a = 5.604$	to ~200 °C	10, 11
talnakhite	$Cu_9Fe_8S_{16}$	(c) $a = 10.591$	to ~186 °C	12, 13
intermediate I	$Cu_9Fe_8S_{16}$?		186 ° to 230 °C	14
intermediate II	$Cu_9Fe_8S_{16}$?		230 ° to 520 °C	14
chalcopyrite	$CuFeS_2$	(t) $a = 5.28$, $c = 10.40$	to 557 °C	15, 16
mooihoekite	$Cu_9Fe_9S_{16}$	(t) $a = 10.585$, $c = 5.383$	to ~167 °C	14, 15
intermediate A	$Cu_9Fe_9S_{16}$		167 to 236 °C	14
putoranite	$Cu_{1.1}Fe_{1.2}S_2$	(c) $a = 5.30$		17
haycockite	$Cu_4Fe_5S_8$	(o) $a \approx b = 10.71$, $c = 31.56$		14, 18
pc			to ~200 °C	14
iss		(c) $a = 5.36$	20-200 ° to 960 °C	19
cubanite	$CuFe_2S_3$	(o) $a = 6.46$, $b = 11.117$, $c = 6.233$	to 200-210 °C	20
isocubanite	$CuFe_2S_3$	(c) $a = 5.303$	>200-210 °C	21

Symmetry: c- cubic, t- tetragonal, h- hexagonal, r- hexagonal-rhombohedral, o- orthorhombic

References: 1. Morimoto and Kullerud (1963); 2. Morimoto and Kullerud (1966); 3. Yund and Kullerud (1966); 4. Morimoto (1970); 5. Wang (1984); 6. Inan and Einaudi (2002); 7. Merwin and Lombard (1937); 8. Clark (1970); 9. Seal et al. (2001); 10. Kajiwara (1969); 11. Bayliss (1989); 12. Cabri and Harris (1971); 13. Hall and Gabe (1972); 14. Cabri (1973); 15. Cabri and Hall (1972); 16. Barton (1973); 17. Filimonova et al. (1980); 18. Hiller and Pobsthain (1956); 19. Kullerud et al. (1969); 20. Cabri et al. (1973); 21. Caye (1988)

Notes: 1. bornite is orthorhombic, pseudo-tetragonal- $a = 10.950$, $b = 21.862$, $c = 10.950$ Å; 2. fukuchilite may be ferroan villamaninite (Bayliss 1989); 3. *pc* is primitive cubic phase; 4. *iss* is intermediate solid solution; isocubanite is mineral form of *iss* of $CuFe_2S_3$ composition

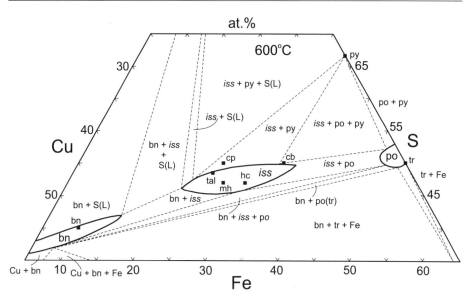

Figure 12. Phase relations in the central portion of the Cu-Fe-S system at 600 °C, highlighting extensive fields of solid solution centered on bornite, the chalcopyrite-derivative phases (*iss*), and pyrrhotite: solid squares are stoichiometric compositions for bornite, chalcopyrite, talnakhite, mooihoekite, haycockite, cubanite, troilite, and pyrite (after Cabri 1973).

talnakhite, mooihoekite, putoranite, and haycockite have not yet been associated with special composition points in the isometric solid solution (*iss*) by reversal of reaction.

Following Kullerud et al. (1969), the immiscible liquid of intermediate composition in the Cu-S system above 1105 °C extends far into the ternary system with increasing temperature, and reaches within 5 wt% of the Fe-S join at 1400 °C. Pyrrhotite appears on the Fe-S join at 1192 °C and high digenite solid solution on the Cu-S join at 1129 °C, quickly becoming a field of high digenite-high bornite solid solution extending far beyond the ideal bornite composition at 900 °C. Sphalerite-structure *iss* initially crystallizes from the central area of the liquid field at about 960 °C, and extends to the cubanite composition point by 700 °C. The 600 °C isothermal section shows the typical phase relations at intermediate temperatures (Fig. 12). The ternary phase diagram is characterized by three extensive solid solutions: (1) high digenite-high bornite; (2) *iss*; and (3) pyrrhotite. The lensoid *iss* field extends largely to metal-excess compositions. Cabri (1973) noted that it may be divided into three zones each characterized by a different quenching behavior, but this feature is not prominently recognized in other studies. Tetragonal chalcopyrite solid solution appears at 557 °C, in the *iss* + pyrite field, and remains isolated from all other Cu-Fe sulfides until temperature is decreased further. Covellite crystallizes at 507 °C and nukundamite at 501 °C. The field of *iss* shifts toward cubanite composition, and that of pyrrhotite shrinks toward the Fe-S join with further decrease in temperature.

Whereas most previous workers had studied the ternary Cu-Fe-S phase relations under dry conditions, using evacuated sealed silica glass tubes, Susaki et al. (1975) used the hydrothermal gradient transport method of Chernyshev and Anfilogov (1968) and Scott and Barnes (1971) at 350 ° and 300 °C. Consistent with previous studies, especially Cabri (1973), they found that the extensive high-temperature field of *iss* in the center of the system separated at low temperature into tetragonal chalcopyrite and a more restricted field of *iss*. At 350 °C, chalcopyrite has a narrow solid solution field extending from stoichiometric $CuFeS_2$ to $Cu_{0.9}Fe_{1.1}S_{2.0}$ along a line with metal/S atomic ratio of approximately 1, while *iss* extended from nearly stoichiometric

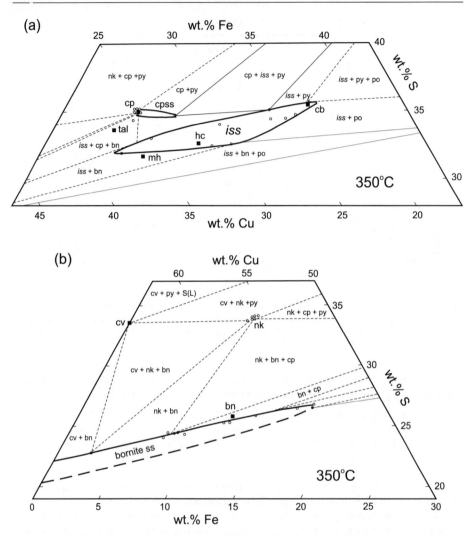

Figure 13. Detailed phase relations in the system Cu-Fe-S at 350 °C around: (a) chalcopyrite and cubanite composition; and (b) bornite composition: nk is nukundamite (idaite of earlier studies) (after Sugaki et al. 1975).

$CuFe_2S_3$ (cubanite composition) to $Cu_{1.2}Fe_{1.1}S_{2.0}$. Digenite-bornite solid solution extended far beyond stoichiometric Cu_5FeS_4 toward an Fe-rich composition, and the Cu/Fe atomic ratio reached 3.2 at 300 °C and 2.9 at 350 °C (Fig. 13a,b): the composition of this solid solution in equilibrium with chalcopyrite extended from about atomic Cu/Fe 4.0 to 7.5. The pyrrhotite solid solution was fairly extensive in ternary space at high temperature, but retreated rapidly toward the Fe-S join with decreasing temperature, so that the maximum Cu in pyrrhotite was only 0.6 wt% at 350 °C and 0.3 wt% at 300 °C. Nukundamite (idaite in Susaki et al. 1975) composition was very close to $Cu_{3.38}Fe_{0.6}S_4$ and exhibited no solid solution. Susaki et al. (1975) established that the nukundamite-chalcopyrite tie-line exists stably at 350 °C and 300 °C under hydrothermal conditions, and that the tie-line change from *iss*-pyrite to chalcopyrite-pyrrhotite occurred at 328 ± 5 °C, in agreement with the dry experiments of Yund and Kullerud (1966).

Chalcopyrite, talnakhite and mooihoekite were synthesized by Cabri (1973), but he was unable to synthesize cubanite and haycockite. Synthetic talnakhite transformed at ~186 °C to intermediate high-temperature phase I which, in turn, transformed at ~230 °C to intermediate phase II, and then to sphalerite-structure *iss* at 520-525 °C (Table 4). Synthetic mooihoekite transformed to intermediate high-temperature phase A at ~167 °C which, in turn, transformed to *iss* at ~236 °C. At haycockite composition, an unquenchable high-temperature phase formed between 20 ° and 200 °C. The transformation of cubanite to isocubanite (formerly *iss* of near-cubanite composition) occurs on heating to 200-210 °C: this reaction has not been reversed.

In other studies, Wang (1984) demonstrated using dry synthesis that a ternary phase similar in composition to idaite reacted to nukundamite + chalcopyrite at 260 °C. The assemblage covellite + nukundamite + chalcopyrite was replaced above 290 °C by nukundamite + pyrite + bornite. Also, there was evidence of an incomplete pyrite-type solid solution [i.e., $(Cu,Fe)S_2$] below 325 °C: note that fukuchilite (Cu_3FeS_8) may be ferroan villamaninite (Bayliss 1989). The limited compositional range of nukundamite (essentially $Cu_{3.38}Fe_{0.62}S_4$) was confirmed by Sugaki et al. (1981) in their hydrothermal synthesis of the mineral. The bahavior of digenite-bornite solid solution below 265 °C is dominated by two time- and temperature-dependent processes: (1) ordering of metal cations and vacancies, which leads to a variety of ordered structures (Morimoto and Kullerud 1966; Putnis and Grace 1976; Kanazawa et al. 1978; Grguric et al. 1998), and (2) rapid exsolution and coarsening which results in the formation of distinctive exsolution microtextures (Grguric and Putnis 1999). Grguric et al. (2000) revised the long-standing pseudobinary phase diagram of Morimoto and Kullerud (1966) using work of Morimoto and Gyobu (1971) and new experimental results based mainly on differential scanning calorimetry over the temperature interval 50-300 °C on a natural digenite and synthetic compositions at 5 mol% intervals along the Cu_9S_5-Cu_5FeS_4 join (Fig. 14). Their phase diagram shows a consolute point at Cu_5FeS_4 and 265 °C, with the temperature corresponding to the tricritical intermediate-high transition in bornite. Robie et al. (1994) measured the heat capacities of synthetic bornite between 5 K and 78 °C and 65 °C and 488 °C, and revised the Gibbs free energy of formation expressions for covellite, anilite, chalcocite, chalcopyrite, bornite, and nukundamite.

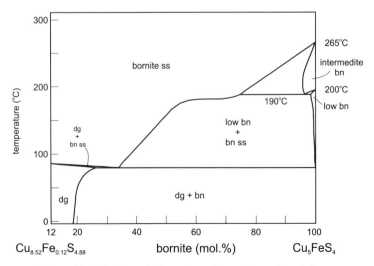

Figure 14. Revised pseudobinary phase diagram for condensed phases in the bornite-digenite solid solution series (after Grguric et al. 2000).

Sulfur fugacity in ternary Cu-Fe-S system

Merwin and Lombard (1937) achieved equilibrium in ternary Cu-Fe-S phase assemblages in the presence of vapor by independently controlling the sulfur vapor pressure. However, Barton and Toulmin (1964a,b) and Toulimin and Barton (1964) were the first to actually measure the sulfur fugacity [$f(S_2)$] in equilibrium with Cu-Fe-S and Fe-S assemblages: they used the electrum tarnish method and noted that the Fe-S assemblages uniquely determined $f(S_2)$. Subsequently, Schneeberg (1973) measured $f(S_2)$ directly, using the Ag/AgI/Ag$_{2+x}$S,$f(S_2)$ electrochemical cell, and investigating equilibrium sulfide assemblages in the systems Cu-Fe-S, Ni-S and Fe-S between 210-445 °C. Recently, Lusk and Bray (2002) used Schneeberg's electrochemical procedures to investigate the following seven reactions between 185-460 °C at 1 bar:

$$\text{pyrite + nukundamite = bornite + sulfur} \tag{4}$$

$$\text{pyrite + bornite = chalcopyrite + sulfur} \tag{5}$$

$$\text{pyrite + chalcopyrite = isocubanite + sulfur} \tag{6}$$

$$\text{pyrrrhotite + chalcopyrite = isocubanite + sulfur} \tag{7}$$

$$\text{pyrite + isocubanite1 = pyrrrhotite + isocubanite2 + sulfur} \tag{8}$$

$$\text{pyrite + chalcopyrite1 = pyrrhotite + chalcopyrite2 + sulfur} \tag{9}$$

$$\text{pyrite = pyrrhotite + sulfur} \tag{10}$$

Their data for these reactions are mapped in $\log f(S_2)$ vs. ($1/T$) space in Figure 15. The sulfur fugacities for Reactions (5) and (10) are in excellent agreement with Schneeberg (1973) at higher temperatures, but are lower than Schneeberg (1973) for Reaction (4) and reveal undetected inflections at ~221 °C and ~237 °C for Reactions (4) and (5), respectively. Also, an inflection at ~291 °C for Reaction (10) was interpreted to imply a low-temperature phase transition in hexagonal pyrrhotite.

Quaternary System Cu-Fe-Zn-S

In natural ores, chalcopyrite and sphalerite occur frequently as intimate intergrowths in the form of: (1) skeletal sphalerite crystals (sphalerite stars) in chalcopyrite, and (2) chalcopyrite dots and blebs in sphalerite. These characteristic intergrowths have been variously interpreted as products of exsolution, epitaxial growth or replacement, and mechanical mixing (see review in Kojima and Sugaki 1985). Naturally, the solubility of CuS in high-temperature sphalerite is of concern as a possible source of error for sphalerite solid solution (Zn,Fe)S geobarometers calibrated in the Fe-Zn-S system (see following section). The Cu-Fe-Zn-S system has been studied by Wiggins and Craig (1980), Kojima and Sugaki (1984, 1985) and Lusk and Calder (2004). The first two studies both investigated the low pressure phase relations from 500-800 °C using the sealed silica glass tube method. Kojima and Sugaki (1984) found that the maximum solubility of Zn in chalcopyrite at 500 °C was less than 0.9 at%, whereas the extensive field of intermediate solid solution (*iss*) dissolves considerable amounts of Zn with maximum values in Fe-rich *iss* from 12.7 at% at 800 °C to 3.3 at% at 500 °C (Fig. 16). Correspondingly, sphalerite solid solution dissolves considerable amounts of Cu, with maximum CuS contents of 10.7, 8.6 and 4.6 mol% at 800, 700 and 600 °C, respectively, in sphalerite with more than 40 mol% FeS. Thus, Kojima and Sugaki (1984) showed that entry of Cu into sphalerite was dependent on both temperature and $f(S_2)$. These phase relations were extended down to 300 °C by Kojima and Sugaki (1985), using both thermal gradient transport and isothermal recrystallization under hydrothermal conditions with an aqueous NH$_4$Cl solution mineralizer. The maximum Zn contents at 300 °C were 0.9 at% in chalcopyrite and 1.2 at% in *iss*. Maximum contents of CuS in sphalerite were about 2.4 mol% at the three temperatures investigated, and bore no relation to FeS^{sp} content and $f(S_2)$.

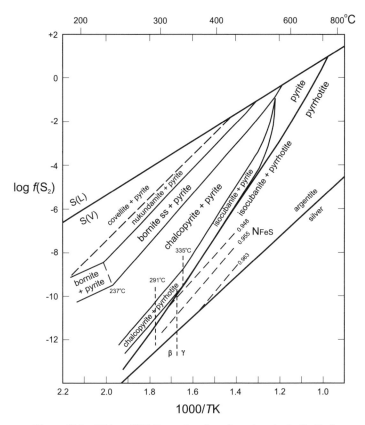

Figure 15. $\log f(S_2)$ vs. $1000/T$ map for selected reactions in the Cu-Fe-S and Fe-S systems (after Lusk and Bray 2002).

Recently, Lusk and Calder (2004) generated temperature-composition data for sphalerites and other sulfides from a sequence of reactions within the Cu-Fe-Zn-S, Fe-Zn-S and Cu-Fe-S systems at 1 bar and 250-535 °C. Two eutectic salt fluxes (NH$_4$Cl-LiCl, 50 mol% NH$_4$Cl, and KCl-LiCl, 41 mol% KCl; introduced by Boorman 1967) were used routinely for crystallizing the sulfide products *in situ*, and the experiments were made in evacuated sealed Pyrex or Vycor glass capsules. Lusk and Calder (2004) superimposed the 1 bar data for equilibrium sphalerite compositions on Lusk and Bray's (2002) $\log(f(S_2))$-$(1/T)$ map for buffer reactions in the Cu-Fe-S and Fe-S systems (see Fig. 15). The FeS contents of sphalerites (mol%) were about 1.7 for pyrite + nukundamite = bornite + S^{vap}, 4.4 for pyrite + bornite = chalcopyrite + S, 14.4 for pyrite + chalcopyrite = isocubanite + S, 22.9 for pyrite + isocubanite1 = pyrrhotite + isocubanite2 + S, and 21.5 for pyrite = pyrrhotite + S. In general agreement with Kojima and Sugaki (1985), they found that Cu contents in sphalerite were low, ranging between about 3 and 0.5 mol% CuS, with highest values for sphalerites associated with bornites. However, the Lusk and Calder (2004) sphalerites for assemblages of high $f(S_2)$ had significantly higher FeS contents than in Czamanske (1974) and Kojima and Sugaki (1985). Lusk and Calder (2004) fitted their results and selected literature data for buffer assemblages of sphalerite with either pyrite or pyrrhotite + pyrite in the Fe-Zn-S and Cu-Fe-Zn-S systems at 250-550 °C and 1 bar to the following Equation for $f(S_2)$:

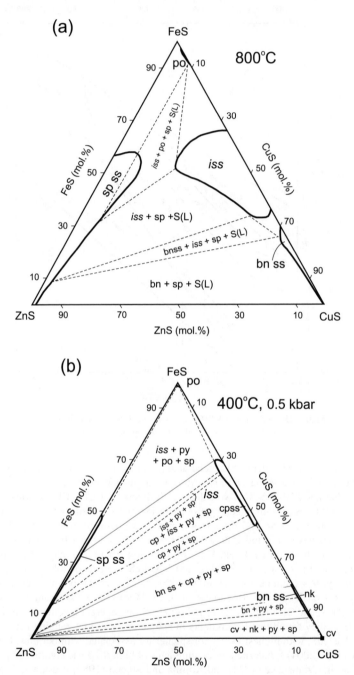

Figure 16. Isothermal phase relations in quaternary Cu-Fe-Zn-S system at: (a) 800 °C and low pressure (after Kojima and Sugaki 1984); and (b) 400 °C and 0.5 kbars, showing marked dependence of solubility of Cu in sphalerite and of Zn in intermediate solid solution (*iss*) with temperature (after Kojima and Sugaki 1985).

$$\log f(S_2) = 11.01 - 9.49\left(\frac{1000}{T}\right) +$$
$$\left[0.187 - 0.252\left(\frac{1000}{T}\right)\right]FeS^{sp} + \left[0.35 - 0.2\left(\frac{1000)}{T}\right)\right]CuS^{sp} \qquad (11)$$

where T is Kelvin, and FeS^{sp} and CuS^{sp} are mol%.

ZnS AND Fe-Zn-S SYSTEM

ZnS

Zinc monosulfide (ZnS) exhibits both polymorphism and polytypism at room pressure. Sphalerite, wurtzite and the wurtzite polytypes all have structures based on the closest packing of their constituent atoms. The Zn and S atoms form interleaved stacking sequences of closest-packed layers, and Zn is in four-fold tetrahedral coordination with S. Interestingly, sphalerite and wurtzite are readily distinguished in petrographic thin section on account of the birefringence of wurtzite and the wurtzite polytypes (e.g., Fleet 1977). Sphalerite [cubic (Zn,Fe)S solid solution] is the dominant zinc sulfide mineral in sulfide ore deposits, occurring in association with iron sulfides (pyrite, pyrrhotite, troilite) ± Cu-Fe sulfides ± galena. Elsewhere, sphalerite is found in a wide variety of rocks and unconsolidated lithologies, including metamorphic, hydrothermally altered and sulfidized rocks, glacial tills, iron meteorites and enstatite chondrites, which collectively represent a very wide range in temperature and pressure of formation or annealing. Zinc monosulfide is a refractory compound. Pure sphalerite structure ZnS sublimes at 1185 °C and melts congruently at 1830 °C and 3.7 atm. At room temperature, sphalerite- and wurtzite-structure ZnS invert to halite structure ZnS with Zn in six-fold octahedral coordination near 15 GPa (e.g., 15 GPa in Zhou et al. 1991; 16.2±0.4 GPa in Yagi et al. 1976). Measurements by Desgreniers et al. (2000) exclude the existence of an intermediate phase of cinnabar type structure. Desgreniers et al. (2000) also found that the halite structure is stable up to 65 GPa, where it inverts to a structure with $Cmcm$ symmetry, an orthorhombic distortion of the halite structure, without significant change in volume. The low-pressure sphalerite/wurtzite equilibrium phase relations remain unclear, and their understanding has not changed substantially since the review of Craig and Scott (1974). Wurtzite is generally regarded as the stable high-temperature phase at low pressure and, indeed, wurtzite and the wurtzite polytypes readily transform to sphalerite in supergene alteration and diagenesis (e.g., Fleet 1977). The early study of Allen et al. (1913) reported a transformation temperature of 1020 °C at 1 atm, albeit based on a sluggish reversal, and this has been widely adopted in the absence of a more systematic study. However, Scott and Barnes (1972) reviewed literature values for the transition temperature ranging from 600 °C to above 1240 °C. Using hydrothermal recrystallization and gas-mixing experiments, they showed that sphalerite and wurtzite can coexist over a range of temperatures well below 1020 °C, as a function of $f(S_2)$. Scott and Barnes (1972) argued that wurtzite was deficient in S relative to sphalerite at the same T, P and $f(S_2)$, and that the sphalerite/wurtzite equilibrium was actually univariant and represented by an equation of the type:

$$ZnS^{sp} = ZnS_{1-x}{}^{wz} + \left(\frac{x}{2}\right)S_2 \qquad (12)$$

or, more generally:

$$\left(\frac{1}{1-y}\right)Zn_{1-y}S^{sp} = ZnS_{1-x}{}^{wz} + \frac{1}{2}\left(x - 1 + \frac{1}{1-y}\right)S_2 \qquad (13)$$

The univariant boundary was located at 500 atm from 465-517 °C over a corresponding calculated $f(S_2)$ range of $10^{-9.5}$ to $10^{-8.7}$ atm. Reversal of the transition was demonstrated from 0.28 to 0.55 kbars and about 450-470 °C in 15 m NaOH solutions. Direct determination of $f(S_2)$ at the transition was made by passing $H_2 + H_2S$ mixtures over ZnS powder, giving values of 10^{-5} atm at 890 °C, $10^{-5.5}$ to $10^{-6.4}$ atm at 800 °C, and $10^{-6.5}$ to $10^{-8.5}$ at 700 °C. Ueno et al. (1996) investigated the sphalerite = wurtzite transition in the system Zn-Fe-Ga-S. They found that, for a sample of composition $(Zn_{0.70}Ga_{0.30})S$, the sphalerite phase inverted to wurtzite structure near 875 °C, but the transition was sluggish in both directions and the transition temperature varied with cation composition and $f(S_2)$.

Ternary Fe-Zn-S system and geobarometry

The Fe content of sphalerite varies with the iron sulfide assemblage [i.e., with $f(S_2)$] and lithostatic pressure, and tends to be preserved during exhumation of metamorphic rocks and orebodies under dry conditions and in the post-formation thermal history of meteorites. Therefore, it is widely used for geobarometry of sphalerite-bearing ore bodies, metamorphic rocks and meteorites. The phase relations of sphalerite in the Fe-Zn-S and Cu-Fe-Zn-S systems have been studied extensively using synthesis experiments, largely to support these geobarometric applications (e.g., Kullerud 1953; Barton and Toulmin 1966; Boorman 1967; Boorman et al. 1971; Scott and Barnes 1971; Scott 1973; Scott and Kissin 1973; Czamanske 1974; Lusk and Ford 1978; Wiggins and Craig 1980; Hutchison and Scott 1983; Kojima and Sugaki 1984, 1985; Bryndzia et al. 1988; Lusk et al. 1993; Balabin and Urusov 1995; Mavrogenes et al. 2001; Lusk and Calder 2004), and thermodynamic and model calculations and field relationships (e.g., Einaudi 1968; Schwarcz et al. 1975; Scott 1976; Hutcheon 1978, 1980; Barker and Parks 1986; Banno 1988; Bryndzia et al. 1990; Toulmin et al. 1991; Balabin and Urusov 1995; Balabin and Sack 2000; Martín and Gil 2005).

The fundamental understanding of the important phase relations close to the FeS-ZnS join of the Fe-Zn-S system was provided by Barton and Toulmin (1966) using evacuated sealed silica glass tube experiments. They investigated isothermal sections at about 580, 600, 640, 700, 740, 800, and 850 °C (e.g., Fig. 17), and clearly demonstrated that the Fe content of sphalerite decreases systematically

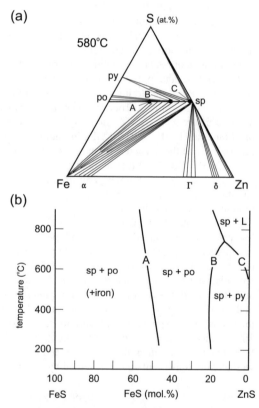

Figure 17. Sphalerite compositions determined by important reactions in the Fe-Zn-S system: (a) 580 °C isothermal section showing ternary phase relations; and (b) temperature-composition projection onto FeS-ZnS join showing sphalerites buffered by reactions A, B and C at 1 bar (after Lusk and Calder 2004).

with increase in $f(S_2)$, consistent with reactions of the type:

$$FeS^{sp} + \tfrac{1}{2}S_2 = FeS_2^{py} \tag{14}$$

for coexisting sphalerite + pyrite. Sphalerite solid solutions lie essentially along the FeS-ZnS join, and neither pyrrhotite nor pyrite takes up appreciable amounts of zinc: the latter is a reflection of the strong tendency of Zn for four-fold tetrahedral coordination with anions in ionic compounds and ligands in covalently bonded compounds. Sphalerite (of various compositions) may be in equilibrium with any of the other phases in the ternary Fe-Zn-S system, and always contains Fe in Fe-bearing bulk compositions. Nevertheless, the join Fe-ZnS is nearly binary; i.e., sphalerite of nearly pure ZnS composition is in equilibrium with α-iron of nearly pure Fe composition. The principal change in the phase relations with increase in temperature is the disappearance of pyrite above 742 °C (at 1 bar).

Barton and Toulmin (1966) identified three univariant reactions, labeled A, B and C in Figure 17 and representing the buffered assemblages sphalerite + troilite (or high-temperature pyrrhotite at or near stoichiometric FeS; presently abbreviated as FeS^{HT}) + α-iron (curve A), sphalerite + hexagonal pyrrhotite (hpo) + pyrite (curve B), and sphalerite + pyrite + liquid sulfur (curve C). These univariant curves are the sphalerite solvi for the three buffered assemblages. Barton and Toulmin (1966) reported that the solvus for sphalerite coexisting with FeS^{HT} is very steep, passing through 56 mol% FeS at 850 °C and 52 mol% at 580 °C. The solvus for sphalerite coexisting with hexagonal pyrrhotite + pyrite is also steep, but concave towards the ZnS composition due mainly to the pinching out of the sphalerite + pyrite field and its replacement by sphalerite + hpo and sphalerite + liquid sulfur assemblages. The composition of sphalerite in equilibrium with hexagonal pyrrhotite + pyrite changes from 13 mol% FeS at 742 °C to 19 mol% at 580 °C. Barton and Toulmin (1966) suggested that the composition of sphalerite in these buffered assemblages might be useful in geothermometry and outlined how the effect of pressure could be estimated by correlating change in the natural logarithm of the activity coefficient for FeS in sphalerite with change in partial molar volume. This pressure correction theory has been adopted in numerous studies using the univariant curve B for sphalerite geobarometry (e.g., Scott 1973) and A for sphalerite cosmobarometry (e.g., Hutchison and Scott 1983; Balabin and Urusov 1995).

Subsequent studies (Boorman 1967; Scott and Barnes 1971) further defined the low-pressure univariant curve for the buffered assemblage sphalerite + pyrrhotite + pyrite using halide fluxes to achieve equilibrium at lower temperatures and 1 bar and 0.25 to 1 kbar, respectively. Scott (1973) extended the experimental calibration of the sphalerite geobarometer to pressures of 2.5, 5 and 7.5 kbars at 325-710 °C using hydrothermal recrystallization in aqueous alkali halide fluxes and cold sealed pressure vessels. This study established the sphalerite geobarometer as a viable method for estimating pressure of formation from assemblages of sphalerite + pyrrhotite + pyrite. However, since the solvus for this buffered assemblage was essentially independent of temperature over the temperature range of most interest (550 °C to below 300 °C; Fig. 18) it was clearly unsuitable for geothermometry. The experimental calibration was extended to about 10 kbars and 420 °C to 700 °C by Lusk and Ford (1978) using an internally heated pressure vessel and various fluxes; i.e., aqueous solutions of 4.5 M KCl or 5 M NH_4Cl or KCl-LiCl and NH_4Cl-LiCl salt mixtures (Fig. 19). The progressive decrease in FeS content of sphalerite with increase in pressure in the temperature independent region was expressed in the equation:

$$FeS^{sp} = 20.53 - 1.313P + 0.0271P^2 \tag{15}$$

where FeS^{sp} is mol% and P is kbars, and the equation was fitted to the preferred experimental results of Boorman (1967), Scott and Barnes (1971), Scott (1973), and Lusk and Ford (1978). Lusk et al. (1993) investigated sphalerite + pyrite + pyrrhotite phase relations to lower temperature (150 °C and 325 °C, and 0.4 to 5.1 kbars), using recrystallization in saturated

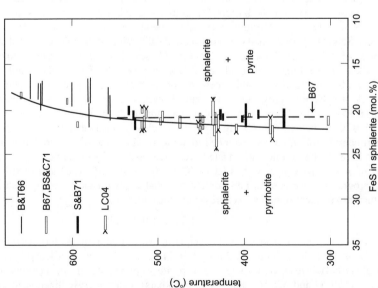

Figure 18. (*left image*) Experimental measurements of temperature vs. FeS content of sphalerite (FeS^{sp}) buffered along univariant curve B (Fig. 17) by the equilibrium assemblage pyrrhotite + pyrite + sulfur(vapor): B&T66 is Barton and Toulmin (1966); B67 is Boorman (1967); BS&C71 is Boorman et al. (1971); S&B71 is Scott and Barnes (1971); and LC04 is Lusk and Calder (2004) (after Lusk and Calder 2004).

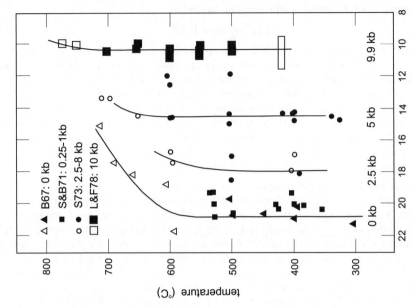

Figure 19. (*right image*) Experimental measurements of FeS content of sphalerite (FeS^{sp}) coexisting with hexagonal pyrrhotite + pyrite as a function of temperature and confining pressure: B67 is Boorman (1967); S&B71 is Scott and Barnes (1971); S73 is Scott (1973); and L&F78 is Lusk and Ford (1978) (after Lusk and Ford 1978).

ammonium iodide solution. Solvus isobars were deflected slightly to lower FeS^{sp} content at about 280 °C. They presented a *P-T-X* calibration map for sphalerite composition coexisting with hexagonal pyrrhotite and pyrite based on the data for equilibrated assemblages in the literature and their study. Although monoclinic pyrrhotite (mpo) most commonly coexists with pyrite under very-low temperature conditions in nature, it was not present in the experimental products of Lusk et al. (1993). It was suggested, therefore, that monoclinic pyrrhotite was stable under isotropic stress conditions only below 150 °C.

Univariant reaction A (Fig. 17) represents the FeS saturation limit of sphalerite coexisting with FeS^{HT} + α-iron (which quenches to troilite + alloy). Although the sphalerite + troilite + alloy assemblage is unknown on Earth, sphalerite does occur as a minor phase in troilite nodules of some iron meteorites as well as in a few enstatite chondrites. Therefore, the composition of sphalerite in this assemblage may be used to estimate the pressures of formation of meteorites. This "cosmobarometer" was calibrated with the 1 bar experimental data of Barton and Toulmin (1966) and additional experiments at 2.5 and 5.0 kbars between 400-800 °C using both aqueous and anhydrous alkali halide flux recrystallization techniques in Hutchison and Scott (1983). They graphed the 1 bar, 2.5 and 5.0 kbar isobars as a function of temperature and sphalerite composition (Fig. 20) and regressed the sphalerite composition data to give an equation for estimating pressure of formation, as follows:

$$P = -3.576 + 0.0551T - 0.0296T \log(\text{FeS}^{sp}) \qquad (16)$$

where *P* is kbars, *T* is Kelvin, and FeS^{sp} is mol%. Unlike the sphalerite + pyrrhotite + pyrite equilibrium, the shifts in the solvus to lower values of FeS^{sp} due to the pressure effect were large (Fig. 20). Also, the measured pressure effect was larger than that calculated by Schwarcz et al. (1975). The composition of sphalerite in equilibrium with troilite + α-iron

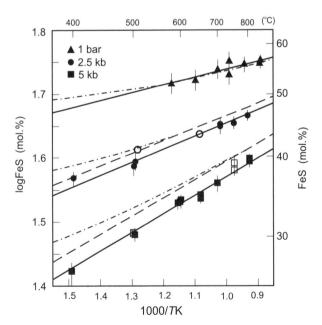

Figure 20. Experimentally determined and calculated isobars along the sphalerite + troilite (FeS^{HT}) + α-iron solvus (curve A in Fig. 17): measurements were made at 1 bar, and 2.5 and 5 kbars; dashed and dot-dashed curves are calculated isobars of Hutchison and Scott (1983) and Schwarcz et al. (1975), respectively (after Hutchison and Scott 1983).

was reinvestigated experimentally from 400-840 °C at 1 bar by Balabin and Urusov (1995). They used evacuated sealed silica glass tubes and dry sintering of pure ZnS and FeS starting materials at 660-840 °C, and an anhydrous halide flux of eutectic NaCl + KCl + PbCl$_2$ + FeCl$_2$ composition at lower temperature. Their experimental data for the FeS^{sp} saturated solvus at 1 bar were approximated by the following equation:

$$FeS^{sp} = 44.09 + 0.0125T \tag{17}$$

where FeS^{sp} is mol% and T is Kelvin. Based on an updated thermodynamic model for the sphalerite [(Zn,Fe)S] solid solution, their new 1 bar solvus (univariant curve A) was in apparent consistency with the high-pressure experimental data of Hutchison and Scott (1983).

The compositions of coexisting hexagonal pyrrhotite + sphalerite are dependent on pressure and temperature and offer the potential of a third sphalerite barometer when appropriately calibrated. This has been investigated by Bryndzia et al. (1988) who recrystallized sphalerite and hexagonal pyrrhotite in a eutectic LiCl-KCl salt flux at 450-750 °C and 1-6 kbars. Plots of the activity of FeS^{hpo} vs. activity of FeS^{sp} were linear for any given pressure, as observed in the earlier study of Barton and Toulmin (1966) at 1 bar. Thus, Bryndzia et al. (1988) concluded that the activity coefficient of FeS^{sp} (i.e., γ_{FeS}^{sp}) is a function of pressure only (Fig. 21) and, following the theory of Barton and Toulmin (1966), used the pressure dependence of γ_{FeS}^{sp} to establish a geobarometer for coexisting sphalerite + hexagonal pyrrhotite, as follows:

$$P = 27.982 \log(\gamma_{FeS}^{sp}) - 8.549 \quad (\pm 0.5 \text{ kbars}) \tag{18}$$

(Bryndzia et al. 1990). This geobarometer resulted in an estimate of 5.8 ± 0.7 kbars for the pressure of metamorphism in the Broken Hill area, Australia, in excellent agreement with 6.0 ± 0.5 kbars for aluminosilicate mineral geobarometers applied to local granulite-grade rocks.

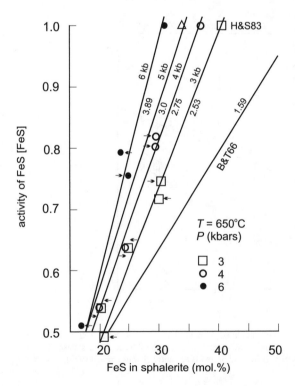

Figure 21. Activity of FeS as a function of sphalerite composition and confining pressure at 650 °C. The activity coefficient in the system FeS-S$_2$ (values on straight line fits) increases with increase in pressure but is essentially independent of sphalerite composition at constant pressure (after Bryndzia et al. 1988).

Calculated phase relations

Overall, the two geobarometers and the cosmobarometer have been well calibrated by the laboratory experiments and give pressure values consistent with independent estimates from nearby aluminosilicate assemblages. However, an integrated and unifying theoretical basis for these various sphalerite barometers is lacking. Thermodynamic theories for the pressure dependence of FeS^{sp} (or γ_{FeS}^{sp}) result in serious discrepancies between calculated and experimental results below 500 °C (e.g., Fig. 22) and, in the absence of a resolution for these discrepancies, Banno (1988) and Toulmin et al. (1991) regarded sphalerite pressures as tentative. Without rigorous thermodynamic support, the low-temperature experimental reversals obtained using seed crystals remain suspect.

The pronounced pressure dependence of the sphalerite solvus results from the large molar volume change for equilibria between sphalerite (Fe^{2+} in tetrahedral coordination) and pyrrhotite (Fe^{2+} in octahedral coordination). Sphalerite- and wurtzite-type structures are inefficient in respect to space filling by their constituent atoms. For hard shell atoms, the filled space in these structures is estimated as only about 50%, compared with about 70% for the corresponding halite structure compounds (Rooymans 1969). Thus, sphalerite-structure compounds are generally expected to have a relatively low-pressure stability. The relatively high-pressure stability of end-member sphalerite (up to about 15 GPa at room temperature) is anomalous for this structure type, and attributable to the very strong preference of divalent Zn for tetrahedral coordination with ligands. On the other hand, sphalerite-structure FeS is expected to behave similarly to sphalerite-structure MnS which inverts to alabandite at 0.3 GPa and room temperature. The equilibrium between sphalerite and troilite (or FeS^{HT}) in the Fe-Zn-S system may be represented by the reaction:

$$FeS^{sp} = FeS^{tr} \tag{19}$$

with an equilibrium constant $K_{sp-tr} = [FeS^{tr}]/[FeS^{sp}]$, where box brackets indicate activity. The molar volume change for this reaction at 1 bar and room temperature (ΔV^{tr-sp}) is

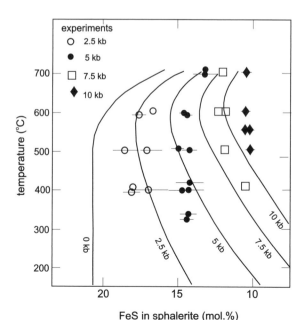

Figure 22. Calculated shifts in univariant curve B [Fig. 17; i.e., sphalerite coexisting with pyrrhotite + pyrite + sulfur(vapor)] for confining pressures of 2.5, 5, 7.5, and 10 kbars, showing marked discrepancies with experimental measurements, particularly in the low-temperature region (after Toulmin et al. 1991).

−5.90 cm^3/mol (or −0.59 J/bar; using unit-cell data of Fleet 1975, and Yund and Hall 1968). As a first approximation, the experimental results of Barton and Toulmin (1966), Boorman (1967) and Balabin and Urusov (1995), for the 1 bar isobar for sphalerite in equilibrium with troilite (or FeS^{HT}) and α-iron, may be extrapolated to higher pressure using the isothermal pressure correction term for deviations from end-member compositions in silicate mineral geobarometers when parameterized as:

$$P = aT + b - RT \ln K/(\Delta V) \tag{20}$$

where T is Kelvin, R is the universal gas constant, K is the equilibrium constant, and the effects of thermal expansion and isothermal compression on the volume change are ignored. The change in pressure due to dilution of end-member component(s) is given by:

$$P - P_0 = -RT \ln K/(\Delta V) \tag{21}$$

where, P_0 is the reference pressure when $K = 1$ at temperature T. For the 1 bar sphalerite isobar of Balabin and Urusov (1995), the composition of sphalerite (in mole fractions of FeS; i.e., X_{FeS}^{sp}) is 0.525 at 700 K and 0.565 at 1000 K. Rearranging equation (21) to:

$$P_0 - P = RT \ln K/(\Delta V) \tag{22}$$

and assuming ideal solid solution in sphalerite of composition (Zn,Fe)S and [FeS^{tr}] = 1, so that $K = 1/X_{FeS}^{sp}$, results in fictive values for the reference pressure (P_0) of −6.4 and −8.0 kbars at 700 K and 1000 K, respectively. Using these fictive values for P_0, we calculate that, at 700 K, the solvus is shifted from $X_{FeS}^{sp} = 0.525$ at 1 bar to 0.41 at 2.5 kbars and 0.32 at 5.0 kbars and, at 1000 K, it is shifted from 0.565 at 1 bar to 0.47 at 2.5 kbars and 0.40 at 5.0 kbars. These pressure corrections shift the sphalerite solvus, respectively, toward the 2.5 and 5.0 kbars experimental isobars of Hutchison and Scott (1983). Agreement with the experimental isobars is improved by accounting for non-ideality of the (Zn,Fe)S solid solution. For example, using the 850 °C activity coefficients of Fleet (1975), the solvus is shifted, at 700 K, to $X_{FeS}^{sp} = 0.39$ at 2.5 kbars and 0.28 at 5.0 kbars and, at 1000 K, to $X_{FeS}^{sp} = 0.46$ at 2.5 kbars and 0.37 at 5.0 kbars. The latter set of calculated values reproduces the experimental isobars of Hutchison and Scott (1983) reasonably well (Fig. 23).

The pressure corrections of Schwarcz et al. (1975) and Hutchison and Scott (1983) for the sphalerite cosmobarometer and of Scott (1973), Hutcheon (1980) and Toulmin et al. (1991) for the sphalerite geobarometer are all based on the equation derived by Barton and Toulmin (1966) for the pressure dependence of X_{FeS}^{sp}:

$$\frac{dX_{FeS^{sp}}}{dP} = -\frac{(\bar{V}_{FeS^{sp}} - V_{FeS^{tr}})X_{FeS^{sp}}}{RT} \tag{23}$$

where $\bar{V}$ is the partial molar volume of FeS^{sp}, V_{FeS}^{tr} is the molar volume of pure troilite, and T is Kelvin. These authors have variously corrected for nonideality of mixing, thermal expansion and isothermal compression of coexisting (Zn,Fe)S and Fe$_{1-x}$S. The most significant discrepancy with the experimental results is the failure to reproduce the curious temperature independence of isobars for sphalerite in equilibrium with pyrrhotite and pyrite (curve B in Fig. 17) below 600 °C (e.g., Figs. 19 and 22). Balabin and Urusov (1995) used a more rigorous equation for pressure dependence of X_{FeS}^{sp} in the sphalerite cosmobarometer (curve B in Fig. 17), and reevaluated the mixing properties of the sphalerite solid solution using the analytical method of Chuang et al. (1985) to solve an expanded Guggenheim-type equation for the excess free energy of mixing. Values for three coefficients and four interaction parameters were obtained by fitting a data set of experimental values for the 1 bar isobar of sphalerite in equilibrium with troilite (or FeS^{HT}) and α-iron, from Barton and Toulmin (1966) and their study. Extrapolation to 2.5 and 5.0 kbars resulted in very good agreement with selected experimental results from Hutchison and Scott (1983) (Fig. 23). Martín and Gil

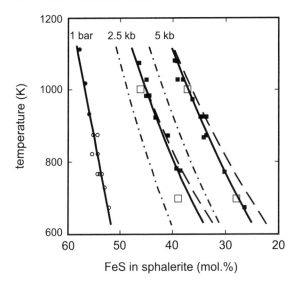

Figure 23. Experimental data of Hutchison and Scott (1983) for the high-pressure sphalerite solvus buffered by FeS^{HT} + α-iron (curve A in Fig. 17; i.e., the cosmobarometer) compared with solvi calculated for different models of the mixing of FeS and ZnS in sphalerite: dot-dashed curves assume ideal mixing behavior; dashed curves are a simple regular solution; solid curves are a regular solution with optimized (fitted) parameters; and open squares are values of pressure-induced shifts from 1 bar isobar calculated in present text; cf. Figure 20 (after Balabin and Urusov 1995).

(2005) followed Hutcheon (1978, 1980) in using asymmetric Margules parameters to describe the mixing in (Zn,Fe) sphalerites: this resulted in a complex analytical expression which they fitted to a database of 279 experiments. As a footnote to these theoretical pressure corrections, Fleet (1975) pointed out that the activity system used in Barton and Toulmin (1966) is a hybrid between that of Lewis and Randall (1961) and the absolute activity system of Guggenheim (1959, p. 220) and does not properly reflect mixing in the crystal lattice of sphalerite. For example, the activity of FeS^{sp} is obtained from the equality $[FeS^{sp}] = [FeS^{po}]$, in the manner of the absolute activities of Guggenheim (1959), rather than from equality of the Gibbs chemical potentials. It follows that reported values for activity coefficients have to be viewed relative to the mixing system being adopted; e.g., using the Lewis and Randall system, Fleet (1975) found that the departures of ZnS and FeS from ideal solution in sphalerite of composition (Zn,Fe)S were not very great, whereas Balabin and Urusov (1995) reported relatively large positive deviations from ideality. More recently, the cluster variation model (CVM) of Balabin and Sack (2000) predicted only moderate deviations from ideality, similar to those calculated by Fleet (1975) at 850 °C.

HALITE STRUCTURE MONOSULFIDES

The monosulfides of Mg, Ca, Mn, and Fe with the cubic halite structure occur naturally as niningerite [(Mg, Fe, Mn)S], alabandite (MnS) and oldhamite (CaS) in EH enstatite chondrite meteorites, where they appear to result from metamorphism under strongly reducing conditions (e.g., Fleet and MacRae 1987). Alabandite also occurs in low-temperature vein deposits in the Earth's crust.

Electronic structure and magnetism

As noted above, the electrical and magnetic properties of metal sulfides are strongly influenced by covalence of the metal-S bond, which results in hybridization of S $3p$ and metal $3d$ bonding states and direct or indirect metal-metal (M-M) bonding interactions in favourable cases. The metal and S atoms in these monosulfides are both in octahedral coordination with nearest neighbors, and there is little direct interaction between MS_6 octahedra. As reviewed in Farrell et al. (2001), the alkaline earth chalcogenides are considered to be insulators or large band gap semiconductors. Divalent Mg differs from Ca, Mn and Fe in that there are no metal $3d$ orbitals in the energy band gap available for hybridization with S $3p$ orbitals. At room temperature and pressure, both MgS and CaS are diamagnetic and classical insulators or large indirect ($\Gamma- X$) band gap semiconductors. The band gap of 2.7 eV in MgS is bounded by S $3p$ bonding orbitals in the upper part of the valence-band (VB) and by Mg $3s$ σ^* (and to a lesser extent, S $3d$) antibonding orbitals in the lower part of the conduction-band (CB). In CaS, the band gap (given variously as 4.4 or 2.1 eV) lies between the S $3p$ bonding orbitals in the VB and predominantly Ca $3d$ (and to a lesser extent Ca $4s$ σ^*) antibonding orbitals in the CB. In MnS, Mn has the high spin t_{2g}^3-e_g^2 electronic configuration, with the majority spin ($\uparrow$) t_{2g}^α and e_g^α bands filled and minority spin ($\downarrow$) t_{2g}^β and e_g^β bands empty. At room temperature, α-MnS (Néel temperature ~152 K) is antiferromagnetic. It is a diluted magnetic semiconductor and has outstanding magnetic and magneto-optical properties derived through interaction of hybridized S sp and Mn $3d$ states (Sato et al. 1997). The band gap lies between occupied Mn $3d$ states in the VB and unoccupied S $3p$ σ^* antibonding states hybridized with Mn $3d(e_g)$ and $3d(t_{2g})$ σ^* antibonding states.

Phase relations

CaS and MgS are refractory compounds. Solid solutions within the system MgS-MnS-CaS-FeS up to about 1100 °C were investigated by Skinner and Luce (1971) using the evacuated sealed silica glass tube method. The extent of solid solution in binary systems is well displayed in their plot of cubic unit-cell edge (a) vs. composition (present Fig. 24). MgS and MnS form a completely miscible solid solution at all temperatures investigated (from 600-1000 °C), and melting was not detected up to 1100 °C. These two sulfides also displayed extensive and strongly temperature-dependent solid solutions toward FeS^{HT}, with about 74 mol% FeS in (Mn,Fe)S and 68 mol% FeS in (Mg,Fe)S at 1000 °C. Correspondingly, the high-temperature FeS phase accommodated 7.4 mol% MnS at the same temperature. However,

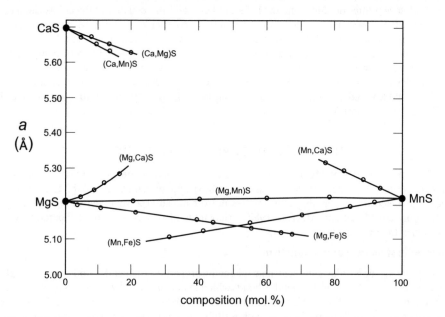

Figure 24. Effect of binary cation substitutions on unit-cell edge (a) of halite-structure monosulfides: measurements were made at room temperature on quenched experimental samples; discontinuous distributions reflect solvus limits at various temperatures investigated (after Skinner and Luce 1971).

solid solution of the ionic monosulfides MgS and CaS in NiAs-type FeS^{HT} was not detected. It is interesting to note at this point that substitution of divalent Fe into MgS results in a decrease in unit-cell edge (Fig. 24), extrapolating to $a = 5.07$ Å for the hypothetical FeS end-member, compared with $a = 5.202$ Å for MgS. Since the radius of Fe^{2+} is some 8% greater than that of Mg^{2+} in oxy structures (e.g., Shannon 1976), this decrease in the cell edge of the (Mg,Fe)S solid solution surely highlights the greater covalence of the Fe-S bond and hybridization of S $3p$ and metal $3d$ bonding states on the Fe atoms, compared with both MgS and Fe oxy compounds. The substitution of Mn into MgS also results in little increase in a, in spite of the significant difference in effective ionic radius for oxy structures (0.83 and 0.72 Å, respectively). However, covalence and p-d hybridization are diminished in MnS compared with halite-structure FeS, because of the stable $3d^5$ high spin electron configuration of Mn^{2+}. The subsolidus regions of the CaS-MnS and CaS-MgS binary phase diagrams are both dominated by fairly symmetrical and temperature-dependent solvi. At 1000 °C, the solvus limits in Skinner and Luce (1971) were 12.7 and 72.4 mol% MnS in the former system and 25.1 and 81.9 mol% MgS in the latter, with complete miscibility anticipated at higher temperatures.

The stability of niningerite (Mg,Fe)S in enstatite chondrites was investigated by Fleet and MacRae (1987) in their experiments on the sulfidation of Mg-rich olivine under reducing conditions at 1100-1200 °C. Experimental products were niningerite, clinoenstatite, and (Fe,Ni)-mss (Fig. 25). Niningerite was present as bubble-like blebs that were interpreted to reflect liquid immiscibility during quenching but are now properly recognized as equilibrium crystals. The average value for the distribution coefficient for Ni/Fe exchange between niningerite solid solution (nss) and mss was 23.7 ± 2.6. On the other hand, the values of the distribution coefficient for coexisting niningerite and troilite in four enstatite chondrites were close to unity and, thus, inconsistent with the equilibration of Ni between these two minerals.

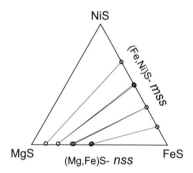

Figure 25. Compositions of coexisting niningerite (nss) and (Fe,Ni) monosulfide (mss) solid solutions from experiments on the sulfidation of Mg-rich olivine at 1200 °C (after Fleet and MacRae 1987).

ACKNOWLEDGMENTS

I thank Maggy Lengke for library work and the Natural Sciences and Engineering Research Council of Canada for financial support.

REFERENCES

Albers W, Schol K (1961) The P-T-X phase diagram of the system Sn-S. Philips Research Reports 16:329-342
Allen ET, Crenshaw JL, Merwin HE (1913) The sulfides of zinc cadmium and mercury; their crystalline forms and genetic conditions. Am J Sci Ser 4 34:341-397
Andresen AF (1960) Magnetic phase transitions in stoichiometric FeS studied by means of neutron diffraction. Acta Chem Scand 14:919-926
Andresen AF, Torbo P (1967) Phase transitions in Fe_xS ($x = 090$-100) studied by neutron diffraction. Acta Chem Scand 21:2841-2848
Anzai S (1997) Effect of correlations on electronic and magnetic properties of transition metal chalcogenides. Physica B237:142-145
Anzai S, Ozawa K (1974) Effect of pressure on the Néel and the ferromagnetic Curie temperature of $FeS_{1-\delta}$. Phys Stat Sol 24:K31-K34
Arnold RG (1962) Equilibrium relations between pyrrhotite and pyrite from 325 ° to 743 °C. Econ Geol 57:72-90
Arnold RG, Malik OP (1975) The NiS-S system above 980 °C: a revision. Econ Geol 70:176-182

Avilov AS, Imamov RM, Muradyan LA (1971) An electron diffraction study of some phases in the Cu-Sb-S system. Trans, Sov Phys Crystallogr 15:616-619
Balabin AI, Sack RO (2000) Thermodynamics of (ZnFe)S sphalerite: A CVM approach with large basis clusters. Mineral Mag 64:923-943
Balabin AI, Urusov VS (1995) Recalibration of the sphalerite cosmobarometer: Experimental and theoretical treatment. Geochim Cosmochim Acta 59:1401-1410
Balabin AI, Osadchiy YeG, Urusov VS, Senin VG (1986) Phase relationships involving daubreelite in the Fe-Cr-S, Mn-Fe-Cr-S, Mg-Fe-Cr-S systems at 840, 745, 660, and 550 °C. Geochem Internat 23:30-43
Banno S (1988) On the sphalerite geobarometer. Geochem J Japan 22:129-131
Barker WW, Parks TC (1986) The thermodynamic properties of pyrrhotite and pyrite: a re-evaluation. Geochim Cosmochim Acta 50:2185-2194
Barnes HL (1973) Polymorphism and polytypism. Final Report U.S. Army Research Office (Durham) Contract No DAHCO4-69-C-0043, 38 p
Barstad J (1959) Phase relations in the system Ag-Sb-S at 400 °C. Acta Chem Scand 13:1703-1708
Barthelemy E, Carcaly C (1987) Phase relations and ageing effects in $Fe_{1-x}Co_xS$ system. J Solid State Chem 66:191-203
Barton PBJr (1969) Thermochemical study of the system Fe-As-S. Geochim Cosmochim Acta 33:841-857
Barton PBJr (1971) The Fe-Sb-S system. Econ Geol 66:121-132
Barton PBJr (1973) Solid solutions in system Cu-Fe-S. Part I: The Cu-S and CuFe-S joins. Econ Geol 68:455-465
Barton PBJr, Toulmin PIII (1964a) The electrum-tarnish method for the determination of the fugacity of sulfur in laboratory sulfide systems. Geochim Cosmochim Acta 28:619-640
Barton PBJr, Toulmin PIII (1964b) Experimental determination of the reaction chalcopyrite + sulfur = pyrite + bornite from 350 °C to 500 °C. Econ Geol 59:747-752
Barton PBJr, Toulmin PIII (1966) Phase relations involving sphalerite in the Fe-Zn-S system. Econ Geol 61:815-849
Baumann IH (1964) Patronite VS_4 und die Mineral-Paragenese der bituminosen schiefes von Minasragra Peru. Neues Jahrb Mineral, Abhdl 101:97-108
Bayliss P (1968) The crystal structure of disordered gersdorffite. Am Mineral 53:290-293
Bayliss P (1969) Isomorphous substitution in synthetic cobaltite and ullmannite. Am Mineral 54:426-430
Bayliss P (1989) Crystal chemistry and crystallography of some minerals within the pyrite group. Am Mineral 74:1168-1176
Beglaryan ML, Abrikasov HK (1959) The Bi_2Te_3-Bi_2Se_3 system. Dokl Akad Nauk USSR 129:135
Bell PM, Kullerud G (1970) High pressure differential thermal analysis. Carnegie Inst Wash, Year Book 68:276-277
Bell PM, El Goresy A, England JL, Kullerud G (1970) Pressure-temperature diagram for Cr_2FeS_4. Carnegie Inst Wash, Year Book 68:277-278
Bell RE, Herfert RE (1957) Preparation and characterization of a new crystalline form of molybdenum disulfide. J Am Chem Soc 79:3351-3355
Berner RA (1962) Tetragonal FeS a new iron sulfide. Science 137:669
Berner RA (1964) Iron sulfides formed from aqueous solution at low temperatures and atmospheric pressure. J Geol 72:293-306
Berry LG (1954) The crystal structure of covellite CuS and klockmannite CuSe. Am Mineral 39:504-509
Bertaut EF (1953) Contribution à l'étude des structures lacunaires: la pyrrhotine. Acta Crystallogr 6:557-561
Bertaut EF (1956) Structure de FeS stoechiométrique. Bulletin Societe Francais Minéralogie et Cristallographie 79:276-292
Bethke PM, Barton PBJr (1971) Subsolidus relations in the system PbS-CdS. Am Mineral 56:2034-2039
Bethke PM, Barton PBJr (1961) Unit-cell dimension versus composition in the systems PbS-CdS, PbS-PbSe, ZnS-ZnSe, and $CuFeS_{190}$-$CuFeSe_{190}$. U.S. Geol Surv Prof Paper 429B:266-270
Bloem J, Kroger FA (1956) The P-T-X phase diagram of the lead-sulfur system. Z Physikalische Chemie (München) 7:1-14
Boehler R (1992) Melting of the Fe-FeO and the Fe-FeS systems at high pressure: Constraints on core temperatures. Earth Planet Sci Lett 111:217-227
Boer KW, Nalesnik NJ (1969) Semiconductivity of CdS as a function of S-vapor pressure during heat treatment between 500 ° and 700 °C. Mat Res Bull 4:153-160
Boorman RS (1967) Subsolidus studies in the ZnS-FeS-FeS_2 system. Econ Geol 62:614-631
Boorman RS, Sutherland JK, Chernyshev LV (1971) New data on the sphalerite-pyrrhotite-pyrite solvus. Econ Geol 66:670-675
Bortnikov NS, Nekrasov IYa, Mozgova NN, Tsepin AI (1981) Phases and phase relations in the central portion of the system Fe-Pb-Sb-S between 300 and 500 °C in relation to lead-antimony sulphosalts. Neues Jahrb Mineral, Abhdl 143:37-60

Bouchard RJ (1968) The preparation of pyrite solid solutions of the type $Fe_xCo_{1-x}S_2$, $Co_xNi_{1-x}S_2$ and $Cu_xNi_{1-x}S_2$. Mat Res Bull 3:563-570
Brett PR (1964) Experimental data from the system Cu-Fe-S and their bearing on exsolution textures in ores. Econ Geol 59:1241-1269
Brett PR, Kullerud G (1967) The Fe-Pb-S system. Econ Geol 62:354-369
Brower WS, Parker HS, Roth RS (1974) Reexamination of synthetic parkerite and shandite. Am Mineral 59: 296-301
Bryndzia LT, Scott SD, Spry PG (1988) Sphalerite and hexagonal pyrrhotite geobarometer: Experimental calibration and application to the metamorphosed sulfide ores of Broken Hill Australia. Econ Geol 83: 1193-1204
Bryndzia LT, Scott SD, Spry PG (1990) Sphalerite and hexagonal pyrrhotite geobarometer: Correction in calibration and application. Econ Geol 85:408-411
Buerger JJ, Buerger MH (1944) Low chalcocite. Am Mineral 29:55-65
Buhlmann E (1965) Untersuchungen im System Cu-Bi-S. PhD Thesis, University of Heidelberg, Germany
Buhlmann E (1971) Untersuchungen im System Bi_2S_3-Cu_2S und geologische Schlussfolgerungen. Neues Jahrb Mineral, Monat 137-141
Bunch TE, Fuchs LH (1969) A new mineral: brezinaite Cr_3S_4 and the Tuscon meteorite Am Mineral 54:1503-1518
Burgmann W, Urbain G, Frohberg MG (1968) Iron-sulfur system in the iron monosulfide (pyrrhotite) region. Memoires Scientifiques de la Revue de Metallurgie 65:567-578
Burkart-Baumann I, Ottemann J, Amstutz GC (1972) The X-ray amorphous sulfides from Cerro de Pasco Peru and the crystalline inclusions. Neues Jahrb Mineral, Monat 433-446
Cabri LJ (1967) A new copper-iron sulfide. Econ Geol 62:910-925
Cabri LJ (1972) The mineralogy of the platinum group elements. Mineral Sci Engin 4:3-29
Cabri LJ (1973) New data on phase relations in the Cu-Fe-S system. Econ Geol 68:443-454
Cabri LJ, Hall SR (1972) Mooihoekite and haycockite two new copper-iron sulfides and their relationship to chalcopyrite and talnakhite. Am Mineral 57:689-708
Cabri LJ, Harris DC (1971) New compositional data on talnakhite. Econ Geol 66:673-675
Cabri LJ, Harris DC, Stewart JM (1970a) Costibite (CoSbS) a new mineral from Broken Hill New South Wales. Am Mineral 55:10-17
Cabri LJ, Harris DC, Stewart JM (1970b) Paracostibite (CoSbS) and nisbite ($NiSb_2$) new minerals from the Red Lake area Ontario Canada. Can Mineral 10:232-246
Cambi L, Elli M (1965) Hydrothermal processes: III Synthesis of thiosalts derived from silver and antimony sulfides. Chimica e l'Industria (Milan) 47:282-290
Canneri G, Fernandes L (1925) Contributo allo studio di alcuni minerali contenenti tallio Analise ternuca dei sistemi Tl_2S-As_2S_3 Tl_2S-PbS. Atti della Accademia Nazionale dei Lincei Classe di Scienze Fisiche Matematiche e Naturali Rendiconti 1:671-676
Carpenter RH, Desborough GA (1964) Range in solid solution and structure of naturally occurring troilite and pyrrhotite. Am Mineral 49:1350-1365
Caye R, Cervelle B, Cesbron F, Oudin E, Picot P, Pillard F (1988) Isocubanite a new definition of the cubic polymorph of cubanite $CuFe_2S_3$. Mineral Mag 52:509-514
Chang LLY (1963) Dimorphic relations in Ag_3SbS_3. Am Mineral 48:429-432
Chang LLY, Bever JE (1973) Lead sulphosalt minerals: Crystal structures stability relations and paragenesis. Mineral Sci Engin 5:181-191
Chang LLY, Knowles CR (1977) Phase relations in the systems PbS-Fe_{1-x}S-Sb_2S_3 and PbS-Fe_{1-x}S-Bi_2S_3. Can Mineral 15:374-379
Chang LLY, Wu D, Knowles CR (1988) Phase relations in the system Ag_2S-Cu_2S-PbS-Bi_2S_3. Econ Geol 83: 405-418
Chernyshev LV, Anfilogov VN (1968) Subsolidus phase relations in the ZnS-FeS-FeS_2 system. Econ Geol 63: 841-844
Chevreton M, Sapet A (1965) Structure de V_3S_4 et de quelques sulfures ternaires isotypes. Comptes Rendus de l'Académie des Sciences, Paris 261:928-930
Chuang Y-Y, Hsieh K-C, Chang YA (1985) Thermodynamics and phase relationships of transition metal-sulfur systems: Part V A reevaluation of the Fe-S system using an associated solution model for the liquid phase. Metall Trans 16B:277-285
Clark AH (1966a) Heating experiments on gudmundite. Mineral Mag 35:1123-1125
Clark AH (1966b) Stability field of monoclinic pyrrhotite. Trans, Inst Mining Metall (London) B75:B232-B235
Clark AH (1970a) An occurrence of the assemblage native sulfur-covellite-"$Cu_{55x}Fe_xS_{65x}$", Aucanquilcha, Chile. Am Mineral 55:913-918
Clark AH (1970b) Compositional differences between hexagonal and rhombohedral molybdenite. Neues Jahrb Mineral, Monat 1:33-38

Clark AH (1970c) Cuprian sphalerite and a probable copper zinc sulfide, Cachiyuyo de Llampos, Copiapo, Chile. Am Mineral 55:1021-1025
Clark AH (1970d) Supergene metastibnite from Mina Alacrán, Pampa Larga, Copiapo, Chile. Am Mineral 55: 2104-2106
Clark AH (1971) Molybdenite $2H_1$ molybdenite 3R and jordisite from Carrizal Alto, Atacama, Chile. Am Mineral 56:1832-1835
Clark AH (1974) Hypogene and supergene cobalt-copper sulfides, Carrizal Alto, Atacama, Chile. Am Mineral 59:302-306
Clark AH, Sillitoe RH (1971) Cuprian galena solid solutions Zapallar Mining district, Atacama, Chile. Am Mineral 56:2142-2145
Clark LA (1960a) The Fe-As-S system: Phase relations and applications. Econ Geol 55:1345-1381, 1631-1652
Clark LA (1960b) The Fe-As-S system: Variations of arsenopyrite composition as functions of T and P. Carnegie Inst Wash, Year Book 59:127-130
Clark LA, Kullerud G (1963) The sulfur-rich portion of the Fe-Ni-S system. Econ Geol 58:853-885
Coey JMD, Roux-Buisson H (1979) Electronic properties of $(Ni_{1-x}Fe_x)S$ solid solutions. Mat Res Bull 14:711-716
Collin G, Gardette MF, Comes R (1987) The $Fe_{1-x}Co_xS$ system (x < 025); transition and the high temperature phase. J Phys Chem Solids 48:791-802
Corrl JA (1964) Recovery of the high pressure phase of cadmium sulfide. J Appl Phys 35:3032-3033
Craig JR (1967) Phase relations and mineral assemblages in the Ag-Bi-Pb-S system. Mineralium Deposita 1: 278-306
Craig JR (1971) Violarite stability relations. Am Mineral 56:1303-1311
Craig JR (1973) Pyrite-pentlandite assemblages and other low temperature relations in the Fe-Ni-S system. Am J Sci 273-A:496-510
Craig JR, Kullerud G (1968) Phase relations and mineral assemblages in the copper-lead-sulfur system. Am Mineral 53:145-161
Craig JR, Kullerud G (1973) The Cu-Zn-S system. Mineralium Deposita 8:81-91
Craig JR, Scott SD (1974) Sulfide phase equilibria. Rev Mineral 1:CS-1–CS-110
Craig JR, Skinner BJ, Francis CA, Luce FD, Makovicky E (1974) Phase relations in the As-Sb-S system. Trans, Am Geophys Union 55:483
Cubicciotti D (1962) The bismuth-sulfur phase diagram. J Phys Chem 66:1205-1206
Cubicciotti D (1963) Thermodynamics of liquid bismuth and sulfur. J Phys Chem 67:118-123
Curlook W, Pidgeon LM (1953) The Co-Fe-S system. Can Mining Metall Bull 493(56):297-301
Czamanske GK (1969) The stability of argentopyrite and sternbergite. Econ Geol 64:459-461
Czamanske GK (1974) The FeS content of sphalerite along the chalcopyrite-pyrite-sulfur fugacity buffer. Econ Geol 69:1328-1334
Czamanske GK, Larson RR (1969) The chemical identity and formula of argentopyrite and sternbergite. Am Mineral 54:1198-1201
Delafosse D, Can Huang Van (1962) Étude du systeme Ni-Co-S a temperature elevee. Comptes Rendus de l'Académie des Sciences, Paris 254:1286-1288
Desgreniers S, Beaulieu L, Lepage I (2000) Pressure-induced structural change in ZnS. Phys Rev B 61:8726-8733
Dickson FW, Tunell G (1959) The stability relations of cinnabar and metacinnabar. Am Mineral 44:471-487
Dickson FW, Radtke AS, Weissberg BG, Heropoulos C (1974) Solid solutions of antimony arsenic and gold in stibnite (Sb_2S_3) orpiment (As_2S_3) and realgar (As_2S_2). Econ Geol 70:591-594
Djurle S (1958) An X-ray study of the system Ag-Cu-S. Acta Chem Scand 12:1427-1436
Dolanski J (1974) Sulvanite from Thorpe Hills, Utah. Am Mineral 59:307-313
D'sugi J, Shimizu K, Nakamura T, Onodera A (1966) High pressure transition in cadmium sulfide. Rev Phys Chem Japan 36:59-73
DuPreez JW (1945) A thermal investigation of the parkerite series. University of Stellenbosch 22:A 97-104
Einaudi MT (1968) Sphalerite-pyrrhotite-pyrite equilibria- A re-evaluation. Econ Geol 63:832-834
Einaudi MT (1987) Phase relations among silicates copper iron sulfides and aqueous solutions at magmatic temperatures-A discussion. Econ Geol 82:497-502
El Goresy A, Kullerud G (1969) Phase relations in the system Cr-Fe-S. In: Meteorite research D. Millman PM (ed) Reidel, p 638-656
Erd RC, Evans HT, Richter HD (1957) Smythite a new iron sulfide and associated pyrrhotite from Indiana. Am Mineral 42:309-333
Etschmann B, Pring A, Putnis A, Grguric BA, Studer A (2004) A kinetic study of the exsolution of pentlandite $(NiFe)_9S_8$ from the monosulfide solid solution (FeNi)S. Am Mineral 89:39-50
Evans HTJr (1971) Crystal structure of low chalcocite. Nature 232:69-70
Farrell SP, Fleet ME (2000) Evolution of local electronic structure in cubic $Mg_{1-x}Fe_xS$ by S K-edge XANES spectroscopy. Solid State Commun 113:69-72

Farrell SP, Fleet ME (2001) Sulfur K-edge XANES study of local electronic structure in ternary monosulfide solid solution [(Fe Co Ni)$_{0.923}$S]. Phys Chem Mineral 28:17-27
Farrell SP, Fleet ME (2002) Phase separation in (FeCo)$_{1-x}$S monosulfide solid solution below 450 °C with consequences for coexisting pyrrhotite and pentlandite in magmatic sulfide deposits. Can Mineral 40: 33-46
Farrell SP, Fleet ME, Stekhin IE, Soldatov A, Liu X (2001) Evolution of local electronic structure in alabandite and niningerite solid solutions [(MnFe)S (MgMn)S (MgFe)S] using sulfur K- and L-edge XANES spectroscopy. Am Mineral 87:1321-1332
Fei Y, Bertka CM, Finger LW (1997) High-pressure iron-sulfur compound Fe$_3$S$_2$ and melting relations in the Fe-FeS system. Science 275:1621-1623
Fei Y, Li J, Bertka CM, Prewitt CT (2000) Structure type and bulk modulus of Fe$_3$S a new iron-sulfur compound. Am Mineral 85:1830-1833
Fei Y, Prewitt CT, Mao HK, Berka CM (1995) Structure and density of FeS at high-pressure and high-temperature and the internal structure of Mars. Science 268:1892-1894
Filimonova AA, Evstigneeva TL, Laputina IP (1980) Putoranite and nickeloan putoranite- new minerals of the chalcopyrite group. Zapiski Vsesoyuznogo Mineralogicheskogo Obshchestva 109:335-341
Fleet ME (1968) On the lattice parameters and superstructures of pyrrhotites. Am Mineral 53:1846-1855
Fleet ME (1971) The crystal structure of a pyrrhotite (Fe$_7$S$_8$). Acta Crystallogr B27:1864-1867
Fleet ME (1972) The crystal structure of α-Ni$_7$S$_6$. Acta Crystallogr B28:1237-1241
Fleet ME (1973) The crystal structure of parkerite (Ni$_3$Bi$_2$S$_2$). Am Mineral 58:435-439
Fleet ME (1975) Thermodynamic properties of (ZnFe)S solid solutions at 850 °C. Am Mineral 60:466-470
Fleet ME (1977) The birefringence-structural state relation in natural zinc sulfides and its application to the schalenblende from Pribram. Can Mineral 15:303-308
Fleet ME (1982) Synthetic smythite and monoclinic Fe$_3$S$_4$. Phys Chem Mineral 8:241-246
Fleet ME (1987) Structure of godlevskite Ni$_9$S$_8$. Acta Crystallogr C43:2255-2257
Fleet ME (1988) Stoichiometry structure and twinning of godlevskite and synthetic low-temperature Ni-excess nickel sulphide. Can Mineral 26:283-291
Fleet ME, MacRae ND (1987) Sulfidation of Mg-rich olivine and the stability of niningerite in enstatite chondrites. Geochim Cosmochim Acta 51:1511-1521
Francis CA, Fleet ME, Misra KC, Craig JR (1976) Orientation of exsolved pentlandite in natural and synthetic nickeliferous pyrrhotite. Am Mineral 61:913-920
Franzen HF, Smeggil J, Conard BR (1967) The group IV di-transition metal sulfides and selenides. Mat Res Bull 2:1087-1092
Fujimori A, Namatame H, Matoba M, Anzai S (1990) Photoemission study of the metal-nonmetal transition in NiS. Phys Rev B42:620-623
Gaines RV (1957) Luzonite famatinite and some related minerals. Am Mineral 42:766-779
Gaines RV, Skinner HCW, Foord EE, Mason B, Rosenzweig A (1997) Dana's New Mineralogy. John Wiley
Gammon JB (1966) Some observations on minerals in the system CoAsS-FeAsS. Norsk Geol Tids 46:405-426
Geller S (1962) Refinement of the crystal structure of Co$_9$S$_8$. Acta Crystallogr 15:1195-1198
Goble RJ (1980) Copper sulfides from Alberta: yarrowite Cu$_9$S$_8$ and spionkopite Cu$_{39}$S$_{28}$. Can Mineral 18: 511-518
Goble RJ, Robinson G (1980) Geerite Cu$_{1.60}$S a new copper sulfide from Dekalb Township, New York. Can Mineral 18:519-523
Godovikov AA, Il'yasheva NA (1971) Phase diagram of a bismuth-bismuth sulfide-bismuth selenide system. Materialy po Geneticheskoi i Eksperimental'noi Mineralogii 6:5-14
Godovikov AA, Ptitsyn AB (1968) Syntheses of Cu-Bi-sulfides under hydrothermal conditions. Eksperimental'noi Issled Mineralogii 6:29-41
Godovikov AA, Kochetkova KV, Lavrent'ev YuG (1970) Bismuth sulfotellurides from the Sokhondo deposit. Geologiya i Geofizika 11:123-127
Goodenough JB (1967) Description of transition metal compounds: application to several sulfides. *In*: Propértés thermodynamiques physiques et structurales des dérivés semi-metalliques. Paris: Centre National de la Recherche Scientifique, p 263-292
Gopalakrishnan J, Murugesan T, Hegde MS, Rao CNR (1979) Study of transition-metal monosulphides by photoelectron spectroscopy. J Phys C: Solid State Phys 12:5255-5261
Gosselin JR, Townsend MG, Tremblay RJ (1976a) Electric anomalies at the phase transition in iron monosulfide. Solid State Commun 19:799-803
Gosselin JR, Townsend MG, Tremblay RJ, Webster AH (1976b) Mössbauer effect in single-crystal Fe$_{1-x}$S. J Solid State Chem 17:43-48
Graeser S (1964) Über Funde der neven rhombohedrichan MoS$_2$ - Modifikation (Molybdanit - 3R) und von Tungstenit in den Alpen Schweiz. Mineral Petrogr Mitt 44:121-128
Graeser S (1967) Ein Vorkommen von Lorandite (TlAsS$_2$) in der Schweiz. Contrib Mineral Petrol 16:45-50

Graf R B (1968) The system Ag_3AuS_2-Ag_2S. Am Mineral 53:496-500
Grguric BA, Putnis A (1998) Compositional controls on phase transition temperatures in bornite: a differential scanning calorimetry study. Can Mineral 36:215-227
Grguric BA, Putnis A (1999) Rapid exsolution behaviour in the bornite-digenite series: implications for natural ore assemblages. Mineral Mag 63:1-14
Grguric BA, Harrison RJ, Putnis P (2000) A revised phase diagram for the bornite-digenite join from in situ neutron diffraction and DSC experiments. Mineral Mag 64:213-231
Grguric BA, Putnis A, Harrison RJ (1998) An investigation of the phase transitions in bornite (Cu_5FeS_4) using neutron diffraction and differential scanning calorimetry. Am Mineral 83:1231-1239
Grønvold F, Haraldsen H (1952) The phase relations of synthetic and natural pyrrhotites ($Fe_{1-x}S$). Acta Chem Scand 6:1452-1469
Grønvold F, Westrum EF (1980) The anilite/low-digenite transition. Am Mineral 65:574-575
Grønvold F, Haraldsen H, Kjekshus A (1960) On the sulfides selenides and tellurides of platinum. Acta Chem Scand 14:1879-1893
Grover B, Moh GH (1966) Experimentelle untersuchungen des quaternaren systems Kupfer-Eisen Molybdan-Schwefel. Referat 44 Jahrestagung der Dentschen Mineralogischen Gesellschaft, München
Grover B, Moh GH (1969) Phasen gleichgewichtsbeziehungen im system Cu-Mo-S in Relation zu naturlichen Mineralien. Neues Jahrb Mineral, Monat 529-544
Grover B, Kullerud G, Moh GH (1975) Phase equilibrium conditions in the ternary iron-molybdenum-sulfur system in relation to natural minerals and ore deposits. Neues Jahrb Mineral, Abhdl 124:246-272
Guggenheim EA (1959) Thermodynamics 4th ed. North-Holland
Hahn H, Harder B, Brockmuller W (1965) Untersuchungen der ternären Chalcogenide X Versuche zur Umsetzung von Titan sulfiden mit sulfiden Zweiwretiger Ubergangsmetalle. Z. Anorganische und Allgemeine Chemie 288:260-268
Hall HT (1966) The systems Ag-Sb-S Ag-As-S and Ag-Bi-S: Phase relations and mineralogical significance. PhD Dissertation, Brown University
Hall HT (1968) Synthesis of two new silver sulfosalts. Econ Geol 63:289-291
Hall SR, Gabe EJ (1972) The crystal structure of talnakhite $Cu_{18}Fe_{16}S_{32}$. Am Mineral 57:368-380
Hansen M, Anderko K (1958) Constitution of Binary Alloys. McGraw-Hill
Haraldsen H (1941) The high-temperature transformations of iron(II) sulphide mixed crystals. Z Anorganische und Allgemeine Chemie 246:195-226
Hayase K (1955) Minerals of bismuthinite-stibnite series with special reference to horobetsuite from the Horobetsu Mine. Mineral J Japan 1:188-197
Hiller JE, Probsthain K (1956) Thermische und röntgenographische Untersuchungen am Kupferkies. Z Kristallogr 108:108-129
Hobbs D, Hafner J (1999) Magnetism and magneto-structural effects in transition-metal sulphides. J Phys: Cond Matt 11:8197-8222
Horwood JL, Townsend MG, Webster AH (1976) Magnetic susceptibility of single-crystal $Fe_{1-x}S$. J Solid State Chem 17:35-42
Hsieh K-C, Chang YA (1987) Thermochemical description of the ternary iron-nickel-sulfur system. Can Metall Quart 26:311-327
Hurlbut CS (1957) The wurtzite-greenockite series. Am Mineral 42:184-190
Hutcheon I (1978) Calculation of metamorphic pressure using the sphalerite-pyrrhotite-pyrite equilibrium. Am Mineral 63:87-95
Hutcheon I (1980) Calculated phase relations for pyrite-pyrrhotite-sphalerite: correction. Am Mineral 65:1063-1064
Hutchison MN, Scott SD (1983) Experimental calibration of the sphalerite cosmobarometer. Geochim Cosmochim Acta 47:101-108
Inan EE, Einaudi MT (2002) Nukundamite ($Cu_{3.38}Fe_{0.62}S_4$)-bearing copper ore in the Bingham porphyry deposit Utah: Result of upflow through quartzite. Econ Geol 97:499-515
Janosi A (1964) La structure du sulfure cuivreux quadratiques. Acta Crystallogr 17:311-312
Jellinek F (1957) The structures of the chromium sulphides. Acta Crystallogr 10:620-628
Jensen E (1942) Pyrrhotite: Melting relations and composition. Am J Sci 240:695-709
Jensen E (1947) Melting relations of chalcocite. Norske Videnskaps– Akademi Oslo Avhandl I Mat-Naturw Klasse 6, 14 p
Kajiwara Y (1969) Fukuchilite Cu_3FeS_8 a new mineral from the Hanawa mine Akita prefecture. Mineral J Japan 5:399-416
Kanazawa Y, Koto K, Morimoto N (1978) Bornite (Cu_5FeS_4): stability and crystal structure of the intermediate form. Can Mineral 16:397-404
Karakhanova MI, Pushinkin AS, Novoselova AV (1966) Phase diagram of the tin-sulfur system. Izvestiya Akademii Nauk USSR Neorganicheskie Materialy 2:991-996

Karunakaran C, Vijayakumar V, Vaidya SN, Kunte NS, Suryanarayana S (1980) Effect of pressure on electrical resistivity and thermoelectric power of FeS. Mat Res Bull 15:201-206

Karup-Møller S (1972) New data on pavonite gustavite and some related sulphosalt minerals. Neues Jahrb Mineral, Monat 19-38

Karup-Møller S (1985) The system Pd-Co-S at 1000 800 600 and 400 °C. Fourth Internat Platinum Symp, Can Mineral 23:306

Karup-Møller S, Makovicky E (1974) Skinnerite Cu_3SbS_3 a new sulfosalt from the Ilimaussag alkaline intrusion, South Greenland. Am Mineral 59:889-895

Karup-Møller S, Makovicky E (1995) The phase system Fe-Ni-S at 725 °C. Neues Jahrb Mineral, Monat 1-10

Karup-Møller S, Makovicky E (1998) The phase system Fe-Ni-S at 900 °C. Neues Jahrb Mineral, Monat 373-384

Karup-Møller S, Makovicky E (2002) The system Fe-Os-S at 1180 °, 1100 ° and 900 °C. Can Mineral 40: 499-507

Kato A (1959) Ikunolite a new bismuth mineral from the Ikuno Mine, Japan. Mineral J Japan 2:397-407

Kavner A, Duffy TS, Shen G (2001) Phase stability and density of FeS at high pressures and temperatures: implications for the interior structure of Mars. Earth Planet Sci Lett 185:25-33

Keighin CW, Honea RM (1969) System Ag-Sb-S from 600 to 200 °C. Mineralium Deposita 4:153-171

Keil K, Brett R (1974) Heiderite $(FeCr)_{1+x}(TiFe)_2S_4$ a new mineral in the Bustee enstatite achondrite. Am Mineral 59:465-470

Keil K, Snetsinger KG (1967) Niningerite a new meteorite sulfide. Science 155:451-453

Keller-Besrest F, Collin G (1990a) Structural aspects of the α transition in off-stoichiometric iron sulfide $(Fe_{1-x}S)$ crystals. J Solid State Chem 84:211-225

Keller-Besrest F, Collin G (1990b) Structural aspects of the α transition in stoichiometric iron sulfide (FeS): identification of the high-temperature phase. J Solid State Chem 84:211-225

King HEJr, Prewitt CT (1982) High-pressure and high-temperature polymorphism of iron sulfide (FeS). Acta Crystallogr B38:1877-1887

Kirkinskiy VA, Ryaposov AP, Yakushev VG (1967) Phase diagram for arsenic tri-sulfide up to 20 kilobars. Isvest Akad Nauk USSR 3:1931-1933

Kissin, SA (1974) Phase relations in a portion of the Fe-S system. PhD Thesis, University of Toronto

Kissin SA, Scott SD (1982) Phase relations involving pyrrhotite below 350 °C. Econ Geol 77:1739-1754

Kitakaze A, Sugaki A (2004) The phase relations between $Fe_{45}Ni_{45}S_8$ and Co_9S_8 in the system Fe-Ni-Co-S at temperatures from 400 ° to 1100 °C. Can Mineral 42:17-42

Kitakaze A, Sugaki A, Shima H (1995) Study of the minerals on the $PbS-Sb_2S_3$ join; Part 1, Phase relations above 400 °C. Mineral J Japan 17:282-289

Klemm DD (1962) Untersuchungen Über die Mischkristall bildung im Dreieck diagramm FeS_2-CoS_2-NiS_2 und ihre Beziehungen zum Aufbau der naturlichen "Bravoite". Neues Jahrb Mineral, Monat 76-91

Klemm DD (1965) Synthesen und Analysen in den Dreickdiagrammen FeAsS-CoAsS-NiAsS und FeS_2-CoS_2-NiS_2. Neues Jahrb Mineral, Abhdl 103:205-255

Klominsky J, Rieder M, Kieft C, L Mraz (1971) Heyrovskyite $6(Pb_{0.86}Bi_{0.08}(AgCu)_{0.04})SBi_2S_3$ from Hurky, Czechoslovakia; new mineral of genetic interest. Mineralium Deposita 6:133-147

Knop O, Ibrahim MA (1961) Chalcogenides of the transition elements. II. Existence of the π phase in the M_9S_8 section of the system Fe-Co-Ni-S. Can J Chem 39:297-317

Kobayashi H, Kamimura T, Ohishi Y, Takeshita N, Mori N (2005) Structural and electrical properties of stoichiometric FeS compounds under high pressure at low temperature. Phys Rev B71:014110/1-014110/7

Kobayashi H, Sato M, Kamimura T, Sakai M, Onodera H, Kuroda N, Yamaguchi Y (1997) The effect of pressure on the electronic states of FeS and Fe_7S_8 studied by Mössbauer spectroscopy. J Phys: Cond Matt 9:515-527

Kojima S, Sugaki A (1984) Phase relations in the central portion of the Cu-Fe-Zn-S system between 800 ° and 500 °C. Mineral J Japan 12:15-28

Kojima S, Sugaki A (1985) Phase relations in the Cu-Fe-Zn-S system between 500 ° and 300 °C under hydrothermal conditions. Econ Geol 80:158-171

Kracek FC (1946) Phase relations in the system sulfur-silver and the transitions in silver sulfide. Trans, Am Geophys Union 27:364-374

Kraft A, Stiller H, Vollstädt H (1982) The monosulfide solid solution in the Fe-Ni-S system; relationship to the Earth's core on the basis of experimental high-pressure investigations. Phys Earth Planet Int 27:255-262

Krestovnikov AN, Mendelevich AY, Glazov VM (1968) Phase equilibria in the Cu_2S-Ag_2S system. Inorganic Mat 4:1047-1048

Kretschmar U, Scott SD (1976) Phase relations involving arsenopyrite in the system Fe-As-S and their application. Can Mineral 14:364-386

Krupp RE (1994) Phase relations and phase transformations between the low-temperature iron sulfides mackinawite, greigite and smythite. Eur J Mineral 6:265-278

Kruse O (1990) Mössbauer and X-ray study of the effects of vacancy concentration in synthetic hexagonal pyrrhotites. Am Mineral 75:755-763

Kruse O (1992) Phase transitions and kinetics in natural FeS measured by X-ray diffraction and Mössbauer spectroscopy at elevated temperatures. Am Mineral 77:391-398

Kullerud G (1953) The FeS-ZnS system a geological thermometer. Norsk Geol Tids 32:61-147

Kullerud G (1961) Two-liquid field in the Fe-S system. Carnegie Inst Wash, Year Book 60:174-176

Kullerud G (1963) Thermal stability of pentlandite. Can Mineral 7:353-366

Kullerud G (1965a) Covellite stability relations in the Cu-S system. Freiberger Forschungshefte B 186C:145-160

Kullerud G (1965b) The mercury-sulfur system. Carnegie Inst Wash, Year Book 64:193-195

Kullerud G (1967a) The Fe-Mo-S system. Carnegie Inst Wash, Year Book 65:337-342

Kullerud G (1967b) Sulfide studies. *In*: Researches in Geochemistry, Vol 2. Abelson PH (ed) John Wiley and Sons, p 286-321

Kullerud G (1969) The lead-sulfur system. Am J Sci 267-A:233-267

Kullerud G, Yoder HS (1959) Pyrite stability relations in the Fe-S system. Econ Geol 54:533-572

Kullerud G, Yund RA (1962) The Ni-S system and related minerals. J Petrol 3:126-175

Kullerud G, Yund RA, Moh GH (1969) Phase relations in the Cu-Fe-S, Cu-Ni-S and Fe-Ni-S systems. *In*: Magmatic ore deposits. Wilson HDB (ed) Econ Geol Mono 4, p 323-343

Kusaba K, Syono Y, Kikegawa T, Simomura O (1997) Structure of FeS under high pressure. J Phys Chem Solids 58:241-246

Kusaba K, Syono Y, Kikegawa T, Simomura O (1998) Structures and phase equilibria of FeS under high pressure and temperature. *In*: Properties of Earth and Planetary Materials at High Pressure and Temperature. AGU, Washington DC, p 297-305

Kutoglu A (1969) Rönt genographicsche und thermische Untersuchungen in Quasibinaren system $PbS-As_2S_3$. Neues Jahrb Mineral, Monat 68-72

Kuznetsov VG, Sokolova MA, Palkina KK, Popova ZV (1965) System cobalt-sulfur. Izvest Akad Nauk USSR Neorgan Mat 1:675-689

Lamprecht G (1978) Phase relations within the Co-Ni-S system below 1000 °C. Neues Jahrb Mineral, Monat 176-191

Lamprecht G, Hanus D (1973) Phase equilibrium relations in the ternary system Co-Ni-S. Neues Jahrb Mineral, Monat 236-240

Lange W, Schlegel H (1951) Die Zustandbilder des Systeme Eisen-Antimon-Schwefel und Kobalt-Antimon-Schwefel. Z Metall 42:257-268

Larimer JW (1968) An experimental investigation of oldhamite CaS and the petrologic significance of oldhamite in meteorites. Geochim Cosmochim Acta 32:965-982

Lawson AC (1972) Lattice instabilities in superconducting ternary molybdenum sulfides. Mat Res Bull 7:773-776

Le Bihan M-Th (1963) Étude structurale de quelques sulfures de plomb et d'arsenic naturels du gisement de Binn. Mineral Soc Am Spec Paper 1:149-152

Learned RE (1966) The solubilities of quartz quartz-cinnabar and cinnabar-stibnite in sodium sulfide solutions and their implications for ore genesis. PhD Thesis, University of California, Riverside

Learned RE, Tunnell G, Dickson FW (1974) Equilibria of cinnabar stibnite and saturated solutions in the system $HgS-Sb_2S_3-Na_2S-H_2O$ from 150 ° to 250 °C at 100 bars with implications concerning ore genesis. J Res U.S. Geol Surv 2:457-466

Lennie AR, Vaughan DJ (1996) Spectroscopic studies of iron sulfide formation and phase relations at low temperatures. *In*: Mineral spectroscopy: A tribute to Roger G Burns. Spec Publ Geochem Soc 5:117-131

Lennie AR, England KER, Vaughan DJ (1995a) Transformation of synthetic mackinawite to hexagonal pyrrhotite: A kinetic study. Am Mineral 80:960-967

Lennie AR, Redfern SAT, Champness PE, Stoddart CP, Schofield PF, Vaughan DJ (1997) Transformation of mackinawite to greigite; an in situ X-ray powder diffraction and transmission electron microscope study. Am Mineral 82:302-309

Lennie AR, Redfern SAT, Schofield PF, Vaughan DJ (1995b) Synthesis and Rietveld crystal structure refinement of mackinawite tetragonal FeS. Mineral Mag 59:677-683

Leonard BF, Desborough GA, Page NJ (1969) Ore microscopy and chemical composition of some lavrites. Am Mineral 54:1330-1346

Lewis GN, Randall M (1961) Thermodynamics, 2nd ed. McGraw-Hill

Li F, Franzen HF (1996) Ordering, incommensuration, and phase transitions in pyrrhotite, Part II: A high-temperature X-ray powder diffraction and thermomagnetic study. J Solid State Chem 126:108-120

Li F, Franzen HF, Kramer MJ (1996) Ordering, incommensuration, and phase transitions in pyrrhotite, Part I: A TEM study of Fe_7S_8. J Solid State Chem 124:264-271

Lin J-F, Fei Y, Sturhahn W, Zhao J, Mao H, Hemley RJ (2004) Magnetic transition and sound velocities of Fe_3S at high pressure: implications for Earth and planetary cores. Earth Planet Sci Lett 226:33-40

Lin RY Hu DC, Chang YA (1978) Thermodynamics and phase relations of transition metal-sulfur systems: II The nickel-sulfur system. Metall Trans B 9B 531-538
Liu H, Chang LLY (1994) Phase relations in the system PbS-PbSe-PbTe. Mineral Mag 58:567-578
Liu M, Fleet ME (2001) Partitioning of siderophile elements (W Mo As Ag Ge Ga and Sn) and Si in the Fe-S system and their fractionation in iron meteorites. Geochim Cosmochim Acta 65:671-682
Lundqvist D (1947) X-ray studies in the ternary system Fe-Ni-S. Arkiv Kemi Mineral Geol 24A, No 22, 12p
Lusk J, Bray DM (2002) Phase relations and the electrochemical determination of sulfur fugacity for selected reactions in the Cu-Fe-S and Fe-S systems at 1 bar and temperatures between 185 and 460 °C. Chem Geol 192:227-248
Lusk J, Calder BOE (2004) The composition of sphalerite and associated sulfides in reactions of the Cu-Fe-Zn-S, Fe-Zn-S and Cu-Fe-S systems at 1 bar and temperatures between 250 and 535 °C. Chem Geol 203:319-345
Lusk J, Ford CE (1978) Experimental extension of the sphalerite geobarometer to 10 kbar. Am Mineral 63:516-519
Lusk J, Scott SD, Ford CE (1993) Phase relations in the Fe-Zn-S system to 5 kbars and temperatures between 325 ° and 150 °C. Econ Geol 88:1880-1903
MacLean WH, Cabri LJ, Gill JE (1972) Exsolution products in heated chalcopyrite. Can J Earth Sci 9:1305-1317
Makovicky E (2002) Ternary and quaternary phase systems with PGE. *In*: The geology geochemistry, mineralogy and mineral beneficiation of platinum-group elements. CIM Special Volume 54. Cabri LJ (ed) Can Inst Mining Metall Petroleum, p 131-175
Mandziuk ZL, Scott SD (1977) Synthesis, stability and phase relations of argentian pentlandite in the system Ag-Fe-Ni-S. Can Mineral 15:349-364
Marfunin AS (1979) Physics of Minerals and Inorganic Compounds. Springer-Verlag
Mariano AN, Warekois EP (1963) High pressure phases of some compounds of groups II-VI. Science 142:672-673
Markham NL (1962) Plumbian ikunolite from Kingsgate, New South Wales. Am Mineral 47:1431-1434
Marshall WG, Nelmes RJ, Loveday JS, Klotz S, Besson JM, Hamel G, Parise JB (2000) High-pressure diffraction study of FeS. Phys Rev B61:11201-11204
Martín JD, Gil ASI (2005) An integrated thermodynamic mixing model for sphalerite geobarometry from 300 to 850 °C and up to 1 GPa. Geochim Cosmochim Acta 69995-1006
Martin P, Price GD, Vočadlo L (2001) An *ab initio* study of the relative stabilities and equations of state of FeS polymorphs. Mineral Mag 5:181-191
Maske S, Skinner BJ (1971) Studies of the sulfosalts of copper. I. Phases and phase relations in the system Cu-As-S. Econ Geol 66:901-918
Matoba M, Anzai S, Fujimori A (1994) Effect of early transition-metal (M=Ti V and Cr) doping on the electronic structure of charge-transfer type compound NiS studied by thermoelectric power and X-ray photoemission measurements. J Phys Soc Japan 63:1429-1440
Mavrogenes JA, MacIntosh IW, Ellis DJ (2001) Partial melting of the Broken Hill galena-sphalerite ore: experimental studies in the system PbS-Fe-S-Zn-S-(Ag_2S). Econ Geol 96:205-210
McCammon CA, Price DC (1982) A Mössbauer effect investigation of the magnetic behaviour of (Fe Co)S_{1+x} solid solutions. J Phys Chem Solids 43:431-437
McKenzie WF, Helgeson HC (1985) Phase relations among silicates copper iron sulfides and aqueous solutions at magmatic temperatures. Econ Geol 80:1965-1973
McKenzie WF, Helgeson HC (1987) Phase relations among silicates copper iron sulfides and aqueous solutions at magmatic temperatures – a reply. Econ Geol 82:501-502
Merwin HE, Lombard RH (1937) The system Cu-Fe-S. Econ Geol 32:203-284
Miller RO, Dachille F, Roy R (1966) High pressure phase equilibrium studies of CdS and MnS by static and dynamic methods. J Appl Phys 37:4913-4918
Misra KC, Fleet ME (1973) The chemical compositions of synthetic and natural pentlandite assemblages. Econ Geol 68:518-539
Misra KC, Fleet ME (1974) Chemical composition and stability of violarite. Econ Geol 69:391-403
Moh GH (1964) Blaubleibender covellite. Carnegie Inst Wash, Year Book 63:208-209
Moh GH (1969) The tin-sulfur system and related minerals. Neues Jahrb Mineral, Abhdl 111:227-263
Moh GH (1971) Blue-remaining covellite and its relations to phases in the sulfur-rich portion of the copper sulfur system at low temperatures. Mineral Soc Japan Spec Paper 1:226-232
Moh GH (1973) Das Cu-W-S system und seine Mineralien sowie ein neues Tungstenit vorkommen in Kipushi/Katanga. Mineralium Deposita 8:291-300
Moh GH (1975) Tin-containing mineral systems: Part II. Phase relations and mineral assemblages in the Cu-Fe-Zn-Sn-S system. Chemie der Erde 34:1-61
Moh GH, Berndt F (1964) Two new natural tin sulfides Sn_2S_3 and SnS_2. Neues Jahrb Mineral, Monat 94-95

Moore DE, Dixon FW (1973) Phases of the system Sb_2S_3-As_2S_3. Trans, Am Geophys Union 54:1223-1224
Mootz D, Puhl H (1967) Crystal structure of Sn_2S_3. Acta Crystallogr 23:471-476
Morimoto N (1962) Djurleite, a new copper sulphide mineral. Mineral J Japan 3:338-344
Morimoto N (1970) Crystal-chemical studies on the Cu-Fe-S system. *In*: Volcanism and ore genesis. Tatsumi T (ed) University of Tokyo Press
Morimoto N, Clark LA (1961) Arsenopyrite crystal-chemical relations. Am Mineral 46:1448-1469
Morimoto N, Gyobu A (1971) The composition and stability of digenite. Am Mineral 56:1889-1909
Morimoto N, Koto K (1970) Phase relations of the Cu-S system at low temperatures: stability of anilite. Am Mineral 55:106-117
Morimoto N, Kullerud G (1962) The Mo-S system. Carnegie Inst Wash, Year Book 61:143-144
Morimoto N, Kullerud G (1963) Polymorphism in digenite. Am Mineral 48:110-123
Morimoto N, Kullerud G (1966) Polymorphism on the Cu_9S_5-Cu_5FeS_4 join. Z Kristallogr 123:235-254
Morimoto N, Gyobu A, Mukaiyama H, Izawa E (1975a) Crystallography and stability of pyrrhotites. Econ Geol 70:824-833
Morimoto N, Gyobu A, Tsukuma K, Koto K (1975b) Superstructure and nonstoichiometry of intermediate pyrrhotite. Am Mineral 60:240-248
Morimoto N, Koto K, Shimazaki Y (1969) Anilite Cu_7S_5 a new mineral. Am Mineral 54:1256-1268
Mukaiyama H, Izawa E (1970) Phase relations in the Cu-Fe-S system the copper-deficient part. *In*: Volcanism and Ore Genesis. Tatsumi T (ed) University of Tokyo Press, p 339-355
Mumme WG, Sparrow GJ, Walker GS (1988) Roxbyite a new copper sulphide mineral from the Olympic Dam deposit, Roxby Downs, South Australia. Mineral Mag 52:323-330
Munson RA (1966) Synthesis of copper disulfide. Inorg Chem 5:1296-1297
Murowchick JB, Barnes HL (1986a) Formation of cubic FeS. Am Mineral 71:1243-1246
Murowchick JB, Barnes HL (1986b) Marcasite precipitation from hydrothermal solutions. Geochim Cosmochim Acta 50:2615-2629
Nagao I (1955) Chemical studies on the hot springs of Nasu. Nippon Kageke Zasshi 76:1071-1073
Nakamura M, Fujimori A, Sacchi M, Fuggle JC, Misu A, Mamori T, Tamura H, Matoba M, Anzai S (1993) Metal-nonmetal transition in NiS induced by Fe and Co substitution: X-ray-absorption spectroscopic study. Phys Rev B48:16942-16947
Nakazawa H, Osaka T, Sakaguchi K (1973) A new cubic iron sulphide prepared by vacuum deposition. Nature, Phys Sci 242:13-14
Nakazawa K, Morimoto N (1971) Phase relations and superstructures of pyrrhotite $Fe_{1-x}S$. Mat Res Bull 6:345-358
Naldrett AJ (1989) Magmatic Sulfide Deposits. Oxford Monographs on Geology and Geophysics, No 14, Clarendon Press
Naldrett AJ, Kullerud G (1966) Limits of the $Fe_{1-x}S$-$Ni_{1-x}S$ solid solution between 600 °C and 250 °C. Carnegie Inst Wash, Year Book 65:320-327
Naldrett AJ, Craig JR, Kullerud G (1967) The central portion of the Fe-Ni-S system and its bearing on pentlandite exsolution in iron-nickel sulfide ores. Econ Geol 62:826-847
Nelmes RJ, McMahon MI, Belmonte SA, Parise JB (1999) Structure of the high-pressure phase III of iron sulphide. Phys Rev B59:9048-9052
Nozaki H, Nakazawi H, Sakaguchi K (1977) Synthesis of mackinawite by vacuum deposition method. Mineral J Japan 8:399-405
Ohtsuki T, Kitakaze A, Sugaki A (1981b) Synthetic minerals with quaternary components in the system Cu-Fe-Sn-S: Synthetic sulfide minerals (X). Science Reports, Tohoku University, Series 3. Mineral Petrol Econ Geol 14:269-282
Ohtsuki T Sugaki A, Kitakaze A (1981a) Three new phases in the system Cu-Fe-Sn-S: Synthetic sulfide minerals, XI. Science Reports, Tohoku University, Series 3. Mineral Petrol Econ Geol 15:79-87
Okazaki A (1959) The variation of superstructure in iron selenide Fe_7Se_8. J Phys Soc Japan 14:112-113.
Okazaki A (1961) The superstructures of iron selenide Fe_7Se_8. J Phys Soc Japan 16:1162-1170
Ontoev DO (1964) Characteristic features of bismuth mineralization in some tungsten deposits of Eastern Transbaikal. Trudy Mineral Muz Akad Nauk USSR 15:134-153
Osadchii EG (1986) Solid solutions and phase relations in the system Cu_2SnS_3-ZnS-CdS at 850 ° and 700 °C. Neues Jahrb Mineral, Abhdl 155:23-38
Otto HH, Strunz H (1968) Zur Kristallchemie synthetischer Blei - Wismut – Spiessglanze. Neues Jahrb Mineral, Abhdl 108:1-19
Ozawa K, Anzai S (1966) Effect of pressure on the α-transition point of iron monosulphide. Phys Stat Sol 17:697-700
Parise JB, Moore FH (1981) A superstructure of α-Ni_7S_6 and its relationship to heazlewoodite (Ni_3S_2) and millerite (NiS). Twelfth Internat Conf Crystallogr, Collected Abstracts, C-182

Peacock MA, McAndrew J (1950) On parkerite and shandite and the crystal structure of $Ni_3Pb_2S_2$. Am Mineral 35:425-439
Pearce CI, Pattrick RAD, Vaughan DJ (2006) Electrical and Magnetic properties of sulfides. Rev Mineral Geochem 61:127-180
Pettit FS (1964) Thermodynamic and electrical investigations on molten antimony sulfide. J Chem Phys 38: 9-20
Pichulo RO (1979) Polymorphism and phase relations in iron(II) sulfide (troilite) at high pressure. PhD Thesis, Columbia University
Plovnick RH, Vlasse M, Wold A (1968) Preparation and structural properties of some ternary chalcogenides of titanium. Inorg Chem 7:127-129
Pósfai M, Buseck PR (1994) Djurleite digenite and chalcocite: intergrowths and transformations. Am Mineral 79:308-315
Pósfai M, Dódony I (1990) Pyrrhotite superstructures. Part I: Fundamental structures of the NC (N = 2, 3, 4 and 5) type. Eur J Mineral 2:525-528
Potter RWII (1973) The systematics of polymorphism in binary sulfides: I. Phase equilibria in the system mercury-sulfur. II. Polymorphism in binary sulfides. PhD Thesis, Pennsylvania State University
Potter RWII (1977) An electrochemical investigation of the system copper-sulfur. Econ Geol 72:1524-142
Potter RWII, Evans HTJr (1976) Definitive X-ray powder data for covellite anilite djurlite and chalcocite. J Res U.S. Geol Surv 4:205-212
Powell AV, Vaqueiro P, Knight KS, Chapon LC, Sánchez RD (2004) Structure and magnetism in synthetic pyrrhotite Fe_7S_8: A powder neutron-diffraction study. Phys Rev B70:014415-1-014415-12
Power LF, Fine HA (1976) The iron-sulphur system part 1: the structures and physical properties of the compounds of the low-temperature phase fields. Mineral Sci Engin 8:106-128
Putnis A (1974) Electron-optical observations on the α-transition in troilite. Science 186:439-440
Putnis A (1976) Observations of transformation behavior in Ni_7S_6 by transmission electron microscopy. Am Mineral 61:322-535
Putnis A, Grace J (1976) The transformation behaviour of bornite. Contrib Mineral Petrol 55:311-355
Radtke AS, Taylor CM, Heropoulos C (1973) Antimony-bearing orpiment, Carlin gold deposit, Nevada. J Res U.S. Geol Surv 1:85-87
Radtke AS, Taylor CM, Dickson FW, Heropoulos C (1974) Thallium bearing orpiment, Carlin Gold deposit, Nevada. J Res U.S. Geol Surv 2:341-342
Radtke AS, Taylor CM, Erd RC, Dickson FW (1974) Occurrence of lorandite $TlAsS_2$ at the Carlin gold deposit, Nevada. Econ Geol 69:121-124
Raghavan V (1988) The Co-Fe-S system. In: Phase diagrams of ternary iron alloys, Part 2. Indian Institute of Metals, Calcutta, p 93-106
Rau H (1965) Thermodynamische Messungen an SnS. Berichte der Bunsen-Gesellschaft 69:731-736
Rau H (1976) Range of homogeneity and defect energetics in $Co_{1-x}S$. J Phys Chem Solids 37:931-934
Raybaud P, Hafner J, Kresse G, Toulhoat H (1997) Ab initio density functional studies of transition-metal sulphides: II. Electronic structure. J Phys: Cond Matt 9:11107-11140
Richardson FD, Jeffes JHE (1952) The thermodynamics of substances of interest in iron and steel making: III. Sulphides. J Iron Steel Inst 171:165-175
Rickard DT (1968) Synthesis of smythite- rhombohedral Fe_3S_4. Nature 218:356-357
Rickard DT (1972) Covellite formation in low temperature aqueous solution. Mineralium Deposia 7:180-188
Ritvo G, White GN, Dixon JB (2003) A new iron sulfide precipitated from saline solutions. Soil Sci Soc Am J 67:1303-1308
Robie RA, Hemingway BS, Fisher JR (1978) Thermodynamic properties of minerals and related substances at 29815 K and 1 bar (10^5 Pascals) pressure and at higher temperatures. Geol Surv Bull 1452, U.S. Govt Printing Office, Washington DC
Robie RA, Seal RRII, Hemingway BS (1994) Heat capacity and entropy of bornite (Cu_5FeS_4) between 6 and 760 K and the thermodynamic properties of phases in the system Cu-Fe-S. Can Mineral 32:945-956
Roland GW (1968) Synthetic trechmannite. Am Mineral 53:1208-1214
Roland GW (1970) Phase relations below 575 °C in the system Ag-As-S. Econ Geol 65:241-252
Rooymans CJM (1969) The behaviour of some groups of chalcogenides under very-high-pressure conditions. In: Advances in High Pressure Research. Vol. 2. Bradley RS (ed) Academic Press, London, p 1-100
Rösch H, Hellner E (1959) Hydrothermale untersuchungen am system $PbS-As_2S_3$. Naturwissenshaften 46:72
Roseboom EH (1962) Djurleite $Cu_{196}S$ a new mineral. Am Mineral 47:1181-1183
Roseboom EH (1966) An investigation of the system Cu-S and some natural copper sulfides between 25 ° and 700 °C. Econ Geol 61:641-672
Roseboom EH, Kullerud G (1958) The solidus in the system Cu-Fe-S between 400 °C and 800 °C. Carnegie Inst Wash, Year Book 57:222-227
Rosenqvist T (1954) A thermodynamic study of the iron cobalt and nickel sulfides. J Iron Steel Inst 176:37-57

Roy R Majumdar AJ, Hulbe CW (1959) The Ag_2S and Ag_2Se transitions as geologic thermometers. Econ Geol 54:1278-1280

Roy-Choudhury K (1974) Experimental investigations on the solid solution series between Cu_2SnS_3 and Cu_2ZnSnS_4 (kesterite). Neues Jahrb Mineral, Monat 432-434

Rueff JP, Kao CC, Struzhkin VV, Badro J, Shu J, Hemley RJ, Mao HK (1999) Pressure-induced high-spin to low-spin transition in FeS evidenced by X-ray emission spectroscopy. Phys Rev Lett 82:3284-3287

Sack RO, Ebel DS (2006) Thermochemistry of sulfide mineral solutions. Rev Mineral Geochem 61:265-364

Sakkopoulos S, Vitoratos E, Argyreas T (1984) Energy-band diagram for pyrrhotite. J Phys Chem Solids 45: 923-928

Salanci B (1965) Untersuchungen am System Bi_2S_3-PbS. Neues Jahrb Mineral, Monat 384-388

Salanci B, Moh GH (1969) Die experimentelle Untersuchung des pseudobinären Schnittes PbS-Bi_2S_3 innerholb des Pb-Bi-S systems in Beziehung zu natürlichen Blei - Bismut – Sulfosalzen. Neues Jahrb Mineral, Abhdl 112:63-95

Sanloup C, Fei Y (2004) Closure of the Fe-S-Si liquid miscibility gap at high pressure. Phys Earth Planet Int 147:57-65

Sanloup C, Guyot F, Gillet P, Fei Y (2002) Physical properties of liquid Fe alloys at high pressure and their bearings on the nature of metallic planetary cores. J Geophys Res 107(B11):2272-2280

Sato H, Mihara T, Furuta A, Tamura M, Mimura K, Happo N, Taniguchi M, Ueda Y (1997) Chemical trend of occupied and unoccupied Mn $3d$ states in MnY (Y=S SeTe). Phys Rev B 56:7222-7231

Schenck R, von der Forst P (1939) Gleichgewichtsstudien an erzbildenchen Sulfiden II. Z Anorganische und Allgemeine Chemie 240:145-157

Schenck R, Hoffman I, Knepper W, Vogler H (1939) Goeichgewichsstudien über erzbildende Sulfide, I. Z Anorganische und Allgemeine Chemie 240:178-197

Schlegel H, Schüller A (1952) Die Schmelz- und Kristallisations-gleischgewichte im System Cu-Fe-S und ihre Bedeutung fur Kupfergewinnung. Freiberger Forschungshefte B No 2:1-32

Schneeberg EP (1973) Sulfur fugacity measurements with the electrochemical cell $Ag/AgI/Ag_{2+x}Sf(S_2)$. Econ Geol 68:507-517

Schüller A, Wholmann E (1955) Betektinite ein neues Blei-Kupfer-Sulfid aus dem Mansfelder Rücken. Geologie 4:535-555

Schwarcz HP, Scott SD, Kissin SA (1975) Pressures of form·.tion of iron meteorites from sphalerite compositions. Geochim Cosmochim Acta 39:1457-1466

Schwarz EJ, Vaughan DJ (1972) Magnetic phase relations of pyrrhotite. J Geomag Geoelect 24:441-458

Scott SD (1973) Sphalerite geothermometry and geobarometry. Econ Geol 66:653-669

Scott SD (1976) Application of the sphalerite geobarometer to regionally metamorphosed terrains. Am Mineral 61:661-670

Scott SD, Barnes HL (1969) Hydrothermal growth of single crystals of cinnabar (red HgS). Mat Res Bull 4: 897-904

Scott SD, Barnes HL (1971) Experimental calibration of the sphalerite geobarometer. Econ Geol 68:466-474

Scott SD, Barnes HL (1972) Sphalerite-wurtzite equilibria and stoichiometry. Geochim Cosmochim Acta 36: 1275-1295

Scott SD, Kissin SA (1973) Sphalerite composition in the Zn-Fe-S system below 300 °C. Econ Geol 68:75-479

Seal RRII, Inan EE, Hemingway BS (2001) The Gibbs free energy of nukundamite ($Cu_{338}Fe_{062}S_4$): A correction and implications for phase equilibria. Can Mineral 39:1635-1640

Secco RA, Rutter MD, Balog SP, Liu H, Rubie DC, Uchida T, Frost D, Wang Y, Rivers M, Sutton SR (2002) Viscosity and density of Fe-S liquids at high pressure. J Phys: Cond Matt 141:1325-11330

Shannon RD (1976) Revised effective ionic radii and systematic studies of interatomic distances in halides and chalcogenides. Acta Crystallogr A32:751-767

Sharma RC, Chang YA (1980) Thermodynamics and phase relationships of transition metal-sulfur systems: IV. Thermodynamic properties of the nickel-sulfur liquid phase and calculation of Ni-S phase diagram. Metall Trans B 11B:139-146

Shelton KL, Merewether PA, Skinner BJ (1981) Phases and phase relations in the system Pd-Pt-Sn. Can Mineral 19:599-605

Sherman DM (1995) Stability of possible Fe-FeS and Fe-FeO alloy phases at high pressure and the composition of the Earth's core. Earth Planet Sci Lett 132:87-98

Shewman RW, Clark LA (1970) Pentlandite phase relations in the Fe-Ni-S system and notes on the monosulfide solid solution. Can J Earth Sci 7:67-85

Shimada K, Mizokawa T, Mamiya K, Saitoh T, Fujimori A, Ono K, Kakizaki A, Ishii T, Shirai M, Kamimura T (1998) Spin-integrated and spin-resolved photoemission study of Fe chalcogenides. Phys Rev B57: 8845-8853

Shunk FA (1969) Constitution of Binary Alloys, Supplement No 2. McGraw-Hill

Silverman MS (1964) High-temperature high-pressure synthesis of a new bismuth sulfide. Inorg Chem 3:1041
Skinner BJ (1966) The system Cu-Ag-S. Econ Geol 61:1-26
Skinner BJ (1970) Stability of the tetragonal polymorph of Cu_2S. Econ Geol 65:724-730
Skinner BJ, Luce FD (1971) Solid solutions of the type (CaMgMnFe)S and their use as geothermometers for the enstatite chondrites. Am Mineral 56:1269-1296
Skinner BJ, Erd RC, Grimaldi FS (1964) Greigite the thio-spinel of iron; a new mineral. Am Mineral 49:543-555
Skinner BJ, Jambor JL, M Ross (1966) Mckinstryite a new copper-silver sulfide. Econ Geol 61:1383-1389
Skinner BJ, Luce FD, Makovicky E (1972) Studies of the sulfosalts of copper. III. Phases and phase relations in the system Cu-Sb-S. Econ Geol 67:924-938
Snetsinger KG (1971) Erlichmanite (OsS_2), a new mineral. Am Mineral 56:1501-1506
Somanchi S (1963) Subsolidus phase relations in the systems Ag-Sb and Ag-Sb-S. MS Thesis, McGill University
Springer G (1969) Naturally occurring compositions in the solid solution series Bi_2S_3-Sb_2S_3. Mineral Mag 37: 295-296
Springer G, LaFlamme JHG (1971) The system Bi_2S_3-Sb_2S_3. Can Mineral 10:847-853
Stemprok M (1967) The Bi-Mo-S system. Carnegie Inst Wash, Year Book 65:336-337
Stemprok M (1971) The Fe-W-S system and its geological application. Mineralium Deposita 6:302-312
Strunz H, Geier BH, Seeliger E (1958) Gallit $CuGaS_2$ das erste selbstandige Gallium-Mineral und seine Paragenese in Tsumeb. Neues Jahrb Mineral, Monat 85-96
Stubbles JR, Birchenall CS (1959) A redetermination of the lead-lead sulfide equilibrium between 585 °C and 920 °C. Trans, AIMME 215:535-538
Sugaki A, Kitakaze A (1998) High form of pentlandite and its thermal stability. Am Mineral 83:133-140
Sugaki A, Shima H (1972) Phase relations of the Cu_2S-Bi_2S_3 system. Technical Reports of Yamaguchi University, p 45-70
Sugaki A, Shima H (1977) On solvuses of solid solutions among troilite hexagonal and monoclinic pyrrhotites below 300 °C - Studies on pyrrhotite group minerals (3)- Science Reports, Tohoku University, Series 3, 13:147-163
Sugaki A, Kitakaze A, Hayashi K (1984) Hydrothermal synthesis and phase relations of the polymetallic sulfide system, especially on the Cu-Fe-Bi-S system. *In*: Materials Science of the Earth's Interior. Terra Science, p 545-583
Sugaki A, Kitakaze A, Hayashi T (1982) High-temperature phase of pentlandite. Annual Meeting of the Mineralogical Society of Japan, Abstracts 22 (in Japanese)
Sugaki A, Kitakaze A, Kitazawa H (1985) Synthesized tin and tin-silver sulfide minerals: Synthetic sulfide minerals (XIII). Science Reports, Tohoku University, Series 3. Mineral Petrol Econ Geol 16:199-211
Sugaki A, Kitakaze A, Shimizu Y (1982) Phase relations in the Cu_3AsS_4-Cu_3SbS_4 join. Science Reports, Tohoku University, Series 3. Mineral Petrol Econ Geol 15:257-271
Sugaki A, Shima H, Kitakaze A (1972) Synthetic sulfide minerals (IV). Technical Reports of Yamaguchi University 1:71-77
Sugaki A, Shima H, Kitakaze A, Harada H (1975) Isothermal phase relations in the system Cu-Fe-S under hydrothermal conditions at 350 °C and 300 °C. Econ Geol 70:806-823
Sugaki A, Shima H, Kitakaze A, Fukuoka M (1977) Hydrothermal synthesis of pyrrhotites and their phase relation at low temperature - Studies on the pyrrhotite group minerals (4)- Science Reports, Tohoku University, Series 3, 13:165-182
Sugaki A, Shima H, Kitakaze A, Mizota T (1981) Hydrothermal synthesis of nukundamite and its crystal structure. Am Mineral 66:398-402
Suhr N (1965) The Ag_2S-Cu_2S system. Econ Geol 50:347-250
Sutarno, Knop O, Reid KIG (1967) Chalcogenides of the transition elements. V. Crystal structures of the disulfides and ditellurides of ruthenium and osmium. Can J Chem 45:1391-1400
Svendsen B, Anderson WW, Ahrens TJ, Bass JD (1989) Ideal Fe-FeS, Fe-FeO phase relations and Earth's core. Phys Earth Planet Int 55:154-186
Takele S, Hearne GR (1999) Electrical transport magnetism and spin-state configurations of high-pressure phases of FeS. Phys Rev B60:4401-4403
Takeno S, Moh GH, Wang N (1982) Dry mackinawite syntheses. Neues Jahrb Mineral, Abhdl 144:297-301
Takeno S, Zôka H, Nihara T (1970) Metastable cubic iron sulfide- with special reference to mackinawite. Am Mineral 55:1639-1649
Tatsuka K, Morimoto N (1973) Composition variation and polymorphism of tetrahedrite in the Cu-Sb-S system below 400 °C. Am Mineral 58:425-434
Taylor LA (1969) The significance of twinning in Ag_2S. Am Mineral 54:961-963
Taylor LA (1970a) The system Ag-Fe-S: Phase equilibria and mineral assemblages. Mineralium Deposita 5: 41-58

Taylor LA (1970b) The system Ag-Fe-S: Phase relations between 1200 ° and 700 °C. Metall Trans 1:2523-2529
Taylor LA (1970c) Low-temperature phase relations in the Fe-S system. Carnegie Inst Wash, Year Book 68: 259-270
Taylor LA, Kullerud G (1971) Pyrite-type compounds. Carnegie Inst Wash, Year Book 69:322-325
Taylor LA, Kullerud G (1972) Phase equilibria associated with the stability of copper disulfide. Neues Jahrb Mineral, Monat 458-463
Taylor P, Rummery TE, Owen DG (1979) On the conversion of mackinawite to greigite. J Inorg Nucl Chem 41:595-596
Tenailleau C, Etschmann B, Wang H, Pring A, Grguric BA, Studer A (2005) Thermal expansion of troilite and pyrrhotite determined by *in situ* cooling (873 to 373 K) neutron powder diffraction measurements. Mineral Mag 69:205-216
Terukov EI, Roth S, Krabbes G, Oppermann H (1981) The magnetic properties of cobalt-doped iron sulphide $Fe_{1-y}Co_yS$ (y ≤ 0.13). Phys Stat Sol A68:233-238
Tokonami M, Nishiguchi K, Morimoto N (1972) Crystal structure of a monoclinic pyrrhotite (Fe_7S_8). Am Mineral 57:1066-1080
Tossell JA, Vaughan DJ (1992) Theoretical Geochemistry: Applications of Quantum Mechanics in the Earth and Mineral Sciences. Oxford University Press
Toulmin PIII (1963) Proustite-pyrargyrite solid solutions. Am Mineral 48:725-736
Toulmin PIII, Barton PBJr (1964) A thermodynamic study of pyrite and pyrrhotite. Geochim Cosmochim Acta 28:641-671
Toulmin PIII Barton PBJr, Wiggins LB (1991) Commentary on the sphalerite geobarometer Am Mineral 76: 1038-1051
Townsend MG Tremblay R Horwood JL, Ripley LJ (1971) Metal-semiconductor transition in single crystal hexagonal nickel sulphide J Physics C: Solid State Phys 4 598-606
Trail RJ, Boyle RW (1955) Hawleyite isometric cadmium sulfide a new mineral Am Mineral 40 555-559
Tremel W, Hoffmann R, Silvestre J (1986) Transitions between NiAs and MnP type phases: An electronically driven distortion of triangular (3^6) nets. J Am Chem Soc 108:5174-5187
Tunnell G (1964) Chemical processes in the formation of mercury ores and ores of mercury and antimony. Geochim Cosmochim Acta 28:1019-1037
Ueno T, Scott SD (1994) Phase relations in the system Ga-Fe-S at 900 °C and 800 °C. Can Mineral 32:203-210
Ueno T, Ito S-I, Nakatsuka S, Nakano K, Harada T, Yamazaki T (2000) Phase equilibria in the system Fe-Ni-S at 500 °C and 400 °C. J Mineral Petrol Sci 95:145-161
Ueno T, Nagasaki K, Horikawa T, Kawakami M, Kondo K (2005) Phase equilibria in the system Cu-Ga-S at 500° and 400 °C. Can Mineral 43:1643-1651
Ueno T, Scott SD, Kojima S (1996) Inversion between sphalerite and wurtzite-type structures in the system Zn-Fe-Ga-S. Can Mineral 34:949-958
Urakawa S, Someya K, Terasaki H, Katsura T, Yokoshi S, Funakoshi K-I, Utsumi W, Katayama Y, Sueda Y-I, Irifune T (2004) Phase relationships and equations of state for FeS at high pressures and temperatures and implications for the internal structure of Mars. Phys Earth Planet Int 143-144:469-479
Urazov GG, Bol'shakov KA, Federov PI, Vasilevskaya II (1960) The ternary bismuth-iron-sulfur system (a contribution to the theory of precipitation smelting of bismuth). Russian J Inorg Chem 5:303-307
Valverde DN (1968) Phase diagram of the Cu-Ag-S system at 300 °C. Z Physikalische Chemie (München) 62: 218-220
van Hook JJ (1960) The ternary system $Ag_2S-Bi_2S_3$-PbS. Econ Geol 55:759-788
Vaughan DJ, Craig JR (1974) The crystal chemistry and magnetic properties of iron in the monosulfide solid solution of the Fe-Ni-S system. Am Mineral 59:926-933
Vaughan DJ, Craig JR (1978) Mineral chemistry of metal sulfides. Cambridge University Press
Vaughan DJ, Rosso KM (2006) Chemical bonding in sulfide minerals. Rev Mineral Geochem 61:231-264
Viaene W (1968) Le systeme Fe-Ge-S a 700 °C. Comptes Rendus de l'Académie des Sciences Paris 266:1543-1545
Viaene W (1972) The Fe-Ge-S system: phase equilibria. Neues Jahrb Mineral, Monat 23-35
Viaene W, Kullerud G (1971) The Ti-S and Fe-Ti-S Systems. Carnegie Inst Wash, Year Book 70:297-299
Vlach KC (1988) A study of sulfur pressures and phase relationships in the Fe-FeS-Co-CoS and Fe-Cr-S systems. PhD Thesis, University of Wisconsin, Madison, Wisconsin
Vogel R (1953) Über das System Blei silber-schwefal. Z Metallkunde 44:133-135
Vogel R (1956) The system bismuth sulfide-copper sulfide and the ternary system $Bi-Bi_2S_3-Cu_2S$-Cu. Z Metallkunde 47:694-699
Vogel R (1968) Zur Entstehung des Daubréelith im meteorischen Eisen. Neues Jahrb Mineral, Monat 453-463

Vollstädt H, Seipold U, Kraft A (1980) Structural and electrical relations of monosulphide solid solution in the Fe-Ni-S system at high pressures and temperatures. Phys Earth Planet Int 22:267-271

Wang H, Pring A, Ngothai Y, O'Neill B (2005) A low-temperature kinetic study of the exsolution of pentlandite from the monosulfide solid solution using a refined Avrami method. Geochim Cosmochim Acta 69:415-425

Wang N (1973) Phases on the pseudobinary join lead sulfide-antimony sulfide. Neues Jahrb Mineral, Monat 79-81

Wang N (1974) The three ternary phases in the system Cu-Sn-S. Neues Jahrb Mineral, Monat 424-431

Wang N (1984) A contribution to the Cu-Fe-S system: the sulfidization of bornite at low temperatures. Neues Jahrb Mineral, Monat 346-352

Wang N (1988) Experimental study of the copper-germanium-sulfur ternary phases and their mutual relations. Neues Jahrb Mineral, Abhdl 159:137-151

Ward JC (1970) The structure and properties of some iron sulphides. Rev Pure Appl Chem 20:175-206

Wehmeier FH, Laudise RA, Shieves JW (1968) The system $Ag_2S-As_2S_3$ and the growth of crystals of proustite smithite and pyrargyrite. Mat Res Bull 3:767-778

Weissberg GA (1965) Getchellite $AsSbS_3$ a new mineral from Humboldt County, Nevada. Am Mineral 50: 1817-1826

Werner A (1965) Investigations on the system Cu-Ag-S. Z Physikalische Chemie (München) 47:267-285

White RM, Mott NF (1971) The metal-non-metal transition in nickel sulphide (NiS). Phil Mag 24:845-856

Wiegers GA, Jellinek F (1970) The system titanium-sulfur II The structure of Ti_3S_4 and Ti_4S_5. J Solid State Chem 1:519-525

Wieser E, Krabbes G, Terukov EI (1982) Detection of the metastable simultaneous occurrence of high- and low-temperature states in $Co_yFe_{1-y}S$ by Mössbauer spectroscopy. Phys Stat Sol A72:695-699

Wiggins LB, Craig JR (1980) Reconnaissance of the Cu-Fe-Zn-S system: sphalerite phase relationships. Econ Geol 75:742-751

Williamson DP, Grimes NW (1974) An X-ray diffraction investigation of sulphide spinels. J Phys Appl Phys 7: 1-6

Wolthers M, Van der Gaast SJ, Rickard D (2003) The structure of disordered mackinawite. Am Mineral 88: 2007-2015

Wuensch BJ (1974) Determination relationships and classification of sulfide mineral structures. Rev Mineral 1: W-1-WW-44

Wuensch BJ, Buerger MJ (1963) The crystal structure of chalcocite Cu_2S. Mineral Soc Am Spec Paper 1:164-170

Wyszomirski P (1976) Experimental studies of the ternary Fe-Co-S system in the temperature range 500-700 °C. Mineralogica Polonica 7:39-49

Wyszomirski P (1977) Phase relations in the Fe-Co-S system at 800 °C. Mineralogica Polonica 8:75-78

Wyszomirski P (1980) The pure dry Fe-Co-S system at 400 °C. Neues Jahrb Mineral, Abhdl 139:131-132

Yagi T, Akimoto S (1976) Pressure fixed points between 100 and 200 kbar based on the compression of sodium chloride. J Appl Phys 47:3350-3354

Ying-chen J, Yu-jen T (1973) Isomorphous system $RuS_2-OsS_2-IrS_2$ and the mineral system PdS-PtS. Geochemia 4:262-270

Yu WD, Gidisse PJ (1971) High pressure polymorphism in CdS CdSe and CdTe. Mat Res Bull 6:621-638

Yund RA (1962) System Ni-As-S: phase relations of mineralogical significance. Am J Sci 260:761-782

Yund RA (1963) Crystal data for synthetic $Cu_{55x}Fe_xS_{65x}$ (idaite). Am Mineral 48:672-676

Yund RA, Hall HT (1968) The miscibility gap between FeS and $Fe_{1-x}S$. Mat Res Bull 3:779-784

Yund RA, Kullerud G (1966) Thermal stability of assemblages in the Cu-Fe-S system. J Petrol 7:454-488

Yusa K, Kitakaze A, Sugaki A (1979) Synthesized bismuth-tellurium-sulfur system: Synthetic sulfide minerals. Science Reports Tohoku University Series 3: Mineral Petrol Econ Geol 14:121-133

Zelikman AN, Belyaevskaya LV (1956) The melting point of molybdenite. Zhurnal Neorganicheskoi Khimii 1: 2239-2244

Zhou Y, Campbell AJ, Heinz DL (1991) Equations of state and optical properties of the high pressure phase of zinc sulfide. J Phys Chem Solids 52:821-825

Metal Sulfide Complexes and Clusters

David Rickard
School of Earth, Ocean and Planetary Sciences
Cardiff University
Cardiff CF103YE, Wales, United Kingdom
e-mail: rickard@cardiff.ac.uk

George W. Luther, III
College of Marine Studies
University of Delaware
Lewes, Delaware, 19958, U.S.A.
e-mail: luther@udel.edu

INTRODUCTION

In this chapter we show that

1. Metal sulfide complexes and clusters enhance the solubility of metal sulfide minerals in natural aqueous systems, explaining the transport of metals in sulfidic solutions and driving the biology and ecology of some systems.

2. There is little or no evidence for the composition or structure of many of the metal sulfide complexes proposed in the geochemical and environmental literature.

3. Voltammetry appears to be a powerful tool in providing additional evidence about the composition of metal sulfide complexes and clusters which complements the increasing use of techniques such as UV-VIS, Raman and IR spectroscopy, EXAFS, XANES and mass spectrometry.

4. Many of the stability constants for metal sulfide complexes are very uncertain because of the lack of independent evidence for their existence.

5. Experimental measurements of metal sulfide complex stability constants is constrained by the lack of knowledge about the composition, structure and behavior of, often nanoparticulate, low temperature metal sulfide precipitates.

6. The competitive kinetics of metal sulfide complex and cluster formation in complicated natural sulfidic systems contributes to the distribution of metals in the environment.

7. The mechanisms of the formation of metal sulfide complexes and clusters provide basic information about the mechanism of formation of metal sulfide minerals and explain the stabilities and compositions of the complexes and clusters.

8. There appears to be a continuum between metal sulfide complexes, metal sulfide clusters and metal sulfide solids.

9. The nature of the first-formed metal sulfide mineral, which is often metastable, can be largely determined by the structure of the metal sulfide cluster in solution.

Background

The metal chemistry of anoxic systems is dominated by reactions with reduced sulfur species. These reduced sulfur systems presently characterize the Earth's subsurface and are occasionally important in marine and freshwater systems. In the first half of Earth history, of course, the surface environments were also anoxic and the reduced sulfur chemistry played an even more widespread role in the geochemistry of base metals (e.g., Canfield 1998; Holland 2004).

In order for metals to be transported within anoxic systems, the metals must be held in solution. Intuitively, this would appear problematical because of the widespread assumption of relative insolubility of metal sulfide minerals. In fact, as shown in Table 1, this assumption is misplaced. Even with the simplest of solubility computations, some 35% of the metals listed are more soluble in sulfidic systems. The system considered in Table 1 is for pure water so that side reactions, such as chloride complexing in seawater or the formation of carbonate solids, are not included. In some of these more complex environments, the solubility of the sulfides may be even more significant. We used +0.5 V for the Eh in oxidized systems here, since this is the average Eh of aqueous environments in contact with the atmosphere, according to the classical studies of Baas Becking et al. (1960). We assumed an Eh of −0.2 V for the sulfide systems and this is a maximum value for microbiological sulfate reduction. Lower Eh values would increase sulfide solubility.

This problem of the solubility of metals in sulfidic environments is of current interest because of its effect on the bioavailability of these metals, all of which are variously critical to fundamental biochemical processes (Williams and Frausto da Silva 1996). Indeed, in the geologic past when life developed, the availability of metals may have been a kinetic inhibitor to key reactions.

In fact, at low temperatures, metastable metal sulfides are kinetically significant and these have enhanced solubilities compared to their more stable counterparts. Furthermore a number

Table 1. Comparison of solubilities of oxides and sulfide solids of metals considered in this chapter. Bold script is used for the most soluble solid. The solubilities are presented in both molal, m, and ppm values of total dissolved metal. The solubilities are calculated for pure water at 25 °C, 1.013 bars total pressure, pH = 7, Eh = 0.5 V (oxide), Eh = −0.2 V, total S(−II) = 10^{-3} m (sulfides). The Davies equation is used for activity computations (see text). Where at least 1 g of solid dissolves in 1000 g H_2O, this is indicated. Data are mainly from the standard *thermo.com. v8.r6* database which in turn derives mostly from Helgeson and his co-workers, modified with pK_{2,H_2S} = 18. FeS data is taken from Rickard (unpublished). Mo and As, which form molybdates and arsenates in oxidized conditions, are excluded.

Oxides and Metals			Sulfides		
solid	m	ppm	solid	m	ppm
Cr_2O_3	2×10^{-9}	1×10^{-4}	CrS	dissolves	
MnO_2	5×10^{-4}	3×10^{1}	MnS	1×10^{-4}	6×10^{0}
FeOOH	3×10^{-12}	2×10^{-7}	FeS	2×10^{-6}	6×10^{-2}
Co_3O_4	5×10^{-5}	3×10^{0}	CoS	1×10^{-10}	5×10^{-3}
NiO	9×10^{-2}	5×10^{3}	NiS	3×10^{-10}	2×10^{-5}
CuO	2×10^{-6}	1×10^{-1}	CuS	5×10^{-19}	3×10^{-14}
ZnO	2×10^{-3}	1×10^{2}	ZnS	1×10^{-13}	8×10^{-9}
CdO	dissolves		CdS	1×10^{-18}	4×10^{-13}
PbO	dissolves		PbS	4×10^{-17}	1×10^{-13}
SnO_2	3×10^{-8}	3×10^{-3}	**SnS_2**	3×10^{-5}	3×10^{-5}
Sb_2O_5	4×10^{-12}	1×10^{-17}	**Sb_2S_3**	4×10^{-8}	5×10^{-3}
Ag	9×10^{-6}	1×10^{0}	Ag_2S	1×10^{-19}	2×10^{-14}
Au	7×10^{-21}	1×10^{-15}	Au	1×10^{-32}	2×10^{-27}
Hg	2×10^{-10}	4×10^{-5}	HgS	3×10^{-41}	6×10^{-36}

of metal sulfide complexes with considerable thermodynamic stabilities has been identified at concentrations which are very low experimentally but significant in natural systems. These are not considered in the calculations listed in Table 1, since they are the major subject of this chapter. These observations provide possible explanations for the mobility and bioavailability of metals in anoxic systems.

In this chapter we consider current knowledge about dissolved metal sulfide clusters. We examine the complexation of S(–II) and S_n(–II) ligands with base metal sulfides. We do not consider the more oxidized sulfur species, such as the sulfur oxyanions. We also limit our discussion to low temperatures (0-100 °C), where these species play a particularly important role, and aqueous solutions, since these are more important geochemically. In fact, from a theoretical or experimental point of view, water is the least suitable medium to consider, and much of the pure chemical literature on complexation is concerned with less polar to non-polar, usually organic, solvents.

Metals considered in this chapter

The formal chemical definition of a metal is all-embracing and of little application to the natural sciences. Here we have focused on those metals which are of significance to environmental science. These are fundamentally the metals which form fairly common sulfide minerals or where sulfide complexes are significant in their (bio)geochemistry. We have also incorporated some metalloids, such as As and Sb because of their close association with metal sulfides. The metals and metalloids considered are conveniently listed in the form of a periodic table (Fig. 1).

The metals discussed form a diverse group of elements. As with all elements in the Periodic Table, their properties can be considered in terms of horizontal rows (the Periods) or vertical columns (the Groups). Both approaches have advantages. The Periods show large numbers of elements whose properties change, often systematically, as the electrons fill a given shell. Elements in individual Groups have related properties, since their electronic configurations are similar, even if these configurations are situated in different shells (Table 2). As can be seen from Table 2, the metals Sc in the first transition series through to Hg in the third, constitute the d-block elements, where chemical properties are influenced by the electron configuration of the nd-electrons.

We consider the metals in both classifications. Thus the largest single homologous series is the metals of the first transition series Sc–Cu. Of these, Cr–Cu have sulfide complexes which

s-block		d-block										p-block					
1 **H**																	2 **He**
3 **Li**	4 **Be**											5 **B**	6 **C**	7 **N**	8 **O**	9 **F**	10 **Ne**
11 **Na**	12 **Mg**											13 **Al**	14 **Si**	15 **P**	16 **S**	17 **Cl**	18 **Ar**
19 **K**	20 **Ca**	21 **Sc**	22 **Ti**	23 **V**	24 **Cr**	25 **Mn**	26 **Fe**	27 **Co**	28 **Ni**	29 **Cu**	30 **Zn**	31 **Ga**	32 **Ge**	33 **As**	34 **Se**	35 **Br**	36 **Kr**
37 **Rb**	38 **Sr**	39 **Y**	40 **Zr**	41 **Nb**	42 **Mo**	43 **Tc**	44 **Ru**	45 **Rh**	46 **Pd**	47 **Ag**	48 **Cd**	49 **In**	50 **Sn**	51 **Sb**	52 **Te**	53 **I**	54 **Xe**
55 **Cs**	56 **Ba**	57 **La**	72 **Hf**	73 **Ta**	74 **W**	75 **Re**	76 **Os**	77 **Ir**	78 **Pt**	79 **Au**	80 **Hg**	81 **Tl**	82 **Pb**	83 **Bi**	84 **Po**	85 **At**	86 **Rn**
87 **Fr**	88 **Ra**	89 **Ac**	90 **Th**	91 **Pa**	92 **U**												

Figure 1. Periodic table of elements (excluding lanthanides and actinides) highlighting metals and metalloids considered in this chapter.

Table 2. Ground state electronic properties of the elements considered in this chapter.

	Mn	Fe	Co	Ni	Cu	Zn		As
[Ar]	$3d^54s^2$	$3d^64s^2$	$3d^74s^2$	$3d^84s^2$	$3d^{10}4s^1$	$3d^{10}4s^2$		$3d^{10}4s^24p^3$
					Ag	Cd	Sn	Sb
[Kr]					$4d^{10}5s^1$	$4d^{10}5s^2$	$4d^{10}5s^25p^2$	$4d^{10}5s^25p^3$
					Au	Hg	Pb	
[Xe]					$4f^{14}5d^{10}6s^1$	$4f^{14}5d^{10}6s^2$	$4f^{14}5d^{10}6s^26p^2$	

are of potential geochemical interest. Other metals are considered most effectively in Groups. We look at Mo as a Group 6 element since its chemistry and biochemistry are becoming increasingly important geochemically. Mo sulfide also forms a number of classical polynuclear forms, known as cages or clusters, which enlighten discussions of these types of complexes in other metals. Mo is part of the group with Cr, which is considered with the first row transition metals. The precious metals include Au and Ag from Group 10, both situated to the right of the d-block where the d-electron orbitals tend to be filled and the elements become resistant to common environmental reactions, such as oxidation. Ag and Au are also related to Cu. We consider the Group 12 metals, Zn, Cd and Hg separately, since with these elements the d orbital configuration becomes less significant and the chemistry is largely determined by the outer ns-electrons. To the right of the d-block elements are the p-block where the ns- and np-electron orbitals determine the chemistry. Sn and Pb are important p-block metals although their properties are really extensions of the non-metals C, Si and Ge in the same group. Finally, we look at the metalloids, As and Sb, from Group 15. These are important elements geochemically and have a significant, and burgeoning, sulfide chemistry.

The metals considered in this chapter display various oxidation state numbers. This is significant in the consideration of metal sulfide complex and cluster chemistry since the sulfide moiety is an effective electron donor which means that only selective metal oxidation states are likely to form stable sulfide species. The oxidation states of the metals considered in this chapter are summarized in Table 3.

Table 3. Oxidation states of metals considered in this chapter. Only compounds are considered. Bold are the most common and [ ] indicate rare oxidation states. Sb also displays a −3 oxidation state which appears not to be significant in natural systems.

Mn	Fe	Co	Ni	Cu	Zn		As
0	0	0	0	[0]			[0]
1	1	1	1	**1**	1		
2	**2**	**2**	**2**	**2**	**2**		
3	**3**	**3**	3	3			**3**
4	4	4	4	[4]			
5							**5**
6	6						
7							

Mo		Ag	Cd		Sn	Sb
0						0
		1				
2		2	**2**			
3		3				3
4					**4**	
5						**5**
6						

	Au	Hg		Pb
	[0]			[0]
	1	1		
		[2]	**2**	**2**
	3			
				4
	5			

Lewis acids and bases

At the same time that Brønsted and Lowry defined acids in terms of the transfer of a proton between species, Lewis (1923) proposed a more general definition A *Lewis acid* is a compound that possesses an empty orbital for the acceptance of a pair of electrons. A *Lewis base* is a substance that acts as an electron pair donor. The fundamental reaction for Lewis acids and bases is complex formation where bonds are formed between the acid and base by sharing the electron pair supplied by the base. In kinetics, equivalent forms would be the *nucleophile* for the donor and the *electrophile* for the acceptor. Any proton is a Lewis acid because it can attach to an electron pair, as HS^-, for example. Thus any Brønsted acid, like H_2S, exhibits Lewis acidity.

Thus H_2O is a weak Lewis base, but the H_2O coordinated to metal ions in the hydration shell (see below) are stronger acids because of the repulsion of the protons in the H_2O molecules by the metal (hydrolysis reactions). Thus the metal cations can be regarded as Lewis acids and their acidity will vary according to their size and charge. Similarly, the sulfide complexes such as $[MeHS]^+$ are Lewis acids because the metal can accept electrons from the Lewis base, HS^-, to form $[Me(HS)_2]$.

Hard A and soft B metals

Ahrland et al. (1958) and Schwarzenbach (1961) divided metal ions into two classes, A and B, based on whether they formed their most stable complexes with the first ligand atom of each periodic group (F,O,N) or with later members (I,S,P). Stumm and Morgan (1970) promulgated this approach in geochemistry. The A-B classification is basically a reflection of the number of outer shell electrons and the deformability (i.e., polarizability) of the electron configuration. Thus, Class A metal ions have inert gas-type electron configurations with essentially spherical symmetries which are not easily deformed. Class A metals are referred to as being *hard*. In contrast, Class B metals have more readily deformable electron configurations and are referred to as *soft*. Pearson (1965) expanded the Class A hard metal classification to include metals that have a tendency to form ion pairs with ligands with low polarizability.

In this classification scheme, the transition metals form an intermediate group. These have between zero and 10 *d*-electrons. Irving and Williams (1953) showed that there is a systematic change in complex stability for these metals with multidentate chelates, known as the *Irving-Williams* order, where the stability increases $Mn^{2+} < Fe^{2+} < Co^{2+} < Ni^{2+} < Cu^{2+} > Zn^{2+}$. Irving and Williams (1953) explained this trend in terms of increased effective nuclear charge and crystal field theory (see below): the stability increases through the increased crystal field stabilization energy (CFSE) resulting from *d*-electrons preferentially occupying lower energy *d*-orbitals. Thus Mn^{2+} (5 *d*-electrons) and Zn^{2+} (10 *d*-electrons) have no CFSE, but the CFSE will increase from Mn^{2+} to the d^9 Cu^{2+} ion.

Pearson (1965) extended the hard and soft classification for metals into acids and bases. This idea, sometimes referred to with the acronym HSAB, classifies F^-, I^-, Cl^-, OH^- and NH_3 as hard bases and S(–II) as the classical example of a soft base. From the definition of hardness it follows that hard acids tend to bind more readily with hard bases and soft bases bind with soft acids. In this classification then, S(–II), HS(–I) and S_n(–II) are soft bases and have a strong tendency to form strong complexes with the class B, or soft, metals (Fig. 2). The hard base-hard acid and soft base–soft acid approach is especially valuable in geochemistry since it explains some parts of Goldschmidt's classification into lithophile and chalcophile elements. The lithophile elements are generally hard cations and are associated with the hard base O^{2-}. The chalcophile elements, which are the main subject of this book tend to be soft and are found in association with the soft base, S(–II).

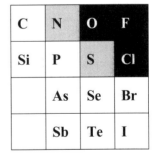

Figure 2. Hard (black), borderline (gray) and soft (white) acids according to Pearson (1963). Sulfur is a borderline acid because species such as SO_3 are hard, SO_2 are borderline and S(–II), HS(–I) and S_n(–II) are soft.

The Irving-Williams order (Irving and Williams 1953) predicts that the transition metals will have a gradational increase in forming sulfide complexes from Mn^{2+} through Cu^{2+}. Interestingly for the sulfide geochemist, Zn^{2+} is defined as borderline whereas Cd^{2+} is soft. That is, the common geochemical assumption that Zn^{2+} and Cd^{2+} would be predicted to behave similarly with respect to sulfide might be questioned by a chemist. However, as shown in Figure 3, the metals considered in this chapter are all soft or borderline and can be expected to have a significant chemistry with the soft base, S(−II).

23	24	25	26	27	28	29	30	31	32	33
		Mn	**Fe**	**Co**	**Ni**	**Cu**	**Zn**			**As**
41	42	43	44	45	46	47	48	49	50	51
	Mo					**Ag**	**Cd**		**Sn**	**Sb**
73	74	75	76	77	78	79	80	81	82	83
						Au	**Hg**		**Pb**	

Figure 3. Hard (black), borderline (gray) and soft (white) classification of the metals considered in this chapter, according to Pearson (1963).

Complexes and clusters

In the standard definition (e.g., Cotton et al. 1999), a complex is identified as a coordination compound, where a central atom or ion, M, unites with one or more ligands, L, to form a species of the form $ML_iL_jL_k$. In these species the metal and the ligand may all bear charges. Cotton et al. (1999) place further restraints on complexes: (1) the central metal ion should be capable of significant existence and (2) the reaction forming the complex can occur in significant conditions. The idea of significance is a subjective one, of course. What Cotton et al. (1999) are addressing is the problem that, statistically, it is probable that all imagined combinations of metals and ligands occur—but the ones that are significant are those that exist for a substantial period of time and contribute a measurable amount to the total dissolved concentrations of metals and/or ligands. This is a typical pragmatic view of an equilibrium chemist; kineticists find that ephemeral complexes, such as the transition state complex in many reactions, are exceptionally important since they determine the rate and direction of the reaction.

This problem becomes apparent when we address clusters. It is obviously possible for complexes to be formed with no central metal atom and with metal atoms which are bonded to each other. These complexes are called clusters or cages. Cotton et al. (1999, p 9) distinguish clusters (or cages) from complexes:

"In each type of structure a set of atoms define the vertices of a polyhedron, but in a complex these atoms are each bound to a central atom and not to each other, whereas in a cage or cluster there need not be a central atom and the essential feature is a system of bonds connecting each atom directly to its neighbors in the polyhedron."

In this definition, a cluster is essentially a polynuclear complex. In contrast, in the surface science and physics literature, clusters are also equated to embryos, the groups of molecules that ultimately develop into the nucleus of the condensed phase. As we demonstrate below, the aqueous iron, copper and zinc sulfide clusters defined and characterized by Buffle et al. (1988), Davison (1980), Davison and Heaney (1980), Theberge and Luther (1997), Theberge (1999), Helz et al. (1992), Luther et al. (1999, 2002) display both properties: they are multinuclear complexes which may develop to form the nuclei of the first condensed phase. As pointed out by Luther and Rickard (2005) this definition is determined to a large extent by the present difficulty in distinguishing between true aqueous clusters and electroactive nanoparticles in *in situ* analyses of natural systems using electrochemical methods.

Care has to be taken in critically assessing literature reports regarding metal sulfide clusters, because of contrasting definitions of exactly what the authors are referring to as clusters. Thus, for example, Sukola et al. (2005) define their "clusters" or "nanoclusters" as something between colloids and truly dissolved species ranging in size from 2–10 nm. In the geochemical literature these forms are usually termed nanoparticles (Banfield and Zhang 2001). Zhang et al. (2003) described 3 nm ZnS nanoparticles, for example, and Ofhuji and Rickard (2006) characterized 4 nm FeS nanoparticles (see also in this volume Pattrick et al. 2006).

As discussed by Luther and Rickard (2005), although there may appear to be an electrochemical operational continuum between clusters and the first condensed solid, theoretically there is an abrupt change of state. A solid can be defined as a state with a surface, although this is often not very helpful practically in low temperature aqueous systems where the first particles are nanometer-sized. Rather more interesting is the sudden increase in density from the aqueous cluster to the solid. We discuss present knowledge about the relationship between dissolved clusters and solids in more detail below.

Coordination numbers and symmetries

In the classical chemical definition there is little difference between complexes and coordination compounds—except that nearly all chemical compounds are coordination compounds. In this view, complexes are a special class of coordination compounds which, as we use the term in this chapter, occur as dissolved species in aqueous solutions. The idea of coordination number and symmetry at a metal center therefore plays a central role in understanding the chemistry and behavior of complexes.

Coordination theory was developed by the Swiss chemist, Alfred Werner, and he received the Nobel Prize for this in 1913. Werner (1904) noted that individual atoms in a chemical species have two different attributes in aqueous solutions: (1) the oxidation number or valence and (2) the coordination number or the number of other atoms directly linked it. The concept of coordination number and the consequent geometry provides a point of divergence for classical equilibrium chemistry. We are no longer considering the state of the system but looking at the real world.

The coordination number reflects the bonding between any atom in a chemical species and its neighbors (see Cotton et al. 1999). In inorganic chemistry, the coordination number is the number of σ-bonds formed between the metal and the ligand, and π-bonds are not included. σ-bonds are the strongest type of covalent bonds. σ-bonds form when (1) the ligand donates a pair of electrons directly to the metal on one of the metal's bond axes defined by the x, y, z axes in Cartesian coordinates or (2) the metal and ligand each share an electron on the metal's bond axis. In the latter case, both atoms give an electron from the s-orbital (or a hybrid orbital) in conjunction with additional electrons from the p- and sometimes d- (and above) orbitals. In contrast, π-bonds are those bonds between two atoms in a molecule that do not have electron density on the bond axis and do not exhibit orbital hybridization. π-bonds directly share electrons between the p-orbitals that are parallel to each other, between a p-orbital and 2 lobes of the d-orbitals, or between 2 lobes of d-orbitals from two different atoms.

There is no simple way of predicting the coordination number of any particular atom in a solution chemical species. Note that this is different from coordination in crystals where Pauling's rules will give a first approximation (Pauling 1960). Although the concept of coordination number applies to main group elements, coordination compounds in the classical sense include mostly transition metals.

Equilibrium constants of complexes

For the reaction between a metal, M, and a ligand, L, to form a complex M_mL_l (Eqn. 1):

$$mM + lL = M_mL_l \tag{1}$$

the state at equilibrium can be defined by an equilibrium constant, K, (Eqn. 2) where:

$$K = \frac{\{M_mL_l\}}{\{M\}^m\{L\}^l} \tag{2}$$

where { } refers to the activities of the species. It is convenient to take logarithms of this relationship (Eqn. 3) since:

$$\log K = \log\{M_mL_l\} - m\log\{M\} - l\log\{L\} \tag{3}$$

and the equibrium constant for an overall reaction which can be represented as a series of simple reactions is then merely the sum of the logarithms of the equilibrium constants for each reaction. For example, the sum of the reactions (Eqns. 4, 5):

$$H_2S = HS^- + H^+ \tag{4}$$

for which the equilibrium constant is K_1 and

$$HS^- = S^{2-} + H^+ \tag{5}$$

with the constant K_2 at equilibrium is (Eqn.6):

$$H_2S = S^{2-} + 2H^+ \tag{6}$$

for which the equilibrium constant K_{12} is given by relationship (Eqn. 7):

$$\log K_{12} = \log K_1 + \log K_2 \tag{7}$$

The logarithmic approach has a further advantage since pH is defined as $-\log\{H+\}$. Then the equilibrium constant for reaction (Eqn. 4) is given by (Eqn. 8):

$$\log K_1 = \log\{HS^-\} - \log\{H^+\} = \log\{HS^-\} + pH \tag{8}$$

This has led to equilibrium constants being listed in terms of pK values (Eqn. 9) where, by analogy with pH,

$$pK = -\log K \tag{9}$$

A further modification of the equilibrium constant nomenclature is the use of β to describe formation constants. The equilibrium constant can be written for the forward or back reaction (e.g., Eqn. 4) and the logarithm of the constants will have opposite signs. The use of formation constants overcomes this possible confusion. The formation constant for a complex is written is the form of a reaction, which results in the production of the complex. Thus, reaction (Eqn. 4) becomes (Eqn. 10):

$$HS^- + H^+ = H_2S \tag{10}$$

and the formation constant, β, (Eqn. 11) is given by

$$\beta = \frac{\{H_2S\}}{\{HS^-\}\{H^+\}} \tag{11}$$

In general,

$$\beta°_{ml} = \frac{\{M_mL_l\}}{\{M\}^m\{L\}^l} \tag{12}$$

In reality, concentrations, [ ], are measured and these are related to the activities through the activity coefficients, γ_i, so that Equation (12) becomes Equation (13)

$$\beta_{ml} = \frac{\gamma_{M_mL_l}[M_mL_l]}{\gamma_M \gamma_L [M]^m [L]^l} \tag{13}$$

Simple inspection shows that β_{ml} only equals $\beta°_{ml}$ where $\gamma_{MmLl} = \gamma_M\gamma_L$, or where the activity coefficients approach 1 in solutions at infinite dilution. So $\beta°_{ml}$ is the thermodynamic equilibrium formation constant or the constant at infinite dilution. Operational equilibrium constants are commonly employed in geochemistry since several of the natural media, such as seawater, can be approximated as having a constant ionic strength. In seawater, for example, the ionic strength is around 0.7 and at this sort of concentration the estimation of individual ion activity constants can be a source of serious uncertainty. Various algorithms for activity coefficients are used (Table 4). Each of these has limited applicability in terms of the ionic strength, I. Several equilibrium computer programmes use the Davies equation, but even here the deviation from measured values becomes more uncertain above $I = 0.5$, which is still less than seawater. The Pitzer approach, which is based on knowledge of a series of coefficients shows excellent agreement in these high ionic strength solutions, but requires *a priori* knowledge of the values of the coefficients for each species under consideration. And these values are commonly not available.

The uncertainties associated with individual activity coefficient estimates can be considerable (Fig. 4). The divergence in the interesting range for natural waters, with ionic strengths between 0.1 and 0.7 M, is apparent from the diagram. The activity coefficient diverges at seawater ionic strengths from around 0.75 to 0.6, and this is a multiplier to the measured concentration. For divalent ions, such as Fe^{2+}_{aq}, the problem is exacerbated by the small value of the activity coefficient ranging from 0.4 at $I = 0.1$ M to 0.2 at 0.7 M, according to the Davies equation, for example. This constitutes a substantial correction to an analytical concentration approaching a factor of 5. For this reason, thermodynamic stability constants are often not cited but the data presented in the form of *conditional stability constants*; that is, a stability constant which is only valid for the conditions stated, such as an ionic strength of 0.7.

For some metals [e.g.; Cu(I,II), Ag(I), Cd(II), Pb(II), Hg(I,II)], the ionic strength is not the key factor in determining the activity of the metal, which can bind strongly to chloride, hydroxide, carbonate or other ligands naturally present. These metal inorganic complexes are

Table 4. Approximations for individual activity coefficient estimations.

Name	Equation	Range (I)
Debye-Huckel	$\log \gamma_i = -\dfrac{A z_i^2 \sqrt{I}}{1 + a_i B \sqrt{I}}$	$<10^{-2.3}$
Davies	$\log \gamma_i = -A z_i^2 \left[\dfrac{\sqrt{I}}{1 + \sqrt{I}} - 0.3 I \right]$	<0.5
B-Dot	$\log \gamma_i = -\dfrac{A z_i^2 \sqrt{I}}{1 + a_i B \sqrt{I}} + \dot{B} I$	$<0.3 - 1$
Pitzer	$\ln \gamma_i = \ln \gamma_i^{dh} + \sum_j D_{ij}(I) m_j + \sum_j \sum_k E_{ijk} m_j m_k$	>6

i,j,k = species; A, B = coefficients; z = electrical charge; I = ionic strength; $å$ = ion size parameter; $\dot{B}$ = coefficient; γ_i^{dh} = Debye-Huckel activity; D_{ij}, E_{ijk} = virial coefficients

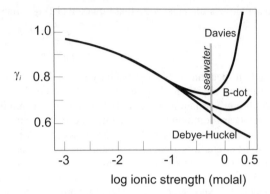

Figure 4. Calculated activity coefficient, γ, for a singly charged ion such as HS⁻ with å = 4 Å using different algorithms. Seawater ionic strength is indicated for reference.

well known. Thus, the free metal $[M^{n+}]$ plus the metal bound to other inorganic ligands, MX_i, equals $[M']$ (Eqn. 14) and

$$[M'] = [M^{n+}] + \sum MX_i \tag{14}$$

and the fraction of free metal, α_M, in the solution without sulfide (or other strong organic ligands) is given by Equations (15) and (16):

$$[M^{n+}] = [M'] \alpha_M \tag{15}$$

where

$$\alpha_M = \frac{1}{(1 + \sum K_{MX_i}[X]_i)} \tag{16}$$

This has also been expressed as the side reaction coefficient for M', $\alpha_{M'}$, (Eqn. 17) which is the reciprocal of α_M:

$$a_{M'} = \frac{[M']}{[M^{n+}]} \tag{17}$$

The side reaction coefficients for inorganic ligands bound to metals in seawater have been tabulated by Turner et al. (1981).

The conditional constant for $M'L$ is related to $M^{n+}L$ by

$$K_{cond\,ML} = \frac{[ML]}{[Mn^{n+}][L']} = K_{cond\,M'L}\,(\alpha_{M'}) \tag{18}$$

Similar equations can be written for sulfide or other anions binding with protons (and common metal ions such as Na, K, Ca and Mg) to give a thermodynamic or pH independent stability constant, K_{therm} (or β_{therm}) (Eqn. 19):

$$K_{therm} = \frac{[ML]}{[M^{n+}][L^{n-}]} = K_{cond\,M'L}\,(\alpha_{M'})\,(\alpha_{L'}) \tag{19}$$

METHODS FOR MEASUREMENT OF METAL SULFIDE STABILITY CONSTANTS

Rickard and Nriagu (1978) commented that if the stability constant for a complex appears to be known to within one logarithmic unit, then it probably has not been measured enough times! Inspection of any non-critical compilation of stability constants (e.g., Sillén and Martell 1964; IUPAC 2006) might suggest that this is case. There are numerous listings of selected values used as a basis for popular equilibrium calculation algorithms. (Smith and Martell 1976; Robie et al. 1978; Lindsay 1979; Wolery 1979; Helgeson et al. 1981; Högfeldt 1982; Wagman et al. 1982; Ball et al 1987; Cox et al. 1989; Delany and Lundeen 1990; Johnson et al. 1991; Robie and Hemingway 1995; Parkhurst 1995; NIST 2005; van der Lee 2005). Lars-Gunnar Sillén himself, when asked how he chose a particular constant, replied that he did this on the basis of his personal knowledge of the laboratory that produced it.

Sillén was being a tad disingenuous: the real work in selecting constants, involves ensuring compatibility between different data sets. We illustrate this with reference to the Fe system. The problem is that popular compilations, such as Wagman et al. (1969, 1982) listed a series of stability constants which were based on the NBS Gibbs free energy of formation for the hexaqua Fe^{2+} ion at 25 °C and 1 atmosphere pressure, $\Delta G°_f(Fe^{2+}_{aq})$, value of -78.9 kJ·mol^{-1}. This value was ultimately derived from the measurements collected by Randall and Frandsen (1932) of 84.9 kJ·mol^{-1}. This value was used in compilations in some very influential textbooks such as Latimer (1952) and Pourbaix (1966) and was later apparently confirmed by the work of Patrick and Thompson (1953) who obtained -78.8 kJ·mol^{-1} and Whittemore and Langmuir (1972) ($\Delta G°_f = -74.3$ kJ·mol^{-1}) which were similar to the value selected by the NBS group. In contrast, Hoar and Hurlen (1958) found -90.0 kJ·mol^{-1}, Larson et al. (1968) found -91.1 kJ·mol^{-1}, Cobble and Murray (1978) -91.5 kJ·mol^{-1}, Sweeton and Baes (1970) -91.8 kJ·mol^{-1}, Tremaine and LeBlanc (1980) -88.92 ± 2 kJ·mol^{-1}. The whole matter was critically reviewed on behalf of the CODATA Task Force on Chemical Thermodynamic Tables by Parker and Khodakovskii (1995) and published in the International Union for Pure and Applied Chemistry (IUPAC) Journal of Physical and Chemical Reference Data. Parker and Khodakovskii (1995) recommended the lower values of -90.53 ± 1 kJ·mol^{-1}. They also reviewed the experimental problems encountered in measurements of this value and showed how the various values had been obtained. The higher values had come about through errors in the measurements of the standard potential of the Fe^{2+}/Fe couple using Fe electrodes. Latimer (1952) had warned about the problems this method involved and Hoar and Hurlen (1958) demonstrated how these problems could be overcome with a kinetic approach. In contrast, Larson et al (1958) used measurements of the specfic heat of hydrous Fe(II) sulfate and Cobble and Murray (1978) measured the specific heat of ferrous chloride. Sweeton and Baes (1970) and Tremaine and LeBlanc (1988) measured the solubility of magnetite to obtain their value.

The significance in the uncertainty in the values for $\Delta G°_f(Fe^{2+}_{aq})$ is that this value is fundamental to all computations based on Fe species in complex natural systems. A system of stability constants, or network, needs to be internally consistent so that relationships between the phases and species can be accurately predicted. The difference between the NBS network $\Delta G°_f(Fe^{2+}_{aq})$ value of -78.9 kJ·mol^{-1} and the modern IUPAC value of -90.53 ± 1 kJ·mol^{-1} is substantial. Fe^{2+}_{aq} is far more stable in computations using the IUPAC value than with the old NBS value. The result is that the relative distribution of dissolved species and solids in Fe-bearing systems based on the older NBS value is erroneous. The problem is more extensive since the compatibility between networks of different cation species is required to determine the relative stabilities of Fe and other cation species. For example, Langmuir (1969) produced an excellent set of Fe stability data which is internally very consistent but which is based on the higher NBS $\Delta G°_f(Fe^{2+}_{aq})$ value. It cannot therefore be used for considerations of the stability of Fe species in systems containing components from other networks.

Since Sillén's time, IUPAC has been steadily producing detailed critical analyses of stability constant data for specific systems. These reports not only recommend a value but also give detailed reasons for why this is done. At the time of writing, an IUPAC team is examining iron sulfide complexes and their results will be an invaluable addition to this area of chemistry.

One approach which partially obviates the $\Delta G°_f$ uncertainty problem is to consider only measured equilibrium constants. Geochemists and environmental chemists like the $\Delta G°_f$ approach because it permits the prediction of the chemical equilibrium state into any system, especially if supported by enthalpy and entropy values. Using only measured equilibrium constants means that the data are not necessarily internally consistent and the application of the data is somewhat restricted to the measured systems. This approach is widely used by solution chemists. It has the advantage that the errors and uncertainties can be minimized: $\Delta G°_f$ is always a derived constant and therefore any measurement errors are promulgated through the derivation and may be relatively signifcant. However, equilibrium constants may also be prone to significant error. For example, using the early Fe electrode measurements of the standard potential of the Fe^{2+}/Fe couple in a system of chemical equations will result in a similar error to that using the NBS $\Delta G°_f(Fe^{2+}_{aq})$ approach. And note that such errors may not be obvious unless the source measurement report is consulted. So to be sure that the results of equilibrium computations—the prediction of the chemical state of environmental or geological systems—are not spurious, it is always necessary to examine the source of the data used. In using the major computer-based equilibrium computation engines, this requirement becomes even more important since it is all too easy to press a button and get what appears to be a meaningful result. In fact, the acronym GIGO of the early computing business applies directly to this area of science: Garbage In, Garbage Out.

Theoretical approaches to the estimation of stability constants

There are two basic approaches to evaluating complex stability constants (a) theoretical and (b) experimental. The theoretical approach in geochemistry was pioneered by R.A. Garrells and established by H. Helgeson and their co-workers in some detail. It is popular in geochemistry since it reaches those parts of the system which experimentation cannot presently reach, such as large ranges of temperature and pressure. Even so attempts to predict standard enthalpies and free energies have not been very successful. The problem is that an error of only 6 kJ·mol^{-1} in reaction energies leads to an error in predicting an equilibrium constant of a factor of 10. Therefore reaction energy computations need to be highly precise in order to be useful. Much of the problem stems from the need to account for the interaction of the solvent with the species of interest. Thus whilst the energetics of gas phase reactions can be computed relatively precisely, the energetics of condensed phase reactions have a considerable uncertainty. Methods for computation of the energetics of metal sulfide complexes have been discussed by Tossell and Vaughan (1992, 1993) and Tossell (1994), but these tend to be semi-empirical rather than strictly *ab initio*.

In the absence of *ab initio* computational methods, straightforward, empirical methods have been used for the prediction of metal sulfide complex stability characteristics. These include correlations based on isovalent-isostructural analogues, ligand valence or number and electrostatic models. These empirical techniques have little grounding in theory. They may work as an approximation for a particular complex or series of complexes but they are not generally applicable. They are useful in checking the consistency of experimental data and strong deviations in behavior of a set of complexes from a similar series requires, at least, some explanation.

Dyrssen (1985, 1988) for example used the isovalent-isostructural analogue approach to estimate stability constants for metal sulfide complexes. Dyrssen's anchor points were experimental measurements of Hg(II) and Cd(II) sulfide complexes and the formation constants of extractable dithizonates (Tables 5-7).

Using these data, Dyrssen (1985) found a relationship for Hg(II) and Cd(II) sulfide stability constants and the dithizone extraction coefficients for Hg(II) and Cd(II) (Eqns. 20-23):

$$M^{2+} + 2H_2S = M(HS)_2 + 2H^+ \quad (K_{22}) \tag{20}$$

and
$$M^+ + 2H_2S = MHS_2^- + 3H^+ \quad (K_{12}) \tag{21}$$
to be
$$\log K_{22} = 0.945 \log K_{ex} - 1.42 \tag{22}$$
and
$$\log K_{12} = 0.948 \log K_{ex} - 7.69 \tag{23}$$

Dyrssen (1988) then produced a complete matrix of stability constants (Table 8) for metal sulfide complexes, based on the same approach.

The problem with the method (apart from the need for correcting these data for errors in pK_{1,H_2S} and pK_{2,H_2S}) is the absence of experimental evidence for the existence of the complexes. Even the data for the Hg(II) and Cd(II) sulfide complexes, on which the scheme is anchored, was based on arithmetic fitting to titration data and lacks direct evidence for the assumed complexes. The Cd(II) sulfide data set in particular appears to include some gross anomalies which are inconsistent with a regular trend. Elliot (1988) also noted problems in the variations in the nature of the aqua ions used in the estimation and in the two order of magnitude spread in the values of K_{ex} for dithizone extraction. Elliot concluded that the resulting estimated metal sulfide stability constants may display errors of several magnitudes. This indeed appears to be the case, as is suggested below in the discussion of metal sulfide stability constants.

Table 5. Equilibrium constants for Hg(II) sulfide complexes by Schwarzenbach and Widmer (1963) assuming $pK_{1,H_2S} = 6.88$ and $pK_{2,H_2S} = 14.15$.

	log$K$
$HgS(s) = Hg^{2+} + S^{2-}$	−50.96
$HgS(s) + H_2S = Hg(HS)_2$	−5.97
$Hg(s) + HS^- = HgHS_2^-$	−5.28
$Hg(s) + HS^- = HgS_2^{2-} + H^+$	−13.58
$Hg^{2+} + 2H_2S = Hg(HS)_2 + 2H^+$	23.96
$Hg^{2+} + 2H_2S = HgHS_2^- + 3H^+$	17.77

Table 6. Equilibrium constants for Cd(II) sulfide complexes suggested by Dyrrsen (1985) to fit the data of Ste-Marie et al. (1964), assuming $pK_{1,H_2S} = 6.9$ and $pK_{2,H_2S} = 13.58$.

	log$K$
$CdS(s) + 2H^+ = Cd^{2+} + H_2S$	−4.64
$CdS(s) + H_2S = Cd(HS)_2$	−4.57
$CdS(s) + HS^- = CdHS_2^-$	−3.93
$Cd^{2+} + 2H_2S = Cd(HS)_2 + 2H+$	0.07
$Cd^{2+} + 2H_2S = CdHS_2^- + 3H^+$	−6.19

Table 7. Selected extraction constants (K_{ex}) for dithizone in carbon tetrachloride, $M^{2+} + 2HD_z(CCl_4) = MD_{z^2}(CCl_4) + 2H^+$ compiled by Dyrrsen (1985).

	log K_{ex} (CCl$_4$)
Mn	−6.5
Fe	3.4
Co	1.59
Ni	1.19
Cu	10.53
Zn	2.26
Cd	1.58
Hg	26.86
Pb	0.38

Table 8. Estimated formation constants (logβ) and solubility products (logK_s) for various metal sulfides and their complexes (Dyrrsen 1988).

	logβ1 (MS)	logβ1 (MHS)	logβ2 (MHS)	logK_s (MS)
Cu^+	23.7	13.3	17.2	—
Ag^+	23.7	13.3	17.2	—
Tl^+	12.4	2.27		—
Mn^{2+}	11.4	−0.5	7.0	1
Fe^{2+}	13.3	1.4	8.9	−4.7
Co^{2+}	16.6	4.7	12.2	−4.6
Ni^{2+}	15.7	3.8	11.4	−3.6 to −7.3
Cu^{2+}	26.0	14.1	21.6	−10
Zn^{2+}	18.5	6.5	14.0	−5.87
Cd^{2+}	18.2	6.4	13.8	−6.85
Hg^{2+}	42.0	30.1	37.7	−9
Sn^{2+}	14.6	2.7	10.2	−11.3
Pb^{2+}	16.9	5.0	12.5	−10.5
Pd^{2+}	56.9	45.0	52.5	—

This computational approach for estimating stability constants was taken to its modern limit by Helgeson (1969) and Helgeson et al. (1978). In these compilations of thermodynamic data, Helgeson used standard molal entropies, heat capacities and volumes derived from correlation algorithms and Clapeyron slope constraints to obtain an internally consistent set of data for around 70 minerals and a large number of soluble species for temperatures between 25 °C and 300 °C. Helgeson's data set is largely based on experimental measurements. Apart from the uncertainties in interpolating these values to other conditions, which was ameliorated to some extent by the iterative nature of the computing process, the Helgeson data set still suffers from the basic uncertainties in the experimental values used as the anchor points. To some extent, assuming that the change in values with temperature and pressure is a continuous function, the computational method allowed selection of a best value for the experimental value too. However, the problem of the nature of the complex used in the data set remains. For example, the Helgeson data set still used $pK_{2,H_2S} \sim 14$ at 25 °C and 1 atm, and did not foresee the experimental data which showed that $pK_{2,H_2S} > 18$. This means that all the sulfide complex data in the Helgeson set are affected by this choice of $pK_{2,H_2S} \sim 14$. The Helgeson data set still provides the basis for the thermodynamic data used in many equilibrium computational programs.

Experimental approaches to the measurement of metal sulfide stability constants.

Titrations. Simple acid-base titrations have been widely used to determine metal sulfide stability constants, especially protonation constants. In this type of approach, acid or alkali is titrated against a solution containing the metal and sulfide, or sulfide is titrated against metal solutions and the results are fitted to model complexes and stabilities. The classical example is the titration of acid against a sulfide solution in water and the determination of pK_{1,H_2S}. This constant forms the basis of all measurements of metal sulfide complex stability constants and thus an accurate assessment of its value is fundamental. And this is not a trivial exercise.

The commonly used modern value for pK_{1,H_2S} at 25 °C is 6.998 and is derived from the work of Suleimenov and Seward (1997). They used a spectrophotometric method which required the measurement of absorbances of dilute sulfide solutions (e.g., 10^{-4} M). Charge-transfer-to-solvent transitions cause intense absorption in the ultraviolet region specific only to the HS$^-$ ion (λ = 231 nm). The preparation and analyses of dilute sulfide solutions is difficult as H_2S is volatile and oxygen sensitive. They also used a spectrophotometric method based on the conversion of sulfide sulphur (H_2S and HS^-) to methylene blue (Gustafsson 1960). This method is not entirely hydrogen sulfide specific as the S(−II) sulphur in polysulfides is measurable but not in a quantitative manner (Luther et al. 1985). Since polysulfides determined by *in situ* techniques (Luther et al. 2001) are not normally a significant fraction of the S(−II) pool, this method can be considered a very precise analytical method for the determination of dilute concentrations of H_2S and HS^-. At ambient temperatures, the pH can be measured with a pH electrode. Even here, there are experimental difficulties since H_2S will react to form metal sulfides which will clog the electrode sinter. Rickard (1989) used an agar-KCl salt bridge to overcome this problem.

Suleimenov and Seward (1997) showed a plot which is familiar to sulfide chemists (Fig. 5) of HS^- versus pH in an aqueous 0.001 M Na^+ matrix. This shows that extremely small changes in HS^- concentration result in extremely large pH changes around pH = 7. This in turn means that the analytical precision for sulfide must be extreme in this area. Suleimenov and Seward (1997) therefore had to buffer the system, and the addition of buffer introduces other complications into the measurements. The nice thing about Suleimenov and Seward's (1997) approach is that they were also able to obtain some information about the structure of the solvated HS^- ion from the spectroscopic data. In particular, they were able to measure the radius of the solvent cavity around the HS^- ion.

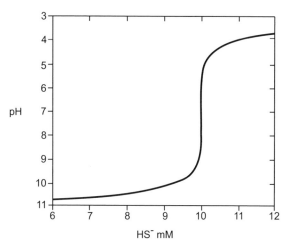

Figure 5. HS^- versus pH in aqueous solution at 25 °C and in the presence of 0.001 M Na^+ (after Suleimenov and Seward 1997).

Voltammetry: general. There has been some discussion about the application of electrochemical methods, especially voltammetry (current, *I* vs. potential, *E* curves), to the measurement of metal sulfide stability constants. To date five methods have been used to determine stability constants. Most methods measure the sulfide (or polysulfide) signal at the mercury electrode which reacts at the Hg electrode according to the following reaction (Eqn. 24) which is expressed in Nernstian form as Equation (25):

$$HS^- + Hg \leftrightarrow HgS + H^+ + 2\ e^- \tag{24}$$

$$E = E^{o\prime} - \left\{\frac{RT}{nF}\ln\left(\frac{[HS^-]}{[H^+]}\right)\right\} \tag{25}$$

where the activity of HgS = 1; $n = -2$; $E = E_p$ (experimentally determined) and $E^{o\prime}$ is the formal potential. Increasing H^+ (decreasing pH) shifts the E_p to positive potentials. Plots of E_p vs. acid/base equivalent give standard "s" shaped curves. Plots of E_p vs. pH from these titrations produce straight line segments of nonzero or zero slope and are related to the number of protons bound to the sulfide (Meites 1965). The operative general reaction for electroactive species, which are not appreciably acidic or basic over the pH range studied, is Equation (26) where R = reduced species and O = oxidized species:

$$RH_q \rightarrow O + qH^+ + ne^- \qquad (26)$$

where $n = -2$ for sulfide electron transfer at the electrode (Eqn. 24). From Meites (1965), q can be evaluated by Equation (27):

$$\frac{dE_p}{d(\text{pH})} = -0.05915 \frac{q}{n} \qquad (27)$$

The slope, $dE_p/d(\text{pH})$, is evaluated from the E_p vs. pH plot over a given pH region. In the neutral to slightly acidic pH region, the uncharged protonated species predominates (for the sulfide case, H_2S exists in this region) and the E_p is independent of pH because no protons are released in the redox reaction (Meites 1965; zero slope for Eqn. 27). The intersection of the basic line segment with the zero slope segment gives pH = pK/q where pK is the dissociation constant of the acid species (e.g.; pK_1 for H_2S; Eqn. 27). The above demonstrates that the stoichiometry of a given sulfur species with H^+ can be determined. Luther et al. (1996) and Chadwell et al. (1999, 2001) used this approach to measure the first pK_a of H_2S and the second pK_a of S_4^{2-} and S_5^{2-}. Their values agree with previous methods used to determine these constants.

There is a possible problem during the determination of the stability constant of a complex in a system where precipitation can occur. For voltammetry this problem has been addressed by Bond and Hefter (1972), who verified that rapid scan voltammetric techniques without a deposition step are amenable to determine stability constants in systems with sparingly soluble salts. They monitored a metal's voltammetric reduction peak while titrating with a known ligand anion. They used the DeFord and Hume (1951) formalism for calculating stability constants which is now discussed.

Voltammetry: titration of sulfide with added metal at constant pH. The first method for determining metal sulfide stability constants is a titration method of sulfide with added metal at constant pH. Here both the decrease in current and the positive shift in sulfide peak potential are monitored. There is no deposition step to preconcentrate the sulfide so that the experiment is performed under diffusion controlled conditions. The DeFord and Hume (1951) equations are used to detect complexes $[M(HS)]^+$, $[M_2(HS)]^-$, $[M_3(HS)]^{-3}$, etc. This method has been verified with known metal ligand complexes (metal thiol and thiosulfate complexes) over a variety of metal concentrations (Luther et al. 2000). For complexes that are labile at the Hg electrode, the method of DeFord and Hume (1951), as modified by Heath and Hefter (1970) determines the successive formation constants of complexes (β_1, β_2, ..., β_n) formed (Eqns. 28 and 29, charges omitted for simplicity):

$$M + L \rightarrow M(L) \qquad \beta_1 = \frac{[ML]}{[M][L]} \qquad (28)$$

$$nM + L \rightarrow [M_n(L)] \qquad \beta_n = \frac{[M_nL]}{[M]^n[L]} \qquad (29)$$

The stability constants can be determined by the relation (Eqn. 30):

$$F_0(X) = \Sigma\beta_n[X]^n = \beta_0 + \beta_1[X] + \beta_2[X]^2 + \ldots + \beta_n[X]^n \qquad (30)$$

where $F_0(X)$ is a polynomial function representing the sum of the $\beta_n[X]^n$ for all complexes, β_n is the overall stability constant of the nth complex, $[X]$ is the analytical or total concentration of the added species (M^{2+} in this case), and $\beta_0 = 1$ for the zeroth complex. $F_0(X)$ is related to the current and potential data by Equation (31):

$$F_0(X) = \text{anti}\log\left\{\left[0.434\frac{nF}{RT}\right][\Delta E_p] + \left[\frac{\log(I_p)_s}{(I_p)_c}\right]\right\} \qquad (31)$$

where $\Delta E_p = (E_p)_s - (E_p)_c$; $n = -2$ for the electrochemical oxidation reaction for sulfide as discussed here (+2 for the electrochemical reduction reaction of divalent cations discussed by DeFord and Hume when the metal concentration is monitored); I_p indicates the peak current; c indicates complexed anion and s indicates free or uncomplexed anion. A plot of $F_0(X)$ versus the metal concentration should give a curve from which the following functions (Eqn. 32) can be evaluated:

$$F_1(X) = \frac{[F_0(X)-1]}{[X]}; \; F_2(X) = \frac{[F_1(X)-\beta_1]}{[X]}; \; \ldots ; \; F_n(X) = \frac{[F_{n-1}(X)-\beta_{n-1}]}{[X]} \quad (32)$$

$F_1(X)$ can be evaluated from $F_0(X)$ and plotted versus ligand concentration. The intercept with the $F_1(X)$ axis is determined by least squares curve fitting and gives β_1. In the original method of DeFord and Hume (1951), the process of calculating $F_n(X)$ from $F_{n-1}(X)$ graphically is repeated until a straight line parallel to the concentration axis (corresponding to the last complex) is obtained. If the $F_0(X)$ plot is a straight line, then the $F_1(X)$ plot is parallel to the concentration axis and only one complex exists. Since the original method, several groups have developed methods to fit the data from the $F_0(X)$ function, but it has become easier to use commercial software to perform (non)linear regression analysis on each $F_n(X)$ vs. $[X]$ curve. In neither DeFord and Hume (1951) nor Heath and Hefter (1977) is there a stipulation that the treatment *must* be performed on either a reduction wave, or a metal ion, and the general form of the equations is presented in Crow (1969).

Klatt and Rouseff (1970) discussed the Lingane formalism (a simple plot of ΔE_p vs. log[X] to determine logβ for a single complex) relative to the DeFord and Hume formalism. They showed that even when the ligand concentration was not in large excess (e.g., $\beta_j C_x^j \sim 1$) all the significant stability constants can be determined by the DeFord and Hume formalism. In the case of $\beta_j C_x^j \sim 1$, a plot of ΔE_p vs. log[X] shows significant curvature when successive formation constants can be determined. The metal-bisulfide system (Luther et al. 1996, 2000) meets these requirements, as do other metal-ligand systems (El-Maali et al. 1989).

Voltammetry: sulfide concentration method. The second method that has been used to determine metal sulfide stability constants is another titration of sulfide with given added metal ion at constant pH (Zhang and Millero 1994; Al-Farawati and van den Berg 1999). The concentration of sulfide (measured as a decrease in current) is monitored during the titration's progress. The sulfide measured is labile and the sulfide not measured is assumed to be tied up in strong, possibly inert complexes but is assumed to be protonated as bisulfide ion (HS^-). There is a deposition step to preconcentrate the sulfide so that the experiment is not performed under diffusion-controlled conditions. To avoid sulfide loss with the Hg pool at the bottom of the cell, Al-Farawati and van den Berg (1999) used a flow analysis method to measure sulfide. A series of simultaneous equations are setup to calculate stability constants for $[M(HS)]^+$, $[M(HS)_2]$, etc. complexes. In the work of Zhang and Millero (1994) only the first two stepwise constants were evaluated by a polynominal regression analysis. Al-Farawati and van den Berg (1999) set up their expression so that higher order HS^- complexes could be measured but, under the conditions of their experiments, only the 1:1 and 1:2 M:HS complexes were reported.

The equations to calculate stability constants are related to the current when sulfide is present, $I_{P,S}$, and when sulfide is not present, I_{max} $[HS]_T$. The ratio, R, is given in Equation (33) and is related to:

$$R = \frac{I_{P,S}}{I_{max}} = \frac{[HS']}{[HS]_T} \quad (33)$$

the sulfide that is measurable in the presence of metal, $[HS']$, with that in the absence of metal, $[HS]_T$. The mass balance for the sulfide (Eqn. 34) is:

$$[HS]_T = [HS'] + [MHS]_T \tag{34}$$

where $[MHS]_T$ (Eqn. 35) is the total concentration of all metal-sulfide species:

$$[MHS]_T = \sum m[M(HS)_m]^{(n-m)} \tag{35}$$

assuming several stepwise complexes can form from the addition of metal to sulfide (Eqn. 36):

$$M^{n+} + mHS^- = [M(HS)_m]^{(n-m)} \tag{36}$$

Combining the mass action expression with these equations gives Equation (37):

$$[HS]_T = [HS'] + \sum m\beta'_m [M^{n+}][HS']^m \tag{37}$$

so that R becomes:

$$R = \frac{[HS']}{[HS'] + \sum m\beta'_m [M^{n+}][HS']^m} = \frac{1}{1 + \sum m\beta'_m [M^{n+}][HS']^{m-1}} \tag{38}$$

Values for β'_m are then obtained by fitting R in Equation (38) using non-linear, least square curve-fitting as a function of $[M^{n+}]$.

Voltammetry: The competitive ligand approach. The third method used by Al-Farawati and van den Berg (1999) is a competitive ligand approach where a metal complex with 8-hydroxyquinoline (or metal-oxine) exhibits a peak that is monitored as sulfide is added to the metal-oxine complex in seawater solutions. The metal-oxine complex current decreases as metal sulfide complexation increases. This method also uses a cathodic stripping experiment with a deposition step to detect the metal in the oxine complex, which is termed labile, as it is not bound to sulfide.

The equations to calculate stability constants are related to the current when sulfide is present, I_S, and when sulfide is not present, I_{max}. This ratio, Q, is also related to the metal that is measurable in the presence of sulfide, $[M^{n+}]_S$, with that in the absence of sulfide, $[M^{n+}]$:

$$Q = \frac{I_S}{I_{max}} = \frac{[M^{n+}]_S}{[M^{n+}]} \tag{39}$$

The mass balance for the metal in the absence of sulfide is given as:

$$[M_T] = [M^{n+}][\alpha_M' + \alpha_{M\text{-oxine}}'] \tag{40}$$

where α_M is the metal side reaction coefficient for binding with the major anions in solution and $\alpha_{M\text{-oxine}}$ is the side reaction coefficient for the metal binding with oxine. In the presence of sulfide, $[M_T]$ is given by:

$$[M_T] = [M^{n+}]_S [\alpha_M' + \alpha_{M\text{-oxine}}' + \alpha_{MHS}'] \tag{41}$$

where α_{MHS} is the side reaction coefficient of metal with sulfide which is related to the stability constant, β'_m, by:

$$\alpha_{MHS}' = \beta'_m [HS']^m \tag{42}$$

so that $[M_T]$ becomes:

$$[M_T] = [M^{n+}]_S \{\alpha_M' + \alpha_{M\text{-oxine}}' + \beta'_m [HS']^m\} \tag{43}$$

Subsituting $[M^{n+}]_S$ and $[M^{n+}]$ into Equation (39), we obtain:

$$Q = \frac{\alpha_M' + \alpha_{M\text{-oxine}}'}{\alpha_M' + \alpha_{M\text{-oxine}}' + \beta'_m [HS']^m} \tag{44}$$

[HS'] is calculated from the sulfide mass balance:

$$[HS'] = [HS]_T - ([M]_T - [M]_{labile}) \quad (45)$$

where $[M]_{labile}$ is the metal that is measurable for each sulfide addition. Values for β'_m are then obtained by fitting Q in Equation (45) using non-linear, least square curve-fitting as a function of sulfide concentration.

Voltammetry: mole ratio method. A fourth method used by Luther et al. (1996, 1999b, 2002) and Luther and Rickard (2005) uses the mole ratio method to determine stability constants for complexes which do not exhibit their own discrete voltammetric wave (peak) and do not dissociate at the electrode (non-labile or inert). The mole ratio method can be used to estimate both the conditional and thermodynamic constants of the complexes. Either the sulfide or metal peak currents can be used to obtain data. Calculations for M_mS_n complexes require that the second dissociation constant for H_2S be known for the calculation. The pK_2 value has changed from 13.78 to 18.5 over the last 40 years and its uncertainty is due to the oxidation of sulfide at high pH to polysulfides (Morse et al. 1987; Schoonen and Barnes 1988). It is possible to calculate the stability constants for metal sulfide complexes without dependence on pK_2. In terms of readily measurable reactants and products, the equations for complex formation and free ligand protonation are Equations (46) and (47), where charges are omitted for simplicity. Equation (46) shows complex formation as a water loss reaction:

$$mM + nHS + nOH \rightarrow [M_mS_n] + nH_2O \quad (46)$$

$$HS + H \rightarrow H_2S \quad (47)$$

Equation (46) is a two component system because the reaction is performed at constant pH. The concentration of a metal is well known from the titration data. Although the sulfide is not readily detected under diffusion control conditions, it can be calculated from titration data. Examples for Zn and Ag are shown in Figure 6 which is plotted as a mole ratio (M/S).

The slopes of the lines in Figure 6 give the stoichiometry of the reaction as the titration progresses. The stability constant, $\beta_{M_mHS_nOH_n}$, for the formation of a metal sulfide species is Equation (48):

$$\beta_{M_mHS_nOH_n} = \frac{[M_mS_n]}{[M]^m[HS]^n[OH]^n} \quad (48)$$

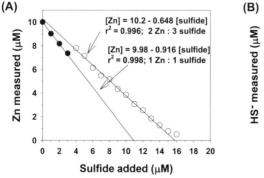
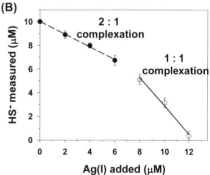

Figure 6. (A) Plot of Zn(II) measured as sulfide is added to a seawater solution with an initial concentration of 10 μM Zn(II). (B) Plot of HS⁻ measured as Ag(I) is added to a sodium nitrate solution containing 10 μM sulfide (from Luther and Rickard 2005).

The total ligand and metal concentrations are given in Equations (49) and (50):

$$c_S = [S] + [HS] + [H_2S] + 3n[M_mS_n] \tag{49}$$

$$c_M = [M] + 3m[M_mS_n] \tag{50}$$

In a typical molar-ratio method, the mole fraction of the complexed metal, α_M (Eqn. 51), is determined experimentally where:

$$\alpha_M = \frac{m[M_mS_n]}{c_M} \tag{51}$$

and $[M]$ is:

$$[M] = (1 - \alpha_M) c_M \tag{52}$$

From Equation (49), the total sulfide concentration ($H_2S + HS^- + S^{2-}$) in terms of bisulfide (HS^-) is:

$$[HS^-]\left(\frac{K_2}{[H]} + 1 + \frac{[H]}{K_1}\right) = c_S - \sum n[M_mS_n] \tag{53}$$

which becomes Equation (54) after substituting with Equation (51):

$$[HS^-]\left(\frac{K_2}{[H]} + 1 + \frac{[H]}{K_1}\right) = c_S - \left(\frac{n}{m}\right)\alpha_M c_M \tag{54}$$

where K_1 and K_2 are the first and second dissociation constants of H_2S. Because K_2 is small relative to K_1, the first term is insignificant whether a value of $10^{-13.78}$ or $10^{-18.5}$ is used. To calculate the thermodynamic constants, only the well documented K_1 value is needed (Morse et al. 1987). Substituting for $[M]$ from Equation (52), $[HS^-]$ from Equation (54) and $[M_mS_n]$ from Equation (51) yields

$$\beta_{M_mHS_nOH_n} = \frac{\left\{\alpha_M\left(\frac{K_2}{[H]} + 1 + \frac{[H]}{K_1}\right)^n c_M^{1-m}\right\}}{\left\{m(1-\alpha_M)^m \left(c_S - \frac{n}{m}\alpha_M c_M\right)^n [OH]^n\right\}} \tag{55}$$

These constants for MS clusters are proton independent as determined by acid-base titrations of the cluster. The metals Zn(II), Cu(II), Pb(II) or Ag(I) with sulfide do not produce an electroactive sulfide signal at circumneutral pH. By adding acid, a sulfide signal was measurable once the the metal sulfide complex or cluster dissociated to produce free sulfide. For AgS clusters, free sulfide only becomes measurable at pH = 2 so the complex is stable to a pH of 2, and that value is used in Equation (55) to calculate $\beta_{M_mHS_nOH_n}$. The values for Zn, Pb and Cu are 6.7, 6.0 and 5.0, respectively (Luther et al. 1996; Rozan et al. 2003).

If pH is kept constant, substitution of the appropriate m and n values must satisfy the relationship (Eqn. 56):

$$K_{COND}^n = \frac{\alpha_M}{m(1-\alpha_M)^m \left(c_S - \frac{n}{m}\alpha_M c_M\right)^n} \tag{56}$$

Normalizing for $m = 1$ gives Equation (57):

$$K_{COND} = \frac{\alpha_M^{1/n}}{m^{1/n}(1-\alpha_M)^{m/n}\left(c_S - \frac{n}{m}\alpha_M c_M\right)} \tag{57}$$

Voltammetry: chelate scale approach. A fifth method employed the chelate scale approach, which Chadwell et al. (1999, 2001) used to measure Zn and Cu polysulfide stability constants. When a metal ligand complex is reduced to a metal amalgam (Eqn. 58):

$$ML + 2e^- \rightarrow M(Hg) + L \tag{58}$$

the half-wave potential of a metal complex, $E_{1/2}'$, or the peak potential, E_p, can be directly related to the thermodynamic stability constant, K_{therm} (Lewis et al. 1995; Croot et al. 1999; Rozan et al. 2003) by Equation (59):

$$E_{1/2}' = E_{1/2} - \frac{2.303\,RT\,\log K_{therm}}{nF} \tag{59}$$

A plot of $E_{1/2}'$ vs. $\log K_{therm}$ for a series of known metal ligand complexes can be constructed from the literature or from experiment to derive information on K_{therm} for newly formed complexes. This particular form of the Lingane equation assumes:

(a) No dependence on the reduced metal since it is an amalgam. Thus the complex is destroyed and this is a measure of the bond strength and K_{therm};

(b) $E_{1/2}'$ is independent of ligand concentration, which can be checked by titrating the metal with ligand until no further change in $E_{1/2}'$ is observed.

This method has not been able to measure metal sulfide stability constants for the metals Cu, Pb, Cd and Zn as no metal sulfide peak was observed. These data indicate that these metal sulfide complexes have stability constants greater than $\log K = 40$.

In summary, the titration studies of Zhang and Millero (1994), Luther et al. (1996) and Al-Farawati and van den Berg (1999) were normally performed only at pH 8 in seawater and at low total sulfide concentrations (<10 μmolar). Zhang and Millero (1994) and Al-Farawati and van den Berg (1999) assumed HS^- complexes for all metals. Luther et al. (1996) observed free HS^- in solution with Mn, Fe, Co and Ni and assigned these as HS^- complexes. However, no free HS^- was observed in titration studies with Cu and Zn (Luther et al. 1996), Pb (Rozan et al. 2003) and Ag (Rozan and Luther 2002; Luther and Rickard 2005) until the pH was lowered. These complexes were assigned as S^{2-} species at pH > 7.

Solubility methods. These methods use pure or synthesized metal sulfide minerals or solids as the starting material. Sulfide, usually at millimolar concentrations, is then added to the solids in sealed tubes over a range of pH values and equilibrated. Filtration is normally used to separate soluble complexes from the solid material after equilibration. Unfortunately earlier work did not always specify the type of filter. Recently, 0.20 μm filters or dialysis membranes have been used for separation. After separation, the total metal and sulfide present in the filtered solution are measured. These data are then modelled to obtain metal sulfide stability constants.

The basic problem of this approach is that curve fitting of a series of supposed complexes with estimated stability constants does not necessarily provide a unique solution. The question of the uniqueness of the solution is rarely addressed although the uncertainty in the reported solution is usually computed. Independent evidence regarding, for example, the degree of protonation or the complex stoichiometry, is required before any reliability can be placed on the computed stability constants. One of the astonishing things the uninitiated reader will discover in this chapter is the large number of sulfide complexes that have been proposed and even modeled with little or no evidence to support their existence in the first place.

For example, a problem with some of the models (Ste-Marie et al. 1964; Gubeli and Ste-Marie 1967; Hayashi et al. 1990; Daskalakis and Helz 1992) is that they assume that mixed complexes with sulfide and hydroxide can exist, e.g., [CdOH(S)]⁻, [Zn(OH)(SH)] and [Zn(OH)(HS)₂]⁻. Dyrssen (1991) pointed out that the stoichiometry of water cannot be determined in aqueous solutions since the activity of water is almost constant; thus, there is limited experimental support for such complexes. Wang and Tessier (1999) also concluded in their experimental study on the Cd-S system that [CdOH(S)]⁻ does not exist.

Most solubility studies model complexes as successive HS⁻ addition to a single metal cation as in [M(HS)]⁻, [M(HS)₂], etc. The problem here is that the proposed stoichiometries, in the absence of independent information, are ambiguous and are not in themselves unique. Therefore, for example, several workers have pointed out that a M(HS)₃⁻ species is indistinguishable from a [M₄S₆] species whose existence is supported by molecular experimental data. Thus, a [Cu₂S(HS)₂]²⁻species (i.e., [Cu₂S₃]) has been suggested by Mountain and Seward (1999) and this species would be analogous to an [M₄S₆] species.

A generalized approach to determining stability constants begins with knowledge of the solubility product (Eqns. 60, 61) of the *MS* solid:

$$MS_{(s)} + H^+ \rightarrow M^{2+} + HS^- \quad (60)$$

$$K_{sp} = \frac{\{M^{2+}\}\{HS^-\}}{\{H^+\}} \quad (61)$$

Equation (61) is combined with equations for stepwise metal bisulfide (Eqn. 62), hydroxide (Eqn. 63) and mixed hydroxide-bisulfide (Eqn. 64) complexes:

$$M^{2+} + nHS^- \rightarrow [M(HS)_n]^{2-n} \qquad K_n \quad (62)$$

$$M^{2+} + qHS^- + rH_2O \rightarrow [M(OH)_r(HS)_q]^{2-r-q} + rH^+ \qquad K_{rq} \quad (63)$$

$$M^{2+} + mH_2O \rightarrow [M(OH)_m]^{2-m} + mH^+ \qquad *K_m \quad (64)$$

The total soluble metal, [*M*], (Eqns. 65, 66) is then a function of the individual metal species:

$$\sum[M] = [M^{2+}] + \sum[M(HS)_n^{2-n}] + \sum[M(OH)_r(HS)_q^{2-r-q}] + \sum[M(OH)_m^{2-m}] \quad (65)$$

$$\sum[M] = f(K_{sp}, K_n, K_{rq}, *K_m, pH, \sum S(-II), \gamma) \quad (66)$$

and multiple-regression analysis of these expressions is used to identify the metal bisulfide complexes that best fit the experimental data. Expressions can also be written that include the S²⁻ ion as a metal ligand.

METHODS USED TO DETERMINE THE MOLECULAR STRUCTURE AND COMPOSITION OF COMPLEXES

There are two distinct aspects to characterizing complexes: (1) measuring their stability and (2) determining their structure and composition. Although, ideally both attributes are described in published reports on complexes, this is not always the case; in fact, it is relatively rare in the geochemical literature. One reason is that natural concentrations of some significant complexes are very small (as is the case in the sulfide complexes) and isolation in sufficient quantities for structural analysis is not possible. Another is that many geochemists live in an equilibrium world where the actual form of the complex is less important than its stability constant, which can be used to predict its distribution. This results in conflicting reports in the published literature about the stability of complexes and indeed about their actual existence in significant concentrations in the real world. This is currently the situation with regard to many aspects of metal sulfide complexes, few of which have been conventionally isolated and characterized.

Commercial programs, such as PEAKFIT®, are widely used for deconvolution of the data obtained by solubility, titration and spectroscopic methods. Such programmes include quite sophisticated fitting engines involving, for example, non-linear peak fitting, and include various data smoothing algorithms. Some groups have developed their own programs for the non-linear treatment of data (e.g., Seward and his co-workers) and these are often based on the same algorithm as the commercial programs. For example, the Marquardt–Levenberg non-linear minimization algorithm is integral to both the PEAKFIT and Seward group approach (Suleimenov and Seward 2000). The problem is that simple titrations or solubility measurements in themselves do not necessarily give a unique solution to complex stability constants (see Suleimenov and Seward 2000). This is because the data are being used to determine the solution to an *inverse* problem. That is, the experimental data provide the result but mathematical analysis is required to determine, or deconvolute, the characteristics of the parameters producing this result. It's the other way around to many mathematical problems where you input the parameters and calculate the result. The parameters to be determined in a stability constant problem usually involve two phenomena: (1) the complexes which are present in the solution and (2) the stability constants for those complexes. Since these two phenomena are interdependent, the problem to be solved is typically non-linear, which usually makes it impossible mathematically to determine unique solutions to the problem. It is important to note that this is not a function of the experimental design but an intrinsic property of the mathematical system. Thus, although the stability algorithms derived may describe the experimental results with apparent precision, the application of these results to the real world may involve large uncertainties.

Chemical synthesis of complexes

The chemical approach to complexes is different to that of the geochemist. The chemist is interested in the nature of the complexes rather than simply in stability constants. Thus the chemical literature on metal sulfide complexes is dominated by syntheses, with most performed in organic solvents (see the mini-review by Rauchfuss 2004). The complexes are then traditionally crystallized as a salt and the structure probed, basically by X-ray analysis. The problem then is to extend the data from the solid phase to information about the complex in solution. It is fairly obvious that the structure and composition of the complex moiety in the crystalline salt is not *a priori* identical to that of the complex in solution because the soluble complex undergoes more molecular motion.

In fact the data obtained on the composition and structure of the complex from the crystal data provides a firm platform from which to go hunting for the complex in solution. The structure and composition of the complex in the solid phase may suggest a number of methods, including extended X-ray absorption fine-structure spectroscopy (EXAFS), X-ray absorption near edge structure (XANES), nuclear magnetic resonance (NMR), Raman, infrared (IR) and ultraviolet-visible (UV-VIS) spectroscopy and mass spectrometry which can provide further information on the complex in solution. In the study of Cu sulfide complexes, for example, we have listed the complexes formed in organic solvents by Achim Müller and his Bielefield group where both crystal chemical and solution spectroscopic evidence are provided. The contrast between this approach and the conflicting and often confused reports in the geochemical literature is marked. On the other hand, the geochemical literature does give stability constants, which is lacking in the chemist's approach.

So you have a choice. You can either chose to compute solution speciation based on a number of complexes that may or may not actually exist—remembering that these are often interdependent; or, you can discuss the chemistry qualitatively in terms of the real entities—but you will not be able to predict the likely solubility in diverse environmental situations. It would be nice if someone were to put both approaches together and, indeed, this must be a primary target for future geochemical research.

LIGAND STABILITIES AND STRUCTURES.

Molecular structures of sulfide species in aqueous solutions

The primary species for sulfide in aqueous solution are H_2S and HS^- (see below). As we show below, HS^- is a Lewis base whereas H_2S can act as a Lewis base or acid.

Qualitative molecular orbital theory provides insights on how electron orbitals interact to control the outcome of reactions. For reactivity the most important orbitals in molecules are the two frontier orbitals: the highest occupied molecular orbital (HOMO) and the lowest unoccupied molecular orbital (LUMO). The LUMO receives electrons donated by the HOMO. The frontier orbitals for the bent molecule H_2S (S-H-S bond angle 92°) are well known (see the compilation of Gimarc 1979). Figure 7 shows the molecular orbital energy level diagram for H_2S which results from the linear combination of the two hydrogen atom's $1s$ orbitals and the sulfur atom's $3s$ and $3p$ orbitals. It also compares the energy level diagrams of HS^- with H_2S. The energies of these orbitals are an important feature of their reactivity.

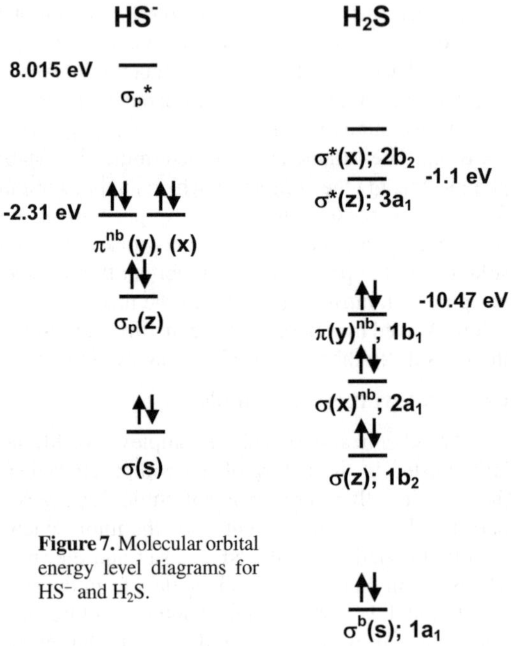

Figure 7. Molecular orbital energy level diagrams for HS^- and H_2S.

In electron-transfer processes the HOMO of the reductant overlaps the LUMO of the oxidant with the same symmetry in order to initiate outer sphere electron transfer. In chemical reactions, a Lewis base HOMO combines with a Lewis acid LUMO. Again the orbitals must have similar symmetries with respect to the bond axis so that they can overlap (Pearson 1976). The reaction is symmetry-allowed if (a) the molecular orbitals are positioned for good overlap (b) the energy of the LUMO is lower than, or less than 6 eV above, that of the HOMO and (c) the bonds thus created or broken are consistent with the expected end-products of the reaction.

The Lowest Unoccupied Molecular Orbital (LUMO) for HS^- was calculated to be +8.015 eV (Rickard and Luther 1997) with no experimental data available for comparison. However, the high positive energy indicates HS^- cannot be an electron acceptor. The Highest Occupied Molecular orbital (HOMO) for HS^- was calculated to be −2.37 eV, which compares well with the experimental value of −2.31 eV (Drzaic et al. 1984; Radzig and Smirnov 1985). The HOMO for HS^- is less stable than that for H_2S (−10.47 eV; see below) indicating that HS^- is more nucleophilic and basic than H_2S, consistent with known reactivity.

The HOMO orbital for H_2S was calculated to be −9.646 eV whereas the experimental value from ionization energy data is −10.47 eV (Drzaic et al. 1984; Radzig and Smirnov

1985). Thus, H_2S is not an excellent electron donor because the HOMO is so stable. This is in accord with known metal sulfide complexes, which have metals bound to HS^- and S^{2-}. At low pH, a ligand field stabilized H_2S complex in water has been documented only for Ru(II) (Kuehn and Taube 1976), which is a low spin metal with a t_{2g}^6 electron configurations. H_2S can act as an electron donor to metals because metal cations have LUMO orbitals of similar energy or more stable energies compared to the HOMO of H_2S or have an empty orbital due to water exchange.

The LUMO orbital for H_2S using the semi-empirical approach was calculated to be +0.509 eV (Rickard and Luther 1997) whereas the experimental value is −1.1 eV based on electron affinity data complied by Radzig and Smirnov (1985). These data indicate that H_2S can be an excellent electron acceptor; in comparison, the LUMO for oxygen is only −0.47 eV. Thus, on energetic considerations alone, H_2S can be an effective electron acceptor. In the reaction of FeS with H_2S to form pyrite, H_2S is the electron acceptor (Rickard 1997; Rickard and Luther 1997).

The LUMO orbital for H_2S (termed $3a_1$) is made from the combination of $1s$ orbitals from each hydrogen and the p_x orbital of sulfur which also mixes with the s orbital of sulfur (Gimarc 1979). The molecular orbital is delocalized across all three atoms since the sign of the wavefunction encompasses all atomic centers. Figure 8 shows the molecular orbital and the charges from the *ab initio* calculations of Trsic and Laidlaw (1980). Because the LUMO is an antibonding orbital in the bent H_2S molecule, it is more destabilized relative to similar molecular orbitals for linear molecules such as BeH_2 (Gimarc 1979). Because of this destabilization, electrons added to this LUMO orbital cause a weakening of both S-H bonds.

Polysulfide ions consist of chains of sulfur atoms. A neat way of illustrating the nature of these chains was developed by Müller and Diemann (1987) and is shown in Figure 9. For polysulfides the dihedral angles vary between 60 and 110°. In Figure 9 the angle is schematically fixed at 90°. The S_3^{2-} ion is necessarily co-planar. Adding a further S atom leads to two possible forms, the *d*- and *l*- isomers. Adding a further S atom to produce S_5^{2-} also provides two possibilities giving rise to *cis* and *trans* forms. The S_6^{2-} ion then can form three enantiomers *cis,cis, trans,trans* and *cis,trans* each *d*- and *l*-isomers respectively (Fig. 9). The *cis*-S_5^{2-} and the *cis,cis*-S_6^{2-} ions effectively constitute fragments of an S_8 ring. In contrast, the *trans*-S_5^{2-} and the *trans,trans*-S_6^{2-} really correspond to parts of an infinite helical chain, as in fibrous sulfur. In complexes, the normal arrangement is *all-trans* conformations, although the *cis*-conformation has been detected in α-Na_2S_5. Interestingly, as shown below, the S_4^{2-}, S_5^{2-} and S_6^{2-} ions are the most abundant in polysulfide solutions at pH > 7. The structures and charges for the S_2, S_3, S_4 and

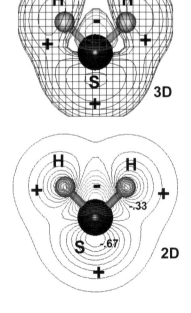

Figure 8. LUMO for H_2S. Upper panel is a three dimensional representation; lower panel is a two dimensional representation with charges from Trsic and Laidlaw (1980). The positive sign indicates the positive part of the orbital's wavefunction and the negative sign the negative part of the orbital's wavefunction.

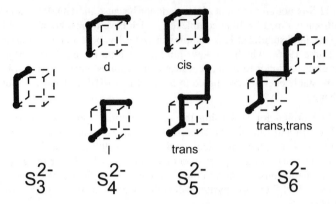

Figure 9. Schematic representation of polysulfide ions S_n^{2-} (where n = 3-6) showing the origin of isomerism in higher chain polysulfides (after Müller and Diemann 1987).

S_5 systems are given in Figure 10. The charges indicated on the S atoms are from the extended Hückel calculations by Meyer et al. (1977), who looked at S_n where n = 2-8. They are in reasonable agreement with the charges from the *ab initio* calculations of Trsic and Laidlaw (1980) who only looked at n = 1-4.

S (–II) equilibria in aqueous solutions

The chemistry of S(–II) in aqueous environmental systems has been well constrained (Morse et al. 1987). pK_{1,H_2S} is close to 7 (e.g., Suleimenov and Seward 1997) which means that H_2S dominates the system at acid pH values and HS^- is the dominant species in alkaline solutions. pK_{2,H_2S} is less precisely constrained but is estimated to be around 18 (Giggenbach 1971; Schoonen and Barnes 1988). This means that the aqueous sulfide ion, S^{2-}, has no significant activity in natural aqueous systems even though MS solids eventually form on precipitation. The problem is that some compilations of stability constants still include older

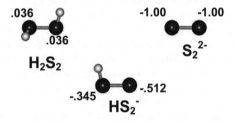

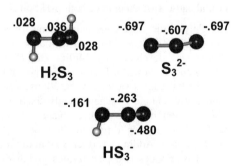

Figure 10. Structures for the S_2, S_3, S_4 and S_5 polysulfide species. Charges are from Meyer et al. (1977).

pK_{2,H_2S} values around 12 or 14, or include sulfide solubility constants which are based on these older values. These still slip readily into the literature since thermodynamic databases may include these intrinsic errors as pointed out originally by Schoonen and Barnes (1988).

In conventional equilibrium diagrams (e.g., Fig. 11), a boundary is often drawn at pH = 7 between areas dominated by H_2S and HS^- since pK_{1,H_2S} is close to 7. At this point the activities of H_2S and HS^- are equal. However, it is important to remember that H_2S exists in quite substantial quantities in solutions at pH > 7, although its relative proportional declines logarithmically-likewise with HS^- in acidic solutions. Also the boundary only refers to

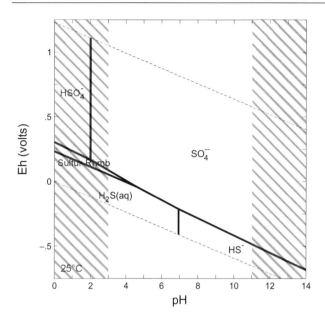

Figure 11. Conventional pH-Eh diagram for stable dissolved sulfur species at 25 °C, 1.013 bars total pressure and a total S activity of 10^{-3}. The hatched areas indicate the limits of the diagram where the activity coefficients of the species become much different from unity. The dashed lines indicate the "stability limit of water" where O_2 and H_2 gas pressures exceed 1.013 bars pressure. See text for discussion.

equality of the activites of the two species. In fact, the empirical Setchenow equation suggests that the activity coefficient for H_2S is close to unity even at seawater ionic strengths (Millero and Schreiber 1982), whereas the Davies equation suggests that the activity coefficient for HS^- approaches 0.6. That is in seawater at pH = 7, there may be equal activities of H_2S and HS^-, but the concentration of HS^- is 40% greater than that of H_2S.

The situation is even more misleading in considering the boundary between aqueous $SO_4(-II)$ and $S(-II)$ species. This boundary is often considered to denote the divide between "oxidizing" and "reducing" environments. In fact, of course, oxidation and reduction is all relative to the species being oxidized or reduced and can occur at any potential. For example, as noted above, H_2S itself is a decent oxidizing agent even when compared with O_2. Another common view is that the boundary marks the limit of oxic systems and below this boundary the conditions are "anoxic". In fact of course, this is not the case, thermodynamically at least. The O_2 partial pressure decreases with decreasing electrode potential but (a) the electrode potential is not in equilibrium with dissolved O_2 in natural aqueous solutions and (b) the calculated O_2 partial pressures are thermodynamic concepts which may have no physical meaning.

The $SO_4^{2-}/S(-II)$ boundary is again a locus where the activities of SO_4^{2-} and $S(-II)$ species are equal. This means that SO_4^{2-} still occurs in substantial quantities below this boundary and, probably more importantly, $S(-II)$ species occur in quite substantial quantities above this boundary—that is, in apparently oxic water. The activity coefficients for the various sulfur species are not equal, as was pointed out for H_2S and HS^- above. In this case we have a divalent species involved and the Davies equation would suggest that the activity coefficient for SO_4^{2-} in seawater approaches 0.2. This means that the concentration of SO_4^{2-} where the activities with $S(-II)$ species are equal is 80% greater that H_2S and 40% greater than HS^-.

Of course, this is all presented in terms of equilibrium thermodynamics, so that it only refers to the state of the system at equilibrium and not to the real world. In the real world, a number of sulfur oxyanions as well as polysulfides (see below) occur and these are not considered in the equilibrium treatment. Equilibrium diagrams can be constructed for these species, but there is some question as to whether the system approaches equilibrium between individual species in the aqueous sulfur system. This together with the intrinsic uncertainty in the stability constants and activity coefficients for these species means that there may be considerable uncertainty in the application of the results of such an approach to natural systems.

Counter intuitively, biologically-mediated processes may help. Microorganisms are intimately involved in many of the transformations involved in sulfur species in natural systems. However, the microorganisms do not produce reactions that are thermodynamically impossible. An example is bacterial sulfate reduction, which produces most of the S(–II) in sedimentary systems. The reduction of SO_4^{2-} to S(–II) is possible inorganically but requires extreme chemical conditions at low temperatures in order to break down the very stable symmetrical SO_4^{2-} molecule. Bacteria bring a very effective enzyme system to bear on the reaction which catalyses an otherwise kinetically hindered process. In this sense, it may well be that biological processes in the sulfur system actually promote the approach to equilibrium rather than complicate it.

Polysulfide stabilities

Much of the published work on the geochemistry of the short chain ($n \leq 5$) polysulfides in low temperature aqueous conditions uses free energy data for these species from Boulegue and Michard (1978), Cloke (1963a,b), Giggenbach (1972), Maronny (1959) and Teder (1971). Rickard and Morse (2005) reviewed the published data and underlined the importance of the report by Kamyshny et al. (2004) which has provided a more secure underpinning for understanding polysulfide geochemistry.

In their classical study, Schwarzenbach and Fischer (1960) titrated HCl against Na_2S_4 and Na_2S_5 solutions. They extrapolated these measurements to S_3(–II) and S_2(–II) species. Kamyshny et al. (2004) trapped aqueous polysulfides with methyl trifluoromethanesulfonate and determined the dimethylpolysulfides formed with HPLC. They used the Schwarzenbach and Fischer (1960) data set in combination with measured data to derive their stability constants. They employed a linear algorithm similar to that originally derived by Cloke (1963a,b), Schoonen and Barnes (1988) and Williamson and Rimstidt (1992) to determine the protonation constants for polysulfides from the original data.

Schwarzenbach and Fischer (1960) only measured protonation constants for S_4^{2-} and S_5^{2-} and their data for S_3^{2-} and S_2^{2-} are in themselves extrapolated. So these linear extrapolations are based on two experimental points. Independent voltammetric measurements of pK_2 for S_4^{2-} and S_5^{2-} were reported by Chadwell et al. (1999, 2001). Chadwell et al. (2001) found a pK_2 for S_4^{2-} of 6.6 and Chadwell et al. (1999) found that $pK_2 = 6.05 \pm 0.5$ for S_5^{2-}. These values agree with Schwarzenbach and Fischer (1960) but are somewhat higher than Kamyshny et al.'s (2004) values. Even so, precise measurements of the protonation constants for the polysulfides are urgently required. The shorter chain polysulfides {S_n(–II) where $n < 4$} and the longer chain polysulfides ($n > 5$) have never been individually isolated in aqueous solutions. Their occurrence is based on an arithmetic analysis of spectroscopic or mass data for total polysulfide solutions under varying conditions.

Stability data for the polysulfides are listed in Tables 9 and 10. Using the Kamyshny et al. (2004) data it is possible to determine polysulfide speciation versus pH (Fig. 12) in the presence of excess S(0). The calculations based on these data show that polysulfides become the dominant species in alkaline solutions relative to S(–II). In the model solution chosen, for example, polysulfides become the dominant species at pH > 9. The most important species in this pH range are S_4^{2-}, S_5^{2-} and S_6^{2-}. Rickard and Morse (2005) commented that one of the features of the Kamyshny et al. (2004) data set is the remarkable relative stability of

Table 9. Thermodynamic constants ($pK_{1,S_n^{2-}}$) for polysulfide formation (from Rickard and Morse (2005): $(n-1)/8\ S_8(s) + HS^- \rightarrow S_n^{2-} + H^+$

n	Maronny (1959)	Cloke (1963a,b)	Teder (1971)	Giggenbach (1972)	Boulegue and Michard (1978)	Kamynshy et al. (2004)
2	12.16	14.43	No data	12.68	12.68	11.46 ± 0.23
3	10.85	13.19	11.75	11.29	12.50	10.44 ± 0.21
4	9.86	9.74	10.07	9.35	9.52	9.70 ± 0.07
5	9.18	9.50	9.41	9.52	9.41	9.47 ± 0.05
6	Does not exsit	9.79	9.43	Does not exsit	9.62	9.6 ± 0.07
7			Does not exist			10.24 ± 0.13
8			Does not exist			10.79 ± 0.16

Table 10. Acid dissociation constants for polysulfides

pK_{1,H_2S_n} $H_2S_n \rightarrow H^+ + HS_n^-$ pK_{2,HS_n^-} $HS_n^- \rightarrow H^+ + S_n^{2-}$

These are corrected values from Rickard and Morse (2005). The Kamyshny data set listed in Rickard and Morse (2005) were uncorrected for ionic strength. The Kamyshny vaues for n = 2–5 are derived from those of Schoonen and Barnes (1998) and both sets are based on the Schwarzenbach and Fisher (1960) data set.

	pK_{1,H_2S_n}		pK_{2,HS_n^-}	
n	Schoonen and Barnes (1988)	Kamyshny et al. (2004)	Schoonen and Barnes (1988)	Kamyshny et al. (2004)
2	5.12	5.11	10.06	10.03
3	4.32	4.31	7.86	7.83
4	3.92	3.91	6.66	6.63
5	3.58	3.61	6.02	6.03
6	—	3.58	—	5.51
7	—	3.48	—	5.18
8	—	3.40	—	4.94

the hydrodisulfide ion, HS_2^-, over the environmentally significant pH range of 6-8. It is the dominant polysulfide at pH < 7 and contributes to ca. 1% of the total dissolved sulfide in much of the system. Rickard and Morse (2005) noted this because of the significance of the disulfide ion in key minerals such as pyrite.

The S(−II) system is highly sensitive to oxidation and the exclusion of air during the sampling and analytical procedures is not entirely possible. The result is that most *ex situ* analyses of environmental polysulfides are probably on the high side, due to artefactual polysulfide production during sampling and analysis. This is particularly true of work on natural sediments as sectioning of sediments could lead to the mixing of oxidized material (including Fe(III) and Mn(III,IV) phases) with reduced sediments. The development of *in situ* analytical methods (Brendel and Luther 1995; Rozan et al. 2000b) has provided a more accurate insight into the distribution of polysulfides in natural aqueous systems (e.g., Luther et al. 2001). Rozan et al. (2000b), for example, found that the presence of polysulfides in estuarine sediments was limited to a thin transition zone between sediments with S(0) dominant and the deeper sulfide zone. The S_x(−II) concentration was 13.9 µM or about 20% of the S(−II) concentration. Some more general implications were derived from examining deep hydrothermal vent systems (Rozan et al. 2000b). Here polysulfides occur in diffuse flow

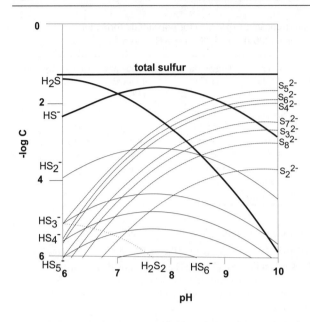

Figure 12. Polysulfide speciation in aqueous solution in terms of log concentration versus pH based on the Kamyshyny et al. (2004) experimental data set for a supersaturated 50 mM total sulfur concentration in the form of a K_2S_5 precursor and $I = 0.3$, with activity coefficients estimated by the Davies equation (from Rickard and Morse 2005).

regions where trace O_2 is present. The $S_n(-II)$ concentration was estimated to be 0.27 μM or around 5% of the S(−II). The implication of these observations is that polysulfides are likely to be present in more oxidized zones in sediments where trace O_2 is a possible constituent.

METAL SULFIDE COMPLEXES

The first transition series: Cr, Mn, Fe, Co, Ni, Cu

The chemistry of the first transition series metals is determined largely by the $3d^n$ electron shell and these can be described as the d-block elements. The chemistry of their complexes has been central to the development of coordination chemistry, as mentioned above. The classical chemistry of the d-block elements developed from the perspective of complexes with a single central metal. Improved techniques for structural determination of the complexes have shown that many d-block complexes have metal-metal bonding, which are described as clusters or cages. In fact, cluster complexes are known throughout the periodic table, but are most numerous in d-block elements.

Chromium. Cr(III) forms a simple $[Cr(H_2O)_5HS]^{2+}$ complex (Ardon and Taube 1967) which was synthesized by the redox reaction of Cr(II) with polysulfide. Because of the stable t_{2g}^3 electron configuration of Cr(III), it is a stable complex to water exchange with a half life of 55 hr at pH 2 and 25 °C, and slowly oxidizes to S_8 in oxygenated waters even in 1 M acid. It has a well-defined Ultraviolet–Visible (UV-VIS) spectrum with peaks at 575 nm (27.5 $M^{-1}cm^{-1}$), 435 nm (43.1 $M^{-1}cm^{-1}$) and 258 nm (6520 $M^{-1}cm^{-1}$). The complex has been precipitated and characterized by total elemental composition (Rasami and Sykes 1976). Al-Farawati and van den Berg (1999) have determined the stability constant, $\log K = 9.5$ (corrected for the side reaction coefficient of Cr(III) in seawater). This complex likely exists in nature and forms directly from the reaction of $[Cr(H_2O)_6]^{3+}$ with H_2S. The reaction of Cr(VI) with excess sulfide to form this complex has not been verified in field or laboratory studies.

Manganese. Manganese occurs in a number of oxidation states in natural systems. However, the Mn(II) ion is the only species which has a significant sulfide chemistry. Mn(II) constitutes a member of the Irving-Williams series of divalent metals from the first transition series. It is therefore classified as a borderline hard-soft metal, but it is placed more towards the harder edge of the group where more ionic bonding is dominant. It thus forms sulfides which are not as stable as subsequent members of this group. For example, Mn(II) forms a number of relatively soluble sulfide phases, including alabandite (α-MnS). The solubility of crystalline, bulk alabandite appears to be well established, but there seems to be a discontinuity between solubility measurements and Mn sulfide complex formation and stabilities. There is a practical problem in controlling the chemistry and structure of the synthetic Mn sulfide precipitates in low temperature aqueous solutions which provides an added uncertainty to solubility measurements. Furthermore, the solubility of alabandite has been measured assuming that the dissolved Mn(II) is entirely in the form of the Mn(II) aqua ion, $[Mn(H_2O)_6]^{2+}$.

Studies of Mn sulfide complexes have mainly been made by voltammetric methods (Zhang and Millero 1994; Luther et al. 1996; Al-Farawati and van den Berg 1999) although Dyrssen (1985, 1988) applied linear free energy estimates to obtain stability constants for the Mn sulfide complexes. These studies have proposed that the $[Mn(HS)]^+$ complex dominates, although the complex has not actually been observed and the results are mainly derived by curve fitting. Some support for the formulation comes from the observation of protons being involved in the complex. Al-Farawati and van den Berg (1999) found evidence for $[Mn(HS)_2]^0$ from curve fitting, but this was not observed by Zhang and Millero, (1994) or Luther et al. (1996). Figure 13 shows structures for these possible complexes. The other metals considered in this chapter with six-coordination would have similar structures.

Polysulfide complexes with compositions $[MnS_4]^0$, $[Mn_2S_4]^{2+}$, $[MnS_5]^0$ and $[Mn_2S_5]^{2+}$ have been described by Chadwell et al. (1999, 2001). They found no evidence for sulfide rich complexes of the form $[Mn(S_n)_2]^{2-}$. Nor did they find any protonated complexes, which is consistent with H^+ outcompeting the metal for the terminal polysulfide site. They found that the $[MnS_4]^0$ and $[MnS_5]^0$ complexes were monodentate (based on similar stability constants to the $[Mn(HS)]^+$ complex) where only one terminal S from the polysulfide binds to one metal (η^1) giving a structure like Mn-S-S-S-S and a formulation $[Mn(\eta^1-S_4)]$. In $[Mn_2S_4]^{2+}$ and $[Mn_2S_5]^{2+}$, the S_n ligand binds two metal centers (μ), with molecular arrangements like Mn-S-S-S-S-Mn and a formulation $[Mn_2(\mu-S_4)]^{2+}$. Figure 14 shows molecular models for these possible complexes with the Mn(II) in octahedral site geometry. The other metals considered in this chapter with six-coordination would have similar structures.

Using organic solvents, Coucouvanis et al. (1985) synthesized and characterized $[Mn(S_6)(S_5)_2]^{2-}$ where the S_n^{2-} species are bidentate. In this complex, the Mn is tetrahedral (Fig. 15) not octahedral. Organic solvents do not significantly solvate or bind to the metal ion, and this permits the polysulfide to complex the metal with both of its terminal sulfur atoms.

A summary of suggested Mn sulfide complexes and their stability complexes is given in Table 11.

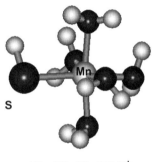

$Mn(H_2O)_5(SH)^+$

$Mn(H_2O)_5(SH)_2^0$

Figure 13. Molecular structures of $[Mn(HS)]^+$ and $[Mn(HS)_2]^0$. There is actually limited evidence for the proposed composition of these species but other metals with 6-coordination would display similar structures.

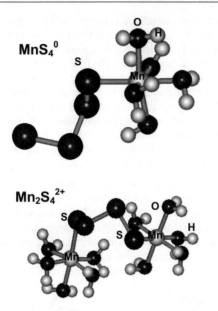

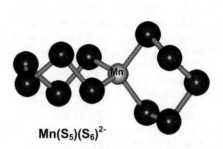

Figure 15. Molecular structure of the tetrahedral Mn complex, [Mn(S$_5$)(S$_6$)]$^{2-}$ which demonstrates the possibility of mixed polysulfide species.

Figure 14. Molecular models for [MnS$_4$]^0 and [Mn$_2$S$_4$]$^{2+}$. S$_4^{2-}$ is a monodentate ligand in these structures.

Table 11. Summary of proposed Mn sulfide complexes and the methods used. The stability of the complex, log$K$, is listed for specific solutions ionic strengths, I, and the method of measurement.

Species	log$K$	I	Method	Reference
[Mn(HS)]$^+$	4.27	0.7	sulfide titration	Al-Farawati & van den Berg (1999)
	6.7	0.7	sulfide titration	Zhang & Millero (1994)
	4.76	0.7	sulfide titration	Luther et al (1996)
[Mn(HS)$_2$]^0	9.9	0.7	sulfide titration	Al-Farawati & van den Berg (1999)
[Mn$_2$(HS)]$^{3+}$	9.67	0.7	sulfide titration	Luther et al. (1996)
[Mn$_3$(HS)]$^{5+}$	15.43	0.7	sulfide titration	Luther et al. (1996)
[Mn(S$_4$)]^0	5.81	0.55	sulfide titration	Chadwell et al. (2001)
[Mn$_2$(S$_4$)]$^{2+}$	11.26	0.55	sulfide titration	Chadwell et al (2001)
[Mn(S$_5$)]^0	5.57	0.55	sulfide titration	Chadwell et al. (1999)
[Mn$_2$(S$_5$)]$^{2+}$	11.54	0.55	sulfide titration	Chadwell et al. (1999)

Iron (II). Iron (II) forms a number of simple Fe sulfide minerals, one of the most important of which at low temperatures is mackinawite, tetragonal FeS. Several studies of sulfide complexation of Fe(II) have been reported using both solubility and voltammetric approaches (e.g., Buffle et al. 1988; Zhang and Millero 1994; Wei and Osseo-Asare 1995; Luther et al. 1996; Theberge and Luther 1997; Davison et al. 1999; Al-Farawati and van den Berg 1999; Chadwell et al. 1999, 2001; Rozan 2000; Luther et al. 2003; Rickard and Morse 2005; Luther and Rickard 2005; Rickard in press). Earlier work was reviewed by Emerson et al. (1983), Davison (1980 1991) and Morse et al. (1987). Somewhat conflicting results have been obtained both as to the form and stability of the complexes.

As with Mn(II) the results of voltammetric titrations provide evidence for [Fe(HS)]$^+$ but no evidence for [Fe(HS)$_2$]^0. The results of the voltammetric experiments are consistent for the stability constant for this complex within one order of magnitude. Wei and Osseo-Asare (1995) also measured a lower stability constant of $\log K = 4.34 \pm 0.15$ at 25 °C ($I = 0$) for $[Fe(HS)]^+$ by using a stopped-flow spectrophotometric technique. They monitored the peak at 500 nm which they attributed to the first formed transient intermediate, $[Fe(HS)]^+$, when Fe(II) and sulfide react at pH > 7. This species is metastable and eventually decomposes to FeS via several possible pathways.

However, curve fitting from solubility studies (Davison et al. 1999) shows that the $[Fe(HS)]^+$ stability constant does not fit the measured solubility. This result has been confirmed by Rickard (in press). Davison et al. (1999) found that the solubility of FeS could be explained by $[Fe(HS)]^+$ using a constant at least two logarithmic units smaller than the measured values and $[Fe(HS)_2]^0$, which is not found with the voltammetric data.

FeS clusters, termed here as FeS_{aq}, are well-known in biochemistry where they constitute the active centers of FeS proteins, such as ferredoxins, and occur in all organisms where they are responsible for basic electron transfer in many key biochemical pathways. Aqueous FeS clusters, in which various numbers of FeS molecules are ligated directly to H_2O molecules, were first observed by Buffle et al. (1988) in lake waters. They were characterized by Theberge and Luther (1997) and Theberge (1999) and are routinely probed electrochemically (Fig. 16). Theberge and Luther (1997) analysed the characteristic wave form from the FeS clusters and showed that the 0.2 V split is consistent with the splitting of Fe(II) in tetrahedral geometries (Fig. 17, modified from data in Theberge and Luther 1997; Theberge 1999).

The stoichiometry of this FeS cluster species is presently unknown, although it has been suggested to be a Fe_2S_2 form (Buffle et al. 1988; Theberge and Luther 1997). However, Theberge and Luther (1997) pointed out that the data could actually fit any FeS phase with a 1:1 stoichiometry. Davison et al. (1988) argued against this species and could not find any evidence in solubility measurements (Davison et al. 1999). However, Rickard (in press) showed that the

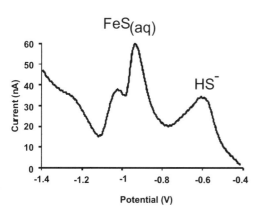
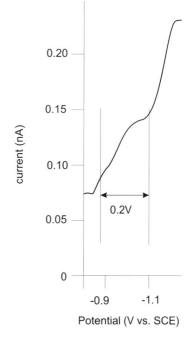

Figure 16 (above). Conventional square wave voltammetric scan of an Fe-S solution showing the typical split peak at around −1.1 V which is assigned to $FeS_{(aq)}$ (from Rickard and Morse 2005).

Figure 17 (at right). Sampled DC polarogram of an FeS cluster showing two waves with 0.2 V center-center distance which reflect two single electron transfers at the Hg electrode: $Fe^{2+} + 2e^- \rightarrow Fe^0$ (after Theberge and Luther 1997 and Theberge 1999).

solutions developing from FeS solubilization in neutral-alkaline systems showed the characteristic voltammetric signature of the aqueous FeS cluster and modeled the solubility using the monomer FeS^0 with a stability constant of $10^{2.2}$ for the acid dissociation reaction (Eqn. 67).

$$FeS^0 + H^+ \rightarrow Fe^{2+} + HS^- \qquad (67)$$

The molecular form of the FeS clusters has been modeled by Luther and Rickard (2005) with the HYPERCHEM™ program, using molecular mechanical calculations with the Polak-Ribiere algorithm where lone pair electrons are considered and the most stable configuration is computed (Fig. 18). The interesting feature of these model structures is that they are very similar in form to the structure of the FeS centers in ferredoxins and show planar and cubane geometries. Since these are neutral species the molecules are liganded directly to water.

Other FeS cluster stoichiometries have been suggested by the work of Rozan (2000) and Luther et al. (2003). These include sulfur rich varieties, such as $[Fe_2S_4]^{4-}$ and metal-rich species like $[Fe_nS_m]^{(n-m)+}$. These are consistent with the sulfide titrations of Luther et al. (1996) (Table 12). It is important to note that these species will probably incorporate a counter ion in natural systems to neutralize the charge. It appears that these counter ions may be organic molecules.

Pyrite, the isometric iron (II) disulfide, is the most common sulfide mineral on the Earth's surface, and thus it is to be expected that Fe(II) should have a significant polysulfide chemistry. Chadwell et al. (1999, 2001) showed Fe polysulfide complexes analogous

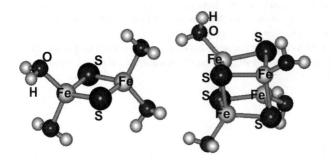

Figure 18. Molecular models of aqueous FeS clusters (modified from Luther and Rickard 2005).

$Fe_2S_2(H_2O)_4$ $Fe_4S_4(H_2O)_4$

Table 12. Summary of stability constants for proposed Fe sulfide complexes and the methods used.

Species	log$K$	I	Method	Reference
$[Fe(HS)]^+$	5.94	0.7	sulfide titration	Al-Farawati and van den Berg (1999)
	5.3	0.7	sulfide titration	Zhang and Millero (1994)
	4.34	0.0	spectrophotometry	Wei and Osseo-Ware (1995)
	5.07	0.7	sulfide titration	Luther et al. (1996)
$[Fe_2(HS)]^{3+}$	10.07	0.7	sulfide titration	Luther et al. (1996)
$[Fe_3(HS)]^{5+}$	16.15	0.7	sulfide titration	Luther et al. (1996)
$[Fe(S_4)]^0$	5.97	0.55	sulfide titration	Chadwell et al. (2001)
$[Fe_2(S_4)]^{2+}$	11.34	0.55	sulfide titration	Chadwell et al. (2001)
$[Fe(S_5)]$	5.69	0.55	sulfide titration	Chadwell et al. (1999)
$[Fe_2(S_5)]^{2+}$	11.30	0.55	sulfide titration	Chadwell et al. (1999)

to the Mn species with compositions [Fe(η^1–S$_4$)], [Fe(η^1–S$_5$)], [Fe$_2$(μ–S$_4$)]$^{2+}$ and [Fe$_2$(μ–S$_5$)]$^{2+}$. The formation of these Fe(II) polysulfide complexes is interesting since they further suggest that non-protonated Fe sulfide complexes could have a significant stability. Molecular models of these species would be similar to those in Figure 14. Coucouvanis et al. (1989) synthesized an interesting bidentate pentasulfido complex, [Fe$_2$S$_2$(S$_5$)]$^{4-}$, which has a Fe$_2$S$_2$ core similar to rubredoxin (Fig. 19). Table 12 lists the iron-sulfide complexes reported in aqueous solution.

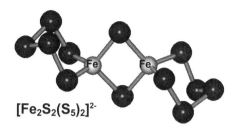

Figure 19. Molecular model for the tetrahedral [Fe$_2$S$_2$(S$_5$)]$^{4-}$ complex. Note the similarity to the structures in Figure 18.

Iron (III). Although an iron(III) sulfide, Fe$_2$S$_3$, appears widely in the earlier literature, the sulfide analog of hematite has not been isolated. However, the sulfide analog of magnetite, the cubic thiospinel greigite, Fe$_3$S$_4$, is a well-established mineral phase, which can be readily synthesized at low temperatures. The synthesis always involves the precursor phase, mackinawite, and proceeds via a solid state transformation (Lennie et al. 1997; Rickard and Morse 2005). The solid state transformation would seem to preclude the formation of Fe(III)-bearing sulfide complexes and no such complexes have been isolated in aqueous solutions. However, the active centers of some FeS proteins are Fe(III) bearing units and the Fe(II)- Fe(III) transition in these moieties are key to the biological electron transfer processes. These clusters have similar cubane forms to the basic structural unit of greigite, and the occurrence of Fe(III)-bearing sulfide clusters in aqueous solutions stabilized by organic ligands is possible.

Cobalt. Cobalt (II) forms a number of sulfide minerals, including CoS. Relatively little is known about the properties of the first formed precipitate from aqueous solutions at low temperatures, and most information derives from studies of well-crystalline bulk material. Recent EXAFS work,(Rickard, Vaughan and coworkers, unpublished data) found that the Co–S distance in the first formed nanoparticulate precipitate is similar to that of cobaltian pentlandite, a cubic phase with a bulk composition given as Co$_9$S$_8$.

Co(II) sulfide complexes have not been observed, but electrochemical titration data suggest that forms like [Co(HS)]$^+$ and [Co(HS)$_2$]^0 fit the observed data. Al-Farawati and van den Berg (1999) found that both [Co(HS)]$^+$ and [Co(HS)$_2$]^0 fitted the titration data whereas Zhang and Millero (1994) and Luther et al. (1996) found no evidence for [Co(HS)$_2$]^0. Luther et al. (1996) showed that the complex was protonated. The values for the stability constants for [Co(HS)]$^+$, which is reported for all three studies, show a range of almost 2 orders of magnitude. Molecular models of these bisulfide species would be similar to those in Figure 13. Chadwell et al. (1999, 2001) showed that Co polysulfide complexes analogous to other members of the iron group, with compositions [Co(η^1–S$_4$)], [Co(η^1–S$_5$)], [Co$_2$(μ–S$_4$)]$^{2+}$ and [Co$_2$(μ–S$_5$)]$^{2+}$. Molecular models of these polysulfides species would be similar to those in Figure 14. Table 13 lists the cobalt sulfide complexes reported in aqueous solution.

Nickel. Nickel (II) precipitates mainly as the mineral millerite, α-NiS, from low temperature aqueous solutions although a number of other phases have been reported and the chemistry of the process is poorly understood. Millerite is an hexagonal phase, and this structural change amongst members of the iron group of sulfides from isometric alabandite, MnS, through tetragonal mackinawite, FeS, and isometric cobaltian pentlandite, CoS, to hexagonal millerite, NiS, continues an unexplained variation in the form of the first precipitated sulfide phases in this homologous group of elements.

Table 13. Summary of stability constants for reported Co sulfide complexes and the methods used.

Species	logK	I	Method	Reference
[Co(HS)]$^+$	6.45	0.7	ligand competition	Al-Farawati and van den Berg (1999)
	5.3	0.7	sulfide titration	Zhang and Millero (1994)
	4.68	0.7	sulfide titration	Luther et al. (1996)
[Co(HS)$_2$]^0	10.15	0.7	ligand competition	Al-Farawati & van den Berg (1999)
[Co$_2$(HS)]$^{3+}$	9.52	0.7	sulfide titration	Luther et al (1996)
[Co$_3$(HS)]$^{5+}$	15.50	0.7	sulfide titration	Luther et al. (1996)
[Co(S$_4$)]^0	5.63	0.55	sulfide titration	Chadwell et al. (2001)
[Co$_2$(S$_4$)]$^{2+}$	11.59	0.55	sulfide titration	Chadwell et al (2001)
[Co(S$_5$)]^0	5.39	0.55	sulfide titration	Chadwell et al. (1999)
[Co$_2$(S$_5$)]$^{2+}$	11.34	0.55	sulfide titration	Chadwell et al. (1999)

In contrast, the Ni(II) sulfide and polysulfide complexes are apparently consistent with the members of the group. Thus Al-Farawati and van den Berg (1999) found that both [Ni(HS)]$^+$ and [Ni(HS)$_2$]^0 fitted their titration data whereas Zhang and Millero (1994) and Luther et al. (1996) found no evidence for [Ni(HS)$_2$]^0. Luther et al. (1996) showed that the [Ni(HS)]$^+$ complex was protonated. The values for the stability constants for [Ni(HS)]$^+$, which is reported for all three studies show a close correlation with a variation of only 0.3 logarithmic units. Chadwell et al. (1999, 2001) reported Ni polysulfide complexes with compositions [Ni(η^1–S$_4$)], [Ni(η^1–S$_5$)], [Ni$_2$(μ–S$_4$)]$^{2+}$ and [Ni$_2$(μ–S$_5$)]$^{2+}$. Using organic solvents, Coucouvanis et al. (1985) synthesized and characterized [Ni(S$_4$)$_2$]$^{2-}$. The molecular structure of this complex is shown in Figure 20 and shows tetrahedral Ni(II) and bidentate polysulfide chelation.

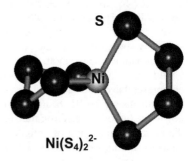

Figure 20. Molecular model for the tetrahedral [Ni(S$_4$)]$^{2-}$ complex.

Reported Ni sulfide complexes and their stability constants are listed in Table 14.

Copper. Cu occurs in oxidation states (I) and (II) in natural systems and these forms have contrasting properties. Thus Cu(II) is a transition metal and a typical borderline hard-soft metal whereas Cu(I) is a soft, B-class metal with a particular predilection for sulfides. The distinction was neatly shown by Luther et al. (2002) where an electron paramagnetic resonance study of the reaction between dissolved S(–II) and Cu(II) in aqueous solution demonstrated that Cu(I) was produced in solution before the formation of the CuS precipitate. That is, that Cu complexed with the soft base S(–II) in solution is soft Cu(I) whereas that complexed with hard H$_2$O in the Cu aqua ion is the relatively hard Cu(II). It appears that in most sulfide minerals Cu occurs as Cu(I). Even the mineral CuS, covellite, is a Cu(I) sulfide (Van der Laan et al. 1992). However, the aqueous Cu(I) species has very limited stability in aqueous solution, and the hexaqua and pentaqua Cu(II) species dominate.

Some debate is on-going about the nature and characteristics of Cu sulfide complexes since the measured solubilities of the first formed phases from aqueous solutions at low temperatures appear to be considerably higher than might be expected from that calculated from the Gibbs free energies of the bulk crystalline equivalents. For this reason there have been many studies of the solubilities of copper sulfides in the geochemical literature which have

Table 14. Summary of stability constants for proposed Ni sulfide complexes and the methods used.

Species	log$K$	I	Method	Reference
[Ni(HS)]$^+$	4.77	0.7	ligand competition	Al-Farawati and van den Berg (1999)
	5.3	0.7	sulfide titration	Zhang and Millero (1994)
	4.97	0.7	sulfide titration	Luther et al. (1996)
[Ni(HS)$_2$]^0	10.47	0.7	ligand competition	Al-Farawati and van den Berg (1999)
[Ni$_2$(HS)]$^{3+}$	9.99	0.7	sulfide titration	Luther et al. (1996)
[Ni$_3$(HS)]$^{5+}$	15.90	0.7	sulfide titration	Luther et al. (1996)
[Ni(S$_4$)]	5.72	0.55	sulfide titration	Chadwell et al. (2001)
[Ni$_2$(S$_4$)]$^{2+}$	11.01	0.55	sulfide titration	Chadwell et al. (2001)
[Ni(S$_5$)]	5.53	0.55	sulfide titration	Chadwell et al. (1999)
[Ni$_2$(S$_5$)]$^{2+}$	11.06	0.55	sulfide titration	Chadwell et al. (1999)

proposed various copper sulfide complexes to explain the solubility characteristics (see Table 15). However, as noted above, simple curve fitting does not give a unique solution and independent evidence for the existence of individual constants is required. Likewise, a number of copper sulfide complexes have been proposed in the chemical literature as examples of unusual coordination or conformation. Much of the data comes from X-ray structural analyses of precipitated salts. However, without independent evidence it is not possible to extrapolate the structure of the moiety in the crystal to the structure and stoichiometry of the species in solution. We have listed a number of complexes determined in this way—with independent corroborative solution data—in Table 15.

Zhang and Millero (1994) and Al-Farawati and van den Berg (1999) reported that [Cu(HS)]$^+$ and [Cu(HS)$_2$]^0 species were consistent with their titration data, whereas Luther et al. (1996) found no evidence for protonation of the Cu sulfide complexes. They reported CuS^0 (i.e., 1:1) and [Cu$_2$S$_3$]$^{2-}$ (i.e., 2:3) species. Mountain and Seward (1999) measured the solubility of Cu$_2$S at 22 °C and argued that the complexes formed were Cu(I) species. They used non-linear curve fitting to suggest [Cu(HS)$_2$]$^-$, [Cu$_2$S(HS)$_2$]$^{2-}$ and CuHS^0 species. They revisited this in Mountain and Seward (2003) for the 35-95 °C temperature range and extrapolated the results to 350 °C.

Luther et al. (2002) used a combination of voltammetric, UV-VIS spectroscopic, mass spectroscopic, ^{63}Cu NMR and EPR spectroscopy to show that the CuS^0 was a polynuclear [Cu$_3$S$_3$] complex and that [Cu$_2$S$_3$]$^{2-}$ was a tetranuclear [Cu$_4$S$_6$]$^{2-}$ species. Both of these stoichiometries are, of course, indistinguishable from the original 1:1 and 2:3 species originally proposed from the results of curve fitting techniques. Luther et al. (2002) calculated molecular models for these complexes (Fig. 21).

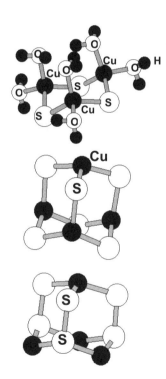

Figure 21. Molecular model for the tetrahedral neutral Cu cluster [Cu$_3$S$_3$(H$_2$O)$_6$] (top). Middle structure is for [Cu$_4$S$_6$]$^{4-}$ where Cu and S only bind to each other. Bottom structure is for [Cu$_4$S$_6$]$^{2-}$ where S-S bonding occurs after reduction of Cu(II) to Cu(I) (from Luther and Rickard 2005).

Table 15. Summary of proposed Cu sulfide complexes and the methods used.

Complex	Method(s)		Reference
$[Cu(HS)]^+$	voltammetric titration	sulfide concentration	Zhang and Millero (1994)
	ligand competition		Farawati and van den Berg (1999)
$[Cu(HS)_2]^0$	voltammetric titration	sulfide concentration	Zhang and Millero (1994)
	ligand competition		Farawati and van den Berg (1999)
$[CuS]^0$, $[Cu_3S_3]$	voltammetry, UV-VIS spectroscopic, mass spectroscopy, ^{63}Cu NMR, EPR spectroscopy	mole ratio	Luther et al (1996) Luther et al. (2002)
$[Cu_2S_3]^{2-}$, $[Cu_4S_6]^{2-}$	voltammetry, UV-VIS spectroscopic, mass spectroscopy, ^{63}Cu NMR, EPR spectroscopy	mole ratio	Luther et al (1996) Luther et al. (2002)
$[Cu(HS)_2]^-$	curve fitting	Cu_2S solubility	Mountain and Seward (1999)
$[Cu_2S(HS)_2]^{2-}$	curve fitting	Cu_2S solubility	Mountain and Seward (1999)
$[CuHS]^0$	curve fitting	Cu_2S solubility	Mountain and Seward (1999)
$[Cu(HS)_3]^{2-}$, $[Cu_3S_4H_2]^{2-}$, etc.	curve fitting	CuS solubility	Thompson and Helz (1994)
$[Cu_2S(HS)_2]^{2-}$, $[Cu_4S_4H_2]^{2-}$, etc.	curve fitting	CuS solubility	Thompson and Helz (1994)
$[Cu_2S_2(HS)_3]^{3-}$	curve fitting	CuS solubility	Shea and Helz (1989) as reinterpreted by Thompson and Helz (1994)
$[Cu_2(S_3)(S_4)]^{2-}$	curve fitting	CuS solubility	Shea and Helz (1989) as reinterpreted by Thompson and Helz (1994)
$[Cu(S_9)(S_{10})]^{3-}$	curve fitting	CuS solubility	Shea and Helz (1989) as reinterpreted by Thompson and Helz (1994)
Multinuclear Cu complexes	EXAFS	CuS in sulfide solutions	Helz et al. (1993)
$[Cu_3(S_4)_3]^{3-}$	Raman, X-ray structure analysis	Synthesis	Müller et al. (1984b)
$[Cu_3(S_6)_3]^{3-}$	IR, Raman, ESCA, UV/VIS spectroscopy, X-ray structure analysis	Synthesis	Müller and Schimanski (1983)
$[Cu_6(S_5)(S_4)_3]^{2-}$	Raman, X-ray structure analysis	Synthesis	Müller et al. (1984a)
$[Cu_4(S_5)_3]^{2-}$	Raman, X-ray structure analysis	Synthesis	Müller et al. (1984b)
$[Cu_4(S_4)(S_5)_2]^{2-}$	Raman, X-ray structure analysis	Synthesis	Müller et al. (1984b)
$[Cu_4(S_4)_2(S_5)]^{2-}$	Raman, X-ray structure analysis	Synthesis	Müller et al. (1984b)
$[Cu(S_4)]_2$	voltammetry	chelate scale	Chadwell et al. (1999)
$[Cu(S_5)]_2$	voltammetry	chelate scale	Chadwell et al. (2000)
$[Cu(S_4)_2]^{3-}$	curve fitting	CuS solubility in polysulfide solution	Cloke (1963) from Höljte and Beckert (1935)
$[Cu(S_5)(S_4)]^{3-}$	curve fitting	CuS solubility in polysulfide solution	Cloke (1963) from Höljte and Beckert (1935)

Ciglenečki et al. (2005) proposed that all the previously reported Cu sulfide complexes were actually nanoparticulate CuS. They stated that, not only the proposal for dissolved Cu sulfide clusters by Luther et al. (2002) was suspect, but also the Cu complexes suggested by the results of Zhang and Millero (1994) and Al-Farawati and van den Berg (1999). They did not cite or comment upon the results of Mountain and Seward (1999, 2003). Earlier, Helz et al. (1993) had proposed multinuclear Cu sulfide complexes on the basis of EXAFS studies. Thompson and Helz (1994) reported that Cu sulfide solubility data suggested the presence of two Cu sulfide complexes, one containing an odd number of Cu atoms (e.g., $[Cu(HS)_3]^{2-}$ or $[Cu_3S_4H_2]^{2-}$ or an even higher multimers) and one containing an even number of Cu atoms (e.g., $[Cu_2S(HS)_2]^{2-}$, $[Cu_4S_4H_2]^-$, etc). Earlier, Shea and Helz (1989) had measured covellite solubility and Thompson and Helz (1994) reinterpreted these results in terms of species $[Cu_2S_2(HS)_3]^{3-}$, $[Cu_2(S_3)(S_4)]^{2-}$ and $[Cu(S_9)(S_{10})]^{3-}$. However, Ciglenečki et al. (2005) then re-reinterpreted these results with the comment that the *"measurement precision was insufficient to exclude mononuclear and dinuclear complexes as predominant species."*

And finally, they appeared to accept that the Cu clusters proposed by Helz et al. (1993) existed on the grounds that *"the measured Cu-S and Cu-Cu distances were similar to those in known $Cu_4(RS)_6^{2-}$ clusters and different from distances in Cu-S solid phases or linear $Cu_2S(HS)_2^{2-}$ complexes"* but then go on to state that the *"existence of clusters as major dissolved Cu species in equilibrium with Cu sulfide minerals is suggested by both studies but remains to be proven."*

Cluster-type complexes of polysulfide with Cu are known and are reviewed by Müller and Diemann (1987). Müller and Schimanski (1983) and Müller et al. (1984a,b) suggested $[(S_6)Cu(S_8)Cu(S_6)]^{4-}$, $[Cu_3(S_4)_3]^{3-}$, $[Cu_3(S_6)_3]^{3-}$, $[Cu_6(S_5)(S_4)_3]^{2-}$ and the series $[Cu_4(S_5)_3]^{2-}$, $[Cu_4(S_4)(S_5)_2]^{2-}$ and $[Cu_4(S_4)_2(S_5)]^2$. Chadwell et al. (1999, 2000) found evidence for the dimers $[Cu_2(S_4)]_2$ and $[Cu_2(S_5)]_2$ in aqueous solutions. Electrochemical evidence demonstrated proton-free 1:1 complexes but the ESR signal was silent. Since Cu(II) is d^9, an ESR signal is expected. Chadwell (1999, 2000) concluded that the complex was a dimer and has an unpaired electron on each of the Cu(II) antiparallel thus producing a silent ESR signal. This type of behavior has been observed for ferredoxins: e.g., complexes of the type $[Fe_2S_2(SR)_4]^{2-,3-}$ and $[Fe_4S_4(SR)_4]^{2-,3-}$ (Holm 1974; Papaefthymiou et al. 1982; Rao and Holm 2004). Chadwell et al. (1999, 2000) also noted some evidence for protonated Cu(II) polysulfide complexes at pH below 6, but the nature of these complexes was not determined. Cloke (1963b) refitted the solubility data from Höljte and Beckert (1935) with covellite in polysulfide solutions and deduced two complexes from the data: $[Cu(S_4)_2]^{-3}$ and $[Cu(S_5)(S_4)]^{-3}$. These data are similar to those of Chadwell et al. (1999, 2001).

The polysulfide data for Cu(II) are consistent with the sulfide results proposed by Luther et al. (2002). It suggests that Cu(II) forms non-protonated complexes in neutral-alkaline solutions but possible protonated species in acid conditions. In addition, Müller and Schimanski (1983) recognized the central role that the Cu_3S_3 (formally $Cu_3(\mu-S)_3$) ring configuration, independently reported by Luther et al. (2002), played in the structure of Cu sulfide clusters. Figure 22 shows the structure for $[Cu_3(S_6)_3]^{3-}$ in which one terminal S atom from S_6^{2-} binds to one Cu and the other terminal S atom binds to two Cu atoms. They described this as a "paradigmatic unit." The polysulfide data also seem similar to the results produced by Ciglenečki et al. (2005) and may provide an alternative explanation for their observations. As pointed out by

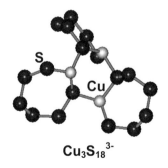

Figure 22. Molecular structure of the Cu polysulfide complex, $[Cu_3(S_6)_3]^{3-}$.

Müller and Diemann (1987) simple reactions of metal ions with H_2S in the presence of oxygen require only small amounts of a "matching ligand" to form a polysulfide complex.

Table 16 lists reported stability data for Cu sulfide complexes. In this table we have separated reported Cu(II) from Cu(I) sulfide complexes. It has been shown that

(i) Cu in a sulfide environment is usually Cu(I) (Van der Laan 1992, Pattrick et al. 1993).

(ii) the reduction of Cu(II) to Cu(I) occurs in solution, within the sulfide complex (Luther et al. 2002)

(iii) the form of the S in minerals such as covellite, CuS, is a disulfide and the composition is better expressed as Cu_2S_2.(Van der Laan 1992, Pattrick et al. 1993, Luther et al. 2002).

Thus the distinction between Cu(II) sulfide complexes and Cu(I) polysulfide complexes may not be as clear as appears from the table.

The confusing situation with regard to Cu sulfide complexes is likely not to be resolved for some time. One contributory factor for the confusion is that most of these studies have been approached from an equilibrium chemistry point of view and the actual species have

Table 16. Reported stability data for Cu sulfide complexes.

Species	log$K$	I	Method	Reference
Cu(II)				
$[Cu(HS)]^+$	11.52	0.7	ligand competition	Al-Farawati and van den Berg (1999)
	7.0	0.7	sulfide titration	Zhang and Millero (1994)
	5.98	0.7	sulfide titration	Luther et al. (1996)[a,b]
$[CuS]^0$	11.2	0.7	sulfide titration	Luther et al. (1996) [a,b]
$[Cu(HS)_2]^0$	18.02	0.7	ligand competition	Al-Farawati and van den Berg (1999)
	13.0	0.7	sulfide titration	Zhang and Millero (1994)
$[Cu_2S_3]^{2-}$	11.68	0.7	sulfide titration	Luther et al (1996) [a,b]
	38.29		corrected for protonation of the sulfide[c]	
$[Cu_2(S_4)_2]^{2-}$	17.81	0.55	sulfide titration	Chadwell et al. (2001)[b]
$[Cu_2(S_5)_2]^{2-}$	20.2	0.55	sulfide titration	Chadwell et al. (1999)[b]
Cu(I)				
$[Cu(HS)]^0$	11.52	0.7	ligand competition	Al-Farawati and van den Berg (1999)
	6.8	0.7	sulfide titration	Zhang and Millero (1994)
	11.8		corrected for metal-chloro complexes	
	13.0	0.0	Cu_2S solubility	Mountain and Seward (1999)[d]
$[Cu(HS)_2]^-$	18.02	0.7	ligand competition	Al-Farawati and van den Berg (1999)
	12.6	0.7	sulfide titration	Zhang and Millero (1994)
	17.6		corrected for metal-chloro complexes	
	17.18	0.0	Cu_2S solubility	Mountain and Seward (1999)[d]
$[Cu_2S(HS)_2]^{2-}$	29.87	0.0	Cu_2S solubility	Mountain and Seward (1999)[d]
$[Cu(S_4)_2]^{3-}$	9.83	0.0	Cu_2S solubility	Cloke (1963)[d]
$[Cu(S_4)(S)_5]^{3-}$	10.56	0.0	Cu_2S solubility	Cloke (1963)[d]

[a] acid-base titrations indicate the species is not protonated. MS species are corrected for protonation of sulfide. Multinuclear clusters are likely.

[b] Reduction of Cu(II) to Cu(I) occurred at some point in the titration so the species are likely Cu(I).

[c] When corrected for the stoichiometry $Cu_2S(HS)_2^{2-}$, which contains protons, and the side reaction coefficient of Cu(I) in seawater the value is 27.44, which compares with the value of 29.87 of Mountain and Seward (1999).

[d] Thermodynamic constants were calculated not conditional constants

not obviously been seen. The exception is the EXAFS data presented by Helz et al. (1993), the protonation data presented by Luther et al. (1996) and the EPR, ^{63}Cu NMR and mass spectroscopic data of Luther et al. (2002). One way around the impasse is to examine the molecular mechanisms involved in the formation of the Cu sulfide species and this, together with supporting Cu isotopic data, is discussed below.

Molybdenum

Molybdenum is a borderline hard-soft member of Group 6. It sits between Cr, a transition metal with characteristic borderline properties and W, a metal which is firmly in the hard category. So Mo has significant oxide and sulfide chemistries with the sulfide MoS_2, molybdenite, being the most important Mo mineral with the molybdate wulfenite, $PbMoO_4$, being also significant, whereas W occurs almost exclusively as tungstates.

MoS_2 occurs in three polytypes, hexagonal molybdenite-2H which is hexagonal, molybdenite-2R, which is rhombohedral, and jordisite, which is apparently amorphous. The term molybdenite usually refers to the 2H polymorph. Even though this is a relatively common mineral, the importance of molybdenum sulfide complexes is related to the use of molybdenum as a potential proxy for redox conditions (Dean et al. 1997) and the well-established molybdenum sulfide cluster chemistry.

The electronic configuration of Mo is $[Kr]5s^14d^5$ and all six outer electrons can be involved in bonding leading to oxidation states in complexes that vary between 0 and +6. Even the -2 state is known in organometallic complexes. Mo forms a variety of sulfide complexes, the thiomolybdates, with a general formula $[MoO_xS_{4-x}]^{n-}$ where x = 0-3: $[MoS_4]^{2-}$, $[MoOS_3]^{2-}$, $[MoO_2S_2]^{2-}$ and $[MoO_3S]^{2-}$. These are bright yellow to red materials and have been the subject of interest since Berzelius's time. The remarkable feature of these complexes is that Mo is in the highest oxidation state in coordination with reductive S(−II) ions. The other notable feature is that the Mo-S distance is consistent at 2.15-2.18 Å for all of these species, and this feature extends to analogous complexes with W, Re, Ta, Nb and V. This suggests some π character in the M_2S bonds—as also suggested from MO calculations (Diemann and Müller 1973). Electron delocalisation through π bonding also explains why these anions do not undergo a spontaneous internal redox reaction. $[MoS_4]^{2-}$ has been the most studied species since it is readily synthesised, easily converts to other dimolybdenum-thiols and has relatively high thermal and hydrolytic stability. It is also central to several key biological processes and has been implicated in geochemical processes. $[MoS_4]^{2-}$ is easily prepared by passing H_2S through an ammoniacal molybdate solution. After 20 minutes $(NH_4)MoS_4$ precipitates as very pure red crystals (Mellor 1943). Formation of $[MoS_4]^{2-}$ from $[MoO_4]^{2-}$ clearly proceeds via successive replacement of oxygen, as evidenced from the colour sequence change in solution: $[MoO_4]^{2-}$(colorless) → $R[MoO_3S]^{2-}$(yellow) → $R[MoO_2S_2]^{2-}$(orange) → $R[MoOS_3]^{2-}$(orange-red) → $R[MoS_4]^{2-}$(red). This can be readily traced with UV-VIS spectroscopy and the kinetics of the process have been established (Harmer and Sykes 1980). Its complete formation in aqueous solution requires high S-Mo ratios and its rate of hydrolysis increases with increasing H^+ concentrations without the formation of the intermediate oxythio ions (Clarke et al. 1987). These data were later used by Erickson and Helz (2000) to suggest a "geochemical switch" in which Mo exists in two distinct regimes in natural environments, as $[MoO_4]^{2-}$ in oxic and low sulfidation (Rickard and Morse 2005) regimes and $[MoS_4]^{2-}$ in high sulfidation environments. The H^+ dependence of the geochemical switch was considered by Tossell (2005) in terms of equilibrium chemistry, and he noted that it would be inhibited at high pH where HS^- dominated.

Earlier claims that $[MoS_4]^{2-}$ polymerises at low pH have not been substantiated (Laurie 2000). Treatment with polysulfides produces dimeric and trimeric species in which the Mo centres have been reduced to Mo(V) and Mo(IV), e.g., $(NH_4)_2[Mo(V)_2(S_2)_6] \cdot 2H_2O$ and $(NH_4)_2[Mo(IV)_3S(S_2)_6]$ (Müller et al. 1978, 1980; Coucouvanis and Hadjikyriacou 1986).

The thiomolybdates form a variety of complexes with ligands where the basic $[MoO_xS_{4-x}]^{n-}$ moiety is retained. The S(–II) ions act as bridging ligands in which they bridge 2, 3 or 4 metal centers. These complexes range from simple linear structures to cubanes and more complex forms (Fig. 23). They have been widely reviewed (Müller and Diemann 1987; Dance and Fischer 1991; Eichhorn 1994; Wu et al. 1996) since they have been used as nonlinear optical materials, industrial catalysts and as models for the active $[Fe_2Mo_2S]$ cluster site in the metalloenzyme nitrogenase. Bostick et al. (2003) showed that when $[MoS_4]^{2-}$ reacts with the surface of pyrite Fe-Mo-S cuboidal clusters form where the original Mo(VI) has been reduced to Mo(IV) and Mo(III) (Mascharak et al. 1983; Osterloh et al. 2000). Vorlicek et al. (2004) noted that the formation of polymeric species of Mo polysulfides (Müller et al. 1978, 1980; Coucouvanis and Hadjikyriacou 1986) is consistent with this observation. The sorption of Mo on pyrite is an important process since Mo has been proposed to be a proxy for anoxic, sulfidic conditions, and much of the Mo in this environment appears to be associated with pyrite (Raiswell and Plant 1980; Coveney 1987; Huerta-Diaz and Morse 1992; Dellwig et al. 2002; Müller 2002).

Silver and gold

Groups 8 through 11 of the periodic table include the noble metals. These are d-block elements whose characteristic property is that they are resistant to oxidation. They are typical soft metals with a strong predilection for forming sulfide complexes. We consider the sulfide complex chemistry of Ag and Au here since these have some geochemical significance. Ag and Au are Group 11 elements together with Cu of the first transition series. They all have a single s electron outside the filled d shell. Even so the properties of these three elements vary more than might be expected from the similar electronic structures. Of course, relativistic effects on the 6s electron might explain some of the differences between Ag and Au. This leads to the relativistic contraction: i.e. Au displays similar, or even smaller, covalent radii in similar compounds to Ag. In contrast, the ionic radius of Au$^+$ is substantially larger than Ag$^+$ (Ishikawa et al. 1995). The result is that in mainly covalent compounds such as sulfides, the stability of the Au$^+$ complexes is similar to or slighty greater than the stability of the Ag$^+$ complexes.

Oxidation states in this group appear erratic. Thus, as we have seen, Cu is commonly Cu(I) or Cu(II), whereas Ag is typically Ag(I) and Au is Au(I) or Au(III). Cu(I), Ag(I) and Au(I) display soft character because of the relatively small energy difference between the frontier orbitals of these ions.

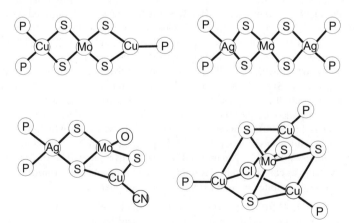

Figure 23. Some simple linear and cubane structures in schematic form illustrating the ability of thiomolybdate anions to act as ligands and as building blocks for more complex structures (P = triphenylphosphane) (adapted from Laurie 2000).

Silver. The black precipitate that occurs through the reaction between S(–II) and Ag salts in aqueous solution at room temperature is monoclinic Ag_2S, acanthite. Above 177 °C this transforms rapidly and reversibly to the isometric polymorph, argentite. It is claimed that Ag_2S is the least soluble of all known silver compounds (e.g., Cotton et al. 1999). However, it was known as long ago as 1949 that the solubility is pH independent in acid solutions but increases markedly with pH in alkaline systems (Treadwell and Hepenstick 1949). Schwarzenbach and Widmer (1966) measured the solubility of Ag_2S and proposed that [AgHS], $[Ag(HS)_2]^-$ and $[Ag_2S(HS)_2]^{2-}$ complexes dominated successively with increasing pH. Sugaki et al. (1987) interpreted the results of similar solubility experiments in terms of binuclear Ag sulfide complexes, $[Ag_2S(H_2S)]^0$, $[Ag_2S(H_2S)(HS)]^-$, $[Ag_2S(H_2S)(HS)_2]^{2-}$ and $[Ag_2S(HS)_2]^{2-}$. Gammons and Barnes (1989) proposed that $[Ag(HS)_2]^-$ was the dominant sulfide complex between pH 5.8 and 7.3. Stefánsson and Seward (2003) studied the solubility of crystalline Ag_2S between 25-400°C and treated the data with a non-linear least squares fitting routine which considered all the hitherto proposed complexes. They found that the data were consistent with [AgHS], $[Ag(HS)_2]^-$ and $[Ag_2S(HS)_2]^{2-}$ being the dominant species.

The result of Stefánsson and Seward's (2003) model for Ag sulfide complexing is shown graphically in Figure 24. This suggests that $[AgHS]^0$ and $[Ag_2S(HS)_2]^{2-}$ are the most important complexes in seawater systems. Zhang and Millero (1994) using voltammetric titrations, provided evidence for $[AgHS]^0$ and $[Ag(HS)_2]^-$ complexes. Al-Farawati and van den Berg

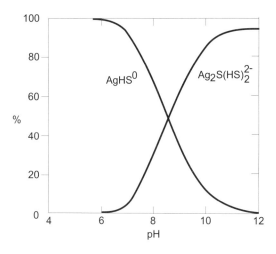

Figure 24. Percentage distribution of Ag(I) species versus pH for a deep anoxic seawater system with total dissolved S(–II) = 5 × 10^{-4} M and total dissolved Cl(–I) = 0.55 at 25 °C from the data in Stefansson and Seward (2003).

Table 17. Comparison of stability constant data for the $[Ag(HS)]^0$ and $[Ag(HS)_2]^-$ species, corrected for Ag side reaction coefficients.

logK_{COND}		Method	Reference
1:1	1:2		
10.8	—	Titration	Luther and Rickard (2005)
≥9.5	≥15.3	Titration	Zhang and Millero (1994)
13.6	17.7	Solubility of Ag_2S	Schwarzenbach and Widmer (1966)
13.49	—	Solubility of Ag_2S	Gammons and Barnes (1989)
11.6	—	Titration	Al-Farawati and van den Berg (1999)

(1999) provided evidence only for [AgHS]^0. Table 17 shows a comparison of stability constant data for the [AgHS]^0 and [Ag(HS)$_2$]$^-$ species, corrected for Ag side reaction coefficients.

The influence of the filled d^{10} shell on Ag chemistry is exemplified by the abundance of Ag cluster compounds. Many compounds show weak Ag-Ag interactions giving rise to the possibility of polynuclear complexes. In many of these, the Ag appears to be zero or even subvalent. Indeed the Ag$_{13}$ unit has been found to be central to a series of superclusters, including both simple carbonyls and more complex mixed Au-Ag species. Rozan and Luther (2002) looked at Ag sulfide complexing in a different way. They found that they could replace Zn and Cu in aqueous ZnS and CuS clusters to produce relatively stable AgS clusters, with suggested [AgS]$^-$ and [Ag$_3$S$_3$]$^{3-}$ stoichiometries. They found that these forms were relatively stable and could explain enhanced dissolved Ag(I) concentrations in fresh water systems. Some support for this idea comes from the voltammetric acid-base titrations reported by Luther and Rickard (2005). These showed evidence for 1:1 and 2:1 Ag-S complexes and also demonstrated that the Ag-S species are not protonated and are likely multi-nuclear clusters. Using the side reaction coefficients for both Ag and sulfide, they obtained stability constants of logK_{therm} = 22.8 for [AgS]$^-$ and logK_{therm} = 29.1 for the [Ag$_2$S]^0 complexes.

Cloke (1963b) investigated the solubility of acanthite in polysulfide solutions and proposed [Ag(S$_4$)$_2$]$^{3-}$, [Ag(S$_4$)(S$_5$)]$^{3-}$ and [Ag(HS)S$_4$]$^{2-}$ complexes. Cloke (1963b) was handicapped by a lack of good polysulfide data but, even so, his work shows the potential importance of polysulfides in Ag solution chemistry. Cloke (1963b) suggested similar stoichiometries for Cu polysulfide species which are now known to be Cu$_2$(S$_4$)$_2$ and Cu$_2$(S$_5$)$_2$ (Chadwell et al. 1999, 2002). It is possible, therefore, that the Ag polysulfide species have similar stoichiometries. It is surprising that it appears that no further work on these Ag polysulfide complexes has been published.

Table 18 lists the stability constants reported for Ag sulfide complexes.

Gold. Gold forms two simple sulfides, Au$_2$S and Au$_2$S$_3$, both of which are metastable and neither of which are known to occur naturally as discrete minerals. The nearest minerals are the rare Au-Ag sulfides. Both Au$_2$S and Au$_2$S$_3$ dissociate readily to form metallic Au, which reflects the extra stability of the Au nuclide in its ground state. This is a major technical problem in experimental studies of Au sulfide complexes (cf. Stefánsson and Seward 2004). Although Au$_2$S can be formed as a pure compound (Gurevich et al. 2004) its synthesis is not simple and the product risks including nanoparticulate Au^0. Since, except for Tossell's (1996) theoretical study, Au sulfide complexes have only been proposed through curve fitting of solubility data, this has led to some uncertainty in their definition.

Au solubility in sulfide solutions is enhanced by the formation of Au sulfide complexes. This has important consequences for the transport of gold in natural systems, especially at elevated temperatures (see Stefánsson and Seward 2004 for a review of this work). Belevantsev et al. (1981) measured the solubility of Au$_2$S in sulfide solutions at 25 °C and proposed that [Au(HS)$_2$]$^-$, [Au$_2$(HS)$_2$S]$^{2-}$ and [Au(HS)(OH)]$^-$ were the dominant complexes. This was a diversion from the classical view that Au existed as [AuS]$^-$ (Latimer 1952). Renders and Seward (1989) measured the solubility of Au$_2$S at 25 °C and proposed that [Au(HS)]^0, [Au(HS)$_2$]$^-$ and [Au$_2$S$_2$]$^{2-}$ complexes dominated. Zotov et al. (1996) proposed [Au(HS)$_2$]$^-$ at near neutral conditions. Baranova and Zotov (1998) fitted [Au(HS)]^0 and [Au(HS)$_2$]$^-$ to Au solubility data in acid sulfidic solutions. Tossell (1996) used quantum mechanical computations to suggest that the species at low pH would be coordinated with H$_2$O and thus be represented as [Au(HS)(H$_2$O)]^0. He confirmed [Au(HS)$_2$]$^-$ at neutral pH but suggested that [Au(HS)(OH)]$^-$ would predominate at high pH rather than dimers like [Au$_2$S$_2$]$^{2-}$.

The results of the Renders and Seward (1989) measurements are shown in Figure 25. The data suggest, even taking into account later strictures on the possibility of Au^0 contamination

Table 18. Reported stability constants for Ag sulfide complexes.

Species	log$K$	I	Method	Reference
[Ag(HS)]^0	6.38	0.7	sulfide titration	Al-Farawati and van den Berg (1999)
	11.6		corrected for metal-ligand complexes in seawater	
	≥9.5	0.7	sulfide titration	Zhang and Millero (1994)
	13.6	1.0	Ag$_2$S solubility	Schwarzenbach and Widmer (1966)[b]
	13.48	0.0	Ag$_2$S solubility	Renders and Seward (1989)[b]
	15.89	0.0	Ag$_2$S solubility	Stephansson and Seward (2003)[b]
	10.8	0.7	sulfide titration	Luther and Rickard (2005)[a]
[AgS]$^-$	22.8	0.7	sulfide titration	Luther and Rickard (2005)[a,b]
[Ag(HS)$_2$]$^-$	≥15.3	0.7	sulfide titration	Zhang and Millero (1994)
	17.17	1.0	Ag$_2$S solubility	Schwarzenbach and Widmer (1966)[b]
	17.28	0.0	Ag$_2$S solubility	Gammons and Barnes (1989)[b]
	17.43	0.0	Ag$_2$S solubility	Renders and Seward (1989)[b]
	17.54	0.0	Ag$_2$S solubility	Stephansson and Seward (2003)[b]
[Ag$_2$S(HS)$_2$]$^{2-}$	72.9	1.0	Ag$_2$S solubility	Schwarzenbach and Widmer (1966)[b]
	31.43	0.0	Ag$_2$S solubility	Renders and Seward (1989)[b]
	31.24	0.0	Ag$_2$S solubility	Stephansson and Seward (2003)[b]
[Ag$_2$S]^0	29.1	0.7	sulfide titration	Luther and Rickard (2005)[a,b]
[Ag(HS)(S$_4$)]$^{2-}$	7.40	0.0	Ag$_2$S solubility	Cloke (1963)[b]
[Ag(S$_4$)$_2$]$^{3-}$	16.46	0.0	Ag$_2$S solubility	Cloke (1963)[b]
[Ag(S$_4$)(S$_5$)]$^{3-}$	16.78	0.0	Ag$_2$S solubility	Cloke (1963)[b]

[a] acid-base titrations indicate the species is not protonated. MS species are corrected for protonation of sulfide. Multi-nuclear clusters are likely.
[b] Thermodynamic constants were calculated not conditional constants.

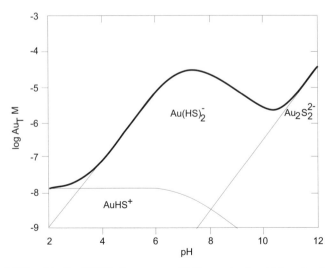

Figure 25. Solubility of Au$_2$S at 25 °C in the presence of 0.01 m total reduced S from Renders and Seward (1989) in terms of the concentration of total dissolved Au (Au$_T$) and pH. The bold curve is the locus of the measured experimental data and the thin lines represent the theoretical species boundaries.

(Stefánsson and Seward 2004), that for all reasonable pH conditions [Au(HS)$_2$]$^-$ is the dominant Au sulfide species.

As with Ag, it is to be expected that Au would form polysulfide complexes. But there are few data about these complexes at low temperatures and even the higher temperature data (Bernt et al. 1994) do not characterize the complexes with any certainty. Bernt et al.'s (1994) data indicate enhanced Au solubility in sulfidic systems saturated with S^0 at temperatures of 100-150 °C. Müller et al. (1984b) have synthesized [Au$_2$(S$_8$)]$^{2-}$ in ethanol. The structure in Figure 26 shows a ten-membered ring of D$_2$ symmetry with possible attraction between the two Au atoms. This structure is also likely for the [Cu$_2$(S$_{4,5}$)] complexes found by Chadwell et al. (1999, 2001) where each terminal S atom in the polysulfide binds to a separate Au or Cu atom.

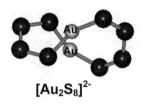

[Au$_2$S$_8$]$^{2-}$

Figure 26. Structure of the Au polysulfide complex, [Au$_2$S$_8$]$^{2-}$.

Zinc, cadmium and mercury

The Group 12 elements continue the trends set by the transition metals. They have ns^2 outer electron configurations but their chemistry is still influenced by the underlying $(n-1)d^{10}$ shell. However, whilst the related elements (Cu, Ag and Au) of Group 11 produce complexes with oxidation states II and III, such compounds are less common for the Group 12 metals. The characteristic oxidation state of the elements is (II) although Hg has a significant (I) chemistry. The (I) oxidation state is of only theoretical significance for Zn and Cd. This is because the Group 12 elements do not lose one or two d electrons like the Group 11 metals. Mercury is the only member of the group to have a significant oxidation state (I) chemistry at normal conditions with the formation of Hg$_2^{2+}$ ions and compounds with Hg$_3^{3+}$ and Hg$_4^{4+}$.

The stability of Group 12 element complexes is relatively high because of their small size and the filled $(n-1)d$ orbitals are relatively easily polarized by ligand electrons producing an effective nuclear charge somewhat greater than the simple +2 charge on the ion. This also means that these metals show mainly soft characteristics although Zn is somewhat different being borderline with a distinctive zinc oxide chemistry, whereas Cd and Hg are typical soft metals.

Zinc. Zinc sulfide occurs as two polymorphs, cubic sphalerite and hexagonal wurtzite. The speciation of Zn sulfide complexes has been of some interest, basically because the measured solubility of Zn(II) in solution in equilibrium with ZnS is far greater than can be accounted for by the simple Zn aqua ion. The problem, as noted by Dyrssen (1991), Hayashi et al. (1991) and Tossell and Vaughan (1993) is that the solubility data do not provide a unique solution to the speciation. In particular, the number of H$_2$O and/or OH$^-$ species cannot be determined. For example, the solubility data can equally well be fitted by species such as [Zn(SH)$_3$(OH)]$^{2-}$, Zn[S(SH)$_2$(OH$_2$)]$^{2-}$, and [ZnS(SH)$_2$]$^{2-}$. Tossell and Vaughan (1993) used computational methods to show that [Zn(HS)$_3$]$^-$ and [Zn(HS)$_3$(OH)]$^{2-}$ would be energetically more stable. Daskalakis and Helz (1993) subsequently modelled solubility data with [Zn(HS)$_4$]$^{2-}$, [Zn(HS)]$^-$ and [ZnS(HS)$_2$]$^{2-}$. However they noted that a complex with the form [Zn$_4$(HS)$_6$(OH)$_4$]$^{2-}$ could equally well fit their data. Zhang and Millero (1994) assumed that the Zn sulfide complexes were similar in form to the Fe, Co, and Ni sulfide complexes and proposed [Zn(HS)]$^+$ and [Zn(HS)$_2$]^0.

Luther et al. (1996) used voltammetric methods together with acid-base titrations to determine the state of protonation of Zn sulfide complexes. They found no evidence for protonation or hydoxy groups and concluded that the major low temperature forms were simple sulfides with 1:1 and 2:3 stoichiometries. They noted that the 2:3 stoichiometry was the equivalent of Daskalakis and Helz (1993) species [Zn$_4$(HS)$_6$(OH)$_4$]$^{2-}$ and [ZnS(HS)$_2$]$^{2-}$. Luther et al. (1996) showed that their data were in good agreement with the ZnS solubility measured by Gubeli and Ste Marie (1967), which had been found to be far too high by Daskalakis and Helz (1993). The key is replacing Gubeli and Ste Marie's (1967) assumed [Zn(OH)(HS)] form with the actual $[Zn_2S_3]^{2-}$. Daskalakis and Helz (1993) had found colloidal ZnS in their systems and had explained the high Gubeli and Ste Marie (1967) numbers by proposing that these investigators were also measuring colloidal suspensions.

Luther et al. (1999) reported that their polymetallic Zn sulfide complexes were essentially clusters according to Cotton et al's (1999) definition cited above. They probed the Zn sulfide complexes with UV-VIS spectroscopy and modelled the species using molecular mechanical calculations. The resulting structures are shown in Figure 27. Luther et al. (1999) agreed with Daskalakis and Helz (1993) that Zn thiolate chemistry (Blower and Dilworth 1987; Prince 1987) suggest that the ZnS cluster with 2:3 stoichiometry is likely to be a tetramer and the acid-base titration data of Luther et al. (1996) suggested a formulation, $Zn_4S_6^{4-}$. In water, this combines with 4 inner shell H_2O molecules to give an overall formulation of $[Zn_4S_6(H_2O)_4]^{4-}$. For the complex with 1:1 stoichiometry they concluded that the probable formulation was $[Zn_3S_3(H_2O)_6]$ based on the structure of known ZnS clusters with thiols. Table 19 lists the evidence for clusters. We list this table, taken from Luther et al. (1999), to demonstrate the sort of independent evidence that can be aquired for the composition of complexes and the methods by which these data can be collected.

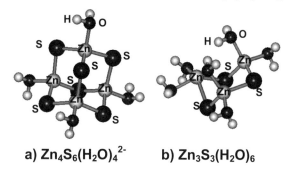

Figure 27. (a) Structure for the cluster $[Zn_4S_6(H_2O)_4]^{4-}$. (b) Structure for the ring $[Zn_3S_3(H_2O)_6]^0$.

As mentioned above, Sukola et al. (2005) defined their "clusters" as what would be called nanoparticles in the geochemical literature. Surprisingly they found no evidence for nanoparticulate ZnS even though it has been characterized in some detail by Zhang et al. (2003). They suggested that the Zn sulfide solutions were a mixture of dissolved Zn sulfide complexes and ZnS colloids.

Chadwell et al. (1999, 2000) characterized Zn polysulfide complexes. They reported monodentate $[Zn(\eta^1-S_4)]$ and $[Zn(\eta^1-S_5)]$ above pH = 6 and their conjugate acids at pH < 6 similar to the forms found for Cu. Again the Zn polysulfide complexes are non-protonated above pH = 6, and this result is consistent with the non-protonated Zn sulfide complexes described above. Using organic solvents, Coucouvanis et al. (1985) synthesized and characterized several bidentate polysulfide complexes including $[Zn(S_4)_2]^{2-}$, $[Zn(S_5)_2]^{2-}$, $[Zn(S_6)_2]^{2-}$.

Proposed Zn sulfide complexes and their reported stability constants are listed in Table 20.

Cadmium. Cadmium (II), in contrast to zinc (II), is classed firmly as a soft metal ion. This is reflected in the relative instability of the cadmiate ion compared to zincates. It is therefore expected that Cd(II) will have a significant sulfide chemistry. Cd(II) is also a toxic metal and thus its sulfide chemistry has some environmental interest. In view of this, it is surprising how little, comparatively speaking, is known about Cd sulfide complexes. None have ever been observed and all the evidence stems from curve fitting of solubility data or voltammetric titrations. Little other evidence for their composition or structure has been reported.

Ste-Marie et al. (1964) explained their solubility data in terms of the complexes $[Cd(HS)]^+$, $[Cd(HS)_2]^0$, $[Cd(HS)_3]^-$ and $[Cd(HS)_4]^{2-}$. However, Dyrssen (1985) re-evaluated

Table 19. Summary of experimental data supporting soluble ZnS cluster species. (From Luther et al. 1999).

Reaction step	$Zn(II) + S(II-) \rightarrow [Zn_3S_3]$	$[Zn_3S_3] \rightarrow [Zn_4S_6]^{4-} + Zn(II)$	$[Zn_4S_6]^{4-} \rightarrow ZnS \downarrow$
Description	Aqueous Zn(II) reacts with aqueous S(–II) to produce an aqueous Zn_3S_3 cluster	Zn_3S_3 clusters react with aqueous S(–II) to produce an aqueous $[Zn_4S_6^{4-}]$ cluster	Aqueous $[Zn_4S_6^{4-}]$ clusters condense with Zn_3S_3 to form solid ZnS with a change of coordination in S from 2 to 4
Cluster MW	$Zn_3S_3(H_2O)_6$ (MW = 400.4 Da)	$[Zn_4S_6(H_2O)_4]^{4-}$ (MW = 526.0 Da)	
Evidence electrochemical	Titrations show an initial 1:1 (S(–II) complex with Zn(II). To be inert the ZnS species must contain S-Zn-S bonds	Further titration shows a 2:3 S(–II) complex with Zn(II). Zn(II) is released into solution and can be accounted for according to stoichiometric calculations.	
spectroscopic	Loss of HS⁻ peak and production of ZnS peaks. Quantum confinement calculations show a size of 1.6 nm (16 Å) or less	Filtration results indicate a size for the initial clusters less than 1000 daltons based on the 200 nm absorption peak.	Bulk ZnS has an absorption at longer λ (380 nm) than the molecular clusters
gel electrophoresis		Negatively charged ZnS cluster gravitates to the positive electrode. Mass is between 350 and 750 daltons	
Molecular modeling	Known ZnS clusters with thiols have Zn_3S_3 configurations. $Zn_3S_3(H_2O)_6$ has a size near 7 Å Calculations show cluster size is smaller than equivalent structured units in the mineral	$Zn_4S_6^{4-}$ is the most stable configuration with a size of 9.7 Å. Molecular orbital (MO) calculations show that clusters form from the overlap of Zn_3S_3 rings (Burdett 1980). Calculations show cluster size is smaller than the mineral since S is 2-coordinated in the cluster and 4-coordinated in the mineral	Zn_3S_3 rings of $Zn_4S_6^{4-}$ have chair configurations as in sphalerite. MO theory shows that additional crosslinking results in mineral formation (Burdett 1980)
Kinetics	Proton loss occurs in the initial reaction (Luther et al. 1996). Observed for Fe(II) (Rickard 1989, 1995; Theberge and Luther 1997) Diffusion controlled reaction related to rate of water loss from Zn(II)	UV-VIS results show on aging that the ZnS absorption at 200 nm shifts to 240, 260 and 290 nm $Zn_4S_6^{4-}$ and Zn_3S_3 are electrochemically inert	Clusters age to form larger particles (Sooklal et al. 1996) UV-VIS results show a very slow increase in absorbance of a ZnS species at 380 nm. This peak appears (0.004 absorbance units) after 2 hours and is similar to bulk semiconductor ZnS (Kortan et al. 1990)
Other work	Significant covalent bonding in Zn-S is in accordance with previous research with thiol capped molecular clusters (Herron et al. 1990; Kortan et al. 1990, Nedeljkovic et al. 1993; Vossmeyer et al. 1995)	Thermodynamic modeling of dissolution experiments shows a Zn:S complex with possible 4:6 stoichiometry (Daskalakis and Helz 1993a) Mass spectrometry of clusters containing thiols show 4 Zn : 6 S stoichiometry (Løver et al. 1997) EXAFS data indicate particles, 200 Å that are not in reversible equilibrium with the solution phase (Helz et al. 1993)	Solid ZnS forms initially as an amorphous precipitate which develops a mainly sphalerite structure (Rickard and Oldroyd, unpub.) Bulk semi-conductor ZnS has an absorption at longer λ (380 nm) than the molecular clusters (Kortan et al. 1990)

Table 20. Proposed Zn sulfide complexes and their reported stability constants.

Species	log$K$	I	Method	Reference
[Zn(SH)(OH)]^0	19.02	1.0	ZnS solubility	Gubelie and Ste-Marie (1967)[b, c]
[Zn(HS)]$^-$	5.78	0.7	ligand competition	Al-Farawati and van den Berg (1999)
	6.0	0.7	sulfide titration	Zhang and Millero (1994)
	6.63	0.7	sulfide titration	Luther et al. (1996)[a]
[ZnS]^0	11.74	0.7	sulfide titration	Luther et al. (1996)[a,b]
[Zn(HS)$_2$]^0	12.9	0.0	ZnS solubility	Dyrssen (1991)[d]
	9.88	0.7	ligand competition	Al-Farawati and van den Berg (1999)
	13.7	0.7	sulfide titration	Zhang and Millero (1994)
[ZnS(HS)]$^-$	13.83	0.0	ZnS solubility	Daskalakis and Helz (1993)
[Zn(HS)$_3$]$^-$	14.9	0.0	ZnS solubility	Dyrssen (1991)[d]
[ZnS(HS)$_2$]$^{2-}$	13.14	0.0	ZnS solubility	Daskalakis and Helz (1993)
[Zn(HS)$_4$]$^{2-}$	14.8	0.0	ZnS solubility	Dyrssen (1991)[d]
	14.64	0.0	ZnS solubility	Daskalakis and Helz (1993)
[Zn$_2$S$_3$]$^{2-}$	13.83	0.7	sulfide titration	Luther et al. (1996)
	41.09		corrected for protonation of the sulfide[c]	
[Zn(S$_4$)]^0	8.37	0.55	sulfide titration	Chadwell et al. (2001)
[Zn(S$_5$)]^0	8.74	0.55	sulfide titration	Chadwell et al. (1999)

[a] acid-base titrations indicate the species is not protonated. MS species are corrected for protonation of sulfide. Multinuclear clusters are likely.
[b] Thermodynamic constants were calculated not conditional constants.
[c] Correcting for Zn$_2$S$_3$$^{2-}$ gives a value of 44.34 which compares with the value of 41.09 of Luther et al (1996).
[d] Dyrssen's recalculation of Hayashi et al (1990).

their experimental measurements and showed that these could be better explained by complexes such as [Cd(HS)$_2$]^0 and [CdHS$_2$]$^-$, by analogy with the Hg complexes proposed by Schwarzenbach and Widmer (1963). Since these were actually derived by curve fitting too, the extrapolation is uncertain. Dyrssen (1988) revisited the Ste-Marie et al. (1964) experimental data in 1988 and proposed that the best fit would be obtained with the complexes, [CdS]^0, [Cd(HS)]$^+$ and [Cd(HS)$_2$]^0. The solubility studies of Wang and Tessier (1999) gave similar results to those of Ste-Marie et al. (1964). However, Daskalakis and Helz (1992) using similar methods found evidence for only [CdOHS]$^-$, [Cd(HS)$_3$]$^-$ and [Cd(HS)$_4$]$^{2-}$. Zhang and Millero (1994) investigated voltammetric titrations of Cd(II) versus S(−II) and proposed [CdHS]$^+$ and [Cd(HS)$_2$]^0 as the dominant species, an idea which was followed by Al-Farawati and van den Berg (1999). If [Cd(HS)$_3$]$^-$ and [Cd(HS)$_4$]$^{2-}$ exist as discrete entities, these would have a tetrahedral Cd(II) center with a structure similar to that in [Cd(HS)$_3$]$^-$ (Fig. 28). Proposed Cd sulfide complexes are listed in Table 21.

The Cd-S system has been studied by several workers using both electrochemical and solubility methods with reasonable agreement for 1:1 and 1:2 Cd-S complexes (Table 22).

By analogy with the Zn-S system, the [Cd(HS)$_3$]$^-$ species is likely a [Cd$_4$S$_6$] cluster species, but Wang and Tessier (1999) as well as Daskalakis and Helz (1992) did not discuss this possibility. Tsang et al. (in press) performed mole

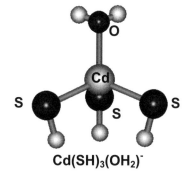

Figure 28. The theoretical structure of an ion with the composition [Cd(SH)$_3$(H$_2$O)]$^-$.

Table 21. Proposed Cd sulfide complexes.

Species	logK	I	Method	Reference
[Cd(HS)]$^+$	6.85	0.7	ligand competition	Al-Farawati and van den Berg (1999)
	6.3	0.7	sulfide titration	Zhang and Millero (1994)
	7.55	1.0	CdS solubility	Ste-Marie et al. (1964)
	7.38	0.0	CdS solubility	Wang and Tessier (1999)
[Cd(HS)$_2$]^0	13.95	0.7	ligand competition	Al-Farawati and van den Berg (1999)
	12.7	0.7	sulfide titration	Zhang and Millero (1994)
	14.61	1.0	CdS solubility	Ste-Marie et al. (1964)
	14.43	0.0	CdS solubility	Wang and Tessier (1999)
[Cd(HS)$_3$]$^-$	16.49	1.0	CdS solubility	Ste-Marie et al. (1964)
	16.44	0.0	CdS solubility	Daskalakis and Helz (1992)
	16.26	0.0	CdS solubility	Wang and Tessier (1999)
[Cd(HS)$_4$]$^{2-}$	18.85	1.0	CdS solubility	Ste-Marie et al. (1964)
	17.89	0.0	CdS solubility	Daskalakis and Helz (1992)
	18.43	0.0	CdS solubility	Wang and Tessier (1999)

Table 22. Comparison of stability constant data for the Cd(HS)$^-$ and Cd(HS)$_2$ species, corrected for Cd side reaction coefficients.

log K_{COND}		Method	Reference
1:1	1:2		
8.7 ± 0.3	14.0 ± 0.4	Titration	Tsang et al (in review)
7.84	14.2	Titration	Zhang and Millero (1994)
7.55 ± 0.16	14.61 ± 0.16	Solubility of CdS	Ste-Marie et al. (1964)
7.38 ± 0.68	14.43 ± 0.01	Solubility of CdS	Wang and Tessier (1999)
9.13 ± 0.02	—	Titration	Al-Farawati and van den Berg (1999)
8.4	15.5	Competitive ligand	Al-Farawati and van den Berg (1999)

ratio titration methodology at pH ~ 7 and showed that 1:1 and 1:2 Cd:S complexes formed. Other titrations at higher pH showed evidence for a [Cd$_2$S$_3$] (or [Cd$_4$S$_6$]) species (Mullaugh and Luther, unpublished). However, the Cd complexes were not protonated based on acid-base titrations. The conditional stability constant for the 1:1 complex agrees with that of the other groups, and correction for the proton side reaction coefficient of sulfide gives a stability constant of logK_{therm} = 23.4 for CdS. As there was no detectable Cd signal for Cd in sulfide solutions, the chelate scale approach indicated that the logK_{therm} is greater than 34.4. Thus a cluster complex [Cd$_4$S$_6$] is likely based on these data. Using organic solvents, Coucouvanis et al. (1985) synthesized and characterized [Cd(S$_6$)$_2$]$^{2-}$.

Mercury. Addition of S(–II) to aqueous Hg^{2+} produces black mercuric sulfide, HgS or metacinnabar, which has a ZnS structure. Metacinnabar is unstable with respect to red, rhombohedral HgS or cinnabar, at temperatures lower than 344°C (Barnes and Seward 1979), and the transformation is kinetically rapid. A third polymorph, hypercinnabar, a dark purple-black phase with a hexagonal structure is also known. Mercury is relatively toxic and its sulfide chemistry is therefore of considerable environmental interest.

Mercury is unique amongst the metals in being a liquid at room temperature. Its filled 4f shell and relativistic effects increase the binding of Hg electrons so that its first ionization potential is greater than any other metal. Relativistic effects are the relative mass increase for a moving particle, and these begin to be significant for the heavier elements, such as Hg. The effect causes the 6s electrons to be more tightly bound and relatively inaccessible to metallic bonding. The relativistic stabilization of the Hg 6s orbital provides an energetic advantage when two Hg$^+$ ions share a pair of 6s electrons which results in the relatively stable Hg$_2^{2+}$ ion, where Hg has a (I) oxidation state, and compounds with Hg$_3^{3+}$ and Hg$_4^{4+}$.

A number of Hg sulfide complexes have been proposed and are listed in Table 23. The first group stems from the original work of Schwarzenbach and Widmer (1963) who measured HgS solubility and showed that [Hg(SH)$_2$], [HgS$_2$H]$^-$, [HgS$_2$]$^{2-}$ fitted the titration curves. Schwarzenbach and Widmer used the black HgS precipitate and therefore were probably examining metacinnabar solubility. Dyrssen (1985) noted that [HgS]^0 would not be detected by the curve fitting of solubility results and Dyrssen and Wedborg (1989, 1991) and Dyrssen (1989) suggested a stability constant for this constant based on the solubility of ZnS and CdS. Benoit et al. (1999) found that the fraction of Hg partitioned into 1-octanol decreased when the sulfide concentration increased and concluded that this was due to the neutral Hg sulfide species, [HgS]^0. Paquette and Helz (1997) also found that cinnabar solubility could be explained without a neutral complex in agreement with the conclusion of Schwarzenbach and Widmer (1963). Tossell (2001) pointed out that isolated molecules with Hg in one-coordination are not known in crystalline solids. Using quantum mechanical computations he found that HgS^0 was probably unstable in aqueous solutions, and probably existed as a neutral [Hg(SH)(OH)] complex hydrated with 4 water molecules. With increasing HS$^-$ concentration, [Hg(SH)$_2$(OH)]$^-$ forms.

There is very limited spectroscopic information on Hg sulfide complexes. Cooney and Hall (1966) used Raman to demonstrate that HgS$_2^{2-}$ was the likely composition in alkali sulfide solutions. Lennie et al. (2003) used EXAFS to demonstrate that Hg was in two coordination with S in alkaline sulfide solutions, which is not inconsistent with this result.

Paquette and Helz (1997) and Jay et al. (2000) showed that the solubility of cinnabar was increased in the presence of elemental sulfur and proposed a series of Hg polysulfide complexes. Paquette and Helz proposed Hg(S$_n$)HS$^-$ and Jay et al. proposed Hg(S$_n$)$_2^{2-}$ in the presence of elemental sulfur and HgS$_n$OH$^-$ at lower sulfide concentrations and high pH. Jay et al. also suggested HgS$_5$ as a possible minor species, which is analogous to the pentasulfide complexes of the transition metals characterized by Chadwell et al. (1999). In fact, Müller et al. (1984c) synthesized [Hg(S$_6$)$_2$]$^{2-}$ in methanol. The structure (Fig. 29) shows a tetrahedral Hg

Table 23. Proposed Hg sulfide complexes.

Complex	Method	Reference
[HgS$_2$]$^{2-}$	HgS solubility	Schwarzenbach and Widmar (1963)
[HgS$_2$H]$^-$	HgS solubility	Schwarzenbach and Widmar (1963)
[Hg(SH)$_2$]^0	HgS solubility	Schwarzenbach and Widmar (1963)
[HgSH]$^+$	HgS solubility	Dyrssen and Wedborg (1991)
[HgS]^0	HgS solubility	Dyrssen (1989)
		Dyrssen and Wedborg (1991)
	Octanol separation	Benoit et al (1999)
		Jay et al.(2000)
[Hg(SH)(OH)]	Quantum mechanical computation	Tossell (2001)
[Hg(SH)$_2$(OH)]$^-$	Quantum mechanical computation	Tossell (2001)
[Hg(S$_n$)HS]$^-$	HgS solubility in presence of S^0	Paquette and Helz (1997)
[Hg(S$_n$)$_2$]$^{2-}$	HgS solubility in presence of S^0	Jay et al. (2000)
[Hg(S$_n$)OH]$^-$	HgS solubility	Jay et al. (2000)
[HgS$_5$]^0	HgS solubility in presence of S^0 and octanol separation	Jay et al. (2000)

bound with two hexasulfido bidentate ligands. Jay et al. (2000) pointed out that the solubility data gave no direct evidence for these species but noted (p. 2199) that *"the indirect evidence provided by the variations in cinnabar solubility with pH and S(−II)$_T$ is fairly strong."*

The potential importance of the Hg polysulfide complexes is illustrated in Figure 30. The Hg polysulfides dominate at the environmentally significant neutral pH area in the presence of S^0. In the absence of S^0, and with total dissolved sulfide > 10^{-10} M, sulfide complexes dominate even in the presence of substantial chloride. A summary of Hg sulfide complexes and their reported stability constants is given in Table 24.

The present data for Hg sulfide complexes are generally displayed in terms of a system being in equilibrium with cinnabar. In fact, in low temperature aqueous systems, this is an unlikely scenario away from Hg ore deposits. The reason is that Hg appears to be equally attracted to thiols as it is to inorganic sulfide species. In fact, the old name for thiol, *mercaptan*, derives from the attraction of this ligand to mercury.

Tin and lead

Tin and lead are members of Group 14 with carbon, silicon and germanium. Although an apparently disparate group, there are clear trends from C (non-metallic), Si (mainly non-metallic), Ge (metalloid) through Sn and Pb which are increasingly metallic. Catenation, which dominates C chemistry, decreases steadily down the group with Sn showing a more marked tendency to produce Sn-Sn bonds than Pb is to produce Pb-Pb bonds. Oxidation states (II) and (IV) dominate in this group with (IV) decreasing in importance with increasing mass. Thus Sn(IV) has a substantial chemistry but Pb(II) is by far the most important Pb state.

Tin. Although in classical geochemistry, tin is considered essentially as a lithophilic element, it does have a significant sulfide chemistry. SnS may be prepared by reaction of S(−II) with Sn(II) salts. Sn_2S_3 with Sn(III) is also known. However, very little appears to be known about Sn sulfide complexes. The main interest has been in polynuclear Sn sulfide clusters, which display complex structures. For example, $[Sn_5S_{12}^{-4}]_\infty$ occurs as a Cs salt and contains both trigonal bipyrimidal and octahedral Sn(IV) (Sheldrick 1988). Similarly, $[Sn_3S_7^{2-}]_\infty$ in Rb salts includes both SnS_4 octahedra and SnS_6 octahedra (Sheldrick and Schaaf 1994). These are all crystalline forms. However, Gu et al. (2005) report $[Sn_2S_6]^{4-}$ anions being produced by reactions between $SnCl_4$, S(−II) and ethylenediamine.

Seby et al. (2001) reviewed all published inorganic thermodynamic data on Sn sulfides and only noted various measurements of the solubility of SnS to produce the $Sn(II)_{aq}$ ion and free S(−II). No data on Sn sulfide complexes were reported. It appears that it is possible that a significant inorganic Sn sulfide complex chemistry exists which has not been widely investigated as yet.

Lead. Lead forms the well-known isometric sulfide phase PbS, galena, which is renowned for its insolubility. In fact in sulfide solutions, Pb displays enhanced solubility over that expected for aqueous Pb^{2+} and a series of sulfide complexes have been assigned. None of these has actually been observed and all are theoretical constructs. Both $[Pb(HS)]^+$ and $[Pb(HS)_2]^0$ have been assumed by a number of authors (e.g., Dyrssen 1985, 1988; Zhang and Millero 1994; Al-Farawati and van den Berg 1999). Earlier suggestions (e.g., Smith and Martell 1976) that higher complexes, such as $[Pb(HS)_3]^-$, might contribute to the total PbS solubility have not been followed up.

Rozan et al. (2003) used mole ratio titration methodology and showed that only a 1:1 complex formed. However, the Pb complexes were not protonated based on acid-base titrations. The logarithm of the PbS conditional stability constant value (corrected for the Pb side reaction coefficient but not for sulfide protonation) was found to be 8.29 by Rozan et al. (2003), 8.0 by Al-Farawati and van den Berg (1999) and 8.64 by Zhang and Millero (1994). These values are

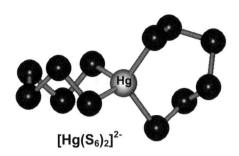

Figure 29. Structure of the Hg polysulfide complex, $[Hg(S_6)_2]^{2-}$.

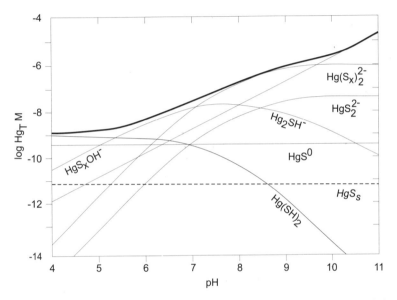

Figure 30. Speciation of total dissolved mercury, Hg_T, in equilibrium with cinnabar, HgS_s, and elemental sulfur at 1 mM total dissolved S(−II) as a function of pH, according to Jay et al. (2000). The bold curve shows the measured concentration of dissolved Hg and the fine lines the species boundaries.

Table 24. Reported stability constants for Hg sulfide complexes.

Complex	log$K$	I	Method	Reference
$[Hg(HS)]^+$	7.8	0.7	sulfide titration	Zhang and Millero (1994)
	20.6		corrected for metal-chloro complexes	
$[Hg(HS)_2]^0$	12.8	0.7	sulfide titration	Zhang and Millero (1994)
	25.6		corrected for metal-chloro complexes	
	37.71	1.0	HgS solubility	Schwarzenbach and Widmer (1963)[a]
$[HgS(HS)]^-$	43.23	1.0	HgS solubility	Schwarzenbach and Widmer (1963)[a]
$[HgS_2]^{2-}$	51.53	1.0	HgS solubility	Schwarzenbach and Widmer (1963)[a]

[a] Thermodynamic constants were calculated not conditional constants.

very similar. Correction for the proton side reaction coefficient of sulfide (Rozan et al. 2003) gives a stability constant of $\log K_{therm} = 16.83$ for PbS. As there was no detectable Pb signal for Pb in sulfide solutions, the chelate scale approach indicated that the $\log K_{therm}$ is greater than 39. Thus a cluster species is likely based on these data. Mass spectrometry data from filtered and freeze dried samples prepared in aqueous solutions indicated a Pb_3S_3 species was present. Pb sulfide complexes and their reported stability constants are listed in Table 25.

Table 25. Proposed Pb sulfide complexes and their reported stability constants.

Complex	log K	I	Method	Reference
$[Pb(HS)]^+$	6.46	0.7	ligand competition	Al-Farawati and van den Berg (1999)
	7.1	0.7	sulfide titration	Zhang and Millero (1994)
	6.75	0.7	sulfide titration	Rozan et al (2003)[a]
$[PbS]^0$	16.83	0.7	sulfide titration	Rozan et al (2003)[a]
$[Pb(HS)_2]^0$	13.86	0.7	ligand competition	Al-Farawati and van den Berg (1999)
	13.5	0.7	sulfide titration	Zhang and Millero (1994)

[a] acid-base titrations indicate the species is not protonated. MS species are corrected for protonation of sulfide. Multinuclear clusters are likely.

Arsenic and antimony

Group 15 elements are known collectively as the pnictides and include N, P, As, Sb and Bi. The shielding effects of the d^{10} and f^{14} electrons, which were noted as being particularly important for the Group 12 elements, are not as marked in Group 15. The trend across the group is for the heavier elements to show quasi-metallic behavior and As, Sb and Bi are described as metalloids. From the point of view of metal sulfide complexes, As and Sb are particularly important geochemically since they are both relatively toxic and their association with sulfidic natural systems has both environmental and economic implications.

The oxidation states are dominated by (III) and (V) configurations. The acceptance of 3 electrons results in the development of the (−III) oxidation state but this configuration is not energetically favorable and of little significance to most environments. As and Sb do not form free aqua cations such as As^{5+} or Sb^{5+} because of the high ionization energies. Free As^{3+} and Sb^{3+} are improbable although the observation that highly acidic solutions of bismuth perchlorate may contain hydrated Bi^{3+} ions suggests that such forms may be theoretically possible in extreme environments.

Arsenic. Arsenic is a metalloid which forms the common sulfide minerals, realgar AsS and orpiment, As_2S_3, apart for being a key constituent of multielement minerals such as arsenopyrite FeAsS. It displays a number of oxidation states in complexes, the most important in the natural environment being As(III) and As(V). The arsenic sulfide complexes have engendered some interest in geochemistry because of the toxicity of As in groundwaters as well as the association of the element with Au.

The arsenic complexes with sulfide are designated thioarsenites if they contain As(III) and thioarsenates if they contain As(V). For convenience, even the O containing species can be generally referred to with the same sobriquets. All of the sulfide complexes with As are thioarsenites, mostly with an As:S ratio of 1:2. Höltje (1929) and Babko and Lisetskaya (1956) proposed a series of monomers. A cyclic trimer was suggested by Angeli and Souchay (1960), Spycher and Reed (1989), Webster (1990) and Eary (1992). However, Mironova et al. (1990) used the same solubility method and concluded that dimeric thioarsenites exist and Krupp (1990) argued for the existence of dimeric thioarsenites by analogy with thioantimonates. Helz et al. (1996) suggested that both monomers and trimers exist. They suggested that the monomers [HAsS$_2$O]$^-$ and [HAsS]$^{3-}$ are present in sulfidic waters undersaturated with respect to orpiment whereas, in nearly saturated solutions, the trimer As$_3$S$_6^{3-}$ dominates. Wilkin et al. (2003) found four arsenic-sulfur species with As:S ratios from 1:1 to 1:4 using ion chromatography with inductively coupled plasma mass spectrometry (IC-ICPMS). The suggested species with ratios from 1:1 to 1:3 are equivalent to the thioarsenites proposed by Höltje (1929) whereas the 1:4 ratio was assigned to the Srivastava and Gosh (1954) species, [H$_4$AsS$_4$]$^-$. Nordstrom (2003 in Stauder et al. 2005) presented a monomeric and a trimeric thioarsenite as the most probable arsenic sulfur complexes in sulfidic waters.

Schwedt and Rieckhoff (1996) detected monothioarsenate by capillary zone electrophoresis and IC-ICPMS. However, they investigated extracts of arsenic slags and minewaters, which probably contained oxygen. Wood et al. (2002) reported up to 8 possible arsenic sulfide complexes in sulfidic solutions and noted that their Raman data suggested the possibility of thioarsenates. However, this possibility was excluded by Stauder et al. (2005) on the basis that the experimental system did not contain oxygen. McCay (1901) showed that in the reaction between H$_2$S and arsenate, thioarsenates are formed by reactions like Equation (68):

$$5[H_3AsO_3] + 3H_2S \rightarrow 2As + 3[H_2AsO_3S]^- + 6H_2O + 3H^+ \qquad (68)$$

The thioarsenates are reduced to arsenites with the separation of sulfur. Rochette et al. (2000) studied the kinetics of this reaction and proposed that thioarsenites are formed as intermediaries in this reaction. Stauder et al. (2005) report that the observations of Spycher and Reed (1989), Webster (1990), Mironova et al. (1990) and Eary (1992) on the solubility products of orpiment dissolution are in error. They found no thioarsenites in 120 samples of hydrogen sulfide with arsenate or arsenide. Only arsenite, arsenate and four thioarsenates, [HAsO$_3$S]$^{2-}$, [HAsO$_2$S$_2$]$^{2-}$, [AsOS$_3$]$^{3-}$, and [AsS$_4$]$^{3-}$, were detected. They conclude that thioarsenates rather than thioarsenites dominate in sulfidic arsenic bearing natural solutions. Table 26 lists the arsenic-sulfide complexes to date. However, we are reluctant, in the absence of any definitive data, to recommend any stability constants at present. The debate over the composition of As sulfide complexes is still unresolved.

Table 26. Proposed compositions for soluble, sulfur containing As species in sulfidic solutions (modified from Staudel et al. 2003).

Composition	Method	Reference
H$_2$AsO$_2$S$^-$, HAsOS$_2$, AsS$_3^{3-}$	orpiment solubility	Höltje (1929)
AsS$_2^-$, H$_2$AsOS$_2^-$, AsS$_3^{3-}$	orpiment solubility	Babko and Lisetskaya (1956)
AsS$_4^{5-}$	—	Srivastava and Gosh (1957)
As$_3$S$_6^{3-}$, As$_2$S$_5^{4-}$	titration freezing point depression	Angeli and Souchay (1960)
HAs$_3$S$_6^{2-}$, As$_3$S$_6^{3-}$	orpiment solubility	Spycher and Reed (1989)
HAs$_2$S^{4-}, As$_2$S$_4^{2-}$, H$_2$As$_2$O$_2$S$_2$	—	Krupp (1989)
H$_2$As$_2$OS$_3$, HAs$_2$S$_4^-$, As$_2$S$_4^{2-}$	titration freezing point depression	Mironova et al. (1990)
H$_2$As$_3$S$_6$–	orpiment solubility	Webster (1991)
H$_2$As$_3$S$_6^-$	orpiment solubility	Eary (1993)
H$_2$As$_3$S$_6^-$, H$_2$AsOS$_2^-$, H$_2$AsS$_3^-$	solubility spectroscopy MO calculations	Helz et al. (1996)
H$_2$AsOS$^-$, HAsOS$_2^{2-}$, HAsS$_3^{2-}$, H$_4$AsS$_4^-$	Ion chromatography ICP-MS analysis	Wilkin et al. (2003)
H$_2$As$_3$S$_6^-$, H$_2$AsOS$_2^-$	—	Nordstrom (in Stauder et al. 2005)

Antimony. Antimony forms a common sulfide mineral stibnite, monoclinic Sb_2S_3, with Sb formally being in the Sb(III) state. The chemistry of antimony in natural waters has been reviewed by Filella et al. (2002). A summary of proposed antimony sulfide complexes, adapted from their study, is shown in Table 27. Their analysis shows the degree of uncertainty in this field at present. Filella et al.'s analysis of the reasons for these discrepancies is instructive in general for both experimentalists and for researchers using metal sulfide complex stability data: *"Discrepancies among published results may be due to:(i) the somewhat wide range of Na_2S concentrations employed in the different studies; (ii) the oxidation in air of Sb(III) species, present in Na_2S solutions in equilibrium with stibnite, to Sb(V) species such as SbS_4^{3-}; (iii) solid phases other than the phases of interest being present in the experimental solutions(i.e., $Sb_2O_3(s)$); (iv) the nature of Sb_2S_3 used in the experiments, crystalline or amorphous (not always identified); and (v) the different experimental pH ranges used (not always given). However, much of the difficulty in determining the speciation may be attributed to the inability of the traditionally used solubility and potentiometric methods to differentiate precisely among species with similar metal/S ratios (case of polymeric species). Stoichiometry often appears to have been simply assumed rather than proved."* (Filella et al. 2002 p. 268).

Krupp (1988) and Spycher and Reed (1989) discussed earlier Sb sulfide solubility data and agreed that when Sb_2S_3 is dissolved in H^+-containing solutions the predominant Sb sulfide species were dimeric, $[Sb_2S_2(SH)_2]$ to $[Sb_2S_4^{2-}]$, depending upon pH. Solubility data for Sb_2S_3 were reported by Krupp (1988). The Raman spectra of species formed by dissolving Sb_2S_3 in alkaline sulfidic solutions appear to be monomeric (Wood 1989). Wood (1989) concluded that polymeric species were unlikely with Sb concentrations <0.1 M. A number of different Sb sulfide species have been proposed by the different investigators studying the solubility of Sb_2S_3 as a function of pH and total sulfide (Krupp 1988). These include many species containing Sb (III) in two- or three-coordination, such as the monomeric species ($[SbS_2]^-$, $[SbS_3]^{3-}$, and their conjugate acids), oligomeric species (e.g., $[Sb_2S_4]^{2-}$, $[Sb_2S_5]^{4-}$, $[Sb_3S_6]^{3-}$, and their conjugate acids), and the high temperature mixed ligand species (e.g., $[Sb_2S_2(OH)_2]$). Four coordinate Sb(III) and Sb(V) sulfide species have not been suggested. The results of Tossell's (1994) semi-empirical quantum mechanical calculations (Fig. 31) were consistent with the experimental observations of Wood (1989) and Krupp (1988).

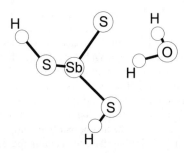

Figure 31. Calculated equilibrium geometry for the $[Sb(SH)_3H_2O]$ complex (after Tossell 1994).

The Eh–pH diagram for the system Sb–S–H_2O at environmentally relevant concentrations might look something like Figure 32. As noted above, these types of diagrams only refer to the equilibrium state of the system and are only valid for the species

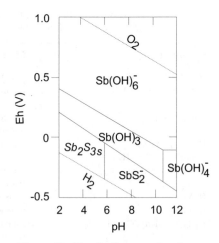

Figure 32. Eh–pH diagram of antimony in the Sb–S–H_2O system at a dissolved antimony concentration of 10^{-8} mol/l and a dissolved sulfur concentration of 10^{-3} mol/l. (adapted from Filella et al. 2002).

Table 27. Proposed Sb sulfide species in aqueous solution at $T < 30$ °C (after Filella et al. 2002).

Species	pH	S (mol/l)	Method	Reference
$SbS_4(H_2O)_2^{3-}$	>12[a]	2.0 $(NH_4)_2S$		Brintzinger and Osswald 1934
SbS^{2-}, SbS_3^{3-}, $Sb_2S_5^{4-}$	>12[a]	0.005–3.0 Na_2S	chemical analysis and microscopy	Fiala and Konopik 1950
SbS^{2-}, SbS_3^{3-}	0.6–12.3	0.005–0.1 H_2S, K_2S	solubility	Akeret 1953
SbS^{2-}	8–9	0.04 H_2S	solubility	Babko and Lisetskaya 1956
$SbS(OH)^{2-}$	10–11	no sulfide	solubility	Babko and Lisetskaya 1956
SbS_3^{3-}	>12[a]	0.005–0.1 Na_2S	solublity	Wei and Saukov 1961
$Sb_2S_4^{2-}$	12.5	0.03–0.06 Na_2S	solubility	Dubey and Ghosh 1962
$Sb_4S_7^{2-}$	>12[a]	0.06–0.92 Na_2S	solubility	Arnston et al. 1966
SbS_3^{3-}	>12[a]	excess Na_2S	solubility	Milyutina et al. 1967
$Sb_2S_4^{2-}$, $H_2Sb_2S_4$, HSb_2S_4	3–9		reinterpretation of Babko and Lisetskaya 1956, Akeret 1953	Kolpakova 1971
SbS_3^{3-}, $Sb_2S_5^{4-}$, $Sb_4S_7^{2-}$	>12[a]	0.25–2.5 Na_2S w/ 0.0–0.12 Sb_2S_3	potentiometry	Shestitko and Demina 1971
SbS_4^{3-}	>12[a]	0.13–1.0 Na_2S w/ 0.02–0.06 Sb_2S_5	potentiometry	Shestitko et al. 1975
SbS_3^{3-}, $SbS_2O_3^-$		excess Na_2S deficiency Na_2S	crioscopy, potentiometry	Chazov 1976
$H_2Sb_2S_4$ (high pH), $HSb_2S_4^-$ (pH 5–9), $Sb_2S_4^{2-}$	3–11	TF Sb = 0.1–0.001	solubility	Krupp 1988
H_2Sb_2S4, $HSb_2S_4^-$, $Sb_2S_4^{2-}$	2–13		reinterpretation of published solubility data	Spycher and Reed 1989
$Sb_2S_4^{2-}$ or $Sb_4S_7^{2-}$ (0.1 m [c] Sb); SbS_2^- or SbS_3^{3-} (< 0.06 m [c] Sb)	12.95	0.95 m [c] Na_2S	Raman	Wood 1989
$SbS_2(SH)^{2-}$, $Sb_2S_2(SH)_2$, $SbS(SH)^{2-}$			*ab initio* quantum mechanical calculations	Tossell 1994
$Sb_2S_4^{2-}$	>12[a]	0.914 m [c] Na_2S_4	Raman	Guschina et al. 2000
SbS_4^{3-}, multimeric species, $Sb_2S_2(SH)_2$	8–14	0.009–2.5 HS$^-$	EXAFS	Mosselmans et al. 2000
$Sb(HS)_4^+$		(a) 1.15 m [c] Na_2S (b) 0.2 m [c] NaHS + 0.06 NaOH (c) 0.2 m [c] NaHS + 0.66 NaOH	EXAFS	Sherman et al. 2000

[a] Not stated by the authors but deduced by Filella et al (2002) from other conditions of the experiments.
[b] TFSb = total free Sb (mol/kg)
[c] mol/kg H_2O

considered. They thus do not describe the real system. However, they can be interesting in pictorially rendering the general trends. Figure 32. reflects the generally accepted view that Sb(V) dominates oxic systems and Sb(III) anoxic systems. The plot suggests that, at environmentally possible Sb and S(−II) concentrations, antimony is present as $[Sb(OH)_6]^-$ in oxic systems and as $[Sb(OH)_3]$ in anoxic ones . Under reducing conditions, and in the presence of sulfur, stibnite, $Sb_2S_3(s)$, is formed at low to intermediate pH values. At higher pH values, the $[SbS]^-$ species replaces stibnite. However, the evidence for the composition of Sb sulfide species remains equivocal and we do not list any selected stability constants.

Selected metal sulfide complex stability constants

The reader will have gathered by now that the evidence for the composition of many complexes in the metal sulfide system is at least uncertain and in some cases absent. The problem is that some complexes have been assumed to occur and little evidence has been forwarded to support their existence. Many of the complexes have been assumed to exist because of curve-fitting of titration or solubility experiments. The problem here is that there is usually no unique solution to the curves obtained. This has often been associated with the proposal of stability constants for the theoretical complexes. So that the experimenter has three unknown variables: (1) composition, in terms of protonation, hydration and the occurrence of other anions, such as oxygen or the halogens; (2) stoichiometry and the actual number of cations and anions involved in the formula, and (3) stability constant, the arithmetic ratio between the product of the undefined products and the reactants. If the experimental data involves three or less variables (e.g., pH, metal concentration, sulfide concentration) then, *a priori*, the solution is not unique. The situation is further compounded by the experimental difficulty in defining the metal sulfide reactants, which are often nanoparticulate materials highly sensitive to the environment of preparation. In the case of FeS, for example, Rickard and Morse (2005) pointed out that the composition of the reactant FeS is rarely determined or reported in the experiments used to measure either its solubility or the Fe sulfide complexes that may be formed. Furthermore the other components in the system are often not well-defined. Because of the sensitivity of sulfide species to oxidation, in particular, experimentation needs to be carried out in very well defined systems—and this has not always been the case.

We present a listing of selective stability constants for metal sulfide complexes in Table 28. This is admittedly a subjective listing. The first thing that will strike the reader about the listing is the relatively small number of complexes compared to the large number of proposed complexes that are discussed under the specific metals above. In fact, many of the complexes for which information does exist for their composition lack data on their stability. This is due to the difference in approach of the chemist and geochemist as noted above. Even in our selective listing there are many constants for which compositional evidence is lacking. And even in some of the complexes for which we have compositional data, it is at best sketchy and often the subject of debate in the literature. In some cases the same constant has been measured in the same medium by more than one group of researchers. However, as seen in Table 28, this does not necessarily mean that there is independent evidence for the composition of the complex. Assumptions often develop a life of their own.

The importance of independent evidence for complexes cannot be overstated. The reported stability constants are only valid for complexes with the stated compositions. If the composition is incorrect, the stability constant is invalid.

The constants we have listed are mainly conditional constants which are only valid for the ionic strength noted. Corrections for other ionic media can be made by the reader but the uncertainties increase as the conditions diverge as discussed above. We list the measured uncertainties where multiple measurements have been made. Where these are absent the uncertainty is at least one log unit.

For [MnHS]$^+$, Zhang and Millero (1994) reported log$K$ = 6.7 at I = 0. This is two logarithmic units higher than the other reported values and is not included in our selective value. The values for the Mn, Fe, Co and Ni monodentate bisulfide complexes, [MnHS]$^+$, are generally similar or have a slight increase. These data are consistent with their known molecular orbital stabilization energies (see below) and their increasing effective nuclear charge. Wei and Osseo-Ware (1995) reported log$K$ = 4.34 for [Fe(SH)]$^+$ at I = 0 and this is probably within the uncertainties of the listed conditional constants at I = 0.7. However, FeS solubility measurements are not consistent with a log$K$ value for [Fe(SH)]$^+$ > 3 (Davison et al. 1999; Rickard unpublished). The reason for the different results for the solubility and titration methods is currently unknown. However, the solubility measurements are not specifically aimed at the determination of complex stability constants or transient intermediates, and this may be a general problem in the use of solubility measurements in investigating the chemistry of dissolved metal sulfide complexes. Also, Wei and Osseo-Ware (1995) noted that [Fe(SH)]$^+$ was an intermediate in their study which decomposed to FeS species.

Al-Farawati and van den Berg (1999) reported log$K$ = 11.52 for [CuHS]$^+$ which is very divergent from other measurements and is not included in our selective values. As noted in footnote to Table 28, Cu appears to be in the form of Cu(I) in most mineral sulfides and there is evidence (Luther et al. 2002) that the reduction takes place in solution (i.e., in the complex). The net result of this is that it is expected that Cu sulfide complexes would have some Cu(I). Likewise, since covellite is effectively a Cu(I) disulfide (i.e., Cu_2S_2) it is probable that the Cu(II) reduction step is accompanied by an effective oxidation of S(–II) to S(–I). The presence of polysulfide in Cu sulfide complexes would explain much of the confusion presently in the literature regarding these species (see above).

Cloke (1963b) lists a table of free energies in his report but this is not properly referred to in his text and thus his Cu and Ag polysulfide complexes are somewhat uncertain. His values for $[Cu(S_4)(S_5)]^{3-}$ or $[Cu(S_4)_2]^{3-}$ are 50% of the Chadwell et al. (1999, 2002) values for $[Cu_2(S_4)_2]^{2-}$ and $[Cu_2(S_5)_2]^{2-}$, which suggests that his complexes have just two Cu-S bonds whereas Chadwell et al. have 4 Cu-S bonds. This is supported by the $[Au_2(S_4)_2]^{2-}$ structure (Fig. 26) found by Müller et al. (1984b).

Ste-Marie et al. (1964) reported log$K$ = 7.55 and 14.61 at I = 1.0 for [CdHS]$^+$ and [Cd(HS)$_2$] respectively and Wang and Tessier (1999) reported 7.38 and 14.43 at I = 0. These values appear to be divergent and are not included in our sected listing. Ste-Marie et al. (1964) also reported log$K$ = 16.49 and 18.85 at I = 1.0 for [Cd(HS)$_3$]$^-$ and [Cd(HS)$_4$]$^{2-}$ respectively. Again, although the numbers are similar, the considerable difference in ionic strengths for these measurements and those listed in out selected values casts some doubt on these values.

Zhang and Millero (1994) reported log$K$ ≥ 9.5 (I = 0.7) for [AgHS]^0 and log$K$ ≥ 15.3 (I = 0.7) for [Ag(HS)$_2$]$^-$ both of which are consistent with our selected values. However, Schwarzenbach and Widmer's (1966) value of 72.9 (I = 1.0) for $[Ag_2S(HS)_2]^{2-}$ appears very divergent from the Seward group's work and has been neglected. Likewise, Schwarzenbach and Widmer's (1963) value of 37.71 (I = 1.0) for [Hg(HS)$_2$] is neglected.

No evidence is reported for the composition of [Fe(HS)$_2$]^0, [Co(HS)$_2$]^0, [Ni(HS)$_2$]^0, [Cu(HS)$_2$]^0, [Zn(HS)$_2$]^0, [Ag(HS)$_2$]$^-$, [Hg(HS)$_2$]^0 and [Pb(HS)$_2$]^0. Luther et al. (1996) could find no evidence for [Fe(HS)$_2$]^0, [Co(HS)$_2$]^0, [Ni(HS)$_2$]^0, [Cu(HS)$_2$]^0 or [Zn(HS)$_2$]^0. It appears that [Fe(HS)$_2$]^0 could be a possible kinetic reaction intermediary in the formation of FeS and FeS$_{aq}$ (Rickard 1995) with limited stability and transient existence.

As mentioned above, As and Sb sulfide stability constants are omitted from the table in the absence of more consistent information about their composition.

Metal sulfide cluster stability. The stability constants for the sulfide complexes are

Table 28. Recommended values for association or stability constants for metal sulfides at 25 °C. Complexes for which there is *some* independent evidence for their composition are in bold. However, this does not mean that the composition has necessarily been entirely proven. Corrections for metal or sulfide side reaction coefficients with other ligands or protons, respectively, are specified. Where more than one source is given the value is the arithmetic mean. The measured uncertainty, $\pm$, covers the range of values cited. In the absence of any duplicate measurements no uncertainty is listed. I is the ionic strength at which the constant is measured.

Metal	Complex	log$K$	$\pm$	I	Reference
Cr(III)	[CrHS]$^+$	3.9		0.7	Al-Farawati and van den Berg (1999)
Mn(II)	[MnHS]$^+$	4.5	0.3	0.7	Al-Farawati and van den Berg (1999)
					Luther et al. (1996)
	[Mn(HS)$_2$]^0	9.9		0.7	Al-Farawati and van den Berg (1999)
	[Mn$_2$(HS)]$^{3+}$	9.7		0.7	Luther et al. (1996)
	[Mn$_3$(HS)]$^{5+}$	15.4		0.7	Luther et al. (1996)
	[MnS$_4$]^0	5.8		0.55	Chadwell et al. (2001)
	[Mn$_2$(S$_4$)]$^{2+}$	11.3		0.55	Chadwell et al. (2001)
	[MnS$_5$]^0	5.6		0.55	Chadwell et al. (1999)
	[Mn$_2$(S$_5$)]$^{2+}$	11.5		0.55	Chadwell et al. (1999)
Fe(II)	**[FeHS]$^+$**	5.4	0.5	0.7	Al-Farawati and van den Berg (1999)
					Zhang and Millero (1994)
					Luther et al. (1996)
	[Fe$_2$(HS)]$^{3+}$	10.07		0.7	Luther et al. (1996)
	[Fe$_3$(HS)]$^{5+}$	16.2		0.7	Luther et al. (1996)
	[FeS$_4$]^0	6.0		0.55	Chadwell et al. (2001)
	[Fe$_2$(S$_4$)]$^{2+}$	11.3		0.55	Chadwell et al. (2001)
	[FeS$_5$]^0	5.7		0.55	Chadwell et al. (1999)
	[Fe$_2$(S$_5$)]$^{2+}$	11.3		0.55	Chadwell et al. (1999)
Co(II)	[CoHS]$^+$	5.5	1.0	0.7	Al-Farawati and van den Berg (1999)
					Zhang and Millero (1994)
					Luther et al. (1996)
	[Co(HS)$_2$]^0	10.2		0.7	Al-Farawati and van den Berg (1999)
	[Co$_2$(HS)]$^{3+}$	9.5		0.7	Luther et al (1996)
	[Co$_3$(HS)]$^{5+}$	15.5		0.7	Luther et al. (1996)
	[CoS$_4$]^0	5.6		0.55	Chadwell et al. (2001)
	[Co$_2$(S$_4$)]$^{2+}$	11.6		0.55	Chadwell et al (2001)
	[CoS$_5$]^0	5.4		0.55	Chadwell et al. (1999)
	[Co$_2$(S$_5$)]$^{2+}$	11.3		0.55	Chadwell et al. (1999)
Ni(II)	[NiHS]$^+$	5.0	0.3	0.7	Al-Farawati and van den Berg (1999)
					Zhang and Millero (1994)
					Luther et al. (1996)
	[Ni(HS)$_2$]^0	10.5		0.7	Al-Farawati and van den Berg (1999)
	[Ni$_2$(HS)]$^{3+}$	10.0		0.7	Luther et al. (1996)
	[Ni$_3$(HS)]$^{5+}$	15.9		0.7	Luther et al. (1996)
	[NiS$_4$]^0	5.7		0.55	Chadwell et al. (2001)
	[Ni$_2$(S$_4$)]$^{2+}$	11.0		0.55	Chadwell et al. (2001)
	[NiS$_5$]^0	5.5		0.55	Chadwell et al. (1999)
	[Ni$_2$(S$_5$)]$^{2+}$	11.1		0.55	Chadwell et al. (1999)
Cu(II)	[CuHS]$^+$	6.5	0.5	0.7	Zhang and Millero (1994)
					Luther et al. (1996)[a,b]
	[CuS]^0	11.2		0.7	Luther et al. (1996)[a,b]
	[Cu(HS)$_2$]^0	15.5	2.5	0.7	Al-Farawati and van den Berg (1999)
					Zhang and Millero (1994)
	[Cu$_2$S$_3$]$^{2-}$	38.3		0.7	Luther et al (1996)[a,b]
	[Cu$_2$(S$_4$)$_2$]$^{2-}$	17.8		0.55	Chadwell et al. (2001)[b]
	[Cu$_2$(S$_5$)$_2$]$^{2-}$	20.2		0.55	Chadwell et al. (1999)[b]
Cu(I)	[CuHS]^0	12.1	0.9	0.7	Al-Farawati and van den Berg (1999)
					Zhang and Millero (1994)[g]
					Mountain and Seward (1999)[d]
	[Cu(HS)$_2$]$^-$	17.6	0.4	0.7	Al-Farawati and van den Berg (1999)
					Zhang and Millero (1994)[g]
					Mountain and Seward (1999)[d]
	[Cu$_2$S(HS)$_2$]$^{2-}$	29.9		0.0	Mountain and Seward (1999)[d]
	[Cu(S$_4$)$_2$]$^{3-}$	9.8		0.0	Cloke (1963)[d]
	[Cu(S$_4$)S$_5$]$^{3-}$	10.6		0.0	Cloke (1963)[d]
Zn(II)	[Zn(SH)(OH)]^0	19.0		1.0	Gubelie and Ste-Marie (1967)[d, e]
	[ZnHS]$^+$	6.1	0.5	0.7	Al-Farawati and van den Berg (1999)
					Zhang and Millero (1994)
					Luther et al. (1996)[a]

Table continued on facing page

Table 28. *continued from facing page*

Metal	Complex	log$K$	±	I	Reference
	[ZnS]^0	11.7		0.7	Luther et al. (1996)[a,d]
	[Zn(HS)$_2$]^0	11.8	1.9	0.7	Al-Farawati and van den Berg (1999)
					Zhang and Millero (1994)
	[ZnS(HS)]$^-$	13.8		0.0	Daskalakis and Helz (1993)
	[Zn(HS)$_3$]$^-$	14.9		0.0	Dyrssen (1991)[f]
	[ZnS(HS)$_2$]$^{2-}$	13.1		0.0	Daskalakis and Helz (1993)
	[Zn(HS)$_4$]$^{2-}$		0.1	0.0	average of Dyrssen (1991)[f]
		14.7			Daskalakis and Helz (1993)
	[Zn$_2$S$_3$]$^{2-}$	41.09		0.7	Luther et al. (1996)
	[ZnS$_4$]^0	8.37		0.55	Chadwell et al. (2001)
	[ZnS$_5$]^0	8.74		0.55	Chadwell et al. (1999)
Mo(VI)	MoO$_3$S^{2-}	4.91	0.24	$I > 0.0$	Harmer and Sykes (1986)
					Brule et al. (1988)
					Helz and Erickson (2000)
	MoO$_2$S$_2$$^{2-}$	10.84	1.21	$I > 0.0$	Brule et al. (1988)
					Helz and Erickson (2000)
	MoOS$_3$$^{2-}$	16.24	2.06	$I > 0.0$	Brule et al. (1988)
					Helz and Erickson (2000)
	MoS$_4$$^{2-}$	22.74	4.3	$I > 0.0$	Brule et al. (1988)
					Helz and Erickson (2000)
Cd(II)	[CdHS]$^+$	6.6	0.3	0.7	Al-Farawati and van den Berg (1999)
					Zhang and Millero (1994)
	[Cd(HS)$_2$]^0	13.3	0.7	0.7	Al-Farawati and van den Berg (1999)
					Zhang and Millero (1994)
	[Cd(HS)$_3$]$^-$	16.3	0.2	0.0	Daskalakis and Helz (1992)
					Wang and Tessier (1999)
	[Cd(HS)$_4$]$^{2-}$	18.2	0.3	0.0	Daskalakis and Helz (1992)
					Wang and Tessier (1999)
Ag(I)	[AgHS]^0	11.2	0.4	0.7	Al-Farawati and van den Berg (1999) [h]
					Luther and Rickard (2005)[a]
		13.5	0.1	0.0	Renders and Seward (1989)[d]
					Schwarzenbach and Widmer (1966)[d]
	[AgS]$^-$	22.8		0.7	Luther and Rickard (2005)[a,d]
	[Ag(HS)$_2$]$^-$	17.35	0.16	1.0	Schwarzenbach and Widmer (1966)[d]
				0.0	Renders and Seward (1989)[d]
					Gammons and Barnes (1989)[d]
					Stephansson and Seward (2003)[d]
	[Ag$_2$S$_3$H$_2$]$^{2-}$	31.33	0.1	0.0	Renders and Seward (1989)[d]
					Stephansson and Seward (2003)[d]
	[Ag$_2$S]^0	29.1		0.7	Luther and Rickard (2005)[a,d]
	[Ag(HS)S$_4$]$^{2-}$	7.4		0.0	Cloke (1963)[d]
	[Ag(S$_4$)$_2$]$^{3-}$	16.5		0.0	Cloke (1963)[d]
	[Ag(S$_4$)(S$_5$)]$^{3-}$	16.8		0.0	Cloke (1963)[d]
Au(I)	[AuHS]^0	24.5		0.0	Renders and Seward (1989)[d]
	[Au(HS)$_2$]$^-$	30.1		0.0	Renders and Seward (1989)[d]
	[Au$_2$S$_2$]$^-$	41.1		0.0	Renders and Seward (1989)[d]
Hg(II)	[HgHS]$^+$	20.6		0.7	Zhang and Millero (1994)[g]
	[Hg(HS)$_2$]^0	25.6		0.7	Zhang and Millero (1994)[g]
	[HgS(HS)]$^-$	43.2		1.0	Schwarzenbach and Widmer (1963)[d]
	[HgS$_2$]$^{2-}$	51.5		1.0	Schwarzenbach and Widmer (1963)[d]
Pb(II)	[PbHS]$^+$	6.8	0.4	0.7	Zhang and Millero (1994)
					Al-Farawati and van den Berg (1999)
					Rozan et al (2003)[a]
	[PbS]^0	16.8		0.7	Rozan et al (2003)[a]
	[Pb(HS)$_2$]	13.8	0.2	0.7	Al-Farawati and van den Berg (1999)
					Zhang and Millero (1994)

[a] acid-base titrations indicate the species is not protonated. *MS* species are corrected for protonation of sulfide. Multinuclear clusters are likely.
[b] Reduction of Cu(II) to Cu(I) occurred at some point in the titration so the species are likely Cu(I).
[c] When corrected for the stoichiometry Cu$_2$S(HS)$_2$$^{2-}$ which contains protons and the side reaction coefficient of Cu(I) in seawater the value is 27.44, which compares with the value of 29.87 of Mountain and Seward (1999).
[d] Thermodynamic constants were calculated not conditional constants.
[e] Correcting for Zn$_2$S$_3$$^{2-}$ gives a value of 44.34 which compares with the value of 41.09 of Luther et al (1996).
[f] Dyrssen's recalculation of Hayashi et al (1990).
[g] Corrected for metal chloro complexes.
[h] corrected for metal-ligand complexes in seawater

significant and increase with clustering (Table 29). Although a voltammetric signal is observed for the destruction of FeS_{aq} clusters, we have not observed a comparable signal for Cu, Pb and Zn sulfide clusters. These divalent cations have measurable reduction peaks as free ions (Cu −0.20 V; Pb −0.45 V; Zn −1.05 V) and with various organic complexes at the mercury electrode. Lewis et al. (1995) demonstrated that Zn-organic complexes give discrete reduction potentials at more negative potentials up to the sodium ion reduction limit. They also showed that the reduction potential for Zn-organic complexes linearly becomes more negative with increasing $logK_{therm}$ of the complexes. The upper limit of detection for Zn-organic complexes is about $logK_{therm}$ = 19. Croot et al. (1999) and Rozan et al. (2003) have performed similar experiments with Cu and Pb, respectively. The upper limit of detection for Cu-organic complexes is $logK_{therm}$ = 47 and for Pb-organic complexes is $logK_{therm}$ = 39. These upper limits are smaller than the stability constants for the M_3S_3 complexes of Cu, Zn and Pb given in Table 29. Thus, discrete metal (Cu, Pb, Zn) sulfide complexes with a cluster stoichiometry of 3:3 or greater will not have a discrete voltammetric signal (Rozan et al. 2000; Rozan et al. 2003) and cannot be detected directly by electrochemical experiments. However, in field samples, comparison of total metal data with either the data obtained from selected acidification or separation experiments (Landing and Lewis 1991; Radford-Knoery and Cutter 1994; Rozan et al. 1999, 2000) or from the amount of metal measured by the pseudovoltammetry method (Lewis et al. 1995; Croot et al. 1999; Rozan et al. 2003) or from the amount of sulfide by other electrochemical or gas chromatographic methods (Kuwabara and Luther 1993; Luther and Tsamakis 1989; Luther and Ferdelman 1993; Radford-Knoery and Cutter 1994) indicate that MS_{aq} complexes can complex up to 90-100% of the metal in sewage treatment plant waters (Rozan et al. 2000; Rozan and Luther 2002) and in oxic waters of rivers, lakes and the ocean.

Table 29. Thermodynamic stability constants calculated for MS clusters using the mole ratio method (after Luther and Rickard 2005).

	M_m	S_n	$\log\beta\ [M_mS_n]$
$[Ag_mS_n]$	2	1	29.1 ± 1.2
	1	1	22.8 ± 0.6
	2	2	50.5 ± 1.1
	3	3	78.3 ± 1.7
	4	4	106.2 ± 2.2
$[Cu_nS_m]$	1	1	11.2 ± 0.78
	3	3	54.7 ± 0.89
	4	6	96.4 ± 1.8
$[Pb_nS_m]$	1	1	16.8 ± 0.33
	2	2	36.4 ± 0.47
	3	3	62.9 ± 0.61
$[Zn_nS_m]$	1	1	11.7 ± 0.14
	3	3	48.5 ± 1.52
	4	6	84.4 ± 1.2

Table 29 also shows that the order of strength for sulfide complexes is $[Ag_nS_m] > [Cu_nS_m] > [Pb_nS_m] > [Zn_nS_m]$. This is consistent with metal replacement reactions as shown in Equations 69-74, all of which have been demonstrated experimentally (Luther and Rickard 2005):

$$Cu^{2+} + [ZnS_{aq}] \rightarrow Zn^{2+} + [CuS_{aq}] \tag{69}$$

$$[Cu(EDTA)]^{2-} + ZnS_{aq} \rightarrow Zn^{2+} + [EDTA]^{4-} + CuS_{aq} \tag{70}$$

$$Ag^+ + ZnS_{aq} \rightarrow Zn^{2+} + AgS^-_{aq} \tag{71}$$

$$[Ag(S_2O_3)]^{2-} + ZnS_{aq} \rightarrow Zn^{2+} + S_2O_3^{2-} + AgS^-_{aq} \tag{72}$$

$$Ag^+ + CuS_{aq} \rightarrow Cu^{2+} + AgS^-_{aq} \tag{73}$$

$$[Ag(S_2O_3)]^{2-} + CuS_{aq} \rightarrow Cu^{2+} + S_2O_3^{2-} + AgS^-_{aq} \tag{74}$$

For example Cu^{2+} replaces Zn in ZnS_{aq} clusters within 2 min (Eqn. 69). The experimental data suggest that the reaction is associative (inner sphere) and that dissociation of the MS_{aq} cluster does not occur because free sulfide is not detected as a dissolved intermediate. For 2 μM $[Cu(EDTA)]^{2-}$ (Eqn. 70), the reaction is slower but complete within 15 min. The copper values in Table 29 are calculated with side reaction coefficients for Cu(II) and are considered approximate as copper reduction to Cu(I) occurs (Luther et al. 2002; Rozan et al. 2002). The side reaction coefficient for Cu(I) is larger than that for Cu(II) and Cu(I)S_{aq} complexes should be stronger than comparable Cu(II)S_{aq} complexes; in addition, S-S bonding also occurs in Cu(I) complexes but not in Cu(II) and the other metal complexes in this study.

The formation of MS_{aq} clusters is of significant importance in different environmental and biological settings. In oxic waters, formation of MS_{aq} clusters is a metal detoxification mechanism. In a recent set of toxicology experiments, Bianichi et al. (2002) showed that silver was not toxic to *Daphnia magna* neonates when free Ag(I) was added to a solution of ZnS_{aq} clusters as the Ag(I) replaced the Zn in the ZnS_{aq} (see Eqn. 71) to form free Zn^{2+}, which is not toxic to the organisms. Samples from a variety of environments including oxygenated waters have measurable sulfide as determined by addition of acid to the sample and subsequent trapping and measurement of the sulfide. Based on the order of stability of MS_{aq} clusters in Table 29 the metals Ag and Cu would be expected to have higher sulfide complexation and thus be less toxic to organisms. Field results confirm the higher percentage of sulfide complexes for Ag (Rozan and Luther 2002) and Cu (Rozan et al. 2000) relative to Zn.

Using *in situ* or real time voltammetry at hydrothermal vents 2500 m below the surface of the ocean, Luther et al. (2001b) demonstrated that *Alvinella pompejana* do not reside in areas where only free sulfide is present but they do reside where FeS_{aq} clusters are the dominant sulfur chemical species. These data indicate that FeS_{aq} clusters are a metal and sulfide detoxification mechanism. Conversely, they found that *Riftia pachyptila* and other organisms dependent on chemosynthesis at hydrothermal vents reside only where free sulfide dominates chemical speciation. Chemosynthetic dependent organisms do not reside where only FeS_{aq} clusters exist, because free sulfide is required by bacteria to perform chemosynthesis.

KINETICS AND MECHANISMS OF METAL SULFIDE FORMATION

Properties of metal aqua complexes

Metal sulfide complexes and clusters are often derived from reactions involving the "free metal ions". In order to consider the kinetics and mechanisms of metal sulfide complex formation, which is an important guide to the probable structure and composition of metal sulfide complexes, we need to address the properties of their building blocks, the "free metal ions" in aqueous solution.

The "free metal ions" in solution are, of course, aqua ions with the metal coordinated to a number of water molecules and their physical-chemical properties are listed in Table 30. The usual way of writing the ferrous ion in aqueous solution, Fe^{2+}, for example, is essentially a shorthand which derives from tiresome balancing of the water molecules in chemical equations. Thus a reaction which might be written in equilibrium shorthand (Eqn. 75) as:

$$Fe^{2+} + HS^- = FeS\downarrow + H^+ \qquad (75)$$

is, in reality, equivalent to the overall reaction (Eqn. 76):

$$[Fe(H_2O)_6]^{2+} + HS^- = FeS\downarrow + H^+ + 6H_2O \qquad (76)$$

ignoring for the moment the fact that the proton is likely to be in the form of the hydronium ion, H_3O^+, and HS^- may be surrounded by coordinated H_2O molecules.

An ion in a solvent like water is solvated; that is, it is surrounded by water molecules (Fig. 33). In equilibrium thermodynamics, the ignoring of the water molecules makes little difference to the resultant state of the system. In contrast, in the real world, the water molecules can determine what products are formed, the rate of formation of the products and

Table 30. Properties of selected metal aqua ions: electronic configuration (config), ligand field stabilization energy (LFSE), molecular orbital stabilization energy (MOSE), negative logarithm of equilibrium constant for hydrolysis reaction $M^{n+}(aq) + H_2O = M(OH)_{aq}^{(n-1)} + H^+$ (pK_{11}), water exchange rate constant $k_{H_2O_{ex}}$ and the activation volume ($\Delta V^{\ddagger}_{H_2O_{ex}}$) (from compilations in Richens et al. 1997 and Lincoln et al. 2003).

Ion	config	LFSE	MOSE	M-OH_2 (pm)	pK_{11} (hydrolysis)	$k_{H_2O_{ex}}$ (s^{-1})	$\Delta V^{\ddagger}_{H_2O_{ex}}$ (cm^3 mol^{-1})
[V(OH$_2$)$_6$]$^{3+}$	t^2_{2g}	$-0.8\Delta_0$	12 βS_σ^2	199	2.26	5.0×10^2	-8.9
[Cr(OH$_2$)$_6$]$^{3+}$	t^3_{2g}	$-1.2\Delta_0$	12 βS_σ^2	198	4.00	2.4×10^{-6}	-9.6
[Mo(OH$_2$)$_6$]$^{3+}$	t^3_{2g}	$-1.2\Delta_0$	12 βS_σ^2	209	?	0.1–1.0	?
[Mn(OH$_2$)$_6$]$^{3+}$	$t^3_{2g}e^1_g$	$-0.6\Delta_0$	9 βS_σ^2	199	0.70	10^3	?
[Fe(OH$_2$)$_6$]$^{3+}$	$t^3_{2g}e^2_g$	$-0.0\Delta_0$	6 βS_σ^2	200	2.16	1.6×10^2	-5.4
[Co(OH$_2$)$_6$]$^{3+}$	$t^4_{2g}e^2_g$	$-2.4\Delta_0$	6 βS_σ^2	187	1.8	?	?
[V(OH$_2$)$_6$]$^{2+}$	t^3_{2g}	$-1.2\Delta_0$	12 βS_σ^2	212		8.7×10^1	-4.1
[Mn(OH$_2$)$_6$]$^{2+}$	$t^3_{2g}e^2_g$	$-0.0\Delta_0$	6 βS_σ^2	218	10.60	2.1×10^7	-5.4
[Fe(OH$_2$)$_6$]$^{2+}$	$t^4_{2g}e^2_g$	$-0.4\Delta_0$	6 βS_σ^2	213	9.50	4.4×10^6	$+3.8$
[Co(OH$_2$)$_6$]$^{2+}$	$t^5_{2g}e^2_g$	$-0.8\Delta_0$	6 βS_σ^2	209	9.65	3.2×10^6	$+6.1$
[Ni(OH$_2$)$_6$]$^{2+}$	$t^6_{2g}e^2_g$	$-1.2\Delta_0$	6 βS_σ^2	207	9.86	3.4×10^4	$+7.2$
[Zn(OH$_2$)$_6$]$^{2+}$	$t^6_{2g}e^4_g$	$-0.0\Delta_0$	0 βS_σ^2	209	8.96	10^7	$+5$
[Cd(OH$_2$)$_6$]$^{2+}$	$t^6_{2g}e^4_g$	$-0.0\Delta_0$	0 βS_σ^2	230	10.08	10^8	-7
[Hg(OH$_2$)$_6$]$^{2+}$	$t^6_{2g}e^4_g$	$-0.0\Delta_0$	0 βS_σ^2	233	3.4	10^9	?
[Cr(OH$_2$)$_5$(OH)]$^{2+}$				199, 230	?	10.8×10^{-4}	$+2.7$
[Cu(OH$_2$)$_6$]$^{2+}$	d^9	$-0.6\Delta_0$	3 βS_σ^2		7.53	4.4×10^9	$+2.0$
[Cu(OH$_2$)$_5$]$^{2+}$	d^9					5.3×10^9	

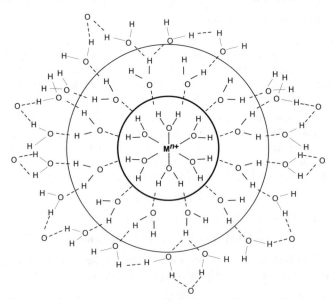

Figure 33. Localized structure of a hydrated metal ion (with coordination number six) in aqueous solution showing primary solvation shell as the inner sphere and secondary solvation shell as the outer sphere. There is still some structure in the outermost tertiary solvation shell which has a diffuse boundary to the bulk where the effects of the metal ion are not significant.

the mechanism of formation. The formation of the metal sulfide complexes therefore basically involves a ligand substitution reaction, where the sulfide species replaces the water in the first coordination sphere of a free metal ion.

Coordination number 6 is common amongst transition metals and usually gives rise to an octahedral geometry. The reason is that six-coordination possesses a very high symmetry. The arrangement also produces a cavity of appropriate size such that the M^{n+} ion can be contained whilst providing enough space for significant M-OH_2 bonding interaction. Ligand-field effects determine the M-OH_2 distance and the lability. In the crystal field model, the electrons in the $d_{x^2-y^2}$ and d_{z^2} orbitals point directly at the electron density on the ligands. They are then regarded as experiencing a greater electrostatic repulsion than those residing in the d_{xy}, d_{yz} and d_{xz} orbitals. Thus in an octahedral $[M(OH)_2)_6]^{n+}$ ion, 3 orbitals with the same energy but different geometries or *degenerate* orbitals (the t_{2g} set) are at a lower energy than the 2 degenerate orbitals in the e_g set. The ligand field splitting parameter, Δ_o, (for an octahedral complex) is the energy required for an electron to jump from one set to the other. For a tetrahedral arrangement, none of the orbitals would be pointing directly at the ligands and the resultant tetrahedral ligand field splitting parameter, Δ_t, is lower than Δ_o so that $\Delta_t \approx 4/9\,\Delta_o$.

Energy is required to pair up electrons in the same orbital in order to overcome the extra repulsion. If $\Delta_o > P$, the pairing energy, a *low-spin* complex results with some of the electrons in pairs in the orbitals. If $P > \Delta_o$, the *high-spin* form is produced, with the electrons unpaired, determined by Hund's rule which states that one electron is added to each of the degenerate orbitals in a subshell before two electrons are added. All first row transition metal $[M(OH)_2)_6]^{2+}$ ions (where M = V to Zn) are high spin. However, all first row transition metal $[M(OH)_2)_6]^{3+}$ ions, excepting Mn^{3+} and Fe^{3+}, are low spin. Ligand field theory uses a molecular orbital approach (MO) to arrive at a similar picture of the bonding orbitals for octahedral complexes except that the repulsive doubly degenerate e_g set is interpreted as *antibonding* (e_g^*). The interesting feaure of the MO approach is that it predicts a change in coordination from six to four when electrons are added into the e_g^* orbitals of an octahedral metal ion. Molecular orbital stabilization energy goes from 12 βS_σ^2 (6 σ bonds predicted to the metal) for the high spin metals up to V^{2+} and begins to decrease towards 0 βS_σ^2 (4 σ bonds predicted to the metal) for Zn^{2+} (Table 30).

However, the LFSE accounts for less than 10% of the total hydration energy of a typical divalent metal ion, despite its effect on M-OH_2 bond distance, hydration number and the kinetic lability of the primary shell. The hydration enthalpy is closely correlated with the degree of hydration or hydration radius. V^{2+} has a large size and the lowest hydration enthalpy of all the first row transition metals and yet its hexa aqua ion, $[V(OH)_2)_6]^{2+}$, is the most inert. This appears to result from a particular kinetic stability of the singly filled t_{2g}^3 set. $[Cr(OH)_2)_6]^{3+}$ is also t_{2g}^3 and is even less labile than $[V(OH)_2)_6]^{2+}$. The Jahn-Teller effect means that the octahedral stuctures of $[Cr(OH)_2)_6]^{2+}$ ($t^3_{2g}e^1_g$) and $[Cu(OH)_2)_6]^{2+}$ ($t^6_{2g}e^3_g$) are tetragonally distorted and very labile. A similar phenomenon is observed for $[Mn(OH)_2)_6]^{3+}$ ($t^3_{2g}e^1_g$) and $[Ag(OH)_2)_6]^{2+}$ ($t^6_{2g}e^3_g$). Although Co(II) forms the octahedral $[Co(OH)_2)_6]^{2+}$ complex at lower temperatures, it tends to form tetrahedral rather than octahedral complexes with other ligands, because of the small differences in LFSE between high spin d^7 octahedral and tetrahedral geometries with the same ligands. Ni(II) forms the paramagnetic $[Ni(OH)_2)_6]^{2+}$ ion because the LFSE is less than the energy required to pair up the two $3d$ electrons in the doubly degenerate e_g set which would lead to distortion.

Mechanisms of formation of metal sulfide complexes

A ligand substitution mechanism is a theoretical construct designed to explain the energetic and stereochemical changes which occur as reactants progress through one or more transition states to the products. Kinetic measurements provide information about the transition state stoichiometry and the enthalpic and entropic changes characterizing the transition state. However, these do not directly provide details of stereochemical changes occurring along the reaction coordinate. The data for metal sulfide complexes are extremely limited, but the fundamental chemical properties of the metal aqua ions permit some predictions to be made with a degree of confidence.

Two extreme mechanistic possibilities arise for the substitution of a water ligand in a metal, M, aqua ion $[M(H_2O)_n]^{m+}$ by a ligand, L^{x-}, and are conveniently discussed using the nomenclature of Langford and Gray (1965). The first occurs when $[M(H_2O)_n]^{m+}$ and L^{x-} pass through a first transition state to form a reactive intermediate, $[M(H_2O)_nL]^{(m-x)}$, in which the coordination number of M^{m+} is *increased* by one:

$$[M(H_2O)_n]^{m+} + L^{x-} \underset{k_{-1}}{\overset{k_1}{\rightleftharpoons}} [M(H_2O)_nL]^{(m-x)+} \underset{k_{-2}}{\overset{k_2}{\rightleftharpoons}} [M(H_2O)_{(n-1)}L]^{(m-x)+} + H_2O \qquad (77)$$

This intermediate survives several molecular collisions before passing through a second transition state to form the product $[M(H_2O)_{(n-1)}L]^{(m-x)}$. Thus, the rate determining step (k_2) is the bond making between L^{x-} with M^{m+} and the mechanism is termed associatively activated and the mechanism *associative*, **A**. The rate of approach to equilibrium in the presence of excess $[L^{x-}]$, characterized by k_{obs} in Equation (78) is dependent on the nature of L^{x-}:

$$k_{obs} = k_2 [M(H_2O)_n]^{m+} [L^{x-}] \qquad (78)$$

The second extreme mechanism operates when $[M(H_2O)_n]^{m+}$ passes through a first transition state to form a reactive intermediate, $[M(H_2O)^{(n-1)}]^{m+}$, in which the coordination number of M^{m+} is *decreased* by one:

$$[M(H_2O)_n]^{m+} + L^{x-} \underset{k_{-1}}{\overset{k_1}{\rightleftharpoons}} [M(H_2O)_{(n-1)}]^{m+} + L^{x-} \underset{k_{-2}}{\overset{k_2}{\rightleftharpoons}} [M(H_2O)_{(n-1)}L]^{(m-x)+} + H_2O \qquad (79)$$

This intermediate also survives several molecular collisions before passing through a second transition state to form the product $[M(H_2O)_{(n-1)}L]^{(m-x)+}$. Thus, the rate determining step (Eqn. 80; k_1) is bond breaking and the mechanism is dissociatively activated and the mechanism is *dissociative*, **D**:

$$k_{obs} = k_1 [M(H_2O)_n]^{m+} \qquad (80)$$

However, the tendency for oppositely charged reactants to form outer-sphere complexes often prevents the observation of either of the rate laws shown in Equations (78) and Equation (80), but this does not preclude the operation of either an **A** or a **D** mechanism within an outer-sphere complex. Between the **A** and **D** mechanistic extremes there exists a continuum of mechanisms in which the entering and leaving ligands make varying contributions to the transition state energetics.

Some indication of the mechanism can be assigned through investigations of the activation volume, $\Delta V^{\ddagger}$. For water exchange on $[M(H_2O)_n]^{m+}$, $\Delta V^{\ddagger}$ is the difference between the partial molar volumes of the ground state and the transition state and is related to the variation of k_{H_2O} with pressure. In simplistic terms, if bond making and bond breaking balance each other in the mechanism, then $\Delta V^{\ddagger} \to 0$; the **A**-type mechanism will dominate where $\Delta V^{\ddagger} < 0$ and the **D**-type mechanism where $\Delta V^{\ddagger} > 0$. Thus inspection of Table 30 suggests that ligand substitution in the divalent aqua metal ions $[Fe(OH_2)_6]^{2+}$, $[Co(OH_2)_6]^{2+}$, $[Ni(OH_2)_6]^{2+}$ and $[Zn(OH_2)_6]^{2+}$ proceeds through a **D** mechanism and the rate of ligand substitution shows little dependence on the ligand. This is an important observation for the formation of sulfide complexes of these metals, for which kinetic information is not available, since it suggests that the rate of formation of the HS$^-$ complexes is similar to the rate for the substitution of other ligands. The limited data available also suggests that there is little variation in the rate of ligand substitution for $[V(OH_2)_6]^{2+}$ (t_{2g}^3) and $[Mn(OH_2)_6]^{2+}$ ($t_{2g}^3e_g^2$). Although these show negative $\Delta V^{\ddagger}$ values, this probably results from a small range of nucleophilicity caused by these ions being borderline hard acids and the substituting ligands are restricted to hard bases.

In contrast, the Cu^{2+} aqua ion displays well-known enhanced lability. The aqua ion exists in water in two forms $[Cu(OH_2)_6]^{2+}$ and $[Cu(OH_2)_5]^{2+}$ (Fig. 34). The rapid interconversion between square pyramidal and trigonal-bipyramidal stereochemistries of these forms leads to the enhanced lability. Cr^{2+}, a $t_{2g}^3 e_g^1$ ion, shows similar Jahn-Teller distortions to the $t_{2g}^6 e_g^3$ Cu^{2+}.

The lability toward water exchange and ligand substitution for the first row divalent transition metal ions then increases in the sequence $Cu^{2+} \sim Cr^{2+} > Zn^{2+} \sim Mn^{2+} > Fe^{2+} > Co^{2+} > Ni^{2+} > V^{2+}$. This sequence results from variations in occupancy of their d orbitals. For Zn^{2+}, Mn^{2+}, Fe^{2+}, Co^{2+}, Ni^{2+} and V^{2+}, the rate of formation of sulfide complexes is probably similar to the rate of water exchange, since the rate is approximately independent of ligand type. *Ab initio* calculations for water exchange on the d^0 to d^{10} first-row hexa-coordinate transition metal ions ranging from Sc^{3+} to Zn^{2+} predict a change from a- to d-activation so that only **A** mechanisms are possible for Sc^{3+}, Ti^{3+}, and V^{3+}, only **D** mechanisms are possible or Ni^{2+}, Cu^{2+}, and Zn^{2+}, while a gradual change from a- to d-activation is predicted for other first-row transition metal ions with d^2 to d^7 electronic configurations. In agreement with these conclusions, density function calculations show that water exchange on $[Zn(H_2O)_6]^{2+}$ occurs through a **D** mechanism (Hartmann et al. 1997). The great labilities of Cu^{2+} and Cr^{2+} probably arise from stereochemical effects induced by their d^9 and d^4 electronic configurations (Jahn-Teller effect).

The practical kinetics of this substitution process can be addressed via the Eigen-Wilkins approach (Eigen and Wilkins 1965) whereby interchange of ligands takes place within a preformed outer sphere complex. Then, a metal aqua ion, $[M(H_2O)_x]^{m+}$ reacts with an incoming ligand, L^{n-}, to form an outer sphere complex, $[M(H_2O)_x L^{(m-n)}]$. The lifetime of this outer sphere complex in water is controlled by the rate of diffusional encounters and is of the order of 1 ns. Therefore, for all reactions taking longer than tens of ns, the formation of the outer sphere complex (Eqn. 81) can be described by the equilibrium constant K_{os} (Eqn. 82) for the reaction:

$$[M(H_2O)_x]^{m+} + L^{n-} = [M(H_2O)_x L^{(m-n)}] \tag{81}$$

$$K_{os} = \frac{[M(H_2O)_x L^{(m-n)}]}{[M(H_2O)_x^{m+}][L^n]} \tag{82}$$

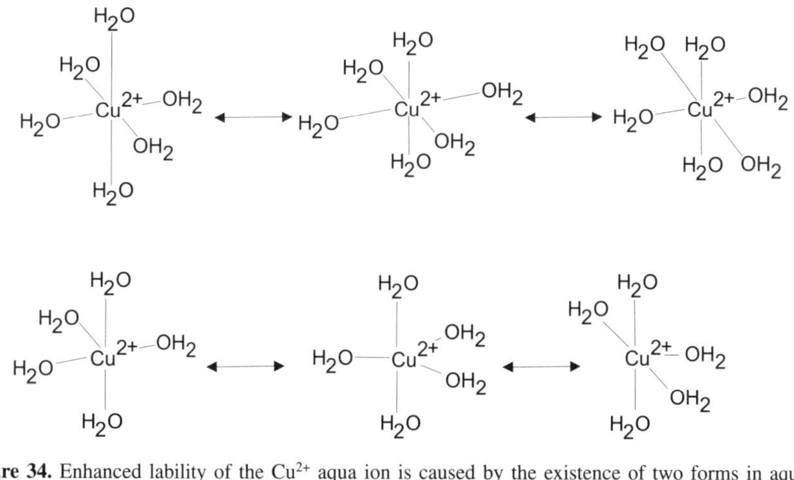

Figure 34. Enhanced lability of the Cu^{2+} aqua ion is caused by the existence of two forms in aqueous solution $[Cu(OH_2)_6]^{2+}$ (top) which shows a bond elongation for each coordinated water and $[Cu(OH_2)_5]^{2+}$ (bottom) in which the square-pyramidal OH_2 exchanges readily with the planar OH_2 via the trigonal-bipyramidal intermediate.

The next step is the exchange of inner sphere water with the outer sphere ligand to produce an inner sphere complex, $[M(H_2O)_{x-1}L]^{(m-n)}$:

$$[M(H_2O)_xL^{(m-n)}] \rightarrow [M(H_2O)_{x-1}L(H_2O)]^{(m-n)} \rightarrow [M(H_2O)_{x-1}L]^{(m-n)} + H_2O \quad (83)$$

The rate of formation of the inner sphere complex, $[M(H_2O)_{x-1}L]^{(m-n)}$, is given by:

$$\frac{d[M(H_2O)_{x-1}L]^{(m-n)}}{dt} = k_f[M(H_2O)_x^{m+}][L^{n-}] \quad (84)$$

where k_f is the rate constant (Eqn. 85) given by:

$$k_f = k_{H_2Oex}K_{os} \quad (85)$$

The kinetics and mechanism of the formation of metal sulfide complexes has only been established in the case of FeS by Rickard (1989, 1995). In this reaction, there are two ligands involved, HS⁻ and H_2S, dependent on pH. Therefore there are two competing reaction pathways and the formation of metal sulfides is pH dependent. The rate laws for both reactions are consistent with Eigen-Wilkins mechanisms (Eigen and Wilkins 1965) where the rate is determined by the exchange between water molecules in hexaqua iron(II) sulfide outer sphere complexes $[Fe(H_2O)_6^{2+}·H_2S]$ and $[Fe(H_2O)_6^{2+}·HS^-]$ and the inner sphere complexes, $[FeH_2S·(H_2O)_5^{2+}]$ and $[Fe(HS)·(H_2O)^{5+}]$. The rates are fast and the rate constant for formation of the inner sphere complexes during the formation of FeS with millimolar solutions of reactants is $>10^7$ M⁻¹ s⁻¹. The subsequent rate of nucleation of FeS is even faster. There is no observable lag phase and, as discussed above, it is probable that aqueous FeS clusters, with the same structures as the fundamental structural elements in mackinawite, are involved. Even so, current EXAFS work (Rickard, Vaughan and coworkers, unpublished data) has shown the co-existence of both Fe-H_2O and Fe-S bonds in the aqueous Fe sulfide solutions at ca. 100 ns reaction time in support of the theory. The process of the formation of the condensed phase, synthetic mackinawite which is the normal product of the reaction, has recently been elucidated and is discussed below.

The implications of the FeS kinetics for the natural environment are interesting. They suggest that the rate of the H_2S pathway becomes equal to and greater than that of the HS⁻ pathway as $\Sigma S(-II)$ reaches 10^{-5} M, or less, under near neutral conditions. In environments with micromolar or greater $\Sigma S(-II)$ concentrations, the rate of sulfide removal is two orders of magnitudes greater in neutral to alkaline solutions than in acid environments, whereas in sulfide-poor systems the rate is greater in neutral to acid conditions.

The importance of the establishment of the Eigen-Wilkins mechanism for iron sulfide complex formation cannot be overestimated. Equation (85) shows that the rate is determined by the rate of exchange of inner sphere H_2O, k_{H_2Oex}, and this is mostly independent of the ligand and dependent on the characteristic of the metal. Therefore the values for k_{H_2Oex} for selected metal aqua ions listed in Table 30 can be used as a first order estimate of the relative rates of formation of the metal sulfide complexes from aqueous solution. Since the rate of formation of the condensed metal sulfide from the inner sphere complex is likely to be very fast, the k_{H_2Oex} values also indicate the relative rate of metal monosulfide mineral formation.

The k_{H_2Oex} values for all the other metals listed in Table 30, except V, Cr and Mo, are larger than that for Fe(II) by one order of magnitude or more and thus the rate of formation of Cu, Zn, Ag and Pb sulfide complexes as discrete [MS] species should be accordingly faster. Based on these data, we calculate a rate constant $>10^8$ M⁻¹ s⁻¹ for the formation of Cu, Zn, Ag and Pb sulfide complexes. This calculation indicates that these reactions are diffusion controlled (Atkins 1978) and much faster than what we can experimentally determine. Unfortunately, comparative experimental kinetic data are unavailable at present to support the theory. However, qualitatively it has been found that the rates are in the order of CuS > ZnS > FeS (Rickard and coworkers, unpublished data), which is consistent with the theory.

Metal isotopic evidence for metal sulfide complexes

Metal isotopes would appear to be potentially powerful probes into metal sulfide complexes and provide additional information about their mechanisms of formation. The widespread availability of multi-collector inductively-coupled plasma mass spectrometers (MC-ICPMS) makes investigations of metal isotope effects possible. However, surprisingly, little experimental work appears to be published in this area.

Butler et al. (2005) demonstrated that kinetic Fe isotope fractionations occur on the formation of FeS by addition of aqueous sulfide to excess Fe(II) solution. The condensed phase Fe(II) is isotopically light relative to its aqueous counterpart. The result is significant since it shows that a significant Fe isotope fractionation can occur in the absence of any redox process in the system. The results support the Eigen-Wilkins mechanism for FeS formation outlined by Rickard (1995). Isotopic fractionation occurs during inner sphere exchange of sulfide and water ligands, with isotopically light Fe reacting fastest and becoming enriched in the condensed phase.

Erlich et al. (2004) investigated the Cu isotopic fractionation between aqueous Cu(II) and CuS, the synthetic equivalent of the mineral covellite. Since CuS is a Cu(I) sulfide the reaction involves a redox process. Erlich et al. (2004) found that the Cu isotopic fractionation in the formation of CuS is of similar magnitude to that of known Cu(II)-Cu(I) redox pairs. The fractionation occurs on the condensation of Cu_3S_3 rings to Cu_4S_5 and Cu_4S_6 clusters which is associated with a release of Cu(II) back into solution (Fig. 35, Luther et al. 2002). During this condensation step the reduction of Cu(II) to Cu(I)—and the oxidation of S(−II) to S(−I)—occurs. The implication of these results, apart from neatly confirming the chemical mechanistic observations, is that the Cu isotopic composition of CuS as well as its structure may be determined in solution.

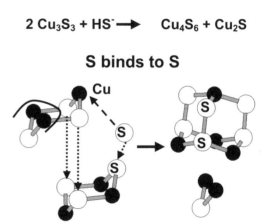

Figure 35. Schematic showing the formation of S-S bonds between two Cu_3S_3 rings. This process implies reduction of all Cu(II) to Cu(I) with formation of S_2^{2-}. S atoms on one ring bind with S atoms on the other with loss of Cu_2S^{2+}. This complex would have the solution stoichiometry of $[Cu_4S_6(H_2O)_7]^{2-}$. Note that three of the Cu atoms are bound to only two S atoms. (after Luther et al. 2002).

Currently work is proceeding on Zn isotope fractionations during the reaction of aqueous Zn(II) with sulfide (Rickard and coworkers, unpublished data). Again Zn isotope fractionations have been measured during this process. As with the Fe system, there is no redox involved in the Zn sulfide formation process. The fractionation must therefore reflect a separation of the nuclides during the ZnS formation process, which is again consistent with the involvement of Zn sulfide cluster complexes in the process.

COMPLEXES, CLUSTERS AND MINERALS

In classical nucleation theory, a cluster is defined as a group of molecules preceding the formation of a nucleus with critical radius. However, it has become apparent that the formation of supernuclei (clusters larger than the critical size) is important in nucleation theory because these are capable of spontaneous growth (e.g., Kashchiev 2000). Recent research has shown that clusters are prevalent in the environment and are possible building blocks in mineral formation (Labrenz et al. 2001; van der Zee et al. 2003; Wolthers et al. 2003).

As discussed above, in chemistry, a cluster is defined as a polynuclear complex; e.g., Fe_4S_4. This has led to some confusion in the literature regarding the nature of sulfide clusters. Luther and Rickard (2005) defined cluster complexes as an operational definition for environmental cluster investigations, since nanoparticles may be electroactive and indistinguishable practically from dissolved species.

As shown by Luther and Rickard (2005), to date we have only defined the boundary between dissolved sulfide species and a nanoparticulate sulfide solid for FeS and ZnS. And even here, there appears to be a possible overlap in the dimensions of aqueous clusters and nanoparticles, such that nanoparticles are small enough to be electroactive and indistinguishable from dissolved species in environmental analyses.

The classical view of the physics of nucleation and crystal growth processes involved in mineral formation is relatively well established (e.g., Adamson 1990; Stumm 1992). In contrast, the chemical or molecular processes involved in the transformation of simple dissolved species to solid products are not as well understood. The schematic diagram (Fig. 36) summarises the processes involved. Thus typical divalent hexaqua complexes of first transition metals are six-coordinate (see above) but they are four-coordinate in the solid phase sulfide, MS, (Krebs 1968; Wells 1986). In addition, the sulfide expands its coordination from 1 or 2 to 4 going from the solution to solid phase. Thus major intra- and inter- molecular arrangements occur during the transformation of the reactants to products in the metal sulfide system. In order to get from one state to the other, a series of reactions or steps must occur between the M(II) and S(−II) species which involve the initial substitution of water by sulfide.

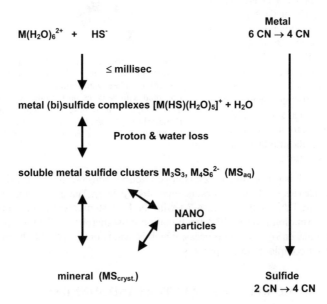

Figure 36. Summary of processes involved in the transformation of dissolved hexaqua metal ions $[M(H_2O)_6]^{2+}$ where the metal is 6 coordinate (CN) and the sulfur 1 or 2 coordinate to the sulfide mineral MS_{cryst} where both the metal and sulfur are both 4-coordinate (from Luther and Rickard 2005).

The kinetics of the process were originally described for iron sulfides by Rickard (1995) and are described above. Reaction intermediates then form that condense, and the coordination of both M and S changes. The process is further complicated by the release of protons, since S^{2-} does not have any meaningful activity in aqueous solutions.

The Ostwald Step Rule or "the rule of stages" postulates that the precipitate with the highest solubility (i.e., the least stable solid phase) will form first in a consecutive precipitation reaction. This rule is well documented in geochemistry (Morse and Casey 1988) and suggests that mineral formation occurs via precursors (intermediates) that can be recognized at the molecular level (Weissbuch et al. 1991). The precipitation sequence results because nucleation of a more soluble phase is kinetically favored over that of a less soluble phase due to the lower solid-solution interfacial tension of the more soluble phase. The classical interpretation of the Ostwald Step Rule is that the metastable phase forms first because it is more soluble than the stable phase. As shown by Luther and Rickard (2005), the formation of aqueous metal sulfide cluster complexes provides an alternative mechanism for the Ostwald Step Rule.

There is a considerable body of information in the biochemical, chemical and semiconductor literature that shows that metal sulfide species yield quantum-size clusters with the metals Fe, Zn, Cu, Cd and Pb. For the FeS system, ferredoxins containing thiols attached to Fe-S clusters are well known (e.g., Huheey et al. 1993). The preparation of other metal-sulfide clusters (e.g., ZnS, CuS, CdS and PbS) has been performed in laboratory studies (Herron et al. 1990; Kortan et al. 1990; Silvester et al. 1991; Nedeljkovic et al. 1993; Vossmeyer et al. 1995; Løver et al. 1997). In these studies, the ion activity product, IAP, would exceed the solubility constant for the bulk metal sulfide, and to prevent precipitation, an organic protective agent (which terminates polymerization or condensation) such as thiols was added to interact with the clusters—sometimes in non-aqueous solvents. The concentration of reactants is often 0.1 mM or higher so that spectroscopic studies could be easily performed for structural elucidation of the clusters. Metal sulfide clusters have also been observed during the dissolution of minerals by molal concentrations of sulfide (Daskalakis and Helz 1993; Helz et al 1993; Pattrick et al. 1997). Methods employed include UV-VIS spectrophotometry, ^{113}Cd NMR, electrospray mass spectrometry, EXAFS, X-ray diffraction, and electron micrographs of powders dispersed in organic solvents. Several of these methods yield information but the interpretation of the data at environmentally relevant concentrations < 2 μM is often ambiguous.

Zinc sulfide clusters

Luther et al. (1999) and Luther and Rickard (2005) tracked the conversion of aqueous Zn(II) and S(−II) to ZnS using UV-VIS spectroscopy. Figure 37 shows spectra of 10 μM Zn(II) and 15 μM HS⁻ alone, as well as the algebraic sum of the two spectra compared with the unfiltered mixture of Zn(II) and HS⁻. For the mixture, the HS⁻ peak at 230 nm decreases, and new peaks at 200 and 290 nm appear and a broad shoulder near 264 nm develops demonstrating that new zinc-sulfide products are produced. The Zn sulfide peaks at 200 and 290 nm shift toward longer wavelengths with time. A peak at 380 nm forms after two hours; this wavelength is similar for semiconductor ZnS (Kortan et al. 1990). The broad shoulder near 264 nm is similar to the shoulder at 264 nm observed by Sooklal et al. (1996) who used a much higher concentration (0.2 mM each) of reactants without the use of thiols and indicated that the ZnS species in solution are molecular clusters. The data in Figure 37 are consistent with the quantum confinement effect (Nedeljkovic et al. 1993; Sooklal et al. 1996) and it is possible to calculate the maximum size of the clusters in solution based on the absorption peaks at 200, 230, 264 and 290 nm. The maximum diameter of the particles is calculated to be 1.6 nm, 2.0 nm, 2.4 nm, and 2.8 nm respectively for the above wavelengths. The 2.4 nm size agrees with the calculations of Sooklal et al. (1996) who detected the 264 nm absorption.

In support of small clusters, we performed filtration experiments. UV-VIS spectroscopic data obtained after filtering through 0.1 μm Nuclepore filters indicated that 95 % or more of the

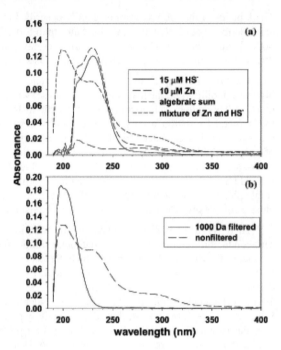

Figure 37. (a) Individual UV–VIS spectra of solutions of 10 μM Zn(II) and 15 μM HS⁻ before reaction, the algebraic sum of these spectra and the chemical mixture of these concentrations of Zn(II) and HS⁻. (b) UV–VIS spectra of the reaction of 10 μM Zn(II) and 15μM HS⁻. The peaks at 230, 264 and 290 nm are not observed in the filtered solution indicating they are due to zinc-sulfide species >1000 Da (from Luther and Rickard 2005).

Zn sulfide species formed passed through the filter. Figure 37A shows distinct peaks at 200, 230, 264 and 290 nm. When 1000 dalton filters were used to filter this stock solution, only the 200 nm peak was observed (Fig. 37B). These data support soluble molecular clusters in solution.

Zn(II) solutions combined with sulfide were also passed through sephadex gel columns which were calibrated with 350 and 750 dalton dyes. A ZnS cluster was observed using UV-VIS spectroscopy and had a retention volume that was intermediate to the two dyes. In addition, gel electrophoresis of ZnS solutions showed that a cluster gravitated to the positive pole and had a retention time intermediate to the 350 and 750 dalton dyes. These data indicate that the stoichiometry is greater than ZnS or ZnS_2 for these solutions and that a negatively charged ion is present in these solutions, which is in agreement with the electrochemical titration data discussed above (Fig. 6). Using anion chromatography on ZnS_{aq} solutions, two peaks were obtained. One occurred near or just after the dead volume peak suggesting a neutral material and the other occurred with a retention time of 14 min. The 200 nm wavelength used for detection indicates that the maximum size of the cluster is 1.6 nm. Luther and Rickard (2005) tracked ZnS particle formation over time with UV-VIS. They showed that the cluster concentrations increased rapidly over 1 day but declined slowly over 30 days. The rapid buildup and decay indicate that aggregation occurs, which removes the ZnS clusters from solution as nanoparticles (Labrenz et al. 2001).

The UV results provide information about the dimensions of the ZnS clusters but not about their mass. The calculated sizes range from 1.6 to 2.8 nm. Banfield and Zhang (2001) reported that the smallest ZnS nanoparticles approach ~1.5 nm diameter, which is closely coincident with the 1.6 nm size calculated from the UV peak at 200 nm. The shape of the ZnS nanoparticles is unknown. Assuming a spherical form, a 1.6 nm ZnS nanoparticle would contain just 13 standard sphalerite unit cells. This indicates 52 ZnS subunits with a molecular weight of 5065 daltons. For wurtzite, the metastable hexagonal ZnS polytype, a 1.6nm spherical nanoparticle would have a mass of 3312 daltons. Although the estimates are based on the dimensions of the bulk crystalline phase and the spherical form is an assumption, the calculation indicates that 1.6 nm spherical ZnS nanoparticles would not pass through a 1000 dalton filter. We thus conclude that at least part of this ZnS material from the UV peak at 200 nm, is in the form of less massive ZnS. The results of dye calibration on sephadex columns indicates that the mass of this fraction is between 350 and 700 daltons. This is equivalent to a maximum between 3 and 7 ZnS subunits.

The electrochemical results show two ZnS cluster species, one neutral with a Zn:S ratio of 1:1 and another, which is anionically charged, with a Zn:S ratio of 2:3 (Fig. 6). Species, which satisfy these parameters, are $[Zn_3S_3]$ (MW = 292) and $[Zn_4S_6]^{4-}$ (MW = 454). In aqueous solutions, both of these species must be coordinated with H_2O giving $[Zn_3S_3(H_2O)_6]$ (MW = 400.4) and $[Zn_4S_6(H_2O)_4]^{4-}$ (MW = 526) and four-coordination is likely as is hydrolysis. Molecular modeling of these structures (Luther et al. 1999) indicates that $[Zn_3S_3(H_2O)_6]$ has a size of 0.7 nm, and $[Zn_4S_6(H_2O)_4]^{4-}$ has a size of 0.95 nm. The masses would permit filtration through 1000 dalton filters, and the sizes are lower that the maximum limit calculated from the 200 nm peak in the UV data.

The essential difference between the aqueous ZnS cluster complexes and ZnS nanoparticles is the density. In $[Zn_3S_3(H_2O)_6]$, there are 3 ZnS monomers in a sphere with a diameter of 0.7nm, giving a molecular density of 0.057 ZnS monomers nm^{-3}. This compares with sphalerite, which has a molecular density of 1.024 ZnS monomers nm^{-3}. In other words, the condensation of the nanoparticulate sphalerite is accompanied by an increase in ZnS molecular density of ~ 200. The density contrast between condensed ZnS and the bulk provides a defined surface which is visible as defining the boundaries of the nanoparticle.

The monomeric unit for converting clusters into higher ordered materials is cyclic six-membered Zn_3S_3 rings with alternating Zn and S atoms (Fig. 27B; Hulliger 1968; Wells 1986). Each metal has two additional water molecules to satisfy four-coordination. For Zn, anion chromatography results suggest that a peak occurs near the dead volume and represents a neutral species. The reaction of this monomeric unit to form higher ordered clusters has been modeled in Luther et al. (1999, 2002). Briefly the overall reaction sequence can be described by Equation (86) (sulfide substitution for water) and Equation (87) (proton and water loss followed by ring formation) with subsequent higher order cluster formation as in Equation (88):

1. *HS⁻ substitution*

$$3[Zn(H_2O)_6]^{2+} + 3HS^- \rightarrow 3[Zn(H_2O)_5(SH)]^+ + 3H_2O \tag{86}$$

2. *Ring formation; H⁺ and H₂O loss*

$$3[Zn(H_2O)_5(SH)]^+ \rightarrow [Zn_3S_3(H_2O)_6] + 3H^+ + 9H_2O \tag{87}$$

3. *Higher order cluster formation*

$$5[Zn_3S_3(H_2O)_6] + 3HS^- \rightarrow 3[Zn_4S_6(H_2O)_4]^{4-} + 3[Zn(H_2O)_6]^{2+} + 3H^+ \tag{88}$$

One result of Equations (86)-(88) is strong covalent *M*-S bond formation. However, the overall effect is a change in metal coordination from 6 to 4 to form the cyclic 6-membered rings, which are found in the mineral sphalerite (Hulliger 1968; Wells 1986). Table 30 shows that the filling of Zn $e_g{}^*$ orbitals should favor four coordination because of loss of molecular orbital stabilization energy. The loss of water molecules (Eqn. 87) results in a substantial increase in entropy and hence Gibbs free energy. These reactions are termed entropy driven.

Iron sulfide clusters

The formation of FeS from solution is the only metal monosulfide system for which the kinetics and mechanism have been determined (Rickard 1995). The process, discussed above, is similar to that proposed for ZnS in that it is entropy driven and characterized by the progressive exclusion of H_2O from the metal complex. FeS clusters were discovered after the Rickard (1995) study. However, their inclusion in the chemical reaction scheme described above would not change the scheme but add more detail to the development of the inner sphere sulfide complexes to the solid phase.

In the case of FeS, the smallest observed particles consist of nanoparticulate mackinawite down to 2.2 nm in size (Wolthers et al. 2003; Michel et al. 2005; Ohfuji and Rickard 2006). Analyses using X-ray Diffraction (XRD; Ohfuji et al. 2006) and High Resolution Transmission Electron Microscopy (HRTEM; Ohfuji and Rickard 2006) show that this material has a well-developed planar structure, defined by sheets of Fe(II) atoms with strong Fe-Fe bonds. A lath-shaped FeS nanoparticle with this structure contains around 150 FeS subunits.

Aqueous Fe_nS_n clusters are electrochemically neutral, and mass spectral data from a field sample indicate that Fe_2S_2 is the likely monomeric unit (Luther and Rickard 2005). The form of this monomeric unit is very similar to the basic structural element of mackinawite (Fig. 38). The calculated Fe-Fe distance in the $Fe_2S_2 \cdot 4H_2O$ cluster complex is 0.283 nm whereas that estimated for the 2 nm mackinawite phase from Wolthers et al.'s (2003) crystallographic parameters is about 0.28 nm and the spacing found by Michel et al. (2005) is 2.61 nm.

The formation of nanoparticulate mackinawite from aqueous FeS cluster complexes is accompanied by a similar discontinuity in density as shown with ZnS. In the case of FeS, this increase in density continues during growth of the nanoparticulate material.

A scheme indicating the development of the nanoparticulate phase is illustrated diagrammatically for FeS in Figure 39. The free energies of the clusters decrease until at 150 FeS subunits condensation occurs. In this model, a spectrum of higher ordered cluster complexes are formed which have relatively short life-times. It is probable, by analogy with atomic cluster theory, that there are certain stoichiometries (or magic numbers) which have greater stability that their neighbors (e.g.; Echt et al. 1981, 1982). We note that, coincidently, 150—the approximate number of FeS subunits in the first observed condensed phase—is close to such a magic number.

The relationship between complexes, clusters and the solid phase

Luther and Rickard (2005) clearly demonstrate that sulfide cluster complexes exist for the metals Fe, Cu, Zn, Ag and Pb. These MS_{aq} clusters form rapidly; e.g., after addition of sulfide to a metal ion the reaction is complete within 5 s or less. Monomeric units of these clusters are essential for the rapid self assembly reactions to form higher ordered materials, such as nanoparticles, if sufficient amounts of both reactants are available.

In both cases discussed above, it appears that there is an overlap in size between cluster complexes and the first-formed nanoparticulate solid phase. The main difference between the clusters and the first condensed phase is a discontinuous increase in density which leads to substantial increases in the mass of the solid metal sulfides compared with the aqueous forms. In environmental analyses based only on voltammetric analyses (Rozan et al. 1999), both nanoparticulate and aqueous forms may react and they cannot be readily distinguished by this means.

Once the cyclic six-membered rings are formed, M-S bonds are not readily dissociated and crosslinking of rings to form higher ordered clusters and nanoparticles is more likely. Thus MS_{aq} clusters of Zn and Cu, as well as Ag and Pb, are not oxidized rapidly and have long half-lives even in oxygenated waters (Luther and Rickard 2005). This is shown in the relatively high stability constants measured for these clusters (Table 30).

In solution, the M_3S_3 rings have two additional water molecules bound to the metal to satisfy four-coordination but are drawn without water attached to the metals. The overlap of sulfur atoms on one M_3S_3 ring with Zn or Cu atoms on another M_3S_3 ring permits formation

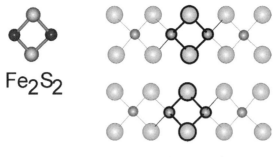

Figure 38. Homology between the structure of the aqueous Fe_2S_2 cluster and mackinawite. Similar structural congruities between aqueous clusters and the first condensed phase were found in the Cu-S and Zn-S systems and led to the theory that the form of the first condensed phase is controlled by the structure of the cluster in solution (from Rickard and Morse 2005).

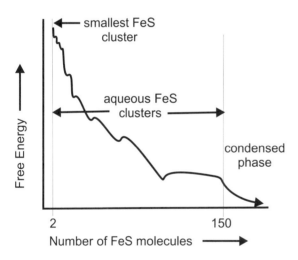

Figure 39. Representation of the relationship between FeS cluster complexes and nanoparticulate FeS solids in terms of energy versus numbers of FeS subunits, from Fe_2S_2 to $Fe_{150}S_{150}$. Certain configurations are more stable than their immediate neighbors and these appear to decrease in frequency with increasing cluster size. A similar diagram could be constructed for ZnS and it may be generally applicable to low temperature aqueous metal sulfide systems.

of Zn_4S_6 or Cu_4S_6 species when concentrations of the reactants are small. The loss of metal in this reaction is consistent with the electrochemical titration results and with the isotopic data as discussed above.

The model seems to be generally applicable to the development of solid phases in low temperature aqueous metal sulfide systems. It implies the lack of any major energy barrier in the nucleation of metal sulfides from aqueous solutions. It provides an alternative explanation of Ostwald's Step Rule—the form of the first solid product is determined by the form of the solution moiety. The process is consistent with the principles of *Chimie Douce* (e.g., Rouxel et al. 1994) applied to environmental chemistry.

ACKNOWLEDGMENTS

Financial support was provided by NERC to DR (NRE/L/S/2000/0061) and NSF to GWL (OCE-0240896, OCE-0326434 and OCE-0424789). We thank the David Vaughan and an anonymous referee for their comments on the draft manuscript.

REFERENCES

Adamson AW (1990) Physical Chemistry of Surfaces. 5th edition. John Wiley
Ahrland S, Chatt J, Davies NR (1958) The relative affinities of ligand atoms for acceptor molecules and ions. Q Rev Chem Soc London 12:265–276
Akeret R (1953) Uber die Löslichkeit von Antimon(3)sulfid. PhD thesis, ETH Zurich
Al-Farawati R, van den Berg CMG (1999) Metal complexation in seawater. Mar Chem 63:331-352
Angeli J, Souchay P (1960) Sur les thioarsenites en solution. C R Acad Sci Paris 250:713-715
Ardon M, Taube H (1967) The thiolchromium (III) ion. J Amer Chem Soc 89:3661-3663
Arntson RH, Dickson FW, Tunell G (1966) Stibnite (Sb_2S_3) solubility in sodium sulfide solutions. Science 153: 1673-1674
Atkins PW (1978) Physical Chemistry. W. H. Freeman and Co.
Baas Becking LGM, Kaplan IR, Moore D (1960) Limits of the natural environment in terms of pH and oxidation-reduction potentials. J Geol 68:243-284
Babko AK, Lisetskaya GS (1956) Equilibrium in reactions of formation of thiosalts of tin, antimony, and arsenic in solution. Russian J Inorg Chem 1:95-107
Ball JW, Nordstrom DK, Zachmann DW (1987) WATEQ4F - A personal computer FORTRAN translation of the geochemical model WATEQ2 with revised data base. US Geological Survey, Water Resources Investigation Report 87-50
Banfield JF, Zhang H (2001) Nanoparticles in the environment. Rev Min Geochem 44:1-58
Baranova NN, Zotov AV (1998) Stability of gold sulphide species ($AuHS°(aq)$) and ($Au(HS)_2^-$) at 300, 350°C and 500 bar: Experimental study. Min Mag 62A:116–117
Barnes HL, Seward TM (1997) Geothermal systems and mercury deposits. *In*: Geochemistry of Hydrothermal Ore Deposits, 3rd ed. Barnes HL (ed) Wiley, p 699-736
Benoit JM, Gilmour CC, Mason RP, Heyes A (1999) Sulfide consitrols on mercury speciation and bioavailability to methylating bacteria in sediment pore waters. Environ Sci Technol 33:951-957
Berndt ME, Buttram T, Earley D, Seyfried WE (1994) The stability of gold polysulfide complexes in aqueous sulfide solutions –100 to 150°C and 100 bars. Geochim Cosmochim Acta 58: 587-594
Bhattacharyya R, Chakrabarty PK, Ghosh PN, Mukherjee AK, Podder D, Mukherjee M (1991) Reaction of MoO_4^{2-} and WO_4^{2-} with aqueous polysulfides: synthesis, structure, and electrochemistry of η^2-polysulfido complexes containng a bridging S,S $\{M_2O_2S_2\}^{2+}$ (M = Mo, W) core. Inorg Chem 30: 3948-3955.
Bianchi AA, Domenech E, Garcia-Espana E, Luis SV (1993) Electrochemical studies on anion coordination chemistry - application of the molar-ratio method to competitive cyclic voltammetry. Anal Chem 65: 3137-3142
Bianchini A, BowlesKC, Brauner CJ, Gorsuch JW, Kramer JR, Wood CM (2002) Evaluation of the effect of reactive sulfide on the acute toxicity of silver(I) to *Daphnia magna*, Part 2. Toxicity results. Environ Toxicol Chem 21:1294-1300
Blower PJ, Dilworth JR (1987) Thiolato-complexes of the transition metals. Coord Chem Rev 76:121-185
Bond AM, Hefter G (1972) Stability constant determination in precipitating solutions by rapid alternating current polarography. J Electroanal Chem 34:227-237
Bostick BC, Fendorf S, Helz G R (2003) Differential adsorption of molybdate and tetrathiomolybdate on pyrite (FeS_2). Environ Sci Technol 37:285-291
Boulegue J, Michard G (1978) Constantes de formation des ions polysulfures S_6^{2-}, S_5^{2-}, et S_4^{2-} en phase aqueuse. J Francais d'Hydrologie 25:27-34
Brendel PJ, Luther III GW (1995) Development of a gold amalgam voltammetric microelectrode for the determination of dissolved Fe, Mn, O^{-2}, and S(-II) in porewaters of marine and fresh-water sediments. Environ Sci Technol 29:751-761
Brintzinger H, Osswald H (1934) Die Anionengewichte einiger Sulfosalze in waßriger Lösung. Z Anorg Allg Chem 220:172–176
Brule JE, Haden YT, Callahan KP, Edwards JO 1988 Equilibrium and rate constants for mononuclear oxythiomolybdate interconversions. Gazzetta Chim Italiana 118: 93-99
Buffle J, de Vitre RR, Perret D, Leppard GG (1988) Combining field measurements for speciation in non perturbable waters. *In*: Metal Speciation: Theory, Analysis and Application. Kramer JR, Allen HE (eds) Lewis Publishers Inc, p 99-124
Butler IB, Archer C, Vance D, Rickard D, Oldroyd A (2005) Fe isotope fractionation on mackinawite formation. Earth Planet Sci Lett 236:430–442
Canfield DE (1998) A new model for Proterozoic ocean chemistry. Nature 396:450-453
Chadwell SJ, Rickard D, Luther III GW (1999) Electrochemical evidence for pentasulfide complexes with Mn^{2+}, Fe^{2+}, Co^{2+}, Ni^{2+}, Cu^{2+} and Zn^{2+}. Aquat Geochem 5:29-57
Chadwell SJ, Rickard D, Luther III GW (2001) Electrochemical evidence for metal polysulfide complexes: Tetrasulfide (S_4^{2-}) reactions with Mn^{2+}, Fe^{2+}, Co^{2+}, Ni^{2+}, Cu^{2+} and Zn^{2+}. Electroanalysis 13:21-29

Chazov VN (1976) Determination of the composition and certain standard potentials of water-soluble sodium thioantimonite. Russian J Phys Chem 50:1793

Ciglenečki I, Krznarić D, Helz G (2005) Voltammetry of copper sulfide nanoparticles: investigation of the cluster hypothesis. Environ Sci Technol 39:7492-7498

Clarke NJ, Laurie SH, Blandamer MJ, Burgess J, Hakin A (1987) Kinetics of the formation and hydrolysis reactions of some thiomolybdate(VI) anions in aqueous solution. Inorg Chim Acta 130:79–83

Cloke PL (1963a) The geologic role of polysulfides - Part I. The distribution of ionic species in aqueous sodium polysulfide solutions. Geochem Cosmochim Acta 27:1265-1298

Cloke PL (1963b) The geologic role of polysulfides - Part II. The solubility of acanthite and covellite in sodium polysulfide solutions. Geochem Cosmochim Acta 27:1299 -1319

Cobble JW, Murray RC Jr. (1978) Unusual ion solvation energies in high temperature water. Faraday Disc Chem Soc 64:144–149

Cooney RPJ, Hall JR (1966) Raman spectrum of thiomercurate(II) ion. Aust J Chem 19:2179–2180

Coucouvanis D, Swenson D, Stremple P, Baenziger NC (1979) Reaction of [Fe(SC$_6$H$_5$)$_4$]$^{2-}$ with organic trisulfides and implications concerning the biosynthesis of ferredoxins. Synthesis and structure of the [(C$_6$H$_5$)$_4$P]$_2$Fe$_2$S$_{12}$ complex. J Amer Chem Soc 101:3392-3394

Coucouvanis D, Patil PR, Kanatzidis MK, Detering B, Baenziger NC (1985) Synthesis and reactions of binary metal sulfides. Structural characterization of the [(S$_4$)$_2$Zn]$^{2-}$, [(S$_4$)$_2$Ni]$^{2-}$, [(S$_5$)Mn(S$_6$)]$^{2-}$, and [(CS$_4$)$_2$Ni]$^{2-}$. Inorg Chem 24:24-31

Coucouvanis D, Hadjikyriacou A (1986) Synthesis and structural characterization of the PH$_4$P+ salts of the [MoS(MoS$_4$)(S$_4$)]$^{2-}$ anion and its [MO$_2$S$_6$]$^{2-}$ and [MO$_2$S$_7$]$^{2-}$ desulfurized derivatives. Inorg Chem 25:4317-4319

Cotton FA, Wilkinson G, Murillo CA, Bockman M (1999) Advanced Inorganic Chemistry. 6th Edition. John Wiley

Coveney Jr RM, Leventhal JS, Glascock MD, Hatch JR (1987) Origins of metals and organic matter in the Mecca Quarry Shale member and stratigraphically equivalent beds across the Midwest. Econ Geol 82: 915–933

Cox JD, Wagman DD, Medvedev VA (1989) CODATA Key Values for Thermodynamics. Hemisphere Publishing Corp

Croot PL, Moffett JW, Luther III GW (1999) Polarographic determination of half-wave potentials for copper-organic complexes in seawater. Mar Chem 67:219-232

Crow DR (1969) Polarography of Metal Complexes. Academic Press Inc

Dance I, Fisher K (1994) Metal chalcogenide cluster chemistry. Prog Inorg Chem 41:637-803

Daskalakis KD, Helz GR (1992) Solubility of CdS greenockite in sulfidic waters at 25 °C. Environ Sci Technol 26:2462–2468

Davison W (1980) A critical comparison of the measured solubilities of ferrous sulphide in natural waters. Geochim Cosmochim Acta 44:803-808

Davison W (1991) The solubility of iron sulphides in synthetic and natural waters at ambient temperature. Aqautic Sci 35:309-329

Davison W, Heaney SI (1980) Determination of the solubility of ferrous sulfide in a seasonally anoxic basin. Limnol Oceanog 25:153-156

Davison W, Phillips N, Tabner BJ (1999) Soluble iron sulfide species in natural waters: Reappraisal of their stoichiometry and stability constants. Aquat Sci 61:23-43

Dean WE, Gardner JV, Piper DZ (1997) Inorganic geochemical indicators of glacial-interglacial changes in productivity and anoxia on the California continental margin. Geochim Cosmochim Acta 61:4507–4518

Delany JM, Lundeen SR (1990) The LLNL Thermodynamic Database. Technical Report UCRL-21658, Lawrence Livermore National Laboratory

Dellwig O, Böttcher ME, Lipinski M, Brumsack H-J (2002) Trace metals in Holocene coastal peats and their relation to pyrite formation (NW Germany). Chem Geol 82:423–442

DeFord DD, Hume DN (1951) The determination of consecutive formation constants of complex ions from polarographic data. J Amer Chem Soc 73: 5321-5322

Draganjac M, Simhon E, Chan LT, Kanatzidis M, Baenziger NC, Coucouvanis D (1982) Synthesis, interconversions, and structural characterization of the [(S$_4$)$_2$MoS][(S$_4$)$_2$MoO]$^{2-}$, [(S$_4$)$_2$MoS][(S$_4$)$_2$MoO]$^{2-}$, (Mo$_2$S$_{10}$)$^{2-}$, and (Mo$_2$S$_{12}$)$^{2-}$ anions. Inorg Chem 21:3321–3332

Drzaic PS, Marks J, Brauman JI (1984) Electron photodetachment from gas phase molecular anions. Gas Phase Ion Chemistry 3:167-211

Dubey KP, Ghosh S (1962) Studies on thiosalts. IV. Formation of thiosalt from antimonous sulphide. Z Anorg Chem 319:204–208

Dyrssen D (1985) Metal complex formation in sulphidic seawater. Mar Chem 15:285-293

Dyrssen D (1988) Sulfide complexation in surface seawater. Mar Chem 24:143-153

Dyrssen D (1991) Comment on "Solubility of sphalerite in aqueous sulfide solutions at temperatures between 25 and 240 °C". Geochim Cosmochim Acta 55: 2682-2684

Dyrssen D, Wedborg M (1989) The state of dissolved trace sulfide in seawater. Mar Chem 26: 289-293
Dyrssen D, Wedborg M (1991) The sulfur-mercury(II) system in natural-waters. Water Air Soil Poll 56:507-519
Eary LE (1992) The solubility of amorphous As_2S_3 from 25 to 90°C. Geochim Cosmochim Acta 56:2267-2280
Echt O, Sattler K, Recknagel (1981) Magic numbers for sphere packings - experimental-verification in free xenon clusters. Phys Rev Let 47:1121-1124
Echt O, Flotte AR, Knapp M, Sattler K, Recknagel E (1982) Magic numbers in mass-spectra of Xe, $C_2F_4Cl_2$ and SF_6 clusters. Ber Bunsen-Gesellschaft-Phys Chem Chem Phys 86:860-865.
Eichhorn BW (1994) Ternary transition-metal sulfides. Prog Inorg Chem 42:139-237
Eigen M, Wilkins RG (1965) The kinetics and mechanism of formation of metal complexes. *In:* Mechanisms of Inorganic Reactions. Advanced Chemistry Series 49:55-60
Elliott S (1988) Linear free energy techniques for estimation of metal sulfide complexation constants. Mar Chem 24:203-213
El-Maali NA, Ghandour MA,Vire J-C, Patriarche GJ (1989) Electrochemical study of cadmium complexes with folic acid and its related compounds. Electroanalysis 1:87-92
Emerson S, Jacobs L, Tebo B (193) The behaviour of trace elements in marine anoxic waters: solubilities at the oxygen-hydrogen sulphide internface. *In:* Trace Metals in Seawater. Wong CS, Boyle E, Bruland KW, Burton JD, Goldberg ED (eds) Plenum Press, p 579-609
Erickson BE, Helz GR (2000) Molybdenum(VI) speciation in sulfidic waters: Stability and lability of thiomolybdates. Geochim Cosmochim Acta 64:1149–1158
Erlich S, Butler IB, Halicz L, Rickard D, Oldroyd A, Matthews A (2004) Experimental study of the copper isotope fractionation between aqueous Cu(II) and covellite, CuS. Chem Geol 209:259-269
Fiala R, Konopik N (1950) Uber das Dreistoffsystem Na_2S–Sb_2S3–H_2O. II. Die auftretenden Bodenkörper und ihre Löslichkeit. Monatsh Chem 81:505–519
Filella M, Belzile N, Chen Y-W (2002) Antimony in the environment: a review focused on natural waters II. Relevant solution chemistry. Earth Sci Rev 59:265–285
Gammons CH, Barnes HL (1989) The solubility of Ag_2S in near-neutral aqueous sulfide solutions at 25 to 300 °C. Geochim Cosmochim Acta 53:279–290
Giggenbach W (1971) Optical spectra of highly alkaline sulfide solutions and the second dissociation constant of hydrogen sulfide. Inorg Chem 10:1333-1338
Giggenbach W (1972) Optical spectra and equilibrium distribution of polysulfide ions in aqueous solutions at 20 °C. Inorg Chem 11:1201-1207
Gimarc BM (1979) Molecular Structure and Bonding: The Qualitative Molecular Orbital Approach. Academic Press
Gobeli AO, Ste-Marie J (1967) Constantes de stabilité de thiocomplexes et produits de solubilité de sulfures de metaux II. Sulfure de zinc. Canadian J Chem 45:2101-2108
Gu XM, Dai J, Jia D X, Zhang Y, ZhuQ-Y (2005) Two-step solvothermal preparation of a coordination polymer containing a transition metal complex fragment and a thiostannate anion: [{Mn(en)$_{(2)}$}$_{(2)}$(μ-en)(μ-Sn$_2$S$_6$)]$_\infty$ (en, ethylenediamine). Crystal Growth Design 5:1845-1848
Gushchina LV, Borovikov AA, Shebanin AP (2000) Formation of antimony(III) complexes in alkali sulfide solutions at high temperatures: an experimental Raman spectroscopic study. Geochem Int 38:510–513
Gurevich VM, Gavrichev KS, Gorbunov VE, Baranova NN, Tagirov BR, Golushina LN, Polyakov VB (2004) The heat capacity of $Au_2S(cr)$ at low temperatures and derived thermodynamic functions. Thermochim Acta 412:85–90
Gustafsson L (1960) Determination of ultramicro amounts of sulphate as methylene blue - I. Talanta 4:227-235.
Harmer MA, Sykes AG (1980) Kinetics of the interconversion of sulfido- and oxomolybdate species $MoO_xS_{4-x}^{2-}$ in aqueous solutions. Inorg Chem 19: 2881-2885
Hartmann M, Clark T, van Eldik R (1997) Hydration and water exchange of zinc (II) ions. Application of density functional theory. J Am Chem Soc 119:7843-7850
Hayashi K, Sugaki A, Kitakaze A (1990). Solubility of sphalerite in aqueous sulfide solutions at temperatures between 25 and 240 °C. Geochim Cosmochim Acta 54:715-725
Heath GA, Hefter G (1977) The use of differential pulse polarography for the determination of stability constants. J Electroanal Chem 84:295-302
Helgeson HC (1964) Thermodynamics of hydrothermal systems at elevated temperatures and pressures. Am J Sci 267:729-804
Helgeson HC, Delany JM, Nesbitt HW, Bird DK (1978) Summary and critique of the thermodynamic properties of rock-forming minerals. Am J Sci 278A:1-229
Helz GR, Charnock JM,Vaughan DJ, Garner CD (1993) Multinuclearity of aqueous copper and zinc bisulfide complexes. An EXAFS investigation. Geochim Cosmochim Acta 57:15-25

Helz GR, Tossell JA, Charnock JM, Pattrick RAD, Vaughan DJ, Garner CD (1996) Oligomerization in As (III) sulfide solutions: Theoretical constraints and spectroscopic evidence. Geochim Cosmochim Acta 59: 4591-4604

Herron N, Wang Y, Eckert H (1990) Synthesis and characterization of surface-capped, sized-quantizied CdS clusters. Chemical control of cluster size. J Amer Chem Soc 112:1322-1326

Hoar TP, Hurlen T (1958) Kinetics of Fe/Fe$_{aq}^{2+}$ electrode. Proc Intern Comm Electrochem Thermodyn Kinetics, 8th Meeting, 445–447

Holland HD (2004) The geologic history of seawater. *In;* Treatise on Geochemistry, Vol. 6. Oceans and Marine Geochemistry. Elderfield H (ed) Academic Press, p 583-625

Högfeldt E (1982) Stability constants of metal ion complexes: Part A. Inorganic ligands. IUPAC Chemical Data Series 21. Pergammon Press, Oxford

Höltje R. (1929) Uber die Löslichkeit von Arsentrisulfid und Arsenpentasulfid. Z Anorg Chem 181:395-407

Höltje R, Beckert J (1935) Die Löslichkeit von Kupfersulfid in Alkalipolysulfidlösungen. Z Anorg Chem 222: 240-244

Holm RH (1974) Equivalence of metal centers in iron-sulfur protein active-site analogs [Fe$_4$S$_4$(SR)$_4$]$_2$. J Amer Chem Soc 96:2644 1974

Huerta-Diaz MG, Morse JW (1992) Pyritization of trace metals in anoxic marine sediments. Geochim Cosmochim Acta 56:2681–2702

Huheey JE, Keiter EA, Keiter RL (1993) Inorganic Chemistry: Principles of Structure and Reactivity. 3rd ed Harper Collins

Irving H, Williams RJP (1953) The stability of transition-metal complexes. J Chem Soc 10:3192-3210

Ishikawa K, Isonaga T, Wakita S, Suzuki Y (1995) Structure and electrical properties of Au$_2$S. Solid State Ionics 79:60-66

IUPAC (2006) The IUPAC Stability Constants Database. www.iupac.org.

Jay JA, Morel FMM, Hemond HF (2000) Mercury speciation in the presence of polysulfides. Environ Sci Technol 34:2196-2200

Johnson JW, Oelkers EH, Helgeson HC (1992) SUPCRT92: A software package for calculating the standard molal thermodynamic properties of minerals, gases and aqueous species from 0 to 1000 °C and 1 to 5000 bar. Comp Geosci 18:799–847

Kamyshny AJ, Goifman A, Gun J, Rizkov D, Lev O (2004). Equilibrium distribution of polysulfide ions in aqueous solutions at 25°C: A new approach for the study of polysulfides equilibria. Environ Sci Technol 38:6633-6644

Kaschiev D (2000) Nucleation: Basic Theory with Applications. Butterworth Heinemann

Klatt LN, Rouseff RL (1970) Analysis of the polarographic method of studying metal complex equilibria. Anal Chem 42:1234-1238

Kolpakova NN (1982) Laboratory and field studies of ionic equilibria in the Sb$_2$S$_3$–H$_2$O–H$_2$S system. Geochem Int 19:46–54

Kortan AR, Hull R, Opila RL, Bawendi MG, Steigerwald ML, Carroll PJ, Brus LE (1990) Nucleation and growth of CdSe on ZnS quantum crystallite seeds, and vice versa, in inverse micelle media. J Amer Chem Soc 112:1327-1332

Krebs H (1968) Fundamentals of Inorganic Crystal Chemistry. McGraw Hill

Krupp RE (1988) Solubility of stibnite in hydrogen sulfide solutions, speciation, and equilibrium constants, from 25 to 350 °C. Geochim Cosmochim Acta 52:3005–3015

Krupp RE (1990) Comment on "As(III) and Sb(III) sulfide complexes: An evaluation of stoichiometry and stability from existing experimental data by N. F. Spycher and M. H. Reed" Geochim Cosmochim Acta 54:3239–3240

Kuehn CG, Taube H (1976) Ammineruthenium complexes of hydrogen sulfide and related sulfur ligands. J Amer Chem Soc 98:689-702

Kuwabara JS, Luther III GW (1993) Dissolved sulfides in the oxic water column of San Francisco Bay, California. Estuaries 16:567-573

Labrenz M, Druschel GK, Thomsen-Ebert T, Gilbert B, Welch SA, Kemner KM, Logan GA, Summons RE, De Stasio G, Bond PL, Lai B, Kelly SD, Banfield JF (2001) Formation of sphalerite (ZnS) deposits in natural biofilms of sulfate-reducing bacteria. Science 290:1744-1747

Landing WM, Lewis BL (1991) Thermodynamic modeling of trace metal speciation in the Black Sea. *In:* Black Sea Oceanography. Izdar E, Murray J (eds) NATO ASI Series p 125-160

Langford CH, Gray HB (1965) Ligand Substitution Processes. W.A. Benjamin Inc.

Langmuir D (1969) The Gibbs free energies of substancies in the system Fe-O$_2$-H$_2$O-CO$_2$ at 25°C. USGS Prof Paper 650-B

Larson JW, Cerruti P, Garber HK, Hepler LG (1968) Electrode potentials and thermodynamic data for aqueous ions. Copper, zinc, cadmium, iron, cobalt and nickel. J Phys Chem 72:2902–2907

Latimer WM (1952) The Oxidation States of the Elements and Their Potentials in Aqueous Solutions. Prentice Hall
Laurie SH (2000) Thiomolybdates—simple but very versatile reagents. Eur J Inorg Chem 2000:2443–2450
Lennie AR, Redfern AT, Champness PE, Stoddart CP, Schofield PF, Vaughan DJ (1997) Transformation of mackinawite to greigite: An in situ X-ray powder diffraction and transmission electron microscope study. Am Mineral 82:302-309
Lennie AR, Charnock JM, Pattrick RAD (2003) Structure of mercury(II)–sulfur complexes by EXAFS spectroscopic measurements. Chem Geol 199:199–207
Lewis GN (1923) Valence and the Structure of Atoms and Molecules. The Chemical Catalog Company
Lewis BL, Luther III GW, Lane H, Church TM (1995) Determination of metal-organic complexation in natural waters by SWASV with pseudopolarograms. Electroanalysis 7:166-177
Lincoln SF, Richens DT, Sykes AG (2003) Metal aqua ions. In: Comprehensive Co-ordination Chemistry II. Vol 1. McCleverty JA, Meyer TJ (eds) Elsevier, p515-555
Lindsay WL(1979) Chemical Equilibria in Soils. Wiley
Løver T, Henderson W, Bowmaker GA, Seakins JM, Cooney RP (1997) Electrospray mass spectrometry of thiophenolate-capped clusters of CdS, CdSe, and ZnS and of cadmium and zinc thiophenolate complexes: Observation of fragmentation and metal, chalcogenide, and ligand exchange processes. Inorg Chem 36: 3711-3723
Luther III GW, Ferdelman TG (1993) Voltammetric characterization of iron (II) sulfide complexes in laboratory solutions and in marine waters and porewaters. Environ Sci Technol 27:1154-1163
Luther III GW, Giblin AE, Varsolona R (1985) Polarographic analysis of sulfur species in marine porewaters. Limnol Oceanogr 30:727-736
Luther III GW, Glazer BT, Hohman L, Popp JI, Taillefert M, Rozan TF, Brendel PJ, Theberge S, Nuzzio DB (2001a) Sulfur speciation monitored in situ with solid state gold amalgam voltammetric microelectrodes: polysulfides as a special case in sediments, microbial mats and hydrothermal vent waters. J Environ Monitoring 3:61-66
Luther III GW, Glazer B, Ma S, Trouwborst R, Shultz BR, Druschel G, Kraiya C (2003) Iron and sulfur chemistry in a stratified lake: evidence for iron rich sulfide complexes. Aquat Geochem 9:87-110
Luther III GW, Rickard D (2005) Metal sulfide cluster complexes and their biogeochemical importance in the environment. J Nano Res 7: 389-407
Luther III GW, Rickard D, Theberge SM, Oldroyd A (1996) Determination of metal (bi)sulfide stability constants of Mn, Fe, Co, Ni, Cu and Zn by voltammetric methods. Environ Sci Technol. 30:671-679
Luther III GW, Rozan TF, Taillefert M, Nuzzio DB, Di Meo C, Shank TM, Lutz RA, Cary SC (2001b) Chemical speciation drives hydrothermal vent ecology. Nature 410:813-816
Luther III GW, Theberge SM, Rickard D (1999) Evidence for aqueous clusters as intermediates during zinc sulfide formation. Geochim Cosmochim Acta 63:3159-3169
Luther III GW, Theberge SM, Rickard D (2000) Determination of stability constants for metal-ligand complexes using the voltammetric oxidation wave of the anion/ligand and the DeFord and Hume formalism. Talanta 5:11-20
Luther III GW, Theberge SM. Rozan TF, Rickard D, Rowlands CC, Oldroyd A (2002) Aqueous copper sulfide clusters as intermediates during copper sulfide formation. Environ Sci Technol 35:94-100
Luther III GW, Tsamakis E (1989) Concentration and form of dissolved sulfide in the oxic water column of the ocean. Mar Chem 27:165-177
Maronny G (1959) Constants de dissociation de l'hydrogene sulfure. Electrochim Acta 1: 58-69
Mascharak PK, Papaefthymiou GC, Armstrong WH, Foner S, Rankel RB, Holm RH (1983) Electronic properties of single and double- $MoFe_3S_4$ cubane type clusters. Inorg Chem 22: 2851–2858
McCay LW (1901) Die Einwirkung von Schwefelwasserstoff auf Arsensäure. Z Anorg Chem 29:36-50
Meites L (1965) Polarographic Techniques 2nd Ed. Wiley Interscience
Mellor JW (1943) A Comprehensive Treatise on Inorganic and Theoretical Chemistry. Vol 11. Longmans
Meyer B, Peter L, Spitzer K (1977) Trends in the charge distribution in sulfanes, sulfanesulfonic acids, sulfanedisulfonic acids, and sulfuous acid. Inorg Chem 16:27-33
Millero FJ, Schreiber DR (1982) Use of the ion pairing model to estimate activity coefficients of the ionic components of natural waters. Amer J Sci 282:1508-1540
Michel FM, Antao SM, Chupas PJ, Lee PL, Parise JB, Schoonen MAA (2005) Short- to medium-range atomic order and crystallite size of the initial FeS precipitate from pair distribution function analysis. Chem Mat 17:6246-6255
Milyutina NA, Polyvyanny IR, Sysoev LN (1967) Solubilities of some sulfides and oxides of some minor metals in aqueous solution of sodium sulfide. Tr Inst Metall Obogashch Akad Nauk Kaz SSR 21: 14–19; CA 76676d
Mironova GD, Zotov AV, Gul'ko NI (1990) The solubilityof orpiment in sulfide solutions at 25-150 °C and the stability of arsenic sulfide complexes. Geochem Intl 27: 61-73

Morse JW, Casey WH (1988) Ostwald processes and mineral paragenesis in sediments. Amer J Sci 288:537-560

Morse JW, Millero J, Cornwell JC, Rickard D (1987) The chemistry of the hydrogen sulfide and iron sulfide systems in natural waters. Earth-Sci Rev 24:1-42

Mosselmans JFW, Helz GR, Pattrick RAD, Charnock JM, Vaughan DJ (2000) A study of speciation of Sb in bisulfide solutions by X-ray absorption spectroscopy. Appl Geochem 15: 879-889

Mountain BW, Seward TM (1999) The hydrosulphide/sulphide complexes of copper(I): Experimental determination of stoichiometry and stability at 22 °C and reassessment of high temperature data. Geochim Cosmochim Acta 63: 11-29

Mountain BW, Seward TM (2003) Hydrosulfide/sulfide complexes of copper(I): Experimental confirmation of the stoichiometry and stability of $Cu(HS)_2$ to elevated temperatures. Geochim Cosmochim Acta 67: 3005-3014

Müller A, Nolte W-O, Krebs B (1978) $[(S_2)_2Mo(S_2)_2Mo(S_2)_2]$, a novel complex containing only S_2^{2-} ligands and a Mo-Mo bond. Angew Chem Int Ed Engl 17:279

Müller A, Diemann E (1987) Polysulfide complexes of metals. Adv Inorg Chem 31:89-122

Müller A, Krickemeyer E, Reinsch U (1980) Reactions of coordinated S_2^{2-} ligands 1. Reaction of $[Mo_2(S_2)_6]^{2-}$ with nucleophiles and a simple preparation of the bis-disulfido complex $[Mo_2O_2S_2(S_2)_2]^{2-}$. Z Anorg Allg Chemie 470:35-38

Müller A, Romer M, Bugge H, Krickemeyer E, Bergmann D (1984a) $[Cu_6S_{17}]^{2-}$, a novel binary discrete polynuclear Cu^I complex with several interesting structural features: the arrangement of the metal atoms and co-ordination of ligands. J Chem Soc Chem Commun 6: 348-349

Müller A, Romer M, Bugge H, Krickemeyer E, Schmitz K (1984b) Novel soluble cyclic and polycyclic polysulfido species: $[Au_2S_8]^{2-}$, $[Cu_4S_x]^{2-}$ (x = 13-15) and other copper clusters. Inorg Chim Acta 85:L39-L41

Müller A, Schimanski J, Schimanski U (1984c) Isolation of a sulphur-rich binary mercury species from a sulphide-containing solution: $[Hg(S_6)_2]^{2-}$, a complex with S_6^{2-} ligands. Angew Chem Int Ed Engl 23: 159-160

Müller A, Schimanski U (1983) $[Cu_3S_{18}]^{3-}$, a novel sulfur rich complex with different kinds of puckered copper sulfur heterocycles, a central Cu_3S_3 and three outer CuS_6 ones. Inorg Chim Acta 77: L187-L188

Nedeljkovic JM, Patel RC, Kaufman P, Joyce-Pruden C, O'Leary N (1993) Synthesis and optical properties of quantum-size metal sulfide particles in aqueous solution. J Chem Ed 70: 342-345

NIST (2005) NIST Critically Selected Stability Constants of Metals Complexes Database. Version 8.0. NIST Standard Reference Database 46. U. S. Department of Commerce

Ohfuji H, Light M, Rickard D, Hursthouse M (2006) Single crystal XRD study on framboidal pyrite. Eur J Min 18:93-98

Ohfuji H, Rickard D (2006) High resolution transmission electron microscopic study on synthetic nanocrystalline mackinawite. Earth Planet Sci Lett 241:227-223

Osterloh F, Segal FM, Achim C, Holm RH (2000) Reduced mono-, di-, and tetracubane-type clusters containing the $[MoFe_3S_4]^z$ core stabilized by tertiary phosphine ligation. Inorg Chem 39:980–989

Papaefthymiou GC, Laskowski EJ, Frotapessoa S, Frankel RB, Holm RH (1982) Anti-ferromagnetic exchange interactions in $[Fe_4S_4(SR)_4]^{2-,3-}$ clusters. Inorg Chem 21:1723-1728

Paquette K, Helz G (1997) Inorganic speciation of mercury in sulfidic waters: the importance of zero-valent sulfur. Environ Sci Technol 31:2148–2153

Parkhurst DL (1995) User's Guide to PHREEQC - A computer program for speciation, reaction path, advective-transport and inverse geochemical calculations', US Geological Survey, USGS 95-4227

Patrick WA, Thompson WE (1953) Standard electrode potential of the iron-ferrous ion couple. J Am Chem Soc 75:1184–1187

Pattrick RAD, van der Laan G, Vaughan DJ, Henderson CMB (1993) Oxidation state and electronic configuration determination of copper in tetrahedrite group minerals by L-edge X-ray absorption spectroscopy. Phys Chem Mineral 20:395-401

Pattrick RAD, Mosselmans JFW, Charnock JM, England KER, Helz GR, Garner CD, Vaughan DJ (1997) The structure of amorphous copper sulfide precipitates: An X-ray absorption study. Geochim Cosmochim Acta 61:2023-2036

Pauling L (1960) The Nature of the Chemical Bond and the Structure of Molecules and Crystals: An Introduction to Modern Structural Chemistry. Oxford University Press

Pearson RG (1968) Hard and soft acids and bases. J Chem Educ 45:581-587

Pourbaix M (1966) Atlas of electrochemical equilibria in aqueous solutions. Pergamon Press

Prince RH (1987) Zinc and cadmium. Comp Coord Chem 5:925-1046

Radford-Knoery J, Cutter GA (1994) Biogeochemistry of dissolved hydrogen sulfide species and carbonyl sulfide in the western North Atlantic Ocean. Geochim Cosmochim Acta 58:5421-5431

Radzic AA, Smirnov BM (1985) Reference Data on Atoms, Molecules, and Ions. Vol. 31, Springer-Verlag

Raiswell R, Plant J (1980) The incorporation of trace elements into pyrite during diagenesis of black shales. Yorkshire, England. Econ Geol 75:684–699

Ramasani T, Sykes AG (1976) Further characterization and aquation of the thiolpentaaquochromium (III), $CrSH^{2+}$, and its equilibration with thiocyanate. Inorg Chem 15:1010-1014

Randall M, Frandsen M (1932) The standard electrode potential ofiron and the activity coefficient of ferrous chloride. J Am Chem Soc 54:47–54

Rao PV, Holm RH (2004) Synthetic analogues of the active sites of iron-sulfur proteins. Chem Rev 104:527-559

Rauchfuss TB (2004) Research in soluble metal sulfides: from polysulfido complexes to functional models for the hydrogenases. Inorg Chem 43:14-26

Renders PJ, Seward TM(1989) The stability of hydrosulphido and sulphido complexes of Au(I) and Ag(I) at 25 °C. Geochim Cosmochim Acta 53:245–253

Richens DT (1997) The Chemistry of Aqua Ions. John Wiley & Sons

Rickard D (1989) Experimental concentration-time curves for the iron(II) sulphide precipitation process in aqueous solutions and their interpretation. Chem Geol 78:315-324

Rickard D (1995) Kinetics of FeS precipitation. 1. Competing reaction mechanisms. Geochem Cosmochim Acta 59:4367-4379

Rickard D (1997) Kinetics of pyrite formation by the H_2S oxidation of iron(II) monosulfide in aqueous solutions between 25 °C and 125 °C: the rate equation. Geochim Cosmochim Acta 61:115-134

Rickard D (in press) The solubility of mackinawite. Geochim Cosmochim Acta

Rickard D, Luther III GW (1997) Kinetics of pyrite formation by the H_2S oxidation of iron(II) monosulfide in aqueous solutions between 25°C and 125°C: the mechanism. Geochim Cosmochim Acta 61:135-147

Rickard D, Morse J (2005) Acid volatile sulfide. Mar Chem 97:141-197

Rickard D, Nriagu JO (1978) Aqueous environmental chemistry of lead. *In*: The Biogeochemistry of Lead in the Environment, Part A. Ecological cycles. Nriagu JO (ed) Elsevier, p 219-284

Robie RA, Hemingway BS, Fisher JR (1978) Thermodynamic properties of minerals and related substances at 298.15 K and one atmosphere presure and at higher temperatures. USGS Bull 1259

Robie RA, Hemingway BS (1995) Thermodynamic properties of minerals and related substances at 298.15 K and 1 Bar (105 Pascals) pressure and at higher temperatures. USGS Bull 2131

Rochette EA, Bostick BC, Li G, Fendorf S (2000) Kinetics of arsenate reduction by dissolved sulfide. Environ Sci Technol 34:4714-4720

Rouxel J, Tournoux M, Brec R (eds) (1994) Soft chemistry routes to new materials. Proceedings of the International Symposium on Soft Chemistry Routes to New Materials, Nantes, France, September 1993. Materials Science Forum, Vols 152 -153

Rozan TF, Benoit G, Luther III GW (1999) Measuring metal sulfide complexes in oxic river waters with square wave voltammetry. Environ Sci Technol 33: 3021-3026

Rozan TF, Luther III GW (2002) Voltammetric evidence suggesting Ag speciation is dominated by sulfide complexation in river water. 2002. *In:* Environmental Electrochemistry: Analyses of Trace Element Biogeochemistry. Taillefert M, Rozan T (eds) American Chemical Society Symposium Series 811: 371-380

Rozan TF, Luther III GW, Ridge D, Robinson S (2003) Determination of Pb complexation in oxic and sulfidic waters using pseudovoltammetry. Environ Sci Technol 37:3845-3852

Rozan TF, Theberge SM, Luther III GW (2000) Quantifying elemental sulfur (S^0), bisulfide (HS^-) and polysulfides (S_x^{2-}) using a voltammetric method. Anal Chim Acta 415:175-184

Schoonen MAA, Barnes HL (1988) An approximation of the second dissociation constant for H_2S. Geochim Cosmochim Acta 52:649-654

Schwarzenbach G (1961) The general, selective and specific formation of complexes by specific metallic cations. Adv Inorg Radiochem 3:265-270

Schwarzenbach G, Fischer A (1960) Die Aciditat der Sulfane und die Zusammensetzung wasseriger Polysulfidlosungen. Helv Chem Acta 43:1365-1390

Schwarzenbach G, Widmer G (1963) Die Löslichkeit von Metallsulfiden. I Schwarzes Quecksilbersulfid. Helv Chim Acta 46:2613-2628

Schwarzenbach G, Widmer G (1966) Die Löslichkeit von Metalsulfiden. II Silbersulfid. Helv Chim Acta 49:111-123

Schwedt G, Rieckhoff M (1996) Separation of thio- and oxothioarsenates by capillary zone electrophoresis and ion chromatography. J Chromatog A736:341-350

Seby F, Potin-Gautier M, Giffaut E, Donard OFX (2001) A critical review of thermodynamic data inorganic tin species. Geochim Cosmochim Acta 65:3041-3053

Shea D, Helz G (1989) Solubility product constants of covellite and a poorly crystalline copper sulfide precipitate at 298 K. Geochim Cosmochim Acta 53:229-236

Sheldrick WS (1988) On CS$_4$SN$_5$S$_{12}$.2H$_2$O, a cesium(I) thiostannate(IV) with 5-fold and 6-fold coordinated tin. Z Anorg Allg Chem 562:23-30

Sheldrick WS, Schaaf B (1994) Preparation and crystal-structure of Rb$_2$Sn$_3$S$_7$.2H$_2$O and Rb$_4$Sn$_2$Se$_6$. Z Anorg Allg Chem 620:1041-1045

Sherman DM, Ragnarsdottir KV, Oelkers EH (2000) Antimony transport in hydrothermal solutions: and EXAFS study of antimony(V) complexation in alkaline sulfide and sulfide -chloride brines at temperatures from 25 °C to 300 °C at P$_{sat}$. Chem Geol 167:161–167

Shestitko VS, Titova AS, Kuzmichev GV (1975) Potentiometric determination of the composition of the thio-anions of antimony(V). Russ J Inorg Chem 20:297–299

Sillén LG, Martell AE (1964) Stability constants of metal-ion complexes. The Chemical Society London, Spec Pub 17

Silvester EJ, Grieser F, Sexton BA, Healy TW (1991) Spectroscopic studies on copper sulfide sols. Langmuir 7:2917-2922

Smith RM, Martell AE (1976). Critical Stability Constants. Vol. 4 Inorganic Complexes. Pergamon Press

Sooklal K, Cullum BS, Angel SM, Murphy CJ (1996) Photophysical properties of ZnS nanoclusters with spatially localized Mn^{2+}. J Phys Chem 100:4551-4555

Spycher NF, Reed MH (1989) As(III) and Sb(III) sulfide complexes: an evaluation of stoichiometry and stability from existing experimental data. Geochim Cosmochim Acta 53: 2185-2194

Stefánsson A, Seward TM (2003) The stability and stoichiometry of sulphide complexes of silver(I) in hydrothermal solutions to 400 °C. Geochim Cosmochim Acta 67:1395–1413

Stefánsson A, Seward TM (2004) Gold(I) complexing in aqueous sulphide solutions to 500 °C at 500 bar. Geochim Cosmochim Acta 68:4121-4143

Stauder S, Raue B, Sacher F (2005) Thioarsenates in sulfidic waters. Environ Sci Technol 39:5933-5939

Ste-Marie J, Torma AE, Gubeli AO (1964) The stability of thiocomplexes and solubility products of metal–sulfides: I. Cadmium sulfide. Can J Chem 42:662–668

Stumm W (1992) Chemistry of the Solid-Water Interface. John Wiley & Sons Inc

Stumm W, Morgan JJ (1996) Aquatic Chemistry. Wiley Interscience

Sugaki A, Scott SD, Hayashi K, Kitakaze A (1987) Ag$_2$S solubility in sulfide solutions up to 250°C. Geochem J 21:291–305

Sukola K, Wang F, Tessier A (2005) Metal-sulfide species in oxic waters. Anal Chim Acta 528:183-195

Suleimenov OM, Seward TM (1997) A spectrophotometric study of hydrogen sulphide ionisation in aqueous solutions to 350°C. Geochim Cosmochim Acta 61:5187-5198

Suleimenov, OM, Seward, TM (2000) Spectrophotometric measurements of metal complex formation at high temperatures: the stability of Mn(II) chloride species. Chem Geol 167:177-192

Sweeton FH, Baes CF Jr. (1970) Solubility of magnetite and hydrolysis of ferrous ion in aqueous solutions at elevated temperatures. J Chem Thermodyn 2:479–500

Teder A (1971) The equilibrium between elemental sulfur and aqueous polysulfide solutions. Acta Chem Scand 25:1722-1728

Theberge SM (1999) Investigations of Metal-Sulfide Complexes and Clusters: A Multimethod Approach. PhD. Dissertation, University of Delaware, Lewes, Delaware

Theberge S, Luther III GW (1997) Determination of the electrochemical properties of a soluble aqueous FeS species present in sulfide solutions. Aquat Geochem 3:191-211

Thompson RA, Helz GR (1994) Copper speciation in sulfidic solutions at low sulfur activity: Further evidence for cluster complexes? Geochim Cosmochim Acta 58:2971–2983

Tossell JA (1994) The speciation of antimony in sulfidic solutions: A theoretical study. Geochim Cosmochim Acta 58:5093-5104

Tossell JA (1994) Calculation of the UV-visible spectra and the stability of Mo and Re oxysulfides in aqueous solution. Geochim Cosmochim Acta 69:2497–2503

Tossel JA (1996) The speciation of gold in aqueous solution: A theoretical study. Geochim Cosmochim Acta 60:17–29

Tossell JA (2001) Calculation of the structures, stabilities, and properties of mercury sulfide species in aqueous solution. J Phys Chem A 105:935–941

Tossell JA, Vaughan DJ (1992) Theoretical Geochemistry: Application of Quantum Mechanics in the Earth and Mineral Sciences. Oxford University Press

Tossell JA, Vaughan DJ (1993) Bisulfide complexes of zinc and cadmium in aqueous solution: Calculation of structure, stability, vibrational, and NMR spectra, and of speciation on sulfide mineral surfaces. Geochim Cosmochim Acta 57:1935-1945

Treadwell WD, Hepenstick H (1949) Uber die Löslichkeit von Silbersulfid. Helv Chim Acta 32:1872–1879

Tremaine PR, LeBlanc JC (1980) The solubility of magnetite and the hydrolysis and oxidation of Fe^{2+} in water to 300°C. J Sol Chem 9:415-442

Trsic M, Laidlaw WG (1980) *Ab initio* Hartree-Fock-Slater calculations of polysulfanes H$_2$S$_n$ (n=1-4) and the ions HS$_n^-$ and S$_n^{2-}$. Internat J Quantum Chem XVII: 969-974

Tsang JJ, Rozan TF, Hsu-Kim H, Mullaugh KM, Luther III GW (in press) Pseudopolarographic determination of Cd^{2+} complexation in freshwater. Envir Sci Technol

Turner DR, Whitfield M, Dickson AG (1981) The equilibrium speciation of dissolved components in freshwater and seawater at 25 °C and 1 atm. Geochim Cosmochim Acta 45:855-881

Van der Laan G, Patrick RAD, Henderson CMB, Vaughan DJ (1992) Oxidation-state variations in copper minerals studied with Cu 2p X-ray absorption-spectroscopy. J Phys Chem Solids 53:1185-1190

Van der Lee J (2005) Common thermodynamic database project. Ecole des Mines de Paris. http://ctdp.ensmp.fr

Van der Zee C, Roberts DR, Rancourt DG, Slomp CP (2003) Nanogoethite is the dominant oxyhydroxide phase in lake and marine sediments. Geology 31:993-996

Vorlicek TP, Kahn MD, Kasuya Y, Helz GR (2004) Capture of molybdenum in pyrite-forming sediments: Role of ligand-induced reduction by polysulfides. Geochim Cosmochim Acta 68:547–556

Vossmeyer T, Reck G, Katsikas L, Haupt ETK, Schulz B, Weller H (1995) A "double-diamond superlattice" built of $Cd_{17}S_4(SCH_2CH_2OH)_{26}$ clusters. Science 267:1476-1479

Wagman DD, Evans WH, Parker VB, Halow I, Bailey SM, Schumm RH (1969) Selected values of chemical thermodynamic properties. Nat Bur Standards Tech Note 270-4

Wagman DD, Evans WH, Parker VB Schumm RH, Halow I, Bailey SM, Churney KL, Nuttall RL (1982) The NBS tables of chemical thermodynamic properties: selected values for inorganic and C_1 and C_2 organic substances in SI units. J Phys Chem Ref Data 11:1-392

Wang F, Tessier A (1999) Cadmium complexation with bisulfide Environ Sci Technol 33:4270–4277

Webster JG (1990) The solubility of As_2S_3 and speciation of As in dilute and sulphide-bearing fluids at 25 and 90 °C. Geochim Cosmochim Acta 54:1009-1017

Wei D, Osseo-Asare K (1995) Formation of iron monosulfide: a spectrophotometric study of the reaction between ferrous and sulfide ions in aqueous solutions. J Coll Interfac Sci 174: 273-282

Wei D-Y, Saukov AA (1961) Physicochemical factors in the genesis of antimony deposits. Geochemistry 6: 510–516.

Weissbuch I, Addadi L, Lahav M, Leiserowitz L (1991) Molecular recognition at crystal interfaces. Science 253:637-645

Wells AF (1986) Structural Inorganic Chemistry, 5th ed. Clarendon Press

Werner A (1904) Lehrbuch der Stereochinie. G. Fischer, Jena.

Whittermore DO, Langmuir D (1972) Standard electrode potential of $Fe^{3+} +e^- =Fe^{2+}$ from 5-35°C. J Chem Engin Data 17:288-290

Wilkin RT, Wallschlager D, Ford RG (2003) Speciation of arsenic in sulfidic waters. Geochem Trans 4:1-7

Williams RJP, Frausto da Silva JJR (1996) The Natural Selection of the Chemical Elements. Clarendon Press

Williamson MA, Rimstidt JD (1992) Correlation between structure and thermodynamic properties of aqueous sulfur species. Geochim CosmochimActa 56:3867-3880

Wolery TJ (1979) Calculation of chemical equilibrium between aqueous solution and minerals: the EQ3/6 software package. Lawrence Livermore Laboratory UCRL-52658

Wolthers M, Van der Gaast SJ, Rickard D (2003) The structure of disordered mackinawite. Am Mineral 88: 2007-2015

Wood SA (1989) Raman spectroscopic determination of the speciation of ore metals in hydrothermal solutions: I. Speciation of antimony in alkaline sulfide solutions at 25 °C. Geochim Cosmochim Acta 53:237–244

Wood SA, Tait CD, Janecky DR (2002) A Raman spectroscopic study of arsenite and thioarsenite species in aqueous solution at 25°C. Geochem Trans 3:31-39

Wu DX, Hong MC, Cao R, Liu HQ (1996) Synthesis and characterization of $[Et_{(4)}N)_{(4)}][MoS_4Cu_{10}Cl_{12}]$: A polynuclear molybdenum-copper cluster containing a central tetrahedral MoS_4 encapsulated by octahedral Cu-6 and tetrahedral Cu-4 arrays. Inorg Chem 35:1080-1082

Zhang J-Z, Millero FJ (1994) Investigation of metal sulfide complexes in sea-water using cathodic stripping square-wave voltammetry. Anal Chim Acta 284:497-504

Zhang H, Gilbert B, Huang F, Banfield JF (2003) Water-driven structure transformation in nanoparticles at room temperature. Nature 424:1025-1029

Zotov AV, Baranova NN, Dar'yina TG, Bannykh LM(1991) The solubility of gold in aqueous chloride fluids at 350-500°C and 500-1500 atm: Thermodynamic parameters of $AuCl_2^-$ up to 750°C and 5000 atm. Geochem Int 28:63–71

Sulfide Mineral Surfaces

Kevin M. Rosso

Chemical Sciences Division
and W.R. Wiley Environmental Molecular Sciences Laboratory
Pacific Northwest National Laboratory
Richland, Washington, U.S.A.
e-mail: kevin.rosso@pnl.gov

David J. Vaughan

School of Earth Atmospheric and Environmental Sciences
and Williamson Research Centre for Molecular Environmental Science
University of Manchester
Manchester, United Kingdom
e-mail: david.vaughan@manchester.ac.uk

INTRODUCTION

The past twenty years or so have seen dramatic development of the experimental and theoretical tools available to study the surfaces of solids at the molecular ("atomic resolution") scale. On the experimental side, two areas of development well illustrate these advances. The first concerns the high intensity photon sources associated with synchrotron radiation; these have both greatly improved the surface sensitivity and spatial resolution of already established surface spectroscopic and diffraction methods, and enabled the development of new methods for studying surfaces. These have included the techniques of X-ray Absorption Spectroscopy (for example, EXAFS and XANES; see Wincott and Vaughan 2006, this volume) which can be used to probe the molecular scale environment of atoms at the surface of a particulate sample in contact with a fluid. The second centers on the scanning probe microscopy (SPM) techniques initially developed in the 1980's with the first scanning tunneling microscope (STM) and atomic force microscope (AFM) experiments. The direct "observation" of individual atoms at surfaces made possible with these methods has truly revolutionized surface science. As well as providing insights into the crystallography and morphology of the pristine surface, the new range of SPM methods has enabled real time and *in situ* observations of surfaces during reaction with gases and fluids. On the theoretical side, the availability of high performance computers coupled with advances in computational modeling has provided powerful new tools to complement the advances in experiment. Particularly important have been the quantum mechanics based computational approaches such as density functional theory (DFT), which can now be easily used to calculate the equilibrium crystal structures of solids and surfaces from first principles, and to provide insights into their electronic structure. There have also been important advances in the use of more empirical or semi-empirical computational methods, both those based on quantum mechanics and on so-called "atomistic" approaches, and in computational tools for the study of reactions at surfaces. All of these computational approaches have a role in gaining greater understanding of surfaces and surface reactivity, and the most sophisticated methods can now produce results in remarkably good agreement with experiment.

These experimental and theoretical developments in surface science have had considerable impact on the study of mineral surfaces. Prior to the 1980's, relatively few studies of the structure and reactivity of mineral surfaces had been undertaken, whereas there is now a substantial and rapidly growing literature in the field. However, certain metal sulfide minerals and synthetic sulfides have been the subject of surface science investigations for much longer. This is for two reasons; one is the importance of a small number of metal sulfides and their synthetic analogs to the electronics industries (examples include GaS, ZnS, PbS), the other is the importance of sulfide minerals as the primary source of many of the metals essential to industry. The processing of "as-mined" ores commonly requires either separation and concentration of the metal-bearing sulfide minerals by froth flotation, or direct extraction of the metals by dissolution (leaching) with acid or alkaline solutions. In either of these processes, the nature and reactivity of the mineral surface is critically important. The metal sulfides continue to be materials of great technological importance, particularly as the raw materials for the metalliferous mining industry. However, the mining of sulfide minerals also leads to the generation of large volumes of waste rock and waste fine materials (tailings) from processing of the mined ores. These sulfide mineral-bearing wastes, commonly containing large amounts of pyrite, are a serious environmental pollutant. On exposure to air and water at the Earth's surface, they can break down, generating sulfuric acid and releasing toxic components (As, Cd, Pb, Hg, etc.) into surface waters and soils. Again, the control exercised by surface structure and chemistry on the behavior of sulfides as pollutants is a key area for the application of research on sulfide surfaces. Of course, this weathering of sulfide wastes created by human activity is just part of the larger scale geochemical cycling of materials at the Earth's surface. In the natural processes of interaction between the Earth's lithosphere, hydrosphere, atmosphere and biosphere, sulfides play an important role; for example, in trapping metals in the sulfidic horizons formed by bacterial reduction of seawater sulfate beneath the sediment surface in many shallow marine environments. The interface between the sulfide mineral surface and biomaterials, whether biofilms, microbes or other organisms, is also a critical part of many important geochemical systems. As discussed elsewhere in this volume, biological agents can be involved in sulfide mineral dissolution or precipitation, and sulfides can fulfill biological functions, for example as magnetic sensors in magnetotactic bacteria.

In this chapter, we review current knowledge of sulfide mineral surfaces, beginning with an overview of the principles relevant to the study of the surfaces of all crystalline solids. This includes the thermodynamics of surfaces, the atomic structure of surfaces (surface crystallography and structural stability, adjustments of atoms at the surface through relaxation or reconstruction, surface defects) and the electronic structure of surfaces. We then discuss examples where specific crystal surfaces have been studied, with the main sulfide minerals organized by structure type (galena, sphalerite, wurtzite, pyrite, pyrrhotite, covellite and molybdenite types). Some examples of more complex phases, where fracture surfaces of unspecified orientation have been studied, are then discussed (millerite, marcasite, chalcopyrite, arsenopyrite, and enargite) before a brief summary of possible future developments in the field. In this chapter, the focus is on the nature of the pristine surface, i.e., the arrangement of atoms at the surface, and the electronic structure of the surface. This is an essential precursor to any fundamental understanding of processes such as dissolution, precipitation, sorption/desorption, or catalytic activity involving the sulfide surface at an interface with a fluid phase. These surface reaction processes form the subject of the following chapter (Rosso and Vaughan 2006).

SURFACE PRINCIPLES

Surface energy

The thermodynamic stability of a surface is characterized by its surface energy (γ). The surface energy is a positive quantity; creation of a surface from an infinite solid must increase the energy of the system because solids do not spontaneously cleave. Because the energy increase is proportional to the surface area (A), surface energy is given in units of energy/area. In this case the internal energy of a system (U) would be given by the following form of the Euler equation (Zangwill 1988).

$$U = TS - PV + \mu N + \gamma A \tag{1}$$

with U being uniquely defined by the temperature (T), entropy (S), pressure (P), volume (V), chemical potential (μ), the number of particles (N), and the surface energy contribution. It is usually on the order of 0.1–1.0 J/m^2; lower values indicate higher stability. Surface energy can also be understood as the work required to form more surface, i.e., to bring molecules from the interior of the phase into the surface region, so it can also be given in terms of force per unit length. The surface free energy for different crystal faces is usually different; the morphology of a crystal grown under near-equilibrium conditions is governed by the minimization of the total surface free energy.

Surface energy is very difficult to measure, so various methods of approximation have been developed. Perhaps the simplest approach is simply to halve the cohesive energy, which is equivalent to taking half of the energy to rupture that number of bonds (of known bond energy) passing through a given area. The most stable surfaces are the ones rupturing the smallest number of bonds. One can imagine that this estimate would work well for solids where there no long range interactions. Such is nominally the case for covalent solids, where the cohesive energy arises primarily from the sharing of electrons in bonds. For ionic solids, where cohesion arises from simple electrostatic attraction between oppositely charged ions, the situation is more complex due to the lattice electrostatic energy (U_{el}). For binary ionic solids, built according to

$$A_a X_{x(s)} \rightarrow aA^{Z_A}_{(g)} + xX^{Z_X}_{(g)} \tag{2}$$

where a and x are stoichiometric numbers and Z indicates the ionic charges, U_{el} can be estimated using

$$U_{el} = -M' \frac{Z_A Z_X (a+x) e^2}{2r} \tag{3}$$

where r is the closest interionic distance and M' is the reduced Madelung constant. Values of M' vary for sulfide minerals; M' for sphalerite is 1.64 for example (Jolly 1991). Given the dependence of U_{el} on the inverse of the first power of r, this contribution to the total internal energy is significantly long-ranged. Sulfide minerals span the range of covalent to ionic interactions; normally we find that end-members do not exist in this regard.

Atomic structure

By the term atomic structure, we mean the arrangement of the atoms at the surface of the solid which is unlikely to be simply represented by a truncation of the structure of the bulk. Generally, atoms at the surface undergo relaxation which involves spontaneous movement of the atoms relative to their bulk positions but such that the symmetry remains the same, or reconstruction, which is a more extensive reorganization leading to a surface symmetry different from that of the bulk. Studies of the structures of pristine surfaces (particularly cleavage surfaces) of minerals and other crystalline solids are an essential precursor to investigations of surface reactivity. In this section, we briefly review elements of surface crystallography and aspects of surface stability and defects.

Crystallography. The "workhorse" technique used to study the crystallography of solid surfaces is low energy electron diffraction (LEED) in which incident electrons from an electron gun are elastically backscattered from a surface and undergo diffraction if the surface atoms are arranged periodically. Since the detected electrons (energies 20-1000 eV) only travel ~5-20 Å into the surface, the diffraction patterns observed are excellent probes of surface structure. It is not appropriate here to describe the LEED technique in detail, and excellent accounts are given by Clarke (1985), Zangwill (1988) and others.

At a growth or cleavage surface of a crystal, periodicity is retained at best in only two dimensions. For a two-dimensional planar periodic structure, every lattice point can be reached from the origin by translation vectors

$$\mathbf{T} = m\mathbf{a}_s + n\mathbf{b}_s \quad (4)$$

where m and n are integers. The primitive vectors a_s and b_s define a unit mesh or surface lattice. There are five possible such lattices in two dimensions as shown in Figure 1 and these are the five Bravais surface lattices. Complete specification of an ordered surface structure requires both the unit mesh and location of the basis atoms, the latter of which must be consistent with certain symmetry restrictions. In two dimensions, the only operations consistent with the five surface lattices that leave one point unmoved are mirror reflections across a line and rotations through an angle of $2\pi / p$ where $p = 1, 2, 3, 4$ or 6. The ten point groups that result, combined with the surface lattices, yields 13 space groups. The addition of glide reflection operations (translation plus mirror reflection) gives a total of 17 two-dimensional space groups.

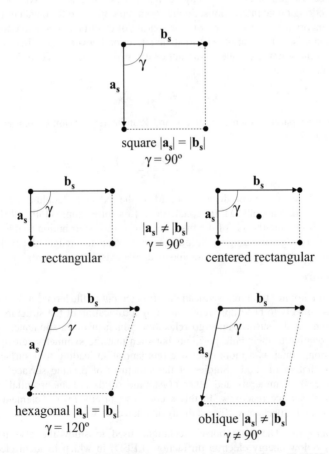

Figure 1. The five Bravais surface lattice types. Adapted from Zangwill (1988).

In labeling and discussing surface diffraction data, atomic resolution images, or model systems the Miller indices familiar to all crystallographers and mineralogists are employed (with the usual convention of crystallographic directions being shown in square brackets and crystallographic planes in curved brackets). Ideal surfaces, or those undergoing simple relaxation, are identified by reference to the bulk plane of the termination e.g., PbS (100). For hexagonal systems, we will use the Miller-Bravais ($hkil$) notation here as opposed to Miller (hkl) notation to be consistent with its predominance in the reviewed literature (note that the Miller-Bravais notation has the property that $h + k + i = 0$). With the surface mesh being the same as the underlying bulk lattice these are therefore 1 × 1 structures. A reconstructed surface would have a primitive surface mesh related to the ideal 1 × 1 structure by $a_s = Na$ and $b_s = Mb$ and, here, the usual nomenclature would be "Formula (hkl) N × M;" e.g., "TiO$_2$ (100) (1 × 3)." If the surface lattice is rotated by an angle φ with respect to the bulk lattice, this angle can be appended (Formula (hkl) N × M – φ).

Stability. The stability of certain arrangements of atoms at a surface may be approached from a variety of perspectives that vary in detail from purely ionic to quantum mechanical. One of the most useful overarching perspectives was developed by Tasker (1979). Tasker envisioned three surface types based on the electrostatic stability of their cation and anion arrangements (Fig. 2). Type I surfaces consist of planes containing both cations and anions, with each plane being charge neutral. Type II surfaces are composed of alternating planes of anions and cations that are arranged in such a way that no net surface dipole moment is possible with some cuts. Type III surfaces are also composed of alternating charged planes, but are arranged in such a way that there will always be a permanent dipole moment normal to the surface. Type III surfaces are termed polar surfaces. Type III surfaces have large surface energy and are therefore very unstable. They are prone to reconstruct to eliminate the surface dipole and thereby lowering the surface energy. Reconstruction leads to a new bonding topology and/or symmetry in the surface lattice relative to the 1 × 1 structure. Polar surfaces have been isolated for study, but often are metastable.

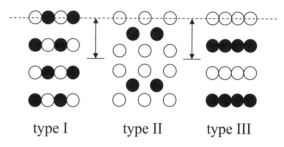

Figure 2. Surface classification according to Tasker (1979) based on the structure of the repeat unit (indicated by arrows) perpendicular to the surface plane (dashed line). Type I and II surfaces are non-polar, charge neutral and stable, whereas a Type III surface is polar and unstable.

Most of the surfaces reviewed here are either type I or II. These surfaces are non-polar and, as a first criterion, are stable with respect to reconstruction. However, these surfaces do undergo relaxation, which is accompanied by lowering of the surface energy. The largest atomic displacements for many of the surfaces reviewed here occur along or approximately along the surface normal direction. Rumpling is an important kind of relaxation where atomic sublattices on the surface move different amounts or in different directions along the surface normal. The displacement of the interplanar spacing between the sublattices is a measure of the surface rumpling (Fig. 3).

Many of the papers covered in this review are applications of molecular modeling techniques to sulfide surfaces. Within this approach, the surface energy can be computed directly, for example, using periodic slab model calculations. Model design is affected by the previously mentioned consideration of the possible role of long-range interactions. The usual

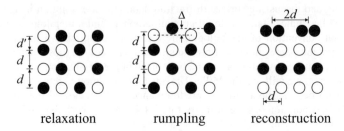

Figure 3. Illustration of differences between surface relaxation, rumpling, and reconstruction. The row of atoms at the top of the diagram represents the surface. Adapted from Noguera (1996).

approach for vacuum-terminated surfaces is to compute

$$\gamma = \frac{U_S - U_B}{A} \quad (5)$$

where subscripts "S" and "B" refer to surface and bulk internal energies for equal numbers of atoms. (For certain "finite" slab models, the denominator is taken as $2A$ to reflect the presence of an upper and lower surface.) To compute surface energies for hydrated surfaces, specifically where associative adsorption occurs, one could use (de Leeuw et al. 2000)

$$\gamma = \frac{U_H - (U_B + U_W)}{A} \quad (6)$$

where "H" refers to the hydrated surface, and U_W is the energy of bulk water. Regardless whether or not this is approached with classical atomistic or quantum mechanical description of the atoms, it is usually rare for these calculations to include all possible energetic contributions to U, especially thermochemical contributions arising from phonon degrees of freedom. Even with the approximations in assessing absolute internal energy values, reasonably good surface free energies can often be predicted this way due to cancellation of errors. Table 1 is a listing of recent theoretical surface energies for non-polar surfaces computed using molecular modeling techniques. Properties underlying the relative stabilities of non-polar surfaces stem in part from the nature of bonding in the solid, which in turn is based on the electronic structures of the constituent atoms. These concepts are discussed in more detail below.

Defects. Surface imperfections are an important aspect of sulfide mineral surfaces because many of these minerals have poor to absent cleavage, yielding very rough fracture topographies. Some macroscopically conchoidal fracture surfaces have been shown to be terraced with a high density of defect structures at microscopic length scales. Defects are also common on surfaces of minerals with very good cleavage. SPM studies have documented a wide range of defects and defect densities occurring on sulfide mineral surfaces. Typical structural or compositional defects are sketched in Figure 4. Steps and kinks have the usual significance as important sites for attachment and detachment of material at the surface during growth or dissolution. Step edges, kinks and corners are reduced coordination sites with increasing numbers of dangling bonds relative to sites on terraces. Outcropping bulk planar and line defects create areas of the surface with strained atomic geometries. All of these kinds of defects can have unique electronic states that are highly reactive, and these will sometimes manifest clearly in STM images. As will be discussed below, substitutional or interstitial impurities are important for modifying the electronic properties of the bulk and surface.

Vacancies can be very common on sulfide mineral surfaces where cleavage quality is poor. Surface or step edge stoichiometry can be altered to attain charge neutrality and stability by forming vacancies. Formation energies for neutral vacancies in sulfide minerals can be

Table 1. Surface energies (J/m^2) for vacuum and hydrated sulfide mineral surfaces computed using computational molecular modeling techniques.

Formula (structure)	Surface	$\gamma_{vac}{}^a$	$\gamma_{water}{}^b$	Reference
FeS$_2$ (pyrite)	(100)	1.23	1.13	de Leeuw et al. (2000)
		1.06		Cai and Philpott (2004)
		0.98		Stirling et al. (2003)
		1.06		Hung et al. (2002a)
		1.23		Guanzhou et al. (2004a)
		1.10		Rosso (2001)
	(100) "crenellated"	3.19	2.60	de Leeuw et al. (2000)
	(110)	2.36	1.66	de Leeuw et al. (2000)
		1.68		Hung et al. (2002a)
	(110) "facetted"	1.97	1.62	de Leeuw et al. (2000)
		1.54		Hung et al. (2002a)
	(111)	1.40		Hung et al. (2002b)
		1.60		Rosso (2001)
	(111) Fe	3.81	2.87	de Leeuw et al. (2000)
	(111) S$_2$	3.92	2.89	de Leeuw et al. (2000)
	(210)	1.50		Hung et al. (2002b)
RuS$_2$ (pyrite)	(100)	1.09		Fréchard and Sautet (1995)
	(111)	3.30		Fréchard and Sautet (1995)
PbS (rocksalt)	(100)	0.34		Wright et al. (1999)
ZnS (sphalerite)	(110)	0.65		Wright et al. (1998)
		0.53		Hamad et al. (2002)
		0.35		Steele et al. (2003)
	(100) Zn	1.25		Wright et al. (1998)
		1.12		Hamad et al. (2002)
	(100) S	1.31		Wright et al. (1998)
		1.30		Hamad et al. (2002)
	(111) Zn	1.82		Wright et al. (1998)
		0.87		Hamad et al. (2002)
	(111) S	1.12		Wright et al. (1998)
	($\bar{1}\bar{1}\bar{1}$) Zn	1.03		Wright et al. (1998)
	($\bar{1}\bar{1}\bar{1}$) S	1.89		Wright et al. (1998)
		1.01		Hamad et al. (2002)
	(210) Zn	1.14		Hamad et al. (2002)
	(210) S	1.08		Hamad et al. (2002)
ZnS (wurtzite)	(11$\bar{2}$0)	0.49		Hamad et al. (2002)
	(10$\bar{1}$0)	0.52		Hamad et al. (2002)
	(12$\bar{3}$0)	0.52		Hamad et al. (2002)
	(0001) Zn	0.90		Hamad et al. (2002)
	(0001) S	0.91		Hamad et al. (2002)
MoS$_2$ (molybdenite)	(0001)	0.28		Fuhr et al. (1999)
		0.25		Weiss and Phillips (1976)
	(10$\bar{1}$0)	1.63		Raybaud et al. (1998)
		1.70		Todorova et al. (2004)
		2.20		Alexiev et al. (2000)

^{a}relaxed surface in vacuum
^{b}relaxed surface with a monolayer of associatively (molecularly) adsorbed water

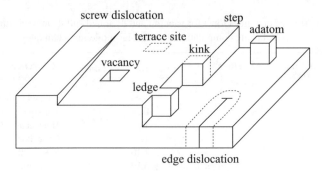

Figure 4. Various types of defects commonly associated with a surface.

addressed using molecular modeling approaches, but they can also be approximated using Van Vechten's (1975) cavity model. This macroscopic model uses covalent and ionic radii, the lattice constant, work function, and surface energies to estimate the energy required to create the cavity for the cationic or anionic vacancy. It was recently shown that vacancy formation energies for sulfide minerals vary over a sizable range between ~1-3 eV (Fiechter 2004). Formation energies for a cationic and anionic pair of vacancies, known as a Schottky pair ranges over 4-9 eV. The energies tend to be smaller for S than for the metal cations and therefore S vacancies tend to be more easily created.

Electronic structure

Surface states. The electronic structure at the surface is related to the bulk electronic structure, but always is different to some degree. For semiconductors, such as the majority of metal sulfides, the surface is a significant perturbation to the electronic structure of the crystal in the surface region. Within the nearly-free electron view, the discontinuation of the periodic potential of the crystal lattice leads to new electronic states that are localized at the surface, which decay both into the bulk and outward from the surface (Fig. 5). From the localized orbital perspective, considering cleavage as the surface forming process for the moment, bond breaking causes orbitals localized on surface atoms to destabilize back towards their free atom

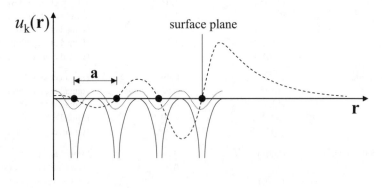

Figure 5. Simplified one-dimensional electronic structure model depicting a surface state in terms of electronic potential energy ($u_k(\mathbf{r})$) versus position ($\mathbf{r}$). Black dots represent the lattice of atoms in a crystal of spacing **a**, forming the periodic potential field for constituent electrons. Electronic states oscillate over this field for an infinite crystal (dotted line). A surface state (dashed line) decays both into the bulk and outward from the surface. Adapted from Zangwill (1988).

orbital character and energies, producing so-called dangling bonds (Fig. 6). From both perspectives, we arrive at the conclusion that unique states are associated with atoms at the surface. Surface states can also arise from surface reconstruction, defects, adsorbates, or discrete surface phases. If these states either reduce the band gap at the surface, or lie fully within the band gap at the surface, they are called surface states.

Because an analogy can be drawn between surface states and the highest-occupied and lowest unoccupied molecular orbitals of frontier molecular orbital theory, surface states tend to dominate the initial stages of surface chemical behavior. The highest occupied states at the surface are energetically predisposed to be the source of electrons, and the lowest unoccupied states are predisposed to accept electrons. The ability of a donor/acceptor to "dock" with a surface atom principally depends on energy and symmetry matching of orbitals between the two. Any study of the chemistry on a clean sulfide mineral surface should consider this idea as central.

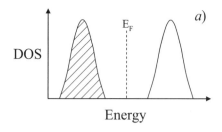
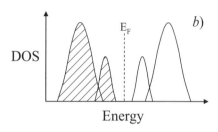

Figure 6. Hypothetical density of states plot for a semiconductor showing *a)* the bulk band gap, and *b)* the introduction of surface states. The Fermi level is designated as E_F. Occupied states are hatched. Surface states are often located in the bulk band gap, causing the surface band gap width to differ from the bulk band gap width.

For semiconducting sulfide minerals with moderate to wide band gaps, there are typically two surface states arising from cleavage, one associated with the dangling bond on the cation and the other with the dangling bond on the anion. Examples include pyrite (100) (Bronold et al. 1994; Rosso et al. 1999a), pyrite (111) (Rosso 2001), greenockite (CdS) (10$\bar{1}$0) (Siemens et al. 1997), sphalerite (111) (Louis and Elices 1975), and sphalerite (110) (Duke 1992). The lateral density of surface states is directly related to the surface energy and therefore surface stability. Surface energy is minimized by rearrangement of either electron density between atoms or by rearrangement of the surface atomic structure itself. It is common for anion dangling bond states to lie energetically below the Fermi level (E_F) while cation states lie above E_F. This can be understood from simple electronegativity arguments, with the cation donating charge to the anion, yielding filled anion dangling bond states in the vicinity of the valence band maximum. The degree of surface stabilization arising from this mechanism depends on the details of the charge transfer. Should the number of electrons available to participate exactly fill all the anion dangling-bonds the driving force for further rearrangement is largely removed. This is another way of stating that the stoichiometric surface is formally charge-neutral with respect to the bulk, and therefore it is a stable surface.

For example, consider extension of this electron-counting argument to the (100) surface of pyrite. In the bulk pyrite structure, each Fe atom contributes two valence electrons to six Fe-S bonds, or 1/3 $e-$ per bond. Each S atom contributes six valence electrons to three Fe-S and one S-S bond. Taking one electron from each S to form the S-S bond, 10 are left for the six Fe bonds around an S_2 group, or 5/3 $e-$ per Fe-S bond. In this way, every bond in pyrite has two electrons and the structure is charge neutral. At the (100) surface, there is a 1:1 ratio of dangling bonds on uppermost Fe and S atoms, arising because only Fe-S bonds are broken. Dangling bonds localized on Fe and S form surface states in the bulk band gap at energies that can approach those for the atomic orbital components of the free atoms. For reasons of relative electronegativity, the 1/3 $e-$ from each Fe dangling bond is transferred to each S dangling bond, completely filling the anion dangling bonds (1/3 + 5/3 = 2 $e-$). Hence, the pyrite (100) surface is said to be "autocompensated;" it will have a low surface energy and should not reconstruct (Harrison 1980; Pashley 1989; Gibson and LaFemina 1996). Hence, all pyrite-type (100), rocksalt-type (100), zinc blende-type (110), and wurtzite-type (10$\bar{1}$0) and (11$\bar{2}$0) mineral surfaces reviewed here are autocompensated.

Charge mobility. For semiconducting sulfides, another factor pertaining to charge redistribution at the surface comes from mobile electrons or holes (see also discussion in Pearce et al. 2006a, this volume). Intrinsic semiconductors have small populations of electrons thermally excited across the band gap into the conduction band, and corresponding holes exist in the valence band (Fig. 7). In extrinsic semiconductors, charge carriers arise principally from charge donating or charge-accepting impurities (dopants) or vacancies scattered throughout the bulk. The position of E_F in the band gap is dependent on the relative availability of charge carriers. In the extrinsic case there is usually an imbalance between the number of charge-carrying electrons and holes, with the more numerous termed the majority carrier. Majority electrons give an n-type semiconductor with E_F displaced up in energy towards the conduction band edge. Majority holes give a p-type semiconductor with E_F displaced to lower energy towards the valence band edge. Majority electrons in the conduction band can populate empty lower-lying surface states to lower their energy. Likewise, electrons from higher-lying filled surface states can "drain" into majority holes in the valence band. The imbalance produces a net charge flow between the surface and bulk, altering the occupation of the surface states. This process is known as charge trapping. Most natural semiconducting minerals are impure and defective to some degree and are therefore more prone to exhibit n- or p-type behavior, depending on the types of impurities and defects present (Pridmore and Shuey 1976). Thus charge trapping due to surface states should not be unexpected.

The net charge exchange between the surface and bulk can strip a wide near-surface region of charge carriers, resulting in the formation of a space charge layer. The space charge

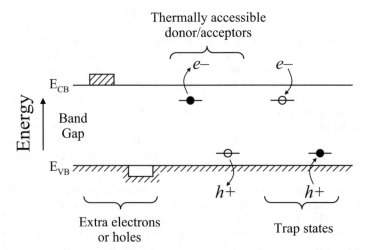

Figure 7. Energy level diagram of possible effects of defects/impurities on the electronic structure of a semiconducting mineral, which include extra electrons or holes in the conduction band (CB) or valence band (VB), respectively, defect levels in the bulk band gap providing electrons or holes by thermal excitation, and defect levels acting as traps for electrons and holes. Adapted from Cox (1987).

layer can penetrate several hundred nanometers from the surface into the bulk. The energies of bands in this region are modified, or "bent;" band-bending is an experimental observable. For example, consider the surface of an *n*-type sample with empty surface states below E_F (Fig. 8). The flow of conduction band electrons into lower-lying empty surface states strips donor atoms in a region below the surface of majority electrons. This causes the surface to become more negative, with a positively charged depletion layer beneath. Viewing surface atoms as being more negatively charged relative to the atoms in the depletion region, the bulk bands bend upwards as they approach the surface. Thermodynamic equilibrium is achieved between the energy gains arising from infilling the surface states and the increasing energy cost to extract electrons from the bulk. This situation yields electron binding energy shifts that can be directly observed in experimental surface core-level photoelectron spectra (SCLS) because the energy of photoejected core electrons is modified by the band bending potential at the surface.

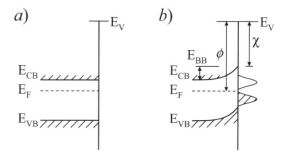

Figure 8. Energy level diagram depicting *a*) *n*-type semiconductor bands, and *b*) upward band-bending due to surface states and the redistribution of charge at the surface. The band bending energy is E_{BB}, the vacuum level energy is E_V, the workfunction (see definition in the following section on Probes) is ϕ, and the electron affinity is χ.

Probes. The energies of electrons in a solid are related to the outside world by the workfunction. The workfunction is defined as the energy difference between an electron at E_F in the solid and an electron in isolation in a vacuum. Although it is defined independent of whether electrons are removed from the bulk versus the surface, there are specific surface contributions to the workfunction because electrons interact with other particles in the solid as they are removed. This interaction is sensitive to the surface structure, charge distribution, and the presence of adsorbates. Therefore, workfunction differences can be thought of as a measure of changes in the affinity of a surface for electrons. For semiconducting sulfide minerals, because states may or may not be available at E_F, the electron affinity is used instead. Electron affinity is defined as the energy to remove an electron from highest occupied state to the vacuum level (Fig. 8). At room temperature in any typical semiconductor, regardless of *n*- or *p*-type behavior, this is the bottom of the conduction band (Sze 1981).

Most of the concepts just presented can be directly or indirectly probed at the atomic scale using STM and STM-based tunneling spectroscopy (TS). For that reason, many sulfide mineral surface studies involve STM. In STM, a bias voltage drives a net electron tunneling current between an atomically sharp metal tip and the sample surface across an insulating gap of a few Ångstroms. For details on the operation of STM, excellent introductory texts are available (Bonnell 1993; Chen 1993). Here we wish to highlight some additional considerations that arise specifically when using STM to interrogate the electronic structure of mineral surfaces. Concerns include the effects of the tip and the electric field on the surface electronic structure.

For STM of semiconducting sulfide minerals, the tunneling junction can be characterized as a "three-phase" metal-insulator-semiconductor interface (Fig. 9). The insulator can be a vacuum gap or another phase such as air or water. The tip is able to induce band bending when in tunneling contact with the surface. For example, as the distance between the tip and an *n*-type semiconductor surface is reduced and electron tunneling contact is established, charge

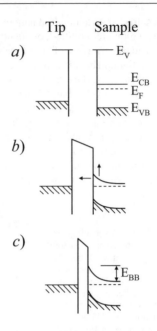

Figure 9. Energy level diagram depicting possible tip-induced bending of *n*-type semiconductor bands at the surface in the operation of the STM. Electrons spontaneously redistribute as the Ångstrom-scale tip-sample distance is decreased from *a*) no tunneling contact, to *b*) initial tunneling contact, to *c*) final tunneling contact. The horizontal arrow shows the direction of electron transfer and the vertical arrow shows the associated direction of band bending for this example case.

carrier electrons can transfer from semiconductor conduction band to the tip conduction band because the semiconductor workfunction will usually be lower than that for the metal. Because this process can be thought of as equivalent to conduction band electrons infilling surface states discussed earlier, the tip can be viewed as becoming increasingly negatively charged with respect to the semiconductor. Likewise, a positive space charge layer then builds up in the semiconductor near-surface that upwardly bends the bulk bands. This spontaneous process continues until the Fermi levels of the tip and semiconductor match (Fig. 9). This tip-induced band bending is effectively an electrostatic coupling of the tip to the sample (Weimer et al. 1988b, 1989; McEllistrem et al. 1993). When the bias voltage is applied across the interface, the band bending is free to be modified by this voltage. This is an unwanted effect in STM that can be a significant source of discrepancies between STM data and other attempts to determine the surface electronic structure. However, if occupied surface states are present, surface state electrons will often be numerous enough to screen the semiconductor bulk states from the buildup of negative charge at the tip. In this case, the semiconductor bands at the interface are less likely to be affected by the presence of the tip.

EXAMPLE SURFACES

Crystallographic surfaces

Rocksalt-type. The rocksalt or halite (NaCl) crystal structure of galena is one of the simplest and best known of all structures, and was one of the first determined by X-ray diffraction. It is based on cubic close packing of sulfur atoms in planes parallel to (111); both lead and sulfur are in regular octahedral (6-fold) coordination. The phases PbSe (clausthalite) and PbTe (altaite) are isostructural with galena. All three exhibit perfect cubic (100) cleavage, and the (100) surface is the dominant growth and cleavage surface of these minerals.

Galena has attracted the interest of materials scientists because of its technological importance (for example in IR detectors, or as nanoparticle catalysts), and mineral technologists because it is the major source of lead. It is also an obvious potential source of pollution from mine wastes. The simple crystal structure of the bulk phase and presence of a perfect cleavage have made galena an attractive substance for surface science experimentation and computer modeling studies.

The first question in regard to the (100) surface of PbS concerns the arrangement of the atoms at the surface; whether and how this differs from a simple truncation of the bulk structure (Fig. 10). This seemingly simple question has proved remarkably controversial. In an earlier publication, Tossell and Vaughan (1987) used a qualitative molecular orbital theory argument to support a proposed lack of displacement of atoms at the surface (relaxation) relative to the bulk. This view has been supported experimentally by Kendelewicz et al. (1998) who studied the surface structure using synchrotron-based X-ray standing wave and photoemission methods. They concluded that neither S nor Pb surface atoms are displaced from their bulk positions and that there is little or no intralayer rumpling or interlayer contraction (see Fig. 3). On the other hand, Becker and Hochella (1996) used periodic Hartree-Fock (HF) calculations to predict a small rumpling at the surface with S atoms displaced into the bulk, although concern has been expressed that the size of the slab (four PbS layers) used in these calculations may not have provided a realistic model. Other computational studies include that of Satta and Gironcoli (2001) using DFT who predicted almost no rumpling in PbS but considerable rumpling in PbTe; the latter result being in good agreement with LEED studies of PbTe by Lazarides et al. (1995) and showing a 7% contraction (0.23 Å) of the top layer of Pb atoms. For both PbS and PbTe, however, a contraction between the top two layers and expansion between the third and fourth layers was calculated. This conclusion, based on similar computational methods, was also reached by Preobrajenski and Chasse (1999) but these authors also found significant rumpling (0.22 Å) of the top layer of atoms, as did Wright et al. (1999).

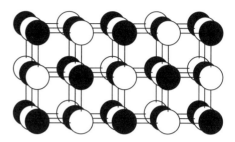

Figure 10. Model of the rocksalt-type (100) surface, which includes PbS (100).

The most recently published computational studies of the "geometric" surface structure of galena do conclude that some relaxation occurs. Muscat and Gale (2003) used a number of different *ab initio* approaches, including both DFT and HF methods, and calculated significant relaxations at the surface (< ~0.2 Å) decaying with depth into the bulk. The magnitude of relaxation was found to oscillate between layers with Pb ions displaced outwards (~0.05 Å) relative to S ions. Ma et al. (2004) using DFT methods to study PbS, PbSe and PbTe found intralayer rumpling and shifts in distances between layers differing between the three chalcogenides. Whereas rumpling of PbSe and PbTe were calculated to be greatest in the outermost layer and to decrease with depth, rumpling of PbS was calculated to increase down to the third layer before then decreasing.

It may seem strange that uncertainty remains over the geometric surface structures of these simple and readily studied materials although, as noted by Kendelewicz et al. (1998), until recently the same has been true of the much studied isostructural phase, MgO (a point that serves to emphasize how difficult it is to acquire definitive information on the surface structure of materials) Although uncertainty does remain as to the surface structure of galena (although the PbTe surface seems certain to undergo relaxation), the weight of argument currently favors some rumpling of the surface and oscillatory relaxation decaying with increasing depth.

These studies of geometric structure have been complemented by experimental and computational investigations of the surface electronic structures of PbS, PbSe and PbTe. The most recent DFT calculations (Ma et al. 2004) suggest that the electronic structures of all three phases are very similar; for PbS, the densities of states (DOS) are calculated to be very similar to those computed for the bulk (Muscat and Gale 2003). Although it was originally believed that there is no significant shift in the core level electron energies of surface atoms in PbS, synchrotron X-ray photoelectron spectroscopy (SXPS) work at low temperatures (Leiro et al. 1998) has identified a shift to lower energy of ~0.3 eV in the S $2p$ states.

Not surprisingly, the electronic structure of the PbS surface has been the subject of a large number of studies (Tossell and Vaughan 1987; Eggleston and Hochella 1993, 1994 Laajalehto et al. 1993; Becker and Hochella 1996; Bottner et al. 1996; Mian et al. 1996; Becker et al. 1997a,b; Eggleston 1997; Becker and Rosso 2001; Muscat and Gale 2003; Ma et al. 2004). These show that the top of the valence band is dominated by combined S $3p$-Pb $6p$ states, where the states with S $3p_z$ character extend out from the surface. Initial oxidation reactions at the PbS surface are likely to be controlled by the shapes and DOS's of orbitals with S $3p_z$ character (see next chapter, Rosso and Vaughan 2006). These orbitals are picked-up in STM images at negative sample bias voltages, whereas STM images at positive sample bias voltages probe the bottom of the conduction band, dominated by states with Pb $6p$ character (Fig. 11) (Becker and Hochella 1996; Becker and Rosso 2001). Angle-resolved spectroscopic studies of the surface electronic structure (Bottner et al. 1996; Cricenti et al. 2001) have been used to define surface electronic states in detail, and combined STM imaging and calculations (using a hybrid DFT and HF approach) employed to explore the differences in electronic structure at steps and kinks on the (100) surface compared with flat terraces (Becker and Rosso 2001). In the latter investigation, an apparent deformation seen in STM images in the vicinity of a step edge (Fig. 11) is found to be mainly an electronic effect rather than a relaxation of atoms (Fig. 12). The implications of these findings for reactivity of the PbS (100) surface are discussed in the next chapter (Rosso and Vaughan 2006).

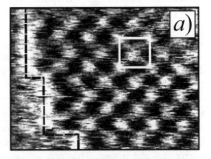

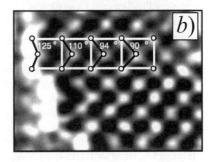

Figure 11. STM images collected in ultra-high vacuum at room temperature showing *a*) raw, and *b*) filtered data for a PbS (100) terrace terminated by a step (at left). The surface was generated by cleaving in vacuum. The square surface lattice unit shown in *a*) is ~5.9 Å on each side. The data was collected at 0.25 V bias voltage and 1 nA setpoint current. Quantum mechanical calculations suggest that the bright spots correspond to Pb atoms, arising mainly from Pb $6p_z$-like states. There is an apparent deformation in the periodicity of the bright spots closer to the step (see also Fig. 12). From Becker and Rosso (2001).

Sphalerite-type. The sphalerite (or "zinc blende") crystal structure of ZnS which is shared with ZnSe (stilleite), ZnTe and CdS (hawleyite) is also one of the classic structure types found in mineralogy, with both cation and anion in regular tetrahedral coordination. It is a structure adopted by a large class of synthetic semiconductor materials which have been of great industrial importance. For this reason, the surface structure and surface chemistry of "zinc blende" phases, in particular those of the (110) and (111) surfaces, have been the subject of much research. Indeed, it has been pointed out by Duke (1992) that these compounds comprise the most thoroughly studied homologous class of semiconductor surface structures, ideal for establishing correlations between bulk and surface properties and universal scaling laws.

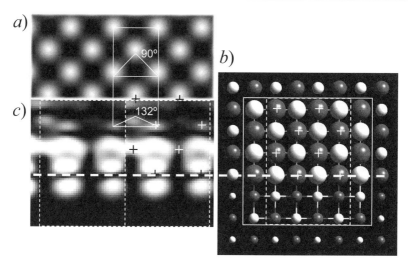

Figure 12. Theoretical STM images and structure model for a step on PbS (100) corresponding to the experimental conditions in Figure 11. (*a*) Calculated constant height STM image for a flat terrace using periodic boundary conditions, (*b*) structure model used for (*c*) the theoretical STM image of a step. The theoretical images in parts a) and c) are aligned such that the lattice is continuous. The calculations suggest the apparent deformation observed in Figure 11 arises mainly from a modified electronic structure of the surface due to the step, as opposed to a relaxation in the atomic structure of the step. From Becker and Rosso (2001).

The (110) surface of sphalerite is a perfect cleavage surface that has been the subject of detailed experimental investigations, and both qualitative arguments and quantitative calculations concerning chemical bonding. In particular, these topics have been addressed in the publications of Duke (1992, 1994) and La Femina (1991, 1992) with a mineralogically-oriented review given by Gibson and La Femina (1996). As can be seen in Figure 13, both Zn and S atoms are 3-fold coordinated at the (110) surface, with bonding to two neighboring atoms in the surface plane and one in the plane directly below the surface. This cleavage surface can achieve autocompensation by having its anion-derived surface states completely filled and cation-derived surface states completely empty (and hence the surface is charge neutral and stable). The simple electron counting arguments are that, in bulk ZnS, each Zn contributes two valence electrons to the four bonds (half electron per bond) and each S contributes six electrons (one and a half per bond) to give a stable "two electron bond." The surface atoms are 3-coordinate with one dangling bond; in the case of the Zn this dangling bond has "one half" electron and in the case of S it has "one and a half" electrons. By transferring the half electron from the Zn dangling bond to the S dangling bond, the latter (and hence anion-derived surface state) can be filled, and the former (cation-derived surface state) can be emptied. Taking this "bond valence" type of approach a stage further, it can be argued that the surface can lower its energy by "rehybridizing the dangling bond charge density," specifically for the surface cations from sp^3 to sp^2, and surface anions to p^3 placing them in a trigonal pyramidal conformation. This change in bonding (electronic structure) at the surface is then an argument for movement of the atom positions in the surface layers to accommodate the change, i.e., for relaxation. Being 3-coordinate at the surface, these atoms can only move in the direction towards or away from the plane of the surface without seriously distorting near-neighbor bond lengths. This leads to the prediction that surface cations move down towards the bulk crystal and surface anions up and away from the surface plane, as illustrated in Figure 14. The main parameter describing this relaxation, the perpendicular shear of the surface ($\Delta_{1,\perp}$) is shown.

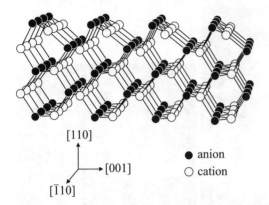

Figure 13. Model of the sphalerite-type (110) surface. [Adapted with permission of the American Vacuum Society from Duke (1992), *J. Vac. Soc. Trans. A*, Vol. 10, Fig. 1, p. 2033.]

Figure 14. Model of the relaxed sphalerite-type (110) surface. The second layer relaxation is exaggerated to render it more visible. The main structural parameter describing the surface relaxation is indicated. [Adapted with permission of the American Vacuum Society from Duke and Wang (1988), *J. Vac. Soc. Trans. B*, Vol. 6, Fig. 1, p. 1440.]

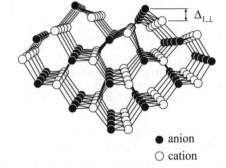

Studies of ZnS (110) using LEED (Lessor et al. 1993) support the predictions of the simple arguments made above, with a values for the parameter $\Delta_{1,\perp}$ in the range 0.55-0.60 Å being determined experimentally. Furthermore, studies of a much wider range of "zinc blende" structure materials confirm that this description of the surface relaxation is universally applicable, and that this principal surface structural parameter scales with the bulk unit cell parameter, as illustrated in Figure 15.

Given the importance of the ZnS (110) surface, it is not surprising that it has attracted the attention of those applying computer modeling techniques both to predict surface geometries and to understand the electronic structure of the surface. Atomistic simulation methods were used by Wright et al. (1998) to investigate surface energies and stabilities. Surfaces of form {110} were calculated to have lowest surface energy, with a marked reduction in energy (~60%) found on comparing unrelaxed with relaxed surface energies ($\gamma = 0.65$ J/m^2, see Table 1). Although the surface relaxation effects described above were predicted by the calculations, they were about half the values described by (Duke et al. 1984) for the topmost layer, although in close agreement for deeper layers. A particular aspect of this study concerned calculating the effects on surface stability of introducing point defects (and hence changing the stoichiometry of the surface). Thus, for example, for Zn-poor surface stoichiometries, the (111) surface is predicted to become the most stable. A later study by Hamad et al. (2002) used both atomistic and quantum mechanical (DFT) approaches to predicting surface structures and crystal morphologies of ZnS, again showing that the (110) surface is the most stable ($\gamma = 0.53$ J/m^2, Table 1). Again, the atomistic simulations yielded surface relaxation values significantly less that experiment, but those from DFT calculations were much closer to experiment.

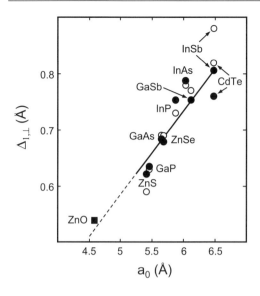

Figure 15. The scaling of the main structural parameter describing the sphalerite-type (110) surface relaxation as a function of the bulk lattice constant. Filled circles show predictions from theory and open circles are data from LEED intensity analyses. [Used with permission of American Physical Society from Skinner and LaFemina (1992), *Phys. Rev. B*, Vol. 45, Fig. 5, p. 3561.]

Among quantum mechanics based computational studies of the ZnS (110) surface, the work of Klepeis et al. (1993) can be mentioned who used a full-potential linear muffin-tin orbital (FPLMTO) method to calculate total energies of the (110) surfaces for the isoelectronic, isostructural series GaP, ZnS, CuCl. With increasing "ionicity" of the bonding across this series, deviation of the surface anion from its ideal bulk-terminated position decreases, although the relaxation phenomena observed are qualitatively the same as described above. Ferraz et al. (1994) used DFT calculations to study the electronic properties of the ZnS, ZnSe and ZnTe (110) surfaces. The relaxed surfaces show good overall agreement with experimental data and dependence of the degree of relaxation on the "ionicity." The energy dispersion of surface states on these zinc chalcogenide (110) surfaces is shown in Figure 16 with the projected bulk band structure shown as hatched regions and the bound surface states as solid lines. The later work of Tutuncu et al. (2000) is broadly in agreement with these results.

As noted above, the (110) surfaces of the isostructural ZnSe and ZnTe phases undergo what are qualitatively the same surface relaxation phenomena as ZnS, varying in degree with their "ionicities" and the same is true for CdS. This phase is also of interest because of the ways in which electronic properties can be manipulated when it occurs as small clusters (Cd_nS_m where $n + m$ lies in the range of 100-10000 atoms). Using cluster calculations based on a simplified LCAO-DFT-LDA scheme, Joswig et al. (2000) suggested that chemical bonds in the surface region are more ionic than in the bulk, and that whereas the HOMO is delocalized over major parts of the nanoparticle, the LUMO is a surface state.

Wurtzite-type. The wurtzite structure, named from the other mineralogically important ZnS polymorph, is another classic structure type found in a significant number of industrially important materials and which forms the basis of a major group of sulfide minerals. There are two cleavage surfaces, ($10\bar{1}0$) and ($11\bar{2}0$), both of which have been studied at the same level of detail and using many of the same techniques as described above for sphalerite-type ZnS. Both of these surfaces have equal numbers of cations and anions in each layer, therefore they are non-polar type I surfaces. In this discussion, we focus on the ($10\bar{1}0$) surface of both wurtzite itself and the isostructural CdS phase (greenockite).

Based upon LEED measurements and associated calculations, as summarized by Duke (1992), the arrangement of atoms at the ($10\bar{1}0$) surface of wurtzite (and the group of materials

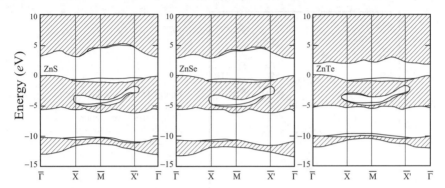

Figure 16. Energy dispersion of surface states on sphalerite-type ZnS, ZnSe, and ZnTe (110) surfaces as calculated using planewave pseudopotential DFT methods. Adapted from Ferraz et al. (1994).

sharing this structure) differs significantly from the (110) surface of sphalerite. The truncated bulk surface is made up from anion-cation dimers, each of which exhibits one bond in the plane of the surface and two into the bulk. The evidence is that on relaxation, these dimers tilt so that the anion is displaced out of the original plane of the surface and the cation into this plane (Fig. 17). These parameters apply to this surface in all wurtzite structure phases; actual values of the parameters derived from LEED intensity analyses were reviewed by Duke (1992) (Table 2). The driving force for this relaxation is similar to that for the sphalerite (110) surface, i.e., the lowering of surface energy that results from the redistribution of the dangling bond charge density.

As in the case of the sphalerite structure sulfide phases, these surfaces have been the subject of detailed studies by those using computational approaches, both to predict surface relaxation effects and to model surface electronic structure. Earlier semi-empirical calculations (Wang and Duke 1988) and later work using *ab initio* methods (Schroer et al. 1994; Rantala et al. 1996) all support the relaxation models of Duke and coworkers (Duke 1992), despite contrary studies (Tsai et al. 1987) that suggested weak surface relaxation in the opposite direction.

The calculations of Schroer et al. (1994) for CdS (10$\bar{1}$0) yielded surface electronic structure information in good agreement with photoemission experiments and showing two occupied surface states at the top of the projected bulk valence bands. The *ab initio* calculations of Pollmann et al. (1996) also led to proposed (anion-derived, *p*-character) dangling bond surface states which could be related to experimental data, although these authors also noted problems which can arise in correct treatment of the *d* electrons in these materials.

Only a limited amount of work has been undertaken in high resolution imaging of the surfaces of these materials. Siemens et al. (1997) reported atomically resolved STM images of CdS (10$\bar{1}$0) cleavage surfaces (and also of CdSe cleavage surfaces) which confirm a 1 × 1 surface unit mesh in agreement with the LEED results described above. As seen in Figure 18, it was possible to image at both negative and positive sample voltages, and hence probe both occupied and empty dangling bond states above anions and cations, respectively. In this and later work (Siemens et al. 1999a,b) it was also possible to use STM to observe features associated with impurity ("dopant") atoms in the CdS lattice; in this case these were indium atoms. As with the CdS nanoparticles noted above, synthetic nanoscale materials based on wurtzite are now being investigated for their potential industrial applications. One example is ZnS (wurtzite) nanoribbons of saw-tooth form produced by a solid-vapor deposition technique (Moore et al. 2004). Fast growth along the *a*-axis forms the body, and slower growth along the *c*-axis the teeth, with the proposed mechanism being self-catalyzed growth of the Zn terminated (0001) surface while the oxygen terminated (000$\bar{1}$) surface is relatively inert.

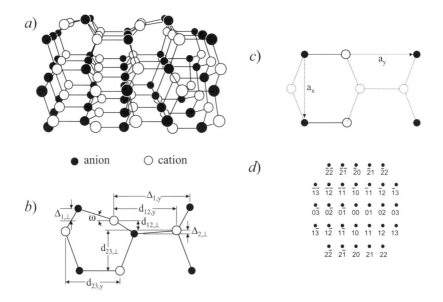

Figure 17. Aspects of wurtzite-type (10$\bar{1}$0) surface structure showing *a*), the relaxed configuration for wurtzite, *b*) side and *c*) top views of structure with independent structural variables defined, and *d*) the associated reciprocal lattice. [Adapted with permission of the American Vacuum Society, from Duke (1992), *J. Vac. Soc. A*, Vol. 10, Figs. 5 and 6a-c, p. 2036.]

Table 2. Structural parameters for the relaxation of wurtzite structure (10$\bar{1}$0) surfaces of ZnS and CdS derived from LEED intensity analyses, from Duke (1992). Distances are given in Å. See also Figure 17.

	a_x	a_y	$\Delta_{1,\perp}$	$\Delta_{1,y}$	$d_{12,y}$	$d_{12,\perp}$	$\Delta_{2,\perp}$	ω
ZnS	3.811	6.234	0.669	4.054	3.513	0.597	0.106	17.0°
CdS	4.135	6.749	0.738	4.410	3.862	0.622	0.081	17.5°

Pyrite-type. The pyrite structure type is closely related to the rocksalt structure type. Each metal atom is 6-fold coordinated to six S atoms in a slightly distorted octahedron, and each S atom is 4-fold coordinated to three metal atoms and one S atom in a distorted tetrahedron. The S-S bond is oriented along body diagonals of the cubic cell. The similarity with the rocksalt structure can be understood by considering that the lattice type is face-centered cubic and that the center of the disulfide group lies at positions equivalent to atomic positions in the rocksalt structure. The pyrite space group is Pa3. Several minerals possess the pyrite structure; here we review surfaces of pyrite (FeS$_2$), vaesite (NiS$_2$), and laurite (RuS$_2$). Major surface planes that occur either as growth or cleavage faces for this structure type are (100), (111), (110), and (210) surfaces, but the prevalence and quality of each depend on the specific mineral. It is noteworthy that despite the similarities in their structures, pyrite has poor (100) and (110) cleavage, while vaesite has good (100) cleavage and laurite has good (100) and (111) cleavage.

Pyrite fracture surfaces, even those where fracture propagation is guided along (100), appear conchoidal to the naked eye. However, STM and LEED data show that true (100) surfaces are present at micron and smaller length scales (Fig. 19) (Eggleston and Hochella

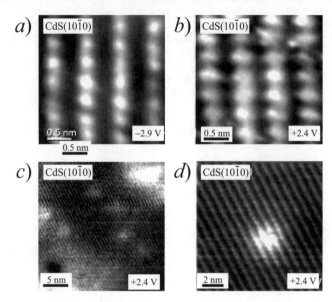

Figure 18. STM images collected at room temperature in ultra-high vacuum of the CdS (10$\bar{1}$0) surface showing *a*) occupied and *b*) empty states arising from dangling bonds on the anions and cations, respectively, and *c*) and *d*) the manifestation of positively charged dopant atoms residing near the surface in the empty state images. [Used with permission of American Physical Society from Siemens (1997), *Phys. Rev. B*, Vol. 56, Figs. 2a,b and 3, pp. 12322-12323.]

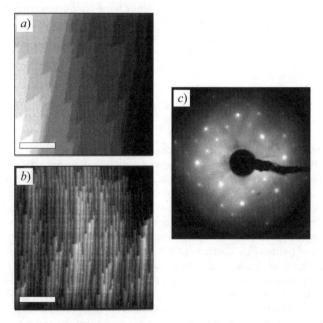

Figure 19. Structural aspects of FeS$_2$ pyrite surfaces produced by fracture in vacuum, showing *a*) the existence of large crystallographic (100) terraces, *b*) a vicinal surface dense with structural imperfections, and *c*) sharp (100) diffraction data assessed by LEED. Pyrite fracture surfaces consist of mixture surface topographic domains shown in *a*) and *b*). The scale bars represent 100 nm. From Rosso et al. (1999a).

1992; Rosso et al. 1999a). The atomic and electronic structure of pyrite (100) surfaces has now been experimentally and theoretically studied in great detail. The charge neutral (100) surface, and therefore the most stable surface (Table 1), is generated by cleavage of Fe-S bonds, leaving S_2 units intact (defects will be discussed later). Little structural relaxation occurs at this surface, as verified by LEED, STM, and computational molecular models of various kinds (Eggleston and Hochella 1990; Pettenkofer et al. 1991; Chaturvedi et al. 1996; Eggleston et al. 1996; Rosso et al. 1999a, 1999b, 2000; de Leeuw et al. 2000; Hung et al. 2002a; Stirling et al. 2003; Cai and Philpott 2004; Philpott et al. 2004; Qiu et al. 2004a,b; von Oertzen et al. 2005), making it close to an ideal termination of the bulk (Fig. 20). Recent planewave pseudopotential DFT calculations (Stirling et al. 2003) are consistent with previous findings that most of the relaxation occurs along the surface normal direction. An oscillatory expansion and contraction of atomic planes is predicted that is concentrated at the upper few atomic planes and rapidly damped within a few atomic layers of the surface (thicker slabs would be required to fully characterize the penetration). Owing to the reduced coordination at the surface, the uppermost S layer expands outward relative to a simple bulk truncation by 17% while the uppermost Fe layer is contracted by 32%. The uppermost Fe-S bond lying closest to the [001] direction shortens by ~0.1 Å. The uppermost S-S bond length is predicted to remain close to unchanged. These trends were confirmed by other workers using similar planewave pseudopotential DFT approaches (Andersson et al. 1994; Cai and Philpott 2004; Qiu et al. 2004a,b), and are largely consistent with localized orbital DFT predictions (Hung et al. 2002a).

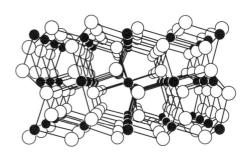

Figure 20. Model of the pyrite-type (100) surface structure. Filled circles are cations and empty circles are anions.

The electronic structure of pyrite (100) has been well studied, but not all issues have been resolved. The many surface spectroscopic studies performed capture a global average of the surface, including information from defects such as steps (which are prevalent on this surface), making characterization of the crystallographic (100) surface itself more difficult. Note also that many studies on the bulk electronic structure of pyrite are available and some of these have been used to estimate the surface electronic structure (Bither et al. 1968; Burns and Vaughan 1970; Li et al. 1974; Ogawa et al. 1974; Schlegel and Wachter 1976; Tossell 1977; van der Heide et al. 1980; Bullett 1982; Folkerts et al. 1987; Folmer et al. 1988; Ferrer et al. 1990; Huang et al. 1993; Zeng and Holzwarth 1994; Mosselmans et al. 1995; Bocquet et al. 1996; Charnock et al. 1996; Fujimori et al. 1996; Raybaud et al. 1997; Eyert et al. 1998; Rosso et al. 1999a; Gerson and Bredow 2000; Rosso 2001; Edelbro et al. 2003; Nesbitt et al. 2004; von Oertzen et al. 2005). Here, we will emphasize direct treatments of the crystallographic surface itself, which include *ab initio* approaches and experimental methods that probe the (100) surface locally rather than the global fracture surface. For details of the bulk electronic structure, see chapter in this volume by Vaughan and Rosso (2006).

The character and localization of the highest occupied and lowest unoccupied states at the surface is a central issue for pyrite (100). In the bulk, the valence band maximum consists primarily of non-bonding Fe $3d$ t_{2g} states, lying energetically above a bonding S $3p$-Fe $3d$ e_g band. The bulk conduction band minimum is thought to consist of a mixture of Fe $3d$ (antibonding e_g states) and S $3p$ states by some workers or exclusively S $3p$ states by others. Using ligand field theory, Bronold et al. (1994) first suggested that surface states of Fe $3d$ character should be localized on the five-fold coordinate Fe atoms at the (100) surface. This suggestion was first confirmed using a combination of dual-bias STM, atomically-resolved tunneling spectroscopy, and local projected densities of states computed at the DFT level of theory (Fig. 21) (Rosso et al. 1999a). That study showed that the highest occupied and lowest unoccupied states at the surface are localized at Fe atoms. The former arise from Fe $3d$ orbitals with an angular momentum component along the surface normal direction, the latter arise from a mixture of Fe $3d$ and S $3p$ orbitals again with an angular momentum component along the surface normal direction. That study also showed direct experimental evidence that the band gap at the uppermost surface is very small, less than 0.04 eV, indicating that while pyrite is a 0.9 eV gap intrinsic semiconductor (at 300 K), the pristine (100) surface is nearly metallic due to the presence of surface states. Local orbital DFT calculations performed by Hung et al. (2002a) are in excellent agreement with all these findings. Planewave DFT calculations performed by Stirling et al. (2003) disagree only on the orbital character of the lowest unoccupied surface state. Stirling et al. (2003), Cai and Philpott (2004), and Qiu et al. (2004a) all differ on the surface gap width, but DFT is incapable of yielding reliable band gaps because it is a ground-state theory and band gaps are an excited state property. Empirical corrections to DFT, such as by including a prescribed amount of exact exchange (Stadele et al. 1999) or by including the self-interaction correction (Perdew and Zunger 1981), are leading to more accurate estimates of band gaps. However, no "improved" DFT methods have been applied to pyrite for this purpose so far.

If we now include the body of work performed on large areas of the pyrite surface, produced either by fracture or ultra-high vacuum (UHV) sputter/annealing techniques, we necessarily incorporate information arising from defects such as steps, kinks, and vacancies where coordination numbers are even lower. These surfaces will often give a (100) diffraction signature but defect densities are likely to vary widely from sample to sample with some being extremely high. Conventional and synchrotron-based XPS and also SCLS studies have shown that monosulfide surface species are present on clean fractured pyrite (Fig. 22) (Bronold et al. 1994; Schaufuss et al. 1998; Nesbitt et al. 2000; Leiro et al. 2003; Kendelewicz et al. 2004). These are consistent with step edges and corners (Rosso

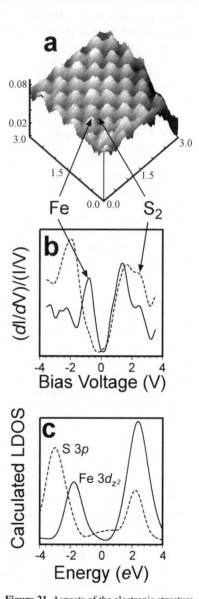

Figure 21. Aspects of the electronic structure of pristine pyrite (100) surfaces at the atomic level, showing *a)* an occupied state STM image (−0.2 V, 2 nA) collected in ultra-high vacuum at room temperature (scale in nm), *b)* experimental atom-resolved tunneling spectra collected at room temperature revealing a surface state localized on surface Fe sites, and *c)* tunneling spectra calculated from first principles showing the surface state to arise from Fe $3d_{z^2}$-like dangling bond orbitals. From Rosso et al. (1999a).

et al. 1999a, 2000; Nesbitt et al. 2000; Rosso 2001; Hung et al. 2003; Leiro et al. 2003). Steps aligned along a variety of directions examined with STM at the atomic scale appear unreconstructed (Fig. 23) (Eggleston and Hochella 1992; Rosso et al. 1999a, 2000). Hence, the presence of broken S-S bonds is to be expected at the very least to maintain charge neutrality at step edges running along directions other than the cubic ones.

However, sulfur vacancies in (100) terraces are also likely. UHV-preparation techniques such as ion-sputtering or annealing readily create S-deficient surfaces (Chaturvedi et al. 1996; Guevremont et al. 1997; Andersson et al. 2004) which can be restored to FeS_2 stoichiometry by, for example, annealing in the presence of an S_2 background gas (Andersson et al. 2004). XPS analysis suggests a presence of Fe^{3+} at the clean pyrite fracture surface based on the high-binding energy tail of the Fe $2p_{3/2}$ spectrum (Nesbitt et al. 1998). It has been hypothesized that the presence of Fe^{3+} is linked with S vacancies in terraces arising from broken S-S bonds through a $Fe^{2+} + S^- \rightarrow Fe^{3+} + S^{2-}$ autooxidation process upon fracture (Nesbitt et al. 2000; Harmer and Nesbitt 2004). The model accounts for an apparent one-to-one ratio of S^{2-} and Fe^{3+} surface constituents from XPS peak-fitting analysis, with surface concentrations thought to be too high to be due to monosulfide at step edges alone (Nesbitt et al. 2000). Much work remains to establish whether monosulfide species on fracture-generated (100) surfaces arise more from steps/kinks or S vacancies on terraces, with microscopic characterization of defect densities being particularly key. In this regard, it is particularly noteworthy that Kendelewicz et al. (2004) report that the percentage of monosulfide defects on fracture surfaces varies from fracture to fracture. Therefore, comparisons from study to study must be done with caution.

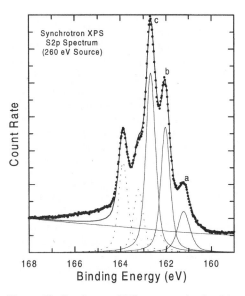

Figure 22. Synchrotron XPS spectrum in the S $2p$ region for pyrite fractured in vacuum collected at ~15 K with the source tuned to 260 eV. Peak "a" is assigned to surface S^{2-} species, peak "b" is assigned to the uppermost S atom of surface S_2^{2-} dimers, and peak "c" is assigned to S atoms of bulk S_2^{2-} dimers. From Nesbitt et al. (2000).

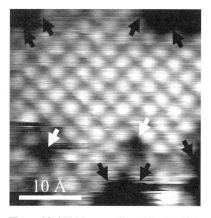

Figure 23. STM image collected in ultra-high vacuum of the pyrite (100) surface at room temperature showing the presence of various surface defects (−0.2 V, 2 nA). Black arrows indicate the location of half-unit cell high steps (stepping down into the dark parts of the image), and white arrows indicate locations of Fe atom vacancies. From Rosso et al. (2000).

The magnetic structure of the (100) surface has been examined from several perspectives. The ligand field analysis of Bronold et al. (1994) suggested the possibility of unpaired d-electrons at 5-fold coordinated Fe^{2+} ions at the pyrite (100) surface. Evidence for spin-polarized Fe^{2+} on fractured pyrite has been obtained from multiplet peak-fitting of Fe $2p_{3/2}$ XPS spectra (Nesbitt et al. 2000). In contrast, all quantum mechanical calculations of the (100) surface electronic structure agree that the (100) surface remains diamagnetic, i.e., the Fe^{2+} ions at the surface remain in a $t_{2g}^6 e_g^0$ $3d^6$ electronic configuration (Rosso et al. 1999a; Hung et al. 2002a; Stirling et al. 2003; Cai and Philpott 2004; Qiu et al. 2004a,b). But the DFT study of Hung et al. (2002a) also showed that pyrite surface sites possessing 4-fold coordinate Fe^{2+} should be spin-polarized. Such sites could be at step edges, but also Fe sites adjacent to S vacancies. Thus it appears plausible that if significant concentrations of S vacancies are present, parts of the surface could be paramagnetic, consistent with the XPS observations.

Comparatively less work by far has been performed on NiS_2 (vaesite) and RuS_2 (laurite) (100) surfaces. The electronic structure of NiS_2 differs from pyrite mainly due to the additional $3d$ electrons partially filling the antibonding e_g states with unpaired spins, but the orbital character of the major bands remains the same (Burns and Vaughan 1970). NiS_2 is an antiferromagnetic semiconductor with a band gap of ~0.3 eV at 300 K (Kautz et al. 1972). As previously mentioned, it shows very good (100) cleavage, in contrast to pyrite, and thus the (100) surface is amenable to clean preparation by cleavage in UHV for study with STM. So far, STM has been performed only on $NiS_{2-x}Se_x$ solid solution phases with x as low as 0.45 (Fig. 24) (Hanaguri et al. 2004; Iwaya et al. 2004). As for pyrite, the NiS_2 (100) surface is autocompensated and does not reconstruct. Interestingly, low temperature STM images clearly show the anion sublattice at positive sample bias, suggesting the lowest unoccupied states at the surface are localized on the anions. Conductivity measurements on cleaved NiS_2 (100) surfaces show evidence for metallic conductivity at the surface at 300 K while the bulk remains semiconducting (Thio and Bennett 1994; Sarma et al. 2003), similar to STM-based observations for pyrite (100) (Rosso et al. 1999a).

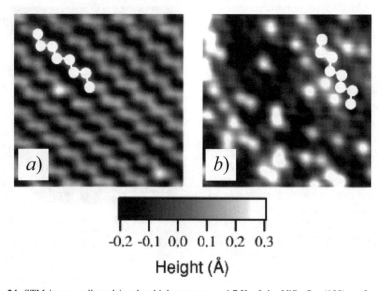

Figure 24. STM image collected in ultra-high vacuum at 4.7 K of the $NiS_{2-x}Se_x$ (100) surface (0.5 V, 0.1 nA), where a) $x = 2.0$, and b) $x = 0.5$. The indicated zig-zag structure is assigned to the uppermost anion plane. The images are 47 Å on each side. [Used with permission of American Physical Society from Iwaya et al. (2004), *Phys. Rev. B*, Vol. 70, Fig. 2a,b, p. 161103-2.]

Ruthenium in RuS_2 is isoelectronic with iron in pyrite, and the band structure of RuS_2 is accordingly very similar to that for pyrite (Holzwarth et al. 1985; Huang and Chen 1988; Frechard and Sautet 1995; Raybaud et al. 1997; Smelyansky et al. 1998). Thus, it is not surprising to find that RuS_2 is a diamagnetic intrinsic semiconductor with a moderate band gap (~1.4 eV) similar to pyrite. Laurite surfaces have been studied in detail to understand their highly efficient hydrodesulfurization (HDS) catalytic properties (Frechard and Sautet 1997); no other pyrite-type metal sulfides have surfaces that are stable under HDS catalytic conditions. The RuS_2 (100) surface is easily generated by cleavage, displaying large (100) terraces hundreds of nanometers across (Colell et al. 1994). LEED and STM imaging at the atomic scale demonstrate that this surface does not reconstruct (Colell et al. 1994; SchaarGabriel et al. 1996; Kelty et al. 1999). HF and DFT slab calculations show that structural relaxation is minimal (Frechard and Sautet 1995; Smelyansky et al. 1998) as is the case for pyrite (100). But in contrast to pyrite, RuS_2 (100) is extremely stable in air, even for periods of many months (Colell et al. 1994; Kelty et al. 1999). STM images in collected in air suggest that both the Ru and S_2 sublattices appear as high tunneling current sites at negative sample bias voltages, suggesting that the highest occupied states at the surface arise from a mixture of Ru $4d$ and S $3p$ states (Colell et al. 1994). This is consistent with periodic HF calculations of the projected densities of states at the surface (Fig. 25) (Frechard and Sautet 1995). Similar STM observations were made by Kelty et al. (1999) over a wider range of bias voltages (Fig. 26), but a more complex partially-oxidized surface structure was suggested to explain the observed periodicity at the surface. Kelty et al. (1999) also show that the band gap at the surface is narrower than in the bulk using tunneling spectroscopy, similar to observations mentioned above for both FeS_2 and NiS_2 (100).

Despite the fact that the (111) surface is a stable, prominent growth face of pyrite, it is not a plane of weakness in the structure that can be isolated by cleavage. This led Guevremont et al. (1998) to develop a low-energy sputter/anneal strategy applied to natural (111) growth faces in UHV for producing clean, stoichiometric (111) surfaces for analysis. Growth surfaces of pyrite, identical to those prepared in this way, have been shown by AFM to be extremely topographically complex (Hochella et al. 1998). While the UHV-cleaning method allowed for surface chemical analysis of pyrite (111) (Guevremont et al. 1998; Elsetinow et al. 2000) no experimental work to our knowledge has directly probed its atomic structure.

Several theoretical discussions of the structure of pyrite (111) are available (de Leeuw et al. 2000; Rosso 2001; Hung et al. 2002b). A central issue is that there is only one stoichiometric 1×1 (111) truncation of pyrite that is charge neutral; all other 1×1 terminations along (111) are polar surfaces (of which there are eight possibilities) (Fig. 27). A large electrostatic driving force is present for the polar surfaces to reconstruct. The polar terminations could be stabilized by introducing vacancies, possibly leading to a superlattice detectable by LEED or STM. The non-polar surface is formed by a planar cut through the midpoint of S-S bonds along the (111) plane. Despite the charge neutrality, geometric relaxation of this surface is more substantial than for pyrite (100), and spin polarization of surface Fe ions due to spin transfer to monosulfide species is predicted (Hung et al. 2002b). Occupied S $3p$ states associated with surface mono- and disulfide groups are displaced above the top of the bulk valence band; surface states are localized at S atoms at the (111) surface in contrast to the (100) surface. The surface energy for the non-polar (111) termination calculated at the DFT level is in the range 1.4-1.6 J/m^2 (Rosso 2001; Hung et al. 2002b), significantly higher than 1.1 J/m^2 calculated on average for pyrite (100) (Table 1). This is consistent with the relative numbers of broken bonds compared over equivalent areas on the two surfaces, which is 14 Fe-S bonds per nm^2 on (100), whereas for (111) it is 10 Fe-S and 7 S-S bonds per nm^2. Some possible polar (111) surfaces of pyrite were modeled by de Leeuw et al. (2000) based on empirical potentials. Half Fe-terminated and half S_2-terminated vacancy structures for (111), built to achieve charge neutrality and to maintain all S-S bonds, yielded relaxed surface energies more than twice as large as that for the relaxed non-polar 1×1 surface (Table 1).

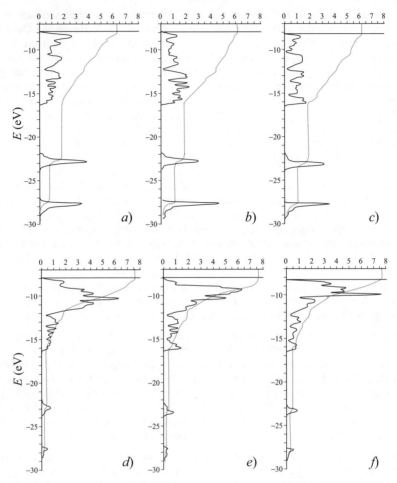

Figure 25. Densities of states (solid curves) and integrated densities of states (dotted curves) for RuS_2 (100) calculated at the Hartree-Fock level of theory projected onto *a)* an upper S atom of the surface S_2 dimer in a slab model, *b)* an internal S atom in the slab model, *c)* an S atom in a bulk model, *d)* an upper Ru atom in a slab model, *e)* an internal Ru atom in a slab model, and *f)* a Ru atom in a bulk model. [Used with permission of Elsevier, from Frechard and Sautet (1995), *Surface Science*, Vol. 336, Fig. 8, p. 158.]

The same structural considerations apply to the RuS_2 (111) surface, which has been studied only by theoretical calculations so far (Frechard and Sautet 1995, 1997; Tan and Harris 1998; Grillo et al. 1999; Grillo and Sautet 2000; Aray et al. 2002). Tan and Harris (1998) model a non-stoichiometric metal termination of RuS_2 (111) using an asymmetric slab (the upper surface does not match the lower surface so as to maintain RuS_2 stoichiometry for the whole slab) and the Fenske-Hall tight-binding band structure approach. Using local orbital periodic HF slab calculations, Frechard and Sautet (1995) show that the charge neutral 1 × 1 termination of RuS_2 (111) has a surface energy three times higher than that for RuS_2 (100) (Table 1). This termination undergoes substantial geometric relaxation as well as electronic relaxation (Frechard and Sautet 1995). Importantly, it was found that the relaxed charge neutral termination is not the lowest energy termination, rather a non-stoichiometric S-rich termination is found instead (Frechard and Sautet 1997). Surface states of S 3*p* character

Sulfide Mineral Surfaces 531

Figure 26. Atomically-resolved images of the RuS$_2$ (100) surface collected under a dry N$_2$ atmosphere at room temperature, where *a*) is an AFM image (~44 Å on each side) showing a square surface unit cell consistent with the size of the bulk unit cell, and *b*) is an STM image (64 Å on each side; −0.7 V, 3.0 nA) showing a similar surface unit cell. [Reprinted with permission from Kelty et al. (1999). © American Chemical Society.]

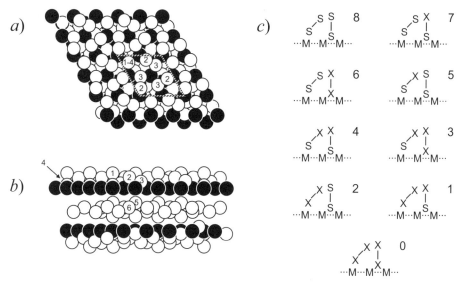

Figure 27. Nine possible (111) 1 × 1 terminations of the pyrite structure type, showing *a*) top-view model of an arrangement of cations (filled circles) and anions (empty circles) with various anion site types labeled, *b*) side-view model showing various possible (111) atomic planes, and *c*) a schematic of the possible terminations with varying stoichiometries. The termination labeled "4" is the charge neutral, stoichiometric termination. [Adapted with permission of Elsevier, from Frechard and Sautet (1997), *J. Catalysis*, Vol. 170, Fig. 2, p. 404.]

associated with surface monosulfide sites are displaced into the bulk band gap (Frechard and Sautet 1995, 1997), consistent with DFT predictions for the equivalent pyrite (111) termination (Rosso 2001; Hung et al. 2002b). These observations suggest that that the initial surface chemical behavior, particular with respect to mechanisms of surface oxidation, could be markedly different for (100) and (111) surfaces of pyrite-type metal sulfide minerals.

A charge neutral stoichiometric (110) termination of pyrite can be generated by a planar cut of Fe-S bonds, exposing 4-fold coordinated Fe (Fig. 28). Using an empirical potential model, de Leeuw et al. (2000) showed that this surface is stable, but significantly less so than pyrite (100) (Table 1). That study also evaluated the stability of an alternative (110) termination derived by "faceting" the surface, exposing alternating (100) planes (Fig. 28). This creates linear channels that are lined with 4-fold coordinate Fe along the ridges, but exposing only 5-fold coordinated Fe everywhere else. By increasing the average coordination of Fe at the surface, faceting effectively decreasing the Fe dangling bond density relative to the planar (110) termination. Both possible (110) terminations show little geometric relaxation; the largest displacements involve reducing the distances of Fe-S bonds opposite missing S ligands at the surface by ~0.1 Å, similar to the behavior of pyrite (100) (de Leeuw et al. 2000). The relaxed faceted (110) surface has a surface energy slightly lower than the planar (110) surface (Table 1). Hung et al. (2002a) calculated the surface energies and the electronic structure for the two (110) terminations described by de Leeuw et al. (2000) at the DFT level of theory and found the same trends in the magnitude of geometric relaxation and relative surface energies (Table 1). Importantly, they also showed that while 5-fold coordinate Fe remains diamagnetic with a low-spin filling of the t_{2g} 3d orbitals, 4-fold coordinate Fe does not. Thus uppermost Fe ions at the planar (110) surface and those along the ridges of faceted (110) surface are spin-polarized. In both cases, surface states of Fe 3d character are predicted to span the bulk band gap, imparting a metallic characteristic to the (110) surface (Hung et al. 2002a).

According to the DFT studies of Hung et al. (2002a,b), the relative surface energies for pyrite follow the order: "faceted (110)" > (210) > (111) > (100). The (210) surface is therefore one of the least stable for pyrite, having nearly the same surface energy as for the faceted (110) surface (Table 1). The charge-neutral stoichiometric (210) pyrite surface is terminated with alternating rows of 4- and 5-fold coordinated Fe atoms, as well as 2- and 3-fold coordinated S atoms (Fig. 29). These uppermost S atoms, which are 2-fold coordinated, undergo displacements of several tens of Ångstroms laterally, along with displacement downward into the surface, relative to bulk positions (Hung et al. 2002b). Both the 4- and 5-fold coordinated Fe atoms contract into the surface, reducing the Fe-S distance underneath by ~0.1 Å, consistent with the behavior of the other pyrite surfaces mentioned above. The 4-fold coordinated Fe atoms are predicted to be spin-polarized, with very minor spin-polarization of the 5-fold coordinated Fe atoms. The magnetic moments of the former are expected to be an order of magnitude larger

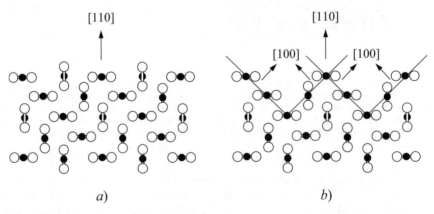

Figure 28. Possible pyrite-type (110) surfaces, showing *a*) a planar termination, and *b*) a micro-facetted termination expressing (100) surfaces. Cations are filled circles and anions are empty circles. [Adapted with permission from de Leeuw et al. (2000). © American Chemical Society.]

(3.2 μ_b) than that of the latter, which are very small (0.1 μ_b). Fe $3d$ states arising from the undercoordinated Fe atoms are predicted to dominate around E_F, forming the highest occupied and lowest unoccupied states at the surface. With regards to the magnetic and electronic structure, the (210) surface shows characteristics very similar to both the faceted (110) and the (100) surfaces of pyrite (Hung et al. 2002a,b).

The (210) surface of RuS_2 shows similar structural and electronic trends as for pyrite (210). Fenske-Hall tight-binding band structure calculations were performed by Tan and Harris (1998) on RuS_2 (210) with the identical atomic topology as the pyrite (210) topology used in Hung et al. (2002b). As was shown for Fe in pyrite (210), 4- and 5-fold coordinated Ru atoms introduce states in the bulk band gap that dominate the highest occupied and lowest unoccupied states near E_F (Tan and Harris 1998).

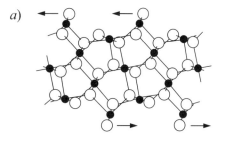

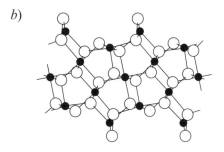

Figure 29. Slab models of the pyrite (210) surface showing the *a*) unrelaxed configuration, and *b*) the relaxed configuration calculated using DFT. Fe atoms are filled circles and S atoms are empty circles. [Adapted with permission of Elsevier, from Hung et al. (2002b), *Surface Science*, Vol. 520, Fig. p. 115.]

Pyrrhotite. As will be evident from other Chapters in this volume (Makovicky 2006; Wincott and Vaughan 2006), pyrrhotite is actually a family of mineral phases with structures based upon the hexagonal NiAs-type structure, and with complex superstructures arising from the ordering of iron atom vacancies in layers parallel to the basal plane. At what is probably the maximum iron deficiency (in compounds of general formula $Fe_{1-x}S$) is the endmember with a superstructure where the vacancies occur on every other iron atom layer and in alternate rows in that layer. As this ordering produces a monoclinic distortion of the structure, this is "monoclinic" pyrrhotite and has a formula Fe_7S_8. The phase of composition FeS also has a superstructure based on the NiAs-type and has its own mineral name, troilite. Intermediate compositions between FeS and Fe_7S_8 exhibit a variety of complex superstructures, although vacancy ordering leads to clustering of phases around compositions such as Fe_9S_{10} and $Fe_{10}S_{11}$ and to structures which are mostly hexagonal (hence the "hexagonal pyrrhotites"). Given the complexity of these materials, it is not surprising that there have been few detailed studies of surface structure and surface chemistry.

Becker et al. (1997c) employed a combination of STM, LEED and quantum mechanical calculations to study the atomic and electronic structure of the (001) surface of a (natural) monoclinic pyrrhotite. A phase transition was observed as a function of temperature in the LEED pattern; above ~300 °C, the pattern showed the periodicity of hexagonal closest-packed S atoms, below that temperature additional spots appear reflecting ordering of Fe vacancies. This parallels the behavior of the bulk material. STM images were obtained at atomic resolution level (Fig. 30) and these showed flat terraces separated by steps with heights of 2.9 Å (or integer multiples thereof). Features on the steps appear as triangles of bright spots with orientations in opposite directions on adjacent layers at 2.9 Å separation; a separation which corresponds to one quarter of the Fe_7S_8 unit cell dimension in the *c*-direction, or the separation between two consecutive Fe or S layers. Although the STM images of single terraces appear to have patterns

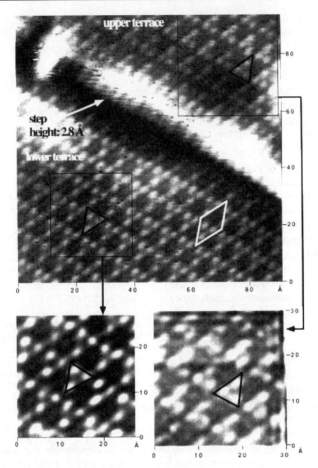

Figure 30. STM images of the (001) surface of monoclinic pyrrhotite collected in ultra-high vacuum at room temperature (−3.5 V bias voltage). The step height is 2.9 Å or one quarter of the 4C Fe_7S_8 unit cell along [001]. Bright spots are assigned to S atoms based on HF electronic structure calculations. Repeating "triangles" assigned to groups of three S atoms alternate directions of orientation on neighboring layers across the 2.9 Å step. [Used with permission of Elsevier, from Becker et al. (1997c), *Surface Science*, Vol. 389, Fig. 4, p. 73.].

reflecting the ordering of vacancies in the Fe atom layers, the bright spots in the images are almost certainly S atoms. The reversed orientation of surface geometry observed is only seen in the S atom layers, and quantum mechanical calculations of the STM images show S $3p$-like states at the top of the valence band. This reversed geometry is seen in the model shown in Figure 31; the vacant surface S sites can be thought of as associated with iron atom vacancies in the (virtual) layer above, hence the mimicking of the vacancy ordering pattern. This is because the S atoms are not equivalent and one S atom per surface unit is drawn into a Fe atom vacancy and, therefore, removed during truncation. Calculated STM images agree well with experimental data and support these interpretations. Further observations from the work of Becker et al. (1997c) include STM images with dislocations and with "etch pits" formed by successive removal of Fe_3S_3 units from the surface. The latter can range from atomic scale to 100 Å across.

In a very different study of monoclinic (Fe_7S_8) and hexagonal ($Fe_{10}S_{11}$) pyrrhotites, using both SXPS and conventional XPS techniques, Nesbitt et al. (2001) probed the nature

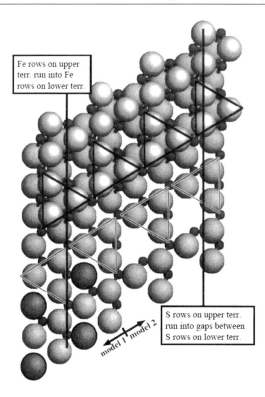

Figure 31. Model for a 2.9 Å step on monoclinic pyrrhotite (001), with possible "triangles" identified corresponding to those observed in experimental STM images (see Fig. 30). [Used with permission of Elsevier, from Becker et al. (1997c), *Surface Science*, Vol. 389, Fig. 7, p. 76.].

of species present at the surface after cleaving in vacuum. The pyrrhotite was fractured "subparallel" to (001), and the spectroscopic data were collected for the S $2p$ peaks. For both pyrrhotite types, the XPS data for the S $2p$ peaks obtained using conventional measurements (employing a monochromatized Al-K$_\alpha$ source) are dominated by one spin-orbit-split doublet but require a second, higher binding energy doublet, to obtain a good fit. This second contribution represents about 20% of the total S signal for both pyrrhotite types. These data can be interpreted in terms of the S atom environments in the bulk structures of these vacancy-containing materials, where the majority of S is in 5-fold coordination (80-85%) and the rest in 6-fold coordination. However, using the synchrotron XPS technique (with the photon source tuned to an energy of 210 eV), a much more surface sensitive spectrum was obtained (Fig. 32). Here, it is estimated that 40-50% of the S $2p$ photoemissions come from the first atomic layer, compared with only 5-15% in the conventional spectrum. Fitting of the SXPS data again indicated the predominance of two doublets, but in this case comparison with the conventional measurements suggested that the second doublet (~40% total intensity) arises from 5-fold coordinate monomeric surface S species. A third feature fitted to the SXPS spectra and contributing ~7% of the intensity is described as of uncertain origin, possibly being derived from the 6-coordinate atoms in the bulk or from surface dimmers (or both).

One other study of the pyrrhotite surface should be mentioned here, and that is the work of Mikhlin et al. (1998) using a range of spectroscopies (XPS, UPS and X-ray emission and absorption) and arguments based upon cluster molecular orbital calculations. Although the emphasis in this study of natural hexagonal pyrrhotite is on the effects of acid leaching on the surface, there are some interesting observations made concerning changes in the electronic structure of the surface following treatment with 1 M HCl. The spectroscopic evidence is of formation of an Fe-deficient layer at the surface with a S:Fe ratio of up to ~5, the presence of

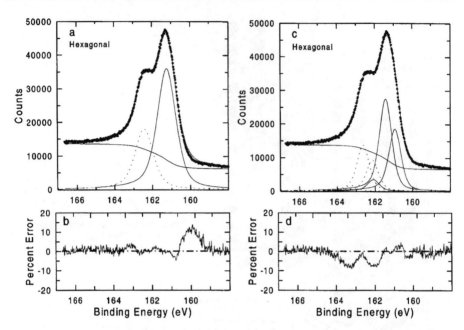

Figure 32. Synchrotron XPS data on hexagonal pyrrhotite fractured in vacuum at room temperature for the S $2p$ region with the beam energy tuned to 210 eV, showing *a*) a spectrum and fit and *b*) residual using only one S $2p$ spin orbit split contribution, and *c*) a spectrum and fit and *d*) residual using two S $2p$ spin orbit split contributions. From Nesbitt et al. (2001).

S-S bonds and a change in the Fe spin state from quintet to singlet (i.e., transition to a low spin state). Readers are referred to the original paper for details of the spectroscopic evidence, but in Figure 33 is shown a summary energy band diagram which contrasts a (one electron model) energy band structure for the original pyrrhotite with that proposed for the Fe-deficient layer.

Covellite. Covellite (CuS) has a layered hexagonal structure in space group $P6_3/mmc$ that cleaves readily along the basal plane to expose large, atomically-flat (0001) surfaces. Each sheet-like unit of covellite consists of three layers of hexagonal close packed S atoms with Cu disposed in the tetrahedral and trigonal interstices (Fig. 34) (Evans and Konnert 1976). The three-layer units are bound together by short S-S bonds (2.07 Å). The atomic structure of the basal surface was studied in detail by Rosso and Hochella (1999). The key difficulties in the determination of the surface atomic structure is that the location of the main plane of weakness perpendicular to the *c*-axis is not obvious, and also that cleavage of Cu-S bonds or cleavage of S-S bonds would lead to very similar surface atomic structures. Bond densities for the two bond types along basal planes are identical. Rosso and Hochella (1999) argued on the basis of an 86% higher electron density in S-S bonds relative to the Cu-S bonds along the *c*-axis that cleaving the Cu-S bonds should require less energy and is therefore more likely. *Ab initio* calculations specifically of the surface atomic structure were not performed at that time due to prohibitive computational expense. With the advent of efficient planewave pseudopotential DFT calculations, progress in this area is possible, and would be valuable for addressing this issue.

Atomic-scale STM images of clean (0001) surfaces in ultra-high vacuum showed a hexagonal periodic array of high tunneling current sites at both negative (CuS valence band) and positive (CuS conduction band) sample bias voltage (Fig. 35) (Rosso and Hochella 1999). The lateral topologies of the surface Cu and S sublattices are identical, irrespective of whether the cleavage plane is through Cu-S or S-S bonds. Bulk electronic structure calculations suggest that

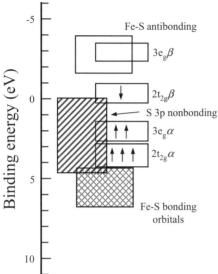

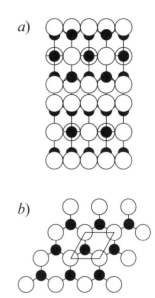

Figure 33. Energy band diagram for hexagonal pyrrhotite, based on X-ray and ultraviolet photoelectron spectroscopy and X-ray emission spectroscopy. [Adapted with permission of Elsevier, from Mkihlin et al. (1998), *Applied Surface Science*, Vol. 125, Fig. 8, p. 81.].

Figure 34. The structure of covellite viewed a) along the c-axis, and b) perpendicular to the c-axis. Cu atoms are filled circles and S atoms are empty circles. [Adapted with permission of Elsevier from Rosso and Hochella (1999), *Surface Science*, Vol 423, Fig. 1, p. 366.]

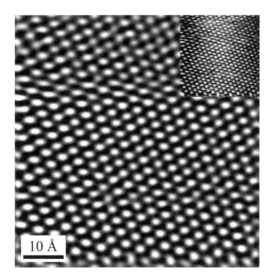

Figure 35. Filtered STM image collected in ultra-high vacuum on the CuS (0001) surface (0.2 V, 1 nA). The inset shows the raw image data. The bright spots are assigned to surface S atoms based on HF electronic structure calculations. [Adapted with permission of Elsevier from Rosso and Hochella (1999), *Surface Science*, Vol 423, Fig. 4, p. 370.]

high-tunneling current sites at negative bias voltage should correspond to surface S atoms due to the dominance of S $3p$ states at the top of the valence band. However the absence of Cu dangling bond surface states could not be confirmed, which is a possible basis for high-tunneling current sites arising from surface Cu atoms. The hexagonal symmetry and lattice constant captured using the STM agreed well with LEED data. The measured surface cell constant of 3.79 Å compared very well with the bulk cell edge of 3.794 Å (Evans and Konnert 1976).

Tunneling spectroscopy was used in an attempt to probe the local densities of states of empty and filled states at various locations in the surface unit cell. In this measurement, the tip position is temporarily fixed over the surface while the bias voltage (V) is ramped and corresponding changes in the tunneling current (I) are recorded. The most pronounced feature was a strong asymmetry in the I(V) curves, where the current rises more rapidly through the negative sample bias part of the ramp relative to that for the positive bias part (Fig. 36). The asymmetry could be explained by tip-induced upward band bending at the covellite (0001) surface. The upward bent

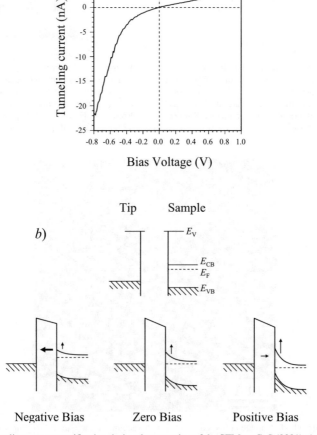

Figure 36. Tunneling current rectification during the operation of the STM on CuS (0001) *a*) as observed by performing current-voltage tunneling spectroscopy, and *b*) as explained by tip-induced formation of a tunneling barrier (Schottky barrier) for positive sample bias voltage. In the schematic diagram, the size and direction of horizontal arrows indicates magnitude and direction of the tunneling current, respectively. [Adapted with permission of Elsevier from Rosso and Hochella (1999), *Surface Science*, Vol. 423, Fig. 7, p. 366.]

bands constitutes a Schottky barrier that impedes tunneling current into the conduction band, leading to the rectifying characteristic observed in the I(V) spectra. The TS measurements also suggested that this surface is metallic, consistent with its bulk metallic character asserted by some (Liang and Whangbo 1993). Covellite has a very low resistivity ($\sim 10^{-7}$ $\Omega \cdot$m), and, after a subtle phase transformation at ~55 K, becomes a superconductor at 1.7 K.

Molybdenite. Molybdenite (MoS_2) has a layered hexagonal structure with space group $P6_3/mmc$ based on the stacking of two-dimensional sheets of MoS_2 units (Fig. 37). Each sheet is comprised of upper and lower close-packed layers of S atoms, with Mo atoms filling the octahedral sites in-between. The 6-fold coordinated Mo sites are trigonal prismatic. S atoms are bonded to three Mo atoms in a trigonal pyramidal fashion. The bond strengths within a layer are covalent and therefore much stronger than the weak van der Waals attraction joining two MoS_2 units together, giving rise to perfect (0001) cleavage. Closest S-S distances are within a sheet, with each S atom surrounded by one S atom along the c-axis at ~3.0 Å and six others in the same (0001) plane each about 3.1 Å away. S-S distances across sheets are ~3.7 Å. Polytypes $2H$ (hexagonal) or $3R$ (rhombohedral) arise from whether or not a two or three vector offset combination, respectively, is found in the stacking sequence of MoS_2 sheets. Edge surfaces of individual MoS_2 sheets follow three directions corresponding to the square

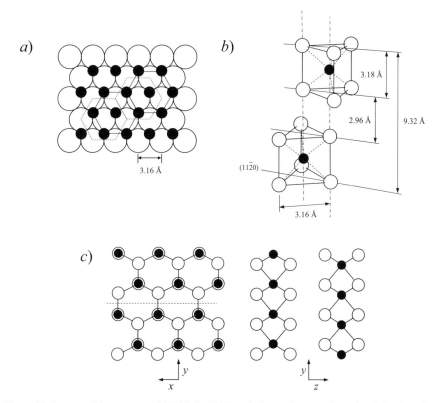

Figure 37. Aspects of the structure of $2H$-MoS_2 (0001) and edge surfaces, as shown by *a*) the view along the c-axis, *b*) trigonal prisms and selected experimental dimensions, and *c*) identification of an edge plane leading to the creation of alternating Mo- and S-edges. Mo atoms are filled circles and S atoms are empty circles. [Parts *a,b* used with permission of the American Physcial Society from Weiss and Phillips (1976), *Phys. Rev. B*, Vol. 14, Figs. 1 and 2, pp. 5392 and 5393, respectively. Part *c* used with permission of Elsevier from Raybaud et al. (1998), *Surf. Sci.*, Vol. 401, Fig. 1, p. 239.]

faces of Mo trigonal prisms projected onto the basal plane (Fig. 37). While the (0001) surface is terminated by coordinatively saturated S atoms, the edge surfaces are terminated by either 2-fold coordinated S atoms (the so-called "S-edge," also referred to as the ($\bar{1}010$) edge), or 4-fold coordinated Mo atoms (the so-called "Mo-edge," also referred to as the ($10\bar{1}0$) edge) (Fig. 37). Thus, it is not surprising that the (0001) surface is a relatively inert surface, giving rise to the well-known excellent lubricating properties of MoS_2, while the edge surfaces have important catalytic properties for HDS reactions.

The MoS_2 (0001) surface has been studied in great detail. LEED intensity-energy (I/V) spectra show for $2H$-MoS_2 that the S plane at the surface is slightly contracted by ~5% (Tong et al. 1976; Mrstik et al. 1977), consistent with the absence of van der Waals attractions at the surface plane (Tan 1996). The surface energy is only ~0.28 J/m^2 (Fuhr et al. 1999) (Table 1), therefore, the strength of the van der Waals attraction is about twice as large (Weiss and Phillips 1976). The underlying van der Waals gap binding the two uppermost MoS_2 sheets together is unchanged (Mrstik et al. 1977). The (0001) surface is extremely stable and relaxation is very subtle, consistent with expectations for a surface where atoms retain their essential bulk coordination. Hence, it follows that the electronic structure at the surface does not depart significantly from the bulk electronic structure. The trigonal prismatic ligand field about Mo atoms splits the $4d$ states into (d_{z^2}, $d_{x^2-y^2}$, d_{xy}), (d_{xz}), and (d_{yz}) groupings; the S $3p$ states are split into (p_z) and (p_x, p_y) groupings (Raybaud et al. 1997). Of the Mo $4d$ states, d_{z^2} has the lowest energy and is completely filled; the rest are unoccupied. Thus, photoelectron spectra and theoretical densities of states show a peak at the top of the valence band corresponding to Mo $4d_{z^2}$ states, which form a ~1 eV wide band that overlaps a 5-7 eV wide S $3p$ band (Fig. 38) (Mattheis 1973a,b; Wertheim et al. 1973; Raybaud et al. 1997). The lowest unoccupied states are derived from a mixture of mostly unoccupied Mo $4d$ orbitals and lesser S $3p$ orbitals (Raybaud et al. 1997).

Because MoS_2 is a moderate band-gap semiconductor (E_g ~ 1.3 eV Mattheis (1973a)), its surfaces have also been examined repeatedly with STM (Weimer et al. 1988a,b; Lieber and Kim 1991; Schlaf et al. 1997; Helveg et al. 2000; Kushmerick et al. 2000; Murata et al. 2001; Becker et al. 2003; Lauritsen et al. 2004a,b). Despite its structural simplicity, interpretation of atomic-scale STM images collected on MoS_2 (0001) has proven to be very challenging. Examples include interpreting image features such as the tunneling current corrugation (e.g., Magonov and Whangbo (1994)), "ring-shaped" defects (e.g., Kushmerick et al. (2000)), vacancies (Caulfield and Fisher 1997), dopant-related features (e.g., Schlaf et al. (1997)), and edge features (e.g., Lauritsen et al. (2004a,b)). In most reports, atomic-scale STM images show a regular hexagonal array of high tunneling current sites with 3.1 Å periodicity and symmetry consistent with the surface lattice expected from the bulk structure (Fig. 39). Because the Mo and S sublattices have the identical spacing and symmetry when projected onto the (0001) plane, assignment of the high tunneling current sites to either sublattice rests entirely upon predictions from electronic structure calculations (Coley et al. 1991; Magonov and Whangbo 1994; Kobayashi and Yamauchi 1995, 1996; Whangbo et al. 1995; Altibelli et al. 1996; Becker et al. 2003). The tunneling current depends essentially on the definite integral (over the bias voltage interval) of the local density of states of the sample at the location of the tip multiplied by the transmission probability (Tersoff and Hamann 1985). This assumes a featureless density of states for the metal tip with a distribution well-approximated as spherical. At positive bias voltage (sampling the unoccupied states at the MoS_2 surface), some theoretical calculations predict that STM images will show the S sublattice because the unoccupied Mo $4d$ states do not have significant angular momentum components normal to the surface plane, and therefore are inaccessible at normal tip-sample separation distances (Kobayashi and Yamauchi 1995; Whangbo et al. 1995). At negative bias voltage (sampling occupied states of the MoS_2 surface), the main issue is the fall-off of the Mo $4d_{z^2}$ states projecting through the uppermost S layer. A two-state matrix element calculation using HF wavefunctions to describe the tip-sample electronic coupling suggested that these states are indeed accessible, and negative bias voltage STM images should be dominated by

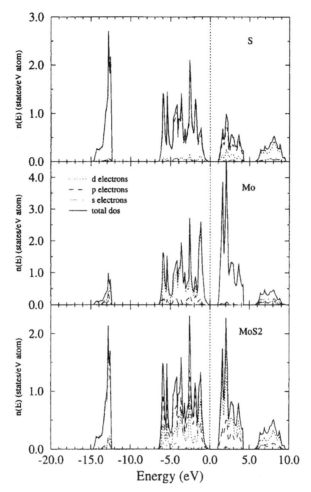

Figure 38. Total, local orbital projected, and partial Mo and S densities of states for bulk MoS$_2$, as calculated using planewave pseudopotential DFT methods. The electronic structure of the (0001) surface is likely to be very similar due to the weak S-S van der Waals interactions binding MoS$_2$ sheets together. [Used with permission of IOP Publishing Limited, from Raybaud et al. (1997), *J. Phys. Condens. Matter*, Vol. 9, Fig. 8, p. 11117.]

the Mo sublattice (Coley et al. 1991). Extended Hückel tight-binding calculations (Magonov and Whangbo 1994), and DFT calculations (Kobayashi and Yamauchi 1995, 1996) suggest the opposite, wherein the S sublattice is imaged via S 3p states on the uppermost S atoms. Evidence for the presence of both Mo and S sublattices in STM images has been given (Weimer et al. 1988a,b). More recent calculations have helped to clarify the negative bias voltage situation, predicting that at smaller tip-sample separations (< ~3 Å), the STM periodicity will arise from a convolution of Mo and S states, whereas at larger tip-sample distances, the S sublattice should dominate the images (Altibelli et al. 1996; Becker et al. 2003).

Edge surfaces of MoS$_2$ have been treated principally by theoretical calculations. The 1 × 1 edge surface of MoS$_2$ includes alternating Mo- and S-edges (Fig. 37). Periodic DFT calculations give surface energies that are significantly higher for this surface than for the (0001) surface (Table 1) (Raybaud et al. 1998; Alexiev et al. 2000; Todorova et al. 2004). Despite the

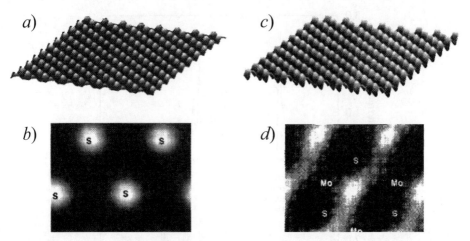

Figure 39. Theoretical STM images for the MoS$_2$ (0001) surface calculated using the elastic scattering quantum chemistry technique, showing in *a*) and *b*) images for tip-surface separation of 4.5 Å and −4.7 V bias voltage, and in *c*) and *d*) images for tip-surface separation of 3.0 Å and −4.8 V bias voltage. [Used with permission of Elsevier, from Altibelli et al. (1996), *Surface Science*, Vol. 367, Fig. 5, p. 214].

undercoordination of Mo and S atoms, these studies showed that surface relaxation displacements are relatively small. Furthermore, using Car-Parrinello molecular dynamics, Raybaud et al. (1998) showed that the 1 × 1 edge surface is stable in vacuum up to 700 K. Mo 4d surface states on the 4-fold coordinate Mo atoms span the bulk band gap, with the most intense states arising from empty d_{xy} and $d_{x^2-y^2}$ orbitals (Fig. 40) (Todorova et al. 2004). Fenske-Hall tight-binding calculations gave a similar prediction (Tan 1996). Sulfur 3p surface states on the 2-fold coordinated S atoms are centered at E_F, and these are thought to create a tendency for subtle S-S pairing across a MoS$_2$ sheet during relaxation (Raybaud et al. 1998; Alexiev et al. 2000; Todorova et al. 2004). Other noteworthy calculations on the structure of MoS$_2$ edge surfaces in vacuum include a periodic DFT study of higher index edge energies (Spirko et al. 2003) and a DFT cluster treatment of a hexagon-shaped Mo$_{27}$S$_{54}$ nanoparticle (Orita et al. 2003).

Fracture surfaces

Many sulfide mineral surfaces have poor or absent cleavage. However, fracturing under vacuum is a useful approach for isolating the clean surface for characterization. These surfaces

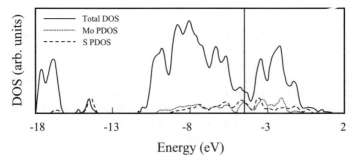

Figure 40. Total and partial Mo and S densities of states for the (10$\bar{1}$0) edge surface of MoS$_2$ calculated at the DFT level of theory using the local density approximation. [Redrawn with permission of The Royal Society of Chemistry, from Todorova et al. (2004), *Phys. Chem. Chem. Phys.*, Vol. 6, Fig. 10, p. 3029.]

are typically very topographically irregular and may involve defects induced by fracture. In cases where planes of weakness exist in the structure, it is possible that the surface would yield good diffraction data revealing the predominant surface crystallographic orientation. In this section, we cover work performed on fracture surfaces with no predominant orientation. This includes surfaces which may possess a predominant orientation but where no attempt to identify the orientation was reported.

Millerite and NiAs-type NiS. Millerite is the low-temperature form of NiS (< 380 °C) possessing a hexagonal structure. In millerite, both Ni and S are in 5-fold coordination. Sulfur atoms coordinating Ni are disposed at the corners of a distorted square pyramid, and vice versa. Millerite is a metallic conductor with electron transport facilitated by a combination of short Ni-Ni distances and Ni $3d$-S $3p$ hybridization (Gibbs et al. 2005). The high-temperature form of NiS has the NiAs (nickeline) hexagonal structure type and typically shows metal deficiency similar to pyrrhotite. Ni is in 6-fold coordination in the high-temperature form of NiS. This NiAs-type NiS has a transition near room temperature from an antiferromagnetic semiconductor below ~263 K to a paramagnetic metallic phase above (Tossell and Vaughan 1992).

Surfaces of both NiS forms have been studied by XPS (Legrand et al. 1998; Nesbitt and Reinke 1999). In both cases, the clean surfaces were produced by fracture in vacuum. Millerite possesses good ($10\bar{1}1$) and ($01\bar{1}2$) cleavage so it is possible that these planes were partly exposed. Narrow scans on the Ni $2p_{3/2}$ photopeaks (~853 eV) for both forms are similar in binding energy and lineshape (Fig. 41). The Ni $2p_{3/2}$ peak is asymmetric toward the higher binding energy side, which arises from the interaction between emitted photoelectrons and conduction band electrons (Doniach and Sunjic 1970; Legrand et al. 1998). There is also a characteristic low-intensity satellite peak at ~860 eV. The S $2p_{3/2}$ photopeak is located at 161.7 eV. The S photopeak was fitted using a doublet, with the doublet attributed to a Ni-monosulfide species (Legrand et al. 1998). No evidence of polysulfide species are observed on these fracture surfaces.

Marcasite. Marcasite (FeS$_2$) is dimorphous with pyrite. These two minerals have identical features of local coordination, but pyrite has a cubic lattice whereas marcasite has an

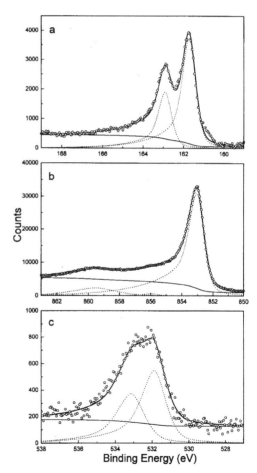

Figure 41. Conventional XPS spectra at room temperature of a millerite NiS surface generated by in-vacuum fracture, showing *a*) the S $2p$ region, *b*) the Ni $2p_{3/2}$ region, and *c*) the O $1s$ region. The small amount of oxygen was attributed to background oxygen in the vacuum chamber. From Legrand et al. (1998).

orthorhombic lattice. The symmetry reduction involves relative cell edge distances that are $b > a > c$, thus the prismatic surfaces are non-equivalent in marcasite. In pyrite FeS$_6$ octahedra share corners, but in marcasite they share edges and form chains parallel to the c-axis. In both minerals, Fe^{2+} is in a $t_{2g}^6 e_g^0$ $3d$ low-spin configuration. Increased distortion of the FeS$_6$ octahedra and stronger Fe-Fe interactions lead to a splitting of the t_{2g} bands and increased stability in marcasite relative to pyrite (Raybaud et al. 1997). Various electronic structure arguments have been put forth to explain the development of the anisotropic lattice of marcasite relative to the isotropic pyrite lattice (Tossell and Vaughan 1992).

Like pyrite, marcasite does not cleave and surfaces generated by fracture are concoidal. Clean fracture surfaces of marcasite made in vacuum have been analyzed with XPS (Pratt et al. 1998; Uhlig et al. 2001; Harmer and Nesbitt 2004). Thin films of marcasite have been synthesized by vacuum deposition and annealing under background S$_2$ gas (Elsetinow et al. 2003). These films have also been characterized with XPS, as well as AFM. Using SXPS, Uhlig et al. (2001) showed that the S $2p$ spectrum for marcasite fracture surfaces consists of four main contributions, assigned to bulk S$_2$ dimers, surface S$_2$ dimers, surface monosulfide, and short chained polysulfides at the surface (Fig. 42). The first three components are also found in the S $2p$ spectrum of fractured pyrite. As for pyrite, a surface monosulfide contribution is expected to arise from S-S bond cleavage during fracture, which is also expected to be coupled to auto-oxidation of Fe^{2+} to Fe^{3+} at the surface. Polysulfide formation on marcasite was postulated as occurring by spontaneous disproportionation of monosulfide and disulfide species at the surface. This process is thought to be more facile for marcasite relative to pyrite owing to the anisotropic nature of the marcasite lattice, which involves interligand distances that are shorter along certain directions than in pyrite (Uhlig et al. 2001; Harmer and Nesbitt 2004).

Chalcopyrite. Chalcopyrite (CuFeS$_2$) is by far the most important ore mineral of copper, so that its surface chemistry is of considerable interest. It has the same crystal structure as sphalerite, but ordering of Cu and Fe atoms results in a tetragonal unit cell essentially double the size of that in sphalerite. Although there has been some confusion in the literature regarding the valence states of the metals in chalcopyrite, to the extent that integral valence states can be assigned in such materials, the appropriate designation is Cu$^+$Fe^{3+}S$_2$ (Pearce et al. 2006b). Chalcopyrite is a narrow band gap semiconductor which is antiferromagnetically ordered at room temperature (see Pearce et al. 2006a, this volume). The absence of any cleavage has meant that studies of the pristine chalcopyrite surface have centered upon photoelectron spectroscopy of material fractured in UHV. Given the limitations for definitive interpretation of XPS data, along with the relative complexity of the electronic structure of chalcopyrite, our understanding of the chalcopyrite surface is currently both incomplete and controversial.

Earlier XPS studies (Buckley and Woods 1984; Yin et al. 1995) drew attention to asymmetry in the Cu, Fe and S $2p$ peaks and, using the very surface sensitive synchrotron radiation XPS technique, Laajalheto et al. (1994) found such asymmetry when studying the S $2p$ spectrum and used it as evidence for more than one surface sulfur species. Klauber (2003) using conventional XPS and UPS methods in conjunction with the semiempirical calculations of Hamajima et al. (1981) has proposed that asymmetry in the S $2p$ spectra can be attributed to a loss feature arising from an interband transition from occupied S levels to unoccupied Fe levels (S $3p \rightarrow$ Fe $3d$) rather than new surface species such as polysulfides (Fig. 43). However, Klauber (2003) does suggest that a second prominent S $2p$ component arises from sulfide dimers (S$_2^{2-}$) forming at the surface. Based primarily on the S $2p$ spectra, but also citing consistency with the Fe $2p$ spectra, this author proposes quite a complex simultaneous surface reconstruction and redox reaction model for the fractured chalcopyrite surface leading to a material comprising roughly the top two layers that has a 50% pyritic content.

Computational studies of both bulk and surface crystal and electronic structures of chalcopyrite are also rather incomplete (Hamajima et al. 1981; Edelbro et al. 2003). For

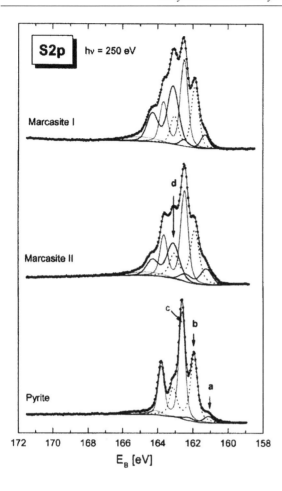

Figure 42. Synchrotron XPS spectra at room temperature comparing the S 2p regions for two different marcasite surfaces and a pyrite surface, all generated by in-vacuum fracture. The beam energy was tuned to 250 eV. Peaks a-d have been assigned to surface S^{2-}, surface S_2^{2-} dimers, bulk S_2^{2-} dimers, and short-chained polysulfides such as sulfur trimers, respectively. [Used with permission of Elsevier, from Uhlig et al. (2001), *Applied Surface Science*, Vol. 179, Fig. 2, p. 224.]

example, Edelbro et al. (2003) have performed *ab initio* full potential calculations using a DFT approach. They note that the calculations show the topmost part of the valence band to be a mixture of Fe 3*d*, Cu 3*d* and S 2*s*,2*p* orbitals (in contrast to pyrite which they also studied where it is mainly non-bonding Fe 3*d* orbitals). However, as noted earlier as a problem with DFT, their calculations predict chalcopyrite to be a metallic conductor not a semiconductor, in conflict with experimental evidence.

Arsenopyrite. As the dominant primary arsenic-containing mineral, arsenopyrite (FeAsS) is of considerable interest as regards its surface chemistry. The crystal structure of arsenopyrite is closely related to that of marcasite (see above); it contains (As-S) dianion units and octahedrally coordinated iron (nominally Fe^{2+}) with edge-sharing octahedra extending along the z-axis. Unlike marcasite, the Fe-Fe distances along the *z* axis direction are alternately short and long (see Makovicky 2006, this volume). Like marcasite (and pyrite) it is a diamagnetic semiconductor.

Beginning with the work of Buckley and Walker (1988) and Richardson and Vaughan (1989) there have been a substantial number of studies of the fresh and altered arsenopyrite surface using XPS methods (Nesbitt et al. 1995; Nesbitt and Muir 1998; Schaufuss et al. 2000; Mikhlin and Tomashevich 2005). Latterly, the interpretation of XPS data for the pristine

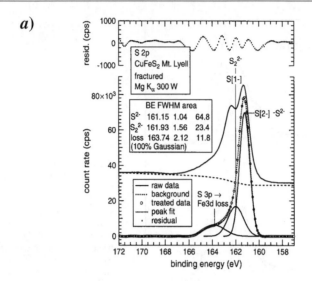

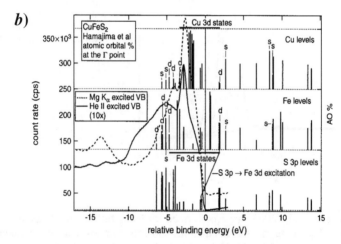

Figure 43. Conventional XPS and UPS spectra at room temperature of a chalcopyrite surface generated by fracturing in N_2, showing the *a)* XPS narrow scan of the S 2p region, and b) XPS and UPS spectra of the valence band overlaid with percentage *s*, *p*, and *d* atomic orbital components calculated using the X_α method. [Used with permission of John Wiley & Sons Ltd., from Klauber (2003), *Surface and Interface Analysis*, Vol. 35, Figs. 2, 8, p. 418, 42. Fig. 8 original source is Hamajima et al. (1981).]

arsenopyrite surface has been facilitated by comparative studies of related dianion species and by the use of SXPS. In particular, Harmer and Nesbitt (2004) have compared Fe 2p, S 2p and As 3d spectra of pyrite, marcasite, arsenopyrite and loellingite and proposed auto-redox and polymerization reactions that stabilize the surfaces of these phases. Thus, these authors argue that whereas the major contribution to the Fe 2p XPS spectrum of arsenopyrite is octahedrally coordinated low-spin Fe^{2+} in the bulk (Fig. 44), there is evidence of additional contributions from under-coordinated high spin surface Fe^{2+} atoms and Fe^{3+} formed by a surface auto-redox reaction of the type

$$S^{1-}_{surf} + Fe^{2+}_{surf} \rightarrow S^{2-}_{surf} + Fe^{3+}_{surf} \tag{7}$$

where S^{1-} surface radicals are formed by rupture of As-S bonds during fracture. The other proposed reactions at the surface involve polymerization of dimers or monomers given by

$$2As_{2\ surf}^{2-} \rightarrow As_{4\ surf}^{2-} + 2e- \qquad (8)$$

and

$$As^{1-} + As_{2\ surf}^{2-} \rightarrow As_{3}^{2-} + e- \qquad (9)$$

Dependant on the bulk structure and electronegativity of the constituent atoms, either the auto-redox or polymerization reactions would dominate; both are suggested to play a role in stabilizing the arsenopyrite surface.

Enargite. Enargite (Cu_3AsS_4) is important in certain ore deposits as a copper source and is the second most significant primary mineral source of arsenic; it causes environmental problems in the vicinity of certain former mining operations. Enargite is a sulfosalt mineral with a structure based on that of the wurtzite form of ZnS, with three-quarters of the Zn replaced by Cu and one-quarter by As. All of the atoms are in tetrahedral coordination, and the AsS_4 tetrahedra share none of their S atoms with one another. Formal oxidation states in this semiconducting mineral are: $Cu^+As^{5+}S^{2-}$. Because of its environmental importance, enargite has recently become the subject of surface science investigations, in particular using XPS to probe surface chemistry (Fullston et al. 1999; Rossi et al. 2001; Velasquez et al. 2002; Pratt 2004). Thus Rossi et al. (2001) examined freshly cleaved samples and "as received" samples of enargite reporting arsenic enrichment at the surface of the latter. Pratt (2004) studied samples fractured in UHV using both conventional and synchrotron radiation XPS. The spectra showed minimal differences in the character of sulfur and of copper in the surface compared with the bulk (and confirm copper as the Cu^+ species). However, the As 3*d* spectra from both conventional and SXPS experiments showed distinct contributions from As at the surface as compared with the bulk (Fig. 45). This led Pratt (2004) to suggest that, following fracture, the enargite surface is reorganized in such a way that the surface is characterized by protrusions of individual arsenic atoms. The results of Rossi et al. (2001) suggest that this surface chemistry may persist in the natural environment.

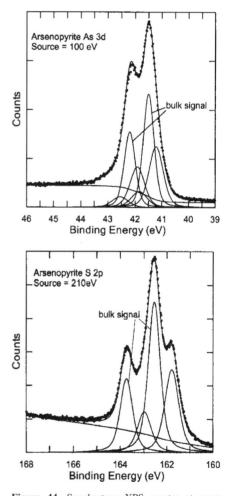

Figure 44. Synchrotron XPS spectra at room temperature of an arsenopyrite surface generated by in-vacuum fracture showing *a*) the As 3*d* region with the beam energy tuned to 100 eV, and *b*) the S 2*p* region with the beam tuned to 210 eV. [Used with permission of Elsevier, from Harmer and Nesbitt (2004), *Surface Science*, Vol. 564, Figs. 2 and 3, p. 41 and 42.]

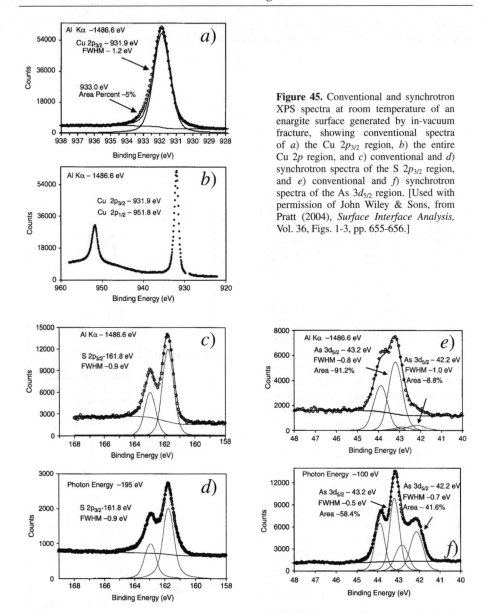

Figure 45. Conventional and synchrotron XPS spectra at room temperature of an enargite surface generated by in-vacuum fracture, showing conventional spectra of *a)* the Cu $2p_{3/2}$ region, *b)* the entire Cu $2p$ region, and *c)* conventional and *d)* synchrotron spectra of the S $2p_{3/2}$ region, and *e)* conventional and *f)* synchrotron spectra of the As $3d_{5/2}$ region. [Used with permission of John Wiley & Sons, from Pratt (2004), *Surface Interface Analysis*, Vol. 36, Figs. 1-3, pp. 655-656.]

FUTURE OUTLOOK

The past two decades have seen significant advances in our understanding of sulfide mineral surfaces. The surface structures of some of the more common sulfide minerals in nature have been imaged with atomic resolution for the first time. The improved energy resolution and surface sensitivity afforded by synchrotron-based spectroscopies have provided new insights into the types and numbers of different species at these surfaces. Highly efficient quantum mechanical modeling codes have enabled theoretical characterization of these surfaces with unprecedented accuracy. Most of these advances have been facilitated by assuming the ideal condition of vacuum termination, a necessary first step for understanding the clean surface properties.

Ultimately, the goal of this work is to provide a foundation for our understanding of sulfide chemical behavior at interfaces with other phases. Much progress has been made in this regard, as highlighted in Rosso and Vaughan (2006, the following chapter in this volume). Despite these advances, many fundamental questions remain for those surfaces already examined, and a range of important surfaces have yet to be examined in detail at all. The harder we look at sulfide mineral surfaces, the more we realize that there is much to be learned. The current state of this field is still very much in a reductionist mode, taking advantage of these new tools to better understand these and other sulfide mineral surfaces at a progressively higher level of detail. Some important near-term focus areas can be easily identified, including correlating surface species determined by XPS with site types and their spatial distribution resolvable by STM, relating tunneling spectroscopic information obtained by STM at the atomic-scale to the local electronic structure, and assessing the origins and structure of defects associated with or emerging at the surface. Beyond these needs, very important broader topics involving realistic surface complexity remain to be addressed. For example, the effect of bulk defects and impurities on the electronic structure and redistribution of charge at these surfaces is poorly understood. The ability of these surfaces to support proximity effects linking the chemical reactivity of spatially separated sites has just been shown. Sustained research over the long-term in these areas is needed if we are to understand the basis for sulfide mineral surface reactivity.

In our assessment the biggest steps forward are going to be made by studying these surfaces using a close coupling of surface spectroscopic and microscopic analysis with molecular modeling calculations. The strength of the combined approach is not in its ability to better "cover" the problem, but in its ability to generate better questions, improving the characterization. More routine coupling between surface analytical spectroscopy with microscopy would allow for an improved understanding of, for example, the site types and spatial distributions of monosulfide defects routinely documented by XPS. More routine coupling between STM imaging and quantum mechanical calculations would allow for a robust description of the surface electronic structure at the local atomic level. A number of recent advances in *ab initio* computation should prove valuable upon more widespread application to sulfide mineral surfaces. These include Car-Parrinello molecular dynamics and so-called *ab initio* thermodynamics for understanding the surface structure and stability, and improved electronic structure methods such as self-interaction corrected DFT and excited state methods applied to the solid state. Given the rapid development of these and other new tools, the future of this field is filled with opportunity.

ACKNOWLEDGMENTS

The authors gratefully acknowledge Mike Hochella for his insightful review of this chapter. KMR acknowledges the support of the U.S. Department of Energy (DOE), Office of Basic Energy Sciences, Geosciences Division, and the Stanford Environmental Molecular Sciences Institute (EMSI) jointly funded by the National Science Foundation and the DOE Office of Biological and Environmental Research (OBER). The W. R. Wiley Environmental Molecular Science Laboratory (EMSL) at Pacific Northwest National Laboratory (PNNL) is a national scientific user facility sponsored by the OBER. PNNL is operated for the DOE by Battelle Memorial Institute under contract DE-AC06-76RLO 1830. DJV acknowledges the support of the Natural Environment Research Council (NERC) in funding the Williamson Research Centre, and both the NERC and the Engineering and Physical Sciences Research Council in funding research in the mineral sciences at Manchester.

REFERENCES

Alexiev V, Prins R, Weber T (2000) Ab initio study of MoS_2 and Li adsorbed on the (10-10) face of MoS_2. Phys Chem Chem Phys 2:1815-1827

Altibelli A, Joachim C, Sautet P (1996) Interpretation of STM images: The MoS_2 surface. Surf Sci 367:209-220

Andersson CM, Brattsand R, Hallberg A, Engman L, Persson J, Moldeus P, Cotgreave I (1994) Diaryl tellurides as inhibitors of lipid peroxidation in biological and chemical systems. Free Radical Res 20:401-410

Andersson K, Nyberg M, Ogasawara H, Nordlund D, Kendelewicz T, Doyle CS, Brown GE, Pettersson LGM, Nilsson A (2004) Experimental and theoretical characterization of the structure of defects at the pyrite FeS_2 (100) surface. Phys Rev B 70:195404

Aray Y, Rodriguez J, Vega D, Coll S, Rodriguez-Arias EN, Rosillo F (2002) Adsorption of thiophene on the RuS_2 (100) and (111) surfaces: A Laplacian of the electronic charge density study. J Phys Chem B 106: 13242-13249

Becker U, Greatbanks SP, Rosso KM, Hillier IH, Vaughan DJ (1997a) An embedding approach for the calculation of STM images: Method development and application to galena (PbS). J Chem Phys 107: 7537-7542

Becker U, Hochella MF (1996) The calculation of STM images, STS spectra, and XPS peak shifts for galena: New tools for understanding mineral surface chemistry. Geochim Cosmochim Acta 60:2413-2426

Becker U, Hochella MF, Vaughan DJ (1997b) The adsorption of gold to galena surfaces: Calculation of adsorption/reduction energies, reaction mechanisms, XPS spectra, and STM images. Geochim Cosmochim Acta 61:3565-3585

Becker U, Munz AW, Lennie AR, Thornton G, Vaughan DJ (1997c) The atomic and electronic structure of the (001) surface of monoclinic pyrrhotite (Fe_7S_8) as studied using STM, LEED and quantum mechanical calculations. Surf Sci 389:66-87

Becker U, Rosso KM (2001) Step edges on galena (100): Probing the basis for defect driven surface reactivity at the atomic scale. Am Mineral 86:862-870

Becker U, Rosso KM, Weaver R, Warren M, Hochella MF (2003) Metal island growth and dynamics on molybdenite surfaces. Geochim Cosmochim Acta 67:923-934

Bither TA, Bouchard RJ, Cloud WH, Donohue PC, Siemons WJ (1968) Transition metal pyrite dichalcogenides. High pressure synthesis and correlation of properties. Inorg Chem 7:2208-2220

Bocquet AE, Mamiya K, Mizokawa T, Fujimori A, Miyadai T, Takahashi H, Mori M, Suga S (1996) Electronic structure of 3d transition metal pyrites MS_2 (M-Fe, Co, or Ni) by analysis of the M 2p core-level photoemission spectra. J Phys Cond Mat 8:2389-2400

Bonnell DA (1993) Scanning Tunneling Microscopy and Spectroscopy: Theory, Techniques, and Applications. VCH

Bottner R, Ratz S, Schroeder N, Marquardt S, Gerhardt U, Gaska R, Vaitkus J (1996) Analysis of angle-resolved photoemission data of PbS (001) surfaces within the direct transition model. Phys Rev B 53: 10336-10343

Bronold M, Tomm Y, Jaegermann W (1994) Surface states on cubic d-band semiconductor pyrite (FeS_2). Surf Sci 314:L931-L936

Buckley AN, Walker GW (1988) The surface composition of arsenopyrite exposed to oxidizing environments. Appl Surf Sci 35:227-240

Buckley AN, Woods R (1984) An X-ray photoelectron spectroscopic study of the oxidation of chalcopyrite. Aust J Chem 37:2403-2413

Bullett DW (1982) Electronic structure of 3d pyrite- and marcasite-type sulfides. J Phys C 15:6163-6174

Burns RG, Vaughan DJ (1970) The interpretation of the reflectivity behavior of ore minerals. Am Mineral 55: 1576-1586

Cai J, Philpott MR (2004) Electronic structure of bulk and (001) surface layers of pyrite FeS_2. Comp Mat Sci 30:358-363

Caulfield JC, Fisher AJ (1997) Electronic structure and scanning tunnelling microscope images of missing-atom defects on MoS_2 and $MoTe_2$ surfaces. J Phys Cond Mat 9:3671-3686

Charnock JM, Henderson CMB, Mosselmans JFW, Pattrick RAD (1996) 3d transition metal L-edge X-ray absorption studies of the dichalcogenides of Fe, Co, and Ni. Phys Chem Mineral 23:403-408

Chaturvedi S, Katz R, Guevremont J, Schoonen MAA, Strongin DR (1996) XPS and LEED study of a single-crystal surface of pyrite. Am Mineral 81:261-264

Chen CJ (1993) Introduction to Scanning Tunneling Microscopy. Oxford University Press

Clarke LJ (1985) Surface Crystallography. An Introduction to Low Energy Electron Diffraction. John Wiley & Sons

Colell H, Bronold M, Fiechter S, Tributsch H (1994) Surface structure of semiconducting ruthenium pyrite (RuS_2) investigated by LEED and STM. Surf Sci 303:L361-L366

Coley TR, Goddard WA, Baldeschwieler JD (1991) Theoretical interpretation of scanning tunneling microscopy images: Application to the molybdenum disulfide family of transition metal dichalcogenides. J Vac Sci Technol B 9:470-474

Cricenti A, Tallarida M, Ottaviani C, Kowalski B, Gutievitz E, Szczerbakow A, Orlowski BA (2001) Differential reflectivity and angle-resolved photoemission of PbS (100). Surf Sci 482:659-663

de Leeuw NH, Parker SC, Sithole HM, Ngoepe PE (2000) Modeling the surface structure and reactivity of pyrite: Introducing a potential model for FeS_2. J Phys Chem B 104:7969-7976

Doniach S, Sunjic M (1970) Many-electron singularity in X-ray photoemission and X-ray line spectra from metals. J Phys C 3:285-291

Duke CB (1992) Structure and bonding of tetrahedrally coordinated compound semiconductor cleavage faces. J Vac Sci Technol A 10:2032-2040

Duke CB (1994) Reconstruction of the cleavage faces of tetrahedrally coordinated compound semiconductors. In Festkorperprobleme - Advances In Solid State Physics 33. 1-36

Duke CB, Paton A, Kahn A (1984) The atomic geometries of GaP (110) and ZnS (110) revisited: A structural ambiguity and its resolution. J Vac Sci Technol A 2:515-518

Duke CB, Wang YR (1988) Surface structure and bonding of the cleavage faces of tetrahedrally coordinated II-VI compounds. J Vac Sci Technol B 6:1440-1443

Edelbro R, Sandstrom A, Paul J (2003) Full potential calculations on the electron bandstructures of sphalerite, pyrite and chalcopyrite. Appl Surf Sci 206:300-313

Eggleston CM (1997) Initial oxidation of sulfide sites on a galena surface: Experimental confirmation of an *ab initio* calculation. Geochim Cosmochim Acta 61:657-660

Eggleston CM, Ehrhardt JJ, Stumm W (1996) Surface structural controls on pyrite oxidation kinetics: An XPS-UPS, STM, and modeling study. Am Mineral 81:1036-1056

Eggleston CM, Hochella MF (1990) Scanning tunneling microscopy of sulfide surfaces. Geochim Cosmochim Acta 54:1511-1517

Eggleston CM, Hochella MF (1992) Scanning tunneling microscopy of pyrite (100): Surface structure and step reconstruction. Am Mineral 77:221-224

Eggleston CM, Hochella MF (1993) Tunneling spectroscopy applied to PbS (001) surfaces: Fresh surfaces, oxidation, and sorption of aqueous Au. Am Mineral 78:877-883

Eggleston CM, Hochella MF (1994) Atomic and electronic structure of PbS (100) surfaces and chemisorption oxidation reactions. *In*: Environmental Geochemistry of Sulfide Oxidation. Alpers CN, Blowes DW (eds) American Chemical Society, p. 201-222

Elsetinow AR, Guevremont JM, Strongin DR, Schoonen MAA, Strongin M (2000) Oxidation of (100) and (111) surfaces of pyrite: Effects of preparation method. Am Mineral 85:623-626

Elsetinow AR, Strongin DR, Borda MJ, Schoonen MA, Rosso KM (2003) Characterization of the structure and the surface reactivity of a marcasite thin film. Geochim Cosmochim Acta 67:807-812

Evans HT, Konnert JA (1976) Crystal structure refinement of covellite. Am Mineral 61:996-1000

Eyert V, Hock K-H, Fiechter S, Tributsch H (1998) Electronic structure of FeS_2: The crucial role of electron-lattice interaction. Phys Rev B 57:6350-6359

Ferraz AC, Watari K, Alves JLA (1994) Surface electronic properties of ZnS, ZnSe and ZnTe (110). Surf Sci 309:959-962

Ferrer IJ, Nevskaia DM, de las Heras C, Sanchez C (1990) About the band gap nature of FeS_2 as determined from optical and photoelectrochemical measurements. Solid State Comm 74:913-916

Fiechter S (2004) Defect formation energies and homogeneity ranges of rock salt-, pyrite-, chalcopyrite- and molybdenite-type compound semiconductors. Solar Energy Mat Solar Cells 83:459-477

Folkerts W, Sawatzky GA, Haas C, de Groot RA, Hillebrecht FU (1987) Electronic structure of some 3d transition metal pyrites. J Phys C 20:4135-4144

Folmer JCW, Jellinek F, Calis GHM (1988) The electronic structure of pyrites, particularly CuS_2, and $Fe_{1-x}Cu_xSe_2$: An XPS and Mossbauer study. J Solid State Chem 72:137-144

Frechard E, Sautet P (1997) RuS_2 (111) surfaces: Theoretical study of various terminations and their interaction with H_2. J Catal 170:402-410

Frechard F, Sautet P (1995) Hartree-Fock *ab initio* study of the geometric and electronic structure of RuS_2 and its (100) and (111) surfaces. Surf Sci 336:149-165

Fuhr JD, Sofo JO, Saul A (1999) Adsorption of Pd on MoS_2 (1000): *Ab initio* electronic structure calculations. Phys Rev B 60:8343-8347

Fujimori A, Mamiya K, Mizokawa T, Miyadai T, Sekiguchi T, Takahashi H, Mori N, Suga S (1996) Resonant photoemission study of pyrite-type NiS_2, CoS_2 and FeS_2. Phys Rev B 54:16329-16332

Fullston D, Fornasiero D, Ralston J (1999) Oxidation of synthetic and natural samples of enargite and tennantite: 2. X-ray photoelectron spectroscopic study. Langmuir 15:4530-4536

Gerson AR, Bredow T (2000) Interpretation of sulphur 2p XPS spectra in sulfide minerals by means of *ab initio* calculations. Surf Int Anal 29:145-150

Gibbs GV, Downs RT, Prewitt CT, Rosso KM, Ross NL, Cox DF (2005) Electron density distributions calculated for the nickel sulfides millerite, vaesite, and heazlewoodite and nickel metal: A case for the importance of Ni-Ni bond paths for electron transport. J Phys Chem B 109:21788-21795

Gibson AS, LaFemina JP (1996) Structure of mineral surfaces. *In*: Physics and Chemistry of Mineral Surfaces. Brady PV (ed) CRC Press, p 1-62

Grillo ME, Sautet P (2000) Density functional study of the structural and electronic properties of RuS_2 (111) II. Hydrogenated surfaces. Surf Sci 457:285-293

Grillo ME, Smelyanski V, Sautet P, Hafner J (1999) Density functional study of the structural and electronic properties of RuS_2 (111) I. Bare surfaces. Surf Sci 439:163-172

Guevremont JM, Elsetinow AR, Strongin DR, Bebie J, Schoonen MAA (1998) Structure sensitivity of pyrite oxidation: Comparison of the (100) and (111) planes. Am Mineral 83:1353-1356

Guevremont JM, Strongin DR, Schoonen MAA (1997) Effects of surface imperfections on the binding of CH_3OH and H_2O on FeS_2 (100): Using adsorbed Xe as a probe of mineral surface structure. Surf Sci 391:109-124

Hamad S, Cristol S, Callow CRA (2002) Surface structures and crystal morphology of ZnS: Computational study. J Phys Chem B 106:11002-11008

Hamajima T, Kambara T, Gondaira KI, Oguchi T (1981) Self-consistent electronic structures of magnetic semiconductors by a discrete variational X_α calculation .3. Chalcopyrite $CuFeS_2$. Phys Rev B 24:3349-3353

Hanaguri T, Kohsaka Y, Iwaya K, Satow S, Kitazawa K, Takagi H, Azuma M, Takano M (2004) STM/STS study of metal-to-Mott insulator transitions. Phys C 408-10:328-329

Harmer SL, Nesbitt HW (2004) Stabilization of pyrite (FeS_2),marcasite (FeS_2), arsenopyrite (FeAsS) and loellingite ($FeAs_2$) surfaces by polymerization and auto-redox reactions. Surf Sci 564:38-52

Harrison WA (1980) Electronic Structure and Properties of Solids. Freeman

Helveg S, Lauritsen JV, Laegsgaard E, Stensgaard I, Norskov JK, Clausen BS, Topsoe H, Besenbacher F (2000) Atomic-scale structure of single layer MoS_2 nanoclusters. Phys Rev Lett 84:951-954

Hochella MF, Rakovan JF, Rosso KM, Bickmore BR, Rufe E (1998) New directions in mineral surface geochemical research using scanning probe microscopes. *In*: Mineral-Water Interfacial Reactions: Kinetics and Mechanisms. Sparks DL, Grundl T (eds) American Chemical Society, 1-22

Holzwarth NAW, Harris S, Liang KS (1985) Electronic structure of RuS_2. Phys Rev B 32:3745-3752

Huang YS, Chen YF (1988) Electronic structure study of RuS_2. Phys Rev B 38:7997-8001

Huang YS, Huang JK, Tsay MY (1993) An electroreflectance study of FeS_2. J Phys Cond Mat 5:7827-7836

Hung A, Muscat J, Yarovsky I, Russo SP (2002a) Density functional theory studies of pyrite FeS_2 (100) and (110) surfaces. Surf Sci 513:511-524

Hung A, Muscat J, Yarovsky I, Russo SP (2002b) Density functional theory studies of pyrite FeS_2 (111) and (210) surfaces. Surf Sci 520:111-119

Hung A, Yarovsky I, Russo SP (2003) Density functional theory studies of xanthate adsorption on the pyrite FeS_2 (110) and (111) surfaces. J Chem Phys 118:6022-6029

Iwaya K, Kohsaka Y, Satow S, Hanaguri T, Miyasaka S, Takagi H (2004) Evolution of local electronic states from a metal to a correlated insulator in a $NiS_{2-x}Se_x$ solid solution. Phys Rev B 70:161103

Jolly WL (1991) Modern Inorganic Chemistry. McGraw-Hill

Joswig JO, Springborg M, Seifert G (2000) Structural and electronic properties of cadmium sulfide clusters. J Phys Chem B 104:2617-2622

Kautz RL, Dresselhaus MS, Adler D, Linz A (1972) Electrical and optical properties of NiS_2. Phys Rev B 6: 2078-2082

Kelty SP, Li J, Chen JG, Chianelli RR, Ren J, Whangbo MH (1999) Characterization of the RuS_2 (100) surface by scanning tunneling microscopy, atomic force microscopy, and near-edge X-ray absorption fine structure measurements and electronic band structure calculations. J Phys Chem B 103:4649-4655

Kendelewicz T, Doyle CS, Bostick BC, Brown GE (2004) Initial oxidation of fractured surfaces of FeS_2 (100) by molecular oxygen, water vapor, and air. Surf Sci 558:80-88

Kendelewicz T, Liu P, Brown GE, Nelson EJ (1998) Atomic geometry of the PbS (100) surface. Surf Sci 395: 229-238

Klauber C (2003) Fracture-induced reconstruction of a chalcopyrite ($CuFeS_2$) surface. Surf Int Anal 35:415-428

Klepeis JE, Mailhiot C, Vanschilfgaarde M, Methfessel M (1993) Role of ionicity in the determination of surface atomic geometries: GaP, ZnS, and CuCl (110) surfaces. J Vac Sci Technol B 11:1463-1466

Kobayashi K, Yamauchi J (1995) Electronic structure and scanning tunneling microscopy image of molybdenum dichalcogenide surfaces. Phys Rev B 51:17085-17095

Kobayashi K, Yamauchi J (1996) Scanning tunneling microscopy image of transition metal dichalcogenide surfaces. Surf Sci 358:317-321

Kushmerick JG, Kandel SA, Han P, Johnson JA, Weiss PS (2000) Atomic-scale insights into hydrodesulfurization. J Phys Chem B 104:2980-2988

Laajalehto K, Kartio I, Nowak P (1994) XPS study of clean metal sulfide surfaces. Appl Surf Sci 81:11-15

Laajalehto K, Smart RS, Ralston J, Suoninen E (1993) STM and XPS investigation of reaction of galena in air. Appl Surf Sci 64:29-39

LaFemina JP (1992) Total-energy calculations of semiconductor surface reconstructions. Surf Sci Rep 16:133-260

LaFemina JP, Duke CB (1991) Dependence of oxide surface structure on surface topology and local chemical bonding. J Vac Sci Technol A 9:1847-1855

Lauritsen JV, Bollinger MV, Laegsgaard E, Jacobsen KW, Norskov JK, Clausen BS, Topsoe H, Besenbacher F (2004a) Atomic-scale insight into structure and morphology changes of MoS_2 nanoclusters in hydrotreating catalysts. J Catal 221:510-522

Lauritsen JV, Nyberg M, Norskov JK, Clausen BS, Topsoe H, Laegsgaard E, Besenbacher F (2004b) Hydrodesulfurization reaction pathways on MoS_2 nanoclusters revealed by scanning tunneling microscopy. J Catal 224:94-106

Lazarides AA, Duke CB, Paton A, Kahn A (1995) Determination of the surface atomic geometry of PbTe (100) by dynamical low energy electron diffraction intensity analysis. Phys Rev B 52:14895-14905

Legrand DL, Nesbitt HW, Bancroft GM (1998) X-ray photoelectron spectroscopic study of a pristine millerite (NiS) surface and the effect of air and water oxidation. Am Mineral 83:1256-1265

Leiro JA, Laajalehto K, Kartio I, Heinonen MH (1998) Surface core-level shift and phonon broadening in PbS (100). Surf Sci 413:L918-L923

Leiro JA, Mattila SS, Laajalehto K (2003) XPS study of the sulphur 2p spectra of pyrite. Surf Sci 547:157-161

Lessor DL, Duke CB, Kahn A, Ford WK (1993) Ionicity dependence of surface bond lengths on the (110) cleavage faces of isoelectronic zincblende structure compound semiconductors: GaP, ZnS, and CuCl. J Vac Sci Technol A 11:2205-2209

Li EK, Johnson KH, Eastman DE, Freeouf JL (1974) Localized and bandlike valence electron states in FeS_2 and NiS_2. Phys Rev Lett 32:470-472

Liang W, Whangbo MH (1993) Conductivity anisotropy and structural phase transition in covellite CuS. Solid State Comm 85:405-408

Lieber CM, Kim Y (1991) Characterization of the structural, electronic and tribological properties of metal dichalcogenides by scanning probe microscopies. Thin Solid Films 206:355-359

Louis E, Elices M (1975) Pseudopotential calculation of surface band structure of (111) diamond and zincblende faces: Ge, α-Sn, GaAs, and ZnS. Phys Rev B 12:618-623

Ma JX, Jia Y, Song YL, Liang EJ, Wu LK, Wang F, Wang XC, Hu X (2004) The geometric and electronic properties of the PbS, PbSe and PbTe (001) surfaces. Surf Sci 551:91-98

Magonov SN, Whangbo MH (1994) Interpreting STM and AFM Images. Adv Mat 6:355-371

Makovicky E (2006) Crystal structures of sulfides and other chalcogenides. Rev Mineral Geochem 61:7-125

Mattheis LF (1973a) Band structures of transition metal dichalcogenide layer compounds. Phys Rev B 8:3719-3740

Mattheis LF (1973b) Energy bands for $2H$-$NbSe_2$ and $2H$-MoS_2. Phys Rev Lett 30:784-787

McEllistrem M, Haase G, Chen D, Hamers RJ (1993) Electrostatic sample tip interactions in the scanning tunneling microscope. Phys Rev Lett 70:2471-2474

Mian M, Harrison NM, Saunders VR, Flavell WR (1996) An *ab initio* Hartree-Fock investigation of galena (PbS). Chem Phys Lett 257:627-632

Mikhlin Y, Tomashevich Y (2005) Pristine and reacted surfaces of pyrrhotite and arsenopyrite as studied by X-ray absorption near-edge structure spectroscopy. Phys Chem Mineral 32:19-27

Mikhlin Y, Tomashevich YV, Pashkov GL, Okotrub AV, Asanov IP, Mazalov LN (1998) Electronic structure of the non-equilibrium iron-deficient layer of hexagonal pyrrhotite. Appl Surf Sci 125:73-84

Moore D, Ronning C, Ma C, Wang ZL (2004) Wurtzite ZnS nanosaws produced by polar surfaces. Chem Phys Lett 385:8-11

Mosselmans JFW, Pattrick RAD, van der Laan G, Charnock JM, Vaughan DJ, Henderson CMB, Garner CD (1995) X-ray absorption near-edge spectra of transition metal disulfides FeS_2 (pyrite and marcasite), CoS_2, NiS_2, and CuS_2, and their isomorphs FeAsS and CoAsS. Phys Chem Mineral 22:311-317

Mrstik BJ, Kaplan R, Reinecke TL, Vanhove M, Tong SY (1977) Surface structure determination of layered compounds MoS_2 and $NbSe_2$ by low energy electron diffraction. Phys Rev B 15:897-900

Murata H, Kataoka K, Koma A (2001) Scanning tunneling microscope images of locally modulated structures in layered materials, MoS_2 (0001) and $MoSe_2$ (0001), induced by impurity atoms. Surf Sci 478:131-144

Muscat J, Gale JD (2003) First principles studies of the surface of galena PbS. Geochim Cosmochim Acta 67:799-805

Nesbitt HW, Bancroft GM, Pratt AR, Scaini MJ (1998) Sulfur and iron surface states on fractured pyrite surfaces. Am Mineral 83:1067-1076

Nesbitt HW, Berlich AG, Harmer SL, Uhlig I, Bancroft GM, Szargan R (2004) Identification of pyrite valence band contributions using synchrotron-excited X-ray photoelectron spectroscopy. Am Mineral 89:382-389

Nesbitt HW, Muir IJ (1998) Oxidation states and speciation of secondary products on pyrite and arsenopyrite reacted with mine waste waters and air. Mineral Pet 62:123-144

Nesbitt HW, Muir LJ, Pratt AR (1995) Oxidation of arsenopyrite by air and air-saturated, distilled water, and implications for mechanism of oxidation. Geochim Cosmochim Acta 59:1773-1786

Nesbitt HW, Reinke M (1999) Properties of As and S at NiAs, NiS, and $Fe_{1-x}S$ surfaces, and reactivity of niccolite in air and water. Am Mineral 84:639-649

Nesbitt HW, Scaini M, Hochst H, Bancroft GM, Schaufuss AG, Szargan R (2000) Synchrotron XPS evidence for Fe^{2+}-S and Fe^{3+}-S surface species on pyrite fracture-surfaces, and their 3d electronic states. Am Mineral 85:850-857

Nesbitt HW, Schaufuss AG, Scaini M, Bancroft GM, Szargan R (2001) XPS measurement of fivefold and sixfold coordinated sulfur in pyrrhotites and evidence for millerite and pyrrhotite surface species. Am Mineral 86:318-326

Noguera C (1996) Physics and Chemistry of Oxide Surfaces. Cambridge University Press

Ogawa S, Waki S, Teranishi T (1974) Magnetic and electrical properties of 3d transition-metal disulfides having the pyrite structure. Int J Mag 5:349-360

Orita H, Uchida K, Itoh N (2003) *Ab initio* density functional study of the structural and electronic properties of an MoS_2 catalyst model: a real size $Mo_{27}S_{54}$ cluster. J Mol Catal A 195:173-180

Pashley MD (1989) Electron counting model and its application to island structures on molecular-beam epitaxy grown GaAs (001) and ZnSe(001). Phys Rev B 40:10481-10487

Pearce CI, Pattrick RAD, Vaughan DJ (2006) Electrical and Magnetic properties of sulfides. Rev Mineral Geochem 61:127-180

Pearce CI, Pattrick RAD, Vaughan DJ, Henderson CMB, van der Laan G (2006b) Copper oxidation state in chalcopyrite: Mixed Cu d^9 and d^{10} characteristics. Geochem Cosmochim Acta *in press*

Perdew JP, Zunger A (1981) Self interaction correction to density functional approximations for many electron systems. Phys Rev B 23:5048-5079

Pettenkofer C, Jaegermann W, Bronold M (1991) Site specific surface interaction of electron donors and acceptors on FeS_2 (100) cleavage planes. Ber Bunsenges Phys Chem 95:560-565

Philpott MR, Goliney IY, Lin TT (2004) Molecular dynamics simulation of water in a contact with an iron pyrite FeS_2 surface. J Chem Phys 120:1943-1950

Pollmann J, Kruger P, Rohlfing M, Sabisch M, Vogel D (1996) *Ab initio* calculations of structural and electronic properties of prototype surfaces of group IV, III-V and II-VI semiconductors. Appl Surf Sci 104:1-16

Pratt A (2004) Photoelectron core levels for enargite, Cu_3AsS_4. Surf Int Anal 36:654-657

Pratt AR, McIntyre NS, Splinter SJ (1998) Deconvolution of pyrite marcasite and arsenopyrite XPS spectra using the maximum entropy method. Surf Sci 396:266-272

Preobrajenski AB, Chasse T (1999) Epitaxial growth and interface structure of PbS on InP (110). Appl Surf Sci 142:394-399

Pridmore DF, Shuey RT (1976) The electrical resistivity of galena, pyrite, and chalcopyrite. Am Mineral 61: 248-259

Qiu GZ, Xiao Q, Hu YH (2004a) First-principles calculation of the electronic structure of the stoichiometric pyrite FeS_2 (100) surface. Comp Mat Sci 29:89-94

Qiu GZ, Xiao Q, Hu YH, Qin WQ, Wang DZ (2004b) Theoretical study of the surface energy and electronic structure of pyrite FeS_2 (100) using a total-energy pseudopotential method, CASTEP. J Coll Int Sci 270: 127-132

Rantala TT, Rantala TS, Lantto V, Vaara J (1996) Surface relaxation of the $(10\bar{1}0)$ face of wurtzite CdS. Surf Sci 352:77-82

Raybaud P, Hafner J, Kresse G, Toulhoat H (1997) *Ab initio* density functional studies of transition metal sulfides: II. Electronic structure. J Phys Cond Mat 9:11107-11140

Raybaud P, Hafner J, Kresse G, Toulhoat H (1998) Structural and electronic properties of the MoS_2 $(10\bar{1}0)$ edge surface. Surf Sci 407:237-250

Richardson S, Vaughan DJ (1989) Arsenopyrite: A spectroscopic investigation of altered surfaces. Mineral Mag 53:223-229

Rossi A, Atzei D, Da Pelo S, Frau F, Lattanzi P, England KER, Vaughan DJ (2001) Quantitative X-ray photoelectron spectroscopy study of enargite (Cu_3AsS_4) surface. Surf Int Anal 31:465-470

Rosso KM (2001) Structure and reactivity of semiconducting mineral surfaces: Convergence of molecular modeling and experiment. Rev Mineral Geochem 42:199-271

Rosso KM, Becker U, Hochella MF (1999a) Atomically resolved electronic structure of pyrite {100} surfaces: An experimental and theoretical investigation with implications for reactivity. Am Mineral 84:1535-1548

Rosso KM, Becker U, Hochella MF (1999b) The interaction of pyrite {100} surfaces with O_2 and H_2O: Fundamental oxidation mechanisms. Am Mineral 84:1549-1561

Rosso KM, Becker U, Hochella MF (2000) Surface defects and self-diffusion on pyrite {100}: An ultra-high vacuum scanning tunneling microscopy and theoretical modeling study. Am Mineral 85:1428-1436

Rosso KM, Hochella MF (1999) A UHV STM/STS and *ab initio* investigation of covellite (001) surfaces. Surf Sci 423:364-374

Rosso KM, Vaughan DJ (2006) Reactivity of sulfide mineral surfaces. Rev Mineral Geochem 61:557-607

Sarma DD, Krishnakumar SR, Weschke E, Schussler-Langeheine C, Mazumdar C, Kilian L, Kaindl G, Mamiya K, Fujimori SI, Fujimori A, Miyadai T (2003) Metal-insulator crossover behavior at the surface of NiS_2. Phys Rev B 67

Satta A, de Gironcoli S (2001) Surface structure and core-level shifts in lead chalcogenide (001) surfaces. Phys Rev B 63:art. no.-033302

SchaarGabriel E, AlonsoVante N, Tributsch H (1996) STM photoeffects mediated by water adsorption on photocatalytic (RuS_2, TiO_2) materials. Surf Sci 366:508-518

Schaufuss AG, Nesbitt HW, Kartio I, Laajalehto K, Bancroft GM, Szargan R (1998) Reactivity of surface chemical states on fractured pyrite. Surf Sci 411:321-328

Schaufuss AG, Nesbitt HW, Scaini MJ, Hoechst H, Bancroft MG, Szargan R (2000) Reactivity of surface sites on fractured arsenopyrite (FeAsS) toward oxygen. Am Mineral 85:1754-1766

Schlaf R, Louder D, Nelson MW, Parkinson BA (1997) Influence of electrostatic forces on the investigation of dopant atoms in layered semiconductors by scanning tunneling microscopy/spectroscopy and atomic force microscopy. J Vac Sci Technol A 15:1466-1472

Schlegel A, Wachter P (1976) Optical properties, phonons, and electronic structure of iron pyrite (FeS_2). J Phys C 9:3363-3369

Schroer P, Kruger P, Pollmann J (1994) Self-consistent electronic structure calculations of the (1010) surfaces of the wurtzite compounds ZnO and CdS. Phys Rev B 49:17092-17101

Siemens B, Domke C, Ebert P, Urban K (1997) Electronic structure of wurtzite II-VI compound semiconductor cleavage surfaces studied by scanning tunneling microscopy. Phys Rev B 56:12321-12326

Siemens B, Domke C, Ebert P, Urban K (1999a) Point defects, dopant atoms, and compensation effects in CdSe and CdS cleavage surfaces. Thin Solid Films 344:537-540

Siemens B, Domke C, Heinrich M, Ebert P, Urban K (1999b) Imaging individual dopant atoms on cleavage surfaces of wurtzite-structure compound semiconductors. Phys Rev B 59:2995-2999

Skinner AJ, LaFemina JP (1992) Surface atomic and electronic structure of ZnO polymorphs. Phys Rev B 45:3557-3564

Smelyansky V, Hafner J, Kresse G (1998) Adsorption of thiophene on RuS_2: An *ab initio* density functional study. Phys Rev B 58:R1782-R1785

Spirko JA, Neiman ML, Oelker AM, Klier K (2003) Electronic structure and reactivity of defect MoS_2 I. Relative stabilities of clusters and edges, and electronic surface states. Surf Sci 542:192-204

Stadele M, Moukara M, Majewski JA, Vogl P, Gorling A (1999) Exact exchange Kohn-Sham formalism applied to semiconductors. Phys Rev B 59:10031-10043

Steele HM, Wright K, Hillier IH (2003) A quantum-mechanical study of the (110) surface of sphalerite (ZnS) and its interaction with Pb^{2+} species. Phys Chem Mineral 30:69-75

Stirling A, Bernasconi M, Parrinello M (2003) *Ab initio* simulation of water interaction with the (100) surface of pyrite. J Chem Phys 118:8917-8926

Sze SM (1981) Physics of Semiconductor Devices, 2nd edition. Wiley

Tan A, Harris S (1998) Electronic structure of Rh_2S_3 and RuS_2, two very active hydrodesulfurization catalysts. Inorg Chem 37:2215-2222

Tan AL (1996) Relaxation in the (001) surface of MoS_2: Band structure calculations. Theo J Mol Struct 363:303-309

Tasker PW (1979) The stability of ionic crystal surfaces. J Phys C 12:4977-4984

Tersoff J, Hamann DR (1985) Theory of the scanning tunneling microscope. Phys Rev B 31:805-813

Thio T, Bennett JW (1994) Hall effect and conductivity in pyrite NiS_2. Phys Rev B 50:10574-10577

Todorova T, Alexiev V, Prins R, Weber T (2004) *Ab initio* study of 2H-MoS_2 using Hay and Wadt effective core pseudo-potentials for modeling the (10-10) surface structure. Phys Chem Chem Phys 6:3023-3030

Tong SY, Vanhove M, Mrstik BJ, Kaplan R, Reinecke T (1976) Surface structure determination of layered compounds MoS_2 and $NbSe_2$ by dynamical LEED approach. J Vac Sci Technol A 13:188-188

Tossell JA (1977) SCF-X_α scattered wave MO studies of the electronic structure of ferrous iron in octahedral coordination with sulfur. J Chem Phys 66:5712-5719

Tossell JA, Vaughan DJ (1987) Electronic structure and the chemical reactivity of the surface of galena. Can Mineral 25:381-392

Tossell JA, Vaughan DJ (1992) Theoretical Geochemistry: Applications of Quantum Mechanics in the Earth and Mineral Sciences. Oxford University Press

Tsai MH, Kasowski RV, Dow JD (1987) Relaxation of the nonpolar (1010) surfaces of wurtzite AlN and ZnS. Sol State Comm 64:231-233

Tutuncu HM, Miotto R, Srivastava GP (2000) Phonons on II-VI (110) semiconductor surfaces. Phys Rev B 62:15797-15805

Uhlig I, Szargan R, Nesbitt HW, Laajalehto K (2001) Surface states and reactivity of pyrite and marcasite. Appl Surf Sci 179:223-230

van der Heide H, Hemmel R, van Bruggen CF, Haas C (1980) X-ray photoelectron spectra of 3d transition metal pyrites. J Sol State Chem 33:17-25

Van Vechten JA (1975) Simple theoretical estimates of schottky constants and virtual enthalpies of single vacancy formation in zincblende and wurtzite type semiconductors. J Electrochem Soc 122:419-422

Rosso KM, Vaughan DJ (2006) Reactivity of sulfide mineral surfaces. Rev Mineral Geochem 61:557-607

Velasquez P, Ramos-Barrado JR, Leinen D (2002) The fractured, polished and Ar^+-sputtered surfaces of natural enargite: An XPS study. Surf Anal 34:280-283

von Oertzen GU, Skinner WM, Nesbitt HW (2005) *Ab initio* and X-ray photoemission spectroscopy study of the bulk and surface electronic structure of pyrite (100) with implications for reactivity. Phys Rev B 72

Wang YR, Duke CB (1988) Cleavage faces of wurtzite CdS and CdSe: Surface relaxation and electronic structure. Phys Rev B 37:6417-6424

Weimer M, Kramar J, Bai C, Baldeschwieler JD (1988a) Tunneling microscopy of $2H-MoS_2$: A compound semiconductor surface. Phys Rev B 37:4292-4295

Weimer M, Kramar J, Bai C, Baldeschwieler JD, Kaiser WJ (1988b) Scanning tunneling microscopy investigation of $2H-MoS_2$: A layered semiconducting transition metal dichalcogenide. J Vac Sci Technol A 6:336-337

Weimer M, Kramar J, Baldeschwieler JD (1989) Band bending and the apparent barrier height in scanning tunneling microscopy. Phys Rev B 39:5572-5575

Weiss K, Phillips JM (1976) Calculated specific surface energy of molybdenite (MoS_2). Phys Rev B 14:5392-5395

Wertheim GK, Disalvo FJ, Buchanan DN (1973) Valence bands of layer structure transition metal chalcogenides. Solid State Comm 13:1225-1228

Whangbo MH, Ren J, Magonov SN, Bengel H, Parkinson BA, Suna A (1995) On the correlation between the scanning tunneling microscopy image imperfections and point defects of layered chalcogenides $2H-MX_2$ (M=Mo, W X=S, Se). Surf Sci 326:311-326

Wincott PL, Vaughan DJ (2006) Spectroscopic studies of sulfides. Rev Mineral Geochem 61:181-229

Wright K, Hillier IH, Vincent MA, Kresse G (1999) Dissociation of water on the surface of galena (PbS): A comparison of periodic and cluster models. J Chem Phys 111:6942-6946

Wright K, Watson GW, Parker SC, Vaughan DJ (1998) Simulation of the structure and stability of sphalerite (ZnS) surfaces. Am Mineral 83:141-146

Yin Q, Kelsall GH, Vaughan DJ, England KER (1995) Atmospheric and electrochemical oxidation of the surface of chalcopyrite ($CuFeS_2$). Geochim Cosmochim Acta 59:1091-1100

Zangwill A (1988) Physics at Surfaces. Cambridge University Press

Zeng Y, Holzwarth NAW (1994) Density functional calculation of the electronic structure and equilibrium geometry of iron pyrite (FeS_2). Phys Rev B 50:8214-8220

Reactivity of Sulfide Mineral Surfaces

Kevin M. Rosso

*Chemical Sciences Division
and W.R. Wiley Environmental Molecular Sciences Laboratory
Pacific Northwest National Laboratory
Richland, Washington, U.S.A.
e-mail: kevin.rosso@pnl.gov*

David J. Vaughan

*School of Earth Atmospheric and Environmental Sciences
and Williamson Research Centre for Molecular Environmental Science
University of Manchester
Manchester, United Kingdom
e-mail: david.vaughan@manchester.ac.uk*

INTRODUCTION

In the preceding chapter, the fundamental nature of sulfide mineral surfaces has been discussed, and the understanding we have of the ways in which the surface differs from a simple truncation of the bulk crystal structure reviewed. This naturally leads on to considering our understanding of sulfide surface chemistry, in the sense of how sulfide surfaces interact and react, particularly with gases and liquids.

As noted elsewhere in this volume, research on sulfide mineral surfaces and surface reactivity is a relatively recent concern of mineralogists and geochemists, partly prompted by the availability of new imaging and spectroscopic methods, powerful computers and new computer algorithms. There has been a significantly longer history of sulfide mineral surface research associated with technologists working with, or within, the mining industry. Here, electrochemical methods, sometimes combined with analytical and spectroscopic techniques, have been used to probe surface chemistry. The motivation for this work has been to gain a better understanding of the controls of leaching reactions used to dissolve out metals from ores, or to understand the chemistry of the froth flotation systems used in concentrating the valuable (usually sulfide) minerals prior to metal extraction.

The need for improved metal extraction technologies is still a major motivation for research on sulfide surfaces, but in the last couple of decades, new concerns have become important drivers for such work. In particular, much greater awareness of the negative environmental impact of acid and toxic metal-bearing waters derived from breakdown of sulfide minerals at former mining operations has prompted research on oxidation reactions, and on sorption of metals at sulfide surfaces. At the interface between fundamental geochemistry and industrial chemistry, the role of sulfide substrates in catalysis, and in the self-assembly and functionalization of organic molecules, has become an area of significant interest. Such work ranges in its application from the development of new industrial processes, to fundamental questions of the possible role of sulfide surfaces in catalyzing the formation of the complex organic molecules leading to the emergence of life on Earth.

In this chapter, we aim to provide an overview of current understanding of sulfide surface chemistry. The size of this research field is already such that it is impossible to discuss all of the published work, but key examples are considered and readers directed to the main literature sources. The chapter begins with some examples of reaction with gaseous species (O_2, H_2O, H_2S, CH_3OH) as these are the most accessible in terms of understanding reactivity at the molecular scale. The very important oxidation and related electron transfer reactions, in both air and aqueous solution, are then considered before considering examples of catalysis and functionalization/self-assembly and interaction with organic molecules. In the final section, sorption of metal ions onto sulfide mineral surfaces is discussed before a few words concerning the future outlook for research in this entire area.

ELEMENTARY ADSORPTION/OXIDATION REACTIONS

The reaction of clean mineral surfaces with pure gaseous molecules in a highly controlled UHV environment can provide important insights into the reactivity of particular surfaces and surface sites. In this way, reaction mechanisms can be explored experimentally for different cleavage or fracture surfaces and types of defect sites, and with close control of temperature and concentration of the reactant molecule. Computer modeling approaches, including *ab initio* computations, are well suited to the theoretical investigation of such reactions. Examples of experimental and computational studies of key sulfide surfaces will now be discussed, before considering the complexities of more advanced surface transformation reactions with air and with aqueous solutions of varying compositions.

Galena

As emphasized in the preceding chapter, the simple crystal structure and perfect cleavage of PbS have made it an attractive substance for surface science studies, and this extends to work on surface reactivity. In particular, the surface breakdown reactions in oxidizing conditions on exposure to water have been a focus of attention because of their obvious environmental relevance. A more fundamental understanding of these reactions is possible through both experimental and computer modeling studies of the interaction of oxygen atoms or molecular species such as H_2O with the dominant cleavage (and growth) surface of galena, the (100) surface.

While a significant body of experimental work has been performed on PbS oxidation in air and aqueous solutions, as reviewed in the next section, very little work has been performed under the controlled conditions of UHV. Therefore, much of what is currently known about the initial adsorption and oxidation reactions due to oxygen exposure comes from a combination of scanning tunneling microscopy (STM) and spectroscopy (STS) studies of the initial stages of oxidation in air and computational molecular modeling. Eggleston and Hochella were the first to show atomic-scale evidence of PbS (100) oxidation due to air exposure (Eggleston and Hochella 1990, 1991, 1993, 1994). Using STM, they showed that nanometer-scale oxidation patches (low tunneling current dark areas) develop on this surface that interrupt the otherwise continuous array of high tunneling current sites (bright spots) corresponding to atomic positions. Becker and Hochella (1996) used *ab initio* computation methods to calculate STM images, STS spectra and X-ray photoelectron spectra (XPS) peak shifts for pristine PbS (100) and for that surface following reaction with oxygen. Experimental STM images of the upper valence band of galena show a periodic array of spots which could arise from either Pb or S; the calculations show that these spots (imaged under conditions where electrons tunnel from occupied states) stem from electronic states with sulfur $3p$ character near the Fermi level. If the galena is exposed to oxygen, more and more of the former bright sulfur spots appear as dark spots. Zyubina et al. (2005) recently demonstrated mechanisms of O_2 dissociation on small PbS (and doped and undoped PbTe) clusters via a peroxide-like surface complex using density functional theory (DFT) cluster calculations. A combination of periodic slab and aperiodic cluster calculations showed that dissociated O_2 yields O adsorbed on top of terrace S atoms, and O migrating beneath the uppermost surface planes at steps and corners (Becker and Hochella 1996; Zyubina et al. 2005).

Computation of STS spectra as well as STM images help in understanding how these reactions would manifest in experimental tunneling data. In STS measurements, the microscope tip is held stationary over a certain atomic position and the bias voltage ramped over a range (e.g., −1 V to +1 V). Under certain conditions and simplifications, the tunneling spectra provide information reflecting the local density of states of the sample at the location of the tip (Tersoff and Hamann 1985). Figure 1 shows a calculated STM image and calculated STS spectra for a sulfur atom with an oxygen adsorbed on top, its second nearest neighbor atom, and a third sulfur atom least affected by the adsorbed oxygen. The depletion of electron density associated with oxygen atom adsorption can be seen in the calculated spectra and accounts for the darkening of regions in the STM image where adsorption has occurred. This effect results from electron density being drawn away from the surface sulfurs by the oxygen atom, depopulating the S $3p$ states (oxidizing the sulfide). This interpretation of STM images was subsequently confirmed experimentally by Eggleston (1997) on PbS (100) surfaces oxidized in air using a dual-bias imaging method which enables simultaneous acquisition of two kinds of images (one of occupied and one of unoccupied states) over a single area of the surface. This showed that S $3p$ states are depopulated by oxidation and so (under negative sample bias) fewer electrons tunnel from these states into the STM tip, giving a darkened area; conversely, when the bias is reversed electrons tunnel from the tip to the sample and oxidized sites appear bright.

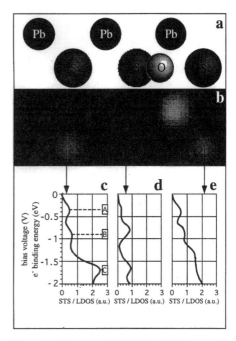

Figure 1. Calculated PbS (100) STM image and STS spectra with adsorbed oxygen; a) the supercell unit for the calculation, b) the calculated image, c-e) calculated spectra at various locations. [Used by permission of Elsevier from Becker and Hochella (1996), *Geochimica et Cosmochimica Acta*, Vol. 60, Fig. 7, p. 2421.]

The PbS (100) surface is particularly amenable to both experimental and computational studies and has therefore become a system for demonstrating new concepts in mineral surface science. The combined experimental and computational work of Becker and Rosso (2001) on geometric and electronic structures of step edges on this surface have been discussed in the previous chapter (Rosso and Vaughan 2006) and clearly have important implications for surface reactivity. Those results show that the density and energy of empty Pb $6p$ states is significantly increased at step edge Pb sites relative to that of terrace Pb sites. This suggests that the edge Pb sites have increased capacity to accept electrons, and are therefore stronger Lewis acid sites. Therefore charge donating adsorbates such as water should bind more strongly to these sites via σ-like interaction between water O sp^3 and surface Pb $6p$ orbitals. This is similar to the earlier findings of Wright et al. (1999a,b) who employed *ab initio* cluster and periodic models to investigate specifically the reaction of water on the PbS surface. They found that the perfect (100) surface is very unreactive, even in the presence of vacancies in the surface layer. However, at step sites (of a single atomic layer), a dissociation reaction described by:

$$PbS + H_2O \rightarrow Pb(OH)^+ + HS^- \qquad (1)$$

was calculated to occur readily, even at ambient temperature. These calculations also provided insights into the reaction mechanism. In Figure 2, the initial configuration, the transition state geometry, and the dissociated end-product structure from the calculations are shown. In a further development of this work, Bryce et al. (2000) used an embedded cluster model adjoined to a dielectric conductor screening model and included the effects of bulk water via a reaction field. The results are in line with the earlier studies and suggest that physisorption occurs at the perfect (100) surface and chemisorption at a step defect. The effect of the polarization and dynamics of surrounding solvent water tends to reduce the driving force and increase the reaction barrier for water dissociation at steps (barrier = 48.6 kcal mol^{-1}).

Another aspect of surface reactivity developed using galena as a model (and also pyrite) is the so-called "proximity effect" on semiconducting mineral surfaces using *ab initio* molecular orbital and planewave DFT calculations, and STM and STS experiments (Becker et al. 2001; Rosso and Becker 2003). Here, the concept is one whereby the chemical reaction of one surface site influences the electronic structure and reactivity of neighboring or nearby sites. The site-to-site interaction of interest here is indirect through the surface, rather than the more well-known direct through-space interadsorbate interaction (such as described by the Fowler adsorption isotherm). The example involving galena which was studied initially was its oxidation via interaction with ferric iron and water. In this case, Fe^{3+} acts as the

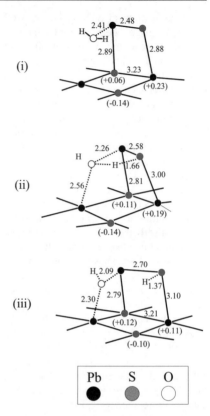

Figure 2. Computational model for H_2O interaction at a PbS (100) step site showing the (i) initial, (ii) transition-state, and (iii) final configurations. [Used by permission of Elsevier from Wright et al. (1999a), *Chemical Physics Letters*, Vol. 299, Fig. 1, p. 529.]

oxidizing agent (electron acceptor) and water promotes the reaction and supplies the oxygen necessary to produce sulfate in the overall reaction (after Rimstidt et al. 1994):

$$PbS + 8Fe^{3+} + 4H_2O \leftrightarrow Pb^{2+} + SO_4^{2-} + 8H^+ + 8Fe^{2+} \qquad (2)$$

It is shown by Becker et al. (2001) that the ferric iron and the water molecule involved in this oxidation do not have to be bonded to the same surface atom. As illustrated in Figure 3, these species may interact at different surface sites including step sites, and electrons may be transported through mineral surface layers or along specific pathways such as steps over distances of several nanometers. Rosso and Becker (2003) used planewave DFT calculations to assess the strength of through-surface interadsorbate interaction between oxygen adsorbed on PbS (100) and also between adsorbed oxygen and vacancies as a function of separation distance. Energy-distance plots indicate that the proximity effect becomes a very strong attractive interaction at separations decreasing below ~5-6 Å, and the proximity effect persists at separations up to 12 Å for these surface species (Fig. 4). Interestingly, the strength of the proximity effect attracting adsorbed O atoms together was found to out-compete their through-space electrostatic repulsion. This finding demonstrated the presence of a strong organizing

Reactivity of Sulfide Mineral Surfaces 561

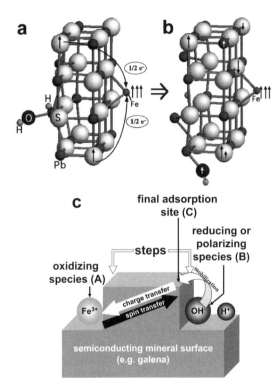

Figure 3. Galena as an example of the proximity effect on semiconducting mineral surfaces, showing a) a cluster model with the initial configuration, b) final configuration, and c) a diagram depicting the relevant processes. [Used by permission of Elsevier from Becker et al. (2001), *Geochimica et Cosmochimica Acta*, Vol. 65, Figs. 1 and 2, pp. 2643 and 2644.]

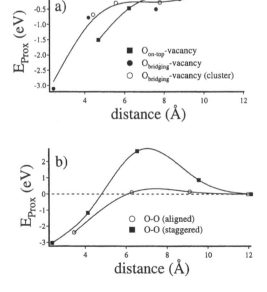

Figure 4. Calculated proximity effect energies as a function of distance of separation for a) adsorbed oxygen-vacancy interaction, and b) oxygen-oxygen interaction on PbS (100). [Used with permission of Elsevier from Rosso et al. (2003), *Geochimica et Cosmochimica Acta*, Vol. 67, Fig. 9, p. 949.]

force on like adatoms that could explain the oxidation of galena and possibly other sulfide surfaces by progressive patchwork growth.

Pyrite

Experimental studies of pyrite surface reactivity are more demanding than those for galena, given that it is more difficult to isolate crystallographically-oriented pyrite surfaces. Pyrite has poor (100) cleavage, and thus fracture leads to macroscopically conchoidal surfaces. However, as reviewed in the previous chapter (Rosso and Vaughan 2006), true pyrite (100) surfaces can be found on fracture surfaces at length scales of several hundred nanometers. Therefore, while crystallographic pyrite surfaces are accessible by local probe methods such as STM, most analytical surface spectroscopic methods necessarily incorporate the overall complicated surface structure. The surface topography and preparation method are particularly key issues for understanding the reactivity of pyrite surfaces because of the wide range of site types and structural environments that occur on differently prepared surfaces.

Using an *in situ* UHV cleaning approach developed by Strongin and co-workers (Chaturvedi et al. 1996), seminal investigations have been undertaken on clean natural pyrite *growth* surfaces for the study of reaction with gaseous molecules in UHV. The cleaning procedure involves cycling between low-energy sputtering, usually with He^+, to remove adventitious material and annealing. These "as-grown" surfaces were shown to be very topographically complex as observed by atomic force microscopy (Hochella et al. 1998), but this cleaning approach gives sharp (100) diffraction as determined by low-energy electron diffraction (LEED). In-UHV fracturing also leads to sharp (100) diffraction, but the topography is distinctly different (Rosso et al. 1999a). The in-UHV cleaned surfaces present a smaller population of monosulfide species than surfaces prepared by in-UHV fracture (Chaturvedi et al. 1996). The cleaning approach has the advantage that it is amenable to preparing pyrite surfaces of orientation other than (100) for UHV study, such as the (111) surface, whereas surfaces of this orientation have not yet been shown to be produced to any degree by fracture. Nevertheless, some differences between the reactivity of pyrite surfaces prepared by in-UHV cleaning and in-UHV fracture are readily apparent in the literature. In general, both sets of results re-enforce the importance of defects, their structure and relative abundance, for controlling at least the initial stages of pyrite surface reactivity.

Guevremont et al. (1997, 1998a,b,c,d) employed a range of UHV surface analytical techniques including XPS and temperature programmed desorption (TPD) to study the interaction of molecular H_2O, O_2 and CH_3OH on cleaned pyrite growth surfaces. The binding of H_2O and CH_3OH at low temperature (90 K) was investigated by techniques including TPD and photoemission of adsorbed xenon (PAX) (Guevremont et al. 1997) which show that both species adsorb molecularly at 90 K and thermally desorb between 170 and 400 K. The PAX data led to the suggestion that adsorption of these molecules occurs preferentially at small populations of defect sites believed to be, at least in part, sulfur vacancies. After defect sites are saturated, it is suggested that adsorbates cluster on the less reactive stoichiometric surface.

Guevremont (1998a) made comparative studies of the oxidation of in-UHV cleaned (100) and (111) growth surfaces exposed to H_2O, O_2 and an H_2O/O_2 mixture in an apparatus allowing surfaces to be reacted at environmentally relevant pressures and then studied using XPS. In these experiments, neither the (100) nor the (111) growth surface exhibited substantial reaction in pure O_2 for exposure pressures up to 1 bar. Exposure to H_2O vapor at similar pressures resulted in significant oxidation of FeS_2 (111) but a much smaller amount of oxidation of the FeS_2 (100) surface; it was suspected by the authors that on the (100) surface, H_2O only reacted at defect sites. Both surfaces showed substantial reaction with the H_2O/O_2 mixture, and this was more than the sum of reactions observed with equivalent amounts of pure H_2O and O_2, leading to the suggestion of a "synergy" between H_2O and O_2 in oxidizing pyrite. Typical XPS data (S $2p$ and Fe $2p$ peaks) for the fresh FeS_2 (100) surface and that exposed to H_2O, O_2 or mixed H_2O/O_2 are shown in Figure 5. In the S $2p$ spectra, features at binding energies of 162.5 and 161.5 eV from the clean surface are assigned to the disulfide group and a monosulfide species; after exposure to H_2O or H_2O/O_2, new features appear at higher binding energies of 163.4 and 168.8 eV whereas equivalent exposure to O_2 has no significant effect. Although these features are clearly products of the oxidation reactions, their precise assignment is uncertain although polysulfide, a sulfur oxide or thiosulfate are suggested as possible reaction products. The growth of a feature at 711 eV in the Fe $2p$ spectrum, following oxidation, was attributed to Fe^{3+}, possibly associated with an iron oxide or hydroxide surface product.

Reactivity studies using precision reactant exposures in UHV with pyrite surfaces generated by fracturing in vacuum demonstrate behavior that is somewhat different than that for cleaned growth surfaces. For example, in contrast to the results of Guevremont et al. (1998a) on cleaned growth surfaces, several studies have shown that fractured pyrite surfaces react with pure O_2. High resolution STM imaging, STS spectra, and ultraviolet photoelectron spectroscopy (UPS) along with quantum mechanical cluster calculations

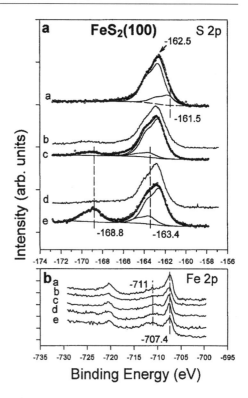

Figure 5. XPS data for pyrite exposed to various amounts of pure or mixtures of O_2 and H_2O, showing changes in the a) S $2p$ and b) Fe $2p$ spectra. From Guevremont et al. (1998b).

were used by Rosso et al. (1999b) to study fundamental oxidation mechanisms when gaseous O_2 and H_2O and their mixtures interact with pyrite (100) surfaces. The UPS data for surfaces exposed to O_2 show that the density of states decreases at the top of the valence band but increases deeper in the valence band, indicating consumption of low binding energy electrons occupying dangling bond surface states localized on surface Fe atoms, consistent with the formation of Fe^{3+}-O^- bonds. This was observable by UPS for O_2 exposures as low as 4 L (e.g., 10^{-7} mbar exposure for 40 seconds); the oxidation process stopped at 20 L (Fig. 6). For exposures to pure H_2O, no oxidation was observed for exposures up to 10 L. Similar to the results of Guevremont et al. (1998a) however, it was found that a mixture of O_2 and H_2O oxidizes the surface much more than does an equivalent concentration of pure O_2 (Fig. 6).

Kendelewicz et al. (2004) have studied FeS_2 fracture surfaces exposed to pure O_2, water vapor, and air using SR-XPS and found behavior largely consistent with Rosso et al. (1999b). Molecular oxygen reacts to a measurable extent with the surface through dissociative chemisorption of O_2 (Fig. 7) but they found this to occur only at much higher O_2 exposures (~10^7 L). These authors went to exposures up to atmospheric O_2 partial pressures. They also found the surface to be unreactive with pure water vapor up to exposures similar to room temperature saturation pressures (~10^{10} L). Consistent with Guevremont et al. (1998a), Kendelewicz et al. (2004) showed the presence of a small percentage of surface monosulfide, which is eradicated by dosing with either O_2 or H_2O. They also identified a range of intermediate oxidation products in their experiments including sulfur oxoanions and zero-valent sulfur.

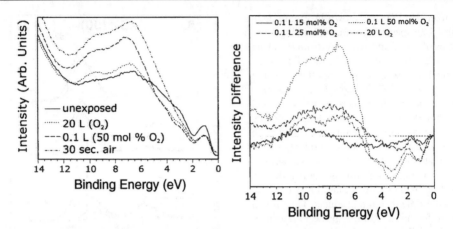

Figure 6. He I UPS valence band spectra (left) and difference spectra (right) comparing pristine pyrite surfaces with those exposed to O_2, O_2-H_2O, and air. From Rosso et al. (1999b).

The progressive oxidation of the (100) surface was experimentally followed at the local atomic level (Rosso et al. 1999b). UHV STM images at atomic resolution were obtained for (100) terraces of the surface exposed to O_2 and to H_2O/O_2 mixtures (Fig. 8). The images show oxidation features in the form of dark "patches" in both the upper valence band and lower conduction band states around the Fermi level. On the basis of *ab initio* calculations discussed below, these features are interpreted as where dissociative chemisorption of O_2 has taken place at Fe sites. At these sites, Fe $3d_{z^2}$ dangling bond surface state density has been lost to bonds with the chemisorbed O species. Consistent with the UPS data, STS spectra show removal of the highest occupied and lowest unoccupied surface state density associated with dangling bond states, in accordance with this interpretation.

The importance of Fe $3d_{z^2}$-like surface states in controlling the initial stages of pyrite (100) reaction with O_2 and H_2O has been re-enforced by various computational molecular modeling studies. The interaction between pyrite (100) surfaces and water molecules is a well-studied example. Hartree-Fock, DFT, and empirical potential models all show that H_2O molecules at various coverages preferentially adsorb by association between their oxygen atoms and surface Fe atoms (Rosso et al. 1999b; de Leeuw et al. 2000; Stirling et al. 2003; Philpott et al. 2004). This result is consistent with experiments (Pettenkofer et al. 1991; Guevremont et al. 1997, 1998a; Rosso et al. 1999b; Kendelewicz et al. 2004). Empirical potential calculations on slab models by de Leeuw et al. (2000) predicted an adsorption energy of −47 kJ mol^{-1}, which compares well with −42 kJ mol^{-1} determined by TPD (Guevremont et al. 1998a). In particular, this interaction has been shown to be mediated by charge transfer from highest occupied O $2p$ states in water to unoccupied Fe $3d_{z^2}$ electronic states near the Fermi level (Rosso et al. 1999b; Stirling et al. 2003). H_2O dissociation is predicted to be strongly unfavorable (Rosso et al. 1999b; de Leeuw et al. 2000; Stirling et al. 2003; Philpott et al. 2004) and partly stabilized at higher coverages by hydrogen bonding interactions (Stirling et al. 2003). Occupied Fe $3d_{z^2}$ states are responsible for binding adsorbed O_2 and providing electrons that destabilize the O_2 molecule towards dissociation by electron transfer from surface Fe into π^* antibonding O_2 molecular orbitals (Rosso et al. 1999b). These authors also showed that H_2O dissociatively sorbs at surface Fe sites only when dissociated O_2 is present at Fe sites nearby, which provides a mechanism explaining the enhanced reactivity of the combined O_2/H_2O gases in causing surface oxidation (Fig. 9). Collectively from the spectroscopic data, STM data, and molecular modeling results to date, the supported synergistic oxidation mechanism is one involving competitive adsorption of O_2 and

H$_2$O at surface Fe sites, oxidation of surface Fe sites by O$_2$, dissociation of co-adsorbed H$_2$O at Fe sites, and charge redistribution in surface S atoms (Rosso et al. 1999b). This is one pathway that allows for production of hydroxyls from dissociated water and subsequent nucleophilic attack of these hydroxyls at surface S sites, consistent with the observations that oxygen in final product sulfate arises predominantly from water molecules (Taylor et al. 1984; Usher et al. 2004, 2005).

AIR/AQUEOUS OXIDATION

The instability of sulfide minerals when exposed to the Earth's atmosphere or to oxygenated surface waters makes redox reactions in general, and oxidation reactions in particular, of great importance in sulfide mineralogy and geochemistry. As seen from the work on clean surfaces and pure gases (O$_2$, H$_2$O) or simple mixtures (H$_2$O/O$_2$) discussed above, progress has been made in studies of "model systems" from which the importance of oxygen and water as oxidants is clear. In natural systems, and in the engineered environments associated with the mineral processing industry, other oxidants are important (notably ferric iron) and other factors commonly make major contributions to the rates of redox reactions. In particular, temperature and pH conditions are significant, and the roles played by a range of bacteria are especially important. In the following section, oxidation in the air and redox reactions under a variety of conditions are discussed with reference to three major binary sulfide minerals (galena, pyrite and pyrrhotite), one mixed cation sulfide (chalcopyrite) and one mixed anion sulfide (arsenopyrite).

Galena

Earlier work on the oxidation of galena in air and in aqueous solution was summarized in Tossell and Vaughan (1987); subsequent work has involved electrochemical methods (e.g., Richardson and Odell 1984; Fornasiero et al. 1994), Raman spectroscopy (Turcotte et al. 1993), thermodynamic methods (Acharya and Paul 1991; Janczuk et al. 1992a,b) and the use of bacteria (Bang et al. 1995). XPS has proved particularly important in providing insights into oxidation processes, more recently through XPS using synchrotron radiation (SR-XPS).

Laajalehto et al. (1997) have applied SR-XPS to air oxidized galena and found that this provides an order of magnitude improvement in surface sensitivity compared with conventional XPS through the ability to tune the energy of the source. After only 10 min

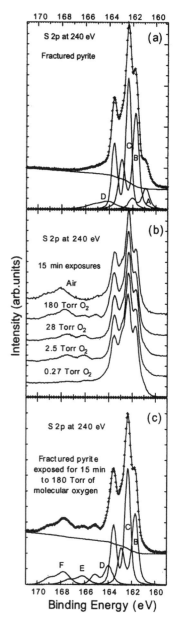

Figure 7. S 2p SR-XPS data for a) fractured pyrite, b) after exposure to different partial pressures of O$_2$ and to ambient air, and c) after exposure to 180 Torr of O$_2$ for 15 min. [Used with permission of Elsevier from Kendelewicz et al. (2004) *Surface Science*, Vol. 558, Fig. 1, p. 82.]

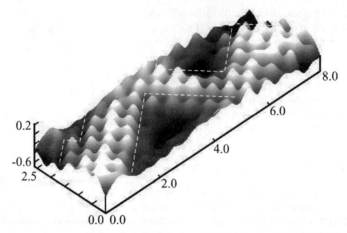

Figure 8. UHV STM image of a pyrite (100) surface exposed to 4 L O_2, showing oxidized patches as low dark areas outlined by white dashed lines. Unoxidized Fe sites affected by neighboring oxidized sites and the still unaffected regions. Scale bars are in nanometers. Adapted from Rosso et al. (1999b).

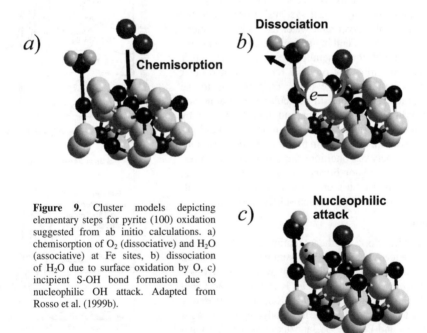

Figure 9. Cluster models depicting elementary steps for pyrite (100) oxidation suggested from ab initio calculations. a) chemisorption of O_2 (dissociative) and H_2O (associative) at Fe sites, b) dissociation of H_2O due to surface oxidation by O, c) incipient S-OH bond formation due to nucleophilic OH attack. Adapted from Rosso et al. (1999b).

exposure to air, oxidized sulfur (both sulfate and metal polysulfide) species can be found, and oxidized sulfur and lead species are observed simultaneously, supporting congruent oxidation of the surface. The kinetics of PbS oxidation contrasts with that of pyrite which was also studied. Whereas a layer of oxidation products forms on pyrite after only a few seconds, it takes several minutes for any evidence of galena oxidation to appear. However, whereas the instantly formed layer on pyrite seems to passivate the surface, on galena, oxidation proceeds continuously with time. Laajalehto et al. (1997) have also drawn attention to the importance of cooling samples of sulfides such as galena before and during XPS experiments (both conventional and SR-XPS) to avoid loss of volatile surface reaction products. Their examination (*ex situ* and over the temperature range 143 – 298 K) of S $2p$ spectra of an electrochemically oxidized PbS (100) surface showed a layer structure of elemental sulfur and metal polysulfides.

Other more recent studies of galena oxidation have also used vibrational spectroscopies to identify oxidation products. Shapter et al. (2000) used Raman spectroscopy to identify oxysulfates (PbO·PbSO$_4$, 3PbO·PbSO$_4$ and 4PbO·PbSO$_4$) produced by laser heating in air. (Unfortunately the temperatures involved, suggested to reach 900°C, could not be exactly determined). Chernyshova (2003) used *in situ* FTIR to study the products of electrochemical oxidation of galena formed at the electrode/electrolyte interface in solution at pH 9.2. From the FTIR measurements it is suggested that under these conditions oxidation of galena starts with the reaction:

$$PbS + 2xh^+ \rightarrow Pb_{1-x}S + xPb^{2+} \tag{3}$$

followed by the reaction:

$$PbS + 2h^+ \rightarrow Pb^{2+} + S^0 \tag{4}$$

where h^+ denotes a hole. Lead sulfate, thiosulfate and polythionate ions are formed from the elementary sulfur formed at the first oxidation stage, and Pb(OH)$_2$ is precipitated (Fig. 10).

Pyrite

The importance of the oxidation of pyrite in air and aqueous media has ensured numerous studies, both experimental and theoretical. Earlier work has been reviewed by Lowson (1982), and amongst many subsequent studies have been those of McKibben and Barnes (1986), Buckley and Woods (1987), Moses et al. (1987), Mishra and Osseo-Asare (1992), Moses and Herman (1991), Raikar and Thurgate (1991), Karthe et al. (1993), Tao et al. (1994), Sasaki (1994), Nesbitt and Muir (1994), Williamson and Rimstidt (1994), Ciminelli and Osseo-Asare (1995), Eggleston et al. (1996), Bonnissel-Gissinger et al. (1998), Evangelou et al. (1998), Schaufuss et al. (1998a,b), England et al. (1999), Kelsall et al. (1999), Chernyshova (2003), Todd et al. (2003), Rimstidt and Vaughan (2003), Abraitis et al. (2004), and Mattila et al. (2004).

More recent work on pyrite oxidation in air has benefited from the application of synchrotron-based spectroscopic methods (see above discussion of Kendelewicz et al. 2004, and also Laajalehto et al. 1997; Schaufuss et al. 1998a,b; Mattila et al. 2004). For example, Schaufuss et al. (1998a,b) using SR-XPS were able to demonstrate the presence of at least three distinct sulfur states at the fractured pyrite surface, each of which oxidizes at a different rate in air. The most reactive surface component is S^{2-} and the second, the surface atom of the first disulfide layer (S_2^{2-}) with sulfur atoms of disulfide groups beneath the surface layer (bulk coordinated sulfurs) being least reactive. A model for the oxidation mechanism based on that originally proposed by Eggleston et al. (1996) was put forward by Schaufuss et al. (1998b) and is illustrated in Figure 11. In this model, Fe^{3+} states are proposed to arise after fracturing of S-S bonds (as suggested by Nesbitt et al. 1998) by electron transfer from Fe^{2+} to this S$^-$ state which then reacts rapidly to sulfate. In turn the sulfate may be displaced from its original site (bonded to Fe^{3+}) as a result of competitive sorption of H$_2$O, O$_2$ and OH$^-$. In Figure 11, the adsorption of O$_2$ as an outer sphere complex is shown. The initial stages in the formation of an Fe$_2$O$_3$ overlayer may begin with the transfer of an electron from an adjacent Fe^{2+} to the Fe^{3+} center, followed by electron transfer to O$_2$ producing O$_2^-$ and the formation of Fe^{3+}-O bonds. This produces a second Fe^{3+}. Based on bulk band edge positions of pyrite and hematite, Eggleston et al. (1996) argued that Fe^{2+} in surface Fe$_2$O$_3$ patches is a better O$_2$ reductant than Fe^{2+} in pyrite (Fig. 12). They proposed electron migration through Fe$_2$O$_3$ "surface states" to

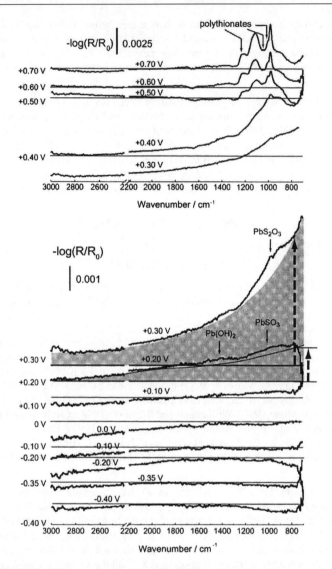

Figure 10. FTIR-ATR difference spectra of the PbS (100) / solution interface measured in situ during an electrochemical positive potential scan at pH 9.2 showing the formation of lead sulfate, thiosulfate, polythionate, and other reaction products. [Used with permission of Elsevier from Chernyshova (2003), *Journal of Electroanalytical Chemistry*, Vol. 558, Fig. 2, p. 87.]

O_2 and facile Fe^{2+}/Fe^{3+} cycling at patch edges by this mechanism. Ultimately, the result is the build-up of islands of Fe^{3+} oxides across the pyrite surface (Eggleston et al. 1996).

The products of pyrite oxidation in aqueous environments are critically dependent upon solution chemistry, with evidence of reaction products formed under different conditions coming from a wide range of (mostly *ex situ*) analytical techniques. For example, England et al. (1999) used synchrotron-based glancing-angle X-ray absorption spectroscopy to probe the surface (~2-3 nm) region of pyrite oxidized at pH 9.2, finding a goethite-like surface species. Todd et al. (2003) used synchrotron-based X-ray absorption spectroscopy (using oxygen *K*- and sulfur and

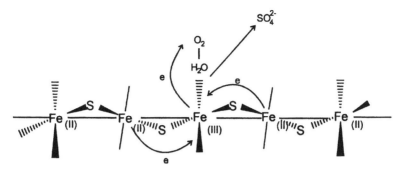

Figure 11. Proposed reaction mechanism of pyrite oxidation in air. [Used with permission of Elsevier from Schaufuss et al. (1998b), *Surface Science*, Vol. 411, Fig. 3, p. 327.]

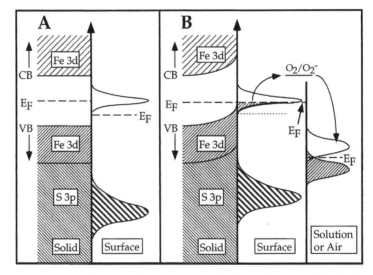

Figure 12. Band structure of the bulk and surface of pyrite a) before and b) after equilibration including the effects of hematite-like conduction band (unfilled) and valence band (bold cross-hatched) surface states. Surface oxidation in air leads to hematite-like surface states that facilitate further electron transfer to O_2. From Eggleston et al. (1996).

iron L-edges) to determine the phases formed by oxidation in aqueous electrolytes at pH 2 to pH 10 in the top 2-5 nm region of the surface. Below pH 4 a ferric (hydroxyl) sulfate is the main product whereas at higher pH an Fe^{3+} oxyhydroxide is also found. The presence of Fe^{3+} in solution autocatalyzes the oxidation, promoting ferric oxyhydroxide formation at low pH. Under the most alkaline conditions, the O K-edge spectrum resembles that of goethite (cf. England et al. 1999). Very similar results were found by Bonnissel-Gissinger et al. (1998) concerning the pH dependence of reaction products. Chernyshova (2003), in an *in situ* FTIR study in aqueous solution, looked at the earlier stages of reaction with the first (monolayer) stage (above 0 V in pH 9.2 solution) producing elemental sulfur and ferric hydroxide and the second (at 0.4 V) producing a sub-monolayer of Fe^{3+}-OH surface species. At higher potentials bulk ferric hydroxide, ferric sulfite and polythionates were reported, accompanied thereafter by sulfate formation. In another FTIR study, Evangelou et al. (1998) explored the role of bicarbonate in solution and suggested the formation of carbonate complexes at the pyrite surface.

Of course, the redox chemistry of pyrite in aqueous solution presents far more complexities than that involving pure gases or air. As noted by Rimstidt and Vaughan (2003), the key controls of mechanisms and hence rates of the oxidation of pyrite remain poorly understood, despite many decades of research, because the processes of aqueous oxidation involve a complex series of elementary reactions. As pointed out by Basolo and Pearson (1967), the elementary steps of redox reactions almost always involve the transfer of only one electron at a time, so that oxidation of a disulfide such as pyrite to release sulfate requires transfer of seven electrons and, hence, up to seven elementary steps. Furthermore, the minerals are semiconductors, so that the reactions are electrochemical in nature and, as already considered above in discussing the proximity effect, various reactions can happen at different sites with electron transfer through the mineral. Following arguments put forward by Kelsall et al. (1999) and Rimstidt and Vaughan (2003), it is suggested that pyrite aqueous oxidation is an electrochemical process with three distinct steps, as illustrated in Figure 13. These are the: (1) cathodic reaction, (2) electron transport, and (3) anodic reaction. The cathodic reaction involves an aqueous species that accepts electrons from an Fe^{2+} site on the mineral surface, such as O_2 in the reaction:

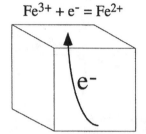

Figure 13. Diagram showing three steps involved in pyrite oxidation in aqueous solution. The cathodic reaction transfers electrons from the surface to the oxidant, which are replenished by electrons collected from anodic surface sites. Water molecules react with sulfur atoms around the anodic surface sites to form sulfoxy species, releasing hydrogen ions into solution. [Used with permission of Elsevier from Rimstidt and Vaughan (2003), *Geochimica et Cosmochimica Acta*, Vol. 67, Fig. 1, p. 874.]

$$FeS_2 + \tfrac{7}{2}O_2 + H_2O \rightarrow Fe^{2+} + 2H^+ + 2SO_4^{2-} \tag{5}$$

It is now clear that the cathodic reaction is the rate-determining step for sulfide mineral oxidation. Williamson and Rimstidt (1994) showed that the pyrite oxidation rate depends on concentration of O_2 or another oxidant (such as Fe^{3+}) and this is the case for other sulfide minerals such as galena, sphalerite, chalcopyrite and arsenopyrite. The exact stages involved in a reaction such as that above are suggested to follow a sequence like that already outlined above for air oxidation. The anodic reaction involves a multistep sulfur atom oxidation (electron removal) process with a sequence of surface reactions of the type:

$$py\text{-S-S} = py\text{-S-S}^+ + e^- \tag{6}$$

$$py\text{-S-S}^+ + H_2O = py\text{-S-S-OH} + H^+ \tag{7}$$

$$py\text{-S-S-OH} = py\text{-S-SO} + e^- + H^+ \tag{8}$$

$$py\text{-S-SO} = py\text{-S-SO}_2 + 2e^- + 2H^+ \tag{9}$$

$$py\text{-S-SO}_2 = py\text{-S-SO}_3 + 2e^- + 2H^+ \tag{10}$$

At this stage there is a tendency for this species to break away from the surface as a thiosulfate complex, which can be written as

$$py\text{-S-S-O}_3 = py + MS_2O_3 \tag{11}$$

The final step is envisaged as being pH dependent so, if the pH is high, the terminal $S\text{-SO}_3$ completely ionizes making the S-S bond stronger than the Fe-S bond, so much of the sulfur is released into solution as $S_2O_3^{2-}$. At low pH, the majority of terminal $S\text{-SO}_3$ groups retain a proton (so are S-SO$_3$H) which encourages transfer of electrons into the S-S bond where they are more easily transferred to the cationic site, leaving the sulfur with a very positive charge. This leads to further nucleophilic attack by a water molecule to produce SO$_4^{2-}$.

Pyrrhotite

A broad review of pyrrhotite oxidation has been published by Belzile et al. (2004), including data on oxidation rates, activation energies and factors controlling rates and oxidation products, as well as discussion of oxidation mechanisms. The air oxidation of pyrrhotite was studied using XPS by Buckley and Woods (1985) and they proposed a mechanism involving diffusion of iron to the surface to form (via an Fe^{2+} oxide) an Fe^{3+} hydrated oxide/hydroxide outer layer. Very detailed XPS and Auger electron spectroscopy (AES) studies of pyrrhotite (Fe$_7$S$_8$) air oxidation by Pratt et al. (1994) and Mycroft et al. (1995b) confirm this outward diffusion of iron. These authors propose the development of a layer structure with (after 50 h of air oxidation) an oxygen-rich and sulfur-depleted outermost layer of <1 nm covering an iron-deficient, sulfur-rich layer which, in turn, shows a gradual decrease in S:Fe ratio with depth. The proposed inward sequence of compositions is FeO$_{1.5}$, FeS$_2$, Fe$_2$S$_3$ and Fe$_7$S$_8$. A mechanism is proposed whereby molecular oxygen is adsorbed onto the pyrrhotite surface and reduced to O^{2-} and then reacts with Fe^{3+} bonded to S to form Fe^{3+}-O bonds and Fe^{3+} oxyhydroxides. Diffusion of iron outwards establishes the outer ferric oxyhydroxide layer, causing the depletion and chemical stratification of the underlying sulfide. Mycroft et al. (1995b) used angle resolved XPS to refine this model, suggesting approximate thicknesses of ~5 Å for the oxidized layer and 30 Å for the S-depleted sulfide layer. In both this work and secondary ion mass spectrometry (SIMS) studies by Smart et al. (2000), S-S bonding was observed and interpreted as due to reorganization of the pyrrhotite structure towards a marcasite-type structure in the proposed FeS$_2$ layer.

Chalcopyrite

Chalcopyrite (CuFeS$_2$) as the most important copper ore mineral, has been the subject of numerous studies of its surface reactivity, particularly of its oxidation in air and redox reactions in aqueous solution. Yin et al. (1995) quantitatively estimated the extent of atmospheric oxidation by determining the charge required to electrochemically reduce chalcopyrite exposed to air for different time periods. The average thickness of the oxidized surface layer was found to develop from ~1.5 nm after 10 min to ~4.5 nm after 100 min. Studies in solution have commonly used electrochemical techniques, and a brief résumé of the results obtained from many of these investigations is provided by Yin et al. (2000).

Typical of the results of such studies is the cyclic voltammogram in Figure 14 which shows a series of peaks arising from oxidation reactions leading to the breakdown of the chalcopyrite in an alkaline solution at pH 9.2, typical of flotation systems (Yin et al. 2000). Although inferences can be drawn directly from the electrochemical data as to the reactions giving rise to peaks A, B and C in the voltammogram, independent information on the phases forming at the chalcopyrite electrode surface can only be obtained from surface analysis. Thus, following transfer of an electrode conditioned under the appropriate potential into the analysis chamber of an XPS instrument, carefully avoiding exposure to atmosphere, O 1s and Fe 2p photoelectron spectra (Fig. 15) can be used to confirm that ferric oxide/hydroxide and some adsorbed O$_2$ are present under the conditions associated with the first oxidation reaction (A in Fig. 14). The S 2p and Cu 2p spectra remain essentially unchanged from those of the fresh surface. Combining electrochemical and spectroscopic data for reaction A, and then for the oxidation reactions at higher potentials (peaks B and C in Fig. 14) enables a sequence of oxidation reactions to be proposed:

$$2CuFeS_2 + 6xOH^- \rightarrow 2CuFe_{1-x}S_2 + xFe_2O_3 + 3xH_2 + 6xe^- \quad (12)$$

$$2CuFeS_2 + 6OH^- \rightarrow 2CuS_2^* + Fe_2O_3 + 3H_2O + 6e^- \quad (13)$$

$$CuFeS_2 + 3OH^- \rightarrow CuS_2^* + Fe(OH)_3 + 3e^- \quad (14)$$

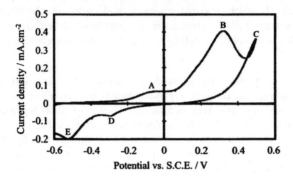

Figure 14. Cyclic voltammogram for oxidation reactions on chalcopyrite in a 0.1 M $Na_2B_4O_7$, pH 9.2 solution at 298 K. The potential was swept at 20 mV/s. See text for discussion of peaks. [Used with permission of The Mineralogical Society from Vaughan et al. (2002), *Mineralogical Magazine*, Vol. 66, Fig. 4a, p. 660.]

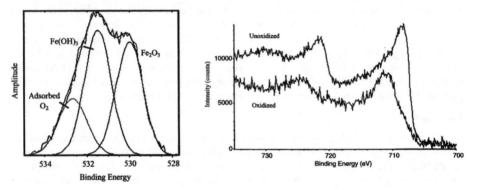

Figure 15. XPS spectra of the O $1s$ (left) and Fe $2p$ (right) regions on an electrochemically oxidized chalcopyrite surface. [Used with permission of The Mineralogical Society from Vaughan et al. (2002), *Mineralogical Magazine*, 66, Vol. Fig. 4b,c, pp. 660-661.]

The implications of these reactions are that at potentials just above the rest potential, a monolayer of $Fe_2O_3/Fe(OH)_3$ is formed leaving Cu and S unoxidized in the original chalcopyrite structure as a metastable phase of CuS_2 stoichiometry (and designated CuS_2^*). Together these phases cause passivation of the surface. A simple model of the development of the surface during oxidation is shown in Figure 16. With increasing potential these reactions continue, removing Fe from deeper within the chalcopyrite and with solid state diffusion being the likely rate-controlling mechanism. Above a critical potential, the metastable CuS_2^* phase decomposes (Yin et al. 2000). The oxidation of chalcopyrite in solution at pH 4 appears to be a similar story to that under alkaline conditions, with an iron oxide/hydroxide "layer" forming at the surface (Farquhar et al. 2003). However, AFM studies of the oxidized surface, both *ex situ* and in the presence of the fluid, show the presence of islands (<0.15 μm wide) of reaction products (Fig. 17). Coverage of the surface with these islands increases with the amount of charge passed. In a study of chalcopyrite oxidation in alkaline solutions, the findings of which were in general agreement with the above, Velasquez et al. (2005) observed similar "islands" using SEM (with EDX analysis facility). They describe the surface as composed of a "very superficial heterogeneous layer of oxidized materials from which protrude islands formed by grains of oxide, hydroxide and sulfate materials."

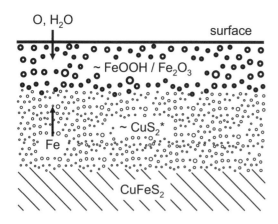

Figure 16. Model for the development of the chalcopyrite surface during oxidation. [Redrawn from Vaughan et al. 2002.]

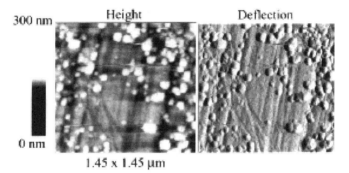

Figure 17. AFM height (left) and deflection (right) images of a chalcopyrite surface electrochemically oxidized at 650 mV in pH 4 solution. [Used with permission of The Mineralogical Society from Vaughan et al. (2002), *Mineralogical Magazine*, Vol. 66, Fig. 4f, p. 661.]

Similar electrochemical and XPS studies of the metal-enriched ("stuffed") derivatives of chalcopyrite, talnakhite ($Cu_9Fe_8S_{16}$), mooihoekite ($Cu_9Fe_9S_{16}$) and haycockite ($Cu_4Fe_5S_8$), show marked differences in oxidation rates compared with the stoichiometric phase (Vaughan et al. 1995). This is to be expected if solid state diffusion is the rate controlling mechanism; the additional interstitial site metals in the stuffed derivatives would be more active and mobile. The relative rates of oxidation are in the order haycockite > mooihoekite > talnakhite > chalcopyrite for reaction in acid solution. In alkaline solutions a similar pattern is observed although, following initial rapid oxidation of the stuffed derivatives, the passivating film which forms retards further oxidation to a greater extent than in chalcopyrite.

Arsenopyrite

A substantial number of studies have been performed to investigate the reactivity of arsenopyrite (FeAsS) in air and in aqueous solutions representative of both environmental systems and mineral processing operations (e.g., Buckley and Walker 1988; Richardson and Vaughan 1989; Nesbitt et al. 1995; Fernandez et al. 1996a,b; Nesbitt and Muir 1998; Schaufuss et al. 2000; Costa et al. 2002; Jones et al. 2003; Mikhlin and Tomashevich 2005). Many of these involve electrochemical methods and XPS studies, although Mikhlin and Tomashevich (2005) used X-ray absorption near-edge spectroscopy (XANES) as a surface probe. Jones et al. (2003) are among the first to investigate microbially mediated surface oxidation reactions, and Schaufuss et al. (2000) further exploit the potential of SR-XPS.

The earlier studies largely employed conventional XPS and AES analysis of fresh surfaces of natural arsenopyrite and of the same surfaces exposed to air, water and a range of acid and alkali aqueous solutions, as well as electrochemical treatments. For example, Buckley and Walker (1988) found that initial oxidation of FeAsS in air occurs rapidly according to the following generalized reaction:

$$FeAsS + (¾x + ½by)O_2 \rightarrow ½xFe_2O_3 + yAsO_b + Fe_{1-x}As_{1-y}S \quad (y > x; b \leq 1.5) \qquad (15)$$

with arsenic being oxidized faster than iron, and subsequent oxidation being slower and leading to the formation of As^{5+} oxides.

The complexities of the pristine FeAsS surface, showing a range of different As and S surface sites as revealed by SR-XPS, have already been discussed elsewhere in this volume (Rosso and Vaughan 2006). The same synchrotron technique was also used by Schaufuss et al. (2000) to study the initial stages of arsenopyrite oxidation on exposure to pure oxygen. The spectra show fast oxidation of As surface sites, and a consecutive reaction scheme for arsenic oxidation involving one-electron transfer steps was indicated by the detection of As^0, As^{2+}, As^{3+} and As^{5+} (Fig. 18). Further oxidation included exposure to the laboratory atmosphere (< 30 min) with the acquired spectra showing As^{5+} to be the final arsenic oxidation product, and with sulfate being the final result of S oxidation (but with numerous intermediate products). These oxidation products reach the surface by diffusion from the bulk. The authors also produced data interpreted in terms of the development of oxidized surface overlayers; an arsenic and iron containing overlayer on exposure to pure O_2 and, following atmospheric oxidation, another layer on top of this with an approximate FeOOH composition. This Fe-O overlayer was estimated to be ~1.8 monolayers thick and formed by the interaction of (atmospheric) water with Fe surface sites. In their electrochemical and XPS study of arsenopyrite oxidation in an acid chloride medium (pH 1.5), Costa et al. (2002) also interpreted their data in terms of the formation of such diffusion-controlled overlayers.

Mikhlin and Tomashevich (2005) studied pristine and air-exposed natural arsenopyrite samples and material which had been subjected to oxidative leaching in acidified ferric chloride, sulfate and nitrate solutions. They recorded Fe L- , S L- and O K-edge XANES spectra. The data from air-exposed samples can be interpreted in terms of iron being initially mainly low-spin Fe^{2+} but with the growth of features associated with high spin Fe^{2+} and Fe^{3+}, the latter becoming predominant after lengthy exposure (1 week). This ferric iron is also interpreted as being associated with As species as reported by others including, as noted above, Schaufuss et al. (2000). Samples leached in the acid ferric chloride and sulfate solutions show spectra consistent with iron and arsenic-depleted, sulfur-enriched surface layers. In contrast, samples leached with acid ferric nitrate show a large concentration of surface Fe^{3+} species. Interesting observations are also made by these authors in regard to the changing electronic structure of the surface following such leaching reactions, and readers are referred to the original publication for further details.

Jones et al. (2003) studied acidic (pH 2.3) oxidative leaching of arsenopyrite in the presence of *Thiobacillus ferrooxidans* and the essential salts required to sustain bacterial growth. Reaction between arsenopyrite and *T. ferrooxidans* in the essential salts solution produced a uniform solid $FePO_4$ overlayer (~0.2 μm thick) within 1 week. The phosphate here is derived from the essential salts (no such layer was seen in a biotic control experiment) and the layer continues to thicken with time, so that reaction continues with oxidation, diffusion, dissolution of arsenopyrite beneath the overlayer. The bacterial cells are separated from the arsenopyrite surface and it appears, therefore, that they do not need to be in contact with the arsenopyrite in order to promote rapid reaction.

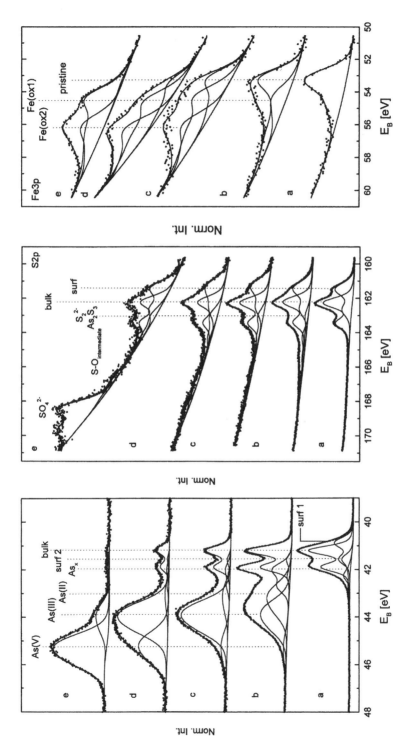

Figure 18. SR-XPS spectra of the As 3*d* (left), S 2*p* (center), and Fe 3*p* (right) regions for an arsenopyrite surface that is a) pristine and after exposure to b) 21,000 L O_2; c) 1 min air, d) 30 min air, and e) 22 hrs air. From Schaufuss et al. (2000).

CATALYSIS

There are very few known cases of true catalytic behavior for transition metal sulfide surfaces, where the surface itself is generally not altered or consumed in the chemical process. Therefore, it is of interest to examine the few cases where this behavior is known to occur. One such case is in hydrodesulfurization (HDS) reactions, which entails selective removal of organosulfur molecules from hydrocarbons produced from oil. Homogeneous HDS is accomplished by reacting these molecules with H_2 to cleave C-S bonds, forming H_2S and hydrocarbons as reaction products. RuS_2 is one of two naturally-occurring transition metal sulfides known to catalyze HDS reactions, the other being MoS_2. Other heteroatom removal reactions are also catalyzed by these surfaces, such as in hydrodenitrogenization. The heterogeneous HDS catalytic conditions are particularly aggressive, involving temperatures of about 600-700 K under a large excess of H_2 gas. Although the standard HDS catalytic material is currently MoS_2, RuS_2 surfaces are 13× more active for HDS of thiophene, a common HDS target molecule, than reference MoS_2 (Frechard and Sautet 1997a,b and references therein). Below, we review the studies of molecular-scale surface reactions underlying the HDS activity of these two materials. Because in both cases the active sites are predominantly believed to be undercoordinated metal atoms associated with sulfur vacancies, this review is also worthwhile as it is an excellent example of defect-driven surface chemistry, which is an important characteristic across the spectrum of possible surface chemical reactions occurring on transition metal sulfides and beyond.

Laurite

The adsorption of H_2 or H_2S gas on laurite (RuS_2) surfaces is relevant because ensuing reactions lead to changes in the surface stoichiometry and the creation of highly undercoordinated Ru. RuS_2 is a pyrite-type structure. As reviewed in the previous chapter (Rosso and Vaughan 2006), the (100) and (111) planes of RuS_2 are good cleavage surfaces involving primarily Ru-S bond rupture. The (100) surface is remarkably inert compared to other pyrite-type metal disulfides. The so-called activation step for preparing RuS_2 surfaces for HDS catalysis involves H_2 adsorption at high temperature to promote sulfur elimination and create coordinatively unsaturated Ru sites. These Ru sites are sufficiently electrophilic to react with electron density-donating organic molecules (Tan and Harris 1998).

The first step toward creating S vacancies is the surface hydrogenation step involving dissociative chemisorption of H_2 or H_2S to form surface -SH groups. The chemisorption of H_2 and H_2S on the stoichiometric RuS_2 (100) surface was modeled by Frechard and Sautet (1997a) using Hartree-Fock periodic slab calculations with post-self-consistent-field (SCF) DFT correlation energy corrections. They found that H_2 associatively (molecularly) chemisorbs preferentially to the 5-fold coordinated surface Ru atoms with a binding energy of 31 kcal mol^{-1}. The H_2 chemisorbs in a "side-on" bidentate configuration to single Ru sites (Fig. 19a). Although H_2 chemisorption at the Ru sites does not lead to dissociation, the H-H bond is greatly weakened by charge transfer from H_2 into p_z and d_{z^2} states of the underlying Ru atom. Dissociation of the H_2 to form Ru-H and an adjacent S-H is less energetically favorable (Fig. 19b,c) by 23 kcal mol^{-1} but is still more favorable than desorption of H_2. Formation of -SH groups only is energetically uphill relative to desorbed H_2, and this is due to the stability of the S_2 unit on this surface. H_2S molecules associatively chemisorb to single surface Ru sites with a binding energy of 35 kcal mol^{-1}. H_2S is electrostatically repelled from binding to surface S sites. As was the case for H_2 chemisorption, dissociation of H_2S at the surface producing Ru-SH and -SH groups is metastable. These observations led Frechard and Sautet (1997a) to conclude that S vacancy creation is likely to be initially slow on (100) terraces. Increasing numbers of these vacancies due to partial surface reduction is predicted to lead to favorable dissociative chemisorption because of enhanced unsaturated character at the Ru sites adjacent to those vacancies.

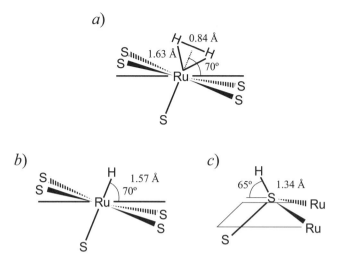

Figure 19. Schematics of the adsorption geometry of a) molecular H_2 and, after a dissociation step, separated hydrogen atoms on b) Ru and c) S surface sites, predicted from first principles calculations for the (100) RuS_2 surface. Redrawn after Frechard and Sautet (1997a).

On the RuS_2 (111) surface, the ability to hydrogenate the surface and create S vacancies is very different than for (100) because monosulfides are predicted which are more reactive with H_2 than S_2 dimers (Frechard and Sautet 1997b). The most stable structure of RuS_2 (111) in vacuum is not the charge neutral one, but rather a non-stoichiometric S-enriched one according to periodic Hartree-Fock calculations with post-SCF DFT, as well as according to planewave pseudopotential DFT calculations (Grillo and Sautet 2000) as reviewed in Rosso and Vaughan (2006). Dissociative chemisorption of H_2 is exothermic leading to a hydrogenated surface termination capped by three tilted –S-SH groups and one perpendicular –SH group. Further surface reduction is endothermic, proceeding through stages that involve decreasing the S:Ru ratio through S removal at protonated S_2 dimers by production of H_2S (costing 66 kcal mol^{-1}), followed by further S removal and the formation of -RuH metal hydride groups (costing an additional 180 kcal mol^{-1}) (Fig. 20). This yields the final (111) surface termination for use in HDS reactions, one which is significantly more catalytically active than the RuS_2 (100) surface.

Interaction of thiophene with hydrogenated and nonhydrogenated (100) and (111) RuS_2 surfaces has been recently modeled at the DFT level of theory (Smelyansky et al. 1998; Aray and Rodriguez 2001; Grillo and Sautet 2001; Aray et al. 2002, 2005). The sulfur atom of thiophene chemisorbs in a tilted η^1 configuration (in the language of organometallic chemistry; monodentate involving one atom in the heterocycle) to 5-fold surface Ru atoms on the stoichiometric nonhydrogenated (100) surface (Fig. 21). This monodentate Ru-S interaction (calculated E_{ads} = 0.43 eV for one thiophene on a 1×1 slab model, E_{ads} = 0.38 eV for one thiophene on a 2×1 supercell model) does not lead to distortions in the thiophene molecule towards cleavage of C-S bonds (Smelyansky et al. 1998). Reduction of the surface by cleaving the uppermost S_2 layer, leaving a 1:1 ratio of surface S vacancies and monosulfides, leads to an additional bond (Ru-C) formed between thiophene and another Ru atom of the surface, yielding a bridging η^2 configuration (Fig. 21). This interaction (E_{ads} = 0.1 eV) is weaker but involves distortion of the charge density distribution and geometry of the thiophene molecule consistent with activation of C-S bonds towards dissociation. The question arises whether hydrogenation of the surface is only important for creating S vacancies

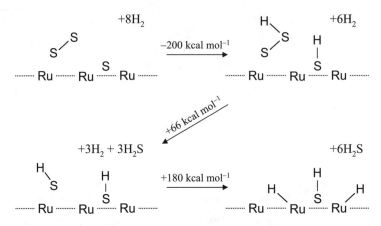

Figure 20. A model for the reduction of the RuS$_2$ (111) surface by molecular hydrogen, based on *ab initio* calculations. Redrawn after Frechard and Sautet (1997b).

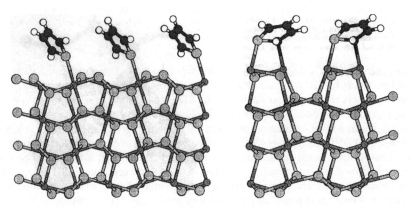

Figure 21. DFT-calculated adsorption geometries for thiophene on stoichiometric nonhydrogenated RuS$_2$ (100) surface (left) and the reduced RuS$_2$ (100) surface (right). [Used with permission of The American Physical Society from Smelyansky et al. (1998), *Physical Review B*, Vol. 58, Figs. 2 and 3, pp. R1783 and R1784.]

at the surface or does it also participate in C-S bond cleavage in thiophene? This question was addressed for a reduced (111) surface consisting of various ratios of -SH and -RuH groups using first principles calculations (Grillo and Sautet 2001). In that study, it was learned that the most favorable adsorption with respect to activation of C-S bonds involves a parallel thiophene η_2 configuration that is stabilized by interaction with the –RuH hydride groups at the surface. The role of H$_2$ therefore appears to be twofold, serving both to promote the formation of undercoordinated Ru by S abstraction and to destabilize C-S bonds by stronger thiophene hybridization to surface hydride groups.

Molybdenite

Considerably more work has been performed to understand the HDS catalytic activity of molybdenite (MoS$_2$) surfaces. As is the case for RuS$_2$, undercoordinated metal atoms are believed to be the catalytic sites for HDS reactions on MoS$_2$. Many of the concepts are the same, H$_2$ reacts with surface Mo and S sites, creating surface groups that lead to modification of the S:Mo ratio under the catalytic conditions. An important difference between the two materials in this regard pertains to the structure of MoS$_2$, which is a hexagonal layered structure type, as reviewed in the previous chapter (Rosso and Vaughan 2006). In contrast to the basal (0001) surface, the hexagonal edge surfaces expose coordinatively unsaturated Mo and S atoms. The Mo-terminated (10$\bar{1}$0) is comprised of 4-fold coordinated Mo atoms, and the S-terminated ($\bar{1}$010) edge is comprised of 2-fold coordinated S atoms. Because of the importance of MoS$_2$ for HDS catalysis, substantial attention has been paid to the structure of the edge surface in equilibrium with H$_2$ and sulfur species such as H$_2$S, sulfhydryl groups, and sulfur-bearing organics (Raybaud et al. 1998; Helveg et al. 2000; Raybaud et al. 2000a,b; Schweiger et al. 2002; Travert et al. 2002; Paul and Payen 2003; Lauritsen et al. 2004a,b). The catalytic activity is known to be correlated with the S:Mo ratio along the edge surfaces, which in turn depends on the background partial pressure of sulfur. DFT studies of S addition to the Mo-edge and S removal from the S-edge under a fixed chemical potential of H$_2$ and H$_2$S have provided thermodynamic stability diagrams for edges with differing S:Mo ratios (Raybaud et al. 2000a; Schweiger et al. 2002). These are partly motivated by STM observations of variable MoS$_2$ nanoparticle morphologies depending on the H$_2$S:H$_2$ ratio during synthesis (Helveg et al. 2000; Lauritsen et al. 2004a). Synthesis in H$_2$S:H$_2$ = 500 is observed to yield a triangular morphology whereas synthesis in H$_2$S:H$_2$ = 0.07 yields a hexagonally truncated morphology (Fig. 22) (Lauritsen et al. 2004a). Raybaud et al. (2000a) used the so-called *ab initio* thermodynamics approach to examine edge structures treated with periodic boundary conditions. Schweiger et al. (2002) used the same approach with aperiodic cluster models to

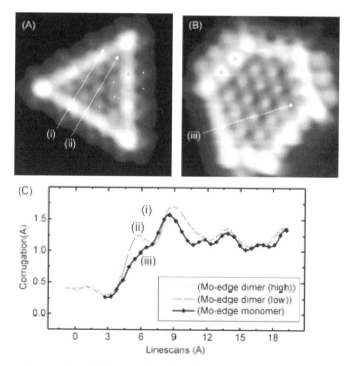

Figure 22. a) Atom-resolved STM image of a triangular single-layer MoS$_2$ nanocluster synthesized under "sulfiding conditions" (45×46 Å^2), and b) atom-resolved STM image of a hexagonal single-layer MoS$_2$ nanocluster synthesized under "sulfo-reductive" conditions (27 × 28 Å^2), and c) STM corrugation measurements across the Mo-edges of the triangle and hexagon. [Used with permission of Elsevier from Lauritsen et al. (2004a), *Journal of Catalysis*, Vol. 221, Fig. 2, p. 512.]

assess edge surface stability and the morphology of single-sheet triangular MoS_2 nanoparticles (for discussion of sulfide nanoparticles see also Pearce et al. 2006). In general, they use the grand canonical potential expression (Schweiger et al. 2002):

$$\Omega(n,\mu_S) = E_{MoS_x} - \mu_{Mo} n_{Mo}^{tot} - \mu_S n_S^{tot} \tag{16}$$

where E_{MoSx} is the DFT total energy of a triangular cluster of size n, and n_{Mo}^{tot} and n_S^{tot} are the total number of Mo and S atoms, respectively. At thermal equilibrium, the chemical potentials of S and Mo satisfy:

$$2\mu_S + \mu_{Mo} \approx E_{MoS_2}^{ref} \tag{17}$$

where $E_{MoS_2}^{ref}$ is the total energy of an infinite MoS_2 monolayer per MoS_2 unit, which gives:

$$\Omega(n,\mu_S) = E_{MoS_x} - n_{Mo}^{tot} E_{MoS_2}^{ref} - \mu_S \left(2n_{Mo}^{tot} - n_S^{tot}\right) \tag{18}$$

where n_S is the excess or lack of S atoms with respect to the MoS_2 stoichiometry. The chemical potential of S is related to the temperature and the partial sulfur pressure by simple thermodynamics as detailed in Raybaud et al. (2000a). The edge surface energy per surface site, $\sigma(\mu)$, is the slope of the grand canonical potential with respect to cluster size, i.e.:

$$d\Omega(\mu) = \sigma(\mu) dn \tag{19}$$

After accounting for the corner energies, the equilibrium morphology of a MoS_2 nanoparticle is determined using:

$$\frac{\sigma_{Mo}}{d_{Mo}} = \frac{\sigma_S}{d_S} \tag{20}$$

where σ's and refer to minima in the surface energies for the Mo and S edge types, and d's refer to distances from the Mo and S edges to the center of mass of the particle. Using this general strategy, Schweiger et al. (2002) were able to predict stable edge structures and nanoparticle morphologies for conditions relevant to HDS, concluding that the stable nanoparticles are deformed hexagons with 60% Mo-edge and 40% S-edge terminations (Fig. 23). They used this same approach to make structure/morphology predictions for the nanoparticles grown under a high partial pressure of H_2S for comparison with those grown for STM observations. This work also showed that S-vacancy formation is easier at the nanoparticle corners under HDS conditions, suggesting that the corner sites may have a higher catalytic activity than the edges. Collectively, this example demonstrates a great deal of utility in using *ab initio* thermodynamics for predicting changes in the structure of a mineral surface due to equilibration with another phase.

FUNCTIONALIZATION AND SELF-ASSEMBLY

In this section, we focus our attention on the chemical functionalization of sulfide mineral surfaces and on their ability to promote self-assembly of organic molecules. Surface functionalization is the introduction of new chemical functional groups to the surface by adsorption of molecules that bear the functional group of interest along with, typically, groups that specifically bind to the sulfide mineral. Functionalization is usually viewed as an intentional activity of the surface chemist performed with the aim of replacing the original chemical interface at the mineral surface (e.g., sulfide/water) with another interface (e.g., sulfide-O-CS_2/water) so as to impart new reactive behavior. Self-assembly is the process of spontaneous matter organization at the surface, usually mediated by a combination of specific adsorbate-surface and adsorbate-adsorbate interactions. It typically involves ordering in one (e.g., along a step) or

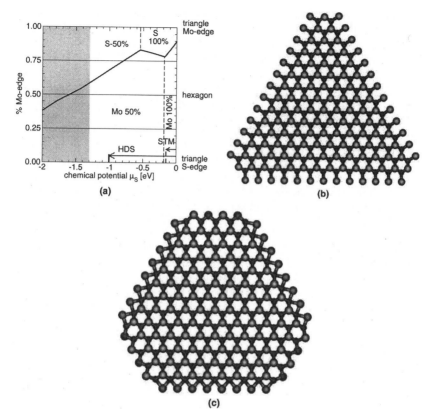

Figure 23. a) Morphology of nanosize MoS_2 particle as a function of the chemical potential of sulfur, based on *ab initio* thermodynamics calculations, and two predicted MoS_2 clusters for b) vacuum conditions during STM imaging, and c) hydrodesulfurization working conditions. [Used with permission of Elsevier from Schweiger et al. (2002), *Journal of Catalysis*, Vol. 207, Fig. 6, p. 83.]

two dimensions along the surface plane. Self-assembly is an important means of synthesizing molecular structures with higher order complexity that would otherwise never spontaneously form. Therefore it is viewed as a possible pathway for construction of, for example, polymeric nucleic acid sequences from monomeric amino acids in prebiotic Earth environments. Functionalization and self-assembly are not mutually exclusive processes; functionalizing adsorbates can also be self-assembled at the surface. Such is possibly the case for surfactant collector molecules as discussed below. Also, it is conceivable that functionalization could depend on self-assembly to attain the desired chemical behavior at the surface. The central theme of specific adsorption at the surface, typically of organic molecules, is an aspect of both of these processes that makes them more convenient to review together than apart.

Collector molecule adsorption

The majority of the work performed on sulfide mineral functionalization so far has been motivated by improving mineral separation technology. Although other applications are emerging, froth flotation is still the most important application for sulfide surface functionalization. Froth flotation is a chemical approach used in the mining industry for achieving mineral separation. A surfactant adsorbate (the collector molecule) is introduced to aqueous mixed-phase mineral suspensions to adsorb specifically to sulfide minerals, usually by chelation of surface metal atoms. The adsorbed collector induces changes in the surface hydrophobicity to facilitate air bubble attachment and sulfide mineral separation into a stable froth. A major research goal is to understand adsorption mechanisms for various collector molecules to improve separation efficiency, selectivity, and to enable the next generation of tailored collector molecules to be developed.

Short chain-length thiol surfactants, such as xanthates (alkyl dithiocarbamates; R-O-CS$_2$), are regularly used as collector molecules. A number of studies have attempted to investigate the mechanistic details of ethylxanthate ($CH_3CH_2OCS_2^-$) and other xanthate anion adsorption onto various sulfide minerals (e.g., Allison et al. 1972; Leppinen and Rastas 1986, 1988, 1989; Ackerman et al. 1987; Leppinen et al. 1989; Nowak 1993; Valli et al. 1994; Valli and Persson 1994a,b; Voigt et al. 1994; Fornasiero et al. 1995; Mielczarski et al. 1995, 1996a,b,c,d, 1997, 1998; Kartio et al. 1999; Pattrick et al. 1999; Larsson et al. 2000; Porento and Hirva 2002, 2003, 2004, 2005; Hung et al. 2003, 2004). Much of the current state of knowledge in this area has been facilitated by vibrational spectroscopic analysis of adsorbate structures and surface speciation, such as by using Fourier transform infrared (FTIR) spectroscopy (sometimes) in the diffuse reflectance (DRIFT) configuration for *ex situ* measurements or the attenuated total reflectance (FTIR-ATR) configuration for *in situ* measurements. For example, Persson and co-workers used *ex situ* DRIFT spectroscopy and XPS to study ethylxanthate interaction with arsenopyrite, millerite, molybdenite, orpiment, and realgar (Valli et al. 1994), and covellite (Valli and Persson 1994a), and chalcopyrite, marcasite, pentlandite, pyrrhotite, and troilite (Valli and Persson 1994b). The experiments were performed on partly oxidized samples by exposing mineral powders made by grinding larger natural crystal specimens in air to ethylxanthate aqueous solutions. A summary of mechanistic findings for these and five other sulfide minerals was given in Valli et al. (1994) and are summarized in Table 1. In general, they found that the interaction could be dominated by: (1) ethylxanthate oxidation

Table 1. Observed surface products from the interaction of sulfide minerals and aqueous solutions of ethylxanthate (EtX = $CH_3CH_2OCS_2$) compared to products expected depending in part on a crystal chemical model for lattice sulfur oxidation potential based on S-S distances. Adapted from Valli et al. (1994).

Mineral	Formula	Shortest S-S distance (Å)	Expected surface product	Observed surface product
Acanthite	Ag$_2$S	4.135	Ag(I)EtX	Ag(I)EtX
Arsenopyrite	FeAsS	3.197	EtX$_2$	EtX$_2$, As(III)EtX$_3$
Chalcocite	Cu$_2$S	3.710	Cu(I)EtX	Cu(I)EtX
Chalcopyrite	CuFeS$_2$	3.685	Cu(I)EtX	Cu(I)EtX
Covellite	CuS	2.084, 3.757	Cu(I)EtX, EtX$_2$	Cu(I)EtX, EtX$_2$
Galena	PbS	4.194	Pb(II)EtX$_2$	Pb(II)EtX$_2$
Marcasite	FeS$_2$	2.223	EtX$_2$	EtX$_2$
Millerite	NiS	3.244	EtX$_2$	EtX$_2$
Molybdenite	MoS$_2$	3.154	EtX$_2$	EtX$_2$
Orpiment	As$_2$S$_3$	3.242	EtX$_2$	As(III)EtX$_3$, EtX$_2$
Pentlandite	(Fe,Ni)$_9$S$_8$	3.362	EtX$_2$	EtX$_2$
Pyrrhotite	Fe$_{1-x}$S	3.390	EtX$_2$	EtX$_2$
Pyrite	FeS$_2$	2.177	EtX$_2$	EtX$_2$
Realgar	AsS	3.295	EtX$_2$	EtX$_2$, As(III)EtX$_3$
Sphalerite	ZnS	3.821	Zn(II)-OCS$_2$	Zn(II)-OCS$_2$
Troilite	FeS	3.348	EtX$_2$	none

to dixanthogen by strongly oxidizing species present at the mineral surface such as $S_2O_7^{2-}$ or $S_2O_8^{2-}$; (2) formation of solid phase metal alkylxanthate salts by dissolution/re-precipitation, and (3) chemisorption of alkylxanthate species. The first two possibilities obscure information pertaining to the formation of alkylxanthate surface complexes. For example, for sulfides where Cu, As, or Pb is the metal atom, their surfaces tend to form the low solubility phases Cu(I)ethylxanthate, As(III)ethylxanthate, and Pb(II)ethylxanthate (Table 1). The relative amounts of $S_2O_x^{2-}$ species present at the surfaces leading in cases of $x > 6$ to the production of dixanthogen was postulated to depend ultimately on S-S distances in the structure, with shorter S-S distances tied to more production through more facile conversion to highly oxidizing $S_2O_x^{2-}$ species. Only in the cases of CdS (greenockite) and ZnS (sphalerite) were chemisorbed alkylxanthate surface complexes not obscured and positively identified (Valli et al. 1994).

The findings of these early studies were further clarified using similar FTIR spectroscopic methods for the minerals pyrrhotite (Fornasiero et al. 1995), chalcopyrite, tetrahedrite, and tennantite (Mielczarski et al. 1996c, 1997). Fornasiero et al. (1995) confirmed the formation of dixanthogen on pyrrhotite surfaces and proposed a model of sequential anion adsorption forming dixanthogen at surface sites. Mielczarski et al. (1996c) confirmed the formation of a cuprous xanthate complex on chalcopyrite surfaces and also found evidence for dixanthogen and the formation of Fe(OH)ethylxanthate due to partial oxidation of the initial chalcopyrite surface. Tetrahedrite and tennantite were found only to form the cuprous xanthate complex. Using a potentiostat controlled cell and a range of overpotentials from the open circuit potential (OCP) to ~200 mV higher than the OCP, Mielczarski et al. (1997) proposed four possible ethylxanthate adsorption mechanisms on chalcopyrite surfaces. These are the formation of dixanthogen according to:

$$EtX^- = 0.5EtX_2 + e^- \tag{21}$$

where EtX = $CH_3CH_2OCS_2$, the formation of iron hydroxyl xanthate complexes by the replacement reaction:

$$Fe(OH)_x + yEtX^- = Fe(OH)_{x-y}EtX_y + yOH^- \tag{22}$$

the formation of cuprous xanthate complex in the reaction:

$$CuFeS_2 + xEtX^- = Cu_{1-x}FeS_2 + xCuEtX + xe^- \tag{23}$$

and the formation of dixanthogen in the reaction:

$$Fe^{3+} + EtX^- = 0.5EtX_2 + Fe^{2+} \tag{24}$$

The availability of oxidized iron at the surface controls iron hydroxyl xanthate complex formation (Eqn. 22). In the case of fresh chalcopyrite surfaces where little or no oxidation has occurred, or for high Cu:EtX$^-$ concentration ratios at the surface, the formation of cuprous xanthate complexes is expected to dominate (Eqn. 23). For low Cu:EtX$^-$ concentration ratios, dixanthogen is produced at the surface. The degree of initial surface oxidation thus strongly influences the nature of the surface products and the degree with which the chalcopyrite surface is functionalized for hydrophobic flotation.

The importance of the metal-xanthate interactions for sulfide mineral flotation led to detailed molecular computational studies being performed. Porento and Hirva (2002) evaluated the energetics of bidentate complex formation between Cu^+, Cu^{2+}, Zn^{2+}, and Pb^{2+} and ethylxanthate, along with several other related prospective collector molecule designs using gas-phase *ab initio* calculations. The ethoxy tail (EtO, where Et = ethyl) adjacent to the thiol groups of the ethylxanthate molecule was systematically replaced with EtS, EtCH$_2$, EtNH, Et$_2$N, Et$_2$P, and (EtO)$_2$P groups to examine the electronic effects on the interaction energies. The types of metal complexes considered are shown in Figure 24, and corresponding complex formation energies (enthalpies under the usual assumption that $\Delta E \sim \Delta H$ where ΔE is the total electronic energy difference) calculated at the DFT level of theory on Hartree-Fock-optimized structures are listed in Table 2. This work showed that metal chelation by the thiol group is strongest for diethyldithiocarbamate anions (Et_2N), suggesting a possible new class of efficient collector molecules.

The success of correlating the behavior of monomeric metal-xanthate complexes to surface complexes is contingent upon factors not treated such as sulfide mineral surface structure and steric effects for collector adsorption. There have been several attempts to directly address this aspect. Using FTIR-ATR, Larsson et al. (2000) studied the adsorption of heptylxanthate on ZnS window elements with undetermined surface structure or crystallographic orientation. They found evidence for a predominantly biatomic bidentate "bridging" surface complex where the sulfur atoms of the xanthate CS_2 group bind to two separate Zn atoms. However, it is known that sphalerite does not float in acid or alkaline media with modest ethylxanthate concentration, possibly due to the high solubility of zinc ethylxanthate. The addition of Cu^{2+} effectively "activates" the sphalerite for flotation by forming copper sulfide on the sphalerite surface via Cu → Zn substitution and subsequently allowing strong copper ethylxanthate complexes to form. Pattrick et al. (1999) studied this system in detail at pH 10 and 12 using fluorescence reflection extended X-ray absorption fine-structure spectroscopy (REFLEXAFS). They found that Cu interacts with surface S atoms and in addition Cu-O bonds are formed.

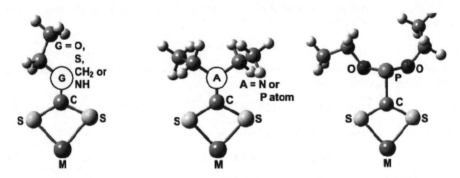

Figure 24. Chelate structure of metal ethyl thiol complexes (left), metal diethyl thiol complexes (center), and metal diethoxyphosphinecarbondithioic acid (right) used in the DFT study of Porento and Hirva (2002); (M = Cu, Zn, or Pb). [Used with permission of Springer-Verlag from Porento and Hirva (2002), *Theoretical Chemistry Accounts*, Vol. 107, Fig. 2, p. 203.]

Table 2. Complex formation energies (kJ/mol) for a range of collector anions and metal cations calculated at the DFT level of theory on HF-optimized structures. From Porento and Hirva (2002).

Collector	Cu^+	Pb^{2+}	Zn^{2+}	Cu^{2+}
Et_2N	−756	−1541	−1765	−1875
EtNH	−751	−1519	−1744	−1845
Et_2P	−704	−1461	−1696	−1785
EtO	−721	−1463	−1685	−1785
$EtCH_2$	−719	−1444	−1682	−1777
$(EtO)_2P$	−691	−1435	−1680	−1767
EtS	−699	−1443	−1670	−1763

S atoms of ethylxanthate were found to efficiently displace these oxygen atoms. The authors proposed a surface complex for ethylxanthate on the Cu-activated sphalerite surface that closely corresponds to covellite Cu-S surface species.

Porento and Hirva (2005) used *ab initio* electronic structure calculations on gas-phase cluster representations of sphalerite (111) with and without Cu substitution to examine mechanisms of forming metal-xanthate surface complexes. Given the focus on alkaline aqueous conditions, their treatment focused on ethylxanthate adsorption by a replacement reaction involving displacement of metal-bound surface hydroxyls. They found a strong correlation between the calculated energy for the replacement reaction and the location of Cu substitution sites in the cluster models (Fig. 25). The replacement reaction is more energetically favorable with hydroxyl bound to a surface Cu substitution site (which binds the hydroxyl less strongly

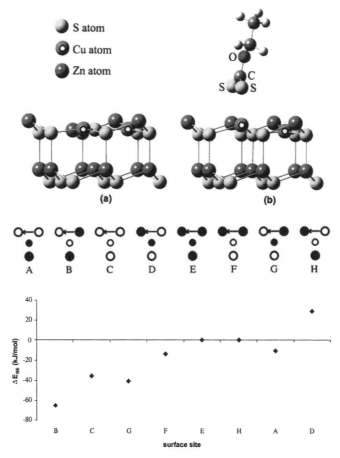

Figure 25. Findings of an *ab initio* calculation study of the influence of Cu substitution for Zn on ethylxanthate adsorption on the ZnS (111) surface. Structure models show example optimized cluster geometries for two Cu→Zn subsitutions a) before and b) after adsorption of ethylxanthate. The center diagram schematically shows the types of Cu-substituted configurations for the metal sublattice considered for co-adsorption energy calculations of OH⁻ and ethylxanthate, where open large circles are Cu atoms and closed large circles are Zn atoms, and the arrow points to the OH⁻ adsorption site and the orientation of the -CS$_2^-$ group. DFT-calculated energies for the various configurations are shown in the graph. [Used with permission of Elsevier from Porento and Hirva (2005), *Surface Science*, Vol. 576, Figs. 2, 4, and 5, pp. 101-103.]

than does a surface Zn site) and with a second Cu substitution present in the adjacent metal layer beneath. Consistent with Larsson et al. (2000), Porento and Hirva (2005) found that the CS_2 group of ethylxanthate binds to the surface by bridging two metal atoms. In an earlier *ab initio* study of the adsorption of ethylxanthate on covellite (001) in the gas-phase, Porento and Hirva (2004) also found that the bridging surface complex was the most stable configuration, in that case involving two surface Cu sites. Collectively these studies suggest that ethylxanthate adsorption onto Cu-activated sphalerite surfaces leading to facile flotation has crystal chemical ties to the structure of covellite.

Ethylxanthate adsorption has also been examined in detail for pyrite and galena. Hung et al. (2003) used planewave pseudopotential DFT to model the gas-phase adsorption of ethylxanthate and its head group moiety $HOCS_2^-$ on periodic slab representations of pyrite (110) and (111) surfaces. They found that $HOCS_2^-$ readily undergoes dissociation at 4-fold Fe sites present on the (110) surface (Fig. 26) and bridging S atoms on the (111) surface (see Rosso and Vaughan 2006). Because of structural similarities between these surface sites and defect sites expected in high densities on other pyrite surfaces, they concluded ethylxanthate would adsorb most strongly, by dissociative chemisorption, to defect sites where, for example, Fe coordination is four or less. In contrast to the bidentate bridging configuration discussed above for sphalerite and covellite, where metal-metal distances at the surface are a good match with the geometry of the xanthate head group, on pyrite (110) the tendency to dissociate may arise from the lack of ideal Fe-Fe distances at the surface. Given the previous discussion above on copper activation, it is noteworthy that pyrite flotation has also been observed to be enhanced by adsorption of copper, mechanisms for which have been examined by XPS (Voigt et al. 1994) and bear similarities to those discussed above for sphalerite (formation of stable copper ethylxanthate surface complexes).

Ethylxanthate adsorption onto galena surfaces has been examined by both macroscopic thermodynamic model fits to adsorption data based on vibrational spectroscopy (Leppinen and Rastas 1988; Nowak 1993), as well as being examined by SR-XPS (Kartio et al. 1999) and most recently by *ab initio* calculations at the DFT level of theory (Porento and Hirva 2003). Motivated by apparent discrepancies in the reported stabilities of surface lead xanthate on

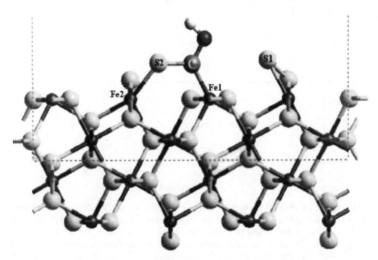

Figure 26. Calculated dissociative adsorption geometry for $HOCS_2^-$ on the FeS_2 (110) surface from DFT calculations. [Used with permission of the American Institute of Physics from Hung et al. (2003), *Journal of Chemical Physics*, Vol. 118, Fig. 2a, p. 6026.]

PbS versus precipitated lead xanthate, Leppinen and Rastas (1988) studied the chemisorption equilbria by analyzing the ethylxanthate and sulfide ion concentrations in the solution phase. It was found that the adsorption data cannot be explained using the thermodynamics of bulk PbS and Pb ethylxanthate phases. Therefore a mixed-phase model was invoked to describe the adsorption in terms of a combination of PbS and (PbEtX)$_2$S species, written as:

$$2EtX^-_{(aq)} + 2PbS_{(sf)} \leftrightarrow (PbEtX)_2S_{(sf)} + S^{2-}_{(aq)} \qquad (25)$$

where "sf" indicates "surface phase". The proposed equilibrium surface phase is characterized by partial surface coverage, as schematically shown in Figure 27. These findings figured heavily in the re-examination of this topic by Nowak (1993), who found in contrast that the adsorption data could be fit under the assumption of an adsorption density of one ethylxanthate molecule for every Pb surface site, and that the adsorption followed a Frumkin isotherm form. Using SR-XPS, Kartio et al. (1999) found that for freshly cleaved PbS exposure to 10^{-4} M ethyl-

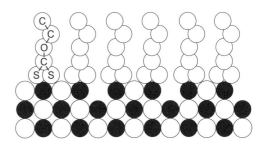

Figure 27. Suggested PbS (100) – ethylxanthate surface phase consisting of partial surface coverage from equilibrium thermodynamic calculations. Redrawn after Leppinen and Rastas (1988).

xanthate at pH 9.2, associated chemical shifts observed in the S 2p photospectra were consistent with the formation of chemisorbed xanthate ions at the surface, as opposed to the formation of bulk lead ethylxanthate (Fig. 28). The chosen exposure conditions had previously been shown by Buckley and Woods (1991) to produce near-monolayer coverage. In a detailed *ab initio* modeling study by Porento and Hirva (2003) at the DFT level using cluster representations of PbS (100), the stability of various possible ethylxanthate surface complex structures were compared (Fig. 29). In that study it was shown that an ethylxanthate molecule preferentially adsorbs as a monodentate complex over a surface Pb atom, where only one of the S atoms is bonded to a surface Pb atom. Collectively, this set of four studies provides a description of PbS functionalization by ethylxanthate that spans a range from the molecular to the macroscopic scales.

While the above discussion focused on ethylxanthate as the main example for sulfide surface functionalization, we note here that many other prospective collector molecules have also been studied. In particular, 2-mercaptobenzothiazole (MBT) has been regularly investigated alongside ethylxanthate since the 1980's. Adsorption mechanisms of MBT on a variety of sulfide minerals have been studied by electrophoresis and microcalorimetry (e.g., Maier and Dobias 1997), SR-XPS (e.g., Szargan et al. 1997), and ATR-FTIR (e.g., Larsson et al. 2001). The reader is referred to these articles and references therein for more information on MBT adsorption.

Self-assembly on sulfide mineral surfaces is a much more limited area of study. Relative to functionalization by collector molecule adsorption, the self-assembly research area can also be described as more exploratory. The molecules chosen for self-assembly studies on sulfide minerals so far have been selected primarily because of their general ability to form a self-assembled monolayer (SAM) on a range of materials. Collector molecule adsorption likely crosses over into this topic of self-assembly, because the collector molecules are surfactants which themselves have a tendency to self-assemble. This is a potentially important aspect of thiol collector molecule adsorption that has not been studied as such, possibly because of the specialized experimental techniques required to address molecular-orientation of sorbed species at surfaces. Other examples can be found where self-assembly was potentially relevant but not specifically addressed, such as in adsorption of 8-hydroxyquinoline (Lan et al. 2002) or

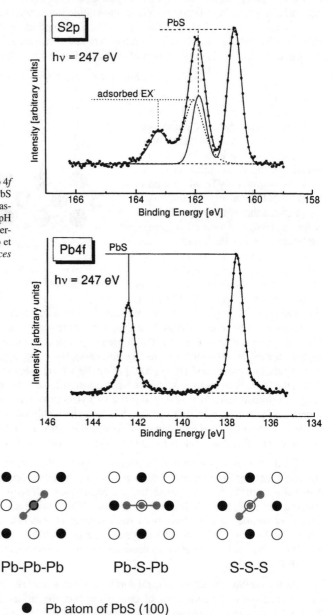

Figure 28. S 2p (top) and Pb 4f (bottom) SR-XPS spectra of PbS after immersion in 10^{-4} M potassium ethylxanthate solution (pH = 9.2) for 45 min. [Used with permission of Elsevier from Kartio et al. (1999), *Colloids and Surfaces A*, Vol. 154, Fig. 1, p. 99.]

Figure 29. Various starting ethylxanthate adsorption configurations on PbS (100) tested in *ab initio* calculations on cluster models at the DFT level of theory. Redrawn after Porento and Hirva (2003).

L-α-phosphatidylcholine hydrogenated lipid and 1,2-*bis*-(10,12-tricosadiynoyl)-sn-glycero-3-phosphocholine lipid (Zhang et al. 2003) to pyrite surfaces (these studies were performed to investigate the formation of thin organic films as pyrite coatings to suppress oxidation). To our knowledge, self-assembly has so far only been specifically examined for pyrite and molybdenite surfaces. We review some of this work below.

The potential for self-assembly of triphenylphosphine, *n*-alkanethiols and *n*-alkylamines on pyrite was examined using XPS, IR spectroscopy, and near-edge X-ray absorption fine structure (NEXAFS) spectroscopy by Himmel et al. (1996). The pyrite was fractured approximately along the (100) plane, polished to 250 nm grit size, and chemically etched to remove oxidation products. XPS showed that the prepared surface was as free of oxidation products as pyrite fractured in inert Ar atmosphere. The authors found that *n*-alkylamines and triphenylphosphine adsorb to the pyrite surface and form dense monolayers. The *n*-alklyamines also displayed a high degree of orientation, with the alkyl chains oriented at a 25° average angle with respect to the surface normal. This is the first known example of a highly oriented molecular film formed on pyrite. The *n*-alkanethiol, which normally forms well defined monolayers on metal substrates, did not adsorb. Molecular modeling has also been used to simulate the formation of self-assembled layers on pyrite. Using MD simulations, Zhang et al. (2005) investigated self-assembly of dilithium phthalocyanine on the pyrite (100) surface. They found a strong dependence of the molecular orientation on packing density, with a preferred stacking axis perpendicular to the surface plane due to strong interactions between Li and surface S atoms.

Molybdenite (0001) surfaces offer the prospect of very flat ordered molecular layers to be formed. The structure of this surface is reviewed in the previous chapter (Rosso and Vaughan 2006). A wide range of organic adsorbates have been examined for their self-assembled film structure on MoS_2 (0001) including metal phthalocyanine (Ludwig et al. 1994; Strohmaier et al. 1996a), naphthalene-1,4,5,8-tetracarboxylic-dianhydride (Strohmaier et al. 1996b), and 4,4′,7,7′-tetrachloro-thioindigo (Petersen et al. 1997) for example. But these and many other studies were performed in vacuum chambers using such techniques as molecular beam epitaxy. Here we focus on the studies performed directly in solution because of their more direct relevance to the natural environment. Highly ordered herringbone-structure monolayers of dotriacontane ($C_{32}H_{66}$) were shown to form on this surface (although primarily on the close analogue phase, $MoSe_2$) using *in situ* STM to directly image the molecular structure of the films in solution (Cincotti and Rabe 1993). The adsorbate lattice is close-packed and oriented relative to the (0001) surface lattice but not simply commensurate. The self-assembly is attributed more to the atomic flatness of the surface and less to specific interactions between the dotriacontane molecules and the surface atoms.

In another set of in situ STM experiments, the influence of the surface lattice on the ordering of self-assembled films was specifically addressed by Giancarlo et al. (1998). They investigated the film structure on MoS_2 (0001) for a series of mono- and disubstituted long-chain hydrocarbons containing –S–, OH, COOH, Br, C=C, and $CONH_2$ functionalities. Specifically, they examined films of octadecyl sulfide, elaidic acid, 12-bromododecanoic acid, 12-bromododecanol, 11-bromoundecanol, octadecanamide. A diverse array of ordering patterns were observed with *in situ* STM; substantial attention was paid to assessing the electronic structure aspects of both the substrate and the film contributing to image contrast by bias dependent imaging (Fig. 30). They compared the ordering patterns obtained on MoS_2 (0001), where the surface lattice repeat is 3.2 Å, with those for the same molecules adsorbed onto graphite basal planes, where the surface lattice repeat is 2.5 Å. Three types of substrate influence on molecular ordering behavior were categorized; from those where there was no detectable influence, to slight influence involving substrate control on the size of ordered molecular domains, to a complete change in molecular ordering induced by the substrate (Table 3). Primary factors contributing to the differences in

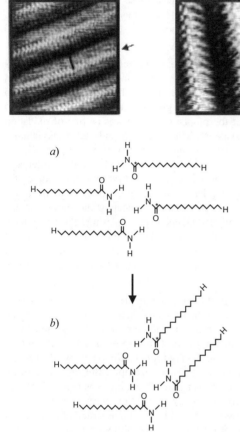

Figure 30. STM images showing a range of self-assembled monolayer structures formed on MoS$_2$ from 12-bromododecanol (left; 12 × 12 nm^2), 11-bromoundecanol (center; 6×6 nm^2), and octadecanamide (right; 10×10 nm^2). Arrows point to bright spot rows that correspond to terminal bromine atoms. The structure diagram shows the a) parallel and b) herringbone configurations of octadecanamide molecules that give rise to different ordering patterns via rotation about a single carbon-carbon bond, denoted with an asterisk. [Used with permission of the American Chemical Society from Giancarlo et al. (1998), *Journal of Physical Chemistry B*, Vol. 102, Figs. 2 and 3, pp. 10258 and 10260.]

Table 3. Comparison between molecular structure parameters and ordering behavior observed by STM for self-assembled thin films on MoS$_2$ compared to graphite (Giancarlo et al. 1998 and references therein). Molecular length is defined as the length of two molecules comprising a dimer pair.

Observed dependence on substrate	Molecule[a]	MoS$_2$ (0001)			Graphite (0001)		
		Molec. length (nm)	Angle (deg.)	Spacing between molec. (nm)	Molec. length (nm)	Angle (deg.)	Spacing between molec. (nm)
None	OS	4.38	90	0.48	4.35	99	0.43
None	EA	2.55	94	0.53	2.43	87	0.43
Slight	12-BA	3.24	115	—	3.70	117	0.42
Significant	12-B	1.86	93	0.45	1.78	—	0.46
None	11-B	1.43	56	0.41	1.51	63	0.49
Significant	O	2.37	117	0.47	2.4-2.6	115-120	0.48

[a]OS = octadecyl sulfide, EA = elaidic acid, 12-BA = 12-bromododecanoic acid, 12-B = 12-bromododecanol, 11-B = 11-bromoundecanol, O = octadecanamide.

behavior include the degree of molecular size-dependent registry between the packed organic film and the substrate repeat unit size, and the strength of adsorbate-substrate and adsorbate-adsorbate interactions, particularly intermolecular hydrogen bonding.

A rapidly growing component of self-assembly studies on mineral surfaces is focused on exploring the potential of this process to construct the biochemical machinery of life. For example, purine-pyrimidine arrays assembled on naturally occurring mineral surfaces might act as possible templates for biomolecular assembly. A number of studies have been performed to examine the monolayer formation mechanisms and structure of purine and pyrimidine bases on MoS_2 (0001), particularly those by Sowerby and co-workers (e.g., Sowerby et al. 1996, 1998a,b; Sowerby and Petersen 1999). Sowerby et al. (1996; 1998b) used STM to investigate the two-dimensional self-assembled monolayer structure for the purine base, adenine ($C_5H_5N_5$), on MoS_2 (0001). The molecules were observed to lie flat on the surface. Although formed from achiral adenine molecules, the observed adsorbate structures were enantiomorphic on molybdenite (Fig. 31). This phenomenon suggests a mechanism for the introduction of a localized chiral symmetry break, and it was concluded that the spontaneous crystallization of these molecules on inorganic surfaces may have some role in the origin of biomolecular optical asymmetry. Molecular mechanics calculations were used to identify possible organizations of the adenine monolayer on the MoS_2 (0001) surface and to determine its registry with the underlying surface atoms. A range of trial mesh and trial packing configurations were evaluated to determine those that are the most energetically stable. The optimal calculated structure involves a rectangular adenine repeat unit at an oblique angle with respect to the substrate lattice vectors. Molecular mechanics calculations were further applied to both adenine and guanine ($C_5H_5N_5O$) adsorption on MoS_2 (0001) and on the graphite basal plane (Sowerby et al. 1998a). The authors found that both self-assembled monolayer structures are similar to each other, adopting a $P2gg$ plane group motif, independent of the structural differences between the two substrates. Central to the adsorbate structures is the formation of centrosymmetric dimers, which neutralizes strong dipole moments intrinsic to the individual

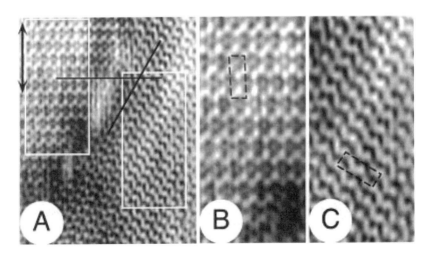

Figure 31. a) STM image of an ordered array of adenine molecules adsorbed to the surface of MoS_2 showing a grain boundary between two domains. Equivalent lattice directions on each of the domains are indicated by the black lines and intersect at an angle of 60°, indicating epitaxial registration with the underlying substrate. b,c) Enlarged portions of each of the domains in a) delineated by the white rectangles. On each image, the adsorbate mesh is described by the black dashed line parallelograms of dimensions 2.28×0.84 nm. [Used with permission of the American Chemical Society from Sowerby et al. (1998b), *Langmuir*, Vol. 14, Fig. 1, p. 5197.]

molecules. Finally, they noted that π-bond cooperativity, which contributes to the stability of cyclic hydrogen bonds between complementary base pairs in nucleic acids, requires that adjacent hydrogen-bonding functional groups to be linked by bonds with π-electron character. This condition is satisfied in the adenine and guanine self-assembled monolayers on MoS_2 (0001).

METAL ION UPTAKE

The uptake of metal ions at the surfaces of sulfide minerals is an important area of study. In nature, it is relevant to understanding the transport and mobility of metals in the Earth's crust and in surface and near-surface environments. Such understanding has applications to processes of importance in ore formation, and in the mobility of pollutants from mine wastes, industrial operations and other sources. Many metals are toxic to life forms, even at low concentrations. Metal uptake processes are also technologically important, for example in mineral extraction.

In highly controlled UHV experimental systems as discussed above, techniques such as XPS and STM can sometimes be used to probe the interaction of individual (gas phase) metal ions with carefully prepared sulfide surfaces. Such studies can provide insights into molecular scale interaction processes. A more direct interest, however, is in the interaction of metal ions in aqueous solution with sulfide mineral surfaces. These interactions can be investigated by performing experiments where sulfides are exposed to metal-containing solutions *ex situ* before transfer into UHV for surface analysis, or by using methods that do not require UHV conditions and where *in situ* analysis of the system is possible.

The uptake of metal ions in solution by mineral surfaces can involve one or more of a number of processes (Fig. 32). The metal ion may retain its hydration sphere and be weakly held to the surface as an *outer sphere complex* (effectively "physisorbed"). It may lose some of its hydration sphere and bond directly and more strongly to the surface as an *inner sphere complex* (effectively "chemisorbed"). It may replace ions at the surface (or even diffuse some distance into the solid) in an exchange reaction, or a more extensive reaction (possibly a redox reaction) with the surface may result in precipitation or co-precipitation of a new phase at the surface and/or wholesale replacement of the substrate.

Galena

Given its simple structure and perfect cleavage, it is not surprising that the uptake of metal ions by galena has been studied both experimentally and computationally. An area of particular interest has been uptake of gold and silver from aqueous solution by galena; here, surface reactions have been suggested as important in ore deposit formation, and also in the control of heavy metal concentrations in reducing aqueous environments. There is now an extensive literature dealing with experimental studies of such interactions with a variety of sulfides, but particularly galena and pyrite. Earlier studies using XPS, AES and (analytical) SEM to characterize surface species have been reviewed by Bancroft and Hyland (1990). These were followed by further investigations using the same range of techniques, in some cases supplemented with electrochemical methods or other spectrometries (Mycroft et al. 1995a; Scaini et al. 1995, 1997). New insights into the molecular scale nature of these interactions were made possible with the application of STM (Eggleston and Hochella 1991, 1993) and of quantum mechanical calculations aimed at exploring the energetics of the processes and simulating STM images, XPS and STS spectra (Becker et al. 1997). Experimental and computational evidence for uptake from solution of species such as $AuCl_4^-$, points to initial adsorption (as an inner sphere complex) at galena surface sites, followed by electron transfer and reduction of Au^{3+} to Au^0 with loss of ligands. This process is calculated

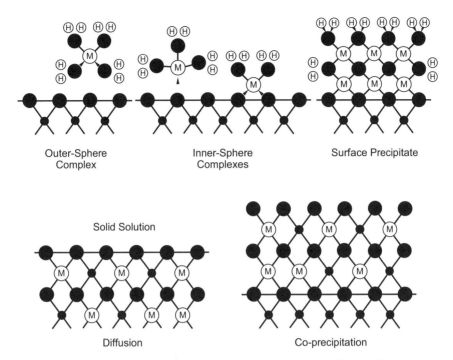

Figure 32. Depiction of various processes of metal ion uptake by mineral surfaces.

to be energetically favorable (Becker et al. 1997) which helps to explain the affinity of gold for surfaces of sulfides such as galena. The Au^0 is then stabilized at the surface by formation of Au-Au bonds to give dimers and islands of the metal.

In two other very different studies of the PbS (100) surface, Kendelewicz et al. (1998) used X-ray standing wave (XSW) and photoemission methods to study interaction with vapor-deposited sodium overlayers, and Genin et al. (2001) used XPS, XAS and reflection high energy electron diffraction (RHEED) to study interaction with mercury vapor. The former was a fundamental study of relaxation effects at the surface and of the possible influence of Na on surface reactivity. There is clear evidence of replacement of surface Pb^{2+} ions by Na$^+$, and the suggestion made is that local charge balance is achieved by having additional Na atoms atop S atoms at the surface. No evidence was found for enhanced reactivity to water, which does not react significantly with the flat PbS (100) surface (see discussion above). In the latter study, Hg^0 was seen to be reduced at the surface to Hg^{2+} and to be adsorbed through the formation of Hg-S bonds. Analysis of XAS data gave a Hg-S bond length of 2.62 Å and a structural model was suggested with Hg atoms adsorbed at 2-coordinate surface sites. At saturation, the adsorbed mercury was estimated to comprise a single monolayer covering the galena surface.

Pyrite

Pyrite, being by far the most abundant natural sulfide mineral, has been the subject of numerous studies of surface interaction with metal ions in solution. For example, a substantial body of work on uptake of precious metals, already referred to above, has been produced. As noted in sections of this and the preceding chapter, much spectroscopic and imaging work on pyrite surfaces has been undertaken to clarify the nature of the pristine surface and the surface reactions involved in pyrite oxidation and breakdown. It has been noted that unsaturated surface groups can form that may react with water (Rosso et al. 1999b), and highly reactive defect sites may occur, partially stabilized by disproportionation (Guevremont et al. 1997; Uhlig et al. 2001). It appears that defect structures can serve as centers for electron transfer and promote disproportionation and cation uptake. For example, the adsorption of Ag^+ on pyrite probably involves sulfur disproportionation and electron transfer (Scaini et al. 1995):

$$FeS_2 + 8Ag^+ + 4H_2O \leftrightarrow Fe^{2+} + Ag_2S + 6Ag^0 + SO_4^{2-} + 8H^+ \qquad (26)$$

The work on precious metal uptake has included studies of Au, Ag and Pd (see review by Bancroft and Hyland 1990) and explored interactions, with gold in particular, in several different oxidation states (Au^{3+}, Au^+) and different solution complexes (chloride, bisulfide). Other variables investigated have included pH and temperature. For example, in uptake of silver as chloride species by pyrite (and galena), changes in metal and chloride concentrations have a significant influence, but not pH variations (Scaini et al. 1997). Reactions of Au^+ bisulfide complexes with pyrite increase at elevated temperature (90 °C vs. 25 °C) and higher pH (6 vs. 3), but far less Au is deposited than from $AuCl^-$ solutions because of the greater stability of $Au(SH)_x^{1-x}$ complexes (Scaini et al. 1998).

The uptake from solution of a range of transition metals and heavy metals by the pyrite surface has also been studied. For example, Parkman et al. (1999) used bulk measurements of uptake as a function of initial concentration combined with X-ray absorption spectroscopy (EXAFS and XANES) to study reactions with Cu^{2+} and Cd^{2+} ions in solution at circum-neutral pH. In these experiments, the bulk uptake of Cd contrasts with that of Cu. The Cd shows decreasing efficiency of uptake with increasing concentration, and the Cu shows constant uptake which is an order of magnitude greater than for Cd. The behavior of Cu is attributed to formation (by precipitation/replacement) of a copper sulfide (of chalcocite or covellite type) at the surface, and this is in line with spectroscopic data showing bonding of Cu to an average of four sulfur atoms at 2.3 Å. The spectroscopic data for Cd indicates bonding to six oxygens at 2.28 Å over all measured concentrations. Combined with the bulk uptake behavior which suggests limited availability of surface sites for "sorption," the simplest interpretation of these data are outer sphere surface complexation. However, Parkman et al. (1999) suggest this is unlikely given the high capacity and relatively strong binding of Cd at low concentrations. The possibility of a more complex interaction process involving oxidation is suggested. Further clarification has come from a study by Bostick et al. (2000) using a range of methods including Raman spectroscopy, electron microscopy and STM. Here, complex reactions involving pyrite surface reconstruction and disproportionation are proposed, leading to a mixture of sulfide and oxide surface products (elemental sulfur, iron hydroxide and CdS).

Other studies of metal uptake by pyrite include investigations of Ni^{2+} (Hacquard et al. 1999), Hg^{2+} (Ehrhardt et al. 2000), As^{3+} and As^{4+} (Farquhar et al. 2002; Bostick and Fendorf 2003) and Mo^{6+} (Bostick et al. 2003). The work on Ni^{2+} "adsorption" was undertaken at pH 10 under conditions where "the mineral surface is rapidly covered by a layer of Fe^{3+} oxides". The first step of the interaction with nickel is then the formation of a hydroxylated surface complex (Hacquard et al. 1999). The "sorption" of Hg^{2+} was investigated as a function of pH and using XPS for analysis of pyrite surfaces exposed to the metal (Ehrhardt et al. 2000). The surfaces studied showed partial oxidation, with islands of Fe^{3+} oxyhydroxides as well as pyrite present. The XPS measurements were interpreted in terms of (inner sphere) surface complexation involving both pyritic functional groups and hydroxyl groups on the oxyhydroxide islands. No evidence was found for the formation of Hg sulfide, sulfate or thiosulfate.

Arsenic and molybdenum are species of particular interest in regard to surface uptake, the former because of its toxicity and the latter because of its importance as a nutrient needed for a variety of biological functions. Farquhar et al. (2002) used EXAFS and XANES to study the mechanisms of uptake of both As^{3+} and As^{5+} from aqueous solution at pH 5.5-6.5.

For both species, the primary coordination of arsenic sorbed at the pyrite surface was to four oxygen atoms (with As-O distances of 1.69 Å for As^{5+} and 1.73 Å for As^{3+} in line with expected trends). Evidence for Fe atoms at 3.35-3.40 Å lends further weight to the argument that both aresenite and arsenate species form outer sphere complexes at the pyrite surface under the conditions studied (Farquhar et al. 2002). In a study of arsenite sorption onto pyrite, Bostick and Fendorf (2003) using bulk sorption measurements, X-ray absorption spectroscopy and XPS, showed patterns of uptake that generally follow the Langmuir isotherms characteristic of interaction with a limited number of surface sites which eventually become saturated (monolayer capacity). Such behavior is in accord with the surface complexation suggested above. However, the spectra reported in this work have been interpreted in terms of formation of an "FeAsS-like surface precipitate." These discrepancies could arise from differences in starting materials and sample handling, although another explanation could be that Farquhar et al. (2002) worked at acid pH, where Bostick and Fendorf (2003) found lower sorption levels and closer adherence to Langmuir isotherm uptake behavior as a function of concentration.

Mo^{6+} uptake on pyrite, both as molybdate (MoO_4^{2-}) and as tetrathiomolybdate (MoS_4^{2-}), was studied by Bostick et al. (2003) using bulk sorption experiments and X-ray absorption spectroscopy. Both forms of Mo partition strongly on to pyrite under a wide range of pH and ionic strength conditions, although molybdate is readily desorbed whereas MoS_4^{2-} is retained even at high pH. The spectroscopic data were interpreted in terms of molybdate forming a bidentate, mononuclear complex on FeS_2 and tetrathiomolybdate forming Mo-Fe-S cubane-type clusters, consistent with its high affinity and resistance to desorption. Structural models for these surface complexes are shown in Figure 33.

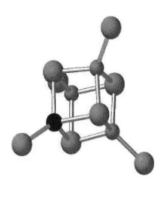

Figure 33. Structural models of (top) MoS_4^{2-} and (bottom) MoO_4^{2-} adsorption on FeS_2. Molybdate forms a bidentate, mononuclear complex with an Fe polyhedron, and MoS_4^{2-} forms a Mo-Fe-S cubane structure. [Used with permission of the American Chemical Society from Bostick and Fendorf (2003), *Environmental Science and Technology*, Vol. 37, Fig. 7, p. 290.]

Pyrrhotite and troilite

Very few studies of the uptake of metals by the surface of pyrrhotite have so far been undertaken. Given the relative complexity of this sulfide, this is not surprising. In a study which further explored the interaction of gold in solution with sulfide surfaces, Widler and Seward (2002) used potentiometric titrations, bulk measurements of uptake, and XPS to examine uptake of gold from Au(I) hydrosulfide solution complexes by pyrrhotite produced by hydrothermal synthesis. Under the conditions of these experiments, the mineral surface is negatively charged from pH 2 to pH 10, and maximum uptake occurs at pH < 3. At alkaline pH, there was no uptake of gold by the pyrrhotite. The process, which was better defined for the pyrite and mackinawite surfaces also studied, involved sorption of an $AuHS^0$ complex.

Bostick and Fendorf (2003), in addition to the work on arsenite sorption on to pyrite discussed above, also studied the reaction of this species with synthetic powdered FeS. Very similar results were reported to those for pyrite, although the absorption affinity increased with pH over the entire range studied rather than becoming constant at higher pH, as in pyrite. Again, the spectroscopic results were interpreted in terms of an FeAsS surface precipitate.

Mackinawite

Mackinawite, the tetragonal FeS (in some cases described as exhibiting a small degree of nonstoichiometry, FeS_{1-x}) is the first iron sulfide to form by precipitation from aqueous solution under a wide range of conditions. The fine-grained black sulfide occurring in the reducing environment just beneath the sediment surface in many marine environments (and resulting from reaction between detritally introduced iron minerals and sulfide generated by bacterial reduction of seawater sulfate) was formerly referred to as "amorphous FeS." However, although this material is poorly crystalline or "X-ray amorphous" due to its fine grain size, on a nanometer scale it is now known to have the mackinawite structure (Vaughan and Craig 1997). This structure, refined by Lennie et al. (1995), is characterized by layers of iron atoms tetrahedrally coordinated to sulfur and with very short Fe-Fe distances within the layers (see Makovicky, 2006 this volume). In sediments, mackinawite is always a very fine particle, transient phase that transforms to other sulfides (particularly pyrite) in the cycle of lithification. However, it is an important phase in estuarine and other near-shore marine environments, and its high level of reactivity (partly due to its small grain size and, hence, large surface area) makes it of considerable environmental interest.

Because mackinawite only occurs (and can only be synthesized) as fine particle material, many of the techniques of surface science cannot be used to study its surface chemistry. However, the EXAFS and XANES methods have been used to probe the interactions of the mackinawite surface with metal ions taken-up from solution, so as to gain understanding of the behavior of pollutants in the marine (and similar related) environments noted above. Here, by acquiring data for an appropriate absorption edge of the pollutant metal of interest, the local coordination environment of that metal (numbers and types of atoms in first, second, and sometimes further surrounding shells) can be established. Such spectroscopic studies have been combined with measurements of the uptake of the metal from an aqueous solution through contact with the mackinawite. Metal uptake has generally been determined as a function of factors such as total available dissolved metal, solution pH, or concentration of other species in solution before separation of "reacted" mackinawite for spectroscopic studies.

Interactions between the mackinawite surface and dissolved toxic heavy metals have been studied for Cu and Cd (Parkman et al. 1999), Co, Ni, Mn (Butler et al. 2006) and for Au (Widler and Seward 2002). The experiments conducted have been similar to those described above for interaction with pyrite. In terms of bulk uptake experiments, the behavior of Cu and Cd involves substantial removal of the metal up to high concentrations, without any apparent limitation, such as would be the case for saturation of surface sites (ie a linear uptake rather than the decreasing sorption associated with a Langmuir isotherm). The results of spectroscopic studies point to formation of a CdS precipitate in the case of cadmium, whereas for copper, a precipitate is again formed but of the ternary sulfide chalcopyrite ($CuFeS_2$). This unexpected result is supported by work on natural Cu-rich sulfidic sediments (Parkman et al. 1999). Preliminary results for reaction of Co, Ni and Mn with mackinawite indicate substantial removal via similar precipitation reactions for Co and Ni as described above for Cu and Cd. However, Mn behaves differently, with limited uptake at pH 7 attributed to surface complexation, and much greater uptake at pH 8 attributed to a precipitation reaction (Butler et al. 2006). In the case of gold, uptake from solutions containing up to 40 mg/kg was 100% below pH 4, whereas no uptake occurred at alkaline pH (9). XPS studies revealed that, in acid solution, a chemisorption reaction occurs at the mackinawite surface leading to reduction of the Au(I) solution complex to Au(0), and formation of surface polysulfides.

The interaction between mackinawite and dissolved arsenic is a topic of particular interest because of concerns over arsenic contamination of rivers and aquifers (see Vaughan 2006 for a general introduction). Farquhar et al. (2002) and Wolthers et al. (2003; 2005) have studied As uptake by mackinawite. Farquhar et al. (2002) used XAS to investigate the mechanisms whereby As^{3+} and As^{5+} in aqueous solution (pH 5.5-6.5) interact with the mackinawite surface and, in line with results described above for pyrite, found evidence for the formation of outer sphere complexes for both species. Wolthers et al. (2005) conducted similar experiments using a "poorly crystalline mackinawite" regarded as more akin to the material found naturally in Recent sediments (and formerly termed "amorphous FeS"). Their results also point to outer sphere complex formation as the mechanism of interaction between arsenic solution species and the mackinawite surface.

The nuclear metals are also of obvious environmental concern, and uranium and neptunium provide contrasting examples of behavior on interaction with the mackinawite surface. Uranium, introduced as the uranyl ion, shows bulk behavior similar to that exhibited by copper, being very effectively removed from solution by mackinawite (Moyes et al. 2000). However, the environment of the metal at the FeS surface, as determined from uranium L-edge EXAFS data, contrasts with that of copper. As illustrated in Figure 34, the uranium is clearly bonded to oxygen with U-O distances characteristic of those expected for the uranyl ion in the first shell, and a further four bonded oxygens at 2.1-2.3 Å. Beyond these shells are further oxygens at 2.9 Å, and possibly iron at ~4 Å (Moyes et al. 2000). Because of the care taken to avoid introduction of oxygen in these experiments, it was suggested that a redox reaction takes place at the FeS surface on interaction with the uranyl ion, leading to precipitation of a U_3O_8-type phase at the surface. Support for this proposal comes from studying the spectra as a function of increasing total uranium concentrations; a systematic decrease in relative intensity of the peak at 1.79 Å which is characteristic of the uranyl species occurs and there is also a systematic decrease in the second shell U-O distance from a value of around 2.2 Å to 2.1 Å, characteristic of reduced uranium (Moyes et al. 2000).

An interesting contrast is provided by studies of the interaction of pentavalent neptunium in solution with mackinawite (Moyes et al. 2002). The uptake of neptunium from solution is relatively low (~10%) and independent of solution concentration over the range 1,000-20,000 ppm. It is also independent of

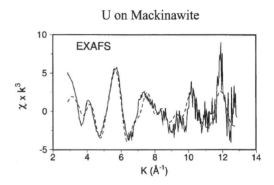

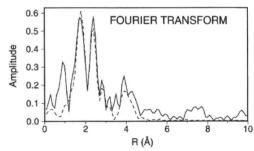

Shell	C.N.	Type	Dist. (Å)	$2\sigma^2 (Å^2)$
1	2	O	1.79	0.012
2	1	O	2.12	0.002
3	3	O	2.32	0.006
4	6	O	2.90	0.040
5	4	Fe	3.96	0.022

Figure 34. EXAFS spectra, Fourier transform, and fit data results for U on mackinawite. Solid lines are the experimental data and dashed lines are best fits. The U-O bond distance of 1.79 Å is characteristic of that expected for the uranyl ion. After Moyes et al. (2000).

equilibration time. The X-ray absorption spectra for the Np L-edge are also essentially the same over this range of concentrations, as may be seen from their Fourier transforms in Figure 35. The best fits to these data involve an environment in which the neptunium is coordinated to four oxygen atoms at around 2.25 Å, two sulfur atoms at ~2.6 Å, and iron atoms at ~3.9 and 4.1 Å. These data suggest that, on interaction with the sulfide surface, a reduction of neptunium (V) to neptunium (IV) occurs, accompanied by a loss of axial oxygen atoms. Then, it appears that neptunium bonds directly to sulfur atoms at the FeS surface as illustrated in Figure 35. This formation of an inner sphere surface complex would explain the limited capacity of FeS to uptake neptunium, in contrast to the behavior of uranium, a result with important implications for the mobility of this element in environments where FeS occurs.

One other example of interaction of a nuclear metal with mackinawite concerns technetium, a very toxic waste product of nuclear fission with a long halflife. Under oxidizing conditions, it forms the mobile pertechnetate (TcO_4^-) anion. The reduced form of technetium, as present in technetium disulfide for example, is immobile but little is known of how technetium in solution as pertechnetate interacts with sulfidic sediments, or what might happen given subsequent re-oxidation. Exploratory work in this area at Manchester (Wharton et al. 2000; Livens et al. 2004) has employed XAS of model systems to study these problems. The technetium K-edge EXAFS for the pertechnetate anion show coordination to four oxygen atoms at ~1.7 Å whereas the disulfide shows coordination to six sulfur atoms at between ~2.1 and 2.5 Å. When pertechnetate is coprecipitated with FeS, the XAS data show coordination to six sulfur atoms at ~2.4 Å indicating formation of a TcS_2-like phase. However, re-oxidation of the material formed in this way does not produce a pertechnetate, but a phase with coordination to six oxygens at ~2 Å, probably a TcO_2 (and hence Tc^{4+}) phase. These observations are confirmed by the Tc K-edge XANES positions which are similar for TcS_2, the coprecipitate, and its re-oxidized equivalent, but clearly differ from the values for pertechnetate.

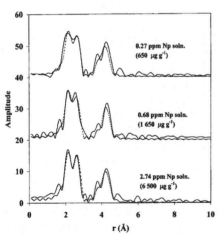
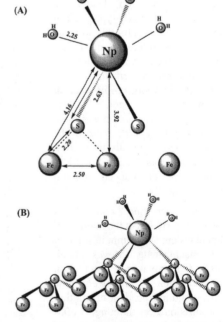

Figure 35. EXAFS spectra for Np on mackinawite and a schematic diagram of the resulting proposed Np binding mechanism to surface S atoms. The models show a) bond distances derived from the spectra, and b) the proposed surface complex. [Used with permission of the American Chemical Society from Moyes et al. (2002), *Environmental Science and Technology*, Vol. 36, Figs. 2 and 4, pp. 181 and 182.]

Sphalerite and wurtzite

Not surprisingly, ZnS has received less attention in regard to its reactions with dissolved metals than the iron sulfides. However, in a very detailed study of the interaction of the wurtzite form of ZnS with Co^{2+}, Persson et al. (1995) made wet-chemical measurements of bulk uptake and used EXAFS to probe the local environment of Co at the ZnS surface. Rather surprisingly, given the existence in nature of Co-bearing ZnS phases (see Wincott and Vaughan 2006 this volume), no evidence was found for exchange between Zn and Co. Analysis of the EXAFS results showed a first shell surrounding Co with both oxygen and sulfur atoms, and a Co-S distance of 2.36 Å which is close to the Zn-S distance in wurtzite of 2.34 Å. Surface complexation (inner sphere) of Co^{2+} to surface sulfide sites is proposed. More detailed EXAFS analysis indicated that the cobalt adsorbs to defect structural positions where Zn atoms are missing at the surface. Maximum surface complex binding capacity was estimated as 2.0 μmol m^{-2} and at increased surface coverages, a structurally disordered $Co(OH)_2$ phase is formed.

A contrasting example is provided by a quantum-mechanical study of the (110) surface of sphalerite and its interaction with Pb^{2+} solution species (Steele et al. 2003). This study was motivated by the problems caused by changes in the surface properties of sphalerite through lead "sorption" during attempts to separate it from other sulfides in mineral processing by froth flotation. These calculations show that replacement of Zn^{2+} by Pb^{2+} in the (110) surface of sphalerite is energetically unfavorable and leads to large distortions of the surface structure; also that Pb^{2+}, $PbOH^+$, $Pb(OH)_2$ and $Pb(H_2O)_2^{2+}$ do not bind to the surface via direct Pb-Zn or Pb-S surface bonds. The calculations do suggest that Pb^{2+} ions bind to the ZnS surface via Pb-O-Zn bonds, and that PbO, $PbOH^+$ and their inner sphere solvated complexes can bind to the surface in this way, a result in agreement with experimental evidence.

FUTURE OUTLOOK

In this chapter we have attempted to provide an overview of the main research areas and the state of understanding of the reactivity of sulfide mineral surfaces. This research area is vast and it crosscuts many scientific, engineering, and industrial domains. It also shares connections with several of the other chapters in this book. By including a wide-range of perspectives, from high vacuum surface science to the mining industry, we have tried to merge traditionally separate research fields. This merging is becoming increasingly necessary as the goal of characterizing and assessing the reactive behavior of sulfides focuses more and more on the molecular-level.

The volume of research in this area is rapidly growing, in part due to significant advances in experimental and computational tools. The availability of techniques using advanced (synchrotron) light sources has provided many new insights into the types and proportions of chemical species that evolve on sulfide surfaces during oxidation and other reactions. Scanning tunneling microscopy and spectroscopy have provided the means to dissect the surface oxidation process, or the structures of self-assembled monolayers, in terms of individual sites at the atomic level. Continual improvements in computational chemistry methods and in their efficiency provide a steady supply of increasingly accurate simulations of many kinds of reactions on sulfide surfaces. Most of these newer tools have yet to be applied across the full range of important sulfide surface chemical problems. Hence the field is currently enjoying a time when well-studied systems are being re-analyzed by application of newer tools.

Given that virtually all of the reactive behavior of sulfide minerals in the natural environment and in industrial systems is controlled by chemical reactions at their surfaces, the need to understand fundamental aspects of these reactions will be ever-present. Our dependence on mineral resources to sustain the raw material and energy needs of a growing and developing population means that sulfide minerals, and therefore their surfaces, will continue to be exposed at the Earth's surface due to mining activities. The need to mitigate acid mine drainage and toxic metal transport, or to improve mineral extraction technologies, whether from base metal ore or coal, have been key drivers for advancing our understanding of sulfide mineral surface chemistry to its current level. While the causes of acid mine drainage are understood now better than ever, the complexity and scale of the problem will continue to provide abundant justification for improving our understanding of, for example, mechanisms and rates of sulfide oxidation reactions at molecular resolution. Similarly, improving mineral separation technologies both in terms of efficiency and affordability by investing in a molecular-level understanding of sulfide surface functionalization reactions could lead to economic and societal benefits of enormous proportions in the long term. Even though motivating factors may gradually change, given the diversity of sulfide minerals and, in some cases, the sheer volumes of the sulfides that we encounter in natural environments, we can expect that sulfide surface chemistry will persist as an important research area for many years to come.

ACKNOWLEDGMENTS

The authors gratefully acknowledge Mike Hochella for his careful review of this manuscript. KMR acknowledges the support of the U.S. Department of Energy (DOE), Office of Basic Energy Sciences, Geosciences Division, and the Stanford Environmental Molecular Sciences Institute (EMSI) jointly funded by the National Science Foundation and the DOE Office of Biological and Environmental Research (OBER). The W. R. Wiley Environmental Molecular Science Laboratory (EMSL) at Pacific Northwest National Laboratory (PNNL) is a national scientific user facility sponsored by the OBER. PNNL is operated for the DOE by Battelle Memorial Institute under contract DE-AC06-76RLO 1830. DJV acknowledges the support of the Natural Environment Research Council (NERC) in funding the Williamson Research Centre, and both the NERC and the Engineering and Physical Sciences Research Council in funding research in the mineral sciences at Manchester.

REFERENCES

Abraitis PK, Pattrick RAD, Kelsall GH, Vaughan DJ (2004) Acid leaching and dissolution of major sulphide ore minerals: processes and galvanic effects in complex systems. Mineral Mag 68:343-351
Acharya HN, Paul A (1991) Thermodynamic analysis of selective reduction of lead-oxide in galena. J Mater Sci Lett 10:257-259
Ackerman PK, Harris GH, Klimpel RR, Aplan FF (1987) Evaluation of flotation collectors for copper sulfides and pyrite .3. Effect of xanthate chain length and branching. Int J Mineral Proc 21:141-156
Allison SA, Granvill.A, Goold LA, Nicol MJ (1972) Determination of products of reaction between various sulfide minerals and aqueous xanthate solution, and a correlation of products with electrode rest potentials. Metall Trans 3:2613
Aray Y, Rodriguez J (2001) Study of hydrodesulfurization by transition metal sulfides by means of the Laplacian of the electronic charge density. Chemphyschem 2:599-604
Aray Y, Rodriguez J, Coll S, Rodriguez-Arias EN, Vega D (2005) Nature of the Lewis acid sites on molybdenum and ruthenium sulfides: An electrostatic potential study. J Phys Chem B 109:23564-23570
Aray Y, Rodriguez J, Vega D, Coll S, Rodriguez-Arias EN, Rosillo F (2002) Adsorption of thiophene on the RuS_2 (100) and (111) surfaces: A Laplacian of the electronic charge density study. J Phys Chem B 106: 13242-13249
Bancroft GM, Hyland MM (1990) Spectroscopic studies of adsorption reduction reactions of aqueous metal complexes on sulfide surfaces. Rev Mineral 23:511-558
Bang SS, Deshpande SS, Han KN (1995) The oxidation of galena using Thiobacillus Ferrooxidans. Hydromet 37:181-192
Basolo F, Pearson RG (1967) Mechanisms of Inorganic Reactions: A Study of Metal Complexes in Solution. Wiley
Becker U, Hochella MF Jr (1996) The calculation of STM images, STS spectra, and XPS peak shifts for galena: New tools for understanding mineral surface chemistry. Geochim Cosmochim Acta 60:2413-2426

Becker U, Hochella MF Jr, Vaughan DJ (1997) The adsorption of gold to galena surfaces: Calculation of adsorption/reduction energies, reaction mechanisms, XPS spectra, and STM images. Geochim Cosmochim Acta 61:3565-3585

Becker U, Rosso KM (2001) Step edges on galena (100): Probing the basis for defect driven surface reactivity at the atomic scale. Am Mineral 86:862-870

Becker U, Rosso KM, Hochella MF Jr (2001) The proximity effect on semiconducting mineral surfaces: A new aspect of mineral surface reactivity and surface complexation theory? Geochim Cosmochim Acta 65: 2641-2649

Belzile N, Chen YW, Cai MF, Li YR (2004) A review on pyrrhotite oxidation. J Geochem Explor 84:65-76

Bonnissel-Gissinger P, Alnot M, Ehrhardt JJ, Behra P (1998) Surface oxidation of pyrite as a function of pH. Environ Sci Technol 32:2839-2845

Bostick BC, Fendorf S (2003) Arsenite sorption on troilite (FeS) and pyrite (FeS_2). Geochim Cosmochim Acta 67:909-921

Bostick BC, Fendorf S, Fendorf M (2000) Disulfide disproportionation and CdS formation upon cadmium sorption on FeS_2. Geochim Cosmochim Acta 64:247-255

Bostick BC, Fendorf S, Helz GR (2003) Differential adsorption of molybdate and tetrathiomolybdate on pyrite (FeS_2). Environ Sci Technol 37:285-291

Bryce RA, Vincent MA, Hillier IH, Hall RJ (2000) Structure and stability of galena (PbS) at the interface with aqueous solution: A combined embedded cluster/reaction field study. J Molec Struct Theochem 500:169-180

Buckley AN, Walker GW (1988) The surface composition of arsenopyrite exposed to oxidizing environments. Appl Surf Sci 35:227-240

Buckley AN, Woods R (1985) X-ray photoelectron spectroscopy of oxidized pyrrhotite surfaces .1. Exposure to air. Appl Surf Sci 22-3:280-287

Buckley AN, Woods R (1987) The surface oxidation of pyrite. Appl Surf Sci 27:437-452

Buckley AN, Woods R (1991) Adsorption of ethyl xanthate on freshly exposed galena surfaces. Colloids Surf 53:33-45

Butler I, Bell R, Bell AMT, Charnock JM, Oldroyd A, Rickard DT, Vaughan DJ (2006) Co^{2+}, Ni^{2+}, Mn^{2+} uptake by mackinawite. Chem Geol (in press)

Chaturvedi S, Katz R, Guevremont J, Schoonen MAA, Strongin DR (1996) XPS and LEED study of a single-crystal surface of pyrite. Am Mineral 81:261-264

Chernyshova IV (2003) An in situ FTIR study of galena and pyrite oxidation in aqueous solution. J Electroanal Chem 558:83-98

Ciminelli VST, OsseoAsare K (1995) Kinetics of pyrite oxidation in sodium carbonate solutions. Metall Mater Trans B 26:209-218

Cincotti S, Rabe JP (1993) Self-assembled alkane monolayers on $MoSe_2$ and MoS_2. Appl Phys Lett 62:3531-3533

Costa MC, do Rego AMB, Abrantes LM (2002) Characterization of a natural and an electro-oxidized arsenopyrite: A study on electrochemical and X-ray photoelectron spectroscopy. Int J Mineral Process 65:83-108

de Leeuw NH, Parker SC, Sithole HM, Ngoepe PE (2000) Modeling the surface structure and reactivity of pyrite: Introducing a potential model for FeS_2. J Phys Chem B 104:7969-7976

Eggleston CM (1997) Initial oxidation of sulfide sites on a galena surface: Experimental confirmation of an ab initio calculation. Geochim Cosmochim Acta 61:657-660

Eggleston CM, Ehrhardt JJ, Stumm W (1996) Surface structural controls on pyrite oxidation kinetics: An XPS-UPS, STM, and modeling study. Am Mineral 81:1036-1056

Eggleston CM, Hochella MF Jr (1990) Scanning tunneling microscopy of sulfide surfaces. Geochim Cosmochim Acta 54:1511-1517

Eggleston CM, Hochella MF Jr (1991) Scanning tunneling microscopy of galena (100) surface oxidation and sorption of aqueous gold. Science 254:983-986

Eggleston CM, Hochella MF Jr (1993) Tunneling spectroscopy applied to PbS (001) surfaces: Fresh surfaces, oxidation, and sorption of aqueous Au. Am Mineral 78:877-883

Eggleston CM, Hochella MF Jr (1994) Atomic and electronic structure of PbS (100) surfaces and chemisorption oxidation reactions. *In*: Environmental Geochemistry of Sulfide Oxidation. Alpers CN, Blowes DW (eds) American Chemical Society, p 201-222

Ehrhardt JJ, Behra P, Bonnissel-Gissinger P, Alnot M (2000) XPS study of the sorption of Hg(II) onto pyrite FeS_2. Surf Interface Anal 30:269-272

England KER, Charnock JM, Pattrick RAD, Vaughan DJ (1999) Surface oxidation studies of chalcopyrite and pyrite by glancing-angle X-ray absorption spectroscopy (REFLEXAFS). Mineral Mag 63:559-566

Evangelou VP, Seta AK, Holt A (1998) Potential role of bicarbonate during pyrite oxidation. Environ Sci Technol 32:2084-2091

Farquhar ML, Charnock JM, Livens FR, Vaughan DJ (2002) Mechanisms of arsenic uptake from aqueous solution by interaction with goethite, lepidocrocite, mackinawite, and pyrite: An X-ray absorption spectroscopy study. Environ Sci Technol 36:1757-1762

Farquhar ML, Wincott PL, Wogelius RA, Vaughan DJ (2003) Electrochemical oxidation of the chalcopyrite surface: an XPS and AFM study in solution at pH 4. Appl Surf Sci 218:34-43

Fernandez PG, Linge HG, Wadsley MW (1996a) Oxidation of arsenopyrite (FeAsS) in acid .1. Reactivity of arsenopyrite. J Appl Electrochem 26:575-583

Fernandez PG, Linge HG, Willing MJ (1996b) Oxidation of arsenopyrite (FeAsS) in acid .2. Stoichiometry and reaction scheme. J Appl Electrochem 26:585-591

Fornasiero D, Li FS, Ralston J (1994) Oxidation of galena .2. Electrokinetic study. J Colloid Interface Sci 164:345-354

Fornasiero D, Montalti M, Ralston J (1995) Kinetics of adsorption of ethyl xanthate on pyrrhotite: In situ UV and infrared spectroscopic studies. J Colloid Interface Sci 172:467-478

Frechard F, Sautet P (1997a) Chemisorption of H_2 and H_2S on the (100) surface of RuS_2: An ab initio theoretical study. Surf Sci 389:131-146

Frechard F, Sautet P (1997b) RuS_2 (111) surfaces: Theoretical study of various terminations and their interaction with H_2. J Catal 170:402-410

Genin F, Alnot M, Ehrhardt JJ (2001) Interaction of vapours of mercury with PbS (001): A study by X-ray photoelectron spectroscopy, RHEED and X-ray absorption spectroscopy. Appl Surf Sci 173:44-53

Giancarlo LC, Fang HB, Rubin SM, Bront AA, Flynn GW (1998) Influence of the substrate on order and image contrast for physisorbed, self-assembled molecular monolayers: STM studies of functionalized hydrocarbons on graphite and MoS_2. J Phys Chem B 102:10255-10263

Grillo ME, Sautet P (2000) Density functional study of the structural and electronic properties of RuS_2 (111) II. Hydrogenated surfaces. Surf Sci 457:285-293

Grillo ME, Sautet P (2001) On the nature of RuS_2 HDS active sites: Insight from ab initio theory. J Mol Catal A 174:239-244

Guevremont JM, Strongin DR, Schoonen MAA (1997) Effects of surface imperfections on the binding of CH_3OH and H_2O on FeS_2 (100): Using adsorbed Xe as a probe of mineral surface structure. Surf Sci 391:109-124

Guevremont JM, Bebie J, Elsetinow AR, Strongin DR, Schoonen MAA (1998a) Reactivity of the (100) plane of pyrite in oxidizing gaseous and aqueous environments: Effects of surface imperfections. Environ Sci Technol 32:3743-3748

Guevremont JM, Elsetinow AR, Strongin DR, Bebie J, Schoonen MAA (1998b) Structure sensitivity of pyrite oxidation: Comparison of the (100) and (111) planes. Am Mineral 83:1353-1356

Guevremont JM, Strongin DR, Schoonen MAA (1998c) Photoemission of adsorbed Xenon, X-ray photoemission spectroscopy, and temperature-programmed desorption studies of H_2O on FeS_2 (100). Langmuir 14:1361-1366

Guevremont JM, Strongin DR, Schoonen MAA (1998d) Thermal chemistry of H_2S and H_2O on the (100) plane of pyrite: Unique reactivity of defect sites. Am Mineral 83:1246-1255

Hacquard E, Bessiere J, Alnot M, Ehrhardt JJ (1999) Surface spectroscopic study of the adsorption of Ni(II) on pyrite and arsenopyrite at pH 10. Surf Interface Anal 27:849-860

Helveg S, Lauritsen JV, Laegsgaard E, Stensgaard I, Norskov JK, Clausen BS, Topsoe H, Besenbacher F (2000) Atomic-scale structure of single layer MoS_2 nanoclusters. Phys Rev Lett 84:951-954

Himmel HJ, Kaschke M, Harder P, Woll C (1996) Adsorption of organic monolayers on pyrite (FeS_2) (100). Thin Solid Films 285:275-280

Hochella MF Jr, Rakovan JF, Rosso KM, Bickmore BR, Rufe E (1998) New directions in mineral surface geochemical research using scanning probe microscopes. *In*: Mineral-Water Interfacial Reactions: Kinetics and Mechanisms. Sparks DL, Grundl T (eds) American Chemical Society, p 1-22

Hung A, Yarovsky I, Russo SP (2003) Density functional theory studies of xanthate adsorption on the pyrite FeS_2 (110) and (111) surfaces. J Chem Phys 118:6022-6029

Hung A, Yarovsky I, Russo SP (2004) Density functional theory of xanthate adsorption on the pyrite FeS_2 (100) surface. Phil Mag Lett 84:175-182

Janczuk B, Wojcik W, Zdziennicka A, Gonzalezcaballero F (1992a) Components of surface free energy of galena. J Mater Sci 27:6447-6451

Janczuk B, Wojcik W, Zdziennicka A, Gonzalezcaballero F (1992b) Determination of the galena surface free energy components from contact angle measurements. Mater Chem Phys 31:235-241

Jones RA, Koval SF, Nesbitt HW (2003) Surface alteration of arsenopyrite (FeAsS) by Thiobacillus ferrooxidans. Geochim Cosmochim Acta 67:955-965

Karthe S, Szargan R, Suoninen E (1993) Oxidation of pyrite surfaces: A photoelectron spectroscopic study. Appl Surf Sci 72:157-170

Kartio I, Laajalehto K, Suoninen E (1999) Characterization of the ethyl xanthate adsorption layer on galena (PbS) by synchrotron radiation excited photoelectron spectroscopy. Colloid Surf A 154:97-101

Kelsall GH, Yin Q, Vaughan DJ, England KER, Brandon NP (1999) Electrochemical oxidation of pyrite (FeS_2) in aqueous electrolytes. J Electroanal Chem 471:116-125

Kendelewicz T, Doyle CS, Bostick BC, Brown GE (2004) Initial oxidation of fractured surfaces of FeS_2 (100) by molecular oxygen, water vapor, and air. Surf Sci 558:80-88

Kendelewicz T, Liu P, Brown GE, Nelson EJ (1998) Interaction of sodium overlayers with the PbS (100) (galena) surface: Evidence for a Na <-> Pb exchange reaction. Surf Sci 411:10-22

Laajalehto K, Kartio I, Suoninen E (1997) XPS and SR-XPS techniques applied to sulphide mineral surfaces. Int J Mineral Proc 51:163-170

Lan Y, Huang X, Deng B (2002) Suppression of pyrite oxidation by iron 8-hydroxyquinoline. Arch Environ Contam Toxicol 43:168-174

Larsson ML, Holmgren A, Forsling W (2000) Xanthate adsorbed on ZnS studied by polarized FTIR-ATR spectroscopy. Langmuir 16:8129-8133

Larsson ML, Holmgren A, Forsling W (2001) Structure and orientation of collectors adsorbed at the ZnS/water interface. J Colloid Interface Sci 242:25-30

Lauritsen JV, Bollinger MV, Laegsgaard E, Jacobsen KW, Norskov JK, Clausen BS, Topsoe H, Besenbacher F (2004a) Atomic-scale insight into structure and morphology changes of MoS_2 nanoclusters in hydrotreating catalysts. J Catal 221:510-522

Lauritsen JV, Nyberg M, Norskov JK, Clausen BS, Topsoe H, Laegsgaard E, Besenbacher F (2004b) Hydrodesulfurization reaction pathways on MoS_2 nanoclusters revealed by scanning tunneling microscopy. J Catal 224:94-106

Lennie AR, Redfern SAT, Scofield PF, Vaughan DJ (1995) Synthesis and Rietveld crystal structure refinement of mackinawite, tetragonal FeS. Mineral Mag 59:677-683

Leppinen JO, Basilio CI, Yoon RH (1988) In situ FTIR spectroscopic study of ethyl xanthate electrosorption on sulfide minerals. J Electrochem Soc 135:C145-C145

Leppinen JO, Basilio CI, Yoon RH (1989) Insitu FTIR study of ethyl xanthate adsorption on sulfide minerals under conditions of controlled potential. Int J Mineral Proc 26:259-274

Leppinen JO, Rastas JK (1986) The Interaction between ethyl xanthate ion and lead sulfide surface. Colloid Surf 20:221-237

Leppinen JO, Rastas JK (1988) Thermodynamics of the system lead sulfide and thiol collector. Colloid Surf 29:205-220

Livens FR, Jones MJ, Hynes AJ, Charnock JM, Mosselmans JFW, Hennig C, Steele H, Collison D, Vaughan DJ, Pattrick RAD, Reed WA, Moyes LN (2004) X-ray absorption spectroscopy studies of reactions of technetium, uranium and neptunium with mackinawite. J Environ Radioact 74:211-219

Lowson RT (1982) Aqueous oxidation of pyrite by molecular oxygen. Chem Rev 82:461-497

Ludwig C, Strohmaier R, Petersen J, Gompf B, Eisenmenger W (1994) Epitaxy and scanning tunneling microscopy image contrast of copper phthalocyanine on graphite and MoS_2. J Vac Sci Technol B 12:1963-1966

Makovicky E (2006) Crystal structures of sulfides and other chalcogenides. Rev Mineral Geochem 61:7-125

Maier GS, Dobias B (1997) 2-mercaptobenzothiazole and derivatives in the flotation of galena, chalcocite and sphalerite: A study of flotation, adsorption and microcalorimetry. Mineral Eng 10:1375-1393

Mattila S, Leiro JA, Heinonen M (2004) XPS study of the oxidized pyrite surface. Surf Sci 566:1097-1101

McKibben MA, Barnes HL (1986) Oxidation of pyrite in low temperature acidic solutions: Rate laws and surface textures. Geochim Cosmochim Acta 50:1509-1520

Mielczarski JA, Mielczarski E, Zachwieja J, Cases JM (1995) In situ and ex situ infrared studies of nature and structure of thiol monolayers adsorbed on cuprous sulfide at controlled potential: Simulation and experimental results. Langmuir 11:2787-2799

Mielczarski JA, Cases JM, Alnot M, Ehrhardt JJ (1996a) XPS characterization of chalcopyrite, tetrahedrite, and tennantite surface products after different conditioning. 1. Amyl xanthate solution at pH 10. Langmuir 12:2531-2543

Mielczarski JA, Cases JM, Alnot M, Ehrhardt JJ (1996b) XPS characterization of chalcopyrite, tetrahedrite, and tennantite surface products after different conditioning. 1. Aqueous solution at pH. Langmuir 12:2519-2530

Mielczarski JA, Cases JM, Barres O (1996c) In situ infrared characterization of surface products of interaction of an aqueous xanthate solution with chalcopyrite, tetrahedrite, and tennantite. J Colloid Interface Sci 178:740-748

Mielczarski JA, Mielczarski E, Cases JM (1996d) Interaction of amyl xanthate with chalcopyrite, tetrahedrite, and tennantite at controlled potentials. Simulation and spectroelectrochemical results for two-component adsorption layers. Langmuir 12:6521-6529

Mielczarski JA, Mielczarski E, Cases JM (1997) Infrared evaluation of composition and structure of ethyl xanthate monolayers produced on chalcopyrite, tetrahedrite, tennantite at controlled potentials. J Colloid Interface Sci 188:150-161

Mielczarski JA, Mielczarski E, Cases JM (1998) Influence of chain length on adsorption of xanthates on chalcopyrite. Int J Mineral Proc 52:215-231

Mikhlin Y, Tomashevich Y (2005) Pristine and reacted surfaces of pyrrhotite and arsenopyrite as studied by X-ray absorption near-edge structure spectroscopy. Phys Chem Mineral 32:19-27

Mishra KK, Osseoasare K (1992) Electroreduction Of Fe^{3+}, O_2, And $Fe(CN)_6^{3-}$ at the n-type pyrite (FeS_2) surface. J Electrochem Soc 139:3116-3120

Moses CO, Nordstrom DK, Herman JS, Mills AL (1987) Aqueous pyrite oxidation by dissolved oxygen and by ferric iron. Geochim Cosmochim Acta 51:1561-1571

Moses CO, Herman JS (1991) Pyrite oxidation at circumneutral pH. Geochim Cosmochim Acta 55:471-482

Moyes LN, Parkman RH, Charnock JM, Vaughan DJ, Livens FR, Hughes CR, Braithwaite A (2000) Uranium uptake from aqueous solution by interaction with goethite, lepidocrocite, muscovite, and mackinawite: An X-ray absorption spectroscopy study. Environ Sci Technol 34:1062-1068

Moyes LN, Jones MJ, Reed WA, Livens FR, Charnock JM, Mosselmans JFW, Hennig C, Vaughan DJ, Pattrick RAD (2002) An X-ray absorption spectroscopy study of neptunium(V) reactions with mackinawite (FeS). Environ Sci Technol 36:179-183

Mycroft JR, Bancroft GM, McIntyre NS, Lorimer JW (1995a) Spontaneous deposition of gold on pyrite from solutions containing Au(III) and Au(I) chlorides. 1. A surface study. Geochim Cosmochim Acta 59:3351-3365

Mycroft JR, Nesbitt HW, Pratt AR (1995b) X-Ray photoelectron and Auger electron spectroscopy of air oxidized pyrrhotite: Distribution of oxidized species with depth. Geochim Cosmochim Acta 59:721-733

Nesbitt HW, Bancroft GM, Pratt AR, Scaini MJ (1998) Sulfur and iron surface states on fractured pyrite surfaces. Am Mineral 83:1067-1076

Nesbitt HW, Muir IJ (1994) X-Ray photoelectron spectroscopic study of a pristine pyrite surface reacted with water vapor and air. Geochim Cosmochim Acta 58:4667-4679

Nesbitt HW, Muir IJ, Pratt AR (1995) Oxidation of arsenopyrite by air and air-saturated, distilled water, and implications for mechanism of oxidation. Geochim Cosmochim Acta 59:1773-1786

Nesbitt HW, Muir IJ (1998) Oxidation states and speciation of secondary products on pyrite and arsenopyrite reacted with mine waste waters and air. Mineral Pet 62:123-144

Nowak P (1993) Xanthate adsorption at PbS surfaces: Molecular model and thermodynamic description. Colloid Surf A 76:65-72

Parkman RH, Charnock JM, Bryan ND, Livens FR, Vaughan DJ (1999) Reactions of copper and cadmium ions in aqueous solution with goethite, lepidocrocite, mackinawite, and pyrite. Am Mineral 84:407-419

Pattrick RAD, England KER, Charnock JM, Mosselmans JFW (1999) Copper activation of sphalerite and its reaction with xanthate in relation to flotation: An X-ray absorption spectroscopy (reflection extended X-ray absorption fine structure) investigation. Int J Mineral Proc 55:247-265

Paul JF, Payen E (2003) Vacancy formation on MoS_2 hydrodesulfurization catalyst: DFT study of the mechanism. J Phys Chem B 107:4057-4064

Pearce CI, Pattrick RAD, Vaughan DJ (2006) Electrical and Magnetic properties of sulfides. Rev Mineral Geochem 61:127-180

Persson P, Parks GA, Brown GE (1995) Adsorption and structural environment of Co(II) at the zinc oxide-aqueous and zinc sulfide-aqueous solution interfaces. Langmuir 11:3782-3794

Petersen J, Strohmaier R, Gompf B, Eisenmenger W (1997) Monolayers of tetrachloro-thioindigo and thioindigo in the STM: Orientational disorder and the absence of photochromism. Surf Sci 389:329-337

Pettenkofer C, Jaegermann W, Bronold M (1991) Site specific surface interaction of electron donors and acceptors on FeS_2 (100) cleavage planes. Ber Bunsenges Phys Chem 95:560-565

Philpott MR, Goliney IY, Lin TT (2004) Molecular dynamics simulation of water in a contact with an iron pyrite FeS_2 surface. J Chem Phys 120:1943-1950

Porento M, Hirva P (2002) Theoretical studies on the interaction of anionic collectors with Cu^+, Cu^{2+}, Zn^{2+} and Pb^{2+} ions. Theor Chem Acc 107:200-205

Porento M, Hirva P (2003) The adsorption interaction of anionic sulfhydryl collectors on different PbS (100) surface sites. Surf Sci 539:137-144

Porento M, Hirva P (2004) A theoretical study on the interaction of sulfhydryl surfactants with a covellite (001) surface. Surf Sci 555:75-82

Porento M, Hirva P (2005) Effect of copper atoms on the adsorption of ethyl xanthate on a sphalerite surface. Surf Sci 576:98-106

Pratt AR, Muir IJ, Nesbitt HW (1994) X-Ray photoelectron and Auger electron spectroscopic studies of pyrrhotite and mechanism of air oxidation. Geochim Cosmochim Acta 58:827-841

Raikar GN, Thurgate SM (1991) An Auger and EELS study of oxygen adsorption on FeS_2. J Phys Cond Mater 3:1931-1939

Raybaud P, Hafner J, Kresse G, Toulhoat H (1998) Structural and electronic properties of the MoS$_2$ (10-10) edge surface. Surf Sci 407:237-250

Raybaud P, Hafner J, Kresse G, Kasztelan S, Toulhoat H (2000a) Ab initio study of the H$_2$-H$_2$S/MoS$_2$ gas-solid interface: The nature of the catalytically active sites. J Catal 189:129-146

Raybaud P, Hafner J, Kresse G, Kasztelan S, Toulhoat H (2000b) Structure, energetics, and electronic properties of the surface of a promoted MoS$_2$ catalyst: An ab initio local density functional study. J Catal 190:128-143

Richardson PE, Odell CS (1984) Semiconducting characteristics of galena electrodes. J Electrochem Soc 131:C99-C99

Richardson S, Vaughan DJ (1989) Arsenopyrite: A spectroscopic investigation of altered surfaces. Mineral Mag 53:223-229

Rimstidt JD, Chermak JA, Gagen PM (1994) Rates of reaction of galena, sphalerite, chalcopyrite, and arsenopyrite with Fe(III) in acidic solutions. In Environmental Geochemistry of Sulfide Oxidation. 2-13

Rimstidt JD, Vaughan DJ (2003) Pyrite oxidation: A state-of-the-art assessment of the reaction mechanism. Geochim Cosmochim Acta 67:873-880

Rosso KM, Becker U, Hochella MF Jr (1999a) Atomically resolved electronic structure of pyrite {100} surfaces: An experimental and theoretical investigation with implications for reactivity. Am Mineral 84:1535-1548

Rosso KM, Becker U, Hochella MF Jr (1999b) The interaction of pyrite {100} surfaces with O$_2$ and H$_2$O: Fundamental oxidation mechanisms. Am Mineral 84:1549-1561

Rosso KM, Becker U (2003) Proximity effects on semiconducting mineral surfaces II: Distance dependence of indirect interactions. Geochim Cosmochim Acta 67:941-953

Rosso KM, Vaughan DJ (2006) Sulfide mineral surfaces. Rev Mineral Geochem 61:505-556

Sasaki K (1994) Effect of grinding on the rate of oxidation of pyrite by oxygen in acid solutions. Geochim Cosmochim Acta 58:4649-4655

Scaini MJ, Bancroft GM, Lorimer JW, Maddox LM (1995) The interaction of aqueous silver species with sulfur-containing minerals as studied by XPS, AES, SEM, and electrochemistry. Geochim Cosmochim Acta 59:2733-2747

Scaini MJ, Bancroft GM, Knipe SW (1997) An XPS, AES, and SEM study of the interactions of gold and silver chloride species with PbS and FeS$_2$: Comparison to natural samples. Geochim Cosmochim Acta 61:1223-1231

Scaini MJ, Bancroft GM, Knipe SW (1998) Reactions of aqueous Au$^+$ sulfide species with pyrite as a function of pH and temperature. Am Mineral 83:316-322

Schaufuss AG, Nesbitt HW, Kartio I, Laajalehto K, Bancroft GM, Szargan R (1998a) Incipient oxidation of fractured pyrite surfaces in air. J Elec Spec Rel Phen 96:69-82

Schaufuss AG, Nesbitt HW, Kartio I, Laajalehto K, Bancroft GM, Szargan R (1998b) Reactivity of surface chemical states on fractured pyrite. Surf Sci 411:321-328

Schaufuss AG, Nesbitt HW, Scaini MJ, Hoechst H, Bancroft MG, Szargan R (2000) Reactivity of surface sites on fractured arsenopyrite (FeAsS) toward oxygen. Am Mineral 85:1754-1766

Schweiger H, Raybaud P, Kresse G, Toulhoat H (2002) Shape and edge sites modifications of MoS$_2$ catalytic nanoparticles induced by working conditions: A theoretical study. J Catal 207:76-87

Shapter JG, Brooker MH, Skinner WM (2000) Observation of the oxidation of galena using Raman spectroscopy. Int J Mineral Proc 60:199-211

Smart RS, Jasieniak M, Prince KE, Skinner WM (2000) SIMS studies of oxidation mechanisms and polysulfide formation in reacted sulfide surfaces. Mineral Eng 13:857-870

Smelyansky V, Hafner J, Kresse G (1998) Adsorption of thiophene on RuS$_2$: An ab initio density functional study. Phys Rev B 58:R1782-R1785

Sowerby SJ, Heckl WM, Petersen GB (1996) Chiral symmetry breaking during the self-assembly of monolayers from achiral purine molecules. J Molec Evol 43:419-424

Sowerby SJ, Edelwirth M, Heckl WM (1998a) Self-assembly at the prebiotic solid-liquid interface: Structures of self-assembled monolayers of adenine and guanine bases formed on inorganic surfaces. J Phys Chem B 102:5914-5922

Sowerby SJ, Edelwirth M, Reiter M, Heckl WM (1998b) Scanning tunneling microscopy image contrast as a function of scan angle in hydrogen-bonded self-assembled monolayers. Langmuir 14:5195-5202

Sowerby SJ, Petersen GB (1999) Scanning tunnelling microscopy and molecular modelling of xanthine monolayers self-assembled at the solid-liquid interface: Relevance to the origin of life. Orig Life Evol Biosph 29:597-614

Steele HM, Wright K, Hillier IH (2003) A quantum-mechanical study of the (110) surface of sphalerite (ZnS) and its interaction with Pb^{2+} species. Phys Chem Mineral 30:69-75

Stirling A, Bernasconi M, Parrinello M (2003) Ab initio simulation of water interaction with the (100) surface of pyrite. J Chem Phys 118:8917-8926

Strohmaier R, Ludwig C, Petersen J, Gompf B, Eisenmenger W (1996a) Scanning tunneling microscope investigations of lead phthalocyanine on MoS$_2$. J Vac Sci Technol B 14:1079-1082

Strohmaier R, Ludwig C, Petersen J, Gompf B, Eisenmenger W (1996b) STM investigations of NTCDA on weakly interacting substrates. Surf Sci 351:292-302

Szargan R, Uhlig I, Wittstock G, Rossbach P (1997) New methods in flotation research: An application of synchrotron radiation to investigation of adsorbates on modified galena surfaces. Int J Mineral Proc 51: 151-161

Tan A, Harris S (1998) Electronic structure of Rh_2S_3 and RuS_2, two very active hydrodesulfurization catalysts. Inorg Chem 37:2215-2222

Tao DP, Li YQ, Richardson PE, Yoon RH (1994) The incipient oxidation of pyrite. Colloid Surf A 93:229-239

Taylor BE, Wheeler MC, Nordstrom DK (1984) Stable isotope geochemistry of acid mine drainage: Experimental oxidation of pyrite. Geochim Cosmochim Acta 48:2669-2678

Tersoff J, Hamann DR (1985) Theory of the scanning tunneling microscope. Phys Rev B 31:805-813

Todd EC, Sherman DM, Purton JA (2003) Surface oxidation of pyrite under ambient atmospheric and aqueous (pH = 2 to 10) conditions: Electronic structure and mineralogy from X-ray absorption spectroscopy. Geochim Cosmochim Acta 67:881-893

Tossell JA, Vaughan DJ (1987) Electronic structure and the chemical reactivity of the surface of galena. Can Mineral 25:381-392

Travert A, Nakamura H, van Santen RA, Cristol S, Paul JF, Payen E (2002) Hydrogen activation on Mo-based sulfide catalysts, a periodic DFT study. J Am Chem Soc 124:7084-7095

Turcotte SB, Benner RE, Riley AM, Li J, Wadsworth ME, Bodily DM (1993) Surface analysis of electrochemically oxidized metal sulfides using Raman spectroscopy. J Electroanal Chem 347:195-205

Uhlig I, Szargan R, Nesbitt HW, Laajalehto K (2001) Surface states and reactivity of pyrite and marcasite. Appl Surf Sci 179:222-229

Usher CR, Cleveland CA, Strongin DR, Schoonen MA (2004) Origin of oxygen in sulfate during pyrite oxidation with water and dissolved oxygen: An in situ horizontal attenuated total reflectance infrared spectroscopy isotope study. Environ Sci Technol 38:5604-5606

Usher CR, Paul KW, Narayansamy J, Kubicki JD, Sparks DL, Schoonen MAA, Strongin DR (2005) Mechanistic aspects of pyrite oxidation in an oxidizing gaseous environment: An in situ HATR-IR isotope study. Environ Sci Technol 39:7576-7584

Valli M, Malmensten B, Persson I (1994) Interactions between sulfide minerals and alkylxanthates .9. A vibration spectroscopic study of the interaction between arsenopyrite, millerite, molybdenite, orpiment and realgar and ethylxanthate and decylxanthate ions in aqueous solution. Colloid Surf A 83:219-225

Valli M, Persson I (1994a) Interactions between sulfide minerals and alkylxanthates. 7. A vibration and X-ray photoelectron spectroscopic study of the interaction between covellite and alkylxanthate ions and the non-sulfide mineral malachite and ethylxanthate ions in aqueous solution. Colloid Surf A 83:199-206

Valli M, Persson I (1994b) Interactions between sulfide minerals and alkylxanthates .8. A vibration and X-ray photoelectron spectroscopic study of the interaction between chalcopyrite, marcasite, pentlandite, pyrrhotite and troilite, and ethylxanthate and decylxanthate ions in aqueous solution. Colloid Surf A 83: 207-217

Vaughan DJ (ed) (2006) Arsenic. Elements 2:71-107

Vaughan DJ, England KER, Kelsall GH, Yin Q (1995) Electrochemical oxidation of chalcopyrite ($CuFeS_2$) and the related metal-enriched derivatives $Cu_4Fe_5S_8$, $Cu_9Fe_9S_{16}$, and $Cu_9Fe_8S_{16}$. Am Mineral 80:725-731

Vaughan DJ, Craig JR (1997) Sulfide ore mineral stabilities, morphologies, and intergrowth textures. In: Geochemistry of Hydrothermal Ore Deposits. Barnes HL (ed) Wiley, p 367-434

Velasquez P, Leinen D, Pascual J, Ramos-Barrado JR, Grez P, Gomez H, Schrebler R, Del Rio R, Cordova R (2005) A chemical, morphological, and electrochemical (XPS, SEM/EDX, CV, and EIS) analysis of electrochemically modified electrode surfaces of natural chalcopyrite ($CuFeS_2$) and pyrite (FeS_2) in alkaline solutions. J Phys Chem B 109:4977-4988

Voigt S, Szargan R, Suoninen E (1994) Interaction of copper(II) ions with pyrite and its influence on ethyl xanthate adsorption. Surf Interface Anal 21:526-536

Wharton MJ, Atkins B, Charnock JM, Livens FR, Pattrick RAD, Collison D (2000) An X-ray absorption spectroscopy study of the coprecipitation of Tc and Re with mackinawite (FeS). Appl Geochem 15:347-354

Widler AM, Seward TM (2002) The adsorption of gold(I) hydrosulphide complexes by iron sulphide surfaces. Geochim Cosmochim Acta 66:383-402

Williamson MA, Rimstidt JD (1994) The kinetics and electrochemical rate determining step of aqueous pyrite oxidation. Geochim Cosmochim Acta 58:5443-5454

Wincott PL, Vaughan DJ (2006) Spectroscopic studies of sulfides. Rev Mineral Geochem 61:181-229

Wolthers M, Charlet L, van der Weijden CH (2003) Arsenic sorption onto disordered mackinawite as a control on the mobility of arsenic in the ambient sulphidic environment. J Phys IV 107:1377-1380

Wolthers M, Charlet L, Van der Weijden CH, Van der Linde PR, Rickard D (2005) Arsenic mobility in the ambient sulfidic environment: Sorption of arsenic(V) and arsenic(III) onto disordered mackinawite. Geochim Cosmochim Acta 69:3483-3492

Wright K, Hillier IH, Vaughan DJ, Vincent MA (1999a) Cluster models of the dissociation of water on the surface of galena (PbS). Chem Phys Lett 299:527-531

Wright K, Hillier IH, Vincent MA, Kresse G (1999b) Dissociation of water on the surface of galena (PbS): A comparison of periodic and cluster models. J Chem Phys 111:6942-6946

Yin Q, Kelsall GH, Vaughan DJ, England KER (1995) Atmospheric and electrochemical oxidation of the surface of chalcopyrite ($CuFeS_2$). Geochim Cosmochim Acta 59:1091-1100

Yin Q, Vaughan DJ, England KER, Kelsall GH, Brandon NP (2000) Surface oxidation of chalcopyrite ($CuFeS_2$) in alkaline solutions. J Electrochem Soc 147:2945-2951

Zhang X, Borda MJ, Schoonen MAA, Strongin DR (2003) Adsorption of phospholipids on pyrite and their effect on surface oxidation. Langmuir 19:8787-8792

Zhang YC, Wang YX, Scanlon LG, Balbuena PB (2005) Ab initio and classical molecular dynamics studies of the dilithium phthalocyanine/pyrite interfacial structure. J Electrochem Soc 152:A1955-A1962

Zyubina TS, Neudachina VS, Yashina LV, Shtanov VI (2005) XPS and ab initio study of the interaction of PbTe with molecular oxygen. Surf Sci 574:52-64

Sulfide Mineral Precipitation from Hydrothermal Fluids

Mark H. Reed and James Palandri

Department of Geological Sciences
University of Oregon
Eugene, Oregon, 97403, U.S.A.
e-mail: mhreed@uoregon.edu

INTRODUCTION

Sulfide minerals precipitate from hydrothermal solution where their constituent metals and sulfide are more stable residing in the solid than in aqueous solution. An understanding of how and why sulfide minerals precipitate means understanding what chemical species hold metals and sulfide in solution, and what fundamental chemical and physical processes may liberate those reactants to precipitate in solids–processes such as pH increase, chemical reduction, dilution or cooling.

In the natural setting, the fundamental chemical and physical drivers of sulfide precipitation are a response to a wide range of processes. For example, pH increases where acidic hydrothermal fluids react with feldspars or carbonates in wall rocks, or where a hydrothermal fluid boils out CO_2 to the gas phase. The pH also increases where a brine is diluted by fresh water while maintaining equilibrium with a silicate wall-rock assemblage. Cooling follows from isoenthalpic boiling, which is caused by pressure drop as a fluid ascends in a hydrothermal system. Cooling also results from dilution by cold water or from heat conduction into cold wall rocks. Chemical reduction occurs where an oxidized fluid reacts with carbon or methane or ferrous iron.

In this chapter, we explore the fundamental chemical and physical processes that drive sulfide mineral dissolution and precipitation as well as a limited set of more complex processes that control the fundamental drivers. The basic goal is to understand the chemistry of the simplest processes so as to improve our ability to deduce what actually happens in the natural world. Our basic approach is to run thermodynamic calculations of whole-system models of the processes causing sulfide precipitation and dissolution—models that take into account simultaneous equilibria involving assemblages of minerals at chemical equilibrium among themselves and with hundreds of aqueous species. We then interpret the models in terms of fundamental effects of pH change, oxidation state change, and changes in temperature and pressure. Such models are always incomplete and approximate, but in their own terms, they do provide a complete view of the processes; i.e., we can pull apart specific precipitation reactions, complexing of metals, pH effects on minerals and metal ligands, temperature effects on mineral solubility and metal complex stability, etc. Through this approach, we may not always pin down definite answers, but we do gain insight into the natural processes that help build our intuitive understanding.

Calculation method

The calculations are executed using computer program CHILLER (Reed 1998) in combination with current thermochemical data in data base SOLTHERM (see below). In such calculations, we simulate processes of temperature or composition change (by mixing, pH change, etc.) by incremental changes of composition or temperature with an overall equilibrium calculation at each increment (Reed 1998). A fundamental basis of the models is that thermodynamic equilibrium applies on appropriate scales in time and space to be useful. We do not assume or define global equilibrium; we explore local chemical equilibria, including metastable equilibria, among minerals and fluid and make judgments case-by-case as to how kinetics or metastability apply to the situation. We do know that local chemical equilibrium among fluids, alteration minerals and vein minerals is a valid assumption in many magmatic and volcanic-hosted hydrothermal systems, as argued explicitly by Reed (1997), and addressed in many other studies (Giggenbach 1980, 1981; Reed 1982, 1998; Arnórsson 1983; Arnórsson et al. 1983; Reed and Spycher 1984; Pang and Reed 1998), and if fluid flow rates are sufficiently slow, in many sedimentary systems to temperatures as low as 75 °C (Bazin et al. 1997a,b; Palandri and Reed 2001), and in ultramafic systems to temperatures as low as 25 °C (Palandri and Reed 2004). Program CHILLER is especially suited for multiphase calculations in hydrothermal and diagenetic systems. Its capabilities overlap those of other programs such as EQ6, Geochemist's Workbench and PHREEQC, among others, reviewed by Zhu and Anderson (2002) and by Bethke (1996).

We do not account for chemical kinetics except in allowing or disallowing metastable equilibria; however, since the temperature range of interest is greater than 70 °C, below which many natural geochemical reactions are slow (Palandri and Reed 2001), we do well without kinetics. In any case, the stable and metastable equilibrium mineral assemblages we compute closely resemble those in natural sulfide mineral deposits, indicating that chemical equilibrium generally applies in natural hydrothermal settings.

Data sources

The thermodynamic data used in the numerical experiments described in this chapter are compiled in data base SOLTHERM (Reed and Palandri 2006), which contains equilibrium constants for minerals, gases, and aqueous species and serves as an input file for program CHILLER. All data sources are referenced within the data base. Most of the equilibrium constants in SOLTHERM are computed by using SUPCRT92 (Johnson et al. 1992), modified to use internally consistent mineral thermodynamic data for silicates, oxides, hydroxides, carbonates, and gases from Holland and Powell (1998; available at URL: *http://www.esc.cam.ac.uk/astaff/holland/ds5/HP98_index.html*), but also including sulfide mineral data from a revised SUPCRT92 data base, SLOP.98 (Shock et al. 1997, revised in 1998), with further data from Bessinger and Apps (unpublished manuscript) and Sack (2000). Thermodynamic data for water and aqueous species are documented by Shock and Helgeson (1988) and by Shock et al. (1997). SOLTHERM also contains equilibrium constants for minerals and aqueous species derived from over 60 additional different sources referenced in the database. All calculations in this chapter apply to pressures and temperatures along the H_2O liquid/vapor curve.

SULFIDE MINERAL SOLUBILITY: BASIC CONTROLS

Consider a basic mineral dissolution-precipitation equilibrium, such as the following for galena:

$$PbS_{(galena)} + H^+_{(aq)} = Pb^{2+}_{(aq)} + HS^-_{(aq)} \qquad (1)$$

and related complexing affecting the concentrations of Pb^{2+} and HS^-, expressed here as dissociation equilibria,

$$PbCl^+_{(aq)} = Pb^{2+}_{(aq)} + Cl^-_{(aq)} \qquad (2)$$

$$Pb(HS)_2^-_{(aq)} = Pb^{2+}_{(aq)} + 2HS^-_{(aq)} \qquad (3)$$

A complexing reaction such as (2) can be combined with galena dissolution (1), to express the effect of complexing within the dissolution reaction as follows:

$$PbS_{(galena)} + H^+_{(aq)} + Cl^-_{(aq)} = PbCl^+_{(aq)} + HS^-_{(aq)} \qquad (4)$$

These reactions illustrate that we can understand the solubility of the mineral in natural systems by examining the ability of the mineral to sequester Pb owing to the mineral's intrinsic thermodynamic stability in competition with the tendency for Pb to reside in the aqueous phase owing to acid attack (Reaction 1) or to complexing of aqueous Pb^{2+} by sulfide or chloride species (Reactions 2, 3). We can examine the dependence of equilibria such as Reactions (1),(2) and (3) on temperature, pressure, pH and Cl^- concentration.

Except for pyrrhotite, sulfide minerals are much less soluble in pure water at temperatures below 100 °C than at high temperature, as shown by a graph of $\log K_{min}$ vs. T (Fig. 1) for reactions such as Reaction (1) with the mineral $\log K$'s normalized to a "per metal atom" basis. At temperatures above 300 °C, the solubility curves are fairly flat, except for pyrite, which becomes more stable (smaller $K_{dissolution}$) at high temperature. The generally smaller $\log K$'s at low T mean that temperature decrease, by itself, tends to drive sulfide mineral precipitation. A graph of $\log K_{dissolution}$ vs. P (Fig. 2) shows the small effect of pressure alone on mineral solubility in pure water; the small decrease in $\log K$ with decreasing pressure favors sulfide precipitation upon pressure decrease. Notice that in both the P and T graphs (Figs. 1, 2) of equilibria such as Reaction (1), the intrinsic mineral stability is convolved with the T and P effects on the properties of pure water, but that effects of pH and complexing are not expressed.

Metal complexing by aqueous Cl^- and HS^- is has a strong effect in determining mineral solubility as a function of temperature in natural systems because the aqueous Cl^- or HS^- ligands tie up the metals, preventing sulfide mineral precipitation, as implied in Reaction (4). The graphs of complex $\log K_{dissocation}$ vs. T (Fig. 3) show that the chloride species of Pb^{2+}, Zn^{2+} and Fe^{2+} are much more stable at high temperature than at low, which means decreasing temperature would tend to drive precipitation of Pb, Zn and Fe sulfides. Cu^+ chloride complex dissociation peaks at around 200 °C, thus cooling to 200 °C tends to drive mineral precipitation by complex dissociation. The temperature dependences of Ag^+ and Au^+ chloride complex stabilities are similar those of Cu^+. In contrast to Pb chlorides, Pb bisulfide dissociation reaches maxima at about 200 °C. Similarly, Cu bisulfides are much more stable at low temperature than at high temperature. Bisulfides of Ag^+ and Au^+ are distinctly more stable at low temperature than at high, which means that with respect to bisulfide complexing, temperature decrease tends to hold Au and Ag in solution, undoubtedly contributing to transport of those metals to the periphery in some natural systems.

The pressure dependence of dissociation for essentially all chloride and bisulfide complexes is such that decreasing pressure makes the complexes more stable (smaller $\log K$'s, Fig. 4). Thus the effect of pressure decrease on complexing, in itself, tends to favor sulfide mineral dissolution by modest amounts.

The effect of decreasing H^+ activity (increased pH) in causing sulfide minerals to precipitate can be inferred from example Reactions (1) and (4), which, whether written to H_2S or HS^-, are displaced toward sulfide precipitation (left side) when H^+ activity decreases, because sulfide is liberated from H_2S and HS^- by decreased H^+ activity. This effect does not necessarily apply for metals complexed by HS^- because the ligand concentration, itself, is a function of pH. These issues are explored further below.

SULFIDE MINERAL NUMERICAL EXPERIMENTS

The controls on sulfide mineral precipitation are explored below in series of numerical experiments where we examine the effects of salinity, pH, temperature change and change of oxi-

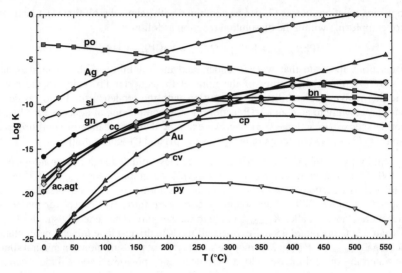

Figure 1. Sulfide mineral dissolution log$K$ vs. T, 25-600 °C, P = 800 bar. Log$K$'s apply to reactions written for dissolution of the mineral (left-hand side; as Reaction 1 in text) to component species, as follows: Cu^+, Pb^{2+}, Zn^{2+}, Fe^{2+}, Ag^+, HS^-, H^+, SO_4^{2-}, H_2O. Log$K$'s are normalized to a "per metal atom" basis to facilitate comparisons; i.e., log$K$ is divided by the number of metal atoms in the formula unit. Notice that a large log$K$ for the reaction written with the mineral on the left side (as Reaction 1, text) indicates a large solubility. For these minerals, the small log$K$'s at low temperature indicate that the minerals are intrinsically insoluble at low T.

Mineral abbreviations for all figures are as follows: ab, albite; ac, acanthite; agt, argentite; Ag, silver; Au, gold; bn, bornite; cc, chalcocite; cp, chalcopyrite; ct, cattierite; cv, covellite; daph, daphnite; gn, galena; hem, hematite; kf, K-feldspar; ln, linnaeite; mt, magnetite; mus, muscovite; pyrol, pyrolusite; sl, sphalerite; po, pyrrhotite; py, pyrite. qz, quartz.

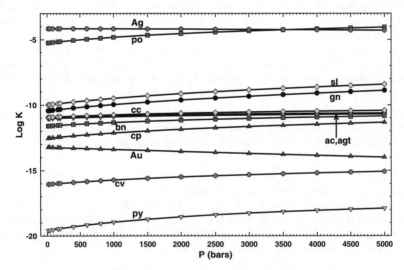

Figure 2. Sulfide mineral dissolution Log$K$ vs. P, 1 to 5000 bar, T = 200 °C. Log$K$'s apply to reactions written for dissolution of the mineral (left-hand side; as Reaction 1, text) to component species, as follows: Cu^+, Pb^{2+}, Zn^{2+}, Fe^{2+}, Ag^+, HS^-, H^+. Log$K$'s are written for a "per metal atom" basis; see caption to Figure 1. Notice that a smaller log$K$ for the reaction written with the mineral on the left side (as Reaction 1, text) indicates a lesser solubility. Mineral abbreviations are given in the caption to Figure 1.

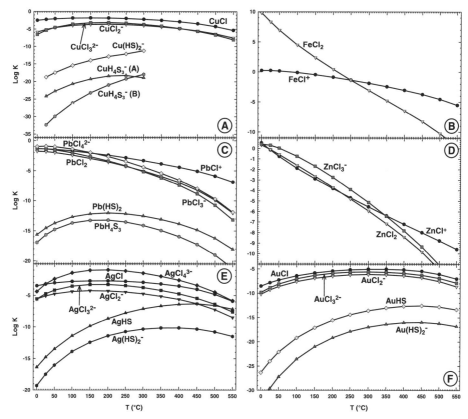

Figure 3. Stability of complexes of base metals, Ag and Au, involving Cl⁻ and HS⁻, expressed as log$K_{dissociation}$ vs. T, for $T = 25$-550 °C at $P = 800$ bar, except Cu-sulfide complexes, $T = 25$-300 °C at P corresponding to liquid-vapor saturation. Log$K$'s apply to dissociation to component species, Cu^+, Pb^{2+}, Zn^{2+}, Fe^{2+}, Ag^+, HS^-, H^+, Cl^-, H_2O.

dation state on the solubility of Pb, Zn, Ag, Cu, and Fe sulfides. In many cases, we hold constant all variables except one, so as to isolate its effects. The starting conditions, fluid compositions, and variable controls in the experiments are listed in Table 1. In all calculations, the pressure is that of H_2O liquid-vapor equilibrium at the stated temperature. In all numerical experiments reported here, there is no path dependence in the experimental run. That is, the experimental variable (T, pH, m_{NaCl}, etc.) can be regarded as changing from high to low or low to high, depending on what view is more meaningful to consider in any given situation. Such symmetry does not apply to runs with fractionation of solids, but no such runs are reported here.

Chloride addition (Run 1)

The effect of chloride concentration on sulfide solubility is explored in Run 1 (Fig. 5). The initial system consists of the sulfide minerals chalcopyrite, galena, sphalerite, pyrite, accompanied by magnetite and electrum, all in contact with one kilogram of water at 300 °C. The electrum is included to provide for monitoring the behavior of Au and Ag, and magnetite provides a redox control. The experiment consists of incrementally adding 4.0 moles of pure NaCl to the initial mixture of minerals and water, while holding pH constant at 5.0, and H_2S essentially constant (Fig. 5A). As NaCl is added, the total aqueous concentrations of metals

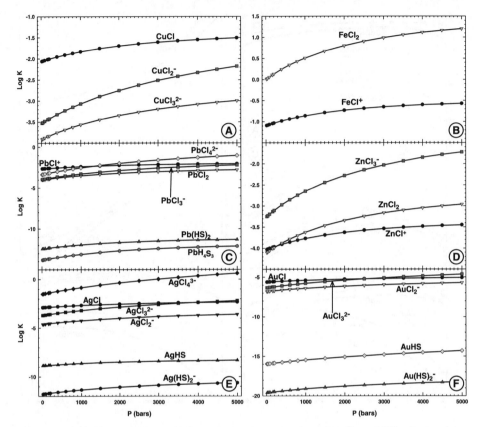

Figure 4. Stability of complexes of base metals, Ag and Au, involving Cl^- and HS^-, expressed as $\log K_{dissociation}$ vs. P, for 1-5000 bar, at $T = 200$ °C. $\log K$'s apply to dissociation to component species, Cu^+, Pb^{2+}, Zn^{2+}, Fe^{2+}, Ag^+, HS^-, H^+, Cl^-, H_2O.

increase by many orders of magnitude (Fig. 5A), as their respective sulfide minerals dissolve. The mineral dissolution is entirely a consequence of progressive and stepwise formation of chloride complexes of the metals, starting from metal hydroxide or bisulfide at the initial small Cl^- concentration (Fig. 5B-E), then progressing through Cl^- ligand numbers up to four. At the run pH of 5.0, Au is held in $Au(HS)_2^-$ throughout (Fig. 5G), but Ag behaves as do the base metals, in shifting from a bisulfide (AgHS) to chlorides (Fig. 5F).

Such chloride complexing of metals is the fundamental reason that natural saline waters carry substantial concentrations of base metals and silver, as articulated by Helgeson (1964), and illustrated by many natural saline waters that have large metal concentrations, e.g., Salton Sea brines (McKibben and Hardie 1997), Mississippi Oil Field Brines, (Carpenter et al. 1974), Red Sea brines (Degens and Ross 1969), Mid-ocean ridge black smokers (Von Damm 1990). The process of precipitation of base metal sulfides from most such natural waters is largely a matter of overcoming the chloride ligand's hold on base metals in solution.

Simple dilution (Run 2)

Isothermal pure water dilution of the 1m NaCl metal-bearing solution from Run 1 at constant pH, 5.0, yields a small amount of galena, and substantial Au°, magnetite and hematite (Fig. 6). The removal of metals from aqueous solution to minerals is revealed by comparing the slopes

Sulfide Mineral Precipitation from Hydrothermal Fluids 615

Table 1. Summary of numerical experiment conditions.

Run #	Fig.	Run Type	Run Action	Notes	T, °C	Initial pH	Final pH
1	5	Cl addition	Add NaCl	a	300	5	5
2	7	Dilution	Add H$_2$O	a,b	300	5	5
3	6	Dilution, buffered	Add H$_2$O	b,c,d	300	5.65	7.8
4	8	Cooling	Cool at constant composition	b,e,f	300-25	5.15	6.5
5	9	pH increase w/gn	Add NaOH		300	1.6	10.6
6	10	pH increase	Add NaOH	b,c	300	5.65	10.2
7	11	pH decrease	Add HCl	b,c	300	5.65	0.6
8	12	pH increase	Add NaOH	g	200	0.8	7
9	13	Dilute w/ cold H$_2$O	Add 25°C H$_2$O		300-25	5.0	5.7
10	14	Reduction	Remove O$_2$	a,h	125	6	6

Run #	Fig.	Total Molality (m), Initial							
		Na	Cl	Fe	Cu	Pb	Zn	Ag	Au	HS$^-$
1	5	0	0	.41e-6	.33e-9	.14e-6	.33e-05	.16e-7	.21e-8	.22e-2
2	7	1.0	1.0	.17e-3	.65e-6	.37e-4	.10e-3	.68e-5	.72e-8	.21e-2
3	6	.94	1.0	.42e-5	.12e-6	.13e-5	.10e-4	.12e-5	.97e-7	.46e-2
4	8	1.0	1.0	.82e-4	.46e-6	.19e-4	.52e-4	.48e-5	.10e-7	.21e-2
5	9	.41	1.0	--	--	.25	--	--	--	.25
6	10	1.0	1.0	.84e-5	.16e-6	.21e-5	.86e-5	.16e-5	.31e-7	.21e-2
7	11	1.0	1.0	.84e-5	.16e-6	.21e-5	.86e-5	.16e-5	.31e-7	.21e-2
8	12	.53	1.0	.56e-1	.17e-3	.41e-3	.11e-2	-	-	.14e-1
9	13	1.0	1.0	.17e-3	.65e-6	.37e-4	.10e-3	.68e-5	.72e-8	.21e-2
10	14	1.0	1.0	.54e-10	.21e-6	.15e-8	.17e-7	.64e-7	.34e-7	.57e-3

a) pH held constant.
b) Initial fluid saturated, but not in contact with py, mt, sl, gn, cp, electrum
c) Neutral initial pH@300 °C = 5.65
d) Dilution pH buffered with ab, kf, mus, qz.
e) Final T = 25.
f) Initial pH = 5.15. pH held 0.5 units lower than neutral at all T.
g) Initial fluid in equilibrium with 0.27 moles py, 0.11 moles cv.
h) Quartz saturated.

of concentration curves in Figure 6B, where plotting of log(kg solvent water) on the abscissa yields a single constant negative slope for all species whose concentrations reflect dilution only (e.g., ΣZn, ΣAg, Fig. 6B) but steeper negative slopes for species removed by mineral precipitation, e.g., ΣFe, ΣAu, which precipitate in iron oxides and electrum, respectively.

Dilution reverses the stepwise complexation series from NaCl addition (Run 1), causing stepwise changes to smaller ligand numbers; e.g., from PbCl$_4^{2-}$ in the saline fluid to PbCl$^+$ upon dilution (Fig. 6E). For Ag$^+$, Au$^+$, and Cu$^+$, dilution increases the absolute concentration of the naked metal ion (e.g., Fig. 6D), but the increases are insufficient to cause metals to precipitate. Iron precipitates in oxides as dilution of Cl$^-$ frees the metal to precipitate, but the dominant effect on Fe is a decreases aqueous H$_2$ concentration (Fig. 6C), driving magnetite and hematite precipitation by oxidation of aqueous Fe^{2+}:

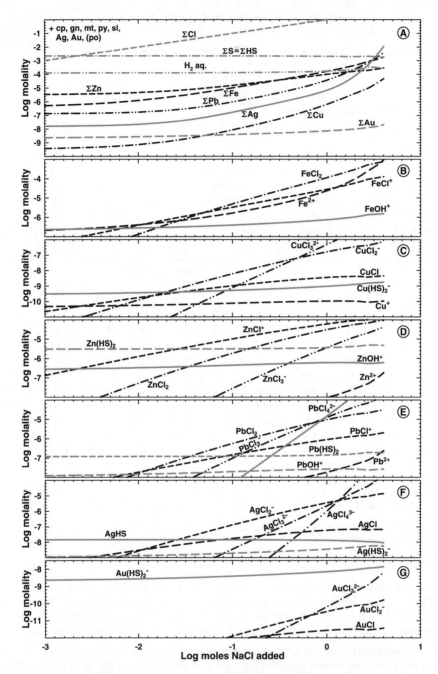

Figure 5. Run 1. Addition of NaCl to an aqueous phase saturated with excess chalcopyrite, galena, magnetite, sphalerite and electrum (pyrrhotite also precipitates near $\log m_{NaCl} = 0.43$) at $T = 300$ °C and pH = 5. The sulfide minerals dissolve as the added chloride progressively complexes metals in a stepwise series of increasing ligand numbers, as Cl$^-$ concentration increases from 0 to 4 m. A) Total concentrations of aqueous components and dissolved H$_2$. B, C, D, E, F) Concentrations of dominant aqueous species: B) Fe, C) Cu, D) Zn, E) Pb, F) Ag and G) Au.

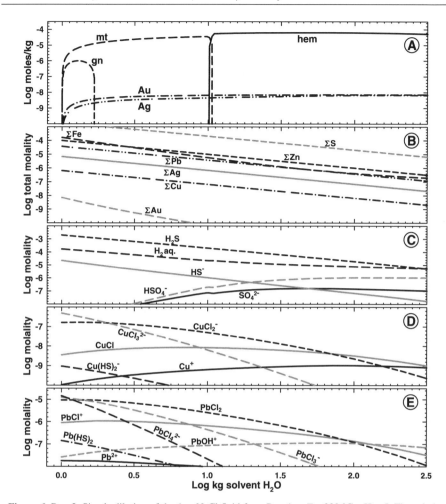

Figure 6. Run 2. Simple dilution of the 1 m NaCl fluid from Run 1 at $T = 300$ °C, pH = 5. The solution is initially saturated with sphalerite, chalcopyrite, galena, pyrite, magnetite and electrum ($X_{Au} = 0.69$) but is not in contact with those minerals. The dilution decreases H_2 concentration and reverses the stepwise complexation series in Run 1 (Fig. 5) causing electrum and iron oxides to precipitate. A) Mineral assemblage. B) Total concentrations of aqueous components. A single constant negative slope shows for all species whose concentrations reflect dilution only, e.g., Cu, Zn. The small amounts of galena and Ag precipitated are not enough to be visible in the aqueous Pb and Ag curves. The steeper negative slopes of Au and Fe reflect precipitation of those metals in electrum and iron oxides. C,D,E) Concentrations of selected aqueous species: C) H_2 and S species, D) Cu complexes, E) Pb complexes.

$$3FeCl^-_{(aq)} + 4H_2O_{(aq)} \rightarrow Fe_3O_{4(magnetite)} + H_{2(aq)} + 6H^+_{(aq)} + 3Cl^-_{(aq)} \quad (5)$$

The simple isothermal dilution results indicate that such dilution, alone, may not drive significant sulfide precipitation in nature—a conclusion that we did not expect, although we need to explore other conditions to understand it better. It also seems likely that isothermal, pure water dilution is rare in nature, in contrast to dilution with cold water with consequent simultaneous cooling, which is probably common (see Run 9, below) and may be compared to natural examples. Dilution by steam-heated, but cooler, waters is also inferred for some systems, as has been explored for the Creede epithermal system by Plumlee et al. (1994).

Silicate-buffered dilution (Run 3)

In contrast to the simple dilution case (Run 2), wherein we held pH constant at 5.0, the pH must increase in a solution that equilibrates with wall-rock alteration minerals in the course of dilution, as explored in Run 3, an isothermal (300 °C) pure-water dilution. The silicate buffered dilution starts with a fluid similar to that in simple dilution (Run 2), except that pH is controlled by the assemblage quartz-K-feldspar-albite-muscovite, in accordance with the following equilibrium (cf. Reed 1997):

$$2H^+_{(aq)} + 2NaAlSi_3O_{8(ab)} + KAlSi_3O_{8(kf)} = KAl_3Si_3O_{10}(OH)_{2(mus)} + 2Na^+_{(aq)} + 6SiO_{2(qz)} \quad (6)$$

for which

$$K = a^2(Na^+)/a^2(H^+) \quad (7)$$

where a refers to activity of the indicated aqueous species. Equation (7) requires that the activity ratio of aqueous Na^+ and H^+ remain constant, thus dilution of Na^+ forces a decrease in $a(H^+)$, and pH increases from 5.65 to 7.8 (Fig. 7B), simply as a result of decrease of Na^+ concentration from 0.94 m (−0.3 log molality) to .0015 m (−2.8 log molality) in this silicate-buffered system.

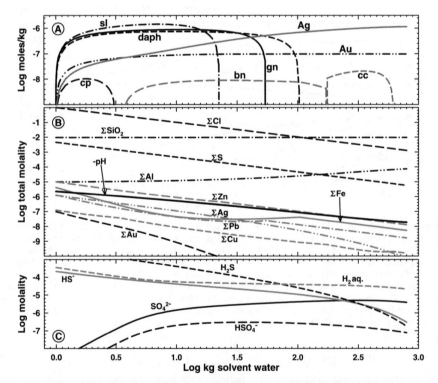

Figure 7. Run 3. Silicate buffered, pure water dilution of the 1 m NaCl fluid from Run 1 at T = 300 °C. The initial solution is saturated with sphalerite, chalcopyrite, galena, pyrite, magnetite and electrum (X_{Au} = 0.69), but is not in contact with those minerals. The solution maintains equilibrium with assemblage quartz-albite-muscovite-K-feldspar, which buffers pH by Reaction (6), between 5.6 and 7.8. Much more sulfide precipitates than in simple dilution (Run2, Fig. 6) because buffered dilution drives pH upward, displacing Reaction (1) to the left (see text). A) Mineral assemblage. B) pH and total concentrations of aqueous components. A single constant negative slope shows for species whose concentrations reflect dilution only, e.g., Cl. C) Concentrations of H_2 and selected S species.

Whereas simple dilution (Run 2) yields little sulfide precipitation, the pH rise in buffered dilution yields moderate amounts of base metal sulfides plus electrum (Fig. 7A). Sulfide precipitates in accordance with the pH effect on equilibria such as Reaction (1), which are displaced toward mineral precipitation by increasing pH. Except for Pb, the maximum quantities of sulfide precipitates are small relative to the initial amounts of metals in solution. Where minerals reach their maximum amounts, 68% of Pb is in galena, 14% of Zn is in sphalerite, and 8%, 36%, and 33% of Cu are in chalcopyrite, bornite and chalcocite, respectively. Most of the aqueous Au and Ag form electrum in response to pH increase and H_2 dilution (Fig. 7C), illustrated by the following reaction for precipitation of Au°:

$$Au(HS)_2^-{}_{(aq)} + 1\tfrac{1}{2}H_2O_{(aq)} = Au° + H_{2(aq)} + 1\tfrac{5}{8}HS^-{}_{(aq)} + \tfrac{3}{8}SO_4^{2-}{}_{(aq)} + 1\tfrac{3}{8}H^+{}_{(aq)} \qquad (8)$$

Wall-rock buffered dilution likely occurs in the periphery of natural systems where fluids percolate slowly through a fracture network, thereby maintaining contact with wall-rock minerals, such as in the propylitic halo of porphyry copper deposits where the dilution effect should be evidenced by a gradient of decreasing salinity in fluid inclusions.

Temperature change (Run 4)

Temperature decrease has long been cited as a likely cause of sulfide precipitation in hydrothermal systems, and there is no doubt that it has a significant controlling effect. We explore the effect of temperature decrease in Run 4 (Fig. 8), where the initial system consists of the 300 °C metal-bearing solution from Run 1 at 1m NaCl concentration, with pH reset to 5.15, and saturated with chalcopyrite, galena, sphalerite, pyrite, magnetite and electrum, but separated from those minerals (i.e., mineral masses reduced to zero). The experiment consists of incrementally decreasing temperature from 300 °C to 25 °C, while holding pH 0.5 unit lower than neutral pH at all temperatures [i.e., $\Delta pH(T) = pH(neutral,T) - 0.5$]. By holding a constant ΔpH, we partially separate the effects of pH change on sulfide precipitation from effects of temperature in controlling mineral solubility and metal complexing.

As temperature decreases from 300 °C, nearly all of the aqueous metal precipitates in pyrite, sphalerite, galena, chalcopyrite and acanthite (Fig. 8A), but at temperatures below 200 °C, chalcopyrite begins to redissolve, yielding Cu^+ back to solution, and at about 175 °C, silver moves substantially from acanthite to electrum. Precipitation of sulfides of Fe, Zn, Pb and Cu (at high T) is driven by breakdown of the respective metal chloride complexes as decreasing temperature renders them less stable (Fig. 8, Fig. 3), combined with the decrease in intrinsic mineral solubility with decreasing temperature below 300 °C (Fig. 1). The dissolution of chalcopyrite results from the increasing formation of Cu sulfide complexes (Fig. 8E), which become more stable at lower temperature (Fig. 3A).

Temperature decrease is one of the most commonly suspected processes in natural systems, leading to sulfide precipitation; however, simple temperature decrease, as opposed to temperature decrease by cold water mixing or boiling, must occur by conduction of heat into the wall-rock, the efficacy of which is limited once wall rock heats-up, thereby eliminating the temperature gradient necessary to conduct heat. Cooling by conduction does apply where high pressure magmatic fluids propagate through fractures into cold rock, as in the early "biotite crackles" in Butte, Montana ("narrow biotitic" veins of Brimhall, 1977). Conductive cooling also applies where a natural "heat exchanger" operates, such as in a seafloor black smoker setting where a hot water conduit is sealed away from cold seawater by anhydrite, but heat from the conduit drives sub-seafloor convection of cold seawater adjacent to the vent conduit.

pH change (Runs 5, 6, 7, 8)

We examined the effect of pH change on mineral precipitation in four ways:

1. Run 5, solubility of galena alone, between pH 1 and 10.5, examining Pb complexing;

2. Run 6, pH increase from 5.65 to 10.2, starting from a 300 °C 1m NaCl fluid initially saturated with sphalerite, chalcopyrite, galena, pyrite, magnetite and electrum [x(Au) = 0.69], but not in contact with those minerals (Table 1);

3. Run 7, pH decrease from 5.65 to 1.0 in the same fluid as in Run 6 at 300 °C (Table 1); no minerals in contact initially;

4. Run 8, pH increase from 0.8 to 7.0, starting with a 200 °C 1 m NaCl fluid in contact with 0.113 mole (10.8 g) of covellite and 0.268 mole (32.1g) of pyrite.

In all four of these pH runs, since the bulk composition is held constant except for the addition of removal of acid or base, the results are the same whether they are regarded as running from high pH to low or low pH to high; i.e., the graphs can be read from left to right or right to left, depending on the process one wishes to consider. Thus, although we choose to describe them as "pH increase" or "pH decrease" that choice is arbitrary.

pH effect on galena solubility and Pb complexing (Run 5). The complicated basic control of pH on sulfide mineral solubility is explored by an examination of a simple system consisting of galena saturated in a 1m NaCl solution at 300 °C at pH values ranging from 1.6 to 10.6 (Fig. 9). We examine Pb, but the same basic chemical effects described below apply to all other sulfides. Under acidic conditions, galena is highly soluble as $PbCl_2$ and $PbCl_4^{2-}$ hold Pb (Fig. 9B), yielding a total aqueous Pb concentration of 0.25m at pH 1.6 (Fig. 9B). The large galena solubility reflects displacement of the following equilibrium to the right by a large H^+ activity at low pH:

$$PbS_{(galena)} + H^+_{(aq)} + 4Cl^-_{(aq)} = PbCl_4^{2-}_{(aq)} + HS^-_{(aq)} \tag{9}$$

As pH increases, Reaction (9) is driven to the left, and essentially all aqueous Pb precipitates in galena as pH rises from 1.6 to 8.1 (Fig. 9A). Galena solubility reverses at pH 8.1, such that aqueous Pb climbs back to nearly 10^{-4} m at pH 10.6. The reversal in galena solubility is caused by the increasing concentration of $Pb(HS)_2$ then $Pb(OH)_2$ and $Pb(OH)_3^-$ (Fig. 9b), as the HS^- and OH^- ligands increase in concentration at high pH (Fig. 9C). This effect is illustrated for OH^- as follows:

$$PbS_{(galena)} + H_2O_{(aq)} + OH^-_{(aq)} = Pb(OH)_{2(aq)} + HS^-_{(aq)} \tag{10}$$

pH increase (Run 6). The same basic pattern of sulfide mineral solubility illustrated in Run 5 for galena applies to a more complex mixture of sulfide minerals, as explored in Run 6, in which the 300 °C 1 m-NaCl solution is initially saturated with sphalerite, galena, chalcopyrite, magnetite and electrum at pH 5.65 (from Run 1), but the saturated sulfides are not in contact with the solution. As pH increases, all sulfides precipitate along with electrum and magnetite (Fig. 10A), in response to reaction (9) for galena, and its analogs, all of which are driven by pH increase. Sphalerite, galena and magnetite redissolve, as aqueous complexing shifts from Cl^- to HS^- and OH^- ligands, as illustrated for Pb in Run 5 (Fig. 9B). Chalcopyrite redissolves as $Cu(HS)_2^-$ forms (Fig. 10C), but chalcopyrite is also partly replaced by bornite, then bornite by chalcocite with increasing pH (Fig. 10A). The latter mineral replacements are driven by pH increase in accordance with a reaction such as the following for replacement of chalcopyrite by bornite:

$$5CuFeS_{2(cp)} + 14H_2O_{(aq)} =$$
$$Cu_5FeS_{4(bn)} + 4Fe(OH)_3^-_{(aq)} + 5\frac{1}{2}HS^-_{(aq)} + \frac{1}{2}SO_4^{2-}_{(aq)} + 10\frac{1}{2}H^+_{(aq)} \tag{11}$$

wherein the iron hydroxide species is one of the two hydroxides that dominate aqueous iron at pH > 7.

pH decrease (Run 7). When pH decreases from 5.65 to 1.0 in the same initial fluid as in Run 6 (above), the initially saturated sulfides and oxides become undersaturated, but saturated

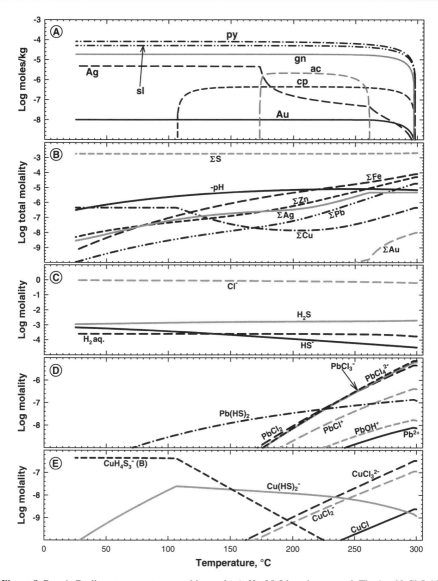

Figure 8. Run 4. Cooling at constant composition and a ΔpH of 0.5 less than neutral. The 1 m NaCl fluid from Run 1 initially saturated with sphalerite, chalcopyrite, galena, pyrite, magnetite and electrum (X_{Au} = 0.69), but not in contact with those minerals, is cooled from 300 to 25 °C. Cooling causes quantitative precipitation of metals (except Cu) in metal sulfides. A) Mineral assemblage. B) pH and total concentrations of aqueous components. C,D,E) Concentrations of selected aqueous species. C) H_2, Cl^- and S species, D) Pb complexes, E) Cu complexes.

electrum precipitates down to pH 4, then dissolves, disappearing at pH 2.8 (Fig. 11A). Sulfides dissolve upon pH decrease as reactions such as (1), (4), and (9) are driven to the right by increased H^+ activity. Electrum precipitates as decreasing pH drives HS^- to H_2S, which drives dissociation of the dominant Au complex, $Au(HS)_2^-$. Electrum dissolves as it is apparently oxidized by H^+ under acidic conditions, forming $AuCl_3^{2-}$ at low pH.

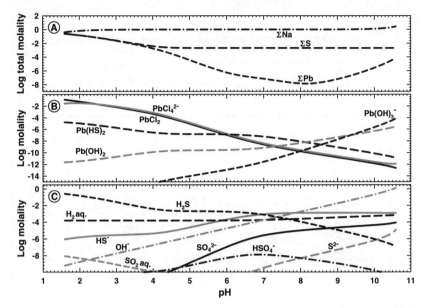

Figure 9. Run 5. Galena solubility as a function of pH in a 1 m NaCl solution at 300 °C. As pH increases (by addition of NaOH) from highly acidic, galena precipitates as Reaction (1) is displaced to the left; at pH > 8, galena dissolves as OH⁻ complexes Pb, displacing Reaction (1) to the right. A) Total concentrations of Na, Pb, and S. B) Concentrations of dominant Pb species. C) Concentrations of H_2, OH⁻, and selected S-bearing species.

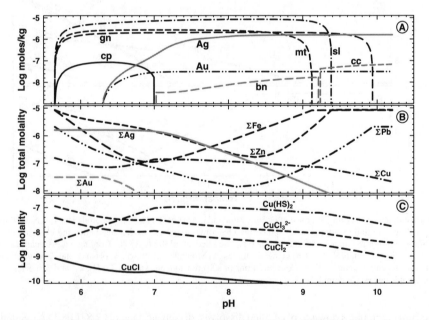

Figure 10. Run 6. Mineral solubilities as a function of increasing pH, resembling Run 5, but with many more metals. The 1 m NaCl fluid from Run 1, saturated with sphalerite, chalcopyrite, galena, pyrite, magnetite and electrum, but not in contact with those minerals, is adjusted to an initial neutral pH of 5.65 at 300 °C. Addition of NaOH causes sulfides, magnetite and electrum to precipitate, as pH increases. Fe, Zn and Pb minerals redissolve at high pH as hydroxide complexes form. A) Minerals. B) Total concentrations of selected aqueous components. C) Concentrations of dominant Cu-bearing species.

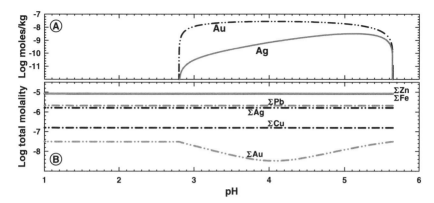

Figure 11. Run 7. pH decrease in a sulfide saturated solution yields electrum. The 1 *m* NaCl fluid from Run 1, saturated with sphalerite, chalcopyrite, galena, pyrite, magnetite and electrum, but not in contact with those minerals, is adjusted to an initial neutral pH of 5.65, then NaOH is removed to decrease pH. $T =$ 300 °C. This pH decrease experiment complements Run 6 (Fig. 10) showing that only electrum saturates, as HS^- is converted to H_2S, driving dissociation of the dominant Au complex, $Au(HS)_2^-$. At pH < 4, electrum dissolves, forming $AuCl_3^{2-}$. A) Minerals. B) Total concentrations of selected aqueous components.

pH change from acidic to neutral: Cu-Fe-S minerals, sphalerite and galena (Run 8). In the last of these simple pH runs, we examine the effect of pH change on the assemblage of Cu-Fe-S minerals and on galena and sphalerite, by increasing pH from 0.8 to 7 in a 200 °C mixed metal 1 *m* NaCl solution (Fig. 12). The starting fluid is in contact with 0.113 mole of covellite and 0.268 mole of pyrite; additional metals in the initial solution include Pb and Zn (Table 1). As pH increases, chalcocite precipitates, replacing covellite completely at pH 1.9 (Fig. 12A). At pH 2.1, bornite precipitates, replacing chalcocite and some pyrite, and at pH 3.5, chalcopyrite precipitates, consuming bornite and pyrite. Sphalerite and galena precipitate at pH 4 to 4.2.

In the system as a whole, the rise in pH quantitatively removes metals from solution as shown in Figure 12B, where metal concentrations decrease by many orders of magnitude as pH enters the neutral range. The pH increase drives sulfide precipitation by reactions such as (1) for galena, and its analogs for other minerals such as the following, which are displaced to the left:

$$ZnS_{(sphalerite)} + H^+_{(aq)} + Cl^- = ZnCl^+_{(aq)} + HS^-_{(aq)} \tag{12}$$

$$CuFeS_{2(cp)} + {}^7/_8H^+_{(aq)} + {}^1/_2H_2O = Fe^{2+}_{(aq)} + Cu^+_{(aq)} + 1{}^7/_8HS^-_{(aq)} + {}^1/_4SO_4^{2-}_{(aq)} \tag{13}$$

The overall mineral zoning sequence, from acidic to neutral, is covellite, chalcocite, bornite, chalcopyrite, galena, sphalerite, all accompanied by pyrite. The pH range of the dominant mineral in the Cu-Fe-S system is clearly affected by the initial metal concentrations and by sulfide concentration; e.g., decreased aqueous H_2S enlarges the bornite span relative to chalcocite and covellite; similarly, larger concentrations of Pb and Zn move galena and sphalerite precipitation to lower pH values.

Among the Cu-Fe-S minerals, acid neutralization yields the progression covellite-chalcocite-bornite-chalcopyrite, all with pyrite, in response to increasing pH as illustrated by the following series of reactions:

covellite to chalcocite:

$$2CuS_{(cv)} + H_2O_{(aq)} = Cu_2S_{(cc)} + {}^1/_4SO_4^{2-}_{(aq)} + {}^1/_2H^+_{(aq)} + {}^3/_4H_2S_{(aq)} \tag{14}$$

chalcocite + pyrite to bornite:

$$Cu_2S_{(cc)} + FeS_{2(py)} + H_2O_{(aq)} = Cu_5FeS_{4(bn)} + {}^1/_4SO_4^{2-}_{(aq)} + {}^1/_2H^+_{(aq)} + {}^3/_4H_2S_{(aq)} \tag{15}$$

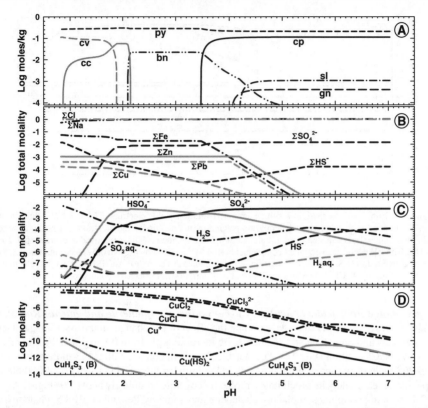

Figure 12. Run 8. Cu-Fe-S minerals as a function of pH showing zonation resulting from pH variation. The initial 1 m NaCl acidic fluid is in contact with covellite (0.11 mole) and pyrite (0.27 mole) and contains additional dissolved Zn and Pb (Table 1), to which NaOH is added to increase pH from 0.8 to 7. $T = 200$ °C. The Cu-Fe-S minerals lie as a function of pH in the classic acidic-to-neutral zoning sequence: cv-cc-bn-cp A) Minerals. B) Total concentrations of aqueous components. C,D) Concentrations of selected aqueous species: C) H_2 and S species, D) Cu complexes.

bornite + pyrite to chalcopyrite:

$$Cu_5FeS_{4(bn)} + 4FeS_{2(py)} + 2H_2O_{(aq)} = 5CuFeS_{2(cp)} + \tfrac{1}{2}SO_4^{2-}{}_{(aq)} + H^+{}_{(aq)} + 1\tfrac{1}{2}H_2S_{(aq)} \quad (16)$$

In each reaction, the removal of H⁺ (pH increase) drives the reaction toward the right hand side, yielding the next mineral assemblage in the series. This mineral progression matches the zoning sequence described for the Main Stage Veins of the Butte District, Montana by Sales (1914) and Meyer et al. (1968), where the Butte Main Stage Horsetail ores consist of covellite-enargite-pyrite immersed in advanced argillic alteration (kaolinite, topaz, sericite, quartz, ± pyrophyllite, ± alunite), passing outward to chalcocite + pyrite, then bornite + pyrite, then chalcopyrite + pyrite. Mineral textures in each zone reveal complex intergrowths and partial replacements including mineral pairs covellite-chalcocite, chalcocite-bornite and bornite-chalcopyrite. The bornite and chalcopyrite zones substantially overlap with the sphalerite zone, which yields outward to added galena and rhodochrosite (Meyer et al. 1968).

The covellite-chalcocite-pyrite (+enargite) end of the series was called the "high-sulfur assemblage" by Meyer and Hemley (1967) in reference to the stoichiometric control whereby addition of S to chalcopyrite yields bornite plus pyrite; S addition to bornite yields chalcocite plus pyrite, and S added to chalcocite yields covellite. Each reaction in the series is also a redox reaction, reflecting the change in nominal oxidation state of sulfur in pyrite from −1 to −2 and to +6 (sulfate), and the changes of Cu oxidation state between +1 and +2 in reactions involving covellite and bornite.

The calculation results (Fig. 12) show that the "high-sulfur assemblage" (covellite-chalcocite-pyrite-enargite) prevails at very acidic pH in the range of 2 and lower, and the mineral progression stoichiometry in the reaction series (Reactions 14, 15, 16) shows the connection of progressive addition of H$^+$ to forming the high-sulfur assemblage at low pH. The role of intense acidity in forming the high-sulfur assemblage is illustrated by the direct coincidence in the field of covellite-chalcocite-pyrite-enargite assemblages with intense advanced argillic wall-rock alteration (including subsets of kaolinite, topaz, quartz, pyrophyllite, zunyite, and alunite), which are formed by acid attack of wall-rocks at pHs' in the range of 1 or 2 (e.g., Reed, 1997). Examples include the classic Butte example (Meyer et al. 1968), the covellite-enargite-pyrite-S° assemblages bounded by quartz-alunite-kaolinite at Summitville, Colorado (Stoffregen 1987) and the Nansatsu deposits of Japan (Izawa 1991), and many other examples in porphyry copper and Natsatsu-Summitville-type deposits worldwide.

pH change: comment. Isothermal pH increase is likely one of the most common causes of sulfide mineral precipitation in natural settings. In most instances where base metal sulfides such as chalcopyrite, sphalerite or galena formed in altered wall rock (distinguish from pyrite and pyrrhotite formed by sulfidation of wall-rock iron), the sulfides precipitate owing to an increase in fluid pH as wall-rock feldspars or carbonates neutralize acidic metal-bearing fluids. Deposits that fall into this category clearly include skarns of all kinds, manto lead-zinc deposits, and the disseminated chalcopyrite in sericitized feldspar sites in alteration envelopes on veinlets in porphyry copper deposits.

Simultaneous dilution and cooling by cold water mixing (Run 9)

In the calculations described above, we have attempted to separate the effects of changes of temperature, pressure, pH, salinity and redox state by varying only one at a time so as to isolate controlling variables that apply to sulfide mineral precipitation in natural systems. It is quite difficult to separate these variables, even in the context of contrived computer models as described above. In the natural setting, simultaneous change in these variables, and many more especially those involving composition, must be the normal circumstance. In Run 9, we examine simultaneous dilution, cooling and pH change by mixing a cold pure water with a hot saline fluid.

Starting with a 300 °C, 1 m NaCl, pH 5 solution in equilibrium with chalcopyrite, sphalerite, galena, pyrite, magnetite and electrum, but removed from these minerals, we add 32 kg of 25 °C pure water, which causes temperature to decrease from 300 °C to 34 °C (Fig. 13, x-axis scale). Upon dilution, sulfide minerals and electrum precipitate abundantly (Fig. 13A), driving aqueous metal concentrations except Fe down by several orders of magnitude. The vast majority of metal precipitation is complete after 1 kg of added cold water (2 kg total) at which point temperature is 177 °C. Further dilution moves Ag from electrum to argentite, and yields substantial pyrite, although not enough to deplete aqueous Fe by much (Fig. 13B).

The pH decreases from 5.0 to 4.1 at 160 °C (~2+ kg solvent water, Fig. 13B), then gradually increases again to 5 (Fig. 13B). Since this solution contains no CO_2 (H_2CO_3), little sulfuric acid (HSO_4^-), and minimal associated HCl at pH 5, the effect of decreasing temperature in driving dissociation of those acid sources could not cause the pH decrease between 300 °C and 160 °C. pH decreases in response to sulfide mineral precipitation, which yields H$^+$, e.g., for sphalerite:

$$ZnCl^+_{(aq)} + H_2S_{(aq)} \rightarrow ZnS_{(sphalerite)} + 2H^+_{(aq)} + Cl^-_{(aq)} \qquad (17)$$

Metal sulfide precipitation is driven largely by the effect of temperature decrease in destabilizing chloride complexes of the metals (Fig. 3), allowing the metals to precipitate in

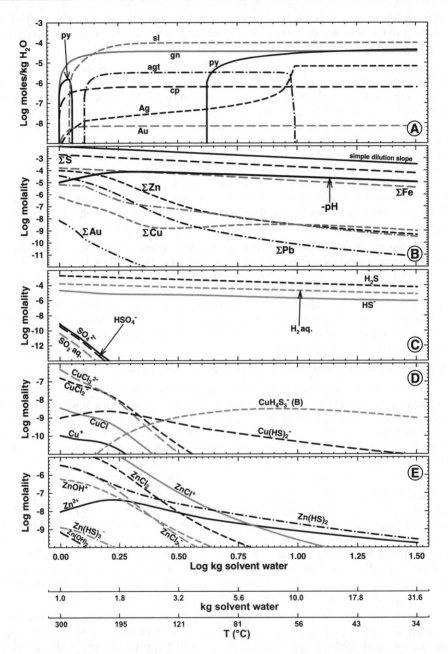

Figure 13. Run 9. Dilution and cooling by cold water mixing with a hydrothermal fluid. Pure 25 °C water is titrated into the 300 °C 1 m NaCl fluid from Run 1, which is saturated with sphalerite, chalcopyrite, galena, pyrite, magnetite and electrum ($X_{Au} = 0.69$), but is not in contact with those minerals. Mixing causes nearly quantitative precipitation of metals into sulfide minerals and electrum because chloride complexes dissociate upon cooling and Ag and Au precipitate owing to the intrinsic stability of the native metals at low temperature. A) Minerals. B) pH and total concentrations of aqueous components. The "simple dilution curve" shows the single constant negative slope for species whose concentrations reflect dilution only. C,D,E) Concentrations of selected aqueous species: C) H_2 and S species, D) Cu complexes, E) Zn complexes.

sulfides, combined with the decrease in mineral solubility with decreasing temperature below 300 °C (Fig. 1), as explained for Run 4 (Fig. 8). The phenomenon of sulfide precipitation driving pH decrease is recognized in the formation of sulfide scale from geothermal waters in Fushime, Japan, where boiling drives temperature decrease, which causes sphalerite and galena to precipitate in the well, yielding H^+ to the water (Akaku et al. 1991).

The transition, upon cooling, from concentrated aqueous metal chlorides to small amounts of metal bisulfides in this run is illustrated for Cu and Zn in Figures 13D and 13E. Apparently in this case (Run 9), in contrast to Run 4 (simple cooling at controlled ΔpH 0.5 unit below neutral) where chalcopyrite dissolves out at $T < 110$ °C, the more acidic pH in Run 9 goes not allow the Cu bisulfides to form as in Run 4 where the bisulfide complexing drives chalcopyrite dissolution. For Au and Ag, despite the dominance of strong bisulfide complexes, electrum precipitate abundantly (Fig. 13A) because the intrinsic stability of the native metals increases substantially at low temperature (Fig. 1).

One of the most interesting conclusions from Run 9 is that, despite substantial dilution of metals and complexing ligands, the effect of the cold water causing cooling dominates in driving sulfide mineral precipitation. This cold water mixing effect applies to the seafloor precipitation of pyrrhotite, pyrite, chalcopyrite and sphalerite in "black smoker" smoke and chimneys (Scott 1997) although, in that instance, the diluting cold water is seawater rather than the pure water used here.

Oxidation and reduction of a sulfide-oxide system (Run 10)

Redox effects play a critical role in sulfide-oxide systems near the Earth's surface in both the origin of metal-bearing fluids and in precipitation of metal sulfides. In Run 10, we examine the effect of reducing an oxygenated metal-oxide-sulfate system, containing Cu, Pb, Zn, Fe, Mn, Co, Ag and Au in a 1m NaCl pH 6 solution at 125 °C in contact with hematite, pyrolusite (MnO_2), cobalt oxide, tenorite (CuO), hemimorphite [$Zn_4Si_2O_7(OH)_2 \cdot H_2O$] and quartz (Table 1, Fig. 14). (The actual calculation was run from reduced to oxidized, but the run direction does not affect the result in this type of calculation.) We view the calculation as starting at a large O_2 fugacity approximating that in a substantially oxidized brine-sediment system, but buried to a depth where temperature is 125 °C, resembling the Olympic Dam rift environment described by Haynes et al. (1994), or approximating parts of the red bed Cu systems such as those described by Oszczepalski (1989) and Hayes et al. (1989).

Upon reduction, no chemical changes occur while pyrolusite persists (Fig. 14A) but, once it is reduced to Mn_2O_3, tenorite and Co_3O_4 dissolve, and electrum, then native copper precipitate, and the composition of the aqueous phase changes radically (Fig. 14C), yielding maxima in aqueous Cu, Co, and Mn and a minimum in Au. Further reduction yields a replacement series among Cu-Fe-S minerals: Cu°, chalcocite, bornite and chalcopyrite (Fig. 14A,B). Other minerals are interspersed in the Cu series, making a complete reduction sequence as follows: Cu°, chalcocite, linnaeite (Co_3S_4), galena, sphalerite, cattierite (CoS_2), bornite, chalcopyrite, pyrite, acanthite. Hematite begins to dissolve where chalcopyrite precipitates, and is fully replaced by pyrite upon further reduction (Fig. 14A). All oxide-to-sulfide mineral replacements reflect the reduction of aqueous sulfate to sulfide, the aqueous concentration of which remains modest at $10^{-6.5}\,m$ (Fig. 14C) until hematite is fully reduced, whereupon it rises to $10^{-3}\,m$.

This chemical system is simple relative to natural analogs, because we omit CO_2 and all rock-forming species except silica, and we hold pH constant at 6. Even so, the interactions among redox-sensitive minerals and aqueous species yields a complex pattern of mineral assemblages and aqueous metal concentrations. Metals such as Fe, Mn, Cu, and Co that form stable oxide minerals have relatively small concentrations under oxidizing conditions while Au, Ag, Zn and Pb concentrations are large because they are not limited by minerals (or only

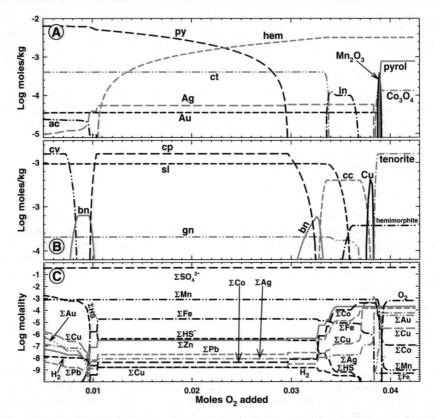

Figure 14. Run 10. Reduction of an oxidized metal-oxide-brine system. Oxygen is removed from an oxygenated metal-oxide-sulfate system containing Cu, Pb, Zn, Fe, Mn, Co, Ag and Au in a 1 m NaCl pH 6 solution at 125 °C in contact with hematite, pyrolusite (MnO_2), cobalt oxide, tenorite (CuO), hemimorphite [$Zn_4Si_2O_7(OH)_2 \cdot H_2O$] and quartz (right-hand side of graphs). Reduction yields a progression from the metal oxide assemblage to a metal sulfide assemblage with concomitant dramatic changes in aqueous metal concentrations. T = 125 °C, pH = 6. A,B) Minerals. C) Total concentrations of aqueous components, and concentrations of aqueous H_2 and O_2.

modestly for Zn). Under moderately reducing conditions (hem-py assemblage), Fe and Mn concentrations are relatively large, but all other metal concentrations are small because their sulfide minerals precipitate (or native metals in electrum), removing metals from solution. Under hematite-absent conditions (Fig. 14A, left side), aqueous sulfide (~60% H_2S, ~40% HS^-) increases, further removing aqueous Fe and Co into sulfide minerals, but increasing aqueous Au, as $Au(HS)_2^-$ and $Au_2H_2S_3^{2-}$ form in subequal amounts.

If we cast these redox results in a natural tectonic rift environment with red beds and evaporites (e.g., Jowett 1989), these model results can be applied to understand metal precipitation in a black shale, or to metal leaching from oxidized source rocks. Where black shales overlie red beds from which oxidized, metaliferous, sulfate-bearing brines flow into the black shale (carbonaceous mud), carbon reduces aqueous metals and reduces sulfate, precipitating sulfide minerals (e.g., Oszczepalski 1989). The sequence of minerals in the natural setting is essentially the same sequence shown upon reduction in Figure 14, starting from native gold, silver and copper. In a more complete model of this process (not shown), we titrate a reductant (C) into fluid advecting through silicate-carbonate buffered sediment, and fractionate minerals as they form, thereby distinctly separating metals into spatial zones in the same order shown in Figure 14, but more distinctly separated.

Carbon or hydrocarbon is one common natural reductant. Another less obvious one is ferrous iron, either in mineral or aqueous form, which reacts with a metal-sulfate solution such as the oxidized solution in Run 10, precipitating hematite or magnetite, and yielding H_2S to react with aqueous metals, e.g.:

$$2Fe^{2+}_{(aq)} + 2H_2O_{(aq)} + \tfrac{1}{4}SO_4^{2-}_{(aq)} = Fe_2O_{3(hem)} + 3\tfrac{1}{2}H^+_{(aq)} + \tfrac{1}{4}H_2S_{(aq)} \qquad (18)$$

Aqueous ferrous iron is postulated by Haynes et al. (1994) as the reductant causing copper sulfide precipitation at Olympic Dam, where the mineral zoning sequence from oxidized to reduced is as in Figure 14, cv-cc-bn-cp, all accompanied by hematite.

We may also apply the Run 10 reduction results to understanding the origin of metalliferous brines formed in an oxidized red bed pile. We interpret the sharp differences in metal concentrations in the aqueous phase (Fig. 14C) as reflecting differences in natural waters emerging from potential source rocks of different oxidation states (cf. Haynes et al. 1994). The oxidation variation in Figure 14 can be regarded as resulting from differences in the intensity of original oxidation of the sediments. The sediment may have more or less reduced Fe in the detrital minerals and rock fragments, or more or less organic debris deposited with the sediment, either of which would result in differences in oxidation state in the buried and heated red bed pile. Depending on the particular source rock oxidation state, Figure 14C shows that the system could produce the following example fluids (from oxidized to reduced): a) Zn-Pb-Ag-Au with modest Cu, b) Cu-Co-Zn-Pb-Mn, c) Zn, Co, Mn, d) Mn, Fe. In the natural setting, the metal ratios depend on both metal-limiting authigenic (diagenetic) minerals such as those computed here (Fig. 14A,B), *and* on ratios of trace metals in the primary source rock detrital grains.

CONCLUSION

We can readily distinguish two categories of sulfide mineral precipitation: a) sulfide precipitation where metal and sulfur, both within a single fluid, combine owing to a change in physical or chemical conditions; b) combining metals in one fluid with sulfur from another source. The separate source category (b) is exemplified by the mixing of a metal-bearing sedimentary brine with an H_2S-bearing phase, as postulated for some Mississippi Valley Type ore deposits (Leach et al. 2006), or the diffusion of aqueous sulfide into wall rock, sulfidizing iron to form pyrite or pyrrhotite, for example (thereby desulfidizing the fluid so that Au° precipitates (e.g., Reed 1997).

In this chapter, we have explored sulfide precipitation from single hydrothermal fluids that contain both sulfur and metals, wherein sulfide precipitates owing to the following effects:

1. cooling, thereby destabilizing metal complexes, thus liberating metals to combine with aqueous sulfide;
2. cooling to a lower temperature where intrinsic sulfide mineral stability is greater;
3. pH increase, liberating sulfide from H_2S or HS^- to precipitate in sulfide minerals;
4. reduction of aqueous sulfate to sulfide, thereby generating sulfide to combine with metals.

These fundamental processes unfold in the context of various natural processes of fluid-fluid mixing, wall-rock reaction, heat conduction and fluid boiling, for example.

Two of the numerical experiments reported above lend themselves to understanding the zoning of Cu-Fe-S sulfides in natural deposits. In Run 8 (Fig. 12) we find that covellite-chalcocite-bornite-chalcopyrite (all with pyrite) form in that order as pH changes from low to high. This order matches the series in numerous magmatic-hydrothermal settings under matching pH conditions indicated by wall-rock alteration assemblages. The same sequence of minerals forms, except without pyrite and with $Cu°$ preceding covellite, in the redox series (Run 10, Fig. 14) from oxidized to reduced. That sequence is also observed in corresponding natural redox settings where oxidation is indicated by distance from fully oxidized rocks (e.g., Olympic Dam, Kupferschiefer). In the first instance, the sequence reflects a pH gradient; in the second, it reflects a redox gradient. The greater complexities of metal zoning involving the full complement of common ore metals, e.g., Mo, Cu, Fe, Zn, Pb, Ag, Au, Hg, that form sulfides, sulfosalts and native elements is beyond the scope if this chapter. However, we find that numerical models of source fluid compositions combined with models of sulfide, sulfosalt, and electrum solubility controls in various settings involving cooling, boiling, fluid-fluid mixing, depressurization, and wall-rock reaction, enable a full understanding of depositional zoning in metal sulfides and kindred minerals.

REFERENCES

Akaku K, Reed MH, Yagi M, Kai K, Yasuda Y (1991) Chemical and physical processes occurring in the Fushime geothermal system, Kyushu, Japan. Geochem J 25:315-334

Arnórsson S (1983) Chemical equilibria in Icelandic geothermal systems—implications for chemical geothermometry investigations. Geothermics 12:119-128

Arnórsson S, Gunnlaugsson E, Hördur S (1983) The chemistry of geothermal waters in Iceland II. Mineral equilibria and independent variables controlling water compositions. Geochim Cosmochim Acta 47: 547-566

Bazin B, Brosse E, Sommer F (1997a) Reconstruction of the chemistry of oil-field formation waters for the purpose of numerical modeling. Bull Soc géol France 168:231-242 (in French).

Bazin B, Brosse É, Sommer F (1997b) Chemistry of oil-field brines in relation to diagenesis of reservoirs 1. Use of mineral stability fields to reconstruct *in situ* water composition. Example of the Mahakam basin. Marine Petrol Geol 14:481-495

Bessinger B, Apps JA (unpublished manuscript) The Hydrothermal Chemistry of Gold, Arsenic, Antimony, Mercury and Silver.

Bethke C (1996) Geochemical Reaction Modeling. Oxford University Press

Brimhall GH (1977) Early fracture-controlled disseminated mineralization at Butte, Montana. Econ Geol 72: 37-59

Carpenter AB, Trout M, Pickett EE (1974) Preliminary report on the origin and chemical evolution of lead-and zinc-rich oil field brines in central Mississippi. Econ Geol 69:1191-1206

Degens ET, Ross DA (eds) (1969) Hot Brines and Recent Heavy Metal Deposits in the Red Sea. Springer-Verlag

Giggenbach WF (1980) Geothermal gas equilibria. Geochim Cosmochim Acta 44:2021-2032

Giggenbach WF (1981) Geothermal mineral equilibria. Geochim Cosmochim Acta 45:393-410

Hayes TS, Rye RO, Whelan JF, Landis GP (1989) Geology and sulphur isotope geothermometry of the Spar Lake stratabound Cu-Ag deposit in the Belt Supergroup, Montana. *In*: Sediment-hosted copper deposits. Geol Assoc of Canada, Spec Paper 36. Boyle R, Beown AC, Jefferson CW, Jowett, EC, Kirkahm, RV (eds) Geol Assoc of Canada, p 319-338

Haynes DW, Cross KC, Bills RT, Reed MH (1994) Olympic Dam ore genesis: a fluid mixing model. Econ Geol 90:281-307

Helgeson HC (1964) Complexing and hydrothermal ore deposition. MacMillan

Holland TJB, Powell R (1998) An internally consistent thermodynamic data set for phases of petrological interest. J Metamorph Geol 16:309-343

Izawa E (1991) Hydrothermal alteration associated with Nansatsu-type gold mineralization in the Kasuga Area, Kagoshima Prefecture. Japan Geol Surv Japan 277:49-52

Johnson J, Oelkers E, Helgeson H (1992) SUPCRT92: a software package for calculating the standard molal thermodynamic properties of minerals, gases, aqueous species, and reactions from 1 to 5000 bar and 0 to 1000 °C. Comp Geosci 18:899-947

Jowett EC (1989) Effects of continental rifting on the location and genesis of stratiform copper-silver deposits. *In*: Sediment-Hosted Copper Deposits. Geol Assoc of Canada, Spec Paper 36. Boyle, RW, Beown, AC, Jefferson, CW, Jowett, EC Kirkahm, RV (eds) Geol Assoc of Canada, p 53-66

Leach D, Macquar J-C, Lagneau V, Leventhal J, Emsbo P, Premo W (2006) Precipitation of lead–zinc ores in the Mississippi Valley-type deposit at Trèves, Cévennes region of southern France. Geofluids 6:24-44

McKibben M, Hardie L (1997) Ore-forming brines in active continental rifts. *In*: Geochemistry of Hydrothermal Ore Deposits, Third Edition. Barnes HL (ed) Wiley, p 877-936

Meyer C, Shea E. Goddard C (1968) Ore deposits at Butte, Montana. *In*: Ore Deposits of the United States, 1933-1967. Brown JS (ed) Am. Inst. Mining, Metallurgical and Petroleum Engineers, p 1373-1416

Meyer C, Hemley J (1968) Wall rock alteration. *In*: Geochemistry of Hydrothermal Ore Deposits, First Edition. Barnes HL (ed) Holt, Rinehart and Winston, p 166-235

Oszczepalski S (1989) Kupferschiefer in Southwestern Poland: Sedimentary environments, metal zoning and ore controls. *In*: Sediment-Hosted Copper Deposits. Geol Assoc of Canada, Spec Paper 36. Boyle RW, Beown AC, Jefferson CW, Jowett EC Kirkahm, RV (eds) Geol Assoc of Canada, p 571-600

Palandri J, Reed MH (2001) Reconstruction of *in situ* composition of sedimentary formation waters. Geochim Cosmochim Acta 65:1741-1767

Palandri J, Reed MH (2004) Geochemical models of metasomatism in ultramafic systems: Serpentinization, related rodingitization and silica-carbonate alteration. Geochim Cosmochim Acta 68:1115-1133

Pang Z, Reed MH (1998) Theoretical chemical thermometry on geothermal waters: Problems and methods. Geochim Cosmochim Acta 62:1082-1091

Plumlee G (1994) Fluid chemistry evolution and mineral deposition in the Main-Stage Creede epithermal system. Econ Geol 89:1860-1882

Reed MH (1982) Calculation of multicomponent chemical equilibria and reaction processes in systems involving minerals, gases and an aqueous phase. Geochim Cosmochim Acta 46:513-528.

Reed M (1997) Hydrothermal Alteration and Its Relationship to Ore Fluid Composition. *In*: Geochemistry of Hydrothermal Ore Deposits, Third Edition. Barnes HL (ed) Wiley, p 303-366

Reed MH (1998) Calculation of simultaneous chemical equilibria in aqueous-mineral-gas systems and its application to modeling hydrothermal processes. *In*: Techniques in Hydrothermal Ore Deposits Geology. Reviews in Economic Geology, Volume 10. Richards J, Larson P (eds) p 109-124

Reed M, Palandri J (2006) SOLTHERM.H06, a data base of equilibrium constants for minerals and aqueous species. Available from the authors, University of Oregon.

Reed MH, Spycher N (1984) Calculation of pH and mineral equilibria in hydrothermal waters with application to geothermometry and studies of boiling and dilution. Geochim Cosmochim Acta 48:1479-1492

Sack RO (2000) Internally consistent database for sulfides and sulfosalts in the system $Ag_2S-Cu_2S-ZnS-Sb_2S_3-As_2S_3$. Geochim Cosmochim Acta 64:3803-3812

Sales RH (1914) Ore Deposits at Butte, Montana. Am Inst Mining Engineers Trans, v 46.

Scott S (1997) Submarine hydrothermal systems and deposits. *In*: Geochemistry of Hydrothermal Ore Deposits, Third Edition. Barnes HL (ed) Wiley, p 797-875

Shock EL, Helgeson HC (1988) Calculation of the thermodynamic and transport properties of aqueous species at high pressures and temperatures: Correlaton algorithms for ionic spesies and equation of state predictons to 5 Kb and 1000 °C. Geochim Cosmochim Acta 52:2009-2036

Shock EL, Sassini DC, Willis M, Sverjensky DA (1997) Inorganic species in geologic fluids: Correlations among standard molal thermodynamic properties of aqueous ions and hydroxide complexes. Geochim Cosmochim Acta 61:907-950

Stoffregen R (1987) Genesis of acid sulfate alteration and Au-Cu-Ag mineralization at Summitville, Colorado. Econ Geol 82:575-1591

Von Damm (1990) Seafloor hydrothermal activity: black smoker chemistry and chimneys. Annu Rev Earth Planet Sci 18:173-204

Zhu C, Anderson G (2002) Environmental Applications of Geochemical Modeling. Cambridge University Press

Sulfur Isotope Geochemistry of Sulfide Minerals

Robert R. Seal, II

U.S. Geological Survey
954 National Center
Reston, Virginia, 20192, U.S.A.
e-mail: rseal@usgs.gov

INTRODUCTION

Sulfur, the 10th most abundant element in the universe and the 14th most abundant element in the Earth's crust, is the defining element of sulfide minerals and provides insights into the origins of these minerals through its stable isotopes. The insights come from variations in the isotopic composition of sulfide minerals and related compounds such as sulfate minerals or aqueous sulfur species, caused by preferential partitioning of isotopes among sulfur-bearing phases, known as fractionation. These variations arise from differences in temperature, or more importantly, oxidation and reduction reactions acting upon the sulfur. The oxidation and reduction reactions can occur at high temperature, such as in igneous systems, at intermediate temperatures, such as in hydrothermal systems, and at low temperature during sedimentary diagenesis. At high temperatures, the reactions tend to occur under equilibrium conditions, whereas at low temperatures, disequilibrium is prevalent. In addition, upper atmospheric processes also lead to isotopic fractionations that locally appear in the geologic record.

Sulfur isotope geochemistry as a subdiscipline of the geological sciences began in the late 1940s and early 1950s with early publications by Thode et al. (1949) and Szabo et al. (1950) on natural variations of sulfur isotopes, and Macnamara and Thode (1950) on the isotopic composition of terrestrial and meteoritic sulfur. Sakai (1957) presented an early scientific summary of sulfur isotope geochemistry, with a particular emphasis on high-temperature processes. Thode et al. (1961) also presented an early summary, but with an emphasis on low-temperature processes. Both of these summaries outlined salient aspects of the global sulfur cycle. Sulfur isotope geochemistry understandably has had a long history of application to the study of sulfide-bearing mineral deposits. Early noteworthy papers include those by Kulp et al. (1956) and Jensen (1957, 1959). Similarly, there is also a legacy of contributions to understanding sedimentary diagenesis and the origin of diagenetic pyrite. The paper by Thode et al. (1951) represents one of the earliest efforts investigating sulfur isotope fractionations associated with bacterial sulfate reduction. Subsequent advances in the field of sulfur isotope geochemistry have been motivated by applications to an increasing variety of geochemical systems and by technological advances in analytical techniques. Noteworthy reviews related to the sulfur isotope geochemistry of sulfide minerals include those of Jensen (1967), Ohmoto and Rye (1979), and Ohmoto and Goldhaber (1997), all of which emphasize mineral deposits, Seal et al. (2000a) which emphasized sulfate minerals and their interactions with sulfides, and Canfield (2001) which emphasized biogeochemical aspects of sulfur isotopes.

A considerable body of knowledge exists on the metal stable isotopic composition of sulfide minerals—a topic that will not be covered in this paper. Recent analytical advances in plasma-source mass spectrometry have enabled precise isotopic measurements of numerous other metals in sulfide minerals including Fe (Johnson et al. 2003; Beard and Johnson 2004), Cu (Maréchal et al. 1999; Zhu et al. 2000; Larson et al. 2003; Albarède 2004), Zn (Maréchal et al. 1999; Albarède 2004) and Mo (Barling et al. 2001; Anbar 2004), among others.

The intent of this chapter is to build upon previous reviews of sulfur isotope geochemistry as they relate to sulfide minerals, summarize landmark studies in the field, resolve, or at least discuss, existing controversies and summarize recent advances for a variety of geochemical settings. The first part of this chapter is designed to provide the reader with a basic understanding of the principles that form the foundations of stable isotope geochemistry. Next, an overview of analytical methods used to determine the isotope composition of sulfide minerals is presented. This overview is followed by a discussion of geochemical processes that determine the isotope characteristics of sulfide minerals and related compounds. The chapter then concludes with an examination of the stable isotope geochemistry of sulfide minerals in a variety of geochemical environments.

FUNDAMENTAL ASPECTS OF SULFUR ISOTOPE GEOCHEMISTRY

An isotope of an element is defined by the total number of protons (Z) and neutrons (N) present, which sum together to give the atomic mass (A). For example, the element sulfur is defined by the presence of 16 protons, but can have either 16, 17, 18, 19, or 20 neutrons, giving atomic masses of 32, 33, 34, 35, and 36 amu, respectively. These isotopes are written as ^{32}S, ^{33}S, ^{34}S, ^{35}S, and ^{36}S. Four of the five naturally occurring sulfur isotopes are stable (^{32}S, ^{33}S, ^{34}S, and ^{36}S) and one (^{35}S) is unstable, or radiogenic. The isotope ^{35}S is formed from cosmic ray spallation of ^{40}Ar in the atmosphere (Peters 1959). It undergoes beta decay with a half-life of 87 days; therefore, it is not important from the perspective of naturally occurring sulfide minerals. The four stable isotopes of sulfur, ^{32}S, ^{33}S, ^{34}S, and ^{36}S, have approximate terrestrial abundances of 95.02, 0.75, 4.21, and 0.02%, respectively (Macnamara and Thode 1950).

Stable isotope geochemistry is concerned primarily with the relative partitioning of stable isotopes among substances (i.e., changes in the ratios of isotopes), rather than their absolute abundances. The difference in the partitioning behavior of various isotopes, otherwise known as fractionation, is due to equilibrium and kinetic effects. In general, heavier isotopes form more stable bonds; molecules of different masses react at different rates (O'Neil 1986). Isotope ratios are usually expressed as the ratio of a minor isotope of an element to a major isotope of the element. For sulfide minerals, the principal ratio of concern is $^{34}S/^{32}S$. However, renewed interest in $^{33}S/^{32}S$ and $^{36}S/^{32}S$ ratios has been generated by the discovery of unexpected variations of these minor isotopes in Precambrian sulfide and sulfate minerals and in Martian meteorites (Farquhar et al. 2000a,b; Farquhar and Wing 2003). Most fractionation processes will typically cause variations in these ratios in the fifth or sixth decimal places. Because we are concerned with variations in isotopic ratios that are relatively small, the isotopic composition of substances is expressed in delta (δ) notation, as parts per thousand variation relative to a reference material. The δ-notation for the $^{34}S/^{32}S$ composition of a substance is defined as:

$$\delta^{34}S = \left(\frac{\left(^{34}S/^{32}S\right)_{sample} - \left(^{34}S/^{32}S\right)_{reference}}{\left(^{34}S/^{32}S\right)_{reference}} \right) \times 1000 \qquad (1)$$

which has units of parts per thousand or permil (‰), also found in the literature spelled "per mil," "per mill," and "per mille." The values for $\delta^{33}S$ and $\delta^{36}S$ are similarly defined for the ratio of $^{33}S/^{32}S$ and $^{36}S/^{32}S$, respectively. The agreed upon reference for sulfur isotopes is Vienna Canyon Diablo Troilite (VCDT) with $\delta^{34}S = 0.0‰$ by definition, which is currently defined relative to a silver sulfide reference material IAEA-S-1 with an assigned value of −0.3‰ because the supply of the Canyon Diablo Troilite reference material has been exhausted (Krouse and Coplen 1997). The reference was originally defined by the isotopic composition of troilite (FeS) from the Canyon Diablo iron meteorite. The absolute $^{34}S/^{32}S$ ratio for Canyon

Diablo Troilite is 4.50045 × 10⁻³ (Ault and Jensen 1963). The selection of a meteoritic sulfide mineral as the reference for sulfur is useful because meteoritic sulfide is thought to represent the primordial sulfur isotopic composition of Earth (Nielsen et al. 1991). Thus, any variations in the isotopic composition of terrestrial sulfur relative to VCDT reflects differentiation since the formation of Earth.

For sulfur, which has more than two stable isotopes, $^{34}S/^{32}S$ is the ratio most commonly measured in studies of terrestrial systems. This ratio was chosen for two main reasons. Firstly, it represents the most abundant isotopes of these elements, which facilitates analysis. Secondly, isotopic fractionation is governed by mass balance such that different isotopic ratios tend to vary systematically with one another in proportions that can be approximated by the mass differences among the isotopes. In other words, the variations in the $^{33}S/^{32}S$ ratio of a sample will be approximately half that of the $^{34}S/^{32}S$ ratio because of the relative differences in masses. Likewise, the variations in the $^{36}S/^{32}S$ ratio of a sample will be approximately twice that of the $^{34}S/^{32}S$ ratio. This linear fractionation trend due to physical and chemical processes is known "mass-dependent fractionation" (Urey 1947; Hulston and Thode 1965a,b), which is in distinct contrast to "mass-independent fractionation." Mass-independent fractionation is reflected by non-linear variations in isotopic fractionation with mass, and will be discussed in more detail below.

Fractionation can be considered in terms of isotopic exchange reactions, which are driven thermodynamically toward equilibrium. Thus, isotopic equilibrium, for example between sphalerite (Sl) and galena (Gn), can be described by an isotopic exchange reaction such as:

$$Pb^{34}S + Zn^{32}S = Pb^{32}S + Zn^{34}S \tag{2}$$

which is written in a form with one exchangeable atom of sulfur. The equilibrium constant (K) for this reaction is equivalent to the isotopic fractionation factor (α):

$$K = \frac{Pb^{32}S \cdot Zn^{34}S}{Pb^{34}S \cdot Zn^{32}S} = \frac{\left(^{34}S/^{32}S\right)_{Sl}}{\left(^{34}S/^{32}S\right)_{Gn}} = \alpha_{Sl\text{-}Gn} \tag{3}$$

where the isotopic species are meant to represent their respective chemical activities. Thus, in a more general form, the partitioning of stable isotopes between two substances, A and B, is quantitatively described by a fractionation factor, which is defined as:

$$\alpha_{A\text{-}B} = \frac{R_A}{R_B} \tag{4}$$

where R is $^{34}S/^{32}S$. This equation can be recast in terms of δ values using Equation (1) as:

$$\alpha_{A\text{-}B} = \frac{1 + \frac{\delta_A}{1000}}{1 + \frac{\delta_B}{1000}} = \frac{1000 + \delta_A}{1000 + \delta_B} \tag{5}$$

Values of α are typically near unity, with variations normally in the third decimal place (1.00X). For example, the equilibrium $^{34}S/^{32}S$ fractionation between sphalerite and galena at 300 °C has been measured to have an $\alpha_{Sl\text{-}Gn}$ value of 1.0022. Thus, sphalerite is enriched in ^{34}S relative to galena by 2.2‰ (i.e., the fractionation equals 2.2‰). For an α value less than unity, such as $\alpha_{Gn\text{-}Sl}$, which equals 0.9978, the galena is depleted in ^{34}S relative to sphalerite by 2.2‰ (i.e., the fractionation equals −2.2‰). In the literature, fractionation factors may be expressed in a variety of ways including α, 1000lnα, and Δ, among others. The value $\Delta_{A\text{-}B}$ is defined as:

$$\Delta_{A\text{-}B} = \delta_A - \delta_B \tag{6}$$

A convenient mathematical relationship is that $1000\ln(1.00X)$ is approximately equal to X, so that:

$$\Delta_{A-B} \approx 1000\ln\alpha_{A-B} \tag{7}$$

Isotopic fractionations may also be defined in terms of an enrichment factor (ε), where:

$$\varepsilon_{A-B} = (\alpha_{A-B} - 1) \times 1000 \tag{8}$$

ANALYTICAL METHODS

Several procedures are available to determine the sulfur isotopic compositions of sulfide minerals. Conventional analyses typically involve mineral separation procedures that may include handpicking or gravimetric techniques (heavy liquids, panning, etc.) or wet chemical techniques. Once a suitable concentration of the desired compound is obtained, the sulfur is extracted and converted to a gaseous form that is amenable to mass spectrometric analysis. For sulfur, the gas is SO_2. Alternatively, the gas SF_6 may be used, which has the advantages of being an inert, non-absorbing gas, and lacking ambiguity in isotopic speciation because fluorine has only one stable isotope. It has the disadvantage of requiring potential hazardous fluorinating reagents. The amount of sample required varies among laboratories, but typically ranges from 5 to 20 mg of pure mineral separate for $\delta^{34}S$ using conventional techniques. Typical analytical uncertainties (1σ) for conventional techniques are ±0.1‰ for $\delta^{34}S$.

For conventional $\delta^{34}S$ analysis of sulfide minerals, SO_2 is produced for analysis by reacting the sulfate mineral with an oxidant (CuO, Cu_2O, or V_2O_5) at elevated temperatures (1000 to 1200 °C) under vacuum (Holt and Engelkemeier 1970; Haur et al. 1973; Coleman and Moore 1978). SF_6 can be prepared using BrF_3, BrF_5, or elemental F as reagents at elevated temperatures (300 °C) in nickel reaction vessels; the SF_6 is then purified cryogenically and through gas chromatography (Hulston and Thode 1965a; Puchelt et al. 1971; Thode and Rees 1971). Other conventional techniques for the $\delta^{34}S$ analysis of sulfide minerals have been summarized by Rees and Holt (1991).

Isotopic analysis is done on a gas-source, sector-type, isotope ratio mass spectrometer. In gas-source mass spectrometers, SO_2 gas molecules are ionized to positively charged particles, such as SO_2^+, which are accelerated through a voltage gradient. The ion beam passes through a magnetic field, which causes separation of various masses such as 64 ($^{32}S^{16}O_2$) and 66 ($^{34}S^{16}O_2$, $^{34}S^{18}O^{16}O$). In conventional dual-inlet mass spectrometers, a sample gas is measured alternately with a reference gas. The beam currents are measured in faraday cups and can be related to the isotopic ratio when the sample and standard gases are compared.

Technological advances over the past decade have opened new frontiers in stable isotope analysis of sulfide minerals. One new area is the *in situ* microanalysis of minerals. For *in situ* analysis, a growing body of sulfur isotope data has been generated from samples of sulfide minerals using the secondary ion mass spectrometer (SIMS) otherwise known as the ion microprobe (Eldridge et al. 1988; Paterson et al. 1997; McKibben and Riciputi 1998). The ion microprobe bombards a sample with a beam of charged Cs or O. The ion beam causes the sample to be ablated as secondary ionic species, which are measured in a mass spectrometer. Spatial resolution less than 20 µm can be achieved with an analytical uncertainty of ±0.25‰ for sulfur isotope analyses using the ion microprobe (Paterson et al. 1997).

Techniques for *in situ* analysis have also been developed using lasers as heat sources to drive reactions producing either SO_2 or SF_6 for isotopic analysis, and have been reviewed by Shanks et al. (1998). Laser-based techniques resulting in SO_2 for isotopic analysis were first developed by Crowe et al. (1990). Spatial resolution can be achieved as good as a spot size of 150 µm having an analytical precision of ±0.3 to 0.6‰. Early development of laser-based sulfur isotope analysis on SF_6 was by Rumble et al. (1993) and Beaudoin and Taylor (1994). Spatial resolution (< 150 μm) and analytical precision (±0.2‰) for *in situ* analysis are routinely similar to those achieved for the analysis of SO_2.

Another recent advance is the development of continuous-flow techniques that use a combination of an elemental analyzer and gas chromatograph for online combustion and purification of gases that are then carried in a He stream directly into the ion source of a mass spectrometer, which allows for the mass production of data from small samples. Continuous-flow systems can measure the sulfur isotopic ratios of sulfide samples in the microgram range, compared to the milligram range for conventional techniques (Giesemann et al. 1994). Sample gases are prepared by on-line peripheral devices such as elemental analyzers that are capable of processing 50 to 100 samples per day in a highly automated fashion. Furthermore, most sulfur isotope measurements can be made without mineral purification, if bulk sulfur data are all that is desired.

REFERENCE RESERVOIRS

Sulfur isotope variations on Earth can be considered relative to geologically important reservoirs. The most common reference reservoirs for sulfur isotopes in terrestrial systems are meteoritic sulfur and seawater. Meteoritic sulfur, such as that in Canyon Diablo troilite, provides a convenient reference because it is generally regarded as approximating the bulk composition of the Earth. The iron meteorites have an average sulfur isotope composition of $\delta^{34}S = 0.2 \pm 0.2$‰ (Kaplan and Hulston 1966), which is indistinguishable from that of pristine mid-ocean ridge basalts ($\delta^{34}S = 0.3 \pm 0.5$‰; Sakai et al. 1984). Geochemical processes, the most notable of which are oxidation and reduction, profoundly fractionate sulfur isotopes away from bulk-Earth values in geological systems (Fig. 1). Oxidation processes produce species that are enriched in ^{34}S relative to the starting material, whereas reduction produces species that are depleted in ^{34}S.

Oxidation-reduction reactions involving reduced sulfur from the interior of the Earth throughout its history have resulted in a $\delta^{34}S$ of 21.0 ± 0.2‰ for dissolved sulfate in modern oceans (Rees et al. 1978). Because of the volume and importance of the ocean in the global sulfur cycle, this composition is another important reference reservoir from which to evaluate sulfur isotope variations in geological systems. The $\delta^{34}S$ of sulfate in ancient oceans as recorded by marine evaporite sequences (Claypool et al. 1980) has varied from a low near 0‰ during

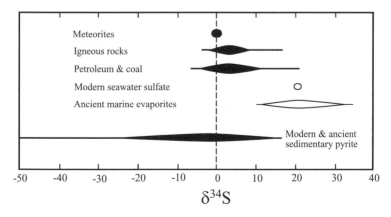

Figure 1. $\delta^{34}S$ of various geologic reservoirs. Modified from Seal et al. (2000a). All isotopic values in permil (VCDT).

Archean time to a high of 35‰ during Cambrian time. The causes and implications of the secular variations in the sulfur isotope composition of seawater are discussed in a later section.

FACTORS THAT CONTROL SULFUR ISOTOPE FRACTIONATION

Most isotopic fractionation is the result of variations in thermodynamic properties of molecules that are dependent on mass. Details of the thermodynamic basis for understanding isotope fractionation have been presented by Urey (1947), Bigeleisen and Mayer (1947), and Bigeleisen (1952). Isotope fractionation may result from equilibrium or kinetically controlled chemical and physical processes. Equilibrium processes include isotopic exchange reactions, which redistribute isotopes among molecules of different substances. Equilibrium isotope effects result from the effect of atomic mass on bonding; molecules containing a heavier isotope are more stable than those containing a lighter isotope. Kinetic processes include irreversible chemical reactions, such as bacterially mediated processes like sulfate reduction and physical processes such as evaporation and diffusion (O'Neil 1986). Kinetic isotope effects are related to greater translational and vibrational velocities associated with lighter isotopes. It is easier to break bonds with lighter isotopes, for example the ^{32}S–O bond, compared with the ^{34}S–O bond, in processes such as bacterially mediated reduction of dissolved sulfate to sulfide.

Among the several factors that influence the magnitude of equilibrium stable isotope fractionations are temperature, chemical composition, crystal structure and pressure (O'Neil 1986). For the present discussion, temperature and chemical composition are the most important. Pressure effects are minimal at upper crustal conditions. The temperature dependence of fractionation factors results from the relative effect of temperature on the vibrational energies of two substances. Theoretical considerations indicate that the stable isotope fractionation between two substances should approach zero at infinite temperature (Bigeleisen and Mayer 1947). These fractionations are generally described well by equations of the form:

$$1000\ln\alpha = \frac{a}{T^2} + \frac{b}{T} + c \qquad (9)$$

where a, b, and c are empirically determined constants.

The dependence of isotopic fractionation can be related to chemical variables such as oxidation state, ionic charge, atomic mass, and the electronic configuration of the isotopic elements and the elements to which they are bound (O'Neil 1986). For sulfur-bearing systems, the effect of the oxidation state of sulfur is especially important. The higher oxidation states of sulfur are enriched in the heavier isotopes relative to lower oxidation states such that ^{34}S enrichment follows the general trend $SO_4^{2-} > SO_3^{2-} > S_x^{\circ} > S^{2-}$ (Sakai 1968; Bachinski 1969). In the geological record, this trend is reflected by the fact that sulfate minerals typically have higher δ^{34}S values than cogenetic sulfide minerals in a variety of geochemical settings.

Cationic substitutions also play an important role in stable isotope fractionations. Heavier elements such as Ba or Pb form stronger bonds than lighter elements such as Ca or Zn. Thus, on a relative basis, the heavier elements are able to bond more effectively with lighter, more energetic stable isotopes such as ^{16}O or ^{32}S. O'Neil et al. (1969) documented a cation-mass dependence of ^{18}O enrichment in divalent metal-carbonate minerals with ^{18}O enrichment following the order $CaCO_3 > SrCO_3 > BaCO_3$. Likewise, the ^{34}S enrichment in divalent sulfide minerals is such that ZnS > PbS.

EQUILIBRIUM FRACTIONATION FACTORS

Equilibrium isotopic fractionation factors are typically derived by one of three methods:

(1) experimental determination, (2) theoretical estimation using calculated bond strengths or statistical mechanical calculations based on data on vibrational frequencies of compounds, or (3) analysis of natural samples for which independent estimates of temperature are available. Each method has advantages and disadvantages. Experimental determination provides a direct measurement of the fractionation, but such efforts are commonly hampered by experimental kinetic limitations and the fact that media used in experiments typically do not approximate natural conditions. Theoretical estimation avoids the kinetic problems of experimental studies, but is limited by the availability and accuracy of data required for the estimation. Fractionation factors derived from the analysis of natural materials provide a means of investigating isotopic fractionations when data from neither of the other methods are available. However, the accuracy of this method can be affected by retrograde isotopic exchange and uncertainties related to whether or not the mineral pairs are cogenetic and to the independent temperature estimate derived from fluid inclusions, for example.

Experimentally determined fractionation factors

Experimental sulfur isotopic fractionation factors for sulfide minerals are limited to a few mineral species, despite the geological importance of numerous sulfides, particularly to ore-forming systems. Ohmoto and Rye (1979) reviewed and critically evaluated the available experimental sulfur isotope fractionation data relative to H_2S, which included temperature-dependent fractionation factors for sulfites, SO_2, H_2S gas, HS^-, S^{2-}, and S, and the minerals pyrite (FeS_2), sphalerite (ZnS), pyrrhotite ($Fe_{1-x}S$), chalcopyrite ($CuFeS_2$), and galena (PbS). Their evaluation and compilation included experimental studies by Grootenboer and Schwarz (1969), Schiller et al. (1969), Grinenko and Thode (1970), Kajiwara and Krouse (1971), Salomons (1971), Thode et al. (1971), Kiyosu (1973), Robinson (1973), and Czamanske and Rye (1974), and estimates following Sakai (1968) and Bachinski (1969). Ohmoto and Lasaga (1982) re-evaluated experimental studies investigating sulfur-isotope fractionations between aqueous sulfate and sulfide (Robinson 1973; Bahr 1976; Igumnov et al. 1977; Sakai and Dickson 1978) and presented a revised equation describing SO_4^{2-}-H_2S sulfur-isotope fractionation as a function of temperature. No further re-evaluation of these data is made in this chapter. Expressions describing the temperature-dependent sulfur isotope fractionation of these compounds relative to H_2S are summarized in Table 1 and Figure 2.

Several other experimental studies of sulfur-isotope fractionation have been published since the compilation of Ohmoto and Rye (1979). Szaran (1996) measured the sulfur isotope fractionation between dissolved and gaseous H_2S from 11 to 30 °C and found that dissolved H_2S is minimally enriched in ^{34}S relative to the gaseous H_2S, ranging from 2.2‰ at 11 °C to 1.1‰ at 30 °C. In comparison, Ohmoto and Rye (1979) reported no fractionation, presumably for all temperatures. A least-squares fit to the data of Szaran (1996) is presented in Table 1.

Hubberten (1980) conducted synthesis experiments investigating sulfur isotope fractionations between 280 and 700 °C for galena, argentite (Ag_2S), covellite (CuS) or digenite (Cu_9S_5) equilibrated with sulfur. Bente and Nielsen (1982) conducted reversed experiments between 150 and 600 °C on isotopic fractionations between bismuthinite (Bi_2S_3) and sulfur. Suvorova and Tenishev (1976) and Suvorova (1978) conducted synthesis experiments investigating sulfur isotope fractionation between 300 and 600 °C between various mineral pairs including sphalerite-molybdenite (Sl-Mb), galena-molybdenite (Gn-Mb), galena-herzenbergite (SnS)(Gn-Hz), tungstenite (WS_2)-molybdenite (Tn-Mb), and stibnite-molybdenite (St-Mb).

The accuracy of these more recent fractionation factors, especially those from the synthesis experiments, warrants evaluation. The rates of solid-state reactions among various sulfides minerals are known to vary by several orders of magnitude. Molybdenite is considered to be one of the most refractory and argentite to be one of the most reactive (Barton and

Table 1. Equilibrium isotopic fractionation factors for sulfide minerals and related compounds described by the equation

$$1000\ln\alpha_{i-H_2S} = \frac{a \times 10^6}{T^2} + \frac{b \times 10^3}{T} + c; \quad (T \text{ in K})$$

Compound or component (i)	a	b	c	T (°C) range*	Data sources
Sulfate minerals and aqueous sulfate	6.463		0.56	200 - 400	(2)
Sulfites	4.12	5.82	−5.0	> 25	(1)
SO$_2$	4.70		−0.5	350 - 1050	(1)
S(=S$_8$)	−0.16			200 - 400	(1)
H$_2$S aqueous-gaseous	0.71		−6.67	11 - 30	(3)
HS$^-$	−0.06		−0.6	50 - 350	(1)
S^{2-}	−0.21	−1.23	−1.23	> 25	(1)
FeS$_2$	0.40			200 - 700	(1)
FeS	0.10			200 - 600	(1)
CuFeS$_2$	−0.05			200 - 600	(1)
PbS	−0.63			50 - 700	(1)
ZnS	0.10			50 - 705	(1)
Ag$_2$S	−0.62			280 - 700	(4)
Cu$_2$S	−0.06			510 - 630	(4)
CuS	0.04			280 - 490	(4)
Bi$_2$S$_3$	−0.67			150 - 600	(5)

* Temperature range refers to the experimental temperature range; note that fractionation factors may extrapolate significantly beyond these ranges (see text).
Data sources: (1) Ohmoto and Rye 1979; (2) Ohmoto and Lasaga 1982; (3) Szaran 1996; (4) Hubberten 1980; (5) Bente and Nielsen 1982.

Skinner 1979). The methodologies and systems can be evaluated critically, in part through comparison with systems evaluated by Ohmoto and Rye (1979).

The experimental data of Hubberten (1980) for sulfur isotope fractionation between galena and sulfur (280 to 700 °C) can be evaluated by comparison with galena-sulfur fractionations derived by the combination of the galena-H$_2$S (50 to 700 °C) and sulfur-H$_2$S (200 to 400 °C) expressions from Ohmoto and Rye (1979). From 280 to 700 °C, the two estimates of galena-sulfur isotope fractionation are identical within analytical uncertainty. No independent comparisons based on experimental results can be made for the argentite, digenite, and covellite data of Hubberten (1980), but the galena-sulfur comparison at least adds confidence to the experimental technique. Nevertheless, the fractionations for argentite, digenite, and covellite are consistent with those predicted following the methods of Sakai (1968) and Bachinski (1969) as summarized by Ohmoto and Rye (1979).

Likewise, no independent comparison of the results of Bente and Nielsen (1982) for bismuthinite-sulfur fractionations can be made, but their results are also consistent with theoretical predictions. Expressions for sulfur isotope fractionation of argentite, digenite, covellite, and bismuthinite with H$_2$S, based on the experimental results of Hubberten (1980) and Bente and Nielsen (1982) combined with the sulfur-H$_2$S fractionations from Ohmoto and Rye (1979) are presented in Table 1.

The experimental results of Suvorova and Tenishev (1976) and Suvorova (1978) for sulfur isotope exchange between molybdenite and a variety of sulfide minerals, and between galena and herzenbergite, are problematic. Derived expressions for fractionation between sphalerite and galena are within 0.4‰ of expressions derived from Ohmoto and Rye (1979). However, fractionations for various sulfide minerals relative to H$_2$S derived on the basis of their results

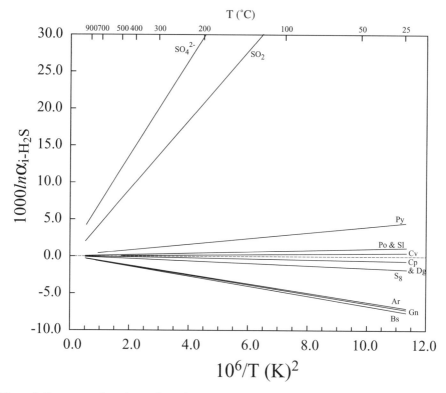

Figure 2. Temperature dependence of experimentally determined equilibrium sulfur isotope fractionation factors relative to H_2S for a variety of sulfur species and sulfide minerals. The dashed line indicates a 0.0‰ $1000ln\alpha$ value. Data from Table 1. Abbreviations: Ar argentite, Bs bismuthinite, Cp chalcopyrite, Cv covellite, Dg digenite, Gn galena, Po pyrrhotite, Py pyrite, Sl sphalerite.

are significantly different from those summarized by Ohmoto and Rye (1979), or those based on theoretical predictions. In fact, the fractionations appear to be the opposite of what would be expected. On the basis of the information provided by Suvorova and Tenishev (1976) and Suvorova (1978), it is unclear whether the discrepancy results from experimental or computational error. Therefore, the results of these studies are not included in Table 1 or Figure 2.

Geothermometry

The temperature-dependence of sulfur isotope fractionation between two phases, typically solids, forms the basis of sulfur isotope geothermometry. Sulfur isotope geothermometry is based on the partitioning of sulfur isotopes between two substances such as sphalerite and galena, or pyrite and barite. Sulfur isotope fractionation between dissolved SO_4^{2-} and H_2S has been used to assess reservoir temperatures in geothermal systems. The use of sulfur isotopes for this type of geothermometry is based on several requirements or assumptions. Firstly, the minerals must have formed contemporaneously and in equilibrium with one another at a single temperature. Secondly, subsequent re-equilibration or alteration of one or both minerals must not have occurred. Thirdly, pure minerals must be separated for isotopic analysis. Fourthly, the temperature dependence of the fractionation factors must be known. In addition, greater precision in the temperature estimate will be achieved from the use of mineral pairs that have the greatest temperature dependence in their fractionations. Kinetic considerations offer both advantages and disadvantages to geothermometry. Rapid kinetics of isotope exchange promotes mineral formation under equilibrium conditions. Unfortunately, rapid exchange kinetics also makes mineral pairs prone to re-equilibration during cooling. In contrast, sluggish kinetics hampers isotopic equilibration between minerals. However, once equilibrated, mineral pairs with sluggish exchange kinetics will tend to record formation conditions without subsequent re-equilibration at lower temperatures.

Because of the relationships expressed in Equations (6) and (7), mineral-mineral fractionation equations can be derived from the equations in Table 1. An equation to calculate the temperature recorded by the coexisting pair of sphalerite (Sl) and galena (Gn) can be derived as follows:

$$1000 \ln\alpha_{Sl-Gn} \approx \Delta_{Sl-Gn} = \delta^{34}S_{Sl} - \delta^{34}S_{Gn} \tag{10}$$

Thus:

$$\Delta_{Sl-Gn} = \Delta_{Sl-H_2S} - \Delta_{Gn-H_2S} = \delta^{34}S_{Sl} - \delta^{34}S_{H_2S} - (\delta^{34}S_{Gn} - \delta^{34}S_{H_2S}) \tag{11}$$

or

$$\Delta_{Sl-Gn} \approx 1000 \ln\alpha_{Sl-H_2S} - 1000 \ln\alpha_{Gn-H_2S} \tag{12}$$

Substituting from Table 1 gives:

$$\Delta_{Sl-Gn} = \left(\frac{0.10 \times 10^6}{T^2}\right) - \left(\frac{-0.63 \times 10^6}{T^2}\right) \tag{13}$$

with T in K, or:

$$\Delta_{Sl-Gn} = \frac{0.73 \times 10^6}{T^2} \tag{14}$$

Solving for T, and converting to °C yields:

$$T(°C) = \sqrt{\frac{0.73 \times 10^6}{\Delta_{Sl-Gn}}} - 273.15 \tag{15}$$

For example, for a sample with $\delta^{34}S_{Sl} = 8.7‰$ and $\delta^{34}S_{Gn} = 6.1‰$, a temperature of 257 °C is calculated using Equation (15). Uncertainties in sulfur isotope temperature estimates generally range between ± 10 and 40 °C (Ohmoto and Rye 1979).

PROCESSES THAT RESULT IN STABLE ISOTOPIC VARIATIONS OF SULFUR

Variations in the stable isotopic composition of natural systems can result from a variety of equilibrium and kinetically controlled processes, which span a continuum. These processes can be further divided into mass-dependent and mass-independent fractionation processes. Mass-dependent fractionation processes are the most common in geochemical systems and cause systematic correlations among the various stable sulfur isotopes (i.e., ^{32}S, ^{33}S, ^{34}S, and ^{36}S) on the basis of their relative mass differences. As the name implies, mass-independent fractionation does not.

Mass-dependent fractionation processes

The most important steps for producing mass-dependent sulfur isotopic variations in sulfide minerals are the geochemical processes that initially produce the sulfide from other sulfur species such as sulfate or sulfite, or transform sulfide to other sulfur species, rather than the actual precipitation of the sulfide mineral from dissolved sulfide. In addition, the low-temperature rates of many of the oxidation and reduction processes are enhanced by bacterial mediation, which can impart distinct isotopic fractionations to these processes. Thus, the complex aqueous geochemistry of sulfur species is a key aspect for understanding the stable isotope geochemistry of sulfate minerals. Ohmoto (1972) developed the principles for application of sulfur isotope systematics to sulfur speciation in hydrothermal ore deposits. Comprehensive reviews of the controls on the sulfur isotope systematics of sulfides in ore deposits have been given by Ohmoto and Rye (1979), Ohmoto (1986), and Ohmoto and Goldhaber (1997).

Significant isotopic variations may be caused by progressive fractionation processes in a setting where the reservoir of sulfur available is finite, especially where the sulfur isotope fractionation factor between the starting and final sulfur species is large. Under these conditions many equilibrium and kinetic processes can be described as Rayleigh distillation processes. Rayleigh processes are described by the equation:

$$R = R_o f^{(\alpha-1)} \tag{16}$$

where R_o is the initial isotopic ratio, R is the isotopic ratio when a fraction (f) of the starting amount remains, and α is the fractionation factor, either equilibrium or kinetic. This equation can be recast in the δ notation for sulfur isotopes as:

$$\delta^{34}S = (\delta^{34}S_o + 1000)f^{(\alpha-1)} - 1000 \tag{17}$$

Rayleigh models accurately describe isotopic variations associated with processes such as the precipitation of minerals from solutions, the precipitation of rain or snow from atmospheric moisture, and the bacterial reduction of seawater sulfate to sulfide, among others. Bacterial reduction of seawater sulfate can be modeled using Equation (17). If $\alpha = 1.0408$ and $\delta^{34}S_o = 21.0‰$, then precipitation of pyrite from H_2S produced from bacterial reduction of sulfate will preferentially remove ^{32}S and the first pyrite formed will have $\delta^{34}S \approx -20‰$. The preferential removal of ^{32}S will cause the $\delta^{34}S$ of the residual aqueous sulfate to increase which, in turn, will lead to an increase in the $\delta^{34}S$ of subsequently formed pyrite (Fig. 3). Under closed-system behavior, after all sulfate has been reduced, the bulk isotopic $\delta^{34}S$ of the precipitated pyrite will equal the $\delta^{34}S$ of the initial sulfate. However, the $\delta^{34}S$ of individual pyrite grains or growth zones can be both lower and higher than the bulk composition, depending on when they formed.

Mixing is another important process that can cause isotopic variations. It can be modeled on the basis of simple mass-balance equations such as:

$$\delta_{mixture} = X_A \delta_A + X_B \delta_B \tag{18}$$

where $\delta_{mixture}$ is the resulting isotopic composition of the mixture, δ_A and δ_B are the isotopic compositions of components A and B, and X_A and X_B are the mole fractions of components A and B.

Kinetics of isotope exchange reactions. The kinetics of isotopic exchange between aqueous sulfate and sulfide at elevated temperatures are important in determining the isotopic composition of sulfide minerals and associated aqueous or solid sulfate. Ohmoto and Lasaga (1982) found that exchange rates between dissolved SO_4^{2-} and H_2S decreased with increasing pH at pH < 3; from pH $\approx$ 4 to 7, the rates remain fairly constant; at pH > 7, the rate also decreases with increasing pH. The reason for these changes in rate as a function of pH is the pH dependence of sulfur speciation. Ohmoto and Lasaga (1982) proposed that the overall rate of exchange is limited by exchange reactions involving intermediate valence thiosulfate species ($S_2O_3^{2-}$), the abundance of which is dependent on pH. The rate-limiting step was postulated to be an intramolecular exchange between non-equivalent sulfur sites in thiosulfate, which has been further investigated by Chu et al. (2004). Ohmoto and Lasaga (1982) calculated the most rapid equilibration rates at high temperature ($T = 350\ °C$) and

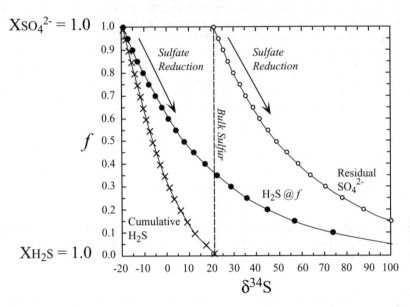

Figure 3. Rayleigh distillation curves for bacterial reduction of seawater sulfate showing the change in $\delta^{34}S$ of resultant H_2S (filled circles), residual sulfate (open circles), and bulk sulfide (X) as a function of reaction progress. Pyrite precipitated from the H_2S would be expected to have a $\delta^{34}S$ that is approximately 4‰ higher than that shown for H_2S assuming equilibrium fractionation between pyrite and H_2S. Modified from Ohmoto and Goldhaber (1997) and Seal et al. (2000a). Isotopic values in permil (VCDT).

low pH (pH ≈ 2) of approximately 4 hours for 90% equilibrium between aqueous sulfate and sulfide; however, at low temperature (T = 25 °C) and near neutral pH (pH = 4-7), the time to attain 90% equilibrium reached 9×10^9 years. Thus, disequilibrium between sulfate and sulfide minerals should be prevalent in hydrothermal and geothermal systems below 350 °C, except under extremely acidic conditions (Fig. 4).

Sulfate reduction. Sulfate reduction in natural systems tends to produce characteristic, kinetically controlled, non-equilibrium sulfur isotope fractionations in both biotic and abiotic environments. Isotopic variations associated with the biogenic reduction of sulfate have been studied by numerous researchers, most of whom have concentrated on the role of dissimilatory sulfate-reducing bacteria such as *Desulfovibrio desulfuricans*. The activity of sulfate-reducing bacteria in marine sediments throughout most of geological time had a profound effect on the sulfur isotope composition of seawater sulfate, which is discussed in a later section.

Sulfate-reducing bacteria are active only in anoxic environments such as below the sediment–water interface, and in anoxic water bodies. Various species of sulfate-reducing bacteria can survive over a range of temperature (0 to 110 °C) and pH (5 to 9.5) conditions, but prefer near-neutral conditions between 20 and 40 °C and can withstand a range of salinities from dilute up to halite saturation (Postgate 1984; Canfield 2001). The metabolism of sulfate-reducing bacteria can be described by the general reaction:

$$2\ CH_2O + SO_4^{2-} \rightarrow H_2S + 2\ HCO_3^- \qquad (19)$$

where CH_2O represents generic organic matter (Berner 1985). The H_2S can be lost to the water column, reoxidized, fixed as iron-sulfide minerals (i.e., pyrite, mackinawite, greigite) or other sulfide minerals if reactive metals are present, or it can be fixed as organic-bound sulfur. In near-surface sediments deposited under normal (oxygenated) marine settings, the

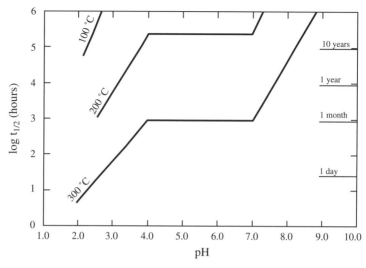

Figure 4. The kinetics of sulfur isotope exchange in terms of pH and log $t_{1/2}$ for a solution with 0.1 m ΣNa and 0.01 m ΣS. The bends in the isotherms are due to changes in the speciation of sulfur as a function of pH. Modified from Ohmoto and Lasaga (1982).

activity of sulfate-reducing bacteria is limited by the supply and reactivity of organic matter; in freshwater and euxinic basins, the activity is limited by sulfate availability (Berner 1985).

The fractionation of sulfur isotopes between sulfate and sulfide during bacterial sulfate reduction is a kinetically controlled process in which ^{34}S is enriched in the sulfate relative to the sulfide, in the same sense as equilibrium fractionation between sulfate and sulfide (Chambers and Trudinger 1979). The sulfate-reducing bacteria more readily metabolize ^{32}S relative to ^{34}S. Thus, the δ^{34}S of the residual aqueous sulfate increases during the reaction progress.

The magnitude of the fractionation has been shown to be a function of the rate of sulfate reduction, which can be related to sedimentation rates. The smaller fractionations correspond to faster rates of sulfate reduction and sedimentation, whereas the larger fractionations correspond to slower rates of sulfate reduction and sedimentation (Goldhaber and Kaplan 1975). In contrast, other compilations fail to show such distinct correlations between isotopic fractionation and sedimentation rate (Canfield and Teske 1996). The fractionation associated with bacterial sulfate reduction (1000ln$\alpha_{SO_4-H_2S}$) typically ranges from 15 to 71‰ (Goldhaber and Kaplan 1975; Canfield and Teske 1996) in marine settings, compared to an equilibrium, abiotic fractionation of approximately 73‰ at 25 °C. However, Canfield and Teske (1996) and Canfield (2001) noted fractionations ranging only between 4 and 46‰ that can be directly attributed to bacterial sulfate reduction. Canfield (2001) and Habicht and Canfield (2001) suggested that the greater amount of fractionation found in nature may result from the near-ubiquitous partial oxidation in marine settings of resultant sulfide, and subsequent isotopic effects associated with disproportionation of intermediate sulfur species.

The isotopic composition of the pyrite resulting from bacterial sulfate reduction depends on how open or closed is the system. In natural settings, evolution of the isotopic system may occur in a closed basin, where the reservoir of sulfate is finite and becoming exhausted, or sulfate availability may be limited due to diagenetic cementation of pore spaces in the sediments which isolates the sulfate undergoing reduction from replenishment. Open systems requires an unlimited source of sulfate and the ability to transport rapidly the sulfate below the sediment-water interface to the site of sulfate reduction.

Closed-system and open-system behavior, and the gradations between these two extremes, will produce distinctive frequency distributions in the resultant $\delta^{34}S$ values of the pyrite. Seal and Wandless (2003) modeled the spectrum of distribution patterns using a combination of Rayleigh and mixing equations in the context of seawater-sulfate reduction during Ordovician time (Fig. 5a). End-member open system behavior produces a sharp mode that corresponds to the $1000\ln\alpha_{SO_4-H_2S}$ value—in their example, 55‰, which was based on the difference between the inferred seawater composition (28‰) and the lowest $\delta^{34}S$ value from pyrites in sedimentary rock near the Bald Mountain massive sulfide deposit in northern Maine (−27‰; Fig. 5b). In contrast, end-member closed system behavior does not produce a mode; instead, a flat distribution results from the continued depletion of the sulfate reservoir (Figs. 3 and 5). When diffusive transport of sulfate is just half the rate of reduction, distributions lacking a mode result. When the rate of diffusive transport and reduction are equal, a skewed distribution with a distinct mode is produced. As advective transport exceeds the rate of reduction and diffusive transport, similar skewed distributions are produced with decreasing ranges of values until end-member open-system conditions are reached.

Abiotic (thermochemical) reduction of aqueous sulfate through high-temperature (200 to 350 °C) interactions with Fe^{2+} (fayalite and magnetite) can be modeled as an equilibrium Rayleigh process (Shanks et al. 1981). The $\delta^{34}S$ of residual aqueous sulfate increases during reduction in accordance with published equilibrium fractionation factors (Ohmoto and Lasaga 1982; Table 1). Similarly, Sakai et al. (1980) found that sulfur isotope fractionations associated with thermochemical reduction of dissolved sulfate through reaction with olivine ($X_{Fo} = 0.90$) at 400 °C produces results consistent with equilibrium exchange between sulfate and sulfide. Orr (1974) and Kiyosu (1980) documented sulfur isotopic effects associated with thermochemical reduction of sulfate because of the interaction with organic matter, and found that sulfate-sulfide kinetic fractionation was less than 10‰.

Sulfide oxidation. The oxidative weathering of sulfide minerals to form sulfate minerals or aqueous sulfate is a quantitative, unidirectional process that produces negligible sulfur-isotope fractionation. The $\delta^{34}S$ of resulting sulfate is indistinguishable from that of the parent sulfide mineral; likewise, the isotopic composition of residual sulfide minerals is unaffected (Gavelin et al. 1960; Nakai and Jensen 1964; Field 1966; Rye et al. 1992). Gavelin et al. (1960) and Field (1966) documented similar sulfur isotope compositions among hypogene sulfide ore minerals and various associated supergene sulfate minerals. A similar conclusion was reached regarding the relationship of aqueous sulfate with sulfide minerals in acid mine drainage settings. Taylor and Wheeler (1994) and Seal (2003) found no discernible difference between $\delta^{34}S$ in the parent sulfides and the associated dissolved sulfate. In contrast, the oxygen

Figure 5 (*on facing page*). Hypothetical sulfur isotope composition of sedimentary pyrites formed under different rates of sulfate transport relative to sulfate reduction. Modified from Seal and Wandless (2003). (a) Sulfur isotope evolution of sedimentary pyrite related to bacterial sulfate reduction and diffusive and advective transport into pore spaces below the sediment-water interface for conditions approximating the inferred seafloor environment of Bald Mountain (Maine). Calculations were made at reduction steps of 0.282 mmol of SO_4. Calculations for no diffusion of sulfate into the system are identical to closed-system Rayleigh behavior. For calculations with the rate of diffusion less than the rate of reduction, the sulfate supply will become exhausted, resulting in the most positive $\delta^{34}S$ values for pyrite near the last reduction steps. Calculations were terminated after 67.4 mmol of SO_4 was reduced, to reflect the amount of organic carbon available for reduction in typical marine sediments. Note that only in cases where the rate of reduction is faster than the rate of diffusion is the $\delta^{34}S$ value for pyrite found to be higher than for coeval seawater. For conditions where the rate of advective transport is greater than the rate of reduction, the isotopic evolution is modeled as mixtures of residual sulfate and pristine seawater sulfate (curves labeled $X_{SW} = 0.0$ to 1.0). (b) Hypothetical histograms for the sulfur isotope composition of sedimentary pyrites for various rates of sulfate reduction and sulfate transport by diffusion and advection, compared to the isotopic composition of pyrites from the graphitic argillite found at the Bald Mountain deposit, Maine (Seal and Wandless 2003). All isotopic values are given in permil (VCDT).

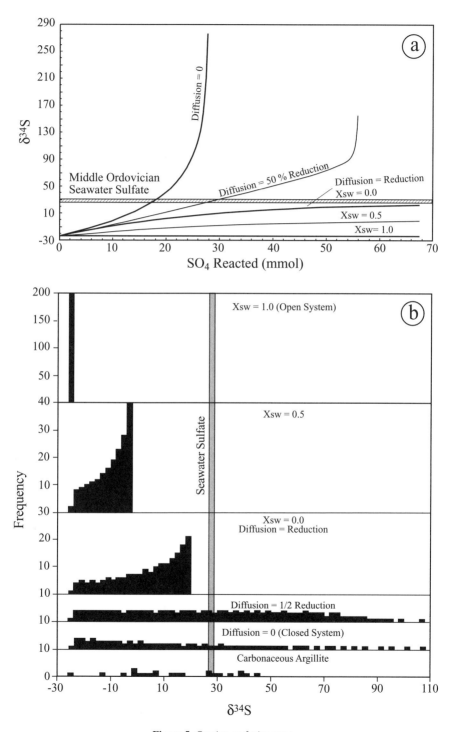

Figure 5. *Caption on facing page.*

isotope composition of sulfate derived from the oxidative weathering of sulfide minerals shows significant variations depending upon the oxygen isotopic composition and pH of the associated water, and the oxidizing agent (i.e., oxygen or ferric iron), among other factors (Taylor and Wheeler 1994; Seal 2003).

Mechanisms of sulfide precipitation. Precipitation mechanisms for sulfide minerals and their associated environment can have significant affects on the sulfur isotopic fractionation between minerals, as discussed by Ohmoto and Goldhaber (1997). For simple sulfides such as ZnS, PbS and $Fe_{1-x}S$, the relative proportions of metal and H_2S are important. These minerals can be precipitated through simple cooling, dilution to destabilize chloride complexes, or acid neutralization. Under conditions where the molalities of dissolved metals exceed that of H_2S, as is commonly found during precipitation of monometallic ores, disequilibrium fractionations are expected where the observed fractionation is less than that expected under equilibrium conditions. This discordance is due to the fact that sulfur needs to be obtained at the site of sulfide deposition. The mineral that reaches saturation first will consume a significant portion of the H_2S reservoir causing a shift through Rayleigh processes in the isotopic composition the residual H_2S available for later minerals. Sulfide minerals from polymetallic ores formed from fluids where the concentration of H_2S greatly exceeds those of the metals, generally show equilibrium fractionation between simple sulfides because of their precipitation resulted from being mutually saturated rather than from one metal becoming exhausted in the fluid so that next metal can reach saturation.

The precipitation of pyrite and chalcopyrite is more complex because it typically requires an oxidation step in addition to other depositional mechanisms (Ohmoto and Goldhaber 1997). For example, pyrite has disulfide (S_2^{2-}) rather than sulfide (S^{2-}) anion units in its structure. Likewise, chalcopyrite precipitation can commonly involve oxidation-reductions reactions of iron and copper. Replacement is another important mechanism for the formation of chalcopyrite where the isotopic composition may be inherited partly, or wholly, from the precursor mineral. Thus, because of the greater complexity of precipitation mechanisms for pyrite and chalcopyrite, equilibrium isotopic fractionations with other sulfide minerals are less likely.

Disproportionation of SO_2. Sulfur dioxide (SO_2) is the most important oxidized sulfur species in high fO_2 magmatic-hydrothermal systems; H_2S is the dominant reduced sulfur species. Upon cooling below ~400 °C, the SO_2 undergoes hydrolysis or disproportionation described by the reaction:

$$4\ H_2O + 4\ SO_2 \rightarrow H_2S + 3\ H^+ + 3\ HSO_4^- \quad (20)$$

producing H_2S and SO_4^{2-} (Holland 1965; Burnham and Ohmoto 1980). The isotopic effects associated with the disproportionation of SO_2 will be discussed below in the sections of porphyry and epithermal mineral deposits.

Mass-independent fractionation processes

Non mass-dependent fractionation or "mass-independent" fractionation refers to processes that cause variations in the abundances of isotopes that are independent of their masses. In mass-dependent fractionation, variations in $^{34}S/^{32}S$ should be approximately twice those of $^{33}S/^{32}S$, and approximately half those of $^{36}S/^{32}S$ (Hulston and Thode 1965a); likewise, the fractionation of $^{17}O/^{16}O$ should be approximately half that of $^{18}O/^{16}O$ (Bigeleisen and Mayer 1947; Urey 1947). On geochemical plots of one isotopic ratio versus another, for example $\delta^{33}S$ versus $\delta^{34}S$, or $\delta^{17}O$ versus $\delta^{18}O$, samples that have experienced mass-dependent fractionation processes would fall along a line known as a *mass-fractionation line*, which has a slope corresponding to the relative mass difference between the two ratios. The Earth-Moon system has characteristic mass-fractionation lines for sulfur and oxygen isotopes because all of the isotopes are well mixed in these bodies. Deviations from these lines, or "isotope anomalies" reflect mass-independent fractionation processes. For sulfur isotopes, these deviations are expressed as non-zero $\Delta^{33}S$ and $\Delta^{36}S$ values, which are defined respectively as:

$$\Delta^{33}S = \delta^{33}S - 1000\left[\left(1 - \frac{\delta^{34}S}{1000}\right)^{0.515} - 1\right] \quad (21)$$

$$\Delta^{36}S = \delta^{36}S - 1000\left[\left(1 - \frac{\delta^{34}S}{1000}\right)^{1.91} - 1\right] \quad (22)$$

where 0.515 and 1.91 are the approximate slopes on the respective δ-δ diagrams, and represent deviations from the terrestrial fractionation line.

Some of the earliest identified isotopic anomalies were found in the oxygen isotope compositions of meteorites, which can be interpreted to reflect heterogeneity in the early history of the solar system (Clayton 1986). In fact, most meteorite types fall off the terrestrial oxygen isotope mass-fractionation line. Sulfur isotopic anomalies in meteorites, discussed below, are less impressive.

Photochemical processes in the upper atmosphere have been found to cause mass-independent fractionations in sulfur and oxygen isotopes (Farquhar and Wing 2003; Rumble 2005). Sulfur isotope anomalies, presumably derived from upper atmospheric ultraviolet radiation-induced photochemical processes, have been identified in pyrite, pyrrhotite, chalcopyrite, and galena, in addition to sulfate minerals, in the Archean geologic record (Farquhar et al. 2000a; Hu et al. 2003; Mojzsis et al. 2003; Ono et al. 2003; Bekker et al. 2004). Prior to 2.4 Ga, sulfide and sulfate values from a variety of geologic settings are variably anomalous, with $\Delta^{33}S$ values for sulfides and sulfates ranging from −2.5 to 8.1‰ (Farquhar et al. 2000a; Ono et al. 2003; Rumble 2005). Since 2.4 Ga, samples have a much more limited range of $\Delta^{33}S$ from −0.5 to 0.7‰ (Fig. 6; Savarino et al. 2003; Bekker et al. 2004). The abrupt change in the magnitude of the anomalous mass-independent fractionations around 2.4 Ga has been interpreted as reflecting the development of an oxygenated atmosphere. The increase in the partial pressure of oxygen would have been conducive to the development of an ozone layer, which would have shielded lower parts of the atmosphere from photochemical processes induced by ultraviolet radiation; in addition, the associated lower abundances of reactive, more reduced sulfur species in the atmosphere may have also contributed to the abrupt end of mass-independent anomalies (Farquhar et al. 2000a; Pavlov and Kasting 2002; Farquhar and Wing 2003; Bekker et al. 2004).

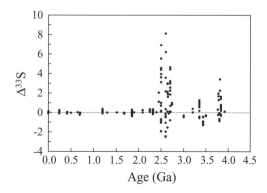

Figure 6. Age distribution of sulfur isotope anomalies in sedimentary rocks. Modified from Rumble (2005), and includes data from Bekker et al. (2004), Farquhar et al. (2000a), Hu et al. (2003), Mojzsis et al. (2003), and Ono et al. (2003).

GEOCHEMICAL ENVIRONMENTS

The scientific literature concerning sulfur isotopes from sulfide minerals is voluminous, and that pertaining to mineral deposits is overwhelming. Therefore, the following sections on specific geochemical environments will focus on examples to illustrate important aspects of sulfur isotope geochemistry from high- and low-temperature settings. An attempt was made to balance coverage of pioneering studies with that of emerging ideas and applications of sulfur isotopes from sulfide minerals.

Meteorites

Sulfur isotope data have provided a variety of insights into the origins of the Earth and the solar system as recorded by meteorites. Sulfur isotope compositions have been determined for a variety of sulfide minerals in meteorites, including troilite, pyrrhotite, pyrite, chalcopyrite, pentlandite, oldhamite (CaS), alabandite (MnS), and daubreelite (FeCr$_2$S$_4$) among others, in addition to sulfate, sulfur in solid solution in native iron, and a variety of species that are extractable with various solvents. The earliest researchers investigating the sulfur isotope composition of meteorites found remarkably constant and fairly homogeneous compositions among all types of meteorites, with the δ^{34}S of most falling between −2.5 and 2.5‰ (Fig. 7; Macnamara and Thode 1950; Hulston and Thode 1965a; Monster et al. 1965; Kaplan and Hulston 1966) which is in stark contrast to the oxygen isotope compositions of meteorites, which varies widely (Clayton 1986). In fact, it was this restricted compositional range, particularly for troilite from iron meteorites, that led to the designation of Canyon Diablo troilite as the basis for the sulfur isotope scale. The isotopic composition of meteoritic sulfur is also used as a reference point for the bulk earth from which to evaluate global scale fractionations in the sulfur cycle.

Local evidence has been found for secondary alteration of the sulfur isotopic composition in a limited number of meteorites that reflects aqueous processes occurring on the parent bodies from which the meteorites came. In the SNC type meteorites, which likely originated on Mars, Shearer et al. (1996) identified vug-filling pyrite with unusually high δ^{34}S between 4.8 and 7.8‰, probably reflecting Martian hydrothermal alteration. Greenwood et al. (2000a)

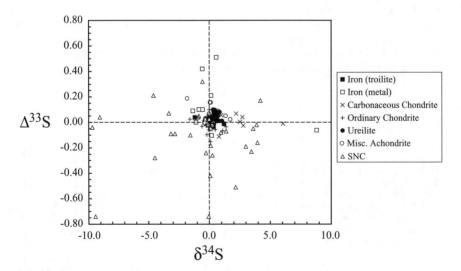

Figure 7. Sulfur isotope composition of meteorites plotted in terms of Δ^{33}S and δ^{34}S. See text for sources of data. Dashed lines indicate 0‰. Isotopic values are given in permil (VCDT).

also found $\delta^{34}S$ values in Martian meteorites ranging from −6.1 to 4.9‰, that were consistent with hydrothermal alteration, in pyrrhotite, pyrite, and chalcopyrite, some of which was vein filling. McSween et al. (1997) found texturally unique pyrrhotite and pentlandite in a carbonaceous chondrite having $\delta^{34}S$ values between −4.2 and 1.1‰, and −5.7 and −3.0‰, respectively, which were interpreted to be the result of alteration on the asteroid on which the chondrites originated. In another carbonaceous chondrite, Monster et al. (1965) found sulfur isotopic compositions for sulfide ($\delta^{34}S = 2.6$‰), native sulfur ($\delta^{34}S = 1.5$‰) and sulfate ($\delta^{34}S = -1.3$‰) that clearly reflect disequilibrium conditions.

Mass-independent sulfur isotopic anomalies have also been identified in meteorites. Such anomalies can be generated by mixing of sulfur from different nucleosynthetic reservoirs (Clayton and Ramadurai 1977), cosmic-ray induced reactions involving iron (Hulston and Thode 1965a; Gao and Thiemens 1991), and photochemical and other chemical reactions (Farquhar et al. 2000b). Evidence for mixing of different nucleosynthetic reservoirs has been elusive in sulfur isotopes. Rees and Thode (1977) found a 1‰ ^{33}S anomaly in the Allende carbonaceous chondrite, but subsequent researchers analyzing Allende were unable to find additional evidence (Gao and Thiemens 1993a). Enstatite chondrites and ordinary chondrites, which come from other primitive asteroids, generally lack sulfur isotopic anomalies (Gao and Thiemens 1993b). The most compelling evidence for nebular sulfur heterogeneity is the small, but distinguishable ^{33}S anomalies ($\Delta^{33}S = 0.042$‰) found in ureilites, a type of achondrite associated with carbonaceous chondrites (Farquhar et al. 2000c). However, Rai et al. (2005) attributed mass-independent anomalies in other achrondritic meteorites to photochemical processes in the early solar nebula. The sulfur dissolved in the metallic phase of iron meteorites, which are the cores of differentiated asteroids, can have $\Delta^{33}S$ and $\Delta^{36}S$ values up to 2.7 and 21.5‰, respectively, that are consistent with cosmic-ray induced spallation reactions, and have a characteristic $\Delta^{36}S/\Delta^{33}S$ ratio of ~8 (Gao and Thiemens 1991).

The largest mass-independent anomalies, not from metallic or organic phases in meteorites, were found in Martian (SNC) meteorites where $\Delta^{33}S$ ranges from −0.302 to 0.071‰ (Fig. 7); $\Delta^{36}S$ values range from 0.0 to 2.6‰ (Farquhar et al. 2000b). Farquhar et al. (2000b) attributed these anomalies to UV-induced photochemical reactions in the Martian atmosphere. Greenwood et al. (2000b) suggested that an additional component of the ^{33}S anomaly may have resulted from inherited nebular material in the Martian regolith that, unlike on Earth, was not homogenized into the bulk planet due to the lack of tectonic processes on Mars.

Marine sediments

The modern oceanic sulfur cycle reflects the mass balance among inputs from the erosion of sulfides and sulfates on the continents, removal through the formation of diagenetic sulfide minerals, and evaporative precipitation of sulfate locally along the margins of the oceans (Claypool et al. 1980). Volcanic outgassing and subduction of sedimentary rocks also play significant roles in the addition and subtraction, respectively, of sulfur relative to the oceanic reservoir, particularly prior to the presence of an oxygenated atmosphere (Canfield 2004). The sulfur isotopic composition of sedimentary sulfide minerals has been intimately linked to the sulfur isotopic composition of dissolved sulfate in the oceans, at least since ~2.4 Ga—the inferred onset of significant partial pressures of oxygen in the atmosphere (Farquhar et al. 2000a; Pavlov and Kasting 2002; Mojzsis et al. 2003). The link between the isotopic compositions of dissolved sulfate and sedimentary sulfides is caused by the activity of sulfate-reducing bacteria as discussed above.

Modern seawater sulfate has a globally homogenous $\delta^{34}S$ of 21.0 ± 0.2‰ (Rees et al. 1978). In contrast, the $\delta^{34}S$ of modern sedimentary sulfide, mostly pyrite, is quite variable, depending upon setting and ranging between −50 and 20‰, although most values are negative (Fig. 8; Chambers 1982; Sælen et al. 1993; Strauss 1997). This range includes anoxic settings

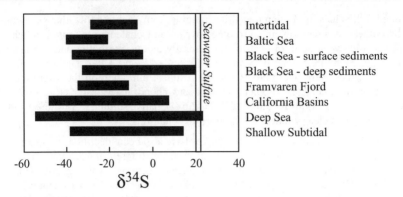

Figure 8. Range of sulfur isotopic values of sedimentary sulfides from a variety of settings compared to seawater sulfate (modified from Strauss 1997). Isotopic values are given in permil (VCDT).

such as the Black Sea and the Framvaren fjord (Norway), and oxic settings ranging from the deep sea to shallow subtidal and intertidal settings (Chambers 1982; Strauss 1997).

Marine sulfate as preserved in evaporite deposits and disseminated in marine sediments provides a robust record of past variations in the isotopic composition of sulfate in the oceans. It has pronounced secular trends in both $\delta^{34}S$ and $\delta^{18}O$, which can be interpreted in terms of these processes (Holser et al. 1979; Claypool et al. 1980). The secular variations in the isotopic compositions of sedimentary sulfide minerals is less distinctive because Rayleigh fractionation of sulfur isotopes during diagenetic bacterial sulfate reduction typically causes a wide range of largely negative $\delta^{34}S$ values within a given sedimentary unit (Hayes et al. 1992; Strauss 1997; Canfield 2004). Numerous studies have investigated or reviewed the sulfur and oxygen isotope systematics of modern and ancient marine sulfate and sulfide (Holser et al. 1979; Claypool et al. 1980; Hayes et al. 1992; Strauss 1997; Seal et al. 2000a; Canfield 2004).

Ancient seawater sulfate had mean $\delta^{34}S$ values that varied from around 4‰ at 3.4 Ga, to a high of 33‰ during Cambrian time, to a Phanerozoic low of about 10‰ during Permian and Triassic time, and ultimately to a modern value around 21‰ (Fig. 9). The mean $\delta^{18}O$ of marine evaporitic sulfate has varied from around 17‰ at 900 Ma to a Phanerozoic low of 10‰ during Permian time, to a modern value of 13‰. The $\delta^{18}O$ of modern seawater sulfate is also homogeneous throughout the oceans with a value of 9.5‰ (Longinelli and Craig 1967; Nriagu et al. 1991). The limited range of sulfur and oxygen isotope compositions for any given time in geologic history results from the rapid mixing time of seawater (~1,000 years) relative to the residence time of sulfate in seawater (8×10^6 years; Holland 1978).

One of the most important milestones in the sulfur cycle of the Earth was the development of an oxic atmosphere around 2.4 Ga (Farquhar et al. 2000a; Pavlov and Kasting 2002; Mojzsis et al. 2003; Bekker et al. 2004). Prior to that time, both sedimentary sulfides and sulfates had limited ranges in $\delta^{34}S$ and clustered near 0‰ (Fig. 9), the inferred value of the bulk Earth because, in the absence of significant concentrations of oxygen in the atmosphere and, therefore, sulfate in seawater, there were few mechanisms available to fractionate sulfur isotopes. As atmospheric concentrations of oxygen increased, the $\delta^{34}S$ of seawater sulfate increased and that of sedimentary sulfides decreased (Fig. 9). Coincident with the divergence of sulfate and sulfide $\delta^{34}S$ values was the abrupt disappearance of mass-independent sulfur isotope anomalies in sedimentary sulfides and sulfates (Fig. 6), presumably because of the development of an ozone layer which had the effect of shielding the atmosphere from UV-induced photochemical reactions known to cause mass-independent anomalies (Farquhar et al. 2000a; Bekker et al. 2004).

Another abrupt change in the sulfur isotope compositions of sedimentary sulfides and seawater sulfate occurred around 0.7 Ga when sedimentary sulfide $\delta^{34}S$ values became more negative and those for seawater sulfate became more variable (Fig. 9). Canfield and Teske (1996) and Canfield (2004) interpreted this change as reflecting a greater level of oxygenation of the oceanic water column, even though episodes of deep water anoxia occurred periodically throughout the Phanerozoic (Leggett 1980).

Coal

The sulfur geochemistry of coal, including its stable isotopes, has been an important research topic because of its use in evaluating coal quality and environmental concerns such as acid rain and acid mine drainage. Sulfur isotope studies have been published for sulfide minerals in coals from the USA (Price and Shieh 1979; Westgate

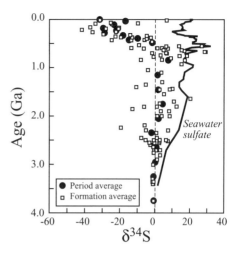

Figure 9. Secular variations of the $\delta^{34}S$ of sulfide in marine sedimentary rocks and in seawater sulfate (modified from Canfield 2004). Heavy line shows the composition of seawater sulfate. All isotopic values are given in permil (VCDT).

and Anderson 1982; Hackley and Anderson 1986; Whelan et al. 1988; Lyons et al. 1989; Spiker et al. 1994), Australia (Smith and Batts 1974; Smith et al. 1982), China (Dai et al. 2002), Japan (Shimoyama et al. 1990) and the Czech Republic (Bouška and Pešek 1999) among others. Isotopic data are available for a variety of sulfur species including pyrite, marcasite, sphalerite, sulfate (mostly secondary), elemental sulfur, and organic sulfur. Collectively, the coals exhibit a wide range of $\delta^{34}S$ values for pyrite (and marcasite), sphalerite, and organic sulfur ranging from −52.6 to 43.1‰, −14.6 to 18.7‰, and −18.7 to 30.6‰, respectively (Fig. 10).

Sulfur in coal is generally interpreted as coming either from sulfur in source plant material or from the bacterial reduction of aqueous sulfate. Sulfides derived from bacterial sulfate reduction may form near the time of original deposition of the peat, during diagenesis, or during coalification (Spiker et al. 1994). Most of the sulfur in low-sulfur coals generally is organically bound sulfur, which has been interpreted to be derived from the original plant material (Price and Shieh 1979; Hackley and Anderson 1986). The isotopic composition of primary plant sulfur should be similar to the isotopic composition of its source—dissolved sulfate—because the assimilation of sulfur by plants during growth only results in minimal fractionation of sulfur isotopes (Chambers and Trudinger 1979). Most modern oxygenated fresh waters have $\delta^{34}S$ values of dissolved sulfate ranging between 0 and 10‰ (Nriagu et al. 1991), which may be analogous to the source waters. Price and Shieh (1979) and Hackley and Anderson (1986) found $\delta^{34}S$ values for organic sulfur from low-sulfur coal were generally between 0 and 10‰, with no correlation with the $\delta^{34}S$ of associated pyrite. However, Price and Shieh (1986) noted a strong correlation between the $\delta^{34}S$ for pyrite and organic sulfur, which they interpreted to reflect the post-depositional mineralization of organic matter associated with the activity of sulfate-reducing bacteria. This correlation between the $\delta^{34}S$ of pyrite and organic sulfur (OS) in coal and oil shale can be described by the equation:

$$\delta^{34}S_{Py} = 1.16\, \delta^{34}S_{OS} - 4.8 \tag{23}$$

which implies that the pyrite-organic sulfur isotope fractionation ($\Delta_{Py\text{-}OS}$) is −4.8‰ (Fig. 11). The data compiled in Figure 11 indicate that this equation provides a reasonable description

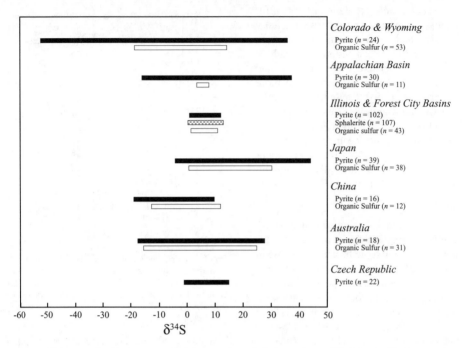

Figure 10. Range of sulfur isotope values from pyrite, sphalerite, and organicly bound sulfur from coals throughout the world. See text for sources of data. Isotopic values are given in permil (VCDT).

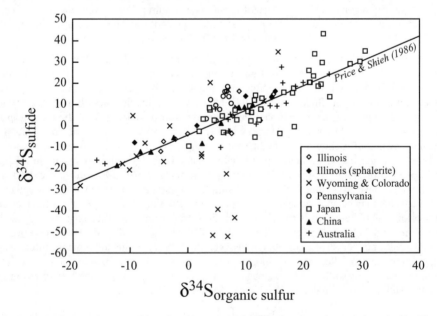

Figure 11. Sulfur isotopic composition of coal in terms of the $\delta^{34}S$ of pyrite and organic bound sulfur. The line describing the covariation of $\delta^{34}S_{Pyrite}$ and $\delta^{34}S_{Orgainc\ S}$ is from Price and Shieh (1986). Isotopic values are given in permil (VCDT).

of sulfide and organic-sulfur sulfur isotopes from coal beds around the world. Shimoyama et al. (1990) identified correlations with slopes ranging between 1.38 and 1.44, which require some additional process beyond Rayleigh fractionation during bacterial sulfate reduction to explain their data from Japanese coals. Mixing between primary sulfur from the original plant material, and sulfide derived from bacterial sulfate reduction, is a possible explanation.

Sulfur isotopes from sulfides are also useful for fingerprinting the incursion of seawater into coal-forming systems. Two general isotopic profiles have been identified, as described by Smith and Batts (1974). In the first case, where the rate of bacterial sulfate reduction is greater than the rate of downward sulfate supply, as documented in the Pelton coal seam (Australia), the total mass of pyrite (~0.8 wt%) and its $\delta^{34}S$ (~25‰) are high near the top of the coal bed, but decrease rapidly with depth (<0.1 wt% and ~ −3‰, respectively). This pattern was interpreted to reflect rapid, quantitative reduction of isotopically heavy seawater sulfate at the top of the section, giving way to plant-derived sulfur at depth. The second case, where the rate of downward supply is greater than that for reduction, the pyrite at top has a lower $\delta^{34}S$ value due to the kinetic fractionation between sulfate and sulfide during bacterial sulfate reduction, but increases progressively with depth due to Rayleigh processes. Smith and Batts (1974) observed this pattern in the Garrick seam (Australia). Lyons et al. (1989) observed a similar pattern in a more detailed data set from the Lower Bakerstown coal bed (Maryland, U.S.A.) (Fig. 12). Whelan et al. (1988) noted a similarity between the isotopic compositions of sphalerite in coal beds in the northern part of the Forest City Basin and those in the nearby Upper Mississippi Valley Zn-Pb deposits, and suggested that some of the sulfur in the coal beds may have been derived from basinal brines.

Mantle and igneous rocks

Insights into the sulfur isotopic composition of the mantle can be obtained from mantle xenoliths, diamonds, and primitive igneous rocks, presumably derived from the mantle.

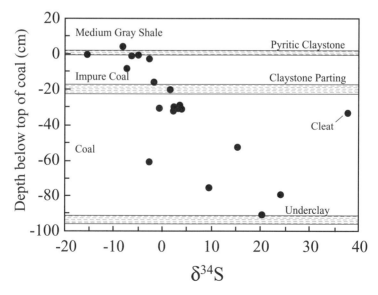

Figure 12. Sulfur isotope composition of pyrite from a cross section through the Bakerstown coal bed, Maryland, USA. (modified from Lyons et al.1989). The section is interpreted to represent the downward incursion of seawater and the associated bacterial sulfate reduction. Compare profile with sulfide curve in Figure 3 at various fractions of reduction (H_2S @ f). Isotopic values are given in permil (VCDT).

Sulfur isotope studies of igneous rocks unrelated to sulfide-bearing mineral deposits are limited, but shed light on the processes of partial melting and assimilation of country rocks. Sulfide minerals that have been analyzed from mantle and other igneous settings unrelated to hydrothermal activity include pyrrhotite, pyrite, pentlandite, chalcopyrite, monosulfide solid solution, and intermediate solid solution.

The sulfur isotopic composition of the mantle has traditionally been considered to be 0 ± 2‰, similar to meteoritic compositions (Thode et al. 1961), but evidence suggests that the sulfur isotope composition is heterogeneous. The $\delta^{34}S$ values of sulfide inclusions in mantle xenoliths ($\delta^{34}S$ = 1.3 ± 3.8‰), sulfide in mid-ocean ridge (MORB; $\delta^{34}S$ = −0.3 ± 2.3‰) and oceanic island basalts (OIB; $\delta^{34}S$ = 1.0 ± 1.9‰), both of which are thought to represent mantle melts, and related gabbros are quite variable (Sakai et al. 1984; Chaussidon et al. 1989; Torssander 1989) but cluster around 0‰ (Fig. 13). For the basalt, some of the variability in sulfide $\delta^{34}S$ values can be attributed to isotopic exchange between sulfide and sulfate and variable sulfide:sulfate ratios in the magmas (Sakai et al. 1984).

The isotopic composition of sulfide inclusions in diamonds are remarkably variable having an average composition of $\delta^{34}S$ = 1.2 ± 5.6‰, and a range from −11 to 14‰ (Fig. 13; Chaussidon et al. 1987; Eldridge et al. 1991; Farquhar et al. 2002). Peridotitic diamonds, generally considered to have a strictly mantle provenance, typically have $\delta^{13}C$ values around −7‰ and sulfide inclusions with $\delta^{34}S$ values clustering between −5 and 5‰ (Eldridge et al. 1995). In contrast, eclogitic diamonds, which have been interpreted to reflect the subduction of biogenic carbon and sulfur into the mantle (Eldridge et al. 1995), have $\delta^{13}C$ values reaching below −30‰ and $\delta^{34}S$ values of sulfide inclusion spanning the entire range observed in diamonds (−11 to 14‰). This interpretation is further supported by Farquhar et al. (2002) who found mass-independent anomalies in the sulfur isotope composition of sulfide inclusions hosted by eclogitic diamonds that suggest that sulfur involved in Archean atmospheric processes has been transferred to the mantle. Thus, the wide range of $\delta^{34}S$ and $\delta^{13}C$ values associated with eclogitic diamonds attests to the inefficiency of mixing processes within the mantle.

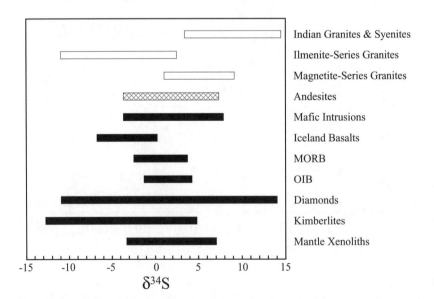

Figure 13. Range of $\delta^{34}S$ values for sulfides from mantle and oceanic and continental igneous settings. See text for sources of data. Isotopic values are given in permil (VCDT).

The sulfur isotope variability found in all igneous rocks, from the most primitive to the most evolved, appears to reflect the global sulfur cycle as moderated by oxidation reactions because of the development of an oxygenated atmosphere, and reduction reactions because of bacterial activity. The sulfur isotope compositions of continental and island arc basalts and gabbros ($\delta^{34}S = 1.0 \pm 3.2‰$) are virtually indistinguishable from those from MORB and OIB (Fig. 13; Ueda and Sakai 1984; Chaussidon et al. 1987). In contrast, andesites have slightly higher $\delta^{34}S$ values (2.6 ± 2.3‰; Rye et al. 1984; Luhr and Logan 2002). Granitoid rocks have an average $\delta^{34}S$ value of 1.0 ± 6.1‰, but range from −11 to 14.5‰, which presumably reflects variable assimilation or partial melting of either pyritic sedimentary rocks with low $\delta^{34}S$ values, or evaporites with high $\delta^{34}S$ values (Sasaki and Ishihara 1979; Ishihara and Sasaki 1989; Santosh and Masuda 1991). Ishihara and Sasaki (1989) found that ilmenite-series granitoids, generally regarded as having formed through partial melting of dominantly sedimentary protoliths, had sulfide $\delta^{34}S$ values less than 0‰, whereas magnetite-series granitoids thought to originate from dominantly igneous protoliths had $\delta^{34}S$ values greater than 0‰ (Fig. 13).

Magmatic sulfide deposits

Magmatic sulfide deposits are generally regarded to be those deposits that form as the result of magmatic crystallization processes, typically prior to saturation with respect to an aqueous phase. This summary focuses on magmatic Ni-Cu and related deposits associated with mafic magmas, which generally formed as immiscible sulfide liquids during the crystallization of a mafic melt. These deposits are important resources for Ni, Cu, and platinum-group elements (PGE). Magmatic sulfide ore-forming systems can be divided into sulfur-poor and sulfur-rich end members where the sulfur-poor systems are the more important sources of PGE and the sulfur-rich systems are the more important sources of Ni and Cu (Ripley and Li 2003). Examples discussed in this section include the Duluth Complex (Minnesota), the Stillwater Complex (Montana), the Bushveld Complex (South Africa), Sudbury (Ontario), Voisey's Bay (Labrador), and Noril'sk (Russia). General aspects of sulfur geochemistry and specific aspects of sulfur isotope geochemistry associated with magmatic sulfide deposits have been reviewed by Ohmoto (1986) and more recently Ripley and Li (2003); their research forms the basis of the following discussion. Ripley and Li (2003) described hypothetical mixing relationships for sulfur isotopes and various metals between mantle-derived magmas and country rocks in the context of magmatic sulfide deposits.

Sulfur isotope data from sulfide minerals are a powerful tool for identifying sulfur contamination of the magma through interactions with the country rocks, if the sulfur isotopic composition of the country rocks was significantly different from the magma. In sulfur-poor systems, such as the Merensky Reef of the Bushveld Complex and the J-M Reef of the Stillwater Complex, in which sulfur requirements were more easily accommodated by solubility of sulfur in mafic magmas, the $\delta^{34}S$ values of sulfide minerals have a limited range and cluster around mantle values (i.e., 0‰; Fig. 14) (Buchanan et al. 1981; Zientek and Ripley 1990; Ripley and Li 2003). In contrast, in sulfur-rich systems, such as the Duluth Complex and Noril'sk, the $\delta^{34}S$ values have a wide range and are significantly positive, suggesting contamination by crustal sources (Fig. 14; Ripley and Al-Jassar 1987; Li et al. 2003; Ripley et al. 2003).

The ability of sulfur isotopes to fingerprint crustal contamination of magmas associated with magmatic sulfide deposits depends upon the isotopic composition of the country rocks. In high-sulfur systems such as Sudbury and Voisey's Bay, the $\delta^{34}S$ values of sulfide minerals have a limited range of near mantle values (Thode et al. 1962; Schwarcz 1973; Ripley et al. 1999; Ripley et al. 2002; Fig. 14). At Sudbury, the $\delta^{34}S$ values of Archean metasedimentary rocks in the footwall of the deposits are near zero (Thode et al. 1962), making sulfur isotope evidence of crustal contamination of the magma equivocal at best. However, Schwarcz (1973) noted small, but discernable differences in the mean $\delta^{34}S$ values of magmatic sulfide deposits near Sudbury. He also documented a general decrease in the $\delta^{34}S$ values of the ore bodies moving

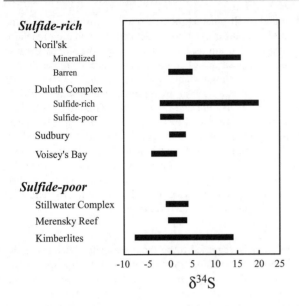

Figure 14. Range of $\delta^{34}S$ values for sulfides from magmatic sulfide deposits (modified from Ripley and Li 2003). Note greater range for sulfide-rich deposits, suggesting crustal contamination of magmas. See text for sources of data. Isotopic values are given in permil (VCDT).

from the country rocks to the intrusion, which supports the idea of crustal contamination of the magma. At Voisey's Bay, Ripley et al. (1999) found that the country rocks had a wide range of $\delta^{34}S$ values, but averaged near zero. Like Schwarcz (1973), in the Sudbury camp they also noted small variations in the isotopic composition of various mineralized zones. With a combination of sulfur, oxygen, and carbon isotope data and Se/S ratios from sulfides, Ripley et al. (1999, 2002) were able to define the role of crustal contamination at Voisey's Bay.

Porphyry and epithermal deposits

Porphyry deposits. Porphyry deposits formed from hydrothermal fluids exsolved from granitic magmas as they cooled with variable involvement of meteoric waters. They typically are large tonnage, low grade deposits. The different classes of porphyry deposits are important sources of Cu, Mo, W, Sn, and Au, as well as other elements. From the perspective of the sulfur isotope composition of sulfide minerals, more significant and interesting isotopic variations can be found in magmatic hydrothermal systems having higher oxidation states, lower pH values, or both, as opposed to near-neutral or more reducing systems. Hydrothermal systems associated with more oxidized magmas, such as I-type granitoids, generally show more sulfur isotopic variations because SO_2 and H_2S are present in the fluids in subequal proportions as opposed to those associated with S-type granitoids which are dominated by H_2S (Burnham and Ohmoto 1980). Crustal contamination of the magmas, as discussed for magmatic sulfide deposits, can also affect the sulfur isotopic composition of magmatic hydrothermal systems. The sulfur isotope geochemistry of magmatic hydrothermal systems have been reviewed by Ohmoto (1986), Rye (1993), Seal et al. (2000a) and, most recently, by Rye (2005).

In high-temperature magmatic hydrothermal settings, such as those for porphyry copper deposits, many of the important processes contributing to sulfur isotope variations of sulfide minerals can be illustrated on diagrams showing the $\delta^{34}S$ of coexisting sulfide and sulfate minerals (Fifarek and Rye 2005; Rye 2005). These diagrams can provide information about the temperature of hydrothermal activity and the SO_4^{2-}/H_2S ratio of the fluid provided that: (1) the SO_4^{2-}/H_2S ratio of the fluid remained constant; (2) the bulk sulfur isotopic composition of the fluid ($\delta^{34}S\Sigma S$) remained constant, and (3) the only cause of isotopic variations in the initial $\delta^{34}S$ of the fluid was exchange between SO_4^{2-} and H_2S in the fluid. Pairs of coexisting sulfate and sulfide minerals should fall along linear arrays having negative slopes ranging from vertical to horizontal, the slope being defined by the SO_4^{2-}/H_2S ratio of the fluid (Fig. 15). The line intersects the line corresponding to zero sulfur isotope fractionation between SO_4^{2-} and H_2S at the bulk isotopic composition of the system ($\delta^{34}S\Sigma S$; Fig. 15). Isotherms plot as lines having positive slopes of unity, the lower temperatures falling down and to the right. When $\delta^{34}S_{sulfide}$ is plotted along the ordinate and $\delta^{34}S_{sulfate}$ plotted along the abscissa, a fluid having H_2S as the only sulfur species would plot as a horizontal line and a fluid with SO_4^{2-} as the only sulfur species would plot as a vertical line. These lines represent limiting conditions for hydrothermal fluids; a line with a slope of −1 would have equal proportions of SO_4^{2-} and H_2S ($SO_4/H_2S = 1$). Natural settings seldom satisfy all of the conditions described above. Interpretation of natural data sets can be complicated by fluctuating fluid compositions, kinetic processes related to isotopic exchange and precipitation, and mixing of multiple sulfur reservoirs, among other processes (Shelton and Rye 1982; Ohmoto 1986; Krouse et al. 1990).

The isotopic characteristics of sulfate and sulfide minerals from porphyry environments suggest a general approach to equilibrium at elevated temperatures. Data from El Salvador, Chile (Field and Gustafson 1976), Gaspé, Quebec (Shelton and Rye 1982), Papua New Guinea (Eastoe 1983), and Butte, Montana (Field et al. 2005) are plotted in Figure 16. In general, the paired data plotted in Figure 16 suggest that equilibrium conditions were approximated by these hydrothermal systems. Collectively, the data suggest that the bulk sulfur isotope composition ($\delta^{34}S\Sigma S$) of porphyry copper hydrothermal systems typically ranges between 1 and 8‰, and that the inferred temperature is between 315 and 730 °C, consistent with

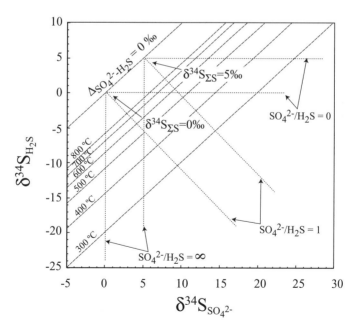

Figure 15. Theoretical aspects of equilibrium fractionation of sulfur isotopes between sulfide and sulfate in hydrothermal systems. Genetically related samples should define linear arrays due to cooling. These arrays should project back to the 0‰ fractionation line at the bulk isotopic composition of the system as represented by the dotted lines. The slope of the dotted lines reflects the SO_4^{2-}/H_2S ratio of the hydrothermal fluid. The range of slopes is limited by a horizontal line indicating all H_2S and no SO_4^{2-}, and a vertical line indicating all SO_4^{2-} and no H_2S; a line with a slope of unity indicates equal proportions of H_2S and SO_4^{2-} (modified from Rye 2005). Isotopic values are given in permil (VCDT).

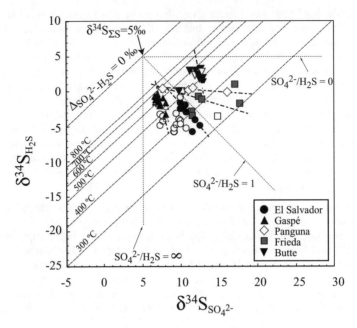

Figure 16. Plot of $\delta^{34}S$ of sulfide versus sulfate for porphyry hydrothermal systems (modified from Rye 2005). The different shapes of symbols are for different deposits. The shading of the symbols is for different sulfide minerals: black and gray, pyrite; white, chalcopyrite. See text for sources of data. Short dashed lines are regressed to the data from individual deposits by Rye (2005). Isotopic values are given in permil (VCDT).

temperature estimates from porphyry copper deposits based on other techniques such as fluid inclusions (Rye 2005). The trends in $\delta^{34}S_{sulfide}$ and $\delta^{34}S_{sulfate}$ in Figure 16 also suggest that porphyry copper hydrothermal systems have a range of oxidation states. Gaspé is one of the more sulfate-rich systems, and Panguna is one of the more sulfide-rich systems.

Many of the data from individual deposits define linear arrays, but some do not. Rye (2005) noted that the linearity of the data varies from mineral to mineral. For example at El Salvador the pyrite-anhydrite pairs define a line having a slope near unity, but the chalcopyrite-anhydrite pairs do not (Fig. 16). This lack of correlation for the chalcopyrite data suggests that chalcopyrite is more prone to retrograde re-equilibration than pyrite, which is consistent with the known reactivities of these two sulfide minerals (Barton and Skinner 1979).

The interpretation of paired sulfide and sulfate data can include additional challenges, Eastoe (1983) questioned the equation of bulk fluid compositions with the composition of magmatic sulfur because of the complexities in the evolution of volatile phases from magmas. In high temperature porphyry environments, Shelton and Rye (1982) suggested that the discrepancies between fluid inclusion temperatures and sulfate–sulfide sulfur isotope temperatures may have resulted from the short time span between the disproportionation of SO_2 to SO_4^{2-} and H_2S, and the subsequent precipitation of sulfate as anhydrite.

Epithermal deposits. Epithermal deposits are hydrothermal mineral deposits that form at shallow levels in the Earth's crust. They form from magmatic water, meteoric water, or mixtures of the two. Epithermal deposits can be divided into two types: acid-sulfate, and adularia-sericite types (Heald et al. 1987). Of these two types, acid-sulfate deposits tend to have more variation in the sulfur isotope composition of sulfide minerals because of the presence of significant quantities of both sulfide and sulfate in the hydrothermal fluids at the time of mineralization. Data for sulfide-sulfate mineral pairs from adularia-sericite type deposits are limited. Accordingly, the following discussion will focus primarily on acid-sulfate deposits. The stable isotope geochemistry of acid-sulfate deposits has been discussed by Rye et al. (1992) and Rye (2005).

Common aspects of the sulfur geochemistry of epithermal deposits can be identified using sulfur isotope data from pairs of sulfide and sulfate minerals. Paired sulfur isotope data from acid-sulfate epithermal deposits are available from Julcani, Peru (Rye et al. 1992), Rodalquilar, Spain (Arribas et al. 1995), Summitville, Colorado (Bethke et al. 2005), Pierina, Peru (Fifarek and Rye 2005), and Tapajós, Brazil (Juliani et al. 2005). In addition, a single pair from the Sunnyside, Colorado adularia-sericite type deposit is available (Casadevall and Ohmoto 1977). Sulfide minerals for which data are available include pyrite, pyrrhotite, and chalcopyrite; sulfate minerals include anhydrite, barite, alunite, and sulfate-bearing apatite. Figure 17 shows the $\delta^{34}S_{sulfide}$ and $\delta^{34}S_{sulfate}$ values for these deposits. Compared to porphyry copper deposits, the mineral pairs from epithermal deposits have a wider range of compositions, which reflects the generally lower temperatures of precipitation. With the exception of the deposits for which data are available from phenocrystic apatite (i.e., Summitville and Julcani), the data tend not to define linear trends. Nevertheless, the range of values is consistent with hydrothermal temperatures determined by other methods such as fluid inclusions, and suggest that the compositions record equilibrium conditions. Thus, the lack of linear trends for data from a given deposit may result from open-system behavior (i.e., boiling), which can alter the bulk composition of the hydrothermal fluid. Acid-sulfate epithermal deposits form at shallow levels in the Earth's crust. Many are thought to be the apical parts of porphyry copper hydrothermal systems (e.g., Lepanto, Philippines; Hedenquist et al. 1998).

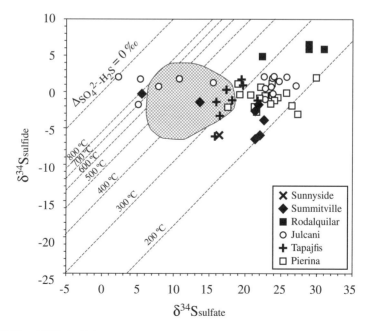

Figure 17. Plot of $\delta^{34}S$ of sulfide versus sulfate for epithermal hydrothermal systems. See text for sources of data. The shaded field encompasses the data from porphyry deposits depicted in Figure 16 for comparison. Note that the data from epithermal deposits imply lower temperatures than those for porphyry deposits, consistent with their inferred genetic relationships to intrusions. Isotopic values are given in permil (VCDT).

Seafloor hydrothermal systems

Modern systems. The stable isotope characteristics of modern seafloor hydrothermal systems from mid-ocean ridges have been summarized recently by Shanks et al. (1995), Herzig et al. (1998), Seal et al. (2000a), and Shanks (2001). Sulfur isotope data are available for a variety of sulfide minerals from seafloor hydrothermal systems including pyrite, marcasite, pyrrhotite, chalcopyrite, bornite, cubanite, sphalerite and wurtzite, in addition to vent fluid H_2S. A summary of the range of $\delta^{34}S$ values from various vent systems is present in Figure 18.

Igneous activity along submarine spreading centers causes hydrothermal convection that instigates a series of geochemical processes defining the sulfur cycle of these settings. In mid-ocean ridge hydrothermal systems, sulfide can be derived from three main sources: (1) leaching from country rocks, both igneous and sedimentary; (2) thermochemical reduction of seawater sulfate due to interaction with ferrous silicates and oxides, or with organic matter; and (3) leaching of sulfide minerals in sediments that were produced by bacterial sulfate reduction. Each of these sources has distinctive sulfur isotope signatures. Mid-ocean ridge basalts have sulfate/sulfide ratios (mass basis) ranging from 0.05 to 0.45 that correlate positively with the water content of the basalts (Sakai et al. 1984). The bulk sulfur isotopic composition averages $0.3 \pm 0.5‰$, and the average $\delta^{34}S$ of the sulfide fraction is $-0.7 \pm 0.8‰$ (Sakai et al. 1984). Sulfate is stripped from seawater by the precipitation of sulfate minerals during heating associated with downwelling because of the retrograde solubility of anhydrite and other sulfate minerals (Bischoff and Seyfried 1978; Seyfried and Bischoff 1981; Shanks et al. 1995). Shanks et al. (1981) and Woodruff and Shanks (1988) proposed that most of the H_2S in vent fluids is derived from monosulfide solid solution. Because pyrite is the main sulfide

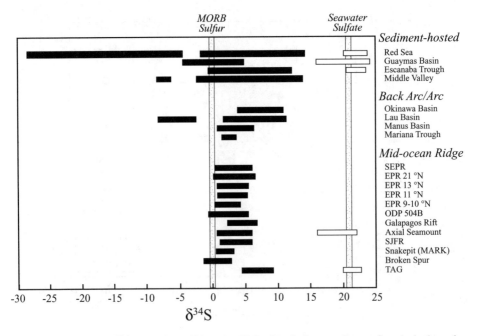

Figure 18. Range of $\delta^{34}S$ values for sulfide and sulfate minerals from modern seafloor hydrothermal systems. Modified from Shanks et al. (1995), Herzig et al. (1998), and Shanks (2001). Note limited variations from largely barren mid-ocean ridge systems and wide variations from sedimented systems where biogenic sulfide may be an important component. Isotopic values are given in permil (VCDT).

mineral in altered oceanic crust, the derivation of H_2S from monosulfide solid solution would require an oxidation step as described by the reaction:

$$8\ FeS + 10\ H^+ + SO_4^{2-} \rightarrow 4\ FeS_2 + H_2S + 4H_2O + 4\ Fe^{2+} \tag{24}$$

which should release H_2S with a $\delta^{34}S$ of 1 to 1.5‰ (Woodruff and Shanks 1988).

The isotopic effects associated with seawater-basalt interactions and associated hydrothermal activity have been modeled by Janecky and Shanks (1988) as two end-member processes: simple adiabatic mixing, and shallow thermochemical reduction. They concluded that simple adiabatic mixing can produce H_2S having a maximum $\delta^{34}S$ of 4.5‰. They also found that thermochemical reduction of seawater sulfate through interactions with ferrous silicates or oxides is more likely to be important at moderate temperature, off-axis settings where the retrograde solubility of sulfate minerals has not removed as much sulfate as in higher temperature settings. Shanks et al. (1981) demonstrated experimentally the effectiveness of sulfate reduction through interactions with olivine and magnetite. According to the model of Janecky and Shanks (1988), thermochemical reduction of modern seawater sulfate through interactions with magnetite can generate H_2S having a $\delta^{34}S$ as high as 15‰. In contrast, hydrogen sulfide derived from the dissolution of biogenic sulfides in sedimentary rocks would be expected to have negative $\delta^{34}S$ values reflecting bacterial sulfate reduction, as described above.

Modern seafloor hydrothermal sulfide minerals and vent fluids reflect the combination of the processes of simple adiabatic mixing, thermochemical reduction, and dissolution of biogenic sulfide minerals. Mid-ocean ridge systems, largely barren of sediments, have $\delta^{34}S$ values that typically range between 0 and 6‰, with the exception of the TAG field along the Mid-Atlantic Ridge (Fig. 18). Compared to the other examples, TAG is a slow-spreading center, which includes a greater component of shallow seawater entrainment, sub-seafloor hydrothermal mineral precipitation and basalt alteration, compared to fast-spreading centers. These processes are more conducive to thermochemical reduction of seawater sulfate, which imparts higher $\delta^{34}S$ values to the resulting sulfide. Sedimented systems have a greater range to both higher and lower $\delta^{34}S$ values. The lower values undoubtedly document the remobilization of sulfide initially precipitated by bacterial sulfate reduction. The sulfur isotopic characteristics of back-arc and arc settings are interesting because some negative $\delta^{34}S$ values have been documented at sites lacking significant sedimentary cover. These sites also have low pH fluids that exceed seawater concentrations of sulfate; disproportionation of SO_2, as described by Equation (20), has been proposed to explain the low $\delta^{34}S$ values (Herzig et al. 1993; Gamo et al. 1997). In essence, these back-arc seafloor hydrothermal systems would represent the modern seafloor equivalents of the terrestrial acid-sulfate epithermal systems discussed above.

Another interesting aspect of the sulfur isotope characteristics of sulfides in seafloor hydrothermal systems is in the isotopic composition of the vent fluid H_2S. Shanks (2001) described how the $\delta^{34}S$ of vent fluid H_2S is commonly 1‰ to more than 4‰ higher than that of sulfide minerals on the inner walls of hydrothermal chimneys. For most of the common sulfide minerals found in seafloor chimneys, the $\delta^{34}S$ of the mineral should be higher than that for the associated H_2S, and in all cases the difference should be less than 1‰ at measured temperatures. Shanks (2001) suggested that local reduction of seawater sulfate in the chimney environment or equilibrium restricted to the minute, innermost layer of sulfide minerals may partially explain this discrepancy. An equally impressive observation on the sulfur isotopic composition of vent fluid H_2S is found in time-series sampling of individual vents on the scale of weeks, months, or a few years. Along the East Pacific Rise, the $\delta^{34}S$ of H_2S from the Aa vent was found to increase by approximately 2‰ over the course of approximately three years, whereas that of the P vent decreased by over 3‰ over a similar period.

Ancient systems. The sulfur isotope compositions of sulfide minerals from ancient seafloor massive sulfide deposits are interpreted in terms of the same geochemical processes as operative in modern systems, but with a few additional complexities. Secular variations in the sulfur isotope composition of seawater, discussed previously, result in one potential component of sulfide sulfur having a composition that varies as a function of time. A second complexity is the periodic occurrence of anoxic bottom waters in the oceans on a global scale (Leggett 1980), which can result in the presence of H_2S in the water column near the seafloor. The following discussion focuses on two general classes of ancient seafloor deposits containing sulfide minerals: volcanic-associated (volcanogenic) massive sulfide deposits, and sedimentary-exhalative (sedex) massive sulfide deposits. Ohmoto (1986) and Huston (1999) have provided reviews of the stable isotope characteristics of ancient volcanic-associated massive sulfide deposits; Seal et al. (2000a) reviewed their isotopic characteristics from the perspective of their associated sulfate minerals.

Volcanic-associated deposits form at active mid-ocean ridge spreading centers, and in arc-volcanic rocks, continental rifts, and Archean greenstone belts, whereas sedex deposits form in failed continental rift settings. Volcanic-associated and sedex deposits are dominated by sulfide minerals, most commonly pyrite, pyrrhotite, chalcopyrite, sphalerite, and galena in varying proportions. They can also have associated sulfate minerals, typically anhydrite, barite, or gypsum. These deposits are major sources of Cu, Pb, Zn, Ag, and Au.

The secular variations observed in the sulfur isotopic composition of sulfide and sulfate minerals in seafloor massive sulfide deposits mimic, to a remarkable degree, the secular variations observed in sedimentary pyrite and marine sulfate and attest to the dominant role that atmospheric oxygen has on the global sulfur cycle (Huston 1999; Figs. 9 and 19). The compilation of Huston (1999) has been expanded to include data from sedimentary exhalative deposits and additional volcanic-associated deposits in Figure 19. The $\delta^{34}S$ values of sphalerite, galena and pyrite from the Mississippian Red Dog deposit, Alaska, range from −45.8 to 12.3‰, with most values between −2.5 and 7.5‰ (Kelley et al. 2004). The lowest values were produced during the earliest stages

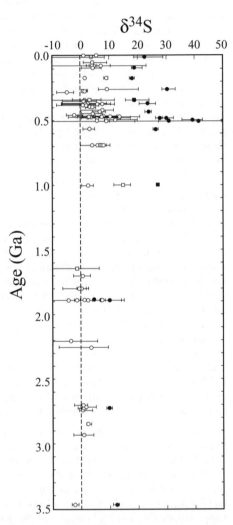

Figure 19. Secular variation of the isotopic composition of sulfide and sulfate minerals from seafloor massive sulfide deposits. Modified from Huston (1999) with data from Whelan et al. (1984), Seal and Wandless (2003), Seal et al. (2000b, 2001), and Kelley et al. (2004). Volcanic-associated deposits are shown by circles; sedimentary exhalative deposits are shown by squares. Sulfides are white symbols; sulfates are black symbols. Isotopic values are given in permil (VCDT).

of hydrothermal activity. Similarly, the $\delta^{34}S$ values of pyrrhotite, pyrite, sphalerite, and galena from the Proterozoic Sullivan and nearby deposits, in the Purcell Supergroup, British Columbia, range between −11 and 6‰ having a distinct mode around −1‰ (Seal et al. 2000b; Taylor and Beaudoin 2000). Taylor and Beaudoin (2000) found significant stratigraphic variations in the $\delta^{34}S$ values of sulfide minerals within the deposit, which they interpreted in terms of variations in the relative proportions of H_2S-derived bacterial sulfate reduction and thermochemical sulfate reduction throughout the period of hydrothermal activity. The Proterozoic Balmat-Edwards Zn-Pb deposits, which experienced amphibolite-facies regional metamorphism, have a limited range of $\delta^{34}S$ values from pyrite, sphalerite, and galena (11.5 to 17.5‰), presumably because of the homogenizing effects of the metamorphism (Whelan et al. 1984).

Prior to 2.4 Ga, the inferred onset of oxygen in the atmosphere, the $\delta^{34}S$ of hydrothermal sulfide and sulfate indicated limited variations and both cluster near 0‰. Between 2.4 and 0.7 Ga, the $\delta^{34}S$ of hydrothermal sulfide and sulfate has a significantly wider range and increasing $\delta^{34}S$ values. Beginning after 0.7 Ga, the time proposed by Canfield and Teske (1996) for the onset of higher oxygen levels in the atmosphere, the $\delta^{34}S$ of hydrothermal sulfides and sulfates indicates a dramatic increase both in range and average value (Fig. 19).

The general correlation between the average $\delta^{34}S$ of a volcanic-associated massive sulfide deposit and coeval seawater was first identified by Sangster (1968). He noted a roughly 17.5‰ difference between seawater and the mean composition of volcanic-associated massive sulfide deposits. For sediment-hosted deposits, which include sedimentary-exhalative deposits among other types, he found a smaller 11.7‰ fractionation between seawater and sulfide. Janecky and Shanks (1988) quantified the relationship between the $\delta^{34}S$ of coeval seawater and sulfide in basaltic seafloor hydrothermal systems using reaction-path geochemical modeling coupled with sulfur-isotope mass-balance equations. They found that for simple adiabatic mixing, as discussed above, the maximum $\delta^{34}S$ of sulfides that can be achieved is 4.5‰, which corresponds to a seawater-sulfide fractionation of 16.5‰, remarkably similar to the fractionation proposed by Sangster (1968). Janecky and Shanks (1988) found a maximum $\delta^{34}S$ of sulfides formed through thermochemical reduction of 15‰, which corresponds to a seawater-sulfide fractionation of 6‰. These maximum compositions of sulfide resulting from these two end-member processes will vary accordingly with secular variations in the $\delta^{34}S$ of seawater sulfate. Despite the predictable relationship between the $\delta^{34}S$ of seawater and hydrothermal sulfides, Janecky and Shanks (1988) found that sulfur isotope disequilibrium best describes sulfide and sulfate in seafloor vent systems, and that the systematic relationship is established at depth in the reaction zone of the seafloor hydrothermal system.

Sulfur isotope studies in the Selwyn Basin (Yukon) by Goodfellow (1987) and Shanks et al. (1987) suggested the significance of H_2S-bearing anoxic bottom waters in determining the isotopic composition of sedimentary-exhalative massive sulfide deposits. Goodfellow and Jonasson (1984) investigated the sulfur isotope composition of sedimentary pyrite and barite within the Cambrian to Mississippian strata of the Selwyn basin. They used the barite data to define a local Selwyn basin sulfate sulfur isotope secular curve that is locally over 20‰ higher than the global marine sulfate curve of Claypool et al. (1980). They used the higher values within the Selwyn basin as evidence that the Selwyn basin had restricted access to the open ocean and that the bottom waters were anoxic and H_2S-bearing. Shanks et al. (1987) extended the Selwyn basin curve farther back into Cambrian time with additional data from the Anvil district. They used data from sedimentary pyrite to define a secular H_2S curve. Goodfellow (1987) used the coincidence of the sulfur isotopic composition of the massive sulfide deposits with that of the sedimentary pyrites to conclude that sulfur for the mineral deposits was dominantly derived from H_2S in an anoxic water column during periods of stagnation in the Selwyn basin. He proposed that these stagnation events may have been global in extent. Goodfellow and Peter (1996) provided additional support for the global extent of these anoxia events from their studies of the sulfur isotope geochemistry of the Brunswick No. 12 deposit in the Bathurst mining camp (New Brunswick), which has sulfur isotope values that fall on the secular pyrite curve for the Selwyn basin.

The role of anoxic bottom waters for the genesis of volcanic-associated massive sulfide deposits can be evaluated by comparing the secular variations in the sulfur isotope composition of seawater and hypothetical hydrothermal sulfide with the sulfur isotope compositions of sulfide and sulfate minerals from massive sulfide deposits. Seal and Wandless (2003) compared secular variations in the sulfur isotopic composition of sulfide and sulfate minerals from Cambrian and Ordovician volcanic-associated massive sulfide deposits from around the world with the global marine sulfate curve, Selwyn basin pyrite curve, and the maximum $\delta^{34}S$ values attainable through simple adiabatic mixing and shallow thermochemical reduction (Fig. 20), as modeled by Janecky and Shanks (1988). Figure 20 has been extended into the latest Proterozoic to include data from the Barite Hill deposit, South Carolina for comparison (Seal et al. 2001). Seal and Wandless (2003) found that the sulfur isotope composition of many of the deposits fell within the permissible range for simple adiabatic mixing, and that all fell within the permissible range for shallow thermochemical reduction (Fig. 20). Thus, the isotopic characteristics of sulfide minerals from all of these deposits can be explained without the need for anoxic H_2S-bearing bottom waters, although their role is not necessarily excluded. The $\delta^{34}S$ values of the associated sulfate minerals provide the most compelling evidence for anoxic waters in the case of the deposits of the Mount Read volcanic belt, Tasmania (Solomon et al. 1969, 1988; Gemmell and Large 1992; McGoldrick and Large 1992), which have $\delta^{34}S$ values for sulfate well in excess of the global marine sulfate curve (Fig. 20).

Mississippi Valley-type deposits

Mississippi Valley-type Pb-Zn deposits typically form in continental settings in low-temperature (<200 °C), near-neutral environments in which sulfur isotope disequilibrium would be expected to dominate (Ohmoto and Lasaga 1982). Thus, stable isotope data from sulfide minerals from Mississippi Valley-type deposits should provide information about the source of sulfide and its geochemical history. Stable isotope studies of Mississippi Valley-type deposits are dominated by sulfur isotope data from both sulfide and sulfate minerals (Ault and Kulp 1960; Sasaki and Krouse 1969; Ohmoto 1986; Kaiser et al. 1987; Richardson et al. 1988; Kesler et al. 1994; Appold et al. 1995; Kesler 1996; Misra et al. 1996; Jones et al. 1996).

Sulfide sulfur in Mississippi Valley-type environments can be derived from a variety of sources including organically bound sulfur, H_2S reservoir gas, evaporites, connate seawater, and diagenetic sulfides. In all cases, these sources are seawater sulfate that has followed various geochemical pathways that impart different isotopic fractionations. The reduction of sulfate occurs either through bacterially mediated processes or abiotic thermochemical processes. Bacterial sulfate reduction, as discussed above, can produce sulfate–sulfide fractionations that typically range from 15 to 60‰ (Goldhaber and Kaplan 1975), whereas those associated with abiotic thermochemical reactions with organic compounds range from zero to as much as 10‰ (Orr 1974; Kiyosu 1980). Bacterial sulfate reduction has been documented at temperatures up to 110 °C (Jørgensen et al. 1992), but the optimum temperature range is between 30 and 40 °C. Ohmoto and Goldhaber (1997) argued that at the site of ore deposition, thermochemical reduction is not effective at $T < 125$ °C because of slow reaction kinetics. For thermochemical reduction to be an important process in Mississippi Valley-type environments, reduction must occur away from the site of ore deposition, in the deeper, hotter parts of the basin. It should be noted that although the kinetic fractionations associated with both reduction processes are distinct, they can produce H_2S of similar compositions if bacterial sulfate reduction occurs quantitatively (or nearly so) in an environment with little or no Fe to sequester the sulfide.

Sulfur isotope data from other Mississippi Valley-type deposits suggest two major sulfide reservoirs, one centered between −5 to 15‰ and one greater than 20‰ (Fig. 21). The higher values of sulfides typically coincide with those of the composition of associated sulfate

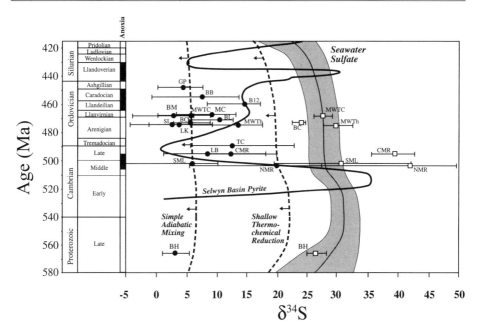

Figure 20. Variation of sulfur isotope composition with age for various Early Paleozoic and Late Proterozoic massive sulfide deposits (modified from Seal and Wandless 2003). The age distribution of anoxic events in the Iapetus Ocean indicated on the figure is reflected by the presence of black shales (after Leggett 1980). The compositions of the deposits are shown as the mean and range. Data are from Huston (1999) unless otherwise noted in text. The seawater sulfate curve (gray field) is modified from Claypool et al. (1980) to account for the 1.65‰ fractionation between evaporitic sulfate minerals and dissolved sulfate (Seal et al. 2000). The Selwyn Basin pyrite curve (heavy black line) is modified from Goodfellow and Jonasson (1984). The upper limits for the composition of sulfide derived from simple adiabatic mixing of vent-fluid H_2S and ambient seawater and of sulfide derived from shallow level thermochemical reduction of seawater sulfate are based on the models of Janecky and Shanks (1988) and are shown as the dashed curves with arrows. Black circles depict data from sulfide minerals; white squares depict data from sulfate minerals. Abbreviations: B12, Brunswick No. 12, Bathurst Mining camp, New Brunswick; BB, Boucher Brook Formation deposits, Bathurst Mining Camp, New Brunswick; BC, Buchans, Newfoundland; BH, Barite Hill, South Carolina; BL, Balcooma, Queensland; CMR, Central Mount Read Volcanics Belt, Tasmania; GP, Gull Pond, Newfoundland; LB, Lush's Bight Ophiolite, Newfoundland; LK, Lokken Ophiolite, Norway; MWTC, Mount Windsor-Trooper Creek, Queensland; MWTh, Mount Windsor-Thalanga, Queensland; NMR, Northern Mount Read Volcanics Belt, Tasmania; SL, Sulitjelma, Norway; SML, Southern Mount Lyell Volcanics Belt, Tasmania TC, Tilt Cove, Newfoundland.

minerals, and have been interpreted to reflect the minimal fractionation associated with abiotic thermochemical reduction (Kesler 1996). However, similar compositions of sulfide could be generated by closed-system, quantitative bacterial reduction of sulfate. A carbonate aquifer is an ideal environment for such a geochemical process due to the lack of reactive Fe to scavenge and fractionate sulfur. The lower values may reflect formation from H_2S derived either directly or indirectly from open-system bacterial reduction of sulfate. Kesler et al. (1994) proposed that low $\delta^{34}S$ H_2S was derived from oil in the deeper parts of the basin for the Central Tennessee and Kentucky Mississippi Valley-type districts. This H_2S ultimately would have been derived from the bacterial reduction of sulfate. The H_2S from both bacterial and abiotic reduction is not in sulfur isotope equilibrium with associated sulfate minerals (Fig. 21).

Stable isotope and fluid-inclusion studies by Richardson et al. (1988) of samples from the Deardorff mine from the Cave-in-Rock fluorspar district, Illinois, indicate mineralization was

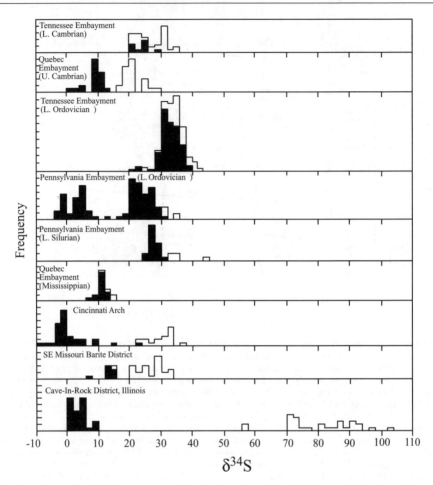

Figure 21. Sulfur isotope histograms for Mississippi Valley-type hydrothermal systems in North America. Data from sphalerite are shown in black; data from sulfate minerals are shown in white. All isotopic values are given in permil (VCDT).

dominated by two formation fluids recharged by meteoric waters, one of which circulated into the basement rocks under low water/rock conditions. Liquid hydrocarbons are present in fluid inclusion in most minerals. The low $\delta^{34}S$ of sulfides (4.0 to 8.9‰ for sphalerite) indicates a significant contribution of H_2S from petroleum sources. The sulfides are completely out of isotopic equilibrium with late stage barites, which have $\delta^{34}S$ values ranging from about 57 to 103‰. These data suggest that the aqueous sulfate was derived from a small fluid reservoir in which the residual seawater sulfate underwent thermal chemical reduction with organic matter. Supporting evidence includes the decrease in the $\delta^{13}C_{CO_2}$ of the fluids during carbonate deposition.

SUMMARY

The Earth is assumed to have a bulk $\delta^{34}S$ value of around 0‰, essentially the same as most meteorites. Some of the most important factors affecting sulfur isotope fractionation throughout the history of the Earth have been oxidation and reduction reactions that ultimately have been facilitated by the progressive development of an oxygenated atmosphere. Prior to 2.4 Ga, the sedimentary record reveals limited variation in the $\delta^{34}S$ of sulfides and sulfates, presumably due to the lack of oxygen. The ancient geologic record also preserves significant mass-independent sulfur isotope anomalies which have been interpreted to be the result of UV-induced photochemical reactions in the atmosphere due to the absence of an ozone layer (Farquhar et al. 2000a). The mass-independent anomalies stop abruptly after 2.4 Ga and the $\delta^{34}S$ of sedimentary sulfides and sulfates begin to show greater variability, consistent with the onset of an oxygenated atmosphere. Beginning at about 0.7 Ga, another major change in the variability of the $\delta^{34}S$ of sedimentary sulfides and sulfates occurs that is indicated by much wider ranges of compositions, which again has been interpreted in terms of increased atmospheric concentrations of oxygen (Canfield and Teske 1996). These same transitions in sedimentary isotopic compositions are also apparent in the isotopic signatures of marine, volcanic-associated massive sulfide deposits.

Throughout much of the history of the Earth, the metabolism of sulfate-reducing bacteria has been important in producing the variability recorded in the geologic record. The profound impact of sulfate-reducing bacteria on the global sulfur cycle may even be discernible in the mantle, where the negative $\delta^{34}S$ values of sulfide inclusions are likely derived from subducted sedimentary sulfides (Chaussidon et al. 1987; Eldridge et al. 1991). Equally impressive is the fact that these mantle heterogeneities may have persisted for billions of years as indicated by the identification of mass-independent anomalies in sulfide inclusions in diamonds, for which the most likely explanation is that the anomalies were locked in the mineral record prior to 2.4 Ga when the atmosphere became oxygenated (Farquhar et al. 2002). The isotopic imprint of sulfate-reducing bacteria can be found in many reaches of the sulfur cycle, from sedimentary sulfides and sulfates, to coal beds, to seafloor hydrothermal mineral deposits, to continental Mississippi Valley-type deposits formed from basinal brines, and to magmas of all compositions that have interacted with crustal rocks.

Sulfur-rich magmatic sulfide deposits associated with mafic igneous rocks commonly record the fingerprint of contamination by crustal sedimentary sulfur. Hydrothermal systems associated with oxidized felsic magmas emplaced into shallow levels of the crust reflect sulfur isotope signatures determined by high-temperature isotopic exchange between reduced and oxidized sulfur species such as H_2S and SO_4^{2-}, which commonly are the result of the disproportionation of SO_2. High-temperature settings associated with porphyry environments tend to record equilibrium sulfur isotope fractionations, whereas moderate temperature settings, such as those for epithermal deposits are more likely to record disequilibrium fractionations, unless the fluids are more acidic when the kinetics of sulfur isotope exchange are more favorable (Ohmoto and Lasaga 1982). In even lower temperature environments, such as those associated with diagenetic sulfides including coal beds, and basinal brines associated with Mississippi Valley-type deposits, kinetic fractionations dominate rather than equilibrium fractionations. In all cases, it is the reaction of a reduced sulfur species to an oxidized sulfur species, the reaction of an oxidized sulfur species to a reduced sulfur species, or isotopic exchange between an oxidized and a reduced sulfur species that causes the most significant sulfur isotope variations.

ACKNOWLEDGMENTS

Discussions with Bob Rye and Pat Shanks have been incredibly rewarding over the years. Don Canfield kindly shared his sulfur-isotope data base. Laboratory assistance and support from Greg Wandless has been indispensable. Assistance with literature searches from Carmen O'Neill and Nadine Piatak is greatly appreciated. Brenda Pierce assisted with Russian translations. The manuscript benefited from reviews by Avery Drake, Jeff Grossman, Nadine Piatak, Pat Shanks, and David Vaughan. The preparation of this chapter has been supported by Kate Johnson, Program Coordinator of the Mineral Resources Program of the U.S. Geological Survey.

REFERENCES

Albarède F (2004) The stable isotope geochemistry of copper and zinc. Rev Mineral Geochem 55:409-427
Anbar AD (2004) Molybdenum stable isotopes: observations, interpretations and directions. Rev Mineral Geochem 55:429-454
Appold MS, Kesler SE, Alt JC (1995) Sulfur isotope and fluid inclusion constraints on the genesis of Mississippi Valley-type mineralization in the Central Appalachians. Econ Geol 90:902-919
Arribas A Jr., Cunningham CG, Rytuba JJ, Rye RO, Kelly WC, Podwysocki MH, McKee EH, Tosdal RM (1995) Geology, geochronology, fluid inclusions, and isotope geochemistry of the Rodalquilar gold-alunite deposit, Spain. Econ Geol 90:795-822
Ault WU, Kulp JL (1960) Sulfur isotopes and ore deposits. Econ Geol 55:73-100
Ault WU, Jensen ML (1963) Summary of sulfur isotope standards. *In:* Biogeochemistry of Sulfur Isotopes. Jensen ML (ed) Nat Sci Found, Symp Proc, Yale University
Bachinski DJ (1969) Bond strength and sulfur isotopic fractionation in coexisting sulfides. Econ Geol 64:56-65
Bahr JR (1976) Sulfur isotopic fractionation between H_2S, S and SO_4^{2-} in aqueous solutions and possible mechanisms controlling isotopic equilibrium in natural systems. M.Sc. thesis, Pennsylvania State University
Barton PB Jr, Skinner BJ (1979) Sulfide mineral stabilities. *In:* Geochemistry of Hydrothermal Ore Deposits. Barnes HL (ed) J Wiley and Sons, 278-403
Barling J, Arnold GL, Anbar AD (2001) Natural mass-dependent variations in the isotopic composition of molybdenum. Earth Planet Sci Lett 193:447-457
Beard BL, Johnson CM (2004) Fe isotope variations in the modern and ancient earth and other planetary bodies. Rev Mineral Geochem 55:319-357
Beaudoin G, Taylor BE (1994) high precision and spatial resolution sulfur isotope analysis using MILES laser microprobe. Geochim Cosmochim Acta 58:5055-5063
Bekker A, Holland HD, Wang P-L, Rumble D, III, Stein HJ, Hannah JL, Coetzee LL, Beukes NJ (2004) Dating the rise of atmospheric oxygen. Nature 427:117-120
Bente K, Nielsen H (1982) Experimental S isotope fractionation studies between coexisting bismuthinite (Bi_2S_3) and sulfur (S°). Earth Planet Sci Lett 59:18-20
Berner RA (1985) Sulphate reduction, organic matter decomposition and pyrite formation. Phil Trans R Soc London A 315:25-38
Bethke PM, Rye RO, Stoffregen RE, Vikre PG (2005) Evolution of the magmatic-hydrothermal acid-sulfate system at Summitville, Colorado: integration of geological, stable-isotope, and fluid-inclusion evidence. Chem Geol 215:281-315
Bigeleisen J (1952) The effects of isotopic substitution on the rates of chemical reactions. J Phys Chem 56:823-828
Bigeleisen J, Mayer MG (1947) Calculation of equilibrium constants for isotopic exchange reactions. J Chem Phys 15:261-267
Bischoff JL, Seyfried WE (1978) Hydrothermal chemistry of seawater from 25 ° to 350 °C. Am J Sci 278:838-860
Bouška V, Pešek J (1999) Quality parameters of lignite of the North Bohemian Basin in the Czech Republic in comparison with world average lignite. Intl J Coal Geol 40:211-235
Buchanan DL, Nolan J, Suddaby P, Rouse JE, Viljoen MJ, Davenport JWJ (1981) The genesis of sulfide mineralization in a portion of the Potgietersrus Limb of the Bushveld Complex. Econ Geol 76:568-579
Burnham CW, Ohmoto H (1980) Late-stage process of felsic magmatism. Soc Mining Geol Japan Spec Issue 8:1-11
Canfield DE (2001) Biogeochemistry of sulfur isotopes. Rev Mineral Geochem 43:607-636
Canfield DE (2004) The evolution of the Earth surface sulfur reservoir. Am J Sci 304:839-861
Canfield DE, Teske A (1996) Late Proterozoic rise in atmospheric oxygen concentration inferred from phylogenetic and sulfur-isotope studies. Nature 382:127-132
Casadevall T, Ohmoto H (1977) Sunnyside mine, Eureka mining district, San Juan County, Colorado: geochemistry of gold and base metal ore deposition in a volcanic environment. Econ Geol 72:1285-1320
Chambers LA (1982) Sulfur isotope study of a modern intertidal environment, and the interpretation of ancient sulfides. Geochim Cosmochim Acta 46:721-728
Chambers LA, Trudinger PA (1979) Microbiological fractionation of stable sulfur isotopes: A review and critique. Geomicrobiol J 1:249-293

Chaussidon M, Albarède F, Sheppard SMF (1987) Sulphur isotope heterogeneity in the mantle from ion microprobe measurements of sulphide inclusions in diamonds. Nature 330:242-244

Chaussidon M, Albarède F, Sheppard SMF (1989) Sulphur isotope variations in the mantle from ion microprobe analyses of micro-sulphide inclusions. Earth Planet Sci Lett 92:144-156

Chu X, Ohmoto H, Cole DR (2004) Kinetics of sulfur isotope exchange between aqueous sulfide and thiosulfate involving intra- and intermolecular reactions at hydrothermal conditions. Chem Geol 211:217-235

Claypool GE, Holser WT, Kaplan IR, Sakai H, Zak I (1980) The age curves of sulfur and oxygen isotopes in marine sulfate and their mutual interpretations. Chem Geol 28:199-260

Clayton DD, Ramadurai S (1977) On presolar meteoritic sulphides. Nature 265:427-428

Clayton RN (1986) High temperature effects in the early solar system. Rev Mineral 16:129-140

Coleman ML, Moore MP (1978) Direct reduction of sulfates to sulfur dioxide for isotopic analysis. Anal Chem 50:1594-1595

Crowe DE, Valley JW, Baker KL (1990) Micro-analysis of sulfur-isotope ratios and zonation by laser microprobe. Geochim Cosmochim Acta 54:2075-2092

Czamanske GK, Rye RO (1974) Experimentally determined sulfur isotope fractionations between sphalerite and galena in the temperature range 600 ° to 275 °C. Econ Geol 69:17-25

Dai S, Ren D, Tang Y, Shao L, Li S (2002) Distribution, isotopic variation and origin of sulfur in coals in the Wuda coalfield, Inner Mongolia, China. Intl J Coal Geol 51:237-250

Eastoe CJ (1983) Sulfur isotope data and the nature of the hydrothermal systems at the Panguna and Frieda porphyry copper deposits, Papua New Guinea. Econ Geol 78:201-213

Eldridge CS, Compston W, Williams IS, Both RA, Walshe JL, Ohmoto H (1988) Sulfur isotope variability in sediment-hosted massive sulfide deposits as determined using the ion microprobe SHRIMP: I. an example from the Rammelsberg orebody. Econ Geol 83:443-449

Eldridge CS, Compston W, Williams IS, Harris JW, Bristow JW (1991) Isotope evidence for the involvement of recycled sediments in diamond formation. Nature 353:649-653

Eldridge CS, Compston W, Williams IS, Harris JW, Bristow JW, Kinny PD (1995) Applications of the SHRIMP I ion microprobe to the understanding of processes and timing of diamond formation. Econ Geol 90:271-280

Farquhar J, Bao H, Thiemens M (2000a) Atmospheric influence of Earth's earliest sulfur cycle. Science 289: 756-758

Farquhar J, Savarino J, Jackson TL, Thiemens MH (2000b) Evidence of atmospheric sulphur in the martian regolith from sulphur isotopes in meteorites. Nature 404:50-52

Farquhar J, Jackson TL, Thiemens MH (2000c) A ^{33}S enrichment in ureilite meteorites: evidence for a nebular sulfur component. Geochim Cosmochim Acta 64:1819-1825

Farquhar J, Wing BA, McKeegan KD, Harris JW, Cartigny P, Thiemens MH (2002) Mass-independent sulfur of inclusions in diamond and sulfur recycling on early Earth. Science 298:2369-2372

Farquhar J, Wing B (2003) Multiple sulfur isotopes and the evolution of the atmosphere. Earth Planet Sci Lett 213:1-13

Field CW (1966) Sulfur isotopic method for discriminating between sulfates of hypogene and supergene origin. Econ Geol 61:1428-1435

Field CW, Gustafson LB (1976) Sulfur isotopes in the porphyry copper deposit at El Salvador, Chile. Econ Geol 71:1533-1548

Field CW, Zhang L, Dilles JH, Rye RO, Reed MH (2005) Sulfur and oxygen isotopic record in sulfate and sulfide minerals of early, deep, pre-Main Stage porphyry Cu-Mo and late Main Stage base-metal mineral deposits, Butte district, Montana. Chem Geol 215:61-93

Fifarek RH, Rye RO (2005) Stable-isotope geochemistry of the Pierina high-sulfidation Au-Ag deposit, Peru: influence of hydrodynamics on SO_4^{2-}-H_2S sulfur isotopic exchange in magmatic-steam and steam-heated environments. Chem Geol 215:253-279

Gamo T, Okamura K, Charlou JL, Urabe T, Auzende JM, Ishibashi J, Shitashima K, Chiba H (1997) Acidic and sulfate-rich hydrothermal fluids from Manus back-arc basin, Papua New Guinea. Geology 25:139-142

Gao X, Thiemens MH (1991) Systematic study of sulfur isotopic composition in iron meteorites and the occurrence of excess ^{33}S and ^{36}S. Geochim Cosmochim Acta 55:2671-2679

Gao X, Thiemens MH (1993a) Isotopic composition and concentration of sulfur in carbonaceous chondrites. Geochim Cosmochim Acta 57:3159-3169

Gao X, Thiemens MH (1993b) Variations in the isotopic composition of sulfur in enstatite and ordinary chondrites. Geochim Cosmochim Acta 57:3171-3176

Gavelin S, Parwel A, Ryhage R (1960) Sulfur isotope fractionation in sulfide mineralization. Econ Geol 55: 510-530

Gemmell JB, Large RR (1992) Stringer system and alteration zones underlying the Hellyer volcanic-hosted massive sulfide deposit, Tasmania, Australia. Econ Geol 87:620-649

Giesemann A, Jäger H-J, Norman AL, Krouse HR, Brand WA (1994) On-line sulfur-isotope determination using an elemental analyzer coupled to a mass spectrometer. Anal Chem 66:2816-2819

Goldhaber MB, Kaplan IR (1975) Controls and consequences of sulfate reduction rates in recent marine sediments. Soil Sci 119:42-55

Goodfellow WD (1987) Anoxic stratified oceans as a source of sulphur in sediment-hosted stratiform Zn-Pb deposits (Selwyn Basin, Yukon, Canada). Chem Geol 65:359-382

Goodfellow WD, Jonasson IR (1984) Ocean stagnation and ventilation defined by $\delta^{34}S$ secular trends in pyrite and barite, Selwyn Basin, Yukon. Geology 12:583-586

Goodfellow WD, Peter JM (1996) Sulphur isotope composition of the Brunswick No. 12 massive sulphide deposit, Bathurst mining camp, New Brunswick: implications for ambient environment, sulphur source, and ore genesis. Can J Earth Sci 33:231-251

Greenwood JP, Riciputi LR, McSween HY Jr., Taylor LA (2000a) Modified sulfur isotopic compositions of sulfides in the nakhlites and Chassigny. Geochim Cosmochim Acta 64:1121-1131

Greenwood JP, Mojzsis SJ, Coath CD (2000b) Sulfur isotopic compositions of individual sulfides in Martian meteorites ALH84001 and Nakhla: implications for crust-regolith exchange on Mars. Earth Planet Sci Lett 184:23-35

Grinenko VA, Thode HG (1970) Sulfur isotope effects in volcanic gas mixtures. Can J Earth Sci 7:1402-1409

Grootenboer J, Schwarz HP (1969) Experimentally determined sulfur isotope fractionation between sulfide minerals. Earth Planet Sci Lett 7:162-166

Habicht KS, Canfield DE (2001) Isotope fractionation by sulfate-reducing natural populations and the isotopic composition of sulfide in marine sediments. Geology 29:555-558

Hackley KC, Anderson TF (1986) Sulfur isotopic variation in low-sulfur coals from the Rocky Mountain region. Geochim Cosmochim Acta 50:1703-1713

Haur A, Hladikova J, Smejkal V (1973) Procedure of direct conversion of sulfates into SO_2 for mass spectrometric analysis of sulfur. Isotopenpraxis 18: 433-436

Hayes JM, Lambert IB, Strauss H (1992) The sulfur-isotopic record. *In:* The Proterozoic Biosphere. Schopf JW, C Klein C (eds) Cambridge Univ Press, p 129-132

Heald P, Foley NK, Hayba DO (1987) Comparative anatomy of volcanic-hosted epithermal deposits: acid-sulfate and adularia-sericite types. Econ Geol 82:1-26

Hedenquist JW, Arribas A Jr, Reynolds TJ (1998) Evolution of an intrusion-centered hydrothermal system: Far-Southeast-Lepanto porphyry-epithermal Cu-Au deposits, Philippines. Econ Geol 93:373-404

Herzig PM, Hannington MD, Fouquet Y, von Stackelberg U, Petersen S (1993) Gold-rich polymetallic sulfides from the Lau back arc and implications for the geochemistry of gold in sea-floor hydrothermal systems of the southwest Pacific. Econ Geol 88:2182-2209

Herzig PM, Petersen S, Hannington MD (1998) Geochemistry and sulfur-isotopic composition of the TAG hydrothermal mound, Mid-Atlantic Ridge 26 °N. Proc ODP, Sci Results 158:47-70

Holland HD (1965) Some applications of thermochemical data to problems of ore deposits, II. Mineral assemblages and the composition of ore-forming fluids. Econ Geol 60:1101-1166

Holland HD (1978) The Chemistry of the Atmosphere and Oceans. J. Wiley and Sons

Holser WT, Kaplan IR, Sakai H, Zak I (1979) Isotope geochemistry of oxygen in the sedimentary sulfate cycle. Chem Geol 25:1-17

Holt BD, Engelkemeier AG (1970) Thermal decomposition of barium sulfate to sulfur dioxide for mass spectrometric analysis. Anal Chem 42:1451-1453

Hu G, Rumble D, Wang P-L (2003) An ultraviolet laser microprobe for the *in situ* analysis of multisulfur isotopes and its use in measuring Archean sulfur isotope mass-independent anomalies. Geochim Cosmochim Acta 67:3101-3118

Hubberten H-W (1980) Sulfur isotope fractionation in the Pb-S, Cu-S and Ag-S systems. Geochem J 14:177-184

Huston DL (1999) Stable isotopes and their significance for understanding the genesis of volcanic-associated massive sulfide deposits: A review. *In:* Volcanic-Associated Massive Sulfide Deposits: Processes and Examples in Modern and Ancient Settings. Reviews in Economic Geology, Vol. 8. Barrie CT, Hannington MD (eds) Soc Econ Geol, p 157-179

Hulston JR, Thode HG (1965a) Variations in the S^{33}, S^{34}, and S^{36} contents of meteorites and their relation to chemical and nuclear effects. J Geophys Res 70:3475-3484

Hulston JR, Thode HG (1965b) Cosmic-ray produced S^{33} and S^{36} in the metallic phase of iron meteorites. J Geophys Res 70:4435-4442

Igumnov SA, Grinenko VA, Poner NB (1977) Temperature dependence of the distribution coefficient of sulfur isotopes between H_2S and dissolved sulfates in the temperature range 260-400 °C. Geokhimiya 7:1085-1087

Ishihara S, Sasaki A (1989) Sulfur isotopic ratios of the magnetite-series and ilmenite-series granitoids of the Sierra Nevada batholith – a reconnaissance study. Geology 17:788-791

Janecky DR, Shanks WC III (1988) Computational modeling of chemical and sulfur isotopic reaction processes in seafloor hydrothermal systems: chimneys, massive sulfides, and subjacent alteration zones. Can Mineral 26:805-825

Jensen ML (1957) Sulfur isotopes and mineral paragenesis. Econ Geol 52:269-281
Jensen ML (1959) Sulfur isotopes and hydrothermal mineral deposits. Econ Geol 54:374-394
Jensen ML (1967) Sulfur isotopes and mineral genesis. In: Geochemistry of Hydrothermal Ore Deposits. Barnes HL (ed) Holt, Rinehart, and Winston, p 143-165
Johnson CM, Beard BL, Beukes NJ, Klein C, O'Leary JM (2003) Ancient geochemical cycling in the earth as inferred from Fe isotope studies of banded iron formations from the Transvaal Craton. Contrib Mineral Petrol 144:523-547
Jones HD, Kesler SE, Furman FC, Kyle JR (1996) Sulfur isotope geochemistry of southern Appalachian Mississippi Valley-type deposits. Econ Geol 91:355-367
Jørgensen BB, Isaksen MF, Jannasch HW (1992) Bacterial sulfate reduction above 100 °C in deep-sea hydrothermal vent sediments. Science 258:1756-1757
Juliani C, Rye RO, Nunes CMD, Snee LW, Corrêa Silva RH, Monteiro LVS, Bettencourt JS, Neumann R, Neto AA (2005) Paleoproterozoic high-sulfidation mineralization in the Tapajós gold province, Amazonian Craton, Brazil: geology, mineralogy, alunite argon age, and stable-isotope constraints. Chem Geol 215:95-125
Kaiser CJ, Kelly WC, Wagner RJ, Shanks WC III (1987) Geologic and geochemical controls on mineralization in the Southeast Missouri Barite district. Econ Geol 82:719-734
Kajiwara Y, Krouse HR (1971) Sulfur isotope partitioning in metallic sulfide systems. Can J Earth Sci 8:1397-1408
Kaplan IR, Hulston JR (1966) The isotopic abundance and content of sulfur in meteorites. Geochim Cosmochim Acta 30:479-496
Kelley KD, Leach DL, Johnson CA, Clark JL, Fayek M, Slack JF, Anderson VM, Ayuso RA, Ridley WI (2004) Textural, compositional, and sulfur isotope variations of sulfide minerals in the Red Dog Zn-Pb-Ag deposits, Brooks Range, Alaska: implications for ore formation. Econ Geol 99:1509-1532
Kesler SE (1996) Appalachian Mississippi Valley-type deposits: paleoaquifers and brine provinces. Soc Econ Geol Spec Pub 4:29-57
Kesler SE, Jones HD, Furman FC, Sassen R, Anderson WH, Kyle JR (1994) Role of crude oil in the genesis of Mississippi Valley-type deposits: Evidence from the Cincinnati Arch. Geology 22:609-612
Kiyosu Y (1973) Sulfur isotopic fractionation among sphalerite, galena and sulfide ions. Geochem J 7:191-199
Kiyosu Y (1980) Chemical reduction and sulfur-isotope effects of sulfate by organic matter under hydrothermal conditions. Chem Geol 30:47-56
Krouse HR, Coplen TB (1997) Reporting of relative sulfur isotope-ratio data. Pure Appl Chem 69:293-295
Krouse HR, Ueda A, Campbell FA (1990) Sulphur isotope abundances in coexisting sulphate and sulphide: Kinetic isotope effects versus exchange phenomena. In: Stable Isotopes and Fluid Processes in Mineralization. Herbert HK, Ho SE (eds) The Univ Western Australia, Univ Extension Pub 23:226-243
Kulp JL, Ault WU, Feely HW (1956) Sulfur isotope abundances in sulfide minerals. Econ Geol 51:139-149
Larson PB, Maher K, Ramos FC, Chang Z, Gaspar M, Meinert LD (2003) Copper isotope ratios in magmatic and hydrothermal ore-forming environments. Chem Geol 201:337-350
Leggett JK (1980) British Lower Paleozoic black shales and their palaeo-oceanographic significance. J Geol Soc London 137:139-156
Li C, Ripley EM, Naldrett AJ (2003) Compositional variations of olivine and sulfur isotopes in the Noril'sk and Talnakh intrusions, Siberia: implications for ore-forming processes in dynamic magma conduits. Econ Geol 98:69-86
Longinelli A, Craig H (1967) Oxygen-18 variations in sulfate ions in sea water and saline lakes. Science 156:56-59
Luhr JF, Logan AV (2002) Sulfur isotope systematics of the 1982 El Chichón trachyandesite: an ion microprobe study. Geochim Cosmochim Acta 66:3303-3316
Lyons PC, Whelan JF, Dulong FT (1989) Marine origin of pyritic sulfur in the Lower Bakerstown coal bed, Castleman coal field, Maryland (U.S.A.). Intl J Coal Geol 12:329-348
Macnamara J, Thode HG (1950) Comparison of the isotopic constitution of terrestrial and meteoritic sulphur. Phys Rev 78:307-308
Maréchal CN, Télouk P, Albarède F (1999) Precise analysis of copper and zinc isotopic compositions by plasma-source mass spectrometry. Chem Geol 156:251-273
McGoldrick PJ, Large RR (1992) Geologic and geochemical controls on gold-rich stringer mineralization in the Que River deposit, Tasmania. Econ Geol 87:667-685
McKibben MA, Riciputi LR (1998) Sulfur isotopes by ion microprobe. In Applications of Microanalytical Techniques to Understanding Mineralizing Processes. Reviews in Economic Geology Vol. 7. McKibben MA, Shanks WC III, Ridley WI (eds) Soc Econ Geol, p 121-139
McSween HY Jr, Riciputi LR, Paterson BA (1997) Fractionated sulfur isotopes in sulfides of the Kaidun meteorite. Meteoritics Planet Sci 32:51-54
Misra KC, Gratz JF, Lu C (1996) Carbonate-hosted Mississippi Valley-type mineralization in the Elmwood-Gordonsville deposits, Central Tennessee zinc district: A synthesis. Soc Econ Geol Spec Pub 4:58-73

Mojzsis SJ, Coath CD, Greenwood JP, McKeegan KD, Harrison TM (2003) Mass-independent isotope effects in Archean (2.5 to 3.8 Ga) sedimentary sulfides determined by ion microprobe analysis. Geochim Cosmochim Acta 67:1635-1658

Monster J, Anders E, Thode HG (1965) $^{34}S/^{32}S$ ratios for the different forms of sulfur in the Orgueil meteorite and their mode of formation. Geochim Cosmochim Acta 29:773-779

Nakai N, Jensen ML (1964) The kinetic isotope effect in the bacterial reduction and oxidation of sulfur. Geochim Cosmochim Acta 28:1893-1912

Nielsen H, Pilot J, Grinenko LN, Grinenko VA, Lein AY, Smith JW, Pankina RG (1991) Lithospheric sources of sulphur. *In:* Stable Isotopes in the Assessment of Natural and Anthropogenic Sulphur in the Environment. Krouse HR, Grinenko VA (eds) SCOPE 43, J Wiley and Sons, p 65-132

Nriagu JO, Ress CE, Mekhtiyeva VL, Lein AY, Fritz P, Drimmie RJ, Pankina RG, Robinson RW, Krouse HR (1991) Hydrosphere. *In:* Stable Isotopes in the Assessment of Natural and Anthropogenic Sulphur in the Environment. Krouse HR, Grinenko VA (eds) SCOPE 43, J Wiley and Sons, p 177-265

Ohmoto H (1972) Systematics of sulfur and carbon isotopes in hydrothermal ore deposits. Econ Geol 67:551-578

Ohmoto H (1986) Stable isotope geochemistry of ore deposits. Rev Mineral 16:185-225

Ohmoto H, Goldhaber MB (1997) Sulfur and carbon isotopes. *In:* Geochemistry of Hydrothermal Ore Deposits. Barnes HL (ed) J Wiley and Sons, p 517-611

Ohmoto H, Lasaga AC (1982) Kinetics of reactions between aqueous sulfates and sulfides in hydrothermal systems. Geochim Cosmochim Acta 46:1727-1745

Ohmoto H, Rye RO (1979) Isotopes of sulfur and carbon. *In:* Geochemistry of Hydrothermal Ore Deposits. Barnes HL (ed) J Wiley and Sons, p 509-567

O'Neil JR (1986) Theoretical and experimental aspects of isotopic fractionation. Rev Mineral 16:1-40

O'Neil JR, Clayton RN, Mayeda TK (1969) Oxygen isotope fractionation in divalent metal carbonates. J Phys Chem 51:5547-5558

Ono S, Eigenbrode JL, Pavlov AA, Kharecha P, Rumble D, III, Kasting JF, Freeman KH (2003) New insights into Archean sulfur cycle from mass-independent sulfur isotope records from Hamersley Basin, Australia. Earth Planet Sci 213;15-30

Orr WL (1974) Changes in sulfur content and isotopic ratios of sulfur during petroleum maturation–study of Big Horn Paleozoic oils. Am Assoc Petrol Geol Bull 58:2295-2318

Paterson BA, Riciputi LR, McSween HY Jr. (1997) A comparison of sulfur isotope ratio measurement using two ion microprobe techniques and application to analysis of troilite in ordinary chondrites. Geochim Cosmochim Acta 61:601-609

Pavlov AA, Kasting JF (2002) Mass-independent fractionation of sulfur isotopes in Archean sediments: strong evidence for an anoxic Archean atmosphere. Astrobiology 2:27-41

Peters B (1959) Cosmic-ray produced radioactive isotopes as tracers for studying large-scale atmospheric circulation. J Atmos Terr Physics 13:351-370

Postgate JR (1984) The Sulfate-Reducing Bacteria. 2nd Ed., Cambridge University Press

Price FT, Shieh YN (1979) The distribution and isotopic composition of sulfur in coals from the Illinois Basin. Econ Geol 74:1445-1461

Price FT, Shieh YN (1986) Correlation between the $\delta^{34}S$ of pyritic and organic sulfur in coal and oil shale. Chem Geol 58:333-337

Puchelt H, Sabels BR, Hoering TC (1971) Preparation of sulfur hexafluoride for isotope geochemical analysis. Geochim Cosmochim Acta 35:625-628

Rai VK, Jackson TL, Thiemens MH (2005) Photochemical mass-independent sulfur isotopes in achondritic meteorites. Science 309:1062-1065

Rees CE, Holt BD (1991) The isotopic analysis of sulphur and oxygen. *In:* Stable Isotopes in the Assessment of Natural and Anthropogenic Sulphur in the Environment. Krouse HR, Grinenko VA (eds) SCOPE 43, J Wiley and Sons, p 43-64

Rees CE, Thode HG (1977) A ^{33}S anomaly in the Allende meteorite. Geochim Cosmochim Acta 41:1679-1682

Rees CE, Jenkins WJ, Monster J (1978) The sulphur isotope geochemistry of ocean water sulphate. Geochim Cosmochim Acta 42:377-382

Richardson CK, Rye RO, Wasserman MD (1988) The chemical and thermal evolution of the fluids in the Cave-in-Rock fluorspar district, Illinois: stable isotope systematics at the Deardorff mine. Econ Geol 83:765-783

Ripley EM, Al-Jassar TJ (1987) Sulfur and oxygen isotope studies of melt-country rock interaction, Babbitt Cu-Ni deposit, Duluth Complex, Minnesota. Econ Geol 82:87-107

Ripley EM, Li C (2003) Sulfur isotope exchange and metal enrichment in the formation of magmatic Cu-Ni-(PGE) deposits. Econ Geol 98:635-641

Ripley EM, Park Y-R, Li C, Naldrett AJ (1999) Sulfur and oxygen isotope evidence of country rock contamination in the Voisey's Bay Ni-Cu-Co deposit, Labrador, Canada. Lithos 47:53-68

Ripley EM, Li C, Shin D (2002) Paragneiss assimilation in the genesis of magmatic Ni-Cu-Co sulfide mineralization at Voisey's Bay, Labrador: $\delta^{34}S$, $\delta^{13}C$, and Se/s evidence. Econ Geol 97:1307-1318

Ripley EM, Lightfoot PC, Li C, Elswick ER (2003) Sulfur isotopic studies of continental flood basalts in the Noril'sk region: implications for the association between lavas and ore-bearing intrusions. Geochim Cosmochim Acta 67:2805-2817

Robinson BW (1973) Sulfur isotope equilibrium during sulfur hydrolysis at high temperatures. Earth Planet Sci Lett 18:443-450

Rumble D (2005) A mineralogical and geochemical record of atmospheric photochemistry. Am Mineral 90: 918-930

Rumble D III, Hoering TC, Palin JM (1993) Preparation of SF_6 for sulfur isotope analysis by laser heating sulfide minerals in the presence of F_2 gas. Geochim Cosmochim Acta 57:4499-4512

Rye RO (1993) The evolution of magmatic fluids in the epithermal environment: the stable isotope perspective. Econ Geol 88:733-753

Rye RO (2005) A review of the stable-isotope geochemistry of sulfate minerals in selected igneous environments and related hydrothermal systems. Chem Geol 215:5-36

Rye RO, Luhr JF, Wasserman MD (1984) Sulfur and oxygen isotope systematics of the 1982 eruptions of El Chichón volcano, Chiapas, Mexico. J Volcanol Geotherm Res 23:109-123

Rye RO, Bethke PM, Wasserman MD (1992) The stable isotope geochemistry of acid sulfate alteration. Econ Geol 87:225-262

Sælen G, Raiswell R, Talbot MR, Skei JM, Bottrell SH (1993) Heavy sedimentary sulfur isotopes as indicators of super-anoxic bottom-water conditions. Geology 21:1091-1094

Sakai H (1957) Fractionation of sulfur isotopes in nature. Geochim Cosmochim Acta 12:150-169

Sakai H (1968) Isotopic properties of sulfur compounds in hydrothermal processes. Geochem J 2:29-49

Sakai H, Dickson FW (1978) Experimental determination of the rate and equilibrium fractionation factors of sulfur isotope exchange between sulfate and sulfide in slightly acid solutions at 300 °C and 1000 bars. Earth Planet Sci Lett 39:151-161

Sakai H, Takenaka T, Kishima N (1980) Experimental study of the rate and isotope effect in sulfate reduction by ferrous oxides and silicates under hydrothermal conditions. Proc Third Internat Symp Water-Rock Interact, Edmonton, Alberta, p 75-76

Sakai H, Des Marais DJ, Ueda A, Moore JG (1984) Concentrations and isotope ratios of carbon, nitrogen, and sulfur in ocean-floor basalts. Geochim Cosmochim Acta 48:2433-2442

Salomons W (1971) Isotope fractionation between galena and pyrite and between pyrite and elemental sulfur. Earth Planet Sci Lett 11:236-238

Sangster DF (1968) Relative sulphur isotope abundances of ancient seas and strata-bound sulphide deposits. Geol Assoc Canada Proc 19:79-91

Santosh M, Masuda H (1991) Reconnaissance oxygen and sulfur isotopic mapping of Pan-African alkali granites and syenites in the southern Indian Shield. Geochem J 25:173-185

Sasaki A, Krouse HR (1969) Sulfur isotopes and the Pine Point Lead-Zinc mineralization. Econ Geol 64:718-730

Sasaki A, Ishihara S (1979) Sulfur isotopic composition of the magnetite-series and ilmenite-series granitoids in Japan. Contrib Mineral Petrol 68:107-115

Savarino J, Romero A, Cole-Dai J, Bekki S, Thiemens MH (2003) UV induced mass-independent sulfur isotope fractionation in stratospheric volcanic sulfate. Geophys Res Lett 30, doi:10.1029/2003GL018134

Schiller WR, von Gehlen K, Nielsen H (1969) Hydrothermal exchange and fractionation of sulfur isotopes in synthesized ZnS and PbS. Econ Geol 64:350-352

Schwarcz HP (1973) Sulfur isotope analyses of some Sudbury, Ontario, Ores. Can J Earth Sci 10:1444-1459

Seal RR II (2003) Stable-isotope geochemistry of mine waters and related solids. In: Environmental Aspects of Mine Wastes. Jambor JL, Blowes DW, Ritchie AIM (eds) Mineralogical Association of Canada Short Series 31:303-334.

Seal RR II, Ayuso RA, Foley NK, Clark SHB (2001) Sulfur and lead isotope geochemistry of hypogene mineralization at the Barite Hill gold deposit, Carolina Slate Belt, southeastern United States: a window into and through regional metamorphism. Min Deposita 36:137-148

Seal RR II, Wandless GA (2003) Sulfur isotope evidence for sea-floor mineralizing processes at the Bald Mountain and Mount Chase massive sulfide deposits, northern Maine. Econ Geol Monogr 11:567-587

Seal RR II, Rye RO, Alpers CN (2000a) Stable isotope systematics of sulfate minerals. Rev Mineral Geochem 40:541-602

Seal RR II, Turner RJW, Leitch CHB (2000b) Reconnaissance sulphur, oxygen, and hydrogen isotope data for mineralization and alteration in the Sullivan-North Star corridor and vicinity, British Columbia, Chapter 24. In: The Geological Environment of the Sullivan Deposit, British Columbia. Lydon JW, Höy T, Slack JF, Knapp ME (eds) Geol Assoc Canada, Min Dep Div, Spec Pub 1:470-481

Seyfried WE, Bischoff JL (1981) Experimental seawater-basalt interaction at 300 °C, 500 bars, chemical exchange, secondary mineral formation and implications for the transport of heavy metals. Geochim Cosmochim Acta 45:135-149

Shanks WC III (2001) Stable isotopes in seafloor hydrothermal systems. Rev Mineralogy Geochem 43:469-525

Shanks WC III, Bischoff JL, Rosenbauer RJ (1981) Seawater sulfate reduction and sulfur isotope fractionation in basaltic systems: interaction of seawater with fayalite and magnetite at 200-350 °C. Geochim Cosmochim Acta 45:1977-1995

Shanks WC III, Woodruff LG, Jilson GA, Jennings DS, Modene JS, Ryan BD (1987) Sulfur and lead isotope studies of stratiform Zn-Pb-Ag deposits, Anvil Range, Yukon: basinal brine exhalation and anoxic bottom-water mixing. Econ Geol 82;600-634

Shanks WC III, Böhlke JK, Seal RR II (1995) Stable isotopes in mid-ocean ridge hydrothermal systems: interactions between fluids, minerals, and organisms. *In:* Seafloor Hydrothermal Systems: Physical, Chemical, Biological, and Geological Interactions. Humphris SE, Zierenberg RA, Mullineaux LS, Thomson RE (eds) Geophys Monogr 91:194-221

Shanks WC III, Crowe DE, Johnson C (1998) Sulfur isotope analyses using the laser microprobe. *In:* Applications of Microanalytical Techniques to Understanding Mineralizing Processes. Reviews in Economic Geology, Vol. 7. McKibben MA, Shanks WC III, Ridley WI (eds) Society of Economic Geologists, p 141-153

Shearer CK, Layne GD, Papike JJ, Spilde MN (1996) Sulfur isotopic systematics in alteration assemblages in martian meteorite Allan Hills 84001. Geochim Cosmochim Acta 60:2921-2926

Shelton KL, Rye DM (1982) Sulfur isotopic compositions of ores from Mines Gaspé, Quebec: an example of sulfate-sulfide isotopic disequilibria in ore-forming fluids with applications to other porphyry type deposits. Econ Geol 77:1688-1709

Shimoyama T, Yamazaki K, Iijima A (1990) Sulphur isotopic composition in the Paleogene coal of Japan. Intl J Coal Geol 15:191-217

Smith JW, Batts BD (1974) The distribution and isotopic composition of sulfur in coal. Geochim Cosmochim Acta 38:121-133

Smith JW, Gould KW, Rigby D (1982) The stable isotope geochemistry of Australian coals. Organic Geochem 3:111-131

Solomon M, Rafter TA, Jensen ML (1969) Isotope studies on the Rosebery, Mount Farrell and Mount Lyell ores, Tasmania. Min Deposita 4:172-199

Solomon M, Eastoe CJ, Walshe JL, Green GR (1988) Mineral deposits and sulfur isotope abundances in the Mount Read Volcanics between Que River and Mount Darwin, Tasmania. Econ Geol 83:1307-1328

Spiker EC, Pierce BS, Bates AL, Stanton RW (1994) Isotopic evidence for the source of sulfur in the Upper Freeport coal bed (west-central Pennsylvania, U.S.A.). Chem Geol 114:115-130

Strauss H (1997) The isotopic composition of sedimentary sulfur through time. Palaeogeogr Palaeoclimat Palaeoecol 132:97-118

Suvorova VA (1978) Sulfur isotopic distribution between Mo, W and Sb sulfides by experimental fractionation. Doklady Akademii Nauk SSSR 243:485-488

Suvorova VA, Tenishev AS (1976) An experimental study of equilibrium distribution of sulfur isotopes between Mo, Pb, Zn, and Sn sulfides. Geokhimiya 11:1739-1742

Szabo A, Tudge A, Macnamara J, Thode HG (1950) The distribution of S^{34} in nature and the sulfur cycle. Science 111:464-465

Szaran J (1996) Experimental investigation of sulphur isotopic fractionation between dissolved and gaseous H_2S. Chem Geol 127:223-228

Taylor BE, Beaudoin G (2000) Sulphur isotope stratigraphy of the Sullivan Pb-Zn-Ag deposit, B.C.: evidence for hydrothermal sulphur, and bacterial and thermochemical sulphate reduction, Chapter 37. *In* The geological environment of the Sullivan deposit, British Columbia. Lydon JW, Höy T, Slack JF, Knapp ME (eds) Geol Assoc Canada, Min Dep Div, Spec Pub 1:696-719

Taylor BE, Wheeler MC (1994) Sulfur- and oxygen-isotope geochemistry of acid mine drainage in the western United States. *In:* Environmental geochemistry of sulfide oxidation. Alpers CN, Blowes DW (eds) Amer Chem Soc Symp Ser 550:481-514

Thode HG, Rees CE (1971) Measurements of sulphur concentrations and the isotope ratios $^{33}S/^{32}S$, $^{34}S/^{32}S$ and $^{36}S/^{32}S$ in Apollo 12 samples. Earth Planet Sci Lett 12:434-438

Thode HG, Macnamara J, Collins CB (1949) Natural variations in the isotopic content of sulphur and their significance. Can J Res B27:361-373

Thode HG, Kleerekoper H, McElcheran D (1951) Isotope fractionation in the bacterial reduction of sulphate. Research, London 4:581-582

Thode HG, Monster J, Dunford HB (1961) Sulfur isotope geochemistry. Geochim Cosmochim Acta 25:159-174

Thode HG, Dunford HB, Shima M (1962) Sulfur isotope abundances of the Sudbury district and their geologic significance. Econ Geol 57:565-578

Thode HG, Cragg CB, Hulston JR, Rees CE (1971) Sulphur isotope exchange between sulphur dioxide and hydrogen sulphide. Geochim Cosmochim Acta 35:35-45

Torssander P (1989) Sulfur isotope ratios of Icelandic rocks. Contrib Mineral Petrol 102:18-23

Ueda A, Sakai H (1984) Sulfur isotope study of Quaternary volcanic rocks from the Japanese island arc. Geochim Cosmochim Acta 44:579-587

Urey HC (1947) The thermodynamic properties of isotopic substances. J Chem Soc 1947:562-581

Westgate LM, Anderson TF (1982) Extraction of various forms of sulfur from coal and shale for stable sulfur isotope analysis. Anal Chem 54:2136-2139

Whelan JF, Cobb, JC, Rye RO (1988) Stable isotope geochemistry of sphalerite and other mineral matter in coal beds of the Illinois and Forest City basins. Econ Geol 83:990-1007

Whelan JF, Rye RO, deLorraine W (1984) The Balmat-Edwards zinc-lead deposits – synsedimentary ore from Mississippi Valley-type fluids. Econ Geol 79: 239-265

Woodruff LG, Shanks WC III (1988) sulfur isotope study of chimney minerals and vent fluids from 21 °N, East Pacific Rise: hydrothermal sulfur sources and disequilibrium sulfate reduction. J Geophys Res 93: 4562-4572

Zhu XK, O'Nions RK, Guo Y, Belshaw NS, Rickard D (2000) Determination of natural Cu-isotope variation by plasma-source mass spectrometry: implications for use as geochemical tracers. Chem Geol 163:139-149

Zientek ML, Ripley EM (1990) Sulfur isotope studies of the Stillwater Complex and associated rocks, Montana. Eco Geol 85:376-391

Sulfides in Biosystems

Mihály Pósfai

Department of Earth and Environmental Sciences
University of Veszprém
Veszprém, Hungary
e-mail: posfaim@almos.vein.hu

Rafal E. Dunin-Borkowski

Department of Materials Science and Metallurgy
University of Cambridge
Cambridge, United Kingdom

INTRODUCTION

Organisms that live on and near the surface of the Earth affect the cycling of sulfur and metals and thus the formation and decomposition of sulfide minerals. Biological mediation of mineral formation can take many forms. Some organisms have evolved to synthesize minerals that are used for a particular function, such as structural support, protection against predators, hardening, or magnetic sensing. In these cases, the organism exerts strict control over the properties and the location of the mineral. The process by which such minerals form is termed biologically controlled mineralization (BCM) (Lowenstam and Weiner 1989).

Biominerals can also form as a byproduct of the metabolism of organisms, or as a consequence of their mere presence. Life can create chemical environments that result in the precipitation of minerals, and biological surfaces can serve as nucleation sites for mineral grains. In such cases, the adventitious deposition of minerals is termed biologically induced mineralization (BIM) (Lowenstam and Weiner 1989). Whereas only a few examples of the formation of sulfide minerals by BCM are known, iron sulfides form in vast quantities by BIM and affect the global cycling of iron, sulfur, oxygen, and carbon (Canfield et al. 2000; Berner 2001).

Organisms are also able to break minerals down. The dissolution of sulfides can be enhanced by biological processes, while some micro-organisms gain their energy by oxidizing the sulfur or the metal in sulfide minerals, thereby converting sulfides into dissolved species or oxides (Kappler and Straub 2005). The biological mediation of both the precipitation and the dissolution of sulfides can be used for practical purposes, such as bioremediation and bioleaching.

Over the past decade, several reviews have been published on biomineralization, many of which include details on sulfides. In the *Reviews in Mineralogy & Geochemistry* series, three volumes have been devoted to interactions between minerals and organisms (Banfield and Nealson 1997; Dove et al. 2003; Banfield et al. 2005). A further short course volume, which includes several chapters on sulfides, was published by the Mineralogical Association of Canada (McIntosh and Groat 1997). A textbook on environmental mineralogy, published by the European Mineralogical Union (Vaughan and Wogelius 2000), also contains material related to biominerals. More recently published general books on BCM include those by Mann (2001) and Baeuerlein (2000, 2004).

The aim of the present chapter is to discuss some aspects of sulfide biomineralization and sulfide bioweathering. In order to avoid repeating the content of recent reviews, this chapter does not provide a comprehensive treatment of interactions between sulfide minerals and organisms. Instead, its primary focus is a description of the properties of biogenic sulfide minerals that distinguish them from their inorganically-formed counterparts. The relationships between mineral properties and biological functions are discussed, some aspects of sulfide formation by BIM are highlighted, and sulfide bioweathering processes are mentioned briefly. Since iron sulfides are by far the most important and abundant sulfide minerals in biosystems, most of this chapter deals with such minerals.

BIOLOGICAL FUNCTION AND MINERAL PROPERTIES: CONTROLLED MINERALIZATION OF IRON SULFIDES

Biologically controlled mineralization is a highly regulated process that results in the formation of minerals that have species-specific physical and chemical properties. These properties include size, morphology, structure, crystallographic orientation, composition, and texture. As discussed by Mann (2001), several levels of regulation combine in BCM to provide distinct mineral properties (Table 1). Chemical control through coordinated ion transport is involved in producing supersaturated solutions in spatially separated spaces such as vesicles or gaps in organic frameworks. Organic surfaces play a crucial role in providing nucleation sites and in selecting the phase and orientation of the nucleating mineral (Weiner and Dove 2003). Chemical, spatial, and morphological regulations combine to shape the growing crystals and to assemble them into complex architectures.

Minerals can serve various functions in living organisms. In association with organic materials, they can form inorganic-organic composites that have favorable mechanical properties. Well-known examples include bones that are used for structural support, teeth that are used for grinding, and shells that are used for mechanical strengthening and protection.

Table 1. Processes and mechanisms that control the properties of minerals formed by biologically controlled mineralization, based on concepts that are described by Mann (2001).

Type of Regulation	Key Factors of Mineral Formation that are Controlled	Means of Control	Result
Chemical	Ion concentration in solution	Coordinated ion transport	Supersaturation and nucleation
	Crystal growth	Promotors and inhibitors	- Controlled crystal morphology - Phase transformations
Spatial	Supersaturation and crystal growth	Vesicles or organic framework	Controlled location, size and shape of the mineral
Structural	Nucleation	Organic surfaces as templates, molecular recognition at organic/inorganic interfaces	- Polymorph selection - Controlled crystallographic orientation
Morphological and constructional	Nucleation and growth	Organic boundaries, vectorial regulation	- Complex morphologies - Time-dependent patterning

Table 2. Sulfide minerals that are formed by biologically controlled mineralization.

Organism	Mineral	Function	References
Magnetotactic bacteria	Greigite, Fe_3S_4	Magnetic sensing	Farina et al. 1990; Mann et al. 1990; Heywood et al. 1990; 1991
	Mackinawite, FeS	Precursor to greigite	Pósfai et al. 1998a,b
	Cubic FeS (identified tentatively)	Precursor to greigite	Pósfai et al. 1998a,b
Scaly-foot gastropod	Pyrite, FeS_2	Mechanical protection	Warén et al. 2003; Suzuki et al. 2006
	Greigite, Fe_3S_4	Mechanical protection	Warén et al. 2003; Suzuki et al. 2006
	Mackinawite, FeS	Precursor to greigite	Suzuki et al. 2006

However, the biological uses of minerals are not only mechanical. Biominerals can also serve as optical, magnetic, or gravity sensing devices, and may be used for the storage of materials such as iron (Mann 2001; Baeuerlein 2004).

In contrast to some mineral groups that are common functional materials in many organisms (e.g., carbonates, phosphates, silica), only a few sulfide minerals are known to serve biological functions (Table 2). Although these sulfide minerals include common species such as pyrite (FeS_2), their formation pathways by BCM were only discovered in the last 15 years. Greigite (Fe_3S_4) is used for magnetic sensing in magnetotactic bacteria (Farina et al. 1990; Mann et al. 1990; Rodgers et al. 1990), and greigite and pyrite both serve as hardening materials on the foot of a deep-sea snail species (Warén et al. 2003; Suzuki et al. 2006). The physical and chemical properties and the apparent functions of these sulfide biominerals are reasonably well known. However, very little is understood about the specific biological control mechanisms that govern crystal nucleation and growth (as listed in Table 1).

Biologically controlled mineralization in magnetotactic bacteria

Magnetotactic bacteria contain intracellular magnetic iron oxide or sulfide minerals that are typically organized in chains. Such cells are aligned by magnetic fields, and as a result the bacteria are constrained to swim parallel to the direction of the geomagnetic field in their natural aquatic environment (Blakemore 1975). This magnetic alignment mechanism enables the bacteria to find their optimal positions in environments that are characterized by vertical chemical gradients (Frankel et al. 1997). Since geomagnetic field lines are inclined with respect to the surface of the Earth (except at the equator), the bacteria do not have to search for their optimal chemical environment in three dimensions, but are guided up and down along the field lines. Nevertheless, several questions remain about the utility of magnetotaxis; neither the benefit of magnetotaxis at the equator, nor the reason for the presence of south-seeking bacteria in the Northern Hemisphere (Simmons et al. 2005) is fully understood.

The term magnetosome refers to an intracellular magnetic mineral grain enclosed by a biological membrane. Such magnetosome membranes were shown to exist in magnetite-producing bacteria (Balkwill et al. 1980), and some of the specific membrane proteins and their encoding genes have been identified (Komeili et al. 2004; Schüler 2004; Fukumori 2000). The magnetosome membrane provides spatial, chemical, structural, and morphological regulation (Table 1) of the nucleation and growth of magnetite crystals. The membrane controls the transport of ions into the magnetosome vesicle, a delimited space in which supersaturation can be achieved (Fig. 1). It is also likely that the membrane provides the organic template for the oriented nucleation of magnetite crystals (Bazylinski and Frankel 2004). The growth of magnetite crystals is controlled by an unknown mechanism to produce well-defined morphologies. Recently, it was found that magnetite particles are assembled into chains by an acidic membrane protein that anchors the magnetosomes to a filamentous structure (Scheffel et al. 2005) (Fig. 1).

The presence of a magnetosome membrane has never been established in sulfide-producing bacteria. Since such bacteria are not yet available in pure culture, it is difficult to determine whether the iron sulfide crystals are enclosed by membranes that are similar to those in magnetite-producing cells. Little is therefore known about the biological regulation of mineral formation in sulfide-bearing bacteria. However, the properties of the inorganic sulfide phases themselves are fairly well understood. These properties can provide indirect information about the mineral-forming process.

The biomineralization of magnetite and sulfides by magnetotactic bacteria, including their micro- and molecular biology and ecology, has been reviewed by Bazylinski and Moskowitz (1997), Baeuerlein (2003), and Bazylinski and Frankel (2003, 2004). Some of the mineralogical aspects of sulfide formation in magnetotactic bacteria are now described, and recent measurements of the magnetic microstructures of chains of greigite magnetosomes in magnetotactic cells are reviewed.

Sulfide-producing magnetotactic bacteria

Sulfide-producing magnetotactic organisms are known to exist in anaerobic marine environments, saltwater ponds, and sulfur-rich marshes (Farina et al. 1990; Mann et al. 1990; Bazylinski and Frankel 2004). The cell morphologies of sulfide-bearing magnetotactic bacteria appear to be very similar in geographically distant locations (Farina et al. 1990; Mann et al. 1990; Bazylinski et al. 1990; Pósfai et al. 1998b; Simmons et al. 2004). One organism is termed the many-celled magnetotactic prokaryote (Rodgers et al. 1990), or alternatively

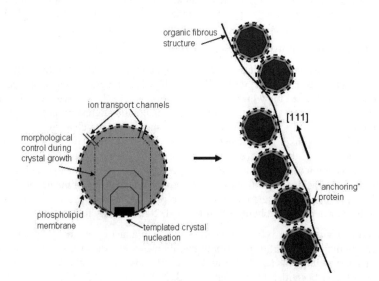

Figure 1. Stages of biologically controlled mineralization in magnetotactic bacteria, as known in the case of magnetite-producing cells. Iron sulfide-producing species may use similar strategies for mineralizaton. The inorganic crystal nucleates and grows inside a magnetosome vesicle, and then the magnetosomes are attached to a filamentous structure by an acidic protein. (Based largely on the model by Scheffel et al. 2005.)

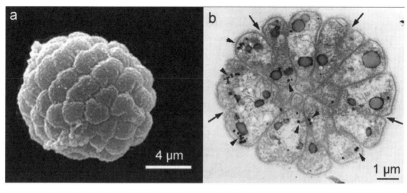

Figure 2. (a) SEM image of the magnetotactic multicellular aggregate (MMA) that consists of many cells and moves as a single unit. (b) Ultrathin section of an MMA. The arrows mark invaginations of the cell wall, indicating the sites of cell division, and the arrowheads mark iron sulfide magnetosomes. [Used with permission of Elsevier, from Keim et al. (2004) *J. Structural Biology*, Vol. 145, Figs. 3c and 5, p. 254-262.]

the magnetotactic multicellular aggregate (MMA) (Lins and Farina 1999). This organism consists of an aggregate of 10 to 30 cells that are arranged in an ordered fashion, enclosing an acellular internal compartment. Each cell contains one or more chains of greigite crystals, which are aligned approximately parallel to each other within the individual cells (Keim et al. 2004) (Fig. 2). The MMA moves as a single unit, guided by Earth's magnetic field. Other common morphological types include rod-shaped cells that may contain single or multiple chains of iron sulfide crystals (Heywood et al. 1991; Bazylinski et al. 1995) (Fig. 3). Although attempts to cultivate sulfide-producing magnetotactic bacteria in pure culture have to date been unsuccessful, fluorescent in situ hybridization studies indicated that the MMA is closely related to

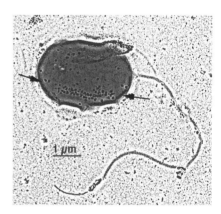

Figure 3. A single, rod-shaped magnetotactic cell that contains a double chain of iron sulfide magnetosomes between the two arrows. (Image from Kasama et al. 2006.)

known sulfate reducers among the δ-proteobacteria (DeLong et al. 1993), whereas a large rod was found to be a member of the γ-proteobacteria and is likely involved in metal cycling (Simmons et al. 2004).

Sulfide-bearing magnetotactic bacteria live below the oxic-anoxic transition zone (OATZ), where H_2S is abundant (Bazylinski and Frankel 2004). MMAs and rod-shaped cells have been observed in distinct zones below the OATZ in Salt Pond, Massachusetts, USA (Simmons et al. 2004). Whereas the concentration of MMAs was largest just below the OATZ, rod-shaped cells appeared to be broadly distributed vertically in a zone that was characterized by the absence of dissolved oxygen and by a high H_2S concentration (Fig. 4). In such an environment, the benefit of possessing an internal compass is unclear. It was speculated that intracellular iron sulfide (and oxide) crystals could serve purposes other than magnetically-assisted navigation (Simmons et al. 2004; Flies et al. 2005). In addition, populations of south-seeking magnetotactic bacteria were recently observed in the Northern Hemisphere (Simmons et al. 2006), challenging the widely-held view about the utility of magnetic navigation for these mi-

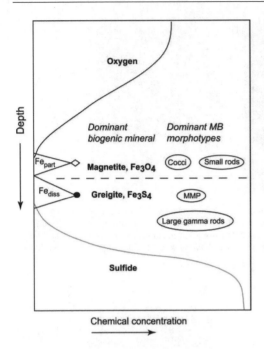

Figure 4. The positions of the types of magnetotactic bacteria (MB) in the water column of Salt Pond, Massachusetts, with respect to depth and the concentrations of oxygen and sulfide. Magnetite-bearing *cocci* and small rods predominate at the oxic-anoxic transition zone, whereas iron sulfide-bearing magnetotactic multicellular prokaryotes (MMP) and large gamma rods predominate below it. Peaks in the concentrations of particulate and dissolved iron (Fe_{part} and Fe_{diss}, respectively) are also shown. [Used with permission of American Society for Microbiology, from Simmons et al. (2004), *Applied and Environmental Microbiology*, Vol. 10, Fig. 7, p. 6230-6239.]

cro-organisms. Further studies on the physiology and ecology of magnetotactic bacteria will be required in order to establish whether the synthesis and presence of magnetosomes serve purposes other than magnetic sensing.

Structures and compositions of iron sulfide magnetosomes

The inorganic part of each iron sulfide magnetosome is typically a crystal of greigite. However, in freshly-collected cells (a few days old), mackinawite (FeS) was identified (Pósfai et al. 1998a). When the samples were stored in air, mackinawite was observed to convert into greigite. This observation suggests that non-magnetic mackinawite precipitates initially, and then converts into magnetic greigite through the loss of ¼ of its iron. Disordered crystals may represent transitional states between the mackinawite and greigite structures and suggest that the transformation takes place in the solid state. Structural similarities between the cubic close-packed sulfur substructures of mackinawite and greigite would allow such a conversion to take place by the diffusion of iron atoms, leaving the sulfur atomic arrangement intact (Fig. 5).

The transformation that was observed in the stored specimens is also thought to take place within living bacteria. The transformation is likely to be faster in living bacteria than in the stored samples, since non-magnetic mackinawite cannot be used for magnetotaxis. In addition to mackinawite, cubic FeS with a sphalerite-type structure was identified tentatively in some magnetotactic cells, based on electron diffraction patterns (Pósfai et al. 1998a). Since this initial identification of cubic FeS, several further attempts to confirm its presence have been unsuccessful. It remains to be established unequivocally that cubic FeS is also a precursor of greigite in magnetotactic bacteria.

The conversions of iron sulfides in bacteria follow similar paths as the well-known phase transformations of authigenic sulfides that form by BIM in anoxic sediments (see Luther and Rickard, 2006; this volume, and the section below on BIM sulfides). However, in marine sediments the final product of iron sulfide formation is commonly pyrite instead of greigite (Schoo-

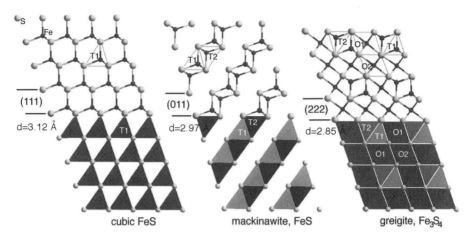

Figure 5. The structural relationships among cubic FeS, mackinawite, and greigite. Light and dark circles represent sulfur and iron atoms, respectively. The lower half of the image shows the same structures in polyhedral respresentation. T1 and T2 mark tetrahedral, and O1 and O2 mark octahedral positions. (Figure from Pósfai et al. 1998b.)

nen 2004). Rickard et al. (2001) found that mackinawite converts to either greigite or pyrite, depending on the presence or absence of carboxylic aldehydes in the solution, respectively. Even though the organic compound was present in very low concentration, it served as a switch that determined the mineral phase. Similar molecular switches have not yet been identified in magnetotactic bacteria, but the concept of a chemical control mechanism over the selection of the mineral phase is consistent with the principles of BCM that are outlined in Table 1.

Greigite magnetosomes typically exhibit patchy contrast in transmission electron microscope (TEM) images (Heywood et al. 1990; Pósfai et al. 1998b). This appearance may be related to the presence of defects that arise from the solid-state transformation of mackinawite into greigite. It may also result from thickness variations. High-resolution TEM images provide evidence that many greigite magnetosomes are aggregates of smaller, flake-like fragments that combine to form a single crystal (Kasama et al. 2006), and that such aggregates can have highly irregular shapes. Synthetic mackinawite was found to precipitate in the form of plate-like nanocrystals with an average size of a few nm (Wolthers et al. 2003; Ohfuji and Rickard 2006). The formation of primary mackinawite in magnetotactic bacteria by a similar mechanism, through the nucleation and aggregation of plate-like nanocrystals, cannot be ruled out.

Although greigite magnetosomes are typically pure iron sulfides, in some samples copper was found to substitute for iron by up to 12 at% (Bazylinski et al. 1993a; Pósfai et al. 1998b). The copper content appeared to be independent of cell type, but was related to geographical location, and therefore presumably to the copper concentration in the environment of the bacteria. When the samples of greigite-containing bacteria are stored in air, the greigite crystals oxidize partially, and an amorphous iron oxide shell forms on them (Lins and Farina 2001; Kasama et al. 2006) (Fig. 6). This phenomenon was observed to reduce the magnetic moments of the magnetosomes (Kasama et al. 2006).

Magnetic sensing with sulfide magnetosomes

Magnetotactic bacteria are the only organisms that are known to make use of the magnetic properties of iron sulfide crystals for navigation. Other organisms that navigate magnetically include algae, protists, bees, ants, fishes, turtles, and birds (Wiltschko and Wiltschko 1995;

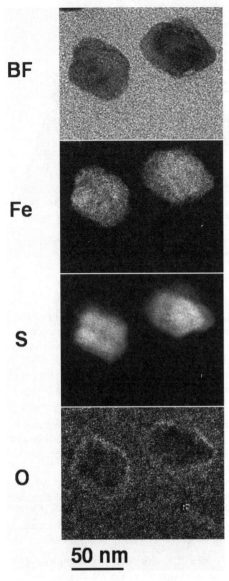

Figure 6. Three-window, background-subtracted elemental maps of two iron sulfide magnetosomes from a magnetotactic bacterium. BF: bright-field image; the images marked Fe, S, and O show the distributions of the respective elements. The magnetosomes have a crystalline iron sulfide core and an amorphous iron oxide shell.

Walker et al. 2002). In the few cases for which the mechanism of magnetic sensing is known, the mineral involved is magnetite (Kirschvink et al. 2001; Winklhofer et al. 2001; Diebel et al. 2000). Magnetite also occurs in the human brain (Kirschvink et al. 1992; Dobson 2001), but it remains to be established whether it has a biological function.

Magnetosomes in magnetotactic bacteria are typically arranged in chains, with each chain behaving as a magnetic dipole (Frankel 1984). The Earth's magnetic field exerts a torque on this dipole, and competes with the effect of Brownian motion that tends to randomize the orientation of the cell. When the magnetic moment of a cell is known, its average orientation with respect to the external magnetic field can be calculated on the basis of the Langevin function, as discussed in detail by Bazylinski and Moskowitz (1997). Both calculations and experiments show that magnetite-producing bacteria typically contain enough magnetosomes to allow their cells to migrate parallel to the small (50 µT) magnetic field of the Earth with a net velocity that is in excess of 90% of their forward velocity (Frankel 1984; Schüler et al. 1995). Since the magnetic induction of greigite (0.16 T) is only about one quarter of that of magnetite (0.60 T) (Dunlop and Özdemir 1997), a cell needs a larger number of greigite than magnetite crystals (of similar size) in order to be magnetotactic (Heywood et al. 1991).

The mechanism of magnetic alignment described above requires the magnetosome crystals to be magnetized approximately parallel to each other at room temperature. The combined effects of their shape and magnetocrystalline anisotropy, as well as interparticle interactions between magnetosomes, determine the magnetic domain state, and therefore the net magnetic dipole moment, of each magnetosome. Based on theoretical considerations, Diaz-Ricci and Kirschvink (1992) calculated the size and shape-dependent magnetic properties of greigite, and determined that the sizes of bacterial magnetosomes place them at the boundary between the superparamagnetic and single magnetic domain size range for isolated crystals. They also reported that crystal shape affects the magnetic properties of greigite significantly. Whereas isolated ~70-nm crystals with prismatic habits were calculated to be single domains, spheroidal particles of similar size were superparamagnetic at room temperature. Experimental results obtained by Chen et al. (2005) also indicate that the magnetic properties of acicular and irregularly-shaped greigite nanocrystals differ.

Measurements of the magnetic properties of greigite-producing magnetotactic bacteria are scarce. As a result of the present inability to grow sulfide-producing magnetotactic organisms in pure culture, it has not been possible to apply bulk magnetic characterization techniques to their study. Recently, Kasama et al. (2006) used off-axis electron holography in the TEM to study the magnetic properties of greigite magnetosomes in rod-shaped cells. Electron holography is a powerful and relatively specialized technique that can be applied to the study of magnetic and electrostatic fields in materials (Dunin-Borkowski et al. 2004). By using electron holography, it is possible to measure parameters such as the magnetic moments and coercivities of individual magnetosomes and their chains quantitatively, as well as to form two-dimensional images of the projected magnetic induction (Dunin-Borkowski et al. 1998, 2001).

The magnetic properties of sulfide magnetosomes were studied in a cell that was at the point of division (Fig. 7a) (Kasama et al. 2006). The structures of some of the magnetosomes in this cell were studied using selected-area electron diffraction and high-resolution TEM, their compositions were determined using energy-filtered TEM, and their three-dimensional morphologies were studied using high-angle annular dark-field electron tomography. The electron holography experiments revealed that the direction of the magnetic field is less uniform within the magnetosome chains, and undulates to a greater degree than in magnetite-containing cells. In addition, some of the greigite crystals (marked by arrows in Fig. 7b) appeared to be only weakly magnetic, with the apparent saturation magnetic induction varying between 0 and 0.16 T for individual crystals in the cell. This behavior could result either from the presence of non-magnetic sulfides other than greigite, or from the fact that some of the greigite crystals may be magnetized in a direction that is almost parallel to that of the electron beam. Since electron holograms are only sensitive to the components of the magnetic field in the plane of the specimen, i.e., perpendicular to the electron beam direction, magnetic crystals with large out-of-plane components of their magnetization would appear to be non-magnetic. Diffraction patterns obtained from several of the apparently non-magnetic crystals were found to be consistent with greigite. The diffraction patterns also showed that the greigite crystals were oriented randomly within the cell, and that their elongation directions appeared to be random. The variable degree of the apparent magnetization of the greigite magnetosomes is therefore likely to be primarily a consequence of their random orientations. Figure 7b also reveals that the magnetic contours within individual crystals are generally parallel to their axes of elongation. These observations are consistent with the calculations of Diaz-Ricci and Kirschvink (1992) that suggest that shape anisotropy has a much larger effect on the magnetization of greigite than magnetocrystalline anisotropy.

Interestingly, the multiple magnetosome chain shown in Figure 7b contains magnetite crystals in addition to the greigite magnetosomes (Kasama et al. 2006). Whereas the greigite grains are equidimensional or only slightly elongated, the iron oxide particles have distinctly elongated shapes, and their axes of elongation are aligned parallel to that of the magnetosome chain (Fig. 7c). Their elongated morphologies constrain their magnetic contours to be parallel to the chain axis (Fig. 7d). In addition, since magnetite is much more strongly magnetic than greigite, the magnetite particles contribute as much as ~30% of the total magnetic moment of the chain, which was measured by electron holography to be 1.8×10^{-15} Am^2. Whereas the randomly-oriented greigite particles produce an undulating magnetic field, the well-aligned magnetite particles provide a distinct "magnetic backbone" to the chain (Fig. 7d). The presence of both greigite and magnetite magnetosomes in the same cells was reported previously by Bazylinski et al. (1993b). The distinct shapes and orientations of these two mineral species suggest that their formation may be regulated by different biological mechanisms.

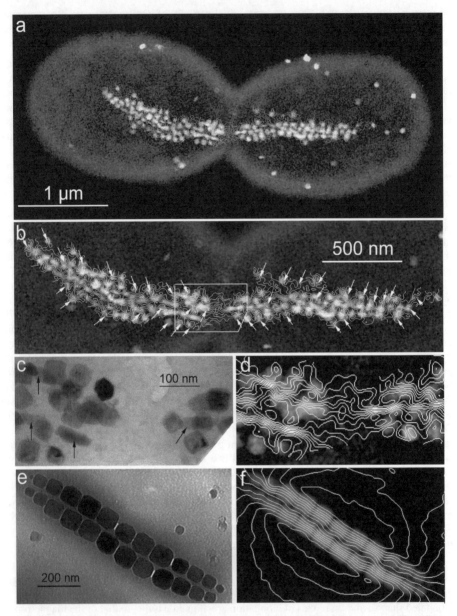

Figure 7. (a) Compositional map of a rod-shaped cell that contains iron sulfide magnetosomes. The cell was caught at the point of cell division. The image was constructed from electron energy-loss maps. (b) Magnetic induction map of the magnetosome chain in (a), obtained from electron holography. The magnitude and the direction of magnetic induction within the crystals is represented by the density and direction of the contour lines, respectively. The arrowed particles appear to be either non-magnetic or weakly magnetic. (c) Bright-field TEM image of the boxed region in (b). The arrowed particles are elongated magnetite crystals. (d) Magnetic induction map from the same area that is shown in (c). The density of the contour lines is much higher in the elongated magnetite crystals than in the equidimensional greigite crystals. (e) Bright-field image and (f) magnetic induction map obtained from a double magnetite chain from a magnetotactic coccus. In contrast to the greigite chain in (b), the magnetic contour lines are straight and their densities uniform within the particles in (f). [Based on images from Kasama et al. 2006.]

As mentioned above, the biological regulation of the nucleation and growth of magnetosomes has only been studied in magnetite-producing bacteria. The use of analogies with magnetite formation to explain control over greigite deposition in bacteria appears to be limited, because there are significant differences between the properties of sulfide and oxide magnetosomes. Some of these differences are illustrated in Figure 7f, which shows a magnetic contour map of a double magnetite chain in a cell of a magnetotactic coccus. The magnetite crystals in this cell have identical morphologies along the entire chain (Fig. 7e), and their [111] axes are aligned with the magnetosome chain within a few degrees, resulting in the same direction of magnetic induction in each crystal (Simpson et al. 2005a). In contrast, the dividing cell in Figure 7a appears to exhibit a lack of control over the shapes and orientations of the greigite crystals. As a result, the magnetic induction is highly variable along the magnetosome chain. The bacterium appears to compensate for the magnetically less efficient assembly of magnetosomes, as well as for the lower magnetization of greigite than magnetite, by forming a multiple chain that contains several times as many crystals as the magnetite chain shown in Figure 7e.

Not only the processes of crystal nuclation and growth, but also the mechanisms of chain assembly appear to be different in the magnetite and greigite producers. Whereas magnetite particles in magnetotactic spirilla were found to be aligned along a filament that runs along the long axis of the cell (Scheffel et al. 2005; Komeili et al. 2006), electron tomography experiments on the dividing cell shown in Figure 7a revealed a three-dimensional arrangement of the crystals in the multiple greigite chain (Kasama et al. 2006).

To date, the magnetic moments of magnetosome chains in three different strains of magnetite-producing bacteria (MS-1 and MV-1, Dunin-Borkowski et al. 1998, 2001; Itaipu-1, McCartney et al. 2001) and in two cells of unnamed sulfide producers (Kasama et al. 2006) have been measured experimentally using electron holography. Remarkably, in the different types of cell the magnetic moments per cell are all the same to within a factor of two. Therefore, even though the biomineralization processes and the properties of magnetosomes may vary between different groups of magnetotactic bacteria, natural selection appears to have favored structures that serve the function of magnetic sensing equally well.

Mechanical protection: iron sulfides on the foot of a deep-sea snail

Hydrothermal vents in mid-ocean ridge systems provide chemical energy for diverse populations of chemoautotrophic bacteria (as reviewed by Jannasch and Mottl 1985). The abundance of micro-organisms at deep-sea vents makes it possible for more complex organisms (such as worms, shrimp, crabs, clams, mussels, gastropods, anemones, barnacles, etc.) to thrive in an environment where no light is available. Thus, entire ecosystems depend on geochemical rather than on solar energy. Based on variations in the species composition of invertebrate communities, faunas at oceanic vents are recognized to belong to six provinces (Van Dover et al. 2001). One of these provinces is the central ridge system in the Indian Ocean, where, among many other animals, the vent fields harbor a snail that bears mineralized scales on its foot (Van Dover et al. 2001).

The sides of this gastropod's foot are covered in a tile-like fashion by black sclerites (Fig. 8). The scales consist of iron sulfide minerals (Warén et al. 2003), making this snail the first known organism that uses sulfide minerals for structural support. Initially, greigite and pyrite were described as the primary mineral phases (Warén et al. 2003; Goffredi et al. 2004), but mackinawite was also subsequently identified (Suzuki et al. 2006). The presence of greigite makes the scales magnetic. As Suzuki et al. (2006) note, "it is rare for animals to produce macroscopic materials that stick to a hand magnet." The only other known organisms that produce such structures are chiton mollusks that have magnetite-bearing radular teeth (Lowenstam 1962).

The spatial distributions, microstructures, magnetic and mechanical properties, and the isotopic compositions of the iron sulfide minerals in this organism were studied by Suzuki et al. (2006). The sulfides were found to be present in three distinct layers, which were defined both by their positions and by their mineral species. An "iron sulfide" layer covers the outer surface of the sclerites and consists primarily of greigite. A "mixed layer" and a "conchiolin layer" occur within the organic matrix, and consist of nanocrystalline pyrite and mackinawite, respectively (Fig. 9). The greigite crystals in the iron sulfide layer are rod-shaped and highly elongated along [110], with average lengths and widths of 118 and 14 nm, respectively. The space between the greigite rods is filled by fibrous mackinawite (Fig. 10). The orientation relationship between the two phases appears to be the same as that described above for sulfides in magnetotactic bacteria, although the boundary plane is different. The pyrite in the mixed layer has an unusual appearance, since it takes the form of nanoparticles that are as small as 3 nm. Remarkably, the nanoparticles have a consistent crystallographic orientation. In the conchiolin layer, mackinawite forms ~3–10 nm particles within amorphous iron sulfide.

Figure 8. Two views of the "scaly-foot gastropod" that has iron sulfide sclerites on its foot. [Used by permission of Elsevier, from Suzuki et al. (2006), *Earth and Planetary Science Letters*, Vol. 242, Fig. 1, p. 40.]

The complex composite of three iron sulfide minerals and organic material results in interesting magnetic and mechanical properties. The presence of ferrimagnetic greigite raises the question of whether the snail uses this mineral for magnetic sensing. Bulk magnetic measurements reveal that most of the greigite crystals are single magnetic domains, but a significant fraction of superparamagnetic greigite is also present (Suzuki et al. 2006). Measurements of anhysteretic remanent magnetization indicate strong interparticle interactions. In addition, the ratio of natural remanent magnetization to isothermal remanent magnetization is consistent with the presence of random orientations of the greigite crystals. All of these observations suggest that the properties of the greigite crystals are not optimized for magnetic sensing, and that the snail does not use the greigite crystals as a magnetic compass (Suzuki et al. 2006).

The mechanical properties of the biomineralized layers are consistent with a hardening function. Nanoindentation studies show that the iron sulfide layer is harder and stiffer than human enamel, and stiffer than molluscan shell nacre (Suzuki et al. 2006). Whereas the minerals provide rigidity, the associated organic material provides toughness. Since the scaly-foot gastropod shares its habitat at the base of black smoker chimneys with predators such as brachyurean crabs (Suzuki et al. 2006) and other gastropods (Warén et al. 2003), it is likely that the hard and tough iron sulfide/organic composite is used for protection.

There is some ambiguity about whether the snail controls the deposition of the sequence of iron sulfide minerals. The iron sulfide layer is known to be covered by bacteria where it is overlain by adjacent sclerites (Warén et al. 2003). The phylogenetic affiliations of these episymbiotic bacteria have been studied by Goffredi et al. (2004), who found a predominance of bacteria belonging to lineages that are involved in sulfur cycling. Similar bacteria were not

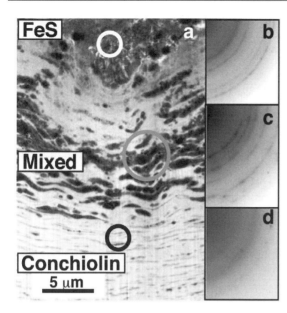

Figure 9. TEM image (a) of a cross-section of the sclerite of the scaly-foot gastropod. Selected-area electron diffraction patterns (b, c, d) obtained from the circled regions in (a), indicating (b) greigite from the FeS layer, (c) pyrite from the mixed layer, and (d) mackinawite from the conchiolin layer. [Used by permission of Elseveir, from Suzuki et al. (2006), *Earth and Planetary Science Letters,* Vol. 242, Fig. 2, p. 42.]

Figure 10. (a) TEM image of rod-shaped crystals from the iron sulfide layer of the sclerite of the scaly-foot gastropod, and (b) electron diffraction pattern from one of the crystals, indicating that it is greigite. The fibrous material next to the rods consists of mackinawite. [Used by permission of Elseveir, from Suzuki et al. (2006), *Earth and Planetary Science Letters,* Vol. 242, Fig. 3, p. 43.]

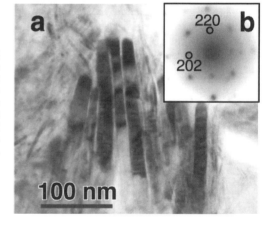

found on other available surfaces or among other gastropods within the same habitat. These observations prompted Goffredi et al. (2004) to speculate that iron sulfide mineralization is a consequence of the metabolism of these symbiotic bacteria. If sulfate-reducing bacteria were the source of sulfur for the sclerites, then a significant enrichment of light isotopes would be expected. However, Suzuki et al. (2006) measured the isotopic compositions of iron and sulfur and found the values to be close to those of the sulfide and iron in the hydrothermally-deposited chimneys. Thus, hydrothermal fluids appear to be a more likely source than episymbiotic bacteria of the iron and sulfur that are involved in sclerite mineralization. The presence of iron sulfides within the conchiolin tissue may also indicate the involvement of the snail in the precipitation of sulfide minerals.

Sulfide mineralization by the scaly-foot snail is the first known case of pyrite formation by BCM. The sulfide mineral assemblage in this organism is also unique in terms of its macroscopically magnetic character and its structural role. Although greigite and mackinawite form in both magnetotactic bacteria and the scaly-foot snail, some of their physical properties and their biological functions are different in the two cases. Much research is still needed to understand the biological control of the deposition of iron sulfides in both types of organisms.

BIOLOGICALLY INDUCED FORMATION OF SULFIDE MINERALS

Biologically induced mineralization is usually considered to be an uncontrolled consequence of metabolic activity, which produces minerals that are characterized by poor crystallinity, a broad particle size distribution, and a lack of well-defined crystal morphology and chemical purity (Frankel and Bazylinski 2003). If the metabolic products diffuse away from the micro-organism, and if the mineral-forming reactions take place in solution or on sediment particles, then the precipitated products may be indistinguishable from minerals that form by purely inorganic processes. However, in many cases bacterial surfaces or extracellular polymeric materials act as passive or active nucleation sites (Fortin et al. 1997; Schultze-Lam et al. 1996; Fortin and Langley 2005). In such cases, the biological material plays a direct role in crystal nucleation, and the minerals that form may have species-specific physical or chemical properties. Thus, BIM encompasses a broad range of mineral-forming processes, many of which are unique to the particular minerals or organisms that are involved in their formation.

Many common sulfide minerals can form by BIM, but the precipitation of iron sulfides is geologically the most important and the most extensively studied problem. Recently, Rickard and Morse (2005) provided a critical review of research into iron sulfide formation, including an assessment of the "myths and facts" that have accumulated over the past 40 years. Sedimentary pyrite formation has also been reviewed by Schoonen (2004), and aspects of the formation of sulfides by BIM are discussed in this volume by Rickard and Luther (2006). Here, the key processes that are involved in BIM are described, including a brief discussion of iron sulfide formation and a review of interesting examples of zinc sulfide mineralization.

Microbial sulfate and metal reduction

The activity of dissimilatory sulfate-reducing prokaryotes (SRP), which supplies reactive sulfide ions, is key to the formation of sulfide minerals by BIM (Frankel and Bazylinski 2003). Bacteria inhabit distinct redox zones according to their physiology (as reviewed in several textbooks of mineralogy and geochemistry, e.g., Nealson and Stahl 1997; Gould et al. 1997; Aplin 2000). Micro-organisms oxidize carbon in organic matter, using a variety of terminal electron acceptors, ranging from O_2 under aerobic conditions to SO_4^{2-} in anoxic environments.

SRP represent a morphologically and phylogenetically heterogeneous group. They are generally strict anaerobes that oxidize simple organic compounds or hydrogen using sulfate ions, as shown for example by the reaction (Tuttle et al. 1969):

$$2CH_2O + SO_4^{2-} \rightarrow 2HCO_3^- + H_2S$$

In this process, the sulfur in the sulfate ion is reduced completely to sulfide, which is released into the environment. Whereas a considerable proportion of the reactive sulfide diffuses upwards and is reoxidized (Jørgensen 1977), part of it combines with metals (primarily iron) to form sulfide minerals (Berner 1970).

Since SRP can use relatively small organic molecules as electron donors, they generally depend on other microbial populations that degrade complex organic compounds. Two major groups of SRP exist, one that incompletely oxidizes organic substrates into acetate (e.g., *Desulfovibrio, Desulfotomaculum, Desulfomonas, Desulfobulbus*), and another that completely oxidizes organic matter to CO_2 (e.g., *Desulfococcus, Desulfosarcina, Desulfonema*) (Gould et al. 1997). Some hyperthermophilic archaea are also dissimilatory sulfate reducers. SRP are ubiquitous in many anaerobic environments, including lakes, swamps, soils, waste ponds, hydrothermal systems, and even within the lithosphere (Lovley and Chapelle 1995). In terms of the amount of sulfide mineralization and its global biogeochemical effect, SRP that occur in marine sediments are most important (Schoonen 2004).

In addition to sulfate reduction, the microbial reduction of metals (such as iron and manganese) is also important in biogenic sulfide mineralization, since it may contribute to the pool of metal ions that are available for mineral formation (Rickard and Morse 2005). Dissimilatory iron-reducing prokaryotes were shown to respire using ferric iron in minerals, and to exert a strong influence on the geochemistry of many environments (Nealson and Saffarini 1994; Methe et al. 2003). Iron reducers are phylogenetically diverse, and include several genera of bacteria (such as *Geobacter* and *Shewanella*) and even archaea (Kappler and Straub 2005). Many of these organisms are phylogenetically closely related to SRP, and include species that can also reduce elemental sulfur. Iron-reducing microorganisms can even use iron from relatively poorly reactive minerals such as magnetite and sheet silicates. The potential role of such micro-organisms in dissolving iron and indirectly affecting the sulfur cycle in sediments is only now beginning to be appreciated (Rickard and Morse 2005).

The role of biological surfaces in mineral nucleation

In general, the heterogeneous nucleation of biominerals is favored kinetically over homogeneous nucleation. Biological surfaces provide excellent nucleation sites for a number of minerals, including sulfides. The properties of different types of mineral-nucleating biological surfaces were reviewed by Schultze-Lam et al. (1996), Fortin et al. (1997), Konhauser (1998), Frankel and Bazylinski (2003), and Gilbert et al. (2005).

The outer surfaces of bacterial cell walls are predominantly negatively charged at near neutral pH, irrespective of whether they belong to gram-positive or gram-negative structural types (Fortin et al. 1997). Therefore, they attract positive ions from solution and thereby initiate the nucleation of metal sulfides. In natural environments, additional biological layers exist on the cell walls. These layers include capsules that usually consist of acidic polysaccharides, S-layers that consist of regular arrays of proteins (Beveridge 1989), sheaths, stalks, and filaments (Gilbert et al. 2005). Many of these surfaces are known to induce the nucleation of metal oxides and sulfides (Fortin et al. 1997; Gilbert et al. 2005).

The ability of bacterial surfaces to bind metal ions is related to the presence of acidic functional groups. As discussed by Gilbert et al. (2005), proteins or polysaccharides that are rich in negatively charged carboxyl (COO^-) groups are the most common and effective cation-binding macromolecules in biomineral nucleation. A general sorption reaction for a metal cation M of charge z (M^{z+}) at a carboxyl binding site, as described by Ferris (1997), results in the release of a proton according to the reaction:

$$B\text{-}COOH + M^{z+} = B\text{-}COOM^{z-1} + H^+$$

Thus, the sorption of metal ions depends not only on the number of reactive chemical groups on the bacterial surface, but also on the pH and on the concentration of dissolved metal ions. The sorption of cationic species is enhanced as the pH increases and as surface groups deprotonate. As a result, the metal binding capacity of natural biofilms is enhanced significantly under circumneutral pH conditions, with respect to that in acidic metal-contaminated waters (Ferris 1997).

Iron sulfides in marine sediments

Iron sulfide minerals are ubiquitous both in modern anoxic sediments and in sedimentary rocks. The primary stages of sedimentary iron sulfide formation were identified by Berner (1970; 1984), and the topic has since been reviewed several times (Morse et al. 1987; Rickard et al. 1995; Schoonen 2004; Rickard and Morse 2005). For the past four decades, the key processes appeared to be well understood. The remaining uncertainties were related to the importance of specific reactions, the physical and chemical properties of transient phases, and the roles of microbes. However, the most recent review by Rickard and Morse (2005) challenged many long-standing views, and identified several areas where more research is necessary. Here, primary attention is paid to the aspects of sedimentary iron sulfide formation that are related to the activities of micro-organisms.

The formation of iron sulfides in sediments is a typical example of BIM. The rate of iron sulfide formation depends primarily on the rate of microbial sulfate reduction (which also depends on the availability of organic carbon), and on the amount of competing electron acceptors including reactive Fe(III)-bearing minerals (Berner 1970) (Fig. 11). When dissolved sulfide produced by SRP reacts with Fe^{2+}, the precipitate that forms is generally termed "amorphous FeS," and appears to correspond to poorly-ordered or nanocrystalline mackinawite (Lennie and Vaughan 1996), or mixtures of mackinawite and greigite. Most earlier literature on sedimentary pyrite formation assumes that pyrite forms by the conversion of mackinawite or greigite (Schoonen 2004). However, according to Rickard and Morse (2005), these precursors are not required for pyrite formation.

Our understanding of the roles of bacteria in each pyrite-forming stage has changed considerably over the past ten years (Donald and Southam 1999; Schoonen 2004; Rickard and Morse 2005). Whereas the role of bacteria had been thought to be restricted to providing sulfide ions, it now appears that micro-organisms affect in many ways the processes that lead to the formation of iron sulfides (Fig. 11).

The mineral species. In addition to pyrite, which is the most abundant species, other iron sulfides that occur in sediments include mackinawite and greigite. The latter minerals (and pyrrhotite ($Fe_{1-x}S$)) are also termed "iron monosulfides." Significantly, mackinawite and greigite have rarely been identified in the field. In most studies, the operationally-defined category of acid volatile sulfides (AVS) is used, and is assumed to include amorphous FeS, mackinawite, and greigite. However, as pointed out by Rickard and Morse (2005), AVS is not

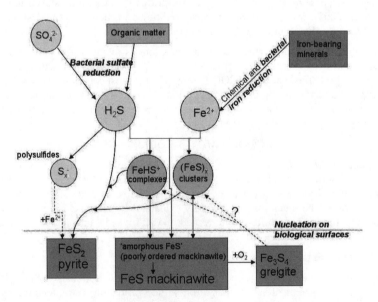

Figure 11. The primary pathways of sedimentary iron sulfide formation, based on Berner (1984) and Rickard and Morse (2005). Circles and rectangles denote dissolved and solid species, respectively. Text in italics refers to processes that involve the activity of bacteria.

equivalent to the sum of solid iron monosulfides but is a complex and variable component of the sediment. AVS likely includes dissolved iron and sulfur species and their complexes, aqueous iron sulfide clusters (FeS$_{aq}$) (see Rickard and Luther, 2006, in this volume), and an unidentified fraction of mackinawite and greigite. In addition, even though pyrite is insoluble in weak acids, commonly used extraction methods may partially dissolve fine-grained pyrite, which may then also contribute to AVS (Rickard and Morse 2005).

The thermodynamic constraints that determine which iron sulfide is stable in an anoxic sediment were discussed by Schoonen (2004). Iron monosulfides are predicted, by equilibrium thermodynamic calculations, to be stable over a very narrow range of pe-pH conditions. Marcasite is metastable with respect to pyrite and forms under acidic conditions (pH < 5) (Murowchick and Barnes 1986). Therefore, in equilibrium, only pyrite would be expected to occur in a low-temperature sedimentary environment. However, many field studies attest to the prevalence of iron monosulfides in modern marine sediments. In euxinic basins, the amount of iron monosulfides exceeds that of pyrite (Hurtgen et al. 1999). In addition, evidence has accumulated over the past 15 years that greigite is the primary carrier of magnetization in many types of sedimentary rock, some of which are as old as Cretaceous (Reynolds et al. 1994; Roberts 1995; Dekkers et al. 2000; Rowan and Roberts 2006; Pearce et al. 2006, in this volume). The presence of metastable iron monosulfides has generally been attributed to the presence of a high nucleation barrier for the formation of pyrite (Schoonen and Barnes 1991; Benning et al. 2000). If pyrite seed crystals are present, then this nucleation barrier can be overcome (Benning et al. 2000).

Availability of iron. The balance between the rate of H$_2$S formation and the availability of reactive iron exerts a controlling factor over FeS formation (Schoonen 2004). Raiswell and Canfield (1998) documented the importance of the mineral phase of iron oxide present in the sediment on the rate of its sulfidation. Highly reactive minerals include ferrihydrite, lepidocrocite, goethite, and hematite, with half-lives of less than a year. Magnetite and "reactive" iron silicates have half-lives on the order of ~10^2 years, whereas the half-lives of poorly reactive minerals (such as ilmenite and some silicates) are in the 10^6-year range. As discussed above, dissimilatory metal-reducing bacteria use oxidized forms of iron as terminal electron acceptors, thereby causing the dissolution of oxide minerals under anaerobic conditions (Frankel and Bazylinski 2003; Kappler and Straub 2005). The released metal ions can participate in various mineral-forming reactions, including those that produce sulfides. Although inorganic and biogenic pathways for metal reduction are not easy to distinguish in most natural systems, bacterial processes are likely to be important for supplying iron for sedimentary iron sulfide formation (Rickard and Morse 2005).

Nucleation and the physical properties of mackinawite and greigite. The role of bacteria in the nucleation of iron monosulfides is uncertain, although there is evidence that FeS nucleates preferentially on the cell envelopes of SRP. Bacterial cells and their remains were found to be prominent nucleation sites for amorphous FeS (and nanocrystalline millerite, NiS) in a contaminated lake sediment (Ferris et al. 1987). Donald and Southam (1999) found that thin layers of FeS coated both the inner and the outer surfaces of cells. Anionic cell surface polymers likely interacted with Fe^{2+}, and the immobilized cations could then react with H$_2$S, forming the films of FeS. Similarly, iron sulfides encrusted the surfaces of SRP in experiments by Watson et al. (2000) (Fig. 12). They formed on the surface of hematite to which SRP were attached, and initiated the precipitation of FeS (Neal et al. 2001). Thus, micro-organisms are important nucleation sites for the formation of iron sulfides.

The initial FeS precipitate is difficult to characterize because of its small grain size and poorly ordered structure. The morphologies and sizes of nanocrystals appear to be strongly affected by experimental conditions. Whereas Wolthers et al. (2003) described FeS precipitates as nanocrystals with an average size of ~4 nm, Herbert et al. (1998) found that platy mackinawite crystals with diameters of 100 to 300 nm precipitated in growth media of SRP, and formed 1 to 2 μm spherical aggregates. Ohfuji and Rickard (2006) showed that mackinawite precipitated as nanocrystalline particles, and presented a list of particle sizes and specific surface areas observed in various studies. Structurally, all of these studies identified the primary phase of the precipitate as "poorly ordered" or nanocrystalline mackinawite, although Wolthers et al. (2003) described two types of crystalline domains ("MkA" and "MkB"), with different d-values that bore little resemblance to those of mackinawite. High-resolution TEM images and electron diffraction patterns were obtained from an FeS precipitate by Ohfuji and Rickard (2006). The diffraction patters contained diffuse rings, indicating that the particles were poorly ordered (Fig. 13). The observed d-spacings suggested that a mackinawite-like short-range order is present, consistent with the high-resolution images.

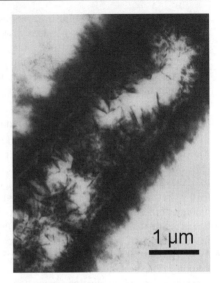

Figure 12. TEM image of a cell of a sulfate-reducing bacterium that is encrusted by iron sulfide minerals. [Used with permission from Elsevier, from Watson et al. (2000) *Journal of Magnetism and Magnetic Materials*, Vol. 214, Fig. 1, p. 13-30.]

In addition to mackinawite, greigite was also identified in several studies in the initial FeS precipitate. Herbert et al. (1998) inferred that the surfaces of the aggregated nanocrystals had a greigite composition, whereas the remaining bulk material consisted of disordered mackinawite. On the basis of magnetic measurements, Watson et al. (2000) found that greigite formed a significant fraction of SRP-precipitated iron sulfide.

Greigite also forms from mackinawite by solid-state transformation. Two basic routes have been suggested, either through iron loss (Lennie et al. 1997) or through sulfur addition

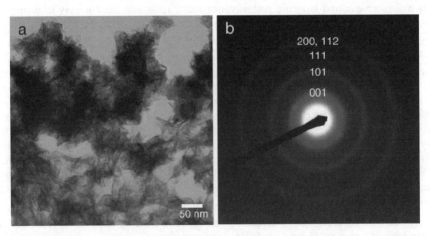

Figure 13. (a) TEM image and (b) electron diffraction pattern of precipitated mackinawite. [Used with permission of Elsevier, from Ohfuji and Rickard (2006), *Earth and Planetary Science Letters*, Vol. 241, Fig. 2a,b, p. 227-233.]

(Horiuchi 1971). It appears that the conversion of mackinawite into either greigite or pyrite can be controlled by the presence of catalytic quantities of organic compounds (Rickard et al. 2001). In the presence of aldehydic carbonyls in the solution, Fe^{2+} in the iron monosulfide is partially oxidized, whereas S^{2-} remains unchanged, forming greigite. In the absence of aldehydic carbonyls, S^{2-} is oxidized and pyrite forms (Rickard et al. 2001). It is not yet known whether similar organic switches operate in natural systems as in the laboratory experiments.

The greigite that forms from mackinawite is also nanocrystalline. This observation has important implications for the magnetic properties of sediments. The magnetic single domain range for greigite is particle-shape-dependent and extends from ~50 nm (Diaz-Ricci and Kirschvink 1992) to a poorly-constrained upper limit of 200–1000 nm (Hoffmann 1992; Diaz-Ricci and Kirschvink 1992). Crystals within this range have a high coercivity and therefore contribute significantly to the remanent magnetism of sediments. Rowan and Roberts (2006) found that single-domain and superparamagnetic greigite populations coexisted in Neogene marine sediments, providing for a complex magnetic behavior. Greigite formed with pyrite in framboids, but a later generation of very fine-grained superparamagnetic greigite appeared to grow on the pyrite crystals. Such late diagenetic changes can complicate paleomagnetic interpretations, since such crystals aquired their remenance > 1 Myr after deposition.

The formation of pyrite. Three primary pathways for pyrite formation are usually considered (Schoonen 2004), including (1) FeS oxidation by a polysulfide species (Luther 1991; Schoonen and Barnes 1991); (2) FeS oxidation by H_2S (Rickard 1997); and (3) conversion of FeS by iron loss through an intermediate greigite phase (Wilkin and Barnes 1996). The reactions are:

(1) $FeS + S_n^{2-} \rightarrow FeS_2 + S_{n-1}^{2-}$

(2) $FeS + H_2S \rightarrow FeS_2 + H_2$

(3) $4 FeS + \frac{1}{2} O_2 + 2 H^+ \rightarrow Fe_3S_4 + Fe^{2+} + H_2O$

$Fe_3S_4 + 2 H^+ \rightarrow FeS_2 + Fe^{2+} + H_2$

Experimental tests by Benning et al. (2000) showed that Reaction (2) does not produce appreciable amounts of pyrite if H_2S is the only reactant in the system with mackinawite. Pyrite formation is induced only if the aqueous sulfur species or the mackinawite is oxidized. However, the importance of Reaction (2) is supported indirectly by the persistence and large proportion of iron monosulfide in euxinic sediments. In such an environment, reactive iron is available in abundance. Consequently, dissolved sulfide is depleted by iron sulfide formation, and the lack of dissolved sulfide prevents it from reacting with FeS and converting it into pyrite (Hurtgen et al. 1999). The conversion of mackinawite into greigite via iron loss (3) was observed by Lennie et al. (1997). On the basis of an analysis of molar volume changes, Furukawa and Barnes (1995) argued that the precursor phase converts to pyrite via the iron loss pathway (Reaction 3 above).

Studies by Luther and coworkers (Theberge et al. 1997; Luther et al. 2001; Luther and Rickard 2005) demonstrated the biogeochemical importance of aqueous metal sulfide complexes (see Rickard and Luther 2006, in this volume). Highly reactive FeS_{aq} clusters appear to be key intermediaries in pyrite formation, as they react with either H_2S or polysulfide species to nucleate pyrite (Rickard and Morse 2005). In light of these results, the conversion of mackinawite or greigite into pyrite cannot be regarded as a solid-state transformation. Instead, these minerals may be partially dissolved, forming aqueous FeS clusters that react to form pyrite (Fig. 11). Since FeS clusters can form by other routes, the presence of mackinawite and greigite is not a necessary condition for pyrite formation (Rickard and Morse 2005).

Experiments by Donald and Southam (1999) indicated that the conversion of FeS to FeS_2 is promoted by the formation of a thin FeS film on the surfaces of bacterial cells. Sulfur-disproportionating bacteria also appeared to play a role in converting organic sulfur into H_2S in experiments by Canfield et al. (1998). Since radiolabeled organic sulfur was incorporated into the final pyrite product in this study, the FeS to pyrite transformation took place via the sulfur addition pathway (reaction (1) above). Fortin and Beveridge (1997) observed the intact remains of SRP encrusted by iron sulfides, while Grimes et al. (2001) found that organic matter provided nucleation sites for the reaction of FeS to FeS_2. It appears that bacterial activity mediates both the initial precipitation of FeS and its conversion to pyrite.

Framboidal pyrite. The interesting morphologies of sedimentary pyrite have long captivated the attention of researchers. A variety of morphological types occurs, including euhedral, irregular, and ooidic pyrite (Hámor 1994). However, the most widespread and characteristic appearance of pyrite is framboidal (Schoonen 2004; Ohfuji and Rickard 2005). The term framboid refers to a spherical structure, which consists of densely-packed pyrite crystals that have similar sizes and morphologies (Fig. 14). In addition to pyrite, greigite has also frequently been found as a component of framboids (Bonev et al. 1989; Wilkin and Barnes 1997; Rowan and Roberts 2006). The diameters of framboids are in the 1–30 μm range (but most are smaller than 10 μm), while the individual constituent crystals range from ~0.1 to 2 μm (Wilkin et al. 1996). Framboids were once thought to be fossilized bacteria. They were then considered to be pyritized organic particles or colloids (Raiswell et al. 1993) or abiotic products of the conversions of magnetic precursor iron sulfides, i.e., greigite (Sweeney and Kaplan 1973; Wilkin and Barnes 1997). However, Butler and Rickard (2000) synthesized pyrite framboids in the absence of magnetic intermediates and biological intervention. They found that the framboidal texture results from rapid nucleation from a strongly supersaturated solution, through the reaction of aqueous FeS cluster complexes with H_2S (see Rickard and Luther 2006, in this volume). Thus, even though the peculiar morphologies of framboids are suggestive of biological processes, the development of framboids may be the least likely of the various stages of sedimentary pyrite formation to be affected by biogenic activity.

Since framboids form either in the water column (in euxinic environments) or during early diagenesis within the top few centimeters of the sediment, their sizes reflect the conditions of the environment of deposition. In a very thorough study of framboid size distributions, Wilkin et al. (1996) established relationships between the size distributions of pyrite framboids and the redox conditions of the depositional environment.

Framboids can have remarkably ordered architectures, forming either cubic or icosahedral close-packed structures (Ohfuji and Akai 2002). In an electron backscatter

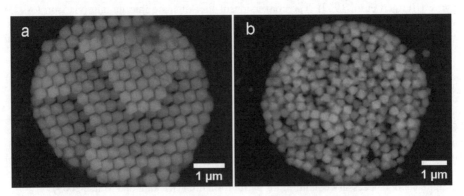

Figure 14. SEM images of synthetic pyrite framboids. (a) Morphologically ordered and (b) disordered framboid. [Used with permission of Elsevier, from Ohfuji and Rickard (2006), *Earth Science Reviews*, Vol. 71, Fig. 1a,c, p. 147-170.]

diffraction study, Ohfuji et al. (2005) distinguished morphologically ordered and disordered framboids (Fig. 14), and determined the morphological and crystallographic orientations of individual nanocrystals. Even in morphologically ordered framboids, low- and high-angle crystallographic misorientations were observed, the latter resulting from the fact that pyrite has only a two-fold axis along <100>. The results suggested that the self-organized structure results from the aggregation and subsequent reorientation of equimorphic nanocrystals.

Biogenic zinc sulfides: from mine-water to deep-sea vents

The biologically mediated precipitation of zinc sulfide has been studied recently in two widely different natural systems, in a flooded lead-zinc mine (Labrenz et al. 2000; Moreau et al. 2004) and in the tubes of a deep-sea vent worm (Zbinden et al. 2001, 2003; Maginn et al. 2002). Remarkably, the ZnS minerals that formed in these distinct environments showed similar morphological and chemical features.

Spherical aggregates of ZnS formed in a biofilm of sulfate-reducing bacteria in the flooded tunnel of a carbonate-hosted Pb-Zn deposit (Labrenz et al. 2000). The spherules were 1 to 5 µm in diameter and consisted of 1 to 5 nm, semi-randomly oriented, crystalline ZnS nanoparticles (Fig. 15). Both sphalerite and wurtzite structures occured within the nanoparticles, and stacking faults, twins, and disordered sequences of close-packed layers were observed to be present in many nanocrystals (Moreau et al. 2004). The ZnS particles were chemically pure, with no measurable iron content, and occured in layers within the biofilm, in close association with bacterial cells or extracellular polymeric material. The bacteria were shown by small-subunit ribosomal RNA gene analyses to belong to the sulfate-reducing family *Desulfobacteriaceae*, and verified to be metabolically active by fluorescence *in situ* hybridization (Labrenz et al. 2000). Some cells were encrusted and fossilized by ZnS spheroids, indicating the intimate association of bacteria and ZnS mineralization. Thus, the ZnS precipitation at this site was wholly attributable to the activity of SRP (Moreau et al. 2004).

The precipitation of pure ZnS consisting of both sphalerite and wurtzite structural elements is an interesting feature of this biomineralization. According to experimentally determined stability fields (Scott and Barnes 1972), sphalerite should form from cold (8-10 °C) groundwater. However, the presence of wurtzite is consistent with a size-dependence of ZnS phase stability, which has been predicted by molecular dynamics simulations (Zhang et al. 2003).

The extreme environment of deep-sea hydrothermal vents of the East Pacific Rise hosts the alvinellid or so-called Pompeii worms (*Alvinella*

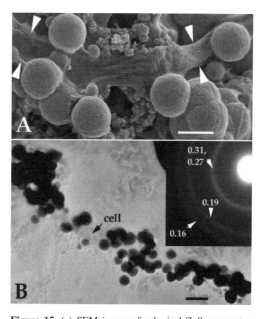

Figure 15. (a) SEM image of spherical ZnS aggregates that are associated with a biofilm (marked by arrowheads) from a flooded lead-zinc mine. (b) TEM image and selected-area electron diffraction pattern, showing that the ZnS spherules are associated with bacterial cells, and that both sphalerite and wurtzite structural elements occur in the spherules. [Reprinted with permission from Labrenz et al., *Science*, Vol. 290, Fig. 2a,b, p. 1744-1747. Copyright (2000) AAAS.]

pompejana and *Alvinella caudata*) (Zbinden et al. 2003). These animals dwell in organic tubes that contain zinc and iron sulfide minerals. The worms live on active sulfide chimney walls, where they are exposed to steep thermal and chemical gradients and intense mineral precipitation (Desbruyeres et al. 1998). Since sulfide-rich hydrothermal fluids and seawater mix within the worm tubes, the inorganic chemical deposition of sulfide minerals is possible. However, whereas hydrothermally-precipitated ZnS grains outside the worm tubes have variable iron contents and crystal sizes, the mineral grains in the tube wall have specific compositions, grain sizes, and positions within the wall structure (Zbinden et al. 2001, 2003). The specific properties of the minerals reflect the direct effects of the biological environment.

The worm tubes consist of concentric layers of fibrous organic material. The inner surface of the tube is covered by filamentous bacteria and ZnS grains. Each time the worm secretes a new layer, the bacteria and the minerals become entombed in the organic matrix of the tube (Zbinden et al. 2001, 2003) (Fig. 16). The spherical ZnS grains are aggregates of 1-5 nm crystals, and have a remarkably uniform composition of $Zn_{0.88}Fe_{0.12}S$. Powder electron diffraction patterns obtained from the round ZnS grains are consistent with the structures of both sphalerite and wurtzite (Zbinden et al. 2001). The sulfide mineral aggregates are attached to sheathed and branching bacterial filaments that occur on the inner tube surfaces (Maginn et al. 2002).

A specific feature of the mineralization associated with the Pompeii worms is that mineralogical gradients are present both from the outside to the inside and from the bottom to the top of the tubes (Zbinden et al. 2003). FeS_2 minerals predominate on the bottom outer surfaces of the tubes, with a marcasite to pyrite ratio of approximately 3:1. The relative proportion of FeS_2 minerals decreases from the bottom to the top of the tubes. Whereas Zbinden et al. (2003) found no iron sulfide in the mineralized layers within the tube wall and on the inner surfaces of the tubes, Maginn et al. (2002) observed pyrite, marcasite and other iron sulfides (presumably mackinawite and greigite) associated with ZnS inside the tubes. Although the particular features of the zinc and iron sulfide distributions may change between worm tubes observed in various sites, the mineralogical gradients indicate that the fluid compositions inside

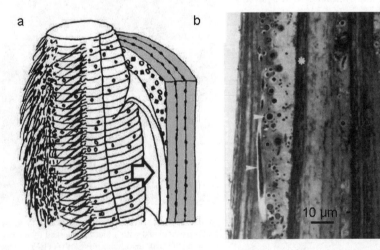

Figure 16. (a) A sketch of an Alvinellid worm and its tube. The worm secretes a new layer on the inside of the tube. The black dots represent ZnS grains and bacteria that are present on the surfaces of the tube layers. (b) Optical micrograph of a cross-section of several layers of an Alvinellid tube. The arrowheads indicate entrapped bacteria. The left side of the figure corresponds to the inside of the tube. [Used with permission of Schweizerbart (*http://www.schweizerbart.de*), from Zbinden et al. (2001), *European Journal of Mineralogy*, Vol. 13, Figs. 1, 5, p. 653-658.]

and outside the tube differ. Either the worm creates special hydrodynamic conditions within the tube, and thereby changes the ratio of vent fluid and seawater with respect to the conditions outside the tube (Desbruyeres et al. 1998), or the metabolic activity of the worm or its epibiotic bacteria causes the exclusive formation of ZnS minerals inside the tube (Zbinden et al. 2003).

There are some remarkable similarities between the ZnS minerals in the worm tubes and in the biofilm that was recovered from the groundwater in the lead-zinc mine. The sizes of the ZnS nanocrystals, and even those of the aggregate spherules, are similar in both cases. It is likely that the nanocrystals in the worm tubes have disordered layer sequences, similar to those in the sphalerite/wurtzite grains from the mine-water. Although the mineral grains from the mine are pure ZnS, while the crystals in the worm tubes contain iron, both types of biogenic ZnS are characterized by constant compositions. All of these features indicate that even minerals that are formed by BIM can have characteristics that distinguish them from inorganically-formed crystals. Although the bacteria that were found on the inner surface of the worm tube were not isolated and cultured, their close association with the ZnS grains suggests that they may be sulfate reducers, and that they could play a role in the detoxification of the fluid surrounding the worm (Zbinden et al. 2001).

BIOLOGICALLY MEDIATED DISSOLUTION OF SULFIDE MINERALS

The dissolution of sulfide minerals has important ramifications, both globally and locally, since it affects global geochemical cycles, generates acid mine drainage (AMD), and is used in industrial metal extraction. There is a long history of research on micro-organisms that oxidize sulfur or iron or both, and thereby enhance the rates of sulfide mineral weathering. Here, a few aspects of sulfide bioweathering are discussed. For comprehensive reviews on the geomicrobiology of sulfide mineral oxidation, the reader is referred to Nordstrom and Southam (1997) and McIntosh et al. (1997). The microbiological cycling of iron was recently reviewed by Kappler and Straub (2005).

Acid mine drainage

Wherever sulfide-bearing rocks or mine tailings are exposed to oxidative conditions, the sulfide minerals dissolve and produce acidic waters (Jambor et al. 2000). AMD is a major environmental concern, since the acidity of the water (generally between pH 2 and 4), as well as the presence of toxic concentrations of metals, are detrimental for many aquatic organisms. Although the oxidation of many sulfide mineral species contributes to AMD, pyrite is usually considered to be the most abundant and important mineral involved in the production of acidic waters (for a review of studies of the oxidation of various sulfides see Nordstrom and Southam 1997, and for the reaction products see Jambor et al. 2000).

Chemical and microbial processes are coupled in the generation of AMD. Under oxidative conditions, sulfide minerals react with oxygen, and the reaction is catalyzed by iron and sulfur-oxidizing bacteria. Members of the bacterial genus *Thiobacillus* are the most widely studied micro-organisms that break down sulfide minerals. Other important genera include *Leptospirillum*, *Sulfobacillus*, and some species of Archaea (Davis 1997; Nordstrom and Southam 1997; Baker and Banfield 2003). Most of these bacteria are acidophilic lithoautotrophs, i.e., they require an environment with a pH < 3 for optimal growth, and they use inorganic compounds as their source of metabolic energy (McIntosh et al. 1997). Microbial processes in the anoxic sections of mine tailings also contribute to the cycling of metals and sulfur, but these processes are still poorly understood (Fortin et al. 2002). Research into microbial communities involved in the production of AMD was recently reviewed by Baker and Banfield (2003).

The mechanism of bacterial catalysis of sulfide mineral oxidation was discussed by Nordstrom and Southam (1997). There has been some controversy over the role of a hypothesized, direct enzymatic reaction induced by bacterial cells that attach to the mineral surface. The existence of such a reaction mechanism was supported by observations of etch pits that reflected the effects of attached cells (Bennett and Tributsch 1978). However, Nordstrom and Southam (1997) provided a comprehensive analysis of the observed rates of specific abiotic and biotic reactions that are involved in pyrite oxidation, and concluded that these data are consistent with the indirect microbial catalysis of sulfide mineral dissolution. The primary role of the bacteria is the oxidation of aqueous Fe^{2+} into Fe^{3+} under acidic conditions, according to the reaction:

$$Fe^{2+} + \frac{1}{4} O_2 + H^+ = Fe^{3+} + \frac{1}{2} H_2O$$

The ferric iron then attacks the mineral surface, and pyrite is oxidized at a rate that is determined by the bacterial oxidation step:

$$FeS_2 + 14 Fe^{3+} + 8 H_2O = 15 Fe^{2+} + SO_4^{2-} + 16 H^+$$

Thus, although bacteria preferentially adhere to mineral surfaces in order to reduce the distance for the diffusion of iron between the mineral and the bacterium, there is no need to invoke an enzymatic reaction for sulfide mineral degradation (Nordstrom and Southam 1997).

The effects of cell attachment on mineral dissolution are the subject of continued interest. Edwards et al. (2001) experimented with reacting pyrite, marcasite, and arsenopyrite with the iron-oxidizing bacterium *Acidithiobacillus ferrooxidans*, the archaeon *Ferroplasma acidarmanus*, and abiotically with Fe^{3+}. Interestingly, in both the biotic and the abiotic experiments, bacillus-sized etch pits developed on pyrite, indicating that the attachment of cells is not necessary for the development of etch pits with characteristic shapes and sizes. However, attached cells of *F. acidarmanus* induced pitting on the more reactive surface of arsenopyrite (Fig. 17). Thus, the reactivity of the mineral may determine whether the surface features that develop during oxidative dissolution are related directly or indirectly to the presence of micro-organisms (see also Rosso and Vaughan 2006, in this volume).

Despite a century of research, the rate of microbially-assisted oxidation of sulfide minerals under natural conditions is still uncertain (Edwards et al. 2000). Laboratory experiments that involved the use of the same strain of *Thiobacillus ferrooxidans*, the same pyrite source, and the same experimental procedures resulted in consistent and reproducible rates, with the rate of microbial iron release being 34 times larger than the abiotic rate (Olson 1991). In contrast, field studies showed a wide variety in the degree of microbial enhancement of chemical processes. For example, in the AMD at the ore body of Iron Mountain, California, microbial

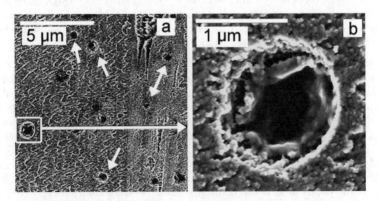

Figure 17. (a) Etch pits on the surface of arsenopyrite that was reacted with *F. acidarmanus*. (b) Magnified image of the area shown within the square in (a). The dehydrated cell in (b) is situated within a cell-sized and -shaped dissolution pit. Other cells are indicated with arrows in (a). [Used with permission of Blackwell Publishing, from Edwards et al. (2001), *FEMS Microbiology Ecology*, Vol. 34, Fig. 10, p. 203.]

enhancement of iron release was found to be much lower (~3× the rate of the abiotic reaction) than in laboratory studies (Edwards et al. 2000). Several poorly-constrained factors, including the surface area that is available for reaction, and the properties of pore fluids (that determine microbial population densities) affect the rates of microbial and abiotic contributions to AMD (Edwards et al. 2000).

In addition to the well-known case of sulfide mineral dissolution under oxidative and acidic conditions, sulfides can be dissolved at higher pH and anoxic conditions. Significant bacterially-assisted dissolution of copper from its sulfide minerals was observed in the slightly alkaline water of a tropical river, downstream of a large copper mine (Simpson et al. 2005b). Although the bacteria that were responsible for the copper release could not be isolated, indirect evidence suggested that they were lithoautotrophs. The fate of sulfide minerals and the cycling of metals in AMD are complicated further by the presence of SRP and iron-reducing bacteria in the anoxic portions of mine tailings (Fortin et al. 2002). The activity of these organisms results in the re-precipitation of sulfide minerals and a reduction in metal concentration in the solution (as discussed below in the section on bioremediation).

Microbial degradation of sulfides in marine environments

Vast quantities of sulfide minerals are present on and beneath the seafloor, in widely varying environments, including marine sediments, hydrothermal systems associated with mid-ocean ridges, and in the bare basaltic rocks on the flanks of mid-ocean ridges (Edwards et al. 2005). Wherever sulfides are in contact with the oxic seawater, the possibility of their dissolution arises. As for AMD generation, both chemical and biogenic processes are involved in the breakdown of sulfides in marine environments. However, there appear to be significant differences between the roles of bacteria in geologically distinct regions of the seafloor.

The pyrite and iron monosulfides that form by BIM within the anoxic sediments can be transported by bioturbation to the surface of the sediment, where they are oxidized chemically by O_2 (Thamdrup et al. 1994; Schippers and Jørgensen 2002). In the process, bacterial involvement may be limited to the oxidation of aqueous sulfur-bearing compounds, intermediates that result from the oxidation of pyrite, into sulfate (Kuenen et al. 1992). The bacterially-assisted oxidation of sulfide minerals under anoxic conditions was also considered by Schippers and Jørgensen (2002), who found that iron monosulfides could be oxidized by Fe^{2+}- or H_2S-oxidizing and NO_3^--reducing bacteria, but pyrite was not attacked by the same processes. Since many metals can be incorporated into iron sulfides, it is of environmental importance to trace the fate of these metals during their oxidation (Bertolin et al. 1995). Since iron monosulfides oxidize more readily than pyrite, they are prone to release incorporated metals (Holmes 1999). In general, there is no uniform behavior of pollutant metals. Whereas some metals remain in the particulate phase, others dissolve, depending on the particular environmental conditions (for a review see Schoonen 2004).

The potentially significant role of micro-organisms in the alteration of seafloor sulfide minerals at hydrothermal vents and in exposed basalt is just beginning to be addressed (as reviewed by Edwards et al. 2005). Hydrothermal metal sulfide deposits support communities of lithoautotrophic bacteria, among which sulfide- and iron- or manganese-oxidizing microbes promote the oxidative surface reaction of sulfide minerals. Verati et al. (1999) observed an external layer on black smoker chimneys that consisted of the oxidation products of sulfides and contained the imprints of bacterial cells. They concluded that bacterial sulfur and iron oxidation are responsible for the weathering of the sulfides. Schrenk et al. (2003) found diverse communities within hot, active black smoker chimney structures. Primarily archaea were found inside the chimneys, whereas in the external, cooler portions of the chimney walls bacteria dominated. In contrast, cold, inactive black smoker chimneys harbored only bacterial communities (Suzuki et al. 2004). Both iron- and sulfur-oxidizing lithoautotrophic bacteria were cultured from extinct chimneys, indicating their likely role in the weathering of black smoker sulfides (Edwards et al. 2005).

In an *in situ* incubation study, several minerals (pyrite, marcasite, chalcopyrite, sphalerite, elemental sulfur) and a fragment of a natural black smoker chimney were left to react for two months in the vicinity of a seafloor hydrothermal system (Edwards et al. 2003). The surfaces of these minerals were colonized by bacteria, likely belonging to iron- and sulfur-oxidizing species. The colonization densities were found to correlate positively with the abiotic reactivity of the minerals, i.e., the more reactive was the mineral, the more cells colonized its surface. The only exception was elemental sulfur, which was both the most heavily colonized and the least reactive. The black smoker chimney fragment was even more heavily colonized and weathered than the individual minerals, suggesting that the weathering of natural, fine-grained sulfide structures is enhanced significantly by micro-organisms (Edwards et al. 2003).

Rock-hosted microbial communities on the seafloor are not restricted to hydrothermal chimney structures. Geographically vast regions exist on the flanks of mid-ocean ridges, which consist of unsedimented basalt and are potential targets for the colonization of lithoautotrophic micro-organisms. Iron-oxidizing bacteria were inferred to be present on the ocean basalt (Thorseth et al. 2001), and uncultured bacteria from deep-sea basalt were genetically similar to known sulfur- and iron-metabolizing bacteria (Lysnes et al. 2004).

Hydrothermal fluids mix with seawater in a shallow, sub-seafloor region, which is inferred to host a "deep biosphere" of endolithic microbial communities (Summit and Baross 2001). Such habitats are not yet accessible for direct sampling, but diffuse vents on ridge flanks are thought to offer a glimpse into the sub-seafloor biota (Edwards et al. 2005). Microbial populations in the diffuse vents were found to be distinct from those in the bottom seawater (Huber et al. 2003). The study of the interactions between microbes and minerals on and below the seafloor is a new field, which will likely bring interesting results concerning the bio-assisted precipitation and alteration of sulfide minerals.

PRACTICAL APPLICATIONS OF INTERACTIONS BETWEEN ORGANISMS AND SULFIDES

Biomimetic materials synthesis

The processes that are involved in biologically controlled mineralization provide valuable insights for materials chemists. "Bio-inspired materials chemistry" makes use of strategies learned from studies on biominerals, and has developed into a large field (Mann 2001). Various types of nanocrystals are synthesized using biomimetic approaches, including several sulfides. Here we mention a few typical examples for the concepts and strategies that are applied—for further reading we recommend the books by Mann (1997, 2001).

In BCM, crystals often precipitate in confined spaces such as phospholipid vesicles or ferritins (Table 1). Similar artificial vesicles can be used to create nanoscale reaction droplets. For example, ferritin is a spherical protein cage with an internal space about 8 nm in diameter that normally contains ferrihydrite (Mann 2001). The supramolecular structure of ferritin is remarkably stable, and the iron oxide core can be removed chemically without affecting the protein shell. Either empty ferritin cages can be used as confined reaction spaces, or the iron oxide core can be transformed chemically. The latter approach was used to produce amorphous FeS particles with controlled sizes ranging from 2 to 7 nm (Meldrum et al. 1991). Semiconducting CdS nanocrystals were synthesized both in reverse micelles and as the cores of artificial ferritin (Wong and Mann 1996). By attaching antibodies and antigens to the proteins, the ferritin cages can be linked together in solution, and a network of preformed, protein-coated inorganic nanoparticles can be engineered.

Functionalized organic structures can be used as epitaxial surfaces for templating the nucleation and growth of inorganic crystals. Oriented crystals of PbS were synthesized on self-assembled surfactant films (Langmuir monolayers) (Belman et al. 2004), while the ordered structures of bacterial S-layers proved to be an efficient template for the nucleation of ordered two-dimensional arrays of 5-nm CdS nanoparticles (Shenton et al. 1997).

Bioremediation

As iron sulfides precipitate in marine sediments, minor and trace amounts of various metals can be incorporated into the structures of iron monosulfides and pyrite. Schoonen's review (2004) contains an extensive compilation of observed metal concentrations in pyrite. Since many of the metal and metalloid impurities are toxic, it is of considerable interest to determine whether the host minerals are stable sinks for these metals. Iron sulfide formation under anoxic conditions in mine tailings and wetlands can be exploited for the immobilization of metal contaminants (Fortin and Beveridge 1997; Paktunc and Dave 2002). A wide range of metals has been co-precipitated with bacterially-produced iron sulfide (Watson and Ellwood 1994). Pesumably, greigite was one of the components of the precipitated sulfide, since the product could be divided into a weakly and a strongly magnetic fraction (Watson et al. 2000). The immobilization of heavy metals in magnetic iron sulfides offers the possibility of removing the contaminants by magnetic separation methods.

In the flooded ZnS mine that was described by Labrenz et al. (2000), coupled geochemical and microbial processes efficiently strip Zn from solutions containing <1 ppm Zn. The biofilm contains Zn in a concentration of about a million times that of the groundwater. As discussed by Moreau et al. (2004), some of the sedimentary sulfide deposits may have formed by similar BIM processes. Since metals such as As, Se, Cd, and Pb can be incorporated into or adsorbed on ZnS minerals, biomineralization may provide a suitable means to control the concentration of toxic metals in groundwater or wetlands. Such bioremediation strategies require that the minerals are relatively stable against dissolution. The solubilities of both sphalerite and wurtzite decrease with coarsening, since the growth of particles reduces the surface-to-volume ratio, decreasing reactivity with respect to oxidative dissolution (Moreau et al. 2004).

There are many possibilities for environmental applications of sulfide mineral precipitation by the mediation of bacteria. As discussed by Lovley (2003), bioremediation has been an empirical practice, but it could transform into a science thanks to new environmental genomic techniques that have become available. Experimental genomic and modelling techniques can be used to understand the physiologies of uncultured micro-organisms, and the resulting biological information, when combined with geochemical models, will be an invaluable tool for designing bioremediation strategies.

Bioleaching of metals

Whereas acidophilic, iron- and sulfur-oxidizing bacteria may be a curse when acid mine drainage is concerned, they are a blessing when used for the leaching of metals from their sulfide ores. Bacterial leaching of metals from low-grade ores is a well-established industrial technology that has been used for centuries (Rawlings and Silver 1995). Primarily mesophilic micro-organisms, such as *Thiobacillus ferrooxidans*, *Thiobacillus thiooxidans*, and *Leptospirillum ferrooxidans,* are applied (Hackl 1997). Historically, the targets of bacterial leaching have been sulfidic copper and refractory gold ores, but bioleaching practices are now used for the solubilization of a wide range of metals from their sulfide minerals (Nemati et al. 1998; Pina et al. 2005).

The use of bacteria in mineral processing may have additional advantageous side-effects. When *T. ferrooxidans* was added to a mixture of sulfide minerals during flotation, the cells adhered preferentially to pyrite and thus suppressed its floatability (Nagaoka et al. 1999). By using the microbes, pyrite could be separated from chalcocite, molybdenite, millerite, and galena. A detailed discussion of bioleaching practices is beyond the scope of this chapter. The interested reader is referred to reviews by Brierley (1978) and Hackl (1997).

IRON SULFIDES AND THE ORIGIN OF LIFE

Any review of the role of sulfide minerals in biosystems would be incomplete without mentioning the hypotheses that implicate sulfides in the origins of life. However, this topic is much more complex than is possible to deal with within the scope of this chapter. Also, an entire issue of the magazine *Elements* (June 2005) was devoted to problems related to the geochemical origin of life. Therefore, here the results of research on the possible roles of sulfides in the emergence of life are discussed only in the most general terms. The reader is referred to reviews by Cody (2005) and Schoonen et al. (2004).

Several origin-of-life hypotheses assume that the first organisms were autotrophs rather than heterotrophs, i.e., that they used small inorganic molecules (such as CO_2, NH_3, H_2S, H_2O, PO_4^{3-}) for building their biomolecules. Since the conversion of the inorganic forms of biogenic elements (C, N, O, S, H, and P) requires energy or the surpassing of an activation barrier (Schoonen et al. 2004), a catalyst is necessary. Modern biocatalysts that promote the formation of organic molecules from small components include protein enzymes that contain clusters of transition metals (Fe, Ni, Co) and sulfur at their active sites (Beinert et al. 1997). The important roles of metal sulfide clusters in microbial biosynthesis inspired two distinct hypotheses by Wächtershäuser (1988, 1990) and Russell and Hall (1997), in both of which it was proposed that sulfide minerals could catalyze the production of the first biomolecules.

According to Wächtershäuser's theory (1988, 1990), the formation of pyrite could have provided the energy source for the first organism, reducing CO_2 in the process, resulting in organic molecules according to the reaction:

$$CO_2(aq) + FeS + H_2S \rightarrow HCOOH + FeS_2 + H_2O$$

The small organic molecules then presumably combined into the larger biomolecules that are necessary for life. Some of the critical points of this hypothesis have been tested both experimentally and theoretically. Iron sulfide minerals were found to promote organic reduction reactions (Blöchl et al. 1992; Kaschke et al. 1994), while Huber and Wächtershäuser (1997) reported that a (Ni,Fe)S compound enhanced reactions that were designed to emulate the carbonyl-inserting reaction in modern microbial enzymes that have key roles in inorganic carbon fixation. However, on the basis of thermodynamical considerations, Schoonen et al. (1999) showed convincingly that the $FeS-H_2S/FeS_2$ redox couple is unlikely to initiate the proposed prebiotic carbon fixation cycle. The key point of their study is that the reducing power of the $FeS-H_2S/FeS_2$ couple diminishes with increasing temperature, whereas the reduction of CO_2 and the formation of carboxylic acids require increasingly higher reducing power with temperature. Nevertheless, transition metal sulfides do act as catalysts for reactions that can form important organic molecules. Cody et al. (2001, 2004) showed that NiS and common minerals (including chalcopyrite, bornite, and chalcocite) have the capacity to convert simple organic molecules into carboxylic acids. These reactions appeared to be surface-catalyzed, since they resulted in a high degree of isomeric selectivity and the reaction yield was correlated with mineral surface area. In another interesting experiment, Bebié and Schoonen (1999) demonstrated that anionic phosphate and phosphorylated organic molecules interacted with the surface of pyrite, and suggested that phosphate could have been concentrated on metal sulfide minerals on a prebiotic Earth, promoting the selective concentration of organic molecules from aqueous solutions.

The origin-of-life hypothesis developed by Russell and Hall (1997) is based on a geochemical consideration of the conditions that may have prevailed on the young Earth.

Soon after the discovery of hot vents on the deep seafloor, it was suggested that they provided the only stable environment where life could have emerged (Corliss et al. 1981). Shock (1992) demonstrated the potential for the formation of various organic molecules when CO_2 and carbonates in seawater mix with hydrothermal solutions. Russell and Hall (1997) and Russell et al. (1998) argued that life started at a redox and pH front where the acidic and warm (~90 °C) water of the Hadean ocean merged with reduced, alkaline, bisulfide-bearing, hot (~150 °C) water emitted at diffuse submarine vents. Under these conditions, colloidal FeS precipitated spontaneously in the form of thin films. According to Russell and Hall (1997), such FeS films could have formed bubbles, creating semipermeable membranes that separated the two fluids with different chemistries. The assimilation of CO_2 was catalyzed by the nickeliferous mackinawite, and amino acids and organic sulfide polymers could be synthesized within the FeS compartments. As the concentrations of carboxylic acids and organic polymers inside the bubbles increased, these organic molecules organized themselves either as coatings on the interiors of the FeS membrane or as micelles, and gradually took over the role of separating the two contrasting fluids. Thus, the first cell-like structures emerged, in which the generation of RNA and DNA may have become possible. The hypotheses by both Wächtershäuser (1988) and Russell and Hall (1997) are great intellectual achievements, and will likely continue to motivate much interesting experimental research in the future.

CONCLUDING THOUGHTS

Sulfide minerals were likely present at the beginning of life, and may even have catalyzed the first metabolic reactions at deep-sea hydrothermal vents. Interactions between organisms and sulfide minerals were important throughout most of Earth's history. Many types of sulfur-metabolizing microbes are rooted deeply in the Tree of Life, including sulfate reducers (Canfield and Raiswell 1999). A large radiation of sulfate reducers accompanied the general radiation of bacterial life. The formation of sulfide minerals by BIM must be at least as old as the first geochemical evidence for sulfate reduction, which is found between 2.7 and 2.5 Ga. Based on the available phylogenies, sulfate reducers may have appeared even earlier, by 3.4 Ga (Canfield and Raiswell 1999). Studies of the isotopic compositions of sedimentary pyrite led to the conclusion that the bottom waters of the oceans became sulfidic around ~1.8 Ga, when increased atmospheric oxygen levels enhanced terrestrial sulfide weathering, supplying sulfate to the oceans and increasing the rate of sulfate reduction (Poulton et al. 2004). Sulfidic conditions may have persisted until between 0.8 and 0.58 Ga ago, when a second major rise in oxygen concentration took place. At this time, the widespread oxidation of marine surface sediments promoted an evolutionary radiation of sulfide oxidizing bacteria (Canfield and Raiswell 1999), setting the stage for interactions between microbes and sulfide minerals as we know them today.

The geological history of sulfide mineral production by BCM is not known with any certainty. Sulfide magnetofossils from magnetotactic bacteria were identified tentatively from Miocene rocks (Pósfai et al. 2001) and soil (Stanjek et al. 1994). Since magnetite magnetofossils were described from Archaean rocks, Kirschvink and Hagadorn (2000) hypothesized that all BCM processes originated from magnetite biomineralization by magnetotactic bacteria.

Since biologically controlled or mediated mineralization produces nanocrystalline sulfide particles, their physical, structural, and chemical characterization will likely remain an exciting and challenging field of mineralogical research. Concerning the iron sulfides that are produced by magnetotactic bacteria, the key problems to be addressed are related to biological control over crystal nucleation and growth. It has still not been established whether greigite crystals in mangnetotactic bacteria are surrounded by a magnetosome membrane, as are the crystals in magnetite-producers, or whether they are deposited in a less controlled manner, perhaps as a consequence of the sulfate-reducing metabolism of the host cell. The mineral phases of the initially formed precipitates, as well as their conversions, also deserve further study. In addition, the possibility exists that additional sulfides with biological functions will be discovered. Given the geological and environmental importance of the bacterially-assisted formation and dissolution of sulfide minerals, interactions between microbes and sulfides will continue to be the subject of intensive research.

ACKNOWLEDGMENTS

We thank Takeshi Kasama for contributing the results of his electron holography work on magnetotatic bacteria, and Ryan Chong for electron tomography of sulfide-bearing bacteria. Samples and discussions with Richard Frankel and discussions with Peter Buseck are gratefully acknowledged. WMP acknowledges support from the Hungarian Science Fund (OTKA-T030186). RDB acknowledges the Royal Society for support.

REFERENCES

Aplin AC (2000) Mineralogy of modern marine sediments: A geochemical framework. *In*: Environmental Mineralogy. EMU Notes in Mineralogy 2. Vaughan DJ, Wogelius RA (eds) Eötvös University Press, p 125-172
Baeuerlein E (2000) Biomineralization: From biology to biotechnology and medical application. Wiley
Baeuerlein E (2003) Biomineralization of unicellular organisms: An unusual membrane biochemistry for the produdction of inorganic nano- and microstructures. Angew Chem Int Ed 42:614-641
Baeuerlein E (2004) Biomineralization: Progress in biology, molecular biology and application. Wiley
Baker BJ, Banfield JF (2003) Microbial communities in acid mine drainage. FEMS Microbiol Ecol 44:139-152
Balkwill DL, Maratea D, Blakemore RP (1980) Ultrastructure of a magnetic spirillum. J Bacteriol 141:1399-1408
Banfield JF, Cervini-Silva J, Nealson KH (eds) (2005) Molecular Geomicrobiology. Rev Mineral Geochem 59. Mineralogical Society of America
Banfield JF, Nealson KH (eds) (1997) Geomicrobiology: Interactions between microbes and minerals. Rev Mineral 35. Mineralogical Society of America
Bazylinski DA, Frankel RB (2003) Biologically controlled mineralization in prokaryotes. Rev Mineral Geochem 54:217-247
Bazylinski DA, Frankel RB (2004) Magnetosome formation in prokaryotes. Nat Rev Microbiol 2:217-230
Bazylinski DA, Frankel RB, Garratt-Reed AJ, Mann S (1990) Biomineralization of iron sulfides in magnetotactic bacteria from sulfidic environments. *In*: Iron Biominerals. Frankel RB, Blakemore RP (eds) Plenum Press, p 239-255
Bazylinski DA, Frankel RB, Heywood BR, Mann S, King JW, Donaghay PL, Hanson AK (1995) Controlled biomineralization of magnetite (Fe_3O_4) and greigite (Fe_3S_4) in a magnetotactic bacterium. Appl Environ Microbiol 61:3232-3239
Bazylinski DA, Garratt-Reed AJ, Abedi A, Frankel RB (1993a) Copper association with iron sulfide magnetosomes in a magnetotactic bacterium. Arch Microbiol 160:35-42
Bazylinski DA, Heywood BR, Mann S, Frankel RB (1993b) Fe_3O_4 and Fe_3S_4 in a bacterium. Nature 366:218
Bazylinski DA, Moskowitz BM (1997) Microbial biomineralization of magnetic iron minerals. Rev Mineral 35:181-223
Bebié J, Schoonen MAA (1999) Pyrite and phosphate in anoxia and an origin-of-life hypothesis. Earth Planet Sci Lett 171:1-5
Beinert H, Holm RH, Munck E (1997) Iron-sulfur clusters: Nature's modular, multipurpose structures. Science 277:653-659
Belman N, Berman A, Ezersky V, Lifshitz Y, Golan Y (2004) Transmission electron microscopy of epitaxial PbS nanocrystals on polydiacetylene Langmuir films. Nanotechnol 15:S316-S321
Bennett JC, Tributsch H (1978) Bacterial leaching patterns on pyrite crystal-surfaces. J Bacteriol 134:310-317
Benning LG, Wilkin RT, Barnes HL (2000) Reaction pathways in the Fe-S system below 100 degrees C. Chem Geol 167:25-51
Berner RA (1970) Sedimentary pyrite formation. Am J Sci 270:1-23
Berner RA (1984) Sedimentary pyrite formation: An update. Geochim Cosmochim Acta 48:605-615

Berner RA (2001) Modeling atmospheric O^{-2} over Phanerozoic time. Geochim Cosmochim Acta 65:685-694
Bertolin A, Frizzo P, Rampazzo G (1995) Sulfide speciation in surface sediments of the Lagoon of Venice - a geochemical and mineralogical study. Mar Geol 123:73-86
Beveridge TJ (1989) Role of cellular design in bacterial metal accumulation and mineralization. Ann Rev Microbiol 43:147-171
Blakemore RP (1975) Magnetotactic bacteria. Science 190:377-379
Blöchl E, Keller M, Wächtershäuser G, Stetter KO (1992) Reactions depending on iron sulfide and linking geochemistry with biochemistry. Proc Nat Acad Sci USA 89:8117-8120
Bonev IK, Khrischev KG, Neikov HN, Georgiev VM (1989) Mackinawite and greigite in iron sulphide concretions from Black Sea sediments. Compt Rend Acad Bulg Sci 42:97-100
Brierley CL (1978) Bacterial leaching. CRC Crit Rev Microbiol 6:207-262
Butler IB, Rickard D (2000) Framboidal pyrite formation via the oxidation of iron (II) monosulfide by hydrogen sulphide. Geochim Cosmochim Acta 64:2665-2672
Canfield DE, Habicht KS, Thamdrup B (2000) The Archean sulfur cycle and the early history of atmospheric oxygen. Science 288:658-661
Canfield DE, Raiswell R (1999) The evolution of the sulfur cycle. Am J Sci 299:697-723
Canfield DE, Thamdrup B, Fleischer S (1998) Isotope fractionation and sulfur metabolism by pure and enrichment cultures of elemental sulfur-disproportionating bacteria. Limnol Oceanogr 43:253-264
Chen X, Zhang X, Wan J, Wang Z, Qian Y (2005) Selective fabrication of metastable greigite (Fe$_3$S$_4$) nanocrystallites and its magnetic properties through a simple solution-based route. Chem Phys Lett 403: 396-399
Cody GD (2005) Geochemical connections to primitive metabolism. Elements 1:139-143
Cody GD, Boctor NZ, Brandes JA, Filley TR, Hazen RM, Yoder HS (2004) Assaying the catalytic potential of transition metal sulfides for abiotic carbon fixation. Geochim Cosmochim Acta 68:2185-2196
Cody GD, Boctor NZ, Hazen RM, Brandes JA, Morowitz HJ, Yoder HS (2001) Geochemical roots of autotrophic carbon fixation: Hydrothermal experiments in the system citric acid, H$_2$O-(±FeS)-(±NiS). Geochim Cosmochim Acta 65:3557-3576
Corliss JB, Baross JA, Hoffman SE (1981) An hypothesis concerning the relationship between submarine hot springs and the origin of life on Earth. Oceanol Acta Suppl 4:59-69
Davis BS (1997) Geomicrobiology of the oxic zone of sulfidic mine tailings. In: Biological-mineralogical interactions. McIntosh JM, Groat LA (eds) Mineralogical Association of Canada, Ottawa, p 93-112
Dekkers MJ, Passier HF, Schoonen MAA (2000) Magnetic properties of hydrothermally synthesized greigite (Fe$_3$S$_4$)— II. High- and low-temperature characteristics. Geophys J Int 141:809-819
DeLong EF, Frankel RB, Bazylinski DA (1993) Multiple evolutionary origins of magnetotaxis in bacteria. Science 259:803-806
Desbruyeres D, Chevaldonne P, Alayse AM, Jollivet D, Lallier FH, Jouin-Toulmond C, Zal F, Sarradin PM, Cosson R, Caprais JC, Arndt C, O'Brien J, Guezennec J, Hourdez S, Riso R, Gaill F, Laubier L, Toulmond A (1998) Biology and ecology of the "Pompeii worm" (*Alvinella pompejana Desbruyeres* and *Laubier*), a normal dweller of an extreme deep-sea environment: A synthesis of current knowledge and recent developments. Deep-Sea Res II 45:383-422
Diaz-Ricci JC, Kirschvink JL (1992) Magnetic domain state and coercivity predictions for biogenic greigite (Fe$_3$S$_4$): A comparison of theory with magnetosome observations. J Geophys Res 97:17309-17315
Diebel CE, Proksch R, Green CR, Neilson P, Walker MM (2000) Magnetite defines a vertebrate magnetoreceptor. Nature 406:299-302
Dobson J (2001) Nanoscale biogenic iron oxides and neurodegenerative disease. FEBS Lett 496:1-5
Donald R, Southam G (1999) Low temperature anaerobic bacterial diagenesis of ferrous monosulfide to pyrite. Geochim Cosmochim Acta 63:2019-2023
Dove PM, De Yoreo JJ, Weiner S (eds) (2003) Biomineralization. Rev Mineral Geochem 54. Mineralogical Society of America
Dunin-Borkowski RE, McCartney MR, Frankel RB, Bazylinski DA, Pósfai M, Buseck PR (1998) Magnetic microstructure of magnetotactic bacteria by electron holography. Science 282:1868-1870
Dunin-Borkowski RE, McCartney MR, Pósfai M, Frankel RB, Bazylinski DA, Buseck PR (2001) Off-axis electron holography of magnetotactic bacteria: Magnetic microstructure of strains MV-1 and MS-1. Eur J Mineral 13:671-684
Dunin-Borkowski RE, McCartney MR, Smith DJ (2004) Electron holography of nanostructured materials. In: Encyclopedia of Nanoscience and Nanotechnology. Nalwa HS (ed) American Scientific Pubs., p 41-100
Dunlop DJ, Özdemir Ö (1997) Rock Magnetism: Fundamentals and Frontiers. Cambridge University Press
Edwards KJ, Bach W, McCollom TM (2005) Geomicrobiology in oceanography: Microbe-mineral interactions at and below the seafloor. Trends Microbiol 13:449-456
Edwards KJ, Bond PL, Druschel GK, McGuire MM, Hamers RJ, Banfield JF (2000) Geochemical and biological aspects of sulfide mineral dissolution: Lessons from Iron Mountain, California. Chem Geol 169:383-397

Edwards KJ, Hu B, Hamers RJ, Banfield JF (2001) A new look at microbial leaching patterns on sulfide minerals. FEMS Microbiol Ecol 34:197-206
Edwards KJ, McCollom TM, Konishi H, Buseck PR (2003) Seafloor bioalteration of sulfide minerals: Results from *in situ* incubation studies. Geochim Cosmochim Acta 67:2843-2856
Farina M, Esquivel DMS, Lins de Barros HGP (1990) Magnetic iron-sulphur crystals from a magnetotactic microorganism. Nature 343:256-258
Ferris FG (1997) Formation of authigenic minerals by bacteria. *In*: Biological-mineralogical interactions. McIntosh JM, Groat LA (eds) Mineralogical Association of Canada, Ottawa, p 187-208
Ferris FG, Fyfe WS, Beveridge TJ (1987) Bacteria as nucleation sites for authigenic minerals in a metal-contaminated lake sediment. Chem Geol 63:225-232
Flies CB, Jonkers HM, de Beer D, Bosselmann K, Bottcher ME, Schüler D (2005) Diversity and vertical distribution of magnetotactic bacteria along chemical gradients in freshwater microcosms. FEMS Microbiol Ecol 52:185-195
Fortin D, Beveridge TJ (1997) Microbial sulfate reduction within sulfidic mine tailings: Formation of diagenetic Fe sulfides. Geomicrobiol J 14:1-21
Fortin D, Ferris FG, Beveridge TJ (1997) Surface-mediated mineral development by bacteria. Rev Mineral 35:161-180
Fortin D, Langley S (2005) Formation and occurrence of iron-rich biogenic minerals. Earth-Sci Rev 72:1-19
Fortin D, Rioux J-P, Roy M (2002) Geochemistry of iron and sulfur in the zone of microbial sulfate reduction in mine tailings. Water Air Soil Pollution 2:37-56
Frankel RB (1984) Magnetic guidance of organisms. Annu Rev Biophys Bioeng 13:85-103
Frankel RB, Bazylinski DA (2003) Biologically induced mineralization by bacteria. Rev Mineral Geochem 54:95-114
Frankel RB, Bazylinski DA, Johnson MS, Taylor BL (1997) Magneto-aerotaxis in marine coccoid bacteria. Biophys J 73:994-1000
Fukumori Y (2000) Enzymes for magnetite synthesis in *Magnetospirillum magnetotacticum*. *In*: Biomineralization. Baeuerlein E (ed) Wiley, p 93-108
Furukawa Y, Barnes HL (1995) Reactions forming pyrite from precipitated amorphous ferrous sulfide. *In*: Geochemical transformations of sedimentary sulfur. Vairavamurthy MA, Schoonen MAA (eds) American Chemical Society, p 194-205
Gilbert PUPA, Abrecht M, Frazer BH (2005) The organic-mineral interface in biominerals. Rev Mineral Geochem 59:157-185
Goffredi SK, Warén A, Orphan VJ, Van Dover CL, Vrijenhoek RC (2004) Novel forms of structural integration between microbes and a hydrothermal vent gastropod from the Indian Ocean. Appl Environ Microbiol 70:3082-3090
Gould WD, Francis M, Blowes DW, Krouse HR (1997) Biomineralization: Microbiological formation of sulfide minerals. *In*: Biological-mineralogical interactions. McIntosh JM, Groat LA (eds) Mineralogical Association of Canada, p 169-186
Grimes ST, Brock F, Rickard D, Davies KL, Edwards D, Briggs DEG, Parkes RJ (2001) Understanding fossilization: Experimental pyritization of plants. Geology 29:123-126
Hackl RP (1997) Commercial applications of bacteria-mineral interactions. *In*: Biological-mineralogical interactions. McIntosh JM, Groat LA (eds) Mineralogical Association of Canada, Ottawa, p 143-167
Hámor T (1994) The occurrence and morphology of sedimentary pyrite. Acta Geol Hung 37:153-181
Herbert RB Jr., Benner SG, Pratt AR, Blowes DW (1998) Surface chemistry and morphology of poorly crystalline iron sulfides precipitated in media containing sulfate-reducing bacteria. Chem Geol 144:87-97
Heywood BR, Bazylinski DA, Garratt-Reed A, Mann S, Frankel RB (1990) Controlled biosynthesis of greigite (Fe_3S_4) in magnetotactic bacteria. Naturwiss 77:536-538
Heywood BR, Mann S, Frankel RB (1991) Structure, morphology and growth of biogenic greigite (Fe_3S_4). Mat Res Soc Symp Proc 218:93-108
Hoffmann V (1992) Greigite (Fe_3S_4): Magnetic properties and first domain observations. Phys Earth Planet Int 70:288-301
Holmes J (1999) Fate of incorporated metals during mackinawite oxidation in sea water. Appl Geochem 14:277-281
Horiuchi S (1971) Zur Umwandlung von Mackinawit (FeS) in Greigit (Fe_3S_4) durch Elektronenstrahlen. Z Anorg Allg Chem 386:208-212
Huber C, Wächtershäuser G (1997) Activated acetic acid by carbon fixation on (Fe,Ni)S under primordial conditions. Science 276:245-247
Huber JA, Butterfield DA, Baross JA (2003) Bacterial diversity in a subseafloor habitat following a deep-sea volcanic eruption. FEMS Microbiol Ecol 43:393-409
Hurtgen MT, Lyons TW, Ingall ED, Cruse AM (1999) Anomalous enrichments of iron monosulfide in euxinic marine sediments and the role of H_2S in iron sulfide transformations: Examples from Effingham Inlet, Orca Basin, and the Black Sea. Am J Sci 299:556-588

Jambor JL, Blowes DW, Ptacek CJ (2000) Mineralogy of mine wastes and strategies of remediation. *In*: Environmental Mineralogy. Vaughan DJ, Wogelius RA (eds) EMU Notes in Mineralogy 2, Eötvös University Press, p 255-290

Jannasch HW, Mottl MJ (1985) Geomicrobiology of deep-sea hydrothermal vents. Science 229:717-725

Jørgensen BB (1977) Sulfur cycle of a coastal marine sediment (Limfjorden, Denmark). Limnol Oceanogr 22: 814-832

Kappler A, Straub KL (2005) Geomicrobiological cycling of iron. Rev Mineral Geochem 59:85-108

Kasama T, Pósfai M, Chong RKK, Finlayson AP, Buseck PR, Frankel RB, Dunin-Borkowski RE (2006) Magnetic properties, microstructure, composition and morphology of greigite nanocrystals in magnetotactic bacteria from electron holography and tomography. Am Mineral (in press)

Kaschke M, Russell MJ, Cole WJ (1994) FeS/FeS$_2$ - a redox system for the origin of life - (some experiments on the pyrite-hypothesis). Orig Life Evol Biosph 24:43-56

Keim CN, Abreu F, Lins U, de Barros HL, Farina M (2004) Cell organization and ultrastructure of a magnetotactic multicellular organism. J Struct Biol 145:254-262

Kirschvink JL, Hagadorn JW (2000) A grand unified theory of biomineralization. *In*: Biomineralization. Baeuerlein E (ed) Wiley, p 139-150

Kirschvink JL, Kobayashi-Kirschvink A, Woodford BJ (1992) Magnetite biomineralization in the human brain. Proc Natl Acad Sci USA 89:7683-7687

Kirschvink JL, Walker MM, Diebel CE (2001) Magnetite-based magnetoreception. Curr Opin Neurobiol 11: 462-467

Komeili A, Li Z, Newman DK, Jensen GJ (2006) Magnetosomes are cell membrane invaginations organized by the actin-like protein MamK. Science 311:242-245

Komeili A, Vali H, Beveridge TJ, Newman DK (2004) Magnetosome vesicles are present before magnetite formation, and MamA is required for their activation. Proc Natl Acad Sci USA 101:3839-3844

Konhauser KO (1998) Diversity of bacterial iron mineralization. Earth-Sci Rev 43:91-121

Kuenen JG, Robertson LA, Tuovinen OH (1992) The genera Thiobacillus, Thiomicrospira, and Thiosphera. *In:* The Prokaryotes. Balows A, Truper HG, Dworkin M, Harder W, Schleifer KH (eds) Springer, p 2638-2657

Labrenz M, Druschel GK, Thomsen-Ebert T, Gilbert B, Welch SA, Kemner KM, Logan GA, Summons RE, De Stasio G, Bond PL, Lai B, Kelly SD, Banfield JF (2000) Formation of sphalerite (ZnS) deposits in natural biofilms of sulfate-reducing bacteria. Science 290:1744-1747

Lennie AR, Redfern SAT, Champness PE, Stoddart CP, Schofield PF, Vaughan DJ (1997) Transformation of mackinawite to greigite: An *in situ* X-ray powder diffraction and transmission electron microscope study. Am Mineral 82:302-309

Lennie AR, Vaughan DJ (1996) Spectroscopic studies of iron sulfide formation and phase relations at low temperatures. Mineral Spectr 5:117-131

Lins U, Farina M (1999) Organization of cells in magnetotactic multicellular aggregates. Microbiol Res 154: 9-13

Lins U, Farina M (2001) Amorphous mineral phases in magnetotactic multicellular aggregates. Arch Microbiol 176:323-328

Lovley DR (2003) Cleaning up with genomics: Applying molecular biology to bioremediation. Nature Rev Microbiol 1:35-44

Lovley DR, Chapelle FH (1995) Deep subsurface microbial processes. Rev Geophys 33:365-381

Lowenstam HA, Weiner S (1989) On Biomineralization. Oxford University Press

Luther GW (1991) Pyrite synthesis via polysulfide compounds. Geochim Cosmochim Acta 55:2839-2849

Luther GW, Rickard D (2005) Metal sulfide cluster complexes and their biogeochemical importance in the environment. J Nanoparticle Res 7:389-407

Luther GW, Rozan TF, Taillefert M, Nuzzio DB, Di Meo C, Shank TM, Lutz RA, Cary SC (2001) Chemical speciation drives hydrothermal vent ecology. Nature 410:813-816

Lysnes K, Thorseth IH, Steinsbu BO, Ovreas L, Torsvik T, Pedersen RB (2004) Microbial community diversity in seafloor basalt from the Arctic spreading ridges. 50:213-230

Maginn EJ, Little CTS, Herrington RJ, Mills RA (2002) Sulphide mineralisation in the deep sea, hydrothermal vent polychaete, Alvinella pompejana: implications for fossil preservation. Mar Geol 181:337-356

Mann S (1997) Biomimetic Materials Chemistry. Wiley-VCH

Mann S (2001) Biomineralization: Principles and Concepts in Bioinorganic Materials Chemistry. Oxford University Press

Mann S, Sparks NHC, Frankel RB, Bazylinski DA, Jannasch HW (1990) Biomineralization of ferrimagnetic greigite (Fe$_3$S$_4$) and iron pyrite (FeS$_2$) in a magnetotactic bacterium. Nature 343:258-261

McCartney MR, Lins U, Farina M, Buseck PR, Frankel RB (2001) Magnetic microstructure of bacterial magnetite by electron holography. Eur J Mineral 13:685-689

McIntosh JM, Groat LA (1997) Biological-mineralogical interactions. Mineralogical Association of Canada

McIntosh JM, Silver M, Groat LA (1997) Bacteria and the breakdown of sulfide minerals. *In*: Biological-mineralogical interactions. McIntosh JM, Groat LA (eds) Mineralogical Association of Canada, Ottawa, p 63-92
Meldrum FC, Wade VJ, Nimmo DL, Heywood BR, Mann S (1991) Synthesis of inorganic nanophase materials in supramolecular protein cages. Nature 349:684-687
Methe BA, Nelson KE, Eisen JA, Paulsen IT, Nelson W, Heidelberg JF, Wu D, Wu M, Ward N, Beanan MJ, Dodson RJ, Madupu R, Brinkac LM, Daugherty SC, DeBoy RT, Durkin AS, Gwinn M, Kolonay JF, Sullivan SA, Haft DH, Selengut J, Davidsen TM, Zafar N, White O, Tran B, Romero C, Forberger HA, Weidman J, Khouri H, Feldblyum TV, Utterback TR, Van Aken SE, Lovley DR, Fraser CM (2003) Genome of Geobacter sulfurreducens: Metal reduction in subsurface environments. Science 302:1967-1969
Moreau JW, Webb RI, Banfield JF (2004) Ultrastructure, aggregation-state, and crystal growth of biogenic nanocrystalline sphalerite and wurtzite. Am Mineral 89:950-960
Morse JW, Millero FJ, Cornwell JC, Rickard D (1987) The chemistry of the hydrogen sulfide and iron sulfide systems in natural waters. Earth-Sci Rev 24:1-42
Murowchick JB, Barnes HL (1986) Marcasite precipitation from hydrothermal solutions. Geochim Cosmochim Acta 50:2615-2629
Nagaoka T, Ohmura N, Saiki H (1999) A novel mineral flotation process using *Thiobacillus ferrooxidans*. Appl Environ Microbiol 65:3588-3593
Neal AL, Techkarnjanaruk S, Dohnalkova A, McCready D, Peyton BM, Geesey GG (2001) Iron sulfides and sulfur species produced at hematite surfaces in the presence of sulfate-reducing bacteria. Geochim Cosmochim Acta 65:223-235
Nealson KH, Saffarini D (1994) Iron and manganese in anaerobic respiration. Annu Rev Microbiol 48:311-343
Nealson KH, Stahl DA (1997) Microorganisms and biogeochemical cycles: What can we learn from layered microbial communities? Rev Mineral 35:5-34
Nemati M, Harrison STL, Hansford GS, Webb C (1998) Biological oxidation of ferrous sulphate by *Thiobacillus ferrooxidans*: a review on the kinetic aspects. Biochem Eng J 1:171-190
Nordstrom DK, Southam G (1997) Geomicrobiology of sulfide mineral oxidation. Rev Mineral 35:361-390
Ohfuji H, Akai J (2002) Icosahedral domain structure of framboidal pyrite. Am Mineral 87:176-180
Ohfuji H, Boyle AP, Prior DJ, Rickard D (2005) Structure of framboidal pyrite: An electron backscatter diffraction study. Am Mineral 90:1693-1704
Ohfuji H, Rickard D (2005) Experimental syntheses of framboids - a review. Earth-Sci Rev 71:147-170
Ohfuji H, Rickard D (2006) High resolution transmission electron microscopic study of synthetic nanocrystalline mackinawite. Earth Planet Sci Lett 241:227-233
Olson GJ (1991) Rate of pyrite bioleaching by *Thiobacillus ferrooxidans* - results of an interlaboratory comparison. Appl Environ Microbiol 57:642-644
Paktunc AD, Dave NK (2002) Formation of secondary pyrite and carbonate minerals in the Lower Williams Lake tailings basin, Elliot Lake, Ontario, Canada. Am Mineral 87:593-602
Pearce CI, Pattrick RAD, Vaughan DJ (2006) Electrical and Magnetic properties of sulfides. Rev Mineral Geochem 61:127-180
Pina PS, Leao VA, Silva CA, Daman D, Frenay J (2005) The effect of ferrous and ferric iron on sphalerite bioleaching with Acidithiobacillus sp. Mineral Eng 18:549-551
Pósfai M, Buseck PR, Bazylinski DA, Frankel RB (1998a) Iron sulfides from magnetotactic bacteria: Structure, composition, and phase transitions. Am Mineral 83:1469-1481
Pósfai M, Buseck PR, Bazylinski DA, Frankel RB (1998b) Reaction sequence of iron sulfide minerals in bacteria and their use as biomarkers. Science 280:880-883
Pósfai M, Cziner K, Márton E, Márton P, Buseck PR, Frankel RB, Bazylinski DA (2001) Crystal-size distributions and possible biogenic origin of Fe sulfides. Eur J Mineral 13:691-703
Poulton SW, Fralick PW, Canfield DE (2004) The transition to a sulphidic ocean ~1.84 billion years ago. Nature 431:173-177
Raiswell R, Canfield DE (1998) Sources of iron for pyrite formation in marine sediments. Am J Sci 298:219-245
Raiswell R, Whaler K, Dean S, Coleman ML, Briggs DEG (1993) A simple 3-dimensional model of diffusion-with-precipitation applied to localized pyrite formation in framboids, fossils and detrital iron minerals. Mar Geol 113:89-100
Rawlings DE, Silver S (1995) Mining with microbes. Bio-Technol 13:773-778
Reynolds RL, Tuttle ML, Rice CA, Fishman NS, Karachewski JA, Sherman DM (1994) Magnetization and geochemistry of greigite-bearing Cretaceous strata, North-Slope Basin, Alaska. Am J Sci 294:485-528
Rickard D (1997) Kinetics of pyrite formation by the H_2S oxidation of iron (II) monosulfide in aqueous solutions between 25 and 125 degrees C: The rate equation. Geochim Cosmochim Acta 61:115-134
Rickard D, Butler IB, Oldroyd A (2001) A novel iron sulphide mineral switch and its implications for Earth and planetary science. Earth Planet Sci Lett 189:85-91
Rickard D, Luther GW (2006) Metal sulfide complexes and clusters. Rev Mineral Geochem XX

Rickard D, Morse JW (2005) Acid volatile sulfide (AVS). Mar Chem 97:141-197

Rickard D, Schoonen MAA, Luther GW (1995) Chemistry of iron sulfides in sedimentary environments. *In*: Geochemical Transformations of Sedimentary Sulfur. ACS Symposium Series 612, p 168-193

Roberts AP (1995) Magnetic properties of sedimentary greigite (Fe_3S_4). Earth Planet Sci Lett 134:227-236

Rodgers FG, Blakemore RP, Blakemore NA, Frankel RB, Bazylinski DA, Maratea D, Rodgers C (1990) Intercellular structure in a many-celled magnetotactic prokaryote. Arch Microbiol 154:18-22

Rosso KM, Vaughan DJ (2006) Reactivity of sulfide mineral surfaces. Rev Mineral Geochem 61:557-607

Rowan CJ, Roberts AP (2006) Magnetite dissolution, diachronous greigite formation, and secondary magnetizations from pyrite oxidation: Unravelling complex magnetizations in Neogene marine sediments from New Zealand. Earth Planet Sci Lett 241:119-137

Russell MJ, Daia DE, Hall AJ (1998) The emergence of life from FeS bubbles at alkaline hot springs in an acid ocean. *In*: Thermophiles: The Keys to Molecular Evolution and the Origin of Life. Adams MWW, Ljungdahl LG, Wiegel J (eds) Taylor and Francis, p 77-121

Russell MJ, Hall AJ (1997) The emergence of life from iron monosulphide bubbles at a submarine hydrothermal redox and pH front. J Geol Soc 154:377-402

Scheffel A, Gruska M, Faivre D, Linaroudis A, Plitzko JM, Schüler D (2005) An acidic protein aligns magnetosomes along a filamentous structure in magnetotactic bacteria. Nature AOP:DOI 10.1038/nature04281

Schippers A, Jørgensen BB (2002) Biogeochemistry of pyrite and iron sulfide oxidation in marine sediments. Geochim Cosmochim Acta 66:85-92

Schoonen M, Smirnov A, Cohn C (2004) A perspective on the role of minerals in prebiotic synthesis. Ambio 33:539-551

Schoonen MAA (2004) Mechanisms of sedimentary pyrite formation. *In*: Sulfur Biogeochemistry - Past and present Geological Society of America Special Paper. Amend JP, Edwards KJ, Lyons TW (eds) Geological Society of America, p 117-134

Schoonen MAA, Barnes HL (1991) Reactions forming pyrite and marcasite from solution: I. Nucleation of FeS_2 below 100 °C. Geochim Cosmochim Acta 55:1495-1504

Schoonen MAA, Xu Y, Bebie J (1999) Energetics and kinetics of the prebiotic synthesis of simple organic acids and amino acids with the $FeS-H_2S/FeS_2$ redox couple as reductant. Orig Life Evol Biosph 29:5-32

Schrenk MO, Kelley DS, Delaney JR, Baross JA (2003) Incidence and diversity of microorganisms within the walls of an active deep-sea sulfide chimney. Appl Environ Microbiol 69:3580-3592

Schultze-Lam S, Fortin D, Davis BS, Beveridge TJ (1996) Mineralization of bacterial surfaces. Chem Geol 132:171-181

Schüler D (2004) Molecular analysis of a subcellular compartment: The magnetosome membrane of Magnetospirillum gryphiswaldense. Arch Microbiol 181:1-7

Schüler DR, Uhl E, Baeuerlein E (1995) A simple light scattering method to assay magnetism in Magnetospirillum gryphiswaldense. FEMS Microbiol Lett 132:139-145

Scott SD, Barnes HL (1972) Sphalerite-wurtzite equilibria and stoichiometry. Geochim Cosmochim Acta 36:1275-1295

Shenton W, Pum D, Sleytr UB, Mann S (1997) Synthesis of cadmium sulphide superlattices using self-assembled bacterial S-layers. Nature 389:585-587

Shock EL (1992) Chemical environments of submarine hydrothermal systems. Orig Life Evol Biosph 22:67-107

Simmons SL, Bazylinski DA, Edwards KJ (2006) South-seeking magnetotactic bacteria in the Northern Hemisphere. Science 311:371-374

Simmons SL, Sievert SM, Frankel RB, Bazylinski DA, Edwards KJ (2004) Spatiotemporal distribution of marine magnetotactic bacteria in a seasonally stratified coastal salt pond. Appl Environ Microbiol 70:6230-6239

Simpson ET, Kasama T, Pósfai M, Buseck PR, Harrison RJ, Dunin-Borkowski RE (2005a) Magnetic induction mapping of magnetite chains in magnetotactic bacteria at room temperature and close to the Verwey transition using electron holography. J Phys Conf Ser 17:108-121

Simpson SL, Apte SC, Davies CM (2005b) Bacterially assisted oxidation of copper sulfide minerals in tropical river waters. Environ Chem 2:49-55

Stanjek H, Fassbinder JWE, Vali H, Wägele H, Graf W (1994) Evidence of biogenic greigite (ferrimagnetic Fe_3S_4) in soil. Eur J Soil Sci 45:97-103

Summit M, Baross JA (2001) A novel microbial habitat in the mid-ocean ridge subseafloor. Proc Natl Acad Sci USA 98:2158-2163

Suzuki Y, Inagaki F, Takai K, Nealson KH, Horikoshi K (2004) Microbial diversity in inactive chimney structures from deep-sea hydrothermal systems. Microbial Ecol 47:186-196

Suzuki Y, Kopp RE, Kogure T, Suga A, Takai K, Tsuchida S, Ozaki N, Endo K, Hashimoto J, Kato Y, Mizota C, Hirata T, Chiba H, Nealson KH, Horikoshi K, Kirschvink JL (2006) Sclerite formation in the hydrothermal-vent "scaly-foot" gastropod - possible control of iron sulfide biomineralization by the animal. Earth Planet Sci Lett 242:39-50

Sweeney RE, Kaplan IR (1973) Pyrite framboid formation: Laboratory synthesis and marine sediments. Geology 68:618-634

Thamdrup B, Fossing H, Jørgensen BB (1994) Manganese, iron, and sulfur cycling in a coastal marine sediment, Aarhus Bay, Denmark. Geochim Cosmochim Acta 58:5115-5129

Theberge SM, Luther GW, Farrenkopf AM (1997) On the existence of free and metal complexed sulfide in the Arabian Sea and its oxygen minimum zone. Deep-Sea Res 44:1381-1390

Thorseth IH, Torsvik T, Torsvik V, Daae FL, Pedersen RB (2001) Diversity of life in ocean floor basalt. Earth Planet Sci Lett 194:31-37

Tuttle JH, Dugan PR, MacMillan CB, Randles CI (1969) Microbial dissimilatory sulfur cycle in acid mine water. J Bacteriol 97:594-602

Van Dover CL, Humphris SE, Fornari D, Cavanaugh CM, Collier R, Goffredi SK, Hashimoto J, Lilley MD, Reysenbach AL, Shank TM, Von Damm KL, Banta A, Gallant RM, Gotz D, Green D, Hall J, Harmer TL, Hurtado LA, Johnson P, McKiness ZP, Meredith C, Olson E, Pan IL, Turnipseed M, Won Y, Young CR, Vrijenhoek RC (2001) Biogeography and ecological setting of Indian Ocean hydrothermal vents. Science 294:818-823

Vaughan DJ, Wogelius RA (2000) Environmental Mineralogy. EMU Notes in Mineralogy 2. Eötvös University Press

Verati C, de Donato P, Prieur D, Lancelot J (1999) Evidence of bacterial activity from micrometer-scale layer analyses of black-smoker sulfide structures (Pito Seamount Site, Easter microplate). Chem Geol 158:257-269

Wächtershäuser G (1990) Evolution of the first metabolic cycles. Proc Natl Acad Sci USA 87:200-204

Wächtershäuser G (1988) Pyrite formation, the first energy source for life: A hypothesis. System Appl Microbiol 10:207-210

Walker MM, Dennis TE, Kirschvink JL (2002) The magnetic sense and its use in long-distance navigation by animals. Curr Opin Neurobiol 12:735-744

Warén A, Bengtson S, Goffredi SK, Van Dover CL (2003) A hot-vent gastropod with iron sulfide dermal sclerites. Science 302:1007-1007

Watson JHP, Cressey BA, Roberts AP, Ellwood DC, Charnock JM, Soper AK (2000) Structural and magnetic studies on heavy-metal-adsorbing iron sulphide nanoparticles produced by sulphate-reducing bacteria. J Magn Magn Mat 214:13-30

Watson JHP, Ellwood DC (1994) Biomagnetic separation and extraction process for heavy-metals from solution. 7:1017-1028

Weiner S, Dove PM (2003) An overview of biomineralization processes and the problem of the vital effect. Rev Mineral Geochem 54:1-29

Wilkin RT, Barnes HL (1997) Formation processes of framboidal pyrite. Geochim Cosmochim Acta 61:323-339

Wilkin RT, Barnes HL, Brantley SL (1996) The size distribution of framboidal pyrite in modern sediments: An indicator of redox conditions. Geochim Cosmochim Acta 60:3897-3912

Wiltschko R, Wiltschko W (1995) Magnetic orientation in animals. Springer

Winklhofer M, Holtkamp-Rötzler E, Hanzlik M, Fleissner G, Petersen N (2001) Clusters of superparamagnetic magnetite particles in the upper-beak skin of homing pigeons: Evidence of a magnetoreceptor? Eur J Mineral 13:659-670

Wolthers M, Van der Gaast SJ, Rickard D (2003) The structure of disordered mackinawite. Am Mineral 88:2007-2015

Wong KKW, Mann S (1996) Biomimetic synthesis of cadmium sulfide-ferritin nanocomposites. Adv Mater 8:928-931

Zbinden M, Le Bris N, Compere P, Martinez I, Guyot F, Gaill F (2003) Mineralogical gradients associated with alvinellids at deep-sea hydrothermal vents. Deep-Sea Res 50:269-280

Zbinden M, Martinez I, Guyot F, Cambon-Bonavita MA, Gaill F (2001) Zinc-iron sulphide mineralization in tubes of hydrothermal vent worms. Eur J Mineral 13:653-659

Zhang HZ, Huang F, Gilbert B, Banfield JF (2003) Molecular dynamics simulations, thermodynamic analysis, and experimental study of phase stability of zinc sulfide nanoparticles. J Phys Chem B 107:13051-13060